Design and Control of Physical and Cyber-Physical Systems

Focusing on basic theory, emerged technologies and underlying engineering solutions, we investigate frontiers of engineering and science in design and control of physical and cyber-physical systems (CPS). Advanced-technology multiphysics systems are designed by seamlessly integrating components and modules, guaranteeing overall specifications and functionality by means of data processing, management and control. Analog and digital controllers and data management are implemented by analog and mixed-signal ASICs, as well as by microcontrollers and field-programmable gate arrays. Cybersecurity and information security have become critical.

This book focuses on engineering solutions and covers frontiers of engineering and science. Basic theory, emerged technologies, advanced software, enabling hardware and computing environments are applied. Recent innovations and discoveries are reported. We demonstrate and apply underlying fundamentals in design of complex systems. This book aims to:

1. Consistently cover topics on engineering science and engineering design pertaining to physical systems and their components;
2. Educate and assist one in development of problem-solving skills and design proficiency;
3. Ensure in-depth pervasive presentation and consistent coverage;
4. Endow the end-user with adequate knowledge in engineering analysis and design.

The emerged technologies and hardware have advanced system organization and enabled capabilities. The recent developments are empowered by new solutions. The major objective of this book is to empower concurrent design, as well as to enable a deep understanding of engineering underpinnings and integrated technologies. Key concepts and paradigms are covered.

Design and Control of Physical and Cyber-Physical Systems

Sergey Edward Lyshevski

CRC Press
Taylor & Francis Group
Boca Raton London New York

CRC Press is an imprint of the
Taylor & Francis Group, an **informa** business

First edition published 2024
by CRC Press
6000 Broken Sound Parkway NW, Suite 300, Boca Raton, FL 33487-2742

and by CRC Press
4 Park Square, Milton Park, Abingdon, Oxon, OX14 4RN

CRC Press is an imprint of Taylor & Francis Group, LLC

ISBN: 978-0-367-36390-1 (hbk)
ISBN: 978-1-032-52678-2 (pbk)

Typeset in Times

Publisher's note: This book has been prepared from camera-ready copy provided by the authors.

Contents

Preface

Focusing on basic theory, emerged technologies and underlying engineering solutions, we investigate frontiers of engineering and science in design and control of physical and cyber-physical systems (CPS). Advanced-technology multi-physics systems are designed by seamlessly integrating components and modules, guaranteeing overall specifications and functionality by means of data processing, management and control. Analog and digital controllers and data management are implemented by analog and mixed-signal ASICs, as well as by microcontrollers and field-programmable gate arrays. Cybersecurity and information security have become critical.

Single- and multi-agent physical systems are comprised of processing units, communication modules, and, integrated multi-physics components, such as transducers, actuators, multi-degree-of-freedom sensors, microelectromechanical systems, microelectronic and electronic components, etc. Systems interact and operate in rapidly changing dynamic environments. Information is measured by networked sensors, and, communication links support secure sharing of data streams. Data fusion and management are considered. *Cloud*, distributed, centralized, decentralized, multi-level hierarchical and adaptively coordinated organizations are developed and commercialized. Multi-input-multi-output systems are analyzed and optimized in physical and information domains. Considering traditional applications, examples of physical systems and CPS are semi-autonomous and autonomous aerial and ground vehicles, *smart* robots, advanced manufacturing, electrical grids, etc. System complexity and challenges pose needs for transformative research. To enable safety, resiliency, security and usability, adequate design schemes are applied. *Cloud* CPS are enabled by emerged advancements in internet of things, machine learning, etc.

With the emergence of new technologies and adverse dynamic environments, system intelligence and information-centric paradigms have become increasingly important in *smart* cars, consumer electronic devices, power systems, robots, medical devices, etc. Physical systems and their components must be controlled. Integration, interfacing, information sharing and data fusion are critical for components and modules which are designed and fabricated using different technologies. Pertained challenges, open problems and adaptive user-configured systems result in needs to advance basic, applied and experimental research. For physical systems and key components, this book covers basic engineering physics, underlying engineering design, emerged hardware solutions and physical implementation. Application-specific findings, experimental results and enabling hardware are reported.

This book focuses on engineering solutions, as well as covers frontiers of engineering and science. Basic theory, emerged technologies, advanced software, enabling hardware and computing environments are applied. Recent innovations and discoveries are reported. We examine underlying fundamentals in design of complex systems by:

1. Consistently covering engineering science and engineering design, pertained to physical systems and their components;
2. Educating one in development of problem-solving skills and design proficiency;
3. Ensuring in-depth pervasive presentation and consistent coverage;
4. Endowing the end-user with adequate knowledge in analysis and design.

The emerged technologies and hardware have being advanced system organization and enabled capabilities. The recent developments are empowered by new solutions. The major objective of this book is to empower concurrent design, as well as to enable a deep understanding of engineering underpinnings and integrated technologies. Key concepts and paradigms are covered.

Educational Objective – This textbook aims to foster adequate learning, technology training, skill advancements, interactive expertise experience, as well as knowledge generation, retention and use. A wide range of worked-out problems, examples and solutions are studied in sufficient depth. This bridges the gap between theory, practical problems and hardware-software co-design. Step-by-step, one is introduced to basics, and guided from theoretical foundations to applications and implementation. To enable analysis and accomplish design tasks, we use MATLAB[R] and SIMULINK[R]. The book demonstrates the MATLAB capabilities, helps one to master MATLAB and investigates solutions. Our objective is to enable the designer's productivity by showing how to apply algorithms and tools. The MATLAB toolboxes offer a set of capabilities to effectively solve a variety of problems. One can modify and refine studied problems, and apply the reported results to solve application-specific problems. Our results provide solutions for various modeling, simulation, control, optimization, implementation, hardware-software co-design and other problems.

Learning and Technological Objectives – Our nation's success and competitiveness in developing practical engineering solutions, discovering novel technologies and deploying systems, requires consistent coverage, education, training, learning retention and knowledge generation. Economic, geopolitical, security, ecological and societal developments depend on the needs of sustainable scientific, engineering and evolving technologies. Discoveries and innovations of national importance can be accomplished by highly-educated, skilled and trained researchers, engineers and practitioners. Many engineering disciplines and technology commercialization depend on solutions of challenging problems. Among these problems are analysis and design systems, which operate in physical and cyber domains.

The author's goal is to contribute and provide a learning platform to some of the aforementioned problems. Innovations, discoveries and technological superiority are enabled by emerged technologies. Safety, resiliency, cost effectiveness, advanced technological solutions and performance metrics are under consideration.

The *contemporary modernism*, hypothetical generalization, and axiomatization of engineering problems may be inadequate to cope with challenges. The needs for technology-centric engineering science and consistent engineering design are strengthened. The author hopes that readers will enjoy this book.

Acknowledgments – The author would like to express sincere acknowledgments and gratitude to colleagues, peers and students who provided valuable suggestions. I would like to express my sincere gratitude and acknowledgements for encouragements, assistance and support from the Department of Electrical and Microelectronic Engineering, and Kate Gleason College of Engineering, Rochester Institute of Technology. The author is very grateful for the hardware support (devices, components and modules) from the Analog Devices, Inc. www.analog.com, Maxon Group www.maxongroup.com, REYAX Technology Corporation http://reyax.com, STMicroelectronics www.st.com and Texas Instruments www.ti.com. The MathWorks, Inc. www.mathworks.com MATLAB® environment is used. With a great pleasure I acknowledge the assistance and help from outstanding Francis and Taylor / CRC Press team, especially Nora Konopka (Acquisitions Editor, Electrical Engineering). Many thanks to all of you.

Sergey Edward Lyshevski, Ph.D., Professor of Electrical Engineering
Department of Electrical and Microelectronic Engineering
Rochester Institute of Technology, Rochester, NY 14623, USA

Author biography

Dr. Lyshevski received his M.S. and Ph.D. in Electrical Engineering from Kiev Polytechnic Institute in 1980 and 1987. From 1980 to 1992 Dr. Lyshevski held research and faculty positions at the Department of Electrical Engineering, National Technical University of Ukraine (Kiev Polytechnic Institute) and the Academy of Sciences of Ukraine. From 1989 to 1992 he was the Microelectronic and Electromechanical Systems Division Head at the Academy of Sciences of Ukraine. From 1993 to 2002 Dr. Lyshevski was an Associate Professor of Electrical and Computer Engineering, Purdue School of Engineering, Indianapolis. In 2002 he joined Rochester Institute of Technology as a professor of Electrical Engineering. Dr. Lyshevski served as a Professor of Electrical and Computer Engineering for the US Department of State, Fulbright program. Dr. Lyshevski is a Full Professor Faculty Fellow at the Air Force Research Laboratories, US Naval Surface Warfare Center, and, US Naval Undersea Warfare Center.

Dr. Lyshevski is the author and co-author of 13 books, 15 handbook chapters, more than 80 journal articles and 300 refereed conference papers. He served as an editor of handbooks. Dr. Lyshevski made more than 100 invited tutorials, workshops and keynote talks. As a principal investigator (project director), he performed contracts and grants for high-technology industry, US Department of Defense (DARPA, ONR and Air Force) and government agencies (DoE, DoT and NSF). Dr. Lyshevski conducts research and technology developments in cyber-physical systems, data security and quality, control, mechatronics, electromechanical systems and microelectromechanical systems. Dr. Lyshevski has made significant contributions in the design of advanced aerial, automotive and naval systems.

1 Systems in Physical and Cyber Domains

1.1. Introduction

Physical and cyber-physical systems (CPS) are designed and commercialized in semi-autonomous and autonomous aerial and ground vehicles, *smart* robots, advanced manufacturing, etc. Autonomous systems possess adaptation, perception, control and execution abilities to adequately operate in interacting environments. These imply seamless integration, organization, data and information sharing, optimization and control in physical and cyber domains. Designs are accomplished for autonomous swarms, industrial control systems, process control systems, distributed control systems, supervisory control and data acquisition systems, etc. Developments and commercialization of advanced physical systems and CPS ensure the United States superiority and technological supremacy, as well as accelerate technological developments guaranteeing security and competitiveness.

Physical Systems – Physical systems are engineered multi-physics systems, designed to guarantee seamless integration of components and modules to ensure functionality in data-driven applications by performing data processing, management and control. Humanoid robots, remote-controlled industrial and mobile robots, supermaneuverable aircraft, radio-controlled and autonomous unmanned aerial and ground vehicles, as well as a spectrum of semi-autonomous and autonomous multi-agent systems have being designed and commercialized. In industrial systems and consumer applications, human interactions, adaptiveness, data quality and seamless secure communication have being achieved. Continuum mechanics, electromagnetics, electrostatics, hydrodynamics and thermodynamics yield continuum energy transductions and processes in actuators, sensors, transducers and microelectronic devices, controlled by analog and digital controllers. Second and third generation jet fighter aircraft, first generation of industrial and mobile robots, as well as other complex systems effectively use analog electronics, sensing, processing and control schemes. Analog ASICs ensure ultra-low power and high efficiency. Unprecedented analog information processing, sensing and control are exhibited by all living systems. Challenges and open problems in analog conditional logic and continuous logic calculus have being resulted in digital design and implementation of digital logic, computing, interfacing and processing. Control, data fusion and processing are accomplished by analog, digital and mixed-signal ASICs, designed using discrete calculus, Boolean algebra, etc. Microprocessors and field-programmable gate arrays (FPGAs) empower autonomy, adaptive perception, control, data management, interfacing, and execution of complex tasks. We investigate analog and digital premises in analysis and control for emerged and legacy systems. Cybersecurity, information security, security awareness and cryptography have being addressed and solved.

Cyber-Physical System Definition – The National Science Foundation offers the following definition: "Cyber-physical systems (CPS) are engineered systems that are built from, and depend upon, the seamless integration of computation and physical components. Advances in CPS will enable capability, adaptability, scalability, resiliency, safety, security, and usability that will expand the horizons of these critical systems" [1]. Cybersecurity, information security, information assurance, security awareness and others critical problems have being emerged in *cloud* computing and *fog*-networked *system of systems*. Systems operate in interacting environments in physical and cyber domains.

Physical domain considerations imply a focus on:

1. Seamless integration of devices, components and modules;
2. Synergy of hardware and software within optimized organization and advanced technologies;
3. High-fidelity analysis, optimization and control;
4. Dynamic data-driven data fusion and management;
5. Physical data quality;
6. Descriptive, predictive and prescriptive data analytics.

Cyber domain considerations imply a focus on:

1. *Cloud* computing and *fog* networking;
2. Information space, researching information confidentiality, integrity, security and vulnerability;
3. Cognitive space, investigating how information is perceived and processed.

Cybernetics, Governance and Control – A word κυβερνητική "kybernetike" was used by Plato (427-347 BC) to describe ship navigation and steering. Most significant chronological achievements in system governance and control are reported in chapter 5. For steam engines, a mechanical *centrifugal governor* with sensing and feedback regulator was commercialized in 1776 by James Watt. Studies in *RLC* circuits, including resonant circuits and control of electromagnetic transducers, were conducted by Alessandro Volta (1745-1827), Félix Savary (1797-1841) and Michael Faraday (1791-1867). In 1834, André-Marie Ampère defined "*la cybernétique*" as "la science du gouvernement des hommes", e.g., "the science of the government of men".

Wireless communication, data fusion and control were developed by Nikola Tesla. In 1898, Nikola Tesla demonstrated a radio-controlled boat, and, secured a U.S. Patent 613,809 *Method of an Apparatus for Controlling Mechanism of Moving Vessels or Vehicles*. In 1912, Arthur Pollen and Frederic Charles Dreyer investigated the fire control systems using mechanical computing apparatuses. Since early 1900 to present days advanced designs, passive *RLC* circuits and analog integrated circuits perform filtering and analog processing. Analog *RLC* filters and regulators are broadly implemented in electromechanical servos, gasoline engines, aircraft and ship control, communication and fire control systems since the early 1900s. In 1948, Norbert Wiener published a book *Cybernetics: Or Control and Communication in the Animal and the Machine*. For many decades, there have being efforts to study information processing, fusion and sensing in living organisms. Moderate progress has being achieved. In contrast, pioneering discoveries by Michael Faraday, Heinrich Hertz,

Nikola Tesla and others in electromagnetics, analog electronics and communication have being empowered astonishing progress in electronics, transducers, etc.

Information security and data quality are critical in CPS and internet of things (IoT) applications. The design taxonomy is depicted in Figures 1.1.a and b. One applies cyber- and informatics tools, data analytics, high-performance computing and microelectronics. Aerodynamically optimized aerial vehicles are comprised of engines, aerodynamic control surfaces, actuators, transducers, multi-physics multi-degree-of-freedom (MDOF) sensors, inertial measurement units (IMU), microprocessing units (MPU), transceivers (transmitters and receivers), input-output devices, power electronics and other components. In addition to conventional requirements and specifications, resiliency to electromagnetic spectrum, data quality, information security and security services are essential.

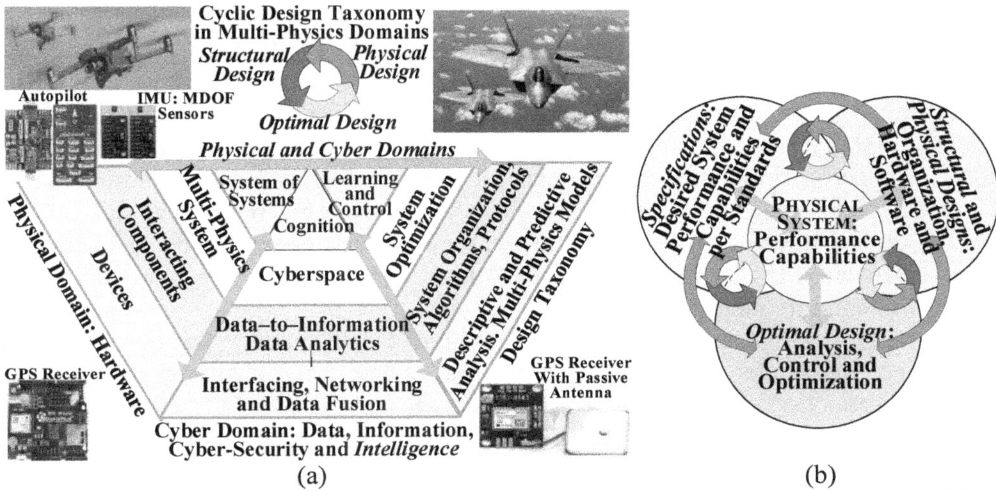

Figure 1.1. (a) Cyclic design taxonomy for multi-physics aerial and autonomous systems. Fifth-generation fighter aircraft and quadrotor helicopters are flying in close-proximity formation, controlled by flight management systems and autopilots. In many commercial multirotor helicopters and fix-wing unmanned aerial systems, the Pixhawk 4® autopilot (44×84×12 mm, 15.8 g) is used. The Pixhawk 4® has a main flight management unit processor STM32F765 (32-bit Arm Cortex-M7, 216 MHz), IO processor STM32F100 (32-bBit Arm Cortex-M3, 24 MHz), MDOF MEMS-technology high resolution sensors with MPUs, SPI and I^2C interfaces (TDK InvenSense IMU ICM-20689, Bosh Sensortec six-axis IMU BMI055, iSentek triaxial magnetometer IST8310, TE Connectivity barometric pressure altimeter sensor MS5611), GPS/GLONASS receiver, RC interfaces, etc. Images of GPS receivers with a passive patch antenna (25×25 mm), and, low-power 26×16 mm system-in-package IMU with triaxial accelerometers, gyroscopes and magnetometers, barometric pressure and temperature sensors; (b) Specification and cyclic design flow in developing system organization, hardware and algorithms to support design tasks ensuring achievable performance and capabilities. The design phases are the conceptual, preliminary and final designs, as well as verification, optimization, etc.

Physical systems interact, function and operate in rapidly changing environments. Physical variables and quantities are measured by sensors. Distributed sensing, control and coordination tasks are considered. Images of aerial vehicles, autopilot and MDOF sensors are shown in Figure 1.1.a. High throughput communication links support secure data streams fusion and management. Flight management systems are deployed on all-weather stealth tactical fighter aircraft as well as in civilian, commercial and military airplanes. Satellite navigation, wind-and-wake sensing and trajectory tracking for extended close-proximity formation flight are implemented. In fixed-wing, vertical take-off and landing, and, rotary-wing rotorcraft vehicles, guidance, navigation and control (GNC) are accomplished by a flight management system with a core processor and auxiliary advanced reduced instruction set computer (RISC) machine (ARM) MPUs.

Access over network, information security, device integration, and information content procurement are critical in CPS. There are complex management, operation and complementarity relationships between devices. Time-varying dynamic relationships between connected devices and interconnect among modules, content creation and other tasks are investigated.

Standards and Design Taxonomy – Quantitative specifications, requirements, guidelines and standards are imposed by the U.S. government and agencies, industry, IEEE, etc. Functionality, safety, interoperability, reliability, integration, scaling, cost-effectiveness and other specifications are procured. As documented in Figure 1.1.b, applying specifications and available technologies, one performs a cyclic design to find:

1. System organization and subsystem configuration by applying modular design and advanced-technology hardware, such as kinematics, transducers, actuators, sensors, MPUs, ASICs, microelectronics, power electronics, transceivers and other components;
2. Algorithms and protocols for spatially-distributed, asynchronous and heterogeneous nodes and interacting components ensuring data quality, information sharing and security;
3. Sensing, data fusion, data management, control and adaptation schemes.

Three tasks, shown in Figure 1.1.b, illustrate a cyclic design flow with alternative solutions, optimal searches, sequential redesign tasks, hardware–software co-design and other refinements upon evaluations. The overall compliance and seamless interoperability can be accomplished by means of modularity with standardized layers, nodes and advanced off-the-shelf components. The main phases and interrelated design activities are:

1. *System Organization Design* – Synthesize integrated organization with interacting nodes considering high-level goals meeting application-specific objectives;
2. *Physical Design* – Develop and validate hardware ensuring overall component compliance, aggregation, connectivity, coordination, interconnectivity and interoperability. The tasks include:
    - Select, design, manufacture and package devices, components, circuit boards, modules, etc.;
    - Select off-the-shelf component with specific volumetric, energy density, efficiency, overloading, accuracy, bandwidth, throughput, data rate, linearity, lifecycle, tolerance, environmental and other requirements;

3. *Optimal Design* – Perform multi-physics analysis, optimization and verification tasks;
4. *Informatics* and *Information Security Design* – Develop a framework and solutions ensuring data quality, integrity, confidentiality and availability.

One starts with a conceptual designs, and progresses to preliminary and final designs. There are cyclic evaluation, characterization, validation, verification and implementation activities.

Software Design – Cyber-physical systems pose challenges in software design due to diversity of hardware platforms, various applications, and, algorithmic complexity. Software, algorithms and protocols should support scheduling, sharing and coordination between services and applications, as well as pervasive networking and real-time cyber capabilities. The *software design* is a subject of extensive research and developments in computer science and software engineering, and, is not emphasized in this book.

Cloud Physical Systems – Cloud, distributed, centralized, decentralized, multi-level hierarchical and adaptively coordinated organizations are developed and deployed. System complexity and needs for improvements pose needs for evolutionary improvements and transformative research [2-6]. To enable functionality, effectiveness, coordination, interconnectivity, resiliency, reliability, interoperability and modularity, adequate design schemes are applied. *Cloud* systems are enabled by the emerged advances in IoT, *machine learning*, data analytics, etc. Due to a wide spectrum of associated research activities, we do not focus on certification and standardization, as well as on safety and reliability. We perform studies on topics associated with secure information sharing and control.

The mobile applications and IoT, e.g., internet networking and connectivity of devices which have an internet protocol address, have being commercialized in automation, energy management, environment, healthcare, media, *smart* home, security, transportation, wearable electronics and other applications. The IoT-support *cloud* computing, data sharing and other services. For medium-complexity applications, the software, middleware and hardware exist. The developed schemes and solutions may be reused and refined. Physical systems and CPS may use the IoT attributes. The scope of our developments and studies also contribute to networking and secure connectivity.

Systems Complexity and Descriptive Design – The intrinsic multi-physics complexity pose design and deployment challenges due to distributed sensing, asynchronous multi-sensor data fusion heterogeneity, latencies, uncertainties, adverse environments, electromagnetic interference, electromagnetic spectrum, high-throughput resilient communication needs, etc. Applying fundamental science, conducting applied research and discovering engineering solutions, new schemes are under developments. System organization design, multi-physics analyses and other problems must be solved to ensure compliance, interoperability and integration with legacy systems. Abstract algebra and graph theory may mathematically define interactions among homogenous and heterogeneous components. Inadequate and *incompatible* models should be avoided. Abstraction and abstract semantics have limited practicality unless supported by engineering and technological solutions. Compatible and compliant components within a modular organization may ensure adequateness. Considering aerial, automotive, consumer electronic, energy, manufacturing, robotic and transportation system applications, we cover core science and engineering aspects pertained to cyclic conceptual, preliminary and final design phases on:

1. *System organization design* by synthesizing application-specific hardware-defined organizations;
2. *Physical design* of advanced-technology transducers, sensors and power electronics components;
3. *Optimal design* with a sequence of analysis, identification, estimation, control and optimization tasks.

Due to complexity, multi-physics and data heterogeneity, control and optimization are critical. Devices, components and modules are designed, fabricated and integrated using different technologies. A spectrum of open problems in design and technology developments must be solved when systems operate in adverse dynamic environments. The aforementioned grand challenges and use of adaptive user-configured systems result in needs of basic, applied and experimental research. Engineering science, underlying engineering design, application-specific findings, physical implementation and experimental studies are investigated.

1.2. Physical Design, System Organization and Optimization

One strives to ensure a synergy of compliant mechanical, electromechanical, electronic, microelectronic, electromagnetic, piezoelectric, magnetoelectronic, optoelectronic, processors and other components within the application-specific specification-defined system organization. Consistently of selected off-the-shelf commercialized or newly-developed components should be guaranteed. Advanced technologies ensure advancements in multidisciplinary engineering underpinnings.

The prevailing specifications are imposed. Physical systems $\mathbf{P}(\mathbf{K}\circ\mathbf{A}\circ\mathbf{S}\circ\mathbf{E}\circ\mathbf{U})$ consist of kinematics $\mathbf{K}$, actuators and transducers $\mathbf{A}$, sensors $\mathbf{S}$, electronics $\mathbf{E}$, and processing units $\mathbf{U}$. The processing units $\mathbf{U}$ perform information management, control, filtering, data fusion and management, encryption/decryption and other tasks. The system is evaluated against multiple criteria procured applying the performance $\mathcal{Q}_P$ and capabilities $\mathcal{Q}_C$ measures $\mathcal{Q}(\mathcal{Q}_P,\mathcal{Q}_C)$ which depend on defined quantities $\boldsymbol{q}$, outputs $\boldsymbol{y}$, system states $\boldsymbol{x}$ and control $\boldsymbol{u}$. A physical system $\mathbf{M}(t,\boldsymbol{q},\boldsymbol{x},\boldsymbol{y},\boldsymbol{u})$ is evaluated using the measured data $\mathcal{D}_{\text{data}}$, and,

$$\mathcal{Q}(\mathcal{Q}_P,\mathcal{Q}_C)=\mathcal{K}(t,\boldsymbol{q},\boldsymbol{x},\boldsymbol{y},\boldsymbol{u}),\ (t,\boldsymbol{q},\boldsymbol{x},\boldsymbol{y},\boldsymbol{u})\in T\times Q\times X\times Y\times U,\ \mathcal{D}_{\text{data}}\subseteq T\times Q\times X\times Y\times U. \quad (1.1)$$

The objectives are to ensure the desired specifications and requirements $(\mathbf{s},\mathbf{r})\in\mathbf{S}\times\mathbf{R}$ such that

$$\mathcal{Q}(\mathcal{Q}_P,\mathcal{Q}_C)\subseteq\mathbf{S}\times\mathbf{R},\ \forall(t,\boldsymbol{q},\boldsymbol{x},\boldsymbol{y},\boldsymbol{u}),\ \forall\mathcal{D}_{\text{data}}. \quad (1.2)$$

Specifications on Aerial Vehicles – Figures 1.1.a and 1.2.a illustrate aerial and underwater vehicles. The Federal Aviation Administration specifications, as well as military standards, define: Levels of flying qualities for the operational and mission flight phases, normal and failure states, flying and handling qualities, pitch-roll-yaw responses, pitch-roll-yaw control and stability, residual oscillations, spiral stability, maneuvering characteristics, control motions in maneuvering flight, longitudinal and lateral flying qualities, longitudinal and lateral control and oscillations, operational and permissible flight envelopes, flight outside the service envelope, maximum and minimum permissible speed, etc. These quantifiable specifications, performance indicators and capabilities metrics are asserted and evaluated using a quadruple $(\boldsymbol{q},\boldsymbol{x},\boldsymbol{y},\boldsymbol{u})\in Q\times X\times Y\times U$.

System- and component-level organizations for commercialized physical systems are different. For example, an underwater vehicle is controlled by steering the fins, shown in Figure 1.2.a. The system organization for electromechanical servosystems, supervised by a GNC system, is illustrated. There are hierarchical top-to-bottom and bottom-to-top designs phases. The synthesis taxonomy is shown in Figures 1.1 and 1.2.b.

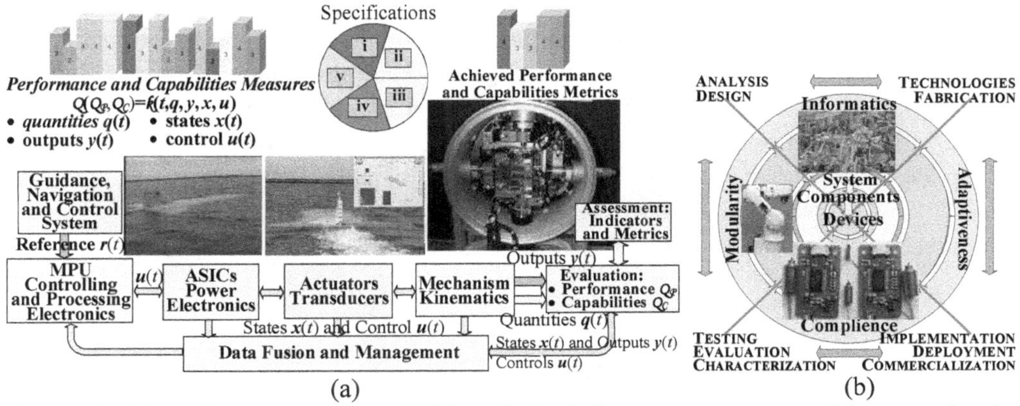

Figure 1.2. (a) Performance and capabilities of physical systems are assessed by a quadruple $(\boldsymbol{q},\boldsymbol{x},\boldsymbol{y},\boldsymbol{u})$. An image of the underwater vehicle, and, a vehicle hull with electronics, servomechanisms and kinematics. To control attitude, fins are steered by servos;
(b) Physical design of hardware–software solutions focusing on Informatics – Modularity – Compliance – Adaptiveness at the device-, component- and system levels. Sensing, data acquisition and control tasks are solved at the component levels: Images of *smart* manufacturing (system-level), robot (subsystem) as well as permanent-magnet synchronous minimotors controlled by high-frequency PWM amplifier with MPU.

Performance and Capabilities: Optimization Problem – One quantitatively evaluates system performance applying metrics, estimates, indicators and measures (1.1) and (1.2). Mathematically, optimization problem may be formulated by considering a tuple $(\boldsymbol{d},\boldsymbol{p},\boldsymbol{c})\in \mathcal{D}\times\mathcal{P}\times C$, where $\boldsymbol{d}$ are the design factors, such as system organization, selected components, cost, effectiveness, volumetric, etc.; $\boldsymbol{p}$ are the performance indicators, such as bandwidth, efficiency, throughput, stability, data quality, trustworthiness, security, etc.; $\boldsymbol{c}$ are the capabilities indicators, such as safety, functionality, reliability, etc.

One finds $\boldsymbol{p}=f_p(\boldsymbol{q},\boldsymbol{x},\boldsymbol{y},\boldsymbol{u})$ and $\boldsymbol{c}=f_c(\boldsymbol{q},\boldsymbol{x},\boldsymbol{y},\boldsymbol{u})$, where $f_p(\cdot):\mathbb{R}^q\times\mathbb{R}^n\times\mathbb{R}^b\times\mathbb{R}^m\to\mathbb{R}^p$ and $f_c(\cdot):\mathbb{R}^q\times\mathbb{R}^n\times\mathbb{R}^b\times\mathbb{R}^m\to\mathbb{R}^c$ are the nonlinear maps. The minimax problem implies optimization of the objective function $W(\boldsymbol{d},\boldsymbol{p},\boldsymbol{c})=p_W(\boldsymbol{d},\boldsymbol{k}(\boldsymbol{q},\boldsymbol{x},\boldsymbol{y},\boldsymbol{u}))$

$$J = \min_{\mathcal{d}\in\mathcal{D},\mathbf{p}\in\mathcal{P},\mathbf{c}\in C} \max W(\mathcal{d}, \mathbf{p}, \mathbf{c}),\ J{:}\mathcal{D}\times\mathcal{P}\times C\to\mathbb{R} \tag{1.3}$$

subject to design, performance and capability levels (l_d,l_p,l_c), constraints $L_{\min}\le L(\mathcal{d},\mathbf{p},\mathbf{c})\le L_{\max}$, and $(\mathbf{s},\mathbf{r})\in\mathbf{S}\times\mathbf{R}$.

Nonlinear optimization and minimax problems (1.3) are evaluated. Heterogeneity and uncertainties are due to computational and algorithmic complexities, as well as problem incompleteness. Optimization problems in multi-dimensional solution spaces are very difficult to solve even at the component-level applying dominant descriptive scaling, space and dimensionality reductions, etc. At system level, the minimax problem (1.3) cannot be solved due to complexity and incompleteness. Using engineering solutions, experienced practitioners solve encountered problems without mathematical considerations. Design phases can be modified, and, one finds solutions to meet objectives, requirement and specifications.

Design and Optimization – Selecting adequate organization and components, the solution space is reduced by applying a real-valued multivariate function $W(\mathbf{p},\mathbf{c})\equiv W(\mathbf{q},\mathbf{x},\mathbf{y},\mathbf{u})$ using the performance and capabilities variables $(\mathbf{p},\mathbf{c})$. For a given organization and chosen components, $\mathcal{d}$ is known. An optimization problem becomes complete and admits a solution. The minimax problem

$$J = \min_{\mathbf{p}\in\mathcal{P},\,\mathbf{c}\in C} \max W(\mathbf{p}, \mathbf{c}),\ J{:}\mathcal{P}\times C\to\mathbb{R},\ W(\mathbf{p},\mathbf{c})=f_W(\mathbf{q},\mathbf{x},\mathbf{y},\mathbf{u}),\ \mathbf{p}=f_p(\mathbf{q},\mathbf{x},\mathbf{y},\mathbf{u}),\ \mathbf{c}=f_c(\mathbf{q},\mathbf{x},\mathbf{y},\mathbf{u}), \tag{1.4}$$

$$L_{p,c\ \min}\le L(\mathbf{p},\mathbf{c})\le L_{p,c\ \max},\ (l_p,l_c)\ge 0,$$

is solved subject to physical system functionality, specifications, evolutions, governing equation, constitutive relationships and constraints.

For physical systems, governing and constitutive equations are

$$\dot{x}(t) = F(t,\mathbf{x},\mathbf{r}) + B(t,\mathbf{x})\mathbf{u},\ \mathbf{y}=H(\mathbf{x}),\ \mathbf{q}=\kappa(\mathbf{x},\mathbf{y},\mathbf{u}),\ \mathbf{u}_{\min}\le\mathbf{u}\le\mathbf{u}_{\max}, \tag{1.5}$$

where $\mathbf{x}\in X\subset\mathbb{R}^n$ is the state vector which evolves in X; $\mathbf{u}\in U\subset\mathbb{R}^m$ is the bounded control which evolves in U; $\mathbf{r}\in R\subset\mathbb{R}^b$ and $\mathbf{y}\in Y\subset\mathbb{R}^b$ are the reference and output vectors; $F(\cdot){:}\mathbb{R}_{\ge 0}\times\mathbb{R}^n\times\mathbb{R}^b\to\mathbb{R}^n$, $B(\cdot){:}\mathbb{R}_{\ge 0}\times\mathbb{R}^n\to\mathbb{R}^{n\times m}$, $H(\cdot){:}\mathbb{R}^n\to\mathbb{R}^b$ and $\kappa(\cdot){:}\mathbb{R}^n\times\mathbb{R}^b\times\mathbb{R}^m\to\mathbb{R}^q$ are the nonlinear maps.

Consider an aerial system. Solving control and dynamic optimization problems, for aerial systems, one considers minimal-time attitude and pitch-roll-yaw responses, minimal tracking error $e(t)$ in maneuvering flight, minimal energy consumption $E(t)$, *admissible* control $u(t)$, maximum endurance, range and efficiency $(\mathcal{E},\mathcal{R},\eta)$, etc. To solve the dynamic minimax optimization, the representative functional

$$J = \min_{t,e,E,u}\ \max_{\mathcal{E},\mathcal{R},\eta} \int_0^\infty \left(t|e| + e^2 + E^2 + u^2\right)dt \tag{1.6}$$

should be minimized subject to the constraints, governing and constitutive equations.

Systems Synergy – Microelectronics, MEMS sensors, massively parallel processing and other advances enable system synergy. There are raising industrial and societal needs with strong growing market and demands in commercialization of cost-effective CPS which guarantee overall superiority. Mechanical and electromagnetic transducers, sensors, power electronics, ASICs and MPUs are the key components. Reflecting a broad spectrum of recent discoveries and technologies, the following problems are emphasized:

1. Selection, design and optimization of devices and components according to their applications within the system organization;
2. Integration of compliant transducers, sensors, power electronics, microelectronics and kinematics emphasizing modularity, matching, compliance, complementarity and completeness.

In aerial vehicles, supervisory control is accomplished by a flight management system implemented in autopilot. Synergistic combination of devices, components and modules is achieved. Analog and digital controllers control fault-tolerant servos of aerodynamic surfaces as illustrated in Figure 1.3. The MDOF sensors measure and process variables (***x***,***y***,***u***). Actuators are controlled by MPUs.

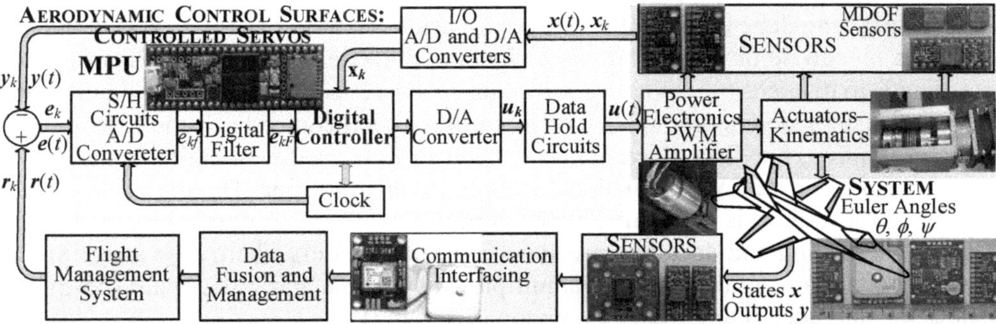

Figure 1.3. High-level organization and functional diagram with hardware components (actuators, amplifiers, MPUs, sensors, and, GPS receiver with a 25×25 mm ceramic passive antenna): Flight actuators and servos are steered using the reference commands ($\mathbf{r}(t)$,$\mathbf{r}_k$) from the flight management system. The aerodynamic control surfaces are displaced. The supervisory control is accomplished by a core flight management system processor, while control of actuators is performed by MPUs. Physical devices are analog. Signal conversions, filtering and data fusion are performed to achieve analog and digital controls.

Aircraft and actuators are regulated by a flight management system and controllers. Analog controllers and MPUs implement control laws ($\mathbf{u}(t)$, $\mathbf{u}_k$) using the state variables($\mathbf{x}(t)$,$\mathbf{x}_k$) and tracking errors ($\mathbf{e}(t)$, $\mathbf{e}_k$), $\mathbf{e}(t)=\mathbf{r}(t)-\mathbf{y}(t)$, $\mathbf{e}_k=\mathbf{r}_k-\mathbf{y}_k$. Actuators displace aerodynamic control surfaces thereby controlling attitude, flight path, pitch-roll-yaw responses, etc. The flight management system computes and coordinates the actuator displacements to execute airplane maneuvers. The tracking errors $\mathbf{e}$ for actuators are the differences between the reference inputs $\mathbf{r}$ and outputs $\mathbf{y}$. Continuous-time transducers, actuators and servos are controlled by applying pulse-width-modulated (PWM) voltage. To implement control laws, analog and digital control schemes are used. System states and output are discretized yielding $\mathbf{x}_k$ and $\mathbf{y}_k$.

For ground vehicles, robots, manipulators, aircraft, submarine and other motional systems, the Euler angles (θ, ϕ, ψ), geographic coordinate system (latitude, longitude and elevation), angle of attack, sideslip angle, acceleration, velocity and other variables are considered as outputs, depending on the problem under consideration. For the output vector $y=[\theta, \phi, \psi]$, the reference inputs are the desired Euler angles r_θ, r_ϕ and r_ψ. The reference vector is $r=[r_\theta, r_\phi, r_\psi]$. To control aircraft, control surfaces are displaced using hydraulic and electromechanical servos. Advanced aircraft is controlled by varying the thrust, as well as displacing rotational and translational actuators.

Using states, outputs and errors (x_k,y_k,e_k), MPUs are aimed to:

1. Implement filters, *estimators*, *observers* and controllers;
2. Perform data acquisition, fusion and data management;
3. Perform decision-making;
4. Generate control signals; etc.

An analog error signal $e(t)=r(t)-y(t)$ may be converted to discrete e_k, perform digital filtering, compute a control function u_k, and, generate the PWM control sequences. If analog $e(t)$ is used, as illustrated in Figure 1.3, the sample-and-hold circuit (S/H circuit) receives the analog signal and holds this signal at the constant value for the specified time depending on the sampling period. Analog-to-digital A/D converter converts a piecewise or continuous-time signal to discrete. Conversion of continuous-time signals to discrete-time is called sampling or discretization. The input signal to the filter is the sampled version of $e(t)$. The input of a digital controller is the filter output. At each sampling, the discretized value of the error e_k is used by a digital controller to compute and generate PWM signals which control an amplifier. The digital-to-analog conversion (decoding) is performed by the digital-to-analog D/A converter and the data hold circuit. Coding and decoding are synchronized by clock. There are various signal conversions, such as multiplexing, demultiplexing, sample and hold, analog-to-digital (quantizing and encoding), digital-to-analog (decoding), etc.

Industrial Solution – Texas Instruments C2000 high-performance microcontrollers are designed and commercialized to control power electronics, actuators and transducers. It ensures digital signal processing in aerial, automotive, consumer electronics, industrial, manufacturing and other applications. With the 32-bit core C28x CPU, the TMS320F280 digital signal processors features 16 PWM outputs, 6 high resolution PWM outputs, 16 channels 12-bit ADC, C/C++ compiler/assembler/linker, Code Composer Studio™, digital motor control and power control software libraries, etc.

Concluding Narrative – Various concepts are applied in design of physical systems. Application of abstract premises and hypothetical consideration under projected assumptions and assumed conjectures are frequently impractical. Axiomatization without use of laws of physics and existing technologies commonly result in inadequacies. Quantitative technology-focused designs may not be accomplished by means of abstract descriptions, *learning*, *knowledge-based* libraries, *virtual designs*, random graphs, etc. Usually, qualitative conjectural tools do not support descriptive features and practical schemes. Experienced practitioners perform near-optimal designs despite ambiguities and complexity. Using specifications, one performs design by finding adequate solutions through a sequence of steps and complementary activities. The device-to-system and system-to-device compliance, complementarity and interoperability is guaranteed. One ensures a consistency between various design steps, phases and tasks. System performance and capabilities are predefined by physical design. The design taxonomy is built on hierarchy, modularity and complementarity. No matter how well individual components perform, the overall performance can be degraded if overall consistency and control adequateness are not achieved. The component-centric *divide-and-solve* approach is applicable in a preliminary design. The overall system interoperability and functionality are addressed to solve pertained problems accomplishing general and specific objectives, specifications, requirements, etc. We study basic design fundamentals and practical solutions across physical, cyber and information domains.

Exploratory and Pragmatic Controls – Model-free, model-following, *learning-enabled* and other design premises have being investigated. These concepts are restricted to open-loop stable systems, assume homogeneous measurements, exceptional superior processing capabilities, etc. Axiomatized formulations, description inadequacy, complexity, data heterogeneities, algorithmic impediments and other obscurities of search and heuristic algorithms may not guarantee multi-physics consistency at device and system levels. These schemes do not admit closed-form unique solutions, and, result in algorithmic complexities, hardware impediments and sensitivity.

In process control and automation, industrial control systems, as well as in other applications, practitioners design and implement pragmatic data-driven control schemes. Designers implement fixed-structure or reconfigurable analog and digital PID control laws $\boldsymbol{u}=\varphi(t, \boldsymbol{e})$ using tracking error $\boldsymbol{e}(t)=[\boldsymbol{r}(t)-\boldsymbol{y}(t)]$

$$u(t)=k_p e(t)+k_i\int e(t)dt+k_d\frac{de(t)}{dt}, \tag{1.7}$$
$$u_n=u_{n-1}+k_{e0}e_n+k_{e1}e_{n-1}+k_{e2}e_{n-2},\ k_{e0}=k_p+T_s k_i+\tfrac{1}{T_s}k_d,\ k_{e1}=-k_p-\tfrac{2}{T_s}k_d,\ k_{e2}=\tfrac{1}{T_s}k_d$$

with constant-coefficient or variable feedback gains. These *minimal complexity* robust, low-latency controllers are implemented by low-power analog electronics and MPUs, ensuring near-optimal performance of commercialized systems. Designers minimize the use of mathematic-centric optimization and design findings, and, focus on adjusting preconfigured reconfigurable controllers on operational simulators, hardware experimental platforms or functioning systems. Examples include MDOF robots, aerial and ground vehicles, fifth and sixth generation of fighter aircraft, etc.

Information Management and Data Quality Control in CPS – The National Institute of Standards and Technology (NIST) Special Publication 800-53, *Security and Privacy Controls for Information Systems and Organizations*, as well as other publications, outline compliance and security controls for information systems, such as general purpose computing systems, CPS, cloud systems, mobile and Internet of Things devices, as well as industrial control systems. The cybersecurity control measures are aimed to ensure resiliency to cyberattacks, enable information assurance, achieve trustworthiness, etc. Some topics on cybersecurity and malware analyses are addressed in chapter 7. We do not examine antivirus software and tools, multifactor authentication tools, and, do not study in details malware and viruses detection and characterization. The focus is on the dynamic-probabilistic analysis of information, physical data quality, information sharing and data management. The dynamic-probabilistic Büchi final state automation, Turing machine and Markov model are examined.

Homework Problems

Homework Problem 1.1.
Provide examples of physical systems and CPS. Report functional diagrams, and, document key interacting components. You may consider aerial and ground vehicles, energy systems, robots, etc.

Homework Problem 1.2.
List key system performance measures and capabilities metrics for a system and modules considered in Homework Problem 1.1.

Homework Problem 1.3.
What are the major differences between physical systems and CPS?

Homework Problem 1.4.
Choose a physical system, module or component. Explicitly define problems to address and solve. Formulate and report typical specifications and requirements. Develop a high-level functional diagram with key components focusing on actuation, sensing, communication, data sharing, data fusion, data management, control and optimization. Report hardware solutions for the selected components, such as actuators, sensors, MPUs, transceivers, etc.

References

1. *Cyber-Physical Systems*, Program Solicitation NSF 21-553, *National Science Foundation*, Washington D.C. 2021.
2. R. Alur, *Principles of Cyber-Physical Systems*, The MIT Press, 2015.
3. M. Engelsberger and T. Greiner, "Dynamic management of cloud- and fog-based resources for cyber-physical production systems with a realistic validation architecture and results," *Proc. Industrial Cyber-Physical Systems Conf.*, pp. 109-114, 2018.
4. E. A. Lee. "Cyber physical systems: Design challenges," *Proc. Symp. Object Oriented Real-Time Distributed Computing*, pp. 363-369, 2008.
5. D. J. Reid, M. J. Pilling and V. G. Ivancevic, *Mathematics of Autonomy: Mathematical Methods For Cyber-Physical-Cognitive Systems,* World Scientific Publishing Co., 2017.
6. H. Noguchi and S. Sugano, "Ephemeral-cyber-physical systems: A cloud-like CPS using shared devices on open IoT," *IEEE Systems Journal*, pp. 1-11, 2020.

2 Analysis and Modeling of Physical Systems

2.1. Physical Systems and Mathematical Models

Physical systems are designed using different paradigms to achieve seamless functionality. The design objective is to ensure the best performance of aerial, automotive, electronic, energy, industrial, power, robotic and other systems under technology limits, cost, volumetric and power constraints, adversarial evens, etc. One investigates designs of process control systems, distributed control systems, security information management systems, physical and information data quality, etc. Design tasks comprise devising modular organization and performing physical designs, applied to a variety of systems, such as supervisory control and data acquisition systems (SCADA), industrial control systems, etc. To meet specifications, the system organization is defined, and, components are selected to ensure complementarity and compliance. Proven technologies and solutions are applied at device, component and system levels. The complexity is increased due stringent specifications, advanced sensing and processing schemes, algorithmic completeness, scalability, etc. Control, sensing, data fusion, data managements and other schemes and technologies affect analysis and design. Multidisciplinary studies are carried out by applying informatics, optimization and signal processing theories. Analysis and modeling of multi-physics systems are accomplished by applying laws of electromagnetics, mechanics and microelectronics. Experienced practitioners effectively design complex systems with minimum level of modeling efforts, and, find near-optimal solutions by choosing adequate components and conducting trial-and-error refinements. Designs have being deployed in complex multi-input multi-output (MIMO) aerial systems, robots, electrical grids, etc. These systems are characterized by complex hierarchical organization. Analyses, control and dynamic optimization problems are solved using physics-consistent descriptions. This chapter focuses on model developments applying laws of physics, model validation, dynamic-probabilistic analysis, descriptive simulations, etc. Technology-proven solutions and commercialized off-the-shelf components are investigated with a focus on testing, characterization and assessment in laboratory and operational environments. Information security, cyber assurance, cryptography and other considerations imply systems analysis in cyberspace.

High-Fidelity Modeling – There is a broad range of dynamic systems. Physical design, application and specificity are considered investigating modeling and analysis tasks [1-3]. Deterministic and probabilistic models, dynamic and stochastic analyses, continuous and *hybrid* domains, information measures as well as other quantitative predicates are investigated to accomplish engineering design. Analysis tasks imply procuring descriptive and governing equations which support information analytics and processing, electromagnetic, thermodynamic and other phenomena using physical and information variables [4-9]. High-fidelity modeling using spatiotemporal three-dimensional Navier-Stokes equations, Maxwell's equations and tensor calculus can be applied at device level. This supports device-level studies. Descriptive lumped-parameter models support descriptive, predictive and prescriptive analyses. Model fidelity, completeness and consistency are investigated. Descriptive models are mathematical description of physical phenomena exhibited, effects observed, transitions and processes, information governance, etc. Mathematical transforms, integral, differential and other operators are applied. Mathematical models are aimed to adequately describe dominant physical phenomena and information management. The Boltzmann equations, Newtonian and Hamiltonian dynamics, Kirchhoff laws, Lagrange equations, Faradays and other laws of physics yield consistent descriptions. Adequate complexity and consistency are ensured by applying nonlinear lumped-parameter models. For robots and aerial systems, illustrated in Figures 2.1, laws of physics and information theory are used at device, components and system levels. Modeling, identification, control, optimization, testing, characterization and other tasks are carried out. Topics in identification, information processing, data acquisition, data management, data quality and security are covered.

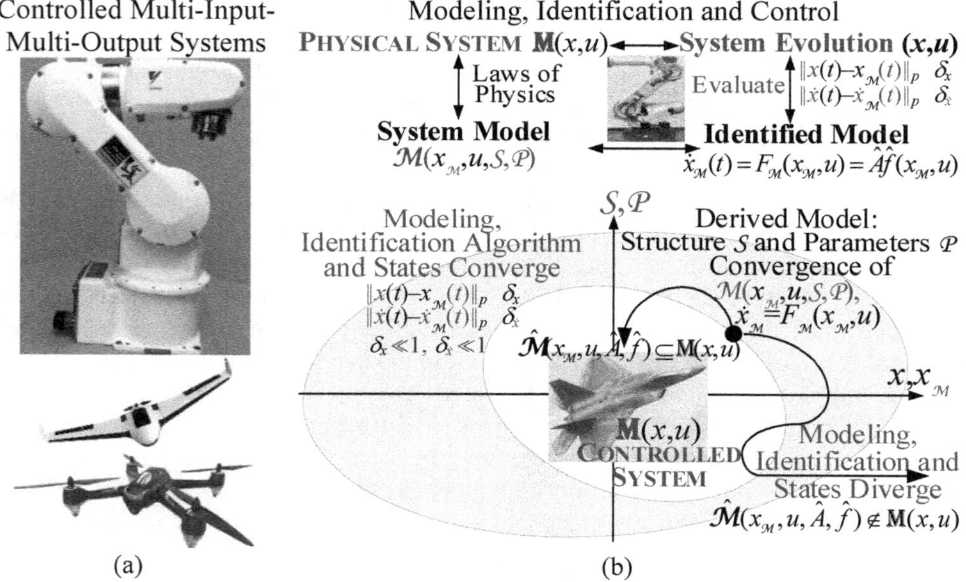

Figure 2.1. (a) Controlled robots and aerial vehicles $\mathbf{M}(x,y,u)$, characterized by the measured states $x\in X$, outputs $y\in Y$ and controls $u\in U$. Systems are controlled by a fixed-structure controller or adaptive control laws which can be reconfigured in near-real-time using estimated parameters; (b) A physical system $\mathbf{M}(x,y,u)$ with finite-dimensional states, outputs and control (x,y,u) evolves in $X\times Y\times U$: A descriptive, predictive and prescriptive model maps $\mathbf{M}(x,y,u)$ by a mathematical model $\mathcal{M}(x_{\mathcal{M}},u,\mathcal{S},\mathcal{P})$, $\dot{x}_{\mathcal{M}}(t)=F_{\mathcal{M}}(x_{\mathcal{M}},u)$, $y_{\mathcal{M}}=H(x_{\mathcal{M}})$. Laws of physics yield model order, dimensionality, structure and parameters $(\mathcal{S},\mathcal{P})$, ensuring sufficient fidelity. An identified model $\dot{x}_{\mathcal{M}}(t)=\hat{A}\hat{f}(x_{\mathcal{M}},u)$ should ensure an explicit representation of phenomena exhibited by $\mathbf{M}(x,y,u)$, guaranteeing $\mathcal{M}\subseteq\mathbf{M}$. Identification, parameter estimation, model validation and fidelity assessment are performed using the dynamic measures on measured states $(x,x_{\mathcal{M}})$ and inputs u.

Modeling Complexity – Fundamental studies, analytic results and numeric computing cannot replace testing, evaluation, characterization and implementation. The *model-free* concepts, *linguistic* models and other conjectural approaches may have limited practicality. Usually, these premises narrowly consider underlying device physics and processing predicates, and, focus on narrative features. Modular system organization, scalability and seamless complementarity of processes yield data sharing, fusion, control and management. Systems may be reconfigured by means of trial-and-error problem-solving schemes. Applying new technologies, physics- and information-centric paradigms guarantee adequate system performance within fixed-structure control and adaptive governance. These schemes and solutions have being successfully implemented in commercialized systems.

Over-simplification, over-complexity and generalization should be avoided. An experienced practitioner accomplishes a near-optimal design avoiding over-complex considerations. Physical controllability, functionality, optimality and other quantities can be achieved avoiding extensive searches and unproven exploratory studies. Selecting advanced off-the-shelf components and proven organizations yield scalable and practical solutions. We examine devices and systems, procure solutions, derive governing equations, design control laws, assess performance, evaluate capabilities, etc. Continuum mechanics, rigid body dynamics, hydrodynamics and continuum electromagnetics yield nonlinear differential-algebraic equations. In digital systems, digital arithmetic and Boolean logic operations are performed by transistors, which are analog electrostatic devices. One investigates analog and digital control and management schemes, data acquisition, data fusion, processing, etc. Challenges arise due to out-of-sequence asynchronous data fusion, sensor heterogeneities, data uncertainties, adverse electromagnetic spectrum, low throughput, latencies, etc. One focuses on data management, high-performance computing, high-throughput resilient communication and other tasks.

Mathematical Models – Lumped-parameter and dynamic-probabilistic models are found using laws of physics and information theory. While model uniqueness and completeness may not be ensured, for physical systems $\mathbf{M}$, one describes physical processes and transductions of states x and outputs y for inputs u. For measured states, outputs and inputs (x,y,u), finds mathematical models $\mathcal{M}(x_{\mathcal{M}},y_{\mathcal{M}},u)$, described by:

1. Ordinary and partial differential equations;
2. Differential-algebraic, integral and integro-differential equations;
3. Functional differential equations;
4. Turing machine;
5. Stochastic Markov chains;
6. etc.

For physical systems, one finds the model order considering (x,u), structure $\mathcal{S}$ (equations class, homogeneity, dependences, etc.) and parameters $\mathcal{P}$. Mathematical transforms and operators are used to describe processes by governing and constitutive equations. For MIMO systems, high-fidelity modeling at device level may be impractical considering system-level governance. For example, the ASICs consist of billions of transistors. It is unrealistic to investigate quantum-mechanical or electrostatics semiconductor device physics, or, conduct device-level analysis, while considering system-level data acquisition and control. Theoretical analysis of quantum communication, elastodynamics, electrodynamics, hydrodynamics and tribology may lead to a quantum-mechanical models in infinite-dimensional abstract variables (*complementary observables*), operators, quantum transforms and operator topological spaces, such as Hilbert $\mathcal{H}=L^2(\mathbb{R}^3)$, Banach $\mathcal{B}$ and L^∞ spaces. While quantum-mechanical model, operators and parameter spaces can be mathematically derived in $\mathbb{R}^\infty\times\mathbb{C}^\infty$, a finite physical continuum yields physics-consistent processing, and control on measurable real-valued variables.

Lumped-parameter models $\mathcal{M}(x_{\mathcal{M}},u,\mathcal{S},\mathcal{P})$ with real space of finite-dimension structure $\mathcal{S}\in\mathbf{S}$ and parameters $\mathcal{P}\in\mathbf{P}$ are adequate to describe multi-physics systems $\mathbf{M}(x,u)$ using the measured states x and inputs u. Physical systems $\mathbf{M}(x,y,u)$ evolve in $X\times Y\times U$, $\mathbf{M}:=\{x\colon x\in X(X_0,U), y=H(x)\}$ with states, outputs and inputs vectors $x\in X\subset\mathbb{R}^n$, $y\in Y\subset\mathbb{R}^b$ and $u\in U\subset\mathbb{R}^m$. While controls, disturbances, perturbations and noise may be referred as inputs, their physics are different. For multi-physics systems $\mathbf{M}(x,y,u)$, assessing dominant governance exhibited, examine:

1. System governance for measured and controlled $(x,y,u)\in X\times Y\times U$, applying laws of physics to define system and model states $(x,x_{\mathcal{M}})\in X\times X_{\mathcal{M}}$, outputs $(y,y_{\mathcal{M}})\in Y\times Y_{\mathcal{M}}$ and inputs $u\in U$;
2. Governing and constitutive equations
   $\mathcal{M}(x_{\mathcal{M}},u,\mathcal{S},\mathcal{P})=\{\dot{x}_{\mathcal{M}}(t)=F_{\mathcal{M}}(x_{\mathcal{M}},u),\ y_{\mathcal{M}}=H(x_{\mathcal{M}})\}$,
   with a finite-dimension structure $\mathcal{S}\in\mathbf{S}$ and parameters $\mathcal{P}\in\mathbf{P}$. A nonlinear map $F(x_{\mathcal{M}},u)$ is defined by $(\mathcal{S},\mathcal{P})$;
3. Identification and parameter estimation by finding a descriptive model $\dot{x}_{\mathcal{M}}(t)=\hat{A}\hat{f}(x_{\mathcal{M}},u)$ using the measured data $(x,y,u)_{\text{data}}\in X\times Y\times U$;
4. Assessment of model fidelity, correspondence, completeness and validity examining system evolutions $(x,y,u)\in X\times Y\times U$ and model dynamics $(x_{\mathcal{M}},y_{\mathcal{M}},u)\in X_{\mathcal{M}}\times Y_{\mathcal{M}}\times U$.

Continuum Physics – Physical laws of continuum electromagnetics, mechanics, thermodynamics and hydrodynamics yield nonlinear differential equations. At device level, one investigates description of plasma and laser kinetics, stimulated and spontaneous emissions, stimulated absorptions, optical resonators and cavity lasers, atomic and subatomic particles transductions, etc. Ultra-fast governance, exhibited by the sensors, photonic transceivers and other components, can be omitted from the governing equations of aerial, robotic and other systems. The quantum-mechanical equations for statistically complete and canonically conjugate variables may be defined using the quantum operators, which are not measurable, not detectable and not observable. The quantum-effect physical transitions and transductions are continuous.

Governing Equations – For a physical system $\mathbf{M}(x,u)$, assert domain and dimensionality of (x,y,u) to yield $(x_{\mathcal{M}},y_{\mathcal{M}},u)$, equations class, order, homogeneity and dependences. The equations of motion with model structure and parameters $(\mathcal{S},\mathcal{P})\in\mathbf{S}\times\mathbf{P}$ to identify are

$$\mathcal{M}(x_{\mathcal{M}},u,\mathcal{S},\mathcal{P})=\{x_{\mathcal{M}}\colon x_{\mathcal{M}}\in X_{\mathcal{M}}(X_{\mathcal{M}0},U)\}, \dot{x}_{\mathcal{M}}(t)=F_{\mathcal{M}}(x_{\mathcal{M}},u), y_{\mathcal{M}}=H(x_{\mathcal{M}}), F_{\mathcal{M}}(\cdot)\colon\mathbb{R}^n\times\mathbb{R}^m\to\mathbb{R}^n. \quad (2.1)$$

Parameters Estimation – Using measured physical system evolutions $(x,u)_{\text{data}} \in X \times U$ and model (2.1) with a given $\mathcal{S}$, for $\dot{x}_{\mathcal{M}}(t)=F_{\mathcal{M}}(x_{\mathcal{M}},u)=A_{\mathcal{M}}(\boldsymbol{a})f_{\mathcal{M}}(x_{\mathcal{M}},u)$, estimate matrix of parameters $\hat{A}$ ensuring convergence, concurrency and validity. Identification and estimation problems are covered in chapter 4, and, illustrated in Figure 2.1.b. The model states $(x_{\mathcal{M}},y_{\mathcal{M}}) \in X_{\mathcal{M}} \times Y_{\mathcal{M}}$ must correspond to the measured system states $(x,y) \in X \times Y$.

System Identification – If $\mathcal{S}$ is not explicitly defined, one considers a class of governing equations which corresponds to the physical system evolutions (x,y,u). For unstructured models, over-structured and over-parameterized map $F_{\mathcal{M}}(x_{\mathcal{M}},u)=A_{\mathcal{M}}(\boldsymbol{a})f_{\mathcal{M}}(x_{\mathcal{M}},u)$ is applied to identify a pair $(\hat{A}, \hat{f})$. For $\dot{x}_{\mathcal{M}}(t)=F_{\mathcal{M}}(x_{\mathcal{M}},u)=A_{\mathcal{M}}(\boldsymbol{a})f_{\mathcal{M}}(x_{\mathcal{M}},u)$, one finds

$$\dot{x}_{\mathcal{M}}(t) = \hat{A}\hat{f}(x_{\mathcal{M}},u),\ (\hat{A}, \hat{f}) \subseteq \mathcal{P} \times \mathcal{S}. \tag{2.2}$$

Using high-dimension structural and parameter spaces, identification is aimed to find parameters and structure $(\hat{A}, \hat{f})$ by truncating terms and applying regularization. As reported in chapter 4, one examines correspondence of model and system dynamics, and, evaluates the quantitative dynamic mismatch measures.

Model Dimensionality and Incompleteness – For physical systems $\mathbf{M}(x,u)$, which are designed to be physically controllable by using the measured states, mathematical models $\mathcal{M}(x_{\mathcal{M}},u,\mathcal{S},\mathcal{P})$ are incomplete and not unique. Model developments, structural identification and parameters estimation are incomplete problems. Due to heterogeneity, uncertainties, parameter variations and secondary phenomena, models $\mathcal{M}(x_{\mathcal{M}},u,\mathcal{S},\mathcal{P})$ are not unique. The model dimensionality and fidelity are defined by the dominant governance of $\mathbf{M}(x,u)$, and, phenomena and transductions exhibited are measured and controlled. Applying laws of physics, adequate models ensure the correspondence of states and outputs. Any mathematical model is a quantitative description of a physical system $\mathbf{M}(x,u)$ in the form of governing and constitutive equations. While not all physical phenomena may be examined and characterized studying transductions and processes, the finite-dimensional model developments problems are solvable.

Lumped-Parameter Models – Mathematical models describe physical phenomena, stochastic processes, etc. The derived and identified models must ensures fidelity, correspondence and completeness, guaranteeing

$$\hat{\mathcal{M}}(x_{\mathcal{M}},u,\hat{A},\hat{f}) \subseteq \mathbf{M}(x,u)\,. \tag{2.3}$$

In the following chapters, except chapter 4, it is assumed that adequate fidelity models are developed, devised and used. Assuming $\hat{\mathcal{M}}(x_{\mathcal{M}},u,\hat{A},\hat{f}) \triangleq \mathbf{M}(x,u)$, conventional state-space notations for dynamic systems are

$$\dot{x}(t)=F(x,u),\ y=H(x),\ x(t_0)=x_0,\ \hat{\mathcal{M}}(x_{\mathcal{M}},u,\hat{A},\hat{f}) \triangleq \mathbf{M}(x,u),\ \begin{cases} x_{\mathcal{M}} \triangleq x \\ y_{\mathcal{M}} \triangleq y \end{cases}, \tag{2.4}$$

where $x \in X \subset \mathbb{R}^n$ is the state vector which evolves in X; $u \in U \subset \mathbb{R}^m$ is the control vector; $y \in Y \subset \mathbb{R}^b$ is the reference and output vectors; $F(\cdot){:}\,\mathbb{R}^n \times \mathbb{R}^m \to \mathbb{R}^n$ and $H(\cdot){:}\,\mathbb{R}^n \to \mathbb{R}^b$ are the nonlinear maps.

2.2. Classical Mechanics and Lumped-Parameter Models

Newtonian, Lagrangian and Hamiltonian mechanics are applied to derive governing equations for rigid body dynamics, multibody dynamics, etc. These paradigms were developed by Sir Isaac Newton (1642–1727), Giuseppe Lagrangia (1736–1813) and Sir William Rowan Hamilton (1805–1865). Study system behavior analyzing the forces which cause motions. Considering one-, two- and three-dimensional motions, the Newton second law, as well as Lagrange and Hamiltonian equations, explicitly define the relationships between dynamics of a body and acting forces.

2.2.1. Newtonian Mechanics and System Energy

Translational Motion – The forces affect system motion and dynamic behavior. The force is the rate of change of the momentum. Using the linear momentum ***p***=*m**v***, the Newton second law of motion is

$$\sum_i \mathbf{F}_i(t,\mathbf{r},\boldsymbol{p}) = \frac{d\boldsymbol{p}}{dt} = \frac{d(m\boldsymbol{v})}{dt}, \tag{2.5}$$

where $\sum_i \mathbf{F}_i(t,\mathbf{r},\boldsymbol{p})$ is the vector sum of all forces acting on the object; $\mathbf{r}$ is the displacement vector; $\boldsymbol{v}$ is the velocity vector; m is the mass.

The object moves uniformly if $\boldsymbol{p}$=const, for which $\frac{d\boldsymbol{p}}{dt}=0$.

If m=const, the second law of motion (2.5) yields

$$\sum_i \mathbf{F}_i(t,\mathbf{r},\boldsymbol{p}) = \frac{d\boldsymbol{p}}{dt} = m\frac{d\boldsymbol{v}}{dt} = m\boldsymbol{a}, \tag{2.6}$$

where $\boldsymbol{a}$ is the vector of accelerations of body with respect to an inertial reference frame due to the acting *net* force.

In (2.6), $m\boldsymbol{a}$ is not a force. A body is at equilibrium, and, the object is at rest or is moving with constant speed and zero acceleration $\boldsymbol{a}$=0 if $\sum_i \mathbf{F}_i(t,\mathbf{r},\boldsymbol{v})=0$.

In the Cartesian system, the equations of motion in the xyz coordinates are

$$\sum_i \mathbf{F}_i(t,\mathbf{r},\boldsymbol{v}) = m\boldsymbol{a} = m\begin{bmatrix}\boldsymbol{a}_x\\ \boldsymbol{a}_y\\ \boldsymbol{a}_z\end{bmatrix},\ \sum_i \mathbf{F}_{x_i} = m\boldsymbol{a}_x,\ \sum_i \mathbf{F}_{y_i} = m\boldsymbol{a}_y,\ \sum_i \mathbf{F}_{z_i} = m\boldsymbol{a}_z,$$

$$m\frac{d\boldsymbol{v}}{dt} = m\begin{bmatrix}\frac{dv_x}{dt}\\ \frac{dv_y}{dt}\\ \frac{dv_z}{dt}\end{bmatrix} = \sum_i \mathbf{F}_i(t,\mathbf{r},\boldsymbol{v}),\quad m\frac{d^2\mathbf{r}}{dt^2} = m\begin{bmatrix}\frac{d^2x}{dt^2}\\ \frac{d^2y}{dt^2}\\ \frac{d^2z}{dt^2}\end{bmatrix} = \sum_i \mathbf{F}_i(t,\mathbf{r},\boldsymbol{v}). \tag{2.7}$$

From (2.5), (2.6) and (2.7), one obtains the ordinary differential equations.

Using the potential function $U(\mathbf{r})$

$$\sum_i \mathbf{F}_i(\mathbf{r}) = -\nabla U(\mathbf{r}). \tag{2.8}$$

The work done per unit time is

$$\frac{dW}{dt} = \sum_i \mathbf{F}_i(\mathbf{r})\frac{d\mathbf{r}}{dt} = -\nabla U(\mathbf{r})\frac{d\mathbf{r}}{dt} = -\frac{dU(\mathbf{r})}{dt}. \tag{2.9}$$

From Newton's second law (2.6), one obtains $\boldsymbol{ma}-\sum_i \mathbf{F}_i(t,\mathbf{r},\boldsymbol{v})=0$.

For a conservative system

$$m\frac{d^2\mathbf{r}}{dt^2}+\nabla U(\mathbf{r})=0. \tag{2.10}$$

Relationships for Work, Energy and Displacement – Gottfried Leibniz determined that the kinetic energy Γ is proportional to squared velocity $\Gamma \propto v^2$, and, velocity is proportional to the squire root of energy, $v\propto\sqrt{\Gamma}$.

The expression for a mechanical work, force and displacement in differential form was found by Gaspard-Gustave de Coriolis as $dW=Fdx$, where $F=ma$.

Hence, $W=\int Fdx=\int madx=\int m\frac{dv}{dt}dx=m\int vdv=\frac{1}{2}mv^2$.

One yields the expression for kinetic energy.

Example 2.1. *Work and Force*

Consider a frictionless movable platform actuated by an actuator. To accelerate a 1 kg mass m=1 kg from v_0=0 m/sec to v_f=1 m/sec, the work required is

$W=\frac{1}{2}(mv_f^2-mv_0^2)=\frac{1}{2}1\times 1^2=0.5$ J.

The work done by the *net* force on a system equals the change in the object's kinetic energy. Hence,

$W_{\text{total}}=\Gamma_2-\Gamma_1=\Delta\Gamma$.

For a varying force, one finds the total work done by the *net* force, $W=\int_{x_1}^{x_2}Fdx$.

Using $a=\frac{dv}{dt}=\frac{dv}{dx}\frac{dx}{dt}=v\frac{dv}{dx}$,

we have

$$W=\int_{x_1}^{x_2}Fdx=\int_{x_1}^{x_2}madx=\int_{x_1}^{x_2}mv\frac{dv}{dx}dx=\int_{v_1}^{v_2}mvdv. \quad \blacksquare$$

Example 2.2. *Mass in the XY Coordinate System*

Consider a body with mass m in the XY coordinate system. The free-body diagram is shown in Figure 2.2.a. The time-varying force $\mathbf{F}_{\text{applied}}$ is applied in the x direction.

Let $\mathbf{F}_{\text{applied}}(t,x)=\text{sech}(x)e^{-t}\sin^2(10t)$, $t\geq 0$.

Assume that the *Coulomb* and static frictions are negligible. Let the viscous friction force is

$F_{\text{friction}}=B_v v$,

where B_v is the viscous friction coefficient.

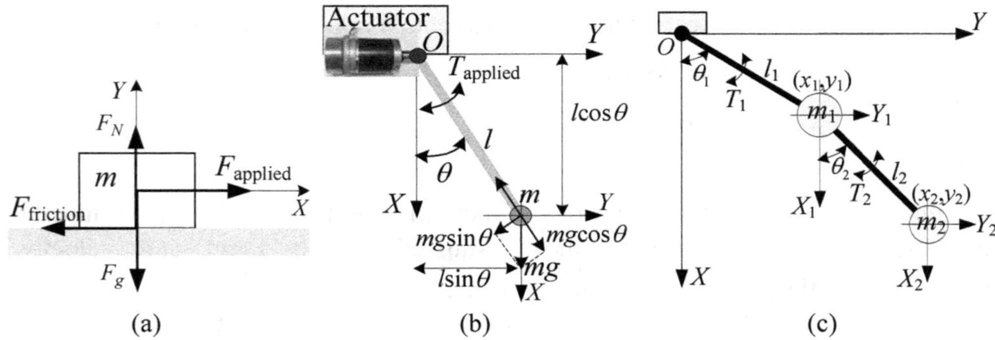

Figure 2.2. (a) Free-body diagram;
(b) Pendulum with a direct-drive actuator, Examples 2.5 and 2.7;
(c) Double pendulum: Two-link rigid robot arm or kinematic chain manipulator with two joints, Example 2.8.

The sum of forces, acting in the *y* direction, is

$$\sum \mathbf{F}_Y = \mathbf{F}_N - \mathbf{F}_g,$$

where $\mathbf{F}_g=mg$ is the gravitational force acting on the mass *m*; $\mathbf{F}_N$ is the normal force which is equal and opposite to the gravitational force.

From (2.6), the equation of motion in the *y* direction is

$$\mathbf{F}_N - \mathbf{F}_g = ma_y = m\frac{d^2y}{dt^2}.$$

From $\mathbf{F}_N=\mathbf{F}_g$, the resulting equation is $m\frac{d^2y}{dt^2}=0$.

The sum of time-invariant or time-varying forces acting in the *x* direction is

$$\sum \mathbf{F}_X = \mathbf{F}_{\text{applied}} - \mathbf{F}_{\text{friction}}.$$

Using (2.6) and (2.7), the equation of motion in the *x* direction is

$$\mathbf{F}_{\text{applied}} - \mathbf{F}_{\text{friction}} = ma_x = m\frac{dv}{dt} = m\frac{d^2x}{dt^2}.$$

One obtains the second-order nonlinear differential equation

$$\frac{d^2x}{dt^2} = \frac{1}{m}\left(F_{\text{applied}} - B_v\frac{dx}{dt}\right) = \frac{1}{m}\left[\operatorname{sech}(x)e^{-t}\sin^2(10t) - B_v\frac{dx}{dt}\right].$$

A set of two first-order linear differential equations is

$$\frac{dv}{dt} = \frac{1}{m}\left[-\operatorname{sech}(x)e^{-t}\sin^2(10t) - B_v v\right],$$

$$\frac{dx}{dt} = v. \qquad \blacksquare$$

Rotational Motion — The Newton second law for a rigid body is

$$\sum_i \boldsymbol{T}_i(t,\boldsymbol{\theta},\boldsymbol{\omega}) = J\boldsymbol{\alpha},\ \boldsymbol{\alpha} = \frac{d}{dt}\frac{d\boldsymbol{\theta}}{dt} = \frac{d^2\boldsymbol{\theta}}{dt^2} = \frac{d\boldsymbol{\omega}}{dt}, \qquad (2.11)$$

where $\sum_i \boldsymbol{T}_i(t,\boldsymbol{\theta},\boldsymbol{\omega})$ is the *net* torque; J is the moment of inertia; $\boldsymbol{\alpha}$ is the angular acceleration vector, $\boldsymbol{\alpha}=\frac{d}{dt}\frac{d\boldsymbol{\theta}}{dt}=\frac{d^2\boldsymbol{\theta}}{dt^2}=\frac{d\boldsymbol{\omega}}{dt}$; $\boldsymbol{\theta}$ and $\boldsymbol{\omega}$ are the angular displacement and velocity.

Using the angular momentum $\boldsymbol{L}_M=\mathbf{R}\times\boldsymbol{p}=\mathbf{R}\times m\boldsymbol{v}$, one has

$$\sum_i \boldsymbol{T}_i=\frac{d\boldsymbol{L}_M}{dt}=\mathbf{R}\times\mathbf{F}, \tag{2.12}$$

where $\mathbf{R}$ is the position vector with respect to the origin.

For a rigid body which rotates around the axis of symmetry, we have $\boldsymbol{L}_M=J\boldsymbol{\omega}$.

For one-dimensional motion, using the *net* moment $\mathbf{M}$, we have $\sum_i \mathbf{M}_i=J\alpha$.

Example 2.3. *Angular Momentum and Torque*
Let the rotor accelerates, and, the angular velocity is $\omega=10t^3$, $t\geq0$. One finds the angular momentum L_M. Assume that there is no load. For a frictionless bearing, $T_{\text{friction}}=0$.

If the equivalent moment of inertia is $J=1$ kg-m^2, the angular momentum is $L_M=J\omega=10t^3$ kg-m^2/s.

The applied torque is $T=\frac{dL_M}{dt}=30t^2$ N-m. ■

Analyze motions using the energy and momentum, which are conserved. The principle of energy conservation states that energy can be only converted from one form to another. Kinetic energy associates with motion, while potential energy is a function of position. For non-conservative system with dissipation D, the sum of energies (Γ,Π,D) yields the total energy of the system, which is conserved.

The total energy remains constant, and,

$$\mathrm{E}=\Gamma+\Pi+D=\text{const}. \tag{2.13}$$

Example 2.4. *Translational and Rotational Motion*
Consider the translational motion of a movable member which is attached to an ideal spring which exerts the spring force described by an ideal Hooke law. An ideal frictionless and massless spring has an equilibrium length, and, the magnitude of the force for stretch or compression Δx is

$$F_{\text{spring}}=k_s|\Delta x|,$$

where k_s is the spring constant.

The translational kinetic energy is $\Gamma=\frac{1}{2}mv^2$.

The elastic potential energy of the spring is $\Pi=\frac{1}{2}k_s x^2$.

Neglecting friction, the total energy is

$$\mathrm{E}=\Gamma+\Pi=\tfrac{1}{2}(mv^2+k_s x^2)=\text{const}.$$

For rotational motion of a rotor with ideal torsional spring, the rotational kinetic energy is $\Gamma=\frac{1}{2}J\omega^2$, while the elastic potential energy is $\Pi=\frac{1}{2}k_s\theta^2$.

Hence, $\mathrm{E}=\Gamma+\Pi=\frac{1}{2}(J\omega^2+k_s\theta^2)=\text{const}$.

If a rigid body exhibits translational and rotational motion, the kinetic energy is

$\Gamma = \frac{1}{2}(mv^2 + J\omega^2)$.

The motion of the rigid body is a combination of translational motion of the center of mass and rotational motion about the axis through the center of mass. The moment of inertia J depends how the mass is distributed with respect to the axis. This J is different for different axes of rotation. If the body is uniform in density, the moment of inertia J is calculated for regular body shapes. For a rigid cylinder of uniformly distributed mass m, radius R, and length l, one has the horizontal and vertical moments of inertia

$J_{\text{horizontal}} = \frac{1}{2}mR^2$, $J_{\text{vertical}} = \frac{1}{4}mR^2 + \frac{1}{12}ml^2$.

The *radius of gyration* can be found for irregularly shaped objects, yielding J.

■

Illustrative Example 2.1. *Total Work*

For a rigid body, if the moment of inertia J is constant, one has

$$\boldsymbol{T}d\boldsymbol{\theta} = J\boldsymbol{\alpha}d\boldsymbol{\theta} = J\frac{d\boldsymbol{\omega}}{dt}d\boldsymbol{\theta} = J\frac{d\boldsymbol{\theta}}{dt}d\boldsymbol{\omega} = J\boldsymbol{\omega}d\boldsymbol{\omega}.$$

The total work W represents the change of kinetic energy,

$$W = \int_{\theta_0}^{\theta_{\text{final}}} \boldsymbol{T}d\boldsymbol{\theta} = \int_{\omega_0}^{\omega_{\text{final}}} J\boldsymbol{\omega}d\boldsymbol{\omega} = \tfrac{1}{2}(J\omega_{\text{final}}^2 - J\omega_0^2).$$ ■

The power is

$$P = \frac{dW}{dt} = \boldsymbol{T}\frac{d\boldsymbol{\theta}}{dt} = \boldsymbol{T}\times\boldsymbol{\omega}. \tag{2.14}$$

Equation (2.14) is an analog of $P=\mathbf{F}\times\boldsymbol{v}$, which is applied for translational motions.

Example 2.5. *Equations of Motion for a Single Pendulum*

Consider a point mass m suspended by a massless unstretchable rigid rod of length l as shown in Figure 2.2.b. The restoring force $-mg\sin\theta$ is the tangential component of the *net* force. The sum of the moments about the pivot point O is

$$\sum M = -mgl\sin\theta + T,$$

where T is the applied torque; l is the length of the pendulum measured from the point of rotation.

Using (2.11), one obtains

$$\frac{d^2\theta}{dt^2} = \frac{1}{J}\left(-mgl\sin\theta + T\right),$$

where J is the moment of inertial of the mass about the point O.

A set of two first-order differential equations is

$$\frac{d\omega}{dt} = \frac{1}{J}\left(-mgl\sin\theta + T\right),\quad \frac{d\theta}{dt} = \omega.$$

The moment of inertia is $J=ml^2$. The equations of motion are

$$\frac{d\omega}{dt} = -\frac{g}{l}\sin\theta + \frac{1}{ml^2}T,$$

$$\frac{d\theta}{dt} = \omega.$$ ■

2.2.2. Lagrange Equations of Motion

Physical systems are comprised of mechanical, electric circuits, electromagnetic, electronic and other components. Using the *generalized coordinates* q_i, which are displacements and electric charges, one finds the total kinetic Γ and potential Π energies. The Lagrangian function L=(Γ–Π) is expressed using a pair (Γ,Π), found using the *generalized coordinates* $(q_1,\ldots,q_n)$ and *generalized velocities* $\left(\frac{dq_1}{dt},\ldots,\frac{dq_n}{dt}\right)$. One has

$$L\left(t,q_1,\ldots,q_n,\frac{dq_1}{dt},\ldots,\frac{dq_n}{dt}\right)=\Gamma\left(t,q_1,\ldots,q_n,\frac{dq_1}{dt},\ldots,\frac{dq_n}{dt}\right)-\Pi\left(t,q_1,\ldots,q_n\right). \quad (2.15)$$

Illustrative Example 2.2. *Dissipating Energy*

Changes in the thermodynamic states due to heat transfer, thermal energy exchange, thermal convection, thermal radiation, electromagnetic energy conversion and other processes are examined. The dissipating energy D is considered. The Rayleigh dissipation function D describes the velocity-dependent frictional forces. The static and dynamic frictions, as well as the dissipation functions, are found using high order multivariate polynomials $D(t,\frac{dq_i}{dt})=\sum_{i,k}B_{i,k}(t)|\dot{q}_i|^k$ and $D(t,\frac{dq_i}{dt})=\sum_{i,k,l}B_{i,k,l}(t)|\dot{q}_i|^{\frac{k}{2l-1}}$. ■

Friction and heat dissipation depend on the velocity, loads, symmetry, eccentricity, contact surfaces wear, corrosion, lubrication, deformation, temperature, etc. The energy dissipation results in irreversible processes in thermodynamic systems. Non-equilibrium thermodynamics may be investigated.

Lagrange Equations of Motion – Using the *total* energies $\Gamma\left(t,q_1,\ldots,q_n,\frac{dq_1}{dt},\ldots,\frac{dq_n}{dt}\right)$, $\Pi\left(t,q_1,\ldots,q_n\right)$ and $D\left(t,q_1,\ldots,q_n,\frac{dq_1}{dt},\ldots,\frac{dq_n}{dt}\right)$, the lumped-parameter models are found using the Lagrange equations of motion

$$\frac{d}{dt}\left(\frac{\partial\Gamma}{\partial\dot{q}_i}\right)-\frac{\partial\Gamma}{\partial q_i}+\frac{\partial D}{\partial\dot{q}_i}+\frac{\partial\Pi}{\partial q_i}=Q_i, \quad (2.16)$$

where q_i and $\dot{q}_i$ are the *generalized coordinates* and *generalized velocities*; Q_i are the *generalized forces*, such as applied, disturbances and perturbation forces, torques, voltages, etc.

For a conservative lossless systems D=0. Equations (2.16) become

$$\frac{d}{dt}\left(\frac{\partial\Gamma}{\partial\dot{q}_i}\right)-\frac{\partial\Gamma}{\partial q_i}+\frac{\partial\Pi}{\partial q_i}=Q_i. \quad (2.17)$$

Euler-Lagrange Equation of Motion – Defining the Lagrangian function as L=(Γ–Π), the Euler-Lagrange equation is $\frac{d}{dt}\left(\frac{\partial L}{\partial\dot{q}_i}\right)=\frac{\partial L}{\partial q_i}$. The *generalized momentum* p_i canonically conjugates to the *generalized coordinate* q_i, and, $p_i=\frac{\partial L}{\partial\dot{q}_i}$.

Example 2.6. *Newton and Euler-Lagrange Equations of One-Dimensional Motion*
Study one-dimensional translation motion of a frictionless mass attached to an ideal spring. The kinetic and potential energies are $\Gamma = \frac{1}{2}mv^2$ and $\Pi = \frac{1}{2}k_s x^2$.

In the Lagrange equations, the *generalized coordinates* $(q_1,\ldots,q_n)$ and *generalized velocities* $\left(\frac{dq_1}{dt},\ldots,\frac{dq_n}{dt}\right)$ are used. The Lagrangian function is $L=(\Gamma-\Pi)$.

For one-dimensional motion, if $D=0$, $F_{\text{applied}}=0$ and m=const, the Newton and Euler-Lagrange equations of motion are

$$m\frac{d^2x}{dt^2} = -k_s x,$$

and $$m\frac{d^2q}{dt^2} = -k_s q,\ q=x.$$

We observe an equivalency of Newton second law and the Lagrange equation. ■

Example 2.7. *Simple Pendulum*
Derive equations of motion for a pendulum, depicted in Figure 2.2.b. The equations were found in Example 2.5 using the Newtonian mechanics. The studied system is lossless if one assumes that friction and aerodynamic drag are negligible, $D=0$. However, systems exhibit frictions and losses.

Using the Cartesian coordinates, x and y describe the location of the mass m. These x and y are dependent. Assume that the length l is constant, $x^2+y^2=l^2$.

Using the angular displacement θ, one has $x=l\cos\theta$, $y=l\sin\theta$.

The kinetic energy is $\Gamma = \frac{1}{2}J\omega^2$, or, $\Gamma = \frac{1}{2}mv^2 = \frac{1}{2}m(l\dot{\theta})^2$.

For $\theta=0$, the potential energy is $\Pi=0$.

One has $x_{\text{height}}=l-l\cos\theta=l(1-\cos\theta)$.

The potential energy is $\Pi=mgl(1-\cos\theta)$.

The angular displacement θ is the *generalized coordinate*, $q=\theta$. Therefore,

$\Gamma = \frac{1}{2}m(l\dot{q})^2$, $\Pi=mgl[1-\cos(q)]$.

The Rayleigh dissipation function expresses the heat energy dissipated due to the viscous friction losses

$$D = \tfrac{1}{2}B_m\dot{\theta}^2 = \tfrac{1}{2}B_m\dot{q}^2,$$

where B_m is the viscous friction coefficient.

With derived (Γ,Π,D), the terms of Lagrange equation (2.17) is found. Consider a *generalized coordinate* $q=\theta$ and time-varying $m(t)$ and $l(t)$.

The product rule is $(fgh)'=f'gh+fg'h+fgh'$. Hence,

$$\frac{\partial\Gamma}{\partial\dot{q}_1} = \frac{\partial\Gamma}{\partial\dot{\theta}} = ml^2\dot{\theta},\ \frac{d}{dt}\left(\frac{\partial\Gamma}{\partial\dot{q}}\right) = \frac{d}{dt}\left(\frac{\partial\Gamma}{\partial\dot{\theta}}\right) = \frac{d}{dt}\left(ml^2\dot{\theta}\right) = ml^2\frac{d^2\theta}{dt^2} + \frac{dm}{dt}l^2\frac{d\theta}{dt} + 2ml\frac{dl}{dt}\frac{d\theta}{dt},$$

$$\frac{\partial\Gamma}{\partial q} = \frac{\partial\Gamma}{\partial\theta} = 0,$$

$$\frac{\partial\Pi}{\partial q} = \frac{\partial\Pi}{\partial\theta} = mgl\sin\theta,$$

$$\frac{\partial D}{\partial\dot{q}} = B_m\dot{\theta}.$$

The *generalized force* Q is the *net* torque ΣT. Consider the actuator torque T_{actuator}, as well as the load and disturbances torques T_{load} and $T_{\text{disturbances}}$. We have

$Q=T_{\text{actuator}}+T_{\text{load}}+T_{\text{disturbances}}$.

If m=const and l=const, the Lagrange equation of motion (2.16) yields

$$ml^2\frac{d^2\theta}{dt^2}+B_v\frac{d\theta}{dt}+mgl\sin\theta=T_{\text{applied}}+T_{\text{load}}+T_{\text{disturbances}}.$$

Therefore,

$$\frac{d^2\theta}{dt^2}=\frac{1}{ml^2}\left(-mgl\sin\theta-B_m\frac{d\theta}{dt}+T_{\text{actuator}}+T_{\text{load}}+T_{\text{disturbances}}\right).$$

We obtain

$$\frac{d\omega}{dt}=\frac{1}{ml^2}\left(-mgl\sin\theta-B_m\omega+T_{\text{actuator}}+T_{\text{load}}+T_{\text{disturbances}}\right),$$

$$\frac{d\theta}{dt}=\omega.$$

There is equivalency of the derived equations of motion found using Newtonian and Lagrangian mechanics. Recall that $J=ml^2$. For time-varying and bounded $m(t)$ and $l(t)$, the derivatives $(\frac{dm}{dt},\frac{dl}{dt})$ exist and bounded. One finds the time-varying governing equations of motion with $m(t)>0$, $l(t)>0$, $\forall t$.

■

Example 2.8. *Two-Link Rigid Robot Arm: Two-Joint Kinematic Chain Manipulator*
Consider a two-degree-of-freedom kinematics, shown in Figure 2.2.c. The angular displacements (θ_1,θ_2) are the independent *generalized coordinates* (q_1,q_2), $q_1=\theta_1$, $q_2=\theta_2$.

In the xy plane, the rectangular coordinates of point masses (m_1,m_2) are (x_1,y_1) and (x_2,y_2). We obtain

$$x_1=l_1\cos\theta_1,\ x_2=l_1\cos\theta_1+l_2\cos\theta_2,$$

$$y_1=l_1\sin\theta_1,\ y_2=l_1\sin\theta_1+l_2\sin\theta_2.$$

Assuming D=0, two Lagrange equations of motion (2.17) are

$$\frac{d}{dt}\left(\frac{\partial\Gamma}{\partial\dot\theta_1}\right)-\frac{\partial\Gamma}{\partial\theta_1}+\frac{\partial\Pi}{\partial\theta_1}=Q_1,$$

$$\frac{d}{dt}\left(\frac{\partial\Gamma}{\partial\dot\theta_2}\right)-\frac{\partial\Gamma}{\partial\theta_2}+\frac{\partial\Pi}{\partial\theta_2}=Q_2.$$

The total kinetic energy Γ is a nonlinear function of displacements (θ_1,θ_2)

$$\Gamma=\tfrac{1}{2}m_1v_1^2+\tfrac{1}{2}m_2v_2^2=\tfrac{1}{2}m_1\left(\dot x_1^2+\dot y_1^2\right)+\tfrac{1}{2}m_2\left(\dot x_2^2+\dot y_2^2\right)$$

$$=\tfrac{1}{2}(m_1+m_2)l_1^2\dot\theta_1^2+m_2l_1l_2\dot\theta_1\dot\theta_2\cos(\theta_2-\theta_1)+\tfrac{1}{2}m_2l_2^2\dot\theta_2^2.$$

For constant (m_1,m_2) and (l_1,l_2), one obtains

$$\frac{\partial\Gamma}{\partial\theta_1}=m_2l_1l_2\sin(\theta_2-\theta_1)\dot\theta_1\dot\theta_2,\ \frac{\partial\Gamma}{\partial\dot\theta_1}=(m_1+m_2)l_1^2\dot\theta_1+m_2l_1l_2\cos(\theta_2-\theta_1)\dot\theta_2,$$

$$\frac{\partial\Gamma}{\partial\theta_2}=-m_2l_1l_2\sin(\theta_1-\theta_2)\dot\theta_1\dot\theta_2,\ \frac{\partial\Gamma}{\partial\dot\theta_2}=m_2l_1l_2\cos(\theta_2-\theta_1)\dot\theta_1+m_2l_1^2\dot\theta_2.$$

For $\theta_1=0$, $\theta_2=0$, the potential energy is $\Pi=0$.

From

$x_{1\text{height}}=l_1(1-\cos\theta_1)$, $x_{2\text{height}}=l_2(1-\cos\theta_2)$,

the total potential energy is

$$\Pi=(m_1+m_2)gl_1(1-\cos\theta_1)+m_2gl_2(1-\cos\theta_2).$$

Hence,

$$\frac{\partial\Pi}{\partial\theta_1}=(m_1+m_2)gl_1\sin\theta_1,\ \frac{\partial\Pi}{\partial\theta_2}=m_2gl_2\sin\theta_2.$$

Let the servos actuate the first and second joints by applying torques (T_1,T_2). Let a time-varying load acts on the second joint $T_{L2}(t)$. Using the *net* torques, the *generalized forces* are $Q_1=T_1$ and $Q_2=T_2+T_{L2}$. The equations of motions are

$$l_1\left[(m_1+m_2)l_1\ddot{\theta}_1+m_2l_2\cos(\theta_2-\theta_1)\ddot{\theta}_2-m_2l_2\sin(\theta_2-\theta_1)\dot{\theta}_2^2+(m_1+m_2)g\sin\theta_1\right]=T_1,$$

$$m_2l_2\left[l_2\ddot{\theta}_2+l_1\cos(\theta_2-\theta_1)\ddot{\theta}_1+l_1\sin(\theta_2-\theta_1)\dot{\theta}_1^2+g\sin\theta_2\right]=T_2+T_{L2}(t).$$

The viscous friction is considered. The dissipating energy is

$$D=\tfrac{1}{2}B_{m1}\dot{\theta}_1^2+\tfrac{1}{2}B_{m2}\dot{\theta}_2^2.$$

Use the Lagrange equations (2.16)

$$\frac{d}{dt}\left(\frac{\partial\Gamma}{\partial\dot{q}_i}\right)-\frac{\partial\Gamma}{\partial q_i}+\frac{\partial D}{\partial\dot{q}_i}+\frac{\partial\Pi}{\partial q_i}=Q_i,\ q_i=(q_1,q_2),$$

where $\dfrac{\partial D}{\partial\dot{q}_1}=B_{m1}\dot{\theta}_1$ and $\dfrac{\partial D}{\partial\dot{q}_2}=B_{m2}\dot{\theta}_2$.

The resulting equations are

$$l_1\left[(m_1+m_2)l_1\ddot{\theta}_1+m_2l_2\cos(\theta_2-\theta_1)\ddot{\theta}_2-m_2l_2\sin(\theta_2-\theta_1)\dot{\theta}_2^2+(m_1+m_2)g\sin\theta_1\right]+B_{m1}\dot{\theta}_1=T_1,$$

$$m_2l_2\left[l_2\ddot{\theta}_2+l_1\cos(\theta_2-\theta_1)\ddot{\theta}_1+l_1\sin(\theta_2-\theta_1)\dot{\theta}_1^2+g\sin\theta_2\right]+B_{m2}\dot{\theta}_2=T_2+T_{L2}(t).$$

Cauchy's form of the governing equations, simulations, nonlinear system analysis, linearization and other pertained problems are reported in section 6.15.3.

■

Example 2.9. *Two-Degree of Freedom Translational-Rotational System*

Consider a two-degree of freedom robotic arm with two actuators to ensure the rotational and translational motions. A robotic arm and lumped-parameter representation are depicted in Figure 2.3.a. Two servos develop the torque T and force F, which are the *generalized forces*. The angular and linear displacements θ and r are the *generalized coordinates* (q_1,q_2). Using the Lagrange equations of motion (2.16), derive nonlinear differential equations.

Denote $\begin{bmatrix}q_1\\q_2\end{bmatrix}=\begin{bmatrix}\theta\\r\end{bmatrix}$ and $\begin{bmatrix}Q_1\\Q_2\end{bmatrix}=\begin{bmatrix}F\\T\end{bmatrix}$.

The Rayleigh dissipation function associates with rotational and translational viscous frictions

$$D=\tfrac{1}{2}(B_m\dot{q}_1^2+B_v\dot{q}_2^2),$$

where B_m and B_v are the viscous friction coefficients.

The translational-rotational system is examined in section 6.15.2.

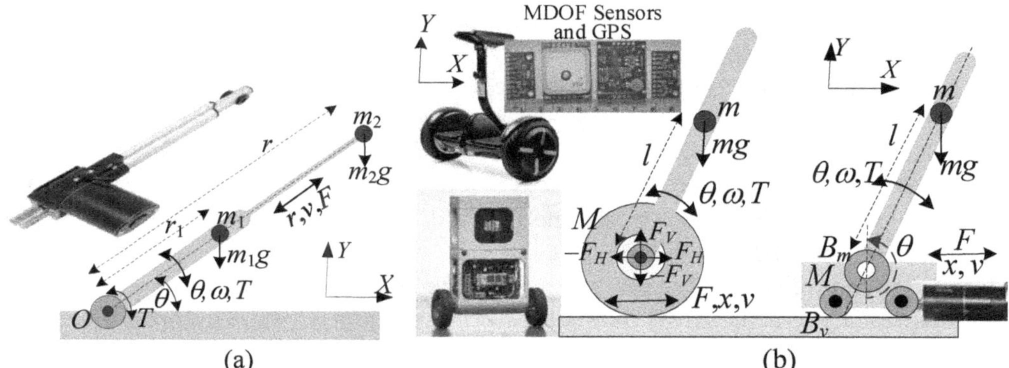

(a) (b)

Figure 2.3. (a) Two-degree-of-freedom robotic arm and a lumped-parameter schematic;
(b) Two- and four-wheel moving platforms with balancing inverted stands. Direct drive servomotors with rotor-wheel and rotor-stand couplings yield translational motion, path and pointing control. Advanced actuators, sensors, microprocessors (MPUs) and communication modules are used. The cart is driven using a direct-drive servomotors. There are acceleration, velocity, displacement and other sensors. System states are measured by MDOF microelectronic and MEMS sensors. The shown images of GPS and inertial sensors are: (1) REYAX RY825 5.5 V, 40 mA, 1575.42 Hz GPS and 1602.5625 Hz GLONASS center frequencies, –167 dBm sensitivity, 32×26×8.3 mm, 14 g global navigation satellite system (GNSS) receiver with embedded GPS/GLONASS antenna, 10 Hz navigation update rate, 2 m position accuracy; (2) TDK InvenSense MPU-9250 inertial measurement unit (IMU) with triaxial accelerometer (±2g, ±4g, ±8g and ±16g measurement range, and, 16-bit ADCs), gyroscope (±250, ±500, ±1000, ±2000 °/sec measurement range, and, 16-bit ADCs) and magnetometer (±4800 μT measurement range, and, 14-bit ADCs ensuring ±0.6 μT resolution), asynchronous data rates (4000, 8000 and 8 Hz for accelerometer, gyroscope and magnetometer), and, 400 kHz peripheral interfacing inter-integrated circuit (I^2C), 25.5×15.4×3 mm package.

■

Example 2.10. *Inverted Stand on the Moving Platform*

Two- and four-wheel self-balancing portable scooters, transporters, pointing systems, manipulators and mobile robots have being commercialized. In two-wheel balancing platforms, there are:

(1) In-wheel direct-drive servomotors with direct rotor-wheel coupling. The left and right motors, mounted on each axle, produce forces $F_L = r_{wheel} T_{Lmotor}$ and $F_R = r_{wheel} T_{Rmotor}$ ensuring steering and curvature path control capabilities;

(2) Direct-drive electromagnetic servo with rotor-arm coupling to steer the mounted stand, $\theta_{min} \le \theta \le \theta_{max}$. Actuator exerts the electromagnetic torque T to rotate the stand.

Depending on applications, one considers:

1. Stabilization of upward arm pointing by controlling F without rotational servo, T=0. Controlled motion of the moving platform, by applying the force F, is a classical control problem to stabilize the arm upwards $\theta = \pi$ despite small initial displacement $|\theta_0| \le (\pi \pm \delta_\theta)$ or perturbations;
2. Stabilization of a standing arm, and, cart repositioning with trajectory and curvature path tracking;
3. Steering of stand as well as cart tracking reposition by controlling T and F.

One may examine three-dimensional kinematic geometry, spatiotemporal distribution of mass, heterogeneous and varying density ρ of kinematics, etc. Using the mass per unit volume ρ, $dm=\rho dV$. The center of mass can be derived for given kinematics, loads, etc.

Consider an arm mounted on a four-wheel moving platform as shown in Figure 2.3.b. For the arm, the center of mass m is at length l. The arm is at stable equilibrium at the downward pointing with $\theta=0$ and $(F,T)=0$. The arm can be stabilized upward, and, steering may be ensured for $\theta_{\min}\le\theta\le\theta_{\max}$. As documented in section 6.15.1, a platform is controlled by regulating F and T. The arm holds by a rotating ball bearing joint. Arm may be actuated by a direct-drive servomotor which develops torque T. A force F is applied to a moving mass M which moves it in the $\pm x$ direction. The mathematical model is found by using the Lagrange equations of motion (2.16).

For a rotating stand in the XY plane, for a point mass m, we have

$x_m=x-l\sin\theta$, $y_m=-l\cos\theta$.

Hence,

$$\dot{x}_m=\dot{x}+l\dot{\theta}\cos\theta,\ \dot{y}_m=l\dot{\theta}\sin\theta.$$

The *total* kinetic energy for a movable cart M and arm m is

$$\Gamma=\Gamma_1+\Gamma_2=\Gamma_M+\Gamma_m=\tfrac{1}{2}M\dot{x}^2+\tfrac{1}{2}m\left(\dot{x}_m^2+\dot{y}_m^2\right)$$

$$=\tfrac{1}{2}M\dot{x}^2+\tfrac{1}{2}m\left[\dot{x}^2+2l\dot{x}\dot{\theta}\cos\theta+l^2\dot{\theta}^2\left(\cos^2\theta+\sin^2\theta\right)\right]$$

$$=\tfrac{1}{2}(M+m)\dot{x}^2+\tfrac{1}{2}ml^2\dot{\theta}^2+ml\dot{x}\dot{\theta}\cos\theta.$$

The potential energy is

$\Pi=mgy_{\text{height}}=mgl(1-\cos\theta)$.

At $\theta=0$ and $\theta=\pi$, $\Pi_{\min}|_{\theta=0}=0$ and $\Pi_{\max}|_{\theta=\pi}=2mgl$.

Hence,

$$\Pi=\Pi_1+\Pi_2=\Pi_M+\Pi_m=0+mgl(1-\cos\theta).$$

Consider the Rayleigh dissipation function, associated with linear and rotational viscous frictions. The dissipating energy is

$$D=D_M+D_m=\tfrac{1}{2}B_v\dot{x}^2+\tfrac{1}{2}B_m\dot{\theta}^2,\ D=D_1+D_2=\tfrac{1}{2}(B_v\dot{q}_1^2+B_m\dot{q}_2^2).$$

Using the *generalized coordinates* ($q_1=x$, $q_2=\theta$) and *generalized forces* ($Q_1=F$, $Q_2=T$), two second-order nonlinear differential equations are found. From (2.16) $\dfrac{d}{dt}\left(\dfrac{\partial\Gamma}{\partial\dot{q}_i}\right)-\dfrac{\partial\Gamma}{\partial q_i}+\dfrac{\partial D}{\partial\dot{q}_i}+\dfrac{\partial\Pi}{\partial q_i}=Q_i$, we have

$$(M+m)\ddot{q}_1+ml\ddot{q}_2\cos q_2-ml\dot{q}_2^2\sin q_2+B_v\dot{q}_1=Q_1,$$

$$ml^2\ddot{q}_2+ml\ddot{q}_1\cos q_2+mgl\sin q_2+B_m\dot{q}_2=Q_2.$$

Therefore

$$(M+m)\ddot{x}+ml\ddot{\theta}\cos\theta-ml\dot{\theta}^2\sin\theta+B_v\dot{x}=F,$$

$$ml^2\ddot{\theta}+ml\ddot{x}\cos\theta+mgl\sin\theta+B_m\dot{\theta}=T.$$

Force Control – To stabilize an arm in upward pointing by moving a platform M, regulate force F, while $T=0$. The resulting equations in the Cauchy's form are

$$\frac{dx}{dt}=v,$$
$$\frac{dv}{dt}=\frac{-lB_v v+B_m\omega\cos\theta+l^2m\omega^2\sin\theta+lmg\sin\theta\cos\theta+lF}{l\left(M+m-m\cos^2\theta\right)},$$
$$\frac{d\theta}{dt}=\omega,$$
$$\frac{d\omega}{dt}=\frac{-B_m(M+m)\omega-(M+m)lmg\sin\theta-l^2m^2\omega^2\sin\theta\cos\theta+lB_vmv\cos\theta-Flm\cos\theta}{l^2m\left(M+m-m\cos^2\theta\right)}.$$

Force and Torque Control – Study two *generalized forces* $Q_1=F$ and $Q_2=T$ to accomplish stand steering and cart repositioning. That is, consider the so-called the force and torque control problem, assuming that F and T are control functions. The Cauchy's form of nonlinear differential equations and linearized model are found using the MATLAB code

```
syms F T M m l g Bv Bm a v x ap om th  % Symbols
% Lagrange equations
eqn1 = (M + m)*a + m*l*ap*cos(th)-m*l*om*om*sin(th)+Bv*v == F;
eqn2 = (m*l^2)*ap+m*l*a*cos(th) + m*g*l*sin(th) + Bm*om == T;
% Cauchy's equations for linear acceleration and angular acceleration
sol = solve([eqn1 eqn2], [a ap]);  acceleration = sol.a; alpha = sol.ap;
pretty(acceleration),pretty(alpha)
```

One finds

$$\frac{dx}{dt}=v,$$
$$\frac{dv}{dt}=\frac{-lB_v v+B_m\omega\cos\theta+l^2m\omega^2\sin\theta+lmg\sin\theta\cos\theta+lF-T\cos\theta}{l\left(M+m-m\cos^2\theta\right)},$$
$$\frac{d\theta}{dt}=\omega,$$
$$\frac{d\omega}{dt}=\frac{-B_m(M+m)\omega-(M+m)lmg\sin\theta-l(lm^2\omega^2\sin\theta+B_vmv)\cos\theta-Flm\cos\theta+(M+m)T}{l^2m\left(M+m-m\cos^2\theta\right)}.$$

We derived a lumped-parameter model not considering the actuator dynamics. An inverted arm on a moving platform is studied in details in section 6.15.1. ■

Example 2.11. *Electric Circuit: Governing Differential Equations*

Consider an electric circuit as depicted in Figure 2.4.a.

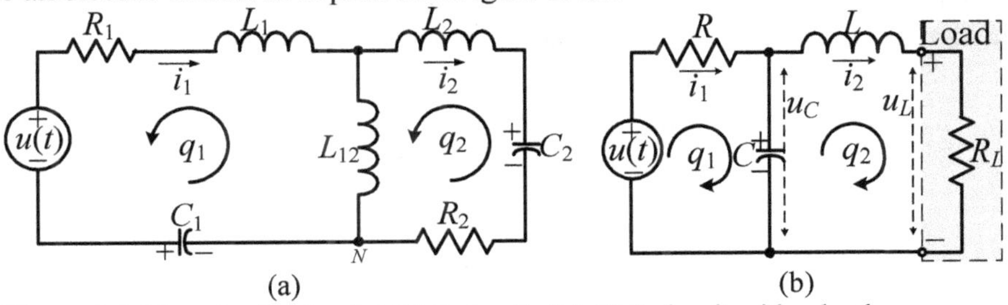

Figure 2.4. Electric circuits: (a) *RLC* circuit; (b) *RLC* circuit with a load.

The electric charges in the first and second loops (q_1,q_2) are the *generalized coordinates* (q_1,q_2). These *generalized coordinates* are related to currents as

$q_1=\int i_1 dt$, $i_1=\dot{q}_1$ and $q_2=\int i_2 dt$, $i_2=\dot{q}_2$.

The supplied voltage u is the *generalized force* Q_1, $u(t)=Q_1$.

The Lagrange equations of motion (2.16) are

$$\frac{d}{dt}\left(\frac{\partial\Gamma}{\partial\dot{q}_1}\right)-\frac{\partial\Gamma}{\partial q_1}+\frac{\partial D}{\partial\dot{q}_1}+\frac{\partial\Pi}{\partial q_1}=Q_1,$$

$$\frac{d}{dt}\left(\frac{\partial\Gamma}{\partial\dot{q}_2}\right)-\frac{\partial\Gamma}{\partial q_2}+\frac{\partial D}{\partial\dot{q}_2}+\frac{\partial\Pi}{\partial q_2}=0.$$

The total magnetic energy is a kinetic energy

$$\Gamma=\tfrac{1}{2}L_1\dot{q}_1^2+\tfrac{1}{2}L_{12}(\dot{q}_1-\dot{q}_2)^2+\tfrac{1}{2}L_2\dot{q}_2^2.$$

Hence,

$$\frac{\partial\Gamma}{\partial q_1}=0,\ \frac{\partial\Gamma}{\partial\dot{q}_1}=(L_1+L_{12})\dot{q}_1-L_{12}\dot{q}_2,\ \frac{\partial\Gamma}{\partial q_2}=0,\ \frac{\partial\Gamma}{\partial\dot{q}_2}=-L_{12}\dot{q}_1+(L_2+L_{12})\dot{q}_2.$$

The total electric energy is a potential energy, $\Pi=\frac{1}{2}\frac{q_1^2}{C_1}+\frac{1}{2}\frac{q_2^2}{C_2}$.

Hence, $\frac{\partial\Pi}{\partial q_1}=\frac{q_1}{C_1}$, $\frac{\partial\Pi}{\partial q_2}=\frac{q_2}{C_2}$.

The total heat energy dissipated is $D=\frac{1}{2}R_1\dot{q}_1^2+\frac{1}{2}R_2\dot{q}_2^2$.

This yields $\frac{\partial D}{\partial\dot{q}_1}=R_1\dot{q}_1$, $\frac{\partial D}{\partial\dot{q}_2}=R_2\dot{q}_2$.

The resulting differential equations are

$$(L_1+L_{12})\ddot{q}_1-L_{12}\ddot{q}_2+R_1\dot{q}_1+\frac{q_1}{C_1}=u,\ -L_{12}\ddot{q}_1+(L_2+L_{12})\ddot{q}_2+R_2\dot{q}_2+\frac{q_2}{C_2}=0.$$

Hence, we have

$$\ddot{q}_1=\frac{1}{L_1+L_{12}}\left(-\frac{q_1}{C_1}-R_1\dot{q}_1+L_{12}\ddot{q}_2+u\right),$$

$$\ddot{q}_2=\frac{1}{L_2+L_{12}}\left(L_{12}\ddot{q}_1-\frac{q_2}{C_2}-R_2\dot{q}_2\right).$$

■

Example 2.12. *Electric Circuit: Equations of Motion*

Consider an electric circuit with a load as shown in Figure 2.4.b.

Study two Lagrange equations

$$\frac{d}{dt}\left(\frac{\partial\Gamma}{\partial\dot{q}_1}\right)-\frac{\partial\Gamma}{\partial q_1}+\frac{\partial D}{\partial\dot{q}_1}+\frac{\partial\Pi}{\partial q_1}=Q_1,$$

$$\frac{d}{dt}\left(\frac{\partial\Gamma}{\partial\dot{q}_2}\right)-\frac{\partial\Gamma}{\partial q_2}+\frac{\partial D}{\partial\dot{q}_2}+\frac{\partial\Pi}{\partial q_2}=0.$$

The total kinetic energy is $\Gamma=\frac{1}{2}L\dot{q}_2^2$.

Therefore,

$$\frac{\partial T}{\partial q_1}=0,\ \frac{\partial T}{\partial \dot q_1}=0,\ \frac{d}{dt}\left(\frac{\partial T}{\partial \ddot q_1}\right)=0,\ \frac{\partial T}{\partial q_2}=0,\ \frac{\partial T}{\partial \ddot q_2}=L\dot q_2,\ \frac{d}{dt}\left(\frac{\partial T}{\partial \ddot q_2}\right)=L\ddot q_2.$$

The total potential energy is $\Pi=\frac{1}{2}\frac{(q_1-q_2)^2}{C}$.

Hence, $\frac{\partial \Pi}{\partial q_1}=\frac{q_1-q_2}{C}$, $\frac{\partial \Pi}{\partial q_2}=\frac{-q_1+q_2}{C}$.

The total dissipating energy is $D=\frac{1}{2}R\dot q_1^{\,2}+\frac{1}{2}R_L\dot q_2^{\,2}$.

Therefore, $\frac{\partial D}{\partial \ddot q_1}=R\dot q_1$, $\frac{\partial D}{\partial \ddot q_2}=R_L\dot q_2$.

From Lagrange equations (2.16)

$$R\dot q_1+\frac{q_1-q_2}{C}=u,$$

$$L\ddot q_2+R_L\dot q_2+\frac{-q_1+q_2}{C}=0,$$

we found a set of differential equations

$$\dot q_1=\frac{1}{R}\left(\frac{-q_1+q_2}{C}+u\right),$$

$$\ddot q_2=\frac{1}{L}\left(-R_L\dot q_2+\frac{q_1-q_2}{C}\right).$$

The equations of motion derived are equivalent the model found using Kirchhoff's law

$$\frac{du_C}{dt}=\frac{1}{C}\left(-\frac{u_C}{R}-i_2+\frac{u}{R}\right),$$

$$\frac{di_2}{dt}=\frac{1}{L}(u_C-R_Li_2).$$

From $i_1=\dot q_1$ and $i_2=\dot q_2$, using $C\frac{du_C}{dt}=i_1-i_2$, we obtain

$$u_C=\frac{q_1-q_2}{C}.$$ ∎

Example 2.13. *Lumped-Parameter Model of a Separately Exited Limited-Angle Electromagnetic Actuator*

Consider a limited-angle electromechanical motion device, widely used to actuate switches, robotic arms, mechanisms, etc. Two independently excited windings on laminated-core stator and rotor form two electromagnets as reported in Figure 2.5.a. The considered separately exited actuator can ensure 360-degree rotation using a commutator to apply a dc voltage to the rotor windings. For a limited angle electromagnetic servo studied, a flexible cable is used to apply voltage u_r. The magnetic coupling between stator and rotor windings, characterized by the mutual magnetizing inductance $L_{sr}(\theta_r)$, results in an electromagnetic torque T_e. The developed T_e is countered by a torsional spring, load torque T_L, weight, etc.

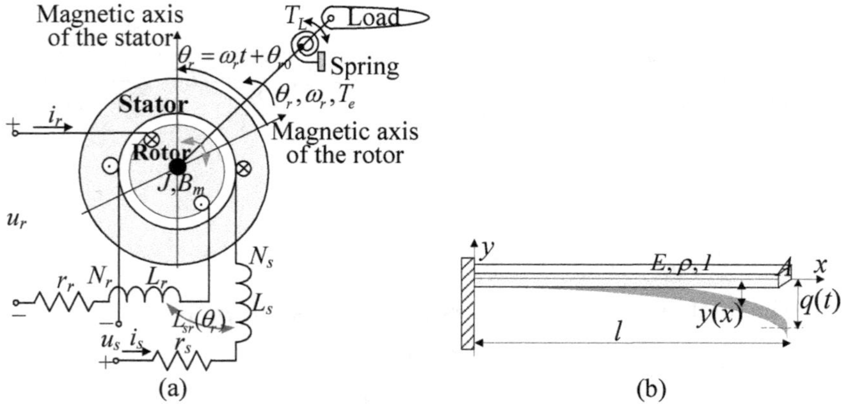

Figure 2.5. (a) Separately exited electromagnetic actuator;
(b) Cantilever beam in the *xy* plane, Example 2.14.

The state variables are the currents in the stator and rotor windings (i_s,i_r), as well as rotor angular velocity and displacement (ω_r, θ_r). Actuator is controlled applying dc voltages to the stator and rotor windings (u_s,u_r). The electromagnetic and load torques are T_e and T_L.

The parameters are the resistances of the stator and rotor windings (r_s,r_r), self-inductances of the stator and rotor windings (L_s,L_r), mutual magnetizing inductance between the stator and rotor windings $L_{sr}(\theta_r)$, number of turns in the stator and rotor windings (N_s,N_r), equivalent moment of inertia of the rotor and attached load J, viscous friction coefficient B_m, and, spring constant k_s.

The mutual magnetizing inductance depends on the displacement of the rotor winding with respect to the stator winding. Recall that for transformers, considering the zero-*electrical*-degree-displaced primary and secondary coils, the mutual inductance is maximum, and, denoted as $\pm M$. If the windings and magnets are orthogonal (displaced by 90-*electrical*-degree), then L_{sr}=0. Rotor rotates. For 360-degree rotation, $L_{sr}(\theta_r)$ is a periodic function of the angular displacement θ_r with a period 2π reaching ($L_{sr\min},L_{sr\max}$).

Hence, $L_{sr\min} \le L_{sr}(\theta_r) \le L_{sr\max}$, $|L_{sr}(\theta_r)| \le L_{sr\max}$.

Assume a sinusoidal magnetic coupling, and, let at $\theta_r = n\pi$, n=0,1,2,…, the mutual inductance $L_{sr}(\theta_r)$ is $L_{sr}(\theta_r)\big|_{\theta_r=0,\pi,2\pi,\ldots} = \pm L_{sr\max} = \pm L_M$.

The mutual inductance is

$$L_{sr}(\theta_r) = L_M \cos\theta_r = L_M \cos q_3,$$

where L_M is the amplitude of the mutual inductance variations.

The independent *generalized coordinates* are the electric charges in the stator and rotor windings (q_1,q_2), as well as the rotor angular displacement q_3. The *generalized forces* are the applied voltages supplied to the stator and rotor windings (Q_1,Q_2), and, the load torque Q_3. We have

$$q_1=\int i_s dt,\ q_2=\int i_r dt,\ q_3=\theta_r,$$

$$\dot{q}_1 = i_s,\ \dot{q}_2 = i_r,\ \dot{q}_3 = \omega_r,$$

$$Q_1=u_s,\ Q_2=u_r,\ Q_3=-T_L.$$

The Lagrange equations (2.16) are expressed in terms of each independent coordinate as

$$\frac{d}{dt}\left(\frac{\partial\Gamma}{\partial\dot{q}_1}\right)-\frac{\partial\Gamma}{\partial q_1}+\frac{\partial D}{\partial\dot{q}_1}+\frac{\partial\Pi}{\partial q_1}=Q_1,$$

$$\frac{d}{dt}\left(\frac{\partial\Gamma}{\partial\dot{q}_2}\right)-\frac{\partial\Gamma}{\partial q_2}+\frac{\partial D}{\partial\dot{q}_2}+\frac{\partial\Pi}{\partial q_2}=Q_2,$$

$$\frac{d}{dt}\left(\frac{\partial\Gamma}{\partial\dot{q}_3}\right)-\frac{\partial\Gamma}{\partial q_3}+\frac{\partial D}{\partial\dot{q}_3}+\frac{\partial\Pi}{\partial q_3}=Q_3.$$

The total kinetic energy of electrical and mechanical systems is a sum of the total electromagnetic $\Gamma_E=\frac{1}{2}L_s\dot{q}_1^2+L_{sr}(q_3)\dot{q}_1\dot{q}_2+\frac{1}{2}L_r\dot{q}_2^2$ and mechanical $\Gamma_M=\frac{1}{2}J\dot{q}_3^2$ energies. We have

$$\Gamma=\tfrac{1}{2}L_s\dot{q}_1^2+L_M\dot{q}_1\dot{q}_2\cos q_3+\tfrac{1}{2}L_r\dot{q}_2^2+\tfrac{1}{2}J\dot{q}_3^2.$$

Hence,

$$\frac{\partial\Gamma}{\partial q_1}=0,\quad \frac{\partial\Gamma}{\partial\dot{q}_1}=L_s\dot{q}_1+L_M\dot{q}_2\cos q_3,$$

$$\frac{\partial\Gamma}{\partial q_2}=0,\quad \frac{\partial\Gamma}{\partial\dot{q}_2}=L_M\dot{q}_1\cos q_3+L_r\dot{q}_2,$$

$$\frac{\partial\Gamma}{\partial q_3}=-L_M\dot{q}_1\dot{q}_2\sin q_3,\quad \frac{\partial\Gamma}{\partial\dot{q}_3}=J\dot{q}_3.$$

The potential energy of the spring is

$$\Pi=\tfrac{1}{2}k_s q_3^2.$$

Thus, $\frac{\partial\Pi}{\partial q_1}=0$, $\frac{\partial\Pi}{\partial q_2}=0$, $\frac{\partial\Pi}{\partial q_3}=k_s q_3$.

The total heat energy dissipated is

$D=D_E+D_M$,

where D_E is the heat energy dissipated in the stator and rotor windings, $D_E=\frac{1}{2}r_s\dot{q}_1^2+\frac{1}{2}r_r\dot{q}_2^2$; D_M is the heat energy dissipated in mechanical system due to the viscous friction losses, $D_M=\frac{1}{2}B_m\dot{q}_3^2$.

From $D=\frac{1}{2}r_s\dot{q}_1^2+\frac{1}{2}r_r\dot{q}_2^2+\frac{1}{2}B_m\dot{q}_3^2$,

we have $\frac{\partial D}{\partial\dot{q}_1}=r_s\dot{q}_1$, $\frac{\partial D}{\partial\dot{q}_2}=r_r\dot{q}_2$, $\frac{\partial D}{\partial\dot{q}_3}=B_m\dot{q}_3$.

For the *generalized coordinates* $q_1=\int i_s dt$, $q_2=\int i_r dt$ and $q_3=\theta_r$, one has $\dot{q}_1=i_s$, $\dot{q}_2=i_r$, $\dot{q}_3=\omega_r$. The *generalized forces* ($Q_1=u_s$, $Q_2=u_r$) are controls, while $Q_3=-T_L$ is the disturbance. Therefore

$$L_s\frac{di_s}{dt}+L_M\cos\theta_r\frac{di_r}{dt}-L_M i_r\sin\theta_r\frac{d\theta_r}{dt}+r_s i_s=u_s,$$

$$L_r\frac{di_r}{dt}+L_M\cos\theta_r\frac{di_s}{dt}-L_M i_s\sin\theta_r\frac{d\theta_r}{dt}+r_r i_r=u_r,$$

$$J\frac{d^2\theta_r}{dt^2}+L_M i_s i_r\sin\theta_r+B_m\frac{d\theta_r}{dt}+k_s\theta_r=-T_L.$$

Analyze device physics and results consistency.

We found that the electromagnetic torque is

$$T_e = -\frac{\partial \Gamma}{\partial q_3} = L_M \dot{q}_1 \dot{q}_2 \sin q_3 = L_M i_s i_r \sin \theta_r ,$$

and, T_e=0 for θ_r=0,π,2π.

One defines the stator-rotor orientation, magnetic axes, mechanical angular displacement, *electrical* angular displacement, etc. The north poles of two electromagnets repel each other. The south poles of the stator and rotor electromagnets are also repelled. The dc voltages of adequate polarity are applied to the stator and rotor windings s and r, and, electromagnet poles are defined by direction of currents i_s and i_r. The north pole of an electromagnet attracts to the south pole of other electromagnet. One specifies the stator-rotor displacement, kinematics, magnetic system, limits $\theta_{r\min} \le \theta_r \le \theta_{r\max}$, etc.

The relevant results, including simulations and linearization, are reported in Practice Problem 2.4 considering $L_{sr}(\theta_r) = L_M \sin\theta_r = L_M \sin q_3$, for which the electromagnetic toque T_e is maximum at θ_r=0,2π,….

■

Example 2.14. *Beam Equations of Motion*

Consider an elastic beam of length l with constant cross-sectional area A, uniform weight per unit volume (density) ρ, Young's modulus of elasticity E, and, moment of inertia of the cross section about its neural axis I. As illustrated in Figure 2.5.b, the vertical beam displacement at free end is $q(t)$.

Forces and loads can be applied at free end, forces may be uniformly or non-uniformly distributed, etc. For the concentrated, uniformly varying and other loads, the expressions for $y(x)$ are found by solving the governing equations or using the experimental data. For the load F applied at the free end, the Euler-Bernoulli beam equation yields a static equilibrium

$$y(x) = \frac{1}{6EI}\left(3lx^2 - x^3\right)F .$$

For uniformly distributed load F_x,

$$y(x) = \frac{1}{24EI}\left(6l^2x^2 - 4lx^3 + x^4\right)F_x .$$

Consider the third degree polynomial for $y(x)$,

$$y(t,x) = \frac{1}{2}\left(3\frac{x^2}{l^2} - \frac{x^3}{l^3}\right)q(t) .$$

Find kinetic and potential energies. For a cubic polynomial, the indefinite integral is $\int \left(3\frac{x^2}{l^2} - \frac{x^3}{l^3}\right)^2 dx = \frac{x^5(5x^2 - 35lx + 63l^2)}{35l^6} + c$.

The kinetic energy is

$$\Gamma(\dot{q}) = \tfrac{1}{2}\int_0^l \dot{y}^2 dm = \tfrac{1}{2}A\rho\int_0^l \tfrac{1}{4}\left(3\frac{x^2}{l^2} - \frac{x^3}{l^3}\right)^2 \dot{q}^2 dx = \tfrac{33}{280}A\rho l\dot{q}^2 .$$

The potential energy of elastic deformation is

$$\Pi(q)=\tfrac{1}{2}\int_0^l EI\left(\frac{\partial^2 y}{\partial x^2}\right)^2 dx=\tfrac{1}{2}EI\int_0^l \frac{3}{2l^3}\left(1-\frac{x}{l}\right)^2 d\left(\frac{x}{l}\right)=\tfrac{3}{2}\frac{EI}{l^3}q^2.$$

The Rayleigh dissipative function is

$$D=\tfrac{1}{2}\int_0^l b_b EI\left(\frac{\partial}{\partial t}\frac{\partial^2 y}{\partial x^2}\right)^2 dx=\tfrac{1}{2}B_d\dot{q}^2,\ B_d=\frac{3b_b EI}{l^3},$$

where b_b is the coefficient of the beam material.

From the Lagrange equation (2.16), the governing equation for a free end beam deflection is

$$\ddot{q}=-\tfrac{140}{11}\frac{EI}{A\rho l^4}q-B_d\dot{q}+F_q(t,x).$$

The potential energy of elastic beam is

$$\Pi=\tfrac{1}{2}\int_{\mathbf{r}}\sigma_{ij}\varepsilon_{ij}d\mathbf{r}+\int_{\mathbf{r}}T(\mathbf{r})w(\mathbf{r})d\mathbf{r}+\int_{\mathbf{r}}F(\mathbf{r})w(\mathbf{r})d\mathbf{r},$$

where $T(\mathbf{r})$ and $F(\mathbf{r})$ are the beam surface traction and force.

The term $\tfrac{1}{2}\sigma_{ij}\varepsilon_{ij}$ gives the strain energy stored. The equations of motion can be derived. For laterally distributed load $T(x)$, the equation for the beam bending is

$$a_b\frac{d^4 w}{dx^4}=T(x),\ a_b=EI.$$

One may solve

$$\xi EI(x)\frac{\partial^5 y(t,x)}{\partial x^4\partial t}+EI(x)\frac{\partial^4 y(t,x)}{\partial x^4}+m_0(x)\frac{\partial^2 y(t,x)}{\partial t^2}+m(x)\frac{d^2\varphi}{dt^2}=F(t,x)$$

or other pertained equations using the distributed force through the beam $F(t,x)$.

■

Illustrative Example 2.3. *Elastic Tensor*

The fourth-order elastic tensor $K_{\alpha\beta\gamma\delta}$ with 81 components is applied to express the force, elastic energy, stress, strain and other quantities of elastic materials and structures. Dynamic aeroelasticity, airframe and wing flexibility, elasticity and stiffness of structures are considered. The expression for elasticity energy Π in orthogonal coordinates is

$$\Pi=\tfrac{1}{2}K_{ijkl}\varepsilon_{ij}\varepsilon_{kl},$$

where K_{ijkl} is the fourth-rank elastic (stiffness) tensor; ε_{ij} and ε_{kl} are the strain tensors.

For a spring, plastics, carbon fiber reinforced polymers and other structural materials, for large deformations using a nonlinear stress–strain curve, the potential energy may be expressed as

$$\Pi(x)=\tfrac{1}{2}k_1x^2+\tfrac{1}{3}k_2\,\mathrm{sgn}(x)x^3+\tfrac{1}{4}k_3x^4.$$

From

$$\Pi(x)=\int F(x)dx,$$

one finds

$$F(x)=k_1x+k_2x|x|+k_3x^3.$$

■

2.2.3. Hamilton Equations of Motion

The Lagrangian function for the conservative systems is a difference between the total kinetic and potential energies, $L=(\Gamma-\Pi)$. Using the *generalized coordinates* q_i and momentum p_i, the total energy Hamiltonian function is expressed using the *canonical coordinates* (q_i,p_i). Hence,

$$H\left(t, q_1,..., q_n, \tfrac{dq_1}{dt},..., \tfrac{dq_n}{dt}\right)=\Gamma\left(t, q_1,..., q_n, \tfrac{dq_1}{dt},..., \tfrac{dq_n}{dt}\right)+\Pi\left(t, q_1,..., q_n\right)+D\left(t, \tfrac{dq_1}{dt},..., \tfrac{dq_n}{dt}\right), \quad (2.18)$$

$$H\left(t, q_1,..., q_n, p_1,..., p_n\right)=\Gamma\left(t, q_1,..., q_n, p_1,..., p_n\right)+\Pi\left(t, q_1,..., q_n\right)+D\left(t, p_1,..., p_n\right).$$

The Hamilton equations of motion for the *canonical coordinates* (q_i,p_i) are

$$\dot{p}_i=-\frac{\partial H}{\partial q_i},\ \dot{q}_i=\frac{\partial H}{\partial p_i}. \quad (2.19)$$

There are the canonical transformation from a $2n$-dimensional pair of *space coordinates* $(\boldsymbol{q},\boldsymbol{p})$ to $(\boldsymbol{Q},\boldsymbol{P})$, $(t,\boldsymbol{q},\boldsymbol{p})\to(t,\boldsymbol{Q},\boldsymbol{P})$. This yields an invariance as $H(t,\boldsymbol{q},\boldsymbol{p})\to K(t,\boldsymbol{Q},\boldsymbol{P})$. One obtains a system of $2n$ first-order differential equations. In contrast, the Lagrange equations of motion (2.16) yield a system of n second-order differential equations of the lumped-parameter models.

Illustrative Example 2.4. *Hamilton-Jacobi Equation*
As documented in section 2.5, the nonlinear differential equation for the Hamiltonian *principal function S* is

$$H\left(t,\boldsymbol{q},\frac{\partial S(t,\boldsymbol{q},\boldsymbol{p}_0)}{\partial \boldsymbol{q}}\right)=-\frac{\partial S(t,\boldsymbol{q},\boldsymbol{p}_0)}{\partial t},$$

where $\boldsymbol{p}_0$ is the initial momentum. ■

Example 2.15. *Harmonic Oscillator*
Consider the unperturbed harmonic oscillator. A frictionless sliding mass m is attached to the ideal spring. The total energy is $\text{E}=\Gamma+\Pi=\frac{1}{2}(mv^2+k_s x^2)$, $D=0$.

Using $q=x$, the Lagrangian is

$$L(x,\tfrac{dx}{dt})=\Gamma-\Pi=\tfrac{1}{2}(mv^2-k_s x^2)=\tfrac{1}{2}(m\dot{x}^2-k_s x^2).$$

The Lagrange equation (2.17) $\frac{d}{dt}\left(\frac{\partial\Gamma}{\partial\dot{q}}\right)-\frac{\partial\Gamma}{\partial q}+\frac{\partial\Pi}{\partial q}=Q$, $q=x$, m=const, yields the second-order differential equation

$$m\frac{d^2x}{dt^2}+k_s x=0,\ \text{or,}\ m\ddot{x}+k_s x=0.$$

The Newton second law (2.6) $ma=\Sigma F$ yields the second-order differential equation

$$m\frac{d^2x}{dt^2}=-k_s x\ \text{or}\ m\ddot{x}=-k_s x.$$

The total energy is

$$\text{E}=\Gamma+\Pi=\tfrac{1}{2}(mv^2+k_s x^2)=\tfrac{1}{2}(\tfrac{1}{m}p^2+k_s x^2).$$

The Hamiltonian function is

$$H(q,p)=\Gamma+\Pi=\tfrac{1}{2}(\tfrac{1}{m}p^2+k_s q^2).$$

From the Hamilton equations of motion (2.19), one obtains

$$\dot{p}=-\frac{\partial H}{\partial x}=-k_s x\,,\ \dot{q}=\dot{x}=\frac{\partial H}{\partial p}=\frac{p}{m}.$$

Recall that $p=mv$. There is an equivalence of the resulting equations of motion. One derives consistent governing equations.

■

2.3. Electromagnetics and Electromechanics

Electromagnetics and mechanics are foundations to devise new devices, analyze device physics, investigate phenomena exhibited, research control schemes, etc. Having covered topics on deriving lumped-parameter models, apply laws of electromagnetic and electromechanics. Electrostatic and electromagnetic actuators are investigated applying laws of electromagnetics.

Ohm's Law – The Ohm law for circuits is $V=rI$. The resistance of the conductor is related to the resistivity ρ and electric conductivity σ,

$$r=\frac{\rho l}{A}=\frac{l}{\sigma A},$$

where l and A are the length and cross-sectional area.

For silver, copper and aluminum at 20°C,

σ=6.21×10^7, σ=5.87×10^7 and σ=3.69×10^7 A/V-m.

The resistivity depends on temperature T, and,

$$\rho(T)=\rho_0[1+\alpha_{\rho 1}(T-T_0)+\alpha_{\rho 2}(T-T_0)^2+\ldots],$$

where $\alpha_{\rho i}$ are the coefficients.

For copper, for $T\leq$180°C which is the case for transducers,

$$\rho(T)=1.7\times 10^{-8}[1+0.0039(T_0-20)],\ T_0=20°\text{C}.$$

Laws of Electromagnetics and Mechanics – The Ohm law relates the current density $\vec{J}$ and electric field intensity $\vec{E}$. Using the conductivity σ and resistivity ρ rank-2 tensors (3×3 matrices), $\vec{J}=\sigma\vec{E}$, $\vec{E}=\rho\vec{J}$. The electrostatic and magnetostatic equations in linear isotropic media are given using vectors of the electric field intensity $\vec{E}$, electric flux density $\vec{D}$, magnetic field intensity $\vec{H}$, and magnetic flux density $\vec{B}$.

Consider the Cartesian coordinate system. The governing electrostatic equations are

$$\nabla\times\vec{E}(x,y,z)=0\,,\ \nabla\cdot\vec{D}(x,y,z)=\rho_v(x,y,z) \tag{2.20}$$

with the constitutive equations $\vec{D}=\varepsilon\vec{E}$.

The magnetostatic equations are

$$\nabla\times\vec{H}(x,y,z)=\vec{J}(x,y,z)\,,\ \nabla\cdot\vec{B}(x,y,z)=0 \tag{2.21}$$

with the constitutive equations $\vec{B}=\mu\vec{H}$.

For steady-state time-invariant fields, electric and magnetic field vectors form separate and independent pairs. That is, ($\vec{E}$, $\vec{D}$) are not related to ($\vec{H}$, $\vec{B}$), and vice versa. For time-varying magnetic fields, the changes of magnetic field affect the electric field, and vice versa. Maxwell's equations in the differential form for time-varying fields are

Faraday's law $\nabla \times \vec{E}(x,y,z,t) = -\frac{\partial \vec{B}(x,y,z,t)}{\partial t}$, (2.22)

Ampere's law $\nabla \times \vec{H}(x,y,z,t) = \vec{J}(x,y,z,t) + \frac{\partial \vec{D}(x,y,z,t)}{\partial t}$, $\vec{J} = \sigma \vec{E}$,

Gauss's law for electric field $\nabla \cdot \vec{D}(x,y,z,t) = \rho_v(x,y,z,t)$,

Gauss's law for magnetic field $\nabla \cdot \vec{B}(x,y,z,t) = 0$.

The constitutive equations are

$$\vec{D} = \varepsilon \vec{E}, \; \vec{D} = \varepsilon \vec{E} + \vec{P}, \tag{2.23}$$

$$\vec{B} = \mu \vec{H}, \; \vec{B} = \mu(\vec{H} + \vec{M}),$$

$$\vec{J} = \sigma \vec{E}, \; \vec{J} = \rho_v \vec{v},$$

where ε, μ and σ are the permittivity, permeability and conductivity tensors; ρ_v is the volume charge density.

Examining transducers, one derives the expressions for force F, torque T, *electromotive* and *magnetomotive* forces, etc. The Lorenz force relates the electromagnetic and mechanical variables

$$\vec{F} = \rho_v(\vec{E} + \vec{v} \times \vec{B}) = \rho_v \vec{E} + \vec{J} \times \vec{B}. \tag{2.24}$$

The total potential energy stored in the electrostatic field is defined using the potential difference V as

$$W_e = \tfrac{1}{2}\int_v \rho_v V dv. \tag{2.25}$$

Using the volume charge density $\rho_v = \vec{\nabla} \cdot \vec{D}$ and $\vec{E} = -\vec{\nabla} V$, the energy stored in the electrostatic field is

$$W_e = \tfrac{1}{2}\int_v \vec{D} \cdot \vec{E} dv. \tag{2.26}$$

The electrostatic volume energy density is $\tfrac{1}{2}\vec{D} \cdot \vec{E}$.

For a linear isotropic medium

$$W_e = \tfrac{1}{2}\int_v \varepsilon \left|\vec{E}\right|^2 dv = \tfrac{1}{2}\int_v \varepsilon^{-1} \left|\vec{D}\right|^2 dv. \tag{2.27}$$

From (2.27), the potential energy stored in the electric field between two surfaces, such as in capacitors, is

$$W_e = \tfrac{1}{2} QV = \tfrac{1}{2} CV^2. \tag{2.28}$$

Using the concept of virtual work, for the lossless conservative system, the differential change of the electrostatic energy dW_e is equal to the differential change of mechanical energy dW_{mec}. That is, $dW_e = dW_{\text{mec}}$.

For translational motion $dW_{\text{mec}} = \vec{F}_e \cdot d\vec{l}$, where $d\vec{l}$ is the differential displacement. From $dW_e = \vec{\nabla} W_e \cdot d\vec{l}$ one concludes that the force is the gradient of the stored electrostatic energy.

That is

$$\vec{F}_e = \vec{\nabla} W_e . \tag{2.29}$$

In the Cartesian coordinates

$$F_{ex} = \frac{\partial W_e}{\partial x},\ F_{ey} = \frac{\partial W_e}{\partial y},\ F_{ez} = \frac{\partial W_e}{\partial z}. \tag{2.30}$$

The stored energy in the magnetostatic field is

$$W_m = \tfrac{1}{2}\int_v \vec{B}\cdot\vec{H}dv,\ W_m = \tfrac{1}{2}\int_v \mu\left|\vec{H}\right|^2 dv = \tfrac{1}{2}\int_v \mu^{-1}\left|\vec{B}\right|^2 dv,\ \mu = \tfrac{dB}{dH}. \tag{2.31}$$

The magnetic energy, stored in the inductor with a single winding, is $W_m = \frac{1}{2}Li^2$. The energy stored in the magnetic field in multi-phase transducers is expressed using the inductance mapping $\boldsymbol{L}$, $W_m = \frac{1}{2}\boldsymbol{i}^T\boldsymbol{L}\boldsymbol{i}$, where $\boldsymbol{i}$ is the current vector $\boldsymbol{i}=[i_1,\ldots,i_n]^T$; T denotes the transpose symbol.

The force is the gradient of the stored magnetic energy. For translational motion

$$\vec{F}_m = \vec{\nabla} W_m,\ F_{mx} = \frac{\partial W_m}{\partial x},\ F_{my} = \frac{\partial W_m}{\partial y},\ F_{mz} = \frac{\partial W_m}{\partial z}. \tag{2.32}$$

For rotational motion, the torque is

$$\vec{T}_e = \vec{\nabla} W_m . \tag{2.33}$$

If the rigid body rotor rotates around the z-axis, $dW_{\text{mec}} = T_e d\theta$, where T_e is the z-component of the torque. For lossless system, the electromagnetic torque is $T_e = \frac{\partial W_m}{\partial \theta}$.

The *electromotive* and *magnetomotive* forces, *emf* and *mmf*, are

$$emf = \oint_l \vec{E}\cdot d\vec{l} = -\int_s \frac{\partial \vec{B}}{\partial t}\cdot d\vec{s},\ emf = \oint_l (\vec{v}\times\vec{B})\cdot d\vec{l}, \tag{2.34}$$

$$mmf = \oint_l \vec{H}\cdot d\vec{l} = \int_s \vec{J}\cdot d\vec{s} + \int_s \frac{\partial \vec{D}}{\partial t}\cdot d\vec{s}. \tag{2.35}$$

Equation (2.34) is the Faraday law of induction. The motional *emf* $\mathscr{E}$ is a function of the velocity and the magnetic flux density. The *emf*, induced in a stationary closed circuit, is equal to the negative rate of increase of the magnetic flux. The magnetomotive force (*mmf*) is the line integral of the time-varying magnetic field intensity $\vec{H}(t)$, and, $\oint_l \vec{H}\cdot d\vec{l} = Ni$. The unit for the *mmf* is amperes or ampere-turns. The induced *mmf* is the sum of the induced current and the rate of change of the flux penetrating the surface bounded by the contour.

According to Faraday's law of induction (2.34)

$$\mathscr{E} = \oint_s \vec{E}(t)\cdot d\vec{l} = -\int_s \frac{\partial \vec{B}(t)}{\partial t}\cdot d\vec{s} = -N\frac{d\Phi}{dt} = -\frac{d\psi}{dt}. \tag{2.36}$$

The total electric flux through a closed surface is given by the Gauss law

$$\Phi = \oint_s \vec{D} \cdot d\vec{s} = \oint_v \rho_v dv, \tag{2.37}$$

where $d\vec{s}$ is the vector surface area, $d\vec{s} = ds\vec{a}_n$; $\vec{a}_n$ is the unit vector, normal to the surface.

The magnetic flux through a closed surface is

$$\Phi = \oint_s \vec{B} \cdot d\vec{s}. \tag{2.38}$$

The current flows in an opposite direction to the flux linkages ψ. The *emf* is a potential difference V in circuits carrying a current. The Kirchhoff voltage law states that around a closed path in an electric circuit, the algebraic sum of the *emf* is equal to the algebraic sum of the voltage drop across the resistances. The algebraic sum of voltages around any closed path is zero. One has

$$u = -ir + \mathscr{E} = -ir - \frac{d\psi}{dt}, \mathscr{E} = -\frac{d\psi}{dt} = -\frac{d(Li)}{dt} = -L\frac{di}{dt} - i\frac{dL}{dt}. \tag{2.39}$$

The Kirchhoff current law states that the algebraic sum of currents at any node is zero.

The inductance L is the ratio of the total flux linkages to the current which they link, while, the reluctance $\Re$ is the ratio of the *mmf* to the total flux. That is

$$L = \frac{N\Phi}{i} = \frac{\int_s \vec{B} \cdot d\vec{s}}{\oint_l \vec{H} \cdot d\vec{l}}, \ \Re = \frac{mmf}{\Phi} = \frac{\oint_l \vec{H} \cdot d\vec{l}}{\int_s \vec{B} \cdot d\vec{s}}. \tag{2.40}$$

The self-inductance is the magnitude of the self-induced *emf* per unit rate of change of current. The inductance variation is used to sense and measure displacement, motions, magnetic fields, frequency, etc. Magnetic pickups and humbuckers use a coil wound around a ferromagnetic core. The inductance varies as a function of frequency and amplitude of relative motions.

The power and torque densities, as well as force- and torque-energy relations in electromagnetic and electrostatic transducers are important. The energy stored in the capacitor is $\frac{1}{2}CV^2$, energy stored in the inductor is $\frac{1}{2}Li^2$, while energy stored in the magnetic field is $\frac{1}{2}\int_v \vec{B} \cdot \vec{H} dv$. The energy in the capacitor is stored in the electric field between plates. The energy in the inductor is stored in the magnetic field. In the variable reluctance electromagnetic transducers (relays, solenoids, magnetic levitation, electromagnets, synchronous motors, rail guns and others), the reluctance and magnetizing inductance vary. One finds the electromagnetic force and torque using the co-energy. In electromagnetic motion devices, the magnetic coupling is between windings that are carrying currents, and, the stationary magnetic field is developed by permanent magnets or electromagnets. The electromagnetic force or torque are

$$\vec{F}_e = -i\oint_l \vec{B}(t) \cdot d\vec{l}, \ \vec{T} = \vec{m} \times \vec{B}. \tag{2.41}$$

Using the spatiotemporal co-energy W_c, one finds the electromagnetic force or torque in translational and rotational devices

$$F_e = -\nabla W_c, \ F_e(i,x) = -\frac{\partial}{\partial x} W_c\left(i, L(x)\right), \tag{2.42}$$

$$T_e = -\nabla W_c, \ T_e(i,x) = -\frac{\partial}{\partial \theta} W_c\left(i, L(\theta)\right).$$

In chapter 3, electrostatic and electromagnetic actuators, generators and transducers are considered in various applications. For example, in all-electric aircraft and electromagnetic control surfaces, high-power and high-torque densities electric machines and actuators are used. Actuators, energy systems and power management systems should ensure seamless functionality. Electrical power systems accomplish generation, distribution, load management and protection. Usually, aircraft's avionics, electronic, radars and auxiliary systems operate at 28 V and 270 V direct current, as well as 115 V alternating current. Two integrated drive generators (120 kVA, 115 V ac, 400 Hz) are deployed in Boeing 777 and Eurofighter Typhoon. Two variable-speed constant-frequency generators (40/45 kVA and 60/65 kVA, 115 V ac, 400 Hz) are on F-18. To control aircraft flight, actuators displace primary aerodynamic control surfaces (ailerons, elevator, flaps, horizontal stabilizers, rudder, slats, spoilers, tails and others), steer gears and nozzle, regulate fuel-injection control valves, etc.

2.4. Rigid Body Dynamics: Aircraft Equations of Motion

To approach analysis, design and optimization problems, this section reports rigid body equations of motion applying conventional notations, avoiding details documented in [10-19]. The translational velocities in body frame (u,v,w), angular rates (p,q,r), Euler angles (ϕ,θ,ψ), translational positions (x,y,z) and other variables are considered.

As illustrated in Figure 2.6, (u,v,w) are the components of the vehicle velocity in the (x,y,z) body axes, e.g., in a body coordinate system. The Euler angles (ϕ,θ,ψ) are the roll, pitch and yaw angles. Studying rigid body aircraft dynamics, one considers coupled longitudinal and lateral dynamics. The aircraft velocity V, angle of attack α and sideslip angle β are expressed using the axial, lateral and normal velocities (u,v,w)

$$\begin{bmatrix} V \\ \alpha \\ \beta \end{bmatrix} = \begin{bmatrix} \sqrt{u^2+v^2+w^2} \\ \tan^{-1}\frac{w}{u} \\ \sin^{-1}\frac{v}{V} \end{bmatrix}, \quad \begin{bmatrix} u \\ v \\ w \end{bmatrix} = \begin{bmatrix} V\cos\alpha\cos\beta \\ V\sin\beta \\ V\sin\alpha\cos\beta \end{bmatrix}, \tag{2.43}$$

$$\begin{bmatrix} \dot{V} \\ \dot{\alpha} \\ \dot{\beta} \end{bmatrix} = \begin{bmatrix} \frac{d}{dt}\sqrt{u^2+v^2+w^2} \\ \frac{d}{dt}\tan^{-1}\frac{w}{u} \\ \frac{d}{dt}\sin^{-1}\frac{v}{V} \end{bmatrix} = \begin{bmatrix} \dot{u}\cos\alpha\cos\beta+\dot{v}\sin\beta+\dot{w}\sin\alpha\cos\beta \\ \frac{1}{V\cos\beta}(-\dot{u}\sin\alpha+\dot{w}\cos\alpha) \\ \frac{1}{V}(-\dot{u}\cos\alpha\sin\beta+\dot{v}\cos\beta-\dot{w}\sin\alpha\sin\beta) \end{bmatrix}.$$

Depending on flight regimes and missions, different states, outputs and variables are used. For example, using the Euler angles and (α,β), the flight path and heading angles (γ,ξ) are $\gamma=(\theta-\alpha)$ and $\xi|_{\phi=0}=(\psi+\beta)$ for a zero-roll angle. For straight and level flight, the aircraft has a non-zero angle of attack α to generate the lift such that the total lift is equal to the total weight. One yields non-zero pitch angle flight with the pertained deflection of aerodynamic control surfaces to generate the lift.

Consider the acting *net* axial, side and normal aerodynamic forces (F_X, F_Y, F_Z), and, *net* rolling, pitching and yawing aerodynamic moments (L, M, N). The aerodynamic forces, moments and acting forces depend on velocity, attitude, ambient density of air at the altitude of flight, etc. The angular rotation vector of the body about the center of mass is $\boldsymbol{\omega}=[p, q, r]$. Figures 2.6 depict the quantities considered.

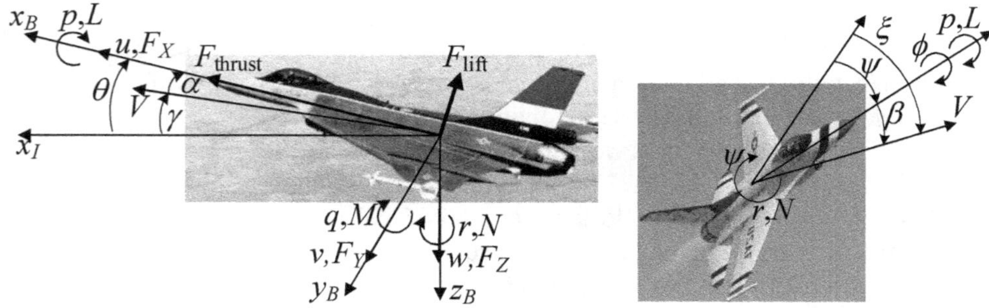

Figure 2.6. Airplane is controlled by deflecting aerodynamic control surfaces, regulating the engine thrust, and, implementing thrust vector control. Aerodynamic forces (F_X, F_Y, F_Z) and moments (L, M, N) depend on engine thrust, deflection of aerodynamic control surfaces δ_i, velocity V, angle of attack α, sideslip angle β, Euler angles (ϕ, θ, ψ), and other quantities. The body coordinate system is vehicle-carried, and, defined on the body of the vehicle. For axes, located at the center of gravity, the X-axis points forward, Y-axis is the starboard right side, and, Z-axis points downward. Coordinate vectors expressed in the body frame are appended with a subscript B. In the body reference frame, the vehicle-carried NED velocity and acceleration are $\boldsymbol{v}_B=[u,v,w]$ and $\boldsymbol{a}_B=[a_X, a_Y, a_Z]$.

Using six-degree-of-freedom longitudinal and lateral variables in the body-axis system, consider a rigid-body dynamics. For different airframes, the coupled translational and rotational dynamics are derived [10-19]. Using the *net* forces $\Sigma\mathbf{F}$, moments $\Sigma\mathbf{M}$ and angular momentums $\Sigma\boldsymbol{L}_M$, Newtonian dynamics in the body-fixed frame is given by the force and moment equations

$$\sum\mathbf{F}=\frac{d}{dt}(m\mathbf{v})+m(\boldsymbol{\omega}\times\mathbf{v}),\ \mathbf{v}=\begin{bmatrix}u\\v\\w\end{bmatrix},\ \sum\mathbf{F}=\begin{bmatrix}F_X\\F_Y\\F_Z\end{bmatrix}, \tag{2.44}$$

$$\sum\mathbf{M}=\frac{d\boldsymbol{L}_M}{dt}+\boldsymbol{\omega}\times\boldsymbol{L}_M=\frac{d}{dt}(\mathbf{I}\boldsymbol{\omega})+(\boldsymbol{\omega}\times\mathbf{I})\boldsymbol{\omega},\ \boldsymbol{\omega}=\begin{bmatrix}p\\q\\r\end{bmatrix},\ \sum\mathbf{M}=\begin{bmatrix}L\\M\\N\end{bmatrix},$$

where $\boldsymbol{v}$ and $\boldsymbol{\omega}$ are the linear and angular velocity vectors, $\mathbf{v}=[u, v, w]$ and $\boldsymbol{\omega}=[p, q, r]$; $\mathbf{F}$ and $\mathbf{M}$ are the aerodynamic forces and moments vectors, and, the control forces (aerodynamic, engine thrust and others) and moments are bounded, $F_{i\,\min}\le F_i\le F_{i\,\max}$ and $M_{i\,\min}\le M_i\le M_{i\,\max}$, $\mathbf{F}=[F_X, F_Y, F_Z]$ and $\mathbf{M}=[L, M, N]$; L, M and N are the rolling, pitching and yawing net moments in the positive p-, q- and r-directions; m is the mass; $\boldsymbol{L}_M$ is the angular momentum vector; $\mathbf{I}$ is the inertia matrix, $\mathbf{I}=\begin{bmatrix}I_{xx} & -I_{xy} & -I_{xz}\\ -I_{yx} & I_{yy} & -I_{yz}\\ -I_{zx} & -I_{zy} & I_{zz}\end{bmatrix}$.

From (2.44), assuming that m and $\mathbf{I}$ are time-invariant, one finds the governing equations of motion

$$\dot{\mathbf{v}} = -\boldsymbol{\omega}\times\mathbf{v} + \tfrac{1}{m}\sum\mathbf{F},\ \ \mathbf{v} = \begin{bmatrix} u \\ v \\ w \end{bmatrix}, \tag{2.45}$$

$$\dot{\boldsymbol{\omega}} = \mathbf{I}^{-1}\left(\sum\mathbf{M} - (\boldsymbol{\omega}\times\mathbf{I})\boldsymbol{\omega}\right),\ \ \boldsymbol{\omega} = \begin{bmatrix} p \\ q \\ r \end{bmatrix}, \boldsymbol{\omega}_B \equiv \boldsymbol{\omega},$$

$$\dot{\boldsymbol{\Theta}} = \begin{bmatrix} 1 & \sin\phi\tan\theta & \cos\phi\tan\theta \\ 0 & \cos\phi & -\sin\phi \\ 0 & \sin\phi\sec\theta & \cos\phi\sec\theta \end{bmatrix}\boldsymbol{\omega},\ \ \boldsymbol{\Theta} = \begin{bmatrix} \phi \\ \theta \\ \psi \end{bmatrix},$$

where $\dot{\boldsymbol{\Theta}} = \begin{bmatrix} \dot\phi \\ \dot\theta \\ \dot\psi \end{bmatrix} = \begin{bmatrix} 1 & \sin\phi\tan\theta & \cos\phi\tan\theta \\ 0 & \cos\phi & -\sin\phi \\ 0 & \sin\phi\sec\theta & \cos\phi\sec\theta \end{bmatrix}\begin{bmatrix} p \\ q \\ r \end{bmatrix} = \mathcal{L}_B^I\boldsymbol{\omega}_B$ is the Euler angle dynamics; $\boldsymbol{\omega}_B$ is the body-axis angular rate vector, $\boldsymbol{\omega}_B = \begin{bmatrix} \omega_x \\ \omega_y \\ \omega_z \end{bmatrix}_B = \begin{bmatrix} p \\ q \\ r \end{bmatrix} = \begin{bmatrix} 1 & 0 & -\sin\theta \\ 0 & \cos\phi & \sin\phi\cos\theta \\ 0 & -\sin\phi & \cos\phi\cos\theta \end{bmatrix}\begin{bmatrix} \dot\phi \\ \dot\theta \\ \dot\psi \end{bmatrix} = \mathcal{L}_I^B\dot{\boldsymbol{\Theta}}$.

Using (2.44) and (2.45) the force and moment equations are

$$\begin{bmatrix} F_x \\ F_y \\ F_z \end{bmatrix} = m\begin{bmatrix} \dot u \\ \dot v \\ \dot w \end{bmatrix} + m\det\begin{bmatrix} \mathbf{i} & \mathbf{j} & \mathbf{k} \\ p & q & r \\ u & v & w \end{bmatrix} = m\begin{bmatrix} \dot u + qw - rv \\ \dot v + ru - pw \\ \dot w + pv - qu \end{bmatrix}, \tag{2.46}$$

$$\begin{bmatrix} L \\ M \\ N \end{bmatrix} = \begin{bmatrix} I_{xx} & -I_{xy} & -I_{xz} \\ -I_{yx} & I_{yy} & -I_{yz} \\ -I_{zx} & -I_{zy} & I_{zz} \end{bmatrix}\begin{bmatrix} \dot p \\ \dot q \\ \dot r \end{bmatrix} + \begin{bmatrix} p \\ q \\ r \end{bmatrix}\times\begin{bmatrix} I_{xx} & -I_{xy} & -I_{xz} \\ -I_{yx} & I_{yy} & -I_{yz} \\ -I_{zx} & -I_{zy} & I_{zz} \end{bmatrix}\begin{bmatrix} p \\ q \\ r \end{bmatrix}$$

$$= \begin{bmatrix} I_{xx}\dot p - I_{xy}\dot q - I_{xz}\dot r + (-I_{xz}p - I_{yz}q + I_{zz}r)q + (I_{xy}p - I_{yy}q + I_{yz}r)r \\ -I_{xy}\dot p + I_{yy}\dot q - I_{yz}\dot r + (I_{xz}p + I_{yz}q - I_{zz}r)p + (I_{xx}p - I_{xy}q - I_{xz}r)r \\ -I_{xz}\dot p - I_{yz}\dot q + I_{zz}\dot r + (-I_{xy}p + I_{yy}q - I_{yz}r)p + (-I_{xx}p + I_{xy}q + I_{xz}r)q \end{bmatrix}.$$

Assuming the left-right symmetry, one has $I_{xy}=I_{yz}=0$, and, $I_{xz}\neq 0$.

Hence, $\begin{bmatrix} L \\ M \\ N \end{bmatrix} = \begin{bmatrix} I_{xx}\dot p - I_{xz}\dot r - I_{xz}pq + (I_{zz} - I_{yy})qr \\ I_{yy}\dot q + I_{xz}p^2 + (I_{xx} - I_{zz})pr - I_{xz}r^2 \\ -I_{xz}\dot p + I_{zz}\dot r + (I_{yy} - I_{xx})pq + I_{xz}qr \end{bmatrix}$.

The gravitational forces are given as

$$\mathbf{F}_g = \begin{bmatrix} F_{Xg} \\ F_{Yg} \\ F_{Zg} \end{bmatrix} = m\mathfrak{R}^T\mathbf{g},\ \ \mathfrak{R} = \begin{bmatrix} 1 & 0 & -\sin\theta \\ 0 & \cos\phi & \sin\phi\cos\theta \\ 0 & -\sin\phi & \cos\phi\cos\theta \end{bmatrix}, \tag{2.47}$$

$$F_{Xg} = -mg\sin\theta,\ F_{Yg} = mg\sin\phi\cos\theta,\ F_{Zg} = mg\cos\phi\cos\theta,$$

where $\mathbf{g}=[0\ \ 0\ \ g]^T$ is the gravity acceleration; $\mathfrak{R}$ is the rotational matrix.

Assume the left-right symmetry, for which $I_{xy}=I_{yz}=0$ and $I_{xz}\neq 0$.
The resulting equations of motion are

$$\dot{u}=rv-qw-g\sin\theta+\tfrac{1}{m}F_X, \tag{2.48}$$
$$\dot{v}=-ru+pw+g\sin\phi\cos\theta+\tfrac{1}{m}F_Y,$$
$$\dot{w}=qu-pv+g\cos\phi\cos\theta+\tfrac{1}{m}F_Z,$$
$$\dot{p}=\tfrac{1}{I_{xx}I_{zz}-I_{xz}^2}\left\{\left[(I_{xx}-I_{yy}+I_{zz})I_{xz}p-(I_{xz}^2-I_{yy}I_{zz}+I_{zz}^2)r\right]q+I_{zz}L+I_{xz}N\right\},$$
$$\dot{q}=\tfrac{1}{I_{yy}}\left[(-I_{xx}+I_{zz})pr-I_{xz}(p^2-r^2)+M\right],$$
$$\dot{r}=\tfrac{1}{I_{xx}I_{zz}-I_{xz}^2}\left\{\left[(I_{xx}-I_{yy}+I_{zz})I_{xz}r-(I_{xx}^2-I_{xx}I_{yy}+I_{xz}^2)p\right]q+I_{xz}L+I_{xx}N\right\},$$
$$\dot{\phi}=p+(q\sin\phi+r\cos\phi)\tan\theta,$$
$$\dot{\theta}=q\cos\phi-r\sin\phi,$$
$$\dot{\psi}=(q\sin\phi+r\cos\phi)\sec\theta,$$
$$\dot{x}=u\cos\theta\cos\psi+v(\sin\phi\sin\theta\cos\psi-\cos\phi\sin\psi)+w(\cos\phi\sin\theta\cos\psi+\sin\phi\sin\psi),$$
$$\dot{y}=u\cos\theta\sin\psi+v(\sin\phi\sin\theta\sin\psi+\cos\phi\cos\psi)+w(\cos\phi\sin\theta\sin\psi-\sin\phi\cos\psi),$$
$$\dot{z}=-u\sin\theta+v\sin\phi\cos\theta+w\cos\phi\cos\theta.$$

Considering the velocities in the body-fixed frame $\mathbf{v}_B$ and in inertial frame $\mathbf{v}_I$, apply the transformations and rotation matrices to find the

Inertial Frame⇒Body-Fixed Frame, and, Body-Fixed Frame⇒Inertial Frame

transformations, as well as differential equations for (x_I, y_I, z_I).

We have

$$\mathbf{v}_I=\mathcal{R}_3(\psi)\mathcal{R}_2(\theta)\mathcal{R}_1(\phi)\mathbf{v}_B=\mathcal{R}_B^I(\phi,\theta,\psi)\mathbf{v}_B, \tag{2.49}$$

$$\mathbf{v}_I=\underbrace{\begin{bmatrix}\cos\psi & -\sin\psi & 0\\ \sin\psi & \cos\psi & 0\\ 0 & 0 & 1\end{bmatrix}}_{\text{Yaw Rotation }\psi,\ \mathcal{R}_3(\psi)}\underbrace{\begin{bmatrix}\cos\theta & 0 & \sin\theta\\ 0 & 1 & 0\\ -\sin\theta & 0 & \cos\theta\end{bmatrix}}_{\text{Pitch Rotation }\theta,\ \mathcal{R}_2(\theta)}\underbrace{\begin{bmatrix}1 & 0 & 0\\ 0 & \cos\phi & -\sin\phi\\ 0 & \sin\phi & \cos\phi\end{bmatrix}}_{\text{Roll Rotation }\phi,\ \mathcal{R}_1(\phi)}\mathbf{v}_B$$

$$=\underbrace{\begin{bmatrix}\cos\theta\cos\psi & \sin\phi\sin\theta\cos\psi-\cos\phi\sin\psi & \cos\phi\sin\theta\cos\psi+\sin\phi\sin\psi\\ \cos\theta\sin\psi & \sin\phi\sin\theta\sin\psi+\cos\phi\cos\psi & \cos\phi\sin\theta\sin\psi-\sin\phi\cos\psi\\ -\sin\theta & \sin\phi\cos\theta & \cos\phi\cos\theta\end{bmatrix}}_{\mathcal{R}_B^I(\phi,\theta,\psi)}\mathbf{v}_B,$$

$$\mathbf{v}_B=\mathcal{R}_1(\phi)\mathcal{R}_2(\theta)\mathcal{R}_3(\psi)\mathbf{v}_I=\mathcal{R}_I^B(\phi,\theta,\psi)\mathbf{v}_I,$$

$$\mathbf{v}_B=\underbrace{\begin{bmatrix}\cos\theta\cos\psi & \cos\theta\sin\psi & -\sin\theta\\ \sin\phi\sin\theta\cos\psi-\cos\phi\sin\psi & \sin\phi\sin\theta\sin\psi+\cos\phi\cos\psi & \sin\phi\cos\theta\\ \cos\phi\sin\theta\cos\psi+\sin\phi\sin\psi & \cos\phi\sin\theta\sin\psi-\sin\phi\cos\psi & \cos\phi\cos\theta\end{bmatrix}}_{\mathcal{R}_I^B(\phi,\theta,\psi)}\mathbf{v}_I,$$

where $\mathcal{R}_B^I(\phi,\theta,\psi)$ is the rotation transformation matrix from body frame to inertial frame; $\mathcal{R}_I^B(\phi,\theta,\psi)$ is the rotation transformation matrix from inertial frame to body frame, $\mathcal{R}_I^B\mathcal{R}_B^I=I$.

Example 2.16. *Angular Momentum and Inertial Matrix*

The angular momentum of a rigid body is $\boldsymbol{L}_M$=**I**$\boldsymbol{\omega}$, $\boldsymbol{\omega}=\begin{bmatrix} p \\ q \\ r \end{bmatrix}$.

For a homogeneous sphere, mass m and radius R, $\mathbf{I}=\frac{2}{5}mR^2\begin{bmatrix} 1 & 0 & 0 \\ 0 & 1 & 0 \\ 0 & 0 & 1 \end{bmatrix}$.

For a ring, one has, $\mathbf{I}=2mR^2\begin{bmatrix} 1 & 0 & 0 \\ 0 & 1 & 0 \\ 0 & 0 & 1 \end{bmatrix}$.

For a homogeneous disk of mass m, radius R and high h, we have

$$\mathbf{I}=\frac{1}{4}mR^2\begin{bmatrix} 1+\frac{h}{3R^2} & 0 & 0 \\ 0 & 1+\frac{h}{3R^2} & 0 \\ 0 & 0 & \frac{1}{2} \end{bmatrix}.$$

In aerial vehicles mass and symmetry vary during flight, engagement scenarios and missions. Hence, $I_{ij}\neq 0$, and, I_{ii} and I_{ij} are time varying. While for symmetric aerial vehicle, I_{xy}=0 and I_{yz}=0, in general, for an asymmetric aircraft, $\mathbf{I}=\begin{bmatrix} I_{xx} & -I_{xy} & -I_{xz} \\ -I_{yx} & I_{yy} & -I_{yz} \\ -I_{zx} & -I_{zy} & I_{zz} \end{bmatrix}$. ■

Illustrative Example 2.5. *Coordinate Systems*

Section 5.12 reports the coordinate systems used in aerial, ground and underwater vehicles, robots, motion platforms, inertial navigation systems, etc. Consider an east fired projectile at the equator. The projectile velocity is V, and, distance to target is d. Ignoring gravity and aerodynamic drag, consider the north-east-up coordinate system. The earth rotates about its axis at angular velocity 2π rad/day, e.g. 7.2722×10^{-5} rad/s. The rotation is positive within the local North axis.

One has $\boldsymbol{\omega}=\begin{bmatrix} p \\ q \\ r \end{bmatrix}=\begin{bmatrix} 7.2722\text{x}10^{-5} \\ 0 \\ 0 \end{bmatrix}$.

From (2.46)

$$\begin{bmatrix} F_x \\ F_y \\ F_z \end{bmatrix}=m\begin{bmatrix} \dot{u}+qw-rv \\ \dot{v}+ru-pw \\ \dot{w}+pv-qu \end{bmatrix}, \begin{bmatrix} 0 \\ 0 \\ 0 \end{bmatrix}=m\begin{bmatrix} \dot{u}+qw-rv \\ \dot{v}+ru-pw \\ \dot{w}+pv-qu \end{bmatrix}=m\begin{bmatrix} \dot{u} \\ \dot{v}-pw \\ \dot{w}+pv \end{bmatrix}, \begin{bmatrix} \dot{u} \\ \dot{v} \\ \dot{w} \end{bmatrix}=\begin{bmatrix} 0 \\ pw \\ -pv \end{bmatrix}.$$

For initial condition $u(0)$=0, $v(0)$=V and $w(0)$=0, one has
$u(t)$=0, $v(t)$=$v(0)\cos(pt)$=$V\cos(pt)$ and $w(t)= -v(0)\sin(pt)= -V\sin(pt)$.
For a small p, $u(t)$=0, $v(t)$=$v(0)$=V, $w(t)= -v(0)pt= -Vpt$.
These yields the displacements for a miss $N(t)$=0, $E(t)$=Vt and $U(t)=-\frac{1}{2}Vpt^2$.
For V=1000 m/s and distance to target d=1000 m, t=1 sec, $E(t)$=Vt=1000 m.
The projectile will miss as $N(t)$=0 m and $U(t)$= –0.07272 m.
For V=1000 m/s and d=2000 m, one has t=2 sec.
Hence, $N(t)$=0 m and $U(t)$= –0.2909 m.
Firing west, for V=1000 m/s and d=2000 m, $N(t)$=0 m and $U(t)$=0.2909 m.

■

Mathematical Models in State-Space Form – The equations of motion (2.48) are given using states

$x=[u\ v\ w\ x\ y\ z\ p\ q\ r\ \phi\ \theta\ \psi]^T \in \mathbb{R}^{12}$.

One applies states and outputs depending on flights phases, regimes, maneuvers, mission, aerial engagement scenarios, etc. Different states are used in analysis, control and optimization. The outputs could be

$y=[x\ \ y\ \ z\ \ h]^T \in \mathbb{R}^4$, $y=[\phi\ \ \theta\ \ \psi]^T \in \mathbb{R}^3$, $y=[V \alpha\ \beta\ h]^T \in \mathbb{R}^4$, $y=[\text{Longitude Latitude}\ \ h]^T \in \mathbb{R}^3$.

The altitude is described by

$\dot{h} = u\sin\theta - v\sin\phi\cos\theta - w\cos\phi\cos\theta$.

The vehicle velocity V, angle of attack α and sideslip angle β are given by (2.43). Differential equations (2.48) can be expanded using the variables of interest for studied MIMO aerial systems. Nonlinear models can be linearized as reported in section 5.11.2 and chapter 6. In some airframes and control schemes, the longitudinal and lateral aerodynamics may be decoupled.

For the longitudinal model, using (V, α, θ),

$\dot{h} = V(\cos\alpha\sin\theta - \sin\alpha\cos\theta)$.

In the state-space form we obtain

$$\dot{x} = \begin{bmatrix} \dot{\mathbf{v}} \\ \dot{\boldsymbol{\omega}} \\ \dot{\boldsymbol{\Theta}} \end{bmatrix} = \begin{bmatrix} \dot{u} \\ \dot{v} \\ \dot{w} \\ \dot{p} \\ \dot{q} \\ \dot{r} \\ \dot{\phi} \\ \dot{\theta} \\ \dot{\psi} \end{bmatrix} = F(\mathbf{v},\boldsymbol{\omega},\boldsymbol{\Theta}) + B(\mathbf{v},\boldsymbol{\omega},\boldsymbol{\Theta}) \begin{bmatrix} \mathbf{F}_{XYZ} \\ \boldsymbol{M}_{pqr} \end{bmatrix} = F(x) + B(x)u. \tag{2.50}$$

The inertial, velocity, body and other frames are considered. As reported in (2.49), frame transformations are applied. For example, the governing equations for translational position and velocities in inertial and body frames are found using

$$\dot{\mathbf{r}}_I(t) = \begin{bmatrix} \dot{x}_I(t) \\ \dot{y}_I(t) \\ \dot{z}_I(t) \end{bmatrix},\ \dot{\mathbf{r}}_I(t) = \mathbf{v}_I,\ \mathbf{v}_I = \mathcal{R}_B^I \mathbf{v}_B,\ \mathbf{v}_B = \mathcal{R}_I^B \mathbf{v}_I,\ \mathbf{v}_B = \begin{bmatrix} u \\ v \\ w \end{bmatrix},\ \mathbf{v}_I = \begin{bmatrix} v_x \\ v_y \\ v_z \end{bmatrix}. \tag{2.51}$$

For an aircraft, considering the drag and side forces (D, S), thrust force by the engine T, as well as the aerodynamic forces and moments due to deflections of the horizontal stabilizers δ_s, flaperons (combined leading-edge and trailing edge flaps, and, ailerons) δ_f, canard δ_c and rudder δ_r.

The longitudinal-lateral aerodynamic model, assuming the left-right symmetry, is found using states $x=[V, \alpha, q, \theta, \beta, p, r, \phi, \psi]$ as

$$\dot{x}(t) = F(x) + B(x)u, \tag{2.52}$$

$$\begin{bmatrix} \dot{V}(t) \\ \dot{\alpha}(t) \\ \dot{q}(t) \\ \dot{\theta}(t) \\ \dot{\beta}(t) \\ \dot{p}(t) \\ \dot{r}(t) \\ \dot{\phi}(t) \\ \dot{\psi}(t) \end{bmatrix} = \begin{bmatrix} \frac{1}{m}[-D\cos\beta+S\sin\beta-mg(\sin\alpha\cos\theta\cos\beta\cos\phi+\cos\theta\sin\beta\sin\phi-\cos\alpha\sin\theta\cos\beta)+T\cos\alpha\cos\beta] \\ q-(p\cos\alpha+r\sin\alpha)\tan\beta+\frac{g}{V\cos\beta}(\cos\alpha\cos\theta\cos\phi+\sin\alpha\sin\theta) \\ \frac{1}{I_{yy}}\left[(-I_{xx}+I_{zz})pr-I_{xz}(p^2-r^2)\right] \\ q\cos\phi-r\sin\phi \\ p\sin\alpha-r\cos\alpha+\frac{g}{V}(\cos\theta\cos\beta\sin\phi+\cos\alpha\sin\theta\sin\beta-\sin\alpha\cos\theta\sin\beta\cos\phi) \\ \frac{1}{I_{xx}I_{zz}-I_{xz}^2}\left[I_{xz}(I_{xx}-I_{yy}+I_{zz})qp-(I_{xz}^2-I_{yy}I_{zz}+I_{zz}^2)qr\right] \\ \frac{1}{I_{xx}I_{zz}-I_{xz}^2}\left[-(I_{xx}^2-I_{xx}I_{yy}+I_{xz}^2)qp+I_{xz}(I_{xx}-I_{yy}+I_{zz})qr\right] \\ p+q\tan\theta\sin\phi+r\tan\theta\cos\phi \\ q\sec\theta\sin\phi+r\sec\theta\cos\phi \end{bmatrix} + B(x)\begin{bmatrix} \delta_s \\ \delta_f \\ \delta_c \\ \delta_r \end{bmatrix}.$$

The control vector is $u=[T, \delta_s, \delta_f, \delta_c, \delta_r]$. In the first equation (2.52), the controlled thrust T is included in the $F(x)$ mapping assuming that the thrust is constant.

Lift and Drag Forces – Recall that $\begin{bmatrix} V \\ \beta \\ \alpha \end{bmatrix} = \begin{bmatrix} \sqrt{u^2+v^2+w^2} \\ \sin^{-1}\frac{v}{V} \\ \tan^{-1}\frac{w}{u} \end{bmatrix}$.

The lift and drag force are estimated as [12, 17]

$$F_{\text{lift}} = \tfrac{1}{2}\rho c_L(\alpha, V)AV^2, \; F_{\text{drag}} = \tfrac{1}{2}\rho c_D(\alpha, V)AV^2,$$

where $c_L(\cdot)$ and $c_D(\cdot)$ are the lift and drag coefficient, which are nonlinear functions of the angle-of-attack, forward velocity V and other variables; ρ is the air density; A is the equivalent area.

The side force S due to fuselage, wing, control surfaces, engine nacelles and other components may not have a significant effect.

Model Fidelity – Researching aerodynamic forces and moments to control aircraft, supersonic and hypersonic aerodynamics, aeroelasticity, unsteady aerodynamics and other phenomena are considered. Engine, actuators, power system, power electronics and other components are key systems. To change aerodynamic forces and moments $\begin{bmatrix} \mathbf{F} \\ \mathbf{M} \end{bmatrix}$, aerodynamic control surfaces are steered, thrust and thrust-vectoring forces are controlled, etc.

Steering of control surfaces is accomplished by rotational or translational servos, regulated by the PWM controllers-drivers which change the voltage applied to actuators. To analyze electromagnetic servos, nonlinear electromagnetics, analytical mechanics and fluid mechanics are applied. While electromagnetic and hydraulic servos are controlled by regulating the applied voltages and pressure, the torque and force control problems are commonly solved using the derived equations (2.44)-(2.52).

Dominant governing dynamics is considered within constitutive relationships, while fast dynamics is neglected. Various concepts and solutions have being applied and verified for different airframes and flight vehicles. Analysis and design problems for aerial systems are reported in illustrative examples, sections 5.11 and 6.8.2, Practice Problems 6.14.6 and 6.14.7, etc.

2.5. Hamiltonian, Dynamic Optimization and Hamilton-Jacobi Equation

The Lagrangian and Hamiltonian functions were defined in section 2.2. A positive definite total energy $\mathbb{E}(t,x)$ yields a real-valued continuous scalar Hamiltonian function $H(t,\boldsymbol{q},\boldsymbol{p})$ (2.18). For dynamic systems

$$H(t,x)\equiv\mathbb{E}(t,x),\ \mathbb{E}=(\Gamma+\Pi+D),\ H>0,\ \mathbb{E}>0,\ \forall(t,x)\in T\times X. \tag{2.53}$$

The governing equations (2.4) and (2.19) are

$$\dot{x}(t)=F(t,x)+B(t,x)u,\ \ x=\begin{bmatrix}\boldsymbol{q}\\ \boldsymbol{p}\end{bmatrix},\ \begin{cases}\dot{q}_i=\dfrac{\partial H}{\partial p_i}\\ \dot{p}_i=-\dfrac{\partial H}{\partial q_i}\end{cases},\ t\geq 0, \tag{2.54}$$

where the control function is $u=\varphi(t,x)$, $\varphi(\cdot):\mathbb{R}_{\geq 0}\times\mathbb{R}^{n}\to\mathbb{R}^{m}$.

The state vector x is found using the *generalized coordinates* $\boldsymbol{q}$ and momentum $\boldsymbol{p}$, and, $x=\begin{bmatrix}\boldsymbol{q}\\ \boldsymbol{p}\end{bmatrix}$. In section 2.2.3, considering the *canonical coordinates* $(\boldsymbol{q},\boldsymbol{p})$ and *principal function S*, the Hamilton-Jacobi equation is

$$H\left(t,\boldsymbol{q},\frac{\partial S(t,\boldsymbol{q},\boldsymbol{p}_0)}{\partial \boldsymbol{q}}\right)=-\frac{\partial S(t,\boldsymbol{q},\boldsymbol{p}_0)}{\partial t}, \tag{2.55}$$

where $\boldsymbol{p}_0$ is the initial momentum.

The *principal function S* is a function, integrated along the path that satisfies the Hamilton equation, and, takes the system from $\boldsymbol{p}_0$ at time t_0 to $\boldsymbol{p}_f$ at t_f.

Solution of (2.55) exists. Positive definite $(\mathbb{E}(t,x),H(t,\boldsymbol{q},\boldsymbol{p}))$ are continuously differentiable. Evolution and definiteness of $\frac{d\mathbb{E}(t,x)}{dt}$ are informative and descriptive considering stability, system dynamics, system governance, etc. A system evolves to the equilibrium state if energy decreases, which implies negative definiteness of $\dfrac{d\mathbb{E}(t,x)}{dt}$.

At equilibrium, $\dfrac{d\mathbb{E}(t,x)}{dt}=0$.

Stability of Dynamic Systems – Conditions for stability of dynamic motion and stable system governance are

$$\left\{\mathbb{E}(t,x)>0,\ \frac{d\mathbb{E}(t,x)}{dt}<0\right\},\ \forall(t,x)\in T\times X. \tag{2.56}$$

Criteria (2.56) are a Lyapunov stability paradigm. The Hamilton-Jacobi equations are applied to solve optimization problems, and, control laws $u=\varphi(t,x)$ are designed in chapters 5 and 6. Section 6.10 covers stability analysis and optimization of nonlinear systems applying the Lyapunov stability theory. Control function $u=\varphi(t,x)$ are found considering system stability, optimality, sensitivity, etc.

Stability and Energy – Consider a continuous and differentiable positive definite energy-dependent function $V(t,x)\equiv\eta[\mathbb{E}(t,x)]$, $V(t,x)>0$. Negative definiteness of $\frac{dV(t,x)}{dt}=\frac{d}{dt}\eta[\mathbb{E}(t,x)]$, $V\in C^k$, $k\geq 2$, defines system stability, optimality, as well as yields closed-form solutions. System dynamics and temporal governance are stable if system evolves to an equilibrium at which for unperturbed system $dV/dt=0$.

Considering a Lyapunov pair $\left\{V(t,x),\frac{dV(t,x)}{dt}\right\}$, and, defining stability and optimality as $\frac{dV}{dt}\leq-\omega(t,x,u)$, one designs a control law $u=\varphi(t,x)$ such as

$$\left\{V(t,x)=\eta[\mathbb{E}(t,x)]>0,\frac{dV(t,x)}{dt}<0\right\},\ \left.\frac{dV}{dt}\leq-\omega(t,x,u)\right|_{u=\varphi(t,x)},\omega(t,x,u)>0,\forall(t,x,u)\in T\times X\times U. \quad (2.57)$$

where $\omega(t,x,u)$ is a continuous positive definite scalar function, and, $\omega(t,x,u)\equiv\gamma\left[\frac{d}{dt}\mathbb{E}(t,x)\right]$.

Function $\omega(t,x,u)$ specifies an evolution of dV/dt, and $\omega(\cdot):\mathbb{R}_{\geq 0}\times\mathbb{R}^n\times\mathbb{R}^m\to\mathbb{R}$ assumes to be locally Hölder continuous, or, has continuous l first derivatives, $\omega\in C^l$, $l\geq 2$. One solves

$$\frac{dV}{dt}=\frac{\partial V}{\partial t}+\left(\frac{\partial V}{\partial x}\right)^T\left[F(t,x)+B(t,x)u\right],\ u=\varphi(t,x). \quad (2.58)$$

Defining $V(t,x)\equiv\eta[\mathbb{E}(t,x)]>0$, specifying stability and optimality as $\frac{dV}{dt}\leq-\omega(t,x,u)$, $\omega(t,x,u)>0$, optimal control function $u^*(t,x)$ is designed by solving an optimization problem. For governing equations of motion (2.54), from (2.56)-(2.58), one has

$$\begin{cases}\text{System Governance: } \dot{x}=F(t,x)+B(t,x)u,\ u=\varphi(t,x),\ u\in U, & (2.59)\\ \text{System Stability: } V(t,x)=\eta[\mathbb{E}(t,x)]>0,\ \frac{dV(t,x)}{dt}<0,\ \forall(t,x)\in T\times X, & \\ \text{Energy Governance: } \frac{dV}{dt}=\frac{\partial V}{\partial t}+\left(\frac{\partial V}{\partial x}\right)^T\left[F(t,x)+B(t,x)u\right],\frac{dV}{dt}\leq-\omega(t,x,u),\ \omega(t,x,u)>0. & \end{cases}$$

From $\frac{dV}{dt}=\frac{\partial V}{\partial t}+\left(\frac{\partial V}{\partial x}\right)^T\left[F(t,x)+B(x)u\right]$, such as $\frac{dV}{dt}\leq-\omega(t,x,u)$, we have

$$-\frac{\partial V}{\partial t}=\omega(t,x,u)+\left(\frac{\partial V}{\partial x}\right)^T\left[F(t,x)+B(t,x)u\right]. \quad (2.60)$$

The scalar Hamiltonian function is

$$H(t,x,u,\tfrac{\partial V}{\partial x})\equiv\omega(t,x,u)+\left(\frac{\partial V}{\partial x}\right)^T\left[F(t,x)+B(t,x)u\right]. \quad (2.61)$$

Derive a control law $u\in U$ by finding a critical point minimizing Hamiltonian (2.61) as

$$\frac{\partial}{\partial u}H(t,x,u,\tfrac{\partial V}{\partial x})=\frac{\partial}{\partial u}\left[\omega(t,x,u)+\left(\frac{\partial V}{\partial x}\right)^T\left[F(t,x)+B(t,x)u\right]\right]=0. \quad (2.62)$$

For $V^*(t,x)\equiv\eta^*[\mathbf{E}(t,x)]$, $V^*\in C^k$, $k\geq 2$, finding a closed-form solution of (2.60), one derives a control function

$$u^* = \varphi\left(t,\frac{\partial V^*}{\partial x}\right),\ V\in C^k,\ k\geq 2. \tag{2.63}$$

Structure of a control function $u^*(t,x)$ depends on $\boldsymbol{\omega}(t,x,u)$ and $V^*(t,x)$.

A scalar function $V^*(t,x)$ should satisfy the Hamilton-Jacobi equation

$$-\frac{\partial V^*}{\partial t} = \boldsymbol{\omega}(t,x,u^*)+\left(\frac{\partial V^*}{\partial x}\right)^T\left[F(t,x)+B(t,x)u^*\right],\ u^* = \varphi\left(t,\frac{\partial V^*}{\partial x}\right),\ \boldsymbol{\omega}\in C^l,\ V\in C^k, \tag{2.64}$$

which admits a solution. A closed-form solution of the Hamilton-Jacobi equation is found with $V\in C^k$ and $\boldsymbol{\omega}\in C^l$. Physics-consistent and design-specific continuous and positive definite admissible $V^*(t,x)$ and $\boldsymbol{\omega}(t,x,u)$ are found yielding solutions of optimization problem

$$\begin{cases} \dot{x} = F(t,x)+B(t,x)u,\ u^* = \varphi\left(t,\frac{\partial V^*}{\partial x}\right),\ u^*\in U, & (2.65)\\ \left\{V(t,x)=\eta\left(\mathbf{E}(t,x)\right)>0,\ \frac{dV}{dt}\leq -\boldsymbol{\omega}(t,x,u)\right\},\ \boldsymbol{\omega}(t,x,u)>0,\ \forall(t,x,u)\in T\times X\times U, \\ \min\limits_{u\in U}\left[H(t,x,u,\frac{\partial V}{\partial x})\right] = \min\limits_{u\in U}\left[\boldsymbol{\omega}(t,x,u)+\frac{\partial V}{\partial x}^T\left[F(t,x)+B(t,x)u\right]\right],\ \frac{\partial}{\partial u}H(t,x,u,\frac{\partial V}{\partial x})=0, \\ -\frac{\partial V^*}{\partial t} = \boldsymbol{\omega}(t,x,u^*)+\left(\frac{\partial V^*}{\partial x}\right)^T\left[F(t,x)+B(t,x)u^*\right]. \end{cases}$$

To solve a dynamic optimization problem, guaranteeing stability and optimality, one:

1. Finds a positive definite total energy $\mathbf{E}(t,x)$, as well as Lebesgue integrable scalar functions (H,V), for which $\left(\partial_u^1 H,\ \partial_u^2 H,\ \partial_x V\right)$ exist;
2. Designs an optimal control law $u^*(t,x)$ as a continuous function;
3. Solves the Hamilton-Jacobi equation.

For an explicit total energy $\mathbf{E}(t,x)$, one finds Lebesgue integrable functions (H,V), such that $\left(\partial_u^1 H,\ \partial_u^2 H,\ \partial_x V\right)$ exist. Consider the *strong derivatives* and *weak partial derivative*. The Sobolev space $W^{k,p}(\boldsymbol{\Omega})$ defines a subset of scalar functions (H,V) in $L^p(\Omega)$, such that (H,V) and their *weak derivatives* up to the order k have finite Lebesgue norms.

For example, define

$$W^{k,p}(\Omega)=\left\{H\in L^p(\Omega):\ \partial_u^d H\in L^p(\Omega),\ \forall d\leq k\right\},\ k\in\mathbb{N},\ p\in[1,\infty],$$

where Ω is a bounded open subset in $\mathbb{R}$.

The Hölder uniformly continuous or locally Hölder continuous real Cantor function $V_{C(\Omega)}$ exists in the Hölder space $C^{k,p}(\Omega)$, $k\in[0,1]$, $p\in[1,\infty]$, and, this Lebesgue singular function is not absolutely continuous. The Cantor functions may are applied in the *minimal complexity* control laws design, as well as considering the Lebesgue singular control functions. The space of Lipschitz continuous functions is $C^{0,1}(\Omega)$, and, $C^{0,1}(\Omega)\cap L^\infty(\Omega)$ is continuously embedded into $W^{1,\infty}(\Omega)$. For defined spaces $\mathcal{F}$ and parameter space $\mathcal{P}$, regularize and parameterize $V^*(t,p,x)$, $p\in\mathcal{P}$, as well as evaluate a closed form solution of (2.64) and (2.65). Solutions, tools and algorithms of dynamic optimization are reported and solved in chapters 5 and 6.

2.6. Numeric and Symbolic Analysis in MATLAB®

For derived equations of motions, found applying the basic laws of physics, solution of analysis, design and optimization problems are supported by high performance interacting computing environments. MATLAB supports a wide spectrum of simulation, analysis, optimization, identification, data acquisition and other problems ensuring data-intensive analysis [21, 22]. MATLAB allows analytic and numeric compiling with high-level programming languages, consistent graphical and interface capabilities, versatility, interactive capabilities, etc. Application-specific toolboxes with build-in libraries, examples and commands are aimed to solve complex problems. An interactive SIMULINK® environment enables MATLAB. A great number of outstanding books and MathWorks resources www.mathworks.com/academia.html are available. This section is not aimed to substitute hundreds of excellent books on MATLAB.

To start MATLAB, double-click the icon . The MATLAB Command Window with Launch Pad and Command History appear on the screen. Typing **ver** or **demo**, the available toolboxes are listed.

Defining Fonts – The MATLAB statements to clear variables and functions from memory, close figures, and, define fonts are

```
clear all; close all; set(gca,'FontName','Times New Roman'); set(gca,'FontWeight','normal');
set(0,'DefaultAxesFontName', 'Times New Roman');
set(0,'DefaultAxesFontSize', 16); set(0,'DefaultAxesFontWeight','normal');
set(0,'DefaultTextFontname', 'Times New Roman');
set(0,'DefaultTextFontSize', 16); set(0,'DefaultTextFontWeight', 'normal');
box on; grid on; grid off;
```

Example 2.17. *Computing Functions, Objective Function Minimization, and, Optimization*
Compute and plot functions

$$y=\sin(x),\ y=e^{-0.25x}\sin(\pi x),\ x\in[0\ \ 4\pi]$$

if x varies from 0 to 4π. Let the increment is 0.05π.

To compute and plot these functions, the MATLAB code is

```
x=0:0.05*pi:4*pi; y1=sin(x); y2=sin(2*x).*cos(x); y3= exp(-0.25*x).*sin(pi*x);
figure(1); plot(x,y1,'k-',x,y1,'ro','Linewidth',2); axis([0 4*pi -1.05 1.05]);
title('Function {\ity}=sin({\itx})','FontSize',22); xlabel('{\itx}','FontSize',20);
figure(2); plot(x,y3,'k-',x,y3,'ro','Linewidth',2); axis([0 4*pi -0.7 0.9]);
title('Function {\ity}={\ite}^{-0.25\itx}sin({\pi}{\itx})','FontSize',22); xlabel('{\itx}','FontSize',20);
```

The resulting plots are shown in Figures 2.7.

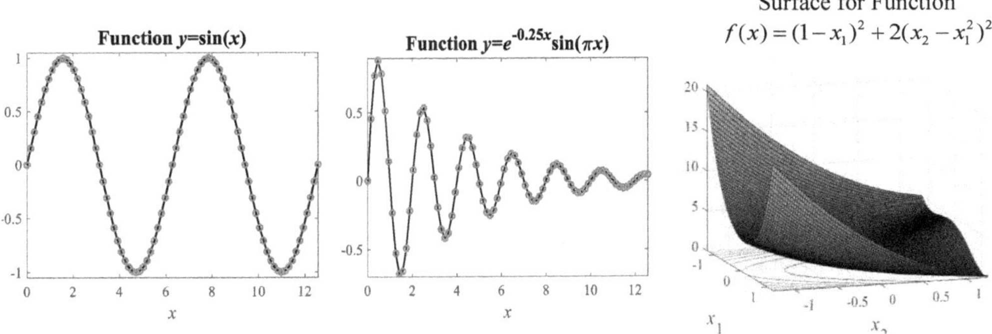

Figure 2.7. Plots for $y=\sin(x)$ and $y=e^{-0.25x}\sin(\pi x)$. Surface of $f(x)=(1-x_1)^2+2(x_2-x_1^2)^2$.

The **fminbnd** and **fminsearch** solvers minimize an objective function $f(x)$. To solve a maximization problem $\max_{x\in X} f(x)$, denote $h(x)=-f(x)$, and define X.

For $y=e^{-0.25x}\sin(\pi x)$, the MATLAB code

```
[x fval]=fminbnd(@(x) -exp(-0.25*x)*sin(pi*x),0,2.5)
```

yields

$\max_{x\in X} f(x)$, which is 0.885 at x=0.4747 for $x\in[0\ \ 2.5]$.

For unimodal, bimodal and multimodal cases, one analyzes and verifies solutions for univariate, bivariate and multivariate functions. Nonlinear constrained optimization problems are solved using the **fmincon** solver.

Consider a non-convex Rosenbrock function

$$f(x)=(a-x_1)^2+b(x_2-x_1^2)^2,\ x\in\mathbb{R}^2,\ (a,b)>0$$

which has a global minimum at $(x_1,x_2)=(a,a^2)$

For a=1 and b=2, a surface for $f(x)=(1-x_1)^2+2(x_2-x_1^2)^2$ is depicted in Figure 2.7. To compute and plot $f(x)=(1-x_1)^2+100(x_2-x_1^2)^2$, the statements are

```
x1=linspace(-1.25,1.25,100); x2=x1; [X1,X2]=meshgrid(x1,x2); a=1; b=2;
f=(a-X1).^2+b*(X2-X1.^2).^2;
axis([-1.25 1.25 -1.25 1.25 min(min(f)) max(max(f))])
figure(1); surfc(x1,x2,f); xlabel('{\itx}_1','FontSize',22'); ylabel('{\itx}_2','FontSize',22');
```

The Optimization Toolbox™ solver considers this function to illustrate and solve constrained nonlinear optimization problems.

For multivariate case, one studies $f(x)=\sum_{i=1}^{n-1}\left[(a-x_i)^2+b(x_{i+1}-x_i^2)^2\right],\ x\in\mathbb{R}^n$. ■

Example 2.18. *Modeling and Analysis of RLC Circuits*

Many nonlinear differential equations cannot be solved analytically. For linear differential equations, analytic solution are found. For parallel and series *RLC* circuits, illustrated in Figures 2.8.a, the differential equations are

$$C\frac{d^2u}{dt^2}+\frac{1}{R}\frac{du}{dt}+\frac{1}{L}u=\frac{di_f}{dt},\quad \frac{d^2u}{dt^2}+\frac{1}{RC}\frac{du}{dt}+\frac{1}{LC}u=\frac{1}{C}\frac{di_f}{dt},$$

and $$L\frac{d^2i}{dt^2}+R\frac{di}{dt}+\frac{1}{C}i=\frac{du_f}{dt},\quad \frac{d^2i}{dt^2}+\frac{R}{L}\frac{di}{dt}+\frac{1}{LC}i=\frac{1}{L}\frac{du_f}{dt}.$$

Evolution of Current $i(t)$

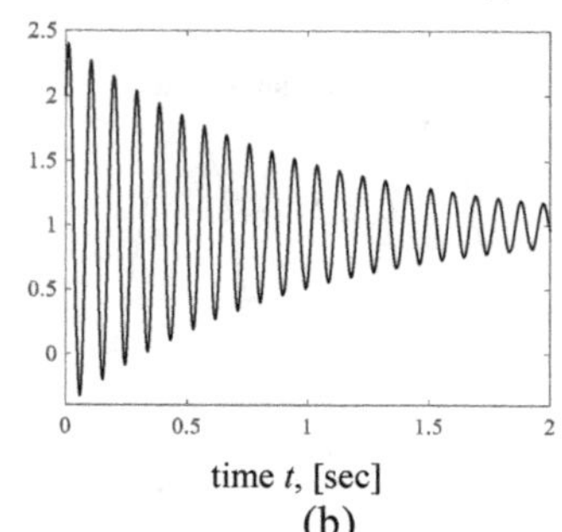

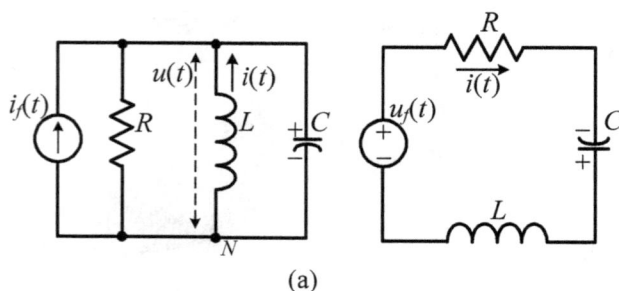

time t, [sec]

(a) (b)

Figure 2.8. (a) Parallel and series *RLC* circuits with inputs $i_f(t)$ and $u_f(t)$; (b) Dynamics due to unit step and initial conditions, Example 2.19.

The general solutions of linear differential equations are found using the characteristic roots of characteristic equation. The transfer functions are found for linear differential equations assuming linear time-invariant system applying the Laplace transform for the output to the input, $G(s)=\frac{Y(s)}{U(s)}$. For a transfer function

$$G(s)=\frac{\omega_0^2}{s^2+2\xi\omega_0 s+\omega_0^2}=\frac{\omega_0^2}{s^2+2\alpha s+\omega_0^2},\ \xi=\frac{\alpha}{\omega_0},$$

the descriptive quantities are the damping coefficient ξ, the *neper* frequency α, and the *natural* frequency ω_0. For the parallel and series *RLC* circuits

$$\xi=\frac{1}{2R}\sqrt{\frac{L}{C}},\ \alpha=\frac{1}{2RC},\ \omega_0=\frac{1}{\sqrt{LC}}$$

and $$\xi=\frac{R}{2}\sqrt{\frac{C}{L}},\ \alpha=\frac{R}{2L},\ \omega_0=\frac{1}{\sqrt{LC}}.$$

One has $\xi=\alpha/\omega_0$.

For the second-order translational motion

$$m\frac{d^2x}{dt^2}+B_v\frac{dx}{dt}+k_s x=F_a(t)$$

with the friction and spring constants (B_v,k_s), $\xi=\frac{B_v}{2\sqrt{k_s m}}$ and $\omega_0=\sqrt{\frac{k_s}{m}}$.

For a linear second-order differential equation

$$\frac{d^2x}{dt^2}+2\alpha\frac{dx}{dt}+\omega_0 x=f(t),$$

using the Laplace operator $s=d/dt$, $s^2=d^2/dt^2$, the characteristic equation is

$$s^2+2\alpha s+\omega_0^2=(s-s_1)(s-s_2)=0.$$

The characteristic roots are $s_{1,2}=-\alpha\pm\sqrt{\alpha^2-\omega_0^2}=-\omega_0\left(\xi\pm\sqrt{\xi^2-1}\right)$.

Case 1. If $\alpha^2>\omega_0^2$, the real distinct characteristic roots are s_1 and s_2.

The general solution is $x(t)=ae^{s_1 t}+be^{s_2 t}+c_f$,

where constants a and b are found using the initial conditions, while c_f is the solution due to the input, which is called the *forcing* function *f*. For the *RLC* circuits, the *forcing function f* is $i_f(t)$ or $u_f(t)$.

Case 2. For $\alpha^2=\omega_0^2$, the characteristic roots are real and identical, $s_1=s_2=-\alpha$.

A solution of the second-order differential equation is

$$x(t)=(a+b)e^{-\alpha t}+c_f.$$

Case 3. If $\alpha^2<\omega_0^2$, the complex distinct characteristic roots are $s_{1,2}=-\alpha\pm j\sqrt{\omega_0^2-\alpha^2}$.

The general solution is

$$x(t)=e^{-\alpha t}\left[a\cos\left(\sqrt{\omega_0^2-\alpha^2}t\right)+b\sin\left(\sqrt{\omega_0^2-\alpha^2}t\right)\right]+c_f=e^{-\alpha t}\sqrt{a^2+b^2}\cos\left[\left(\sqrt{\omega_0^2-\alpha^2}t\right)-\tan^{-1}\left(\frac{b}{a}\right)\right]+c_f.$$

For the unit step input, from $G(s)=\dfrac{X(s)}{U(s)}=\dfrac{\omega_0^2}{s^2+2\xi\omega_0 s+\omega_0^2}$, the output $X(s)$ is

$$X(s)=\frac{1}{s}\frac{\omega_0^2}{s^2+2\xi\omega_0 s+\omega_0^2}=\frac{K_1}{s}+\frac{K_2 s+K_3}{s^2+2\xi\omega_0 s+\omega_0^2}.$$

The partial fractioning yields (K_1,K_2,K_3), and, we have

$$X(s)=\frac{1}{s}-\frac{(s+\xi\omega_0)+\frac{\xi}{\sqrt{1-\xi^2}}\omega_0\sqrt{1-\xi^2}}{(s+\xi\omega_0)^2+\omega_0^2(1-\xi^2)}.$$

The underdamped response for a unit step is

$$x(t)=1-e^{-\xi\omega_0 t}\left[\cos\left(\omega_0\sqrt{1-\xi^2}t\right)+\frac{\xi}{\sqrt{1-\xi^2}}\sin\left(\omega_0\sqrt{1-\xi^2}t\right)\right]$$

$$=1-\frac{1}{\sqrt{1-\xi^2}}e^{-\xi\omega_0 t}\cos\left[\omega_0\sqrt{1-\xi^2}t-\tan^{-1}\left(\frac{\xi}{\sqrt{1-\xi^2}}\right)\right],$$

or

$$x(t)=1-e^{-\alpha t}\left[\cos\left(\sqrt{\omega_0^2-\alpha^2}t\right)+\frac{\alpha}{\sqrt{\omega_0^2-\alpha^2}}\sin\left(\sqrt{\omega_0^2-\alpha^2}t\right)\right]$$

$$=1-\frac{\alpha}{\sqrt{\omega_0^2-\alpha^2}}e^{-\alpha t}\cos\left[\sqrt{\omega_0^2-\alpha^2}t-\tan^{-1}\left(\frac{\alpha}{\sqrt{\omega_0^2-\alpha^2}}\right)\right].$$

Having found analytic solutions and explicit expressions, time evolutions can be computed and plotted in MATLAB as documented in Example 2.19. ■

Example 2.19. *RLC Circuits: Analysis of Second-Order Differential Equations*
Consider the series *RLC* circuit with R=0.5 ohm, L=1 H and C=2 F, see Figure 2.8.a. Compute the transient response if the input $u_f(t)$ is the unit step, $u_f(t)=1(t)$.

The series *RLC* circuit is described by $\dfrac{d^2i}{dt^2}+\dfrac{R}{L}\dfrac{di}{dt}+\dfrac{1}{LC}i=\dfrac{1}{L}\dfrac{du_f}{dt}$.

The characteristic equation is $s^2+\dfrac{R}{L}s+\dfrac{1}{LC}=0$.

The characteristic eigenvalues are $s_{1,2}=-\dfrac{R}{2L}\pm\sqrt{\left(\dfrac{R}{2L}\right)^2-\dfrac{1}{LC}}=-\alpha\pm\sqrt{\alpha^2-\omega_0^2}$.

If $\left(\dfrac{R}{2L}\right)^2>\dfrac{1}{LC}$, the characteristic eigenvalues are real and distinct.

For $\left(\dfrac{R}{2L}\right)^2=\dfrac{1}{LC}$, the eigenvalues are real and identical.

If $\left(\dfrac{R}{2L}\right)^2<\dfrac{1}{LC}$, the eigenvalues are complex.

For R=1 ohm, L=0.47 H and C=470 μF, the characteristic eigenvalues are complex, which implies an underdamped dynamics

$$i(t)=e^{-\alpha t}\left[a\cos\left(\sqrt{\omega_0^2-\alpha^2}t\right)+b\sin\left(\sqrt{\omega_0^2-\alpha^2}t\right)\right]+c_f,\ \alpha=\frac{R}{2L}=1.064,\ \omega_0=\frac{1}{\sqrt{LC}}=67.28.$$

Using initial conditions, one finds (a,b). Figure 2.8.b shows $i(t)$ for a=1, b=1 and c_f=1. MATLAB statements to compute and plot $i(t)$ are

```
R=1; L=0.47; C=4.7e-4; a=1; b=1; cf=1; l=R/(2*L); w0=1/sqrt(L*C);
t=0:1e-5:2.5; x=exp(-l*t).*(a*cos(sqrt(w0^2-l^2)*t)+b*sin(sqrt(w0^2-l^2)*t))+cf;
plot(t,x,'k','Linewidth',1.5); axis([0 2 -0.4 2.5]);
```

■

Example 2.20. *Analytic Solution of Differential Equations Using MATLAB*
Analytically solve the third-order differential equation

$$\frac{d^3x}{dt^3}+2\frac{dx}{dt}+3x=10f\ .$$

Applying the **dsolve** command

```
x=dsolve('D3x+2*Dx+3*x=10*f')
```

the solution is

```
x = (10*f)/3 + C3*exp(-t) + C1*exp(t/2)*cos((11^(1/2)*t)/2) + C2*exp(t/2)*sin((11^(1/2)*t)/2)
```

Using the **pretty** command

```
pretty(x)
```

we have

```
10 f                                    / sqrt(11) t \                         /sqrt(11) t \
---- + C3 exp(-t) + C1 exp(t/2) cos| ---------- | + C2 exp(t/2)sin|---------- |
  3                                     \     2      /                         \     2     /
```

Hence, $x(t)=\frac{10}{3}f+c_3e^{-t}+c_1e^{0.5t}\cos\left(\frac{1}{2}\sqrt{11}t\right)+c_2e^{0.5t}\sin\left(\frac{1}{2}\sqrt{11}t\right)$.

Let the initial conditions are given, and, $\left(\frac{d^2x}{dt^2}\right)_0=5,\left(\frac{dx}{dt}\right)_0=15$ and $x_0=-20$.

Hence, the code is

```
x=dsolve('D3x+2*Dx+3*x=10*f','D2x(0)=5','Dx(0)=15','x(0)=-20'); pretty(x)
```

The resulting solution with derived c_1, c_2 and c_3 yields

$$x(t)=\frac{10}{3}f-e^{-t}(2f+14)-e^{0.5t}\cos\left(\frac{1}{2}\sqrt{11}t\right)\left(\frac{4}{3}f+6\right)-\frac{8\sqrt{11}}{33}e^{0.5t}\sin\left(\frac{1}{2}\sqrt{11}t\right)(f-3)\ .$$

Consider a forcing function $f(t)$=50cos(10t), and, solve

$$\frac{d^3x}{dt^3}+2\frac{dx}{dt}+3x=f(t),\ f(t)=50\cos(10t).$$

Using

```
x=dsolve('D3x+2*Dx+3*x=50*cos(5*t)','D2x(0)=5','Dx(0)=15','x(0)=-20'); pretty(x)
```

one finds a solution for $x(t)$ as

$$x(t)=-\frac{2875}{6617}\sin 5t+\frac{75}{6617}\cos 5t-\frac{187}{13}e^{-t}+\frac{5702\sqrt{11}}{5599}e^{0.5t}\sin\left(\frac{1}{2}\sqrt{11}t\right)-\frac{2864}{509}e^{0.5t}\cos\left(\frac{1}{2}\sqrt{11}t\right)\ .\ ■$$

Example 2.21. *Numeric Solution of Differential Equations*
For many nonlinear differential equations, analytic solution cannot be derived. Apply the **ode45** MATLAB solver to numerically solve differential equations using a variable step Runge-Kutta method. Consider a system which do not admit an analytic solution

$$\frac{dx_1(t)}{dt}=-20x_1+e^{-x_1}\,|\,x_2x_3\,|+10x_1x_2x_3,\ \ x_1(t_0)=x_{10},$$

$$\frac{dx_2(t)}{dt}=-5x_2-10\sin x_1\cos x_3+|x_3|,\ \ x_2(t_0)=x_{20},$$

$$\frac{dx_3(t)}{dt}=-25x_3-5x_1x_2+50\sin x_1\cos x_2,\ \ x_3(t_0)=x_{30}\ .$$

The initial conditions are $x_0=\begin{bmatrix}x_{10}\\x_{20}\\x_{30}\end{bmatrix}=\begin{bmatrix}2\\-1\\-2\end{bmatrix}$. Two m-files `ch21.m` and `ch22.m` are developed. Dynamics of states $x_1(t)$, $x_2(t)$ and $x_3(t)$ are plotted using the **plot** and **plot3** commands.

The first file is

```
% MATLAB file ch21.m
t0=0; tfinal=0.75; tspan=[t0 tfinal]; % Initial and final time
y0=[2 -1 -2]';                        % Initial conditions for state variables
[t,y]=ode45('ch22',tspan,y0);         % ode45 MATLAB solver
% Plot transient dynamics of the state variables solving differential equations specified in ch22.m
figure(1); plot(t,y(:,1),'k-',t,y(:,2),'k--',t,y(:,3),'b:','Linewidth',2.5); axis([0 0.75 -2.1 2.1]);
% xlabel('time [sec]','FontSize',16); title('Dynamics of {\itx}_1(t), {\itx}_2(t), {\itx}_3(t)','FontSize',18);
legend('{\itx}_1({\itt})','{\itx}_2({\itt})','{\itx}_3({\itt})','linewidth',2,'FontSize',22); legend boxoff;
% 3D plot for x1, x2 and x3
figure(2); plot3(y(:,1),y(:,2),y(:,3),'Linewidth',2.5); xlabel('{\itx}_1','FontSize',20);
ylabel('{\itx}_2','FontSize',20); zlabel('{\itx}_3','FontSize',20);
% title('States Evolutions {\itx}_1({\itt}), {\itx}_2({\itt}), {\itx}_3({\itt})','FontSize',18);
```

The second MATLAB file `ch22.m` with a set of differential equations is

```
% MATLAB file ch22.m    Simulation of the third-order differential equations
function yprime = difer(t,y);
% Differential equations coefficients
a11=-20; a12=1; a13=10; a21=-5; a22=-10; a23=1; a31=-25; a32=-5; a33=50;
% Three differential equations: System of three first-order differential equations
yprime=[a11*y(1,:)+a12*(exp(-y(1,:)))*abs(y(2,:)*y(3,:))+a13*y(1,:)*y(2,:)*y(3,:);... % first equation
a21*y(2,:)+a22*sin(y(1,:))*cos(y(3,:))+a23*abs(y(3,:));...                            % second equation
a31*y(3,:)+a32*y(1,:)*y(2,:)+a33*sin(y(1,:))*cos(y(2,:))];                            % third equation
```

The resulting transients are documented in Figure 2.9.a. Three-dimensional evolution of the state variables is illustrates in Figure 2.9.b.

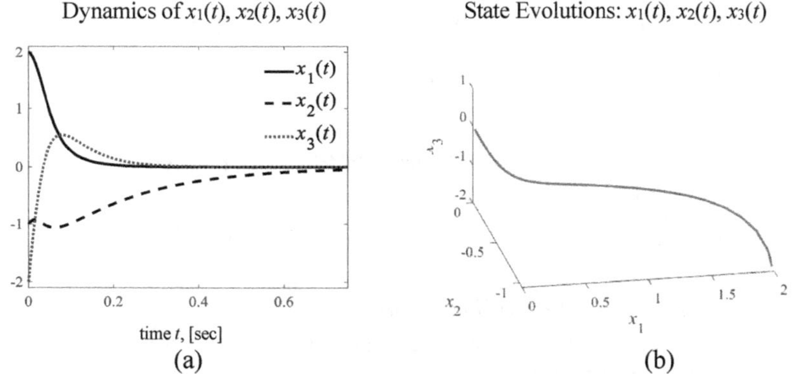

Figure 2.9. (a) Dynamics of states $x_1(t)$,$x_2(t)$,$x_3(t)$ depicted by solid, dashed and dotted lines; (b) Evolution of (x_1,x_2,x_3). ■

SIMULINK is an interactive computing environment to simulate linear, nonlinear, continuous-time, discrete-time, multivariable, multirate and hybrid systems. SIMILINK Blocksets are built-in blocks which is a library of different components from the Library Browser. A C-code from diagrams is generated using the Real-Time Workshop Toolbox. The HDL Coder generates Verilog® and VHDL® codes.

Using a mouse-driven interface, SIMULINK diagrams are built. A library of signal sources, linear and nonlinear functions, connectors and customized S-functions provide flexibility and interactability. SIMULINK menu ensures interactive simulations and visualization. To assess various SIMULINK examples and documentation, illustrated in Figure 2.10, one types in the Command Window **demo Simulink**

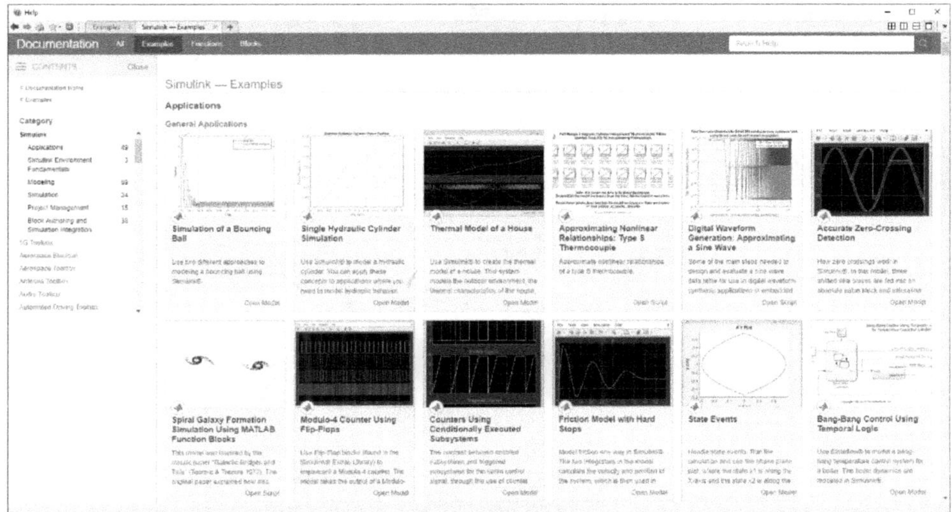

Figure 2.10. SIMULINK demo window **>> demo Simulink**

We provide supplementary coverage and educate the reader on how to solve practical engineering problems. SIMULINK is introduced with step-by-step instructions and practical examples not rewriting user manuals, documentations and books. Figures 2.11 report the SIMULINK demo features with various simple, medium complexity and advanced examples. The van der Pol oscillator and friction model, studied in SIMULINK, are covered in this section providing alternative solutions.

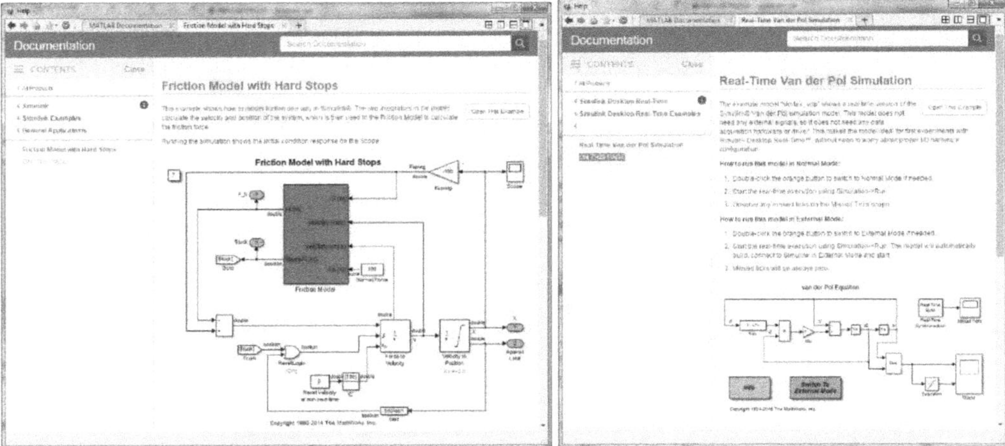

Figure 2.11. SIMULINK demonstrations and examples. The Search icon can be used.

Using differential and difference equations, as well as constitutive relationships, one builds SIMULINK diagrams using blocks from the Library Browser, some of which are reported in Figures 2.12. Nonlinear deterministic and stochastic systems can be simulated. Various examples in aerospace, automotive, electronic, mechanical and robotic systems are available. The designer evaluates the fitness, practicality, applicability and validity of MATLAB toolboxes and environments. For example, to analyze unsteady acoustics, aerodynamics, elasticity, tribology, vibration, vorticity and other fields, the van der Pol and Lorenz equations are used.

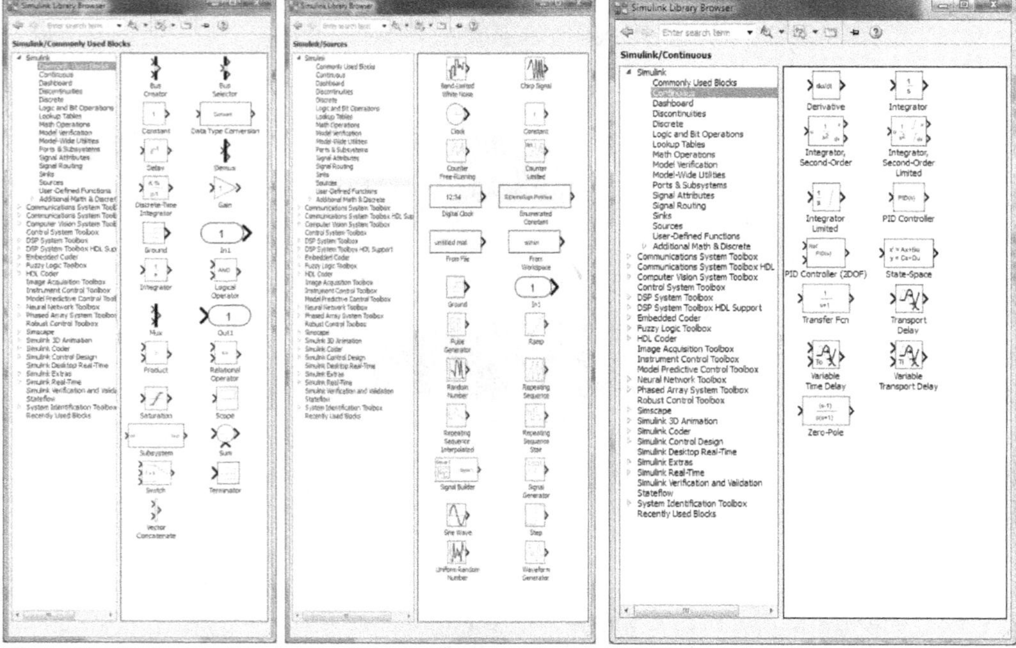

Figure 2.12. Blocks, Sources and Continuous components from the Library Browser

Example 2.22. *Van der Pol Differential Equations*

The van der Pol oscillator, which exhibits limit cycles, is described by the second-order nonlinear differential equation

$$\frac{d^2x}{dt^2}-k\left(1-x^2\right)\frac{dx}{dt}+x=0,\ k\geq 0.$$

The unperturbed equations exhibit limit cycles. The differential equations for the unperturbed van der Pol oscillator are

$$\frac{dx_1(t)}{dt}=x_2,\ (x_{10},x_{20})\neq 0,$$

$$\frac{dx_2(t)}{dt}=-x_1+kx_2(1-x_1^2).$$

For the perturbed system,

$$\frac{d^2x}{dt^2}-k\left(1-x^2\right)\frac{dx}{dt}+x=d(t),\ k\geq 0,$$

where $d(t)$ is the forcing function.

The differential equation is rewritten as a system of two first-order differential equations

$$\frac{dx_1(t)}{dt}=x_2,$$

$$\frac{dx_2(t)}{dt}=-x_1+kx_2\left(1-x_1^2\right)+d(t).$$

The SIMULINK diagram is built using the Function, Integrator, Mux, Scope and other blocks as reported in Figure 2.13.a. Simulations are performed for $k=1$, $d=0$, and initial conditions $x_0=\begin{bmatrix}x_{10}\\x_{20}\end{bmatrix}=\begin{bmatrix}1\\-1\end{bmatrix}$. If $d(t)\neq 0$, depending on input d, one may use the signal sources, such as Signal Generator, White Noise, etc. By double-clicking the Signal Generator block, one selects the waveform, amplitude, frequency, etc. The initial conditions are set by double-clicking the Integrators and setting x_{10} and x_{20}. Specifying the simulation time to be 25 seconds, the results are illustrated in Figure 2.13.b. Saving the Scope data in the array format as a variable x, the plot command is used. The resulting plots are illustrated in Figure 2.13.b.

```
plot(x(:,1),x(:,2),'k-',x(:,1),x(:,3),'k--','LineWidth',2.5); axis([0 25 -3 3]);
legend('{\itx}_1({\itt})','{\itx}_2({\itt})','linewidth',2,'FontSize',18); legend boxoff;
xlabel('time, [sec]','FontSize',18);
title('Van der Pol Oscillator Dynamics {\itx}_1({\itt}), {\itx}_2({\itt})','Fontsize',18);
```

The van der Pol equations may be modified. Consider

$$\frac{dx_1(t)}{dt} = x_2,$$

$$\frac{dx_2(t)}{dt} = -x_1 + kx_2\left(1 - |x_1| - x_1^2 - |x_1^3| - x_1^4\right).$$

The SIMULINK diagram is depicted in Figure 2.13.a. The system evolution with periodic orbits is reported in Figure 2.13.c.

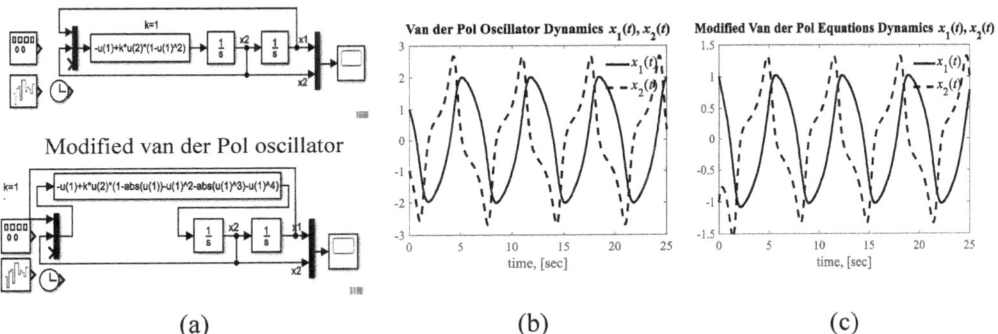

(a) (b) (c)

Figure 2.13. (a) SIMULINK diagrams to simulate *standard* and *modified* van der Pol oscillators;
(b) Transient dynamics of the van der Pol oscillator;
(c) Dynamics of the states $x_1(t)$ and $x_2(t)$ for the *modified* van der Pol equations of motion.

■

Example 2.23. *Lotka-Volterra Differential Equation*

Consider an unperturbed deterministic Lotka-Volterra system, described by

$$\frac{dx_1}{dt} = a_{11}x_1 - a_{12}x_1x_2,\ x_1(t_0)=x_{10},\ x_{10}>0,\ \forall(a_{ii},a_{ij})>0,$$

$$\frac{dx_2}{dt} = -a_{21}x_2 + a_{22}x_1x_2,\ x_2(t_0)=x_{20},\ x_{20}>0$$

with positive-definite coefficients, $(a_{ii},a_{ij})>0$. These equations admits evolutions $(x_1,x_2)>0$ for $(x_{10},x_{20})>0$. The governing dynamics of a system with states (x_1,x_2) are studied for given equilibriums and bifurcations with critical points, saddle points, orbits and limit cycles.

We have

$$\frac{dx_2}{dx_1} = \frac{x_2\left(-a_{21} + a_{22}x_1\right)}{x_1\left(a_{11} - a_{12}x_2\right)}.$$

Separating the variables, one finds $\int \frac{-a_{21} + a_{22}x_1}{x_1} dx_1 = \int \frac{a_{11} - a_{12}x_2}{x_2} dx_2$.

Hence, the phase plane equation is $-a_{21}\ln x_1 + a_{22}x_1 = a_{11}\ln x_2 - a_{12}x_2 + c_1$.

From $\begin{cases} x_1\left(a_{11} - a_{12}x_2\right) = 0 \\ x_2\left(-a_{21} + a_{22}x_1\right) = 0 \end{cases}$, the stationary point is $\left(\dfrac{a_{11}}{a_{12}}, \dfrac{a_{21}}{a_{22}}\right)$.

Using the ode45 solver, simulations are reported in Example 6.6. The SIMULINK diagram is documented in Figure 2.14.a. The system evolution is affected by $(a_{ii},a_{ij})>0$ and initial conditions $(x_{10},x_{20})>0$. For $a_{11}=1$, $a_{12}=1$, $a_{21}=10$ and $a_{22}=2$, Figures 2.14.b and 2.14.c report the transient dynamics and orbits for (x_1,x_2) with $x_{10}=3$ and $x_{20}=3$.

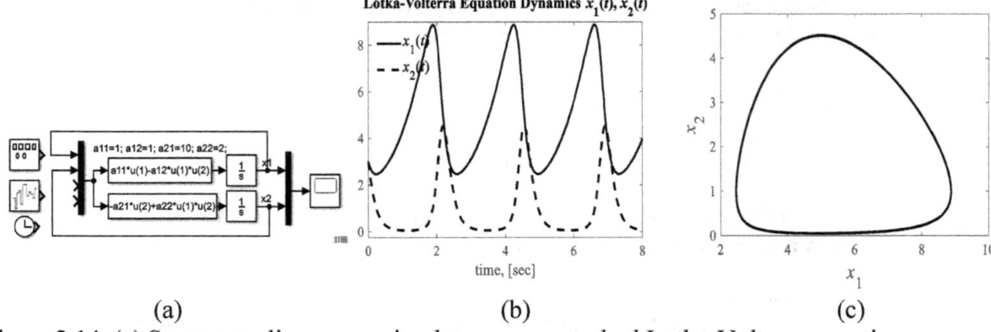

(a) (b) (c)

Figure 2.14. (a) SIMULINK diagram to simulate an unperturbed Lotka-Volterra equation;
(b) Dynamics and periodic evolutions of $x_1(t)$ and $x_2(t)$;
(c) Orbits of states (x_1,x_2).

The considered second-order system is equivalent to

$$\frac{dx_1}{dt} = -a_{11}x_1 + a_{12}x_1x_2,\ (x_{10},x_{20})>0,\ \forall(a_{ii},a_{ij})>0,$$

$$\frac{dx_2}{dt} = a_{21}x_2 - a_{22}x_1x_2.$$

Complimentary studies and results are reported in Example 6.6. ■

Example 2.24. *Stochastic Perturbed Lotka-Volterra Equation*

Dynamic systems may be perturbed by deterministic and stochastic disturbances. Consider

$$\frac{dx_1}{dt} = a_{11}x_1 - a_{12}x_1x_2 + b_{11}\xi_1,\ x_1(t_0)=x_{10},\ x_{10}>0,\ \forall(a_{ii},a_{ij})>0,$$

$$\frac{dx_2}{dt} = -a_{21}x_2 + a_{22}x_1x_2 + b_{21}\xi_2,\ x_2(t_0)=x_{20},\ x_{20}>0.$$

Two independent and identically distributed random variables, which are white noise ξ_1 and ξ_2, are characterized by the finite variances (σ_1^2,σ_2^2) and covariance $E[\xi_1(t_1)\cdot\xi_1(t_2)]=0$, $E[\xi_2(t_1)\cdot\xi_2(t_2)]=0$, $\forall t_1\neq t_2$. The samples are statistically uncorrelated and identically distributed with σ^2. The covariance matrix is $C_{xx} = \begin{bmatrix} \sigma^2 & \cdots & 0 \\ \vdots & \ddots & \vdots \\ 0 & \cdots & \sigma^2 \end{bmatrix} = \sigma^2 I$.

The probability density function is $f_X(x) = \dfrac{1}{\sigma\sqrt{2\pi}} e^{-\frac{1}{2\sigma^2}x^2}$.

Figure 2.15.a reports a SIMULINK diagram. Let $a_{11}=1$, $a_{12}=1$, $a_{21}=10$, $a_{22}=2$, $b_{11}=1$ and $b_{21}=1$. The stochastic dynamics of $x_1(t)$ and $x_2(t)$ are documented in Figures 2.15.b. For stochastic processes (ξ_1,ξ_2), the variances are $\sigma_1^2=0.1$ and $\sigma_2^2=0.1$. The perturbed orbits

(x_1,x_2) are depicted in Figure 2.15.c. For a considered class of differential equations, $(x_1,x_2)>0$, $\forall t \in T$, $\forall \xi \in \Xi$. The numeric method, tolerance, noise power, sample time and other specifications affect the consistency of simulations. The variance of ξ_1 and ξ_2 can be computed using the **var** command. We have $\sigma_1^2=0.1$, $\sigma_2^2=0.0975$. For a Lotka-Volterra class of equations, $x_1(t)>0$ and $x_2(t)>0$ must be guaranteed for any initial conditions and disturbances (ξ_1,ξ_2). The conditional statements can be used ensuring $(x_1,x_2)>0$.

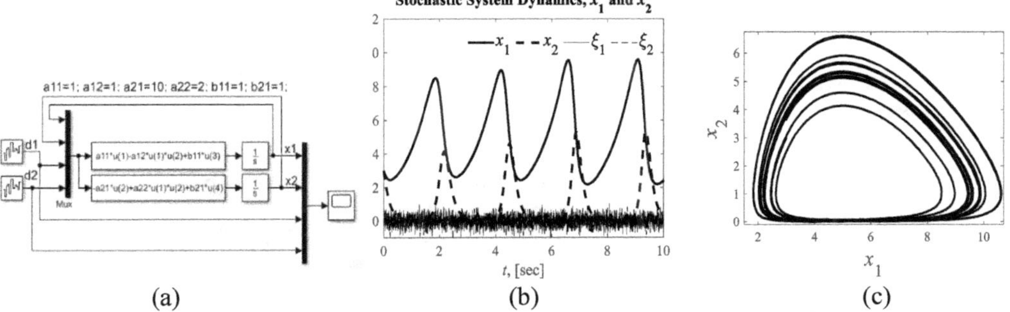

(a) (b) (c)

Figure 2.15. (a) SIMULINK diagram; (b) Dynamics of $x_1(t)$ and $x_2(t)$, $x_{10}=3$ and $x_{20}=3$;
(c) Phase portrait (x_1,x_2) of perturbed periodic solutions for $t\in[0\ \ 30]$ sec. ■

Example 2. 25. *Lorenz Equation*

The unperturbed Lorenz system of differential equations is

$$\frac{dx_1}{dt} = a_{11}\left(-x_1 + x_2\right),$$

$$\frac{dx_2}{dt} = a_{21}x_1 + a_{22}x_2 - a_{23}x_1x_3,$$

$$\frac{dx_3}{dt} = -a_{31}x_3 + a_{32}x_1x_2.$$

The SIMULINK diagram is reported in Figure 2.16.a. Depending on $(a_{ii},a_{ij})\in\mathbb{R}$, states (x_1,x_2,x_3) asymptotically converges to equilibriums, (x_1,x_2,x_3) may exhibit underdamped dynamics, bifurcations with quantified orbits, or, limit cycles. Let $a_{11}=5$, $a_{21}=12.5$, $a_{22}=2$, $a_{23}=2$, $a_{31}=2$ and $a_{32}=1$. Consider the unperturbed system dynamics due to initial conditions $x_{10}=1$, $x_{20}=0.5$ and $x_{30}=0$. System transients and three-dimensional evolution (x_1,x_2,x_3) are depicted in Figures 2.16.b and 2.16.c.

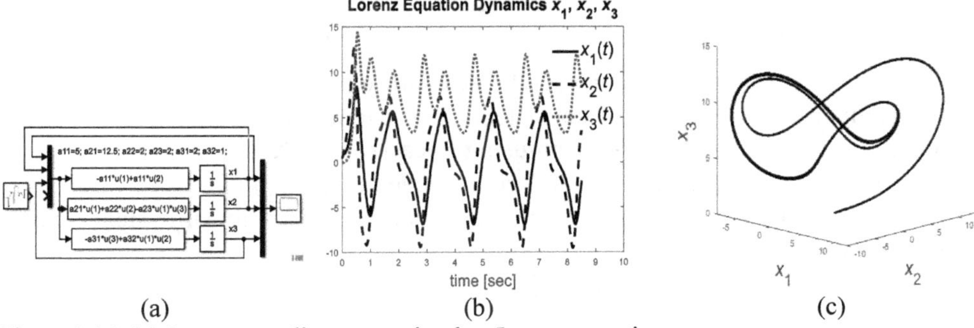

(a) (b) (c)

Figure 2.16. (a) SIMULILNK diagram to simulate Lorenz equations;
(b) Dynamics of $x_1(t)$, $x_2(t)$ and $x_3(t)$ depicted by solid, dashed and dotted lines;
(c) Three-dimensional evolution of states (x_1,x_2,x_3). ■

Example 2.26. *Lorenz Equation and Stability Analysis*
The Lorenz equations are used to model simplified momentum, heat, acoustic, fluids flow (Navier-Stokes) and other partial differential equations. Consider the third-order differential equations covered in Example 2.25 assuming $a_{22}=-1$, $a_{23}=1$ and $a_{32}=1$.

$$\frac{dx_1}{dt}=a_{11}\left(-x_1+x_2\right),\ (a_{11},a_{21},a_{31})>0,$$

$$\frac{dx_2}{dt}=a_{21}x_1-x_2-x_1x_3,$$

$$\frac{dx_3}{dt}=-a_{31}x_3+x_1x_2\,.$$

In the state-space form we have $\dot{x}=F(x)$, $\begin{bmatrix}\dot{x}_1\\ \dot{x}_2\\ \dot{x}_3\end{bmatrix}=\begin{bmatrix}a_{11}(-x_1+x_2)\\ a_{21}x_1-x_2-x_1x_3\\ -a_{31}x_3+x_1x_2\end{bmatrix}$.

The linearization, covered in chapter 6, yields the Jacobian matrix

$$J=\begin{bmatrix}\frac{\partial f_1}{\partial x_1} & \frac{\partial f_1}{\partial x_2} & \frac{\partial f_1}{\partial x_3}\\ \frac{\partial f_2}{\partial x_1} & \frac{\partial f_2}{\partial x_2} & \frac{\partial f_2}{\partial x_3}\\ \frac{\partial f_3}{\partial x_1} & \frac{\partial f_3}{\partial x_2} & \frac{\partial f_3}{\partial x_3}\end{bmatrix}=\begin{bmatrix}-a_{11} & a_{11} & 0\\ a_{21}-x_3 & -1 & -x_1\\ x_2 & x_1 & -a_{31}\end{bmatrix}.$$

The linearized system is $\dot{\delta}_x(t)=\left.\frac{\partial F}{\partial x}\right|_{x=\bar{x}}\delta_x=J|_{x=\bar{x}}\,\delta_x=\left.\begin{bmatrix}\frac{\partial f_1}{\partial x_1} & \frac{\partial f_1}{\partial x_2} & \frac{\partial f_1}{\partial x_3}\\ \frac{\partial f_2}{\partial x_1} & \frac{\partial f_2}{\partial x_2} & \frac{\partial f_2}{\partial x_3}\\ \frac{\partial f_3}{\partial x_1} & \frac{\partial f_3}{\partial x_2} & \frac{\partial f_3}{\partial x_3}\end{bmatrix}\right|_{x=\bar{x}}\delta_x=A\delta_x$.

For the *equilibrium state* $x_e=\bar{x}$, select $\bar{x}=\begin{bmatrix}\bar{x}_1\\ \bar{x}_2\\ \bar{x}_3\end{bmatrix}=\begin{bmatrix}0\\ 0\\ 0\end{bmatrix}$.

We have $A=\begin{bmatrix}-a_{11} & a_{11} & 0\\ a_{21} & -1 & 0\\ 0 & 0 & -a_{31}\end{bmatrix}$.

The characteristic equation $\det(\lambda I-A)=0$ yields the eigenvalues
$\lambda_1=-a_{31}$, $\lambda_{2,3}=\frac{1}{2}\left(-1-a_{11}\pm\sqrt{a_{11}^2-2a_{11}+4a_{11}a_{21}+1}\right)$.

Choice of $x_e=\bar{x}$ affects matrix A, characteristic equation and eigenvalues λ_i. The linearized dynamic analysis is not consistent. One cannot conclude on stability and dynamic transients of a nonlinear system using linearized models.

Dynamic transients and periodic orbits depend on $(a_{ii},a_{ij})\in\mathbb{R}$.

Apply the Lyapunov theory to examine stability of nonlinear systems. For a positive-definite function

$$V(x_1,x_2,x_3)=\tfrac{1}{2}\left(x_1^2+x_2^2+x_3^2\right),\ V>0,$$

the total derivative is

$$\frac{dV}{dt}=\frac{dV}{dx_1}\frac{dx_1}{dt}+\frac{dV}{dx_2}\frac{dx_2}{dt}+\frac{dV}{dx_3}\frac{dx_3}{dt}=x_1\frac{dx_1}{dt}+x_2\frac{dx_2}{dt}+x_3\frac{dx_3}{dt}=-a_{11}x_1^2+(a_{11}+a_{21})x_1x_2-x_2^2-a_{31}x_3^2\,.$$

The definiteness of dV/dt in small depends on (a_{11},a_{21}) and $x\in X$. From $(x_1+x_2)^2\ge 0$, $\forall(x_1,x_2)$, we have $x_1^2+x_2^2\ge 2x_1x_2$. For $a_{21}\gg 1$, the system may exhibit instabilities in small $x\in X]_{dV/dt>0}$, and, stability in large for $x\in X]_{dV/dt<0}$, $X_{dV/dt>0}\subset X_{dV/dt<0}$. Depending on (a_{11},a_{21},a_{31}), the unperturbed system exhibits overdamped and underdamped transients, as well as periodic or aperiodic asymmetric limit cycles.

The SIMULINK diagram is reported in Figure 2.17.a. Figures 2.17.b and c illustrate dynamics for two cases: (1) $a_{11}=10$, $a_{21}=10$, $a_{31}=1$; (2) $a_{11}=10$, $a_{21}=100$, $a_{31}=1$.

The initial conditions are $x_{10}=5$, $x_{20}=-5$ and $x_{30}=0$. Simulation results support fundamental studies derived applying the Lyapunov theory.

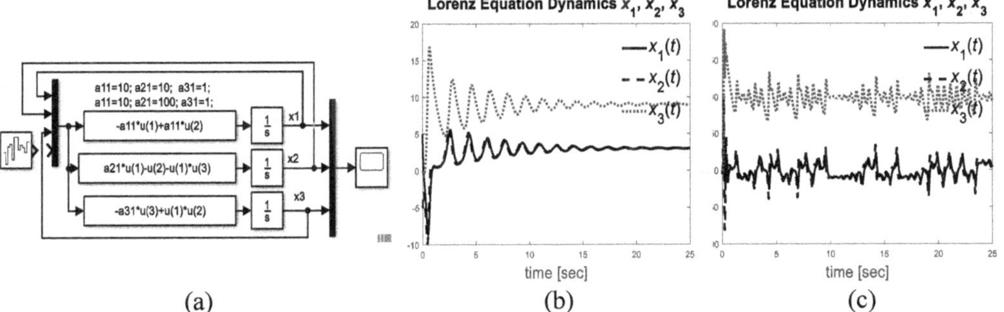

(a) (b) (c)

Figure 2.17. (a) SIMULILNK diagram for an unperturbed Lorenz system;
(b) Transients of $x_1(t)$, $x_2(t)$ and $x_3(t)$ if $a_{11}=10$, $a_{21}=10$, $a_{31}=1$;
(c) Evolutions with aperiodic orbits of $x_1(t)$, $x_2(t)$ and $x_3(t)$, $a_{11}=10$, $a_{21}=100$, $a_{31}=1$. ■

Example 2.27. Nonlinear Interpolation, Parametrization and Parameter Estimations
The MATLAB **lsqnonlin** solver solves nonlinear least-squares interpolation and parametrization problems.

Let a function
$y=f(x)=2e^{-0.5x}(\cos 5x+\sin 2x)=p_1e^{-p_2x}(\cos p_3x+\sin p_4x)$, $p_1=2$, $p_2=0.5$, $p_3=5$, $p_4=2$
is superimposed with zero-mean, high-variance σ^2 noise ξ.

Using a stochastic perturbation ξ, compute a dataset

$$y_{\text{data}}=2e^{-0.5x}(\cos 5x+\sin 2x)+\xi.$$

Using a nonlinear least-squares on y_{data}, estimate the unknown parameters $\hat{p}=(\hat{p}_1,\hat{p}_2,\hat{p}_3,\hat{p}_4)$, and, obtain the parameterized

$$y_p=f_p(\hat{p},x)=\hat{p}_1e^{-\hat{p}_2x}(\cos\hat{p}_3x+\sin\hat{p}_4x),\ (\hat{p}_1,\hat{p}_2,\hat{p}_3,\hat{p}_4)\in\mathbb{R}.$$

The sampled y_{data} is shown in Figure 2.18 by dots. Coefficients $\hat{p}=(\hat{p}_1,\hat{p}_2,\hat{p}_3,\hat{p}_4)$ are found assigning the initial values p_{i0}. Due to high-variance perturbations on y by normally distributed pseudorandom ξ, modelled using the **randn** command, $\hat{p}=(\hat{p}_1,\hat{p}_2,\hat{p}_3,\hat{p}_4)$ vary for different data sets y_{data}. Accurate estimation of (p_1,p_2,p_3,p_4) in most cases is ensured. Plots for a function $y=f(x)$ (dotted line), y_{data} (dots) used for parametrization, as well as parametrized $y_p=f_p(\hat{p},x)$ (solid line) are depicted in Figures 2.18. Least-squares parametrization of $y_{\text{data}}=2e^{-0.5x}(\cos 5x+\sin 2x)+\xi$ by $y_p=f_p(\hat{p},x)=\hat{p}_1e^{-\hat{p}_2x}(\cos\hat{p}_3x+\sin\hat{p}_4x)$ is accomplished for different p_{i0} and σ^2, yielding $(\hat{p}_1,\hat{p}_2,\hat{p}_3,\hat{p}_4)\in\mathbb{R}$.

The resulting $(\hat{p}_1, \hat{p}_2, \hat{p}_3, \hat{p}_4)$ are reported in the Figures 2.18 annotations.

For initial values p_{10}=1, p_{20}=1, p_{30}=10, p_{40}=1, if σ^2=0.25, the MATLAB statements are

```
x=linspace(0,10,100); y=2.*exp(-0.5*x) .*(cos(5*x)+sin(2*x)); Xi=0.5*randn(size(x)); ydata=y+Xi;
variance=var(Xi) ;   % Variance of the normal distribution
fun=@(p) p(1).*exp(-p(2)*x).*(cos(p(3)*x)+sin(p(4)*x)) - ydata;
p0=[1 1 10 1]; p=lsqnonlin(fun,p0), yint=p(1).*exp(-p(2)*x) .*(cos(p(3)*x)+sin(p(4)*x));
plot(x,y,'b:',x,ydata,'ko',x,yint,'k','LineWidth',3); xlabel('{\itx}','Fontsize',20);
legend('Function {\ity}={\itf}({\itx})','Data {\ity}_d_a_t_a ','Parameterized
{\ity_p}={\itf_p}({\itp},{\itx})','linewidth',2,'FontSize',18); legend boxoff;
```

Function $y = f(x) = 2e^{-0.5x}(\cos 5x + \sin 2x)$, Perturbed $y_{\text{data}} = 2e^{-0.5x}(\cos 5x + \sin 2x) + \xi$, and, Parameterized $y_p = f_p(\hat{p}, x) = \hat{p}_1 e^{-\hat{p}_2 x}(\cos \hat{p}_3 x + \sin \hat{p}_4 x)$

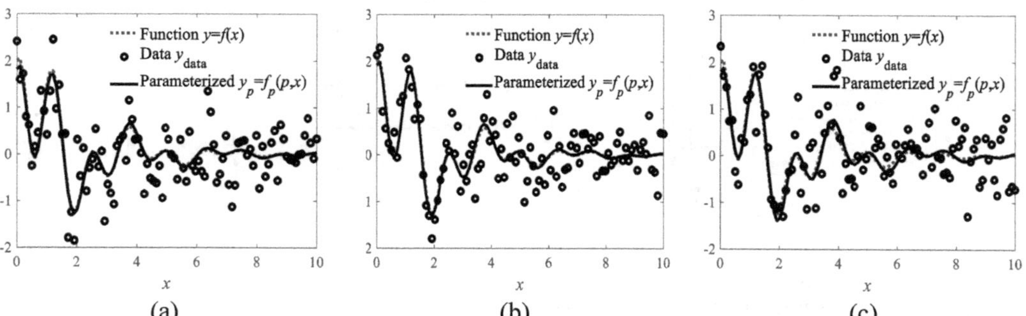

Figure 2.18. Plots for $y = f(x) = 2e^{-0.5x}(\cos 5x + \sin 2x)$, perturbed evolution $y_{\text{data}} = 2e^{-0.5x}(\cos 5x + \sin 2x) + \xi$ and parameterized $y_p = f_p(\hat{p}, x) = \hat{p}_1 e^{-\hat{p}_2 x}(\cos \hat{p}_3 x + \sin \hat{p}_4 x)$. The estimated parameter $\hat{p} = (\hat{p}_1, \hat{p}_2, \hat{p}_3, \hat{p}_4)$ are: (a) $\hat{p}_1$=1.8274, $\hat{p}_2$=0.4506, $\hat{p}_3$=4.9912, $\hat{p}_4$=2.0219; (b) $\hat{p}_1$=1.9224, $\hat{p}_2$=0.4642, $\hat{p}_3$=5.0094, $\hat{p}_4$=2.0178;(c) $\hat{p}_1$=1.8071, $\hat{p}_2$=0.4008, $\hat{p}_3$=4.9294, $\hat{p}_4$=2.0092. ■

Example 2.28. *Bearing Friction in Motion Devices*

Bearing friction is a complex nonlinear phenomenon which depends on angular velocity, loading, sliding surface, fluid layers, wearing, temperature, lubrication, etc. The *Coulomb*, viscous, static and viscous friction torques are exhibited by all motion devices, such as depicted in Figure 2.19.a. A friction force exist between the bearing surface and the shaft surface. The *Coulomb* friction is a retarding force or torque which changes its sign with the reversal of the direction of motion. The amplitude of force or torque remains the same.

For *Coulomb* friction, illustrated in Figure 2.19.b, consider

$$F_{\text{Coulomb}} = k_{Fc}\,\text{sgn}(v) = k_{Fc}\,\text{sgn}\left(\frac{dx}{dt}\right),\ T_{\text{Coulomb}} = k_{Tc}\,\text{sgn}(\omega) = k_{Tc}\,\text{sgn}\left(\frac{d\theta}{dt}\right).$$

The viscous friction is a retarding force or torque, which are nonlinear functions of velocity as depicted in Figure 2.19.c. The linear relationships to describe the viscous friction for translational and rotational motions are

$$F_{\text{viscous}}=B_v v,\ T_{\text{viscous}}=B_m \omega,$$

where B_v and B_m are the viscous friction coefficients.

The static friction, illustrated in Figure 2.19.d, exists when the body is stationary and vanishes as motion begins,

$$F_{\text{static}} = \pm F_{st}\big|_{v=\frac{dx}{dt}=0},\ T_{\text{static}} = \pm T_{st}\big|_{\omega=\frac{d\theta}{dt}=0}.$$

Friction force and torque are modeled using frictional memory, presliding conditions, etc. Equations commonly used are

$$F_{\text{friction}} = \left(k_1 - k_2 e^{-k_4|v|} + k_3|v|\right)\operatorname{sgn}(v),\ T_{\text{friction}} = \left(k_1 - k_2 e^{-k_4|\omega|} + k_3|\omega|\right)\operatorname{sgn}(\omega).$$

The typifying plots for F_{friction} and T_{friction} are shown in Figure 2.19.e. Friction can be experimentally measured, described and parameterized.

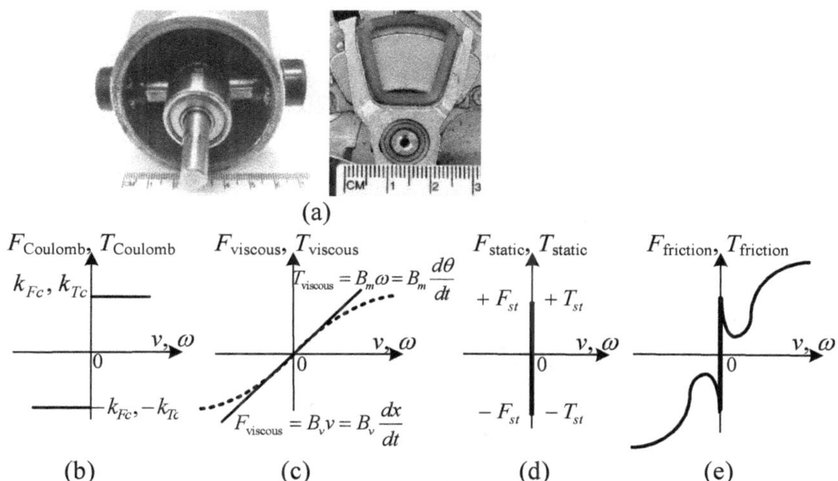

Figure 2.19. (a) Friction is an inherent phenomenon in rotational devices;
(b) *Coulomb* friction; (c) Viscous friction; (d) Static friction; (e) Friction force and torque.

Time-varying dependences, load-dependent asymmetry, eccentricity, lubrication, surface nonuniformity, wearing, temperature, surface roughness and other phenomena affect friction. Consider [22]

$$T_{\text{friction}} = \sum_i B_{m1_i} \operatorname{sgn}(\omega_r)|\omega_r|^{\frac{i}{1+2a_1}} + \sum_j B_{m2_j}\omega_r^{j+2a_2}$$
$$+\operatorname{sgn}(\omega_r)\left[\sum_k b_k\left(1 - e^{-c_k \operatorname{sgn}(\omega_r)|\omega_r|^{\frac{k}{1+2a_3}}}\right) + \sum_l b_l\left(1 - e^{-c_l|\omega_r|^{l+2a_4}}\right)\right],$$

where B_{m1i}, B_{m2j}, b_k, b_l, c_k and c_l are the constants, $(B_{m1i},B_{m2j},b_k,b_l,c_k,c_l)>0$; a_i are the non-negative integers, a_i=0,1,... .

Using experimental data, the nonlinear friction relationship are applied and parameterized finding unknown parameters. Expression for F_{friction} or T_{friction} are parametrized by solving a nonlinear mixed polynomial-exponential interpolation and least-squares parametrization.

For an actuator, illustrated in Figure 2.19.a, at ω_r=[90 187 284 384] rad/sec, the measured friction torque is

$T_{\text{friction data}}$=[3.28×10^{-2} 3.9×10^{-2} 4.37×10^{-2} 4.68×10^{-2}] N-m.

The experimentally measured $T_{\text{friction data}}(\omega_r)$ is documented in Figure 2.20.a by dots.

Consider

$$T_{\text{friction}} = B_m(\omega_r)\omega_r = p_1 \tanh(p_2\omega_r) + p_3\omega_r \text{ and } T_{\text{friction}} = B_m(\omega_r)\omega_r = \frac{p_1 + p_2 e^{-p_3\omega_r}}{1+p_4\omega_r}\omega_r.$$

The unknown coefficients p_i, found by solving a nonlinear least-squares parameterization problem, are

```
CoefficientsT1 = 3.2160e-02  1.7015e-02  3.8778e-05
CoefficientsT2 = 9.0471e-04  1.0000e+00  1.0000e+00  1.7087e-02
```

Hence,

$T_{\text{friction}}=0.0322\tanh(0.017\omega_r)+3.88\times10^{-5}\omega_r$ and $T_{\text{friction}}=\dfrac{9.05\times10^{-4}+e^{-\omega_r}}{1+0.0171\omega_r}\omega_r$.

Applying a linear assumption $T_{\text{viscous}}=B_m\omega_r$, at ω_r=[90 187 284 384] rad/sec, $B_m\big|_{\omega_r=90\,\frac{\text{rad}}{\text{sec}}}=3.64\times10^{-4}$, $B_m\big|_{\omega_r=187\,\frac{\text{rad}}{\text{sec}}}=2.09\times10^{-4}$, $B_m\big|_{\omega_r=284\,\frac{\text{rad}}{\text{sec}}}=1.54\times10^{-4}$, $B_m\big|_{\omega_r=384\,\frac{\text{rad}}{\text{sec}}}=1.22\times10^{-4}$ N-m-sec/rad.

These $B_m(\omega_r)$ are reported in Figures 2.20.b by four dots. For two applied expressions for $T_{\text{friction}}(\omega_r)$, one has

$$B_m(\omega_r)=\frac{p_1\tanh(p_2\omega_r)+p_3\omega_r}{\omega_r},\quad B_m(\omega_r)=\frac{p_1+p_2e^{-p_3\omega_r}}{1+p_4\omega_r}.$$

The unknown coefficients p_i are found by using the **nlinfit** solver.

```
% Parameterize and Deriving the Approximations for Te(wr)
format short e
wr=[90 187 284 384]; Tfriction=[3.28e-02 3.9e-02 4.37e-02 4.68e-02]; Bm=Tfriction./wr;
figure(1); plot(wr,Tfriction,'bo','linewidth',4);
xlabel('Angular Velocity, {\it\omega_r} [rad/sec]','FontSize',20);
ylabel('Measured and Interpolated {\itT}_{friction}, [N-m]','FontSize',16);
title('Viscous Friction {\itT}_{friction}({\it\omega_r})' ,'FontSize',20);
ModelFunctionT1=@(p,x) p(1).*tanh(p(2).*x)+p(3).*x; StartingValuesT1=[0.1 0.1 0];
CoefficientsT1=nlinfit(wr,Tfriction,ModelFunctionT1,StartingValuesT1), xgrid=linspace(0,400,100);
line(xgrid,ModelFunctionT1(CoefficientsT1,xgrid),'Color','k','linewidth',2);
ModelFunctionT2=@(p,x) x.*(p(1)+p(2).*exp(-p(3).*x))./(1+p(4).*x);
StartingValuesT2=[1 1 1 1]; CoefficientsT2=nlinfit(wr, Tfriction,ModelFunctionT2,StartingValuesT2)
line(xgrid,ModelFunctionT2(CoefficientsT2,xgrid),'Color','b','linewidth',3);
% Parameterize and Deriving the Approximations for Bm
figure(2); plot(wr,Bm,'o','linewidth',4); axis([10 400 1e-4 6e-4]);
xlabel('Angular Velocity, {\it\omega_r} [rad/sec]','FontSize',20);
ylabel('Measured and Interpolated {\itB_m}, [N-m-s/rad]','FontSize',16);
title('{\itB_m}({\it\omega_r}), [N-m-sec/rad]' ,'FontSize',20);
ModelFunctionB1=@(p,x) (p(1).*tanh(p(2).*x)+p(3).*x)./x; StartingValuesB1=[0.1 0.1 0];
CoefficientsB1=nlinfit(wr,Bm,ModelFunctionB1,StartingValuesB1), xgrid=linspace(0,400,100);
line(xgrid,ModelFunctionB1(CoefficientsB1,xgrid),'Color','k','linewidth',2);
ModelFunctionB2=@(p,x) (p(1)+p(2).*exp(-p(3).*x))./(1+p(4).*x); StartingValuesB2=[1 1 1 1];
CoefficientsB2=nlinfit(wr,Bm,ModelFunctionB2,StartingValuesB2), xgrid=linspace(0,400,100);
line(xgrid,ModelFunctionB2(CoefficientsB2,xgrid),'Color','b','linewidth',3);
```

The unknown p_i for interpolated $B_m(\omega_r)$ are

```
CoefficientsB1 = 3.1564e-02  1.7843e-02  4.0878e-05
CoefficientsB2 = 1.0170e-03  1.0000e+00  1.0000e+00  2.0005e-02
```

Hence,

$$B_m(\omega_r)=\frac{0.0316\tanh(0.0178\omega_r)+4.09\times10^{-5}\omega_r}{\omega_r}\ \text{and}\ B_m(\omega_r)=\frac{0.001+e^{-\omega_r}}{1+0.02\omega_r}.$$

Plots for $B_m(\omega_r)$ are documented in Figure 2.20.b. Interpolation and parameterization errors and convergence depend on the applied expressions and numeric algorithms. One may apply the bicubic, trilinear (three variable) and multivariate interpolations. Physics-consistent rational, trigonometric and exponential expressions for $T_{\text{friction}}(\omega_r)$ and $B_m(\omega_r)$ are found.

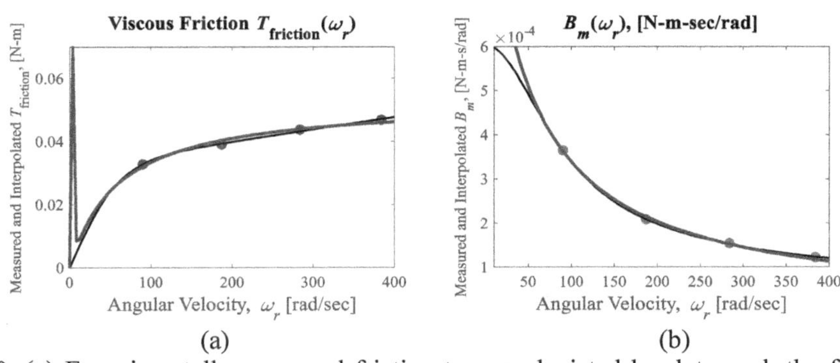

Figure 2.20. (a) Experimentally measured friction torque, depicted by dots, and, the friction torques described by $T_{\text{friction}} = 0.0322\tanh(0.017\omega_r) + 3.88\times10^{-5}\omega_r$ and $T_{\text{friction}} = \dfrac{9.05\times10^{-4} + e^{-\omega_r}}{1+0.0171\omega_r}\omega_r$;

(b) Measured B_m shown by dots, and interpolated expressions for $B_m(\omega_r)$. ■

Example 2.29. *Experimental Studies of Friction in Servos*

For a servo, illustrated in Figure 2.19.a, the experimentally measured $T_{\text{friction data}}$ for $\omega_r \in [-20\ 20]$ rad/sec is reported in Figures 2.21.

The expression for nonlinear friction torque is

$$T_{\text{friction}}(\omega_r) = B_{m10}\omega_r + B_{m11}\,\text{sgn}(\omega_r)|\omega_r|^{\frac{1}{3}} + b\,\text{sgn}(\omega_r)\left(1 - e^{-c\omega_r^2}\right).$$

The parameterization yields $B_{m10}=6.1\times10^{-5}$, $B_{m11}=1.92\times10^{-4}$, $b=5.8\times10^{-4}$ and $c=19.5$. The experimentally measured $T_{\text{friction data}}$ and plots for $T_{\text{friction}}(\omega_r)$ are documented in Figures 2.21.

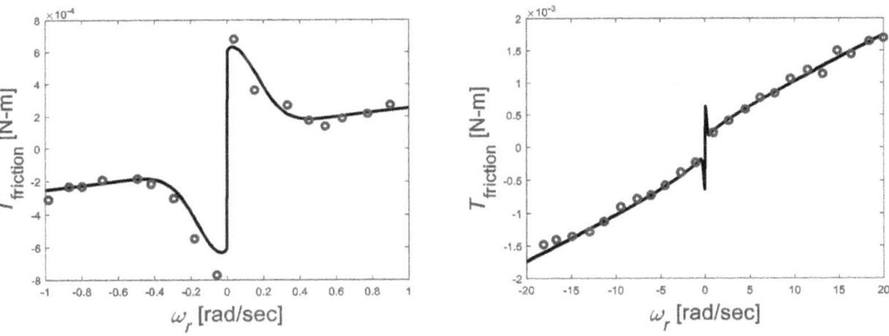

Figure 2.21. Experimental data for the measured $T_{\text{friction data}}$, depicted by dots, and, nonlinear friction torque $T_{\text{friction}} = B_{m10}\omega_r + B_{m11}\omega_r^{\frac{1}{3}} + b\,\text{sgn}(\omega_r)\left(1 - e^{-c\omega_r^2}\right)$, shown by solid lines. ■

Example 2.30. *Nonlinear Magnetization With a Hysteresis Loop*

Ferrites, alloys and ferromagnetic cores are used in inductors, transformers, electromagnetic devices, etc. These materials, e.g., ceramic, composite, thin and ultra-thin grain oriented and non-grain oriented silicon iron, electrical steel lamination as well as others, are characterized by nonlinear magnetization, *BH* curve, remanence B_r, magnetic coercivity H_c, maximum energy product $(BH)_{\max}$, hysteresis loop, etc.

The *mmf* is given by (2.35), $mmf = \oint_l \vec{H}(t)\cdot d\vec{l} = Ni$.

One obtains the relationship between electric field intensity H and current i.

Using the *effective* length l_e, we have $H = \frac{N}{l_e}i$ [A/m] or [A-turn/m].

From $B=\mu H=\mu_0\mu_r H$, one has $\mu_r(H,\frac{dH}{dt})=\frac{1}{\mu_0}\frac{dB(H,\frac{dH}{dt})}{dH}$.

For the ferromagnetic cores with nonlinear BH curves and hysteresis, apply a consistent approximation

$$B(H,\tfrac{dH}{dt})=B_{\max}\tanh\left(aH-\tfrac{a}{b}H_c\,\mathrm{sgn}(\tfrac{dH}{dt})\right),\ (B_{\max},H_c,a,b,)>0,$$

where a and b are the constants, found performing the parameterization using the experimental data.

From $\mu_r(H,\frac{dH}{dt})=\frac{1}{\mu_0}\frac{dB(H,\frac{dH}{dt})}{dH}$, we have

$$\mu_r(H,\tfrac{dH}{dt})=\tfrac{1}{\mu_0}B_{\max}a\,\mathrm{sech}^2\left(aH-\tfrac{a}{b}H_c\,\mathrm{sgn}(\tfrac{dH}{dt})\right),\ H=\tfrac{N}{l_e}i.$$

From (2.39), the inductance and reluctance are $L=\frac{\int_s \vec{B}\cdot d\vec{s}}{\oint_l \vec{H}\cdot d\vec{l}}$ and $\Re=\frac{\oint_l \vec{H}\cdot d\vec{l}}{\int_s \vec{B}\cdot d\vec{s}}$.

One obtains a nonlinear expression for the magnetizing inductance

$$L(i,\tfrac{di}{dt})\equiv\tfrac{N^2A}{l_e}\mu_0\mu_r(i,\tfrac{di}{dt}).$$

For the high-permeability electrical steel lamination core, N=100 and l=10 m. A fed current is $i\in[-75\ \ 75]$ A yields $H\in[-750\ \ 750]$ A/m. At a low frequency $f\lesssim100$ Hz, one measures $B_{\max}$=0.75 T and H_c=150 A/m. The coefficients are computed to be a=0.005 and b=2. Figure 2.22.a reports a SIMULINK diagram to compute and plot the BH curve and $\mu_r(i,\frac{di}{dt})$. Recall that $\mu_0=4\pi\times10^{-7}$ H/m. Figure 2.22.b documents a BH curve and varying $\mu_r(i,\frac{di}{dt})$. One observes a minor inconsistency for evolutions of the magnetic field to the saturation $B_{\max}$ from $(B_0,H_0)=0$. This discrepancy can be overcome by applying the conditional statement, and, is not critical because devices operate at $(B,H)\neq0$.

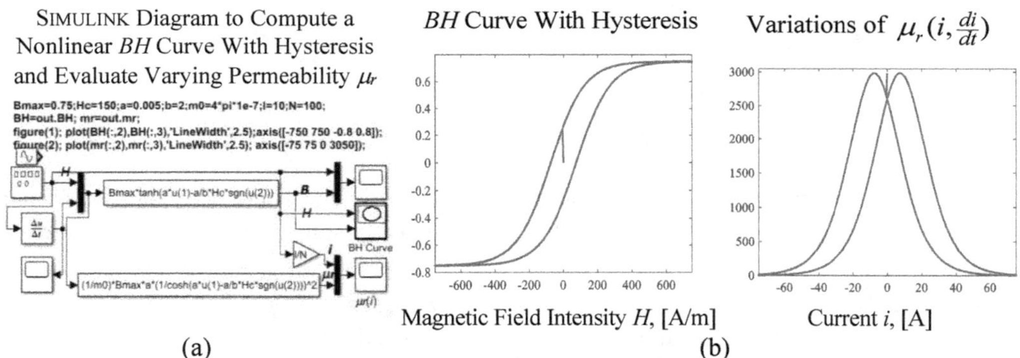

(a) (b)

Figure 2.22. (a) SIMULINK diagram to compute nonlinear steady-state BH curve with a hysteresis loop, as well as investigate the varying relative permeability $\mu_r(i,\frac{di}{dt})$;

(b) BH curve with a hysteresis, and, nonlinear relative permeability $\mu_r(i,\frac{di}{dt})$. The media-dependent parameters and constants are $B_{\max}$=0.75 T, H_c=150 A/m, $\mu_0=4\pi\times10^{-7}$ H/m, a=0.005, b=2, l=10 m, N=100. ■

2.7. Practice Problems

Practice Problem 2.1. *Two-Mass System*
Study a two-mass system (m_1,m_2) depicted in Figure 2.23.a. The damper (absorber) and viscous friction coefficients are B_v, B_{v1} and B_{v2}. The spring constant is k_s. Find differential equations in the x direction using the Newton second law. With the applied forces F_1 and F_2, the equations (2.6) are

$$m_1\frac{d^2x_1}{dt^2}=-(B_v+B_{v1})\frac{dx_1}{dt}-k_s(x_1-x_2)+F_1,$$

$$m_2\frac{d^2x_2}{dt^2}=-B_{v2}\frac{dx_2}{dt}+k_s(x_1-x_2)+F_2.$$

The resulting set of four first-order differential equations is

$$\frac{dx_1}{dt}=v_1,$$

$$\frac{dv_1}{dt}=\frac{1}{m_1}\left[-(B_v+B_{v1})v_1-k_s(x_1-x_2)+F_1\right],$$

$$\frac{dx_2}{dt}=v_2,$$

$$\frac{dv_2}{dt}=\frac{1}{m_2}\left[-B_{v2}v_2+k_s(x_1-x_2)+F_2\right].$$

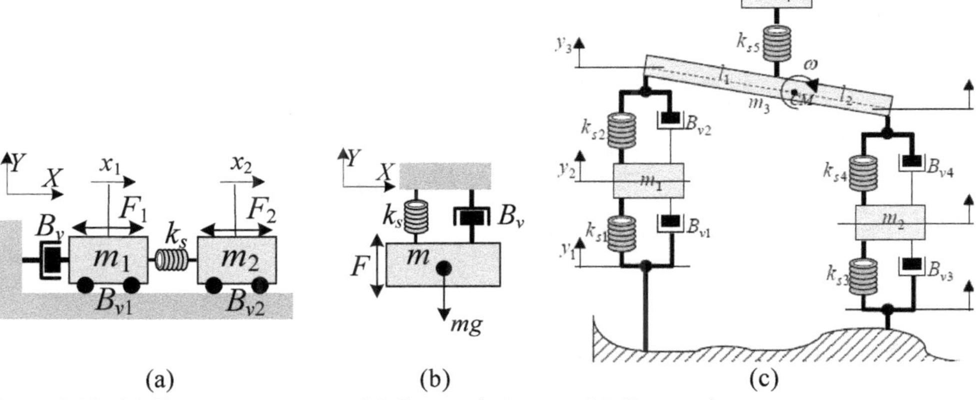

Figure 2.23. (a) Two-mass system; (b) Suspended mass; (c) Suspension system. ∎

Practice Problem 2.2.
A body is suspended with a spring with coefficient k_s, see Figure 2.23.b. The shock absorber has a damping coefficient B_v. Using Newton's second law for translational motion (2.6), one finds the second-order differential equation

$$m\frac{d^2y}{dt^2}=-B_v\frac{dy}{dt}-k_s y+F-mg.$$

The total kinetic Γ, dissipating D and potential Π energies are

$$\Gamma=\tfrac{1}{2}m\left(\frac{dy}{dt}\right)^2,\ D=\tfrac{1}{2}B_v\left(\frac{dy}{dt}\right)^2,\ \Pi=\tfrac{1}{2}k_s y^2.$$

From (2.16) $\frac{d}{dt}\left(\frac{\partial\Gamma}{\partial\dot{y}}\right)-\frac{\partial\Gamma}{\partial y}+\frac{\partial D}{\partial\dot{y}}+\frac{\partial\Pi}{\partial y}=Q$ with q=y and Q=($F_{\text{appplied}}-mg$),

$$\frac{d}{dt}\left(m\frac{dy}{dt}\right)+B_v\frac{dy}{dt}+k_s y=F-mg.$$

For the constant mass m, one finds a time-invariant differential equation

$$m\frac{d^2y}{dt^2}+B_v\frac{dy}{dt}+k_s y=F-mg.$$

■

Practice Problem 2.3. *Modeling of a Suspension System*

Study a suspension system illustrated in Figure 2.23.c. Find differential equations of motions within the y direction. The equivalent masses of the front and rear wheels, frame and seat are m_1, m_2, m_3 and m_4. The spring and damping coefficients are ($k_{s1},k_{s2},k_{s3},k_{s4},k_{s5}$) and ($B_{v1}$, B_{v2}, B_{v3}, B_{v4}). The distances from the left and right ends to the center of mass are denoted as l_1 and l_2.

The Newton's translational law (2.6) results in differential equations

$$m_1\frac{d^2y_2}{dt^2}=k_{s1}(y_1-y_2)+B_{v1}\left(\frac{dy_1}{dt}-\frac{dy_2}{dt}\right)+k_{s2}(y_3-y_2)+B_{v2}\left(\frac{dy_3}{dt}-\frac{dy_2}{dt}\right),$$

and $$m_2\frac{d^2y_6}{dt^2}=k_{s3}(y_5-y_6)+B_{v3}\left(\frac{dy_5}{dt}-\frac{dy_6}{dt}\right)+k_{s4}(y_7-y_6)+B_{v4}\left(\frac{dy_7}{dt}-\frac{dy_6}{dt}\right).$$

From $y_{CM}=y_3+\frac{l_1}{l_1+l_2}(y_7-y_3)$, we have

$$m_3\frac{d^2y_{CM}}{dt^2}=k_{s2}(y_2-y_3)+B_{v2}\left(\frac{dy_2}{dt}-\frac{dy_3}{dt}\right)+k_{s4}(y_6-y_7)$$
$$+B_{v4}\left(\frac{dy_6}{dt}-\frac{dy_7}{dt}\right)+k_{s5}\left(y_4-y_3-\frac{l_2}{l_1+l_2}(y_7-y_3)\right).$$

The Newtonian dynamics yields

$$m_4\frac{d^2y_4}{dt^2}=k_{s5}\left(y_3+\frac{l_2}{l_1+l_2}(y_7-y_3)-y_4\right).$$

Newton's rotational law gives

$$J\frac{d^2\theta}{dt^2}=-l_1k_{s2}(y_2-y_3)-l_1B_{v2}\left(\frac{dy_2}{dt}-\frac{dy_3}{dt}\right)+l_2k_{s4}(y_6-y_7)+l_2B_{v4}\left(\frac{dy_6}{dt}-\frac{dy_7}{dt}\right)$$
$$-(l_1-l_2)k_{s5}\left(y_4-y_3-\frac{l_2}{l_1+l_2}(y_7-y_3)\right).$$

A set of differential equations is derived. ■

Practice Problem 2.4. *Separately Exited Limited-Angle Electromagnetic Actuator*

The lumped-parameter model of a separately exited electromagnetic actuator was developed in Example 2.13. Consider the case when the rotor and stator windings are orthogonal. The mechanical, magnetic axes and *electrical* angular displacements are θ_r.

For an actuator schematics documented in Figure 2.5.a, consider the mutual inductance variations as

$$L_{sr}(\theta_r)=L_M\sin\theta_r=L_M\sin q_3,$$

where L_M is the amplitude of mutual inductance. Using descriptions and notations reported in Example 2.13, the *generalized coordinates* and *generalized forces* are

$q_1=\int i_s dt$, $q_2=\int i_r dt$, $q_3=\theta_r$, $\dot{q}_1=i_s$, $\dot{q}_2=i_r$, $\dot{q}_3=\omega_r$,

$Q_1=u_s$, $Q_2=u_r$, $Q_3=-T_L$.

The Lagrange equations (2.16) yield

$$\frac{d}{dt}\left(\frac{\partial\Gamma}{\partial\dot{q}_1}\right)-\frac{\partial\Gamma}{\partial q_1}+\frac{\partial D}{\partial\dot{q}_1}+\frac{\partial\Pi}{\partial q_1}=Q_1,$$

$$\frac{d}{dt}\left(\frac{\partial\Gamma}{\partial\dot{q}_2}\right)-\frac{\partial\Gamma}{\partial q_2}+\frac{\partial D}{\partial\dot{q}_2}+\frac{\partial\Pi}{\partial q_2}=Q_2,$$

$$\frac{d}{dt}\left(\frac{\partial\Gamma}{\partial\dot{q}_3}\right)-\frac{\partial\Gamma}{\partial q_3}+\frac{\partial D}{\partial\dot{q}_3}+\frac{\partial\Pi}{\partial q_3}=Q_3.$$

The total kinetic, potential and dissipating energy are

$$\Gamma=\tfrac{1}{2}L_s\dot{q}_1^2+L_M\dot{q}_1\dot{q}_2\sin q_3+\tfrac{1}{2}L_r\dot{q}_2^2+\tfrac{1}{2}J\dot{q}_3^2,$$

$$\Pi=\tfrac{1}{2}k_s q_3^2,$$

$$D=\tfrac{1}{2}r_s\dot{q}_1^2+\tfrac{1}{2}r_r\dot{q}_2^2+\tfrac{1}{2}B_m\dot{q}_3^2.$$

We have

$$L_s\frac{di_s}{dt}+L_M\sin\theta_r\frac{di_r}{dt}+L_M i_r\cos\theta_r\frac{d\theta_r}{dt}+r_s i_s=u_s,$$

$$L_r\frac{di_r}{dt}+L_M\sin\frac{di_s}{dt}+L_M i_s\cos\theta_r\frac{d\theta_r}{dt}+r_r i_r=u_r,$$

$$J\frac{d^2\theta_r}{dt^2}-L_M i_s i_r\cos\theta_r+B_m\frac{d\theta_r}{dt}+k_s\theta_r=-T_L.$$

Differential equations in Cauchy's form are found applying the Symbolic Math Toolbox. Using the first and second non-Cauchy equations

```
syms rs rr Ls Lr LM is ir dis dir wr th us ur
% Lagrange equations
eqn1=Ls*dis+LM*sin(th)*dir+LM*ir*cos(th)*wr+rs*is==us;
eqn2=Lr*dir+LM*sin(th)*dis+LM*is*cos(th)*wr+rr*ir==ur;
% Cauchy's equations for stator and rotor currents
sol=solve([eqn1 eqn2], [dis dir]); Dis=sol.dis; Dir=sol.dir;
pretty(Dis), pretty(Dir)
```

The resulting nonlinear differential equations are

$$\frac{di_s}{dt}=\frac{-r_sL_ri_s+L_M^2i_s\omega_r\sin\theta_r\cos\theta_r+r_rL_Mi_r\sin\theta_r-L_rL_Mi_r\omega_r\cos\theta_r+L_ru_s-L_M\sin\theta_ru_r}{L_sL_r-L_M^2\sin^2\theta_r},$$

$$\frac{di_r}{dt}=\frac{-r_rL_si_r+r_sL_Mi_s\sin\theta_r-L_sL_Mi_s\omega_r\cos\theta_r+L_M^2i_r\omega_r\sin\theta_r\cos\theta_r-L_M\sin\theta_ru_s+L_su_r}{L_sL_r-L_M^2\sin^2\theta_r},$$

$$\frac{d\omega_r}{dt}=\frac{1}{J}\left(L_Mi_si_r\cos\theta_r-B_m\omega_r-k_s\theta_r-T_L\right),$$

$$\frac{d\theta_r}{dt}=\omega_r.$$

Simulations are performed for the equations in non-Cauchy's form. The actuator parameters are r_s=2 ohm, r_r=2 ohm, L_s=0.05 H, L_r=0.05 H, L_M=0.1 H, B_m=0.0001 N-m-sec/rad, k_s=10 N-m/rad and J=0.005 kg-m^2.

Figures 2.24 report a SIMULINK diagram and simulations for $\theta_r(t)$ with initial conditions θ_{r0}=0.5 rad as well as θ_{r0}= –0.5 rad. The stator and rotor voltages are u_s=10 V and u_r=10 V. The torsional spring opposes the electromagnetic torque T_e. At steady-state if T_L=0, one has

T_e=T_{spring},

where $T_e = L_M i_s i_r \cos\theta_r$ and T_{spring}=$k_s\theta_r$.

As the dc voltages u_s=10 V and u_r=10 V are applied, the rotor rotates and assumes the equilibrium displacement θ_{re}=0.2427 rad.

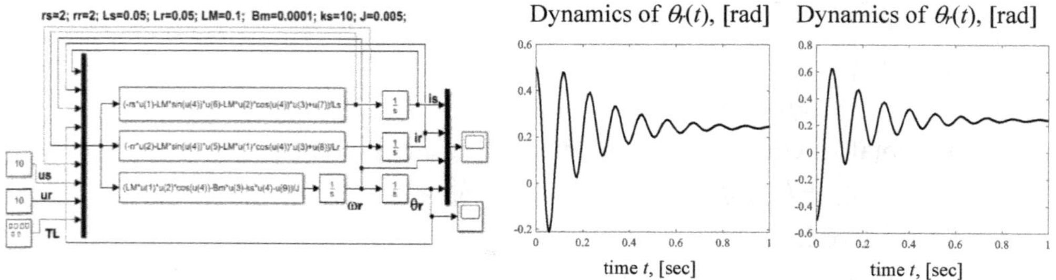

Figure 2.24. SIMULINK diagram, and, dynamics of angular displacement $\theta_r(t)$ if initial conditions are θ_{r0}=0.5 rad, θ_{r0}= –0.5 rad. The applied voltages are u_s=10 V, u_r=10 V.

The linearization of a nonlinear system is performed using a method, covered in section 6.2. The linearized model in the state-space form

$$\dot{\delta}_x(t) = A\delta_x + B\delta_u,\ \dot{x}(t) = Ax + Bu,\ x \equiv \delta_x,\ u \equiv \delta_u,$$

is found by deriving the Jacobian matrices (J, J_u).

Compute an *equilibrium inputs* $\bar{u} \in \mathbb{R}^2$ for chosen *equilibrium states* $x_e = \bar{x}$, $\bar{x} \in \mathbb{R}^4$, such as

$$\dot{x}_e = F(x_e,\bar{u}) = F(\bar{x},\bar{u}) = 0,\ F(\bar{x},\bar{u}) = 0.$$

We obtain

$$\dot{\delta}_x(t) = \left.\frac{\partial F}{\partial x}\right|_{\substack{x=\bar{x}\\ u=\bar{u}}} \delta_x + \left.\frac{\partial F}{\partial u}\right|_{\substack{x=\bar{x}\\ u=\bar{u}}} \delta_u = J|_{\substack{x=\bar{x}\\ u=\bar{u}}}\, \delta_x + J_u|_{\substack{x=\bar{x}\\ u=\bar{u}}}\, \delta_u = \left.\begin{bmatrix} \frac{\partial f_1}{\partial x_1} & \frac{\partial f_1}{\partial x_2} & \frac{\partial f_1}{\partial x_3} & \frac{\partial f_1}{\partial x_4} \\ \frac{\partial f_2}{\partial x_1} & \frac{\partial f_2}{\partial x_2} & \frac{\partial f_2}{\partial x_3} & \frac{\partial f_2}{\partial x_4} \\ \frac{\partial f_3}{\partial x_1} & \frac{\partial f_3}{\partial x_2} & \frac{\partial f_3}{\partial x_3} & \frac{\partial f_3}{\partial x_4} \\ \frac{\partial f_4}{\partial x_1} & \frac{\partial f_4}{\partial x_2} & \frac{\partial f_4}{\partial x_3} & \frac{\partial f_4}{\partial x_4} \end{bmatrix}\right|_{\substack{x=\bar{x}\\ u=\bar{u}}} \delta_x + \left.\begin{bmatrix} \frac{\partial f_1}{\partial u_1} & \frac{\partial f_1}{\partial u_2} \\ \frac{\partial f_2}{\partial u_1} & \frac{\partial f_2}{\partial u_2} \\ \frac{\partial f_3}{\partial u_1} & \frac{\partial f_3}{\partial u_2} \\ \frac{\partial f_4}{\partial u_1} & \frac{\partial f_4}{\partial u_2} \end{bmatrix}\right|_{\substack{x=\bar{x}\\ u=\bar{u}}} \delta_u.$$

At no disturbances T_L=0, for $\bar{x} = \begin{bmatrix} \bar{i}_s \\ \bar{i}_r \\ \bar{\omega}_r \\ \bar{\theta}_r \end{bmatrix} = \begin{bmatrix} 2 \\ 2 \\ 0 \\ 0.2427 \end{bmatrix}$, the equilibria for inputs are found to be $\bar{u} = \begin{bmatrix} \bar{u}_s \\ \bar{u}_r \end{bmatrix} = \begin{bmatrix} 10 \\ 10 \end{bmatrix}$.

Analytic and numeric solutions are found and computed as

```
syms rs rr Ls Lr LM Bm ks J is ir dis dir dwr wr th us ur TL
% Lagrange equations
eqn1=Ls*dis+LM*sin(th)*dir+LM*ir*cos(th)*wr+rs*is==us;
eqn2=Lr*dir+LM*sin(th)*dis+LM*is*cos(th)*wr+rr*ir==ur;
eqn3=J*dwr-LM*is*ir*cos(th)+Bm*wr+ks*th==-TL;
% Cauchy's equations
sol=solve([eqn1 eqn2 eqn3], [dis dir dwr]);
Dis=sol.dis; Dir=sol.dir; Dwr=sol.dwr;
pretty(Dis), pretty(Dir), pretty(Dwr)
rs=2; rr=2; Ls=0.05; Lr=0.05; LM=0.1; Bm=0.0001; ks=10; J=0.005;
x1=is; x2=ir; x3=wr; x4=th; % States
% Derivative of states
dx1=subs(Dis); dx2=subs(Dir); dx3=subs(Dwr); dx4=x3;
% Jacobian matrices
Jx=[diff(dx1,x1) diff(dx1, x2) diff(dx1, x3) diff(dx1, x4);
    diff(dx2,x1) diff(dx2, x2) diff(dx2, x3) diff(dx2, x4);
    diff(dx3,x1) diff(dx3, x2) diff(dx3, x3) diff(dx3, x4);
    diff(dx4,x1) diff(dx4, x2) diff(dx4, x3) diff(dx4, x4)];
Jb=[diff(dx1,us) diff(dx1,ur);diff(dx2,us) diff(dx2,ur);diff(dx3,us) diff(dx3,ur);diff(dx4,us) diff(dx4,ur)];
is=2; ir=2; wr=0; th=0.2427; us=10; ur=10; TL=0;  % Equilibrium
% Linearized system dx/dt=Ax+Bu
S1=subs(Jx); S2=subs(Jb);
A=double(S1), B=double(S2), H=[0 0 0 1]; D=[0 0; 0 0];
```

The MATLAB yields the resulting matrices A and B

```
A =
 -5.2017e+01  2.5002e+01 -2.6223e+00 -1.0626e+02
  2.5002e+01 -5.2017e+01 -2.6223e+00 -1.0626e+02
  3.8828e+01  3.8828e+01 -2.0000e-02 -2.0192e+03
           0           0  1.0000e+00           0
B =
  2.6009e+01 -1.2501e+01
 -1.2501e+01  2.6009e+01
           0           0
           0           0
```

We have

$$\dot{x} = \begin{bmatrix} \dot{x}_1 \\ \dot{x}_2 \\ \dot{x}_3 \\ \dot{x}_4 \end{bmatrix} = \begin{bmatrix} \frac{di_s}{dt} \\ \frac{di_r}{dt} \\ \frac{d\omega_r}{dt} \\ \frac{d\theta_r}{dt} \end{bmatrix} = Ax + Bu = \begin{bmatrix} -52 & 25 & -2.62 & -106.2 \\ 25 & -52 & -2.62 & -106.2 \\ 38.8 & 38.8 & -0.02 & -2019.2 \\ 0 & 0 & 1 & 0 \end{bmatrix} x + \begin{bmatrix} 26 & -12.5 \\ -12.5 & 26 \\ 0 & 0 \\ 0 & 0 \end{bmatrix} u,\ u = \begin{bmatrix} u_s \\ u_r \end{bmatrix}.$$

■

Practice Problem 2.5. *Modeling of a Two-Phase Induction Motor Using Lagrange Equations of Motion*

Develop a mathematical model of a two-phase induction motors using the Lagrange equations of motions. *Practice Problems* 3.8 and 3.9 report model developments. Simulations and dynamic analysis are conducted in *Practice Problem* 3.10. These problems can be solved by applying methodologies covered in this chapter. ■

Homework Problems

Homework Problem 2.1. Three-Degree-of-Freedom Manipulator
Consider a three-degree-of-freedom manipulator, shown in Figure 2.23.a. With angular displacements (θ_1,θ_2,θ_3), the independent *generalized coordinates* are (q_1,q_2,q_1). The *generalized forces* are the actuators' torques (T_{e1},T_{e2},T_{e3}) and load torques (T_{L1},T_{L2},T_{L3}). There are viscous frictions in three joints, $T_{\text{friction}i}=B_{mi}\omega_i$.
Problems to solve:

1. Using the Lagrange equations of motion (2.16), develop a mathematical model;
2. Perform simulations if m_1=1 kg, m_2=0.75 kg, m_3=0.5 kg, l_1=0.25 m, l_2=0.25 m, l_3=0.25 m, B_{m1}=0.1 N-m-sec/rad, B_{m2}=0.1 N-m-sec/rad, B_{m3}=0.1 N-m-sec/rad. Let $T_{e1}=0.1e^{-t}$ N-m, $T_{e2}=-0.1e^{-t}$ N-m, $T_{e3}=0.1e^{-t}$ N-m, while the load torques are T_{L1}=0, $T_{L2}=\mp0.05$ N-m, $T_{L3}=\pm0.05$ N-m at frequency f=0.25 Hz. The initial conditions are θ_{10}=0, $\theta_{20}=-1$ rad and θ_{30}=1 rad;
3. Plot dynamics of $\omega_1(t)$, $\theta_1(t)$, $\omega_2(t)$, $\theta_2(t)$, $\omega_3(t)$ and $\theta_3(t)$.

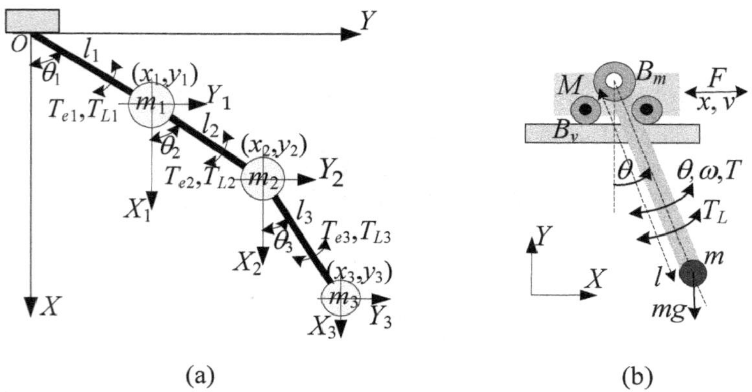

Figure 2.23. (a) Three-degree-of-freedom manipulator;
(b) Four-wheel moving platform with a downward arm: Lumped-parameter schematic.

Homework Problem 2.2. Translational-Rotational System
Consider an open-loop stable non-inverted two-degree-of-freedom translational-rotational system, described in Example 2.9. This problem is studied in section 6.15.2. Let the arm is at downward position, $\theta_0=0.25\pi$ rad.
Problems to solve:

1. Develop a mathematical model. The arm displacement r is $0.25\le r\le0.5$ m. Let m_1=1 kg, m_2=0.25 kg, r_1=0.25 m, r_0=0.375 m (spring *free length*), B_m=0.01 N-m-sec/rad, B_v=0.01 N-sec/m and k_s=10 N/m. Mechanical limits on the angular and linear displacements are $-\frac{1}{2}\pi\le\theta<\frac{1}{2}\pi$ rad and $0.25\le r\le0.5$ m;
2. Perform nonlinear simulations for $T_L=0.1e^{-t}\cos(t)$ N-m and $F_L=0.1e^{-t}\sin(t)$ N. Report transient dynamics for state variables if applied (T, F) are T=0 and F=0.

Homework Problem 2.3. Inverted Translational-Rotational System

Example 2.10 studies the problem when the arm is pointed upward. The relevant mathematical model is developed and simulated in section 6.15.1.

Problems to solve:

1. Develop a mathematical model for open-loop stable four-wheel non-inverted system if the arm is pointed downward, see Figure 2.23.b;
2. Simulate a system, and, report simulation results. Let M=1 kg, m=0.25 kg, l=1 m, B_v=0.01 N-sec/m and B_m=0.01 N-m-sec/rad. The linear and angular displacements are constrained as $-1\le x\le 1$ m and $-\frac{1}{2}\pi\le\theta<\frac{1}{2}\pi$ rad. Consider:
(i) Actuators' forces and torques are F=0 and T=0. The loads are $F_L=e^{-t}$ N and $T_L=0.25e^{-t}$ N-m. The initial conditions are x_0=0 m and $\theta_0=0.25\pi$ rad;
(ii) Actuators' forces and torques are $F=e^{-t}\cos(t)$ N and $T=0.25e^{-t}\sin(t)$ N-m. The loads are F_L=0 and T_L=0. The initial conditions are x_0=0 and θ_0=0. Report transient dynamics for $v(t)$, $x(t)$, $\omega(t)$ and $\theta(t)$.

Homework Problem 2.4. Separately Exited Limited-Angle Electromagnetic Actuator

Consider an actuator studied in Example 2.13 and Practice Problem 2.4.

Problems to solve:

1. Develop a mathematical model using the Lagrange equations of motion if a mutual inductance is $L_{sr}(\theta_r)=L_M\sin^3\theta_r$;
2. Find the resulting equations in Cauchy's form;
3. Perform simulations applying dc voltages (u_s, u_r), $|u_s, u_r|\le 20$ V, $u_{s\max}=\pm 20$ V and $u_{r\max}=\pm 20$ V. The parameters are r_s=2 ohm, r_r=2 ohm, L_s=0.05 H, L_r=0.05 H, L_M=0.2 H, B_m=0.0001 N-m-sec/rad, k_s=5 N-m/rad and J=0.01 kg-m^2. The electromagnetic torque T_e depends on the amplitude and polarity of the stator and rotor voltages (u_s, u_r). For $|\theta_r|\le 0.5$ rad, initial angular displacements are $\theta_{r0}\neq 0$ rad. The load torque is $T_L=\pm 1$ N-m, f=1 Hz.

Homework Problem 2.5. RLC Circuit

For the circuit reported in Figure 2.4.a, the differential equations are

$$\ddot{q}_1=\frac{1}{L_1+L_{12}}\left(-\frac{q_1}{C_1}-R_1\dot{q}_1+L_{12}\ddot{q}_2+u\right),$$

$$\ddot{q}_2=\frac{1}{L_2+L_{12}}\left(L_{12}\ddot{q}_1-\frac{q_2}{C_2}-R_2\dot{q}_2\right).$$

Problems to solve:

1. Develop the SIMULINK diagram, and, perform simulation if L_1=0.01 H, L_2=0.005 H, L_{12}=0.0025 H, C_1=0.02 F, C_2=0.1 F, R_1=10 ohm, R_2=5 ohm and $u=100\sin(200t)$ V;
2. Report dynamics of *generalized coordinates* (q_1,q_2) and currents (i_1,i_2);
3. Derive the Cauchy form of differential equations.

Homework Problem 2.6. Modified Lotka-Volterra Equation

Consider governing differential equations

$$\frac{dx_1}{dt} = a_{11}x_1 - a_{12}x_1x_2 + a_{13}x_1e^{a_{14}\sin x_2} = x_1\left(a_{11} - a_{12}x_2 + a_{13}e^{a_{14}\sin x_2}\right),\ x_1(t_0)=x_{10},\ \forall(a_{ii},a_{ij})>0,$$

$$\frac{dx_2}{dt} = -a_{21}x_2 + a_{22}x_1x_2 - a_{23}x_2e^{a_{24}\cos x_1} = x_2\left(-a_{21} + a_{22}x_1 - a_{23}e^{a_{24}\cos x_1}\right),\ x_2(t_0)=x_{20}.$$

Here, $(x_1,x_2)>0$, and, initial conditions are $(x_{10},x_{20})>0$.

Problems to solve:

1. Separating the variables, analytically derive the phase plane. Analyze stability and evolutions with different initial conditions $(x_{10},x_{20})>0$. Plot the phase planes;
2. Perform simulations in SIMULINK if $a_{11}=1$, $a_{12}=1$, $a_{13}=1$, $a_{14}=1$, $a_{21}=1$, $a_{22}=1$, $a_{23}=1$ and $a_{24}=1$. Use different initial conditions $(x_{10},x_{20})>0$.

Homework Problem 2.7. Deterministic and Stochastic Differential Equations

Consider deterministic and stochastic (white noise perturbation) nonlinear differential equations

$$\frac{dx_1}{dt} = a_{11}(-x_1 + x_2) + \sin(x_1x_2) + b_{11}d(t),$$

$$\frac{dx_2}{dt} = a_{21}x_1 + a_{22}x_2 + \cos(x_1x_3) - a_{23}x_1x_3,$$

$$\frac{dx_3}{dt} = -a_{31}x_3 + x_1x_2.$$

Problems to solve:

1. Develop a SIMULINK diagram, and, perform simulations if $a_{11}=5$, $a_{21}=12.5$, $a_{22}=2$, $a_{23}=1$, $a_{31}=2$ and $b_{11}=1$. The initial conditions are $x_{10}=5$, $x_{20}=0$ and $x_{30}=-5$. A deterministic perturbations are $d(t)=0$, $d(t)=\sin(t)$ and $d(t)=e^{-\sin t}$;
2. Simulate stochastic system, perturbed by $d(t)$, which is the white noise ξ characterized by the finite variances $\sigma^2=0.1$ and $\sigma^2=1$.

References

1. P. Derler, E. A. Lee and A. S. Vincentelli, "Modeling cyber-physical systems," *Proc. IEEE*, vol. 100, issue 1, pp. 13-28, 2012.
2. I. Akkaya, P. Derler, S. Emoto and E. A. Lee, "Systems engineering for industrial cyber-physical systems," *Proc. IEEE*, vol. 104, issue 5, pp. 997-1012, 2016.
3. P. Novak, P. Kadera and M. Wimmer, "Agent-based modeling and simulation of hybrid cyber-physical systems," *Proc. Cybernetics Conf.*, pp. 1-8, 2017.
4. V. Giurgiutiu and S. E. Lyshevski, *Micromechatronics: Modeling, Analysis, and Design With MATLAB®*, CRC Press, Boca Raton, FL, 2009.
5. P. C. Krause and O. Wasynczuk, *Electromechanical Motion Devices*, McGraw-Hill Book Company, New York, 1989.
6. P. C. Krause, O. Wasynczuk and S. D. Sudhoff, *Analysis of Electric Machinery*, IEEE Press, New York, 1994.
7. S. E. Lyshevski, *MEMS and NEMS: Systems, Devices, and Structures*, CRC Press, Boca Raton, FL, 2002.
8. S. E. Lyshevski, *Mechatronics and Control of Electromechanical Systems*, CRC Press, Boca Raton, FL, 2017.
9. D. C. White and H. H. Woodson, *Electromechanical Energy Conversion*, Wiley, New York, 1959.
10. J. Anderson, *Aircraft Performance and Design*, McGraw-Hill, NY, 1999.
11. A. E. Bryson, *Control of Spacecraft and Aircraft.* Princeton University Press, Princeton, NJ, 1993.
12. B. Etkin and L. D. Reid, *Dynamics of Flight: Stability and Control*, John Wiley and Sons, NY, 1995.
13. D. McRuer, I. Ashkenas and D. Graham, *Aircraft Dynamics and Automatic Control.* Princeton University Press, Princeton, NJ, 1976.
14. A. Miele, *Flight Mechanics. Volume 1: Theory of Flight Path,* Addison-Wesley, Reading, MA, 1962.
15. R. C. Nelson, *Flight Stability and Automatic Control*, McGraw-Hill, NY, 2007.
16. E. Seckel, *Stability and Control of Airplanes and Helicopters*, Academic Press, NY, 1964.
17. R. F. Stengel, *Flight Dynamics*, Princeton University Press, NJ, 2022.
18. B. L. Stevens and F. L. Lewis, *Aircraft Control and Simulation*, John Wiley and Sons, NY, 2003.
19. B. L. Stevens, F. L. Lewis and E. N. Johnson, *Aircraft Control and Simulation Dynamics, Controls Design, and Autonomous Systems*, Wiley, NY, 2015.
20. S. E. Lyshevski, *Engineering and Scientific Computations Using MATLAB®*, Wiley, Hoboken, NJ, 2003.
21. *MATLAB R*2023, www.mathworks.com, MathWorks, Inc., 2023.
22. S. E. Lyshevski, "Control of high-precision direct-drive mechatronic servos: Tracking control with adaptive friction estimation and compensation," *Mechatronics,* vol. 43, no. 1, pp. 1-5, 2017.

3 Actuators, Sensors and Electronics

3.1. Introduction

Energy conversion devices, power electronics, sensors, MPUs and other modules are key components of physical systems. Electromagnetic, electrostatic, electroacoustic, piezoelectric, photoelectric, hydraulic, thermoelectric, ultrasonic and other transducers convert electromagnetic, mechanical, thermal and chemical energy from one form to another. Actuators, sensors and transducers ensure energy transductions yielding motions and measurements of physical quantities. Examples of transducers are *smart* piezoelectric ceramic, electrostatic and electromagnetic devices used for actuation and sensing. There are specifics in design and selection of actuators, sensors and transducers within a modular system organization. *Active* and *passive* rotational and translational transducers are characterized by the source of energy, phenomena exhibited, physics of energy conversion and transduction, etc. Traducers are characterized by:

1. Force and torque densities;
2. Energy density;
3. Efficiency and losses;
4. Repeatability, linearity and hysteresis;
5. Stability and sensitivity;
6. Accuracy, precision, resolution, static error and dynamic error;
7. Dynamic range and bandwidth;
8. Dimensionality, weight and volumetric;
9. Failure rate, lifetime, etc.

Control of energy conversion in actuators and sensors are accomplished by regulating schemes accomplished by power amplifiers, dc-dc regulators, inverters, power management electronics, etc. Using pulse-width modulation (PWM) schemes, analog controllers and MPUs regulate switching activities of power MOSFETs. Examine electrostatic, electromagnetic and microelectromechanical actuators and sensors [1-11]. Low-power microelectronics and high-power-density power electronics empower signal conditioning and signal processing supporting control, filtering and interfacing [3, 9, 12-15]. High torque and energy densities electromagnetic actuators guarantee efficiency and effectiveness. For sensing and microactuation, electrostatic devices ensure adequate performance [1-3, 8-10]. Mini and micro scale transducers, actuators and sensors are fabricated using micromachining, 3D printing, additive manufacturing, precision machining, pressure bonding and injection molding of permanent magnets, piezoelectric ceramic, etc. Examples are inertial sensors, magnetic field and pressure sensors, photodetectors, receivers, MEMS strain gauges, proximity sensors, etc. Microfabrication and computer-integrated manufacturing guarantee affordability, high yield and superior capabilities. Carbon, silicon and III-V semiconductor microelectronics, optoelectronics, photonics and MEMS sensors measure various quantities, such as:

1. Electromagnetic quantities, such as capacitance, inductance, current, voltage, charge, power, field intensity and density, etc.;
2. Mechanical and thermal quantities, such as the acceleration, velocity, displacement, force, torque, strain, stress, temperature, temperature gradient, etc.;
3. Optical and optoelectronic quantities;
4. Chemicals and environmental characteristics, such as the mass and molar concentrations, humidity, pressure, temperature, atmospheric concentrations, etc.

Commercialized Microtransducers – To actuate rotational and translational microassemblies, actuators develop the electrostatic or electromagnetic force and torque. The loads and disturbances are evaluated. In many applications, electrostatic devices are used [1-3]. For example, more than 1,000,000 electrostatic 10×10 μm micromirrors are repositioned in the Texas Instruments digital light processing (DLP) module, shown in Figure 3.1.

Texas Instrument digital light processing (DLP) module with more than 1,000,000 individually controlled electrostatic torsional micromirrors

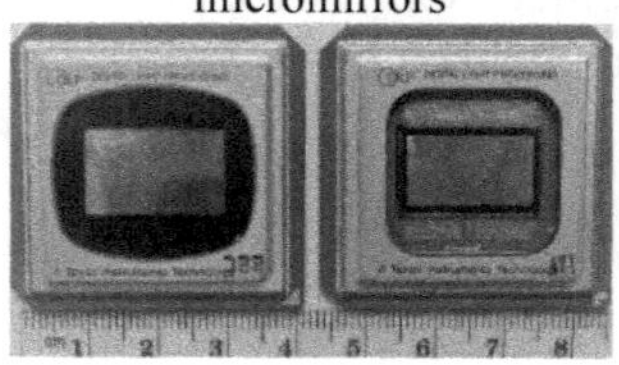

RF transceiver module XBee PRO (200 Kbps, 902 to 928 MHz, 24 dBm output power, 215 mA), and, REYAX RYLR896 transceiver module for IoT long range communication: 15 dBm, –148 dBm sensitivity, 820 to 1020 MHz frequency, 15 km range, 3.6 V, 43 mA transmitter, 17 mA receiver

Analog Devices MEMS technology gyroscopes on evaluation boards

MDOF sensors: Inertial navigation system (INS) and inertial measurement unit (IMU). Nine-degree-of-freedom IMU with triaxial accelerometer, gyroscope and magnetometer is used in system-in-package INS BNO055 and MPU-6500

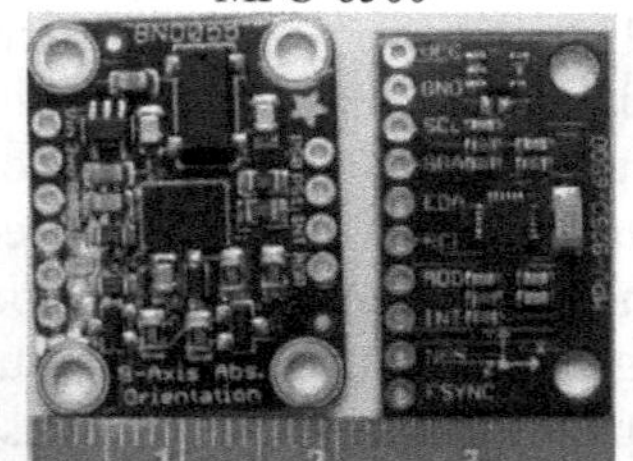

Texas Instruments OPT101 monolithic photodiode with on-chip amplifier

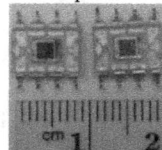

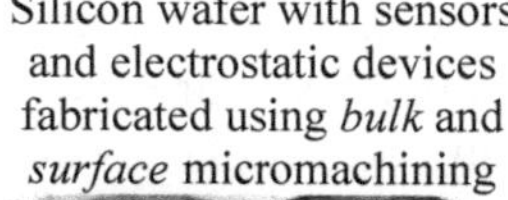

Silicon wafer with sensors and electrostatic devices fabricated using *bulk* and *surface* micromachining

Diced accelerometers, pressure and temperature sensors

Wire-bonded packaged MEMS actuators and sensors

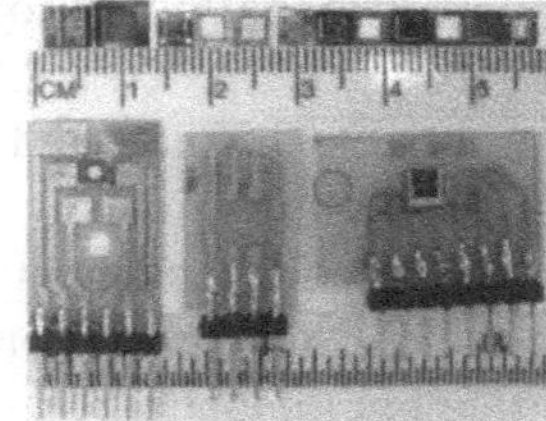

Figure 3.1. Microsystem technology sensors, actuators and transceivers.

In the Texas Instruments DLP470TP, 10×10 μm torsional mirrors within 3840×2160 array are steered by μN electrostatic force. These DLPs are used in high-definition displays and projection systems. Electrostatic microactuators are individually controlled ensuring 8000 Hz bandwidth, which implies millisecond repositioning. Multi-degree-of-freedom (MDOF) microaccelerometers and gyroscopes are designed and commercialized by Analog Devices, Bosch Sensortec, InvenSense, STMicroelectronics and other companies. Using microelectronic and MEMS technologies, electrostatic actuators, sensors and transceivers have being fabricated, diced, bonded and packaged.

Commercialized High-Power and High-Energy Densities Drives and Servos – Having emphasized mini- and micro transducers, in aerial, ground, surface and underwater vehicle propulsion systems, high-torque density ac electromagnetic actuators are used. For example, in propulsion systems of unmanned aerial vehicles, electric ground vehicle drives and ship integrated electric propulsion, 40 to 400 W, 40 to 400 kW and 4 to 40 MW synchronous and induction motors have being commercialized.

Energy Estimates, Force and Torque Analyses – Electromagnetic devices ensure high force, torque and energy densities. There are variable reluctance (solenoids, relays, electromagnets, magnetic levitation, switches, etc.), permanent magnet and other electromagnetic transducers. The stored electric and magnetic volume energy densities ρ_{W_e} and ρ_{W_m} for electrostatic and electromagnetic transducers are

$$\rho_{W_e} = \tfrac{1}{2}\varepsilon E^2,\ \rho_{W_m} = \tfrac{1}{2}\mu^{-1}B^2 = \tfrac{1}{2}\mu H^2, \tag{3.1}$$

where ε is the permittivity, $\varepsilon=\varepsilon_0\varepsilon_r$; ε_0 and ε_r are the permittivity of free space and relative permittivity, $\varepsilon_0=8.85\times10^{-12}$ F/m; E is the electric field intensity; μ is the permeability, $\mu=\mu_0\mu_r$; μ_0 and μ_r are the permeability of free space and relative permeability, $\mu_0=4\pi\times10^{-7}$ T-m/A; B and H are the magnetic field density and intensity, see section 2.3 and Example 2.30 which analyze the BH curve with a hysteresis and nonlinear $\mu = \frac{dB}{dH}$, investigating the nonlinear relative permeability μ_r.

Electrostatic and Electromagnetic Actuators – The maximum energy density of electrostatic actuators is limited by the maximum electric field E which can be applied before electrostatic breakdown occurs. In mini and micro structures, the maximum electric field cannot exceed $E_{\max}$.

The maximum energy density is $\rho_{W_e\max} = \tfrac{1}{2}\varepsilon_0\varepsilon_r E_{\max}^2$.

The maximum electric field is estimated as $E_{\max}=V_{\max}/x$. In 10×10 μm to millimeter size structures with a few micrometers air gap, $E_{\max}$ is 1×10^6 V/m. This yields $\rho_{W_e\max}$ to be 10 J/m^3.

The relative permittivity ε_r of media vary, for example, $\varepsilon_{r\,\text{polytetrafluoroethylene}}=2.1$, $\varepsilon_{r\,\text{polyimide}}=3.4$, $\varepsilon_{r\,\text{Si}}=11.68$, $\varepsilon_{r\,\text{methanol}}=30$, $\varepsilon_{r\max\,\text{lead zirconate titanate}}=5500$.

For electromagnetic actuators, the maximum energy density $\rho_{W_m\max}$ depends on the BH curve, and, limited by the saturation flux density B_{sat}. These topics were covered in section 2.3 and Example 2.30. For many general purpose industrial transducers with electromagnets and permanent magnets, $0.5<B_{\text{sat max}}<2.5$ T. For permanent magnets, $B_{\text{sat max}}\approx1$ T.

The maximum energy product $(BH)_{max}$ in ferrite, alnico, $SmCo_5$ and $Nd_2Fe_{14}B$ sintered magnets vary from 30 to 400 kJ/m^3. The media permeability μ_r varies. For example, the μ_r estimates for electrical steel laminations and superpermalloy $Ni_{79\%}Fe_{16\%}Mo_{5\%}$ reach 1000 to 100,000.

One has $\rho_{W_m} \gg \rho_{W_e}$. In electrostatic and electromagnetic devices, the energy is

$$\mathbf{E}_{\text{electrostatic}} = \tfrac{1}{2}\int_v E \cdot D dv,\ \mathbf{E}_{\text{magnetic}} = \tfrac{1}{2}\int_v H \cdot B dv. \tag{3.2}$$

Energy Conversion in Electrostatic and Electromagnetic Transducers – In electrostatic transducers, the energy is stored in electric field between two surfaces in capacitors. The co-energy is

$$W_{c\ \text{electrostatic}} = \tfrac{1}{2}QV = \tfrac{1}{2}CV^2.$$

In electromagnetic transducers, using the magnetizing inductance L and flux linkages ψ,

$$W_{c\ \text{magnetic}} = \tfrac{1}{2}Li^2 = \tfrac{1}{2}\psi i.$$

One examines the energy variations due to varying displacement-dependent capacitance C or magnetizing inductance L. The electrostatic and electromagnetic forces and torques are given below.

For the electrostatic devices, we have

$$F_e = \frac{\partial W_c}{\partial x} = \frac{\partial}{\partial x}\tfrac{1}{2}C(x)V^2 = \tfrac{1}{2}\frac{\partial C(x)}{\partial x}V^2,\ T_e = \frac{\partial W_c}{\partial \theta_r} = \frac{\partial}{\partial \theta_r}\tfrac{1}{2}C(\theta_r)V^2 = \tfrac{1}{2}\frac{\partial C(\theta_r)}{\partial \theta_r}V^2, \tag{3.3.1}$$

For electromagnetic devices, one has

$$F_e = \frac{\partial W_c}{\partial x} = \frac{\partial}{\partial x}\tfrac{1}{2}L(x)i^2 = \tfrac{1}{2}\frac{\partial L(x)}{\partial x}i^2,\ T_e = \frac{\partial W_c}{\partial \theta_r} = \frac{\partial}{\partial \theta_r}\tfrac{1}{2}L(\theta_r)i^2 = \tfrac{1}{2}\frac{\partial L(\theta_r)}{\partial \theta_r}i^2. \tag{3.3.2}$$

Design Flow – The following sequential steps are carried out:

1. For a given application, specifications and requirements, define technological limits and constraints;
2. Evaluate implement and advance existing solutions by examining components, operating principles, integration, compatibility, etc. Select or design transducers, electronic and control solutions. This task is partitioned into many subtasks related to actuators, sensors, filters, regulators, data fusion and sharing, input-output devices, actuator-sensor-electronics integration, communication, etc.;
3. Perform electromagnetic, energy conversion, mechanical, thermal and sizing estimates;
4. Define materials, processes and technologies to fabricate structures, design assemblies and package components. Integrate components, devices and modules in a functional physical system;
5. Conduct data-intensive electromagnetic, mechanical, vibroacoustic and thermodynamic analyses;
6. Control information sharing and optimize information management;
7. Test, characterize, evaluate and substantiate devices, components, modules and system;
8. Optimize and re-design system ensuring best performance and capabilities.

We investigate advanced-technology transducers and pertained electronics. Basic, applied and experimental studies are conducted and reported.

3.2. Electrostatic Transducers

Consider the translational and rotational electrostatic actuators and sensors fabricated using cost-effective high-yield micromachining, thin film deposition, electroplating and other technologies. The images of electrostatic MEMS are depicted in Figure 3.1. High-performance electrostatic transducers and integrated MEMS are mass-fabricated and commercialized by Analog Devices, Bosch Sensortec, Knowles, STMicroelectronics, Texas Instruments and other companies. Electrostatic transducers are used in audio, communication, healthcare, imaging, microscopy, navigation, photonic, positioning, resonators, robotic, scanners, video and other systems. These transducers are integrated with microelectronics.

Microfabrication yields nonuniform surface- and bulk-micromachined structures and devices. Fabrication imperfections, temperature sensitivity and parameter variations may result in adverse outcomes. Devices exhibit heterogeneity. Controlled microactuators ensure specified accuracy, precision, resolution, dynamic range and bandwidth in ultra-high-definition video systems, precision microservos, microscopy and other applications.

For electrostatic multi-axis transducers, the governing equations of movable structures (proof mass, cantilever beam, diaphragm, plate, etc.) are found using rigid-body and multibody dynamics paradigms. The Euler–Bernoulli beam theory, nonlinear elasticity and Timoshenko beam theory are applied.

The governing equations of Newtonian and translational-rotational dynamics are considered in section 2.4. Considering the linear acceleration and velocity (***a***, ***v***), angular accelerations and velocity ($\boldsymbol{\alpha}$, $\boldsymbol{\omega}$), *net* force and torque ($\Sigma\mathbf{F}$, $\Sigma\mathbf{T}$), we have

$$\sum \mathbf{F} = \frac{d}{dt}(m\boldsymbol{v}) + m(\boldsymbol{\omega}\times\boldsymbol{v}), \tag{3.4.1}$$

$$\sum \mathbf{T} = \frac{d}{dt}(\mathbf{I}\boldsymbol{\omega}) + (\boldsymbol{\omega}\times\mathbf{I})\boldsymbol{\omega},$$

which for time-invariant mass and inertia matrix (m, $\mathbf{I}$),

$$\frac{d\boldsymbol{v}}{dt} = \begin{bmatrix} \dot{v}_x \\ \dot{v}_y \\ \dot{v}_z \end{bmatrix} = -\boldsymbol{\omega}\times\boldsymbol{v} + \mathcal{R}^T\mathbf{g} + \tfrac{1}{m}\sum\mathbf{F}, \tag{3.4.2}$$

$$\frac{d\boldsymbol{\omega}}{dt} = \begin{bmatrix} \dot{\omega}_\phi \\ \dot{\omega}_\theta \\ \dot{\omega}_\psi \end{bmatrix} = \mathbf{I}^{-1}\left(-(\boldsymbol{\omega}\times\mathbf{I})\boldsymbol{\omega} - A_\Omega\mathcal{R}^T\boldsymbol{v} - B_\Omega\boldsymbol{\omega} - \boldsymbol{\tau}_g + \sum\mathbf{T}\right),$$

$$\frac{d\boldsymbol{\Theta}}{dt} = \begin{bmatrix} \dot{\phi} \\ \dot{\theta} \\ \dot{\psi} \end{bmatrix} = \begin{bmatrix} 1 & \sin\phi\tan\theta & \cos\phi\tan\theta \\ 0 & \cos\phi & -\sin\phi \\ 0 & \sin\phi\sec\theta & \cos\phi\sec\theta \end{bmatrix} \begin{bmatrix} \omega_\phi \\ \omega_\theta \\ \omega_\psi \end{bmatrix},$$

where $\boldsymbol{\Theta}=[\phi\ \theta\ \psi]^T$ is the Euler angles vector; $\Sigma\mathbf{F}=[F_X\ F_Y\ F_Z]^T$ and $\Sigma\mathbf{T}=[T_\phi\ T_\theta\ T_\psi]^T$ are the force and torque vectors; $\mathbf{g}=[0\ 0\ g]^T$ is the gravity acceleration; $\boldsymbol{\tau}_g$ is gyroscope torque vector; $\mathcal{R}$ is the rotational matrix; A_Ω and B_Ω are the rotational multi-axis coupling mappings.

The electrostatic actuators are open-loop stable. The single-axis one-dimensional equations of motion for translational and rotational actuators are

$$\frac{dv}{dt}=\frac{1}{m}\left(F_e-F_{\text{drag}}-F_{\text{elastic}}-F_\xi\right),\qquad(3.5.1)$$

$$\frac{dx}{dt}=v,$$

and

$$\frac{d\omega}{dt}=\frac{1}{J}\left(T_e-T_{\text{drag}}-T_{\text{elastic}}-T_\xi\right),\qquad(3.5.2)$$

$$\frac{d\theta}{dt}=\omega,$$

where v and ω are the linear and angular velocities; x and θ are the linear and angular displacements; F_e and T_e are the electrostatic force and torque; F_{drag} and T_{drag} are the drag force and torque, such as air resistance and viscous frictions; F_{elastic} and T_{elastic} are the elastic force and torque; F_ξ and T_ξ are the stochastic perturbations such as cross-axis coupling forces and random disturbances for which the statistical model may exist and multivariate distributions may be derived, see section 5.12.

Electrostatic Force – The electrostatic force is found using the co-energy $W_c=\frac{1}{2}C(x)\mathcal{V}^2$. The applied voltage $\mathcal{V}$ is limited, $\mathcal{V}_{\min}\le\mathcal{V}\le\mathcal{V}_{\max}$. For the parallel-plate capacitor, the capacitance $C(x)$ is a function of the plate displacement x,

$$C(x)=\varepsilon_0\varepsilon_r\frac{A}{x_0-x},$$

where A is the *effective* area; x is the plate relative displacement with equilibrium $x|_{\mathcal{V}=0}=0$, $x\ge0$; x_0 is the equilibrium plates separation at $\mathcal{V}=0$, $x_0>0$, $(x_0-x)>0$.

Translational electrostatic actuators develop one-directional forces to minimize the air gap. The electrostatic force to minimize the plates separation (x_0-x) is

$$F_e=-\frac{\partial W_c}{\partial x}=-\frac{\partial}{\partial x}\left(\tfrac{1}{2}C(x)\mathcal{V}^2\right)=\varepsilon_0\varepsilon_r\frac{A}{2(x_0-x)^2}\mathcal{V}^2,\ x<x_0,\ \mathcal{V}_{\min}\le\mathcal{V}\le\mathcal{V}_{\max}.\qquad(3.6)$$

Example 3.1. *Electrostatic Energy and Force*

Consider the parallel-plate capacitor with the separation between two plates (x_0-x). The dielectric permittivity is ε. Neglect the fringing effect at the edges, and, assume that the electric field is uniform, such that the electric field intensity is $E=\frac{\mathcal{V}}{x_0-x}$.

The stored electrostatic energy is

$$W_c=\tfrac{1}{2}\int_v\varepsilon\left|\vec{E}\right|^2dv=\tfrac{1}{2}\int_v\varepsilon\left(\tfrac{\mathcal{V}}{x_0-x}\right)^2dv=\tfrac{1}{2}\varepsilon\frac{\mathcal{V}^2}{(x_0-x)^2}A(x_0-x)=\tfrac{1}{2}\varepsilon\frac{A}{x_0-x}\mathcal{V}^2=\tfrac{1}{2}C(x)\mathcal{V}^2,$$

where the capacitance $C(x)$ is a function of the plate displacement x, $C(x)=\varepsilon\frac{A}{x_0-x}$.

One-directional electrostatic force is a nonlinear function of the voltage $\mathcal{V}$ and plates separation (x_0-x), and, $F_e=-\frac{\partial W_c}{\partial x}=-\frac{1}{2}\frac{\partial C(x)}{\partial x}\mathcal{V}^2=\frac{1}{2}\varepsilon\frac{A}{(x_0-x)^2}\mathcal{V}^2$.

Considering the absolute plates separation x, the electrostatic force is $F_e=\frac{1}{2}\varepsilon\frac{A}{x^2}\mathcal{V}^2$.■

Elastic and Drag Forces – The elastic and drag forces (F_{elastic},F_{drag}) are component of the *net* force in (3.4). As reported in Example 2.14, using the kinetic, potential and dissipating energies (Γ,Π,D) one finds the equations of motion using Lagrange equations (2.16). The solutions are device and design specific.

The elastic force counters the one-directional electrostatic force ensuring device functionality. The movable member (cantilever beam, diaphragms or membrane) is suspended or fixed using polysilicon and polymer springs or flexible couplings, see Figure 3.2.a. At steady-state, $F_e=F_{\text{elastic}}$.

One analyzes (F_e,F_{elastic}) and displacement, sensing schemes, signal conditioning and electronics, etc. It is important to derive a consistent F_{elastic} which restores the moving member to an equilibrium position. The fourth-order 81-components elastic tensor K_{ijkl} is applied to express the force, elastic energy, stress, strain and other quantities of elastic materials and structures. In orthogonal coordinates, the elasticity energy is

$$\Pi=\tfrac{1}{2}K_{ijkl}\varepsilon_{ij}\varepsilon_{kl},$$

where K_{ijkl} is the fourth-rank elastic (stiffness) tensor; ε_{ij} and ε_{kl} are the strain tensors.

In electrostatic transducers, polysilicon and polymer springs, elastic beams and diaphragms, as well as for linear and progressive variable stiffness springs are used. Consider a one-dimensional problem. For large deformations with a nonlinear stress-strain curve, the potential energy $\Pi(x)$ is a high-order polynomial. Nonlinear elasticity theory is applied to find $\Pi(\cdot)$, see Example 2.14.

Consider a quartic polynomial $\Pi(x)=\frac{1}{2}k_1x^2+\frac{1}{3}k_2\,\text{sgn}(x)x^3+\frac{1}{4}k_3x^4$, $(k_1,k_2,k_3)>0$.

From $\Pi(x)=\int F(x)dx$, one finds the elastic force $F_{\text{elastic}}(x)=k_1x+k_2x|x|+k_3x^3$.

The dissipating energy $D(\cdot)$ may be found as a multivariate polynomial of velocity, displacement and other variables which depend on viscosity, lubrication, temperature, etc.

Consider a univariate quartic polynomial to express the dissipating energy as

$$D(v)=\tfrac{1}{2}c_1v^2+\tfrac{1}{3}c_2\,\text{sgn}(v)v^3+\tfrac{1}{4}c_3v^4,\ (c_1,c_2,c_3)>0.$$

Hence, $F_{\text{drag}}=c_1v+c_2v|v|+c_3v^3$.

The considered F_{elastic} and F_{drag} can be experimentally measured or estimated.

Considering univariate quartic $\Pi(x)$ and $D(v)$, the nonlinear drag and elastic forces are

$$F_{\text{drag}}=c_1v+c_2v|v|+c_3v^3,$$

$$F_{\text{elastic}}=k_1x+k_2x|x|+k_3x^3. \tag{3.7}$$

For coefficients (c_1,c_2,c_3) and (k_1,k_2,k_3), one has $c_i\in[c_{i\min}\ c_{i\max}]$, $c_{i\min}\le c_i\le c_{i\max}$, $c_i=c_{i0}\pm\Delta c_i$, $c_{i\min}>0$, $k_i\in[k_{i\min}\ k_{i\max}]$, $k_{i\,\min}\le k_i\le k_{i\,\max}$, $k_i=k_{i0}\pm\Delta k_i$, $k_{i\min}>0$.

From (3.5), (3.6) and (3.7), we have

$$\frac{dv}{dt}=\frac{1}{m}\left[\left(-c_1v-c_2v|v|-c_3v^3-k_1x-k_2x|x|-k_3x^3\right)+\frac{\varepsilon_0\varepsilon_rA}{2(x_0-x)^2}\mathcal{V}^2\right],\ x<x_0, x\ge 0, \mathcal{V}_{\min}\le\mathcal{V}\le\mathcal{V}_{\max}, \tag{3.8}$$

$$\frac{dx}{dt}=v.$$

Example 3.2. *Single-Axis Electrostatic Microactuator*

The electrostatic microphones and speakers with ~1 mm suspended circular diaphragm were commercialized by Analog Devices, InvenSense, Knowles and STMicroelectronics. Consider a proof-of-concept electrostatic actuator. Figures 3.2.a illustrate a single-axis microactuator with a suspended 270×270 μm, 30 μm-thickness silicon movable plate above conducting stationary plate. Thin-film 0.5 μm aluminum is deposited on the movable and substrate plates to form a parallel plate capacitor and interconnect. In the commercialized Texas Instruments DLPs, torsional mirrors are steered ensuring angular repositioning. The TI DLP® technology and DLP product are reported on the www.ti.com.

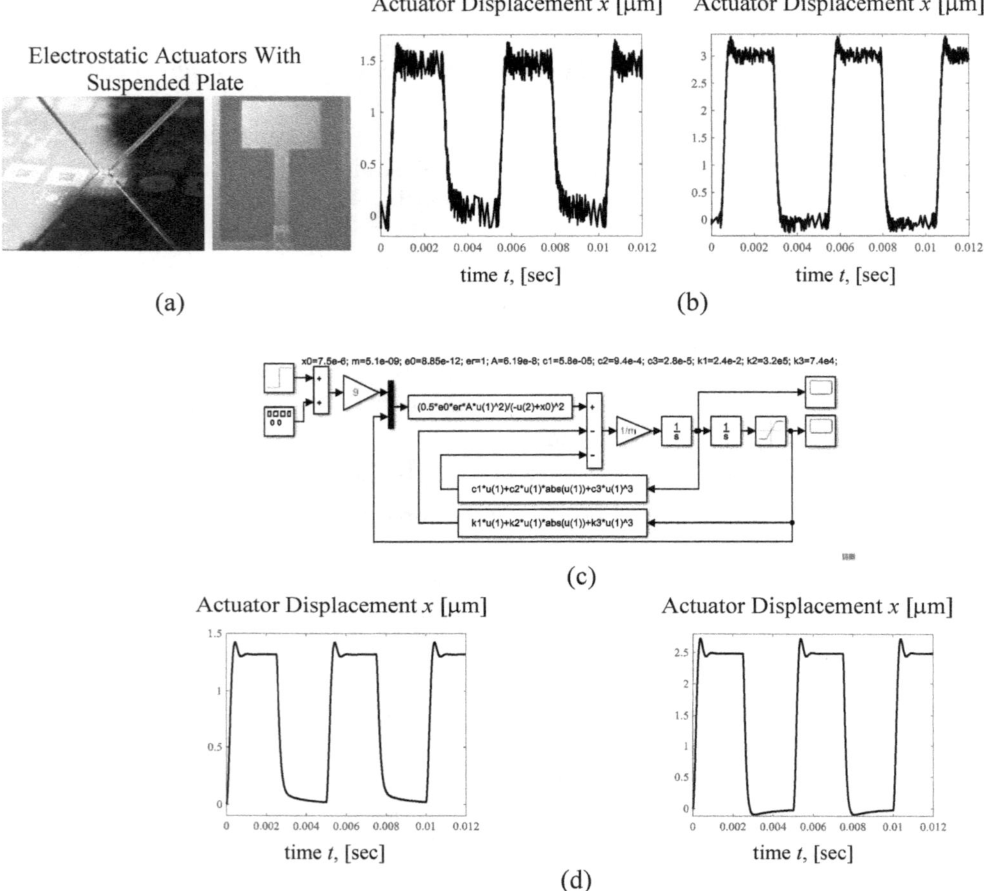

Figure 3.2. (a) Electrostatic microactuators: Bending plate, supported at the fixed end, repositions in the vertical axis x with bending $x(y)$. The displacement sensor, formed by four piezoresistors with a balanced Wheatstone bridge, is calibrated to measure x. The plates separation at $\mathcal{V}=0$ is $(x_0-x)=7.5$ μm, $x_0=7.5$ μm with a free end displacement $x|_{\mathcal{V}=0}=0$;
(b) Experimental results: Open-loop actuator free-end displacement $x(t)$ for 200 Hz voltage pulses $\mathcal{V}(t)=[9\ \ 0]$ and $\mathcal{V}(t)=[13.6\ \ 0]$ V. The free-end of the bending plate is displaced by 1.46 and 3 μm;
(c) SIMULINK diagram to simulate electrostatic actuator with estimated parameters;
(d) Simulation results: Dynamics of $x(t)$ for 200 Hz voltage pulses 9 and 13.6 V which yield 1.32 and 2.49 μm displacements.

For many electrostatic actuators, $\mathcal{V}_{\min}=0$, and, $0\le\mathcal{V}\le\mathcal{V}_{\max}$. Study repositioning from an equilibrium with $x=0$ when $\mathcal{V}=0$ to the specific vertical displacement of a plate free end $x\neq0$ are plate bending $x(y)$ by applying voltage $\mathcal{V}$. The movable plate is restored to equilibrium $x|_{\mathcal{V}=0}=0$ due to the elastic force. In a proof-of-concept, the plate is fixed to the stationary structure, and, plate free end is displaced. The static equilibrium of the bended plate is described by a high-order polynomial $x(y)$, see Example 2.14.

Consider the tip vertical repositioning in the x direction. The applied voltage is $0\leq\mathcal{V}\leq 17.5$ V. For $\mathcal{V}=0$, the upper plate tip is at $x|_{\mathcal{V}=0}=0$. The plates separation is $(x_0-x|_{\mathcal{V}=0})=x_0=7.5$ μm.

If $\mathcal{V}\neq 0$, the plate bends downward, such that $0<(x_0-x|_{\mathcal{V}\neq 0})<7.5$ μm. The plate separation reduces as voltage is applied. The actuator is open-loop stable. Figure 3.2.b reports the experimental studies for the 1.46 μm and 3 μm repositioning for the voltage pulses $\mathcal{V}(t)=\begin{cases}9\text{ V}\\0\end{cases}$ and $\mathcal{V}(t)=\begin{cases}13.6\text{ V}\\0\end{cases}$ applied at frequency $f=200$ Hz. The one-directional electrostatic force reduces the plates separation (x_0-x), while elastic force opposes F_e. At steady-state, $F_e=F_{\text{elastic}}$. Using the experimental data, the unknown parameters of model (3.8) are estimated.

To measure displacement, piezoresistive sensing is considered. Four piezoresistors, embedded in a stationary silicon structure and movable suspended plate, form a Wheatstone bridge. The integrated Wheatstone bridge circuit consists a patterned conducting polysilicon and p-type boron doped piezoresistors with 5×10^{14} atoms/cm^2 dose, 80 keV and 2 μm depth. Due to axial loadings, piezoresistors change resistance which depends on geometric changes in piezoresistive media, such as longitudinal, transverse and other deflections, deformations and distortions. The studied sensing mechanism is used to measure displacement, deformation, force, torque, stress and other quantities in accelerometers, gyroscopes, tactile force, pressure, flow and other sensors. For homogeneous media and uniform geometry, the resistance is

$$R=\rho L/A,\ A=WH,$$

where A, L, W and H are the cross-sectional area, length, width and thickness; ρ is the resistivity, which is a function of strain ε, temperature, etc.

The piezoresistive tensor $\boldsymbol{\pi}$ and stress tensor $\boldsymbol{\sigma}$ yield

$$\frac{\Delta\rho}{\rho}=\boldsymbol{\pi\sigma},\ \sigma_{ij}=C_{ijkl}\varepsilon_{kl},\ \varepsilon_{ij}=S_{ijkl}\sigma_{kl},$$

where C_{ijkl} and S_{ijkl} are the elastic stiffness and inverse compliance matrices.

The Poisson's ratio is $\nu=-\dfrac{\varepsilon_{\text{lateral}}}{\varepsilon_{\text{axial}}}=\left|\dfrac{\varepsilon_y}{\varepsilon_x}\right|=\left|\dfrac{\varepsilon_z}{\varepsilon_x}\right|$.

The resistance change is $\dfrac{\Delta R}{R}=\dfrac{\Delta L}{L}-\dfrac{\Delta A}{A}+\dfrac{\Delta\rho}{\rho}$.

From $\dfrac{\Delta A}{A}=-2\nu\dfrac{\Delta L}{L}$, one obtains $\dfrac{\Delta R}{R}=(1+2\nu)\dfrac{\Delta L}{L}+\dfrac{\Delta\rho}{\rho}$.

For a one-dimensional problem, $\dfrac{\Delta R}{R}=(1+2\nu+\pi C)\varepsilon$.

Using the *gauge factor* G, we have

$$\frac{\Delta R}{R}=\left(\frac{\Delta R}{R}\right)_{\text{longitudinal}}+\left(\frac{\Delta R}{R}\right)_{\text{transverse}}=G_{\text{longitudinal}}\varepsilon_{\text{longitudinal}}+G_{\text{transverse}}\varepsilon_{\text{transverse}},\ G=(1+2\nu+\pi C).$$

Nonuniformity, heterogeneity and environmental factors affect accuracy, precision, linearity, sensitivity, range, etc. Due to variations of the normal and shear stresses tensors ($\boldsymbol{\sigma}$,$\boldsymbol{\tau}$), piezoresistive tensor $\boldsymbol{\pi}$, nonlinearities and aforementioned dependences, the displacement sensor is calibrated using the optical proximity probe which ensures submicron accuracy. One measures the voltage Δu of the balanced Wheatstone bridge with four piezoresistors R_i which experience longitudinal and transverse elongations.

The output voltage depends on the exhibited variations in R_i=$(R_0 \pm \Delta R)_i$.

One has $\Delta u = \frac{1}{4R_0}\left(\Delta R_1 + \Delta R_3 - \Delta R_2 - \Delta R_4\right)V_0$.

The voltage $\Delta u(t)$ is filtered by the fourth-order notch filter with gain 1000 and cutoff frequency 25000 Hz. For $x \in [x_{\min}\ x_{\max}]$ at steady-state, the filter output u_f is calibrated using the optical proximity probe yielding $x=f(u_f)$. Despite static errors in calibration and displacement estimates $x=f(u_f)$, which is ±0.35 μm, this error may not be critical in proof-of-concept analysis.

For the studied actuator, the governing equations are (3.8), $\varepsilon_0=8.85\times10^{-12}$, $\varepsilon_r=1$, $A=6.19\times10^{-8}$ m^2. The control law design is reported in Example 6.49.

From the experimental data and dynamics for $\mathcal{V}(t) = \begin{cases} 9\text{ V} \\ 0 \end{cases}$, the parameters are estimated, yielding $m=5.1\times10^{-9}$ kg, $c_1=5.8\times10^{-5}$ N-s/m, $c_2=9.4\times10^{-4}$ N-s^2/m^2, $c_3=2.8\times10^{-5}$ N-s^3/m^3, $k_1=2.4\times10^{-2}$ N/m, $k_2=3.2\times10^{5}$ N/m^2 and $k_3=7.4\times10^{4}$ N/m^3.

The actuator repositions as voltage applied, and, returns to an equilibrium x_0 when $\mathcal{V}=0$ due to restoring elastic force. For voltage pulses $\mathcal{V}(t) = \begin{cases} 9\text{ V} \\ 0 \end{cases}$, $\mathcal{V}(t) = \begin{cases} 13.6\text{ V} \\ 0 \end{cases}$, f=200 Hz, the settling time is 0.001 sec. Depending on the voltage applied and displacement, the parameters (c_i,k_i) vary by ±28%. Due to open-loop stability, the parameter variations do not considerably affect analyses. Stability analysis and control design are reported in Example 6.49. The SIMULINK diagram and simulated dynamics are illustrated in Figures 3.2.c and d. The model and simulations are in agreement with the experimental results. ■

Example 3.3. *Single-Axis Electrostatic Inertial Sensors*

A movable proof-mass is suspended within the stationary member assembly. The external forces and torques yield linear or angular displacement of proof masses. Considering translational motion of a suspended proof-mass on polysilicon springs, one has $ma=F_{\text{drag}}+F_{\text{elastic}}$ with $F_{\text{drag}}=0$ in vacuum.

The capacitance $C(x)$ or $C(\theta)$ of parallel plates, placed on stationary and movable members, vary. The capacitance difference ΔC is measured. For the parallel-plate capacitors $C=\varepsilon_0\varepsilon_r\dfrac{A}{x}$. For the moving member which rotates with respect to stator, the capacitance of two plates is

$$C=\varepsilon_0\varepsilon_r\frac{A}{g}=\varepsilon_0\varepsilon_r\frac{WL(\theta)}{g},$$

where A is the effective overlapping area of the plates, $A=WL(\theta)$; W and $L(\theta)$ are the width and length of the plates; g is the air gap between the plates.

For small translational and rotational displacement in μm or μrad, assuming linear elasticity of polysilicon or other micromachined springs, $F_{\text{elastic}}=k_s x$ and $T_{\text{elastic}}=k_s\theta$, where k_s is the constant.

From $ma=k_s x$, $J\alpha=k_s\theta$, we have $a=\frac{1}{m}k_s x$ and $\alpha=\frac{1}{J}k_s\theta$.

By measuring the varying capacitance C or ΔC between the plates, the displacement x or θ is estimated. With known (m,J) and k_s, the acceleration is found. The images of the MEMS-technology sensors are reported in Figures 3.1 and 3.3. ■

Example 3.4. *Integrated MEMS Technology MDOF Sensors*
Inertial sensors, such as IMUs and INSs, are used in aerial, ground and underwater vehicles, robots, wearable electronics, etc. Triaxial accelerometers and gyroscopes with ASICs and MPUs comprise IMUs and INSs. The MPUs and digital motion processors perform filtering, processing, interfacing, etc.
Sensors measure linear accelerations and angular rates $\mathbf{y}=[(a_x,a_y,a_z),(\omega_\phi,\omega_\theta,\omega_\psi)]$.

The direct measurements, data filtering and processing yield the IMU outputs $\hat{\mathbf{y}}=[(\hat{a}_x,\hat{a}_y,\hat{a}_z),(\hat{\omega}_\phi,\hat{\omega}_\theta,\hat{\omega}_\psi)]$.

The accuracy, linearity, cross-axis coupling, noise power, data rate and other factors affect the selection of IMUs. Depending on applications, single-, dual- and triaxial accelerometers and gyroscopes are used. The images of InvenSense nine-degree-of-freedom IMU and Bosch Sensortec INS are shown in Figures 3.3.a and b.

The device physics is described in Example 3.3 for a single-axis devices. Figure 3.3.c documents a triaxial gyroscope with four movable masses (M_1,M_2,M_3,M_4) suspended on the polysilicon springs within the stationary assembly. One estimates the roll, pitch and yaw motions using the varying capacitances C_i. To analyze the axis coupling, one may apply the translational-rotational dynamics (3.4), nonlinear elasticity, etc. The commonly applied equations of motion are

$$m\boldsymbol{a}_j=\Sigma\mathbf{F}_j,\ J\boldsymbol{\alpha}_j=\Sigma\mathbf{T}_j.$$

Due to the Coriolis force acting on the suspended movable masses, they are in motion relative to a rotating reference frame. As shown in Figure 3.3.c, the suspended masses (M_1,M_3) move upward and downward out of the plane sensing ω_ϕ. Masses (M_2,M_4) move upward and downward out of the plane sensing ω_θ. The proof masses (M_2,M_4) move in the same horizontal plane sensing ω_ψ. The cross-coupling

$$\mathcal{B}_{Mi}=\mathcal{R}_\theta\mathcal{R}_\psi\mathcal{R}_\phi\mathcal{B}_{IMU}$$

is examined in the stationary and movable members orthogonal bases $\mathcal{B}_{IMU}$ and $\mathcal{B}_{Mi}$.

The rotational mappings are

$$\mathcal{R}_\theta=\begin{bmatrix}\cos\theta & \sin\theta & 0\\ -\sin\theta & \cos\theta & 0\\ 0 & 0 & 1\end{bmatrix},\ \mathcal{R}_\psi=\begin{bmatrix}\cos\psi & 0 & -\sin\psi\\ 0 & 1 & 0\\ \sin\psi & 0 & \cos\psi\end{bmatrix},\ \mathcal{R}_\phi=\begin{bmatrix}1 & 0 & 0\\ 0 & \cos\phi & \sin\phi\\ 0 & -\sin\phi & \cos\phi\end{bmatrix}.$$

To control physical systems, one measures and estimates linear and angular displacements $(\mathbf{x},\boldsymbol{\theta})$, velocities $(\mathbf{v},\boldsymbol{\omega})$, accelerations $(\boldsymbol{a},\boldsymbol{\alpha})$, etc. Using the IMU outputs, $\hat{\mathbf{y}}=[\hat{\omega}_\varphi,\hat{\omega}_\theta,\hat{\omega}_\psi]$ performing filtering, interpolation and mathematical operations, the angular rates yield estimates of the Euler angles by solving the Euler equation

$$\frac{d\mathbf{\Theta}}{dt}=\begin{bmatrix}\dot{\phi}\\ \dot{\theta}\\ \dot{\psi}\end{bmatrix}=\begin{bmatrix}1 & \sin\phi\tan\theta & \cos\phi\tan\theta\\ 0 & \cos\phi & -\sin\phi\\ 0 & \sin\phi\sec\theta & \cos\phi\sec\theta\end{bmatrix}\begin{bmatrix}\omega_\phi\\ \omega_\theta\\ \omega_\psi\end{bmatrix},\ \mathbf{\Theta}=\begin{bmatrix}\phi\\ \theta\\ \psi\end{bmatrix},\ \boldsymbol{\omega}=\begin{bmatrix}\omega_\phi\\ \omega_\theta\\ \omega_\psi\end{bmatrix}=\begin{bmatrix}p\\ q\\ r\end{bmatrix},$$

where $\omega_\phi\equiv p$, $\omega_\theta\equiv q$ and $\omega_\psi\equiv r$ are the angular velocities; $\mathbf{\Theta}$ is the Euler angles vector.

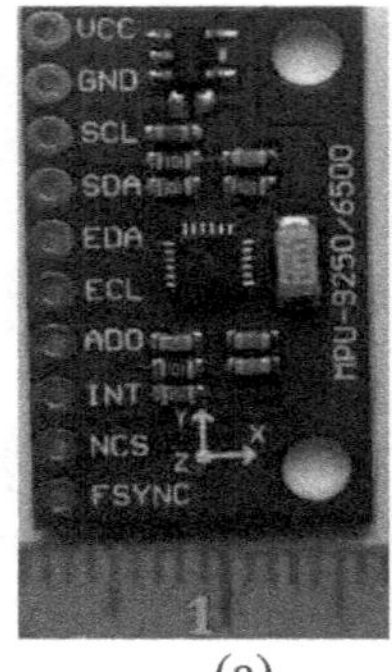

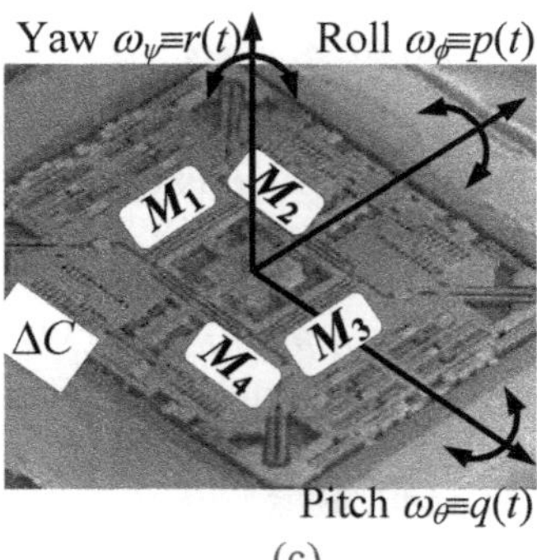

(a) (b) (c)

Figure 3.3. (a) TDK InvenSense IMU with triaxial accelerometer and gyroscope sensors to measure $\mathbf{y}=[(a_x,a_y,a_z),(\omega_\phi,\omega_\theta,\omega_\psi)]$, and, process measurements on-chip yielding filtered outputs $\hat{\mathbf{y}}=[(\hat{a}_x,\hat{a}_y,\hat{a}_z),(\hat{\omega}_\phi,\hat{\omega}_\theta,\hat{\omega}_\psi)]$;

(b) Bosch Sensortec MDOF INS measures accelerations, angular rates, magnetic field and barometric pressure $\mathbf{y}=[(a_x,a_y,a_z),(\omega_\phi,\omega_\theta,\omega_\psi),(B_x,B_y,B_z),p]$, and, calculates orientation, position, motions, latitude, longitude, etc. The filters outputs are $\hat{\mathbf{y}}=[(\hat{a}_x,\hat{a}_y,\hat{a}_z),(\hat{\omega}_\phi,\hat{\omega}_\theta,\hat{\omega}_\psi),(\hat{B}_x,\hat{B}_y,\hat{B}_z),\hat{p}]$;

(c) Triaxial gyroscope computes the angular rates $[\omega_\phi,\omega_\theta,\omega_\psi]$ by measuring orientation of four suspended masses M_i applying capacitive sensing ΔC, and, applying the translational-to-rotational mappings. ■

3.3. Electromagnetic Transducers

Device physics of electromagnetic transducers is based upon three key principles:

1. Variable reluctance electromagnetics. The electromagnetic force (torque) is produced to minimize the reluctance (minimize air gap) of the electromagnetic system in solenoids, translational and rotational relays, magnetic levitation systems, electromagnets, reluctance motors, etc.;
2. Electromagnetic Induction. The electromagnetic torque (force) results due to varying spatiotemporal electromagnetic fields. The phase voltages are induced as *motional emf* in the rotor windings due to time-varying stator magnetic field and relative motion of rotor with respect to stator. Induction motors are covered in the Practice Problems 3.8, 3.9, 3.10, 3.11 and 5.9;
3. Synchronous Electromagnetics. The electromagnetic torque (force) results due to varying spatiotemporal magnetic field established by the stator windings and stationary magnetic field established by the permanent magnets or electromagnets on the rotor. In sensors and generators, the voltage is induced as the *motional emf* if rotor is rotated by the external source of energy.

Fidelity and Complexity – Device physics affect analysis and control schemes. There are numerous commercialized and deployed electromagnetic transducers. One may apply three-dimensional Maxwell's equations and tensor calculus carrying out high-fidelity data-intensive analyses. High-fidelity consideration may yield needless control complexity, as well as algorithmic, computational and numeric unsolvability and undecidability. To solve device- and system-level problems, consider consistently designed and electromagnetically optimized transducers. Experimental results, characterization and evaluation validate common practices and solutions covered.

Performance and Fabrication – Electromagnetic transducers are fabricated using precision computer numerical control (CNC) machining, thin-gauge grain-oriented electrical steel laminations, high energy permanent magnets, etc. The copper coils are coated with a uniform high-temperature, high dielectric constant insulating enamel. Permanent magnet electromagnetic actuators, servos, machines and transducers ensure high torque, energy and power densities, as well as superior performance and exceptional capabilities. Variable reluctance electromechanical motion devices are also used. The images of variable reluctance, permanent magnet direct current (dc), and, permanent magnet synchronous transducers are illustrated in Figure 3.4.

Figure 3.4. Images of variable reluctance device (solenoid), permanent magnet motors with gearhs and angular velocity sensors, and, permanent magnet synchronous machines with controller-drivers and Hall-effect sensors to measure rotor displacement

3.3.1. Variable Reluctance Electromagnetic Actuators: Solenoids, Relays and Magnetic Levitation Systems

Solenoids, relays, electromagnets and magnetic levitation consist of a movable (plunger or rotor) and stationary members made from high-permeability ferromagnetic materials. The windings wound within a helical pattern. The electromagnetic force is developed due to the magnetic path reluctance which varies. Performance of variable reluctance devices depends on magnetic system, ferromagnetic materials, BH curve, energy product $(BH)_{\max}$, permeability $\mu = \mu_0 \mu_r = \frac{dB}{dH}$, friction, lubrication, etc. Nonlinear magnetization and descriptive characteristics are reported in Example 2.30.

Solenoids are shown in Figure 3.5.a. The magnetic levitation devices with the suspended ferromagnetic ball are depicted in Figures 3.5.b and c. The *closed* magnetic system is formed by the magnetic flux linkages paths in stationary and movable members. When the voltage is applied, current flows in the winding, magnetic flux is produced within the established paths, and, the electromagnetic force F_e is exerted to a movable member to minimize the reluctance and air gap.

In solenoids, when the applied voltage becomes zero, the plunger resumes an equilibrium position due to the restoring spring force, $F_e = F_{\text{spring}}$. Lubrication and low friction materials for the central guide (nonmagnetic sleeve) and plunger coating are used to minimize friction and wearing. Glass filled nylon and brass (for the guide), copper, aluminum, tungsten, platinum and other low friction coatings are used. Depending on the surface roughness and material composition, the estimated viscous friction coefficients B_v varies from 0.04 to 0.1 N-sec/m for lubricated (solid film and oil) tungsten on tungsten, copper on copper, aluminum on aluminum, titanium on titanium. For unlubricated materials, such as tungsten on tungsten, copper on copper, aluminum on aluminum, and, titanium on titanium, B_v varies from 0.3 to 1.2 N-sec/m.

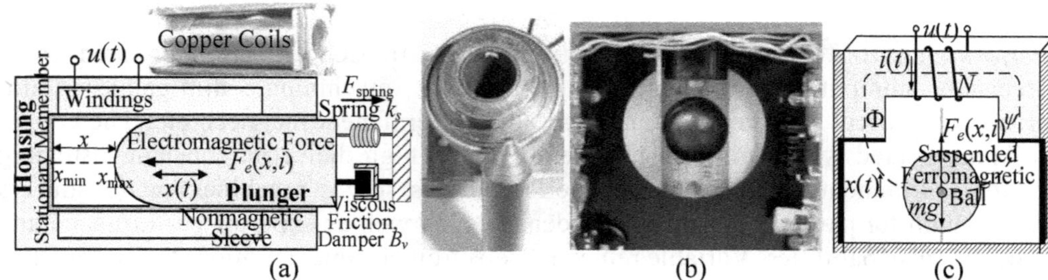

Figure 3.5. (a) Schematic and images of solenoids, used as actuators, relays, mechanical switches or positioners. The one-directional electromagnetic force F_e is countered by the restoring spring force F_{spring}. At equilibrium F_e=F_{spring}. The helical springs are designed for compression and tension. The tension spring operates with a tension load, and, the spring stretches as the force is applied. The compression spring operates with a compression load, and, the spring contracts as the force is applied. Depending on solenoid applications and kinematics, linear and progressive springs have helical coils, *flat* or V-spring designs;
(b) Magnetic levitation system with the suspended ferromagnetic ball. At equilibrium, F_e=mg;
(c) Schematics of a magnetic levitation system with the suspended ferromagnetic mass.

Nonlinear Magnetic System – Armature current i in N coils produces the flux Φ, and, the flux linkages are ψ=$N\Phi$. The displacement of movable member changes the magnetic energy W_m stored in an air gap.

From $W_m=\frac{1}{2}\int_v \mu\left|\vec{H}\right|^2 dv=\frac{1}{2}\int_v \mu^{-1}\left|\vec{B}\right|^2 dv$, we have

$$dW_m=dW_{m\,\mathrm{airgap}}=2\frac{B^2}{2\mu_0}Adx=\frac{\Phi^2}{\mu_0 A}dx,$$

where A is the cross-sectional area; dx is the virtual displacement.

The flux Φ is constant if i=const. The increase of the air gap dx leads to increase of the stored magnetic energy.

Equation $F_e=-\nabla W_m=-\frac{\partial W_m}{\partial x}$ yields the electromagnetic force $\vec{F}_e=-\vec{a}_x\frac{\Phi^2}{\mu_0 A}$.

The inductance and reluctance are (2.39) $L=\frac{\int_s \vec{B}\cdot d\vec{s}}{\oint_l \vec{H}\cdot d\vec{l}}$, $\Re=\frac{\oint_l \vec{H}\cdot d\vec{l}}{\int_s \vec{B}\cdot d\vec{s}}$.

Considering a nonlinear BH curve with hysteresis, the varying relative permeability $\mu_r=\frac{1}{\mu_0}\frac{dB(H)}{dH}$ yields $\mu_r(i,\frac{di}{dt})$.

For example, in Example 2.30

$$\mu_r(H,\tfrac{dH}{dt})=\frac{1}{\mu_0}B_{\max}a\,\mathrm{sech}^2\left(aH-\tfrac{a}{b}H_c\,\mathrm{sgn}\left(\tfrac{dH}{dt}\right)\right),\ H=\tfrac{N}{l_e}i.$$

Explicit expressions for $\mu_r(i,\frac{di}{dt})$ and $L(i,\frac{di}{dt},x)$ are derived for the ferromagnetic materials used, characterized by the BH curves.

Linear Magnetic System – As voltage $u(t)$ is applied, the *motional emf* opposes the applied voltage. Assume a linear magnetic system and μ_r=const. These imply that the reluctance and magnetizing inductance are functions of the varying magnetic path lengths.

The Kirchhoff voltage law

$$u = ri + \frac{d\psi}{dt},\ \psi = L(x)i$$

yields $u = ri + L(x)\frac{di}{dt} + i\frac{dL(x)}{dx}\frac{dx}{dt}$.

The electromagnetic force tends to reduce the air gap minimizing the reluctance.

The magnetizing inductance is $L(x) = \frac{N^2}{\Re_{\text{total}}(x)}$, and, $\psi = L(x)i$.

To derive the electromagnetic force, use the co-energy $W_c = \frac{1}{2}L(x)i^2$.

The one-directional electromagnetic force is $F_e = -\nabla W_c = -\frac{\partial W_c}{\partial x} = -\frac{1}{2}i^2\frac{\partial}{\partial x}L(x)$.

Using the Kirchhoff and Newton second laws

$$\frac{di}{dt} = \frac{1}{L(x)}\left[-ri - \frac{\partial L(x)}{\partial x}iv + u\right],\ L(x) = \frac{N^2}{\Re_{\text{total}}(x)}, \tag{3.9}$$

$$\frac{dv}{dt} = \frac{1}{m}\left[\frac{1}{2}\frac{\partial L(x)}{\partial x}i^2 - F_{\text{friction}} - F_{\text{spring}} - F_L\right],$$

$$\frac{dx}{dt} = v,\ x_{min} \le x \le x_{max},\ x_{min} \ge 0.$$

Explicit equations for $L(x)$ and $\Re_{\text{total}}(x)$ must be derived. These $L(x)$ and $\Re_{\text{total}}(x)$ can be experimentally measured. The force, acceleration, velocity and displacement are vectors. As voltage u is applied, F_e is developed to minimize the air gap x. Nonlinear magnetic system, hysteresis, magnetic losses, cross-section area variations, magnetic field nonuniformity and other effects may be considered. Using experimental data, one verifies model fidelity and makes refinements. The electromagnetic force $F_e(x,i)$ can be experimentally measured and described as studied in Example 3.7.

The friction and spring forces ($F_{\text{friction}}, F_{\text{spring}}$) oppose F_e. The stretched tension spring exhibits the force, and, an ideal Hook's law is $F_{\text{spring}} = k_s x_{\text{spring}}$. At the steady-state if $F_L = 0$, $F_e = F_{\text{spring}}$. In equations (3.9), the movable member displacement x is used, and $x \in [x_{min}\ \ x_{max}]$, $x_{min} \ge 0$. The stroke is $\Delta x = (x_{max} - x_{min})$. In Figure 3.5.a, the electromagnetic force is developed in the $-x$ direction, while the spring force in the $+x$ direction.

Spring Force – A tension spring exhibits zero force at *zero-length* when spring is relaxed. Using the plunger displacement $x(t)$, linear Hook's law yields an expression for spring force $F_{\text{spring}} = k_s(x_0 - x)$.

The maximum force is exerted at $x_{\text{min airgap}}$.

We have $F_{\text{spring max}} = k_s(x_0 - x_{\text{min airgap}})$ and $F_{\text{spring min}} = k_s(x_0 - x_{\text{max airgap}})$.

At *zero-length*, the spring is relaxed exhibiting zero force, for example, $F_{\text{spring min}} = k_s(x_0 - x_{\text{max airgap}}) = 0$.

Within the plunger displacement $x_{min} \le x \le x_{max}$, for, $x_{max} = x_{\text{max airgap}}$, $x_0 = x_{max}$.

There are mechanical limits on the plunger displacement $x_{min} \le x \le x_{max}$, $x_{min} \ge 0$, $x_{min} = x_{\text{min airgap}}$.

As the voltage u is applied, the plunger moves to reduce the air gap.

The maximum reluctance and force occur at zero-air gap, and, $F_e|_{x=0} = F_{e\ max}$.

Hence, $F_{\text{spring max}}=k_s x_0$ at the plunger zero-airgap position, $x_{\text{min airgap}}=0$. For u=0 V, the return spring restores the plunger to the position $x_{\max}$ at which, one may assume $F_{\text{spring}}=k_s(x_0-x_{\max})=0$, yielding $x_0=x_{\max}$. One selects a spring depending on a stroke Δx, $i_{\max}$, $F_{e\max}$, F_L, etc.

Example 3.5. *Modeling of Electromagnet and Magnetic Levitation Device*
Consider a magnetic levitation system or electromagnet as reported in Figure 3.6.a. The relative permeabilities of stationary and movable members are μ_{r1} and μ_{r2}. The equivalent *magnetic circuit* of the *closed* magnetic system with reluctances $(\Re_1,\Re_2,\Re_x)$ is depicted in Figure 3.6.b.

Assume a linear magnetic system for which μ_r=const, and, reluctance and magnetizing inductance are functions of the magnetic path lengths. For μ_r=const, the reluctances of the ferromagnetic materials of stationary and movable members, as well as the reluctance of two air gaps, are

$$\Re_1=\frac{l_1}{\mu_0\mu_{r1}A},\ \Re_2=\frac{l_2}{\mu_0\mu_{r2}A},\ \Re_x=\frac{2x}{\mu_0 A},\ A=l_l l_w.$$

Considering the *fringing* effect, the air gap reluctance is $\Re_x(x)=\dfrac{2x}{\mu_0\left(k_{g1}A+k_{g2}x^2\right)}$,

where k_{g1} and k_{g2} are the nonlinear functions of the ferromagnetic material, cross-sectional area $A=l_l l_w$, l_l/l_w ratio, BH curve, electromagnetic load, etc.

The magnetizing inductance is $L(x)=\dfrac{N^2}{\Re_{\text{total}}(x)}=\dfrac{N^2}{\Re_x(x)+\Re_1+\Re_2}$.

Form the co-energy $W_c=\frac{1}{2}L(x)i^2$, one finds the electromagnetic force

$$F_e=-\frac{\partial W_c}{\partial x}=-\tfrac{1}{2}i^2\frac{\partial}{\partial x}L(x).$$

At equilibrium, we have $F_e=mg$.

Alternatively, one considers a nonlinear BH curve with hysteresis using $\mu_r(i,\frac{di}{dt})$ and $L(i,\frac{di}{dt},x)$ in the pertained analysis, see Example 2.30.

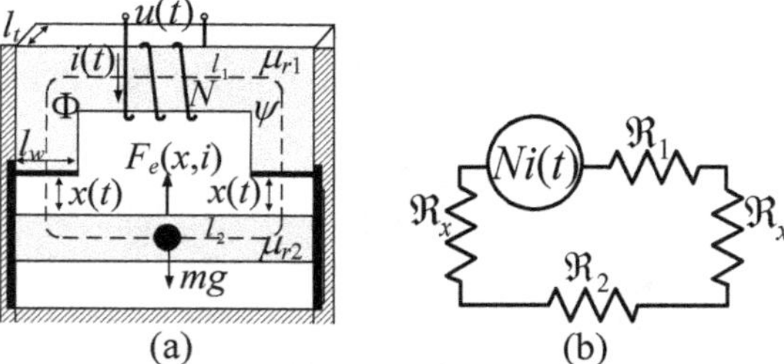

Figure 3.6. (a) Schematic of a variable-reluctance actuator;
(b) Equivalent *magnetic circuit*.

The source of flux linkages in the ferromagnetic members and air gaps is the *magnetomotive force*

$$mmf=\oint_l\vec{H}\cdot d\vec{l}=\sum_j H_j l_j=H_1l_1+H_2l_2+2H_x x.$$

The total reluctance is $\Re_{\text{total}}=(\Re_1+\Re_2+2\Re_x)$, $\Re_x=\dfrac{x(t)}{\mu_0 A}$.

The flux linkages are $\psi = N\Phi = \dfrac{N^2 i(t)}{\dfrac{l_1}{\mu_0\mu_{r1}A} + \dfrac{l_2}{\mu_0\mu_{r2}A} + \dfrac{2x}{\mu_0 A}} = L(x)i$.

The magnetizing inductance $L(x)$ is a nonlinear function of the displacement x

$$L(x) = \frac{N^2\mu_0\mu_{r1}\mu_{r2}A}{\mu_{r2}l_1 + \mu_{r1}l_2 + 2\mu_{r1}\mu_{r2}x}.$$

From $F_e = -\nabla W_c = -\dfrac{\partial W_c}{\partial x} = -\tfrac{1}{2}i^2(t)\dfrac{\partial}{\partial x}L(x)$, we find the force exerted on the movable member as a function of the current $i(t)$ and displacement $x(t)$.

One-directional electromagnetic force minimizes the air gap x, and

$$F_e = \frac{N^2\mu_0\mu_{r1}^2\mu_{r2}^2A}{(\mu_{r2}l_1 + \mu_{r1}l_2 + 2\mu_{r1}\mu_{r2}x)^2}i^2.$$

Using the Kirchhoff voltage law $u = ri + \dfrac{d\psi}{dt} = ri + L(x)\dfrac{di}{dt} + i\dfrac{dL(x)}{dx}\dfrac{dx}{dt}$ and Newtonian dynamics, one yields

$$\frac{di}{dt} = \frac{\mu_{r2}l_1 + \mu_{r1}l_2 + 2\mu_{r1}\mu_{r2}x}{N^2\mu_0\mu_{r1}\mu_{r2}A}\left[-ri - \frac{2N^2\mu_0\mu_{r1}^2\mu_{r2}^2A}{(\mu_{r2}l_1 + \mu_{r1}l_2 + 2\mu_{r1}\mu_{r2}x)^2}iv + u\right],$$

$$\frac{dv}{dt} = \frac{1}{m}\left[\frac{N^2\mu_0\mu_{r1}^2\mu_{r2}^2A}{(\mu_{r2}l_1 + \mu_{r1}l_2 + 2\mu_{r1}\mu_{r2}x)^2}i^2 - B_v v - k_s(x_0 - x) - F_L\right],$$

$$\frac{dx}{dt} = v,\ x_{\min}\le x\le x_{\max},\ x_{\min}\ge 0.$$ ■

Example 3.6. *Modeling of Solenoid*

Consider a solenoid depicted in Figure 3.7.a with the stroke 5 cm, $x_{\min}\le x\le x_{\max}$, $x\in[0.0005\ \ 0.05]$ m. The relative permeabilities of the stationary member and plunger are μ_{rs} and μ_{rp}. The solenoid parameters and coefficients are r=8.5 ohm, leakage inductance L_l=0.001 H, N=700, m=0.095 kg, l_p=0.055 m, l_s=0.095 m, $A=A_s=A_p=A_x$=0.00025 m^2, μ_{rs}=4500, μ_{rp}=5000, B_v=0.06 N-sec/m, k_s=10 N/m and x_0=0.05 m. The subscripts p and s stand for the plunger and stationary member. The length $l_p(x)$ varies. The assumptions l_p=const and d=0 results in error less than 1% because $\mu_r \gg 1$ and $d\approx 0$. Let $l_p=\tfrac{1}{2}(l_{p\,\min}+l_{p\,\max})$.

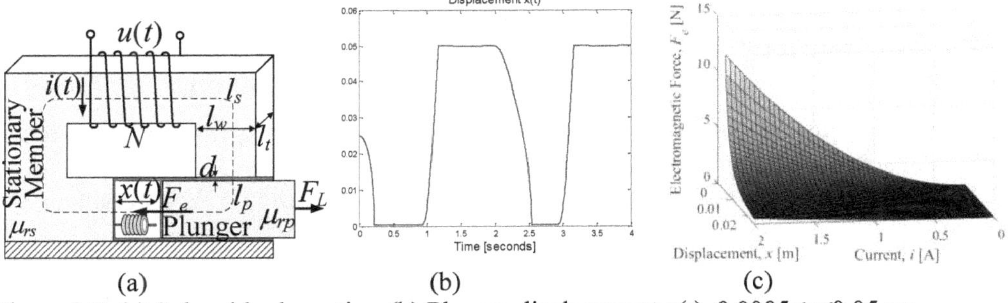

Figure 3.7. (a) Solenoid schematics; (b) Plunger displacement $x(t)$, $0.0005\le x\le 0.05$ m; (c) Surface for the electromagnetic force $F_e(x,i)$.

The magnetizing inductance is

$$L(x)=\frac{N^2}{\Re_s+\Re_p+\Re_x}=\frac{N^2}{\dfrac{l_s}{\mu_0\mu_{rs}A}+\dfrac{l_p}{\mu_0\mu_{rp}A}+\dfrac{x}{\mu_0 A}}=\frac{N^2\mu_0\mu_{rs}\mu_{rp}A}{\mu_{rp}l_s+\mu_{rs}l_p+\mu_{rs}\mu_{rp}x}.$$

The electromagnetic force is

$$F_e=-\tfrac{1}{2}i^2\frac{\partial}{\partial x}L(x)=\frac{N^2\mu_0\mu_{rs}^2\mu_{rp}^2A}{2\left(\mu_{rp}l_s+\mu_{rs}l_p+\mu_{rs}\mu_{rp}x\right)^2}i^2.$$

The maximum force F_e is at $x=x_{\min}$.

One has $F_{e\max}\big|_{x=x_{\min}}=\dfrac{N^2\mu_0\mu_{rs}^2\mu_{rp}^2A}{2\left(\mu_{rp}l_s+\mu_{rs}l_p+\mu_{rs}\mu_{rp}x_{\min}\right)^2}i^2.$

The resulting nonlinear differential equations are

$$\frac{di}{dt}=\frac{1}{L(x)+L_l}\left[-ri-\frac{N^2\mu_0\mu_{rs}^2\mu_{rp}^2A}{2\left(\mu_{rp}l_s+\mu_{rs}l_p+\mu_{rs}\mu_{rp}x\right)^2}iv+u\right],\ L(x)=\frac{N^2\mu_0\mu_{rs}\mu_{rp}A}{\mu_{rp}l_s+\mu_{rs}l_p+\mu_{rs}\mu_{rp}x},$$

$$\frac{dv}{dt}=\frac{1}{m}\left[\frac{N^2\mu_0\mu_{rs}^2\mu_{rp}^2A}{2\left(\mu_{rp}l_s+\mu_{rs}l_p+\mu_{rs}\mu_{rp}x\right)^2}i^2-B_v v-k_s(x_0-x)-F_L\right],$$

$$\frac{dx}{dt}=v,\ x_{\min}\le x\le x_{\max}.$$

The leakage inductance L_l is small, and, one may let $L_l=0$. Figures 3.7.b and c illustrate dynamics of $x(t)$ and three-dimensional plot $F_e(x,i)$ for $x\in[0.0005\ \ 0.05]$ m and $i\in[0\ \ 2]$ A. For $u=0$, the return spring restores the plunger to the *zero-length* spring position at which $F_{\text{spring}}=k_s(x_0-x)=0$.

For $u\neq 0$, at the plunger's steady-state equilibrium position $F_e(x,i)=F_{\text{spring}}=k_s(x_0-x)$. ■

Example 3.7. Magnetizing Inductance

The magnetizing inductance $L(x)$ and electromagnetic force F_e can be experimentally measured. Catalogs report plots for $F_e(x)$ for different currents, voltages and duty cycles. The electromagnetic force depends on the BH curve, and, permeability depends on the electromagnetic loads $\mu_r=\frac{1}{\mu_0}\frac{dB}{dH}$, see Example 2.30.

The Ampere law is $\oint_l H\cdot dl=i$.

In a full operating envelope, nonlinear magnetization is considered, and, $\mu_r(i,\frac{di}{dt})$ is not constant. The $F_e(x)$ are measured by loading a device, or, using the calibrated readouts of the stretch or compression of a calibrating spring with the known spring constant k_s finding the exerted force $F_{\text{spring}}(x)$.

The magnetizing inductance $L(x)$ can be measured. We obtained the expressions for the magnetizing inductance as $L(x)=\dfrac{a_1}{b+cx}$ by assuming μ_r=const and applying other assumptions. The electromagnetic force is

$$F_e=-\frac{\partial}{\partial x}\left(\tfrac{1}{2}L(x)i^2\right)=\frac{a_1c}{(b+cx)^2}i^2=\frac{a}{(b+cx)^2}i^2,\ a=a_1c,\ (a_1,b,c)>0.$$

The unknown coefficients (a_1,b,c) can be found using parameters such as lengths, air gap, movable to stationary members face-to-face geometry, area, μ_r, N, etc. For a solenoid, shown in Figures 3.5.a, the force $F_e(x,i)$ is experimentally measured for i=1 A at x=[0.005 0.01 0.015 0.021 0.025 0.03 0.035 0.04] m.

The experimentally measured $F_e(x)_{\text{experimental data}}$ is depicted in Figures 3.8 by dots. The minimum and maximum plunger displacements are x_{min}=0.05 and x_{max}=5 cm.

We find the unknown constants (a,b,c,d)>0 for

$$F_e = \frac{a}{(b+cx)^2}i^2,\ F_e = \frac{ae^{-dx}}{(b+cx)^2}i^2,\ F_e = ae^{-bx}i^2.$$

Note that the approximation $F_e(x,i)$ as $F_e = ae^{-bx}i^2$ is not physics consistent, and, used to practice simplistic studies. The physics-consistent equations are

$$F_e = \frac{a}{(b+cx)^2}i^2,\ F_e = \frac{ae^{-dx}}{(b+cx)^2}i^2,\ F_e(x,i) = \frac{ae^{-dx}e^{-2f|i|}}{\left(b+ce^{-g|i|}x\right)^2}i^2, F_e(x,i) = \frac{ae^{-dx}e^{-2f|i|}}{\left(be^{-h|i|}+ce^{-g|i|}x\right)^2}i^2.$$

Consider $F_e = \frac{a}{(b+cx)^2}i^2$. The unknowns (a,b,c) are found by nonlinear regression using the **nlinfit** command. The initial values are a_0=100, b_0=0.1 and c_0=100.

```
x=[0.005 0.01 0.015 0.021 0.025 0.03 0.035 0.04]; Fe1=[35 20 12 7 4.5 2.5 1.5 1]; % Measured data
plot(x,Fe1,'ko','linewidth',3); xlabel('Displacement, {\itx} [m]','FontSize',20);
title('Electromagnetic Force {\itF_e}({\itx}), [N]','FontSize',20);
modelFun=@(p,x) p(1)./((p(2)+p(3).*x).^2); startingValues=[100 0.1 100];       % First Fe(x)
modelFun=@(p,x) (p(1).*exp(-p(2).*x)); startingValues=[10 10];                 % Second Fe(x)
modelFun=@(p,x) (p(1).*exp(-p(4).*x))./((p(2)+p(3).*x).^2);startingValues=[100 0.1 100 1]; %Third Fe(x)
CoefEsts=nlinfit(x, Fe1, modelFun, startingValues)
xgrid=linspace(0,0.05,100); line(xgrid, modelFun(CoefEsts,xgrid),'Color','b','linewidth',2.5);
legend boxoff; legend('{\itF_e}({\itx})_{experimental data}', 'Parameterized {\itF_e}({\itx})','Fontsize',22);
```

The computed coefficients (a,b,c) are

CoefEsts = 5.2773e+01 7.3694e-01 9.6106e+01

For a=52.77, b=0.737 and c=96.1, $F_e = \frac{a}{(b+cx)^2}i^2 = \frac{52.77}{(0.737+96.1x)^2}i^2$.

Experimental $F_e(x)_{\text{experimental data}}$ and parameterized $F_e(x)$ are documented in Figure 3.8.a.

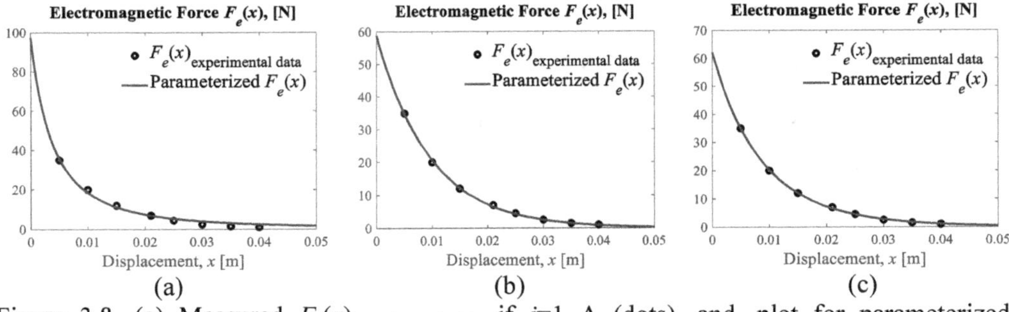

Figure 3.8. (a) Measured $F_e(x)_{\text{experimental data}}$ if i=1 A (dots), and, plot for parameterized $F_e = \frac{a}{(b+cx)^2}i^2$, a=52.77, b=0.737, c=96.1 (solid line);

(b) Plots for the measured $F_e(x)_{\text{experimental data}}$, and, $F_e = ae^{-bx}i^2$, a=58.7, b=105;

(c) Measured $F_e(x)_{\text{experimental data}}$ and $F_e = \frac{ae^{-dx}}{(b+cx)^2}i^2$, a=62.5, b=1, c=27.1, d=64.3.

Approximating the measured $F_e(x,i)$ by not physics-consistent $F_e = ae^{-bx}i^2$, the unknown a and b are found with a_0=10 and b_0=10. One obtains a=58.7 and b=105.

A plot for $F_e = ae^{-bx}i^2 = 58.7e^{-105x}i^2$ is depicted in Figure 3.8.b.

For $F_e = \dfrac{ae^{-dx}}{(b+cx)^2}i^2$, the unknown coefficients are a=62.5, b=1, c=27.1 and d=64.3. A plot for the parameterized $F_e = \dfrac{62.5e^{-64.3x}}{(1+27.1x)^2}i^2$ is reported in Figure 3.8.c. ■

Example 3.8. *Illustrative Simplified Modeling and Simulations*

Consider a solenoid, shown in Figures 3.5.a. The measured parameters and coefficients are r=15 ohm, m=0.1 kg, B_v=0.06 N-sec/m and k_s=10 N/m. Consider an approximation of the electromagnetic force as $F_e = ae^{-bx}i^2$, a=58.7, b=105, $0.0005 \le x \le 0.05$ m, computed in Example 3.7.

From $F_e(i,x) = -\dfrac{\partial W_c(i,x)}{\partial x} = -\tfrac{1}{2}i^2\dfrac{\partial L(x)}{\partial x}$, for a given $F_e = ae^{-bx}i^2$, we find a positive-definite magnetizing inductance

$$L(x) = -\int 2ae^{-bx}dx = 2\frac{a}{b}e^{-bx}.$$

The Kirchhoff and Newton laws yield governing equations, and, a SIMULINK diagram.

$$\frac{di}{dt} = \frac{b}{2ae^{-bx}}\left[-ri - 2ae^{-bx}iv + u\right],$$

$$\frac{dv}{dt} = \frac{1}{m}\left[ae^{-bx}i^2 - B_v v - k_s(x_0 - x) - F_L\right]$$

$$\frac{dx}{dt} = v,\ x_{min} \le x \le x_{max}.$$

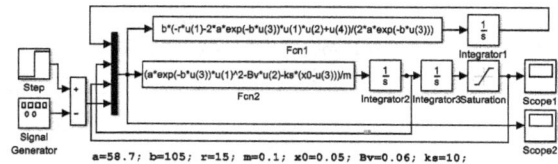

The mechanical limits on the plunger displacement are $0.0005 \le x \le 0.05$ m. During repositioning, dynamics of current $i(t)$, velocity $v(t)$ and position $x(t)$ are reported in Figure 3.9. For the voltage pulses $u = \begin{cases} 3\text{ V} \\ 2\text{ V} \end{cases}$, the plunger is repositioned to the equilibrium positions x_e at which $F_e = F_{spring} = k_s(x_0 - x)$.

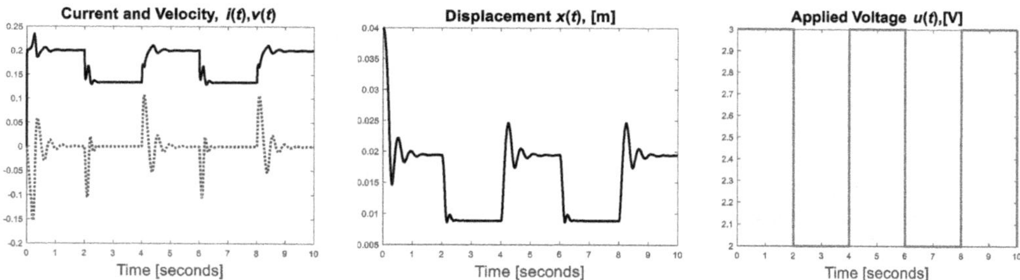

Figure 3.9. (a) SIMULIMK diagram to simulate open-loop solenoid;
(b) Dynamics of current $i(t)$, velocity $v(t)$ and displacement $x(t)$. Initial displacement is $x|_{t=0}$=0.04 m, and, voltage pulses are $u = \begin{cases} 3\text{ V} \\ 2\text{ V} \end{cases}$ with f=0.25 Hz. ■

Example 3.9. *Solenoid Modeling*
Using the equivalent *magnetic circuit*, the magnetizing inductance is

$$L(x)=\frac{ae^{-bx}}{(c+dx)},\ (a,b,c,d)>0.$$

One applies $F_e(i,x)=-\frac{\partial W_c(i,x)}{\partial x}=-\frac{1}{2}\frac{\partial L(x)}{\partial x}i^2$.

For a function f/g, using the quotient rule, the derivative is $(f'g-g'f)/g^2$.

Hence, $F_e=\frac{1}{2}ae^{-bx}\frac{b(c+dx)+d}{(c+dx)^2}i^2$.

The maximum electromagnetic force is at $x=0$,

$$F_{e\max}\Big|_{x=0}=\frac{1}{2}ae^{-bx}\frac{b(c+dx)+d}{(c+dx)^2}i^2\Bigg|_{x=0}=\frac{1}{2}a\frac{bc+d}{c^2}i^2.$$

Using the Faradays law of inductance, the total *emf* is

$$-\frac{d\psi}{dt}=-\frac{d}{dt}\big(L(x)i\big)=-\frac{di}{dt}\frac{ae^{-bx}}{(c+dx)}-ae^{-bx}\frac{b(c+dx)+d}{(c+dx)^2}iv.$$

The resulting nonlinear differential equations are

$$\frac{di}{dt}=\frac{c+dx}{ae^{-bx}}\left[-ri-ae^{-bx}\frac{b(c+dx)+d}{(c+dx)^2}iv+u\right],$$

$$\frac{dv}{dt}=\frac{1}{m}\left[\frac{1}{2}ae^{-bx}\frac{b(c+dx)+d}{(c+dx)^2}i^2-B_v v-k_s(x_0-x)-F_L\right],$$

$$\frac{dx}{dt}=v,\ x_{\min}\le x\le x_{\max}.$$ ■

3.3.2. Experimental Analysis and Control of a Solenoid

Solenoid schematics and equivalent *magnetic circuit* are depicted in Figures 3.10.

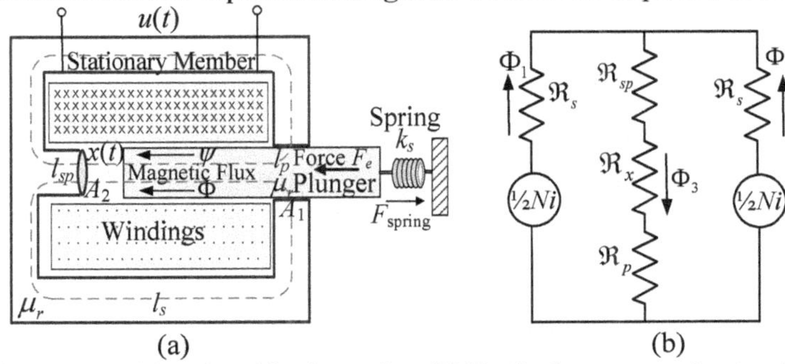

Figure 3.10. (a) Solenoid schematics; (b) Equivalent *magnetic circuit*.

The reluctances of the stationary member, stationary member which faces the plunger, plunger and air gap are

$$\Re_s=\frac{l_s}{\mu_0\mu_r A_1},\ \Re_{sp}=\frac{l_{sp}}{\mu_0\mu_r A_2},\ \Re_p=\frac{l_p}{\mu_0\mu_r A_2},\ l_p=\tfrac{1}{2}(l_{p\,\min}+l_{p\,\max}),\ \Re_x=\frac{x}{\mu_0 A_2}.$$

The equivalent *magnetic circuit* yields

$$\tfrac{1}{2}Ni=\Re_s\Phi_1+(\Re_{sp}+\Re_x+\Re_p)\Phi_3,\ \tfrac{1}{2}Ni=\Re_s\Phi_2+(\Re_{sp}+\Re_x+\Re_p)\Phi_3.$$

From $Ni = \Re_s(\Phi_1 + \Phi_2) + 2(\Re_{sp} + \Re_x + \Re_p)\Phi_3$, $\Phi_1+\Phi_2=\Phi_3$, we obtain $Ni = (\Re_s + 2\Re_{sp} + 2\Re_x + 2\Re_p)\Phi_3$.

The magnetic flux Φ_3 and flux linkages $\psi = N\Phi_3$ are

$$\Phi_3 = \frac{Ni}{\Re_s + 2\Re_{sp} + 2\Re_x + 2\Re_p},\quad \psi = \frac{N^2 i}{\Re_s + 2\Re_{sp} + 2\Re_x + 2\Re_p}.$$

The magnetizing inductance is $L(x) = \dfrac{N^2}{\Re_s + 2\Re_{sp} + 2\Re_x + 2\Re_p} = \dfrac{N^2 \mu_0 \mu_r A_1 A_2}{l_s A_2 + 2l_{sp} A_1 + 2A_1 \mu_r x + 2l_p A_1}$.

Hence, the electromagnetic force is $F_e = -\frac{1}{2} i^2 \dfrac{\partial}{\partial x} L(x) = \dfrac{N^2 \mu_0 \mu_r^2 A_1^2 A_2}{\left(l_s A_2 + 2l_{sp} A_1 + 2A_1 \mu_r x + 2l_p A_1\right)^2} i^2$.

The Kirchhoff voltage and Newton laws yield

$$\frac{di}{dt} = \frac{1}{L(x) + L_l}\left[-ri - \frac{2N^2 \mu_0 \mu_r^2 A_1^2 A_2}{\left(l_s A_2 + 2l_{sp} A_1 + 2A_1 \mu_r x + 2l_p A_1\right)^2} iv + u\right],\ L(x) = \frac{N^2 \mu_0 \mu_r A_1 A_2}{l_s A_2 + 2l_{sp} A_1 + 2A_1 \mu_r x + 2l_p A_1},$$

$$\frac{dv}{dt} = \frac{1}{m}\left[\frac{N^2 \mu_0 \mu_r^2 A_1^2 A_2}{\left(l_s A_2 + 2l_{sp} A_1 + 2A_1 \mu_r x + 2l_p A_1\right)^2} i^2 - B_v v - k_s\left(x_0 - x\right) - F_L\right],$$

$$\frac{dx}{dt} = v,\ x_{min} \le x \le x_{max}.$$

Examine a Ledex B11M-254 solenoid, driving circuit, and, proportional-integral controller

$$u_c = k_p e + k_i \int e dt.$$

For the considered solenoid, he catalog data is: u_{max}=12, 17, 24 and 38 V for 100, 50, 25 and 10% duty cycle operation; 15.5 N holding force; A-class coil insulation; 105°C maximum temperature; 17 g plunger weight. The maximum stroke is 2.2 cm.

The solenoid parameters are r=17.3 ohm, L_l=0.001 H, N=1780, m=0.017 kg, μ_r=5500, l_s=0.048 m, l_p=0.02 m, l_{sp}=0.08 m, $A_1=A_2=2\times10^{-4}$ m^2 and B_v=0.25 N-sec/m. The spring constant is k_s=58 N/m.

A closed-loop system is comprised of solenoid, position sensor, one-quadrant PWM amplifier, filters and proportional-integral analog controller. The solenoid displacement $x(t)$ is measured and compared with the reference displacement $r(t)$ to obtain the tracking error $e(t)=r(t)-x(t)$.

Using pulse-width-modulation (PWM), we control the duty cycle of the MOSFET transistor, thereby changing the *average* voltage $u(t)$ applied to a winding. The control voltage $u_c(t)$ is compared with a periodic triangular or sinusoidal signal u_t. The circuit schematics and images of the system are depicted in Figures 3.11.a and b. Notations, definitions, components and signals are labeled and accentuated.

The output voltage of the low-power, general purpose instrumentation amplifier INA128 is proportional to the tracking error $e(t)$. The error amplifier adds the reference voltage signal $r(t)$ with the inverted linear potentiometers output which corresponds to the measured plunger displacement $x(t)$. The error circuit is implemented using a unit-gain instrumentation amplifier INA128. The tracking error $e(t)=r(t)-x(t)$ is used in the proportional-integral analog controller $u_c = k_p e + k_i \int e dt$. The proportional term k_pe is implemented by using an operational amplifier TLC277 in an inverting configuration. The proportional gain k_p can be changed by the input and feedback resistors, $k_p=-R_{P2}/R_{P1}$. The integral feedback $k_i\int edt$ is implemented using an integrator. An inverted operational amplifier has an input resistor R_I and a feedback capacitor C_I. The integral feedback gain is $k_i= -1/R_IC_I$. Due to the inverting configuration for proportional and integral terms, the outputs are inverted and summed to yield the control voltage u_c. Instrumentation amplifiers are used due to their low noise, linearity, etc.

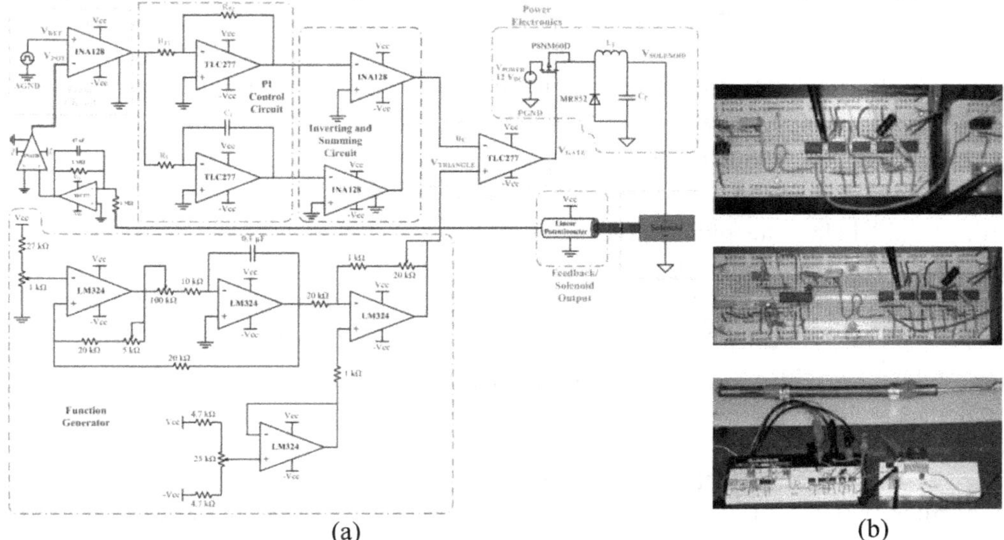

(a) (b)

Figure 3.11. (a) Solenoid with power electronics, sensor, controller and filters; (b) Images of closed-loop solenoid and electronics on prototyping boards.

The control signal u_c is supplied to a comparator which compares two inputs to produce a PWM output, see Figure 3.11.a. The comparator is implemented using an operational amplifier with the control signal u_c supplied to the inverting terminal and a periodic near-triangular waveform u_t applied to the noninverting terminal. The comparator outputs a positive rail voltage V_{CC} for the duration of time when a positive input is greater than the negative input. The comparator outputs $-V_{CC}$ when the negative input is greater than the positive input. The comparator develops the PWM waveform changing the duty cycle of the MOSFET. The periodic triangular waveform u_t is established by a function generator. As depicted in Figure 3.11.a, the first LM324 operational amplifier in the function generator circuit is the comparator. The positive input of the comparator is the output of the second operational amplifier in the circuit which is an integrator. The oscillation of the comparator produces a square waveform which is integrated by the second operational amplifier with a capacitor in the negative feedback to produce a near-triangular waveform u_t. The frequency of u_t is determined by the time constant defined by the input resistance and feedback capacitor of the integrator. The remaining two LM324 operational amplifiers in the function generator circuit are the buffer amplifiers for u_t and dc offset. The oscillating frequency of the n-stage ring pulse-oscillator is $f=\dfrac{1}{n(\tau_1+\tau_2)}$, where τ_1 and τ_2 are the intrinsic propagation delays; n is the number of inverters, n is odd. The maximum frequency is 6.8 kHz. The oscillators, wave generation and shaping circuit can be design using a Schmitt trigger, Wien-bridge oscillator, Colpitts oscillator, Hartley oscillator, RC phase-shift concept and other solutions.

The PWM waveform, produced by comparing the control signal u_c and near-triangular waveform u_t, controls the *average* voltage u applied to the solenoid. The TLC277 operational amplifier, which produces the PWM waveform, can output a maximum current ~30 mA and voltage 18 V. For the solenoid, the rated current and voltage are 2 A and 40 V. We use a one-quadrant stage as reported in Figure 3.11.a. A power MOSFET is controlled by applying the voltage to a gate. The comparator output controls the MOSFET switching. A low-pass filter is implemented using an inductor L_F and capacitor C_F. The filter frequency is $f_c = \frac{1}{2\pi\sqrt{L_F C_F}}$. For L_F=500 μH and C_F=4.7 μF, f_c=3.28 kHz. The plunger of solenoid is connected to a linear potentiometer to measure $x(t)$. One finds $e(t)$. The second-order low-pass filter is implemented to attenuate noise. The low-pass filter is between the potentiometer and error amplifier input. The operating solenoid bandwidth, reciprocal of settling time, is 1 Hz. For an unit-gain filter with the cutoff frequency $f = \frac{1}{2\pi R_f C_f}$, f=3.4 Hz, R_f=1×10^6 ohm and C_f=47 nF.

The control voltage u_c, which is the output of the proportional-integral controller, and near-triangular signal u_t are illustrated in Figure 3.12.a. These u_c and u_t are compared by the comparator, and, a PWM signal drives a MOSFET. The MOSFET gate voltage and the voltage applied to the solenoid are shown in Figure 3.12.b.

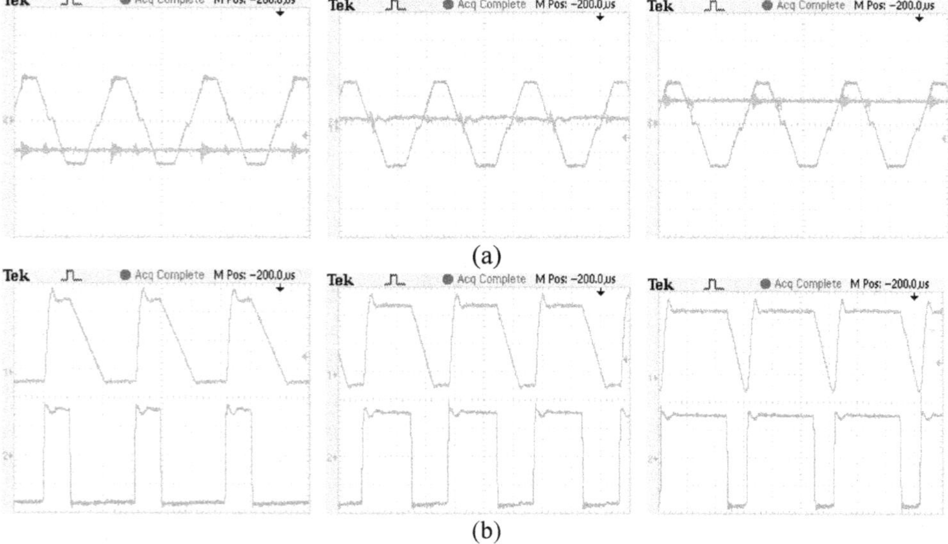

Figure 3.12. (a) Near-triangular waveform u_t and control voltage u_c for three references, which correspond to the plunger positions as r=0.5 cm, r=1 cm and r=1.5 cm;
(b) MOSFET gate voltage and PWM voltage applied to a solenoid u when r=1.5 cm, r=1 cm and r=0.5 cm. By changing the duty cycle d_D, the average voltage, applied to the winding, varies. The bus voltage is 12 V. For r=0.5 cm, we have d_D=0.28 and $u_{average}$=3.43 V, while, for a large displacement with r=1.5 cm, d_D=0.77 and $u_{average}$=9.2 V.

The transient dynamics of the closed-loop system are reported in Figure 3.13. The comparison of the reference $r(t)$ and plunger displacements $x(t)$ provide an evidence that: (1) Fast repositioning is accomplished with the settling time less than 0.8 sec; (2) The steady-state tracking error is minimized; (3) The loads F_L and disturbances are attenuated; (4) Robustness to parameter variations is accomplished in the full operating envelope. One-directional control of the displacement $x(t)$ is accomplished using the proportional-integral control law.

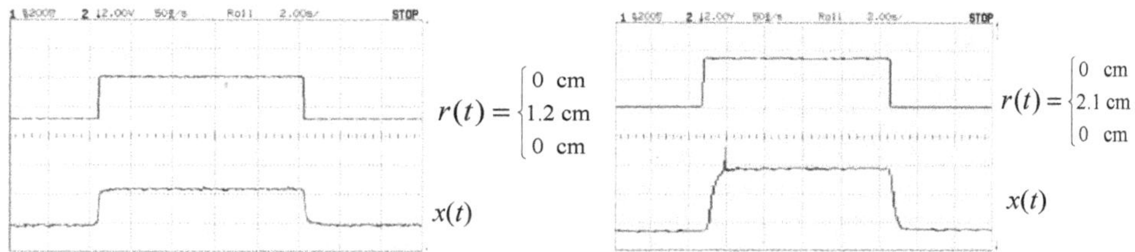

Figure 3.13. For references $r(t)=\begin{cases}0\text{ cm}\\1.2\text{ cm}\\0\text{ cm}\end{cases}$ and $r(t)=\begin{cases}0\text{ cm}\\2.1\text{ cm}\\0\text{ cm}\end{cases}$ (top plots), the plunger consistently repositions with bidirectional load $F_L=\pm5$ N. Reference $r=2.1$ cm corresponds to 2.1 cm stroke. For solenoid, $F_{e\max}|_{x=0,\ r=0.022}=19$ N and $F_{e\min}|_{x=0.022,\ r=0}=0$ N for $u=12$ V. The assigned displacements $x(t)$, illustrated by bottom plots, are ensured. The full stroke repositioning $0\le x\le 2.2$ cm is accomplished. The bidirectional loads $F_L(t)$ cause distortions and variations of $x(t)$, attenuated by the controller.

3.3.3. Variable Reluctance Rotational Actuators and Stepper Motors

Radial topology variable reluctance actuators and stepper motors are used in many applications. These actuators are used as limited angle rotational relays, switches, positioners, etc. The variable reluctance and *hybrid* stepper motors are depicted in Figure 3.14.a. A single-phase synchronous reluctance actuator with a *closed* magnetic system is illustrated in Figure 3.14.b. The path for flux linkages ψ_{as} is illustrated.

If rotor rotates, the positive-definite reluctance $\Re(\theta_r)>0$ and magnetizing inductance $L(\theta_r)>0$ vary with a period π, and, $\Re_{\min}\le\Re(\theta_r)\le\Re_{\max}$, $L_{\min}\le L(\theta_r)\le L_{\max}$.

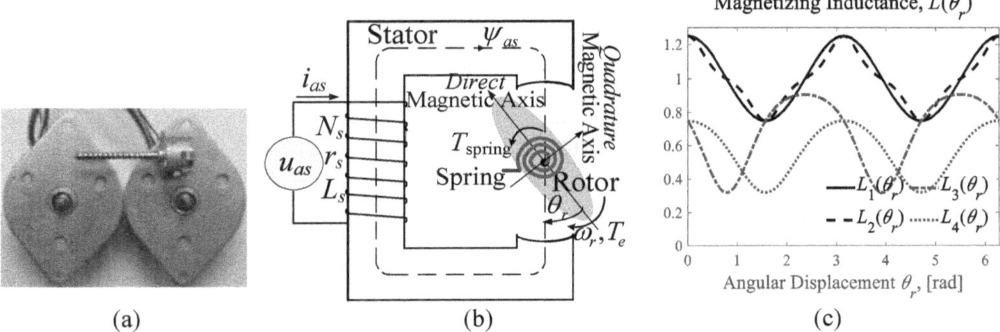

Figure 3.14. (a) Variable reluctance and *hybrid* stepper motors;
(b) Radial topology limited angle variable reluctance actuator with a torsional spring. Rotor and stator are made from electrical steel lamination. To ensure uniform moment of inertia J, a round-rotor has polymer-filled cavities, yielding varying inductance in the *quadrature* and *direct* magnetic axes;
(c) Magnetizing inductances $L_1(\theta_r)=\bar{L}_m+L_{\Delta m}\cos(2\theta_r)$, $L_2(\theta_r)=\bar{L}_m+L_{\Delta m}\cos(2\theta_r)e^{-\sin^2 2\theta_r}$, $L_3(\theta_r)=\bar{L}_m-L_{\Delta m}e^{\sin 2\theta_r}$ and $L_4(\theta_r)=\bar{L}_m-L_{\Delta m}e^{\sin^2\theta_r}$, plotted by solid, dashed, dash-dotted and dotted lines, Example 3.10.

Quadrature and Direct Magnetic Axes – For the *quadrature* (q) and *direct* (d) magnetic axes, reported in Figure 3.14.b, the corresponding reluctances and inductances are ($\Re_{mq}$,$\Re_{md}$) and (L_{mq},L_{md}), such as $\Re_{mq}$>$\Re_{md}$, L_{mq}<L_{md}. For the *direct* axis, the reluctance $\Re_{md}$ is minimum because the air gap is minimum. The reluctance in the *quadrature* axis $\Re_{mq}$ is maximum due to the maximum air gap.

The *quadrature*- and *direct*-axis inductances are L_{mq}=1/$\Re_{mq}$ and L_{md}=1/$\Re_{md}$.

The average inductance is $\bar{L}_m \neq 0$.

For symmetric geometry and magnetic system, $\bar{L}_m = \frac{1}{2}\left(L_{mq} + L_{md}\right)$.

For linear magnetic system assume μ_r=const and $\bar{L}_m$=const.

The magnitude of inductance variations $L_{\Delta m}$ is $L_{\Delta m} = \frac{1}{2}\left(L_{md} - L_{mq}\right)$.

In general, the magnetizing inductance is

$$L(\theta_r) = \bar{L}_m - L_{\Delta m}\phi(\theta_r),$$

where $\phi(\theta_r)$ is the periodic function which depends on the stator and rotor geometry, magnetic system, etc.

Using co-energy $W_c(i_{as}, \theta_r) = \frac{1}{2}L(\theta_r)i_{as}^2$, the electromagnetic torque T_e is

$$T_e = \frac{\partial W_c(i_{as}, \theta_r)}{\partial \theta_r} = \frac{\partial}{\partial \theta_r}\frac{1}{2}L(\theta_r)i_{as}^2 = \frac{1}{2}i_{as}^2\frac{\partial}{\partial \theta_r}L(\theta_r). \tag{3.10}$$

The Kirchhoff voltage law yields $u_{as} = r_s i_{as} + \frac{d\psi_{as}}{dt} = r_s i_{as} + L(\theta_r)\frac{di_{as}}{dt} + i_{as}\frac{dL(\theta_r)}{d\theta_r}\frac{d\theta_r}{dt}$.

Using circuitry-electromagnetic and Newtonian dynamics

$$\frac{di_{as}}{dt} = \frac{1}{L(\theta_r)}\left[-r_s i_{as} - \frac{dL(\theta_r)}{d\theta_r}i_{as}\omega_r + u_{as}\right], \tag{3.11}$$

$$\frac{d\omega_r}{dt} = \frac{1}{J}\left[\frac{1}{2}\frac{dL(\theta_r)}{d\theta_r}i_{as}^2 - F_{\text{friction}} - F_{\text{spring}} - F_L\right],$$

$$\frac{d\theta_r}{dt} = \omega_r, \ \theta_{r\,\min} \le \theta_r \le \theta_{r\,\max},$$

where $L(\theta_r) = \bar{L}_m - L_{\Delta m}\phi(\theta_r)$, $\bar{L}_m = \frac{1}{2}\left(L_{mq} + L_{md}\right)$ and $L_{\Delta m} = \frac{1}{2}\left(L_{md} - L_{mq}\right)$.

As the phase voltage u_{as} is applied, T_e is developed to minimize the air gap. The friction and spring torques (T_{friction},T_{spring}) oppose T_e. A linear Hooks law, applied to the single- and double-coil torsional springs (axial, radial, tangential, etc.), yields

$T_{\text{spring}}=k_s\theta_r$ or $T_{\text{spring}}=k_s|\theta_r|$

for initially unloaded spring with $T_{\text{spring}}\big|_{\theta_r=0} = 0$.

Nonlinear elasticity theory and *zero-angular-displacement* θ_{s0} can be used to explicitly define $T_{\text{spring}}(\theta_r)=f_s(\theta_r,\theta_{s0})$.

Example 3.10. Magnetizing Inductance of Single-Phase Two-Pole-Rotor Variable Reluctance Actuators

One analytically derives, measures and calculates the magnetizing inductance $L(\theta_r)$, which is a periodic function. Due to nonlinear magnetic system, nonlinear BH curve with hysteresis loop, varying cross-section areas on opposite ends of stator and rotor, unequal air gaps, nonuniformity and other affects, $L(\theta_r)$ could be symmetric or asymmetric.

For a symmetric two-pole actuator, depicted in Figure 3.14.b, the typifying magnetizing inductances are

$$L(\theta_r)=\bar{L}_m+L_{\Delta m}\cos(2\theta_r),\; L(\theta_r)=\bar{L}_m+\sum_n L_{\Delta m_n}\cos^{2n-1}(2\theta_r),$$

$$L(\theta_r)=\bar{L}_m+L_{\Delta m}\cos(2\theta_r)e^{-\sin^2 2\theta_r},\; L(\theta_r)=\bar{L}_m+\sum_{n,p} L_{\Delta m_{n,p}}\cos^{2n-1}(2\theta_r)e^{-\sin^{2p} 2\theta_r}.$$

These $L(\theta_r)$ are periodic functions with a period π.

Consider the representative inductances as

$$L_1(\theta_r)=\bar{L}_m+L_{\Delta m}\cos(2\theta_r),\; L_2(\theta_r)=\bar{L}_m+L_{\Delta m}\cos(2\theta_r)e^{-\sin^2 2\theta_r},$$

$$L_3(\theta_r)=\bar{L}_m-L_{\Delta m}e^{\sin 2\theta_r},\; L_4(\theta_r)=\bar{L}_m-L_{\Delta m}e^{\sin^2\theta_r}.$$

Let $\bar{L}_m$ =1 H and $L_{\Delta m}$ =0.25 H. Plots for $L(\theta_r)$ are depicted in Figure 3.14.c. ■

Example 3.11. *Limited Angle Variable Reluctance Actuator*

Consider a magnetic system and geometry with (L_{mq},L_{md}), illustrated in Figure 3.14.b. Let at for the initial stator-rotor displacement, the rotor is displaced such that the air gap reluctances are maximum. Assume the sinusoidal variations of the magnetizing inductance, such as

$$L(\theta_r)=\bar{L}_m-L_{\Delta m}\sin(2\theta_r).$$

The co-energy is $W_c(i_{as},\theta_r)=\frac{1}{2}\left(\bar{L}_m-L_{\Delta m}\sin(2\theta_r)\right)i_{as}^2$.

Hence, the electromagnetic torque T_e is

$$T_e=\frac{\partial W_c(i_{as},\theta_r)}{\partial\theta_r}=\frac{\partial}{\partial\theta_r}\frac{1}{2}\left(\bar{L}_m-L_{\Delta m}\sin(2\theta_r)\right)i_{as}^2=L_{\Delta m}\cos(2\theta_r)i_{as}^2.$$

The electromagnetic torque is developed, and, actuator operates within a limited angle $\theta_{r\,\min}\le\theta_r\le\theta_{r\,\max}$. At θ_{r0}=0 rad, the electromagnetic torque is T_e=0 N-m.

If voltage u_{as} is applied, the current i_{as} yields flux linkages ψ_{as} through the *closed* magnetic system as shown in Figure3.14.b. Apply the Kirchhoff law

$$u_{as}=r_{as}i_{as}+\frac{d\psi_{as}}{dt},$$

where the total *emf* is

$$-\frac{d\psi_{as}}{dt}=-\frac{d}{dt}L_m(\theta_r)i_{as}=-\frac{d}{dt}\left(\bar{L}_m-L_{\Delta m}\sin(2\theta_r)\right)i_{as}=-\left(\bar{L}_m-L_{\Delta m}\sin(2\theta_r)\right)\frac{di_{as}}{dt}+2L_{\Delta m}\cos(2\theta_r)i_{as}\omega_r.$$

One obtains a set of three first-order nonlinear differential equations (3.11) with a motional *emf* $2L_{\Delta m}\cos(2\theta_r)i_{as}\omega_r$ which opposes the applied voltage u_{as}

$$\frac{di_{as}}{dt}=\frac{1}{\bar{L}_m-L_{\Delta m}\sin(2\theta_r)}\left[-r_si_{as}-2L_{\Delta m}\cos(2\theta_r)i_{as}\omega_r+u_{as}\right],$$

$$\frac{d\omega_r}{dt}=\frac{1}{J}\left[L_{\Delta m}\cos(2\theta_r)i_{as}^2-B_m\omega_r-k_s(\theta_r-\theta_{s0})-T_L\right],$$

$$\frac{d\theta_r}{dt}=\omega_r,\; \theta_{r\,\min}\le\theta\le\theta_{r\,\max}.$$

Let the actuator parameters are r_s=10 ohm, L_{md}=0.25 H, L_{mq}=0.05 H, B_m=0.01 N-m-sec/rad, k_s=0.2 N-m/rad, θ_{s0}=0 rad and J=0.001 kg-m^2. The SIMULINK diagram is depicted in Figure 3.15.a. Dynamics of the phase current $i_{as}(t)$, angular velocity $\omega_r(t)$ and displacement $\theta_r(t)$ are reported in Figure 3.15.b if T_L=0. For the phase voltage, applied as pulses $u_{as}=\begin{cases}0\\10\text{ V}\end{cases}$, the rotor repositions to the steady-state equilibrium positions at which $T_e=T_{\text{spring}}+T_L$. The initial conditions are $[i_{as},\ \omega_r,\ \theta_r]_0=[0\ \ 0\ \ 0.0873]$.

For u_{as}=0 V and u_{as}=10 V, the rotor repositions from θ_r=0 to θ_r=0.37 rad.

```
rs=10; Lmd=0.25; Lmq=0.05; J=0.001; Bm=0.01; ks=0.2; ths0=0;
Lmb=(Lmq+Lmd)/2;Ldm=(Lmd-Lmq)/2;
```

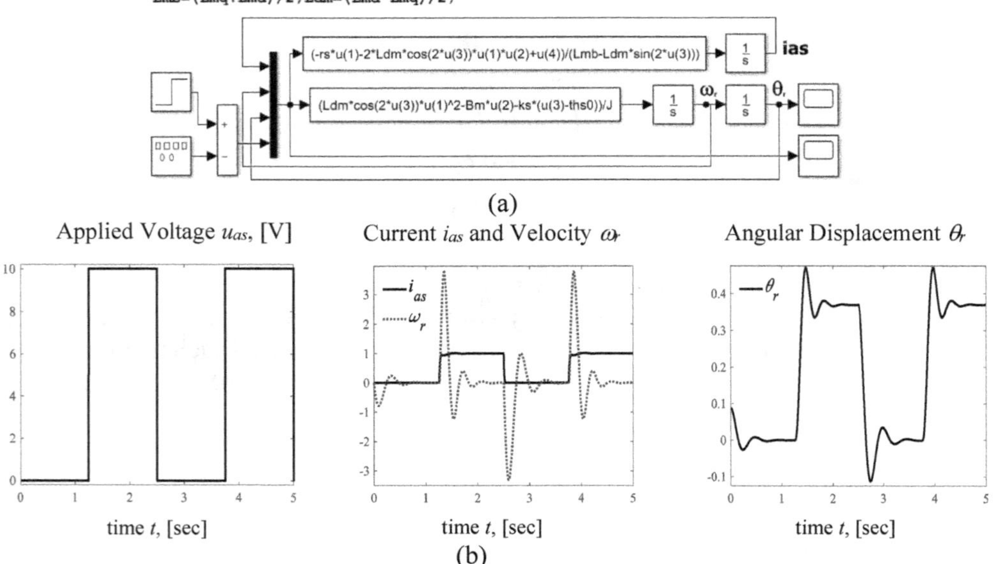

Figure 3.15. (a) SIMULINK diagram to simulate an actuator;
(b) Dynamics of $i_{as}(t)$, $\omega_r(t)$ and $\theta_r(t)$ for $u_{as}=\begin{cases}0\\10\text{ V}\end{cases}$ with f=0.4 Hz, θ_{r0}=0.0873 rad. At steady-state, rotor is displaced to $\theta_r=\begin{cases}0\text{ rad}\\0.37\text{ rad}\end{cases}$. ■

Example 3.12. *Variable Reluctance 360-Degree Rotation Motors and Stepper Actuators*
The magnetizing inductance for actuators, shown in Figures 3.14.a and b with no spring, can be expressed as

$$L(\theta_r)=\bar{L}_m-\sum_n L_{\Delta m_n}\cos^{2n-1}(2\theta_r),\quad L(\theta_r)=\bar{L}_m-\sum_p L_{\Delta m_p}e^{-\sin^{2p}2\theta_r},$$

$$L(\theta_r)=\bar{L}_m-\sum_{n,p} L_{\Delta m_{n,p}}\cos^{2n-1}(2\theta_r)e^{-\sin^{2p}2\theta_r},\ \text{etc.}$$

For $L(\theta_r)=\bar{L}_m-\sum_n L_{\Delta m_n}\cos^{2n-1}(2\theta_r)$, we have

$$T_e=\frac{\partial W_c(i_{as},\theta_r)}{\partial\theta_r}=i_{as}^2\sum_n(2n-1)L_{\Delta m_n}\sin(2\theta_r)\cos^{2n-2}(2\theta_r).$$

Using the derived expression for T_e, one finds the phase current and voltage to ensure continuous rotation such that $T_{e\,\text{average}}\neq 0$.

For n=2, $L(\theta_r)=\bar{L}_m-L_{\Delta m2}\cos^3(2\theta_r)$, we have

$$T_e=3L_{\Delta m2}\sin(2\theta_r)\cos^2(2\theta_r)i_{as}^2.$$

For the dc phase current $i_{as}=i_M$, $T_{e\text{average}}=\frac{1}{\pi}\int_0^{\pi}3L_{\Delta m2}\sin(2\theta_r)\cos^2(2\theta_r)i_{as}^2(\theta_r)d\theta_r=0$.

To ensure 360-degree rotation, the phase current and voltage are

$$i_{as}=\begin{cases}\dfrac{i_M}{\sqrt{\sin 2\theta_r \cos 2\theta_r}}, & \sin 2\theta_r>0\\ 0, & \sin 2\theta_r\le 0\end{cases},\quad |i_{as}|\le i_{\max},$$

$$u_{as}=\begin{cases}\dfrac{u_M}{\sqrt{\sin 2\theta_r \cos 2\theta_r}}, & \sin 2\theta_r>0\\ 0, & \sin 2\theta_r\le 0\end{cases},\quad |u_{as}|\le u_{\max}.$$

Assuming no bounds on i_{as}, $T_{e\,\text{average}}=\frac{1}{\pi}\int_0^{\pi}3L_{\Delta m2}\sin(2\theta_r)\cos^2(2\theta_r)i_{as}^2(\theta_r)d\theta_r$.

For a given $i_{as}(\theta_r)$, $T_{e\,\text{average}}=\frac{3}{2\pi}L_{\Delta m2}$. Hence, $T_{e\text{average}}\neq 0$.

The considered example describes operation of a single-phase variable reluctance stepper and synchronous reluctance motors. The variable reluctance and synchronous alternating current (ac) motors are usually two- and three-phase. ■

3.4. Permanent Magnet Direct Current Transducers and Power Electronics

Study high-performance translational and rotational dc transducers with permanent magnets. These transducers operate due to the spatiotemporal electromagnetic field interaction. The torque tends to align the magnetic moment $\vec{m}$ with field $\vec{B}$, and

$$\vec{T}=\vec{m}\times\vec{B}=i\vec{s}\times\vec{B}. \tag{3.12}$$

For uniform magnetic flux density, the electromagnetic force and torque are

$$\vec{F}=-i\vec{B}\times\oint_l d\vec{l},\ \vec{T}=\vec{R}\times\vec{F}. \tag{3.13}$$

The permanent magnets on stator or moving member establish a stationary magnetic field. The voltage is applied to the rotor windings using brushes and commutator, or, a flexible cord for limited-displacement transducers. Images of permanent magnet rotational and translational transducers are depicted in Figures 3.16.a. These actuators are controlled using power electronics which outputs PWM voltage applied to the winding. As the transducer is selected for given assemblies, the actuator parameters $(u_a,i_a,\omega_r,T_e,P,t_{\text{settling}},\eta)_{\text{rated}}$ and $(u_a,i_a,\omega_r,T_e,P,t_{\text{settling}},\eta)_{\text{peak}}$, are applied to select the matching power electronics. The off-the-shelf PWM controllers-drivers integrate filters, controllers and H-bridge drivers as covered in the application notes by Motorola, Texas Instruments, STMicroelectronics, Linear Technologies and other companies. One uses high power density power operational amplifiers, adequate topologies and other solutions as illustrated in Figures 3.16.b and c.

The schematics of high switching frequency MC33030 controller-driver is documented in Figure 3.17.a. This PWM controller-driver integrates on-chip operational amplifiers, comparators, driving logics, four-quadrant driver, etc. The rated output voltage and current are 36 V and 1 A. One may use MC33030 for ~10 W dc actuators. For a short period of time, dc actuators can operate at high voltage and current. The ratio $T_{e\text{peak}}/T_{e\text{rated}}$, $i_{\text{peak}}/i_{\text{rated}}$ and $P_{\text{peak}}/P_{\text{rated}}$ could reach 10. For power electronics, the $i_{\text{peak}}/i_{\text{rated}}$ ratio may vary from 1 to 2. The MC33030 servomotor driver contains 119 active transistors. The difference between the reference and actual angular velocity or displacement is compared by the error amplifier. Two comparators are used as shown in Figure 3.17.a. A *pnp* differential output power stage ensures driving and braking capabilities. The four-quadrant H-bridge driver guarantees high performance.

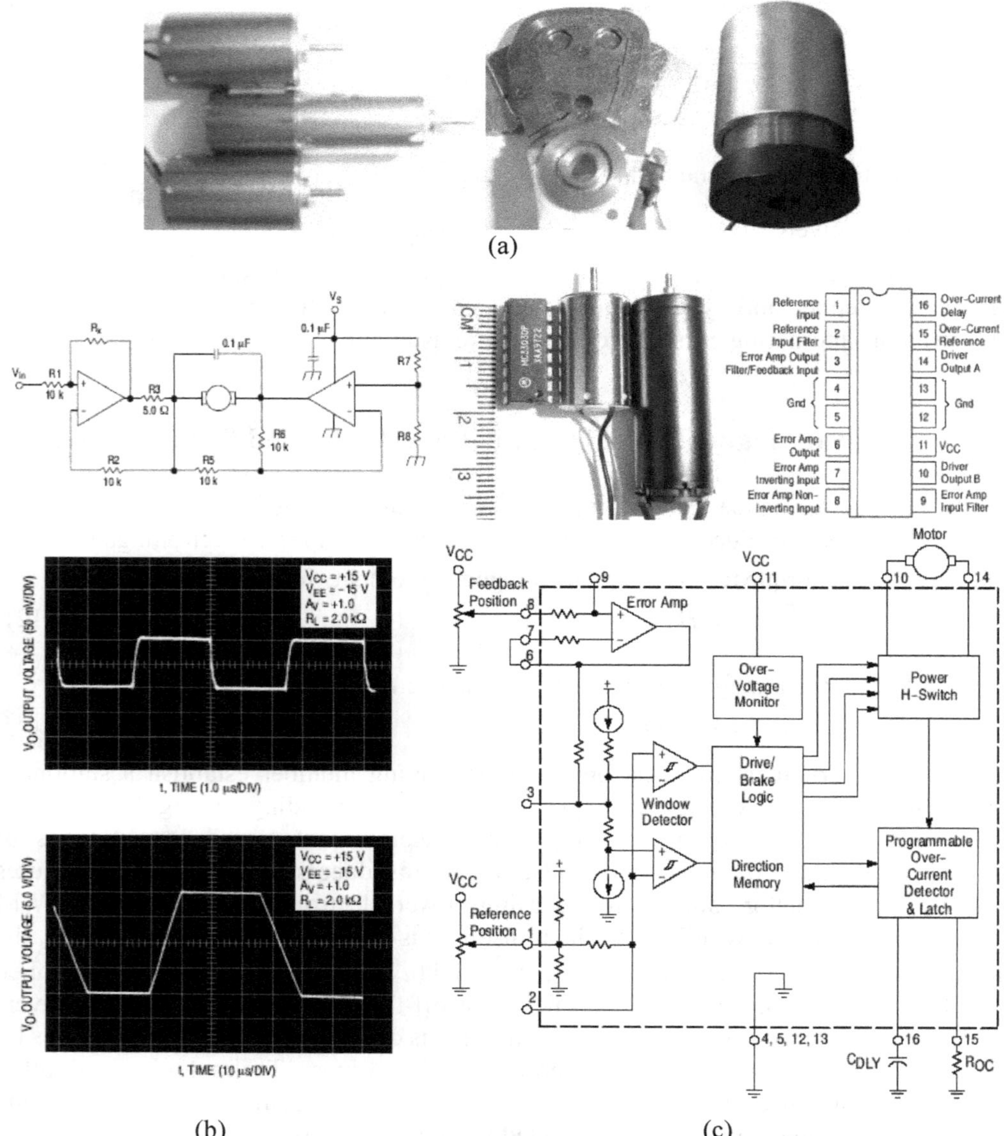

(b) (c)

Figure 3.16. (a) Permanent magnet rotational and translational servo actuators;
(b) Application of a dual amplifier to control permanent magnet dc minimotors: Bidirectional rotation of minimotors using TCA0372 dual power operational amplifier, 40 V and 1 A. The TCA0372 transient responses for the output voltage at light and peak loads are illustrated, copyright of Motorola, used with permission [7];
(c) Image of the MC33030 controller-driver, ~3 W permanent magnet dc minimotor, and, high-level MC33030 diagram, copyright of Motorola, used with permission [7].

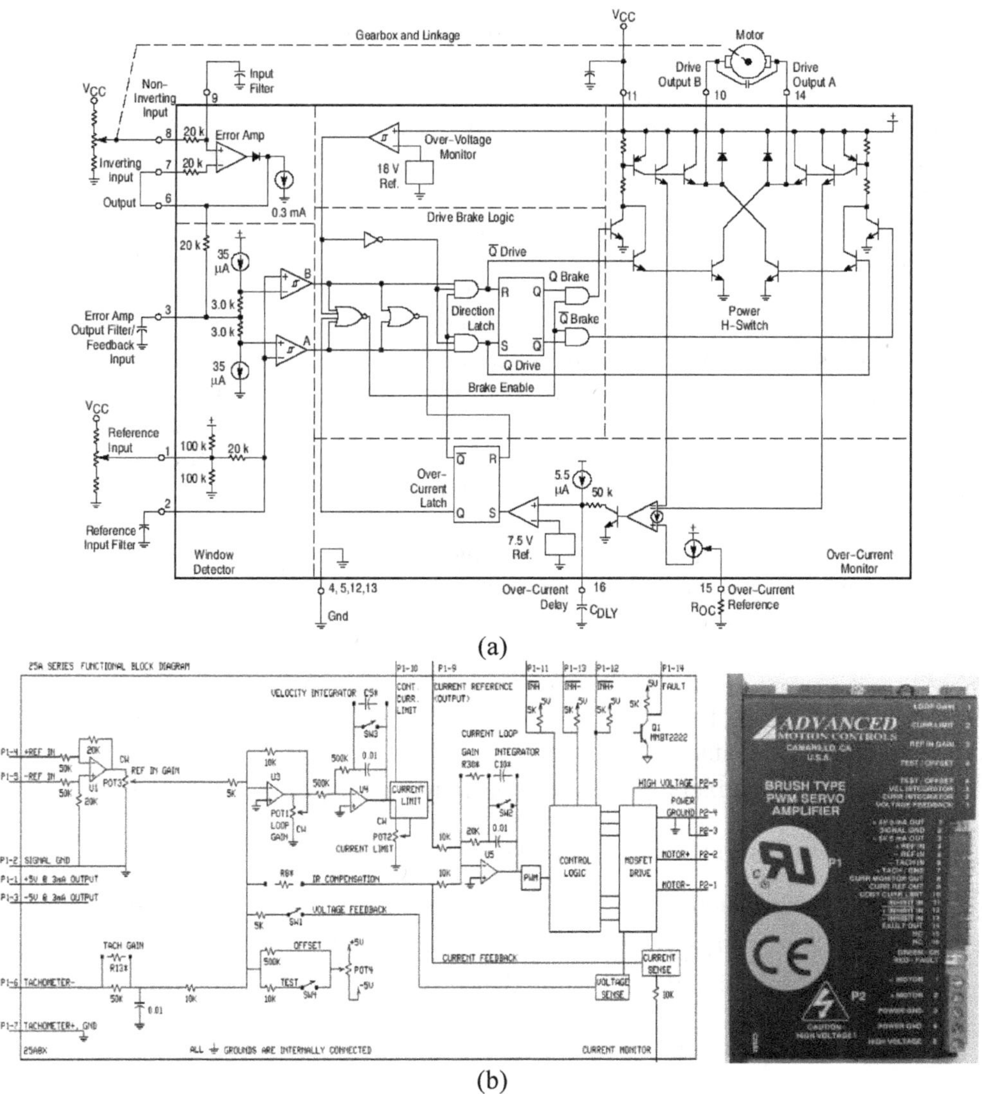

(b)

Figure 3.17. (a) Schematics of a servosystem with MC33030 controller-driver, copyright of Motorola, used with permission [7];
(b) 25A PWM amplifier, courtesy of Advanced Motion Controls, www.a-m-c.com, [7].

To regulate the angular velocity, one changes u_a. To rotate motor clockwise and counterclockwise, the bipolar voltage u_a is applied to the armature winding. For ~500 W dc actuators, the schematics of a four-quadrant 25A PWM servo amplifier (20-80 V, ±12.5 A continuous current, ±25 A peak current, 22 kHz, 129×76×25 mm dimensions) is documented in Figures 3.17.b. The motor armature winding is connected to P2-1 and P2-2. To control the angular velocity, one supplies the reference voltage to P1-4, and the voltage induced by the tachogenerator (proportional to the motor angular velocity) is supplied to P1-6. This amplifier can be used in servosystems, and, the angular or linear displacement should be supplied to P1-6. The proportional-integral analog controller is integrated. One can change the proportional and integral feedback gains adjusting the resistors. Various PWM amplifiers are available.

3.4.1. Radial Topology Permanent Magnet Direct Current Transducers

Permanent magnet alternating current (ac) and dc actuators convert electrical energy to mechanical energy, while generators convert mechanical energy to electrical energy. The output power of actuators is $P=\omega_r T_e$, while for generators $P=\mathscr{E}_a i_a$, where $\mathscr{E}_a$ is the generator induced voltage, which is the motional *emf* $\mathscr{E}$.

Radial topology transducers have stationary and rotating members, separated by an air gap. The armature windings, placed in the rotor slots, connected to a rotating commutator as shown in Figures 3.18.a and b. One applies the armature voltage u_a to the armature windings using graphite brushes and copper commutator. The armature winding consists of uniformly distributed coils. The excitation magnetic field is produced by the ark permanent magnets. The armature windings and permanent magnets produce stationary *mmfs* which are displaced by 90 electrical degrees. The armature magnetic force is along the rotor-fixed *quadrature* magnetic axis, while the *direct* axis corresponds to a permanent magnet magnetic axis.

The electromagnetic torque is produced due to interaction of the magnetic dipole moment and stationary magnetic field. Consistent electromagnetic design results in near-optimal electromagnetic system, uniform magnetic field, etc. From (3.13)

$$T_e=R_\perp F_e= R_\perp N l_{\text{eff}} B_{\text{eff}} i_a = k_a i_a, \qquad (3.14)$$

where i_a is the armature current; k_a is the torque constant; N is the number of turns; l_{eff} is the effective winding length; B_{eff} is the effective field which depends on magnets, air gap, winding topology and pattern, etc.

The electromagnetic torque can be found by using the coenergy $W_c = \int_i \psi di$, $T_e(i,\theta) = \frac{\partial W_c(i,\theta)}{\partial \theta}$. The magnetic flux crossing a surface is $\Phi = \oint_s \vec{B}\cdot d\vec{s}$.

The expression for T_e agrees with the equation (3.12) $\vec{T} = \vec{m}\times\vec{B} = i\vec{s}\times\vec{B} = NiA\vec{a}_m \times \vec{B}$. Assume that the stationary near-uniform magnetic field is produced by permanent magnets, and, the coils experience the same magnetic field.

Hence, $T_e=k_a i_a$.

For motors, in the Kirchhoff law

$$u_a = r_a i_a + \frac{d\psi}{dt} \qquad (3.15)$$

one should explicitly define the motional *back emf* $\mathscr{E}$.

Steady-State Motors and Generator Analyses – Using the armature voltage u_a, motional *emf* $\mathscr{E}$ and voltage drop $r_a i_a$, from (3.15), the steady-state equation for dc motors is

$$u_a - \mathscr{E} = r_a i_a.$$

The difference between the applied voltage u_a and motional *emf* $\mathscr{E}$ is the voltage drop across the armature resistance r_a. The motor rotates at an angular velocity ω_r at which the *emf* $\mathscr{E}$, induced in the armature winding, which *balances* the armature voltage u_a.

For motors, the induced *emf* is less than the voltage applied to the windings. For generators, the armature current i_a is in the same direction as the induced *emf*, and the terminal voltage is $(\mathscr{E} - r_a i_a)$.

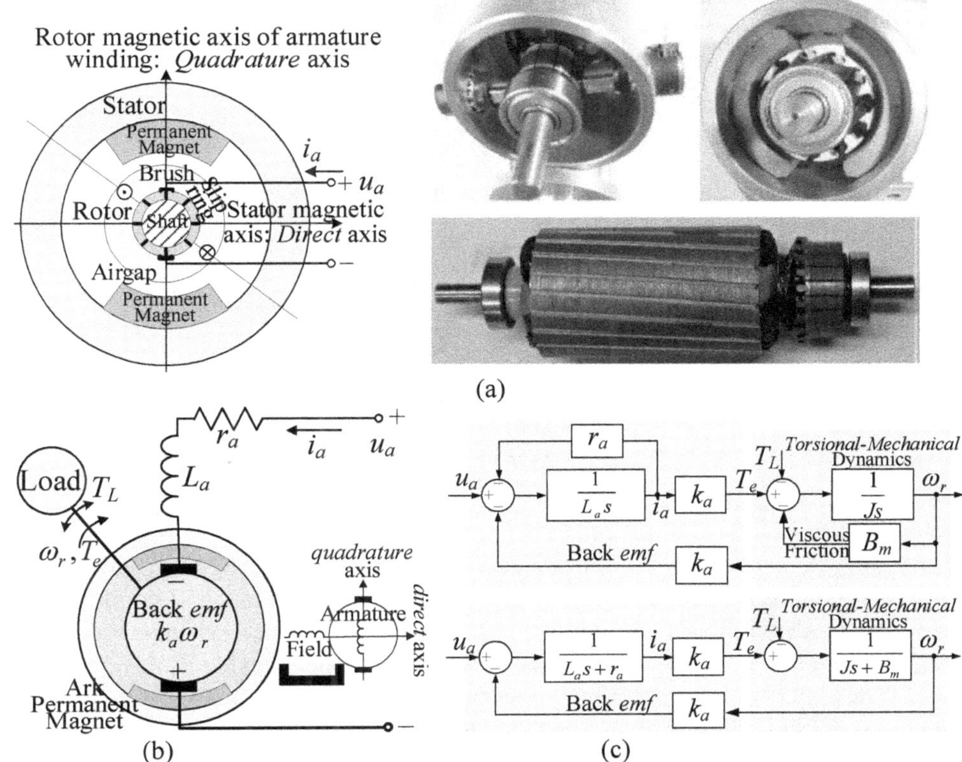

Figure 3.18. (a) Schematics and images of a radial topology permanent magnet dc transducer. The armature windings, connected to a rotating commutator, are placed in the rotor slots. Rotor is made from electrical steel lamination. The voltage is supplied using graphite brushes which slide on the copper commutator. There are two magnetized-through thickness ark magnets on the stator assembly. The bearing hubs are on both sides of the stationary member;
(b) Schematic diagram of a permanent magnet dc actuator. Voltage u_a is applied to the windings on a rotor using carbon or metal-graphite (usually graphite with copper) brushes and brass or copper segments on the commutator;
(c) The *s*-domain diagrams of permanent magnet dc actuators found using the laws of physics and derived equations of motion.

The Faraday law of induction defines the *emf* as

$$emf = \oint_l \vec{E}\cdot d\vec{l} = -\oint_s \frac{\partial \vec{B}}{\partial t}\cdot d\vec{s}\,,\quad emf = \oint_l (\vec{v}\times\vec{B})\cdot d\vec{l}\,. \tag{3.16}$$

The *mmf* is

$$mmf = \oint_l \vec{H}\cdot d\vec{l} = \oint_s \vec{J}\cdot d\vec{s} + \oint_s \frac{\partial \vec{D}}{\partial t}\cdot d\vec{s}\,. \tag{3.17}$$

Assume a uniform magnetic field, linear magnetic system and constant magnetic *susceptibility*. From (3.16), the motional *back emf* is

$$\mathscr{E}=k_a\omega_r, \tag{3.18}$$

where k_a is the *back emf* constant.

Using Kirchhoff's voltage law (3.15) and Newton's second law of motion $\Sigma T=J\alpha$, the resulting differential equations are

$$\frac{di_a}{dt}=\frac{1}{L_a}\left(-r_a i_a-k_a\omega_r+u_a\right),\qquad (3.19)$$
$$\frac{d\omega_r}{dt}=\frac{1}{J}\left(k_a i_a-B_m\omega_r-T_L\right).$$

The s-domain diagrams are illustrated in Figure 3.18.c.

In the state-space form, denote $x=\begin{bmatrix}x_1\\x_2\end{bmatrix}=\begin{bmatrix}i_a\\\omega_r\end{bmatrix}$, $u=u_a$ and $d=T_L$.

From (3.19) we have

$$\frac{dx}{dt}=Ax+Bu+B_d d,\quad \begin{bmatrix}\frac{di_a}{dt}\\\frac{d\omega_r}{dt}\end{bmatrix}=\begin{bmatrix}-\frac{r_a}{L_a} & -\frac{k_a}{L_a}\\\frac{k_a}{J} & -\frac{B_m}{J}\end{bmatrix}\begin{bmatrix}i_a\\\omega_r\end{bmatrix}+\begin{bmatrix}\frac{1}{L_a}\\0\end{bmatrix}u_a-\begin{bmatrix}0\\\frac{1}{J}\end{bmatrix}T_L,\qquad (3.20)$$

where $A\in\mathbb{R}^{2\times2}, B\in\mathbb{R}^{2\times1}, B_d\in\mathbb{R}^{2\times1}$.

Actuator Nonlinearities – The armature resistance r_a and *susceptibility*, which affects k_a, vary as functions of temperature. At peak loads, the magnetic field nonuniformity, magnetic system nonlinearities, nonlinear BH curve with hysteresis, magnetic losses, friction nonlinearities, as well as other effects, must be considered. High fidelity analysis is needed in entire operating envelope. The use of linear models is adequate for the rated operating conditions and loads.

Example 3.13. *Experimental Studies of a Permanent Magnet dc Motor*
Consider a 250 W, 70 V Torquemaster 2620 permanent magnet dc motor. The parameters are r_a=3.15 ohm, L_a=0.0066 H, k_a=0.156 V-sec/rad, B_m=0.0001 N-m-sec/rad and J=0.00015 kg-m^2.
Figure 3.19.a depicts the experimental results if motor starts at stall, ω_{r0}=0. The voltage u_a=60 V is applied at 0.04 sec. The motor accelerates at no load, reaches the angular velocity ω_r=384.62 rad/sec. Figure 3.19.b illustrates the motor dynamics if u_a=60 V is applied at 0.1 sec. Motor is loaded with T_L=0.5 N-m at t=0.4 sec.

For linear systems (3.20) $\frac{dx}{dt}=Ax+Bu$, $y=Hx+Du$, apply the **lsim** command. The angular velocity is the system output. The output equation matrices are H=[0 1] and D=[0]. The initial conditions are zero, and, voltage u_a=60 V is applied at 0.1 sec, while T_L=0.5 N-m at t=0.4 sec.

MATLAB statements, where the matrix B is refined to perform simulations with input u_a and disturbance T_L, are

```
ra=3.15; La=0.0066; ka=0.156; Bm=0.0001; J=0.00015;
A=[-ra/La -ka/La; ka/J -Bm/J]; B=[1/La 0; 0 -1/J]; H=[0 1]; D=[0];
t0=0.1; dT=0.001; tf=0.6; tL=0.4; x10=0; x20=0; t=t0:dT:tf; x0=[x10 x20];
Ua=60; U=Ua*ones(size(t)); TL=0.5; Tl=TL*[zeros(size(t0:dT:tL)) ones(size(tL+dT:dT:tf))]; u=[U; Tl];
[y,x]=lsim(A,B,H,D,u,t,x0);
plot(t,x(:,1),'k :',t,x(:,2),'k-','LineWidth',3); axis([0 0.6 0 400]); xlabel('time {\itt}, [sec]','FontSize',18);
title('Angular Velocity {\it\omega_r} [rad/sec] and Current {\iti_a} [A]','FontSize',18,'FontWeight','Normal');
legend('{\iti_a_s}','{\it\omega_r}','Location','northwest','NumColumns',1,'FontSize',24);
legend boxoff;
```

For u_a=60 V and load T_L=0.5 N-m at t=0.4 sec, the experimental and numeric results for states $i_a(t)$ and $\omega_r(t)$ are depicted in Figures 3.19.b and c. The experiments and simulations illustrate that a motor reaches the angular velocity 384.62 rad/sec within 0.1 sec. Consistency and correspondence of analytic, experimental, modeling and simulation results are achieved.

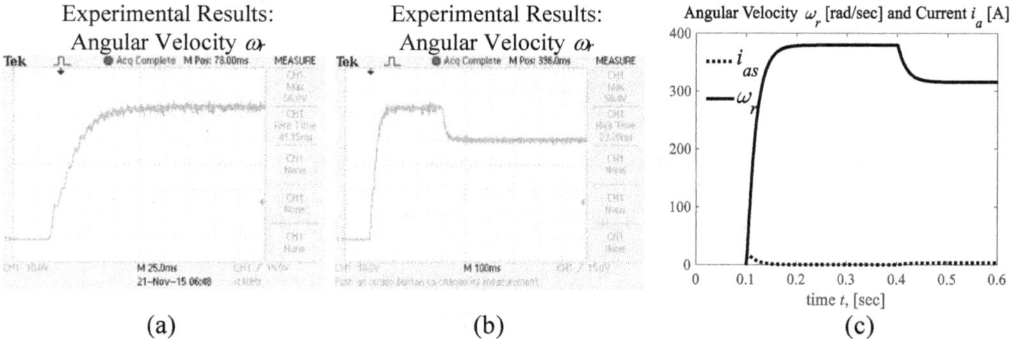

Figure 3.19. (a) Experimental results for u_a=60 V: Motor starts at stall ω_{r0}=0 with no loads, and, reaches speed 384.6 rad/sec;
(b) Experimental results for u_a=60 V: Motor starts at stall ω_{r0}=0 with no loads. The load T_L=0.5 N-m is applied at t=0.4 sec. The angular velocity is measured by a tachometer, $\omega_{r\text{measured}}=k_{\text{tachometer}}u_{\text{tachometer}}$, $k_{\text{tachometer}}$=6.77 V-rad/sec;
(b) Simulation results: Dynamics of $i_a(t)$ and $\omega_r(t)$, u_a=60 V, T_L=0.5 N-m at 0.4 sec. ■

Steady-State Analysis and Torque-Speed Characteristic – Using the first differential equation in (3.19) $\frac{di_a}{dt}=\frac{1}{L_a}(-r_a i_a - k_a\omega_r + u_a)$, at steady-state, $0=-r_a i_a - k_a\omega_r + u_a$.

Hence, $\omega_r = \frac{u_a - r_a i_a}{k_a}$.

The electromagnetic torque is $T_e=k_a i_a$.

In steady-state $T_e=T_{\text{friction}}+T_L$. Hence, $T_e=T_L$ if T_{friction}=0, and, other torques due to losses are negligible. Hence, the torque-speed characteristic $\omega_r=f(u_a,T)$ is

$$\omega_r = \frac{u_a - r_a i_a}{k_a} = \frac{u_a}{k_a} - \frac{r_a}{k_a^2}T. \tag{3.21}$$

One changes the applied voltage u_a to vary the angular velocity. If the load is applied, the angular velocity reduces. The slope is $-r_a/k_a^2$. The torque-speed characteristics are illustrated in Figure 3.20.a for different u_a, $|u_a|\le u_{a\max}$. To reduce the angular velocity, one decreases u_a. The angular velocity at which motor rotates is the intersection of the torque-speed and load characteristics.

From the Newton second law, neglecting electrical, magnetic and mechanical losses, $\frac{d\omega_r}{dt}=\frac{1}{J}(T_e - T_L)$. At $T_e=T_L$ motor rotates at the constant angular velocity. At no load, from (3.21) $\omega_r=u_a/k_a$ assuming that there are no losses. The angular velocity is reversed by changing the polarity of u_a.

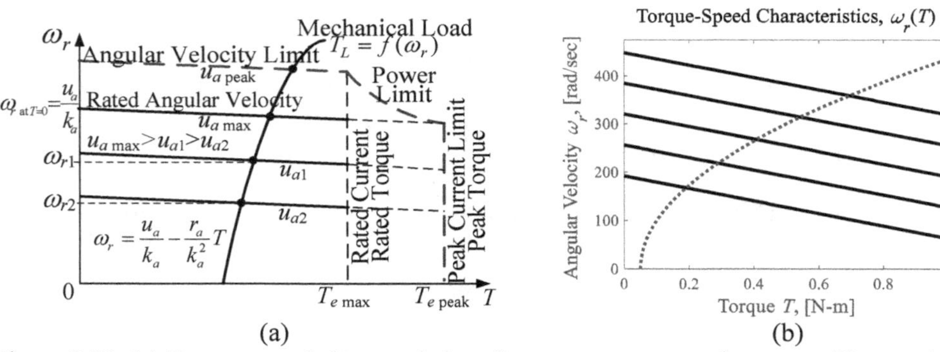

Figure 3.20. (a) Torque-speed characteristics of permanent magnet dc motors. The overloading capabilities are specified by the manufacturer. For a short time, $T_{e\,\text{peak}}/T_{e\,\max}$ could reach ~5; (b) Torque-speed characteristic for a permanent magnet motor and load, Example 3.14.

Example 3.14. *Computing Torque-Speed Characteristics*

Calculate and plot the torque-speed characteristics for a 70 V (rated) permanent magnet dc motor studied in Example 3.13. The motor parameters are r_a=3.15 ohm and k_a=0.156 V-sec/rad.

The load is a nonlinear function of angular velocity. For the aero- and hydrodynamic loads, $T_L=f(\omega_r)=a+b\omega_r^2$. Let $T_L=f(\omega_r)=0.05+5\times10^{-6}\omega_r^2$ N-m.

The torque-speed characteristics are described by equation (3.21). Using different values for the armature voltage u_a, the steady-state characteristics are calculated and plotted as depicted in Figure 3.20.b. The load $T_L=f(\omega_r)$ is illustrated by the dotted line.

```
ra=3.15; ka=0.156; T=0:0.01:1; % Motor parameters and torque
for ua=30:10:70;              % Applied voltage
wr=ua/ka-(ra/ka^2)*T;    % Torque-speed characteristic: Angular velocity for different voltages and loads
wrl=0:10:500; Tl=0.05+5e-6*wrl.^2; % Load torque at different velocities
plot(T,wr,'k-',Tl,wrl,'b:','LineWidth',3); hold on; axis([0 1 0 475]); end; hold off;
title('Torque-Speed Characteristics, {\it\omega_r}({\itT})','FontSize',20,'FontWeight','Normal');
xlabel('Torque {\itT}, [N-m]','FontSize',20'); ylabel('Angular Velocity {\it\omega_r}, [rad/sec] ','FontSize',20);
```

■

Efficiency and Losses – There are aerodynamic, hydrodynamic, electrical, magnetic, mechanical and other losses. The electrical losses P_E in copper winding are $P_E=r_a i_a^2$.

The mechanical losses P_M are due to the friction between the bearings and shaft, friction between the brushes and commutator, aerodynamic and hydrodynamic drag losses, etc. Under many assumptions, the mechanical losses due to viscous friction are assumed to be $P_M=B_m\omega_r^2$.

There are the magnetic losses P_m, such as the hysteresis electrical steel losses, eddy-current losses, etc.

Assuming that magnetic, aerodynamic drag and other losses are negligible, with nonlinear viscous friction

$$P_{\text{losses}}=P_E+P_M==r_a i_a^2+p(B_{m\,i},\omega_r)=r_a i_a^2+B_{m1}(\cdot)\omega_r^{2/3}+B_{m2}(\cdot)\omega_r^{4/3}+B_{m3}(\cdot)\omega_r^2+\ldots, \quad (3.22)$$

where $p(\omega_r)$ is a high-degree fractional polynomial which depend on the electromagnetic and mechanical loads, lubrication, wearing, bending, temperature, instabilities, etc.; B_{mi} are the real-valued coefficients, which depend on the loads, axes eccentricity, balancing, vibration, weight-bearing asymmetry, etc.

Assuming linear viscous friction, $P_{\text{losses}}=P_E+P_M=r_a i_a^2+B_m\omega_r^2$.

The efficiency η is experimentally found for the steady-state operation. For actuators and generators, using the input and output powers P_{input} and P_{output}, we have

$$\eta_{\text{actuator}}=\frac{P_{\text{output}}}{P_{\text{input}}}\times 100\%=\frac{T_L\Omega_r}{U_aI_a}\times 100\%,\ \eta_{\text{generator}}=\frac{P_{\text{output}}}{P_{\text{input}}}\times 100\%=\frac{U_aI_a}{T_L\Omega_r}\times 100\%. \quad (3.23)$$

The load torque T_L may or may not be directly measured. The efficiency is estimated as

$$\eta=\frac{P_{\text{output}}}{P_{\text{input}}}=\frac{P_{\text{input}}-P_{\text{losses}}}{P_{\text{input}}},\ \eta\approx\frac{U_aI_a-\left[r_aI_a^2+p(\Omega_r)\right]}{U_aI_a}. \quad (3.24)$$

For $P_{\text{losses}}=P_E+P_M=r_ai_a^2+B_m\omega_r^2$, we have

$$\eta=\frac{P_{\text{output}}}{P_{\text{input}}}=\frac{P_{\text{input}}-P_{\text{losses}}}{P_{\text{input}}}=\frac{U_aI_a-\left(r_aI_a^2+B_m\Omega_r^2\right)}{U_aI_a}.$$

Example 3.15. *Mechanical Losses*

Using the experimental data and tribology concept, one examines interacting surfaces, lubrication, wear, friction, etc. The physics-consistent expressions for $T_{\text{viscous}}(\omega_r)$ are derived. For example

$$T_{\text{viscous}}=B_{m1}\,\text{sgn}(\omega_r)e^{-b_1|\omega_r|}+B_{m2}\tanh(b_2\omega_r)+B_{m3}\omega_r^{1/3}+B_{m4}\omega_r,\ B_{mi}>0,\ (b_1,b_2)>0.$$

The mechanical losses are estimated as

$$P_M=\left(B_{m1}\,\text{sgn}(\omega_r)e^{-b_1|\omega_r|}+B_{m2}\tanh(b_2\omega_r)+B_{m3}\omega_r^{1/3}+B_{m4}\omega_r\right)\omega_r.$$ ■

3.4.2. Axial Topology Permanent Magnet Direct Current Actuators

This section examines the axial topology transducers used as drives and servos in wearable electronics, hard disk actuators, rotating stages, robots, etc. The wedge-shaped segmented permanent magnets or a single multi-pole magnet are placed on the stator, while the planar windings are on the rotor. Brushes and commutator, or flexible cord, are used to supply armature voltage to the winding on a rotating rotor. The advantages of the axial topology are:

1. Affordability and simplicity to fabricate and assemble transducers with wedge-shaped magnets and windings on the planar non-magnetic rotor;
2. There are relaxed shape, geometry and sizing requirements imposed on magnets and windings. However, device performance is affected by topology, magnets $(BH)_{\max}$, magnet-coil coupling, air gap, spacing, symmetry, volumetric, etc.;
3. There is no back ferromagnetic material required. The fiber-reinforced plastic, carbon fiber reinforced polymers, fiberglass and other non-magnetic materials are used.

The axial topology permanent magnet dc transducer is schematically depicted in Figure 3.21.a. The limited angle axial topology servo is shown in Figure 3.21.b. The electromagnetic force and torque are developed to ensure clockwise or counterclockwise rotation of a rotor.

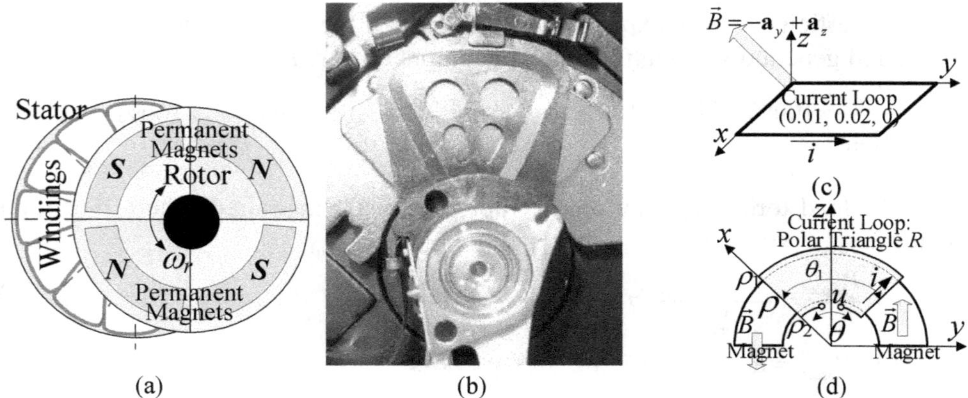

Figure 3.21. (a) Axial topology permanent magnet machine schematics with axially magnetized through thickness wedge-shaped magnets or multi-pole magnets;
(b) Limited angle axial topology servo to reposition a pointer. Planar winding coils are on polymer (polyethylene) rotor. Two axially magnetized through thickness wedge-shaped neodymium-iron-boron magnets are on stator below rotor. A planar winding, with a number of copper coils *N*, is on a rotor;
(c) Rectangular planar current loop in an uniform magnetic field *B*;
(d) Two wedge-shaped magnets and a planar winding in a cylindrical coordinate system. Permanent magnets are magnetized through thickness, producing a near-sigmoid field.

Consider a current loop in the magnetic field, produced by a magnet. The representative drawings are shown in Figures 3.21.c and d. The torque on a planar current loop of any size and shape in the uniform magnetic field is given by (3.12).

The magnetic dipole moment of the coils is $\vec{m} = Ni\vec{s}$ [A-m^2].

In the Cartesian coordinate system, the cross product of $\vec{m} = \begin{bmatrix} m_i \\ m_j \\ m_k \end{bmatrix}$ and $\vec{B} = \begin{bmatrix} B_i \\ B_j \\ B_k \end{bmatrix}$ yields the torque

$$\vec{T} = \vec{m} \times \vec{B} = Ni\vec{s} \times \vec{B},\ \vec{T} = \vec{m} \times \vec{B} = \begin{vmatrix} \vec{i} & \vec{j} & \vec{k} \\ m_i & m_j & m_k \\ B_i & B_j & B_k \end{vmatrix} = \vec{i}\begin{vmatrix} m_j & m_k \\ B_j & B_k \end{vmatrix} - \vec{j}\begin{vmatrix} m_i & m_k \\ B_i & B_k \end{vmatrix} + \vec{k}\begin{vmatrix} m_i & m_j \\ B_i & B_j \end{vmatrix}. \quad (3.25)$$

The cross product of two vectors is a vector which is orthogonal to both vectors. The solution is verified by $\vec{m} \cdot (\vec{m} \times \vec{B}) = 0$ and $\vec{B} \cdot (\vec{m} \times \vec{B}) = 0$.

The torque may be found using (3.13) as $\vec{T} = \vec{R} \times \vec{F}$, $\vec{F} = -i\oint_l \vec{B} \times d\vec{l}$.

In the differential form, $d\vec{F} = id\vec{l} \times \vec{B}$.

For an uniform magnetic field, $\vec{F} = -i\vec{B} \times \oint_l d\vec{l}$. The interaction of the magnetic dipole moment and magnet field results in the electromagnetic force and torque. The torque on a current loop tends to turn the loop to align the magnetic field produced by the loop with magnet.

Illustrative Example 3.1.
A polar triangle for a winding is $R = \{(\rho, \theta) |\ \rho_1 \le \rho \le \rho_2, -\frac{1}{2}\theta_1 \le \theta \le \frac{1}{2}\theta_1\}$.

The transformations from rectangular to cylindrical coordinates, and, from cylindrical to rectangular coordinates, are

$$\begin{cases} x=\rho\cos\theta \\ y=\rho\sin\theta \\ z=z \end{cases}, \begin{cases} \rho=\sqrt{x^2+y^2} \\ \theta=\arctan\frac{y}{x} \\ z=z \end{cases}, 0\le\rho<\infty, -\pi<\theta\le\pi, -\infty<z<\infty.$$

The unit vectors are $\begin{cases} \mathbf{a}_x=\mathbf{a}_\rho\cos\theta-\mathbf{a}_\theta\sin\theta \\ \mathbf{a}_y=\mathbf{a}_\rho\sin\theta+\mathbf{a}_\theta\cos\theta \\ \mathbf{a}_z=\mathbf{a}_z \end{cases}, \begin{cases} \mathbf{a}_\rho=\mathbf{a}_x\cos\theta+\mathbf{a}_y\sin\theta \\ \mathbf{a}_\theta=-\mathbf{a}_x\sin\theta+\mathbf{a}_y\cos\theta \\ \mathbf{a}_z=\mathbf{a}_z \end{cases}.$

For vector components,

$$\begin{bmatrix} A_\rho \\ A_\theta \\ A_z \end{bmatrix}=\begin{bmatrix} \cos\theta & \sin\theta & 0 \\ -\sin\theta & \cos\theta & 0 \\ 0 & 0 & 1 \end{bmatrix}\begin{bmatrix} A_x \\ A_y \\ A_z \end{bmatrix}, \begin{bmatrix} A_x \\ A_y \\ A_z \end{bmatrix}=\begin{bmatrix} \cos\theta & \sin\theta & 0 \\ -\sin\theta & \cos\theta & 0 \\ 0 & 0 & 1 \end{bmatrix}^{-1}\begin{bmatrix} A_\rho \\ A_\theta \\ A_z \end{bmatrix}=\begin{bmatrix} \cos\theta & -\sin\theta & 0 \\ \sin\theta & \cos\theta & 0 \\ 0 & 0 & 1 \end{bmatrix}\begin{bmatrix} A_\rho \\ A_\theta \\ A_z \end{bmatrix}.$$

■

Magnets – Arc, disc, cylinder and ring magnets can be axially and diametrically magnetized. Arcs, blocks (cubes and rectangles) and rings can be magnetized through thickness or width. The segments are magnetized ensuring north (or south) on the outside face, magnetized through circumference, or, magnetized through thickness. The wedge-shaped magnets are used in axial topology transducers. Multi-pole magnets, shown in Figures 3.21, are the axially-magnetized by segmented magnetization, or, segmented arrays are used. The skewed and angular field orientations are applied.

Example 3.16. *Analysis of Axial Actuators and Coordinate Systems*

Consider a 1×2 cm current loop with N coils in an uniform field $\vec{B}=-\mathbf{a}_y+\mathbf{a}_z$ as illustrated in Figure 3.21.c. Using the Cartesian coordinate system, the torque (3.25) is

$$\vec{T}=\vec{m}\times\vec{B}=i\vec{s}\times\vec{B}=Ni\begin{bmatrix} \mathbf{a}_x & \mathbf{a}_y & \mathbf{a}_z \\ 0.01 & 0.02 & 0 \\ 0 & -1 & 1 \end{bmatrix}=Ni\left(0.02\mathbf{a}_x-0.01\mathbf{a}_y-0.01\mathbf{a}_z\right) \text{ N-m.}$$

MATLAB computes the cross product. For N=1 and i=1 A,

```
N=1; i=1; m=[0.01 0.02 0]; B=[0 -1 1]; T=N*i*cross(m, B); T
```

we have

```
T = 2.0000e-02 -1.0000e-02 -1.0000e-02
```

In the Cartesian coordinates, the equation for a rectangle and squire are

$\left|\frac{1}{b}x+\frac{1}{c}y\right|+\left|\frac{1}{b}x-\frac{1}{c}y\right|=2$, $|x-y|+|x+y|=a$.

Using $x=\rho\cos\theta$ and $y=\rho\sin\theta$, one finds

$|\rho\cos\theta-\rho\sin\theta|+|\rho\cos\theta+\rho\sin\theta|=a$.

For a circle with a radius R, $x^2+y^2=R^2$, $\rho=2R\cos\theta$.

Recall that $R=\sqrt{x^2+y^2}$, $\tan\theta=\frac{y}{x}$ and $\theta=\arctan\frac{y}{x}$.

Consider a polar equation for a circle $\rho=-4\cos\theta$. Hence $\rho^2=-4\rho\cos\theta$.

From $\rho^2=x^2+y^2$ and $x=\rho\cos\theta$, we have $x^2+y^2=-4x$.

The resulting equation $(x+2)^2+y^2=4$ is a circle with radius R=2 and center at (–2,0).

Study a four-pole (P=4) axial-topology transducers, depicted in Figures 3.21, with polar triangle planar coils and axially magnetized through thickness wedge-shaped magnets, $\rho_{1\text{coil}}\le\rho_{\text{coil}}\le\rho_{2\text{coil}}$ and $\rho_{1\text{magnet}}\le\rho_{\text{magnet}}\le\rho_{2\text{magnet}}$. Let a winding topology is defined in $(\mathbf{a}_\rho,\mathbf{a}_\theta)$, no spacing between magnets, and, coils are under a uniform filed such that $\vec{s}=0.05\mathbf{a}_\rho+0.01\mathbf{a}_\theta$, and, $\vec{B}=0.25\mathbf{a}_z$.

From $\vec{T}=\vec{m}\times\vec{B}=PNi\vec{s}\times\vec{B}$, under assumptions stated, one yields the estimation for the electromagnetic torque. For P=4 and N=50

$$\vec{T}=\vec{m}\times\vec{B}=PNi\vec{s}\times\vec{B}=200i\begin{bmatrix}\mathbf{a}_\rho & \mathbf{a}_\theta & \mathbf{a}_z\\ 0.05 & 0.01 & 0\\ 0 & 0 & 0.25\end{bmatrix}=i(0.5\mathbf{a}_\rho-2.5\mathbf{a}_\theta)\text{ N-m.}$$

As reported in the Practice Problem 3.13, the unit vectors ($\mathbf{a}_\rho,\mathbf{a}_\theta,\mathbf{a}_z$) of the cylindrical coordinate system are functions of position, while ($\mathbf{a}_x,\mathbf{a}_y,\mathbf{a}_z$) are not functions of position. In particular,

$\mathbf{a}_\rho=\mathbf{a}_x\cos\theta+\mathbf{a}_y\sin\theta$, $\mathbf{a}_\theta=-\mathbf{a}_x\sin\theta+\mathbf{a}_y\cos\theta$, $\mathbf{a}_z=\mathbf{a}_z$.

Clockwise and counterclockwise rotation is accomplished changing the polarity of voltage applied. ■

Axial Topology 360-Degree-Rotation Transducers — As illustrated in Figure 3.21.a, the planar coils are above the axially magnetized through thickness wedge-shaped magnets, or, a multi-pole skewed magnet. The *effective* flux density $B(\theta_r)$ varies because rotor with windings displaces relative to the stator with magnets which produce the stationary field. The voltage u_a is applied to coils using brushes and copper commutator. Depending on coils topology, magnet magnetization, magnet magnetic field orientation, and coil-and-magnet placement, one finds a magnetic coupling. The flux density $B(\theta_r)$, as viewed from windings, is a periodic function of θ_r. Let magnets or poles produces a uniform magnetic field, and, there is a number of magnets or poles $N_m=2m$, m is the integer. Depending on spacing between magnets, segmented array skewed or non-skewed designs, magnets shape and thickness, magnetization variation, topology and pattern, as well as other factors, one may find

$$B(\theta_r)=B_{\max}\operatorname{sgn}\left(\sin(\tfrac{1}{2}N_m\theta_r)\right),\ B(\theta_r)=\sum_n B_n\sin^{2n-1}(\tfrac{1}{2}N_m\theta_r),\qquad(3.26)$$

where $B_{\max}$ is the amplitude of the effective flux density produced by magnets as viewed from coils, $B_{\max}$ depends on the magnets used, magnet-coils separation, temperature, etc.; n is the integer which is a function of the magnet magnetization, geometry, shape, width, thickness, etc.

Example 3.17. *Axial Topology Permanent Magnet dc Servo*

Consider the following $B(\theta_r)$

$B(\theta_r)=B_{\max}\sin(\tfrac{1}{2}N_m\theta_r)$, $B(\theta_r)=B_{\max}\operatorname{sgn}\left(\sin(\tfrac{1}{2}N_m\theta_r)\right)$ and $B(\theta_r)=B_{\max}\sin^5(\tfrac{1}{2}N_m\theta_r)$.

The resulting plots for $B(\theta_r)$ are reported in Figures 3.22 for $B_{\max}$=1 T and N_m=4.

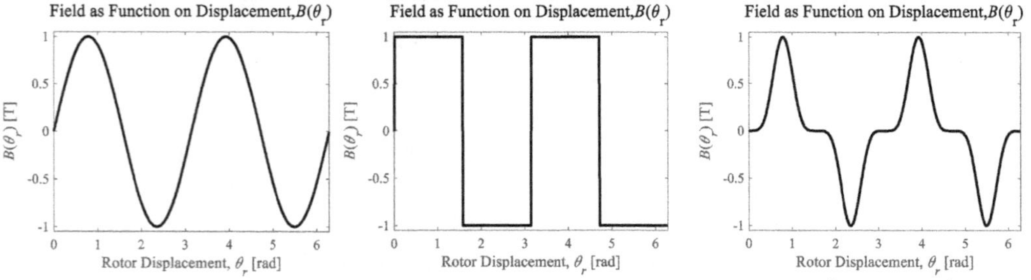

Figure 3.22. Plots of $B(\theta_r)=B_{\max}\sin(\tfrac{1}{2}N_m\theta_r)$, $B(\theta_r)=B_{\max}\operatorname{sgn}\left(\sin(\tfrac{1}{2}N_m\theta_r)\right)$ and $B(\theta_r)=B_{\max}\sin^5(\tfrac{1}{2}N_m\theta_r)$, $B_{\max}$=1 T, N_m=4. ■

Periodic $B(\theta_r)$ can be experimentally measured. The analytic expressions for $B(\theta_r)$ are used to derive the motional *emf* and T_e. Using Kirchhoff's voltage law $u_a = r_a i_a + \frac{d\psi}{dt}$ and Newton's second law of motion $\frac{d\omega_r}{dt} = \frac{1}{J}\left(T_e - T_{\text{friction}} - T_L\right)$, applying (3.13) and (3.16), the differential equations for the axial topology permanent magnet dc servo are

$$\frac{di_a}{dt} = \frac{1}{L_a}\left(-r_a i_a - \frac{d}{dt}\oint_s B(\theta_r)ds + u_a\right), \tag{3.27}$$

$$\frac{d\omega_r}{dt} = \frac{1}{J}\left(R_\perp i_a \oint_l B(\theta_r)dl - T_{\text{friction}} - T_L\right),$$

$$\frac{d\theta_r}{dt} = \omega_r .$$

Winding topology and pattern, magnet geometry and magnetization, *BH* curve, electromagnetic loading and other factors affect modeling and analysis tasks. For the electromagnetic toque

$$T_e = R_\perp N l_{eq} B(\theta_r) i_a, \tag{3.28}$$

where l_{eq} is the *effective* coil filament length.

For $B(\theta_r) = B_{\max}\,\text{sgn}\left(\sin(\tfrac{1}{2}N_m\theta_r)\right)$ and adequately designed electromagnetic system, winding topologies and magnets, we obtain

$$T_e = R_\perp N l_{eq} B_{\max} i_a = k_a i_a,\ k_a = R_\perp N l_{eq} B_{\max}. \tag{3.29}$$

Equations of motion (3.27) and (3.29), yield

$$\frac{di_a}{dt} = \frac{1}{L_a}\left(-r_a i_a - k_a\omega_r + u_a\right), \tag{3.30}$$

$$\frac{d\omega_r}{dt} = \frac{1}{J}\left(k_a i_a - B_m\omega_r - T_L\right),$$

$$\frac{d\theta_r}{dt} = \omega_r .$$

3.4.3. Axial Topology Permanent Magnet Direct Current Servos

Consider limited angle axial topology actuators with wedge-shaped segmented magnets. These servos are used in robots, repositioning servos, rotating platforms, and other applications. Actuators with two nickel-plated NdFeB magnets and rotor planar coils are depicted in Figures 3.23.a. For limited angle actuators, the commutator may not be required. The voltage is supplied by using a flexible cord. To rotate rotor clockwise or counterclockwise, the voltage polarity is changed. The top and bottom winding segments are not experience magnetic field. With axially magnetized through thickness magnets, the electromagnetic torque on the left and right side filaments is developed in the same direction, and, $T_e = T_{eL} + T_{eR}$.

The magnetic field by magnets $B(\theta_r)$ and magnetic dipole moment of coil with current i_a result in F_{eL} and F_{eR}, $\vec{F} = -i\vec{B} \times \oint_l d\vec{l}$. The F_{eL} and F_{eR} depend on $B_L(\theta_r)$ and $B_R(\theta_r)$ established by two nickel-plated NdFeB magnets.

The electromagnetic torques are $T_{eL} = R_\perp F_{eL}$ and $T_{eR} = R_\perp F_{eR}$.

The magnetic field $B(\theta_r)$ can be measured with the μT accuracy. Depending on displacement sensors, the rotor displacement measurement accuracy reaches milliradians. The adequate expressions, consistent with magnets magnetization and measured $B(\theta_r)$, are found. For example

$$B(\theta_r)=B_{\max}\tanh^{2q-1}(a\theta_r),\ a\gg 1,\ q=1,2,3,\dots, \tag{3.31}$$

$$B_3(\theta_r)=B_{\max}\frac{\theta_r}{\sqrt{b+\theta_r^2}},\ b\ll 1.$$

Example 3.18. *Magnetic Field in Axial Topology Servo*

Magnetic system significantly affects performance. Study variations of $B(\theta_r)$ as viewed from a winding. Stationary member with magnets, movable members and rotor with coils are assembled as documented in Figures 3.23.a.

The experimentally measured $B(\theta_r)_{\text{measured}}$, illustrated in Figure 3.23.b by dots, can be described and parameterized using expressions (3.31). We have

$B_1(\theta_r)=B_{\max}\tanh(a\theta_r)$, $B_2(\theta_r)=B_{\max}\tanh^{13}(a\theta_r)$, $B_{\max}$=1.07 T, a=98,

and, $B_3(\theta_r)=B_{\max}\dfrac{\theta_r}{\sqrt{b+\theta_r^2}}$, b=0.000091.

For $-0.175\le\theta_r\le 0.175$ rad, the plots for $B(\theta_r)$ are depicted in Figure 3.23.b.

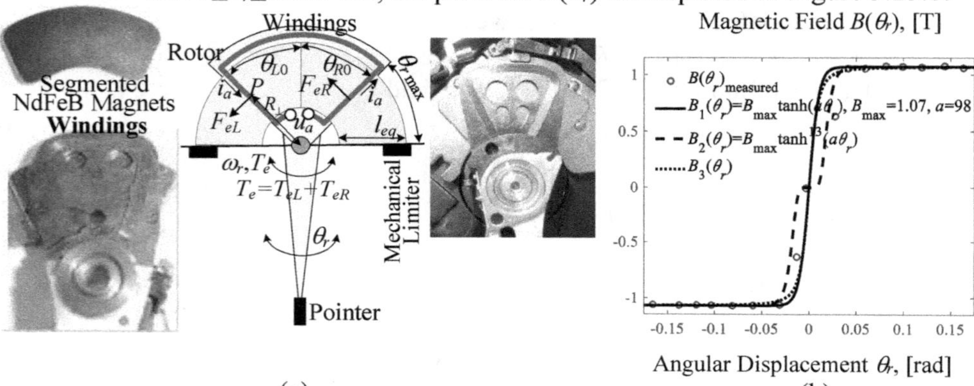

Figure 3.23. (a) Limited angle axial topology servo to reposition a pointer: Planar coils are above nickel-plated axially magnetized through thickness wedge-shaped NdFeB magnets. The mechanical limit on the rotor angular displacement is $-\theta_{r\max}\le\theta_r\le\theta_{r\max}$;
(b) Plots for measured $B(\theta_r)$ (dots), and, $B_1(\theta_r)=B_{\max}\tanh(a\theta_r)$, $B_2(\theta_r)=B_{\max}\tanh^{13}(a\theta_r)$, $B_{\max}$=1.07 T, a=98, $B_3(\theta_r)=B_{\max}\dfrac{\theta_r}{\sqrt{b+\theta_r^2}}$, b=0.000091. ■

The Kirchhoff's voltage law is $u_a=r_ai_a+\dfrac{d\psi}{dt}$.

Using the motional *emf* (3.16) $\mathscr{E}=-\oint_s\dfrac{\partial\vec{B}}{\partial t}\cdot d\vec{s}$, one finds

$$\frac{di_a}{dt}=\frac{1}{L_a}\left(-r_ai_a-\frac{d}{dt}\oint_s B(\theta_r)ds+u_a\right). \tag{3.32}$$

The torque is found using (3.12) or (3.13)

$$\vec{T}=\vec{R}\times\left(\vec{F}_{eL}+\vec{F}_{eR}\right)=-i\vec{R}\times\left(\oint_l \vec{B}_L\times d\vec{l}+\oint_l \vec{B}_R\times d\vec{l}\right). \quad (3.33)$$

Consistent assembly, magnet magnetization as well as optimal mechanical and electromagnetic designs yield an uniform magnetic field, symmetric balanced kinematics, uniform planar circular motion, active planar coil filaments point to the center of rotation, T_z torque component, etc. Let the electromagnetic force is applied at a point P, whose position is $R_\perp$ relative to point of rotation. Using the perpendicular radius (*lever arm*) $R_\perp$, the total electromagnetic torque is

$T_e=(T_{eL}+T_{eR})=R_\perp(F_{eL}+F_{eL})$.

Denote the relative angular filaments displacements as (θ_L,θ_R), at which the filed density of axially magnetized through thickness wedge-shaped magnets is $(B_L(\theta_r),B_R(\theta_r))$. We have $T_{eL}=R_\perp Nl_{eq}B_L(\theta_r)i_a$ and $T_{eR}=R_\perp Nl_{eq}B_R(\theta_r)i_a$.

For $\theta_{r0}=0$, $\theta_{L0}=|\theta_{r\max}|$ and $\theta_{R0}=|\theta_{r\max}|$. For a symmetric kinematics,

$\theta_L(t)=\theta_{L0}\mp\theta_r(t)$, $\theta_R(t)=\theta_{R0}\pm\theta_r(t)$, $-\theta_{r\max}\le\theta_r\le\theta_{r\max}$, $|\theta_{\min}|=|\theta_{r\max}|$.

The Newton second law of motion yields

$$\frac{d\omega_r}{dt}=\frac{1}{J}\left[T_e-T_{\text{friction}}-T_L\right]=\frac{1}{J}\left[R_\perp Nl_{eq}\left(B_L(\theta_r)+B_R(\theta_r)\right)i_a-T_{\text{friction}}-T_L\right], \quad (3.34)$$

$$\frac{d\theta_r}{dt}=\omega_r,\ -\theta_{r\max}\le\theta_r\le\theta_{r\max}.$$

Using the device-specific $B(\theta_r)$, one finds the motional *emf* $\mathscr{E}$, electromagnetic torque T_e, as well as the explicit governing equations (3.32) and (3.34).

Example 3.19. *Axial Topology Actuators*

Axially magnetized through thickness wedge-shaped NdFeB magnets exhibit the stationary magnetic field approximated as $B(\theta_r)=B_{\max}\tanh(a\theta_r)$. In general,

$B(\theta_r)=B_{\max}\tanh^{2q-1}(a\theta_r)$, $a\gg1$, $q=1,2,3,\ldots$.

For $B(\theta_r)=B_{\max}\tanh(a\theta_r)$, the motional *emf* (3.16) is $\mathscr{E}=-N\dfrac{d}{dt}\displaystyle\int_{r_{in}}^{r_{out}}\int_{\theta_R}^{\theta_L}B_{\max}\tanh(a\theta_r)rdrd\theta$.

For a continuous real-valued function f, defined in a closed interval $[b, c]$, let F is an antiderivative of f. For a definite integral, $\int_b^c f(x)dx=F(c)-F(b)$. Thus,

$$\mathscr{E}=-\frac{r_{\text{out}}^2-r_{\text{in}}^2}{2}NB_{\max}\left(\tanh a\theta_L-\tanh a\theta_R\right)\omega_r=-\frac{r_{\text{out}}^2-r_{\text{in}}^2}{2}NB_{\max}\left[\tanh a(\theta_{L0}-\theta_r)-\tanh a(\theta_{R0}+\theta_r)\right]\omega_r.$$

Using (3.33), the electromagnetic torque is $T_e=R_\perp Nl_{eq}B_{\max}\left(\tanh(a\theta_L)+\tanh(a\theta_R)\right)i_a$.

Hence, $T_e=R_\perp Nl_{eq}B_{\max}\left[\tanh a(\theta_{L0}-\theta_r)+\tanh a(\theta_{R0}+\theta_r)\right]i_a$.

Let $T_{\text{friction}}=B_m\omega_r$, and, the elastic restoring force of the flexible cord is $T_{\text{elastic}}=k_s\theta_r$. The Kirchhoff and Newton laws (3.32) and (3.34) yield

$$\frac{di_a}{dt}=\frac{1}{L_a}\left[-r_ai_a-\frac{r_{\text{out}}^2-r_{\text{in}}^2}{2}NB_{\max}\left[\tanh a(\theta_{L0}-\theta_r)-\tanh a(\theta_{R0}+\theta_r)\right]\omega_r+u_a\right],$$

$$\frac{d\omega_r}{dt}=\frac{1}{J}\left[R_\perp Nl_{eq}B_{\max}\left[\tanh a(\theta_{L0}-\theta_r)+\tanh a(\theta_{R0}+\theta_r)\right]i_a-B_m\omega_r-k_s\theta_r-T_L\right],$$

$$\frac{d\theta_r}{dt}=\omega_r,\ -\theta_{r\max}\le\theta_r\le\theta_{r\max}.$$

Consider a servo with mechanical limits on deflection $-10 \le \theta_r \le 10$ deg, $-0.175 \le \theta_r \le 0.175$ rad, $\theta_{L0}=\theta_{R0}=0.175$ rad. Assume the pointer is at center, $\theta_{r0}=0$.

The parameters are $B_{max}=1$ T, $a=100$, $r_a=35$, $L_a=4.1\times10^{-3}$ H, $R_\perp=0.02$ m, $N=125$, $l_{eq}=0.01$ m, $B_m=5\times10^{-4}$ N-m-sec/rad, $k_s=0.05$ N-m/rad, $J=1.5\times10^{-6}$ kg-m^2, $r_{in}=0.015$ m and $r_{out}=0.025$ m.

For the derived equations of motion, the corresponding SIMULINK diagram is reported in Figure 3.24.a. The transient dynamics for $\theta_r(t)$ is depicted in Figure 3.24.b if u_a are the steps ±5 V with $f=10$ Hz, and, $T_L=0$. At steady state, actuator is displaced by $\theta_r(t)=\pm0.143$ rad, and, the settling time is 0.025 sec. Fast repositioning is ensured.

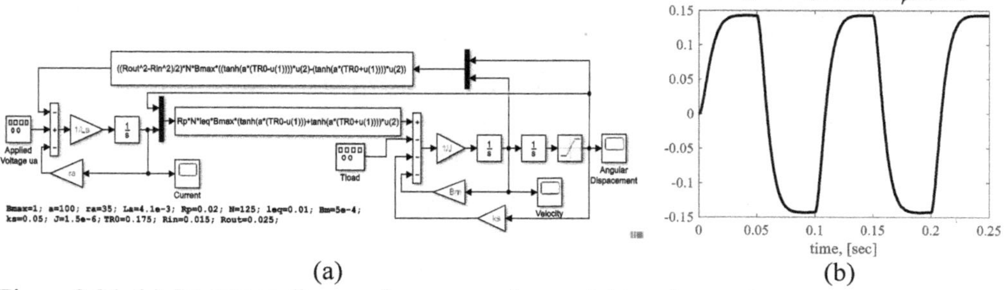

(a) (b)

Figure 3.24. (a) SIMULINK diagram for an open-loop axial topology actuator;
(b) Transient dynamics of the angular displacement $\theta_r(t)$, $u_a(t)=\begin{cases}5\text{ V}\\-5\text{ V}\end{cases}$, $f=10$Hz. ■

Example 3.20. *Axial Topology Actuators With Near-Sigmoid Magnetic Field*
Consider a high-performance actuator with $B(\theta_r)=B_{max}\tanh(a\theta_r)$, $a\gg1$.

We apply the results obtained in Example 3.19. For $a\gg1$, $\tanh(a\theta_i)\approx\pm1$, $\theta_i\neq0$.

Therefore, $T_e=T_{eL}+T_{eR}=2R_\perp Nl_{eq}B_{max}i_a$.

One finds the resulting differential equations

$$\frac{di_a}{dt}=\frac{1}{L_a}\left(-r_ai_a+u_a\right),$$

$$\frac{d\omega_r}{dt}=\frac{1}{J}\left(2R_\perp l_{eq}NB_{max}i_a-B_m\omega_r-k_s\theta_r-T_L\right),$$

$$\frac{d\theta_r}{dt}=\omega_r,\ -\theta_{rmax}\le\theta_r\le\theta_{rmax}.$$

One may neglect the armature transient dynamics because the armature inductance is small. For $L_a\approx0$, one finds $i_a=u_a/r_a$. Hence,

$$\frac{d\omega_r}{dt}=\frac{1}{J}\left(\frac{2}{r_a}R_\perp l_{eq}NB_{max}u_a-B_m\omega_r-k_s\theta_r-T_L\right),$$

$$\frac{d\theta_r}{dt}=\omega_r,\ -\theta_{rmax}\le\theta_r\le\theta_{rmax}.$$

Due to magnetic field and multi-coil nonuniformity and spacing between two magnets, one cannot assume $B(\theta_r)=B_{max}\mathrm{sgn}(\theta_r)$. The results are applicable for (3.31) $B(\theta_r)=B_{max}\tanh^{2q-1}(a\theta_r)$, $a\gg1$, $q>1$. ■

3.5. Translational Permanent Magnet Devices

The translational devices are used in high-bandwidth linear repositioning servos in aerial and ground vehicles, robots, medical devices, etc. The examples are cylindrical housed and frameless linear *voice coil* transducers, loudspeakers, microphones, etc. The images of ironless loudspeakers and linear *voice coil* transducers with radially-magnetized ring magnets are illustrated in Figures 3.25.a. Ceramic (ferrite), alnico and rear-earth magnets are used.

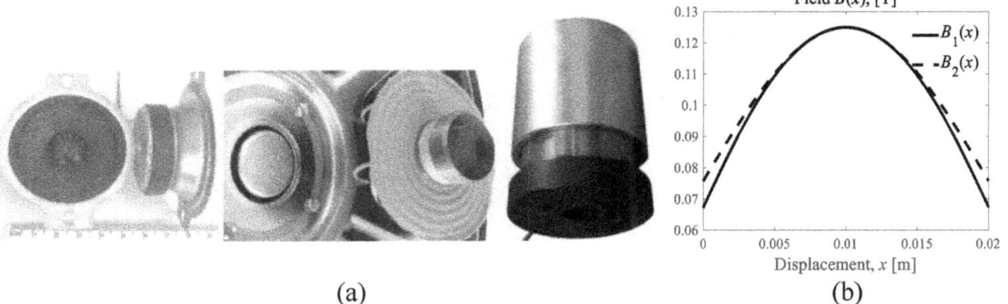

(a) (b)

Figure 3.25. (a) Ironless loudspeakers and long-stroke cylindrical linear actuator;
(b) Plots for $B_1(x) = B_{\max}\left[1 - \frac{1}{1+e^{-a[x-\frac{1}{2}(l_{\max}-l_{\min})]^2}}\right]$, $B_{\max}$=0.25 T, $l_{\min}\leq x\leq l_{\max}$, $0\leq x\leq 0.02$ m, a=10000, as well as, $B_2(x) = B_{\max}e^{-a(x-x_0)^2}$, $B_{\max}$=0.125 T, a=5000, x_0=0.01 m.

A movable member is secured to a stationary frame using different kinematic schemes. An N-turn winding (*voice coils*) is under the magnetic field established by radially-magnetized through thickness permanent magnet as illustrated in Figures 3.25.a. To bidirectionally displace a movable member, one changes the polarity of varying voltage applied to a winding. The bidirectional electromagnetic force is produced. In loudspeakers, a lightweight cone is secured to a rigid frame using a flexible suspension. A variety of different materials are used, such as polymers, composites, etc.

The suspension system maintains core with centered coil ensuring uniform air gap, and, suspension exhibits an elastic force F_{elastic} to make the cone return to an equilibrium position if voltage is not applied. The insulated enameled copper wire may have a circular, rectangular or hexagonal cross-section. The coils are oriented coaxially inside the air gap.

Applying the Kirchhoff and Newton laws, one obtains the resulting equations which are device dependent. For the magnetic system, magnets and kinematics under consideration, one finds the electromagnetic force by applying (3.13) $\vec{F} = \oint_l i d\vec{l} \times \vec{B} = -i\oint_l \vec{B}\times d\vec{l}$, as well as the motional *emf* (3.16) $\mathscr{E} = \oint_l \vec{E}\cdot d\vec{l} = -\oint_s \frac{\partial \vec{B}}{\partial t}\cdot d\vec{s}$.

We obtain

$$\frac{di_a}{dt} = \frac{1}{L_a}\left(-r_a i_a - \frac{d}{dt}\oint_s B(x)ds + u_a\right), \tag{3.35}$$

$$\frac{dv}{dt} = \frac{1}{m}\left(i_a\oint_l B(x)dl - F_{\text{drag}} - F_{\text{elastic}} - F_L\right),$$

$$\frac{dx}{dt} = v,\ x_{\min}\leq x\leq x_{\max},$$

where F_{drag} is the *net* drag force, such as aerodynamic drag, fluid forces, friction, etc.; F_{elastic} is the elastic restoring force. In a preliminary design, one may apply the approximations $F_{\text{drag}}=k_1v+k_2v|v|$ and $F_{\text{elastic}}=k_{\text{elastic}}|x|$.

Example 3.21. *Magnetic Field in Translational Permanent Magnet Actuators*
Study a translational actuator with a radially-magnetized ring magnet with width $l_{w\,\text{magnet}}$, as depicted in Figures 3.25.a. The field $B(x)$, as viewed from the winding coils, significantly affects the overall performance. One can measure $B(x)$, $x_{\min}\le x\le x_{\max}$, and express $B(x)$ by real-valued, continuous and differentiable functions. Trigonometric, exponential and other functions are used depending on the magnet magnetization, magnet shape, coil displacement with respect to magnet, magnet and coil orientation, magnet and coil widths ($l_{w\,\text{magnet}}$,$l_{w\,\text{coil}}$), symmetry, air gap uniformity, asymmetry, etc. For example,
$B(x)=B_{\max}e^{-a|x-x_0|}$, $B(x)=B_{\max}e^{-a(x-x_0)^2}$, $B(x)=B_{\max}e^{-a|(x-x_0)^3|}$, $B(x)=B_{\max}e^{-a(x-x_0)^4}$,$a>1$.

Using the experimental data, one finds $B(x)=\sum_n B_n e^{-a_n|(x-x_0)^n|}$.

At rest $\Sigma F=0$, and, equilibrium is characterized by $F_e=F_{\text{elastic}}$. Within the mechanical limits on the core displacement $x_{\min}\le x\le x_{\max}$, using ($l_{w\,\text{coil}}$,$l_{w\,\text{magnet}}$) and $B(x)$, the electromagnetic force is $F_e(x)=-i\oint_l B(x)dl$. The bidirectional electromagnetic force yields the displacement of the movable member. There are different electromagnetic systems, magnetic field distribution, kinematics, etc. The coils and magnet widths can be $l_{w\,\text{coil}}>l_{w\,\text{magnet}}$, $l_{w\,\text{coil}}=l_{w\,\text{magnet}}$ or $l_{w\,\text{coil}}<l_{w\,\text{magnet}}$.

The plot for $B_1(x,l)=B_{\max}\left[1-\dfrac{1}{1+e^{-a[x-\frac{1}{2}(l_{\max}-l_{\min})]^2}}\right]$, $B_{\max}$=0.25 T, a=10000, $0\le x\le 0.02$ m, $0\le l\le 0.02$ m, $l_{\min}=0$, $l_{\max}=0.02$ is depicted in Figure 3.25.b, $x\in[0\ 0.02]$ m.

Consider $B_2(x)=B_{\max}e^{-a(x-x_0)^2}$, $B_{\max}$=0.125 T, a=5000, x_0=0.01 m, $0\le x\le 0.02$ m. Figure 3.25.b illustrates $B_2(x)$. ■

3.6. Alternating Current Electromagnetic Motion Devices: Permanent Magnet Synchronous Machines

The alternating current (ac) electromagnetic motion devices, such as multi-phase induction and synchronous motors, were invented, demonstrated and implemented by Nikola Tesla. In 1888, Mr. Tesla secured the following US patents on two-phase induction and synchronous motors: (1) 381,968 and 382,279 *Electro Magnetic Motor*; (2) 381,969 *Electro Magnetic Motor*. Hundreds of patents on ac electric machines, apparatuses and devices were granted to Nikola Tesla during 1880 to 1920. In 1895, Nikola Tesla and Westinghouse Electric and Manufacturing Company, led by George Westinghouse, designed and built the first 37 MW hydro-electric power plant on Niagara Falls using 3.7 MW polyphase synchronous generators. These Tesla designed and Westinghouse built generators were operational until 1961. Modern energy systems, power plants and electrical grid are critical infrastructure systems, connected to the Internet, utility systems and IoT services. We focus on industrial control systems, instrumentation, robotics, high power and high torque densities actuators, and, servos. Permanent magnet synchronous machines, called *brushless dc motors*, are emphasized.

Induction Motors – Induction motors are examined in Practice Problems 3.8, 3.9, 3.10, 3.11 and 5.9. Three-phase induction motors [4-7] are used in high-power transportation and propulsion systems, various industrial applications, appliances, etc.

Permanent Magnet Electromagnetic Motion Devices – Permanent magnet synchronous machines guarantee superior performance and capabilities surpassing other actuators, motors and generators. These machines are used in aerial and ground vehicles, energy systems, robots, servos, consumer electronics, etc. The considered mutiphase ac permanent magnet synchronous machines, controlled by changing the magnitude and frequency of phase voltages, are also called *brushless dc motors*.

The electromagnetic torque results due to the interaction of time-varying magnetic field established by the stator windings and magnetic field produced by arc-shaped magnets on the rotor, as depicted in Figure 3.26.a. As shown in Figure 3.26.b, if the considered machine is used as a generators, the voltages are induced in the stator windings. The generator is rotated by the external source. The angular velocity of synchronous motors is fixed with the frequency f of the phase voltages (u_{as},u_{bs},u_{cs}) applied to the windings. These (u_{as},u_{bs},u_{cs}) are applied as functions of the rotor angular displacement θ_r. The steady-state torque-speed characteristics are the horizontal lines as depicted in Figure 3.26.c.

For P-pole machines, the electrical angular velocity ω_r is equal to the synchronous angular velocity ω_e. The synchronous angular velocity is

$$\omega_e = \frac{4\pi f}{P},$$

where P is the number of poles; f is the frequency of the phase voltages.

The mechanical angular velocity ω_{rm} and torque on shaft T_{em} are

$$\omega_{rm} = \frac{2}{P}\omega_r \text{ and } T_{em} = \frac{P}{2}T_e.$$

The maximum load should not exceed a peak electromagnetic torque, $T_{e\,\mathrm{peak}} > T_{L\max}$.

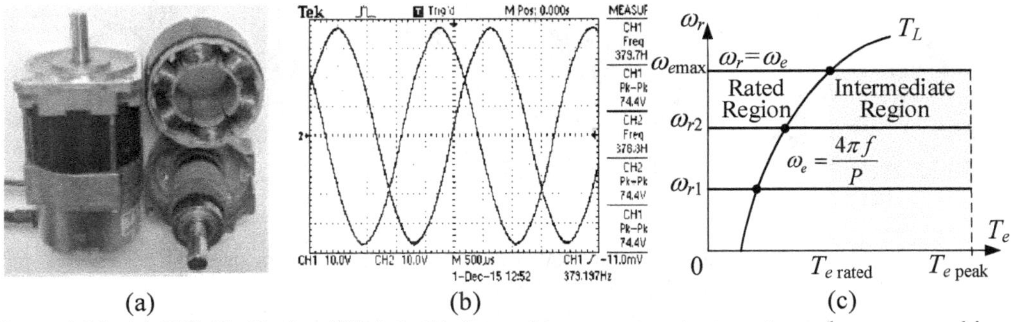

(a) (b) (c)

Figure 3.26. (a) NEMA 23 size (∅2.3 inch) three-phase permanent magnet synchronous machine which can be used as motor and generator: ∅58.4 mm, 67.5 mm length, four-pole (P=4), 40 W rated, 24 V, 2.5 A, 300 rad/sec rated and 500 rad/sec maximum, 0.127 N-m. Three-phase windings (*as*,*bs*,*cs*) are placed in the stator slots. Stator is made from electrical steel lamination. The arc-shaped skewed SmCo magnets are on the rotor. The phase resistance and inductance are 1.18 ohm and 0.0044 H, and, the *back emf* constant is 0.05 V-sec/rad;
(b) Experimental studies: 40W NEMA 23 size three-phase permanent magnet synchronous machine is used as a generator: The induced phase voltages (u_{as},u_{bs},u_{cs}) are displaced by 120 electrical degrees;
(c) Torque-speed characteristics: Synchronous machines rotate at synchronous angular velocity, $\omega_r = \omega_e$, $\omega_e = 4\pi f/P$.

For a short period of time, one may overload permanent magnet synchronous machines. The ratio $T_{e\,\text{peak}}/T_{e\,\text{rated}}$ depends on many factors, and, may reach 5. The limits on electromagnetic torque and power are due to physical limits, such as maximum energy density, nonlinear magnetic system, *BH* curve, maximum current density, thermodynamic limits, maximum insulation temperature, bearing type and design, etc. The motors are controlled by the PWM amplifiers which may admit up to 2 overloading capabilities for very short time. Motors, actuators, sensors, electronics and PWM controllers-drivers should be consistently chosen.

The assumptions are: (1) Ideal electromagnetic symmetry; (2) Magnetization and *BH* curve are *linear*, and, no hysteresis; (3) Magnetic field, established by permanent magnets, is uniform; (4) All coils under the same homogeneous uniform field; (5) Permanent magnet *susceptibility* is constant, despite the fact that the Curie's constant varies as a function of temperature; (6) The temperature-dependent resistivity of the copper winding is constant; (7) Cogging torque is negligible, (8) Magnetic losses, eddy-current losses, and fringing effect are neglected; (9) Bearing symmetry with no wear; etc.

The operating envelope (power, torque, force, load, load profile, angular velocity, etc.) defines T_e and ω_r resulting in the motor dimensionality, characteristics and parameters. Acceleration, settling time and repositioning rate depend on the ratio $(T_e - T_L)/J$. High torque density, energy and power densities, as well as angular velocity are ensured by optimal magnetic system design, magnets used, precision bearing, sensing and control schemes, power electronics, etc. For preliminary estimates, one may assume the power density estimate ~1 W/cm^3. Figures 3.27 document images of permanent magnet synchronous motors with the PWM drivers. Stepper motors are synchronous machines. Optimal torque generation, electromagnetic symmetry, losses and vibration minimization, as well as other key attributes and improvements can be achieved by means of consistent control and electronic solutions.

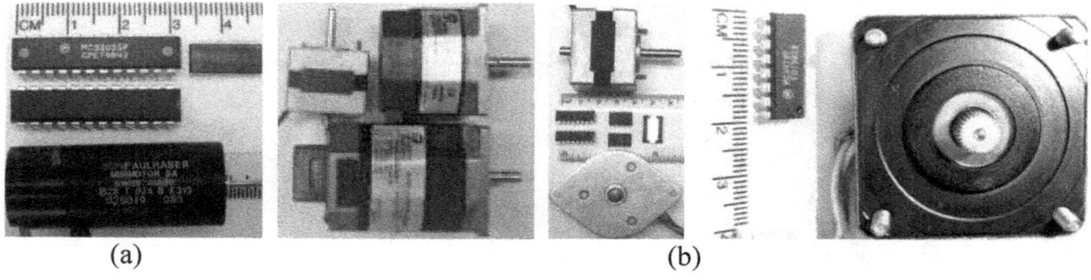

(a) (b)

Figure 3.27. Permanent magnet synchronous machines and controllers-drivers:
(a) Motorola MC33035P PWM driver (~30 V and 1.5 A), STMicroelectronics L6235N driver (~50 V, 2.8 A rated, 5.6 A peak, 100 kHz) and Texas Instruments DRV8312DDWR driver (~50 V, 3 A rated, 6 A peak, 100 to 500 kHz, ~90% to 97% efficiency). These controlled drivers can be used for a Faulhaber 1628 024B permanent magnet synchronous motor with SmCo magnets, P=2, 17 W, 24 V, 0.5 A rated, 1.5 A peak, 3000 rad/sec, 7000 rad/sec maximum, 3.3 mN-m rated, 11 mN-m peak, 15.2 ohm, 0.517 mH, 0.000000054 kg-m^2, up to 70% efficiency;
(b) Permanent magnet stepper motors, and, Texas Instrument DRV8432, 50 V, 6 A peak controllers-driver.

3.6.1. Two-Phase Permanent Magnet Synchronous Machines

Consider two-phase permanent magnet synchronous machines. The excitation field is produced by permanent magnets. Figure 3.28.a illustrates synchronous machines used in drives and servos for wearable electronics, computer hard disk drives, robots, etc.

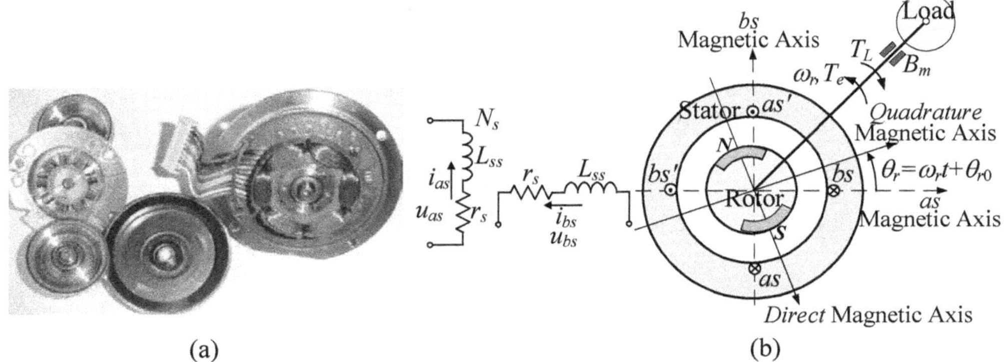

Figure 3.28. (a) Images of P-pole two- and three-phase permanent magnet machines;
(b) Two-phase two-pole round-rotor permanent magnet synchronous machine. For a round-rotor design, one attempts to minimize the torque ripple by ensuring $L_{mq}=L_{md}$, $L_{\Delta m}=0$. Recall that $\vec{T}=\vec{m}\times\vec{B}=N i\vec{s}\times\vec{B}$. For coils, ⊗ indicates that the phase current flows into the plane, while ⊙ indicates that a current flows out the plane.

Using Kirchhoff's voltage law

$$u_{as}=r_s i_{as}+\frac{d\psi_{as}}{dt},\ \psi_{as}=L_{asas}i_{as}+L_{asbs}i_{bs}+\psi_{asm}=L_{ss}i_{as}+\psi_{asm},\tag{3.36}$$

$$u_{bs}=r_s i_{bs}+\frac{d\psi_{bs}}{dt},\ \psi_{bs}=L_{bsas}i_{as}+L_{bsbs}i_{bs}+\psi_{bsm}=L_{ss}i_{bs}+\psi_{bsm},$$

where u_{as} and u_{bs} are the phase voltages applied to the stator windings as and bs; i_{as} and i_{bs} are the phase currents; ψ_{as} and ψ_{bs} are the stator flux linkages; r_s is the resistances of the stator windings; L_{asas} and L_{bsbs} are the self-inductances, $L_{ss}=L_{asas}=L_{bsbs}$; L_{asbs} and L_{bsas} are the mutual inductances of the stator windings, which are displaced by 90 electrical degrees, $L_{asbs}=L_{bsas}=0$.

The flux linkages ψ_{asm} and ψ_{bsm} are periodic functions of the rotor angular displacement θ_r with respect stator. Assume a linear not saturated *BH* curve, sinusoidal winding distribution, sinusoidal *mmf* waveforms, etc.

Let the rotor with magnets is position with respect to stator such that

$$\psi_{asm}=\psi_m\sin\theta_r,\tag{3.37}$$

$$\psi_{bsm}=-\psi_m\cos\theta_r.$$

With a round-rotor design, one minimizes the torque ripple by ensuring $L_{mq}=L_{md}$, $L_{\Delta m}=0$. For synchronous reluctance and *hybrid* stepper motors, which operate due to varying reluctance $\Re(\theta_r)$, $L_{mq}\neq L_{md}$ and $L_{\Delta m}\neq 0$. From (3.36) and (3.37), with $L_{mq}=L_{md}$,

$$u_{as}=r_s i_{as}+\frac{d\left(L_{ss}i_{as}+\psi_m\sin\theta_r\right)}{dt}=r_s i_{as}+L_{ss}\frac{di_{as}}{dt}+\psi_m\cos\theta_r\omega_r,\tag{3.38}$$

$$u_{bs}=r_s i_{bs}+\frac{d\left(L_{ss}i_{bs}-\psi_m\cos\theta_r\right)}{dt}=r_s i_{bs}+L_{ss}\frac{di_{bs}}{dt}+\psi_m\sin\theta_r\omega_r.$$

Using the mechanical angular velocity ω_{rm} and displacement θ_{rm}, the Newton second law yields

$$\frac{d\omega_{rm}}{dt}=\frac{1}{J}\left(T_e-B_m\omega_{rm}-T_L\right), \tag{3.39}$$
$$\frac{d\theta_{rm}}{dt}=\omega_{rm}\,.$$

For P-pole transducers, to guarantee the overall generality of results, use the electrical angular velocity and displacement (ω_r,θ_r). The mechanical angular velocity and displacement of the shaft (ω_{rm},θ_{rm}) are

$$\omega_{rm}=\frac{2}{P}\omega_r,\ \theta_{rm}=\frac{2}{P}\theta_r\,. \tag{3.40}$$

The electromagnetic torque is found by using the coenergy

$$\begin{aligned}W_c&=\tfrac{1}{2}\left(L_{ss}i_{as}^2+L_{ss}i_{bs}^2\right)+\begin{bmatrix}i_{as} & i_{bs}\end{bmatrix}\begin{bmatrix}\psi_{asm}\\ \psi_{bsm}\end{bmatrix}+W_{PM}\\ &=\tfrac{1}{2}\left(L_{ss}i_{as}^2+L_{ss}i_{bs}^2\right)+\left(i_{as}\psi_m\sin\theta_r-i_{bs}\psi_m\cos\theta_r\right)+W_{\text{permanent magnet}}\,.\end{aligned} \tag{3.41}$$

The self-inductance L_{ss} and energy stored in permanent magnets $W_{\text{permanent magnet}}$ are not functions of θ_r. Hence

$$T_e=\frac{P}{2}\frac{\partial W_c}{\partial\theta_r}=\frac{P\psi_m}{2}\left(i_{as}\cos\theta_r+i_{bs}\sin\theta_r\right). \tag{3.42}$$

Using equations (3.38), *torsional-mechanical* dynamics (3.39), as well as relationships (3.40) and (3.42), we obtain

$$\frac{di_{as}}{dt}=\frac{1}{L_{ss}}\left(-r_si_{as}-\psi_m\cos\theta_r\omega_r+u_{as}\right), \tag{3.43}$$
$$\frac{di_{bs}}{dt}=\frac{1}{L_{ss}}\left(-r_si_{bs}-\psi_m\sin\theta_r\omega_r+u_{bs}\right),$$
$$\frac{d\omega_r}{dt}=\frac{1}{J}\left[\frac{P^2\psi_m}{4}\left(i_{as}\cos\theta_r+i_{bs}\sin\theta_r\right)-B_m\omega_r-\frac{P}{2}T_L\right],$$
$$\frac{d\theta_r}{dt}=\omega_r.$$

From (3.42), to guarantee the balanced operation for which T_e is not a function of the angular displacement θ_r, one finds the balanced current set

$$i_{as}=\sqrt{2}i_M\cos\theta_r,\ i_{bs}=\sqrt{2}i_M\sin\theta_r\,. \tag{3.44}$$

The balanced phase voltages are

$$u_{as}=\sqrt{2}u_M\cos\theta_r,\ u_{bs}=\sqrt{2}u_M\sin\theta_r\,. \tag{3.45}$$

Using the balanced current set (3.44), the electromagnetic torque (3.42) is

$$T_e=\frac{P\psi_m}{2}\sqrt{2}i_M\left(\cos^2\theta_r+\sin^2\theta_r\right)=\frac{P\psi_m}{\sqrt{2}}i_M\,. \tag{3.46}$$

To implement the displacement-dependent current or voltage sets (3.44) and (3.45), one measures θ_r by using the Hall or inductive sensors. The estimation of displacement θ_r in the so-called *sensorless control* may not be practical due to hardware and algorithmic complexity, sensitivity, instabilities, insufficient accuracy, etc.

Permanent magnet synchronous motors with high number of poles P develop high electromagnetic torque, while the mechanical angular velocity is low because $\omega_{rm}=2\omega_r/P$. These motors are effectively used as direct drives and servos. The direct actuator-kinematic connection without gears ensures high efficiency, reliability and performance. Equations derived can be refined using ω_{rm} and θ_{rm}. Recall that $\omega_{rm}=\frac{2}{P}\omega_r$ and $\theta_{rm}=\frac{2}{P}\theta_r$. The mechanical angular velocity and displacement (ω_{rm},θ_{rm}) is commonly used in analysis of stepper motors which may operate as open-loop devices without measurements of θ_r.

3.6.2. Permanent Magnet Stepper Motors

Variable reluctance torque $T_{e_{L_{\Delta m}(\theta_r)}}$ and permanent magnet torque $T_e=m\times B$ are used in *hybrid* stepper motors, $T_e=T_{e\ \text{reluctance}}+T_{e\ \text{magnet}}$. Images of permanent magnet stepper motors are depicted in Figure 3.29. *Hybrid* stepper motors ensure the microstepping capabilities due to the magnetizing inductance variations $L(\theta_r)$ and electromagnetic coupling between stator winding and rotor magnets. As illustrated in Figures 3.28.b, for the *quadrature*- and *direct*-axis inductances, $L_{mq}\neq L_{md}$, and, $T_{e\ \text{reluctance}}=T_{e_{L_{\Delta m}(\theta_r)}}\neq 0$. Low-cost two-phase variable reluctance stepper motors are used. Incremental microstepping can be performed at low loads because the holding torque is low. Consider bipolar and unipolar permanent magnet stepper motors used in high-performance applications. Depending on winding topologies, stepper motors are two- or four-phase synchronous machines. A two-phase P-pole machines were studied in section 3.6.1.

Figures 3.29 document unipolar 8-lead two-stack-rotor stepper motors which may be configured as two- and four-phase synchronous machines. The two-phase schematics is realized by using two coils per phase. Example 3.22 documents the experimental studies for the 8-lead permanent magnet stepper motor.

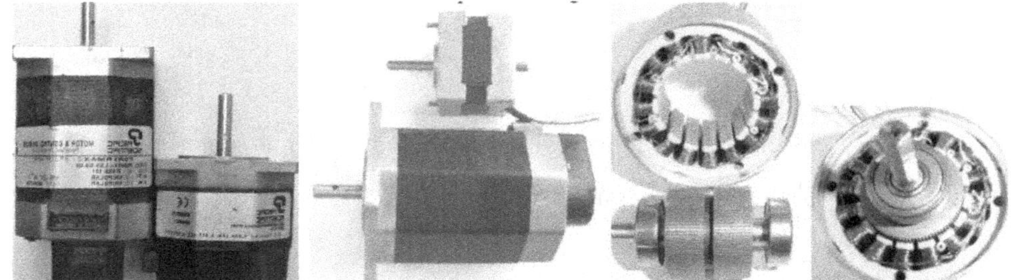

Figure 3.29. NEMA 23 size (2.3×2.3 inch faceplate) stepper motors: There are various stepper motors with different designs, topologies, and, winding configurations. In two-stack-rotor, there are two stacks with the N poles P_N and S poles P_S, $P=P_N+P_S$, $P_N=P_S$.

For stepper motors with two independent orthogonal *as* and *bs* stator windings, one *energizes* windings by adequately applying phase voltages u_{as} and u_{bs}. The direction of T_e can be changed, and, the rotor rotates *counterclockwise* or *clockwise*. Applying u_{as} and u_{bs}, one achieves the incremental rotor displacement equal to a full or half step. The rotor repositioning rate (number of steps per second) is regulated by changing the frequency of phase voltages (u_{as}, u_{bs}). If stepper motors operate in an open-loop mode, the motor can miss the step if: (1) Electromagnetic torque T_e is not sufficient, or, $T_e<T_L$; (2) Phase voltages u_{as} and u_{bs} are supplied at high frequency f, and, f depends on L_{ss} and equivalent moment of inertia J.

The stepper motor can pass the step if: (1) $T_e>>T_L$; (2) Load T_L is bidirectional; (3) High kinetic energy is stored in the moving rotor with the attached kinematics and high J. Other factors contribute to missing or passing steps such as varying J, fast-varying bidirectional T_L, parameter variations, etc.

Open-loop stepper motors are used if $T_L\approx$const and $J\approx$const. The phase voltages switching frequency f is found by examining the system dynamics, loads, etc. Stepper motors are designed with a high number of rotor teeth RT which are the magnets, $RT\equiv P$ for a single stack, and, $P\equiv 2RT$ in two-stack-rotor. To achieve 1.8 degree rotation using the full-step operation, RT=100.

Consider two-stack-rotor stepper motors with the number of rotor teeth RT. The electrical angular velocity and displacement are $\omega_r=RT\omega_{rm}$ and $\theta_r=RT\theta_{rm}$. The flux linkages are the functions of the number of RT and displacement. Let $\psi_{asm}=\psi_m\cos(RT\theta_{rm})$ and $\psi_{bsm}=\psi_m\sin(RT\theta_{rm})$. For a round-rotor with $L_{md}=L_{mq}$, $L_{\Delta m}=0$.

The Kirchhoff law for the *as* and *bs* phases yields

$$u_{as}=r_si_{as}+\frac{d\psi_{as}}{dt},\ \psi_{as}=L_{ss}i_{as}+\psi_{asm},\ \psi_{asm}=\psi_m\cos(RT\theta_{rm}), \tag{3.47}$$

$$u_{bs}=r_si_{bs}+\frac{d\psi_{bs}}{dt},\ \psi_{bs}=L_{ss}i_{bs}+\psi_{bsm},\ \psi_{bsm}=\psi_m\sin(RT\theta_{rm}).$$

One finds

$$\frac{di_{as}}{dt}=\frac{1}{L_{ss}}\left[-r_si_{as}+RT\psi_m\sin(RT\theta_{rm})\omega_{rm}+u_{as}\right],$$

$$\frac{di_{bs}}{dt}=\frac{1}{L_{ss}}\left[-r_si_{bs}-RT\psi_m\cos(RT\theta_{rm})\omega_{rm}+u_{bs}\right]. \tag{3.48}$$

The coenergy is $W_c=\frac{1}{2}\left(L_{ss}i_{as}^2+L_{ss}i_{bs}^2\right)+\psi_m\cos(RT\theta_{rm})i_{as}+\psi_m\sin(RT\theta_{rm})i_{bs}+W_{\text{permanent magnet}}$.

Hence,

$$T_e=\frac{\partial W_c}{\partial\theta_{rm}}=RT\psi_m\left[-\sin(RT\theta_{rm})i_{as}+\cos(RT\theta_{rm})i_{bs}\right]. \tag{3.49}$$

From Newton's second law, using (3.48) and (3.49), we have

$$\frac{di_{as}}{dt}=\frac{1}{L_{ss}}\left[-r_si_{as}+RT\psi_m\sin(RT\theta_{rm})\omega_{rm}+u_{as}\right], \tag{3.50}$$

$$\frac{di_{bs}}{dt}=\frac{1}{L_{ss}}\left[-r_si_{bs}-RT\psi_m\cos(RT\theta_{rm})\omega_{rm}+u_{bs}\right],$$

$$\frac{d\omega_{rm}}{dt}=\frac{1}{J}\left[RT\psi_m\left[-\sin(RT\theta_{rm})i_{as}+\cos(RT\theta_{rm})i_{bs}\right]-B_m\omega_{rm}-T_L\right],$$

$$\frac{d\theta_{rm}}{dt}=\omega_{rm}.$$

Having found T_e (3.49), the stepper motors can be rotated sequentially supplying the phase voltages u_{as} and u_{bs} without measuring θ_{rm}. The step-by-step rotor repositioning is accomplished by sequentially applying phase voltages (u_{as},u_{bs}).

Considering a continuous rotation, measuring θ_{rm}, the balanced current and voltage sets are

$$i_{as} = -\sqrt{2}i_M \sin(RT\theta_{rm}),\ i_{bs} = \sqrt{2}i_M \cos(RT\theta_{rm}), \qquad (3.51)$$
$$u_{as} = -\sqrt{2}u_M \sin(RT\theta_{rm}),\ u_{bs} = \sqrt{2}u_M \cos(RT\theta_{rm})$$

for which $T_e = \sqrt{2}RT\psi_m i_M$.

The Motorola MC3479 as well as other controllers-drivers can be used in various applications such as positioning tables, disk drives, robots, servos, etc. These controllers can drive stepper motors bidirectionally. The representative block diagrams, circuitry, timing diagrams and connections are reported in Figure 3.30. High-performance drivers ensure the full-, half- and mini-stepping operations.

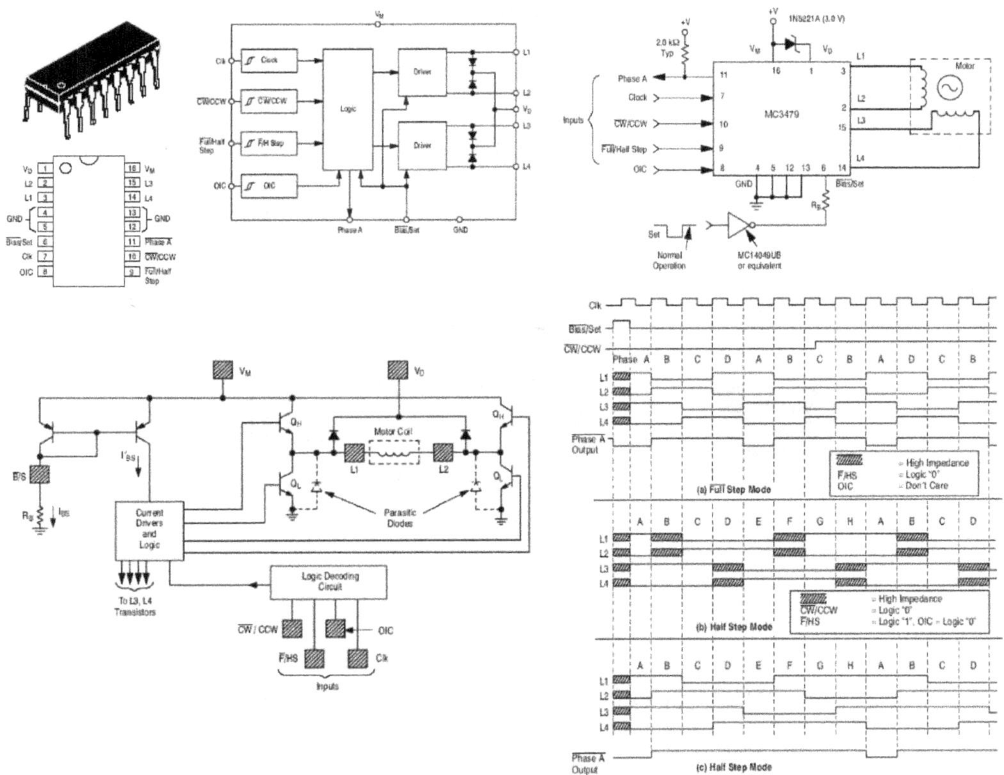

Figure 3.30. Motorola MC3479 stepper motor driver: The output voltage sequences and timing diagrams, as well as the controller-driver schematics. Copyright of Motorola, used with permission [7]

As illustrated in Figure 3.30, the H-bridge topology driver stage supplies the phase voltages u_{as} and u_{bs} to the windings. The terminals are (L1,L2) and (L3,L4). The applied voltage polarity depends on which transistor pair (Q_H, Q_L) is *on*. These transistors are driven by the signal-level voltages from the logic circuit. The maximum sink current is a function of the resistor between pin 6 and ground. When the outputs are in a high impedance state, both transistors (Q_H, Q_L) are *off*. The pin V_D provides a current path for the phase winding ("motor coil") during transients to reject the *back emf* voltage. Pin V_D is connected to V_M (pin 16) through a diode, resistor, or directly.

The instantaneous peak voltage at the outputs must not exceed V_M, which is 6 V. The diodes across Q_L of each output provide a circuit path for the current. When the input is at a Logic "0" (less than 0.8 V), the output corresponds to a full step operation with each clock cycle. The direction depends on the CW/CCW input. There are four switching phases for each cycle of the sequencing logic. As the phase voltage is applied, there are current i_{as} or i_{bs} in the motor windings. For a Logic "1" (higher than 2 V), the outputs change a half step during each clock cycle. Eight switching phases correspond to complete cycles of the sequencing logic.

Example 3.22. *Control of a Stepper Motor*

Experiments are performed for a 8-lead P22NSXA Pacific Scientific permanent magnet stepper motor with two-stack-rotor, 1.8° full-step, 2.7 V and 4.6 A (unipolar) rated, 1 N-m (rated), r_s=0.5 ohm, L_{ss}=7.5×10^{-4} H, ψ_m=4.9×10^{-3} N-m/A, B_m=9.2×10^{-4} N-m-sec/rad and J=1×10^{-4} kg-m^2. A shaft-mount incremental encoder measures θ_{rm}. The angular velocity ω_{rm} can be estimated. The schematics and hardware are reported in Figure 3.31. The clockwise and counterclockwise rotation and precision repositioning should be guarantee despite bidirectional load $T_L(t)$. A servo consists of a MPU, amplifier, stepper motor and kinematics. Using the reference displacement $r=\theta_{ref}(t)$ and the measured angular displacements θ_{rm}, the controller develops the PWM signals which drive high-frequency MOSFETs. The PWM phase voltages $(u_{as}, u_{bs}, \bar{u}_{as}, \bar{u}_{bs})$ are controlled by a full-H-bridge topology 6A MOSFET driver with 200 kHz switching frequency. The *driver* is controlled by a TMS320F28035 C2000 32-bit fixed-point 60 MHz MPU with processing, control, interfacing, peripheral, programming and other capabilities.

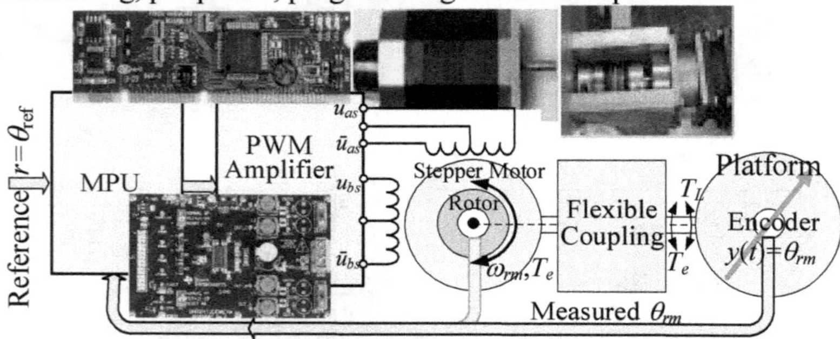

Figure 3.31. Bidirectional repositioning closed-loop servo with a stepper motor

The full-, half-, quarter- and micro stepping can be ensured [17]. The full-step displacement is $2\pi/RT$. For a full step repositioning in the open-loop bipolar winding configuration without measurements of θ_{rm}, the phase voltages u_{as} and u_{bs} are sequentially applied within the allowed frequency as

$$u_{as} = \begin{cases} u_M, \forall\theta_{rm} \in [0 \ \frac{\pi}{RT}] \\ 0, \ \forall\theta_{rm} \in [\frac{\pi}{RT} \ \frac{2\pi}{RT}] \end{cases}, \quad u_{bs} = \begin{cases} u_M, \forall\theta_{rm} \in [\frac{\pi}{RT} \ \frac{2\pi}{RT}] \\ 0, \ \forall\theta_{rm} \in [0 \ \frac{\pi}{RT}] \end{cases}.$$

The maximum switching frequency f of phase voltage pulses (u_{as}, u_{bs}) is defined by the motor transients, affected by L_{ss} and J.

For the unipolar motor in full-, half- and micro stepping, one finds the applied voltages $(u_{as}, u_{bs}, \bar{u}_{as}, \bar{u}_{bs})$. For a full-step, voltages applied as functions of the angular displacement θ_{rm}. In particular, one has

$$u_{as}=\begin{cases}u_M,\forall\theta_{rm}\in[0\ \frac{\pi}{2RT}]\\ 0,\forall\theta_{rm}\in[\frac{\pi}{2RT}\ \frac{2\pi}{RT}]\end{cases},\quad u_{bs}=\begin{cases}u_M,\forall\theta_{rm}\in[\frac{\pi}{2RT}\ \frac{\pi}{RT}]\\ 0,\forall\theta_{rm}\in[0\ \frac{\pi}{2RT}],[\frac{\pi}{RT}\ \frac{2\pi}{RT}]\end{cases},$$

$$\bar{u}_{as}=\begin{cases}u_M,\forall\theta_{rm}\in[\frac{\pi}{RT}\ \frac{3\pi}{2RT}]\\ 0,\forall\theta_{rm}\in[0\ \frac{\pi}{RT}],[\frac{3\pi}{2RT}\ \frac{2\pi}{RT}]\end{cases},\quad \bar{u}_{bs}=\begin{cases}u_M,\forall\theta_{rm}\in[\frac{3\pi}{2RT}\ \frac{2\pi}{RT}]\\ 0,\forall\theta_{rm}\in[0\ \frac{3\pi}{2RT}]\end{cases}.$$

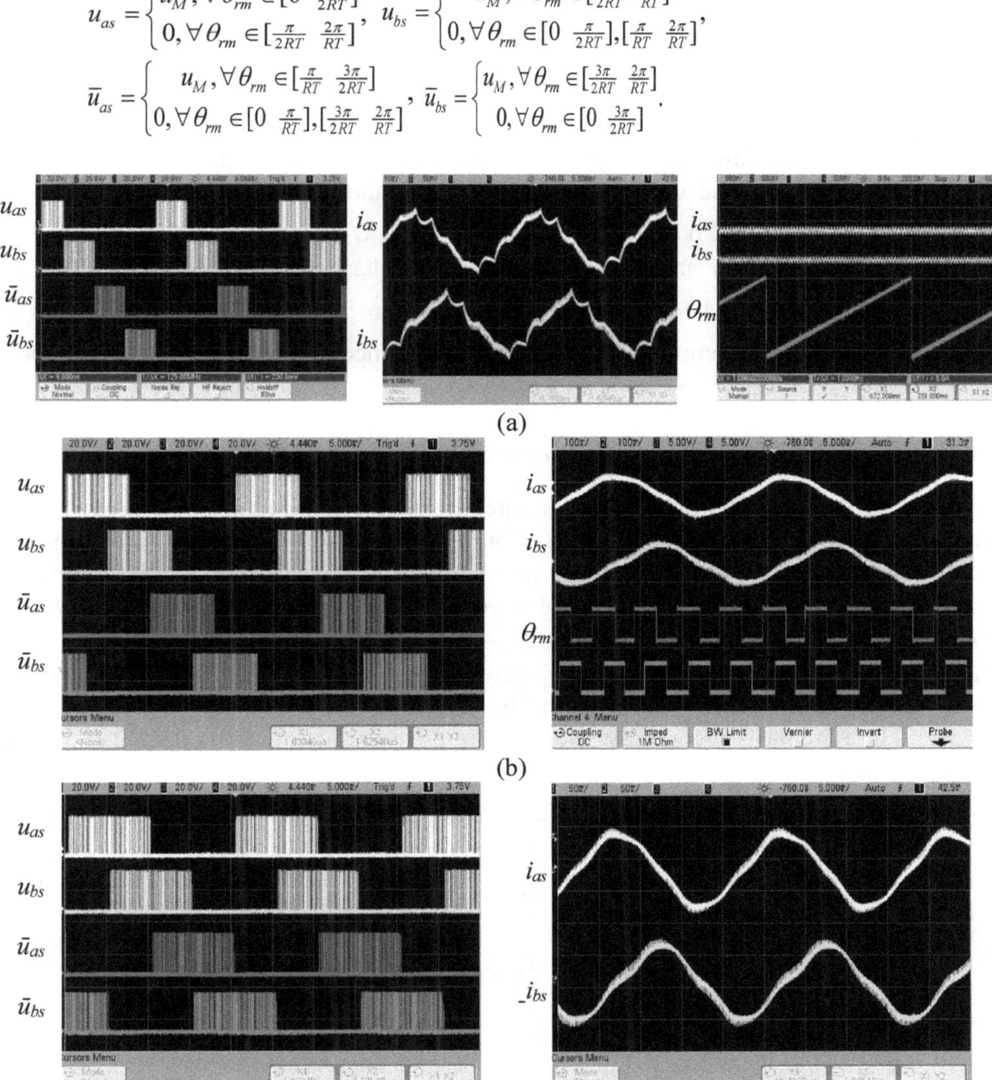

Figure 3.32. (a) Repositioning for the ramp reference $\theta_{ref}(t)$ with a full-step operation. The PWM phase voltages $(u_{as},u_{bs},\bar{u}_{as},\bar{u}_{bs})$ are applied, $-u_{max}\le u\le u_{max}$. Phase currents (i_{as},i_{bs}) and measured θ_{rm} are depicted;
(b) Repositioning with a half-step operation. The phase voltages $(u_{as},u_{bs},\bar{u}_{as},\bar{u}_{bs})$ and currents (i_{as},i_{bs}) are illustrated. The encoder measurements are reported;
(c) 1/128 microstepping: PWM phase voltages and near-sinusoidal phase currents.

The closed-loop system is designed to prevent inadequacy in microstepping and guarantee accuracy in positioning for varying bidirectional loads. With the limits on the phase voltages $-u_{max}\le u\le u_{max}$, u_{max}=2.7 V, a proportional-integral control law and algorithm with feedback functions $\varphi_{as}(\cdot)$, $\varphi_{bs}(\cdot)$, $\bar{\varphi}_{as}(\cdot)$, $\bar{\varphi}_{bs}(\cdot)$ are

$$u=\text{sat}_{-u_{\max}}^{u_{\max}}\left(k_pe+k_i\int edt\right),\ e(t)=(\theta_{\text{ref}}-\theta_{rm}),\ \begin{cases}u_{as}=u(e)\varphi_{as}(e,\theta_{rm}),\forall\theta_{rm}\in\Theta_{as}\\u_{bs}=u(e)\varphi_{bs}(e,\theta_{rm}),\forall\theta_{rm}\in\Theta_{bs}\\\bar{u}_{as}=u(e)\bar{\varphi}_{as}(e,\theta_{rm}),\forall\theta_{rm}\in\bar{\Theta}_{as}\\\bar{u}_{bs}=u(e)\bar{\varphi}_{bs}(e,\theta_{rm}),\forall\theta_{rm}\in\bar{\Theta}_{bs}\end{cases}.$$

For the full-step repositioning, the PWM phase voltages and currents are documented in Figures 3.32.a. The phase voltages $(u_{as},u_{bs},\bar{u}_{as},\bar{u}_{bs})$ and phase and currents (i_{as},i_{bs}) are illustrated in Figures 3.32.b and c for the half-step and 1/128 microstepping. The derived $\varphi_{as}(\cdot),\varphi_{bs}(\cdot),\bar{\varphi}_{as}(\cdot),\bar{\varphi}_{bs}(\cdot)$ are functions of θ_{rm}. The phase voltages $(u_{as},u_{bs},\bar{u}_{as},\bar{u}_{bs})$ are regulated using the PWM concept, $-u_{\max}\le u\le u_{\max}$. Near-sinusoidal currents $(i_{as},i_{bs},\bar{i}_{as},\bar{i}_{bs})$ ensure near-balanced operation, optimizing performance and robust repositioning. ■

3.6.3. Radial Topology Three-Phase Synchronous Machines

Permanent magnet synchronous motors, rated from mW to hundreds of kW, are used in wearable electronics, rotating and positioning stages, servos, robots, medium- and heavy-duty traction drives, etc. A three-phase two-pole permanent magnet synchronous machine is depicted in Figure 3.33.a. For round-rotor machines, the *quadrature-* and *direct-*axis inductances are equal, and, $L_{mq}=L_{md}$, $L_{\Delta m}=0$. Different permeability, stator and field nonuniformity and asymmetry may yields, $L_{mq}\neq L_{md}$ and $L_{\Delta m}\neq 0$. The images of the synchronous transducer is shown in Figure 3.33.

The Kirchhoff second law yields three differential equations for the *as*, *bs* and *cs* stator phases

$$u_{as}=r_si_{as}+\frac{d\psi_{as}}{dt},\ \psi_{as}=L_{asas}i_{as}+L_{asbs}i_{bs}+L_{ascs}i_{cs}+\psi_{asm},\tag{3.52}$$

$$u_{bs}=r_si_{bs}+\frac{d\psi_{bs}}{dt},\ \psi_{bs}=L_{bsas}i_{as}+L_{bsbs}i_{bs}+L_{bscs}i_{cs}+\psi_{bsm},$$

$$u_{cs}=r_si_{cs}+\frac{d\psi_{cs}}{dt},\ \psi_{cs}=L_{csas}i_{as}+L_{csbs}i_{bs}+L_{cscs}i_{cs}+\psi_{csm},$$

$$\mathbf{u}_{abcs}=\mathbf{r}_s\mathbf{i}_{abcs}+\frac{d\boldsymbol{\psi}_{abcs}}{dt},\ \begin{bmatrix}u_{as}\\u_{bs}\\u_{cs}\end{bmatrix}=\begin{bmatrix}r_s&0&0\\0&r_s&0\\0&0&r_s\end{bmatrix}\begin{bmatrix}i_{as}\\i_{bs}\\i_{cs}\end{bmatrix}+\begin{bmatrix}\frac{d\psi_{as}}{dt}\\\frac{d\psi_{bs}}{dt}\\\frac{d\psi_{cs}}{dt}\end{bmatrix},$$

$$\begin{bmatrix}\psi_{as}\\\psi_{bs}\\\psi_{cs}\end{bmatrix}=\begin{bmatrix}L_{asas}i_{as}+L_{asbs}i_{bs}+L_{ascs}i_{cs}+\psi_{asm}\\L_{bsas}i_{as}+L_{bsbs}i_{bs}+L_{bscs}i_{cs}+\psi_{bsm}\\L_{csas}i_{as}+L_{csbs}i_{bs}+L_{cscs}i_{cs}+\psi_{csm}\end{bmatrix},\ \begin{bmatrix}\psi_{asm}\\\psi_{bsm}\\\psi_{csm}\end{bmatrix}=\begin{bmatrix}f(\theta_r)\\f(\theta_r-\frac{2}{3}\pi)\\f(\theta_r+\frac{2}{3}\pi)\end{bmatrix},$$

where ψ_{asm}, ψ_{bsm} and ψ_{csm} are the periodic functions of θ_r with a period 2π.

The stator windings are displaced by 120 electrical degrees.

Denoting the magnitude of flux linkages as ψ_m, let

$$\psi_{asm}=\psi_m\sin\theta_r,\tag{3.53}$$

$$\psi_{bsm}=\psi_m\sin\left(\theta_r-\tfrac{2}{3}\pi\right),$$

$$\psi_{csm}=\psi_m\sin\left(\theta_r+\tfrac{2}{3}\pi\right).$$

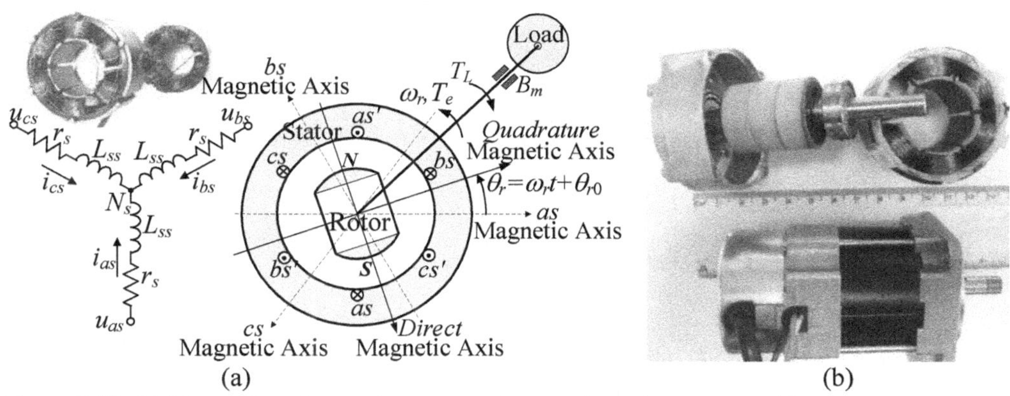

Figure 3.33. (a) Three-phase two-pole permanent magnet synchronous machine;
(b) Image of a ∅58.4 mm four-pole (P=4) permanent magnet synchronous machine.

For a transducer reported in Figure 3.33.a, the *quadrature* and *direct* axes reluctances are examined. For a uniform-permeability round rotor $\Re_{mq}=\Re_{md}$. Consider $\Re_{mq}\neq\Re_{md}$, $\Re_{mq}>\Re_{md}$ due to varying permeability and non-uniformity. The *quadrature* and *direct* axes magnetizing inductances are

$$L_{mq}=\frac{N_s^2}{\Re_{mq}} \text{ and } L_{md}=\frac{N_s^2}{\Re_{md}},\ L_{mq}<L_{md},\ L_{ls}+L_{mq}\le L_{asas}\le L_{ls}+L_{md}.$$

The phase inductance $L_{asas}(\theta_r)$ is a periodic function of θ_r

$$L_{asas}=L_{ls}+\overline{L}_m-L_{\Delta m}\cos 2\theta_r\,, \tag{3.54}$$

$$\overline{L}_m=\tfrac{1}{3}\left(L_{mq}+L_{md}\right)=\tfrac{1}{3}\left(\frac{N_s^2}{\Re_{mq}}+\frac{N_s^2}{\Re_{md}}\right),\ L_{\Delta m}=\tfrac{1}{3}\left(L_{md}-L_{mq}\right)=\tfrac{1}{3}\left(\frac{N_s^2}{\Re_{md}}-\frac{N_s^2}{\Re_{mq}}\right),$$

where $\overline{L}_m$ is the average value of the magnetizing inductance; $L_{\Delta m}$ is the half of amplitude of the sinusoidal variation of the magnetizing inductance.

The equation for flux linkages is

$$\boldsymbol{\psi}_{abcs}=\mathbf{L}_s\mathbf{i}_{abcs}+\boldsymbol{\psi}_m \tag{3.55}$$

$$=\begin{bmatrix} L_{ls}+\overline{L}_m-L_{\Delta m}\cos 2\theta_r & -\frac{1}{2}\overline{L}_m-L_{\Delta m}\cos 2\left(\theta_r-\frac{1}{3}\pi\right) & -\frac{1}{2}\overline{L}_m-L_{\Delta m}\cos 2\left(\theta_r+\frac{1}{3}\pi\right)\\ -\frac{1}{2}\overline{L}_m-L_{\Delta m}\cos 2\left(\theta_r-\frac{1}{3}\pi\right) & L_{ls}+\overline{L}_m-L_{\Delta m}\cos 2\left(\theta_r-\frac{2}{3}\pi\right) & -\frac{1}{2}\overline{L}_m-L_{\Delta m}\cos 2\theta_r\\ -\frac{1}{2}\overline{L}_m-L_{\Delta m}\cos 2\left(\theta_r+\frac{1}{3}\pi\right) & -\frac{1}{2}\overline{L}_m-L_{\Delta m}\cos 2\theta_r & L_{ls}+\overline{L}_m-L_{\Delta m}\cos 2\left(\theta_r+\frac{2}{3}\pi\right) \end{bmatrix}\begin{bmatrix} i_{as}\\ i_{bs}\\ i_{cs}\end{bmatrix}+\psi_m\begin{bmatrix}\sin\theta_r\\ \sin\left(\theta_r-\frac{2}{3}\pi\right)\\ \sin\left(\theta_r+\frac{2}{3}\pi\right)\end{bmatrix}.$$

For round-rotor machines, $\Re_{mq}=\Re_{md}$ and $L_{mq}=L_{md}$. Thus, $\overline{L}_m=\dfrac{2N_s^2}{3\Re_{mq}}=\dfrac{2N_s^2}{3\Re_{md}}$, $L_{\Delta m}=0$.

Denoting $L_{ss}=L_{ls}+\overline{L}_m$, the inductance matrix is

$$\mathbf{L}_s=\begin{bmatrix} L_{ls}+\overline{L}_m & -\frac{1}{2}\overline{L}_m & -\frac{1}{2}\overline{L}_m\\ -\frac{1}{2}\overline{L}_m & L_{ls}+\overline{L}_m & -\frac{1}{2}\overline{L}_m\\ -\frac{1}{2}\overline{L}_m & -\frac{1}{2}\overline{L}_m & L_{ls}+\overline{L}_m\end{bmatrix}=\begin{bmatrix} L_{ss} & -\frac{1}{2}\overline{L}_m & -\frac{1}{2}\overline{L}_m\\ -\frac{1}{2}\overline{L}_m & L_{ss} & -\frac{1}{2}\overline{L}_m\\ -\frac{1}{2}\overline{L}_m & -\frac{1}{2}\overline{L}_m & L_{ss}\end{bmatrix}. \tag{3.56}$$

The expressions for flux linkages are

$$\psi_{as}=\left(L_{ls}+\bar{L}_m\right)i_{as}-\tfrac{1}{2}\bar{L}_m i_{bs}-\tfrac{1}{2}\bar{L}_m i_{cs}+\psi_m\sin\theta_r, \quad (3.57)$$
$$\psi_{bs}=-\tfrac{1}{2}\bar{L}_m i_{as}+\left(L_{ls}+\bar{L}_m\right)i_{bs}-\tfrac{1}{2}\bar{L}_m i_{cs}+\psi_m\sin\left(\theta_r-\tfrac{2}{3}\pi\right),$$
$$\psi_{cs}=-\tfrac{1}{2}\bar{L}_m i_{as}-\tfrac{1}{2}\bar{L}_m i_{bs}+\left(L_{ls}+\bar{L}_m\right)i_{cs}+\psi_m\sin\left(\theta_r+\tfrac{2}{3}\pi\right),$$
$$\boldsymbol{\psi}_{abcs}=\mathbf{L}_s\mathbf{i}_{abcs}+\boldsymbol{\psi}_m=\begin{bmatrix} L_{ls}+\bar{L}_m & -\tfrac{1}{2}\bar{L}_m & -\tfrac{1}{2}\bar{L}_m \\ -\tfrac{1}{2}\bar{L}_m & L_{ls}+\bar{L}_m & -\tfrac{1}{2}\bar{L}_m \\ -\tfrac{1}{2}\bar{L}_m & -\tfrac{1}{2}\bar{L}_m & L_{ls}+\bar{L}_m \end{bmatrix}\begin{bmatrix} i_{as} \\ i_{bs} \\ i_{cs}\end{bmatrix}+\psi_m\begin{bmatrix}\sin\theta_r \\ \sin(\theta_r-\tfrac{2}{3}\pi) \\ \sin(\theta_r+\tfrac{2}{3}\pi)\end{bmatrix}.$$

Using (3.52), (3.57) and $\mathbf{L}_s=\begin{bmatrix} L_{ss} & -\tfrac{1}{2}\bar{L}_m & -\tfrac{1}{2}\bar{L}_m \\ -\tfrac{1}{2}\bar{L}_m & L_{ss} & -\tfrac{1}{2}\bar{L}_m \\ -\tfrac{1}{2}\bar{L}_m & -\tfrac{1}{2}\bar{L}_m & L_{ss}\end{bmatrix}$, we have

$$\mathbf{u}_{abcs}=\mathbf{r}_s\mathbf{i}_{abcs}+\frac{d\boldsymbol{\psi}_{abcs}}{dt}=\mathbf{r}_s\mathbf{i}_{abcs}+\mathbf{L}_s\frac{d\mathbf{i}_{abcs}}{dt}+\frac{d\boldsymbol{\psi}_m}{dt},\quad \frac{d\boldsymbol{\psi}_m}{dt}=\psi_m\begin{bmatrix}\cos\theta_r\omega_r \\ \cos\left(\theta_r-\tfrac{2}{3}\pi\right)\omega_r \\ \cos\left(\theta_r+\tfrac{2}{3}\pi\right)\omega_r\end{bmatrix}. \quad (3.58)$$

Cauchy's form of differential equations are found using $\mathbf{L}_s^{-1}$ as

$$\frac{d\mathbf{i}_{abcs}}{dt}=-\mathbf{L}_s^{-1}\mathbf{r}_s\mathbf{i}_{abcs}-\mathbf{L}_s^{-1}\frac{d\boldsymbol{\psi}_m}{dt}+\mathbf{L}_s^{-1}\mathbf{u}_{abcs}. \quad (3.59)$$

The circuitry-electromagnetic dynamics is

$$\frac{di_{as}}{dt}=-\frac{r_s\left(2L_{ss}-\bar{L}_m\right)}{2L_{ss}^2-L_{ss}\bar{L}_m-\bar{L}_m^2}i_{as}-\frac{r_s\bar{L}_m}{2L_{ss}^2-L_{ss}\bar{L}_m-\bar{L}_m^2}i_{bs}-\frac{r_s\bar{L}_m}{2L_{ss}^2-L_{ss}\bar{L}_m-\bar{L}_m^2}i_{cs} \quad (3.60)$$
$$-\frac{\psi_m\left(2L_{ss}-\bar{L}_m\right)}{2L_{ss}^2-L_{ss}\bar{L}_m-\bar{L}_m^2}\omega_r\cos\theta_r-\frac{\psi_m\bar{L}_m}{2L_{ss}^2-L_{ss}\bar{L}_m-\bar{L}_m^2}\omega_r\cos\left(\theta_r-\tfrac{2}{3}\pi\right)-\frac{\psi_m\bar{L}_m}{2L_{ss}^2-L_{ss}\bar{L}_m-\bar{L}_m^2}\omega_r\cos\left(\theta_r+\tfrac{2}{3}\pi\right)$$
$$+\frac{2L_{ss}-\bar{L}_m}{2L_{ss}^2-L_{ss}\bar{L}_m-\bar{L}_m^2}u_{as}+\frac{\bar{L}_m}{2L_{ss}^2-L_{ss}\bar{L}_m-\bar{L}_m^2}u_{bs}+\frac{\bar{L}_m}{2L_{ss}^2-L_{ss}\bar{L}_m-\bar{L}_m^2}u_{cs},$$
$$\frac{di_{bs}}{dt}=-\frac{r_s\bar{L}_m}{2L_{ss}^2-L_{ss}\bar{L}_m-\bar{L}_m^2}i_{as}-\frac{r_s\left(2L_{ss}-\bar{L}_m\right)}{2L_{ss}^2-L_{ss}\bar{L}_m-\bar{L}_m^2}i_{bs}-\frac{r_s\bar{L}_m}{2L_{ss}^2-L_{ss}\bar{L}_m-\bar{L}_m^2}i_{cs}$$
$$-\frac{\psi_m\bar{L}_m}{2L_{ss}^2-L_{ss}\bar{L}_m-\bar{L}_m^2}\omega_r\cos\theta_r-\frac{\psi_m\left(2L_{ss}-\bar{L}_m\right)}{2L_{ss}^2-L_{ss}\bar{L}_m-\bar{L}_m^2}\omega_r\cos\left(\theta_r-\tfrac{2}{3}\pi\right)-\frac{\psi_m\bar{L}_m}{2L_{ss}^2-L_{ss}\bar{L}_m-\bar{L}_m^2}\omega_r\cos\left(\theta_r+\tfrac{2}{3}\pi\right)$$
$$+\frac{\bar{L}_m}{2L_{ss}^2-L_{ss}\bar{L}_m-\bar{L}_m^2}u_{as}+\frac{2L_{ss}-\bar{L}_m}{2L_{ss}^2-L_{ss}\bar{L}_m-\bar{L}_m^2}u_{bs}+\frac{\bar{L}_m}{2L_{ss}^2-L_{ss}\bar{L}_m-\bar{L}_m^2}u_{cs},$$
$$\frac{di_{cs}}{dt}=-\frac{r_s\bar{L}_m}{2L_{ss}^2-L_{ss}\bar{L}_m-\bar{L}_m^2}i_{as}-\frac{r_s\bar{L}_m}{2L_{ss}^2-L_{ss}\bar{L}_m-\bar{L}_m^2}i_{bs}-\frac{r_s\left(2L_{ss}-\bar{L}_m\right)}{2L_{ss}^2-L_{ss}\bar{L}_m-\bar{L}_m^2}i_{cs}$$
$$-\frac{\psi_m\bar{L}_m}{2L_{ss}^2-L_{ss}\bar{L}_m-\bar{L}_m^2}\omega_r\cos\theta_r-\frac{\psi_m\bar{L}_m}{2L_{ss}^2-L_{ss}\bar{L}_m-\bar{L}_m^2}\omega_r\cos\left(\theta_r-\tfrac{2}{3}\pi\right)-\frac{\psi_m\left(2L_{ss}-\bar{L}_m\right)}{2L_{ss}^2-L_{ss}\bar{L}_m-\bar{L}_m^2}\omega_r\cos\left(\theta_r+\tfrac{2}{3}\pi\right)$$
$$+\frac{\bar{L}_m}{2L_{ss}^2-L_{ss}\bar{L}_m-\bar{L}_m^2}u_{as}+\frac{\bar{L}_m}{2L_{ss}^2-L_{ss}\bar{L}_m-\bar{L}_m^2}u_{bs}+\frac{2L_{ss}-\bar{L}_m}{2L_{ss}^2-L_{ss}\bar{L}_m-\bar{L}_m^2}u_{cs}.$$

The electromagnetic torque is found by using the coenergy

$$W_c = \tfrac{1}{2}\begin{bmatrix} i_{as} & i_{bs} & i_{cs} \end{bmatrix} \mathbf{L}_s \begin{bmatrix} i_{as} \\ i_{bs} \\ i_{cs} \end{bmatrix} + \begin{bmatrix} i_{as} & i_{bs} & i_{cs} \end{bmatrix} \begin{bmatrix} \psi_m \sin\theta_r \\ \psi_m \sin\left(\theta_r - \tfrac{2}{3}\pi\right) \\ \psi_m \sin\left(\theta_r + \tfrac{2}{3}\pi\right) \end{bmatrix} + W_{\text{permanent magnet}} . \quad (3.61)$$

For round-rotor synchronous machines, matrix $\mathbf{L}_s$ and energy stored in permanent magnets $W_{\text{permanent magnet}}$ are not functions of θ_r. One obtains the electromagnetic torque for *P*-pole three-phase permanent magnet synchronous motors

$$T_e = \frac{P}{2}\frac{\partial W_c}{\partial \theta_r} = \frac{P\psi_m}{2}\left[i_{as}\cos\theta_r + i_{bs}\cos\left(\theta_r - \tfrac{2}{3}\pi\right) + i_{cs}\cos\left(\theta_r + \tfrac{2}{3}\pi\right)\right]. \quad (3.62)$$

With the derived T_e (3.62), the Newton second law yields

$$\frac{d\omega_{rm}}{dt} = \frac{1}{J}\left[\frac{P\psi_m}{2}\left[i_{as}\cos\theta_r + i_{bs}\cos\left(\theta_r - \tfrac{2}{3}\pi\right) + i_{cs}\cos\left(\theta_r + \tfrac{2}{3}\pi\right)\right] - B_m\omega_{rm} - T_L\right], \quad (3.63)$$

$$\frac{d\theta_{rm}}{dt} = \omega_{rm} .$$

The electrical angular velocity ω_r and displacement θ_r are related to the mechanical angular velocity and displacement (3.40), $\omega_{rm} = \frac{2}{P}\omega_r$ and $\theta_{rm} = \frac{2}{P}\theta_r$. The differential equations of the *torsional-mechanical* dynamics are

$$\frac{d\omega_r}{dt} = \frac{1}{J}\left[\frac{P^2\psi_m}{4}\left[i_{as}\cos\theta_r + i_{bs}\cos\left(\theta_r - \tfrac{2}{3}\pi\right) + i_{cs}\cos\left(\theta_r + \tfrac{2}{3}\pi\right)\right] - B_m\omega_r - \frac{P}{2}T_L\right], \quad (3.64)$$

$$\frac{d\theta_r}{dt} = \omega_r .$$

A nonlinear mathematical model in Cauchy's form is given by a system of five differential equations (3.60) and (3.64). To control motors, one regulates the *abc* phase currents or voltages. To regulate motion devices, the electromagnetic torque (3.62) must be changed. A balanced three-phase current set is

$$i_{as} = \sqrt{2}i_M\cos\theta_r,\ i_{bs} = \sqrt{2}i_M\cos\left(\theta_r - \tfrac{2}{3}\pi\right),\ i_{cs} = \sqrt{2}i_M\cos\left(\theta_r + \tfrac{2}{3}\pi\right). \quad (3.65)$$

A trigonometric identity is $\cos^2\theta_r + \cos^2\left(\theta_r - \tfrac{2}{3}\pi\right) + \cos^2\left(\theta_r + \tfrac{2}{3}\pi\right) = \tfrac{3}{2}$.

One yields

$$T_e = \frac{P\psi_m}{2}\sqrt{2}i_M\left[\cos^2\theta_r + \cos^2\left(\theta_r - \tfrac{2}{3}\pi\right) + \cos^2\left(\theta_r + \tfrac{2}{3}\pi\right)\right] = \frac{3P\psi_m}{2\sqrt{2}}i_M . \quad (3.66)$$

The angular displacement θ_r is measured by the Hall-effect sensors. The PWM amplifiers change the magnitude u_M of the phase voltages (u_{as},u_{bs},u_{cs}) of the balanced voltage set

$$u_{as} = \sqrt{2}u_M\cos\theta_r,\ u_{bs} = \sqrt{2}u_M\cos\left(\theta_r - \tfrac{2}{3}\pi\right),\ u_{cs} = \sqrt{2}u_M\cos\left(\theta_r + \tfrac{2}{3}\pi\right). \quad (3.67)$$

The balanced phase voltages (3.67) are the functions of θ_r. Synchronous transducers are characterized by the rated power, voltage, current, torque, speed, etc. The matching PWM power amplifiers are used. For a 100 W rated synchronous motors, Figure 3.34.a reports the schematics of a B15A8 amplifier, 20 to 80 V, 7.5 A continuous, 15 A peak, 2.5 kHz bandwidth, 129×76×25 mm dimensions. The motor phase windings are connected to P2-1, P2-2 and P2-3. One connects the Hall sensor outputs to P1-12, P1-13 and P1-14. The "control logic" uses the measured rotor angular displacement to generate the PWM phase voltages (u_{as},u_{bs},u_{cs}) by driving the power MOSFETs. The proportional-integral analog controller is used. The reference voltage is supplied to P1-4. The tachometer voltage, proportional to the angular velocity of rotation, is supplied to P1-6. The reference and measured angular velocities are compared to the tracking error $e(t)$. The analog proportional-integral controller develops the control signals which turn *on* and *off* MOSFETs. The proportional and integral feedback gains are regulated by adjusting the resistors. Figure 3.34.b documents high-performance machines with the PWM drivers-controllers. The evaluation boards may be effectively used. The EVAL6235 with the L6235 driver is shown in Figure 3.34.c.

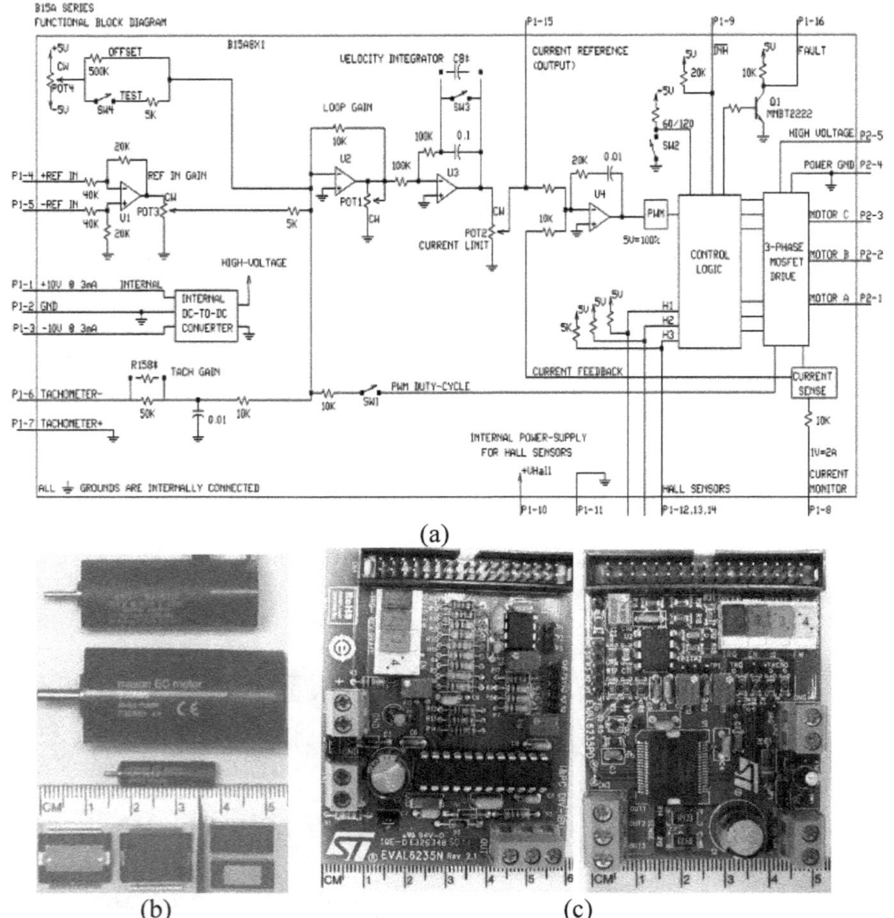

(a)

(b) (c)

Figure 3.34. (a) B15A8 PWM amplifier schematics, courtesy of Advanced Motion Controls, www.a-m-c.com [7];
(b) Images of radial topology permanent magnet synchronous motors;
(c) STMicroelectronics EVAL6235N and EVAL6235PD evaluation boards.

Figures 3.34.b documents images of high-performance radial topology permanent magnet synchronous motors with Hall effect sensors. In particular:

1. Faulhaber 1628 024B motor with SmCo magnets, ∅15 mm, two-pole, 17 W, 24 V, 0.5 A rated, 1.5 A peak, 3000 rad/sec, 7000 rad/sec maximum, 3.3 mN-m rated, 11

mN-m peak, 15.2 ohm, 0.517 mH, 5.4×10^{-8} kg-m^2, up to 70% efficiency. Motor is equipped with the IE2-512 encoder with 512 pulses per revolution;

2. Maxon EC 200685 motor, ∅22 mm, two-pole, 12 V, 2500 rad/sec, 5.75 A and 21 mN-m rated, ~85% efficiency;

3. Maxon EC motor, ∅6 mm, two-pole, 12 V, 4500 rad/sec, 0.27 A and 0.41 mN-m rated, ~65% efficiency. The PWM drivers are: (1) DRV8312DDWR, 52.5 V and 3.5 A rated, 70 V and 6.5 A pea; (2) DRV8332DKDR, 52.5 V and 8 A rated.

Figures 3.34.c reports the STMicroelectronics EVAL6235N and EVAL6235PD evaluation boards with the L6235 DMOS driver for three-phase synchronous motors, 52 V, 2.8 A rated, 5.6 A peak, 100 KHz. The double-diffused metal-oxide-semiconductor (DMOS) power FETs are used. The bipolar-CMOS-DMOS combines isolated DMOS power transistors with CMOS and bipolar circuits on the same chip. The L6235 has a three-phase DMOS bridge and current sensors. The Hall sensors measure displacements implementing a balanced voltage set.

Permanent magnet 10 W rated synchronous motors can be controlled by the 30 V, 1 A MC33035 controller and other controllers-drivers. The phase voltages (u_{as},u_{bs},u_{cs}) are obtained by using the rotor angular displacement measured by the Hall sensors as reported in Figures 3.35. The MC33035 can be used to drive power MOSFETs, see Figure 3.35.b. The representative block diagrams provide the functional system schematics. Three-phase, six-step full-wave converter topology are implemented. Many closed-loop systems are designed with proportional or proportional-integral control laws. The "error amplifier" yields the error $e(t)$. The reader is referred to the application notes for detail information.

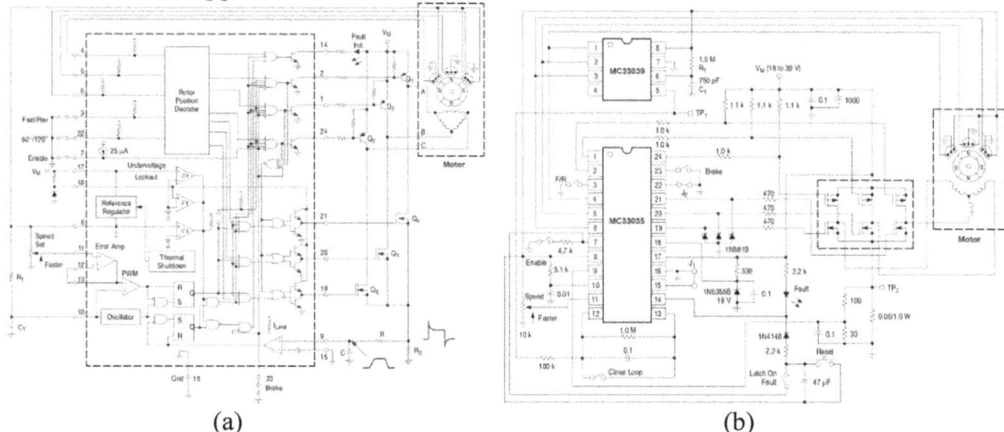

Figure 3.35. (a) Schematics of the MC33035 *Brushless DC Motor Controller*, copyright of Motorola, used with permission [7];
(b) Schematics of the MC33039 and MC33035 controllers to drive the power MOSFETs controlling synchronous motors, copyright of Motorola, used with permission [7].

3.6.4. Permanent Magnet Synchronous Motors and Lagrange Equations

Apply the Lagrange equations

$$\frac{d}{dt}\left(\frac{\partial\Gamma}{\partial\dot{q}_i}\right)-\frac{\partial\Gamma}{\partial q_i}+\frac{\partial D}{\partial\dot{q}_i}+\frac{\partial\Pi}{\partial q_i}=Q_i, \qquad (3.68)$$

where the kinetic Γ, potential Π and dissipating D energies are found by using the *generalized coordinates* q_i. The *generalized coordinates* are the electric charges in the *abc* stator windings $q_1=\int i_{as}dt,\ \dot q_1=i_{as}$, $q_2=\int i_{bs}dt,\ \dot q_2=i_{bs}$, $q_3=\int i_{cs}dt,\ \dot q_3=i_{cs}$ and angular displacement $q_4=\theta_r$, $\dot q_4=\omega_r$. The *generalized forces* are the applied voltages to the *abc* windings $Q_1=u_{as}$, $Q_2=u_{bs}$, $Q_3=u_{cs}$ and the load torque $Q_4=-T_L$.

One yields

$$\frac{d}{dt}\left(\frac{\partial\Gamma}{\partial\dot q_1}\right)-\frac{\partial\Gamma}{\partial q_1}+\frac{\partial D}{\partial\dot q_1}+\frac{\partial\Pi}{\partial q_1}=Q_1,\ \frac{d}{dt}\left(\frac{\partial\Gamma}{\partial\dot q_2}\right)-\frac{\partial\Gamma}{\partial q_2}+\frac{\partial D}{\partial\dot q_2}+\frac{\partial\Pi}{\partial q_2}=Q_2, \quad (3.69)$$

$$\frac{d}{dt}\left(\frac{\partial\Gamma}{\partial\dot q_3}\right)-\frac{\partial\Gamma}{\partial q_3}+\frac{\partial D}{\partial\dot q_3}+\frac{\partial\Pi}{\partial q_3}=Q_3,\ \frac{d}{dt}\left(\frac{\partial\Gamma}{\partial\dot q_4}\right)-\frac{\partial\Gamma}{\partial q_4}+\frac{\partial D}{\partial\dot q_4}+\frac{\partial\Pi}{\partial q_4}=Q_4.$$

The total kinetic energy is kinetic energies of electrical and mechanical systems

$$\Gamma=\Gamma_E+\Gamma_M=\tfrac{1}{2}L_{asas}\dot q_1^2+\tfrac{1}{2}\left(L_{asbs}+L_{bsas}\right)\dot q_1\dot q_2+\tfrac{1}{2}\left(L_{ascs}+L_{csas}\right)\dot q_1\dot q_3+\tfrac{1}{2}L_{bsbs}\dot q_2^2 \quad (3.70)$$

$$+\tfrac{1}{2}\left(L_{bscs}+L_{csbs}\right)\dot q_2\dot q_3+\tfrac{1}{2}L_{cscs}\dot q_3^2+\psi_m\dot q_1\sin q_4+\psi_m\dot q_2\sin\left(q_4-\tfrac{2}{3}\pi\right)+\psi_m\dot q_3\sin\left(q_4+\tfrac{2}{3}\pi\right)+\tfrac{1}{2}J\dot q_4^2.$$

Therefore,

$$\frac{\partial\Gamma}{\partial q_1}=0,\ \frac{\partial\Gamma}{\partial\dot q_1}=L_{asas}\dot q_1+\tfrac{1}{2}\left(L_{asbs}+L_{bsas}\right)\dot q_2+\tfrac{1}{2}\left(L_{ascs}+L_{csas}\right)\dot q_3+\psi_m\sin q_4, \quad (3.71)$$

$$\frac{\partial\Gamma}{\partial q_2}=0,\ \frac{\partial\Gamma}{\partial\dot q_2}=\tfrac{1}{2}\left(L_{asbs}+L_{bsas}\right)\dot q_1+L_{bsbs}\dot q_2+\tfrac{1}{2}\left(L_{bscs}+L_{csbs}\right)\dot q_3+\psi_m\sin\left(q_4-\tfrac{2}{3}\pi\right),$$

$$\frac{\partial\Gamma}{\partial q_3}=0,\ \frac{\partial\Gamma}{\partial\dot q_3}=\tfrac{1}{2}\left(L_{ascs}+L_{csas}\right)\dot q_1+\tfrac{1}{2}\left(L_{bscs}+L_{csbs}\right)\dot q_2+L_{cscs}\dot q_3+\psi_m\sin\left(q_4+\tfrac{2}{3}\pi\right),$$

$$\frac{\partial\Gamma}{\partial q_4}=\psi_m\dot q_1\cos q_4+\psi_m\dot q_2\cos\left(q_4-\tfrac{2}{3}\pi\right)+\psi_m\dot q_3\cos\left(q_4+\tfrac{2}{3}\pi\right),\ \frac{\partial\Gamma}{\partial\dot q_4}=J\dot q_4.$$

The total potential energy is Π=0.

The total dissipating energy is a sum of heat energy dissipated by the electrical system and heat energy dissipated by the mechanical system. That is,

$$D=D_E+D_M=\tfrac{1}{2}\left(r_s\dot q_1^2+r_s\dot q_2^2+r_s\dot q_3^2+B_m\dot q_4^2\right). \quad (3.72)$$

Differentiate D with respect to the *generalized coordinates*

$$\frac{\partial D}{\partial\dot q_1}=r_s\dot q_1,\ \frac{\partial D}{\partial\dot q_2}=r_s\dot q_2,\ \frac{\partial D}{\partial\dot q_3}=r_s\dot q_3,\ \frac{\partial D}{\partial\dot q_4}=B_m\dot q_4. \quad (3.73)$$

The Lagrange equations (3.68) yield

$$L_{asas}\frac{di_{as}}{dt}+\tfrac{1}{2}\left(L_{asbs}+L_{bsas}\right)\frac{di_{bs}}{dt}+\tfrac{1}{2}\left(L_{ascs}+L_{csas}\right)\frac{di_{cs}}{dt}+\psi_m\omega_r\cos\theta_r+r_si_{as}=u_{as}, \quad (3.74)$$

$$\tfrac{1}{2}\left(L_{asbs}+L_{bsas}\right)\frac{di_{as}}{dt}+L_{bsbs}\frac{di_{bs}}{dt}+\tfrac{1}{2}\left(L_{bscs}+L_{csbs}\right)\frac{di_{cs}}{dt}+\psi_m\omega_r\cos\left(\theta_r-\tfrac{2}{3}\pi\right)+r_si_{bs}=u_{bs},$$

$$\tfrac{1}{2}\left(L_{ascs}+L_{csas}\right)\frac{di_{as}}{dt}+\tfrac{1}{2}\left(L_{bscs}+L_{csbs}\right)\frac{di_{bs}}{dt}+L_{cscs}\frac{di_{cs}}{dt}+\psi_m\omega_r\cos\left(\theta_r+\tfrac{2}{3}\pi\right)+r_si_{cs}=u_{cs},$$

$$J\frac{d^2\theta_r}{dt^2}-\psi_mi_{as}\cos\theta_r-\psi_mi_{bs}\cos\left(\theta_r-\tfrac{2}{3}\pi\right)-\psi_mi_{cs}\cos\left(\theta_r+\tfrac{2}{3}\pi\right)+B_m\frac{d\theta_r}{dt}=-T_L.$$

Analyze inductances and flux linkages as reported in section 3.6.3.

For round-rotor transducers, $\Re_{mq}=\Re_{md}$. Hence, $L_{mq}=L_{md}$, $\bar L_m=\dfrac{2N_s^2}{3\Re_{mq}}=\dfrac{2N_s^2}{3\Re_{md}}$, $L_{\Delta m}=0$.

Denoting $L_{ss} = L_{ls} + \bar{L}_m$, we obtain the resulting equations

$$\left(L_{ls} + \bar{L}_m\right)\frac{di_{as}}{dt} - \tfrac{1}{2}\bar{L}_m\frac{di_{bs}}{dt} - \tfrac{1}{2}\bar{L}_m\frac{di_{cs}}{dt} + \psi_m\omega_r\cos\theta_r + r_s i_{as} = u_{as}, \tag{3.75}$$

$$-\tfrac{1}{2}\bar{L}_m\frac{di_{as}}{dt} + \left(L_{ls} + \bar{L}_m\right)\frac{di_{bs}}{dt} - \tfrac{1}{2}\bar{L}_m\frac{di_{cs}}{dt} + \psi_m\omega_r\cos\left(\theta_r - \tfrac{2}{3}\pi\right) + r_s i_{bs} = u_{bs},$$

$$-\tfrac{1}{2}\bar{L}_m\frac{di_{as}}{dt} - \tfrac{1}{2}\bar{L}_m\frac{di_{bs}}{dt} + \left(L_{ls} + \bar{L}_m\right)\frac{di_{cs}}{dt} + \psi_m\omega_r\cos\left(\theta_r + \tfrac{2}{3}\pi\right) + r_s i_{cs} = u_{cs},$$

$$J\frac{d\omega_r}{dt} + B_m\omega_r - \psi_m\left[i_{as}\cos\theta_r + i_{bs}\cos\left(\theta_r - \tfrac{2}{3}\pi\right) + i_{cs}\cos\left(\theta_r + \tfrac{2}{3}\pi\right)\right] = -T_L,$$

$$\frac{d\theta_r}{dt} = \omega_r .$$

The expression $\dfrac{\partial\Gamma}{\partial q_4} = \psi_m\dot{q}_1\cos q_4 + \psi_m\dot{q}_2\cos\left(q_4 - \tfrac{2}{3}\pi\right) + \psi_m\dot{q}_3\cos\left(q_4 + \tfrac{2}{3}\pi\right)$ yields the electromagnetic torque

$$T_e = \frac{\partial W_c}{\partial\theta_r} = \frac{\partial\Gamma}{\partial q_4} = \psi_m\left[i_{as}\cos\theta_r + i_{bs}\cos\left(\theta_r - \tfrac{2}{3}\pi\right) + i_{cs}\cos\left(\theta_r + \tfrac{2}{3}\pi\right)\right]. \tag{3.76}$$

The derived governing equations correspond to equations found in section 3.6.3.

3.6.5. Mathematical Models of Permanent Magnet Synchronous Machines Using the *Quadrature* and *Direct* Operators

Synchronous and induction motors are controlled applying the phase voltages (u_{as}, u_{bs}, u_{cs}). Analysis of ac transducers can be performed using the operators. The measurable machine variable tuple ($\mathbf{u}_{abcs}, \mathbf{i}_{abcs}, \boldsymbol{\psi}_{abcs}$) can be expressed using the *quadrature*, *direct*, and *zero* (*qd*0) operators for stator currents, voltages and flux linkages ($\mathbf{u}_{qd0s}, \mathbf{i}_{qd0s}, \boldsymbol{\psi}_{qd0s}$) [4-7, 16]. This concept was developed to simplify mathematical analysis of ac transducers using a number of assumptions and simplifications. The Park transformation assumes a linear magnetic system, sinusoidal *mmf*, symmetry, etc. The pertained nonlinear analyses is obscured, and, balanced current and voltage sets cannot be derived.

To control motors, if algorithms are derived in the *qd*0 operators, the *inverse* Park transformation must be performed in real time to compute phase voltages (u_{as}, u_{bs}, u_{cs}). Implementation inconsistencies, discrepancies and control inadequateness in high-performance applications are the drawbacks of the *qd*0 analyses and control schemes. Analyses of ac transducers in the machine variables ($\mathbf{u}_{abcs}, \mathbf{i}_{abcs}, \boldsymbol{\psi}_{abcs}$) are performed, and, control and hardware solutions are consistent and straightforward.

Park Transformation – Derive the governing equations using the *qd*0 components of stator currents, voltages and flux linkages, which are not measurable and not observable operators. We find a mathematical model in the *arbitrary* reference frame when the frame angular velocity ω is not specified. The Park transformations and Park mapping are

$$\begin{cases}\mathbf{u}_{qd0s} = \mathbf{K}_s\mathbf{u}_{abcs},\\ \mathbf{i}_{qd0s} = \mathbf{K}_s\mathbf{i}_{abcs},\\ \boldsymbol{\psi}_{qd0s} = \mathbf{K}_s\boldsymbol{\psi}_{abcs},\end{cases} \quad \begin{cases}\mathbf{u}_{abcs} = \mathbf{K}_s^{-1}\mathbf{u}_{qd0s},\\ \mathbf{i}_{abcs} = \mathbf{K}_s^{-1}\mathbf{i}_{qd0s},\\ \boldsymbol{\psi}_{abcs} = \mathbf{K}_s^{-1}\boldsymbol{\psi}_{qd0s},\end{cases} \quad \mathbf{K}_s = \tfrac{2}{3}\begin{bmatrix}\cos\theta & \cos\left(\theta - \tfrac{2}{3}\pi\right) & \cos\left(\theta + \tfrac{2}{3}\pi\right)\\ \sin\theta & \sin\left(\theta - \tfrac{2}{3}\pi\right) & \sin\left(\theta + \tfrac{2}{3}\pi\right)\\ \tfrac{1}{2} & \tfrac{1}{2} & \tfrac{1}{2}\end{bmatrix}. \tag{3.77}$$

The $qd0$ operators of voltages $\mathbf{u}_{qd0s}$, currents $\mathbf{i}_{qd0s}$ and flux linkages $\boldsymbol{\psi}_{qd0s}$ are computed using the *machine* variables and $\mathbf{K}_s$.

From (3.52) $\mathbf{u}_{abcs} = \mathbf{r}_s\mathbf{i}_{abcs} + \frac{d\boldsymbol{\psi}_{abcs}}{dt}$ and (3.77), we have

$$\mathbf{K}_s^{-1}\mathbf{u}_{qd0s} = \mathbf{r}_s\mathbf{K}_s^{-1}\mathbf{i}_{qd0s} + \frac{d\left(\mathbf{K}_s^{-1}\boldsymbol{\psi}_{qd0s}\right)}{dt}, \mathbf{K}_s^{-1} = \begin{bmatrix} \cos\theta & \sin\theta & 1 \\ \cos\left(\theta - \frac{2}{3}\pi\right) & \sin\left(\theta - \frac{2}{3}\pi\right) & 1 \\ \cos\left(\theta + \frac{2}{3}\pi\right) & \sin\left(\theta + \frac{2}{3}\pi\right) & 1 \end{bmatrix}. \quad (3.78)$$

Multiplication of the left and right sides in (3.78) by $\mathbf{K}_s$ yields

$$\mathbf{K}_s\mathbf{K}_s^{-1}\mathbf{u}_{qdos} = \mathbf{K}_s\mathbf{r}_s\mathbf{K}_s^{-1}\mathbf{i}_{qd0s} + \mathbf{K}_s\frac{d\mathbf{K}_s^{-1}}{dt}\boldsymbol{\psi}_{qd0s} + \mathbf{K}_s\mathbf{K}_s^{-1}\frac{d\boldsymbol{\psi}_{qd0s}}{dt}. \quad (3.79)$$

In (3.79), the matrix $\mathbf{r}_s$ is diagonal, and, $\mathbf{K}_s\mathbf{r}_s\mathbf{K}_s^{-1} = \mathbf{r}_s$.

From $\frac{d\mathbf{K}_s^{-1}}{dt} = \omega\begin{bmatrix} -\sin\theta & \cos\theta & 0 \\ -\sin\left(\theta - \frac{2}{3}\pi\right) & \cos\left(\theta - \frac{2}{3}\pi\right) & 0 \\ -\sin\left(\theta + \frac{2}{3}\pi\right) & \cos\left(\theta + \frac{2}{3}\pi\right) & 0 \end{bmatrix}$, we have $\mathbf{K}_s\frac{d\mathbf{K}_s^{-1}}{dt} = \omega\begin{bmatrix} 0 & 1 & 0 \\ -1 & 0 & 0 \\ 0 & 0 & 0 \end{bmatrix}$.

Equation (3.79) yields

$$\mathbf{u}_{qd0s} = \mathbf{r}_s\mathbf{i}_{qd0s} + \omega\begin{bmatrix} \psi_{ds} \\ -\psi_{qs} \\ 0 \end{bmatrix} + \frac{d\boldsymbol{\psi}_{qd0s}}{dt}, \quad (3.80)$$

where $\boldsymbol{\psi}_{qd0s} = \mathbf{K}_s\boldsymbol{\psi}_{abcs}, \boldsymbol{\psi}_{abcs} = \mathbf{L}_s\mathbf{i}_{abcs} + \boldsymbol{\psi}_m = \begin{bmatrix} L_{ls} + \bar{L}_m & -\frac{1}{2}\bar{L}_m & -\frac{1}{2}\bar{L}_m \\ -\frac{1}{2}\bar{L}_m & L_{ls} + \bar{L}_m & -\frac{1}{2}\bar{L}_m \\ -\frac{1}{2}\bar{L}_m & -\frac{1}{2}\bar{L}_m & L_{ls} + \bar{L}_m \end{bmatrix}\begin{bmatrix} i_{as} \\ i_{bs} \\ i_{cs} \end{bmatrix} + \psi_m\begin{bmatrix} \sin\theta_r \\ \sin\left(\theta_r - \frac{2}{3}\pi\right) \\ \sin\left(\theta_r + \frac{2}{3}\pi\right) \end{bmatrix}$.

Hence,

$$\boldsymbol{\psi}_{qd0s} = \mathbf{K}_s\mathbf{L}_s\mathbf{K}_s^{-1}\mathbf{i}_{qd0s} + \mathbf{K}_s\boldsymbol{\psi}_m = \begin{bmatrix} L_{ls} + \frac{3}{2}\bar{L}_m & 0 & 0 \\ 0 & L_{ls} + \frac{3}{2}\bar{L}_m & 0 \\ 0 & 0 & L_{ls} \end{bmatrix}\mathbf{i}_{qd0s} + \psi_m\begin{bmatrix} -\sin\left(\theta - \theta_r\right) \\ \cos\left(\theta - \theta_r\right) \\ 0 \end{bmatrix}. \quad (3.81)$$

Using (3.80) and (3.81), in the *arbitrary* reference frame with not specified θ, one finds

$$\mathbf{u}_{qd0s} = \mathbf{r}_s\mathbf{i}_{qd0s} + \omega\begin{bmatrix} \psi_{ds} \\ -\psi_{qs} \\ 0 \end{bmatrix} + \begin{bmatrix} L_{ls} + \frac{3}{2}\bar{L}_m & 0 & 0 \\ 0 & L_{ls} + \frac{3}{2}\bar{L}_m & 0 \\ 0 & 0 & L_{ls} \end{bmatrix}\frac{d\mathbf{i}_{qd0s}}{dt} + \psi_m\frac{d}{dt}\begin{bmatrix} -\sin\left(\theta - \theta_r\right) \\ \cos\left(\theta - \theta_r\right) \\ 0 \end{bmatrix}. \quad (3.82)$$

Stationary, Rotor and Synchronous Reference Frames – Specify the frame angular velocity and displacement (ω,θ). The stationary $\omega=0$, rotor $\omega=\omega_r$, and synchronous $\omega=\omega_e$ reference frames are commonly used. For synchronous transducers, the electrical angular velocity ω_r is equal to synchronous angular velocity ω_e. Let the angular velocity of the reference frame is $\omega=\omega_e=\omega_r$. From $\theta=\theta_e=\theta_r$, the Park transformations matrix is

$$\mathbf{K}_s^r = \mathbf{K}_s^e = \frac{2}{3}\begin{bmatrix} \cos\theta_r & \cos\left(\theta_r - \frac{2}{3}\pi\right) & \cos\left(\theta_r + \frac{2}{3}\pi\right) \\ \sin\theta_r & \sin\left(\theta_r - \frac{2}{3}\pi\right) & \sin\left(\theta_r + \frac{2}{3}\pi\right) \\ \frac{1}{2} & \frac{1}{2} & \frac{1}{2} \end{bmatrix}, \quad (3.83)$$

where the superscripts r and e correspond to the rotor and synchronous reference frames.

Using $\omega=\omega_r$ and $\theta=\theta_r$ in (3.82), in the rotor and synchronous reference frames, one yields the governing equations for the *quadrature*, *direct* and *zero* current operators

$$\frac{di_{qs}^r}{dt}=-\frac{r_s}{L_{ls}+\frac{3}{2}\bar{L}_m}i_{qs}^r-\frac{\psi_m}{L_{ls}+\frac{3}{2}\bar{L}_m}\omega_r-i_{ds}^r\omega_r+\frac{1}{L_{ls}+\frac{3}{2}\bar{L}_m}u_{qs}^r, \tag{3.84}$$

$$\frac{di_{ds}^r}{dt}=-\frac{r_s}{L_{ls}+\frac{3}{2}\bar{L}_m}i_{ds}^r+i_{qs}^r\omega_r+\frac{1}{L_{ls}+\frac{3}{2}\bar{L}_m}u_{ds}^r,$$

$$\frac{di_{0s}^r}{dt}=-\frac{r_s}{L_{ls}}i_{0s}^r+\frac{1}{L_{ls}}u_{0s}^r.$$

To find T_e, apply the Park transformation $\mathbf{i}_{abcs}=(\mathbf{K}_s^r)^{-1}\mathbf{i}_{qd0s}^r$ to the phase current in the electromagnetic torque equation (3.62). Substitute $i_{as}=\cos\theta_r i_{qs}^r+\sin\theta_r i_{ds}^r+i_{0s}^r$, $i_{bs}=\cos\left(\theta_r-\frac{2}{3}\pi\right)i_{qs}^r+\sin\left(\theta_r-\frac{2}{3}\pi\right)i_{ds}^r+i_{0s}^r$ and $i_{cs}=\cos\left(\theta_r+\frac{2}{3}\pi\right)i_{qs}^r+\sin\left(\theta_r+\frac{2}{3}\pi\right)i_{ds}^r+i_{0s}^r$ in

$$T_e=\frac{P\psi_m}{2}\left[i_{as}\cos\theta_r+i_{bs}\cos\left(\theta_r-\tfrac{2}{3}\pi\right)+i_{cs}\cos\left(\theta_r+\tfrac{2}{3}\pi\right)\right].$$

In the *quadrature* current operator, the electromagnetic torque is

$$T_e=\frac{3P\psi_m}{4}i_{qs}^r. \tag{3.85}$$

Alternatively, from $W_c=\frac{1}{2}\mathbf{i}_{abcs}^T\mathbf{L}_s\mathbf{i}_{abcs}+\mathbf{i}_{abcs}^T\boldsymbol{\psi}_m+W_{\text{permanet magnet}}$, using the Park transformations $\mathbf{i}_{abcs}=(\mathbf{K}_s^r)^{-1}\mathbf{i}_{qd0s}^r$ and $\boldsymbol{\psi}_{abcs}=(\mathbf{K}_s^r)^{-1}\boldsymbol{\psi}_{qd0s}^r$ one finds (3.85).

The electromagnetic and *torsional-mechanical* dynamics are described by

$$\frac{di_{qs}^r}{dt}=-\frac{r_s}{L_{ls}+\frac{3}{2}\bar{L}_m}i_{qs}^r-\frac{\psi_m}{L_{ls}+\frac{3}{2}\bar{L}_m}\omega_r-i_{ds}^r\omega_r+\frac{1}{L_{ls}+\frac{3}{2}\bar{L}_m}u_{qs}^r, \tag{3.86}$$

$$\frac{di_{ds}^r}{dt}=-\frac{r_s}{L_{ls}+\frac{3}{2}\bar{L}_m}i_{ds}^r+i_{qs}^r\omega_r+\frac{1}{L_{ls}+\frac{3}{2}\bar{L}_m}u_{ds}^r,$$

$$\frac{di_{0s}^r}{dt}=-\frac{r_s}{L_{ls}}i_{0s}^r+\frac{1}{L_{ls}}u_{0s}^r.$$

$$\frac{d\omega_r}{dt}=\frac{3P^2\psi_m}{8J}i_{qs}^r-\frac{B_m}{J}\omega_r-\frac{P}{2J}T_L,$$

$$\frac{d\theta_r}{dt}=\omega_r.$$

A balanced current set (3.65) is

$$i_{as}(t)=\sqrt{2}i_M\cos\theta_r,\ i_{bs}(t)=\sqrt{2}i_M\cos\left(\theta_r-\tfrac{2}{3}\pi\right),\ i_{cs}(t)=\sqrt{2}i_M\cos\left(\theta_r+\tfrac{2}{3}\pi\right).$$

Using $\mathbf{i}_{qd0s}^r=\mathbf{K}_s^r\mathbf{i}_{abcs}$, one obtains

$$\begin{bmatrix} i_{qs}^r \\ i_{ds}^r \\ i_{0s}^r \end{bmatrix}=\frac{2}{3}\begin{bmatrix} \cos\theta_r & \cos\left(\theta_r-\frac{2}{3}\pi\right) & \cos\left(\theta_r+\frac{2}{3}\pi\right) \\ \sin\theta_r & \sin\left(\theta_r-\frac{2}{3}\pi\right) & \sin\left(\theta_r+\frac{2}{3}\pi\right) \\ \frac{1}{2} & \frac{1}{2} & \frac{1}{2} \end{bmatrix}\begin{bmatrix} \sqrt{2}i_M\cos\theta_r \\ \sqrt{2}i_M\cos\left(\theta_r-\frac{2}{3}\pi\right) \\ \sqrt{2}i_M\cos\left(\theta_r+\frac{2}{3}\pi\right) \end{bmatrix}=\begin{bmatrix} \sqrt{2}i_M \\ 0 \\ 0 \end{bmatrix}.$$

Furthermore,

$$\begin{bmatrix} u_{qs}^r \\ u_{ds}^r \\ u_{0s}^r \end{bmatrix} = \frac{2}{3}\begin{bmatrix} \cos\theta_r & \cos\left(\theta_r - \frac{2}{3}\pi\right) & \cos\left(\theta_r + \frac{2}{3}\pi\right) \\ \sin\theta_r & \sin\left(\theta_r - \frac{2}{3}\pi\right) & \sin\left(\theta_r + \frac{2}{3}\pi\right) \\ \frac{1}{2} & \frac{1}{2} & \frac{1}{2} \end{bmatrix}\begin{bmatrix} \sqrt{2}u_M \cos\theta_r \\ \sqrt{2}u_M \cos\left(\theta_r - \frac{2}{3}\pi\right) \\ \sqrt{2}u_M \cos\left(\theta_r + \frac{2}{3}\pi\right) \end{bmatrix} = \begin{bmatrix} \sqrt{2}u_M \\ 0 \\ 0 \end{bmatrix}.$$

To ensure balanced operation, the *quadrature*, *direct* and *zero* current operators are

$$i_{qs}^r(t) = \sqrt{2}i_M,\ i_{ds}^r(t) = 0,\ i_{0s}^r(t) = 0. \tag{3.87}$$

From (3.67), the $qd0$ voltage operators are

$$u_{qs}^r(t) = \sqrt{2}u_M,\ u_{ds}^r(t) = 0,\ u_{0s}^r(t) = 0. \tag{3.88}$$

For the synchronous reference frame, the model is identical to the rotor reference frame because

$$\mathbf{u}_{qd0s}^e = \mathbf{u}_{qd0s}^r,\ \mathbf{i}_{qd0s}^e = \mathbf{i}_{qd0s}^r,\ \boldsymbol{\psi}_{qd0s}^e = \boldsymbol{\psi}_{qd0s}^r.$$

From $\omega=\omega_e=\omega_r$ and $\theta=\theta_e=\theta_r$, differential equations (3.86) result in the state-space model in the synchronous reference frame

$$\begin{bmatrix} \frac{di_{qs}^e}{dt} \\ \frac{di_{ds}^e}{dt} \\ \frac{di_{0s}^e}{dt} \\ \frac{d\omega_r}{dt} \\ \frac{d\theta_r}{dt} \end{bmatrix} = \begin{bmatrix} -\frac{r_s}{L_{ls}+\frac{3}{2}\bar{L}_m} & 0 & 0 & -\frac{\psi_m}{L_{ls}+\frac{3}{2}\bar{L}_m} & 0 \\ 0 & -\frac{r_s}{L_{ls}+\frac{3}{2}\bar{L}_m} & 0 & 0 & 0 \\ 0 & 0 & -\frac{r_s}{L_{ls}} & 0 & 0 \\ \frac{3P^2\psi_m}{8J} & 0 & 0 & -\frac{B_m}{J} & 0 \\ 0 & 0 & 0 & 1 & 0 \end{bmatrix}\begin{bmatrix} i_{qs}^e \\ i_{ds}^e \\ i_{0s}^e \\ \omega_r \\ \theta_r \end{bmatrix} + \begin{bmatrix} -i_{ds}^e\omega_r \\ i_{qs}^e\omega_r \\ 0 \\ 0 \\ 0 \end{bmatrix}$$

$$+\begin{bmatrix} \frac{1}{L_{ls}+\frac{3}{2}\bar{L}_m} & 0 & 0 \\ 0 & \frac{1}{L_{ls}+\frac{3}{2}\bar{L}_m} & 0 \\ 0 & 0 & \frac{1}{L_{ls}} \\ 0 & 0 & 0 \\ 0 & 0 & 0 \end{bmatrix}\begin{bmatrix} u_{qs}^e \\ u_{ds}^e \\ u_{0s}^e \end{bmatrix} - \begin{bmatrix} 0 \\ 0 \\ 0 \\ \frac{P}{2J} \\ 0 \end{bmatrix} T_L. \tag{3.89}$$

The balanced operations is ensured by

$$i_{qs}^e(t) = \sqrt{2}i_M,\ i_{ds}^e(t) = 0,\ i_{0s}^e(t) = 0, \text{ and } u_{qs}^e(t) = \sqrt{2}u_M,\ u_{ds}^e(t) = 0,\ u_{0s}^e(t) = 0.$$

Example 3.23. Simulation of the Closed-Loop Permanent Magnet Synchronous Motor Using a Mathematical Model in the qd0 Operators

Using differential equations (3.86), simulate a four-pole 50 W, 50 V (rms) permanent magnet motor with the following parameters: r_s=1.5 ohm, $L_{SS}=L_{ls}+\frac{3}{2}\bar{L}_m$=0.005 H, ψ_m=0.1 V-sec/rad, B_m=0.0001 N-m-sec/rad and J=0.002 kg-m^2.

The bounded proportional-integral control law ensures the balanced operation, and $u_{qs}^r = \text{sat}_{u_{moi}}^{u_{max}}(k_p e + k_i \int e dt)$, $-u_{max} \le |u_{qs}^r| \le u_{max}$, $u_{max} = \sqrt{2}50$ V, k_p=10, k_i=10, $e(t)=\omega_{ref}-\omega_r$.

The SIMULINK diagram for a closed-loop system is reported in Figure 3.36.a. Figure 3.36.b documents the angular velocity $\omega_r(t)$ and currents (i_{qs}^r, i_{ds}^r). The voltage u_{qs}^r is bounded as $-u_{\max} \le \left|u_{qs}^r\right| \le u_{\max}$, $u_{\max} = \sqrt{2}50$ V. The bidirectional rotation of the motor is achieved. The reference is specified to be $\omega_{\text{ref}}=\pm 200$ rad/sec. The motor is controlled by applying the phase voltages (u_{as}, u_{bs}, u_{cs}). These phase voltages must be computed in real-time by measuring θ_r and computing the *inverse* Park transformation.

The voltages of the PWM amplifier to be applied to *abc* phases are

$$\mathbf{u}_{abcs} = \begin{bmatrix} u_{as} \\ u_{bs} \\ u_{cs} \end{bmatrix} = \mathbf{K}_s^{-1}\mathbf{u}_{qd0s}^r = \begin{bmatrix} \cos\theta_r & \sin\theta_r & 1 \\ \cos\left(\theta_r - \frac{2}{3}\pi\right) & \sin\left(\theta_r - \frac{2}{3}\pi\right) & 1 \\ \cos\left(\theta_r + \frac{2}{3}\pi\right) & \sin\left(\theta_r + \frac{2}{3}\pi\right) & 1 \end{bmatrix} \begin{bmatrix} u_{qs}^r \\ u_{ds}^r \\ u_{0s}^r \end{bmatrix}.$$

The *quadrature* voltage operator u_{qs}^r and phase voltages (u_{as}, u_{bs}, u_{cs}) are illustrated in Figure 3.36.c. The *direct* and *zero* voltage operators are $u_{ds}^r = 0$, $u_{0s}^r = 0$. The motor starts at stall and reaches the reference angular velocity ±200 rad/sec. The applied load is $T_L(t)=\pm 0.25$ N-m.

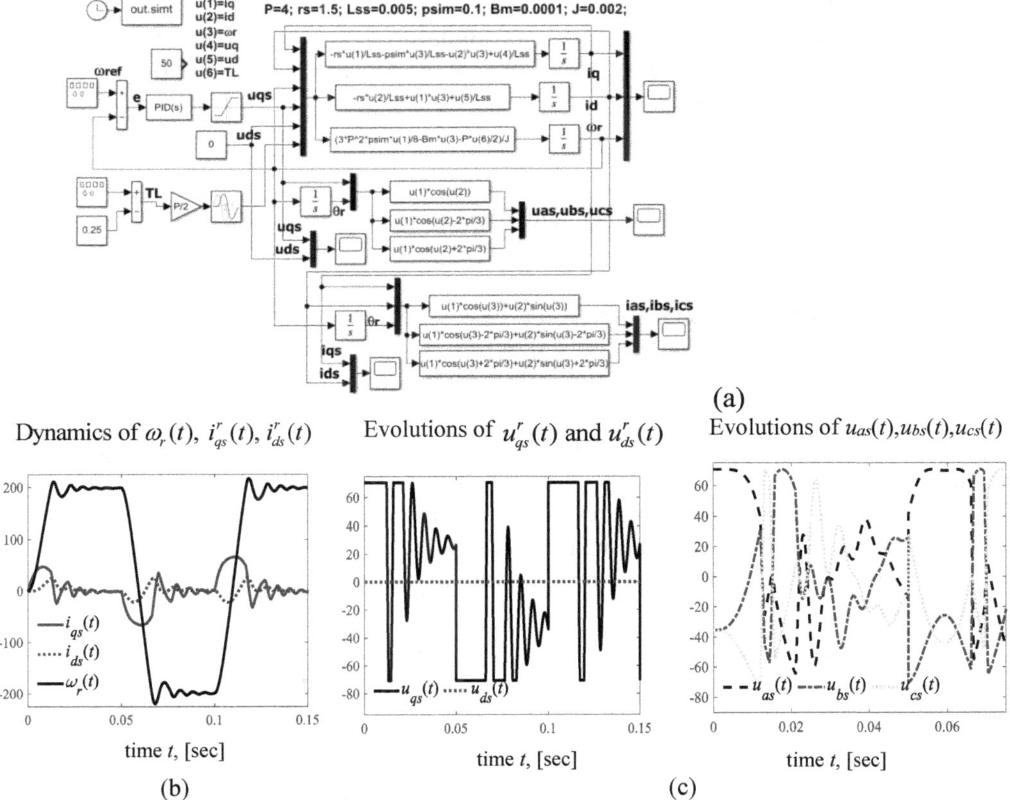

Figure 3.36.(a) SIMULINK diagram to simulate a closed-loop electric drive with bounds on control $u_{qs}^r = \text{sat}_{u_{\min}}^{u_{\max}}(k_p e + k_i \int e dt)$, $u_{ds}^r = 0$, $u_{0s}^r = 0$, $e(t)=\omega_{\text{ref}}-\omega_r$, $-u_{\max} \le \left|u_{qs}^r\right| \le u_{\max}$, $u_{\max} = \sqrt{2}50$ V;

(b) Closed-loop dynamics $\omega_r(t)$, $i_{qs}^r(t)$ and $i_{ds}^r(t)$, $\omega_{\text{ref}}=\pm 200$ rad/sec, $T_L(t)=\pm 0.25$ N-m;

(c) Evolutions of a *quadrature* voltage operator u_{qs}^r and phase voltages (u_{as}, u_{bs}, u_{cs}). ■

Example 3.24. *Experimental Studies: Control of Permanent Magnet Synchronous Motors Using the qd0 Operators*

Electric drives with permanent magnet dc and ac motors are open-loop stable. The Lyapunov stability theory is used in stability analysis. A positive definite function is

$$V(i_{qs}^r, i_{ds}^r, \omega_r) = \tfrac{1}{2}(i_{qs}^{r\,2} + i_{ds}^{r\,2} + \omega_r^2),\ V>0.$$

Using (3.86), one finds $\dfrac{dV(i_{qs}^r, i_{ds}^r, \omega_r)}{dt} < 0$. Hence, an open-loop system is stable. Details are reported in Example 6.42.

Physical implementation, characterization and deployment of closed-loop systems are important. The Hall-effect, inductive, optical encoders, resolvers and other sensors measure the rotor displacement θ_r to implement the balanced voltage sets (u_{as},u_{bs},u_{cs}). There are sensing, interfacing and processing delays and latencies in measurements, filtering and fetching θ_r. Considering control in the *machine variables*, one avoids inadequacies and impediments despite of microseconds latencies. The PWM controllers-drivers with proportional-integral controllers are commercialized and implemented controlling the balanced voltage sets (u_{as},u_{bs},u_{cs}). Control laws use the tracking error $e(t)$.

For electric drives $e(t)=\omega_{\text{ref}}(t)-\omega_r(t)$, and, for servos $e(t)=\theta_{\text{ref}}(t)-\theta_r(t)$.

For open-loop stable systems, one may implement the proportional-integral control law. For electric drives

$$u = k_p e + k_i \int e\,dt,\ e(t)=\omega_{\text{ref}}(t)-\omega_r(t),$$

where k_p and k_i are the proportional and integral feedback gains.

The rotor angular velocity and displacement (ω_r,θ_r) are measured, and, the specified accuracy and precision can be accomplished using off-the-shelf sensors and PWM amplifiers.

Some solutions use the *torque command*, *vector control*, *field oriented control*, *sensorless control* and other concepts attempting to use the mathematical operators for fluxes, currents and voltages in the *qd*0 reference frames. The *qd*0 reference frame yields modeling simplicity assuming ideal sinusoidal *mmf*, linear magnetization, linear *BH* curve, field uniformity, ideal electromagnetic design, assembly homogeneity and other assumptions. In physical implementation, the *direct* and *inverse* Park transformations must be computed in real-time under constrains and axiomatized postulates. Correspondingly, inconsistencies and impediments are observed.

The steady-state operation is characterized by the equilibrium when the *net* torque is $\Sigma T=0$.

For negligible friction and drag torques, $\Sigma T=0$ yields $T_e=T_L$.

At $\Sigma T=0$, motor operates at the constant angular velocity ω_r=const.

From (3.85), the electromagnetic toque is $T_e = \dfrac{3P\psi_m}{4} i_{qs}^r$.

For a known load T_L, assuming the model validity, one may find the *quadrature* current and voltage operators (i_{qs}^r, u_{qs}^r). The steady-state analysis is performed by using the first and second differential equations (3.86). The applied voltages are balanced by the motional *emf*. From

$$-r_s i_{qs}^r - \psi_m\omega_r - (L_{ls} + \tfrac{3}{2}\bar{L}_m) i_{ds}^r \omega_r + u_{qs}^r = 0 \text{ and } -r_s i_{ds}^r + (L_{ls} + \tfrac{3}{2}\bar{L}_m) i_{qs}^r \omega_r + u_{ds}^r = 0,$$

one finds the relationship between ($u_{qs}^r \neq 0, u_{ds}^r = 0$), ($i_{qs}^r, i_{ds}^r$) and steady-state ω_r.

The *quadrature* voltage operator u_{qs}^r is balanced by $\left[-\psi_m\omega_r - (L_{ls} + \frac{3}{2}\bar{L}_m)i_{ds}^r\omega_r\right]$ and voltage drop $r_s i_{qs}^r$.

Furthermore, from $-r_s i_{ds}^r + (L_{ls} + \frac{3}{2}\bar{L}_m)i_{qs}^r\omega_r = 0$ one yields

$i_{ds}^r = \frac{1}{r_s}(L_{ls} + \frac{3}{2}\bar{L}_m)i_{qs}^r\omega_r$.

The aforementioned concept may suit the steady-state analysis assuming the measured load T_L, which is a challenging problem. Other assumptions are the validity of the *qd*0 models, found under many assumptions. Parameters ($r_s, \psi_m, (L_{ls} + \frac{3}{2}\bar{L}_m), B_m$), which are assumed to be constant, vary. Measurements, errors and data heterogeneities are other impediments.

As an education exercise and demonstration, use the STMicroelectronics STM32 Development Board. The *torque* and *quadrature current commands* are specified, and, the *sensorless control* is implemented estimating the angular displacement θ_r. Figure 3.37 documents the transient responses for different current feedback gains (k_p,k_i) in control

$$u_{qs}^r = k_p \boldsymbol{e}_{i_{qs}^r} + k_i \int \boldsymbol{e}_{i_{qs}^r} dt,$$

where the *quadrature current command* difference is $\boldsymbol{e}_{i_{qs}^r}(t) = i_{qs\ \text{command}}^r - i_{qs}^r(t)$.

Here, $\boldsymbol{e}_{i_{qs}^r}$ is not the tracking error because the input-output pair ($r=\omega_{\text{ref}}(t)$, $y=\omega_r(t)$) is not used.

It is observed that:

1. The *quadrature current command* should be adjusted for different ω_r, load T_L and (k_p,k_i) to ensure the steady-state angular velocity ω_r=const;
2. If motor is not loaded, T_L=0. For ω_r=const, assuming time-invariant ψ_m and linear viscous friction coefficient B_m, at steady-state $T_e=T_{\text{viscous friction}}$. Equations $T_e = \frac{3P\psi_m}{4} i_{qs}^r$ and $T_{\text{viscous friction}}=B_m\omega_r$ yield i_{qs}^r at the steady-state;
3. System dynamics are affected by the state observers, filter transfer function, filter gain and various processing tasks associated with the Park transformations and estimation of $\hat{\theta}_r(t)$. The latencies due to Park transformations and state estimations alter statics and dynamics;
4. There are inconsistencies of *sensorless* control due to inadequate estimation of rotor displacement $\hat{\theta}_r(t)$, while the Hall-effect and inductive sensors directly measure θ_r;
5. High gains (k_p,k_i) result in saturation and nonlinearities even at no loads;
6. Time-varying load significantly affect the steady-state and dynamic performance, and, causes significant challenges in finding the corresponding *quadrature current command*;
7. There are hardware and algorithmic complexities and vulnerabilities;
8. etc.

As illustrated in Figures 3.37, specifying $\omega_{\text{ref}}(t)$=750 rad/sec, desired stability margins and robustness are not achieved. The closed-loop system is sensitive to feedback gains and *torque commands*. Control of ac transducers in the *machine variables*, using consistent control laws and classical schemes, are achieved with minimum level of assumptions and postulates minimizing complexity and vulnerability.

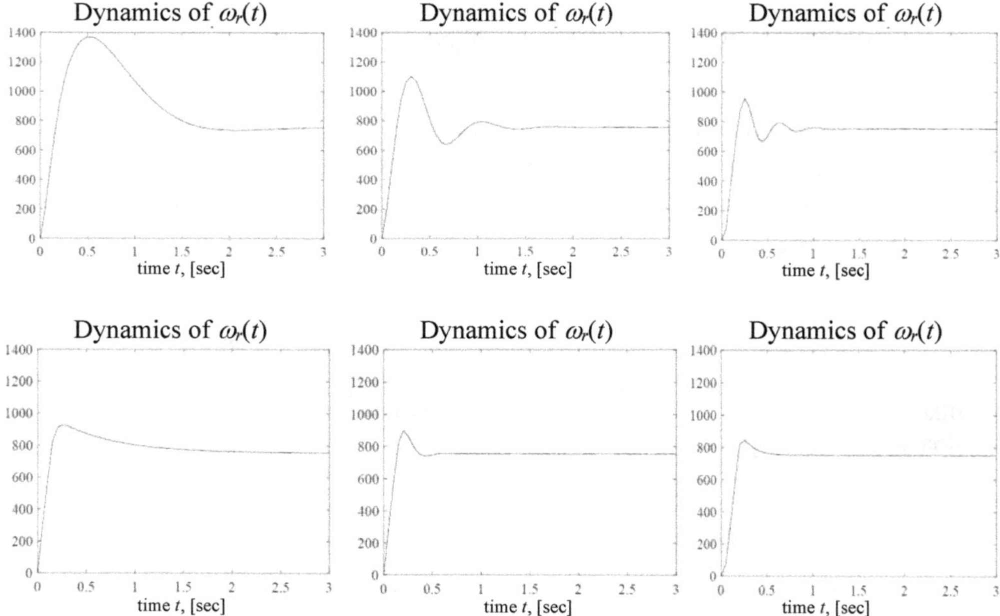

Figure 3.37. Experimental result using the STMicroelectronics STM32 Development Board. The current feedback gains of a control law, affected by the implemented filter and estimator gains, are: (1) k_p=3000, k_i=500; (2) k_p=4000, k_i=4000; (3) k_p=6000, k_i=8000; (4) k_p=18000, k_i=2000; (5) k_p=20000, k_i=80000; (6) k_p=30000, k_i=80000. ∎

3.6.6. High-Fidelity Analysis of Permanent Magnet Motion Devices

Nonlinear magnetic system, magnetic field nonuniformity, nonlinear *BH* curve with a hysteresis loop, spacing between magnets, winding designs, eccentricity and other effects significantly affect transducer capabilities. Using high-accuracy magnetic field sensors, one finds the field produced by the permanent magnets. The magnetic system nonlinearities are experimentally found in a full operating envelope. One measures the induced phase voltages by permanent magnet generator.

For optimal design and not saturated magnetic system, the induced motional *emfs* ($\mathscr{E}_{as}$,$\mathscr{E}_{bs}$,$\mathscr{E}_{cs}$) are sinusoidal such as

$$\mathscr{E}=\frac{d\boldsymbol{\psi}_m}{dt}=\psi_m\begin{bmatrix}\cos\theta_r\omega_r\\ \cos\left(\theta_r-\frac{2}{3}\pi\right)\omega_r\\ \cos\left(\theta_r+\frac{2}{3}\pi\right)\omega_r\end{bmatrix}.$$

As reported, the assumed $\boldsymbol{\psi}_m=\begin{bmatrix}\psi_{asm}\\ \psi_{bsm}\\ \psi_{csm}\end{bmatrix}=\psi_m\begin{bmatrix}\sin\theta_r\\ \sin\left(\theta_r-\frac{2}{3}\pi\right)\\ \sin\left(\theta_r+\frac{2}{3}\pi\right)\end{bmatrix}$ may not be ensured in full operating envelope, defined by loads, angular velocity, etc. One may find that the flux linkages $\boldsymbol{\psi}_m$ in a full operating envelope as

$$\boldsymbol{\psi}_m = \begin{bmatrix} \psi_{asm} \\ \psi_{bsm} \\ \psi_{csm} \end{bmatrix},$$

$$\psi_{asm} = \psi_m \left[\sum_n a_n \sin^{2n-1}\theta_r + \sum_{l,k} a_{l,k}\,\mathrm{sgn}(\sin\theta_r) \left| \sin^{\frac{2l-1}{2k-1}}\theta_r \right| \right] \circ \left[1 + \sum_{p,q} a_{p,q} e^{\sum_{p,q}\left(a_p \sin^p \theta_r + a_q \cos^q \theta_r\right)} \right], \quad (3.90)$$

$$\psi_{bsm} = \psi_m \left[\sum_n a_n \sin^{2n-1}(\theta_r - \tfrac{2}{3}\pi) + \sum_{l,k} a_{l,k}\,\mathrm{sgn}(\sin(\theta_r - \tfrac{2}{3}\pi)) \left| \sin^{\frac{2l-1}{2k-1}}\theta_r \right| \right]$$

$$\circ \left[1 + \sum_{p,q} a_{p,q} e^{\sum_{p,q}\left(a_p \sin^p (\theta_r - \frac{2}{3}\pi) + a_q \cos^q (\theta_r - \frac{2}{3}\pi)\right)} \right],$$

$$\psi_{csm} = \psi_m \left[\sum_n a_n \sin^{2n-1}(\theta_r + \tfrac{2}{3}\pi) + \sum_{l,k} a_{l,k}\,\mathrm{sgn}(\sin(\theta_r + \tfrac{2}{3}\pi)) \left| \sin^{\frac{2l-1}{2k-1}}\theta_r \right| \right]$$

$$\circ \left[1 + \sum_{p,q} a_{p,q} e^{\sum_{p,q}\left(a_p \sin^p (\theta_r + \frac{2}{3}\pi) + a_q \cos^q (\theta_r + \frac{2}{3}\pi)\right)} \right],$$

where a_i and $a_{i,j}$ are the coefficients which depend on the electromagnetic system, operating envelope, electromagnetic loading, electrical steel lamination, magnets, transducer assembly, fabrication technology and other factors, $a_{i,j}(E,D,B,H,\varepsilon,\mu,i_{abcs},\omega_r)$.

Example 3.25. *Symmetric Flux Linkages*
Using (3.90), consider flux linkages, observed in two-phase permanent magnet synchronous machines as:

1. $\psi_{asm} = \sin\theta_r + \sin^5\theta_r$, $\psi_{asm} = \sin^3\theta_r + \sin^5\theta_r$;
2. $\psi_{asm} = \frac{1}{2}\sin\theta_r + \mathrm{sgn}(\sin\theta_r)\left|\sin^{1/3}\theta_r\right|$, $\psi_{asm} = \frac{1}{2}\sin\theta_r + \mathrm{sgn}(\sin\theta_r)\left|\sin^{1/9}\theta_r\right|$;
3. $\boldsymbol{\psi}_m = \begin{bmatrix} \psi_{asm} \\ \psi_{bsm} \end{bmatrix} = \begin{bmatrix} \sin\theta_r e^{-a\cos^2\theta_r} \\ \cos\theta_r e^{-a\sin^2\theta_r} \end{bmatrix}$, e.g., $\psi_{asm} = \sin\theta_r e^{-a\cos^2\theta_r}$ and $\psi_{bsm} = \cos\theta_r e^{-a\sin^2\theta_r}$.

   Figures 3.38 document the resulting plots.

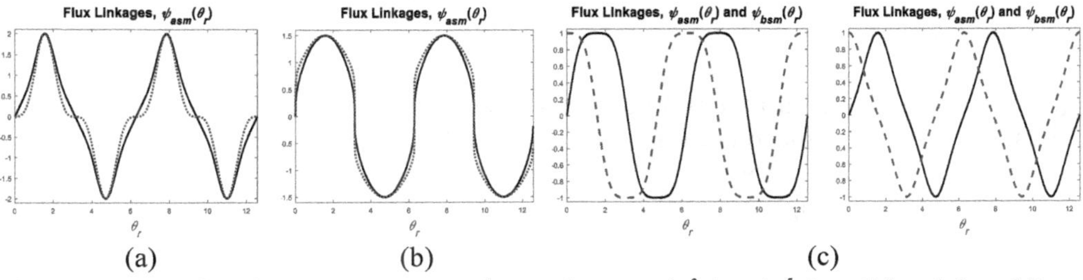

(a) (b) (c)

Figure 3.38. (a) Plots for $\psi_{asm} = \sin\theta_r + \sin^5\theta_r$ and $\psi_{asm} = \sin^3\theta_r + \sin^5\theta_r$, solid and dotted lines; (b) $\psi_{asm} = \frac{1}{2}\sin\theta_r + \mathrm{sgn}(\sin\theta_r)\left|\sin^{1/3}\theta_r\right|$ and $\psi_{asm} = \frac{1}{2}\sin\theta_r + \mathrm{sgn}(\sin\theta_r)\left|\sin^{1/9}\theta_r\right|$;

(c) $\boldsymbol{\psi}_m = \begin{bmatrix} \psi_{asm} \\ \psi_{bsm} \end{bmatrix} = \begin{bmatrix} \sin\theta_r e^{-a\cos^2\theta_r} \\ \cos\theta_r e^{-a\sin^2\theta_r} \end{bmatrix}$, a=0.5, and, $\boldsymbol{\psi}_m = \begin{bmatrix} \psi_{asm} \\ \psi_{bsm} \end{bmatrix} = \begin{bmatrix} \sin\theta_r e^{-a\cos^2\theta_r} \\ \cos\theta_r e^{-a\sin^2\theta_r} \end{bmatrix}$, a=–0.5. ■

Flux Linkages, Electromagnetic Torque, and, Balanced Current and Voltage Sets – For adequately-designed transducers, expressions (3.90) are simplified to

$$\psi_{asm}=\psi_m\sum_n a_n\sin^{2n-1}\theta_r,\qquad (3.91)$$

$$\psi_{bsm}=\psi_m\sum_n a_n\sin^{2n-1}(\theta_r-\tfrac{2}{3}\pi),$$

$$\psi_{csm}=\psi_m\sum_n a_n\sin^{2n-1}(\theta_r+\tfrac{2}{3}\pi).$$

From (3.52), one yields

$$u_{as}=r_si_{as}+\frac{d\psi_{as}}{dt},\quad u_{bs}=r_si_{bs}+\frac{d\psi_{bs}}{dt},\quad u_{cs}=r_si_{cs}+\frac{d\psi_{cs}}{dt},$$

$$\mathbf{u}_{abcs}=\mathbf{r}_s\mathbf{i}_{abcs}+\frac{d\boldsymbol{\psi}_{abcs}}{dt},$$

$$\boldsymbol{\psi}_{abcs}=\begin{bmatrix}\psi_{as}\\ \psi_{bs}\\ \psi_{cs}\end{bmatrix}=\begin{bmatrix}L_{ls}+\bar{L}_m & -\frac{1}{2}\bar{L}_m & -\frac{1}{2}\bar{L}_m\\ -\frac{1}{2}\bar{L}_m & L_{ls}+\bar{L}_m & -\frac{1}{2}\bar{L}_m\\ -\frac{1}{2}\bar{L}_m & -\frac{1}{2}\bar{L}_m & L_{ls}+\bar{L}_m\end{bmatrix}\begin{bmatrix}i_{as}\\ i_{bs}\\ i_{cs}\end{bmatrix}+\psi_m\begin{bmatrix}\sum_n a_n\sin^{2n-1}\theta_r\\ \sum_n a_n\sin^{2n-1}(\theta_r-\frac{2}{3}\pi)\\ \sum_n a_n\sin^{2n-1}(\theta_r+\frac{2}{3}\pi)\end{bmatrix}.\qquad (3.92)$$

One finds the total *emfs* $(\frac{d\psi_{as}}{dt},\frac{d\psi_{bs}}{dt},\frac{d\psi_{cs}}{dt})$ and derives the circuitry-electromagnetic equations. The electromagnetic torque is

$$T_e=\frac{\partial W_c}{\partial\theta_r}=\frac{P\psi_m}{2}\Big[i_{as}\sum_n(2n-1)a_n\cos\theta_r\sin^{2n-2}\theta_r+i_{bs}\sum_n(2n-1)a_n\cos\left(\theta_r-\tfrac{2}{3}\pi\right)\sin^{2n-2}\left(\theta_r-\tfrac{2}{3}\pi\right)$$

$$+i_{cs}\sum_n(2n-1)a_n\cos\left(\theta_r+\tfrac{2}{3}\pi\right)\sin^{2n-2}\left(\theta_r+\tfrac{2}{3}\pi\right)\Big].\qquad (3.93)$$

The Newton second law yields

$$\frac{d\omega_r}{dt}=\frac{P^2\psi_m}{4J}\Big[i_{as}\sum_n(2n-1)a_n\cos\theta_r\sin^{2n-2}\theta_r+i_{bs}\sum_n(2n-1)a_n\cos\left(\theta_r-\tfrac{2}{3}\pi\right)\sin^{2n-2}\left(\theta_r-\tfrac{2}{3}\pi\right)$$

$$+i_{cs}\sum_n(2n-1)a_n\cos\left(\theta_r+\tfrac{2}{3}\pi\right)\sin^{2n-2}\left(\theta_r+\tfrac{2}{3}\pi\right)\Big]-\frac{B_m}{J}\omega_r-\frac{P}{2J}T_L,$$

$$\frac{d\theta_r}{dt}=\omega_r.\qquad (3.94)$$

The balanced current and voltage sets are derived using the expression for T_e (3.93). We have a balanced current set

$$i_{as}=\sqrt{2}i_M\frac{\cos\theta_r}{\sum_n(2n-1)a_n\sin^{2n-2}\theta_r},\quad |i_{as},i_{bs},i_{cs}|\le i_{\max},\qquad (3.95)$$

$$i_{bs}=\sqrt{2}i_M\frac{\cos\left(\theta_r-\frac{2}{3}\pi\right)}{\sum_n(2n-1)a_n\sin^{2n-2}\left(\theta_r-\frac{2}{3}\pi\right)},$$

$$i_{cs}=\sqrt{2}i_M\frac{\cos\left(\theta_r+\frac{2}{3}\pi\right)}{\sum_n(2n-1)a_n\sin^{2n-2}\left(\theta_r+\frac{2}{3}\pi\right)}.$$

The balanced voltage set is

$$u_{as}=\sqrt{2}u_M\frac{\cos\theta_r}{\sum_n(2n-1)a_n\sin^{2n-2}\theta_r},\ |u_{as},u_{bs},u_{cs}|\le u_{\max},\tag{3.96}$$

$$u_{bs}=\sqrt{2}u_M\frac{\cos\left(\theta_r-\frac{2}{3}\pi\right)}{\sum_n(2n-1)a_n\sin^{2n-2}\left(\theta_r-\frac{2}{3}\pi\right)},$$

$$u_{cs}=\sqrt{2}u_M\frac{\cos\left(\theta_r+\frac{2}{3}\pi\right)}{\sum_n(2n-1)a_n\sin^{2n-2}\left(\theta_r+\frac{2}{3}\pi\right)}.$$

The balanced current and voltage sets (3.95) and (3.96) should be implemented. The phase currents and voltages are constrained. Using power MOSFET output stages (usually six- or twelve-step) with hard or soft (*passive* or *active*) switching, one strives to ensure efficiency and consistency. The hardware solutions define the voltage waveforms. Advanced MPUs are used to estimate the unknown a_i and $a_{i,j}$ applying robust algorithms. Nonlinear analysis, conditional logics, look-up tables and other computing tasks are implemented. Within the existing converter topologies it is not always possible to ensure ideal sinusoidal voltage waveforms. The rated solid-state device voltage, current, switching frequency, nonlinearities and latencies affect current and voltage waveforms. We considered nonlinear electromagnetics, control paradigms and hardware solutions to ensure optimal performance.

Example 3.26. *Two-Phase Permanent Magnet Synchronous Motor*

For an optimally designed two-phase permanent magnet synchronous machines,

$$\boldsymbol{\psi}_m=\begin{bmatrix}\psi_{asm}\\ \psi_{bsm}\end{bmatrix}=\begin{bmatrix}\psi_m\sin\theta_r\\ \psi_m\cos\theta_r\end{bmatrix}.$$

The electromagnetic torque is $T_e=\frac{1}{2}P\psi_m\left(\cos\theta_r i_{as}-\sin\theta_r i_{bs}\right)$.

The balanced current set is $i_{as}=i_M\cos\theta_r$, $i_{bs}=-i_M\sin\theta_r$.

Let the flux linkages, established by permanent magnets as viewed from the (*as*,*bs*) windings, are

$$\boldsymbol{\psi}_m=\begin{bmatrix}\psi_{asm}\\ \psi_{bsm}\end{bmatrix}=\begin{bmatrix}\psi_m\sum_n a_n\sin^{2n-1}\theta_r\\ \psi_m\sum_n a_n\cos^{2n-1}\theta_r\end{bmatrix}.$$

The electromagnetic torque is

$$T_e=\frac{1}{2}P\psi_m\left[i_{as}\sum_n(2n-1)a_n\cos\theta_r\sin^{2n-2}\theta_r-i_{bs}\sum_n(2n-1)a_n\sin\theta_r\cos^{2n-2}\theta_r\right].$$

For $a_1\neq0$, $a_2\neq0$ and $\forall a_n=0$, $n>2$, we have

$\psi_{asm}=\psi_m(a_1\sin\theta_r+a_2\sin^3\theta_r)$, $\psi_{bsm}=\psi_m(a_1\cos\theta_r+a_2\cos^3\theta_r)$.

Hence, $T_e=\frac{1}{2}P\psi_m\left[i_{as}\cos\theta_r\left(a_1+3a_2\sin^2\theta_r\right)-i_{bs}\sin\theta_r\left(a_1+3a_2\cos^2\theta_r\right)\right]$.

The phase voltages (u_{as},u_{bs}), which ensure the near-balanced operating conditions, are

$$u_{as}=u_M\frac{\cos\theta_r}{a_1+3a_2\sin^2\theta_r},\ |u_{as}|\le u_{\max},$$

$$u_{bs}=-u_M\frac{\sin\theta_r}{a_1+3a_2\cos^2\theta_r},\ |u_{bs}|\le u_{\max}.$$

If $a_1\gg a_2$, $u_{as}=u_M\cos\theta_r$, $u_{bs}=-u_M\sin\theta_r$. ■

3.6.7. Axial Topology Permanent Magnet Synchronous Motion Devices

In aerial, automotive, biotechnology, consumer electronics, energy, robotics and other applications, axial topology permanent magnet actuators, servos and generators are used. As shown in Figure 3.39.a, the stationary magnetic field is established by magnets placed on the rotor. In motors, the ac phase voltages (u_{as},u_{bs},u_{cs}) are applied to the stator windings as functions of θ_r. In generators, the *emfs* ($\mathscr{E}_{as}$, $\mathscr{E}_{bs}$, $\mathscr{E}_{cs}$) are induced.

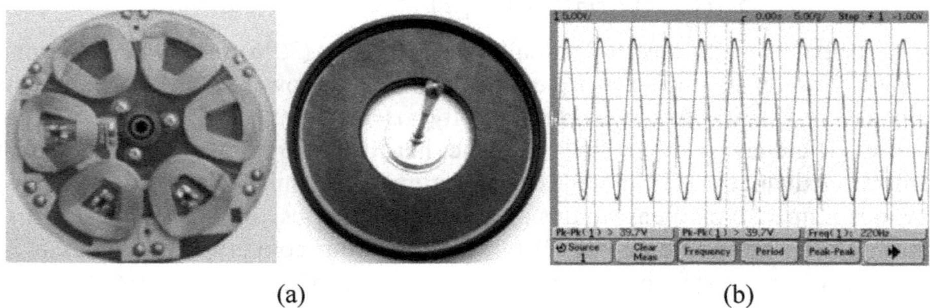

(a) (b)

Figure 3.39. (a) Three-phase axial topology permanent magnet synchronous transducer: Stator with planar windings, and, rotor with axially magnetized through thickness wedge-shaped magnets; (b) Sinusoidal *emf* is induced in the *as* phase which implies a sinusoidal *mmf*.

The advantages of axial topology transducers are affordability, robustness, assembly and packaging simplicity, etc. These benefits are due to: (1) Ferromagnetic materials are not required; (2) No cogging torque; (3) Optimal mechanical and thermal designs, thereby ensuring robustness, ruggedness, favorable heat dissipation, cooling, etc. However, the radial topology transducers ensure higher power and torque densities.

The phase flux linkages (ψ_{as}, ψ_{bs}, ψ_{cs}) vary as a function of θ_r due to the angular displacement of rotor with magnets relative to stator windings. Depending on topology, windings pattern, magnet magnetization and shape, one finds ($B_{as}(\theta_r)$,$B_{bs}(\theta_r)$,$B_{cs}(\theta_r)$) and (ψ_{as}, ψ_{bs}, ψ_{cs}) which are the periodic functions of θ_r, θ_{rm}=2θ_r/P. The number of poles P corresponds to number of magnets N_m.

For an optimal electromagnetic design

$$B_{as}(\theta_r)=B_{\max}\sin(\theta_r),\ B_{bs}(\theta_r)=B_{\max}\sin(\theta_r-\tfrac{2}{3}\pi),\ B_{cs}(\theta_r)=B_{\max}\sin(\theta_r+\tfrac{2}{3}\pi),\quad (3.97)$$

where $B_{\max}$ is the *effective* magnetic field density produced by magnets as viewed from windings, and, $B_{\max}$ depends on the magnets used, magnet-winding separation, number of coil layers, temperature, etc.

The device-dependent *effective* phase flux densities vary, for example, $B_{as}(\theta_r)=B_{\max}\,\mathrm{sgn}(\sin\theta_r)$, $B_{bs}(\theta_r)=B_{\max}\,\mathrm{sgn}\big(\sin(\theta_r-\tfrac{2}{3}\pi)\big)$, $B_{cs}(\theta_r)=B_{\max}\,\mathrm{sgn}\big(\sin(\theta_r+\tfrac{2}{3}\pi)\big)$. Considering nonlinear electromagnetics and exhibited heterogeneities,

$$B_{as}=B_{\max}\sum_n a_{B_n}\sin^{2n-1}\theta_r,\ B_{bs}=B_{\max}\sum_n a_{B_n}\sin^{2n-1}(\theta_r-\tfrac{2}{3}\pi),\ B_{cs}=B_{\max}\sum_n a_{B_n}\sin^{2n-1}(\theta_r+\tfrac{2}{3}\pi).\quad (3.98)$$

The stator is made from non-ferromagnetic materials. The mutual inductances between the planar windings are negligibly small. The magnetic flux crossing a surface is $\Phi=\oint_s \vec{B}\cdot d\vec{s}$. The flux linkages are $\psi=N\oint_s \vec{B}\cdot d\vec{s}$.

The phase flux linkages (ψ_{as}, ψ_{bs}, ψ_{cs}) can be experimentally measured. Figure 3.39.b depicts the motional *emf* $\mathscr{E}_{as}$ generated by a three-phase axial topology generator.

For an ideal sinusoidal *mmf*, from (3.91) $\boldsymbol{\psi}_m = \begin{bmatrix} \psi_{asm} \\ \psi_{bsm} \\ \psi_{csm} \end{bmatrix} = \psi_m \begin{bmatrix} \sin\theta_r \\ \sin\left(\theta_r - \frac{2}{3}\pi\right) \\ \sin\left(\theta_r + \frac{2}{3}\pi\right) \end{bmatrix}$.

In general, consider

$$\begin{bmatrix} \psi_{as} \\ \psi_{bs} \\ \psi_{cs} \end{bmatrix} = \begin{bmatrix} L_{ss} & 0 & 0 \\ 0 & L_{ss} & 0 \\ 0 & 0 & L_{ss} \end{bmatrix} \begin{bmatrix} i_{as} \\ i_{bs} \\ i_{cs} \end{bmatrix} + \psi_m \begin{bmatrix} \sum_n a_n \sin^{2n-1}\theta_r \\ \sum_n a_n \sin^{2n-1}(\theta_r - \frac{2}{3}\pi) \\ \sum_n a_n \sin^{2n-1}(\theta_r + \frac{2}{3}\pi) \end{bmatrix}. \quad (3.99)$$

Using the Kirchhoff second law, the differential equations are

$$u_{as} = r_s i_{as} + \frac{d\psi_{as}}{dt},\ u_{bs} = r_s i_{bs} + \frac{d\psi_{bs}}{dt},\ u_{cs} = r_s i_{cs} + \frac{d\psi_{cs}}{dt}, \quad (3.100)$$

$$\mathbf{u}_{abcs} = \mathbf{r}_s \mathbf{i}_{abcs} + \frac{d\boldsymbol{\psi}_{abcs}}{dt},\ \begin{bmatrix} u_{as} \\ u_{bs} \\ u_{cs} \end{bmatrix} = \begin{bmatrix} r_s & 0 & 0 \\ 0 & r_s & 0 \\ 0 & 0 & r_s \end{bmatrix} \begin{bmatrix} i_{as} \\ i_{bs} \\ i_{cs} \end{bmatrix} + \begin{bmatrix} \frac{d\psi_{as}}{dt} \\ \frac{d\psi_{bs}}{dt} \\ \frac{d\psi_{cs}}{dt} \end{bmatrix}.$$

The electromagnetic torque is

$$T_e = \frac{\partial W_c}{\partial \theta_r} = \frac{P\psi_m}{2}\left[i_{as}\sum_n (2n-1)a_n \cos\theta_r \sin^{2n-2}\theta_r + i_{bs}\sum_n (2n-1)a_n \cos\left(\theta_r - \frac{2}{3}\pi\right)\sin^{2n-2}\left(\theta_r - \frac{2}{3}\pi\right) \right.$$
$$\left. + i_{cs}\sum_n (2n-1)a_n \cos\left(\theta_r + \frac{2}{3}\pi\right)\sin^{2n-2}\left(\theta_r + \frac{2}{3}\pi\right) \right]. \quad (3.101)$$

The *torsional-mechanical* dynamics is given by

$$\frac{d\omega_r}{dt} = \frac{P^2\psi_m}{4J}\left[i_{as}\sum_n (2n-1)a_n \cos\theta_r \sin^{2n-2}\theta_r + i_{bs}\sum_n (2n-1)a_n \cos\left(\theta_r - \frac{2}{3}\pi\right)\sin^{2n-2}\left(\theta_r - \frac{2}{3}\pi\right) \right.$$
$$\left. + i_{cs}\sum_n (2n-1)a_n \cos\left(\theta_r + \frac{2}{3}\pi\right)\sin^{2n-2}\left(\theta_r + \frac{2}{3}\pi\right) \right] - \frac{B_m}{J}\omega_r - \frac{P}{2J}T_L,$$
$$\frac{d\theta_r}{dt} = \omega_r. \quad (3.102)$$

The differential equations are found using (3.100) and (3.102). The balanced current and voltage sets can be derived.

Example 3.27. *Two-Phase Axial Topology Permanent Magnet Synchronous Actuator*

Let $\boldsymbol{\psi}_m = \begin{bmatrix} \psi_{asm} \\ \psi_{bsm} \end{bmatrix} = \begin{bmatrix} \psi_m \sin^5\theta_r \\ \psi_m \cos^5\theta_r \end{bmatrix}$.

The considered $\psi_{asm}=\psi_m\sin^5\theta_r$, $\psi_{bsm}=\psi_m\cos^5\theta_r$ imply $a_3=1$, and, all other $\forall a_n=0$, $a_n\neq 3$ in (3.99). The electromagnetic torque (3.101) is

$$T_e = \frac{\partial W_c}{\partial \theta_r} = \frac{5}{2}P\psi_m\left(i_{as}\cos\theta_r\sin^4\theta_r - i_{bs}\sin\theta_r\cos^4\theta_r\right).$$

Let ψ_m=0.1 N-m/A, P=8 and i_M=2 A.

Consider the phase currents set

$i_{as}=i_M\cos\theta_r$, $i_{bs}=-i_M\sin\theta_r$, i_M=2 A.

Differentiation, computing and plotting are performed using the Symbolic Toolbox. In particular,

```
x=sym('x');   % Symbolic variable
psim=0.1; P=8; iM=2;
psias=psim*sin(x)^5; psibs=psim*cos(x)^5;  % Flux linkages
% Differentiate y1 and y2 using the diff command
dpsias=diff(psias); dpsibs=diff(psibs);
ias=iM*cos(x); ibs=-iM*sin(x);                    % Phase currents
% Derive and plot the electromagnetic torque
Te=P*(dpsias*ias+dpsibs*ibs)/2, Te=simplify(Te)
figure=ezplot(Te,[0 2*pi -0.01 1.01]); set(figure,'LineWidth',3); xlabel('{\it\theta_r}, [rad]','FontSize',20);
legend('{\itT_e}({\it\theta_r})','Location','northeast','FontSize',22);  legend boxoff;
```

The results are

```
Te = 4*cos(x)^2*sin(x)^4 + 4*cos(x)^4*sin(x)^2
Te = 1/2 - cos(4*x)/2
```

The electromagnetic torque is

$$T_e = \tfrac{1}{2}\left(1-\cos 4\theta_r\right) \text{ N-m},$$

or, $$T_e = \tfrac{5}{2} P\psi_m i_M \sin^2\theta_r \cos^2\theta_r \text{ N-m}.$$

Plot for T_e is illustrated in Figure 3.40.a. The electromagnetic torque varies. The torque ripple is undesirable because it results in losses, noise, vibration, etc.

From $T_e = \tfrac{5}{2} P\psi_m\left(i_{as}\cos\theta_r \sin^4\theta_r - i_{bs}\sin\theta_r\cos^4\theta_r\right)$,
the balanced current set is

$$i_{as} = i_M \frac{\cos\theta_r}{\sin^4\theta_r}, |i_{as}| \le i_{\max},$$

$$i_{bs} = -i_M \frac{\sin\theta_r}{\cos^4\theta_r}, |i_{bs}| \le i_{\max}.$$

If the current limits are not considered, one has

```
ias=iM*cos(x)/sin(x)^4; ibs=-iM*sin(x)/cos(x)^4;  % Phase currents
% Derive and plot the electromagnetic torque
Te=P*(dpsias*ias+dpsibs*ibs)/2, Te=simplify(Te)
figure=ezplot(Te,[0 2*pi 0 5]); set(figure,'LineWidth',3); xlabel('{\it\theta_r}, [rad]','FontSize',20);
legend('{\itT_e}({\it\theta_r})','Location','northeast','FontSize',22);  legend boxoff;
```

We obtain

```
Te = 4*cos(x)^2 + 4*sin(x)^2
Te = 4
```

That is, T_e=4 N-m as plotted in Figure 3.40.b. Due to singularity and limits, it is not possible to implement the balanced current and voltage sets. For

$$i_{as} = i_M \frac{\cos\theta_r}{\sin^4\theta_r},\ i_{bs} = -i_M \frac{\sin\theta_r}{\cos^4\theta_r},$$

the inherent PWM controllers-drivers limits on the phase currents are imposed as $|i_{as}| \le i_{\max}$, $|i_{bs}| \le i_{\max}$. Hence, there will be torque ripple, vibration and losses.

Optimally designed transducers should be used because the electromagnetic and control solutions may not always overcome the device selection and hardware limits.

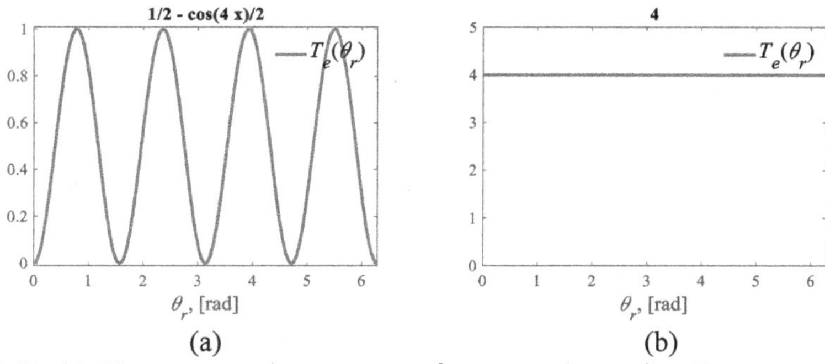

Figure 3.40. (a) Electromagnetic torque $T_e = \frac{5}{2} P\psi_m i_M \sin^2\theta_r \cos^2\theta_r$ N-m;
(b) Electromagnetic torque for a balanced current set is T_e=4 N-m, assuming no limits on the phase currents. ■

3.7. Operational Amplifiers and Analog Electronics

3.7.1. Analog Signal Processing

Physical systems, components and devices are continuous. Sensors measure physical quantifies to be fused into data acquisition, management and control systems. The fused information supports coordination, control and functionality. The laws of physics yield differential equations for continuum mechanics, electromagnetics, thermodynamics, etc. Analog signal processing ensures consistency, accuracy and low power [3, 14, 17, 18]. Complex physical systems necessitate digital systems due to simplicity in integration and implementation of digital processing and computing schemes, interface compliance, etc. The analog-to-digital and digital-to-analog conversions are aimed to ensure digital control and data acquisition. Digital signal processing tasks are performed by MPUs and FPGAs. In high-degree-of-freedom aerial systems, the flight control computer and flight management system are digital due to compliance, interconnectivity and interoperability of various components and modules.

In many systems, analog signal processing, filtering and control are performed using analog signals. Addition, multiplication, scaling and other operations, as well as various integral transforms (Fourier, Hankel, Hilbert, Laplace and others) are performed in time and frequency domains. One implements analog controllers and filters using operational amplifiers. Low power electronics ensures compliancy, effectiveness and simplicity. Analog computers and processing circuits are implemented using multipliers, summers, comparators, integrators, differentiators and signal generators, realized by operational amplifiers.

Consider operational amplifies to implement analog controllers and filters. The signal-level sensor outputs are filtered to attenuate noise and minimize measurement heterogeneities. Analog filters and controllers implementation, signal processing and filter design are considered. Various physical quantities (displacement, velocity, acceleration, force, torque, stress, strain, pressure, temperature, field intensity and density, and others) can be measured. The sensors usually convert the physical quantities to voltage, current or charge. There are numerous high-performance on a single-chip, *smart* and multi-chip sensors commercialized by Analog Devices, Bosh, Honeywell, Emerson, Infineon Technologies, Siemens, Sony, STMicroelectronics, Texas Instruments, etc.

A single operational amplifier has *noninverting* and *inverting* inputs (pins 3 and 2) as well as an output terminal (pin 6), see Figure 3.41.a. The dc voltage is supplied. The terminal 7 is connected to a positive voltage u_{+}, while terminal 4 is connected to a negative voltage u_{-} or ground. The pin connections of the single, dual and quad low-power operational amplifiers MC33171, MC33172 and MC33174 are reported in Figure 3.41.a. There are various packages and configurations. Operational amplifiers, which consist of dozens of FETs, are fabricated using the CMOS and biCMOS technologies. Figure 3.41.a depicts the representative schematics and transient responses. There are general-purpose, instrumental, precision, high-speed, differential, power and other operational amplifiers shown in Figure 3.41.b. Using analog semiconductor devices and operational amplifiers, one may perform summation, subtraction, multiplication and division of input signals. Four-quadrant multipliers-dividers AD534JDZ and AD734 are shown in Figure 3.41.c.

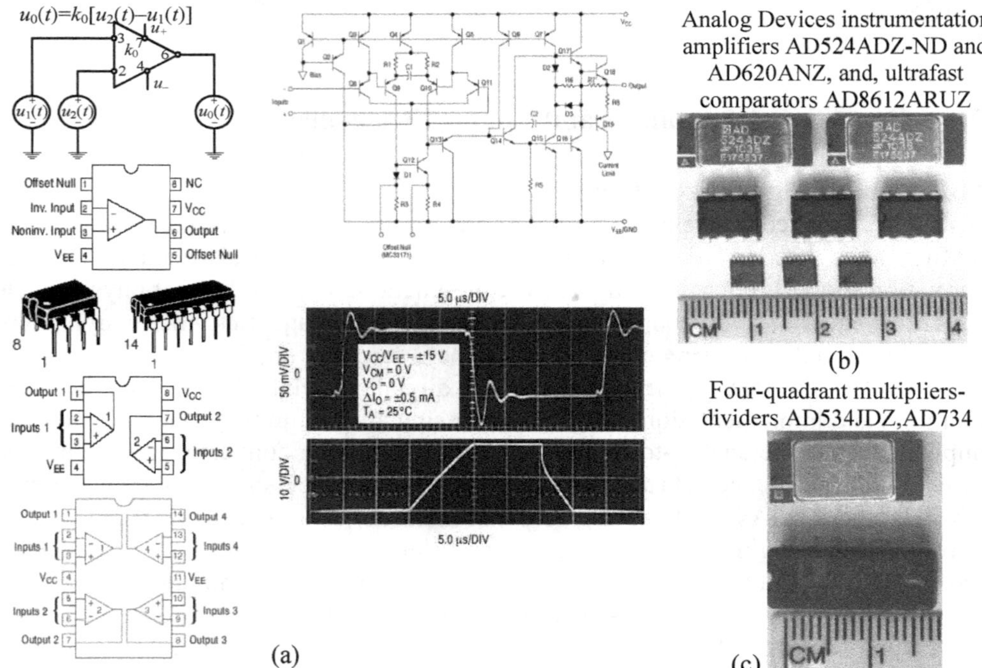

Figure 3.41. (a) General-purpose operational amplifiers, pin connections, packages, schematics, and transient responses, copyright of Motorola, used with permission [7];
(b) Images of Analog Devices 16-CDIP package instrumentation amplifiers AD524ADZ-ND, low-power instrumentation amplifiers AD620ANZ, and, 14-lead ultrafast 4 nsec single supply comparators AD8612ARUZ;
(c) Images of four-quadrant multipliers-dividers AD534JDZ and AD734 with a multiplication error ±0.25%. Summation, subtraction, division and multiplication ($XY+Z$) operations can be implemented. The AD734AQ is a high speed (10 MHz bandwidth, 200 nsec settling time) four-quadrant analog multiplier which performs a division XY/Z with –80 dB distortion. Low capacitance (X,Y,Z) inputs are differential.

The operational amplifier output is the difference between two input voltages $[u_1(t)-u_2(t)]$ applied to the *inverting* input and the *noninverting* input, multiplied by the differential open-loop coefficient k_0. The output voltage is $u_0(t)=k_0[u_2(t)-u_1(t)]$, the gain k_0 is positive, $k_0 \in [1\times10^5\ 1\times10^7]$. The general purpose operational amplifiers have input and output resistances $[1\times10^5\ 1\times10^{12}]$ and [10 1000] ohm, respectively. The *inverting* and *noninverting* input terminals are denoted as "−" and "+", see Figure 3.42.a. Supplying the signal-level input voltage $u_1(t)$ to the *inverting* input terminal using external resistor R_1, grounding the *noninverting* input terminal, and, using the feedback with the external resistor R_2, the output is $u_0 = -\frac{R_2}{R_1}u_1$. The inverting weighted summer is shown in Figure 3.42.b. The instrumentational amplifiers are a preferable choice. The image of the instrumentation amplifier AD524ADZ-ND is documented in Figure 3.42.c.

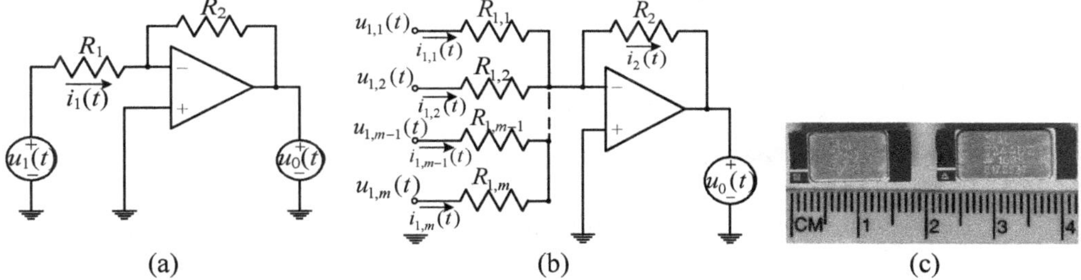

Figure 3.42. (a) Inverting configuration of the operational amplifier;
(b) Summing amplifier with m inputs. The current in the feedback path is $i_2(t)=i_{1,1}(t)+\ldots+i_{1,m}(t)$, $i_{1,1}(t)=u_{1,1}(t)/R_{1,1}$, …, $i_{1,m}(t)=u_{1,m}(t)/R_{1,m}$. Amplifier output is $u_0(t) = -\left(\frac{R_2}{R_{1,1}}u_{1,1}(t)+\ldots+\frac{R_2}{R_{1,m}}u_{1,m}(t)\right)$;
(c) Analog Devices 16-CDIP package instrumentation amplifier AD524ADZ-ND.

The input impedance $Z_1(s)$ and feedback impedance $Z_2(s)$ are illustrated in Figure 3.43.a. Using the frequency domain, $Z_1(s) = Z_1(j\omega)$ and $Z_2(s) = Z_2(j\omega)$. The impedance is the ratio of the phasor voltage to the phasor current.

The impedances of the resistor, capacitor and inductor are:

1. Resistor $Z_R(s) = R,\ Z_R(j\omega) = R$;
2. Capacitor $Z_C(s) = \frac{1}{sC},\ Z_C(j\omega) = \frac{1}{j\omega C} = -\frac{j}{\omega C}$;
3. Inductor $Z_L(s) = sL,\ Z_L(j\omega) = j\omega L$.

The transfer function of the closed-loop amplifier configuration is

$$G(s) = \frac{U_0(s)}{U_1(s)} = -\frac{Z_2(s)}{Z_1(s)}. \tag{3.103}$$

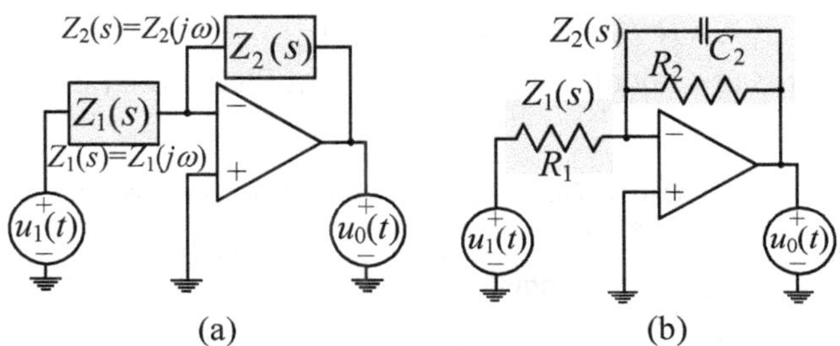

Figure 3.43. (a) Operational amplifier with $Z_1(s)$ and $Z_2(s)$, $G(s) = -\frac{Z_2(s)}{Z_1(s)}$;
(b) Inverting operational amplifier with input and feedback impedances $Z_1(s) = R_1$ and $Z_2(s) = \frac{R_2}{R_2C_2s+1}$, and, transfer function $G(s) = -\frac{Z_2(s)}{Z_1(s)} = -\frac{R_2}{R_1}\frac{1}{(R_2C_2s+1)}$.

Example 3.28. *First- and Second-Order Low-Pass Filters*
For the operational amplifier, shown in Figure 3.43.b, the transfer function is found.

From $Z_1(s)=R_1$ and $Z_2(s)=\dfrac{R_2}{R_2C_2s+1}$, we have

$$G(s)=\frac{U_0(s)}{U_1(s)}=-\frac{Z_2(s)}{Z_1(s)}=-\frac{R_2/R_1}{R_2C_2s+1}.$$

The gain is R_2/R_1, while the time constant is R_2C_2.

In the frequency domain, substituting $s=j\omega$, we have

$$G(j\omega)=\frac{U_0(j\omega)}{U_1(j\omega)}=-\frac{Z_2(j\omega)}{Z_1(j\omega)}=-\frac{R_2}{R_1}\frac{1}{(j\omega R_2C_2+1)}.$$

Consider a low-pass filter $G(j\omega)=\dfrac{R_2}{R_1}\dfrac{1}{(j\omega R_2C_2+1)}$.

Let the filter low-frequency gain and cut-off frequency are specified to be $k=1$ and $\omega_c=1000$ rad/s.

For frequency $\omega>\omega_c$, the attenuation of a sinusoidal input is found using $|G|_{\text{dB}}$.

For the first-order low-pass filter, $k=|G|_{dB}\big|_{\omega<\omega_c}$, $\Delta|G|_{dB}\big|_{\omega=\omega_c}=-3$ dB.

The slope of $|G|_{\text{dB}}$ is –20 dB/dec for $\omega>\omega_c$.

The assigned pass band gain $k=1$ implies $R_1=R_2$. For $\omega_c=1000$ rad/s, we have $\omega_c=1/R_2C_2=1000$ rad/s.

For $R_1=R_2$, $R_1=1\times10^5$ ohm, $R_2=1\times10^5$ ohm, one calculates $C_2=1\times10^{-8}$ F. The high-frequency noise is attenuated.

The Bode plots are calculated and plotted in Figure 3.44.a using MATLAB statements

```
R1=1e5; R2=1e5; C2=1e-8; num=[R2/R1]; den=[R2*C2 1]; bode(num,den,'k')
```

There are multi-frequency multisource noise to be attenuated. For a series configuration of two first-order filers

$$G(s)=\frac{R_2^2}{R_1^2}\frac{1}{(R_2C_2s+1)}\frac{1}{(R_2C_2s+1)}.$$

For the second-order low-pass filter, the slope of $|G|_{\text{dB}}$ is –40 dB/dec for $\omega>\omega_c$.

For the gain $k=1$ and cut-off frequency $\omega_c=1000$ rad/s, $R_1=R_2=1\times10^5$ ohm and $C_2=1\times10^{-8}$ F. From $G(j\omega)=\dfrac{R_2^2}{R_1^2}\dfrac{1}{(j\omega R_2C_2+1)}\dfrac{1}{(j\omega R_2C_2+1)}$, one finds the magnitude $|G|_{\text{dB}}$ and phase ϕ. The Bode plots are calculated and depicted in Figure 6.44.b.

```
R1=1e5; R2=1e5; C2=1e-8;
G=@(s) ((R1/R2)./(R2*C2*s+1)).*((R1/R2)./(R2*C2*s+1)); % Second-order low-pass filter, G(s)
omega=2*pi*logspace(0,4,1000); % Frequency
magG=abs(G(j*omega)); magGDB=20*log10(magG); % Computing magnitude
phaseGdeg=rad2deg(angle(G(j*omega))); % Computing phase
BodePlot=figure(1);MG=subplot(2,1,1,'Parent',BodePlot);PG=subplot(2,1,2,'Parent',BodePlot);%Plots
semilogx(MG, omega/2/pi, magGDB), semilogx(PG, omega/2/pi, phaseGdeg)
MG.XGrid='on'; MG.YGrid='on'; PG.XGrid='on'; PG.YGrid='on';
MG.XLabel.String='Frequency {\it\omega}, [rad/s]'; MG.YLabel.String='Magnitude [dB]';
PG.XLabel.String='Frequency {\it\omega}, [rad/s]'; PG.YLabel.String='Phase [°]';
```

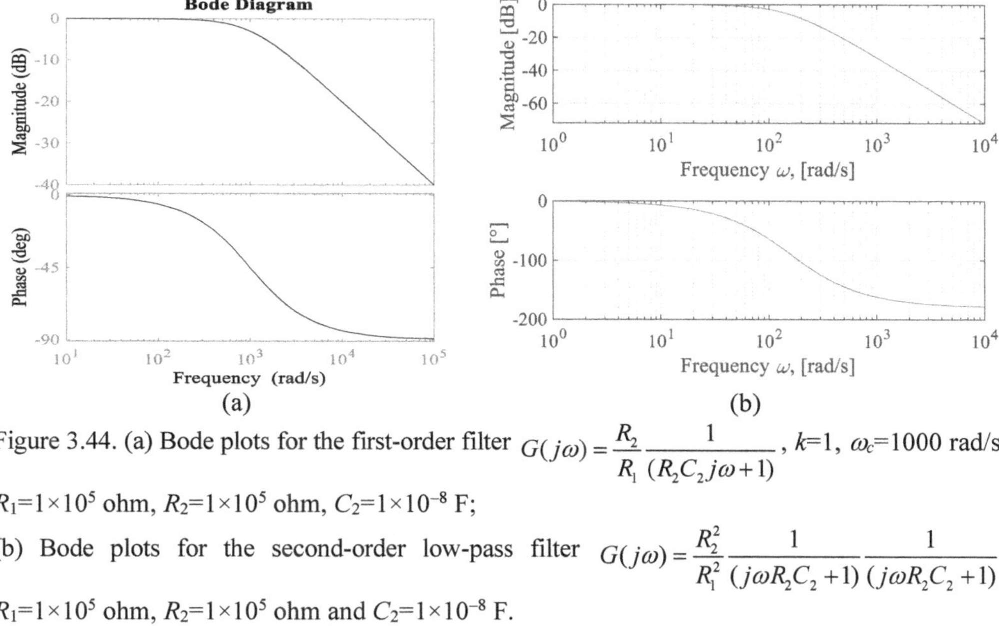

(a) (b)

Figure 3.44. (a) Bode plots for the first-order filter $G(j\omega)=\frac{R_2}{R_1}\frac{1}{(R_2C_2j\omega+1)}$, k=1, ω_c=1000 rad/s, $R_1=1\times10^5$ ohm, $R_2=1\times10^5$ ohm, $C_2=1\times10^{-8}$ F;

(b) Bode plots for the second-order low-pass filter $G(j\omega)=\frac{R_2^2}{R_1^2}\frac{1}{(j\omega R_2C_2+1)}\frac{1}{(j\omega R_2C_2+1)}$, $R_1=1\times10^5$ ohm, $R_2=1\times10^5$ ohm and $C_2=1\times10^{-8}$ F.

■

Analog filters and controllers are implemented using operational amplifiers. The transfer functions of filters and control laws are found, and, physical implementation is considered. For commonly used input and feedback impedances $Z_1(s)$ and $Z_2(s)$, Table 3.1 reports transfer functions of the inverting operational amplifier configurations illustrated in Figure 3.43.a. High-order filters can be implemented using a single operational amplifier as studied in Example 3.33.

3.7.2. Analog Filters: Design and Implementation

Operational amplifiers perform summation, integration, differentiation and other operations in time and frequency domains. One physically implements the *error amplifiers*, filters, controllers, signal conditioning operations, etc. Signals are perturbed by noise of different origins due to device-physics heterogeneity, thermal fluctuations, shot noise, electromagnetic interferences, etc. Low, medium and high frequency noise can be attenuated by filters. System bandwidth, frequency envelope, noise characteristics, robustness, immunity, compliance and other quantities are considered [14, 18, 19]. The pertained studies are covered in sections 5.12, 5.13 and 5.14.

Using the system bandwidth, noise power spectral density, sensitivity and physical controllability, find system dynamic modes and frequencies ($\mathbf{M}$,Ω) to preserve, and, noise modalities and frequencies ($\mathbf{M}_n$,Ω_n) to attenuate. Finding the pass and stop band frequencies, attenuation and admissible ripple, elliptical, Butterworth, Chebyshev, Bessel, Cauer, notch and other filters can be used.

Table 3.1. Transfer Functions of the Inverting Operational Amplifier Configurations

Input impedance $Z_1(s)$	Feedback impedance $Z_2(s)$	Transfer function
$Z_1(s)=R_1$ 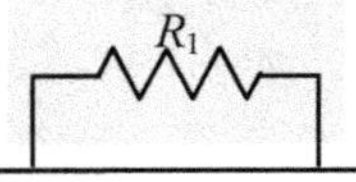	$Z_2(s)=R_2$ 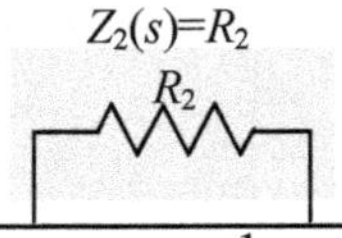	Gain $G(s)=\frac{U_0(s)}{U_1(s)}=-\frac{R_2}{R_1}$
$Z_1(s)=R_1$ 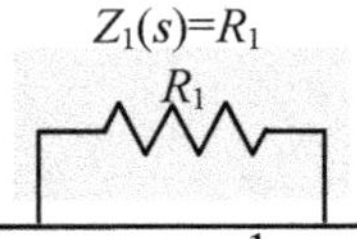	$Z_2(s)=\frac{1}{C_2 s}$ 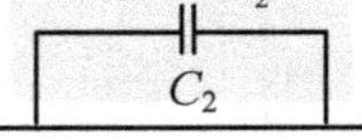	Integrator $G(s)=\frac{U_0(s)}{U_1(s)}=-\frac{1}{R_1 C_2 s}$
$Z_1(s)=\frac{1}{C_1 s}$ 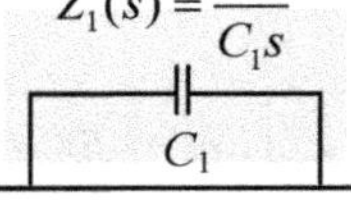	$Z_2(s)=R_2$ 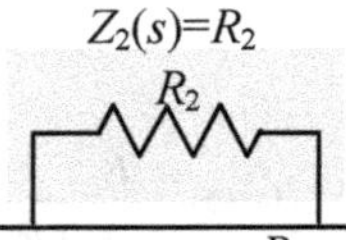	Differentiator $G(s)=\frac{U_0(s)}{U_1(s)}=-R_2 C_1 s$
$Z_1(s)=R_1$ 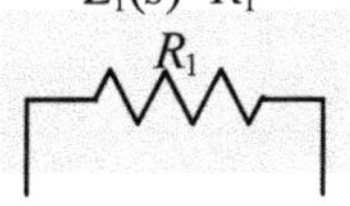	$Z_2(s)=\frac{R_2}{R_2 C_2 s+1}$ 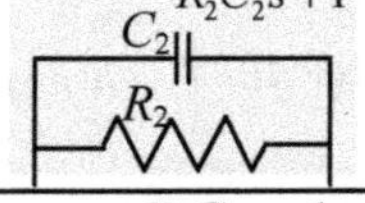	First-order low-pass filter $G(s)=\frac{U_0(s)}{U_1(s)}=-\frac{R_2}{R_1}\frac{1}{(R_2 C_2 s+1)}$
$Z_1(s)=R_1$ 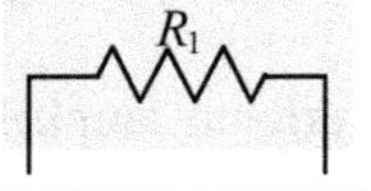	$Z_2(s)=\frac{R_2 C_2 s+1}{C_2 s}$ 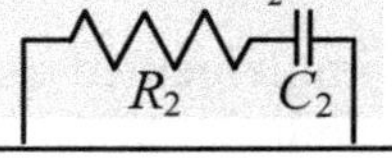	Proportional-integral controller $G(s)=\frac{U_0(s)}{U_1(s)}=-\frac{R_2 C_2 s+1}{R_1 C_2 s}$
$Z_1(s)=\frac{R_1}{R_1 C_1 s+1}$ 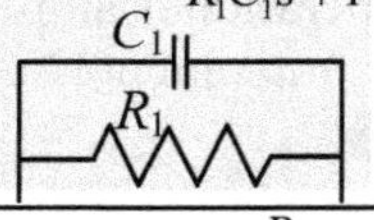	$Z_2(s)=R_2$ 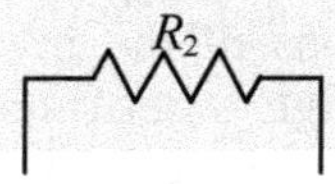	Proportional-derivative controller $G(s)=\frac{U_0(s)}{U_1(s)}=-\frac{R_2(R_1 C_1 s+1)}{R_1}$
$Z_1(s)=\frac{R_1}{R_1 C_1 s+1}$ 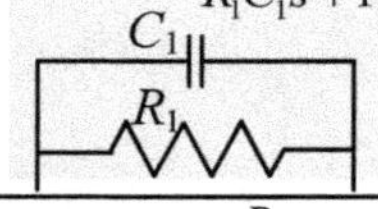	$Z_2(s)=\frac{R_2 C_2 s+1}{C_2 s}$ 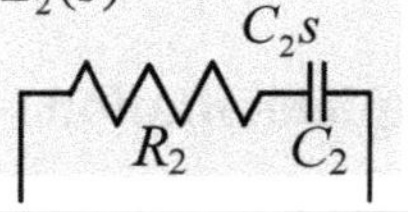	Proportional-integral-derivative controller $G(s)=\frac{U_0(s)}{U_1(s)}=-\frac{(R_1 C_1 s+1)(R_2 C_2 s+1)}{R_1 C_2 s}$
$Z_1(s)=\frac{R_1}{R_1 C_1 s+1}$ 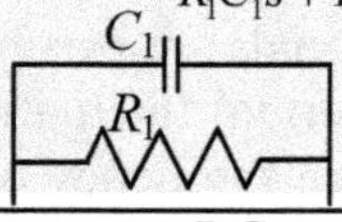	$Z_2(s)=\frac{R_2}{R_2 C_2 s+1}$ 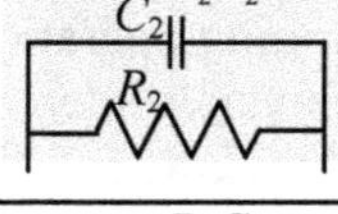	First-order notch filter $G(s)=\frac{U_0(s)}{U_1(s)}=-\frac{R_2}{R_1}\frac{(R_1 C_1 s+1)}{(R_2 C_2 s+1)}$
$Z_1(s)=\frac{R_1 C_1 s+1}{C_1 s}$ 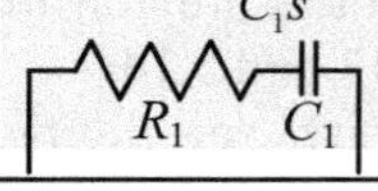	$Z_2(s)=\frac{R_2 C_2 s+1}{C_2 s}$ 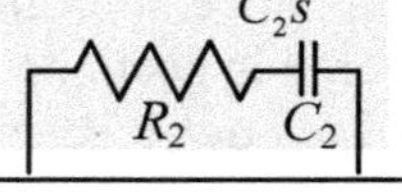	First-order notch filter $G(s)=\frac{U_0(s)}{U_1(s)}=-\frac{C_1}{C_2}\frac{(R_2 C_2 s+1)}{(R_1 C_1 s+1)}$

Design Considerations – Dynamic systems are studied in time and frequency domains, considering dynamic $\mathbf{M}(t)$ and frequency Ω modes $(\mathbf{M},\Omega)$. The frequency domain analysis is applied to accomplish linear filters design, while time domain is applied to analyze nonlinear systems. Filters are designed to attenuate complex multisource random noise, deterministic disturbances, interferences and electromagnetic spectrum of different origins. The measured, transmitted and fused physical quantities $\mathbf{y}(t)$ are distorted by multisource distortions $\boldsymbol{n}(t)=f(\boldsymbol{n}_i)$, characterized by the dynamic and frequency modalities $(\mathbf{M}_n,\Omega_n)$. One designs filters evaluating:

- System frequencies and system dynamic modes to preserve;
- Noise and distortions frequencies and dynamic modalities to attenuate;
- Measured and fused signals in sensors and ASICs, characterized by noise power spectral density, frequency, signal-to-noise ratio, etc.

Problem Formulation – The overall objective is to enable physical data quality, enhance data homogeneity, minimize data heterogeneity and reduce information losses. Among various solutions and schemes, filters design is an important task. For a given physical system, we design filters $\mathcal{F}_{SC}(\boldsymbol{S},\boldsymbol{C})\in\mathbf{C}_{\mathcal{F}}$ on a specified filter class $\mathbf{C}_{\mathcal{F}}$ (analog, finite impulse response, infinite impulse response digital, autoregressive and others), finding the filter structure $\boldsymbol{S}$ (order, dynamically reconfigurable, adaptive feedforward and feedback compensation schemes, etc.) and coefficients $\boldsymbol{C}$ (time invariant or varying). Explicitly defining homogeneity and heterogeneity measures $(\mathcal{M}_{\mathcal{H}},\mathcal{M}_{\mathcal{L}})$, the problem is to maximize $\mathcal{M}_{\mathcal{H}}$ and minimize $\mathcal{M}_{\mathcal{L}}$ by using the *design variables* $(\mathbf{C}_{\mathcal{F}},\boldsymbol{S},\boldsymbol{C})$.

Frequency spectrum $\mathrm{N}(\omega)=\int_{-\infty}^{\infty}\boldsymbol{n}(t)e^{-j\omega t}dt$, $\mathrm{N}[k]=\sum_{n=0}^{N-1}\boldsymbol{n}[n]e^{-j\frac{2\pi}{N}kn}$, as well as power spectral density of continuous $\boldsymbol{n}(t)$ and discrete $\boldsymbol{n}[n]$, are considered.

In the frequency domain, filters

$$\mathcal{F}_{SC}(\boldsymbol{S},\boldsymbol{C})\equiv G_{M,N,a,b}(s),\ s=j\omega,\ G(s)=\frac{N(s)}{D(s)}=\frac{\sum_{i=0}^{m}b_i s^i}{\sum_{i=0}^{n}a_i s^i}$$

are characterized by:

1. Transfer function order, and, degree of numerator $N(s)$ and denominator $D(s)$, $\deg(D(s))\geq\deg(N(s))$, characterized by real and complex conjugate pairs of zeros and characteristic poles;
2. Magnitude of frequency response, characterized by pass band gain k and slope $|G(j\omega)|$;
3. Pass- and stop-band edge frequencies ω_p and ω_s with $|G(j\omega)|_{\omega_p\in\Omega_p}$, $|G(j\omega)|_{\omega_s\in\Omega_s}$, including the admissible ripples, attenuations and distortions of $|G(j\omega)|$;
4. Cut-off frequency ω_c, allowed $|G(j\omega)|_{\omega_c\in\Omega_c}$, and, $|G(j\omega)|$ attenuation at ω_c;
5. Phase $\phi(\omega)$ delay and distortions.

The cut-off frequency ω_c, as well as the pass- and stop-band edge frequencies (ω_p,ω_s), are found examining the system and noise frequency envelopes investigating $(\omega_p,\omega_c,\omega_s)\in\Omega_n\times\Omega$. Low-pass, high-pass, band-pass and band-stop filters with corresponding transfer functions are designed.

The elliptical, Chebyshev, Bessel, Cauer and others filters are used [19]. The low-pass Butterworth and notch filters guarantee no pass-band and no stop-band ripples. These filters ensure the preferable overdamped transient response. In the Butterworth and notch filters, the magnitude $|G(j\omega)|$ and phase $\phi(\omega)$ are the monotonically decreasing or monotonically decreasing functions of ω at all frequencies.

Butterworth and Chebyshev Filters – The n-degree Butterworth and Chebyshev filter transfer functions are

$$G(s)=\frac{k_0}{B_n(\frac{1}{\omega_c}s)},\quad G(s)=\frac{k_0}{\varepsilon C_n(\frac{1}{\omega_c}s)},$$

where B_n and C_n are the n-degree Butterworth and Chebyshev polynomials, normalized for ω_c=1; ε is the ripple parameter.

Hence, $|G(j\omega)|^2=\frac{1}{1+B_n^2(\frac{\omega}{\omega_c})^{2n}}$ and $|G(j\omega)|^2=\frac{1}{1+\varepsilon^2 C_n^2(\frac{\omega}{\omega_c})^{2n}}$, k_0=1.

For low-pass filters, the pass band is $0\le\omega\le\omega_p$.

For the Butterworth filter, $1+\left(\frac{\omega_p}{\omega_c}\right)^{2n}=10^{-\frac{1}{10}|G_p|}$, $1+\left(\frac{\omega_s}{\omega_c}\right)^{2n}=10^{-\frac{1}{10}|G_s|}$.

The filter order M is derived by computing $m=\frac{1}{2\ln\left(\frac{\omega_s}{\omega_p}\right)}\ln\left(\frac{10^{-\frac{1}{10}|G_s|}-1}{10^{-\frac{1}{10}|G_p|}-1}\right)$.

One finds the integer M, $M\ge m$.

For a Chebyshev filter, $m=\frac{1}{\cosh^{-1}\left(\frac{\omega_s}{\omega_p}\right)}\cosh^{-1}\sqrt{\frac{10^{-\frac{1}{10}|G_s|}-1}{10^{-\frac{1}{10}|G_p|}-1}}$.

The specifications yield the filters order. The tabulated transfer functions are found.

Example 3.29. *Butterworth Filters*

For notch and Butterworth filters, the magnitude $|G(j\omega)|$ and phase $\phi(\omega)$ are monotonically decreasing or monotonically increasing functions at all frequencies. One specifies: (1) The pass band gain $|G|_{\text{dB max}}$ at the pass band frequency ω_p; (2) The minimum stop band gain $|G|_{\text{dB min}}$ at the stop band frequency ω_s.

The gain, attenuations and frequencies ($\omega_p, \omega_c, \omega_s$) define the filter order. Using the n-degree Butterworth polynomials

$$B_n(s)=\begin{cases}\prod_{k=1}^{\frac{n}{2}}\left[s^2-2s\cos\left(\frac{2k+n-1}{2n}\pi\right)+1\right], & n \text{ even}\\ (s+1)\prod_{k=1}^{\frac{n-1}{2}}\left[s^2-2s\cos\left(\frac{2k+n-1}{2n}\pi\right)+1\right], & n \text{ odd}\end{cases},$$

the filter transfer function, normalized for ω_c=1, is $G(s)=\frac{k_0}{B_n(\frac{1}{\omega_c}s)}$.

The high-order filters are formed by cascading in series the first-, second- and high-order stages. For example, three second-order low-pass filters can be cascaded to physically implement a sixth-order filter. The normalized n-degree Butterworth polynomials are

$B_1(s)=(s+1)$, $B_2(s)=(s^2+1.414s+1)$, $B_3(s)=(s+1)(s^2+s+1)$,
$B_4(s)=(s^2+0.765s+1)(s^2+1.848s+1)$, $B_5(s)=(s+1)(s^2+0.618s+1)(s^2+1.618s+1)$,
$B_6(s)=(s^2+0.518s+1)(s^2+1.414s+1)(s^2+1.932s+1)$,
$B_7(s)=(s+1)(s^2+0.445s+1)(s^2+1.247s+1)(s^2+1.802s+1)$,
$B_8(s)=(s^2+0.39s+1)(s^2+1.111s+1)(s^2+1.663s+1)(s^2+1.962s+1)$, etc.

The Bode plots for $G(s)$ with $B_n(s)$, $n=1,\ldots,6$ of the normalized Butterworth filters with $k_0=100$ are documented in Figure 3.45.a.

Specify the filter order $n=2$ and cut-off frequency $f_c=1000$ Hz, $\omega_c=2\pi f_c$. Filter is designed using the MATLAB commands.

```
n=2; f=1000; [num,den]=butter(n,2*pi*f,'low','s'); filter=tf(num,den), bode(num,den,'k')
```

The transfer function is $G(s)=\dfrac{3.95\times10^7}{s^2+8886s+3.95\times10^7}$.

The Bode plots are depicted in Figure 3.45.b.

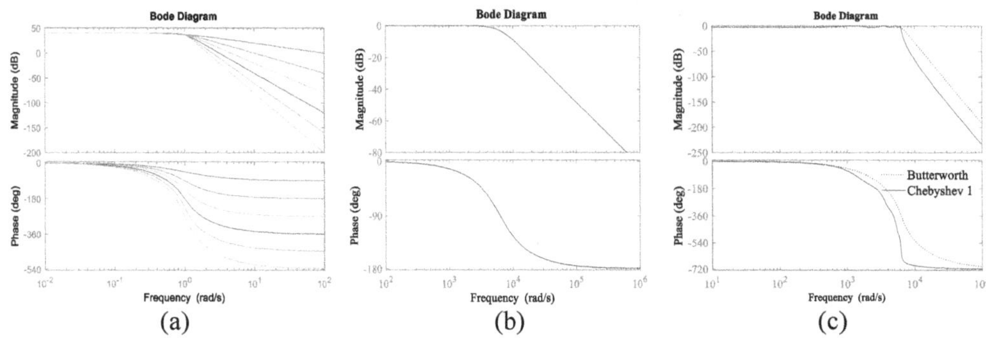

(a) (b) (c)

Figure 3.45. (a) Bode plots for the Butterworth filters with $G(s)=\dfrac{k_0}{B_n(\frac{1}{\omega_c}s)}$, $k_0=100$, $\omega_c=1$ rad/s, $n=1$ to $n=6$;
(b) Bode plots for the second-order filter, $G(s)=\dfrac{3.95\times10^7}{s^2+8886s+3.95\times10^7}$;
(c) Bode plots for the Butterworth and Chebyshev filters, $M=8$ and $f_c=1000$ Hz, Example 3.30.■

Example 3.30. *Filters Design and Analysis*

Consider the M-order unit-gain low-pass Butterworth filter

$$|G(j\omega)|^2=\frac{1}{1+(\omega/\omega_c)^{2M}}=\frac{1}{1+\Delta_p^2(\omega/\omega_p)^{2M}}.$$

At the pass- and stop-band frequencies (ω_p,ω_s), the specified attenuations, characterized by the real-valued (Δ_p,Δ_s), are $|G(j\omega_p)|^2=\dfrac{1}{1+\Delta_p^2}$ and $|G(j\omega_s)|^2=\dfrac{1}{\Delta_s^2}$.

For $(\omega_p,\omega_c,\omega_s)$, $(|G(j\omega_p)|^2, |G(j\omega_s)|^2)$ and (Δ_p,Δ_s), one finds the filter order $M\geq m$ as

$$(\omega_p/\omega_c)^{2M}=\Delta_p^2=\Delta_s^2-1,\ m=\frac{\log(\Delta_p/\sqrt{\Delta_s^2-1})}{\log(\omega_p/\omega_s)}.$$

The specifications are the latency, phase lag, distortion, complexity and sensitivity. The Bode plots for the Butterworth and Chebyshev filters, designed in MATLAB by specifying $M=8$ and cut-off frequency 1000 Hz, are depicted in Figure 3.45.c.

```
M=8; f=1000; [zb,pb,kb]=butter(M,2*pi*f,'s'); % Butterworth filter
[Nb,Db]=zp2tf(zb,pb,kb); bode(Nb,Db,'b:'); hold on
[zc,pc,kc]=cheby1(M,3,2*pi*f,'s'); % Chebyshev type 1 with 3 dB pass band ripple
[Nc,Dc]=zp2tf(zc,pc,kc); bode(Nc,Dc,'k'); hold off;
legend('Butterworth','Chebyshev 1', 'Fontsize',16); legend boxoff;
```

■

Example 3.31. *First-Order Notch Filters*

Consider a schematics, depicted in Figure 3.46.a.

The impedances are $Z_1(s)=\dfrac{R_1}{R_1C_1s+1}$ and $Z_2(s)=\dfrac{R_2}{R_2C_2s+1}$.

One finds a transfer function of the first-order notch filter as

$G(s)=\dfrac{U_0(s)}{U_1(s)}=-\dfrac{Z_2(s)}{Z_1(s)}$, $G(s)=\dfrac{N(s)}{D(s)}=\dfrac{R_2}{R_1}\dfrac{(R_1C_1s+1)}{(R_2C_2s+1)}$, deg($D(s)$)=1, deg($N(s)$)=1.

With a band-stop phase $\phi\to 0$ as $\omega\to\infty$. The filter has a zero $s=-1/(R_1C_1)$ and pole $s=-1/(R_2C_2)$. Let the system bandwidth is 1000 Hz, which implies 0.001 sec settling time. In engineering design, the bandwidth is reciprocal of settling time. The 100000 Hz noise should be attenuated at least 10 times. At frequencies within the system bandwidth, a unit gain k=1 should be ensured. Hence, $k=R_2/R_1=1$.

From $\omega=2\pi f$, find the cut-off frequencies $\omega_{ci}=1/R_iC_i$.

For numerator and denominator, $\omega_{cN}=\dfrac{1}{R_1C_1}$ and $\omega_{cD}=\dfrac{1}{R_2C_2}$.

At the corner frequencies ω_{ci}, the attenuation is $\Delta|G|_{\text{dB}}=-3$ dB of the nominal passband value. To ensure k=1 up to f=1000 Hz, $R_1=R_2$, $\omega_{cD}=1\times10^4$ rad/s, and, $\omega_{cN}=1\times10^7$ rad/s. Hence, R_1=1000 ohm, R_2=1000 ohm, $C_1=1\times10^{-10}$ F and $C_2=1\times10^{-7}$ F. The Bode plots are illustrated in Figure 3.46.a, and, design specifications are met.

For a filter, reported in Figure 3.46.b, $Z_1(s)=\dfrac{R_1C_1s+1}{C_1s}$ and $Z_2(s)=\dfrac{R_2C_2s+1}{C_2s}$.

The first-order notch filter transfer function is $G(s)=\dfrac{N(s)}{D(s)}=\dfrac{C_1}{C_2}\dfrac{(R_2C_2s+1)}{(R_1C_1s+1)}$.

One finds (R_i,C_i) values to meet the design specifications.

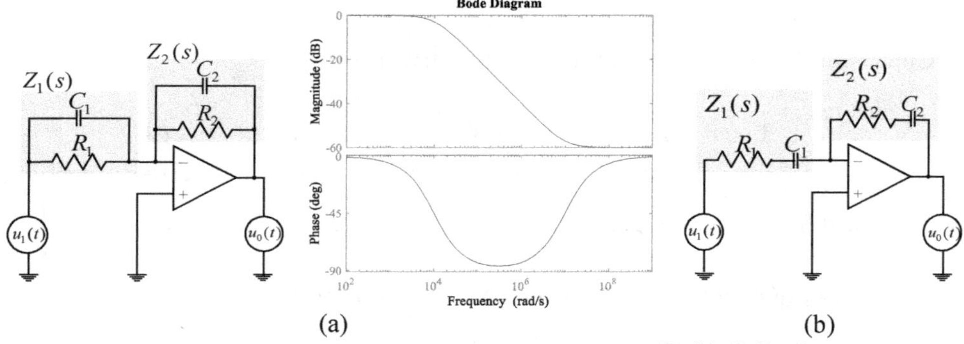

Figure 3.46. (a) Analog notch filter: Transfer function is $G(j\omega)=\dfrac{R_2}{R_1}\dfrac{(j\omega R_1C_1+1)}{(j\omega R_2C_2+1)}$. Bode plots for a notch filter for R_1=1000 ohm, R_2=1000 ohm, $C_1=1\times10^{-10}$ F, $C_2=1\times10^{-7}$ F;
(c) Notch filter, $G(j\omega)=\dfrac{C_1}{C_2}\dfrac{(j\omega R_2C_2+1)}{(j\omega R_1C_1+1)}$. ■

Example 3.32. *Second-Order Filter*

Consider the notch filter as a series configuration of two inverting operational amplifiers with input and feedback impedances as reported in Figure 3.47.a.

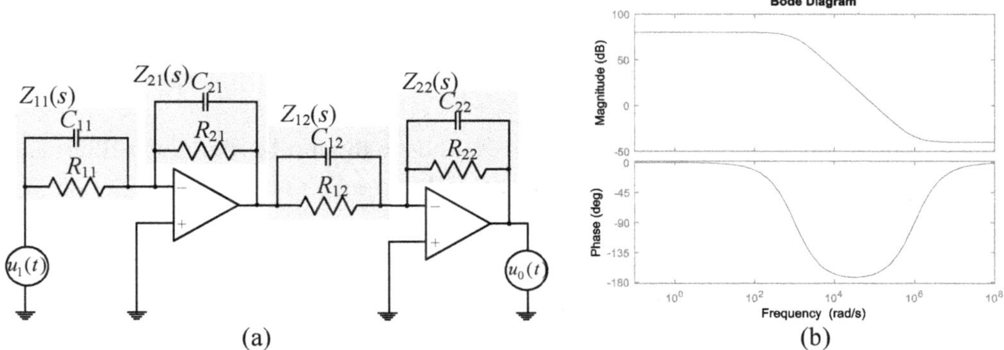

(a) (b)

Figure 3.47. (a) Notch filter schematics. A transfer function is $G(s)=\frac{Z_{21}(s)\,Z_{22}(s)}{Z_{11}(s)\,Z_{12}(s)}=\frac{R_{21}R_{22}}{R_{11}R_{12}}\frac{(R_{11}C_{11}s+1)(R_{12}C_{12}s+1)}{(R_{21}C_{21}s+1)(R_{22}C_{22}s+1)}$;

(b) Bode plots for the second-order $G(s)$ with R_{11}=100 ohm, R_{12}=100 ohm, R_{21}=10000 ohm, R_{22}=10000 ohm, C_{11}=10 nF, C_{12}=10 nF, C_{21}=100 nF and C_{22}=100 nF.

The input and feedback impedances are

$$Z_{11}(s)=\frac{R_{11}}{R_{11}C_{11}s+1},\ Z_{12}(s)=\frac{R_{12}}{R_{12}C_{12}s+1},\ Z_{21}(s)=\frac{R_{21}}{R_{21}C_{21}s+1}\ \text{and}\ Z_{22}(s)=\frac{R_{22}}{R_{22}C_{22}s+1}.$$

The transfer function of the cascaded notch filter is

$$G(s)=\frac{Z_{21}(s)\,Z_{22}(s)}{Z_{11}(s)\,Z_{12}(s)}=\frac{R_{21}R_{22}}{R_{11}R_{12}}\frac{(R_{11}C_{11}s+1)(R_{12}C_{12}s+1)}{(R_{21}C_{21}s+1)(R_{22}C_{22}s+1)}.$$

The specified gain, cut-off frequency and attenuation at the specified frequencies are used to find values of resistors and capacitors. Let the 10 Hz system bandwidth should be preserved. The multi-frequency noise with frequencies from 10000 to 100000 Hz, should be attenuated at least 100 times. The filter gain at low frequency should be k=10000, or, 80 dB. The corner frequencies are $1/(R_{ij}C_{ij})$.

Two zeros are $\omega_{cN1}=\frac{1}{R_{11}C_{11}}$, $\omega_{cN2}=\frac{1}{R_{12}C_{12}}$.

Two poles are $\omega_{cD1}=\frac{1}{R_{21}C_{21}}$, $\omega_{cD2}=\frac{1}{R_{22}C_{22}}$.

Let $\omega_{cD1}=\omega_{cD2}$=1000 rad/s and $\omega_{cN1}=\omega_{cN2}=1\times10^6$ rad/s. Recall that $\omega=2\pi f$.

The system bandwidth f=10 Hz gives $2\pi f$=62.8 rad/s. At the corner frequencies $1/(R_{ij}C_{ij})$, one has $\Delta|G|_{\text{dB}}=-3$ dB attenuation at the nominal passband value.

Using the specified low-frequency gain 80 dB, one has $k=\frac{R_{21}R_{22}}{R_{11}R_{12}}$=10000.

With the cut-off frequencies $1/(R_{ij}C_{ij})$, we have R_{11}=100 ohm, R_{12}=100 ohm, R_{21}=10000 ohm, R_{22}=10000 ohm, C_{11}=10 nF, C_{12}=10 nF, C_{21}=100 nF and C_{22}=100 nF. The Bode plots, documented in Figure 3.47.b, are calculated using MATLAB.

```
R11=100; R12=100; R21=10000; R22=10000; C11=10e-9; C12=10e-9; C21=100e-9; C22=100e-9;
num1=[R11*C11 1]; num2=[R12*C12 1]; num=(R21*R22)/(R11*R12)*conv(num1,num2);
den1=[R21*C21 1]; den2=[R22*C22 1]; den=conv(den1, den2); bode(num,den,{0.1,1e8});
```
■

Example 3.33. *Second- and Third-Order Filters*

Consider the filters reported in Figures 3.48. For a schematic in Figure 3.48.a,

$$Z_1(s)=\frac{\left(R_{11}+\frac{1}{C_{11}s}\right)\left(R_{12}+\frac{1}{C_{12}s}\right)}{R_{11}+\frac{1}{C_{11}s}+R_{12}+\frac{1}{C_{12}s}}=\frac{(R_{11}C_{11}s+1)(R_{12}C_{12}s+1)}{s\left[(R_{11}C_{11}C_{12}+R_{12}C_{11}C_{12})s+C_{11}+C_{12}\right]},$$

$$Z_2(s)=\frac{(R_{21}C_{21}s+1)(R_{22}C_{22}s+1)}{s\left[(R_{21}C_{21}C_{22}+R_{22}C_{21}C_{22})s+C_{21}+C_{22}\right]}.$$

The resulting transfer function, corresponds to the third-order notch filter, and

$$G(s)=\frac{(R_{21}C_{21}s+1)(R_{22}C_{22}s+1)\left[(R_{11}+R_{12})C_{11}C_{12}s+C_{11}+C_{12}\right]}{(R_{11}C_{11}s+1)(R_{12}C_{12}s+1)\left[(R_{21}+R_{22})C_{21}C_{22}s+C_{21}+C_{22}\right]}.$$

Let the low-frequency filter gain should be k=10, and, the system bandwidth is 25 Hz. The 10000 Hz noise should be attenuated by 50 times. The gain, numerator zeros and denominator poles depend on R_i and C_i. The band-pass, cut-off and stop-band frequencies are functions of (R_i,C_i). For a low-pass filter, to meet specifications, $R_{11}=1\times10^7$ ohm, $R_{12}=1\times10^7$ ohm, $R_{21}=1\times10^5$ ohm, $R_{22}=1\times10^5$ ohm, $C_{11}=1\times10^{-10}$ F, $C_{12}=1\times10^{-10}$ F, $C_{21}=1\times10^{-11}$ F and $C_{22}=1\times10^{-11}$ F. The Bode plots are illustrated in Figure 3.48.b. Design specifications are met. Computing of Bode plots is performed using a code

```
R11=1e7; R12=1e7; R21=1e5; R22=1e5; C11=1e-10; C12=1e-10; C21=1e-11; C22=1e-11;
N1=[R21*C21 1]; N2=[R22*C22 1]; N3=[(R11+R12)*C11*C12 (C11+C12)];
N=conv(N1,N2); num=conv(N,N3);
D1=[R11*C11 1]; D2=[R12*C12 1]; D3=[(R21+R22)*C21*C22 (C21+C22)]; D=conv(D1,D2);
den=conv(D,D3); bode(num,den,{0.1,1e8});
```

The Sallen-Key filter is reported in Figure 3.48.c. Using the node analysis

$$G(s)=\frac{U_0(s)}{U_1(s)}=\frac{Z_3(s)Z_4(s)}{Z_1(s)Z_2(s)+\left[Z_1(s)+Z_2(s)\right]Z_3(s)+Z_3(s)Z_4(s)}.$$

Hence, $G(s)=\dfrac{U_0(s)}{U_1(s)}=\dfrac{1}{R_1R_2C_1C_2s^2+(R_1+R_2)C_2s+1}$.

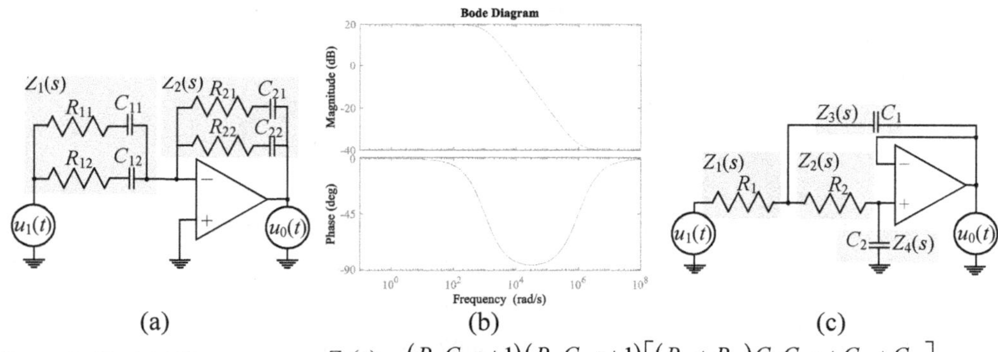

(a) (b) (c)

Figure 3.48. (a) Filter, $G(s)=-\dfrac{Z_2(s)}{Z_1(s)}=\dfrac{(R_{21}C_{21}s+1)(R_{22}C_{22}s+1)\left[(R_{11}+R_{12})C_{11}C_{12}s+C_{11}+C_{12}\right]}{(R_{11}C_{11}s+1)(R_{12}C_{12}s+1)\left[(R_{21}+R_{22})C_{21}C_{22}s+C_{21}+C_{22}\right]}$;

(b) Bode plots for the third-order notch filter, $R_{11}=1\times10^7$ ohm, $R_{12}=1\times10^7$ ohm, $R_{21}=1\times10^5$ ohm, $R_{22}=1\times10^5$ ohm, $C_{11}=1\times10^{-10}$ F, $C_{12}=1\times10^{-10}$ F, $C_{21}=1\times10^{-11}$ F and $C_{22}=1\times10^{-11}$ F;

(c) Analog Sallen-Key low-pass filter, $G(s)=\dfrac{1}{R_1R_2C_1C_2s^2+(R_1+R_2)C_2s+1}$. ■

Example 3.34. *Implementation of Analog Filters With a Single Operational Amplifier*
There are single-amplifier active filter schemes which implement the second-, third- and high-order filters. Using the system bandwidth, noise frequency, frequencies ($\omega_p, \omega_c, \omega_s$), gain, attenuation, sensitivity and complexity, one designs a filter. Input and feedback impedances $Z(s)=Z(j\omega)$ yield low-, high- and band-pass high-order filters.

The admittance $Y(s)=Y(j\omega)$ is reciprocal of impedance $Z(s)=Z(j\omega)$.

For a capacitor and resistor, $Y_C(s)=sC$ and $Y_R(s)=1/R$.

For a filter, depicted in Figure 3.49.a, the node equations using the Kirchhoff current law at node 1 and summing node 2, are

$(Y_1+Y_2+Y_3+Y_4)V_1-Y_1U_1-Y_4U_0=0, \quad -Y_3V_1-Y_5U_0=0.$

Eliminating V_1 and grouping the terms yield

$$G(s)=\frac{U_0(s)}{U_1(s)}=\frac{-Y_1Y_3}{(Y_1+Y_2+Y_3+Y_4)Y_5+Y_3Y_4}.$$

For a filter, reported in Figure 3.49.b, $Y_1=1/R_1$, $Y_2=sC_2$, $Y_3=1/R_3$, $Y_4=1/R_2$ and $Y_5=sC_1$.

The transfer function is

$$G(s)=\frac{U_0(s)}{U_1(s)}=\frac{Z_3(s)Z_4(s)}{Z_1(s)Z_2(s)+\left[Z_1(s)+Z_2(s)\right]Z_3(s)+Z_3(s)Z_4(s)}$$

$$=\frac{R_2}{R_1}\frac{1}{R_2R_3C_1C_2}\frac{1}{\left[s^2+\frac{1}{C_2}\left(\frac{1}{R_1}+\frac{1}{R_2}+\frac{1}{R_3}\right)s+\frac{1}{R_2R_3C_1C_2}\right]}.$$

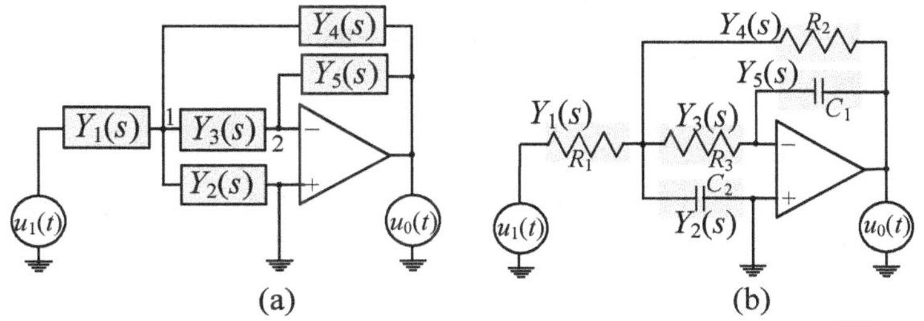

Figure 3.49. (a) Multiple input and feedback filter, $G(s)=\frac{-Y_1Y_3}{(Y_1+Y_2+Y_3+Y_4)Y_5+Y_3Y_4}$;

(b) The second-order low-pass filter, $G(s)=\frac{R_2}{R_1}\frac{1}{R_2R_3C_1C_2}\frac{1}{\left[s^2+\frac{1}{C_2}\left(\frac{1}{R_1}+\frac{1}{R_2}+\frac{1}{R_3}\right)s+\frac{1}{R_2R_3C_1C_2}\right]}$. ■

3.8. High-Switching Frequency dc-dc Regulators

The commercialized high switching frequency PWM amplifiers, regulators and integrated controllers-drivers are the application specific. Using PWM switching, the voltage at the load terminal is regulated [6, 7, 12-14]. Various dc-dc switching regulator topologies have being designed and commercialized in various applications, such as energy systems, power management, robotics, wearable electronics, etc. Consider one-quadrant dc-dc regulators used in consumer electronics, ground and unmanned aerial vehicles, as well as other applications. The dc-dc regulators are open-loop stable. The output voltage should be stabilized or regulated to guarantee the specified profile and governance. Closed-loop systems are designed to ensure adequate dynamics. The *step-down*, *step-up*, *buck-boost* and other regulators are considered.

3.8.1. *Step-Down* Switching Regulators

Step-down topologies of dc-dc switching regulators have being developed, and, these converters are open-lop stable. The commercialized of-the-shelf converters are controlled using the integrated analog and digital proportional-integral controllers. The schematics for a high switching frequency dc-dc *step-down* (*buck*) regulator is shown in Figure 3.50.a. The regulator components are the MOSFET, diode *D*, inductors and capacitor. The *LC* output filter attenuates the voltage ripple at the *RL* load with R_L and L_L. The contact resistances, parasitic resistances, ohmic losses, as well as inherent transistor and inductor resistances are r_s, r_L and r_C. The images of the Texas Instruments TPS5410D regulators are illustrated in Figure 3.50.b.

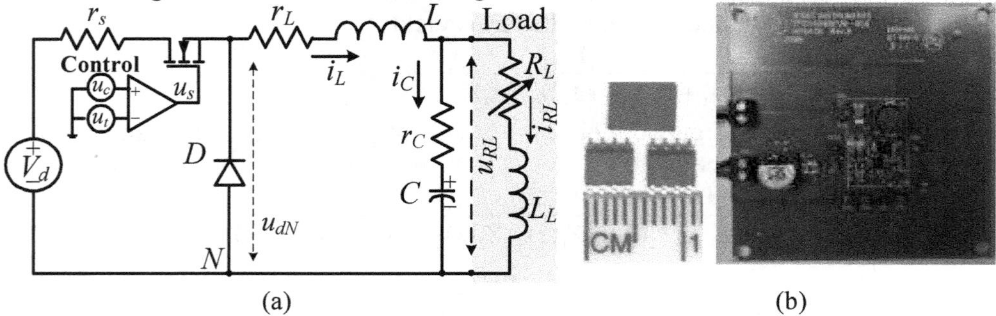

(a) (b)

Figure 3.50. (a) Controlled *step-down* PWM regulator schematics;
(b) TPS5410D *buck* regulator, and, TPS54040EVM evaluation board, 5.5 to 36 V input voltage, 1.23 to 31 V output voltage, 1 A output current, 500 kHz switching frequency.

The MOSFET is open or closed. The switching frequency is $f = \frac{1}{t_{on} + t_{off}}$, where t_{on} and t_{off} are the switching *on* and *off* durations. For a lossless MOSFET, the voltage u_{dN} is equal to the supplied voltage V_d when the transistor is closed. The output voltage is zero if the transistor is open, see Figure 3.51.a. Voltage u_{dN} and voltage applied to load u_{RL} are regulated by controlling the switching *on* and *off* durations t_{on} and t_{off}. The average voltage, applied to the *RL* load, depends on t_{on} and t_{off}. In steady-state,

$$u_{dN\,\text{average}} = \frac{t_{on}}{t_{on} + t_{off}} V_d = d_D V_d,\; d_D = \frac{t_{on}}{t_{on} + t_{off}},\; d_D \in [0\ \ 1], \tag{3.104}$$

where d_D is the duty ratio, also called the duty cycle.

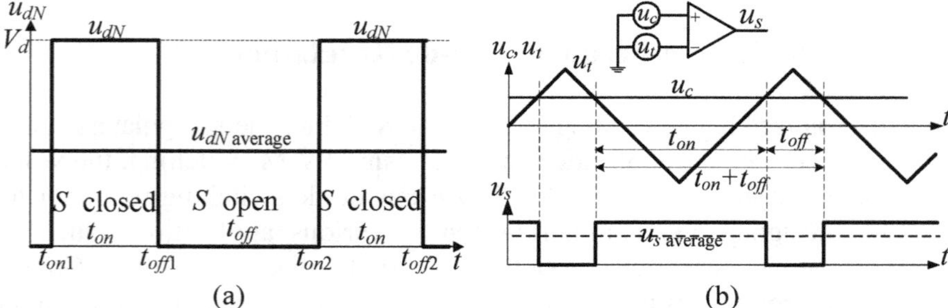

(a) (b)

Figure 3.51. (a) PWM voltage waveform;
(b) Comparator ensures the transistor switching: The control, triangular and driving signal voltages (u_c,u_t,u_s) are documented.

The duty ratio d_D is a function of the switching frequency f of u_t and time t_{on}. If f=const, $d_D \in [0 \quad 1]$, and, d_D=0 if t_{on}=0, d_D=1 if t_{off} =0. As shown in Figures 3.50.a and 3.51.b, changing (u_s,d_D), one controls the transistor switching activity and the average voltage supplied to the load u_{RL}. To establish the PWM switching, a control-triangle concept is used. The switching signal u_s is generated by comparing a signal-level control voltage u_c with a repetitive triangular u_t. The duration of the output pulses u_s represents the weighted value between the triangular voltage u_t with the assigned switching frequency f and control signal u_c. The output voltage of the comparator u_s drives MOSFET. The *on* and *off* switching is accomplished by comparing u_c and u_t. The Motorola dual comparator MC3405 is reported in Figure 3.52. Figures 3.52 illustrate the voltage waveforms.

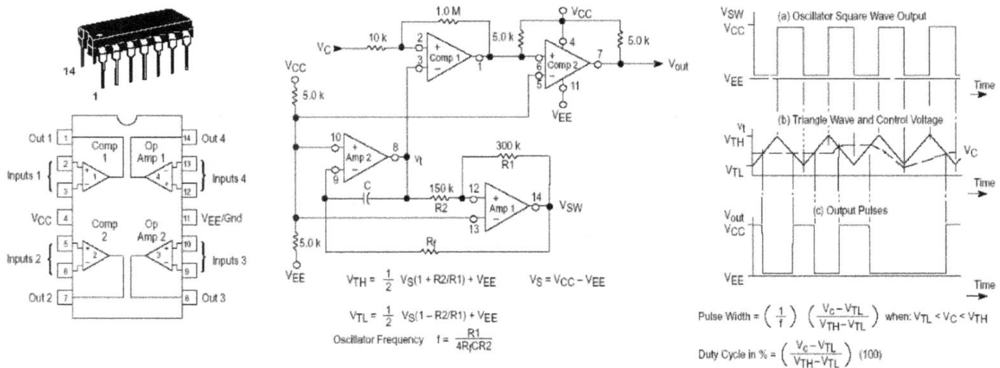

Figure 3.52. MC3405 comparator pin connections, schematics and waveform, copyright of Motorola, used with permission [7]

For a dc-dc regulator, illustrated in Figure 3.50.a, a low-pass first-order LC filter is characterized by the transfer function $G(s)=\dfrac{1}{LCs+1}$ for r_L=0 and r_C=0.

Considering the RL load, one has $G(s)=\dfrac{L_L s+R_L}{L_L LCs^3+R_L LCs^2+(L_L+L)s+R_L}$.

The inductor L and capacitor C ensure the specified voltage ripple. Neglecting small resistances r_s, r_L and r_C, the voltage ripple is $\dfrac{\Delta u_a}{u_a}=\dfrac{1-d_D}{8LCf^2}$.

The minimum value for the LC filter inductor is $L_{\min}=\dfrac{(1-d_D)R_L}{2f}$.

Two circuits for the closed and open MOSFET are illustrated in Figures 3.53.a and b.

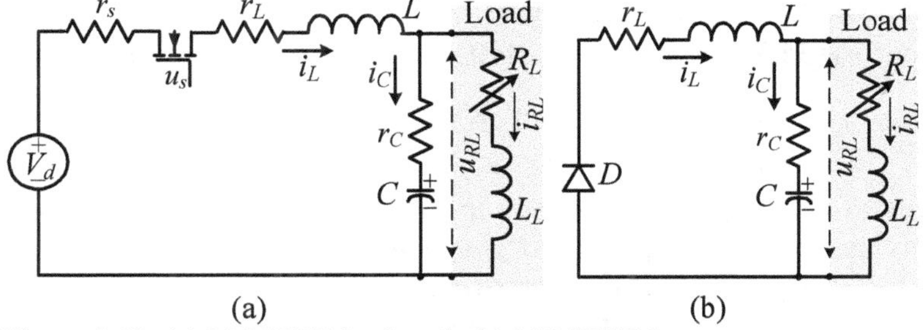

Figure 3.53. (a) MOSFET is closed; (b) MOSFET is open

If the transistor is closed, the diode D is reverse biased. For the circuit, shown in Figure 3.53.a, using Kirchhoff's laws

$$\frac{du_C}{dt}=\frac{1}{C}(i_L-i_a), \tag{3.105}$$
$$\frac{di_L}{dt}=\frac{1}{L}\left[-u_C-(r_L+r_C)i_L+r_Ci_{RL}-r_si_L+V_d\right],$$
$$\frac{di_{RL}}{dt}=\frac{1}{L_L}\left[u_C+r_Ci_L-(R_L+r_C)i_{RL}\right].$$

If transistor is open, the diode D is forward biased, and, $i_d=i_L$, see Figure 3.53.b. One obtains

$$\frac{du_C}{dt}=\frac{1}{C}(i_L-i_a), \tag{3.106}$$
$$\frac{di_L}{dt}=\frac{1}{L}\left[-u_C-(r_L+r_C)i_L+r_Ci_{RL}\right],$$
$$\frac{di_{RL}}{dt}=\frac{1}{L_L}\left[u_C+r_Ci_L-(R_L+r_C)i_{RL}\right].$$

When MOSFET is closed, the duty ratio is $d_D=1$. If MOSFET is open, the duty ratio is $d_D=0$. Using the *averaging* concept, from two sets of differential equations (3.105) and (3.106), we find the resulting nonlinear differential equations

$$\frac{du_C}{dt}=\frac{1}{C}(i_L-i_{RL}), \tag{3.107}$$
$$\frac{di_L}{dt}=\frac{1}{L}\left[-u_C-(r_L+r_C)i_L+r_Ci_{RL}-r_si_Ld_D+V_dd_D\right],\ 0\le d_D\le 1,$$
$$\frac{di_{RL}}{dt}=\frac{1}{L_L}\left[u_C+r_Ci_L-(R_L+r_C)i_{RL}\right].$$

The duty ratio is regulated by the signal-level control voltage u_c

$$d_D=\frac{u_c}{u_{t\max}}\in[0\quad 1],\ u_c\in[0\quad u_{c\max}],\ u_{c\max}=u_{t\max}. \tag{3.108}$$

Neglecting small r_s, r_L and r_C, the analysis of the steady-state performance yields $\frac{u_{RL\ \text{average}}}{V_d}=d_D$. The regulator output is the voltage across the capacitor (filter output) u_C or the voltage applied to the load u_{RL}.

The control limit is $0\le u_c\le u_{c\max}$, $u_c\in[0\ \ u_{c\max}]$. Consider d_D as a control, $0\le d_D\le 1$.

Example 3.35. *Simulation and Proportional-Integral Control of a Step-Down Regulator*
The parameters of a *step-down* regulator, experimentally studied in Example 6.30, are $r_s=0.01$ ohm, $r_L=0.093$ ohm, $r_C=0.065$ ohm, $C=1\times10^{-4}$ F, $L=5\times10^{-6}$ H and $L_L=0.007$ H. For the constant $V_d=35$ V, the load is $R_L\in[10_{\text{peak}}\ \infty_{\text{open}})$ ohm. For open-loop analysis, let the duty ratio is $d_D=\begin{cases}0.5, & 0\le t<0.005 \text{ sec}\\ 0.25, & 0.005\le t<0.01 \text{ sec}\\ 0.75, & 0.01\le t\le 0.015 \text{ sec}\end{cases}$. For a model, given by differential equations (3.107), the SIMULINK diagram is reported in Figure 3.54.a. The steady-state value of the output voltage u_{RL} corresponds to d_D. The voltage at the load terminal is $u_{RL}=(u_C+r_Ci_L-r_Ci_{RL})$. Dynamics of the states (u_C,i_L,i_{RL}) are depicted in Figure 3.54.b for $V_d=35$ V, $R_L=10$ ohm. The settling time is 3.1×10^{-3} sec.

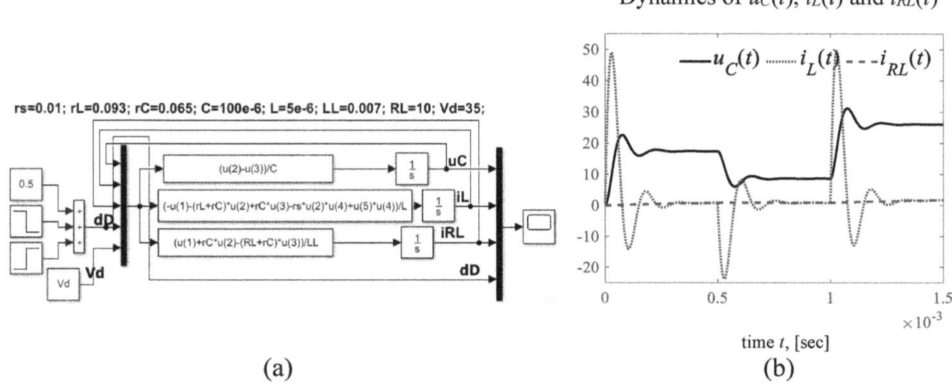

Figure 3.54. (a) Open-loop system: SIMULINK diagram to simulate a dc-dc regulator;

(b) Dynamics of voltage $u_C(t)$, and, currents $i_L(t)$ and $i_{RL}(t)$, $d_D=\begin{cases}0.5, & 0\le t<0.005 \text{ sec}\\ 0.25, & 0.005\le t<0.01 \text{ sec}\\ 0.75, & 0.01\le t\le 0.015 \text{ sec}\end{cases}$.

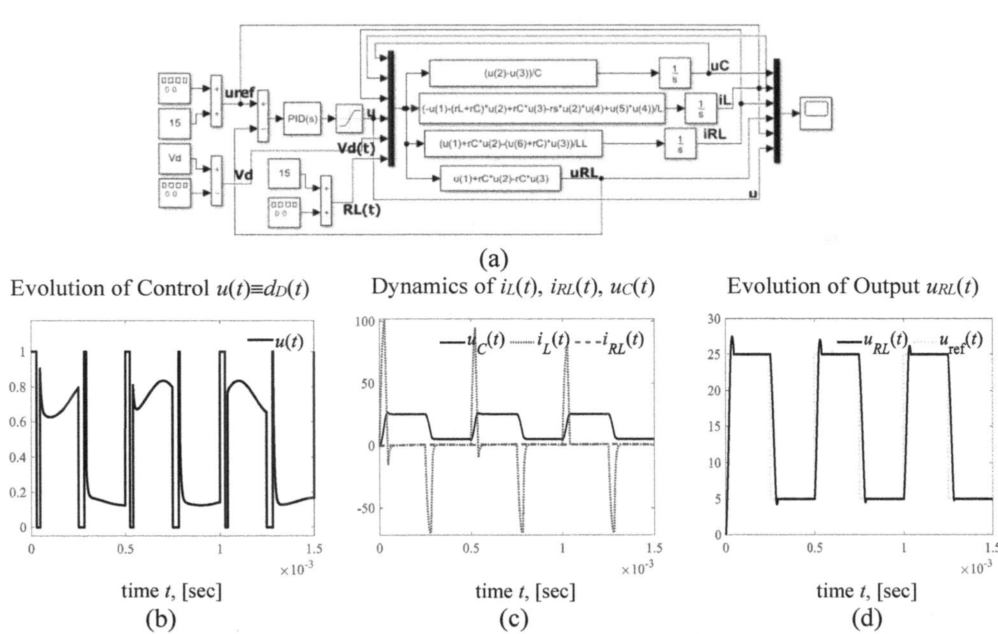

Figure 3.55. (a) Closed-loop regulator: SIMULINK diagram to simulate a closed-loop dc-dc regulator with bounded proportional-integral control law $u=\text{sat}_0^1\left(50e+5\int edt\right)$;

(b) Evolution of control, $0\le u\le 1$;

(c) Transient dynamics of $u_C(t), i_L(t), i_{RL}(t)$ for different references $u_{\text{ref}}(t)$ and loads $R_L(t)$;

(d) Output voltage $u_{RL}(t)$ tracks the specified reference $u_{\text{ref}}(t)=\begin{cases}25 \text{ V}\\ 5 \text{ V}\end{cases}$ for varying load $R_L(t)$=15±5 ohm, f=5000 Hz and voltage source $V_d(t)=35\pm 5\sin(2\pi\times 2500t)$ V.

Tracking of the specified output voltage at the load $u_{RL}=(u_C+r_Ci_L-r_Ci_{RL})$, or other states, can be ensured by using control laws. Consider the proportional-integral control

$$u=\text{sat}_0^1\left(k_pe+k_i\int edt\right),\ e(t)=u_{\text{ref}}-u_{RL},\ u\equiv d_D,\ (k_p,k_i)>0.$$

The amplitude of the control function u is bounded due to the limit $0 \le d_D \le 1$. For the closed-loop system with a proportional-integral-derivative control law $u = \mathrm{sat}_0^1\left(50e + 5\int edt\right)$, implemented using the PID(s) block, the SIMULINK diagram is depicted in Figure 3.55.a. In the closed-loop system, the voltage $u_{RL}(t)$ tracks the reference $u_{\mathrm{ref}} = \begin{cases} 25\ \mathrm{V} \\ 5\ \mathrm{V} \end{cases}$, f=2000 Hz for the varying load $R_L = \begin{cases} 30\ \mathrm{ohm} \\ 10\ \mathrm{ohm} \end{cases}$, f=5000 Hz and varying voltage source V_d, $V_d = V_{d0} + 5\sin(2\pi \times 2500t)$, V_{d0}=35 V. The evolution of the control function $u(t)$, $u(t) \equiv d_D(t)$ is depicted in Figure 3.55.b.

Control law $u = \mathrm{sat}_0^1\left(50e + 5\int edt\right)$ is bounded as $0 \le u \le 1$. Dynamics of states (u_C, i_L, i_{RL}) are illustrated in Figures 3.55.c. Figure 3.55.d depicts the output $u_{RL}(t)$. The experimental studies are performed, and, findings are reported in Example 6.30.

For the output voltage $u_{RL} = (u_C + r_C i_L - r_C i_{RL})$, the settling time is 7.4×10^{-4} sec. ■

3.8.2. *Step-Up* dc-dc Switching Regulators

The *step-up* (*boost*) dc-dc switching regulators are used in the energy systems, such as photovoltaic, hydro- and wind turbine auxiliary units. A typical configuration of a one-quadrant *boost* regulator is depicted in Figure 3.56.

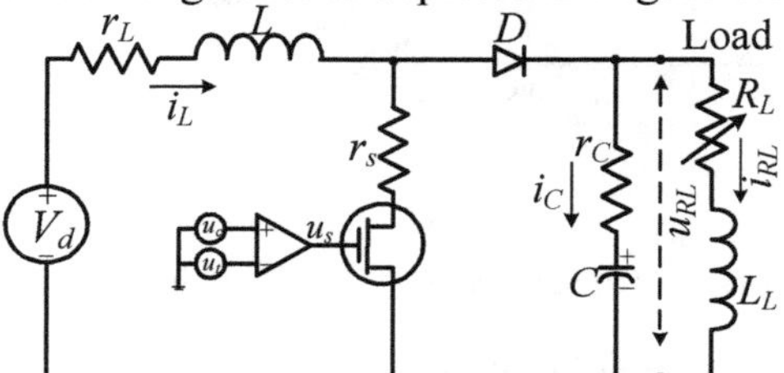

Figure 3.56. High-frequency *boost* regulator.

When the MOSFET is closed, the diode D is reverse biased. One finds

$$\frac{du_C}{dt} = -\frac{1}{C} i_{RL}, \tag{3.109}$$

$$\frac{di_L}{dt} = \frac{1}{L}\left[-(r_L + r_s)i_L + V_d\right],$$

$$\frac{di_{RL}}{dt} = \frac{1}{L_L}\left[u_C - (R_L + r_C)i_{RL}\right].$$

If transistor is open, the diode is forward biased because the direction of the current in the inductor i_L does not change instantly. Hence,

$$\frac{du_C}{dt} = \frac{1}{C}(i_L - i_{RL}), \tag{3.110}$$

$$\frac{di_L}{dt} = \frac{1}{L}\left[-u_C - (r_L + r_C)i_L + r_C i_{RL} + V_d\right],$$

$$\frac{di_{RL}}{dt} = \frac{1}{L_L}\left[u_C + r_C i_L - (R_L + r_C)i_{RL}\right].$$

Applying the *averaging* concept, using d_D, one finds

$$\frac{du_C}{dt}=\frac{1}{C}\left(i_L-i_{RL}-i_L d_D\right),\tag{3.111}$$
$$\frac{di_L}{dt}=\frac{1}{L}\left[-u_C-\left(r_L+r_C\right)i_L+r_C i_{RL}+u_C d_D+\left(r_C-r_s\right)i_L d_D-r_C i_{RL} d_D+V_d\right],$$
$$\frac{di_{RL}}{dt}=\frac{1}{L_L}\left[u_C+r_C i_L-\left(R_L+r_C\right)i_{RL}-r_C i_L d_D\right].$$

The compliance, matching and compatibility of power electronics, energy sources, power electronics and energy storage devices must be ensured. The steady-state analysis results in $\frac{u_{RL\ \text{average}}}{V_d}=\frac{1}{1-d_D}$.

The voltage ripple is $\frac{\Delta u_{RL}}{u_{RL}}=\frac{d_D}{R_L C f^2}$.

The minimum value of the inductance depends on the switching frequency and load resistance, $L_{\min}=\frac{d_D(1-d_D)^2 R_L}{2f}$.

Example 3.36. *Lagrange Equations of Motion to Model a Step-Up Switching Regulator*
Apply the Lagrange equation to derive the mathematical model for a one-quadrant *boost* dc-dc regulator illustrated in Figure 3.56. The Lagrange equations of motion are

$$\frac{d}{dt}\left(\frac{\partial\Gamma}{\partial\dot{q}_1}\right)-\frac{\partial\Gamma}{\partial q_1}+\frac{\partial D}{\partial\dot{q}_1}+\frac{\partial\Pi}{\partial q_1}=Q_1,$$
$$\frac{d}{dt}\left(\frac{\partial\Gamma}{\partial\dot{q}_2}\right)-\frac{\partial\Gamma}{\partial q_2}+\frac{\partial D}{\partial\dot{q}_2}+\frac{\partial\Pi}{\partial q_2}=Q_2.$$

The electric charges in the first and the second loops are q_1 and q_2.

One has $i_L=\dot{q}_1$ and $i_{RL}=\dot{q}_2$.

The generalized forces are $Q_1=V_d$ and $Q_2=0$.

When transistor is closed, the total kinetic Γ, potential Π, and dissipating D energies are

$$\Gamma=\tfrac{1}{2}\left(L\dot{q}_1^{\,2}+L_L\dot{q}_2^{\,2}\right),\ \Pi=\frac{1}{2}\frac{q_2^2}{C},\ D=\tfrac{1}{2}\left(\left(r_L+r_s\right)\dot{q}_1^{\,2}+\left(r_C+R_L\right)\dot{q}_2^{\,2}\right).$$

Assume that the resistances, inductances, and capacitance are constant. We have

$$\frac{\partial\Gamma}{\partial q_1}=0,\ \frac{\partial\Gamma}{\partial q_2}=0,\ \frac{\partial\Gamma}{\partial\dot{q}_1}=L\dot{q}_1,\ \frac{\partial\Gamma}{\partial\dot{q}_2}=L_L\dot{q}_2,\ \frac{d}{dt}\left(\frac{\partial\Gamma}{\partial\dot{q}_1}\right)=L\ddot{q}_1,\ \frac{d}{dt}\left(\frac{\partial\Gamma}{\partial\dot{q}_2}\right)=L_L\ddot{q}_2,$$
$$\frac{\partial\Pi}{\partial q_1}=0,\ \frac{\partial\Pi}{\partial q_2}=\frac{q_2}{C},\ \frac{\partial D}{\partial\dot{q}_1}=\left(r_L+r_s\right)\dot{q}_1,\ \frac{\partial D}{\partial\dot{q}_2}=\left(r_C+R_L\right)\dot{q}_2.$$

The Lagrange equations of motion yield

$$L\ddot{q}_1+\left(r_L+r_s\right)\dot{q}_1=Q_1,\ L_L\ddot{q}_2+\left(r_C+R_L\right)\dot{q}_2+\frac{1}{C}q_2=Q_2.$$

One obtains

$$\ddot{q}_1 = \frac{1}{L}\left(-(r_L + r_s)\dot{q}_1 + Q_1\right),$$

$$\ddot{q}_2 = \frac{1}{L_L}\left(-(r_C + R_L)\dot{q}_2 - \frac{1}{C}q_2 + Q_2\right),$$

If transistor is open, one has

$$\Gamma = \tfrac{1}{2}\left(L\dot{q}_1^2 + L_L\dot{q}_2^2\right),\ \Pi = \tfrac{1}{2}\frac{(q_1 - q_2)^2}{C},\ D = \tfrac{1}{2}\left(r_L\dot{q}_1^2 + r_C(\dot{q}_1 - \dot{q}_2)^2 + R_L\dot{q}_2^2\right).$$

Hence,

$$\frac{\partial\Gamma}{\partial q_1} = 0,\ \frac{\partial\Gamma}{\partial q_2} = 0,\ \frac{\partial\Gamma}{\partial \dot{q}_1} = L\dot{q}_1,\ \frac{\partial\Gamma}{\partial \dot{q}_2} = L_L\dot{q}_2,\ \frac{d}{dt}\left(\frac{\partial\Gamma}{\partial \dot{q}_1}\right) = L\ddot{q}_1,\ \frac{d}{dt}\left(\frac{\partial\Gamma}{\partial \dot{q}_2}\right) = L_L\ddot{q}_2,$$

$$\frac{\partial\Pi}{\partial q_1} = \frac{q_1 - q_2}{C},\ \frac{\partial\Pi}{\partial q_2} = -\frac{q_1 - q_2}{C} \text{ and } \frac{\partial D}{\partial \dot{q}_1} = (r_L + r_C)\dot{q}_1 - r_C\dot{q}_2,\ \frac{\partial D}{\partial \dot{q}_2} = -r_C\dot{q}_1 + (r_C + R_L)\dot{q}_2.$$

The resulting equations are

$$L\ddot{q}_1 + (r_L + r_C)\dot{q}_1 - r_C\dot{q}_2 + \frac{q_1 - q_2}{C} = Q_1,$$

$$L_L\ddot{q}_2 - r_C\dot{q}_1 + (r_C + R_L)\dot{q}_2 - \frac{q_1 - q_2}{C} = Q_2.$$

Therefore,

$$\ddot{q}_1 = \frac{1}{L}\left(-(r_L + r_C)\dot{q}_1 + r_C\dot{q}_2 - \frac{q_1 - q_2}{C} + Q_1\right),$$

$$\ddot{q}_2 = \frac{1}{L_L}\left(r_C\dot{q}_1 - (r_C + R_L)\dot{q}_2 + \frac{q_1 - q_2}{C} + Q_2\right).$$

Cauchy's form of differential equations (3.111) are found using

$$\frac{dq_1}{dt} = i_L,\ \frac{dq_2}{dt} = i_{RL}.$$

The voltage across the capacitor u_C is expressed using the charges. When transistor is closed $u_C = -\frac{q_2}{C}$, while if transistor is open $u_C = \frac{q_1 - q_2}{C}$.

The differential equations found using Kirchhoff's voltage law and the Lagrange equations result in an identical model. ■

3.8.3. *Buck-Boost* Switching Regulators

Different topologies of dc-dc *buck-boost* regulators exist. The *buck-boost* switching regulator is illustrated in Figure 3.57.a. Figures 3.57.b depict images of the TPS63060DSCR regulator and dc-dc converter on evaluation board. The voltage at the load u_{RL} is regulated by controlling the switching *on* and *off* durations t_{on} and t_{off}. The duty cycle $d_D = \frac{t_{on}}{t_{on} + t_{off}} \in [0\ 1]$ varies, $0 \le d_D \le 1$.

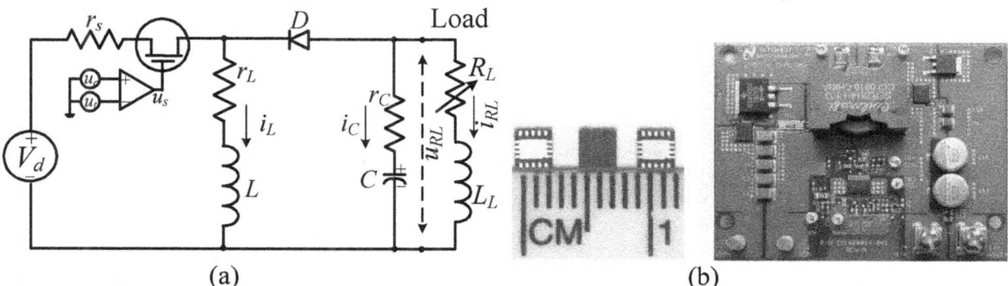

Figure 3.57. (a) High-frequency *buck-boost* switching regulator with the *RL* load;
(b) Texas Instruments TPS63060DSCR *buck-boost* regulator (2.5 to 12 V input voltage, 2.5 to 8 V output voltage, 2 A output current in the *buck* mode and 1.3 A in the *boost* mode, up to 93% efficiency). The dc-dc regulator on the evaluation board.

If MOSFET is closed, the diode is reverse biased. When MOSFET is open, the diode is forward biased. Derive a set of differential equations.

Using Kirchhoff's laws, if MOSFET is closed and open, we have

$$\frac{du_C}{dt}=-\frac{1}{C}i_{RL}, \quad (3.112)$$

$$\frac{di_L}{dt}=\frac{1}{L}\left(-\left(r_L+r_s\right)i_L+V_d\right),$$

$$\frac{di_{RL}}{dt}=\frac{1}{L_L}\left(u_C-\left(R_L+r_C\right)i_{RL}\right),$$

and

$$\frac{du_C}{dt}=-\frac{1}{C}\left(i_L+i_{RL}\right), \quad (3.113)$$

$$\frac{di_L}{dt}=\frac{1}{L}\left(u_C-\left(r_L+r_C\right)i_L-r_Ci_{RL}\right),$$

$$\frac{di_{RL}}{dt}=\frac{1}{L_L}\left(u_C-r_Ci_L-\left(R_L+r_C\right)i_{RL}\right).$$

Applying the *averaging* concept and using d_D, from (3.112) and (3.113)

$$\frac{du_C}{dt}=\frac{1}{C}\left(-i_L-i_{RL}+i_Ld_D\right), \quad (3.114)$$

$$\frac{di_L}{dt}=\frac{1}{L}\left(u_C-(r_L+r_C)i_L-r_Ci_{RL}-u_Cd_D-(r_s-r_C)i_Ld_D+r_Ci_{RL}d_D+V_dd_D\right),$$

$$\frac{di_{RL}}{dt}=\frac{1}{L_L}\left(u_C-r_Ci_L-(R_L+r_C)i_{RL}+r_Ci_Ld_D\right).$$

The steady-state relationship for the supplied and terminal voltages is

$$\frac{u_{RL\ \text{average}}}{V_d}=\frac{-d_D}{1-d_D}.$$

The expressions for the voltage ripple and minimum inductance are

$$\frac{\Delta u_{RL}}{u_{RL}}=\frac{d_D}{R_LCf},\ L_{\min}=\frac{(1-d_D)^2R_L}{2f}.$$

Example 3.37. *Simulations and Dynamic Analysis*
Consider a regulator, described by differential equations (3.114). The parameters are r_s=0.01 ohm, r_L=0.093 ohm, r_C=0.065 ohm, $C=1\times10^{-4}$ F, $L=5\times10^{-6}$ H and L_L=0.007 H.

For V_d= –25 V and R_L=10 ohm, the simulations are performed for $d_D=\begin{cases}0.5, & 0\le t<0.005 \text{ sec}\\ 0.25, & 0.005\le t<0.01 \text{ sec}\\ 0.5, & 0.01\le t\le 0.015 \text{ sec}\end{cases}$. The SIMULINK diagram is reported in Figure 3.58.a.

Dynamics of (u_C, i_L, i_{RL}) are illustrated in Figure 3.58.b. The system is open-loop stable. The settling time is 0.0045 sec. To stabilize the output voltage for the varying $V_d(t)$ and loads $R_L(t)$, or, ensure tracking of the specified command, the closed-loop systems are designed and control laws are implemented.

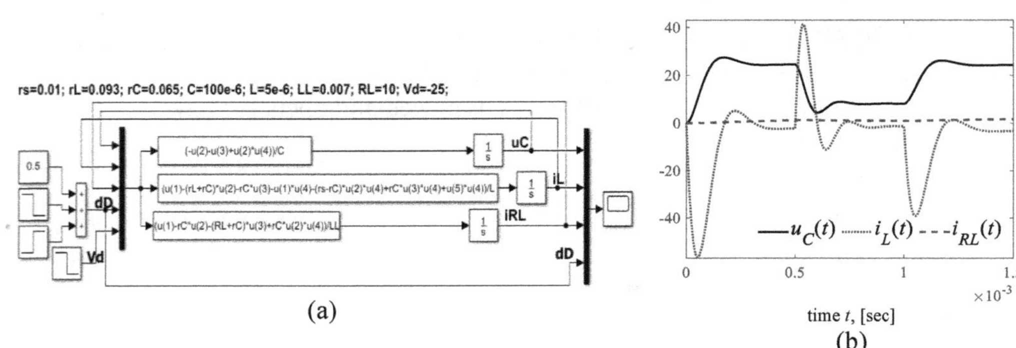

(a) (b)

Figure 3.58. (a) SIMULINK diagram to simulate a dc-dc regulator;
(b) Dynamics of voltage $u_C(t)$, and, currents $(i_L(t), i_{RL}(t))$, $d_D=\begin{cases}0.5, & 0\le t<0.005 \text{ sec}\\ 0.25, & 0.005\le t<0.01 \text{ sec}\\ 0.5, & 0.01\le t\le 0.015 \text{ sec}\end{cases}$.

■

3.8.4. *Cuk* Switching Regulator

The *buck*, many *buck-boost*, *boost* and *flyback* regulators are based on the inductive energy transfer. The Cuk regulator is based on a capacitive-inductive energy transfer. The Cuk regulator is an high-power-factor inverting *buck-boost* dc-dc converter. If MOSFET is *on* or *off*, the currents in the input and output inductors L_1 and L are continuous, see Figure 3.59.a. The output voltage, applied to the load can be smaller or greater than the supplied voltage V_d. When MOSFET is turned *off*, the diode is forward biased. The voltage V_d is supplied, and, capacitor C_1 is charged through the inductor L_1. Assume that the MOSFET is *on*. The current through the inductor L_1 rises, the voltage of capacitor C_1 reverses bias the diode D, and, turns the diode *off*. The capacitor C_1 discharges the stored potential energy through the circuit formed by capacitors C_1, C, the load $R_L L_L$, and the inductor L. If MOSFET is turned *off*, the voltage V_d is applied. The capacitor C_1 charges. The energy, stored in the inductor L, transfers to the load. The diode and transistor are switching synchronously. The capacitor C_1 is a key element for transferring energy from the energy source to the load. Figure 3.59.b reports the images of the Texas Instruments LM2611AMF/NOPB Cuk regulators.

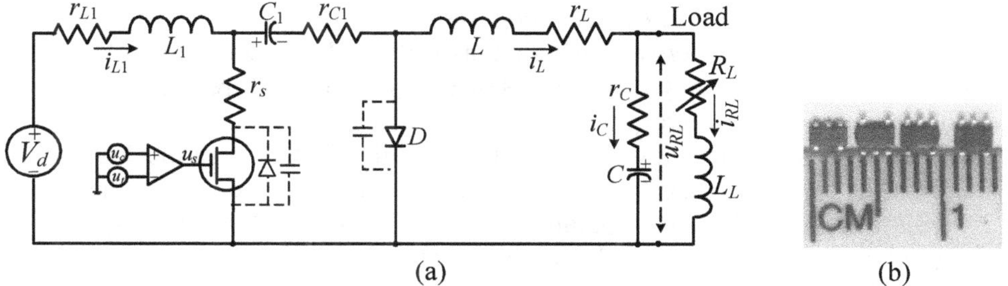

(a) (b)

Figure 3.59. (a) High frequency *soft* switching inverting Cuk regulator;
(b) Texas Instruments LM2611AMF/NOPB SOT-23-5 package Cuk regulator, 2.7 to 14 V input voltage, –1.2 to –27 V output voltage, 900 mA, 1.4 MHz.

Examining when the MOSFET is opened and closed, differential equations are found

$$\frac{du_{C1}}{dt}=\frac{1}{C_1}\left(i_{L1}-i_{L1}d_D+i_L d_D\right), \tag{3.115}$$

$$\frac{du_C}{dt}=\frac{1}{C}\left(i_L-i_{RL}\right),$$

$$\frac{di_{L1}}{dt}=\frac{1}{L_1}\left[-u_{C1}-\left(r_{L1}+r_{C1}\right)i_{L1}+u_{C1}d_D+\left(r_{C1}-r_s\right)i_{L1}d_D+r_s i_L d_D+V_d\right],$$

$$\frac{di_L}{dt}=\frac{1}{L}\left[-u_C-\left(r_L+r_C\right)i_L+r_C i_{RL}-u_{C1}d_D+r_s i_{L1}d_D-\left(r_{C1}+r_s\right)i_L d_D\right],$$

$$\frac{di_{RL}}{dt}=\frac{1}{L_L}\left[u_C+r_C i_L-\left(r_C+R_L\right)i_{RL}\right].$$

The steady-state equations, which are applied to ensure the overall regulator application and regulator-load matching, are

$$\frac{u_{RL\ \mathrm{average}}}{V_d}=-\frac{d_D}{1-d_D},\ \frac{\Delta u_{RL}}{u_{RL}}=\frac{1-d_D}{8LCf^2},\ L_{1\min}=\frac{(1-d_D)^2R_L}{2d_D f},\ L_{\min}=\frac{(1-d_D)R_L}{2f}. \tag{3.116}$$

There is a great variety of high-performance PWM switching and resonant regulators. The inverting Cuk regulators accomplish inductive-capacitive energy conversion and transfer. The dc-dc regulators ensure high efficiency, minimal losses and overloading capabilities due to the use of high-performance MOSFETs, high-Q ceramic capacitors and low-loss inductors. Depending on applications and operating envelopes, the input voltage $V_d(t)$ can be time-invariant or time-varying. The output voltage at the load terminal u_{RL} should be stabilized at the specified value. Or, the voltage u_{RL} should be regulated. Stability, robustness, minimal dynamic error and zero steady-state error must be guaranteed. When the MOSFET is *on* or *off*, it is desirable that the currents in the input and output inductors L_1 and L are smooth. Depending on applications, voltage and current, the MOSFET switching frequency f=1/(t_{on}+t_{off}) may vary from ~100 kHz to 10 GHz. Passive *soft* switching schemes may be used. *Soft* switching regulators use zero-current inductors L_{ZC} which enable zero-current turn-on of MOSFET, zero-voltage capacitors C_{ZV} which ensure zero-voltage turn-off of switches, as well as snubber inductors, voltage-storage capacitors and diodes. In *soft* switching regulators, clamp circuit ensures MOSFET operation at near-zero-voltage switching turn-on. This ensures high frequency operation, thereby reducing the size of components. The clamp capacitor-diode circuit is connected in parallel with the switch. The MOSFET turns on at near-zero-voltage.

Example 3.38. *Cuk Regulator: Closed-Loop Systems*
The kinetic and potential energy conversion, as well as (L_1,C_1) and (L,C) dependencies on the *RL* load and frequency *f*, result in sensitivity of the Cuk regulator and defined operating envelope. There are physical and technological limits on the rated values $(L,C)_{\text{rated}}$ of high-frequency, low-loss and high-quality-factors ceramic inductors and capacitors. The low-inductance *RL* loads cause challenges, and, consider small L_L.

Simulate a regulator with parameters r_s=0.01 ohm, r_L=0.093 ohm, r_{L1}=0.1 ohm, r_C=0.065 ohm, r_{C1}=0.07 ohm, C=1×10^{-7} F, C_1=1×10^{-7} F, L=5×10^{-5} H, L_1=5×10^{-7} H, L_L=0.0001 H and V_d= −30 V. The open-loop regulator dynamics for different d_D is documented in Figure 3.60.a.

Study closed-loop system. The output voltage $u_{RL}(t)$ should track the reference $u_{\text{ref}}(t)$. Using the tracking error $e(t)$, the proportional-integral control law is

$$u = \text{sat}_0^1\left(0.001e + 100\int edt\right),\ e(t)=u_{\text{ref}}-u_{RL},\ u\equiv d_D,\ 0\le u\le 1.$$

Study closed-loop system governance. Let $u_{\text{ref}}(t)$=20±10 V with f=500 Hz.

The load varies as $R_L(t)$=20±5 ohm with f=750 Hz.

Figures 3.60.b depict the closed-loop state dynamics and output voltage $u_{RL}(t)$, which tracks the reference $u_{\text{ref}}(t)$. Despite the varying load $R_L(t)$ and small L_L, the output voltage maintains the specified values.

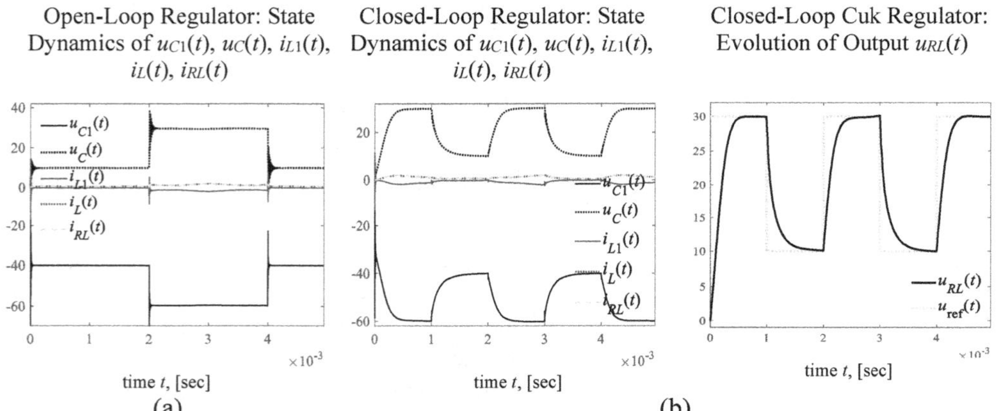

Figure 3.60. (a) Open-loop regulator states governance $u_{C1}(t)$, $u_C(t)$, $i_{L1}(t)$, $i_L(t)$ and $i_{RL}(t)$ for duty ratio $d_D=\begin{cases}0.25, & 0\le t<0.002 \text{ sec}\\ 0.5, & 0.002\le t<0.004 \text{ sec}\\ 0.25, & 0.004\le t\le 0.005 \text{ sec}\end{cases}$;

(b) Closed-loop dc-dc regulator dynamics and output evolution for $u_{\text{ref}}(t)$=20±10 V, f=500 Hz. Time varying load is $R_L(t)$=20±5 ohm, f=750 Hz. Control law is $u = \text{sat}_0^1(0.001e + 100\int edt)$. ■

3.9. Energy Sources: Photovoltaic Modules and Solar Cells

Permanent magnet dc and synchronous generators are covered. These generators are widely used in light- and medium-duty energy systems, auxiliary power systems, airplane electrical power systems, etc. [4-7, 20, 21]. Hydro, gas, turbofan or wind turbine rotates generators, and, voltages are induced. Depending on the prime mover torque and angular velocity of rotation, the generated voltage and frequency vary. In many applications, the induced voltage should be stabilized using dc-dc switching regulators. Inverters convert the dc voltage to ac voltage. For example, in aircraft, the power management provide 5 V, 28 V and 270 V dc voltages, as well as 115 or 235 V 400 Hz ac voltage.

Photovoltaic modules and solar cells are used in renewable energy and power systems. Energy harvesting, energy storage, energy management and distribution are key problems in aerial vehicles, biomedical devices, portable electronics, communication, robots, security, etc. Solar cells and photovoltaic modules may ensure power and energy specifications empowering autonomy, endurance, etc. Commercial single- and multi-junction solar cells are fabricated using amorphous, polycrystalline and monocrystalline silicon, cadmium telluride, copper indium gallium selenide, gallium arsenide and other materials. Solar cells with CdS, CdSe, PbS and Sb_2S_3 quantum dots, as well as organic and polymer solar cells, are commercialized.

The absorption of photons with energy $E=Nhf$ results in generation of electrons, and, subsequent separation of the photo-generated charge carriers. In crystalline silicon solar cells, there are $n+$ doped top and $p+$ back layers. The performance of solar cells depends on many factors, such as: (1) Concentrations N_D and N_A of doping atoms, and, the width of a *p-n* junction space-charge region where the donor atoms donate free electrons, while acceptor atoms accept electrons; (2) Temperature-dependent carriers mobility of electrons μ_n and holes μ_p; (3) Diffusion coefficient D which characterizes the charge carriers transport due to drift and diffusion; (4) Lifetime τ and diffusion length L of the excess carriers which characterize carriers recombination-generation equilibrium and transductions; (5) Photon absorption which depends on the band gap energy E_g, absorption coefficient α, and refractive index R.

Usually, the output voltage of an individual solar cell is ~0.5V. The images of the examined crystalline silicon solar cells are reported in Figure 3.61.a. The experimental and analytic *I–V* and *P–V* characteristics are studied. The solar cell equivalent circuit is depicted at Figure 3.61.b [22, 23].

Figure 3.61. (a) Images of 156×156 mm Si solar cells and photovoltaic module with: (i) STMicroelectronics STEVAL-ISV008V1 300 W, 100 kHz *step-up* dc-dc converter with power MOSFETs board; (ii) High-efficiency *buck-boost* dc-dc switching regulator with input voltage up to 36 V, output voltage 0.8 to 24 V, 5 A, 200 to 400 kHz, controlled by a 16-bit RISC architecture ultra-low-power MPU M430F5438A;
(b) Equivalent circuit to model solar cells and photovoltaic modules.

The experimental *I–V* and *P–V* characteristics are measured to determine the solar cell parameters. These *I–V* and *P–V* characteristics, measured under the standard temperature and irradiance G, may yield (R_s,R_p). Performance and capabilities of solar cells are technology- and design dependent. The efficiency of solar cell is $\eta=P_{max}/GA$, where P_{max} is the maximum power point; G is the irradiance in W/m^2; A is the surface area in m^2. Using an equivalent circuit with two diodes, we have [22, 23]

$$I = I_{ph} - \sum_i I_{D_i} - \frac{V + IR_s}{R_p}, \; I_{D_i} = I_{0_i}(e^{\frac{V+IR_s}{a_{D_i}V_{ta}}} - 1), \tag{3.117}$$

where I_{ph} is the photo current due to irradiation; I_{D_i} are the diode currents; R_s and R_p are the series and shunt resistances, $R_p \gg R_s$; I_{0_i} are the diode leakage currents in the absence of light; a_D is the diode quality coefficient, $1 < a_D < 2$, for a large forward-bias voltage, $a_D \approx 1$, or, $a_D \approx 2$ when diffusion or recombination dominate; V_{ta} is the thermal voltage, $V_{ta} = \frac{1}{q} N_s kT$; N_s is the number of cells connected in series; k is the Boltzmann constant, $k=1.381\times10^{-23}$ J/K; $q=1.602\times10^{-19}$ C; T is the ambient temperature, in Kelvins.

The unknown $(I_{ph}, R_s, R_p, I_0, a_D)$ are estimated using data fitting. The least squares is applied for the experimental I–V characteristics. The residual function is

$$r = y_i - \hat{y}_i, \; r = y_i - f(\alpha, x_i), \tag{3.118}$$

with the model function $f(\cdot,\cdot)$. The least-squares estimator is

$$\hat{\alpha} = \left(X^T X\right)^{-1} X^T y, \; X_{ij} = \frac{\partial}{\partial \alpha_j} f(\alpha, x_i). \tag{3.119}$$

The initial parameters are

$$R_{s_0} = \left.\frac{dV}{dI}\right|_{V_{oc}}, \; R_{p_0} = \left.\frac{dV}{dI}\right|_{I_{sc}}, \; I_{0_0} = \frac{I_{sc}}{2e^{q\frac{V_{oc}}{akT}} - 1} \text{ and } I_{ph_0} = I_{sc}.$$

Here, V_{oc} is the open-circuit voltage; I_{sc} is the short-circuit current.

For nonhomogeneous semiconductors, using the resistivity and conductivity tensors (**ρ**,**σ**), the current density and electric field are **J**=**σE** and **E**=**ρJ**. The electrical resistivity is $\rho = E/J$, where E and J are the electric field and current density.

The current density is a function of the electron and hole mobilities (μ_n, μ_p) and concentrations (n,p),

$$J = q(n\mu_n + p\mu_p)E = \sigma E. \tag{3.120}$$

The resistivity ρ is the inverse of conductivity σ. For intrinsic semiconductors, the resistivity decreases if temperature increases because electrons exhibit the conduction energy band due to thermal energy, while holes fill the valence band.

One has $\rho = \rho_0^{-\alpha T}$, where the Steinhart-Hart coefficient is $T^{-1} = k_1 + k_2\rho + k_3(\ln\rho)^3$.

For extrinsic doped semiconductors, $\sigma = Ae^{-\frac{E_g}{2kT}}$, $E_{g\,\text{Si}}=1.11$ eV.

For one-dimensional case, due to the temperature gradient,

$$J = \sigma\left(\frac{1}{q}\frac{dE_F}{dx} - P_T \frac{dT}{dx}\right), \tag{3.121}$$

where E_F is the Fermi energy level.

Denote the effective density of states in the conducting and valance bands as N_C and N_V. The thermoelectric power is [24]

$$P_T = -\frac{k}{q}\left[\frac{(\frac{5}{2} - s + \ln\frac{N_C}{n})n\mu_n - (\frac{5}{2} - s - \ln\frac{N_V}{p})n\mu_p}{n\mu_n + p\mu_p}\right]. \tag{3.122}$$

Hence,

$$I = I_{ph} - \sum_i I_{D_1} - I_{D_2} - \frac{V + IR_s + Ir_s c^{aI_{pv}}}{R_p + r_p d^{bI_{pv}}}, \tag{3.123}$$

where r_s and r_p are the current dependent resistances; a and b are the constants.

The experimental *I–V* and *P–V* characteristics are examined. For the 156×156 mm solar cell illustrated in Figure 3.61.a, the open-circuit voltage is V_{oc}=0.618 V. At G=1000 W/m^2, short circuit current is I_{sc}=5.49 A, and, $P_{\max}$=2.76 W at 0.532 V and 5.19 A, T=25+T_K, T_K=273.15K. Applying the data fitting procedure, the parameters in (3.123) are found to be R_s=0.00012 ohm, R_p=24.3 ohm, I_{ph}=5.68 A, I_0=1.24×10^{-7} A, a_{D1}=1, a_{D2}=1.34, r_s=0.000035 ohm, r_p=0.28 ohm, a=2.35, b=2.08, c=0.19, d=2.61. The comparison of the experimental and parameterized *I-V* and *P-V* characteristics are reported in Figures 3.62.

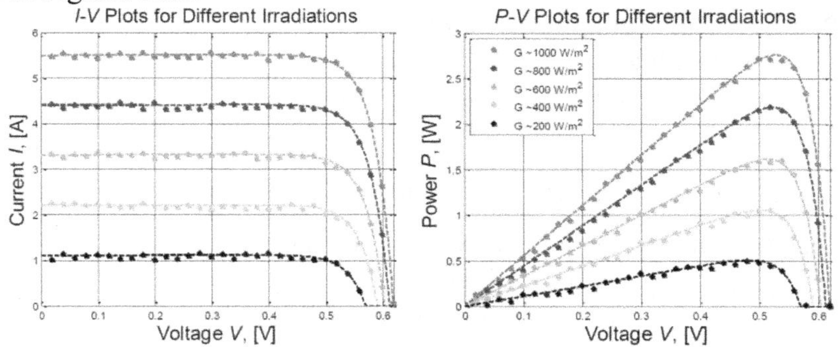

Figure 3.62. Experimental (dots) and parametrized model (solid lines) *I–V* and *P–V* characteristics for different irradiation *G*

3.10. Practice Problems

Practice Problem 3.1. *Model of a Magnetic Levitation System*

Derive the electromagnetic force and equations of motion for a magnetic levitation system if $L(x)=\frac{a}{b+cx+dx^2}$, $(a,b,c,d)>0$.

For the electromagnetic force, $F_e(i,x)=-\frac{\partial W_c(i,x)}{\partial x}=-\frac{1}{2}i^2\frac{dL(x)}{dx}$.

Using the chain rule $\frac{df(u)}{dx}=\frac{df}{du}\frac{du}{dx}$, denote $u=b+cx+dx^2$. Thus, $\frac{d}{du}u^{-1}=-\frac{1}{u^2}$.

From $\frac{d}{dx}(b+cx+dx^2)=c+2dx$, one finds

$$\frac{d}{dx}L(x)=a\left(-\frac{1}{u^2}\right)(c+2dx)=-\frac{a(c+2dx)}{(b+cx+dx^2)^2}.$$

Therefore, $F_e=\frac{a(c+2dx)}{2(b+cx+dx^2)^2}i^2$.

At x=0, the electromagnetic force is maximum, and

$$F_{e\max}=F_e(i,x)\Big|_{x=0}=\frac{a(c+2dx)}{2(b+cx+dx^2)^2}i^2\Bigg|_{x=0}=\frac{ac}{2b^2}i^2.$$

Using the Faradays law of inductance, the total *emf* is

$$\frac{d\psi}{dt}=\frac{d}{dt}\big(L(x)i\big)=\frac{di}{dt}\frac{a}{(b+cx+dx^2)}-\frac{a(c+2dx)}{(b+cx+dx^2)^2}iv.$$

The resulting nonlinear differential equations are

$$\frac{di}{dt}=\frac{b+cx+dx^2}{a}\left[-ri-\frac{a(c+2dx)}{(b+cx+dx^2)^2}iv+u\right],$$

$$\frac{dv}{dt}=\frac{1}{m}\left[\frac{a(c+2dx)}{2(b+cx+dx^2)^2}i^2-B_v v-mg-F_L\right],$$

$$\frac{dx}{dt}=v,\ x_{\min}\le x\le x_{\max}.$$

Practice Problem 3.2. *Analysis of Solenoid*

Consider a solenoid with a magnetizing inductance $L(x)=a\,\mathrm{csch}(bx)$, $(a,b)>0$, $x>0$.

The domain for this $L(x)$ is $\{x\in\mathbb{R}: b\neq 0, x\neq 0, x>0\}$.

The electromagnetic force is

$$F_e(i,x)=-\frac{\partial W_c(i,x)}{\partial x}=-\tfrac{1}{2}i^2\frac{dL(x)}{dx}=\tfrac{1}{2}ab\coth(bx)\mathrm{csch}(bx)i^2.$$

Using the Kirchhoff and Faradays laws, we have

$$u=ri+\frac{d\psi}{dt}=ri+L(x)\frac{di}{dt}+i\frac{dL(x)}{dx}\frac{dx}{dt}=ri+a\,\mathrm{csch}(bx)\frac{di}{dt}-ab\coth(bx)\mathrm{csch}(bx)iv.$$

Applying the laws of electromagnetics and Newtonian mechanics, the governing equations are

$$\frac{di}{dt}=\frac{1}{a\,\mathrm{csch}(bx)}\left[-ri-ab\coth(bx)\mathrm{csch}(bx)iv+u\right],$$

$$\frac{dv}{dt}=\frac{1}{m}\left[\tfrac{1}{2}ab\coth(bx)\mathrm{csch}(bx)i^2-B_v v-k_s(x_0-x)-F_L\right],$$

$$\frac{dx}{dt}=v,\ x_{\min}\le x\le x_{\max},\ x_{\min}>0,\ x_{\min}\neq 0.$$

Let the solenoid stroke is Δx, $\Delta x=(x_{\max}-x_{\min})$. At zero-airgap $x=0$, the reluctance and electromagnetic force are maximum, $F_e|_{x=0}=F_{e\max}$. Let the restoring spring was chosen such as $F_{\text{spring max}}=F_{e\max}$. The spring stretches by δx. For many solenoids $\delta x\ge\Delta x$ or $\delta x=\Delta x$. For an ideal Hooks law, $F_{\text{spring max}}=F_{e\max}=k_s x_0$.

At *zero-length*, $F_{\text{spring min}}=0$ when $F_{e\min}=0$. From $F_{\text{spring}}=k_s(x_0-x)$, $x_0=x_{\max}$.

A spring is selected using $k_s=F_{e\max}/x_0$.

Practice Problem 3.3. *Permanent Magnet dc Generator*

Consider a permanent magnet dc generator which is rotated by a prime mover with ω_r=100 rad/sec, and, $T_{\text{prime mover}}$=1 N-m. The armature voltage at the generator terminal $u_{a\text{terminal}}$ is 10 V, while the current i_a is 8 A. The armature resistance r_a is 0.1 ohm.

The efficiency is

$$\eta=\frac{P_{\text{output}}}{P_{\text{input}}}=\frac{U_a I_a}{T_{\text{prime mover}}\Omega_r}=\frac{10\times 8}{100\times 1}\times 100\%=80\%.$$

For generators, the armature current i_a is in the same direction as the induced *emf*, and, the terminal voltage is $(E_a-r_a i_a)$.

At steady-state, from the Kirchhoff voltage law,

$u_{a\,terminal}=(E_a-r_a i_a)=(k_a\omega_r-r_a i_a)$.

The *back emf* constant is k_a. $k_a=(u_{a\,\text{terminal}}+r_a i_a)/\omega_r=(10+0.1\times 8)/100=0.108$ V-sec/rad.

Practice Problem 3.4. *Efficiency Analysis*

Viscous friction torque $T_{\text{viscous}}(\omega_r)$ and $B_m(\omega_r)$ may be experimentally found in an operating envelope. Let in $\omega_r\in[\omega_{r\min}\ \omega_{r\max}]$, $B_m(\omega_r)=ae^{-b\omega_r}$, $a>0$, $b>0$.

The mechanical and *total* losses are $P_M=B_m(\omega_r)\omega_r^2=ae^{-b\omega_r}\omega_r^2$, $P_{\text{losses}}=r_a i_a^2+ae^{-b\omega_r}\omega_r^2$.

If T_L is not measured, one uses the *total* power losses estimate.

The efficiency estimate $\eta(I_a,\Omega_r)$ is $\eta=\dfrac{P_{\text{output}}}{P_{\text{input}}}=\dfrac{P_{\text{input}}-P_{\text{losses}}}{P_{\text{input}}}=\dfrac{U_aI_a-\left[r_aI_a^2+ae^{-b\Omega_r}\Omega_r^2\right]}{U_aI_a}$.

Practice Problem 3.5. *Axial Topology Limited Angle Actuator*

Model an axial topology limited angle actuator if $B(\theta_r)=B_{\max}k\theta_r$. The angular displacement is limited as $-0.175\le\theta_r\le 0.175$ rad, $\theta_{L0}=\theta_{R0}=0.175$ rad.

The parameters are $B_{\max}=1$ T, $k=5$, $r_a=35$ ohm, $L_a=4.1\times10^{-3}$ H, $R_\perp=0.02$ m, $N=100$, $l_{eq}=0.0125$ m, $B_m=5\times10^{-4}$ N-m-sec/rad, $k_s=0.05$ N-m/rad, $J=1.5\times10^{-6}$ kg-m^2, $\theta_{R0}=0.175$ rad, $r_{\text{in}}=0.015$ m, $r_{\text{out}}=0.025$ m, $A_{eq}=\frac{1}{2}(r_{\text{out}}^2-r_{\text{in}}^2)$.

For $B(\theta_r)=B_{\max}k\theta_r$, one finds

$$\mathscr{E}=-N\frac{d}{dt}\int_{r_{\text{in}}}^{r_{\text{out}}}\int_{\theta_R}^{\theta_L}B_{\max}k\theta_r r dr d\theta_i=-\frac{r_{\text{out}}^2-r_{\text{in}}^2}{2}NB_{\max}k(\theta_L-\theta_R)\omega_r,\ \theta_L(t)=\theta_{L0}-\theta_r(t),\theta_R(t)=\theta_{R0}+\theta_r(t).$$

$$\text{Thus, }\mathscr{E}=-\frac{r_{\text{out}}^2-r_{\text{in}}^2}{2}NB_{\max}k[(\theta_{L0}-\theta_r)-(\theta_{R0}+\theta_r)]\omega_r=(r_{\text{out}}^2-r_{\text{in}}^2)NB_{\max}k\theta_r\omega_r,\ \theta_{L0}=\theta_{R0}.$$

The electromagnetic torque is $T_e=R_\perp Nl_{eq}B_{\max}k(\theta_L+\theta_R)i_a=2R_\perp Nl_{eq}B_{\max}k\theta_{R0}i_a$.

From the Kirchhoff and Newton second laws, one has

$$\frac{di_a}{dt}=\frac{1}{L_a}\left[-r_ai_a-(r_{\text{out}}^2-r_{\text{in}}^2)NB_{\max}k\theta_r\omega_r+u_a\right],$$

$$\frac{d\omega_r}{dt}=\frac{1}{J}\left[2R_\perp Nl_{eq}B_{\max}k\theta_{R0}i_a-T_{\text{friction}}-T_L\right],\ T_{\text{friction}}=B_m\omega_r,\ T_{\text{elastic}}=k_s\theta_r,$$

$$\frac{d\theta_r}{dt}=\omega_r,\ -\theta_{r\max}\le\theta_r\le\theta_{r\max}.$$

The parameters and constants are upload as

```
Bmax=1; k=5; ra=35; La=4.1e-3; Rp=0.02; N=100; leq=1.25e-2; Bm=5e-4; ks=0.05; J=1.5e-6;
TR0=0.175; Rin=0.015; Rou=0.025; Aeq=(Rou^2-Rin^2)/2;
```

The SIMULINK diagram is reported in Figure 3.63.a. The transient dynamics for $i_a(t)$ and $\theta_r(t)$ is depicted in Figure 3.63.b if u_a is applied as steps ±5 V with $f=10$ Hz.

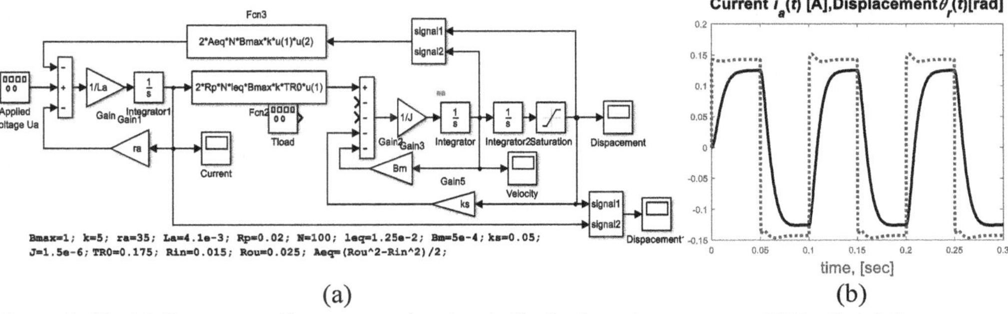

(a) (b)

Figure 3.63. (a) SIMULINK diagram to simulate a limited angle actuator, $B(\theta_r)=B_{\max}k\theta_r$; (b) Transient dynamics for $i_a(t)$ and $\theta_r(t)$ illustrated by dotted and solid lines.

Practice Problem 3.6. *Hard-Disc Drive Actuator*
Examples 3.19 and 3.20 report results for different magnetic fileds $B(\theta_r)$.

The motional *emf*, flux and force are found as

$$\mathscr{E}=\oint_l \vec{E}(t)\cdot d\vec{l} = -\oint_s \frac{\partial \vec{B}}{\partial t}\cdot d\vec{s}\,,\ \Phi=\oint_s \vec{B}\cdot d\vec{s}\,,\ \vec{F}=\oint_l i d\vec{l}\times\vec{B}=-i\oint_l \vec{B}\times d\vec{l}\,.$$

The magnetic system nonlinearities are described by real-valued continuous and integrable functions. Study an axial topology limited angle actuator,

$$B(\theta_r)=B_{\max}\frac{\theta_r}{\sqrt{a+\theta_r^2}},\ a>0,\ a\ll 1.$$

This continuous $B(\theta_r)$ describes the assumed discontinuous field $B(\theta_r)=B_{\max}\mathrm{sgn}(\theta_r)$.

For $f(\theta)=\dfrac{b\theta}{\sqrt{a+\theta^2}}$, $\displaystyle\int\frac{b\theta}{\sqrt{a+\theta^2}}d\theta=\tfrac{1}{2}b\sqrt{a+\theta^2}+C$.

Using results of Section 3.4.3, for $T_{\text{friction}}=B_m\omega_r$ and $T_{\text{elastic}}=k_s\theta_r$, one finds

$$\frac{di_a}{dt}=\frac{1}{L_a}\left[-r_a i_a-\frac{r_{out}^2-r_{in}^2}{2}NB_{\max}\left[\frac{\theta_{L0}-\theta_r}{\sqrt{a+(\theta_{L0}-\theta_r)^2}}-\frac{\theta_{R0}+\theta_r}{\sqrt{a+(\theta_{R0}+\theta_r)^2}}\right]\omega_r+u_a\right],$$

$$\frac{d\omega_r}{dt}=\frac{1}{J}\left[R_\perp N l_{eq}B_{\max}\left[\frac{\theta_{L0}-\theta_r}{\sqrt{a+(\theta_{L0}-\theta_r)^2}}+\frac{\theta_{R0}+\theta_r}{\sqrt{a+(\theta_{R0}+\theta_r)^2}}\right]i_a-B_m\omega_r-k_s\theta_r\right],$$

$$\frac{d\theta_r}{dt}=\omega_r,\ -\theta_{r\max}\le\theta_r\le\theta_{r\max}.$$

Practice Problem 3.7. *Control of a Permanent Magnet Synchronous Motor*
The model developments are covered in section 3.6.1. For a 50 W motor, the parameters are $P=4$, $r_s=1.5$ ohm, $L_{ss}=0.005$ H, $\psi_m=0.1$ V-sec/rad, $B_m=0.0001$ N-m-sec/rad, $J=0.0002$ kg-m^2. Using the Kirchhoff voltage law and Newtonian dynamics, for

$$\psi_{asm}=\psi_m\sin\theta_r,\ \psi_{bsm}=-\psi_m\cos\theta_r,$$

we obtained the resulting model (3.43)

$$\frac{di_{as}}{dt}=\frac{1}{L_{ss}}(-r_s i_{as}-\psi_m\cos\theta_r\omega_r+u_{as}),\ \frac{di_{bs}}{dt}=\frac{1}{L_{ss}}(-r_s i_{bs}-\psi_m\sin\theta_r\omega_r+u_{bs}),$$

$$\frac{d\omega_r}{dt}=\frac{1}{J}\left[\frac{P^2\psi_m}{4}(i_{as}\cos\theta_r+i_{bs}\sin\theta_r)-B_m\omega_r-\frac{P}{2}T_L\right],\ \frac{d\theta_r}{dt}=\omega_r.$$

The SIMULINK diagrams for open-loop and closed-loop systems are reported in Figures 3.64.a. Figure 3.64.b documents the angular velocity $\omega_r(t)$ of the open-loop drive for the balanced voltage set (3.45)

$$u_{as}=\sqrt{2}u_M\cos\theta_r,\ u_{bs}=\sqrt{2}u_M\sin\theta_r,\ u_M=\pm 50\text{ V}$$

if $T_L=0$ N-m. The bidirectional rotation of motor is achieved.

The proportional-integral control law yields the voltage magnitude for the phase voltages

$$u_{as}=u\cos\theta_r,\ u=10e+\textstyle\int e dt,\ e(t)=(\omega_{\text{ref}}-\omega_r),$$

$$u_{bs}=u\sin\theta_r,$$

ensuring the balanced operation when (u_{as},u_{bs}) are not reaching limits. The tracking error is $e(t)=(\omega_{\text{ref}}-\omega_r)$

The amplitude of the voltage is limited as $-\sqrt{2}u_M \le u \le \sqrt{2}u_M$, u_M=50 V.

The limits on phase voltages yield bounds on control

$u_{as} = u\cos\theta_r$, $u_{bs} = u\sin\theta_r$, $u = \mathrm{sat}_{-\sqrt{2}u_M}^{\sqrt{2}u_M}\left(10e + \int edt\right)$.

These limits affect system dynamics and stability. The phase voltage limits reach for the specified reference ω_{ref}. For low $\omega_{\mathrm{ref}}(t)$ and rated $T_L(t)$, the balanced operation is ensured. The closed-loop electric drive dynamics are depicted in Figure 3.64.c. Tracking and disturbance attenuation are accomplished. The motor rotates clockwise and counterclockwise with ω_r=±300 rad/sec at peak loads T_L=±1 N-m.

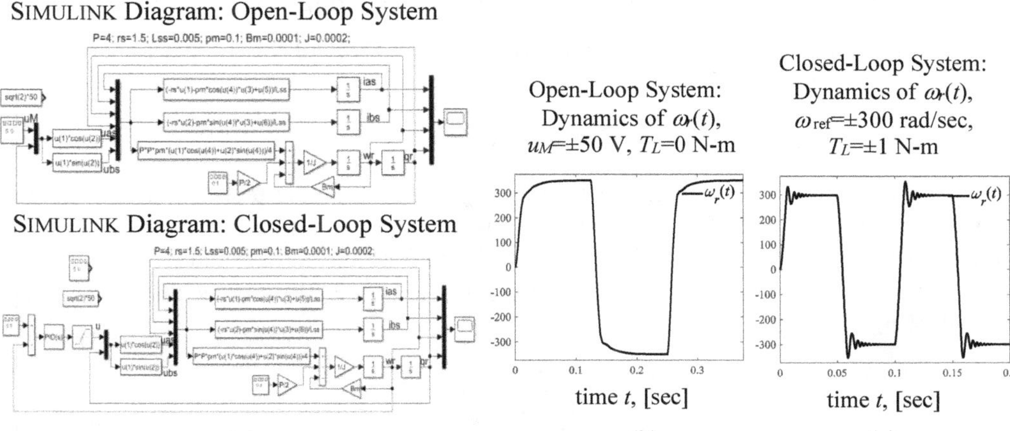

(a) (b) (c)

Figure 3.64. (a) SIMULINK diagrams to simulate a two-phase permanent magnet synchronous motor as open-loop and closed-loop systems;
(b) Open-loop motor accelerations for the balanced voltage set $u_{as} = \sqrt{2}u_M\cos\theta_r$, $u_{bs} = \sqrt{2}u_M\sin\theta_r$, u_M=±50 V. The clockwise and counterclockwise rotation is achieved;
(c) Closed-loop motor velocity $\omega_r(t)$ with the proportional-integral tracking control law which ensures the balanced motor operation with phase voltages $u_{as} = u\cos\theta_r$, $u_{bs} = u\sin\theta_r$, $u = \mathrm{sat}_{-\sqrt{2}u_M}^{\sqrt{2}u_M}\left(10e + \int edt\right)$, $e(t)=\omega_{\mathrm{ref}}-\omega_r$, $-\sqrt{2}u_M \le u \le \sqrt{2}u_M$. The reference is $\omega_{\mathrm{ref}}(t)$=±300 rad/sec, and, peak loads T_L=±1 N-m are applied with frequency 20 Hz .

As reported in section 3.6.6 and Example 3.26, consider

$\psi_{asm} = -\psi_m(a_1\sin\theta_r + a_2\sin^3\theta_r)$, $\psi_{bsm} = \psi_m(a_1\cos\theta_r + a_2\cos^3\theta_r)$.

The governing equations are

$$\frac{di_{as}}{dt} = \frac{1}{L_{ss}}\left[-r_s i_{as} - \psi_m\cos\theta_r(a_1 + 3a_2\sin^2\theta_r)\omega_r + u_{as}\right],$$

$$\frac{di_{bs}}{dt} = \frac{1}{L_{ss}}\left[-r_s i_{bs} - \psi_m\sin\theta_r(a_1 + 3a_2\cos^2\theta_r)\omega_r + u_{bs}\right],$$

$$\frac{d\omega_r}{dt} = \frac{1}{J}\left[\frac{P^2\psi_m}{4}\left[i_{as}\cos\theta_r\left(a_1 + 3a_2\sin^2\theta_r\right) + i_{bs}\sin\theta_r\left(a_1 + 3a_2\cos^2\theta_r\right)\right] - B_m\omega_r - \frac{P}{2}T_L\right],$$

$$\frac{d\theta_r}{dt} = \omega_r.$$

From expression for the electromagnetic torque

$$T_e = \frac{P\psi_m}{2}\left[i_{as}\cos\theta_r\left(a_1 + 3a_2\sin^2\theta_r\right) + i_{bs}\sin\theta_r\left(a_1 + 3a_2\cos^2\theta_r\right)\right],$$

the balanced voltage set is

$$u_{as} = u_M \frac{\cos\theta_r}{a_1 + 3a_2\sin^2\theta_r}, \; |u_{as}| \le u_{\max},$$

$$u_{bs} = u_M \frac{\sin\theta_r}{a_1 + 3a_2\cos^2\theta_r}, \; |u_{bs}| \le u_{\max}.$$

For a_1=1 and a_2=0.1, the simulations are performed. The SIMULINK diagram is depicted in Figure 3.65.a. We investigate the closed-loop electric drive dynamics and tracking with a proportional-integral control law considering voltage limits. We have

$$u_{as} = u\frac{\cos\theta_r}{a_1 + 3a_2\sin^2\theta_r}, \; |u_{as}| \le u_{\max}, \; u = \operatorname{sat}_{-\sqrt{2}u_M}^{\sqrt{2}u_M}\left(10e + \int e dt\right), -\sqrt{2}u_M \le u \le \sqrt{2}u_M,$$

$$u_{bs} = u\frac{\sin\theta_r}{a_1 + 3a_2\cos^2\theta_r}, \; |u_{bs}| \le u_{\max}.$$

For $\omega_{\text{ref}}(t)$=±300 rad/sec at peak load T_L=±1 N-m, the phase voltages (u_{as},u_{bs}) reach the limits as depicted in Figure 3.65.b. Dynamics of phase currents (i_{as},i_{bs}) are reported in Figure 3.65.c. As illustrated in Figure 3.65.c, motor rotates clockwise and counterclockwise ensuring ω_r=±300 rad/sec for $\omega_{\text{ref}}(t)$=±300 rad/sec if the loads are T_L=±1 N-m, f=20 Hz.

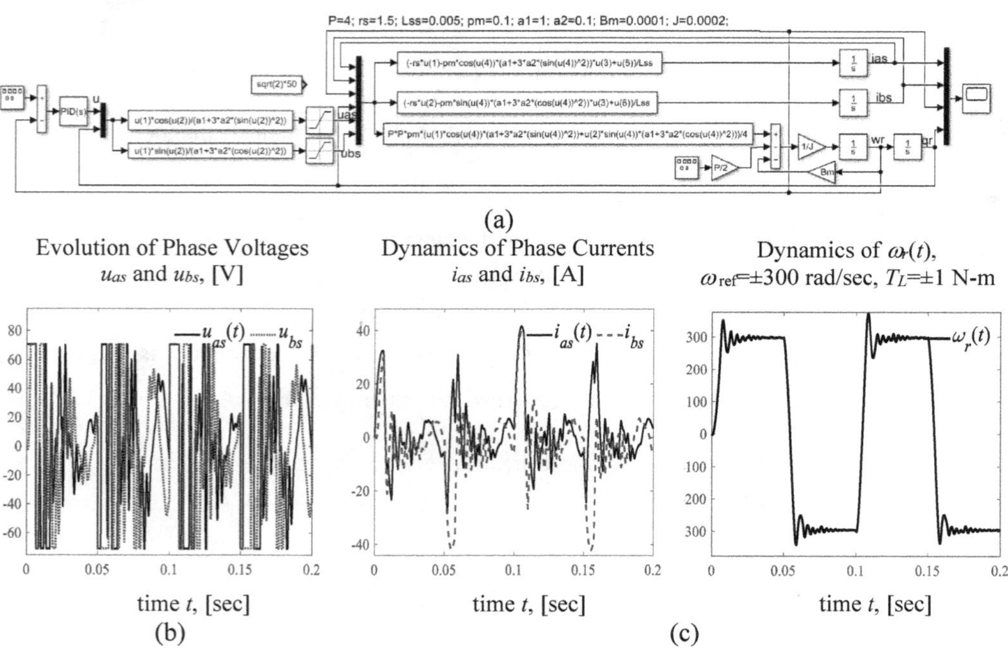

(b) (c)

Figure 3.65. (a) SIMULINK diagrams to simulate a closed-loop electric drive;
(b) Evolution of phase voltages (u_{as},u_{bs}),

$$u_{as} = u\frac{\cos\theta_r}{a_1 + 3a_2\sin^2\theta_r}, \; |u_{as}| \le u_{\max}, \; u_{bs} = u\frac{\sin\theta_r}{a_1 + 3a_2\cos^2\theta_r}, \; |u_{bs}| \le u_{\max}, -\sqrt{2}u_M \le u \le \sqrt{2}u_M, \; u_M\text{=50 V},$$

$$u = \operatorname{sat}_{-\sqrt{2}u_M}^{\sqrt{2}u_M}\left(k_p e + k_i \int e dt\right), \; k_p\text{=10}, \; k_i\text{=1};$$

(c) Dynamics of phase currents (i_{as},i_{bs}) and velocity $\omega_r(t)$, $\omega_{\text{ref}}(t)$=±300 rad/sec, T_L=±1 N-m.

Practice Problem 3.8. *Two-Phase Induction Motors*
Two-phase induction motors were invented and commercialized by Nikola Tesla. In 1888, a US patent 381,968 *Electro Magnetic Motor* was secured by Tesla. Three-phase induction motors have being broadly used. Michail Dolivo-Dobrowolsky acquired the US patent 422746 *Electrical Induction Apparatus or Transformer*, issued in 1890. Excellent books are written on induction motors. In practice problems 3.8, 3.9, 3.10, 3.11 and 5.9, we cover key basics on adequate analysis and control of induction motors.

Study an alternating current (ac) two-phase induction motor, illustrated in Figure 3.66.a. The stator (*as*,*bs*) and rotor (*ar*,*br*) windings, as well as the stator-rotor magnetic coupling, are depicted. To rotate and control squirrel-cage induction motors, one varies the frequency and magnitude of the ac phase voltages (u_{as},u_{bs}) supplied to the stator windings. In squirrel-cage induction motors, due to a time-varying magnetic field coupling and motion of a rotor with respect to stator, the ac voltages ($\mathscr{E}_{ar}$,$\mathscr{E}_{br}$) are induced as motional *emfs*. The ac rotor currents (i_{ar},i_{br}) yield flux linkages (ψ_{ar}, ψ_{br}) within the close magnetic path. The stator-rotor mutual inductances (L_{asar},L_{asbr},L_{bsar},L_{bsbr}) are used to derive the governing equations, as well as to examine the electromagnetic torque T_e. The electromagnetic torque results due to spatiotemporal interacting of electromagnetic fields.

The ac applied voltages to the motor windings cannot exceed the rated voltage, $u_{M\min} \leq u_M \leq u_{M\max}$. The synchronous angular velocity of the *mmf* is a function of the frequency of the applied voltages f. The synchronous angular velocity is given as

$$\omega_e = \frac{4\pi f}{P}.$$

Induction motors, also called *asynchronous motors*, rotate at the electrical angular velocity $\omega_r \leq \omega_e$.

A difference between the synchronous velocity ω_e and the *electrical* angular velocity ω_r is the *slip*, and

$$slip = \frac{\omega_e - \omega_r}{\omega_e}.$$

At no load conditions, assuming no friction and drag, one has $\omega_r = \omega_e$, and, *slip*=0. The limit $f_{\min} \leq f \leq f_{\max}$ is due to the power electronics limits to support sinusoidal PWM voltage waveforms with $f \in [f_{\min}\ f_{\max}]$. There are mechanical limits on the maximum angular velocity $\omega_{r\max}$. To vary ω_r, one changes the magnitude of the applied voltages u_M and f. Control concepts are application specific, depend on induction motor class, power electronics and loads. The torque-speed characteristics $\omega_r = \phi(T)$ are documented in Figure 3.66.b.

Frequency and Voltage-Frequency Control – A key principle to control industrial induction motors is the frequency control. Regulating the frequency f of phase voltages (u_{as},u_{bs}), one also may change the voltage magnitude u_M specifying the ratio u_M/f within the operating envelope as illustrated in Figure 3.66.b.

The Kirchhoff voltage law yields equations for the phase voltages (u_{as},u_{bs},u_{ar},u_{br}), currents (i_{as},i_{bs},i_{ar},i_{br}) and flux linkages (ψ_{as}, ψ_{bs}, ψ_{ar}, ψ_{br}) for the coupled stator and rotor circuitry with the *motional emf* terms.

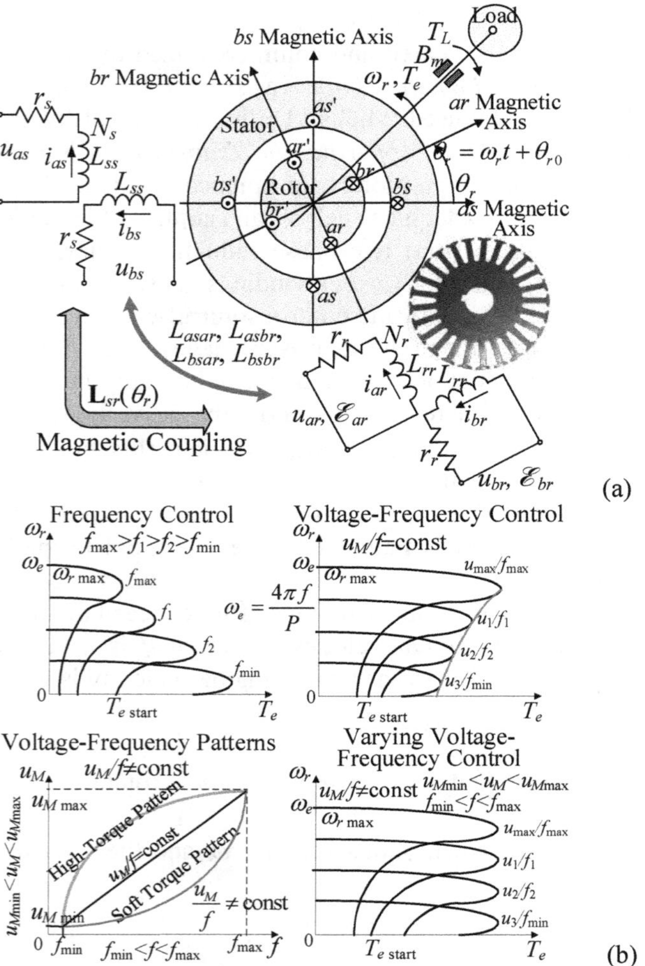

Figure 3.66. (a) Two-phase symmetric induction motor;
(b) For constant magnitude of the supplied phase voltages u_M=const, the angular velocity is regulated by varying the voltage frequency f, $f_{\min} \leq f \leq f_{\max}$, which defines the synchronous angular velocity $\omega_e = \dfrac{4\pi f}{P}$. Frequency control guarantees superior capabilities due to high starting torque $T_{e\ \text{start}}$ and consistent regulation within torque-speed characteristics $\omega_r = \phi(T)$. To minimize losses, the voltage magnitude u_M and frequency f are regulated. The *constant-volts-per-hertz* control is guaranteed if u_M/f=const. Acceleration capabilities are ensured by soft- and stringent torque patterns.

Consider the Kirchhoff voltage law equations, which yield:

Stator Kirchhoff voltage law equations as

$$u_{as} = r_s i_{as} + \frac{d\psi_{as}}{dt},$$

$$u_{bs} = r_s i_{bs} + \frac{d\psi_{bs}}{dt},$$

Rotor Kirchhoff voltage law equations as

$$u_{ar} = r_r i_{ar} + \frac{d\psi_{ar}}{dt},$$

$$u_{br} = r_r i_{br} + \frac{d\psi_{br}}{dt},$$

where u_{as} and u_{bs} are the ac phase voltages applied to the stator windings *as* and *bs*; u_{ar} and u_{br} are the phase voltages in the rotor windings; i_{as}, i_{bs}, i_{ar} and i_{br} are the phase currents in the stator and rotor windings; ψ_{as}, ψ_{bs}, ψ_{ar} and ψ_{br} are the stator and rotor flux linkages; r_s and r_r are the resistances of the stator and rotor windings.

The flux linkages are expressed using the self- and mutual inductances as:

Flux linkages in stator $\psi_{as} = L_{asas}i_{as} + L_{asbs}i_{bs} + L_{asar}i_{ar} + L_{asbr}i_{br}$,

$\psi_{bs} = L_{bsas}i_{as} + L_{bsbs}i_{bs} + L_{bsar}i_{ar} + L_{bsbr}i_{br}$,

Flux linkages in rotor $\psi_{ar} = L_{aras}i_{as} + L_{arbs}i_{bs} + L_{arar}i_{ar} + L_{arbr}i_{br}$,

$\psi_{br} = L_{bras}i_{as} + L_{brbs}i_{bs} + L_{brar}i_{ar} + L_{brbr}i_{br}$,

where L_{asas}, L_{bsbs}, L_{arar} and L_{brbr} are the self-inductances of the stator and rotor windings; L_{asbs}, L_{asar}, L_{asbr}, L_{bsar}, L_{bsbr} and L_{arbr} are the mutual inductances between the stator and rotor windings, $L_{asbs}=L_{bsas}$, $L_{asar}=L_{aras}$, $L_{asbr}=L_{bras}$, $L_{bsar}=L_{arbs}$, $L_{bsbr}=L_{brbs}$, $L_{arbr}=L_{brar}$.

Assume that the magnetic system is linear with an ideal winding design and symmetric magnetic system. The stator and rotor self-inductances are $L_{ss}=L_{asas}=L_{bsbs}$ and $L_{rr}=L_{arar}=L_{brbr}$. The stator and rotor windings are orthogonal. There are no magnetic coupling between the *as* and *bs*, as well as between *ar* and *br* windings. Hence, $L_{asbs}=L_{bsas}=0$ and $L_{arbr}=L_{brar}=0$.

Rotor rotates at the *electrical* angular velocity ω_r. The mutual inductances between stator and rotor windings are periodic functions of the electrical angular displacement θ_r. The stator-rotor mutual inductances have minimum and maximum values. The phases magnetic couplings are periodic with a period 2π. The maximum coupling occurs when the *ar* rotor winding faces the *as* stator winding. The zero coupling occurs when the *ar* rotor winding is orthogonal to the *as* stator winding. Ideally, the induction motor electromagnetic system design yields the sinusoidal mutual inductances between stator and rotor windings. The mutual inductances depend on the initial displacement of rotor windings (*ar*,*br*) with respect to stator windings (*as*,*bs*). The resulting governing equations are identical for any initial orientation of rotor with respect to stator. Let the (*as*,*bs*) windings are displaced with respect to the (*ar*,*br*) windings such that

$L_{asar}=L_{aras}=L_{sr}\cos\theta_r$, $L_{asbr}=L_{bras}=-L_{sr}\sin\theta_r$, $L_{bsar}=L_{arbs}=L_{sr}\sin\theta_r$, $L_{bsbr}=L_{brbs}=L_{sr}\cos\theta_r$.

Consider magnetically coupled windings. Flux linkages expressions for an assumed initial (*as*,*bs*) to (*ar*,*br*) orientation are

$$\psi_{as} = L_{ss}i_{as} + L_{sr}\cos\theta_r i_{ar} - L_{sr}\sin\theta_r i_{br},\ \psi_{bs} = L_{ss}i_{bs} + L_{sr}\sin\theta_r i_{ar} + L_{sr}\cos\theta_r i_{br},$$

$$\psi_{ar} = L_{sr}\cos\theta_r i_{as} + L_{sr}\sin\theta_r i_{bs} + L_{rr}i_{ar},\ \psi_{br} = -L_{sr}\sin\theta_r i_{as} + L_{sr}\cos\theta_r i_{bs} + L_{rr}i_{br}.$$

The flux linkages and current vectors are related using the inductance matrices ($\mathbf{L}_s$,$\mathbf{L}_r$) and inductance mapping $\mathbf{L}_{sr}(\theta_r)$. One has

$$\begin{bmatrix} \boldsymbol{\psi}_{abs} \\ \boldsymbol{\psi}_{abr} \end{bmatrix} = \begin{bmatrix} \mathbf{L}_s & \mathbf{L}_{sr}(\theta_r) \\ \mathbf{L}_{sr}^T(\theta_r) & \mathbf{L}_r \end{bmatrix} \begin{bmatrix} \mathbf{i}_{abs} \\ \mathbf{i}_{abr} \end{bmatrix},$$

$$\mathbf{L}_s = \begin{bmatrix} L_{ss} & 0 \\ 0 & L_{ss} \end{bmatrix},\ \mathbf{L}_r = \begin{bmatrix} L_{rr} & 0 \\ 0 & L_{rr} \end{bmatrix},\ \mathbf{L}_{sr}(\theta_r) = \begin{bmatrix} L_{sr}\cos\theta_r & -L_{sr}\sin\theta_r \\ L_{sr}\sin\theta_r & L_{sr}\cos\theta_r \end{bmatrix},$$

where $\mathbf{L}_s$ is the matrix of the stator self-inductances, $L_{ss}=L_{ls}+L_{ms}$, $L_{ms}=\dfrac{N_s^2}{\Re_m}$; $\mathbf{L}_r$ is the matrix of the rotor self-inductances, $L_{rr}=L_{lr}+L_{mr}$, $L_{mr}=\dfrac{N_r^2}{\Re_m}$; $\mathbf{L}_{sr}(\theta_r)$ is the stator-rotor mutual inductance mapping, $\mathbf{L}_{sr}(\theta_r)=\begin{bmatrix} L_{sr}\cos\theta_r & -L_{sr}\sin\theta_r \\ L_{sr}\sin\theta_r & L_{sr}\cos\theta_r \end{bmatrix}$, $L_{sr}=\dfrac{N_s N_r}{\Re_m}$; L_{ms} and L_{mr} are the stator and rotor magnetizing inductances; L_{ls} and L_{lr} are the stator and rotor leakage inductances; N_s and N_r are the number of turns of the stator and rotor windings; $\Re_m$ is the magnetizing reluctance.

Using the number of turns of stator and rotor windings, define

$$\mathbf{i}'_{abr}=\frac{N_r}{N_s}\mathbf{i}_{abr},\ \mathbf{u}'_{abr}=\frac{N_s}{N_r}\mathbf{u}_{abr},\ \boldsymbol{\psi}'_{abr}=\frac{N_s}{N_r}\boldsymbol{\psi}_{abr}.$$

Applying the turn ratio, the flux linkages are

$$\begin{bmatrix}\boldsymbol{\psi}_{abs}\\ \boldsymbol{\psi}'_{abr}\end{bmatrix}=\begin{bmatrix}\mathbf{L}_s & \mathbf{L}'_{sr}(\theta_r)\\ \mathbf{L}'^{T}_{sr}(\theta_r) & \mathbf{L}'_r\end{bmatrix}\begin{bmatrix}\mathbf{i}_{abs}\\ \mathbf{i}'_{abr}\end{bmatrix},$$

$$\mathbf{L}'_r=\left(\frac{N_s}{N_r}\right)^2\mathbf{L}_r=\begin{bmatrix}L'_{rr} & 0\\ 0 & L'_{rr}\end{bmatrix},\ \mathbf{L}'_{sr}(\theta_r)=\left(\frac{N_s}{N_r}\right)\mathbf{L}_{sr}(\theta_r)=L_{ms}\begin{bmatrix}\cos\theta_r & -\sin\theta_r\\ \sin\theta_r & \cos\theta_r\end{bmatrix},$$

where $L'_{rr}=L'_{lr}+L'_{mr}$, $L_{ms}=\dfrac{N_s}{N_r}L_{sr}$, $L'_{mr}=\dfrac{N_s^2}{N_r^2}L_{mr}$, $L'_{mr}=L_{ms}=\dfrac{N_s}{N_r}L_{sr}$, $L'_{rr}=L'_{lr}+L_{ms}$.

Hence,

$$\begin{bmatrix}\boldsymbol{\psi}_{abs}\\ \boldsymbol{\psi}'_{abr}\end{bmatrix}=\begin{bmatrix}\mathbf{L}_s & \mathbf{L}'_{sr}(\theta_r)\\ \mathbf{L}'^{T}_{sr}(\theta_r) & \mathbf{L}'_r\end{bmatrix}\begin{bmatrix}\mathbf{i}_{abs}\\ \mathbf{i}'_{abr}\end{bmatrix},$$

$$\begin{bmatrix}\psi_{as}\\ \psi_{bs}\\ \psi'_{ar}\\ \psi'_{br}\end{bmatrix}=\begin{bmatrix}L_{ss} & 0 & L_{ms}\cos\theta_r & -L_{ms}\sin\theta_r\\ 0 & L_{ss} & L_{ms}\sin\theta_r & L_{ms}\cos\theta_r\\ L_{ms}\cos\theta_r & L_{ms}\sin\theta_r & L'_{rr} & 0\\ -L_{ms}\sin\theta_r & L_{ms}\cos\theta_r & 0 & L'_{rr}\end{bmatrix}\begin{bmatrix}i_{as}\\ i_{bs}\\ i'_{ar}\\ i'_{br}\end{bmatrix}.$$

In the vector-matrix form

$$\mathbf{u}_{abs}=\mathbf{r}_s\mathbf{i}_{abs}+\frac{d\boldsymbol{\psi}_{abs}}{dt},\ \mathbf{r}_s=\begin{bmatrix}r_s & 0\\ 0 & r_s\end{bmatrix},$$

$$\mathbf{u}'_{abr}=\mathbf{r}'_r\mathbf{i}'_{abr}+\frac{d\boldsymbol{\psi}'_{abr}}{dt},\ \mathbf{r}'_r=\frac{N_s^2}{N_r^2}\mathbf{r}_r=\frac{N_s^2}{N_r^2}\begin{bmatrix}r'_r & 0\\ 0 & r'_r\end{bmatrix}.$$

Assume that the magnetization BH is linear without hysteresis. Therefore, μ_r, L_{ms} and self-inductances (L_{ss}, L'_{rr}) are constants.

We obtain a set of four nonlinear differential equations

$$L_{ss}\frac{di_{as}}{dt}+L_{ms}\frac{d\left(i_{ar}^{'}\cos\theta_r\right)}{dt}-L_{ms}\frac{d\left(i_{br}^{'}\sin\theta_r\right)}{dt}=-r_si_{as}+u_{as},$$

$$L_{ss}\frac{di_{bs}}{dt}+L_{ms}\frac{d\left(i_{ar}^{'}\sin\theta_r\right)}{dt}+L_{ms}\frac{d\left(i_{br}^{'}\cos\theta_r\right)}{dt}=-r_si_{bs}+u_{bs},$$

$$L_{ms}\frac{d\left(i_{as}\cos\theta_r\right)}{dt}+L_{ms}\frac{d\left(i_{bs}\sin\theta_r\right)}{dt}+L_{rr}^{'}\frac{di_{ar}^{'}}{dt}=-r_r^{'}i_{ar}^{'}+u_{ar}^{'},$$

$$-L_{ms}\frac{d\left(i_{as}\sin\theta_r\right)}{dt}+L_{ms}\frac{d\left(i_{bs}\cos\theta_r\right)}{dt}+L_{rr}^{'}\frac{di_{br}^{'}}{dt}=-r_r^{'}i_{br}^{'}+u_{br}^{'}.$$

One finds

$$\frac{di_{as}}{dt}=\frac{1}{L_{ss}}\left[-r_si_{as}-L_{ms}\left(\frac{di_{ar}^{'}}{dt}\cos\theta_r-i_{ar}^{'}\sin\theta_r\omega_r\right)+L_{ms}\left(\frac{di_{br}^{'}}{dt}\sin\theta_r+i_{br}^{'}\cos\theta_r\omega_r\right)+u_{as}\right],$$

$$\frac{di_{bs}}{dt}=\frac{1}{L_{ss}}\left[-r_si_{bs}-L_{ms}\left(\frac{di_{ar}^{'}}{dt}\sin\theta_r+i_{ar}^{'}\cos\theta_r\omega_r\right)-L_{ms}\left(\frac{di_{br}^{'}}{dt}\cos\theta_r-i_{br}^{'}\sin\theta_r\omega_r\right)+u_{bs}\right],$$

$$\frac{di_{ar}^{'}}{dt}=\frac{1}{L_{rr}^{'}}\left[-r_r^{'}i_{ar}^{'}-L_{ms}\left(\frac{di_{as}}{dt}\cos\theta_r-i_{as}\sin\theta_r\omega_r\right)-L_{ms}\left(\frac{di_{bs}}{dt}\sin\theta_r+i_{bs}\cos\theta_r\omega_r\right)+u_{ar}^{'}\right],$$

$$\frac{di_{br}^{'}}{dt}=\frac{1}{L_{rr}^{'}}\left[-r_r^{'}i_{br}^{'}+L_{ms}\left(\frac{di_{as}}{dt}\sin\theta_r+i_{as}\cos\theta_r\omega_r\right)-L_{ms}\left(\frac{di_{bs}}{dt}\cos\theta_r-i_{bs}\sin\theta_r\omega_r\right)+u_{br}^{'}\right].$$

The Faraday law of induction is $\mathscr{E}=\oint_l \vec{E}(t)\cdot d\vec{l}=-N\frac{d\Phi}{dt}=-\frac{d\psi}{dt}$.

In the squirrel-cage induction motors, the rotor voltages are induced as the *motional emfs* $\mathscr{E}_{ar}$ and $\mathscr{E}_{br}$. The *total emfs* in rotor windings are

$$\mathscr{E}_{ar}=-L_{ms}\frac{d\left(i_{as}\cos\theta_r\right)}{dt}-L_{ms}\frac{d\left(i_{bs}\sin\theta_r\right)}{dt}-L_{rr}^{'}\frac{di_{ar}^{'}}{dt},$$

$$\mathscr{E}_{br}=L_{ms}\frac{d\left(i_{as}\sin\theta_r\right)}{dt}-L_{ms}\frac{d\left(i_{bs}\cos\theta_r\right)}{dt}-L_{rr}^{'}\frac{di_{br}^{'}}{dt}.$$

The induced voltages in the (ar,br) windings are the *motional emfs*

$$\mathscr{E}_{ar\omega}=L_{ms}\left(i_{as}\sin\theta_r-i_{bs}\cos\theta_r\right)\omega_r,$$

$$\mathscr{E}_{br\omega}=L_{ms}\left(i_{as}\cos\theta_r+i_{bs}\sin\theta_r\right)\omega_r.$$

The *torsional-mechanical* governing equations are

$$\frac{d\omega_{rm}}{dt}=\frac{1}{J}(T_e-B_m\omega_{rm}-T_L),$$

$$\frac{d\theta_{rm}}{dt}=\omega_{rm}.$$

The mechanical angular velocity and displacement of the rotor $(\omega_{rm},\theta_{rm})$ are expressed by using the *electrical* angular velocity and displacement (ω_r,θ_r).

Using the number of poles P, $\omega_{rm}=\frac{2}{P}\omega_r$ and $\theta_{rm}=\frac{2}{P}\theta_r$.

Using (ω_r, θ_r), we have

$$\frac{d\omega_r}{dt}=\frac{1}{J}\left(\frac{P}{2}T_e-B_m\omega_r-\frac{P}{2}T_L\right),$$

$$\frac{d\theta_r}{dt}=\omega_r\,.$$

The electromagnetic torque T_e is found using the co-energy W_c.

$$T_e=\frac{P}{2}\frac{\partial W_c\left(\mathbf{i}_{abs},\mathbf{i}'_{abr},\theta_r\right)}{\partial\theta_r},$$

$$W_c=\tfrac{1}{2}\mathbf{i}_{abs}^T\left(\mathbf{L}_s-L_{ls}\mathbf{I}\right)\mathbf{i}_{abs}+\mathbf{i}_{abs}^T\mathbf{L}'_{sr}(\theta_r)\mathbf{i}'_{abr}+\tfrac{1}{2}\mathbf{i}'^T_{abr}\left(\mathbf{L}'_r-L'_{lr}\mathbf{I}\right)\mathbf{i}'_{abr},\ \mathbf{I}=\begin{bmatrix}1&0\\0&1\end{bmatrix}.$$

The self-inductances (L_{ss}, L'_{rr}) and leakage inductances (L_{ls}, L'_{lr}) are not functions of the angular displacement θ_r. From $\mathbf{L}'_{sr}(\theta_r)=L_{ms}\begin{bmatrix}\cos\theta_r&-\sin\theta_r\\\sin\theta_r&\cos\theta_r\end{bmatrix}$, the electromagnetic torque of P-pole induction motors is

$$T_e=\frac{P}{2}\frac{\partial W_c\left(\mathbf{i}_{abs},\mathbf{i}'_{abr},\theta_r\right)}{\partial\theta_r}=\frac{P}{2}\mathbf{i}_{abs}^T\frac{\partial\mathbf{L}'_{sr}(\theta_r)}{\partial\theta_r}\mathbf{i}'_{abr}=\frac{P}{2}L_{ms}\begin{bmatrix}i_{as}&i_{bs}\end{bmatrix}\begin{bmatrix}-\sin\theta_r&-\cos\theta_r\\\cos\theta_r&-\sin\theta_r\end{bmatrix}\begin{bmatrix}i'_{ar}\\i'_{br}\end{bmatrix}$$

$$=\frac{P}{2}L_{ms}\left[\left(-i_{as}\sin\theta_r+i_{bs}\cos\theta_r\right)i'_{ar}-\left(i_{as}\cos\theta_r+i_{bs}\sin\theta_r\right)i'_{br}\right]$$

$$=-\tfrac{P}{2}L_{ms}\left[\left(i_{as}i'_{ar}+i_{bs}i'_{br}\right)\sin\theta_r+\left(i_{as}i'_{br}-i_{bs}i'_{ar}\right)\cos\theta_r\right].$$

The *torsional-mechanical* equations are

$$\frac{d\omega_r}{dt}=\frac{1}{J}\left[-\tfrac{P^2}{4}L_{ms}\left[\left(i_{as}i'_{ar}+i_{bs}i'_{br}\right)\sin\theta_r+\left(i_{as}i'_{br}-i_{bs}i'_{ar}\right)\cos\theta_r\right]-B_m\omega_r-\tfrac{P}{2}T_L\right],$$

$$\frac{d\theta_r}{dt}=\omega_r\,.$$

We obtains the resulting governing equations in Cauchy's form, given by six differential equations

$$\frac{di_{as}}{dt}=-\frac{L'_{rr}r_s}{L_\Sigma}i_{as}+\frac{L_{ms}^2}{L_\Sigma}i_{bs}\omega_r+\frac{L_{ms}L'_{rr}}{L_\Sigma}i'_{ar}\left(\omega_r\sin\theta_r+\frac{r'_r}{L'_{rr}}\cos\theta_r\right)$$

$$+\frac{L_{ms}L'_{rr}}{L_\Sigma}i'_{br}\left(\omega_r\cos\theta_r-\frac{r'_r}{L'_{rr}}\sin\theta_r\right)+\frac{L'_{rr}}{L_\Sigma}u_{as}-\frac{L_{ms}}{L_\Sigma}\cos\theta_r u'_{ar}+\frac{L_{ms}}{L_\Sigma}\sin\theta_r u'_{br},$$

$$\frac{di_{bs}}{dt}=-\frac{L'_{rr}r_s}{L_\Sigma}i_{bs}-\frac{L_{ms}^2}{L_\Sigma}i_{as}\omega_r-\frac{L_{ms}L'_{rr}}{L_\Sigma}i'_{ar}\left(\omega_r\cos\theta_r-\frac{r'_r}{L'_{rr}}\sin\theta_r\right)$$

$$+\frac{L_{ms}L'_{rr}}{L_\Sigma}i'_{br}\left(\omega_r\sin\theta_r+\frac{r'_r}{L'_{rr}}\cos\theta_r\right)+\frac{L'_{rr}}{L_\Sigma}u_{bs}-\frac{L_{ms}}{L_\Sigma}\sin\theta_r u'_{ar}-\frac{L_{ms}}{L_\Sigma}\cos\theta_r u'_{br},$$

$$\frac{di'_{ar}}{dt}=-\frac{L_{ss}r'_r}{L_\Sigma}i'_{ar}+\frac{L_{ms}L_{ss}}{L_\Sigma}i_{as}\left(\omega_r\sin\theta_r+\frac{r_s}{L_{ss}}\cos\theta_r\right)-\frac{L_{ms}L_{ss}}{L_\Sigma}i_{bs}\left(\omega_r\cos\theta_r-\frac{r_s}{L_{ss}}\sin\theta_r\right)$$
$$-\frac{L_{ms}^2}{L_\Sigma}i'_{br}\omega_r-\frac{L_{ms}}{L_\Sigma}\cos\theta_r u_{as}-\frac{L_{ms}}{L_\Sigma}\sin\theta_r u_{bs}+\frac{L_{ss}}{L_\Sigma}u'_{ar},$$
$$\frac{di'_{br}}{dt}=-\frac{L_{ss}r'_r}{L_\Sigma}i'_{br}+\frac{L_{ms}L_{ss}}{L_\Sigma}i_{as}\left(\omega_r\cos\theta_r-\frac{r_s}{L_{ss}}\sin\theta_r\right)+\frac{L_{ms}L_{ss}}{L_\Sigma}i_{bs}\left(\omega_r\sin\theta_r+\frac{r_s}{L_{ss}}\cos\theta_r\right)$$
$$+\frac{L_{ms}^2}{L_\Sigma}i'_{ar}\omega_r+\frac{L_{ms}}{L_\Sigma}\sin\theta_r u_{as}-\frac{L_{ms}}{L_\Sigma}\cos\theta_r u_{bs}+\frac{L_{ss}}{L_\Sigma}u'_{br},$$
$$\frac{d\omega_r}{dt}=\frac{1}{J}\left[-\tfrac{P^2}{4}L_{ms}\left[\left(i_{as}i'_{ar}+i_{bs}i'_{br}\right)\sin\theta_r+\left(i_{as}i'_{br}-i_{bs}i'_{ar}\right)\cos\theta_r\right]-B_m\omega_r-\tfrac{P}{2}T_L\right],$$
$$\frac{d\theta_r}{dt}=\omega_r,$$

where $L_\Sigma=L_{ss}L'_{rr}-L_{ms}^2$.

Applying the ac phase voltages u_{as} and u_{bs}, one rotates the motor clockwise or counterclockwise. In squirrel-cage motors, $(u'_{ar},u'_{br})=0$, while the *motional emfs* $(\mathscr{E}_{ar},\mathscr{E}_{br})$ are induced in (ar,br).

Three-Phase Induction Motors – Principles of operation, control and torque-speed characteristics of three-phase induction motor are almost identical to two-phase induction motors. In industrial application, three-phase induction motors are used. Three-phase induction motors were designed and commercialized by Michail Dolivo-Dobrowolsky. He secured the US Patent 422746 *Electrical Induction Apparatus or Transformer*, issued on March 4, 1890.

Practice Problem 3.9. Two-Phase Induction Motor: Lagrange Equations of Motion
A mathematical model of an induction motor is derived using the Lagrange equations of motion (2.16)

$$\frac{d}{dt}\left(\frac{\partial\Gamma}{\partial\dot q_i}\right)-\frac{\partial\Gamma}{\partial q_i}+\frac{\partial D}{\partial\dot q_i}+\frac{\partial\Pi}{\partial q_i}=Q_i.$$

The *generalized* independent coordinates q_i are the charges and rotor angular displacement. That is, $q_1=\int i_{as}dt$, $q_2=\int i_{bs}dt$, $q_3=\int i'_{ar}dt$, $q_4=\int i'_{br}dt$, $q_5=\theta_r$.

The *generalized forces* Q_i are the voltages and load torque,

$Q_1=u_{as}$, $Q_2=u_{bs}$, $Q_3=u'_{ar}$, $Q_4=u'_{br}$, $Q_5=-T_L$.

The resulting Lagrange equations are

$$\frac{d}{dt}\left(\frac{\partial\Gamma}{\partial\dot q_1}\right)-\frac{\partial\Gamma}{\partial q_1}+\frac{\partial D}{\partial\dot q_1}+\frac{\partial\Pi}{\partial q_1}=Q_1,\quad \frac{d}{dt}\left(\frac{\partial\Gamma}{\partial\dot q_2}\right)-\frac{\partial\Gamma}{\partial q_2}+\frac{\partial D}{\partial\dot q_2}+\frac{\partial\Pi}{\partial q_2}=Q_2,$$
$$\frac{d}{dt}\left(\frac{\partial\Gamma}{\partial\dot q_3}\right)-\frac{\partial\Gamma}{\partial q_3}+\frac{\partial D}{\partial\dot q_3}+\frac{\partial\Pi}{\partial q_3}=Q_3,\quad \frac{d}{dt}\left(\frac{\partial\Gamma}{\partial\dot q_4}\right)-\frac{\partial\Gamma}{\partial q_4}+\frac{\partial D}{\partial\dot q_4}+\frac{\partial\Pi}{\partial q_4}=Q_4,$$
$$\frac{d}{dt}\left(\frac{\partial\Gamma}{\partial\dot q_5}\right)-\frac{\partial\Gamma}{\partial q_5}+\frac{\partial D}{\partial\dot q_5}+\frac{\partial\Pi}{\partial q_5}=Q_5.$$

The total kinetic, potential and dissipating energies (Γ,Π,D) are

$$\Gamma=\tfrac{1}{2}L_{ss}\dot{q}_1^2+L_{ms}\dot{q}_1\dot{q}_3\cos q_5-L_{ms}\dot{q}_1\dot{q}_4\sin q_5+\tfrac{1}{2}L_{ss}\dot{q}_2^2+L_{ms}\dot{q}_2\dot{q}_3\sin q_5+L_{ms}\dot{q}_2\dot{q}_4\cos q_5+\tfrac{1}{2}L_{rr}'\dot{q}_3^2+\tfrac{1}{2}L_{rr}'\dot{q}_4^2+\tfrac{1}{2}J\dot{q}_5^2,$$

$$\Pi=0,\quad D=\tfrac{1}{2}\left(r_s\dot{q}_1^2+r_s\dot{q}_2^2+r_r'\dot{q}_3^2+r_r'\dot{q}_4^2+B_m\dot{q}_5^2\right).$$

Hence, we have

$$\frac{\partial\Gamma}{\partial q_1}=0,\ \frac{\partial\Gamma}{\partial\dot{q}_1}=L_{ss}\dot{q}_1+L_{ms}\dot{q}_3\cos q_5-L_{ms}\dot{q}_4\sin q_5,$$

$$\frac{\partial\Gamma}{\partial q_2}=0,\ \frac{\partial\Gamma}{\partial\dot{q}_2}=L_{ss}\dot{q}_2+L_{ms}\dot{q}_3\sin q_5+L_{ms}\dot{q}_4\cos q_5,$$

$$\frac{\partial\Gamma}{\partial q_3}=0,\ \frac{\partial\Gamma}{\partial\dot{q}_3}=L_{rr}'\dot{q}_3+L_{ms}\dot{q}_1\cos q_5+L_{ms}\dot{q}_2\sin q_5,$$

$$\frac{\partial\Gamma}{\partial q_4}=0,\ \frac{\partial\Gamma}{\partial\dot{q}_4}=L_{rr}'\dot{q}_4-L_{ms}\dot{q}_1\sin q_5+L_{ms}\dot{q}_2\cos q_5,$$

$$\frac{\partial\Gamma}{\partial q_5}=-L_{ms}\dot{q}_1\dot{q}_3\sin q_5-L_{ms}\dot{q}_1\dot{q}_4\cos q_5+L_{ms}\dot{q}_2\dot{q}_3\cos q_5-L_{ms}\dot{q}_2\dot{q}_4\sin q_5,$$

$$\frac{\partial\Gamma}{\partial q_5}=-L_{ms}\left[(\dot{q}_1\dot{q}_3+\dot{q}_2\dot{q}_4)\sin q_5+(\dot{q}_1\dot{q}_4-\dot{q}_2\dot{q}_3)\cos q_5\right],\qquad \frac{\partial\Gamma}{\partial\dot{q}_5}=J\dot{q}_5,$$

$$\frac{\partial\Pi}{\partial q_1}=0,\ \frac{\partial\Pi}{\partial q_2}=0,\ \frac{\partial\Pi}{\partial q_3}=0,\ \frac{\partial\Pi}{\partial q_4}=0,\ \frac{\partial\Pi}{\partial q_5}=0,$$

$$\frac{\partial D}{\partial\dot{q}_1}=r_s\dot{q}_1,\ \frac{\partial D}{\partial\dot{q}_2}=r_s\dot{q}_2,\ \frac{\partial D}{\partial\dot{q}_3}=r_r'\dot{q}_3,\ \frac{\partial D}{\partial\dot{q}_4}=r_r'\dot{q}_4,\ \frac{\partial D}{\partial\dot{q}_5}=B_m\dot{q}_5.$$

The machine variables correspond to the *generalized* coordinates as

$$\dot{q}_1=i_{as},\ \dot{q}_2=i_{bs},\ \dot{q}_3=i_{ar}',\ \dot{q}_4=i_{br}',\ \dot{q}_5=\omega_r,$$

$$Q_1=u_{as},\ Q_2=u_{bs},\ Q_3=u_{ar}',\ Q_4=u_{br}',\ Q_5=-T_L.$$

One obtains the differential equations

$$L_{ss}\frac{di_{as}}{dt}+L_{ms}\frac{d\left(i_{ar}'\cos\theta_r\right)}{dt}-L_{ms}\frac{d\left(i_{br}'\sin\theta_r\right)}{dt}+r_si_{as}=u_{as},$$

$$L_{ss}\frac{di_{bs}}{dt}+L_{ms}\frac{d\left(i_{ar}'\sin\theta_r\right)}{dt}+L_{ms}\frac{d\left(i_{br}'\cos\theta_r\right)}{dt}+r_si_{bs}=u_{bs},$$

$$L_{ms}\frac{d\left(i_{as}\cos\theta_r\right)}{dt}+L_{ms}\frac{d\left(i_{bs}\sin\theta_r\right)}{dt}+L_{rr}'\frac{di_{ar}'}{dt}+r_r'i_{ar}'=u_{ar}',$$

$$-L_{ms}\frac{d\left(i_{as}\sin\theta_r\right)}{dt}+L_{ms}\frac{d\left(i_{bs}\cos\theta_r\right)}{dt}+L_{rr}'\frac{di_{br}'}{dt}+r_r'i_{br}'=u_{br}',$$

$$J\frac{d^2\theta_r}{dt^2}+L_{ms}\left[\left(i_{as}i_{ar}'+i_{bs}i_{br}'\right)\sin\theta_r+\left(i_{as}i_{br}'-i_{bs}i_{ar}'\right)\cos\theta_r\right]+B_m\frac{d\theta_r}{dt}=-T_L.$$

The electromagnetic torque is

$$T_e=\frac{\partial\Gamma(\dot{q}_1,\dot{q}_2,\dot{q}_3,\dot{q}_4,q_5)}{\partial q_5}=-L_{ms}\left[(\dot{q}_1\dot{q}_3+\dot{q}_2\dot{q}_4)\sin q_5+(\dot{q}_1\dot{q}_4-\dot{q}_2\dot{q}_3)\cos q_5\right].$$

For P-pole induction motors, six differential equations result as derived in Practice Problem 3.8. The Lagrange equations provide a general and consistent procedure.

Practice Problem 3.10. *Simulation of a Two-Phase Induction Motor*
Simulations are performed for a 7.5 kW, 380 V (rms), 60 Hz, four-pole (P=4) A-class induction motor. A SIMULINK diagram, documented in Figure 3.67, corresponds to the derived circuitry-electromagnetic dynamics described by differential equations in not Cauchy form, see Practice Problems 3.8 and 3.9.

Examine acceleration capabilities and settling time for a motor with parameters r_s=0.742 ohm, r_r=0.495 ohm, L_{ms}=0.128 H, L_{ls}=0.004 H, $L_{ss}=L_{ls}+L_{ms}$, L_{lr}=0.0065 H, $L_{rr}=L_{lr}+L_{ms}$, B_m=0.038 N-m-sec/rad and J=0.19 kg-m^2. For the considered motor, the number of turns of the stator and rotor windings are equal, $N_s=N_r$.

To guarantee the balanced operation, the phase voltages are

$$u_{as}(t)=\sqrt{2}u_M\cos(\omega_f t),\; u_{bs}(t)=\sqrt{2}u_M\sin(\omega_f t).$$

For the rated operation, u_M=380 V and ω_f=377 1/s.

The frequency of the applied phase voltages $\omega_f=2\pi f$ defines the synchronous angular velocity $\omega_e=\frac{4\pi}{P}f$. The electrical angular velocity is $\omega_r \le \omega_e$. Only assuming no friction, no drag, and no load torque T_L, one has $\omega_r=\omega_e$. Otherwise $\omega_r<\omega_e$.

The mechanical angular velocity of the shaft is $\omega_{rm}=\frac{2}{P}\omega_r$.

Dynamics of the stator and rotor currents $i_{as}(t)$, $i_{bs}(t)$, $i_{ar}(t)$ and $i_{br}(t)$, as well as angular velocity $\omega_r(t)$, are potted in Figures 3.68 if the load torque is T_L=0 and T_L=20 N-m. The motor accelerates from stall, ω_{r0}=0 rad/sec. The acceleration capabilities are documented. The A-class motors exhibit high efficiency and low losses.

The torque-speed characteristics $\omega_r(t)=\phi(T(t))$ can be plotted. Physics-consistent and hardware-compliant control laws must be designed and implemented. Practice Problem 5.9 reports the proportional-integral control of the induction motor considering tracking control with load attenuation.

SIMULINK diagram to model two-phase induction motors;

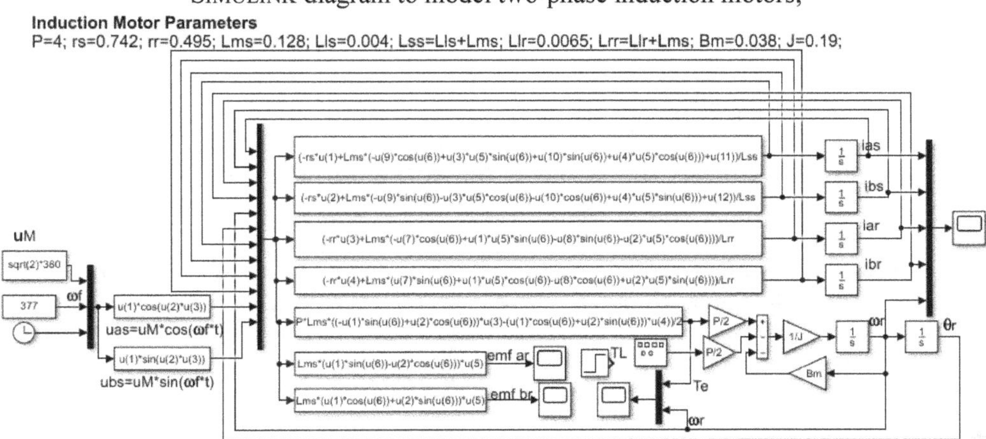

Figure 3.67. SIMULINK diagram to model two-phase induction motors.

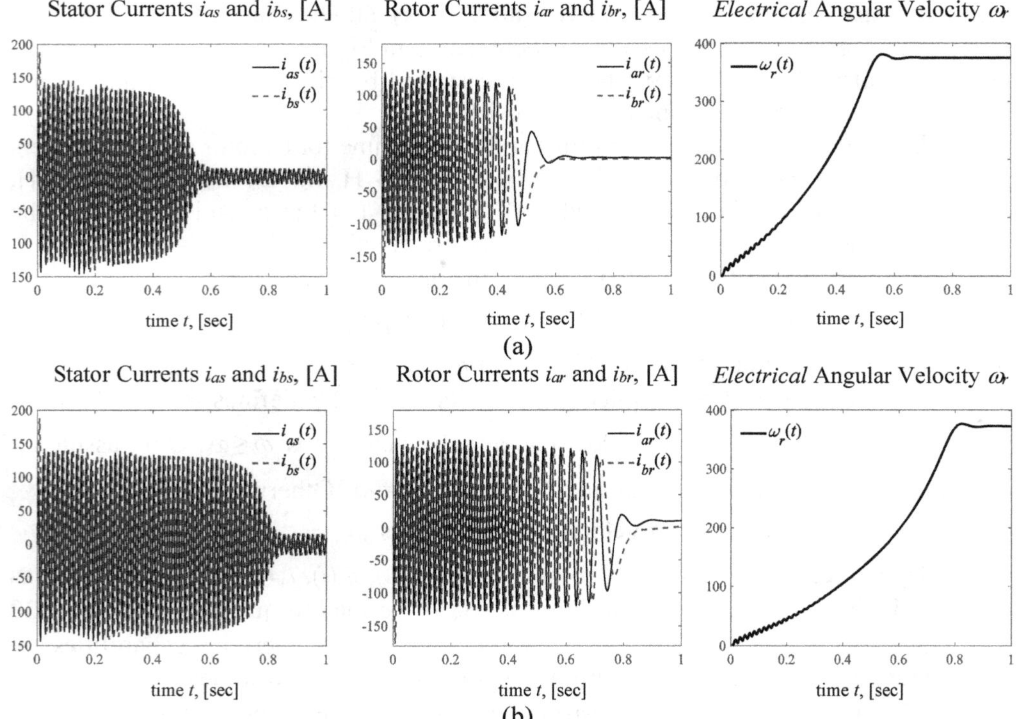

Figure 3.68. Dynamics of an induction motor. The applied phase voltages with the constant frequency are $u_{as}(t)=\sqrt{2}u_M\cos(\omega_f t)$ and $u_{bs}(t)=\sqrt{2}u_M\sin(\omega_f t)$, u_M=380 V, ω_f=377 1/sec.
(a) Motor accelerates with no load, T_L=0 N-m; (b) Motor acceleration with T_L=20 N-m.

Practice Problem 3.11.*Torque-Speed Characteristics: Three-Phase Induction Motors*
Industrial induction motors are three-phase motors. One controls the frequency and magnitude of the phase voltages (u_{as},u_{bs},u_{cs}). Under assumptions and simplifications, the steady-state torque-speed characteristics $\omega_r=\phi(T)$ can be obtained by using the equivalent *magnetic* circuit.

Reactance and Impedance – The inductive reactance is $X=\omega L$ [ohm], $\omega=2\pi f$. For a capacitor and inductor in series, using the inductive and capacitive reactances, $X=\omega L-\dfrac{1}{\omega C}$. The complex impedance is $Z=r+jX$.

Assigning different values for the phase voltages magnitude u_M and frequency f, one evaluates the torque-speed characteristics $\omega_r=\phi(T)$ for three-phase induction motors as

$$T_e=\frac{3\left(u_M\dfrac{X_M}{X_s+X_M}\right)^2\dfrac{r_r'}{slip}}{\omega_e\left[\left(r_s\left(\dfrac{X_M}{X_s+X_M}\right)^2+\dfrac{r_r'}{slip}\right)^2+\left(X_s+X_r'\right)^2\right]},\quad slip=\frac{\omega_e-\omega_r}{\omega_e},\quad \omega_e=\frac{4\pi f}{P},$$

where X_s and X_r' are the stator and rotor reactances, $X_s=\omega L_{ss}$ and $X_r'=\omega L_{rr}'$; X_M is the magnetizing reactance, $X_M=\omega L_{ms}$.

The electrical steel and silicon electrical steel is an iron alloys laminations. These laminations are manufactured and assembled to ensure specific magnetic properties, such as high permeability, low hysteresis and power losses, etc. The *BH* curves and pertained analyses are reported in sections 2.3, 3.1 and 3.3. Parameters of induction motors depend on electrical steels, characterized by the *BH* curve, permeability, magnetostriction, losses, air gap, winding design, etc. One uses the experimentally measured and computed reluctances, inductances and reactances.

Consider two four-pole industrial induction motors with identical windings and electromagnetic system, and, different electrical steel characteristics. At the rated operation, the parameters of 0.75 kW, 220 V, 60 Hz induction motors are:

1. P=4, r_s=4.95 ohm, $r_r^{'}$=3.2 ohm, X_{ss}=19.5 ohm, $X_{rr}^{'}=93$ ohm, X_M=58.1 ohm;
2. P=4, r_s=4.95 ohm, $r_r^{'}$=3.2 ohm, X_{ss}=12.1 ohm, $X_{rr}^{'}=48$ ohm, X_M=32.5 ohm.

The rated voltage is $u_{M\ \max}$=220 V. The phase voltages are supplied with frequencies 20, 40 and 60 Hz. For each frequency f, the synchronous angular velocity is $\omega_e=\frac{4\pi}{P}f$. The *low-slip* torque-speed characteristics of the A/B class induction motors, operated using the frequency control, are found for different frequencies f of the phase voltages (u_{as},u_{bs},u_{cs}), u_M=const. MATLAB is used to compute and plot the torques-speed characteristics.

```
uM=220; P=4; rs=4.95; rr=3.2; Xss=19.5; Xrr=93; XM=58.1; % Induction motor parameters
uM=220; P=4; rs=4.95; rr=3.2; Xss=12.1; Xrr=48; XM=32.5; % Induction motor parameters
for f=20:20:60; we=4*pi*f/P;      % Applied voltage frequency f is 20, 40, 60 Hz
% Compute and plot the torque-speed characteristic
for wr=[1:0.05:4*pi*f/P]; % Angular velocity
slip=(we-wr)/we;          % Slip
Te=3*(uM*XM/(Xss+XM))^2*(rr/slip)/(we*((rs*(XM/(Xss+XM))^2+rr/slip)^2+(Xss+Xrr)^2));
plot(Te,wr,'o'); % title('Torque-Speed Characteristics','FontSize',18);
hold on; end; end;
legend('{\itf}=60 Hz','{\itf}=40 Hz','{\itf}=20 Hz','Location','northeast','NumColumns',1,'FontSize',22); legend boxoff;
```

Frequency Control and Torque-Speed Characteristics – The torque-speed characterristics are documented in Figures 3.69. One assesses the starting torque, acceleration capabilities, control principles, *slip*, etc. The frequency control is a key principle in industrial electric drives. The starting $T_{e\ \mathrm{start}}$ is maximum at minimal frequency $f_{\min}$. By changing the frequency f, one controls ω_r and T_e. Practice Problem 5.9 reports the proportional-integral control of induction motors applying a frequency control concept.

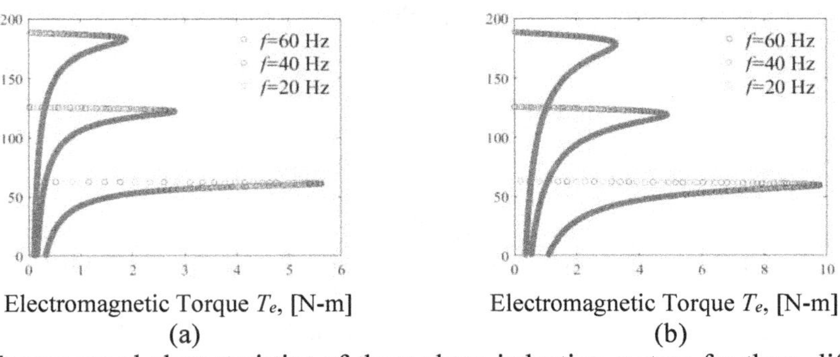

Figure 3.69. Torque-speed characteristics of three-phase induction motors for three different frequencies f of phase voltages (u_{as},u_{bs},u_{cs}), u_M=220 V, f=[20, 40, 60] Hz:

(a) Motor parameters are r_s=4.95 ohm, $r_r^{'}$=3.2 ohm, X_s=19.5 ohm, $X_r^{'}=93$ ohm, X_M=58.1 ohm;

(b) Motor parameters are r_s=4.95 ohm, $r_r^{'}$=3.2 ohm, X_s=12.1 ohm, $X_r^{'}=48$ ohm, X_M=32.5 ohm.

Practice Problem 3.12. *Flyback and Forward dc-dc Regulators*

To decouple the input and output stages, *flyback* and *forward* dc-dc regulators are used. These regulators magnetically isolate the input and output stages using transformers which support switching schemes. Transformers increase the size and cost, reduce efficiency and bandwidth, yields nonlinearities and hysteresis pertained to a magnetic system, high current, etc. The energy is stored in the inductor when transistor is closed. The stored energy is transformed to the load when transistor is open. The *flyback* and *forward* magnetically coupled dc-dc regulators are illustrated in Figures 3.70.

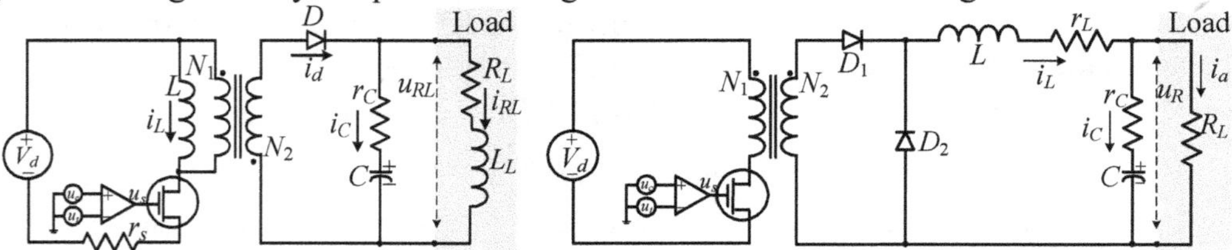

Figure 3.70. *Flyback* dc-dc regulator, and, the *forward* dc-dc regulator

In the *flyback* dc-dc regulator, when MOSFET is closed, the diode is reverse biased. If MOSFET is open, the diode is forward biased. The transistor is closed for time $\frac{d_D}{f}$ and open for $\frac{1-d_D}{f}$. For open transistor $i_d = \frac{N_1}{N_2} i_L$. The differential equations for the *forward* regulator when MOSFET is closed are

$$\frac{du_C}{dt} = \frac{1}{C(r_C + R_L)}(-u_C + R_L i_L),$$

$$\frac{di_L}{dt} = \frac{1}{L}\left(-\frac{R_L}{r_C + R_L}u_C + \left(\frac{R_L^2}{r_C + R_L} - r_L - R_L\right)i_L + \frac{N_2}{N_1}V_d\right).$$

If MOSFET is open

$$\frac{du_C}{dt} = \frac{1}{C(r_C + R_L)}(-u_C + R_L i_L),$$

$$\frac{di_L}{dt} = \frac{1}{L}\left(-\frac{R_L}{r_C + R_L}u_C + \left(\frac{R_L^2}{r_C + R_L} - r_L - R_L\right)i_L\right).$$

Using the duty ratio d_D, the resulting differential equations are

$$\frac{du_C}{dt} = \frac{1}{C(r_C + R_L)}(-u_C + R_L i_L),$$

$$\frac{di_L}{dt} = \frac{1}{L}\left(-\frac{R_L}{r_C + R_L}u_C + \left(\frac{R_L^2}{r_C + R_L} - r_L - R_L\right)i_L + \frac{N_2}{N_1}V_d d_D\right).$$

Using a set of differential equations, simulate the *forward* regulator if V_d=50 V and d_D=0.5. The parameters are r_L=0.02 ohm, r_C=0.01 ohm, R_L=3 ohm, L=0.000005 H, C=0.003 F and N_2/N_1=1. The transient dynamics for the states $u_C(t)$ and $i_L(t)$ are documented in Figures 3.71. The settling time is 0.0015 sec. The steady-state voltage, applied to the load terminal is 25 V.

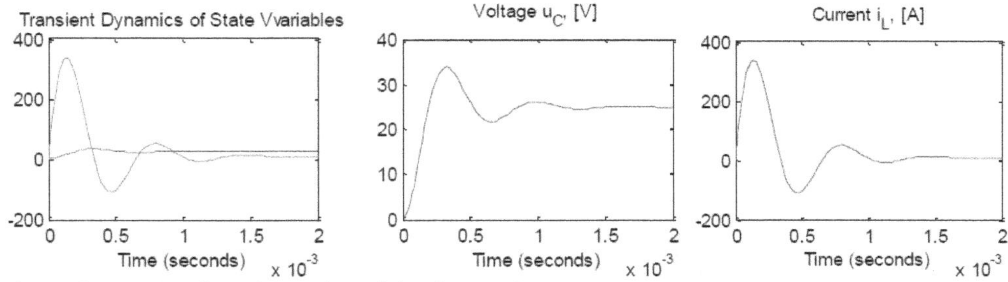

Figure 3.71. Transient dynamics of the *forward* converter.

Practice Problem 3.13. *Coordinate Systems*

In aerial and ground vehicles, autonomous mobile robots, automation, assembly, manufacturing and other systems, path and motion planning, tracking, spatiotemporal navigation and control are accomplished using different coordinate systems. The Cartesin, spherical and cylindrical coordinate systems are used in communication, electromagnetics, mechanics and transducers design, etc.

Consider a vector $\mathbf{F}=a\mathbf{a}_x+b\mathbf{a}_y+c\mathbf{a}_z$.

The equations which express the transformations between rectangular to cylindrical coordinates are

$$\begin{cases} x=\rho\cos\theta \\ y=\rho\sin\theta \\ z=z \end{cases}, \quad \begin{cases} \rho=\sqrt{x^2+y^2} \\ \theta=\arctan\frac{y}{x} \\ z=z \end{cases}, \quad 0\le\rho<\infty,\ -\pi<\theta\le\pi,\ -\infty<z<\infty.$$

The dot products of unit vectors in cylindrical and rectangular coordinate systems are

	$\mathbf{a}_\rho$	$\mathbf{a}_\theta$	$\mathbf{a}_z$
$\mathbf{a}_x$	$\cos\theta$	$-\sin\theta$	0 .
$\mathbf{a}_y$	$\sin\theta$	$\cos\theta$	0
$\mathbf{a}_z$	0	0	1

In the cylindrical coordinates, the components of the vector are

$F_\rho=F\cdot\mathbf{a}_\rho=a(\mathbf{a}_x\cdot\mathbf{a}_\rho)+b(\mathbf{a}_y\cdot\mathbf{a}_\rho)+c(\mathbf{a}_z\cdot\mathbf{a}_\rho)=a\cos\theta+b\sin\theta+c\times0=a\cos\theta+b\sin\theta$,

$F_\theta=F\cdot\mathbf{a}_\theta=a(\mathbf{a}_x\cdot\mathbf{a}_\theta)+b(\mathbf{a}_y\cdot\mathbf{a}_\theta)+c(\mathbf{a}_z\cdot\mathbf{a}_\theta)=a(-\sin\theta)+b\cos\theta+c\times0=-a\sin\theta+b\cos\theta$,

$F_z=F\cdot\mathbf{a}_z=a(\mathbf{a}_x\cdot\mathbf{a}_z)+b(\mathbf{a}_y\cdot\mathbf{a}_z)+c(\mathbf{a}_z\cdot\mathbf{a}_z)=a\times0+b\times0+c\times1=c$.

Practice Problem 3.14. *Circuits and Passive Filters*

Various filters, compensators and regulators can be implemented. Consider circuits reported in Figures 3.72 which implement the first- and second-order low-pass filters. For these circuits, the transfer functions are

$$G(s)=\frac{U_0(s)}{U_1(s)}=\frac{R}{Ls+R}=\frac{1}{\frac{L}{R}s+1},\quad G(s)=\frac{U_0(s)}{U_1(s)}=\frac{\frac{1}{Cs}}{R+\frac{1}{Cs}}=\frac{1}{RCs+1},$$

$$G(s)=\frac{U_0(s)}{U_1(s)}=\frac{\frac{1}{Cs}}{R+Ls+\frac{1}{Cs}}=\frac{1}{LCs^2+RCs+1},$$

$$G(s)=\frac{U_0(s)}{U_1(s)}=\frac{\frac{R}{1+RCs}}{Ls+\frac{R}{1+RCs}}=\frac{R}{RLCs^2+Ls+R}=\frac{1}{LCs^2+\frac{L}{R}s+1}.$$

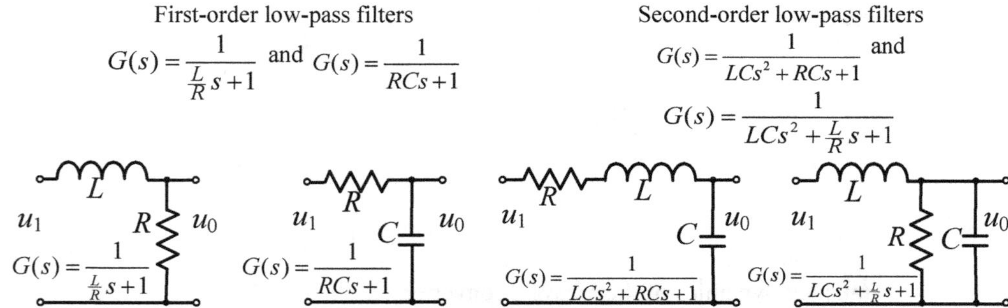

First-order low-pass filters
$G(s)=\frac{1}{\frac{L}{R}s+1}$ and $G(s)=\frac{1}{RCs+1}$
Second-order low-pass filters
$G(s)=\frac{1}{LCs^2+RCs+1}$ and
$G(s)=\frac{1}{LCs^2+\frac{L}{R}s+1}$

Figure 3.72. First- and second-order low pass filters.

Practice Problem 3.15. *Second-Order Passive Filter*

Passive filters, studied in Practice Problem 3.14, ensure a unit dc gain. For the second-order notch filter, documents Figure 3.73.a, the transfer function is

$$G(s)=\frac{U_0(s)}{U_1(s)}=\frac{R_3\left[LCs^2+(R_1+R_2)Cs+1\right]}{(R_2+R_3)LCs^2+\left[(R_1+R_2)R_3C+R_1R_2C+L\right]s+R_1+R_3}.$$

The notch filter yields attenuation, and, the gain is $\frac{R_3}{R_1+R_3}$.

Design a passive filter specifying near-unit dc gain with attenuation up to –5 dB for the system frequency from 1 to 500 Hz. The noise, with frequency higher than 5000 Hz, should be attenuated at least by 10 times. We obtain R_1=1 ohm, R_2=100 ohm, R_3=1 ohm, L=0.01 H and C=0.01 F. The Bode plots, shown in Figure 3.73.b, are computed and plotted.

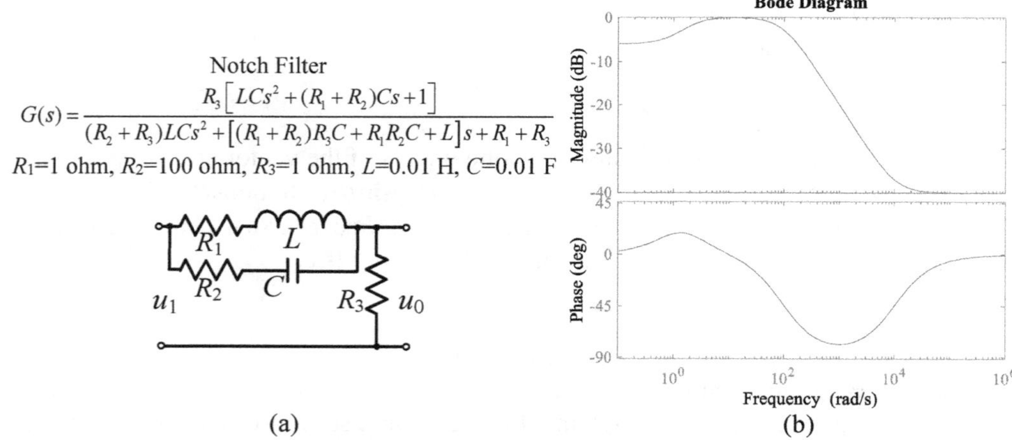

Figure 3.73. (a) Notch passive filter,

$$G(s)=\frac{R_3\left[LCs^2+(R_1+R_2)Cs+1\right]}{(R_2+R_3)LCs^2+\left[(R_1+R_2)R_3C+R_1R_2C+L\right]s+R_1+R_3};$$

(b) Bode plots for the notch filter, R_1=1 ohm, R_2=100 ohm, R_3=1 ohm, L=0.01 H and C=0.01 F.

```
R1=1; R2=100; R3=1; L=0.01; C=0.01;
num=R3*[L*C (R1+R2)*C 1]; den=[(R2+R3)*L*C (R3*(R1+R2)*C*R1*R2*C+L) (R1+R3)];
bode(num,den,{0.1 1e6})
```

Practice Problem 3.16. *Circuits and High-Order Passive Filters*

The band-pass filters are illustrated in Figures 3.74.a with the transfer functions

$$G(s)=\frac{U_0(s)}{U_1(s)}=\frac{RCs}{LCs^2+RCs+1},\quad G(s)=G_1(s)G_2(s)=\frac{U_0(s)}{U_1(s)}=\frac{R_2C_2s}{(R_1C_1s+1)(R_2C_2s+1)}.$$

The passive filter is depicted in Figure 3.74.b. A transfer function is $G(s)=\dfrac{s(a_2s^2+a_1s+a_0)}{b_4s^4+b_3s^3+b_2s^2+b_1s+b_0}$, where $(a_0,a_1,a_2,b_0,b_1,b_2,b_3,b_4)$ are straightforwardly found, and, depend on the circuit parameters $(R_1,R_2,R_3,C_1,C_2,L_1,L_2)$.

The passive filter, reported in Figure 6.74.c, is described by

$$G(s)=\frac{2R_2C_2L_1L_3s^3}{b_6s^6+b_5s^5+b_4s^4+b_3s^3+b_2s^2+b_1s+b_0},$$

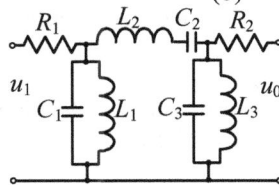

where the $(R_1,R_2,C_1,C_2,L_1,L_2)$-dependent coefficients $(b_0,b_1,b_2,b_3,b_4,b_5,b_6)$ are

$b_6=C_1C_2C_3L_1L_2L_3R_1R_2$, $b_5=C_2L_1L_2L_3(C_1R_1+C_3R_2)$,

$b_4=C_2L_1L_2L_3+C_1C_2L_1(L_2+L_3)R_1R_2+(C_1L_1+C_2L_1+C_2L_2)C_3L_3R_1R_2$,

$b_3=(C_1L_1+C_2L_1+C_2L_2)L_3R_1+C_2L_1(L_2+L_3)R_2+C_3L_1L_3R_2$,

$b_2=L_1L_3+\left[(C_1+C_2)L_1+C_2(L_2+L_3)+C_3L_3\right]R_1R_2$, $b_1=L_3R_1+L_1R_2$ and $b_0=R_1R_2$.

Band-pass filters $G(s)=\dfrac{RCs}{LCs^2+RCs+1}$ and $G(s)=\dfrac{R_2C_2s}{(R_1C_1s+1)(R_2C_2s+1)}$

L C R u_1 u_0 R_1 C_2 C_1 R_2

(a)

Passive filter

$G(s)=\dfrac{s(a_2s^2+a_1s+a_0)}{b_4s^4+b_3s^3+b_2s^2+b_1s+b_0}$

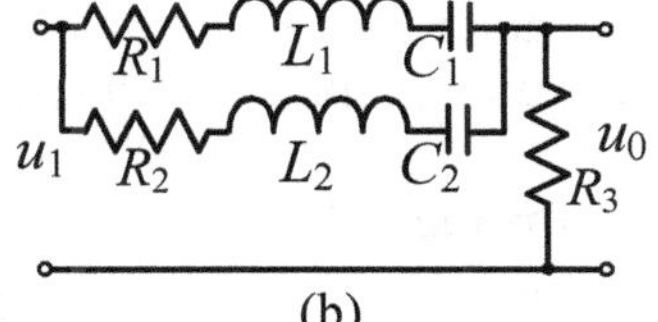

(b)

Passive band-pass filter

$G(s)=\dfrac{2R_2C_2L_1L_3s^3}{b_6s^6+b_5s^5+b_4s^4+b_3s^3+b_2s^2+b_1s+b_0}$

R_1 L_2 C_2 R_2 u_1 u_0 C_1 L_1 C_3 L_3

(c)

Figure 3.74. (a) Band-pass filters, $G(s)=\dfrac{RCs}{LCs^2+RCs+1}$ and $G(s)=\dfrac{R_2C_2s}{(R_1C_1s+1)(R_2C_2s+1)}$;

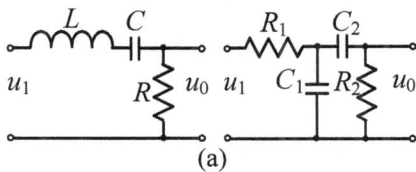

(b) Passive filter with a transfer function $G(s)=\dfrac{s(a_2s^2+a_1s+a_0)}{b_4s^4+b_3s^3+b_2s^2+b_1s+b_0}$;

(c) Passive band-pass filter, $G(s)=\dfrac{2R_2C_2L_1L_3s^3}{b_6s^6+b_5s^5+b_4s^4+b_3s^3+b_2s^2+b_1s+b_0}$.

Homework Problems

Homework Problem 3.1.

A force $\vec{F}=3\vec{i}+2\vec{j}+4\vec{k}$ [N] acts through the point with a position vector $\vec{R}=2\vec{i}+\vec{j}+3\vec{k}$ [m]. Derive a torque $\vec{T}=\vec{R}\times\vec{F}$ [N-m].

Homework Problem 3.2.

For variable reluctance electromagnetic actuators, section 3.3.1 documents the application of laws of physics to derive physics-consistent magnetizing inductance $L(x)$ and electromagnetic force $F_e(x)$.

As a mathematical exercise, considered Example 3.7.

The magnetizing inductance is $L(x)=ae^{-bx}$, $x_{min}\leq x\leq x_{max}$, $x\geq 0$, $a>0$, $b\gg 1$.

Problems to solve:

1. Derive and report an equation for the electromagnetic force F_e. Find an expression for F_{emax}. Let $x_{min}=0$, $a=0.1$ and $b=500$. For $i=1$ A, calculate F_{emax}.
2. In MATLAB, compute and plot $F_{emax}(x)$. Using the **surf** command, compute and plot a surface for $F_{emax}(i,x)$ if $0\leq x\leq 0.01$ m and $1\leq i\leq 10$ A;
3. Derive and report equations of motion;
4. Estimate the spring constant k_s if $i_{max}=10$ A;
5. For the voltage pulses from 10 to 50 V, perform simulations in SIMULINK. Let $r=5$ ohm, $m=0.02$ kg and $B_v=0.1$ N-sec/m. Use the derived k_s.

Homework Problem 3.3.

Consider a solenoid. The magnetizing inductance is

$L(x)=cb^{ax}$, $x\geq 0$, $a=1,2,3,\ldots$, $b=1/n$, $n=2,3,4,\ldots$, $c>0$.

Problems to solve:

1. Derive and report an explicit equation for the electromagnetic force F_e;
2. Find and report equations for the *total* and *motional emfs*;
3. Derive and report a mathematical model for the considered solenoid.

Homework Problem 3.4.

Consider a solenoid. The magnetizing inductance is $L(x)=e^{-ax}\text{csch}(bx)$, $a\gg 1$, $b>0$. Derive equations for the electromagnetic force F_e and *motional emf.*

Homework Problem 3.5.

Consider an axial-topology hard disk drive permanent magnet dc actuator, illustrated in Figures 3.21.b and d. Problems to solve:

1. Derive equations for a magnetic dipole moment for two axially magnetized through thickness wedge-shaped magnets. For the current loop, find $\vec{s}$, $\rho_{min}\leq\rho\leq\rho_{max}$, $\theta_{min}\leq\theta\leq\theta_{max}$. One may assume that magnets establish a uniform sigmoidal field as viewed from the current loop. Polar and cylindrical coordinates can be used to find the torque $\vec{T}=\vec{m}\times\vec{B}=Ni\vec{s}\times\vec{B}$, exerted on rotor;
2. Derive the resulting equations of motion. Note that optimal design yields only the T_z component of the electromagnetic torque;
3. Prove your results analytically and numerically. Use the parameter values reported in Example 3.19. Simulate an actuator, and, examine dynamics.

Homework Problem 3.6.

Using operational amplifiers, develop schematics to implement analog proportional, proportional-integral and proportional-integral-derivative controllers. Report alternative schematics for analog controllers. Derive the transfer functions $G(s)$ and report equations for the feedback gains (k_p,k_i,k_d) using the resistors and capacitors.

Homework Problem 3.7.

Study the PWM switching concept in dc-dc regulators. The switching signal u_s, which drives a MOSFET, is generated by comparing a signal-level control voltage u_c with a high-frequency repetitive periodic signal u_t. Using operational amplifiers, propose the schematics to generate sinusoidal and triangular u_t using oscillator schemes and circuits. Report the voltage waveforms.

Homework Problem 3.8.

Consider the third-order low-pass Butterworth filter, shown in Figure 3.75.a.

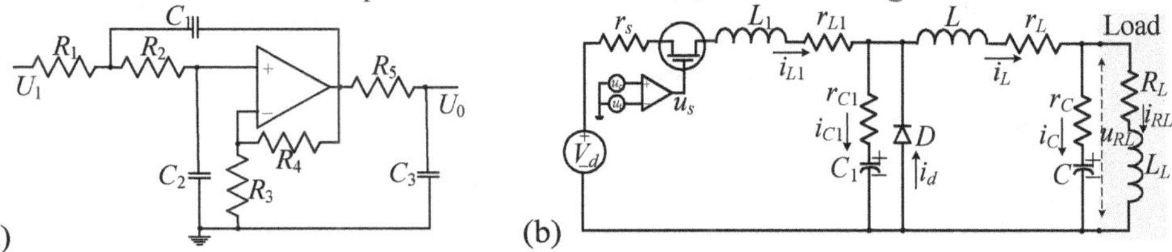

Figure 3.75. (a) Third-order low-pass Butterworth filter;
(b) Resonant dc-dc regulator with zero-current switching.

Derive the transfer function. For a unit gain and cut-off frequency 10000 Hz, compute the values of resistors and capacitors. In MATLAB, compute the Bode plots.

Homework Problem 3.9.

Conduct search on dc-dc regulators and inverters for a chosen application and system. Estimate power, voltage, currents, switching frequency, efficiency and volumetric needed for a chosen system, such as aerial vehicles, mobile robot, auxiliary power system, servo or wearable device. Using the catalog data reported, by the Texas Instruments or STMicroelectronics, consider the particular dc-dc swathing regulators

Different topologies and schemes have being developed and commercialized. For example, zero-voltage and zero-current switching improve efficiency, reliability, switching frequency, stability, robustness, etc.

A variety of resonant regulator topologies and filters are used, including documented in Figure 3.74.b. The output voltage at the *RL* load terminal is regulated by controlling the switching *on* and *off* durations. When MOSFET is open, the diode *D* is forward biased to carry the output inductor current i_L. The voltage across the capacitor C_1 is zero. When MOSFET is closed, the diode remains forward biased while $i_{L1}<i_L$. As i_{L1} reaches i_L, the diode turns *off*. Hence, transistor turns *off* and *on* at zero-current, $i_{L1}=0$. A comparator with inputs (u_c,u_t) implements the PWM switching. The switching signal u_s, which drives MOSFET, is generated by comparing a signal-level control voltage u_c with a high-frequency triangular signal u_t.

For a regulator, reported in Figure 3.75.b, solve the following problems:

1. Develop a set of differential equations to model a regulator;
2. Perform simulations in SUMULINK. The parameters are r_s=0.01 ohm, r_L=0.093 ohm, r_{L1}=0.1 ohm, r_C=0.065 ohm, r_{C1}=0.07 ohm, $C=1\times10^{-5}$ F, $C_1=4.7\times10^{-6}$ F, $L=5\times10^{-5}$ H, $L_1=5\times10^{-6}$ H and L_L=0.0001 H. Let V_d=30 V, and, for an open-loop system for different d_D, $0<d_D<1$.;
3. Using a control law $u=\text{sat}_0^1\left(k_p e+k_i\int e dt\right)$, by trial and error find feedback gains $(k_p,k_i)>0$ which ensure adequate dynamic performance. Perform simulation of the closed-loop system for different k_p and k_i.

References

1. L. J. Hornbeck, “Combining digital optical MEMS, CMOS and algorithms for unique display solutions”, *Proc. Electron Devices Meeting*, pp. 17-24, 2007.
2. V. Giurgiutiu and S. E. Lyshevski, *Micromechatronics: Modeling, Analysis, and Design With MATLAB®*, CRC Press, Boca Raton, FL, 2009.
3. *International Technology Roadmap for Semiconductors,Micro-Electromechanical Systems (MEMS)*, Semiconductor Industry Association, Austin, TX, USA, 2020.
4. P. C. Krause and O. Wasynczuk, *Electromechanical Motion Devices*, McGraw-Hill Book Company, New York, 1989.
5. P. C. Krause, O. Wasynczuk and S. D. Sudhoff, *Analysis of Electric Machinery*, IEEE Press, New York, 1994.
6. S. E. Lyshevski, *Electromechanical Systems and Devices*, CRC Press, Boca Raton, FL, 2008.
7. S. E. Lyshevski, *Electromechanical Systems, Electric Machines, and Applied Mechatronics*, CRC Press, Boca Raton, FL, 1999.
8. S. E. Lyshevski, *MEMS and NEMS: Systems, Devices, and Structures*, CRC Press, Boca Raton, FL, 2002.
9. S. E. Lyshevski, *Mechatronics and Control of Electromechanical Systems*, CRC Press, Boca Raton, FL, 2017.
10. S. D. Senturia, Microsystem Design. Kluwer, Norwell, MA, 2001.
11. D. C. White and H. H. Woodson, *Electromechanical Energy Conversion*, Wiley, New York, 1959.
12. D. W. Hart, *Power Electronics*, McGraw Hill, NY, 2010.
13. S. E. Lyshevski, “Resonant converters: Nonlinear analysis and control,” *IEEE Trans. on Industrial Electronics*, vol. 47, no. 4, pp. 751-758, 2000.
14. A. S. Sedra and K. C. Smith, *Microelectronic Circuit*, Oxford University Press, New York, 2022.
15. R. H. Park, “Two-reaction theory of synchronous machines: Generalized method of analysis – Part I,” *Trans. American Inst. Electrical Engineering*, vol. 48, no. 2, pp. 716-730, 1929.
16. S. E. Lyshevski, “Nonlinear control of servo-systems actuated by permanent-magnet synchronous motors,” *Automatica*, vol. 34, no. 10, pp. 1231-1238, 1998.
17. S. E. Lyshevski, “Microstepping and high-performance control of permanent-magnet stepper motors,” *Journal Energy Conversion and Management*, vol. 85, pp. 245-253, 2014.
18. *Nonlinear Circuits Handbook*, Ed. D. Sheingold, Analog Devices, Boston, MA, 1976.
19. H. Zumbahlen, *Linear Circuit Design Handbook*, Analog Devices, MA, 2008.
20. S. E. Lyshevski, “High-power density mini-scale power generation and energy harvesting systems,” *Energy Conversion and Management*, vol. 52, pp. 46-52, 2011.
21. C. *Wiegand*, B. A. Bullick, J. A. Catt, J. W. Hamstra, G. P. Walker and S. Wurth, “F-35 air vehicle technology overview,” *Proc. Aviation Technology, Integration, and Operation Conf.*, pp. 1-28, 2018.
22. J. A. Gow and C. D. Manning, "Development of a photovoltaic array model for use in power-electronics simulation studies." *Proc. Electric Power Applications Conf.*, vol. 146, no. 2, pp. 193-200, 1999.
23. R. Messenger and J. Ventre, *Photovoltaic Systems Engineering*, CRC Press, Boca Raton, FL, 2010.
24. S. M. Sze and K. K. Ng, *Physics of Semiconductor Devices*, Wiley-Interscience, Hoboken, NJ, 2007.

4 Identification of Dynamic Systems

4.1. Introduction to Identification and Estimation Problems

Identification and parameters estimation are studied to solve analysis, design, optimization and control problems for dynamically evolving physical systems $\mathbf{M}(t,x,y,u)$ [1-10]. Real-time identification is aimed to adaptively control systems, and, optimize system governance by reconfiguring controllers. The outputs $y(t)$ are measured to ensuring functionality, and, states x and inputs u (controls, disturbances and perturbations) may be measured.

Using laws of physics, one finds a class and order of finite-dimension governing differential and constitutive equations which describe mathematical models $\mathcal{M}(t,x_{\mathcal{M}},y,u,\mathcal{S},\mathcal{P})$. Invariant or time-varying structures $(\mathcal{S},\mathcal{S}(t))$ and parameters $(\mathcal{P},\mathcal{P}(t))$ should be identified and estimated. Supporting engineering science and engineering design, adequate models $\mathcal{M}(t,x_{\mathcal{M}},u,\mathcal{S},\mathcal{P})$ are found ensuring the system-to-model and model-to-system complementarity using states and inputs. Different systems have being investigated, and, finite-dimensional models, described by nonlinear differential equations $\dot{x}(t)=F(x,u)$, were developed and analyzed in chapters 2 and 3 assuming $\hat{\mathcal{M}}(x_{\mathcal{M}},u,\hat{A},\hat{f})\triangleq\mathbf{M}(x,u)$, $\begin{cases} x_{\mathcal{M}}\triangleq x \\ y_{\mathcal{M}}\triangleq y\end{cases}$. In this section, the system and model states are denoted as $(x,x_{\mathcal{M}})$. Section 2.1 reports introduction to identification. Key steps and solution hierarchy are illustrated in Figure 4.1.

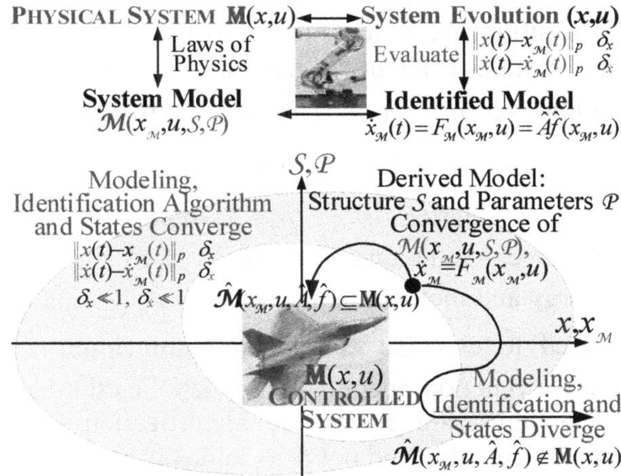

Figure 4.1. For a dynamic physical system $\mathbf{M}(t,x,u)$, find a finite-dimensional mathematical model $\mathcal{M}(t,x_{\mathcal{M}},u,\mathcal{S},\mathcal{P})$ as $\dot{x}_{\mathcal{M}}(t)=F_{\mathcal{M}}(x_{\mathcal{M}},u)$ with identified $\hat{\mathcal{M}}(x_{\mathcal{M}},u,\hat{A},\hat{f})$, $\dot{x}_{\mathcal{M}}(t)=\hat{A}\hat{f}(x_{\mathcal{M}},u)$. The objective is to ensure a descriptive and predictive representation of dynamic governance and phenomena exhibited by $\mathbf{M}(x,u)$, e.g., find $\hat{\mathcal{M}}(x_{\mathcal{M}},u,\hat{A},\hat{f})$ such that $\hat{\mathcal{M}}(x_{\mathcal{M}},u,\hat{A},\hat{f})\subseteq\mathbf{M}(x,u)$.

Modeling and Identification Problems – Laws of physics and observed dynamics of $\mathbf{M}(t,x,u)$ yield system dimensionality for *dominant governance*, yielding structure $\mathcal{S}$ and parameters $\mathcal{P}$. Identification, model validation and fidelity assessment are performed using the measured states $(x,x_{\mathcal{M}})$ and inputs u. Solution of identification problem implies finding a model, such that $\hat{\mathcal{M}}(x_{\mathcal{M}},u,\hat{A},\hat{f})\subseteq\mathbf{M}(x,u)$.

Modeling, analysis, identification, control and optimization depend on:

1. System complexity and descriptive fidelity;
2. Physical system $\mathbf{M}(x,u)$ controllability and governance with input-to-state and input-to-output mappings;
3. Measurability of system variables (x,u) and physical data quality.

Infinite-Dimensional and Finite-Dimensional Models – Continuum mechanics and electrodynamics were covered using laws of physics and measured physical quantities. Abstract models in infinite-dimensional Hilbert, Banach and other spaces can be found applying transforms $\mathcal{T}$ and operators $\mathcal{O}$ to model real-valued transductions using a pair $(\mathcal{T},\mathcal{O})\in\mathbb{R}^\infty\times\mathbb{C}^\infty$.

For example, a time-dependent Schrödinger equation $i\hbar\frac{\partial}{\partial t}\Psi(t,\mathbf{r})=\hat{H}\Psi(t,\mathbf{r})$ is written using a wave function operator $\Psi(t,\mathbf{r})$, and, the Hamiltonian energy operator $\hat{H}$. Abstract models support analysis pertained to theoretical elasticity, elastodynamics, hydrodynamics, tribology and quantum mechanics. To investigate theoretical quantum informatics, *magnetic flux* quantization, theoretical quantum computing and other abstract modeling tasks, one may investigate quantum-mechanical premises in infinite-dimensional spaces using not measurable operators and operator-associated *complementary observables*. For example, the wave function $\psi\in\mathbb{R}^\infty\times\mathbb{C}^\infty$ is characterized by coefficients $c\in\mathbb{C}\times\mathbb{R}$. Computable *commuting quantum operators* and *commuting observables* are not detectable, not observable and cannot not be measured. Using the standard deviation of position and momentum (σ_x,σ_p), as well as energy and time (σ_E,σ_t), the Heisenberg uncertainties principle is given in the position-momentum and energy-time standard deviations, $(\sigma_x\sigma_p\geq\frac{1}{2}\hbar,\sigma_E\sigma_t\geq\frac{1}{2}\hbar)$. For the operators $(\bar{A},\bar{B})$, one has $\sigma_{\bar{A}}^2\sigma_{\bar{B}}^2\geq\left(\frac{1}{2i}\langle[\bar{A},\bar{B}]\rangle\right)^2$. With not detectable position and momentum operators $(\bar{x},\bar{p})$, the canonical commutation relation is $[\bar{x},\bar{p}]=\bar{x}\bar{p}-\bar{p}\bar{x}=i\hbar$. In classical mechanics, the observable displacement and momentum pair is (x, p).

The quantum kinetic energy and momentum operators $-\frac{\hbar^2}{2m}\nabla^2$ and $-i\hbar\frac{\partial}{\partial\mathbf{r}}$ are fundamentally different compared to classical energy and momentum. Using the Hamiltonian operator $\hat{H}$, the wave function operator $\Psi(t,\mathbf{r})$ satisfies the Schrödinger equation, and, the energy eigenvalues are computed. Identification may not be performed on not measurable, not observable and not detectable operators.

For a photon, the detected and measured energy, kinetic energy and momentum are given by the Plank and de Broglie equations, $E=hf=hc/\lambda$, $\Gamma=pc$ and $p=h/\lambda$. The continuum mechanics and electrodynamics explicitly define the kinetic energy, potential energy, dissipating energy and momentum $(\Gamma,\Pi,D,\mathbf{p})$, which are detectable and measurable. Control and identification of systems are performed using measurable physical variables. Microscopic and macroscopic systems operate and exhibit real-valued state transductions, described by the finite-dimensional equations. In commercialized photonic, optoelectronic and electronic systems, which may operate on quantum phenomena, the real-valued quantitative states and inputs are detectable and measurable. These devices are lasers, light emitting diodes, photodiodes, optocouplers, photomultipliers, optical transmitters and receivers, optical gyroscope, interferometric sensor, etc.

Physical Systems, Descriptive Models and Multi-Physics Characterization – Physical systems $\mathbb{M}(x,u)$ are characterized by real-valued, detectable, observable and measurable states $x \in X \subset \mathbb{R}^n$ and inputs $u \in U \subset \mathbb{R}^m$. These (x, u) pertain to physical phenomena exhibited. The prevailing device physics is known. In previous chapters we found the n-dimensional lumped-parameter descriptive models $\mathcal{M}(x_{\mathcal{M}},u,\mathcal{S},\mathcal{P})$, given by the governing differential and constitutive equations, characterized by:

1. Finite-dimensional model structure $\mathcal{S}$ with *a priori* known class and order;
2. Real-valued parameters $\mathcal{P}$.

A physics-defined tuple of real-valued (x,y,u) is used to observe, control and supervise physical systems $\mathbb{M}(x,y,u)$ which evolve in $X \times Y \times U$,

$\mathbb{M}:=\{(x,y,u): x \in X(X_0,U), y=H(x)\}$.

States, outputs and inputs (x,y,u) are measured by sensors. To characterize, optimize and control multi-physics systems $\mathbb{M}(x,u)$, consider its descriptive model $\mathcal{M}(x_{\mathcal{M}},u,\mathcal{S},\mathcal{P})$, modelled by finite-dimensional nonlinear differential equations $\dot{x}_{\mathcal{M}}(t)=F_{\mathcal{M}}(x_{\mathcal{M}},u)$ with:

1. *A priori* known as well as unknown parameters to estimate;
2. *A priori* known governing equations, model structure $\mathcal{S}$ and nonlinear mapping $F_{\mathcal{M}}(x_{\mathcal{M}},u)$;
3. Incomplete model with partially known dimensionality and unspecified structure $\mathcal{S}$.

The identification and estimation problems to solve are:

1. Identification of a class, order, dimension, nonlinearities and state-input dependencies of governing equations by using multivariate polynomials pertained to model structures $\mathcal{S}$;
2. Estimation of unknown parameters $\mathcal{P}$ under uncertainties and heterogeneities.

Effective algorithms are needed to perform identification and parameters estimation. Physical systems exhibit multi-physics phenomena with oscillations, resonance, limit cycles (stable, semi-stable or unstable), aperiodic and irregular cycles, fluctuations, asymmetric phases and other instabilities. The examples are:

1. Nonlinear resonance in electric circuits and mechanical structures;
2. Unsteady velocity fields and flows;
3. Elastic instabilities;
4. Unsteady thermodynamics and thermo-fluidics;
5. Wave turbulence;
6. Reaction–diffusion systems and diffusion-control reaction;
7. Membrane action potentials evolutions in neurons;
8. Cell and pharmacokinetics cycles;

9. Ecological and environmental instabilities;
10. Cyclic and oscillatory population dynamics, predator–prey interaction (*paradox enrichment*) and evolutionary dynamics in bio and ecosystems;
11. Infectious disease transmission cycles and epidemiological evolutions;
12. Market fluctuations and economic cycles.

Structured, Partially-Structured and *Unstructured Models* – Models $\mathcal{M}(x_{\mathcal{M}},u)$ may be derived using laws of physics, which yield governing differential equations and constitutive equations for states, outputs and inputs $(x_{\mathcal{M}},y,u)$. A time-varying or time-invariant model structure $\mathcal{S}$ may be known *a priori*, or, $\mathcal{S}$ may be unknown. For example, multi-physics of nonlinear friction, unsteady flow and micro-tribology may not yield explicit $\mathcal{S}$.

The known and over-structured $\mathcal{S}$ can be expressed by high-degree multivariate polynomials with unknown coefficients to estimate. Applying polynomial calculus, a model $\mathcal{M}(x_{\mathcal{M}},u,\mathcal{S},\mathcal{P})$ is defined by $(x_{\mathcal{M}},u)$ and a pair $(\mathcal{S},\mathcal{P})$. A quantitative fidelity validation may be performed despite unknown structure and parameters of physical systems $\mathbb{M}(x,u)$, $x\in X\subset\mathbb{R}^n$, $u\in U\subset\mathbb{R}^m$. Identification algorithms are sensitive as measurements are perturbed by noise, errors, input-output nonlinearities, data fusion heterogeneities, etc. Using interpolation, splines and kernel estimators, robustness and convergence may be ensured despite impediments and uncertainties [11, 12]. Consider:

1. *Structured* models $\mathcal{M}(x_{\mathcal{M}},u,\mathcal{S},\mathcal{P})$, described by
   $\dot{x}_{\mathcal{M}}(t)=F_{\mathcal{M}}(x_{\mathcal{M}},u)$, $x_{\mathcal{M}}\in X_{\mathcal{M}}\subset\mathbb{R}^n$, $u\in U\subset\mathbb{R}^m$,
   with a known structure $\mathcal{S}$ and nonlinear map $F_{\mathcal{M}}(\cdot)$;
2. *Partially-structured* and *unstructured* models $\mathcal{M}(x_{\mathcal{M}},u,\mathcal{S},\mathcal{P})$ if $\mathcal{S}$ is unknown, and, the model *predictor* for $F_{\mathcal{M}}(\cdot)$ are high-order and fractional polynomials.

4.2. Physical Systems and Mathematical Models

Identification and parameters estimation were performed for the block structured, grey-box and other models [2, 5]. Algorithms should be adequate for physical systems, descriptive fidelity, data heterogeneity, measurement errors, etc. Models of electronic, electromechanical, mechanical and other physical systems are derived by applying laws of electromagnetics, mechanics, thermodynamics, etc. The *generalized coordinates* q_i and *generalized momentum* p_i are used to find the kinetic, potential and dissipating energies (Γ,Π,D). These (q_i,p_i) and *generalized forces* Q_i are related to system states and inputs.

Notations – The Lagrange equations (2.16) $\frac{d}{dt}\left(\frac{\partial\Gamma}{\partial\dot{q}_i}\right)-\frac{\partial\Gamma}{\partial q_i}+\frac{\partial D}{\partial\dot{q}_i}+\frac{\partial\Pi}{\partial q_i}=Q_i$, Newtonian mechanics, Kirchhoff laws, Maxwell's equations and other laws of physics yield the governing and constitutive equations for a model $\mathcal{M}(x_{\mathcal{M}},u,\mathcal{S},\mathcal{P})$, defined as $\dot{x}_{\mathcal{M}}(t)=F_{\mathcal{M}}(x_{\mathcal{M}},u)$. One finds finite-dimensional model structures and real-valued parameters. The input-to-state and structure-to-parameters complementarity of $(x,u,\mathcal{S},\mathcal{P})$ should be guaranteed. Assuming a sufficient model fidelity and explicit system description, $\mathcal{M}(x_{\mathcal{M}},u,\mathcal{S},\mathcal{P})\triangleq\mathbb{M}(x,u)$, $x_{\mathcal{M}}\triangleq x$.

Model Fidelity – Model fidelity is system-specific and application-defined. For example, electronic circuit simulation and design tools use models to examine governance of semiconductor devices and circuits. Various electronic design automation software exist. The examples are analog simulation of complex integrated circuits with a schematic editor in SPICE, or, design and verification of digital circuits in Verilog and VHDL. A netlist is generated from circuit schematics. The microarchitecture simulations of data path, central processing units and ASICs are performed by applying the computer-aided design tools. However, to control aerial or automotive vehicles, there is no need to model fast dynamics of billions of transistors in ASICs, MPUs, FPGAs, etc. System-level generalizations and considerations are applied despite complexity, multi-physics-phenomena, multi-level organization, etc.

Physical System Characterization – System governance is evaluated using the measured states, outputs and inputs (x,y,u). *Dominant governance* of dynamic physical systems $\mathbf{M}(x,y,u)$ assumes a *standard* description, which implies system evolutions as $\dot{x}(t)=F(x,u)=Af(x,u)$ with implicit mappings $(F(x,u),f(x,u))$ and unknown parameters matrix A. Analyzing multi-physics fidelity, characterize a system by implicitly-defined mappings $(F(\cdot),f(\cdot))$ and matrix A. With measured $x\in X\subset\mathbb{R}^n$, $y\in Y\subset\mathbb{R}^m$ and $u\in U\subset\mathbb{R}^m$, the observed physical system governance is

$$\mathbf{M}:\equiv\left\{\underset{\text{Measured}}{(x,y,u)}:\ x\in X(X_0,U),\ \underset{\text{Implicit }(F,A,f)}{\dot{x}(t)=F(x,u)=Af(x,u)},\ x=\begin{bmatrix}x_{\text{measured}}\\ x_{\text{observed}}\end{bmatrix},\ y=H(x)\right\}. \quad (4.1)$$

Dimensionality and Measurability – One strives to avoid an excessive fidelity in analysis and design. Despite model incompleteness and not unique measured pair (x,u), model structure and mapping $F_{\mathcal{M}}(x_{\mathcal{M}},u)$ depend on fidelity, measurability, etc. Assume the system states are detectable and observable yielding the measured finite-dimensional (x,u) considering a *dominant governance*.

Model Description, Model Selection and Governing Equations – The governing equations for multi-physics models are given using the finite-dimensional states and inputs $(x_{\mathcal{M}},u)$ within a model structure and parameters pair $(\mathcal{S},\mathcal{P})$. Despite overall incompleteness, undecidability and axiomatization, sufficient fidelity and reducibility are achieved applying *predictors* $(F_{\mathcal{M}}(\cdot),f_{\mathcal{M}}(\cdot))$ and lumped-parameter equations for a given dimensionality (x,u) as

$$\mathcal{M}=\{x_{\mathcal{M}}:x_{\mathcal{M}}\in X_{\mathcal{M}}(X_{\mathcal{M}0},U),\ \dot{x}_{\mathcal{M}}(t)=F_{\mathcal{M}}(x_{\mathcal{M}},u)=A_{\mathcal{M}}f_{\mathcal{M}}(x_{\mathcal{M}},u),(x_{\mathcal{M}},u)\in X_{\mathcal{M}}\times U,(A_{\mathcal{M}},f_{\mathcal{M}})\subseteq\mathcal{P}\times\mathcal{S}\}, \quad (4.2)$$

where $F_{\mathcal{M}}(x_{\mathcal{M}},u):\mathbb{R}^n\times\mathbb{R}^m\rightarrow\mathbb{R}^n$ is the physics-defined mapping; $A_{\mathcal{M}}(\cdot):\mathbf{P}\rightarrow\mathbb{R}^{n\times z}$ is the matrix of unknown parameters; $f_{\mathcal{M}}(\cdot):X_{\mathcal{M}}\times U\times\mathbf{S}\times\mathbf{P}\rightarrow\mathbb{R}^z$ is the structure- and parameter-dependent nonlinear mapping.

Fidelity and measurability define dimensionality of (x,u), structure and parameter spaces $(\mathcal{S},\mathcal{P})\in\mathbf{S}\times\mathbf{P}$. Despite incompleteness and undecidability, physics-dependent $(\mathcal{S},\mathcal{P})$ can be known *a priori* or deducible from system governance (4.1), mapped by particular classes and order of differential equations (4.2), constitutive relationships, dependences, etc. The goal is to identify $(\mathcal{S},\mathcal{P})$ using (x,u), and, validate a model by assessing dynamic mismatch measures for $(x,x_{\mathcal{M}})$.

Model Predictors and Multivariate Polynomials – In (4.2), the *predictor* maps $F_{\mathcal{M}}(x_{\mathcal{M}},u)$ and $f_{\mathcal{M}}(x_{\mathcal{M}},u)$ are expressed as multivariate polynomials $p_{F_i}(\boldsymbol{a},x_{\mathcal{M}},u)$, defined by number of terms, degree, coefficients and constants. For *fully-structured* models with *a priori* known $f_{\mathcal{M}}(x_{\mathcal{M}},u)$, a matrix of unknown parameters $A_{\mathcal{M}}$ is estimated by solving a parameters estimation problem and finding $\hat{A}$. For $f(x_{\mathcal{M}},u)\equiv p_{F_i}(\boldsymbol{a},x_{\mathcal{M}},u)$, there is a mutual dependences between a structure $f_{\mathcal{M}}(x_{\mathcal{M}},u)$ and matrix $A_{\mathcal{M}}$. If the model structure is unknown, consider finite-dimensional $(\mathcal{S},\mathcal{P})$ with over-structured $\mathcal{S}$ and over-parameterized $\mathcal{P}$. High-degree multivariate polynomials describe over-structured $f_{\mathcal{M}}(x_{\mathcal{M}},u)$ with pertained over-parameterized $A_{\mathcal{M}}$. Regularization and degree reduction yields the resulting $(\hat{A},\hat{f})$. Model selection, structure and parameter dimensionality assessment, model validation and other problems are considered. These tasks are of an importance dealing with under- and over-fitted model.

4.3. Identification and Parameters Estimation

Problem Formulation – For a physical system (4.1) $\mathbf{M}(x,u)$, $(x,u)\in X\times U$, $x\in\mathbb{R}^n$, $u\in\mathbb{R}^m$, and finite-dimensional model $\mathcal{M}(x_{\mathcal{M}},u,\mathcal{P},\mathcal{S})$ (4.2), identify the unknown parameters and model structure by finding

$$\hat{\mathcal{M}}(x_{\mathcal{M}},u,\hat{A},\hat{f})=\left\{x_{\mathcal{M}}:x_{\mathcal{M}}\in X_{\mathcal{M}}(X_{\mathcal{M}0},U),\ \dot{x}_{\mathcal{M}}(t)=\hat{A}\hat{f}(x_{\mathcal{M}},u),\ (\hat{A},\hat{f})\subseteq\mathcal{P}\times\mathcal{S}\right\}, \tag{4.3}$$

$\hat{A}\in\mathbb{R}^{n\times z}$, $\hat{f}(\cdot):X_{\mathcal{M}}\times U\times\mathcal{P}\times\mathcal{S}\to\mathbb{R}^z$.

Consider descriptive systems governance as

$$\begin{bmatrix}\mathbf{M}\\ \mathcal{M}\\ \hat{\mathcal{M}}\end{bmatrix}:=\begin{cases}\text{Physical System } \mathbf{M}:\equiv\left\{\underset{\text{Measured}}{(x,y,u)}:\dot{x}(t)=\underset{\text{Implicit }(F,f,A)}{F(x,u)=Af(x,u)}\right\}\\ \Updownarrow\\ \text{System Model to Identify}\quad \dot{x}_{\mathcal{M}}(t)=F_{\mathcal{M}}(x,u)=A_{\mathcal{M}}f_{\mathcal{M}}(x_{\mathcal{M}},u)\\ \Updownarrow\\ \text{Identified Model}\quad \dot{x}_{\mathcal{M}}(t)=\underset{\hat{A}=\left[(\Phi^T\Phi)^{-1}\Phi^T\dot{x}\right]^T}{\hat{A}\hat{f}(x_{\mathcal{M}},u)}\end{cases}. \tag{4.4}$$

Fit the model parameters by the least-squares estimator $\hat{A}\in\mathbb{R}^{n\times z}$ as

$$\hat{A}=\left[(\Phi^T\Phi)^{-1}\Phi^T\dot{x}\right]^T,\ \hat{A}=[\hat{a}_{ij}]=\begin{bmatrix}\hat{a}_{11} & \cdots & \hat{a}_{1z}\\ \vdots & \ddots & \vdots\\ \hat{a}_{n1} & \cdots & \hat{a}_{nz}\end{bmatrix},\ \hat{A}\in\mathbb{R}^{n\times z},\ \dot{x}_{\text{data}}\in\mathbb{R}^{q\times n},\ \Phi\in\mathbb{R}^{q\times z},\ z=n+m+\text{N}+\text{M}+\text{K},$$

$$\Phi=\begin{bmatrix}f(x(t_1),u(t_1))\\ \vdots\\ f(x(t_q),u(t_q))\end{bmatrix}=\begin{bmatrix}x_{t_1} & u_{t_1} & f_x(x_{t_1}) & f_u(u_{t_1}) & f_{xu}(x_{t_1},u_{t_1})\\ \vdots & \vdots & \vdots & \vdots & \vdots\\ x_{t_q} & u_{t_q} & f_x(x_{t_q}) & f_u(u_{t_q}) & f_{xu}(x_{t_q},u_{t_q})\end{bmatrix},\ f(x,u)=\begin{bmatrix}x\\ u\\ f_x(x)\\ f_u(u)\\ f_{xu}(x,u)\end{bmatrix}, \tag{4.5}$$

$f_x(\cdot):\mathbb{R}^n\to\mathbb{R}^N$, $f_u(\cdot):\mathbb{R}^m\to\mathbb{R}^M$, $f_{xu}(\cdot):\mathbb{R}^n\times\mathbb{R}^m\to\mathbb{R}^K$,

where $A_{\mathcal{M}}(\cdot):\mathbf{P}\to\mathbb{R}$ is the matrix of model parameters to estimate; $f_{\mathcal{M}}(\cdot):X_{\mathcal{M}}\times U\times\mathbf{S}\times\mathbf{P}\to\mathbb{R}$ is the model structure mapping to be identified using the over-structured and over-parameterized $(f_{\mathcal{M}},A_{\mathcal{M}})$ found by using specified classes of physics-consistent differential equations and overdetermined *predictors* given by *d*-degree regular or fractional multivariate polynomials $p_{F_i}(a, x_{\mathcal{M}}, u) \in \mathcal{P}^{d,(n+m)}$, and, $f(x,u) \triangleq f_{\mathcal{M}}(x_{\mathcal{M}}, u)$, $f(x,u) = \begin{bmatrix} x \\ u \\ f_x(x) \\ f_u(u) \\ f_{xu}(x,u) \end{bmatrix}$;

$f_x(\cdot):\mathbb{R}^n \to \mathbb{R}^N$, $f_u(\cdot):\mathbb{R}^m \to \mathbb{R}^M$ and $f_{xu}(\cdot):\mathbb{R}^n \times \mathbb{R}^m \to \mathbb{R}^K$ are the structure-dependent mappings; $\hat{A} \in \mathbb{R}^{n\times z}$ is the matrix of identified parameters; $\hat{f}(\cdot):X_{\mathcal{M}} \times U \to \mathbb{R}^z$ is the identified structure- and parameter-dependent mapping; Φ is the data mapping, $\det(\Phi^T\Phi) \neq 0$, $\Phi \in \mathbb{R}^{q\times z}$.

In (4.5) and least-squares estimators, $\dot{x}$ denotes the measured or observed $\dot{x}_{\text{data}}$. Identification assumes measurements of $(x, \dot{x}, u) \in X \times U$ on $t \in [t_0, t_f]$ to explicitly find $\Phi(x,u)$. The datasets $(x, \dot{x}, u)_{\text{data}}$ are measured with uncertainties.

The objective is to minimize a dynamic mismatch measure pairs $[m_x(x, x_{\mathcal{M}}), m_{\dot{x}}(\dot{x}, \dot{x}_{\mathcal{M}})]$ and $[\Delta x(t), \Delta \dot{x}(t)]$

$$\min_{\substack{m_x, m_{\dot{x}} \\ \Delta x, \Delta \dot{x} \\ (\hat{A}, \hat{f})|_{P \cap S}}} \left\{ \begin{bmatrix} m_x(x, x_{\mathcal{M}}) \\ m_{\dot{x}}(\dot{x}, \dot{x}_{\mathcal{M}}) \end{bmatrix}, \begin{bmatrix} \Delta x(t) = \|x(t) - x_{\mathcal{M}}(t)\|_p \\ \Delta \dot{x}(t) = \|\dot{x}(t) - \dot{x}_{\mathcal{M}}(t)\|_p \end{bmatrix} \right\}, \quad \begin{cases} m_x(x, x_{\mathcal{M}}) = \int_{t_0}^{t_f} \frac{1}{\|x\|_p} \left(\|x(t) - x_{\mathcal{M}}(t)\|_p \right) dt \\ m_{\dot{x}}(\dot{x}, \dot{x}_{\mathcal{M}}) = \int_{t_0}^{t_f} \frac{1}{\|\dot{x}\|_p} \left(\|\dot{x}(t) - \dot{x}_{\mathcal{M}}(t)\|_p \right) dt \end{cases}, \tag{4.6}$$

such as

$$\begin{cases} m_x(x, x_{\mathcal{M}}) \le \delta_{m_x},\ m_{\dot{x}}(\dot{x}, \dot{x}_{\mathcal{M}}) \le \delta_{m_{\dot{x}}},\ 0 < (\delta_{m_x}, \delta_{m_{\dot{x}}}) \ll 1 \\ \Delta x(t) \le \delta_x,\ \Delta \dot{x}(t) \le \delta_{\dot{x}},\ 0 < (\delta_x, \delta_{\dot{x}}) \ll 1 \end{cases}, \tag{4.7}$$

and, validate a model

$$\hat{\mathcal{M}}(x_{\mathcal{M}}, u, \hat{A}, \hat{f}) \subseteq \mathbf{M}(x,u). \tag{4.8}$$

A notation $\|\cdot\|_p$ denotes the p-norm, $p \ge 1$, $\|\Delta x\|_p = \left(\sum_{i=1}^{n} |\Delta x_i|^p \right)^{\frac{1}{p}}$.

Identification is accomplished using measured $(x, \dot{x}, u) \in X \times U$ at t_k on $t \in [t_1, \ldots, t_q]$, $k \in [1, \ldots, q]$. Assuming adequate fidelity with $f(x,u) \triangleq f_{\mathcal{M}}(x_{\mathcal{M}}, u)$, the autoregressive moving average exogenous (ARMAX) algorithms apply the least-squares estimates for $\hat{A}(\cdot)$, given by (4.5).

Using the positive definite diagonal scaling matrix $W \in \mathbb{R}^{q\times q}$, from (4.5),

$$\hat{A} = \left[(\Phi^T W \Phi)^{-1} \Phi^T W \dot{x} \right]^T,\ \hat{A} \in \mathbb{R}^{n\times z}. \tag{4.9}$$

It is challenging to find a matrix W because one evaluates variations, scaling and significance of time-varying mapping $\Phi(\cdot)$.

Singular Value Decomposition in System Identification and Parameters Estimation – Model incompleteness, data heterogeneity, sensitivity and singularity obscure the ARMAX solutions and may yield erroneous results. The singular value decompositions (SVD) for Φ and $\Phi^T\Phi$ exist, and

$$\Phi = u_\Phi \sigma_\Phi v_\Phi^T,\ u_\Phi \in \mathbb{R}^{q\times q},\ \sigma_\Phi \in \mathbb{R}^{q\times z},\ v_\Phi \in \mathbb{R}^{z\times z}, \tag{4.10}$$

$$\Phi^T \Phi = v_\Phi \sigma_\Phi^T \sigma_\Phi v_\Phi^T = U_\Phi \Sigma_\Phi V_\Phi^T,\ U_\Phi \in \mathbb{R}^{z\times z},\ \Sigma_\Phi \in \mathbb{R}^{z\times z},\ V_\Phi \in \mathbb{R}^{z\times z},$$

where (u_Φ, v_Φ) and (U_Φ, V_Φ) are the square orthogonal matrices, whose columns form orthonormal bases; σ_Φ and Σ_Φ are the matrices of singular values, $\forall \sigma_{ii} \ge 0$.

Note: An extensive coverage on the SVD is reported section 7.5.

Existence of SVD for near-singular and matrices and mappings support data homogeneity, ensure uniqueness, regularization and convergence of solutions. For example, in (4.5), one computes $(\Phi^T\Phi)^{-1}$ as $(\Phi^T\Phi)^{-1}=(U_\Phi\Sigma_\Phi V_\Phi^T)^{-1}=V_\Phi\Sigma_\Phi^{-1}U_\Phi^T$.

A computational complexity of SVD for a $l\times k$ matrix is $O(l^3+l^2k+k^2l)$, $l>k$.

One finds $\hat{A}$ by applying the SVDs for mappings $(\Phi,\ \Phi^T\Phi,\ (\Phi^T\Phi)^{-1},\ \hat{x})$.

The corresponding SVDs for $(\Phi,\ \Phi^T\Phi,\ (\Phi^T\Phi)^{-1},\ \hat{x})$ exist for:

1. Non-square and non-invertible matrices and mappings;
2. Near-singular, ill-conditioned and badly scaled matrices and mappings;
3. Matrices and mappings with zero-valued blocks and uncertainties in $(x,\dot{x},u)_{\text{data}}$ and $\Phi(x,u)$.

For (4.5), using (4.10),

$$\hat{A}=\left[(\Phi^T\Phi)^{-1}\Phi^T\dot{x}\right]^T=\left[(v_\Phi\sigma_\Phi^T\sigma_\Phi v_\Phi^T)^{-1}v_\Phi\sigma_\Phi^T u_\Phi^T\dot{x}\right]^T,\ \hat{A}\in\mathbb{R}^{n\times z}, \tag{4.11}$$

$$\hat{A}=\Bigg[\underbrace{(\Phi^T\Phi)^{-1}}_{\Phi^T\Phi=v_\Phi\sigma_\Phi^T\sigma_\Phi v_\Phi^T=U_\Phi\Sigma_\Phi V_\Phi^T}\ \underbrace{\Phi^T}_{\Phi=u_\Phi\sigma_\Phi v_\Phi^T}\ \dot{x}\Bigg]^T=\left[V_\Phi\Sigma_\Phi^{-1}U_\Phi^T v_\Phi\sigma_\Phi^T u_\Phi^T\dot{x}\right]^T.$$

Identification Algorithm: Dynamic Mismatches, Parameter and Structural Differences

For a physical systems $\mathbb{M}$ and model $\mathcal{M}$, solving (4.5) or (4.11), find $(\hat{A},\hat{f})\big|_{\mathsf{P}\cap\mathsf{S}}$ and truncated $(\tilde{A},\tilde{f})\big|_{\mathsf{P}\cap\mathsf{S}}$ by minimizing dynamic mismatch measures $[(m_x, m_{\dot{x}}),(\Delta x,\Delta\dot{x})]$ (4.6), ensuring (4.7), guaranteeing (4.8).

The structural difference Δf and parameter difference ΔA are

$$\Delta f(x,x_\mathcal{M},u)=[f(x,u)-\hat{f}(x_\mathcal{M},u)],\ \hat{f}\equiv f_\mathcal{M}, \tag{4.12}$$
$$\Delta A(t)=[A(t)-\hat{A}(t)],\ \hat{A}\equiv A_\mathcal{M}.$$

A pair $(\Delta A,\Delta f)$ cannot be evaluated because the system $\mathbb{M}(x,u)$ structure and parameters (F,f,A) are unknown. For measured system transients and model evolutions $(x(t),x_\mathcal{M}(t))$, quantitative mismatch evolutions are expressed and assessed by dynamic measured $[\Delta x(t),\Delta\dot{x}(t)]$ (4.6). Furthermore,

$$\Delta\dot{x}(t)=[\dot{x}(t)-\dot{x}_\mathcal{M}(t)]=[Af(x,u)-\hat{A}\hat{f}(x_\mathcal{M},u)]=\Delta A(t)f(x,u)+\hat{A}(t)\Delta f(x,x_\mathcal{M},u),\ \hat{A}\equiv A_\mathcal{M}. \tag{4.13}$$

The state mismatch measure $(\Delta x,\Delta\dot{x})$ (4.13) depends on structure-parametric mismatch pair $(\Delta A,\Delta f(x,x_\mathcal{M},u))$ (4.12). These imply

$$\begin{cases}\left|A-\hat{A}\right|\to 0,\ \forall\left|a_{ij}-\hat{a}_{ij}\right|\to 0\\ [F(\cdot)-\hat{F}(\cdot)]\to 0,\ [f(\cdot)-\hat{f}(\cdot)]\to 0\end{cases}\ \text{if}\ \begin{cases}\Delta\dot{x}(t)=[\dot{x}(t)-\dot{x}_\mathcal{M}(t)]\to 0\\ \Delta x(t)=[x(t)-x_\mathcal{M}(t)]\to 0\end{cases}. \tag{4.14}$$

One also concludes that $\begin{cases}\Delta\dot{x}(t)=[\dot{x}(t)-\dot{x}_\mathcal{M}(t)]\to 0\\ \Delta x(t)=[x(t)-x_\mathcal{M}(t)]\to 0\end{cases}$ if $\begin{cases}\left|A-\hat{A}\right|\to 0,\ \forall\left|a_{ij}-\hat{a}_{ij}\right|\to 0\\ [F(\cdot)-\hat{F}(\cdot)]\to 0,\ [f(\cdot)-\hat{f}(\cdot)]\to 0\end{cases}$.

The derived (4.14) correspond, and, in agreement, with (4.6)-(4.7).

Illustrative Example 4.1.

Let $f(\boldsymbol{a},x)=a_lx^l+a_{l-1}x^{l-1}+\ldots+a_1x+a_0$ and $\hat{f}(\hat{\boldsymbol{a}},x)=\hat{a}_qx^q+\hat{a}_{q-1}x^{q-1}+\ldots+\hat{a}_1x+\hat{a}_0$ are the polynomials with real coefficients, $\forall(a_i,\hat{a}_i)\in\mathbb{R}$.

If $l=q$ and $a_i=\hat{a}_i$, $\forall(a_i,\hat{a}_i)$, we have $f(\boldsymbol{a},x)=\hat{f}(\hat{\boldsymbol{a}},x)$, $\forall x\in\mathbb{R}$. ■

While a tuple (F,f,A) is unknown, the behavioral and descriptive convergences of a model to physical system governance is asserted by (4.13) and (4.14). Using the measured states, the model can be validated by evaluating dynamic mismatch measures (4.6)-(4.7) for $(x,x_{\mathcal{M}})$ and $(\dot{x},\dot{x}_{\mathcal{M}})$. For the measured $(x,x_{\mathcal{M}})$ and $(\dot{x},\dot{x}_{\mathcal{M}})$, or computed $(\dot{x}=\frac{dx}{dt},\ \dot{x}_{\mathcal{M}}=\frac{dx_{\mathcal{M}}}{dt})$ on $t\in[t_0,\ t_f]$, algorithm (4.5) or (4.11) should prevail with convergence to $\hat{A}=[\hat{a}_{ij}]$. Considering the bounded dynamic mismatches (4.7) $\begin{cases} m_x(x,x_{\mathcal{M}}),\ m_{\dot{x}}(\dot{x},\dot{x}_{\mathcal{M}}) \\ \Delta x(t),\ \Delta\dot{x}(t) \end{cases}$ and (4.14) $\begin{cases} \Delta\dot{x}(t)=[\dot{x}(t)-\dot{x}_{\mathcal{M}}(t)]\to 0 \\ \Delta x(t)=[x(t)-x_{\mathcal{M}}(t)]\to 0 \end{cases}$, one has

$$\hat{\mathcal{M}}=\left\{\dot{x}_{\mathcal{M}}(t)=\hat{A}\hat{f}(x_{\mathcal{M}},u),\left[\hat{A}=[\hat{a}_{ij}],\hat{f},\ p_{\hat{F}_i}(\hat{a}_{ij},x_{\mathcal{M}},u)\right]\Big|_{\mathbf{P}\cap\mathbf{S}}\subseteq\mathcal{P}\times\mathcal{S}\right\},\ \hat{\mathcal{M}}(x_{\mathcal{M}},u,\hat{A},\hat{f})\subseteq\mathbf{M}(x,u)\ .\ (4.15)$$

Structure and Parameters Reduction and Regularization – Solving (4.5) or (4.11), we found $\hat{A}$, which admits a SVD as

$$\hat{A}=[\hat{a}_{ij}]=\begin{bmatrix}\hat{a}_{11} & \cdots & \hat{a}_{1z}\\ \vdots & \ddots & \vdots\\ \hat{a}_{n1} & \cdots & \hat{a}_{nz}\end{bmatrix},\ \hat{A}=U_{\hat{A}}\Sigma_{\hat{A}}V_{\hat{A}}^T,\ \hat{A}\in\mathbb{R}^{n\times z},\ U_{\hat{A}}\in\mathbb{R}^{n\times n},\ \Sigma_{\hat{A}}\in\mathbb{R}^{n\times z},\ V_{\hat{A}}\in\mathbb{R}^{z\times z}.(4.16)$$

In structural identification, for $\mathbf{M}(x,u)$, $x\in\mathbb{R}^n$, $u\in\mathbb{R}^m$, using the overdetermined *predictors* $\left(A_{\mathcal{M}},f_{\mathcal{M}},p_{F_i}(\boldsymbol{a}_i,x_{\mathcal{M}},u)\right)\subseteq\mathcal{P}\times\mathcal{S}$, initialize a search in $\mathcal{S}\times\mathcal{P}\in\mathbf{S}\times\mathbf{P}$ with $\left(A_{\mathcal{M}0},f_{\mathcal{M}0},p_{F_i0}(\boldsymbol{a}_{i0},x_{\mathcal{M}},u)\right)$ and $p_{F_i0}(\boldsymbol{a}_{i0},x_{\mathcal{M}},u)\in\mathcal{P}^{d_0,(n+m)}$. Compute (4.11) $\hat{A}=[\hat{a}_{ij}]$, such that (4.15) $\hat{\mathcal{M}}(x_{\mathcal{M}},u,\hat{A},\hat{f})\subseteq\mathbf{M}(x,u)$ is guaranteed, and, regularize matrix of parameters and model structure by determining truncated $(\tilde{A},\tilde{f})\big|_{\mathbf{P}\cap\mathbf{S}}$.

The algorithm to find $\hat{\mathcal{M}}(x_{\mathcal{M}},u,\hat{A},\hat{f})$ and $\tilde{\mathcal{M}}(x_{\mathcal{M}},u,\tilde{A},\tilde{f})$ is:

1. Compute $\hat{A}$ using (4.5) or (4.11), and, solve the minimization problem (4.6)-(4.7) for *admissible* mismatches $[m_x(x,x_{\mathcal{M}}),m_{\dot{x}}(\dot{x},\dot{x}_{\mathcal{M}})]$ and $[\Delta x(t),\Delta\dot{x}(t)]$, yielding $(\hat{A},\hat{f})$ to ensure (4.15), such that $\hat{\mathcal{M}}(x_{\mathcal{M}},u,\hat{A},\hat{f})\subseteq\mathbf{M}(x,u)$ on $[t_0,\ t_f]$;
2. Using $\hat{A}=[\hat{a}_{ij}]$ and $(\hat{A},\hat{f})\big|_{\mathbf{P}\cap\mathbf{S}}$, solving (4.6)-(4.7) and (4.11), compute a regularized matrix of parameters $\tilde{A}$ evaluating entries as $\begin{cases}\forall\hat{a}_{ij}\neq 0 \text{ if } |\hat{a}_{ij}|>\varepsilon_a\\ \forall\hat{a}_{ij}=0 \text{ if } |\hat{a}_{ij}|\leq\varepsilon_a\end{cases}$, $0<\varepsilon_a\ll 1$ with the specified entry-specific tolerances $\varepsilon_a\in\mathbb{R}$. Derive the resulting $(\tilde{A},\tilde{f})\big|_{\mathbf{P}\cap\mathbf{S}}$ and $\tilde{\mathcal{M}}(x_{\mathcal{M}},u,\tilde{A},\tilde{f})$ as

$$\tilde{A}=[\tilde{a}_{kl}]\Big|_{\begin{cases}\hat{a}_{kl}\neq 0,\,|\hat{a}_{ij}|>\varepsilon_a,\,0<\varepsilon_a\ll 1\\ \hat{a}_{kl}=0,\,|\hat{a}_{ij}|\leq\varepsilon_a\end{cases}}=\begin{bmatrix}\tilde{a}_{11} & \cdots & \tilde{a}_{1z}\\ \vdots & \ddots & \vdots\\ \tilde{a}_{n1} & \cdots & \tilde{a}_{nz}\end{bmatrix},\ \tilde{A}\in\mathbb{R}^{n\times z}; \quad (4.17)$$

$$\tilde{\mathcal{M}}=\left\{\dot{x}_{\mathcal{M}}(t)=\tilde{A}\tilde{f}(x_{\mathcal{M}},u),\left[\tilde{A}=[\tilde{a}_{ij}],\tilde{f},\ p_{\tilde{F}_i}(\tilde{a}_{ij},x_{\mathcal{M}},u)\right]\Big|_{\mathbf{P}\cap\mathbf{S}}\subseteq\mathcal{P}\times\mathcal{S}\right\},\ \tilde{\mathcal{M}}(x_{\mathcal{M}},u,\tilde{A},\tilde{f})\subseteq\mathbf{M}(x,u)\,.$$

With computed $(\hat{A},\tilde{A})$, obtain a truncated structure $\tilde{f}(\cdot)$, $(\tilde{A},\tilde{f})\big|_{\mathbf{P}\cap\mathbf{S}}$ and $\tilde{\mathcal{M}}(x_{\mathcal{M}},u,\tilde{A},\tilde{f})$, solving (4.6)-(4.7), such as $\tilde{\mathcal{M}}(x_{\mathcal{M}},u,\tilde{A},\tilde{f})\subseteq\mathbf{M}(x,u)$. The entries of matrices $\hat{A}=[\hat{a}_{ij}]$ and $\tilde{A}=[\tilde{a}_{kl}]$ are the coefficients which associate with states and inputs (controls, disturbances and perturbations). States and inputs specific tolerances ε_a are applied to assess negligibility of nondominant matrix entries, thereby reducing $\hat{f}(\cdot)$ to $\tilde{f}(\cdot)$, and, assessing $\tilde{\mathcal{M}}(x_{\mathcal{M}},u,\tilde{A},\tilde{f})\subseteq\mathbf{M}(x,u)$. Dominant and negligible dependencies, phenomena (stability, sensitivity, robustness, etc.), and other factors are considered.

Singular Value Decomposition and Its Application to Solve Identification Problems – The eigendecomposition factorization empowers matrix and mapping factorizations yielding robustness and convergence despite uncertainties and heterogeneities. The SVDs for (Φ, $\Phi^T\Phi$, $(\Phi^T\Phi)^{-1}$, $\hat{\dot{x}}$) are used to compute $\hat{A}$ as (4.11) and (4.22). The factorization schemes are effectively applied and verified to overcome data uncertainties, heterogeneities, sensitivity and convergence impediments. However, one may not ensure consistency using the *full*, reduced and truncated SVDs in dimensionality reduction, low-rank approximation, as well as correlation and model correspondence on $(\hat{A},\tilde{A})$, $\hat{A}=U_{\hat{A}}\Sigma_{\hat{A}}V_{\hat{A}}^T$ and $\tilde{A}=\tilde{U}_{\tilde{A}}\tilde{\Sigma}_{\tilde{A}}\tilde{V}_{\tilde{A}}^T$. Tuples $(U_{\hat{A}},\Sigma_{\hat{A}},V_{\hat{A}})$ and $(U_{\tilde{A}},\Sigma_{\tilde{A}},V_{\tilde{A}})$ may not provide the overall parametric-structural dependencies, may not yield regularization and preserve descriptive quantitative features. Consistent analysis ensures explicit characterization and descriptive measures.

Model Regularization – Solving (4.5), (4.11) and (4.17), applying (4.6)-(4.7), validate a model (4.15) $\hat{\mathcal{M}}(x_{\mathcal{M}},u,\hat{A},\hat{f})\subseteq\mathbf{M}(x,u)$ with $(\hat{A},\hat{f})\big|_{\mathbf{P}\cap\mathbf{S}}$, as well as $\tilde{\mathcal{M}}(x_{\mathcal{M}},u,\tilde{A},\tilde{f})\subseteq\mathbf{M}(x,u)$ with identified $(\tilde{A},\tilde{f})\big|_{\mathbf{P}\cap\mathbf{S}}$ on $[t_0,t_f]$. The number of columns z in $\hat{A}\in\mathbb{R}^{n\times z}$ and $\tilde{A}\in\mathbb{R}^{n\times z}$, computed using (4.11) and (4.17), may differ because $\hat{f}(x_{\mathcal{M}},u)$ is altered to $\tilde{f}(x_{\mathcal{M}},u)$. Terms and degrees of multivariate polynomials $p_{F_i}(\boldsymbol{a},x_{\mathcal{M}},u)\in\mathcal{P}^{d,(n+m)}$ are defined by multi-physic phenomena in $\mathbf{M}(x,u)$, and, affected by data quality. Assessing states and input dependencies, evaluates sizes, partitioning and blocks in $(\hat{A},\tilde{A})$. If $\dim(\hat{A})>\dim(\tilde{A})$, there is a likelihood of reducability of $(\hat{A},\hat{f})\big|_{\mathbf{P}\cap\mathbf{S}}$ to $(\tilde{A},\tilde{f})\big|_{\mathbf{P}\cap\mathbf{S}}$.

4.4. Multivariate Polynomials and Identification

Mathematical models are conjectural descriptions of physical phenomena in form of descriptive governing and constitutive equations. Multivariate polynomials are used to define $f_{\mathcal{M}}(x_{\mathcal{M}},u)$. Express a model (4.2) as

$$\mathcal{M}=\{x_{\mathcal{M}}\colon x_{\mathcal{M}}\in X_{\mathcal{M}}(X_{\mathcal{M}0},U)\},\ \dot{x}_{\mathcal{M}}(t)=F_{\mathcal{M}}(\boldsymbol{a},x_{\mathcal{M}})+B_{\mathcal{M}}(\boldsymbol{b},x_{\mathcal{M}},u),\ (\boldsymbol{a},\boldsymbol{b},F_{\mathcal{M}},B_{\mathcal{M}})\subseteq\mathcal{P}\times\mathcal{S}, \tag{4.18}$$

$$\dot{x}_{\mathcal{M}}=\begin{bmatrix}\dot{x}_{\mathcal{M}1}\\ \vdots\\ \dot{x}_{\mathcal{M}i}\\ \vdots\\ \dot{x}_{\mathcal{M}n}\end{bmatrix}=\begin{bmatrix}p_{F_1}(\boldsymbol{a},x_{\mathcal{M}})+p_{B_1}(\boldsymbol{b},x_{\mathcal{M}},u)\\ \vdots\\ p_{F_i}(\boldsymbol{a},x_{\mathcal{M}})+p_{B_i}(\boldsymbol{b},x_{\mathcal{M}},u)\\ \vdots\\ p_{F_n}(\boldsymbol{a},x_{\mathcal{M}})+p_{B_n}(\boldsymbol{b},x_{\mathcal{M}},u)\end{bmatrix}.$$

In structure-parametric maps $F_{\mathcal{M}}(\cdot)$ and $B_{\mathcal{M}}(\cdot)$, which are the *predictors*, parameters $(\boldsymbol{a},\boldsymbol{b})$ are the polynomial coefficients and constants. Polynomial calculus supports identification tasks. In a finite-dimensional linear space, model structure $\mathcal{S}$ and unknown parameters $\mathcal{P}(\boldsymbol{a},\boldsymbol{b})$ are described by multivariate polynomials $p_{F_i}(\boldsymbol{a},x_{\mathcal{M}})\in\mathcal{P}_F^{d_1,n}$ and $p_{B_i}(\boldsymbol{b},x_{\mathcal{M}},u)\in\mathcal{P}_B^{d_2,(n+m)}$.

Example 4.1. *Nonlinear Systems With Limit Cycles*
Consider physical systems $\mathbb{M}(x,u)$, $x\in\mathbb{R}^n$, $u\in\mathbb{R}^m$ which exhibit symmetric and asymmetric oscillations, switching, limit cycles and instabilities. For given dimensionalities of (x,u), a model $\mathcal{M}(\cdot)$ should be found using a class of nonlinear differential equations which admit periodic, quasiperiodic or non-periodic solutions. For systems which exhibit instabilities *in small* with $x\in\mathbb{R}^n$ $(n\geq 2)$, if $\mathcal{S}$ is unknown, consider the *standard* and modified Lotka-Volterra, van der Pol, Lorenz and other equations, examined in Examples 2.22 to 2.26. These equations describe mechanical waves, thermodynamic and unsteady aerodynamic periodic symmetric and aperiodic asymmetric oscillations, fluctuations, etc.

The *standard* Lotka-Volterra equations, for which $(x_1,x_2)>0$, are

$\dot{x}_1 = a_{11}x_1 - a_{12}x_1x_2$, $(a_{ii},a_{ij})\in\mathbb{R}$, $\forall a_{ii}>0$,

$\dot{x}_2 = -a_{21}x_2 + a_{22}x_1x_2$.

The Lorenz differential equations are

$\dot{x}_1 = a_{11}x_1 + a_{12}x_2$, $(a_{ii},a_{ij})\in\mathbb{R}$,

$\dot{x}_2 = a_{21}x_1 + a_{22}x_2 + a_{23}x_1x_3$,

$\dot{x}_3 = a_{31}x_3 + a_{32}x_1x_2$.

These differential equations were studied in chapter 2. ■

Example 4.2. *Polynomials*
A monomial polynomial is $p(\boldsymbol{a},x)=a_nx^n + a_{n-1}x^{n-1} + \ldots + a_1x + a_0$, $\forall a_i\in\mathbb{R}$.

A coefficient a_{n-1} is called the degree $(n-1)$ coefficient. If $a_n\neq 0$, then, n is the degree of $p(\boldsymbol{a},x)$, a_n is the leading coefficient of $p(\boldsymbol{a},x)$, and, a_nx^n is the leading term. Degrees 1, 2, 3, 4 and 5 correspond to the linear, quadratic, cubic, quartic and quantic polynomials. For example, the linear and quantic polynomials are

$p(\boldsymbol{a},x)=x+1$, $p(\boldsymbol{a},x)=x^5+2x^3+3x+4$.

A polynomial in one indeterminate is a univariate polynomial. The laws of physics yield multivariate polynomials with many indeterminates. Polynomials with two and three indeterminates are called bivariate and trivariate polynomials. Multivariate polynomials describe nonlinear friction, losses, magnetization, etc. ■

Non-Homogeneous and Homogeneous Polynomials – The n- and $(n+m)$-variate non-homogeneous polynomials $p_{F_i}(\boldsymbol{a},x)\in\mathcal{P}_F^{d_1,n}$ and $p_{B_i}(\boldsymbol{b},x,u)\in\mathcal{P}_B^{d_2,(n+m)}$ with degrees (d_1,d_2) in spaces $(\mathcal{P}_F^{d_1,n},\mathcal{P}_B^{d_2,(n+m)})$, $\dim\mathcal{P}_F^{d_1,n}=\begin{bmatrix} d_1+n+1 \\ n+1\end{bmatrix}$, $\dim\mathcal{P}_B^{d_2,(n+m)}=\begin{bmatrix} d_2+n+m \\ n+m\end{bmatrix}$ are

$$p_{F_i}(\boldsymbol{a},x)=\boldsymbol{a}_0+\sum_{0\leq i_1+i_2+\ldots+i_{n-1}+i_n\leq d_1}\boldsymbol{a}_{i_1i_2\ldots i_{n-1}i_n}x_1^{i_1}x_2^{i_2}\ldots x_{n-1}^{i_{n-1}}x_n^{i_n}, \tag{4.19}$$

$$p_{B_i}(\boldsymbol{b},x,u)=\boldsymbol{b}_1u+\sum_{0\leq i_1+i_2+\ldots+i_{n-1}+i_n\leq d_2}\boldsymbol{b}_{i_1i_2\ldots i_{n-1}i_n}(x_1^{i_1}x_2^{i_2}\ldots x_{n-1}^{i_{n-1}}x_n^{i_n})u.$$

For homogeneous polynomials $\mathcal{H}_F^{d_1,n}$ and $\mathcal{H}_B^{d_2,(n+m)}$,

$$p_{F_i}^{\mathcal{H}}(\boldsymbol{a},x)=\boldsymbol{a}_0+\sum_{i_1+i_2+\ldots+i_{n-1}+i_n=d_1}\boldsymbol{a}_{i_1i_2\ldots i_{n-1}i_n}x_1^{i_1}x_2^{i_2}\ldots x_{n-1}^{i_{n-1}}x_n^{i_n},\ \dim\mathcal{H}_F^{d_1,n}=\begin{bmatrix} d_1+n \\ n\end{bmatrix}, \tag{4.20}$$

$$p_{B_i}^{\mathcal{H}}(\boldsymbol{b},x,u)=\boldsymbol{b}_1u+\sum_{i_1+i_2+\ldots+i_{n-1}+i_n=d_2}\boldsymbol{b}_{i_1i_2\ldots i_{n-1}i_n}(x_1^{i_1}x_2^{i_2}\ldots x_{n-1}^{i_{n-1}}x_n^{i_n})u,\ \dim\mathcal{H}_B^{d_2,(n+m)}=\begin{bmatrix} d_2+n+m-1 \\ n+m-1\end{bmatrix}.$$

Overdetermined Predictors – If the model structure is unknown, the search algorithm refines the degrees of polynomials (4.19)-(4.20) (d_1,d_2) with ($d_{10}\geq 1$,$d_{20}\geq 1$), and, finds $(\hat{A},\hat{f})$. Adequate model *predictors* with $p_{F_i}(\boldsymbol{a},x)$ and $p_{B_i}(\boldsymbol{b},x,u)$ are applied. For systems with limit cycles, $d_1\geq 2$, see Examples 2.22 to 2.26, and 4.1. To describe multivariate polynomials, develop computationally efficient algorithms and support polynomial calculus, apply matrices, row-by-row product and other operations.

Example 4.3.
Consider a polynomial $p(\boldsymbol{a},x)$ of degree d with n variables, $x=[x_1,\ldots,x_n]$.
The number of terms and coefficients is

$$\binom{n+d}{d}=\frac{(n+d)(n+d-1)\ldots(n+1)}{d(d-1)\ldots 1}=\frac{(n+d)!}{d!n!},\ (n+d)\geq d.$$ ■

Example 4.4.
For n=2 and d=3, if $\forall a_i$=1, we have

$$p(x_1,x_2)=x_1^3+x_2^3+x_1^2x_2+x_2^2x_1+x_1^2+x_1x_2+x_2^2+x_1+x_2+1 \text{ with } \binom{2+3}{3}=10.$$ ■

Example 4.5. *Matrices and Polynomials*
For the second degree polynomial,

$$p_{\mathcal{H}}=a_{0i}+\sum_{i=1}^{n}a_ix_i+\sum_{i=1}^{n}\sum_{j=1}^{n}a_{ij}x_ix_j,\ a_{ij}\neq a_{ji}.$$

A quadratic polynomial is a polynomial of degree 2. In vector-matrix notations, $p_{\mathcal{H}}=x^TAx$, $A\in\mathbb{R}^{n\times n}$.
For linear and quadratic $p_{F_i}(\boldsymbol{a},x)$, and, linear $p_{B_i}(\boldsymbol{b},x,u)$, we have

$$\dot{x}(t)=\begin{bmatrix}a_{01}\\ \vdots\\ a_{0n}\end{bmatrix}+A_1x+\begin{bmatrix}x^TA_{21}x\\ \vdots\\ x^TA_{2n}x\end{bmatrix}+\left[\begin{bmatrix}b_{11}\\ \vdots\\ b_{1n}\end{bmatrix}+B_1x\right]u,\ x\in\mathbb{R}^n,\ u\in\mathbb{R}^m,$$

where $a_{0i}\in\mathbb{R}$ and $b_{1i}\in\mathbb{R}$ are the scalars; $A_1\in\mathbb{R}^{n\times n}$, $A_{2i}\in\mathbb{R}^{n\times n}$ and $B_1\in\mathbb{R}^{n\times n}$ are the matrices of coefficients of the first- and second-degree polynomials $p_{F_i}(\boldsymbol{a},x)$ and $p_{B_i}(\boldsymbol{b},x,u)$. ■

Example 4.6. *High-Degree and Fractional Polynomials*
Polynomials of degree m with powers w=(0,1,…,m) and w=(w_1,…,w_m) are

$$p(\boldsymbol{a},x)=a_0+a_1x^1+a_2x^2+\ldots+a_mx^m,$$

$$p(\boldsymbol{a},x)=a_1x^{w_1}+a_2x^{w_2}+\ldots+a_mx^{w_m}.$$

For $w\in\Omega$, use nonnegative integers (i,j,k,l), and,

$$\Omega=\left\{\sum_i\sum_j\left(-\tfrac{2i-1}{2j-1}\right),\ 0,\ \sum_k\sum_l\left(\tfrac{2k-1}{2l-1}\right)\right\}.$$

For $\Omega=\left\{0,\ \sum_k\sum_l\left(\tfrac{2k-1}{2l-1}\right)\right\}$, k=1,2,3,4,5 and l=2, the fractional polynomial is

$$p(\boldsymbol{a},x)=a_{02}+a_{12}x^{1/3}+a_{22}x+a_{32}x^{5/3}+a_{42}x^{7/3}+a_{52}x^3.$$

Polynomials are expressed as the product of basis power vector to a square matrix A.

$$p(\boldsymbol{a},x)=(x^{\langle m\rangle})^TAx^{\langle m\rangle}=\sum_{i,j=1}^{n}a_{ij}x_i^{\langle m\rangle}x_j^{\langle m\rangle}.$$

For $p(\boldsymbol{a},x)=1-x_1-2x_2+x_1^2+2x_2^2+x_1(x_2-2x_1^2-3x_2^2)+x_2(x_1^2+2x_2^2)+x_1^4+2x_1^2x_2^2+x_2^4$,

$$p(\boldsymbol{a},x)=\begin{bmatrix}1 & x_1 & x_2 & x_1^2 & x_2^2\end{bmatrix}\begin{bmatrix}1 & 0 & 0 & 0 & 0\\ -1 & 0 & 0 & 0 & 0\\ -2 & 1 & 0 & 0 & 0\\ 1 & -2 & 1 & 1 & 0\\ 2 & -3 & 2 & 2 & 1\end{bmatrix}\begin{bmatrix}1\\ x_1\\ x_2\\ x_1^2\\ x_2^2\end{bmatrix}.$$

Using symmetric matrices A_1 and A_2, one has

$p(x)=(x^{\langle m\rangle})^T\left(A_1+A_2\right)x^{\langle m\rangle}$.

Fractional and regular high-degree polynomials as *predictors* may be found as

$$p(\boldsymbol{a},x)=(x^{\langle m_1\rangle})^T Ax^{\langle m_2\rangle}=\sum_{i,j=1}^{n}a_{ij}x_i^{\langle m_1\rangle}x_j^{\langle m_2\rangle}. \qquad \blacksquare$$

Example 4.7.

The Lotka-Volterra equations are

$\dot{x}_1=a_{11}x_1-a_{12}x_1x_2$, $x_1(t_0)=x_{10}$, $\forall(a_{ii},a_{ij})>0$,

$\dot{x}_2=-a_{21}x_2+a_{22}x_1x_2$, $x_2(t_0)=x_{20}$.

System dynamics and evolutions of states $x_1(t)$ and $x_2(t)$ depend on initial conditions $(x_{10},x_{20})>0$. A not-perturbed second-order system with $u=0$ exhibits symmetric periodic orbits $x\geq 0$, $x_0\neq 0$, $x\in\mathbb{R}^2$, $x=[x_1\ x_2]^T$. For a second-order system, which exhibits limit cycles with evolutions which depend on x_0 and $x\geq 0$, assume a known structure, described by the *standard* Lotka-Volterra equations. We have

$$\dot{x}_{\mathcal{M}}(t)=F_{\mathcal{M}}(\boldsymbol{a},x_{\mathcal{M}})=A_{\mathcal{M}}f_{\mathcal{M}}(x_{\mathcal{M}})=\begin{bmatrix}a_{11} & 0 & -a_{12}\\ 0 & -a_{22} & a_{23}\end{bmatrix}\begin{bmatrix}x_{\mathcal{M}1}\\ x_{\mathcal{M}2}\\ x_{\mathcal{M}1}x_{\mathcal{M}2}\end{bmatrix},\ (A_{\mathcal{M}},f_{\mathcal{M}})\subseteq\mathcal{P}\times\mathcal{S}.$$

If the model structure $\mathcal{S}$ is unknown, one may find a *predictor* by modifying the Lotka-Volterra equation. Apply the second degree polynomials to define over-structured $\mathcal{S}$ and over-parameterized $\mathcal{P}$. The corresponding $F_{\mathcal{M}}(\boldsymbol{a},x_{\mathcal{M}})$, $A_{\mathcal{M}}$ and $f_{\mathcal{M}}(x_{\mathcal{M}})$ are

$$\dot{x}_{\mathcal{M}}(t)=A_1x_{\mathcal{M}}+\begin{bmatrix}x_{\mathcal{M}}^TA_{21}x_{\mathcal{M}}\\ x_{\mathcal{M}}^TA_{22}x_{\mathcal{M}}\end{bmatrix}=\begin{bmatrix}a_{11} & a_{12}\\ a_{21} & a_{22}\end{bmatrix}\begin{bmatrix}x_{\mathcal{M}1}\\ x_{\mathcal{M}2}\end{bmatrix}+\begin{bmatrix}\begin{bmatrix}x_{\mathcal{M}1} & x_{\mathcal{M}2}\end{bmatrix}\begin{bmatrix}a_{21(11)} & a_{21(12)}\\ a_{21(21)} & a_{21(22)}\end{bmatrix}\begin{bmatrix}x_{\mathcal{M}1}\\ x_{\mathcal{M}2}\end{bmatrix}\\ \begin{bmatrix}x_{\mathcal{M}1} & x_{\mathcal{M}2}\end{bmatrix}\begin{bmatrix}a_{22(11)} & a_{22(12)}\\ a_{22(21)} & a_{22(22)}\end{bmatrix}\begin{bmatrix}x_{\mathcal{M}1}\\ x_{\mathcal{M}2}\end{bmatrix}\end{bmatrix}$$

$$=\begin{bmatrix}a_{11}x_{\mathcal{M}1}+a_{12}x_{\mathcal{M}2}\\ a_{21}x_{\mathcal{M}1}+a_{22}x_{\mathcal{M}2}\end{bmatrix}+\begin{bmatrix}a_{21(11)}x_{\mathcal{M}1}^2+(a_{21(21)}+a_{21(12)})x_{\mathcal{M}1}x_{\mathcal{M}2}+a_{21(22)}x_{\mathcal{M}2}^2\\ a_{22(11)}x_{\mathcal{M}1}^2+(a_{22(21)}+a_{22(12)})x_{\mathcal{M}1}x_{\mathcal{M}2}+a_{22(22)}x_{\mathcal{M}2}^2\end{bmatrix},$$

and

$$\dot{x}_{\mathcal{M}}(t)=F_{\mathcal{M}}(\boldsymbol{a},x_{\mathcal{M}})=A_{\mathcal{M}}f(x_{\mathcal{M}})=\begin{bmatrix}a_{11} & a_{12} & a_{13} & a_{14} & a_{15}\\ a_{21} & a_{22} & a_{23} & a_{24} & a_{25}\end{bmatrix}\begin{bmatrix}x_{\mathcal{M}1}\\ x_{\mathcal{M}2}\\ x_{\mathcal{M}1}^2\\ x_{\mathcal{M}1}x_{\mathcal{M}2}\\ x_{\mathcal{M}2}^2\end{bmatrix}. \qquad \blacksquare$$

Example 4.8.

Consider the fourth-order system with linear and quadratic polynomials. Using the matrices and column vector (column matrix) of coefficients $A_1\in\mathbb{R}^{4\times4}$, $A_{2i}\in\mathbb{R}^{4\times4}$, $\boldsymbol{b}_1\in\mathbb{R}^{4\times1}$,

$$\dot{x}_{\mathcal{M}}(t) = A_1 x_{\mathcal{M}} + \begin{bmatrix} x_{\mathcal{M}}^T A_{21} x_{\mathcal{M}} \\ x_{\mathcal{M}}^T A_{22} x_{\mathcal{M}} \\ x_{\mathcal{M}}^T A_{23} x_{\mathcal{M}} \\ x_{\mathcal{M}}^T A_{24} x_{\mathcal{M}} \end{bmatrix} + \boldsymbol{b}_1 u,\ x_{\mathcal{M}} = \begin{bmatrix} x_{\mathcal{M}1} \\ x_{\mathcal{M}2} \\ x_{\mathcal{M}3} \\ x_{\mathcal{M}4} \end{bmatrix},$$

$$A_1 = \begin{bmatrix} a_{11} & a_{12} & a_{13} & a_{14} \\ a_{21} & a_{22} & a_{23} & a_{24} \\ a_{31} & a_{32} & a_{33} & a_{34} \\ a_{41} & a_{42} & a_{43} & a_{44} \end{bmatrix},\ A_{2i} = \begin{bmatrix} a_{2i11} & a_{2i12} & a_{2i13} & a_{2i14} \\ a_{2i21} & a_{2i22} & a_{2i23} & a_{2i24} \\ a_{2i31} & a_{2i32} & a_{2i33} & a_{2i34} \\ a_{2i41} & a_{2i42} & a_{2i43} & a_{2i44} \end{bmatrix},\ \boldsymbol{b}_1 = \begin{bmatrix} b_{11} \\ b_{12} \\ b_{13} \\ b_{14} \end{bmatrix},$$

For $A_{\mathcal{M}}(\boldsymbol{a},\boldsymbol{b})\in\mathbb{R}^{4\times 15}$ and $f_{\mathcal{M}}(x_{\mathcal{M}},u)\in\mathbb{R}^{15}$, we have

$$\dot{x}_{\mathcal{M}}(t) = F_{\mathcal{M}}(\boldsymbol{a},x_{\mathcal{M}}) + \boldsymbol{b}_1 u = A_{\mathcal{M}}(\boldsymbol{a},\boldsymbol{b}) f_{\mathcal{M}}(x_{\mathcal{M}},u),\ f_{\mathcal{M}}(x_{\mathcal{M}},u) = \begin{bmatrix} x_{\mathcal{M}1} \\ x_{\mathcal{M}2} \\ x_{\mathcal{M}3} \\ x_{\mathcal{M}4} \\ x_{\mathcal{M}1}^2 \\ x_{\mathcal{M}1}x_{\mathcal{M}2} \\ x_{\mathcal{M}1}x_{\mathcal{M}3} \\ x_{\mathcal{M}1}x_{\mathcal{M}4} \\ x_{\mathcal{M}2}^2 \\ x_{\mathcal{M}2}x_{\mathcal{M}3} \\ x_{\mathcal{M}2}x_{\mathcal{M}4} \\ x_{\mathcal{M}3}^2 \\ x_{\mathcal{M}3}x_{\mathcal{M}4} \\ x_{\mathcal{M}4}^2 \\ u \end{bmatrix}.$$ ■

4.5. Data Heterogeneity and Kernel Estimators

Sensitivity and deficiencies of identification algorithms are due to over-structured $\mathcal{S}$ and over-parameterized $\mathcal{P}$ models, as well as heterogeneity of measured or observed states [11-13]. Particular challenges arise in measurements and estimations of $\dot{x}(t)$. High order filters attenuate noise ξ and minimize distortions δ in measurements. Asynchronous data fusion, data gaps, latencies and other factors affect convergence and fidelity of computed $\hat{A}$ because $(x,\dot{x},u)_{\text{data}}$ is perturbed by $(x,\dot{x},u)_{\text{filer}}$. Statistical models for (ξ,δ) are unknown, vary, and, depend on sensor linearity, environmental sensitivity, delays, ASICs errors, data sharing and processing latencies, interference, noise, etc. The weighted least-squares and variance-covariance observation dependences may not ensure expected results and numeric stability. In addition to singular value factorizations to deal with data uncertainties and heterogeneities in computing $(\hat{A},\tilde{A})$, other schemes are investigated to enable robustness.

Filtering, interpolation, factorization and fusion schemes are addressed to enable homogeneity. The following factors impose challenges:

1. Dynamics, delays and latencies in measurements;
2. Interferences and distortions;
3. Multisource noise and measurement errors;
4. Sensor linearity, accuracy, dynamic precision and environmental sensitivity;
5. Digital-to-analog and analog-to-digital conversion errors;
6. Missed data and data gaps;
7. Asynchronous sampling and fusion from distributed sensors.

The measured $(x,\dot{x},u)$ are fused with uncertainties ζ_i as

$$\begin{bmatrix} x(t) \\ \dot{x}(t) \\ u(t) \end{bmatrix}_{\substack{\text{data}\\\text{fused}}} = \begin{bmatrix} x(t) \\ \dot{x}(t) \\ u(t) \end{bmatrix}_{\substack{\text{physical}\\\text{quantities}}} + \begin{bmatrix} \zeta_x(t) \\ \zeta_{\dot{x}}(t) \\ \zeta_u(t) \end{bmatrix}. \tag{4.21}$$

Using $(\widehat{x}, \widehat{\dot{x}}, u)_{\text{data}}$, compute $\hat{A}$ and $\tilde{A}$ as

$$\hat{A} = \begin{bmatrix} \underbrace{(\hat{\Phi}^T\hat{\Phi})^{-1}}_{\hat{\Phi}^T\hat{\Phi}=U_{\hat{\Phi}}\Sigma_{\hat{\Phi}}V_{\hat{\Phi}}^T} & \underbrace{\hat{\Phi}^T}_{\Phi=u_{\hat{\Phi}}\sigma_{\hat{\Phi}}v_{\hat{\Phi}}^T} & \underbrace{\widehat{\dot{x}}}_{\widehat{\dot{x}}=U_x\Sigma_xV_x^T} \end{bmatrix}^T = \begin{bmatrix} V_{\hat{\Phi}}\Sigma_{\hat{\Phi}}^{-1}U_{\hat{\Phi}}^T v_{\Phi}\sigma_{\Phi}^T u_{\Phi}^T U_x\Sigma_x V_x^T \end{bmatrix}^T, \tag{4.22}$$

$$\hat{\Phi} = \begin{bmatrix} f(\widehat{x}(t_1), u(t_1)) \\ \vdots \\ f(\widehat{x}(t_q), u(t_q)) \end{bmatrix} = \begin{bmatrix} \widehat{x}_{t_1} & u_{t_1} & f_x(\widehat{x}_{t_1}) & f_u(u_{t_1}) & f_{xu}(\widehat{x}_{t_1}, u_{t_1}) \\ \vdots & \vdots & \vdots & \vdots & \vdots \\ \widehat{x}_{t_q} & u_{t_q} & f_x(\widehat{x}_{t_q}) & f_u(u_{t_q}) & f_{xu}(\widehat{x}_{t_q}, u_{t_q}) \end{bmatrix} = u_{\hat{\Phi}}\sigma_{\hat{\Phi}}v_{\hat{\Phi}}^T,$$

$$\tilde{A} = [\tilde{a}_{kl}]\Big|\begin{cases} \hat{a}_{kl}\neq 0, |\hat{a}_{ij}|>\varepsilon_a, 0<\varepsilon_a\ll 1 \\ \hat{a}_{kl}=0, |\hat{a}_{ij}|\leq\varepsilon_a \end{cases} = \begin{bmatrix} \tilde{a}_{11} & \cdots & \tilde{a}_{1z} \\ \vdots & \ddots & \vdots \\ \tilde{a}_{n1} & \cdots & \tilde{a}_{nz} \end{bmatrix}, \ \tilde{A}\in\mathbb{R}^{n\times z}.$$

Identification Time – Hardware-, algorithms- and distortions-dependent probabilistic models of l-tuple $\zeta_i(\cdot)$ are unknown. To minimize noise, distortions and perturbations, filters are designed and implemented as covered in sections 5.14. Digital filters process $(x,\dot{x},u)$ yielding $(\widehat{x}, \widehat{\dot{x}}, u)$. Non-parametric kernel *estimators* minimize data uncertainties, support truncated data matching and enable data homogeneity. However, the latency is increased. For $(x,\dot{x},u)_{\text{data}}$, find $(\widehat{x}, \widehat{\dot{x}}, u)$ to compute $(\hat{A},\tilde{A})$. Data analytic time for measurements, interfacing, fusion, filtering, estimations and computing of $(\hat{A},\tilde{A})$ by microcontroller of FPGA yields time for identification $t_{\text{identification}}=(t_{\text{data logging}} + t_{\text{processing}})$. Implemented filters, kernel density *estimators*, least-squares, SVD and other algorithms are aimed to ensure robustness and convergence.

Kernel Density Estimation – Despite associated latency with the kernel density estimation, consider this approach. The input-output data (Y_i,X_i) is measured with $\zeta_i(\cdot)$. Apply statistical estimates for a real-valued function using the measured or observed data. For the independent, identically distributed errors ε_i with zero mean and finite variance [14, 15]

$$Y_i=m(X_i)+\varepsilon_i,\ \varepsilon_i\equiv\zeta_i,\ i=1,\ldots,n, \tag{4.23}$$

where $m(\cdot)$ is the nonparametric conditional expectation of Y with respect to X, $m(x) = \mathbb{E}(Y|X = x) = \frac{1}{g(x)}\int yg(x,y)dy$.

The kernel *estimators* are [14, 15]

$$\hat{m}(x,h) = \frac{\sum_{i=1}^{n} K(X_i - x)Y_i}{\sum_{i=1}^{n} K(X_i - x)} = \sum_{i=1}^{n} w_{ni}(x)Y_i, \tag{4.24}$$

$$\hat{m}(x,h) = \frac{1}{h}\sum_{i=1}^{n}(X_i - X_{i-1})K\left(\tfrac{1}{h}(X_i - x)\right)Y_i,$$

where h is the kernel radius, which is a parameter, $h>0$.

For the kernel function $K(\cdot)$, $\int K(z)^2dz<\infty$ and $\int z^2K(z)dz=\sigma_K^2<\infty$. In near-real time and offline identification, one computes $(\hat{x}, \hat{\dot{x}}, \hat{u})$ to find $(\hat{A},\tilde{A})$ using (4.22).

Illustrative studies in applying the reported results are documented in Example 4.10 with experimental verification.

4.6. Structure-Parametric Regularization on Partitioned Datasets

For over-structured and over-parametrized models $\mathcal{M}(\mathcal{S},\mathcal{P})$, high-dimensional structure and parameter spaces of high-order multivariate polynomials define $f_{\mathcal{M}}(x_{\mathcal{M}},u)$. Unknown $\hat{A}\in\mathbb{R}^{n\times z}$ and $\tilde{A}\in\mathbb{R}^{n\times z}$ are computed by solving (4.5), (4.11), (4.17) and (4.22). Measured $(x,\dot{x},u)\in X\times U$ on $t\in T$ can be synchronously or asynchronously retrieved and updated as partitioned datasets $(x,\dot{x},u)_{h=1,\ldots,H}$.

For $(x,\dot{x},u)_{h=1,\ldots,H}$, one yields Φ_h and computes $(\hat{A}_h,\tilde{A}_h)$ as

$$\hat{A}_h=\left[(\Phi_h^T\Phi_h)^{-1}\Phi_h^T\dot{x}\right]_h^T,\ \hat{A}_h=[\hat{a}_{ij}]_h=\begin{bmatrix}\hat{a}_{11} & \cdots & \hat{a}_{1z}\\ \vdots & \ddots & \vdots\\ \hat{a}_{n1} & \cdots & \hat{a}_{nz}\end{bmatrix}_h,\ \hat{A}_h\in\mathbb{R}^{n\times z},\ h=1,\ldots,H, \quad (4.25)$$

$$\tilde{A}_h=[\tilde{a}_{kl}]_h\Big|\begin{cases}\hat{a}_{kl}\neq 0,\ |\hat{a}_{ij}|>\varepsilon_a,\ 0<\varepsilon_a\ll 1\\ \hat{a}_{kl}=0,\ |\hat{a}_{ij}|\le\varepsilon_a\end{cases}=\begin{bmatrix}\tilde{a}_{11} & \cdots & \tilde{a}_{1z}\\ \vdots & \ddots & \vdots\\ \tilde{a}_{n1} & \cdots & \tilde{a}_{nz}\end{bmatrix}_h,\ \tilde{A}_h\in\mathbb{R}^{n\times z}.$$

Using partitioned $(x,\dot{x},u)_h\in X\times U$ on $t\in T$, one finds $\tilde{A}_h$, and, $\tilde{A}_i\neq\tilde{A}_j$.

Variations of $\tilde{A}_h$ are due to:

1. Time varying parameters and structures;
2. Algorithmic instabilities, sensitivity, numeric errors, etc.;
3. Electromagnetic spectrum, data modification cyberattacks, etc.

It is difficult to assume time-varying and statistical dependences in $\hat{A}_h$ and $\tilde{A}_h$, evaluate parameters estimation errors and uncertainties, and, find multivariate probability distributions. Hence, the Fisher, Riemannian, entropy and likelihood statistical metrics may not be effectively applied to identify parameters, estimate dimensions, validate models, etc. Matrices $(\hat{A},\tilde{A})_h$ can be clustered, correlated, regularized and evaluating (4.6)-(4.8) and (4.13). The algorithm implies shifting and sorting identified matrices and mappings $(\hat{A},\hat{f})_h\big|_{\mathbf{P}\cap\mathbf{S}}$, finding $(\tilde{A},\tilde{f})_h\big|_{\mathbf{P}\cap\mathbf{S}}$ by evaluating mismatches, *admissible* tolerance, errors, etc. Using equally or unequally spaced and partitioned data $(x,\dot{x},u)_{h\ \text{data}}$, identification may yield time-varying $(\tilde{A}(t),\tilde{f}(t))_h\big|_{\mathbf{P}\cap\mathbf{S}}$.

Example 4.9.

Matrix factorization concepts may not assert computed $(\hat{A},\tilde{A})_h$, while very effective dealing with data uncertainties and heterogeneities to compute $(\hat{A},\tilde{A})_h$. The Eckart-Young theorem formulates the t-rank approximation for P, yielding a truncated P_t using the $t\times t$ truncated matrix Σ_t with $\sigma_{i,i}\in[\sigma_t\ \sigma_{0\text{null}}]$. For the r-rank matrix Σ, the factorized P is $P=U\Sigma V^T$. One solves

$$\arg\min_{P_t}\|\ P-P_t\ \|_p,\ P_t\approx U_t\Sigma_t V_t^T$$

and examines dominant and negligible dependences. The matrix of *factor scores* yields projections of observations on principle components. The t-rank approximation yields P_t. Approximation and truncation algorithms may not ensure a consistency asserting computed $(\hat{A},\tilde{A})_h$. While low-rank matrix approximation may be investigated, this concept may cause inadequateness and erroneous results. ■

Example 4.10. *Series RLC Circuit: Kernel Estimator, and, Parameters Identification*
Many electronic, electromechanical and mechanical systems exhibit underdamped dynamic responses with measurements impediments and data heterogeneities. For the series *RLC* circuit, the governing equation is

$$Ri + L\frac{di}{dt} + \frac{1}{C}\int_0^t idt + u_0 = u,$$

where *R*, *L* and *C* are the constant or time-varying and current-dependent parameters; u_C is the voltage across the capacitor, $u_C = \frac{1}{C}\int_0^t idt$.

For time-invariant (*R*,*L*,*C*), the characteristic equation is $s^2 + \frac{R}{L}s + \frac{1}{LC} = 0$.

The eigenvalues are $s_{1,2} = -\frac{R}{2L} \pm \sqrt{\left(\frac{R}{2L}\right)^2 - \frac{1}{LC}}$.

The damping coefficient ξ, *neper* frequency α and resonant *natural* frequency ω_0 are

$$\xi = \frac{R}{2}\sqrt{\frac{C}{L}},\ \alpha = \frac{R}{2L},\ \omega_0 = \frac{1}{\sqrt{LC}}.$$

The underdamped not perturbed dynamics is

$$x(t) = e^{-\alpha t}\left\lfloor a\cos\left(\sqrt{\omega_0^2 - \alpha^2 t}\right) + b\sin\left(\sqrt{\omega_0^2 - \alpha^2 t}\right)\right\rfloor,$$

$$x(t) = e^{-\alpha t}\left[a\cos(\omega_d t) + b\sin(\omega_d t)\right],\ \omega_d = \sqrt{\omega_0^2 - \alpha^2}.$$

For a circuit with R=10 ohm, L=47 mH and C=100 μF, ξ=0.2306, α=106.38 Np/s, ω_0=461.27 1/s and ω_d=448.83. The high-Q ceramic resistors, capacitors and inductors exhibit specified tolerances. These (*R*,*L*,*C*) may be temperature dependent.

Perform parameter estimations. The state vector is $x=[x_1\ x_2]^T$, $x_1=i$ and $x_2=u_C$. The circuit is described as

$$\dot{x} = \begin{bmatrix}\dot{x}_1\\ \dot{x}_2\end{bmatrix} = \begin{bmatrix}\frac{d}{dt}i\\ \frac{d}{dt}u_C\end{bmatrix} = \begin{bmatrix}-\frac{R}{L} & -\frac{1}{L}\\ \frac{1}{C} & 0\end{bmatrix}\begin{bmatrix}x_1\\ x_2\end{bmatrix} + \begin{bmatrix}\frac{1}{L}\\ 0\end{bmatrix}u,\ x = \begin{bmatrix}x_1\\ x_2\end{bmatrix} = \begin{bmatrix}i\\ u_C\end{bmatrix},$$

$$\dot{x} = Ax + Bu,\ A = \begin{bmatrix}-\frac{R}{L} & -\frac{1}{L}\\ \frac{1}{C} & 0\end{bmatrix} = \begin{bmatrix}a_{11} & a_{12}\\ a_{21} & 0\end{bmatrix},\ B = \begin{bmatrix}\frac{1}{L}\\ 0\end{bmatrix} = \begin{bmatrix}b_{11}\\ 0\end{bmatrix}.$$

Contactless measurements of low current represent challenges due to sensor error, noise and interference. Using a contactless Hall-effect current sensor and voltage sensors (optically isolated operational amplifier), experiments were performed. Current $i(t)$ is measured with distortions $\zeta(\cdot)$. Figures 4.2 repot the measured current $x_{\text{data}}(t)\equiv i_{\text{data}}(t)$, as well as the kernel estimation $\hat{x}(t)$ using (4.24), h=1.62. Current is measured with the data rate 10,000 Hz. A non-parametric kernel *estimator* estimates current, and, $(\hat{x},\hat{\dot{x}})_h$ are found with monotonically smooth $(\hat{x}(t),\hat{\dot{x}}(t))_h$ on intervals t_i to be used in identification.

A model is given as

$$\dot{x}_{\mathcal{M}}(t) = A_{\mathcal{M}}(\boldsymbol{a},\boldsymbol{b})f_{\mathcal{M}}(x_{\mathcal{M}},u),\ f(x,u) \triangleq f_{\mathcal{M}}(x_{\mathcal{M}},u),\ f(x,u) = \begin{bmatrix}x_1\\ x_2\\ u\end{bmatrix},\ f(\cdot){:}X\times U\to\mathbb{R}^3.$$

Estimate parameters by finding unknown matrices $\hat{A}_i \neq \hat{A}_j$ and $\tilde{A}_i \neq \tilde{A}_j$, thereby obtaining $A_{\mathcal{M}} \in \mathbb{R}^{2\times 3}$ and $f_{\mathcal{M}}(x_{\mathcal{M}}, u)$. For the estimated current and measured voltage $u_C(t)$, derivatives $(\dot{x}_1, \dot{x}_2)$ are estimated, yielding $(\hat{\dot{x}}_1, \hat{\dot{x}}_2)$.

The measured $(x,u) \in X \times U$ on $t \in T$ is partitioned as $(x,u)_{h=1,\ldots,H}$, and, updated as $(\hat{x}, \hat{\dot{x}}, u)_i$. Matrices of parameters $\hat{A}_i \in \mathbb{R}^{2\times 3}$ and $\tilde{A}_i \in \mathbb{R}^{2\times 3}$ are computed using (4.25). Convergence is guaranteed for different initial conditions and applied voltage (input) waveforms. While $\hat{A}_i \neq \hat{A}_j$ and $\tilde{A}_i \neq \tilde{A}_j$, variations of $\tilde{A}_i$ are not exceeding 2.93% for consistency estimated $\hat{\dot{x}}$ on intervals t_i, where continuous $(\hat{x}(t), \hat{\dot{x}}(t))_i$ are monotonically smooth and infinitely differentiable.

To identify $\hat{A}_i \in \mathbb{R}^{2\times 3}$, a mapping $\Phi \in \mathbb{R}^{q\times 3}$, $q \geq 10$ yields consistent results, ensuring adequate parameter estimation accuracy.

From $\hat{A}_i = \begin{bmatrix} \hat{a}_{11} & \hat{a}_{12} & \hat{a}_{13} \\ \hat{a}_{21} & \hat{a}_{22} & \hat{a}_{23} \end{bmatrix}_i$, find the regularized matrix of parameters.

We have

$$\tilde{A}_1 = \begin{bmatrix} -211.6 & -21.5 & 21.5 \\ 9940.3 & 0 & 0 \end{bmatrix}.$$

The estimated parameters are $\tilde{R}$=9.84 ohm, $\tilde{L}$=46.5 mH and $\tilde{C}$=100.6 μF.

For a model with estimated $(\tilde{R}, \tilde{L}, \tilde{C})$, the evolution of current is found as $i(t) = e^{-\tilde{\alpha}t}\left[a\cos(\tilde{\omega}_d t) + b\sin(\tilde{\omega}_d t)\right]$, $\tilde{\omega}_d = \sqrt{\tilde{\omega}_0^2 - \tilde{\alpha}^2}$, or, computed by solving the state-space differential equation. Constants (a,b) depend on initial conditions, while $(\tilde{\alpha}, \tilde{\omega}_d)$ are the (R,L,C)-dependent. For $\tilde{R}$=9.84 ohm, $\tilde{L}$=46.5 mH and $\tilde{C}$=100.6 μF, one has $\tilde{\alpha} = 105.8$ Np/s and $\tilde{\omega}_d = 450.1$. With $(\tilde{R}, \tilde{L}, \tilde{C})$, the current dynamics are depicted in Figures 4.2 by solid lines. For three different initial conditions, the constants (a,b) for $x(t)$ are $(2.94\times10^{-4}, 7.88\times10^{-4})$, $(5.84\times10^{-4}, 0.31\times10^{-4})$, $(1.78\times10^{-4}, 5.49\times10^{-4})$.

Data fusion, data management, kernel estimation and parameter identification take less than 0.01 sec.

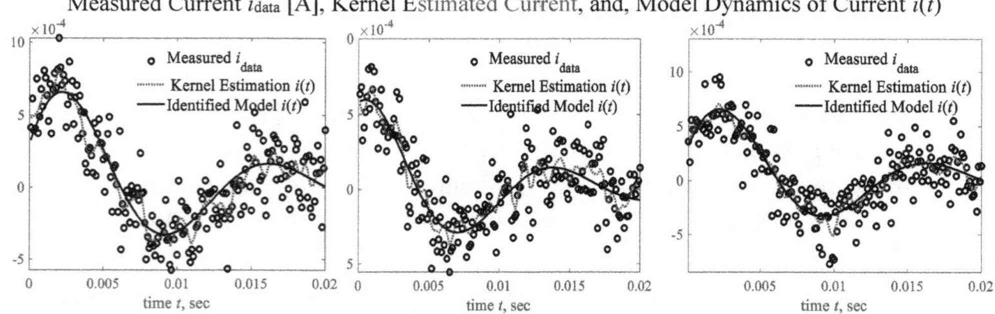

Figure 4.2. Kernel estimation of current (dashed line) for a measured current $x_{data}(t)=x(t)+\zeta$, $x_{data} \equiv i_{data}$ (circles), kernel estimation (dashed line), and, model evolution of $i(t)$ with identified $(\tilde{R}, \tilde{L}, \tilde{C})$ for three different initial conditions (solid line). ■

4.7. Examples in Identification of Nonlinear Systems

Challenges in identification arise when physical systems exhibit instabilities and measurement heterogeneities. If the model structure and $F_{\mathcal{M}}(\cdot)$ are unknown, one applies a consistent class of differential equations and *predictors* $f_{\mathcal{M}}(x_{\mathcal{M}},u)$ with corresponding dimensionality, defined by $\mathbb{M}(x,u)$. For example, for stable-in-large and unstable-in-small systems $\mathbb{M}(x,u)$, $x\in\mathbb{R}^2$, which exhibit bounded periodic evolutions due to initial conditions with $u=0$, there is a class of nonlinear models such as van der Pol oscillator with $|x_1,x_2|\le M$, or, Lotka-Volterra equations with $(x_1,x_2)>0$. If a system $\mathbb{M}(x,u)$, $x\in\mathbb{R}^n$ $(n\ge 3)$ exhibits the asymmetric aperiodic limit cycles, one considers the Lorenz and other equations with specific state and input dimensionalities (n,m). Physics-defined states $x\in\mathbb{R}^n$ and inputs $u\in\mathbb{R}^m$, as well as *predictors*, consistent with laws of physics, are used. The *standard* van der Pol, Lotka-Volterra and other differential equations can be modified refining order, dimensionality and structure-parametric mappings.

4.7.1. Identification of a Second-Order System

Consider a nonlinear system $\mathbf{M}(x,u)$, $x\in\mathbb{R}^2$, $(x_1,x_2)>0$, $u\in\mathbb{R}^1$, which exhibits limit cycles. To obtain datasets to conduct identification, let a system $\mathbf{M}(x,u)$ is described by differential equations to be simulated with the fixed step size 0.01 sec. One obtains datasets $[(x_1,x_2),(\dot{x}_1,\dot{x}_2),u]\in X\times U$ for identification. As an input, use the discrete white noise. Hence, $u=\xi[n]$ is a sequence of uncorrelated random $\xi[n]$ with zero mean and a finite variance σ^2. The model structure is unknown. As the overdetermined *predictor* use the differential equations for a model $\mathcal{M}(x_{\mathcal{M}},u,\mathcal{S},\mathcal{P})$, $x_{\mathcal{M}}\in\mathbb{R}^2$ which admit limit cycles.

High-Degree Polynomials – The governing equations for $\mathbf{M}(x,u)$ are

$$\dot{x}_1 = x_1 - x_1x_2 - x_1^2x_2 + u,\ u=\xi[n],\ x_1(t_0)=x_{10},\ \sigma^2=1\times 10^{-4},$$

$$\dot{x}_2 = -x_2 + x_1x_2 + x_1x_2^2 + u,\ u=\xi[n],\ x_2(t_0)=x_{20}.$$

The overdetermined *predictor* is

$$\dot{x}_{\mathcal{M}1} = a_{11}x_{\mathcal{M}1} + a_{12}x_{\mathcal{M}2} + a_{13}x_{\mathcal{M}1}x_{\mathcal{M}2} + a_{14}x_{\mathcal{M}1}^2x_{\mathcal{M}2} + a_{15}x_{\mathcal{M}1}x_{\mathcal{M}2}^2 + b_{11}u,\ (\boldsymbol{a},\boldsymbol{b})\in\mathbb{R},$$

$$\dot{x}_{\mathcal{M}2} = a_{21}x_{\mathcal{M}1} + a_{22}x_{\mathcal{M}2} + a_{23}x_{\mathcal{M}1}x_{\mathcal{M}2} + a_{24}x_{\mathcal{M}1}^2x_{\mathcal{M}2} + a_{25}x_{\mathcal{M}1}x_{\mathcal{M}2}^2 + b_{12}u\,.$$

Hence,

$$\dot{x}_{\mathcal{M}}(t) = A_{\mathcal{M}}(\boldsymbol{a},\boldsymbol{b})f_{\mathcal{M}}(x_{\mathcal{M}},u),\ f(x,u)\triangleq f_{\mathcal{M}}(x_{\mathcal{M}},u),\ f(x,u)=\begin{bmatrix} x_1\\ x_2\\ x_1x_2\\ x_1^2x_2\\ x_1x_2^2\\ u\end{bmatrix},\ f(\cdot){:}X\times U\to\mathbb{R}^6,$$

where $A_{\mathcal{M}}$ is the matrix of unknown parameters, $A_{\mathcal{M}}\in\mathbb{R}^{2\times 6}$.

Estimate parameters and identify the structure. The SIMULINK diagram is reported in Figure 4.3.a. Using the simulation results, applying the fix size differential equations solver, obtain $[(x_1,x_2),(\dot{x}_1,\dot{x}_2),u]_{\text{data}}$ and compute (4.5) $\hat{A}\in\mathbb{R}^{2\times 6}$, finding $(\hat{A},\hat{f})$, and, identifying models as $\hat{\mathcal{M}}(x_{\mathcal{M}},u,\hat{A},\hat{f})$ and $\tilde{\mathcal{M}}(x_{\mathcal{M}},u,\tilde{A},\tilde{f})$.

For computed $\hat{A}=\begin{bmatrix}\hat{a}_{11} & \hat{a}_{12} & \hat{a}_{13} & \hat{a}_{14} & \hat{a}_{15} & \hat{a}_{16}\\ \hat{a}_{21} & \hat{a}_{22} & \hat{a}_{23} & \hat{a}_{24} & \hat{a}_{25} & \hat{a}_{26}\end{bmatrix}$, the convergence is guaranteed for any initial conditions, datasets $((x_1,x_2),(\dot{x}_1,\dot{x}_2),u)$, and, mapping $\Phi\in\mathbb{R}^{q\times z}$, $q\geq z$. For a given $[(x_1,x_2),(\dot{x}_1,\dot{x}_2),u]_{\text{data}}$, an identification algorithms takes less than 0.005 sec. The computed $\hat{A}\in\mathbb{R}^{2\times 6}$ for $\Phi\in\mathbb{R}^{50\times 6}$ and $\Phi\in\mathbb{R}^{100\times 6}$ are

$\hat{A}=$ 1.0000e+00 -9.1729e-07 -1.0000e+00 -1.0000e+00 8.8604e-07 9.9984e-01
9.9336e-07 -1.0000e+00 1.0000e+00 -1.6687e-06 1.0000e+00 9.9888e-01

and

$\hat{A}=$ 1.0000e+00 -9.4925e-11 -1.0000e+00 -1.0000e+00 -2.4309e-09 1.0000e+00
-1.8066e-08 -1.0000e+00 1.0000e+00 5.0387e-08 1.0000e+00 1.0000e+00

The MATLAB code to estimate parameters using (4.5) and (4.11) is reported in Figure 4.3.a. For $\Phi\in\mathbb{R}^{100\times 6}$, applying (4.5) $\hat{A}=\left[(\Phi^T\Phi)^{-1}\Phi^T\dot{x}\right]^T$, we have

$\hat{A}=$ 1.0000e+00 -9.4925e-11 -1.0000e+00 -1.0000e+00 -2.4309e-09 1.0000e+00
-1.8066e-08 -1.0000e+00 1.0000e+00 5.0387e-08 1.0000e+00 1.0000e+00

Compute $\hat{A}$ using SVD as (4.11) $\hat{A}=\left[V_\Phi\Sigma_\Phi^{-1}U_\Phi^T v_\Phi\sigma_\Phi^T u_\Phi^T\dot{x}\right]^T$.

One finds $\hat{A}$ and $\hat{\mathcal{M}}(x_{\mathcal{M}},u,\hat{A},\hat{f})$,

$\hat{A}=$ 1.0000e+00 -3.7476e-11 -1.0000e+00 -1.0000e+00 -5.2319e-10 1.0000e+00
-1.9926e-08 -1.0000e+00 1.0000e+00 6.3712e-08 1.0000e+00 1.0000e+00

Computing $\hat{A}$ as $\hat{A}=\left[(v_\Phi\sigma_\Phi^T\sigma_\Phi v_\Phi^T)^{-1}v_\Phi\sigma_\Phi^T u_\Phi^T\dot{x}\right]^T$, we have

$\hat{A}=$ 1.0000e+00 1.7528e-12 -1.0000e+00 -1.0000e+00 2.3327e-10 1.0000e+00
-1.6223e-09 -1.0000e+00 1.0000e+00 8.9129e-10 1.0000e+00 1.0000e+00

Note: In these and following studies, the fixed-step Euler, Runge-Kutta and other solvers are used. There are standard MATLAB command and custom algorithms to compute SVDs and perform matrix factorizations. Commands and solvers used, algorithms, step size, initial conditions and other factors result in some differences for $\hat{A}$, $\tilde{A}$, $(m_x,m_{\dot{x}})$, etc. Despite discrepancies, overall consistency is guaranteed.

For $(x,x_{\mathcal{M}})$, the model is quantitatively examined using the p-norms for the dynamic mismatch measures. For $\hat{A}_1$, we have $m_x\leq 0.001$ and $m_{\dot{x}}\leq 0.005$, $p=1$. The dynamic mismatch measures $[(m_x,m_{\dot{x}}),(\Delta x,\Delta\dot{x})]$ are evaluated on a specified $t\in[t_0,t_f]$.

Having found $\hat{A}$, apply (4.17) to obtain a regularized $\tilde{A}$ and truncated $\tilde{f}(\cdot)$.

For computed $\hat{A}=\begin{bmatrix}\hat{a}_{11} & \hat{a}_{12} & \hat{a}_{13} & \hat{a}_{14} & \hat{a}_{15} & \hat{a}_{16}\\ \hat{a}_{21} & \hat{a}_{22} & \hat{a}_{23} & \hat{a}_{24} & \hat{a}_{25} & \hat{a}_{26}\end{bmatrix}$, null negligibly small entries, $\varepsilon_a=1\times 10^{-6}$.

The regularized matrix of parameters is $\tilde{A}=\begin{bmatrix}1 & 0 & -1 & -1 & 0 & 1\\ 0 & -1 & 1 & 0 & 1 & 1\end{bmatrix}$.

Finding $\tilde{A}$, we obtain $\tilde{f}(\cdot)$. The identified model with $(\tilde{A},\tilde{f})\big|_{\mathrm{P}\cap\mathrm{S}}$ is

$$\dot{x}_{\mathcal{M}1}=x_{\mathcal{M}1}-x_{\mathcal{M}1}x_{\mathcal{M}2}-x_{\mathcal{M}1}^2x_{\mathcal{M}2}+u,$$

$$\dot{x}_{\mathcal{M}2}=-x_{\mathcal{M}2}+x_{\mathcal{M}1}x_{\mathcal{M}2}+x_{\mathcal{M}1}x_{\mathcal{M}2}^2+u.$$

Criteria (4.6)-(4.7) on dynamic mismatch measures $[m_x(x,x_{\mathcal{M}}),m_{\dot{x}}(\dot{x},\dot{x}_{\mathcal{M}})]$, $[\Delta x(t),\Delta\dot{x}(t)]$ are ensured. A model $\tilde{\mathcal{M}}(x_{\mathcal{M}},u,\tilde{A},\tilde{f})$ is the same as $\mathbf{M}(x,u)$.

Evolutions of $x(t)$ and $x_{\mathcal{M}}(t)$ are identical, and, reported in Figure 4.3.b.

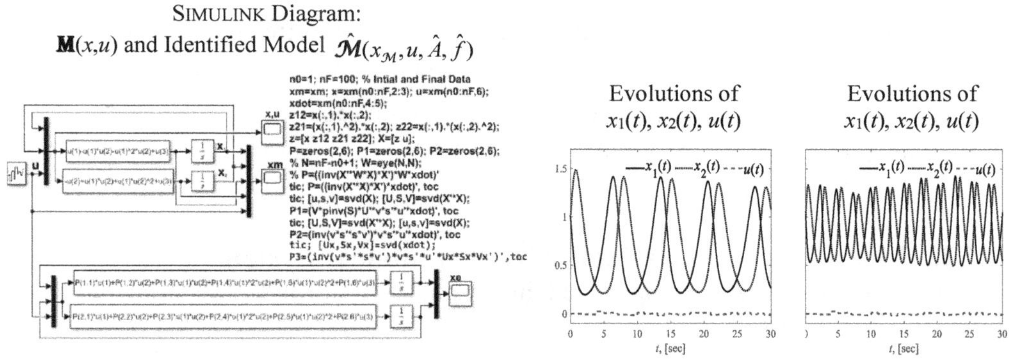

SIMULINK Diagram:
$\mathbf{M}(x,u)$ and Identified Model $\hat{\mathcal{M}}(x_{\mathcal{M}}, u, \hat{A}, \hat{f})$

```
n0=1; nF=100; % Intial and Final Data
xm=xm; x=xm(n0:nF,2:3); u=xm(n0:nF,6);
xdot=xm(n0:nF,4:5);
z12=x(:,1).*x(:,2);
z21=(x(:,1).^2).*x(:,2); z22=x(:,1).*(x(:,2).^2);
z=[x z12 z21 z22]; X=[z u];
P=zeros(2,6); P1=zeros(2,6); P2=zeros(2,6);
% N=nF-n0+1; W=eye(N,N);
% P=((inv(X'*W*X)*X')*W*xdot)'
tic; P=((inv(X'*X)*X')*xdot)', toc
tic; [u,s,v]=svd(X); [U,S,V]=svd(X'*X);
P1=(V*pinv(S)*U'*v*s'*u'*xdot)', toc
tic; [U,S,V]=svd(X'*X); [u,s,v]=svd(X);
P2=(inv(v*s'*s*v')*v*s'*u'*xdot)', toc
tic; [Ux,Sx,Vx]=svd(xdot);
P3=(inv(v*s'*s*v')*v*s'*u'*Ux*Sx*Vx')',toc
```

Evolutions of $x_1(t)$, $x_2(t)$, $u(t)$

Evolutions of $x_1(t)$, $x_2(t)$, $u(t)$

(a) (b) (c)

Figure 4.3. (a) SIMULINK diagram to simulate $\mathbf{M}(x,u)$ and $\hat{\mathcal{M}}(x_{\mathcal{M}}, u, \hat{A}, \hat{f})$ with computed $\hat{A}$. A MATLAB code to perform parameter estimation using (4.5) and (4.11) is reported;
(b) System $\mathbf{M}(x,u)$ with high-degree polynomials: Dynamics and model evolutions, $[x_{10}, x_{20}]=[1, 1]$;
(c) System $\mathbf{M}(x,u)$ with fractional polynomials: Dynamics and model evolutions, $[x_{10}, x_{20}]=[1, 1]$.

Fractional Polynomials – The governing equations for $\mathbf{M}(x,u)$ with the factional polynomials are

$$\dot{x}_1 = x_1 + x_1^{\frac{1}{3}} + x_1^{\frac{4}{3}} - x_1x_2 - 2.5x_1^2x_2 - x_1^3x_2^3 + u,\ u=\xi[n],\ x_1(t_0)=x_{10},$$

$$\dot{x}_2 = -x_2 - x_2^{\frac{1}{3}} - x_2^{\frac{4}{3}} + x_1x_2 + 2.5x_1x_2^2 + x_1^3x_2^3 + u,\ u=\xi[n],\ x_2(t_0)=x_{20}.$$

With fractional multivariate polynomials, state-space equations, $A_{\mathcal{M}}\in\mathbb{R}^{2\times11}$ and $f_{\mathcal{M}}(\cdot)\in\mathbb{R}^{11}$ are

$$\dot{x}_{\mathcal{M}}(t) = A_{\mathcal{M}}(\boldsymbol{a},\boldsymbol{b})f_{\mathcal{M}}(x_{\mathcal{M}}) = \begin{bmatrix} a_{\mathcal{M}11} & a_{\mathcal{M}12} & a_{\mathcal{M}13} & a_{\mathcal{M}14} & a_{\mathcal{M}15} & a_{\mathcal{M}16} & a_{\mathcal{M}17} & a_{\mathcal{M}18} & a_{\mathcal{M}19} & a_{\mathcal{M}110} & b_{\mathcal{M}01} \\ a_{\mathcal{M}21} & a_{\mathcal{M}22} & a_{\mathcal{M}23} & a_{\mathcal{M}24} & a_{\mathcal{M}25} & a_{\mathcal{M}26} & a_{\mathcal{M}27} & a_{\mathcal{M}28} & a_{\mathcal{M}29} & a_{\mathcal{M}210} & b_{\mathcal{M}02} \end{bmatrix} \begin{bmatrix} x_{\mathcal{M}1} \\ x_{\mathcal{M}2} \\ x_{\mathcal{M}1}^{\frac{1}{3}} \\ x_{\mathcal{M}1}^{\frac{4}{3}} \\ x_{\mathcal{M}2}^{\frac{1}{3}} \\ x_{\mathcal{M}2}^{\frac{4}{3}} \\ x_{\mathcal{M}1}x_{\mathcal{M}2} \\ x_{\mathcal{M}1}^2x_{\mathcal{M}2} \\ x_{\mathcal{M}1}x_{\mathcal{M}2}^2 \\ x_{\mathcal{M}1}^3x_{\mathcal{M}2}^3 \\ u \end{bmatrix}.$$

Compute $\hat{A}\in\mathbb{R}^{2\times11}$ using (4.5) with $\Phi\in\mathbb{R}^{250\times6}$ and $\Phi\in\mathbb{R}^{500\times6}$. We have

$\hat{A}_1$=

1.0000e+00 -7.0677e-07 1.0000e+00 1.0000e+00 6.6768e-07 3.1437e-07 -1.000e+00 -2.5000e+00 5.3316e-08 -1.000e+00 1.000e+00
-3.6842e-07 -1.0000e+00 5.5701e-07 2.2074e-07 -1.0000e+00 -1.0000e+00 1.000e+00 5.9153e-08 2.5000e+00 1.000e+00 1.000e+00

and

$\hat{A}_2$=

1.0000e+00 3.2500e-08 1.0000e+00 1.0000e+00 -2.2182e-08 -1.3746e-08 -1.000e+00 -2.5000e+00 2.9759e-09 -1.000e+00 1.000e+00
-2.0136e-07 -1.0000e+00 1.6383e-07 9.9759e-08 -1.0000e+00 -1.0000e+00 1.000e+00 -6.8997e-09 2.5000e+00 1.000e+00 1.000e+00

The regularized matrix of parameters is computed applying (4.17)

$$\tilde{A} = \begin{bmatrix} 1 & 0 & 1 & 1 & 0 & 0 & -1 & -2.5 & 0 & -1 & 1 \\ 0 & -1 & 0 & 0 & -1 & -1 & 1 & 0 & 2.5 & 1 & 1 \end{bmatrix}.$$

We find $(\tilde{A},\tilde{f})$, and, obtain $\tilde{\mathcal{M}}(x_{\mathcal{M}},u,\tilde{A},\tilde{f})\subseteq\mathbf{M}(x,u)$.

To typify physical systems, for simulated system of differential equations, a tuple (F,f,A) is assymed to be unknown. The identified models $\hat{\mathcal{M}}(x_{\mathcal{M}},u,\hat{A},\hat{f})$ and $\tilde{\mathcal{M}}(x_{\mathcal{M}},u,\tilde{A},\tilde{f})$ correspond to $\mathbf{M}(x,u)$. Figure 4.3.c documents dynamics of $x(t)$. Transients for $x_{\mathcal{M}}(t)$ are identical to $x(t)$.

4.7.2. Identification of a Fourth-Order Nonlinear System

The Lorenz equations were analyzed in Examples 2.25 and 2.26. The third-order differential equations

$$\dot{x}_1 = a_{11}\left(-x_1 + x_2\right),$$
$$\dot{x}_2 = a_{21}x_1 + a_{22}x_2 - a_{23}x_1x_3,$$
$$\dot{x}_3 = a_{31}x_3 + x_1x_2,$$

were simulated if a_{11}=5, a_{21}=12.5, a_{22}=2, a_{23}=250 and a_{31}=2.

The *standard* Lorenz equation can be refined meeting system order and dimensionality $\mathbf{M}(x,u)$. In particular, $x\in\mathbb{R}^n$, $n\geq 3$ and $u\in\mathbb{R}^m$, $m\geq 1$. To describe thermodynamics, nonuniform thermofluidics, turbulence, unsteady aerodynamic periodic fluctuations, atmospheric convection, non-equilibrium cycles and other phenomena, consider systems $\mathbf{M}(x,u)$ and models $\mathcal{M}(x_{\mathcal{M}},u,\mathcal{S},\mathcal{P})$ with $x\in\mathbb{R}^n$ and $u\in\mathbb{R}^m$. Consider a system of differential equations, which typifies a physical system $\mathbf{M}(x,u)$, $x\in\mathbb{R}^4$, $u\in\mathbb{R}^1$

$$\dot{x}_1 = -5x_1 + 4x_4 + x_1x_2 - 0.25x_2x_3 + u,$$
$$\dot{x}_2 = -3x_2 + x_1x_4,$$
$$\dot{x}_3 = -3x_3 + x_1x_2 + 2u,$$
$$\dot{x}_4 = 5x_1 + 2x_4 - 2x_1x_2 - x_2x_3 + u.$$

With the fix step-size differential equations solver, simulate governing equations to obtain $(x,\dot{x},u)\in X\times U$. Identify $\mathcal{M}(x_{\mathcal{M}},u,\mathcal{S},\mathcal{P})$ using the over-structured and over-parameterized $(\mathcal{S},\mathcal{P})$. As model $\mathcal{M}(x_{\mathcal{M}},u,A_{\mathcal{M}}(\boldsymbol{a},\boldsymbol{b}),f_{\mathcal{M}}(x_{\mathcal{M}},u))$, consider a *predictor* which admit aperiodic asymmetric oscillations

$$\dot{x}_{\mathcal{M}}(t) = A_{\mathcal{M}}(\boldsymbol{a},\boldsymbol{b})f_{\mathcal{M}}(x_{\mathcal{M}},u),\ A_{\mathcal{M}}\in\mathbb{R}^{4\times 11}, f_{\mathcal{M}}(\cdot)\in\mathbb{R}^{11}.$$

In particular,

$$\begin{aligned}
\dot{x}_{\mathcal{M}1} &= a_{11}x_{\mathcal{M}1} + a_{12}x_{\mathcal{M}2} + a_{13}x_{\mathcal{M}3} + a_{14}x_{\mathcal{M}4} + a_{15}x_{\mathcal{M}1}x_{\mathcal{M}2} + a_{16}x_{\mathcal{M}1}x_{\mathcal{M}3} + a_{17}x_{\mathcal{M}1}x_{\mathcal{M}4}\\
&\quad + a_{18}x_{\mathcal{M}2}x_{\mathcal{M}3} + a_{19}x_{\mathcal{M}2}x_{\mathcal{M}4} + a_{110}x_{\mathcal{M}3}x_{\mathcal{M}4} + b_{11}u,\\
\dot{x}_{\mathcal{M}2} &= a_{21}x_{\mathcal{M}1} + a_{22}x_{\mathcal{M}2} + a_{23}x_{\mathcal{M}3} + a_{24}x_{\mathcal{M}4} + a_{25}x_{\mathcal{M}1}x_{\mathcal{M}2} + a_{26}x_{\mathcal{M}1}x_{\mathcal{M}3} + a_{27}x_{\mathcal{M}1}x_{\mathcal{M}4}\\
&\quad + a_{28}x_{\mathcal{M}2}x_{\mathcal{M}3} + a_{29}x_{\mathcal{M}2}x_{\mathcal{M}4} + a_{210}x_{\mathcal{M}3}x_{\mathcal{M}4} + b_{12}u,\\
\dot{x}_{\mathcal{M}3} &= a_{31}x_{\mathcal{M}1} + a_{32}x_{\mathcal{M}2} + a_{33}x_{\mathcal{M}3} + a_{34}x_{\mathcal{M}4} + a_{35}x_{\mathcal{M}1}x_{\mathcal{M}2} + a_{36}x_{\mathcal{M}1}x_{\mathcal{M}3} + a_{37}x_{\mathcal{M}1}x_{\mathcal{M}4}\\
&\quad + a_{38}x_{\mathcal{M}2}x_{\mathcal{M}3} + a_{39}x_{\mathcal{M}2}x_{\mathcal{M}4} + a_{310}x_{\mathcal{M}3}x_{\mathcal{M}4} + b_{13}u,\\
\dot{x}_{\mathcal{M}4} &= a_{41}x_{\mathcal{M}1} + a_{42}x_{\mathcal{M}2} + a_{43}x_{\mathcal{M}3} + a_{44}x_{\mathcal{M}4} + a_{45}x_{\mathcal{M}1}x_{\mathcal{M}2} + a_{46}x_{\mathcal{M}1}x_{\mathcal{M}3} + a_{47}x_{\mathcal{M}1}x_{\mathcal{M}4}\\
&\quad + a_{48}x_{\mathcal{M}2}x_{\mathcal{M}3} + a_{49}x_{\mathcal{M}2}x_{\mathcal{M}4} + a_{410}x_{\mathcal{M}3}x_{\mathcal{M}4} + b_{14}u,
\end{aligned}
\qquad
f(x,u) = \begin{bmatrix} x_1 \\ x_2 \\ x_3 \\ x_4 \\ x_1x_2 \\ x_1x_3 \\ x_1x_4 \\ x_2x_3 \\ x_2x_4 \\ x_3x_3 \\ u \end{bmatrix}.$$

The SIMULINK diagram is reported in Figure 4.4.a. Use the simulated datasets for $(x,\dot{x},u)$ to solve (4.5) and find $\hat{A}\in\mathbb{R}^{4\times 11}$, which yields $\hat{f}(\cdot)$. We obtain $\hat{\mathcal{M}}(x_{\mathcal{M}},u,\hat{A},\hat{f})$. Convergence is guaranteed for any initial conditions and datasets $((x_1,x_2,x_3,x_4),(\dot{x}_1,\dot{x}_2,\dot{x}_3,\dot{x}_4),u)_{\text{data}}$, for mapping $\Phi\in\mathbb{R}^{q\times z}$, $q>z$.

Compute $\hat{A}\in\mathbb{R}^{4\times 11}$ for $\Phi\in\mathbb{R}^{500\times 11}$ using (4.5) and (4.11). The MATLAB code to estimate parameters is reported in Figure 4.4.a.

Using (4.5) $\hat{A}=\left[(\Phi^T\Phi)^{-1}\Phi^T\dot{x}\right]^T$, one obtains

$\hat{A}=$

```
-5.0000e+00  1.8566e-11 -1.1658e-10  4.0000e+00  1.000e+00  2.0463e-11  5.0484e-12 -2.5000e-01 -2.3956e-12  4.0973e-12  1.000e+00
 1.1610e-10 -3.0000e+00  8.0306e-11  4.3645e-12  1.458e-12 -1.7577e-11  1.0000e+00  5.4183e-12  1.2546e-12 -2.4998e-12  6.0619e-12
-5.3705e-11  1.9415e-11 -3.0000e+00 -4.7490e-12  1.000e+00  1.1939e-11  4.8682e-12  1.4677e-13 -5.3424e-13  2.0617e-12  2.000e+00
 5.0000e+00 -4.2451e-11  1.0665e-10  2.0000e+00 -2.000e+00 -2.9406e-11 -8.8876e-12 -1.0000e+00  2.4953e-12 -5.7341e-12  1.000e+00
```

Applying (4.11), for $\hat{A}=\left[V_\Phi\Sigma_\Phi^{-1}U_\Phi^T v_\Phi\sigma_\Phi^T u_\Phi^T\dot{x}\right]^T$ and $\hat{A}=\left[(v_\Phi\sigma_\Phi^T\sigma_\Phi v_\Phi^T)^{-1}v_\Phi\sigma_\Phi^T u_\Phi^T\dot{x}\right]^T$ we computed

$\hat{A}=$

```
-5.0000e+00 -1.1395e-11  8.1459e-10  4.0000e+00  1.0000e+00 -5.2199e-11  5.1927e-12 -2.5000e-01  1.3383e-11 -1.5570e-11  1.000e+00
 1.3063e-10 -3.0000e+00  1.2931e-09  8.4345e-11 -6.2821e-12 -1.5309e-10  1.0000e+00  9.6696e-11  2.1461e-11 -3.0628e-11  9.2449e-11
-1.0148e-10 -1.5652e-11 -3.0000e+00  3.0454e-11  1.0000e+00 -3.7063e-11  2.0840e-12  5.2221e-11  9.7373e-12 -1.0713e-11  2.000e+00
 5.0000e+00 -2.3195e-11 -4.3700e-10  2.0000e+00 -2.000e+00  -1.0194e-11 -1.9353e-11 -1.000e+00  -9.1778e-12  4.7757e-12  1.000e+00
```

and

$\hat{A}=$

```
-5.0000e+00 -5.3451e-12  2.8383e-11  4.0000e+00  1.0000e+00 -5.0790e-12 -1.4078e-12 -2.5000e-01  1.0814e-12 -9.3237e-13  1.000e+00
-1.3099e-11 -3.0000e+00 -1.7415e-10 -1.1990e-11  7.5850e-13  2.3592e-11  1.0000e+00 -1.2172e-11 -2.9687e-12  4.0317e-12 -9.029e-12
 5.2083e-12 -2.9301e-12 -3.0000e+00  4.6455e-12  1.0000e+00 -6.9198e-12  1.2557e-13  4.7459e-12  1.2332e-12 -8.9950e-13  2.000e+00
 5.0000e+00  3.6149e-12 -8.7965e-11  2.0000e+00 -2.0000e+00  1.5145e-11 -1.4669e-12 -1.0000e+00 -2.6052e-12  1.8092e-12  1.000e+00
```

Using the measured states $(x,x_{\mathcal{M}})$, the quantitative measures are defined using the p-norms. We have $m_x\le 0.001$ and $m_{\dot{x}}\le 0.01$ for $p=1$.

Apply (4.17) to find $\tilde{A}$. The regularized parameter matrix is

$$\tilde{A}=\begin{bmatrix}-5 & 0 & 0 & 4 & 1 & 0 & 0 & -0.25 & 0 & 0 & 1\\ 0 & -3 & 0 & 0 & 0 & 0 & 1 & 0 & 0 & 0 & 0\\ 0 & 0 & -3 & 0 & 1 & 0 & 0 & 0 & 0 & 0 & 2\\ 5 & 0 & 0 & 2. & -2 & 0 & 0 & -1 & 0 & 0 & 1\end{bmatrix}.$$

This $\tilde{A}$ can be refined because we have zero columns. However, $\tilde{A}$ is reported keeping the same dimensionality as $\hat{A}$. One obtains the resulting truncated structure for $\tilde{f}(\cdot)$. An identified model $\tilde{\mathcal{M}}(x_{\mathcal{M}},u,\tilde{A},\tilde{f})$ is

$$\dot{x}_{\mathcal{M}1}=-5x_{\mathcal{M}1}+4x_{\mathcal{M}4}+x_{\mathcal{M}1}x_{\mathcal{M}2}-0.25x_{\mathcal{M}2}x_{\mathcal{M}3}+u,$$
$$\dot{x}_{\mathcal{M}2}=-3x_{\mathcal{M}2}+x_{\mathcal{M}1}x_{\mathcal{M}4},$$
$$\dot{x}_{\mathcal{M}3}=-3x_{\mathcal{M}3}+x_{\mathcal{M}1}x_{\mathcal{M}2}+2u,$$
$$\dot{x}_{\mathcal{M}4}=5x_{\mathcal{M}1}+2x_{\mathcal{M}4}-2x_{\mathcal{M}1}x_{\mathcal{M}2}-x_{\mathcal{M}2}x_{\mathcal{M}3}+u.$$

Matching of a model to system is guaranteed. That is, $\tilde{\mathcal{M}}(x_{\mathcal{M}},u,\tilde{A},\tilde{f})\subseteq\mathbf{M}(x,u)$. Evolutions of $x(t)$ and $x_{\mathcal{M}}(t)$ are identical, and, documented in Figures 4.4.b.

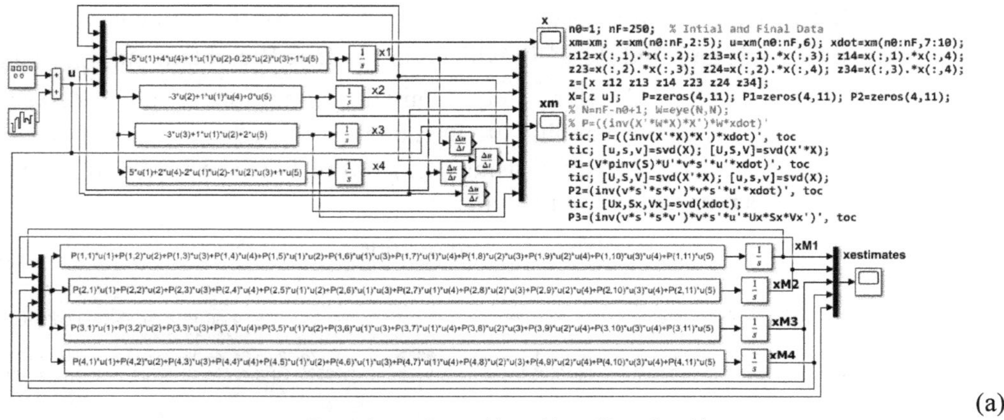

(a)

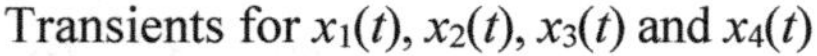

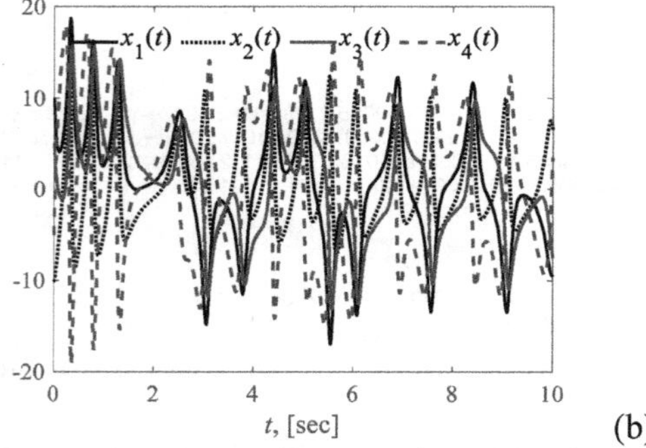

(b)

Figure 4.4. (a) SIMULINK diagram to model $\mathbf{M}(x,u)$ and $\hat{\mathcal{M}}(x_{\mathcal{M}}, u, \hat{A}, \hat{f})$;
(b) System and model dynamics, x_0=[10, –10, 5, –5].

Derivatives $(\dot{x}_1, \dot{x}_2, \dot{x}_3, \dot{x}_4)$ may not be directly measured. In SIMULINK, the derivatives of states are numerically computed using the derivative blocks which yield $(\hat{\dot{x}}_1, \hat{\dot{x}}_2, \hat{\dot{x}}_3, \hat{\dot{x}}_4)$ with numeric errors and latency, which depend on the differential equation solver, step size, etc. The SIMULINK diagram is shown in Figure 4.4.a.

Solving (4.22) $\hat{A} = \left[(\hat{\Phi}^T \hat{\Phi})^{-1} \hat{\Phi}^T \hat{\dot{x}} \right]^T$, $\hat{A} \in \mathbb{R}^{4\times 11}$, we have

$\hat{A}$=

-4.7591	-0.0479	-0.3133	3.7899	0.9864	0.0009	0.0028	-0.1743	-0.0228	-0.0021	0.9853
-0.1521	-2.2549	0.1008	-0.0989	0.0236	0.0323	0.9373	-0.0163	-0.0074	0.1136	-1.0412
0.4692	-0.0413	-3.3385	-0.2884	0.9430	0.0012	0.0020	0.0599	-0.0353	-0.0011	1.8827
3.4941	0.0601	0.9532	2.6455	-1.7798	0.0051	-0.0031	-1.1681	0.0870	0.0131	0.9490

Computed $\hat{A} = \begin{bmatrix} -4.76 & -0.048 & -0.313 & 3.79 & 0.986 & 0.001 & 0.003 & -0.174 & -0.023 & -0.002 & 0.985 \\ -0.152 & -2.24 & 0.101 & -0.1 & 0.023 & 0.032 & 0.937 & -0.016 & -0.007 & 0.114 & -1.04 \\ 0.469 & -0.041 & -3.34 & -0.288 & 0.943 & 0.001 & 0.002 & 0.06 & -0.035 & -0.001 & 1.88 \\ 3.49 & 0.06 & 0.953 & 2.65 & -1.78 & 0.005 & -0.003 & -1.17 & 0.087 & 0.013 & 0.949 \end{bmatrix}$ does not yield a seamless system description. However, the derived $\hat{\mathcal{M}}(x_{\mathcal{M}}, u, \hat{A}, \hat{f})$ typifies $\mathbf{M}(x,u)$. Identification convergences, and, accuracy depends on many factors, such as the ability to measure $(x, \dot{x}, u)$, estimated $(\hat{x}, \hat{\dot{x}} = \frac{d\hat{x}}{dt})$, dataset size, *predictors* used, heterogeneities, etc.

4.8. Identification of Dynamic Environmental System

Various spatiotemporal climate models were developed by government agencies and laboratories. Many research teams worldwide have being applied concepts of atmospheric thermodynamics, biogeochemistry and fluid dynamics to investigate climate, atmospheric, circulation and other models.

Zero- and One-Dimensional Energy Balance Models – The energy balance models analyze equilibrium states of a body which absorbs, dissipates and reemits energies. A model, applied to the Earth, is aimed to find an average temperature of the Earth's surface T as a function of the sun irradiation, dissipation, etc. An energy balance equation $E_{in}=E_{out}$ at the thermal equilibrium is found by assuming that all bodies radiate energy as the electromagnetic radiation. The *black body* radiation temperature dependence is satisfied by the Stefan-Boltzmann law with the *radiant emittance*

$$J=\sigma T^4,$$

where σ is the Stefan-Boltzmann constant, $\sigma=5.6703744\times10^{-8}$ W/m^{2}K^4.

The radius of the Earth is $R=6.378\times10^6$ m. Assume that an Earth surface area, which absorbs electromagnetic radiation is a disk, facing the sun. The disk area is πR^2. The area of the Earth's surface is $4\pi R^2$.

From $E_{in}=\pi R^2 S$, $E_{out}=4\pi R^2\sigma T^4$, we have

$$\pi R^2 S=4\pi R^2\sigma T^4,$$

where S is the energy flux density per unit area constant, $S=1367.6$ W/m^2.

Hence, $Q=\sigma T^4$, $Q=S/4$, and, $T=\left(\frac{S}{4\sigma}\right)^{\frac{1}{4}}=\left(\frac{Q}{\sigma}\right)^{\frac{1}{4}}$.

One computes $T=278.7$K, or, 5.5°C, which is a reasonable estimate with not accounted surface curvature and shape nonuniformity, varying surface composition and density, absorption heterogeneity, varying thermal conductivity, assumption that the total solar irradiance is measured on a surface perpendicular to the source which yields constant S, topographic irregularities (elevation, slope angle and aspect), rainfall, water vapor, relative humidity, soil-moisture content, evaporative cooling effects near the surface, air gases, and, other factors.

The average temperature of the Earth is ~14°C or 287.15K. Using the global mean planetary albedo constant α and greenhouse factor ε, one has

$$\pi R^2 S(1-\alpha)=4\pi R^2\varepsilon\sigma T^4,$$

For $\alpha=0.3$ and $\varepsilon=0.62$, we obtain

$$T=\left(\frac{S(1-\alpha)}{4\varepsilon\sigma}\right)^{\frac{1}{4}}=\left(\frac{Q(1-\alpha)}{\varepsilon\sigma}\right)^{\frac{1}{4}}=287.25\text{K}.$$

For land and ice, α is ~0.3 and ~0.6. The temperature-dependent $\alpha(T)$ and $\varepsilon(T)$ are used. Models and parameters are parameterized to experimental data, and, fidelity is enabled by considering multi-physics phenomena.

The first law of thermodynamics states that the change in the internal energy of a closed system ΔE is equal to the difference of the heat energy supplied to the system E_{in} and the thermodynamic work W done by the system on surrounding, $\Delta E=(E_{in}-W)$. To model temperature evolutions, assume that the temperature varies at a rate proportional to the energy difference

$$dT/dt=k(E_{in}-E_{out}).$$

The evolution of temperature is described by a nonlinear differential equation

$$\frac{dT}{dt}=\frac{1}{c}\left[-\varepsilon\sigma T^4+(1-\alpha)Q\right],$$

where c is the *effective* constant, [J/K-m^2].

Experimental Data – Large datasets were collected from the ice core samples in Antarctica, Arctic and Greenland with entrapped air and gas inclusions in the ice sheet. A dataset for a period of 420,000 years is collected by drilling the ice core to the depth of 3623 meters on Princess Elizabeth Land in Antarctica by an international team, which yields measured changes in atmospheric gas traces, gases concertation, chemical compositions, temperature and solar insolation [16-18]. These broadly cited and applied findings document a data on concentration of carbon dioxide CO_2 [parts per million by volume (pPv)], temperature variations T [°C], concentration of methane CH_4 [pPv], level of oxygen isotope ^{18}O [%], and, the solar insolation u [Joules measured at 65^0N]. Any model is incomplete. Simplified descriptions are found applying laws of chemistry, physics, etc. Ensuring variables complementarity, standard and non-standard systems of units and quantities are considered.

Model Fidelity – Multi-physics phenomena and effects result in complex spatiotemporal dynamics, described by a system of high-dimension nonlinear partial differential equations and constitutive relationships. There are debates and ongoing efforts aimed to enable model fidelity and completeness under uncertainties [17, 21, 22]. In addition to atmospheric thermodynamics and convection, analysis is affected by nonuniform absorption, emission, atmospheric and oceanic circulations, spatiotemporal coupled dynamics, heterogeneities, statistical variabilities, etc. Even solution of the standard three-dimensional heat conduction in heterogeneous media

$$\rho c\frac{\partial T}{\partial t}-\nabla\cdot(k\nabla T)=\dot{Q}_V+\xi_{\dot{Q}_V}$$

is not a trivial problem. Here, $T(t,x,y,z)$ is the spatiotemporal temperature; ρ, c and k are the media density, specific heat and thermal conductivity; $\dot{Q}_V$ and $\xi_{\dot{Q}_V}$ are the volumetric heat source and perturbations.

Exploratory Model and Identification – There are debates on validity and accuracy of climate models which are simplified and parametrized to admit analytic and numeric solutions and analyses. Incompleteness, complexity and heterogeneity of thermodynamics, chemophysical and other phenomena affect fidelity. The system structure and parameters are unknown. With limited available data and variables, investigate a highly simplified exploratory model which does not include many phenomena and effects. A system $\mathbf{M}(x,u)$ exhibits evolutions with aperiodic asymmetric limit cycles on states [16-18]. Adequate classes of nonlinear differential equations are considered.

States and Inputs for Illustrative and Exploratory Studies – Use the *arbitrary units* for the obfuscated measured data to obtain quantitative results under discrepancies and model incompleteness stated above. Use the measured aperiodic and asymmetric variations of four variables

$$x=[x_1, x_2, x_3, x_4],\ x_1\equiv CO_2,\ x_2\equiv T,\ x_3\equiv CH_4,\ x_4\equiv {}^{18}O$$

as an environmental system states. There is a state-input-structure-parameter complementarity. Climate cycles and atmospheric variations are due to state coupling and inputs. Inputs u are the variations of solar insolation, sunspot activity cycles, solar activity, solar magnetic field, Earth's orbital variations and eccentricity (axial tilt), volcano irruptions, pollution, burning fossil fuels, agricultural burning, wildfire, deforestation, contamination, etc. The considered inputs (*forcing functions*) $u\in\mathbb{R}^m$, $m\geq 1$ in our studies are

$$u=[u_1, u_2, u_3, u_4, u_5],$$

where u_1 is the solar insolation; u_2 is the eccentricity; u_3 is the longitude of perihelion; u_4 is the precession index; u_5 is the axial tilt.

The Earth's axial tilt with respect to the orbital plane varies from 22.1° to 24.5° with a period 41,000 years [19, 20]. The period of axial variation of the Earth's axis of rotation relative to the fixed stars is 25,772 years. As inputs, one may also consider the apsidal precision, orbital inclination, orbital eccentricity, etc.

For $\mathbf{M}(x,u)$, $x\in\mathbb{R}^4$, $u\in\mathbb{R}^m$, perform identification using linear and truncated quadratic multivariate polynomials. There are incompleteness and heterogeneities in (x,u), unknown states and inputs dependencies, unit inconsistencies, etc. Perform identification by assuming a state-input-structure-parameters $(x,u,S,\boldsymbol{a})$ complementarity, scaling and applying *arbitrary units* for (x,u).

Identification For a Single Input – A system with four nonlinear differential equation, studied in section 4.7.2, yields evolutions with asymmetric aperiodic limit cycles. Using the results of section 4.7.2, for a system with $x=[x_1, x_2, x_3, x_4]$, $x\in\mathbb{R}^4$, and, a solar insolation as an input $u\in\mathbb{R}^1$, apply the linear and truncated quadratic polynomials as a *predictor* for $f_{\mathcal{M}}(x_{\mathcal{M}},u)$. Study

$$\dot{x}_{\mathcal{M}}(t)=A_{\mathcal{M}}(\boldsymbol{a},\boldsymbol{b})f_{\mathcal{M}}(x_{\mathcal{M}},u),\ f_{\mathcal{M}}(x_{\mathcal{M}},u)\triangleq f(x,u),$$

$$\begin{aligned}\dot{x}_{\mathcal{M}1}&=a_{11}x_{\mathcal{M}1}+a_{12}x_{\mathcal{M}2}+a_{13}x_{\mathcal{M}3}+a_{14}x_{\mathcal{M}4}+a_{15}x_{\mathcal{M}1}x_{\mathcal{M}2}+a_{16}x_{\mathcal{M}1}x_{\mathcal{M}3}+a_{17}x_{\mathcal{M}1}x_{\mathcal{M}4}\\&\quad+a_{18}x_{\mathcal{M}2}x_{\mathcal{M}3}+a_{19}x_{\mathcal{M}2}x_{\mathcal{M}4}+a_{110}x_{\mathcal{M}3}x_{\mathcal{M}4}+b_{11}u,\\\dot{x}_{\mathcal{M}2}&=a_{21}x_{\mathcal{M}1}+a_{22}x_{\mathcal{M}2}+a_{23}x_{\mathcal{M}3}+a_{24}x_{\mathcal{M}4}+a_{25}x_{\mathcal{M}1}x_{\mathcal{M}2}+a_{26}x_{\mathcal{M}1}x_{\mathcal{M}3}+a_{27}x_{\mathcal{M}1}x_{\mathcal{M}4}\\&\quad+a_{28}x_{\mathcal{M}2}x_{\mathcal{M}3}+a_{29}x_{\mathcal{M}2}x_{\mathcal{M}4}+a_{210}x_{\mathcal{M}3}x_{\mathcal{M}4}+b_{12}u,\\\dot{x}_{\mathcal{M}3}&=a_{31}x_{\mathcal{M}1}+a_{32}x_{\mathcal{M}2}+a_{33}x_{\mathcal{M}3}+a_{34}x_{\mathcal{M}4}+a_{35}x_{\mathcal{M}1}x_{\mathcal{M}2}+a_{36}x_{\mathcal{M}1}x_{\mathcal{M}3}+a_{37}x_{\mathcal{M}1}x_{\mathcal{M}4}\\&\quad+a_{38}x_{\mathcal{M}2}x_{\mathcal{M}3}+a_{39}x_{\mathcal{M}2}x_{\mathcal{M}4}+a_{310}x_{\mathcal{M}3}x_{\mathcal{M}4}+b_{13}u,\\\dot{x}_{\mathcal{M}4}&=a_{41}x_{\mathcal{M}1}+a_{42}x_{\mathcal{M}2}+a_{43}x_{\mathcal{M}3}+a_{44}x_{\mathcal{M}4}+a_{45}x_{\mathcal{M}1}x_{\mathcal{M}2}+a_{46}x_{\mathcal{M}1}x_{\mathcal{M}3}+a_{47}x_{\mathcal{M}1}x_{\mathcal{M}4}\\&\quad+a_{48}x_{\mathcal{M}2}x_{\mathcal{M}3}+a_{49}x_{\mathcal{M}2}x_{\mathcal{M}4}+a_{410}x_{\mathcal{M}3}x_{\mathcal{M}4}+b_{14}u,\end{aligned}\qquad f(x,u)=\begin{bmatrix}x_1\\x_2\\x_3\\x_4\\x_1x_2\\x_1x_3\\x_1x_4\\x_2x_3\\x_2x_4\\x_3x_4\\u\end{bmatrix},$$

where $A_{\mathcal{M}}\in\mathbb{R}^{4\times 11}$ and $f_{\mathcal{M}}(\cdot):X\times U\to\mathbb{R}^{11}$.

Using the datasets [16-18], perform identification for partitioned $(x,u)_h$, $(x,u)\in X\times U$, $t\in[0\ 400{,}000]$ years. States $x=[x_1, x_2, x_3, x_4]$ and input u are reported in Figure 4.5.a. Note that evolutions of (x,u) are given backward.

Scale (x,u) with zero mean and *arbitrary units* as $\bar{x}=0, \bar{u}=0$, $|(x,u)_{\text{mean}=0}|\leq 1$.

While the quantitative analysis could be ensured, for nonlinear systems, scaling and unit discrepancies are ambiguous. Using (4.24), the kernel density estimation for asynchronously spaced (x_1, x_2, x_3, x_4) yields $(\hat{x}_1,\hat{x}_2,\hat{x}_3,\hat{x}_4)$.

The derivatives $(\hat{\dot{x}}_1=\frac{d\hat{x}_1}{dt},\hat{\dot{x}}_2=\frac{d\hat{x}_2}{dt},\hat{\dot{x}}_3=\frac{d\hat{x}_3}{dt},\hat{\dot{x}}_4=\frac{d\hat{x}_4}{dt})$ are computed.

Using $(\hat{x}_1,\hat{x}_2,\hat{x}_3,\hat{x}_4)$ and $(\hat{\dot{x}}_1,\hat{\dot{x}}_2,\hat{\dot{x}}_3,\hat{\dot{x}}_4)$, solve (4.25) and compute matrices of parameters $\hat{A}_h \in \mathbb{R}^{4\times 11}$, $\hat{A}_h = [\hat{a}_{ij}]_h$ for partitioned datasets.

Evaluate the system and model dynamics $\dot{x}_{\mathcal{M}} = \hat{A}_h \hat{f}_h(x_{\mathcal{M}}, u)$ by applying measures (4.6)-(4.7) for $(x, x_{\mathcal{M}}, u) \in X \times X_{\mathcal{M}} \times U$ with computed $\hat{A}_h$. For partitioned $x=[x_1, x_2, x_3, x_4]_h$, $t \in [t_0\ t_{\text{final}})_h$, compute $\hat{A}_h \in \mathbb{R}^{4\times 11}$. For a time-varying system $\mathbf{M}(x,u)$, data heterogeneities and other factors, $\hat{A}_h$ are distinct. However, computed $\hat{A}_h$ validate the applied $f_{\mathcal{M}}(\cdot)$, and, $\hat{f}_h(\cdot)$ are found. Consider $t_1 \in [0\ 150{,}000)$, $t_2 \in [150{,}000\ 300{,}000)$ and $t_3 \in [300{,}000\ 400{,}000]$ years. We have

$$\hat{A}_1 = \begin{bmatrix} -0.1014 & 0.2685 & -0.0664 & 0.0718 & -0.0038 & 0.0388 & 0.2161 & -0.1619 & -0.0621 & -0.2769 & 0.0014 \\ -0.0739 & 0.1055 & 0.0221 & 0.0684 & -0.0705 & 0.1762 & 0.1249 & -0.2065 & 0.0697 & -0.2172 & 0.0165 \\ -0.0315 & 0.1090 & -0.0226 & 0.0824 & -0.0052 & -0.0886 & 0.1163 & 0.1403 & 0.0096 & -0.0171 & 0.0032 \\ 0.0556 & 0.0669 & -0.2436 & -0.0656 & -0.0095 & 0.2278 & -0.0721 & -0.4359 & -0.1839 & 0.0003 & -0.0489 \end{bmatrix},$$

$$\hat{A}_2 = \begin{bmatrix} -0.1012 & 0.2643 & -0.0652 & 0.0749 & -0.0324 & 0.0488 & 0.2244 & -0.1322 & -0.0581 & -0.2617 & 0.0029 \\ -0.0730 & 0.0984 & 0.0236 & 0.0732 & -0.1109 & 0.1918 & 0.1340 & -0.1641 & 0.0809 & -0.1909 & 0.0177 \\ -0.0300 & 0.0985 & -0.0206 & 0.0896 & -0.0631 & -0.0657 & 0.1299 & 0.2009 & 0.0258 & 0.0218 & 0.0046 \\ 0.0510 & 0.0696 & -0.2412 & -0.0658 & -0.0412 & 0.2289 & -0.0566 & -0.4045 & -0.2050 & -0.0090 & -0.0417 \end{bmatrix},$$

$$\hat{A}_3 = \begin{bmatrix} -0.1143 & 0.3058 & -0.0766 & 0.0712 & -0.0111 & 0.0893 & 0.3029 & -0.2114 & -0.0078 & -0.3549 & 0.0106 \\ -0.0877 & 0.0945 & 0.0411 & 0.0600 & -0.0809 & 0.2142 & 0.1864 & -0.1582 & 0.0249 & -0.1043 & 0.0348 \\ -0.0275 & 0.0956 & -0.0245 & 0.0775 & -0.1579 & 0.0008 & 0.1799 & 0.2706 & -0.0922 & 0.1899 & 0.0204 \\ 0.0480 & 0.0865 & -0.2419 & -0.0239 & -0.0450 & 0.2504 & -0.2611 & -0.3538 & -0.3272 & 0.2431 & -0.0148 \end{bmatrix}.$$

Time-Varying $\hat{A}(t)$ – For not equal intervals $[t_0\ t_{\text{final}})_h \in T$ on $t \in T$, solutions $x_{\mathcal{M}}(t)$ with distinct initial and final conditions $(x_{\mathcal{M}0}, x_{\mathcal{M}\text{final}})_h$ are aggregated finding evolutions for $x_{\mathcal{M}}(t)$, $t \in T$. The overall adequateness of the identified model governance, state and input dependences, estimated parameters $[\hat{a}_{ij}]_h$, as well as structure-parametric regularization of identified $(\hat{A}_h, \hat{f}_h)$ with $\forall \hat{f}_h \equiv f_{\mathcal{M}}(\cdot)$ are guaranteed despite model incompleteness, inconsistencies, time-varying $\hat{A}(t)$, data heterogeneity, scaling and unit discrepancies, measurement uncertainties, estimation errors, etc.

The model is validated, and, for $(x, x_{\mathcal{M}})_{t\in[0\ 400000]}$, $m_x \leq 0.062$, $m_x \leq 0.133$, $p=1$. A system $\mathbf{M}(x,u)$ and model $\hat{\mathcal{M}}(x_{\mathcal{M}}, u, \hat{A}_h, \hat{f}_h)$ exhibit identical dynamics for $(x, x_{\mathcal{M}})$, documented in Figure 4.5.b. That is, time-varying $\hat{A}(t) = \sum_h \hat{A}_h$ ensures consistency with minor jump discontinuities at $(t_{(i-1)\text{final}}, t_{i0})$.

Time-Invariant System – Matrices $\hat{A}_h$ are found. Evolutions of $x_{\mathcal{M}}(t)$ using three $\hat{A}_h$, $h=1,2,3$ are simulated on an entire time horizon $t \in [0\ 400000]$ years. Results are illustrated in Figures 4.5.c. Assuming a time-invariant $\hat{A}$, the overall objective is achieved for computed $\hat{A}_h$, and, $\hat{\mathcal{M}}(x_{\mathcal{M}}, u, \hat{A}_h, \hat{f}_h)$ typifies $\mathbf{M}(x,u)$. Convergence of an identification algorithm and adequacy of results are ensured despite discrepancies and challenges. There is an overall consistency between the measured $(CO_2, T, CH_4, {}^{18}O, u)$ of $\mathbf{M}(x,u)$ and evolutions of an identified model.

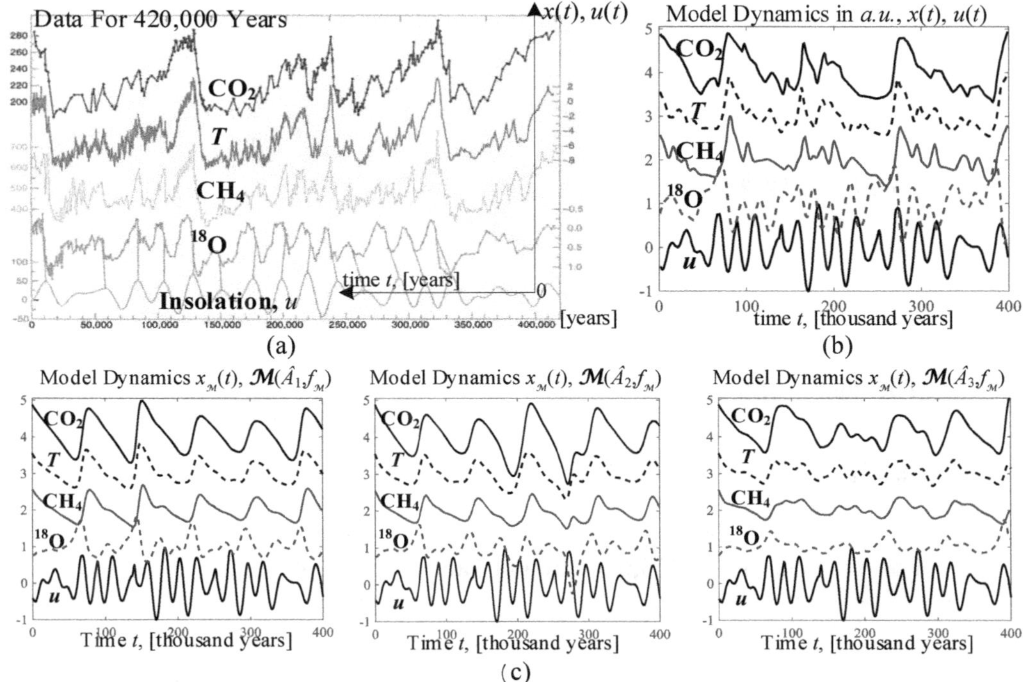

Figure 4.5. (a) Measured evolutions of $x_1 \equiv CO_2$, $x_2 \equiv T$, $x_3 \equiv CH_4$, $x_4 \equiv {}^{18}O$ and $u \equiv$solar insolation. The data is reported backward [16-18];

(b) Identified model $\hat{\mathcal{M}}(x_{\mathcal{M}},u,\hat{A}_h,\hat{f}_h)$: Dynamics for the model states $x_{\mathcal{M}}(t)$ with $(\hat{A}_h,\hat{f}_h)_{h=1,2,3}$, $\forall \hat{f}_h \equiv f_{\mathcal{M}}(\cdot)$. A model with time-varying $\hat{A}(t)=\sum_h \hat{A}_h$ exhibits identical dynamics to a *scaled* system $\mathbf{M}(x,u)$. Evolutions of $(x_{\mathcal{M}},u)_{\bar{x}=0,\bar{u}=0}$ are in the *arbitrary units*, and, reported with offsets to display $|(x_{\mathcal{M}},u)| \le 1$ on the same plot, typifying experimental documented data [16-18];

(c) Dynamics of $x_{\mathcal{M}}(t)$ for $\hat{\mathcal{M}}(x_{\mathcal{M}},u,\hat{A}_h,\hat{f}_h)$ with identified $(\hat{A}_h,\hat{f}_h)$, $\forall \hat{f}_h \equiv f_{\mathcal{M}}(\cdot)$ for three time-invariant $\hat{A}_h$, h=1,2,3 on an entire time horizon $t \in [0\ 400{,}000]$ years.

Identification For a Multi-Input Multi-Output System – Consider $\mathbf{M}(x,u)$, $x \in \mathbb{R}^4$, $\forall x>0$ with input vector $u \in \mathbb{R}^5$,

$$u=[u_1, u_2, u_3, u_4, u_5],$$

where u_1 is the solar insolation; u_2, u_3, u_4 and u_5 are the eccentricity, longitude of perihelion $\sin(\varpi)$, precession index $e\sin(\varpi)$ and axial tilt ε, which are the Milankovitch cycles affecting the Earth motions [19, 20].

The *arbitrary units* are used. The *scaled* (x,u) are

$$0 \le (x_1,x_2,x_3,x_4) \le 1,\ 0 \le (u_1,u_2,u_5) \le 1,\ |u_3| \le 1,\ |u_4| \le 1,\ |(u_3,u_4)_{\text{mean}=0}| \le 1.$$

The state variations are reported in Figures 4.4.a and 4.5.a [16-20, 23]. Identification is performed using a model

$$\dot{x}_{\mathcal{M}}(t)=A_{\mathcal{M}}(\boldsymbol{a},\boldsymbol{b})f_{\mathcal{M}}(x_{\mathcal{M}},u),\ f_{\mathcal{M}}(x_{\mathcal{M}},u)\triangleq f(x,u),\ f(x,u)=\begin{bmatrix}x_1\\x_2\\x_3\\x_4\\x_1x_2\\x_1x_3\\x_1x_4\\x_2x_3\\x_2x_4\\x_3x_4\\u_1\\u_2\\u_3\\u_4\\u_5\end{bmatrix}.$$

The matrix of parameters $\hat{A}_h \in \mathbb{R}^{4\times 15}$ is

$$\hat{A}_h=\left[\hat{a}_{ij}\right]_h=\begin{bmatrix}\hat{a}_{11}&\hat{a}_{12}&\hat{a}_{13}&\hat{a}_{14}&\hat{a}_{15}&\hat{a}_{16}&\hat{a}_{17}&\hat{a}_{18}&\hat{a}_{19}&\hat{a}_{110}&\hat{a}_{111}&\hat{a}_{112}&\hat{a}_{113}&\hat{a}_{114}&\hat{a}_{115}\\\hat{a}_{21}&\hat{a}_{22}&\hat{a}_{23}&\hat{a}_{24}&\hat{a}_{25}&\hat{a}_{26}&\hat{a}_{27}&\hat{a}_{28}&\hat{a}_{29}&\hat{a}_{210}&\hat{a}_{211}&\hat{a}_{212}&\hat{a}_{213}&\hat{a}_{214}&\hat{a}_{215}\\\hat{a}_{31}&\hat{a}_{32}&\hat{a}_{33}&\hat{a}_{34}&\hat{a}_{35}&\hat{a}_{36}&\hat{a}_{37}&\hat{a}_{38}&\hat{a}_{39}&\hat{a}_{310}&\hat{a}_{311}&\hat{a}_{312}&\hat{a}_{313}&\hat{a}_{314}&\hat{a}_{315}\\\hat{a}_{41}&\hat{a}_{42}&\hat{a}_{43}&\hat{a}_{44}&\hat{a}_{45}&\hat{a}_{46}&\hat{a}_{47}&\hat{a}_{48}&\hat{a}_{49}&\hat{a}_{410}&\hat{a}_{411}&\hat{a}_{412}&\hat{a}_{413}&\hat{a}_{414}&\hat{a}_{415}\end{bmatrix}_h.$$

For two datasets $t\in[225{,}000\ 350{,}000]$ and $t\in[250{,}000\ 375{,}000]$, we find

$$\hat{A}_1=\begin{bmatrix}-0.2302&0.0062&0.1947&0.0267&-0.0334&0.1177&0.2822&-0.1164&0.0437&-0.2755&-0.0418&-0.0672&-0.069&0.0688&0.0558\\-0.1093&0.1115&0.1093&0.1037&-0.5453&0.2283&0.33&-0.0345&-0.1177&-0.3657&-0.014&-0.0026&0.3705&-0.3682&-0.0419\\0.1743&0.3189&-0.2562&-0.1286&-0.0288&-0.1725&-0.1474&-0.1321&-0.1&0.399&0.0434&0.02559&0.2066&-0.2254&-0.0063\\-0.3264&0.0251&0.2599&0.0046&0.3385&0.0436&0.256&-0.257&-0.1594&-0.2537&0.0233&0.0107&-0.1242&0.1222&0.0209\end{bmatrix},$$

$$\hat{A}_2=\begin{bmatrix}-0.2422&0.0082&0.1857&0.0194&-0.0283&0.12946&0.2904&-0.1129&0.0471&-0.2736&-0.0368&-0.0604&-0.0809&0.0811&0.056\\-0.0873&0.1079&0.1256&0.117&-0.5547&0.2068&0.315&-0.0408&-0.1238&-0.3691&-0.0231&-0.0151&0.3922&-0.3907&-0.0423\\0.1739&0.3189&-0.2565&-0.1288&-0.0287&-0.1721&-0.1471&-0.132&-0.1&0.3991&0.0436&0.0258&0.2063&-0.225&-0.0063\\-0.3278&0.0253&0.2587&0.0037&0.3391&0.045&0.2569&-0.2566&-0.159&-0.2535&0.0239&0.0115&-0.1255&0.1237&0.0209\end{bmatrix}.$$

System $\mathbf{M}(x,u)$ and model $\hat{\mathcal{M}}(x_{\mathcal{M}},u,\hat{A}_h,\hat{f}_h)$, $\forall\hat{f}_h\equiv f_{\mathcal{M}}(\cdot)$ governances are reported in Figures 4.6. Dynamics of a model resemble evolutions of $\mathbf{M}(x,u)$. Convergence of an identification algorithm and consistency are ensured. Model incompleteness, nonlinear state-input dependencies, unit inconsistencies and other factors impair the model to system matching. Fidelity, under- and over-fitted mapping, state-input and structure-parameter complementarity, as well as other problems should be addressed.

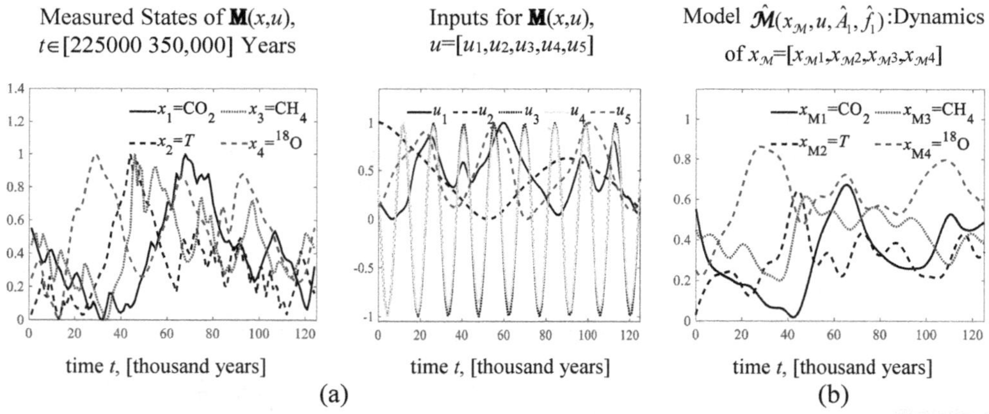

(a) (b)

Figure 4.6. (a) System $\mathbf{M}(x,u)$ evolutions for scaled states $x_1\equiv CO_2$, $x_2\equiv T$, $x_3\equiv CH_4$, $x_4\equiv{}^{18}O$, $\forall x>0$, and inputs, $u_1\equiv$solar insolation, $u_2\equiv$eccentricity, $u_3\equiv$longitude of perihelion, $u_4\equiv$precession index, $u_5\equiv$axial tilt, $t\in[225{,}000\ 350{,}000]$ years;
(b) Dynamics of model states $(x_{\mathcal{M}1},x_{\mathcal{M}2},x_{\mathcal{M}3},x_{\mathcal{M}4})$ with identified $(\hat{A}_1,\hat{f}_1)$, $\hat{f}_1=f_{\mathcal{M}}(\cdot)$.

Homework Problems

Homework Problem 4.1.

Consider unperturbed van der Pol differential equations, studied in Example 2.22.

Let a system $\mathbf{M}(x,u)$ is described as

$\dot{x}_1 = x_2$, $(x_{10},x_{20})\neq 0$,

$\dot{x}_2 = -x_1 + x_2(1-x_1^2)$,

Perform simulations in SIMULINK to obtain datasets for identification $(x,\dot{x})\in X$, and, identify a model $\mathcal{M}(x_{\mathcal{M}},u,\mathcal{S},\mathcal{P})$ with over-structured and over-parameterized $(f_{\mathcal{M}},A_{\mathcal{M}})$

$$\dot{x}_{\mathcal{M}1} = a_{11}x_{\mathcal{M}1} + a_{12}x_{\mathcal{M}2} + a_{13}x_{\mathcal{M}1}x_{\mathcal{M}2} + a_{14}x_{\mathcal{M}1}^2 x_{\mathcal{M}2} + a_{15}x_{\mathcal{M}1}x_{\mathcal{M}2}^2,$$

$$\dot{x}_{\mathcal{M}2} = a_{21}x_{\mathcal{M}1} + a_{22}x_{\mathcal{M}2} + a_{23}x_{\mathcal{M}1}x_{\mathcal{M}2} + a_{24}x_{\mathcal{M}1}^2 x_{\mathcal{M}2} + a_{25}x_{\mathcal{M}1}x_{\mathcal{M}2}^2.$$

Compute $\hat{A}$ and $\tilde{A}$ using (4.5), (4.11) and (4.17). Find $(\tilde{A},\tilde{f})$.

Compute $[(m_x,m_{\dot{x}}),(\Delta x,\Delta\dot{x})]$ for $p=1$ and $p=2$.

Validate $\hat{\mathcal{M}}(x_{\mathcal{M}},u,\hat{A},\hat{f})\subseteq \mathbf{M}(x,u)$ and $\hat{\mathcal{M}}(x_{\mathcal{M}},u,\tilde{A},\tilde{f})\subseteq \mathbf{M}(x,u)$.

Homework Problem 4.2.

Examples 2.23 and 4.7 report the *standard* Lotka-Volterra equations.

Consider a system $\mathbf{M}(x,u)$

$\dot{x}_1 = a_{11}x_1 - a_{12}x_1x_2 - a_{13}x_1^3x_2^3$, $x_1(t_0)=x_{10}$, $\forall(a_{ii},a_{ij})>0$,

$\dot{x}_2 = -a_{21}x_2 + a_{22}x_1x_2 + a_{23}x_1^3x_2^3$, $x_2(t_0)=x_{20}$, $(x_{10},x_{20})>0$.

Assign $(a_{ii},a_{ij})>0$, and, perform simulations in SIMULINK. For datasets $(x,\dot{x})\in X$, perform identification specifying an over-structured and over-parameterized $(\mathcal{S},\mathcal{P})$ for a model $\mathcal{M}(x_{\mathcal{M}},u,\mathcal{S},\mathcal{P})$ using an adequate class of differential equations. Compute $\hat{A}$ using (4.5), and, find $\tilde{A}$ applying (4.17). Evaluate your results by computing $[(m_x,m_{\dot{x}}),(\Delta x,\Delta\dot{x})]$ for $p=1$.

Homework Problem 4.3.

Consider a perturbed system $\mathbf{M}(x,u)$, see Example 2.24

$\dot{x}_1 = a_{11}x_1 - a_{12}x_1x_2 - a_{13}x_1^3x_2^3 + b_{11}\xi_1$, $x_1(t_0)=x_{10}$, $\forall(a_{ii},a_{ij})>0$,

$\dot{x}_2 = -a_{21}x_2 + a_{22}x_1x_2 + a_{23}x_1^3x_2^3 + b_{21}\xi_2$, $x_2(t_0)=x_{20}$, $(x_{10},x_{20})>0$.

Assign $(a_{ii},a_{ij})>0$, and, perform simulations in SIMULINK. For datasets $(x,\dot{x})\in X$, perform identification for a known model structure, as well as for an over-structured and over-parameterized $(\mathcal{S},\mathcal{P})$ problem using consistent *predictors*. Validate identification results proving $\hat{\mathcal{M}}(x_{\mathcal{M}},u,\hat{A},\hat{f})\subseteq \mathbf{M}(x,u)$ with computed $\hat{A}$ and $\tilde{A}$. Evaluate measures $[(m_x,m_{\dot{x}}),(\Delta x,\Delta\dot{x})]$.

Homework Problem 4.4.

Consider perturbed differential equations which describe a system $\mathbf{M}(x,u)$

$\dot{x}_1 = x_2$,

$\dot{x}_2 = -x_1 + x_2(1-x_1^2-x_1^4) + x_3 + u$, $u\equiv\xi$,

$\dot{x}_3 = -x_1x_2 - x_3$.

Let the initial conditions are $(x_{10},x_{20},x_{30})=0$ and $(x_{10},x_{20},x_{30})\neq 0$.

1. Having found $(x,\dot{x},u)\in X\times U$ from simulations for different initial conditions, identify $\mathcal{M}(x_{\mathcal{M}},u,\mathcal{S},\mathcal{P})$ using an adequate class of differential equations. Use as overdetermined model *predictors* with high-degree polynomials. Investigate convergence of an identification algorithm. Validate your model. Prove $\hat{\mathcal{M}}(x_{\mathcal{M}},u,\hat{A},\hat{f})\subseteq \mathbf{M}(x,u)$ with computed $\hat{A}$ and $\tilde{A}$.
2. Perform simulations in SIMULINK to obtain datasets $(x,\dot{x},u)\in X\times U$. Let $\dot{x}$ are not directly measured. Evaluate the derivatives as $\hat{\dot{x}}=\frac{dx}{dt}$. Identify a system, finding $\mathcal{M}(x_{\mathcal{M}},u,\mathcal{S},\mathcal{P})$. Evaluate algorithm convergence and consistency.

Homework Problem 4.5.

Perform experiments and collect data for a series or parallel *RLC* circuit, described by the second-order linear differential equations, see Examples 2.18, 2.19 and 4.10. Let the model structure is known, and, parameters (R,L,C) are unknown. Measure the state dynamics (x_1,x_2) and input $u(t)$. Evaluate $(\dot{x}_1,\dot{x}_2)$ as $\hat{\dot{x}}_1=\frac{dx_1}{dt}$, $\hat{\dot{x}}_2=\frac{dx_2}{dt}$, and, obtain $(x,\hat{\dot{x}},u)$ or $(\hat{x},\hat{\dot{x}},u)\in X\times U$. Perform parameters estimation, compute $(\hat{A},\tilde{A})$, and, examine model fidelity. The circuit parameters (R,L,C) are known with a specific tolerance, and, may be measured. Note that (R,L,C) are temperature dependent, and, may vary. Validate your results.

References

1. V. Adetola, M. Guay and D. Lehrer, "Adaptive estimation for a class of nonlinearly parameterized dynamic systems," *IEEE Trans. Automatic Control*, vol. 59, no. 10, pp. 2818-2824, 2017.
2. S. A. Billings, *Nonlinear System Identification: NARMAX Methods in the Time, Frequency, and Spatio-Temporal Domains*, Wiley, Hoboken, NJ, 2013.
3. W. R. Jacobs, T. Baldacchino, T. Dodd and S. R. Anderson, "Sparse Bayesian nonlinear system identification using variational inference," *IEEE Trans. Automatic Control*, vol. 99, pp. 1-16, 2018.
4. I. Karafyllis and M. Krstic, "Adaptive certainly-equivalence control with regulation-triggered finite-time least-square identification," *IEEE Trans. Automatic Control*, vol. 99, pp. 1-16, 2018.
5. L. Ljung, *System Identification: Theory for the User*, Prentice-Hall, Englewood Cliffs, NJ, 1999.
6. S. E. Lyshevski, *Control Systems Theory With Engineering Applications*, Birkhauser, Boston, MA, 2001.
7. S. E. Lyshevski, "Identification of nonlinear flight dynamics: Theory and practice," *IEEE Trans. Aerospace and Electronic Systems,* vol. 36, no. 2, pp. 383-392, 2000.
8. S. E. Lyshevski, "State-space model identification of deterministic nonlinear systems: Nonlinear mapping technology and application of Lyapunov theory," *Automatica,* vol. 34, no. 5, pp. 659-664, 1998.
9. H. Rios, D. Efimov, J. A. Moreno, W. Perruquetti and J. G. Rueda-Escobedo, "Time-varying parameter identification algorithms: Finite and fixed-time convergence," *IEEE Trans. Automatic Control*, vol. 62, no. 7, pp. 3671-3678, 2017.
10. M. M. Tobenkin, I. R. Manchester and A. Megretski, "Convex parameterization and fidelity bounds for nonlinear identification and reduced-order modeling," *IEEE Trans. Automatic Control*, vol. 62, no. 7, pp. 3679-3686, 2017.
11. B. Mu, T. Chen and L. Ljung, "On the input design for kernel-based regularized LTI system identification: Power-constrained inputs," *Proc. Conf. Decision and Control*, pp. 5262-5267, 2017.
12. W. Pan, Y. Yuan, L. Ljung, J. Goncalves and G.-B. Stan, "Identification of nonlinear state-space systems from heterogeneous datasets," *IEEE Trans. Control Network Systems*, vol. 5, no. 2, pp. 737-747, 2018.
13. T. Strutz, *Data Fitting and Uncertainty: A Practical Introduction to Weighted Least Squares and Beyond*, Springer Vieweg, NY, 2016.
14. J. Arenas-García, K. B. Petersen, G. Camps-Valls and L. K. Hansen, "Kernel multivariable analysis framework for supervised subspace learning: A tutorial on linear and kernel multivariate methods," *IEEE Signal Processing Magazine*, vol. 30, issue 4, pp. 16-29, 2013.
15. K.-R. Müller, S. Mika, G. Rätsch, K. Tsuda and B. Schölkopf, "An introduction to kernel-based learning algorithms," *IEEE Trans. Neural Networks*, vol. 12, no. 2, pp. 181-201, 2001.
16. J. R. Petit, J. Jouzel, D. Raynaud, N. I. Barkov, J.-M. Barnola, I. Basile, M. Bender, J. Chappellaz, M. Davis, G. Delaygue, M. Delmotte, V. M. Kotlyakov, M. Legrand, V. Y. Lipenkov, C. Lorius, L. Pepin, C. Ritz, E. Saltzman and M.

Stievenard, "Climate and atmospheric history of the past 420,000 years from the Vostok ice core, Antarctica," *Nature*, vol. 399, issue 6735, pp. 429-436, 1999.

17. J. R. Petit, *Vostok – Le dernier secret de l'Antarctique*, Paulsen, 2017.
18. A. N. Salamatin, E. A. Tsyganova, V. Y. Lipenkov and J. R. Petit, "Vostok (Antarctica) ice-core time-scale from datings of different origins," *Annals Glaciology*, vol. 39, pp. 283-292, 2004.
19. M. M. Milankovic, *Canon of Insolation and the Ice-Age Problem*, Textbook Publishing Company, Belgrade, 1998.
20. M. M. Milankovitch, *Theorie Matheematique des Phenomenes Thermiques Produits par la Radiation Solaire*, Gauthier-Villars, Paris, 1920.
21. K. E. Trenberth, *Climate System Modeling*, Cambridge University Press, 1993.
22. W. M. Washington and C. L. Parkinson, *Introduction to Three-Dimensional Climate Modeling*, University Science, NY, 2005.
23. N. C. Wells, *The Atmosphere and Ocean: A Physical Introduction*, Wiley and Sons, West Sussex, UK, 2012.

5 Multi-Input Multi-Output Systems: Control and Optimization

Narration on Engineering Discoveries – Control and measurement apparatuses for water flow, hydraulic gravity control, water levels regulation and water distribution, hydraulic and pneumatic devices, steam mechanisms, mining and metallurgy machinery, mills, torsion catapults, gimbals, and other inventions were originated and implemented in Roman Empire, Greece and Ptolemaic Egypt. Inventions are dated back to the fourth and fifth century BC by Archimedes of Syracuse, Ctesibius of Alexandria, Heron of Alexandria, Philo of Byzantium and others. In various applications and broad use, empirical designs, measurements, regulation and analysis of systems governance were applied and pragmatic. Medieval European and Byzantine scientists developed various original concepts, and, discovered various apparatuses.

Founding studies in nonlinear dynamics and analysis were pioneered and accomplished by Johann Bernoulli (1667-1742), Leonard Euler (1707-1783), Sir William Hamilton (1805-1865), Carl Jacobi (1804-1851), Joseph Lagrange (1736-1813), Gottfried Leibniz (1646-1716), Sir Isaac Newton (1642-1726) and others. Physicists and mathematicians Blaise Pascal and Daniel Bernoulli developed foundations of hydrodynamics and statics, as well as investigated and demonstrated hydraulic devices. Joseph Bramah applied and secured the Great Britain patent #2045 Hydraulic Press in 1795. The first controlled rotary hydraulic motors was designed and deployed in 1876 by William Armstrong for a swing bridge over the River Tyne. In the controlled rotational and translational electro-hydraulic actuators, controlled electric motor rotates a pump, and, regulated flow and hydraulic pressure are converted into torque and force, thereby yielding rotational or translational motion. Hydraulic cylinders yield a translational motion. Industrial automation and control solutions were commercialized in the 18th century. For steam engines, a centrifugal governor with angular velocity measurements and feedback control was commercialized in 1776 by James Watt. Stability of the governor was analytically studied by James Maxwell, Edward Roth and Ivan Vyshnegradsky in the 1860th. Stability theory for linear and nonlinear dynamic systems was developed by Alexander Lyapunov in 1892 [1].

Industrial electromagnetic actuators and transducers were invented and demonstrated by Michael Faraday, who discovered, pioneered and founded electromagnetics and electrochemistry in numerous applications. In the early 1830th, Moritz von Jacobi designed and commercialized direct current electric motors for ship propulsion with an electrochemical battery as energy source. Nikola Tesla demonstrated wireless communication, *RLC* filters and compensators (regulators), electrochemical battery energy source, as well as controlled direct current motors to actuate a propeller and steer rudder in radio-controlled boats. Mr. Tesla secured the U.S. Patent 613,809 *Method of an Apparatus for Controlling Mechanism of Moving Vessels or Vehicles*, 1898. Nikola Tesla invented and commercialized various controlled energy, electromechanical, electromagnetic and communication systems, securing hundreds of patents during 1886 to 1922. Mr. Tesla's patents were commercialized, and, his pioneering discoveries have being empowered industrial progress, striving economic developments, welfare, as well as improved standards of living and quality of life.

Physical systems should be controlled. Passive *RLC* electric circuits were used to implement analog filters, compensators, regulators and controllers. These systems were implemented to ensure vehicle guidance and control, as well as control of electromechanical servos in battleships and airplane starting early 1910. Resonators, filters, compensators and regulators were implemented using the *RLC* circuits, as reported in Practice Problems 3.15, 3.16, 3.17 and 5.10. In 1912, with a limited success, Arthur Pollen and Frederic Dreyer conducted the feasibility studies on fire control systems using mechanical computing apparatuses. In early 1900, Elmer Sperry pioneered revolutionary developments in gyroscopes and magnetic compass with implementation and deployment capabilities on battleships and airplanes. Practical stabilizing and steering control schemes with aforementioned sensors, researched and demonstrated since early 1900, use analog electric circuits to implement filters, estimators, compensators, regulators, controllers, etc. The *RLC* circuits, electron tubes (vacuum tubes) and solid-state electronic devices implemented analog control and processing schemes. Proportional-integral-derivative (PID) control for linear and nonlinear systems was reported in 1922 by Nikolai Minorsky [2].

Minimax optimization to design control laws and optimize dynamic systems were proposed by Richard Bellman [3], Rudolf Kalman [4, 5] and others in the late 1950 and thereafter. Textbooks [6-14] report key concepts and methods in design and optimization of dynamic systems. While many control premises have being investigated in last decades, industrial systems, consumer electronics and other platforms predominantly implement practical state feedback and PID control schemes using the tracking error. These control algorithm are supported by practical algorithmic and hardware solutions with consistent digital and analog implementation schemes.

5.1. System Performance and Optimality

Quantifiable specifications and requirements are imposed by industry, security and IEEE standards, guidance and regulations. Systems should ensure functionality, safety, controllability, adaptiveness, etc. One analyzes underlying system organization, and, performs a physical design tasks for industrial control systems (ICS), process control systems, distributed control systems, supervisory control and data acquisition systems (SCADA), and, other platforms. The designer investigates:

1. System organization with advanced-technology compatible hardware, such as compliant kinematics, transducers, actuators, sensors, ASICs, MPUs, FPGAs, power electronics, receivers, transmitters and other components;
2. Software, algorithms and protocols for spatially distributed, asynchronous and heterogeneous interacting components ensuring sensing, data fusion, information management, diagnostics, etc.;
3. Seamless integration, information sharing, control and optimization schemes;
4. Reliability, interoperability, modularity and scalability.

System Complexity – There are challenges in design of high-performance physical systems, such as aerial and ground manned and unmanned vehicles, semi-autonomous robots, autonomous swarms, etc. Designers encounter and overcome various challenges. Complexity is due to components integration, asynchronous data fusion, decentralized and centralized controls at module and system levels, etc. Impediments are due to fundamental and technological limits which are addressed in prevailing analyses and concurrent designs. Despite challenges, complex command, control, communications, computers and intelligence platforms were designed and deployed. To achieve desirable performance and capabilities, hardware matching and hardware-software co-design tasks are performed ensuring:

- Explicit articulation and application of advanced solutions and technologies;
- Component complementarity, compatibility, interconnectivity, interfacing, networking, etc.

One solves optimization, control, identification, estimation and other problems at the component and system levels. Paradigms and methods in design of systems have being emerged by addressing and solving problems within a domain-specific context.

Multiple-Level Hierarchical Systems – Closed-loop systems must ensure functionality and specified performance. Complex physical systems can be decomposed into subsystems and aggregated using hierarchical decomposition and aggregation. Physical system design implies organization consistency, modular integrity, compliance, connectivity, modularity, etc. Hardware-consistent consideration and aggregation are critical. Software developments and physical hardware designs should guarantee compliance and matching. One controls physical and information processes, mathematically described by nonlinear governing equations, constitutive relationships and statistical models. Multi-physics, spatial dependencies, high dimensionality and multiple domains are considered using a compliant aggregation. Optimization and control problems are considered at the component, subsystem and system levels. Advanced technology hardware and modular design stimulate studies for multi-input multi-output (MIMO) systems examining:

1. System aggregation and integration;
2. Fidelity analysis and multidimensional scaling;
3. Dimensionality reduction and decoupling;
4. Control and optimization schemes.

Control and Dynamic Optimization – Multi-physics high-fidelity analyses are aimed to ensure practicality and guarantee adequacy of engineering solutions. Dynamic optimization implies design and implementation of centralized, decentralized and distributed control schemes to ensure optimality, defined by performance functionals, indexes, stability criteria, etc. Control algorithms must guarantee adequate dynamics, stability and robustness. Departing from axiomatized formulations and conjectural paradigms, investigate design methods and implementation schemes to ensure:

- Safety and functionality;
- Effectiveness and productivity;
- Adaptiveness and resiliency;
- Mechanical and electromagnetic integrity;
- Stability, robustness, dynamic shaping, accuracy and precision.

There are many components in physical systems. Due to fast transients of microelectronic devices and sensors, dominant dynamics of mechanical and electromechanical aggregated subsystems are considered. While latencies and dynamics of key system components (MPUs, ASICs, receiver, transmitter and others) are not considered in dynamic optimization, these modules are critical. Sensor nonlinearities, data uncertainties, information heterogeneity, ASICs errors and other impediments are considered. Using laws of physics, informatics and probability theory, mathematical models of physical and information processes are found in form of governing and constitutive equations. Nonlinear time-invariant and time-varying equations adequately model physical systems. These equations are applied to solve analysis and optimization problems. Analytic design of control laws may assume that all states are measured or observed. Not all physical states can be measured, estimated and observed. Physical control should be consistent with device physics and compliant with sensing and electronic hardware, algorithms and protocols. *Minimal complexity* control laws should guarantee simplicity and adequateness.

Design Objectives – Closed-loop physical systems $\mathbb{M}(\boldsymbol{x},\boldsymbol{y},\boldsymbol{u})$ should ensure adequate performance, measured against specifications. System organization and hardware predefine system capabilities. In aerial, automotive and robotic applications, high-performance *smart* servos are comprised of permanent magnet actuators, soft-switching PWM amplifiers, microsystem-technology multi-degree-of-freedom (MDOF) sensors, MPUs and FPGAs, transceivers, etc. Abstract formulations and hypothetical solutions may result in chattering, oscillatory dynamics, losses, low efficiency, noise, vibration, wearing and other adverse phenomena. Axiomatized control algorithms may destabilize open-loop stable systems. Consistent control schemes ensure implementation practicality and near-optimal overall performance. Investigate physical control schemes, hardware, as well as synthesize and implement adequate control laws to guarantee:

- Effective energy conversion and multi-physics transductions in transducers, actuators, sensors, etc.;
- Stability with specified stability margins in the full operating envelope;
- Robustness to parameter variations, disturbances, structural variations and kinematic changes;
- Electrical, mechanical and thermal efficiency,
- Noise, vibration and losses minimization;
- Disturbances and perturbations rejection;
- Acceptable steady-state, static and dynamic tracking errors with specified accuracy and precision;
- Transient response specifications, such as settling times, overshoot, etc.

System performance is evaluated against multiple criteria. Performance and capabilities can be assigned and assessed using high-dimensional estimates, measures and metrics. Control laws $u(t)$ are found applying defined-by-specifications performance functionals, optimality and stability criteria, etc. Controlled dynamic systems are characterized by physical quantities $\boldsymbol{q}$, outputs $\boldsymbol{y}(t)$, states $\boldsymbol{x}(t)$ and controls $\boldsymbol{u}(t)$, which form high-dimensional manifold in high-dimensional $Q\times Y\times X\times U$ space.

As documented in chapter 1 and Figure 5.1, performance and capabilities measures $(\mathcal{Q}_{\mathcal{P}},\mathcal{Q}_{C})$ depend on real-valued variable-dependent quantities $\boldsymbol{q}(\boldsymbol{x},\boldsymbol{y},\boldsymbol{u})$.

In an operating envelope $Y\times X\times U$, one:

1. Examines $\mathcal{Q}(\mathcal{Q}_{\mathcal{P}},\mathcal{Q}_{C})=\boldsymbol{k}(\boldsymbol{q},\boldsymbol{x},\boldsymbol{y},\boldsymbol{u})$ and assesses performance evaluating variables $(\boldsymbol{p},\boldsymbol{c})\in\mathcal{P}\times C$;
2. Design a real-valued multivariate function $W(\boldsymbol{p},\boldsymbol{c})=f_W(\boldsymbol{q},\boldsymbol{x},\boldsymbol{y},\boldsymbol{u})$, and, solve the minimax problem (1.4)

$$J = \min_{p\in\mathcal{P},\, c\in C}\max W(\boldsymbol{p},\boldsymbol{c}),\; J{:}\mathcal{P}\times C\to\mathbb{R},\; \boldsymbol{p}=f_p(\boldsymbol{q},\boldsymbol{x},\boldsymbol{y},\boldsymbol{u}),\; \boldsymbol{c}=f_c(\boldsymbol{q},\boldsymbol{x},\boldsymbol{y},\boldsymbol{u}), \tag{5.1}$$

$$(\boldsymbol{q},\boldsymbol{x},\boldsymbol{y},\boldsymbol{u})\in \mathcal{Q}\times X\times Y\times U,\; L_{\min}\le L(\boldsymbol{p},\boldsymbol{c})\le L_{\max},\; (l_p,l_c)\ge 0.$$

Complexity and high dimensionality may not yield a solution of the minimax problem. Applying consistent governing equations and constitutive relationships, research and substantiate design methods minimize functionals with perceptive performance integrands, finding closed-form solutions.

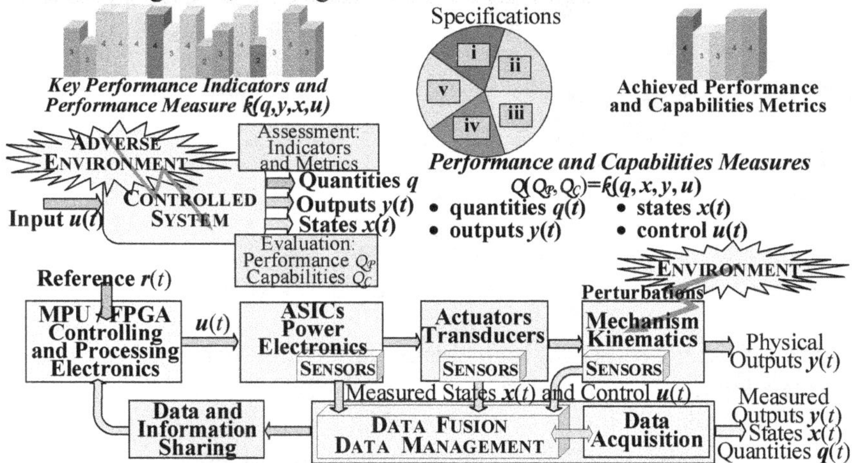

Figure 5.1. Performance and capabilities are assessed by state-dependent physical quantities $\boldsymbol{q}$, outputs $\boldsymbol{y}(t)$, states $\boldsymbol{x}(t)$ and controls $\boldsymbol{u}(t)$. Quantifiable specifications, performance indicators and capability metrics are asserted and evaluated. The physical quantities $\boldsymbol{q}$ are stability, stability margins, sensitivity, robustness, disturbance rejection, vulnerability, efficiency, bandwidth, physical data quality, information losses, accuracy, precision, etc.

Figure 5.2 reports a functional diagram of a closed-loop electromechanical servo used in high-accuracy pointing systems, flight servos, precision steering mechanisms, robots, etc. Considering a physical domain, information domain and cyberspace, Figure 5.3 documents components and integrated modules.

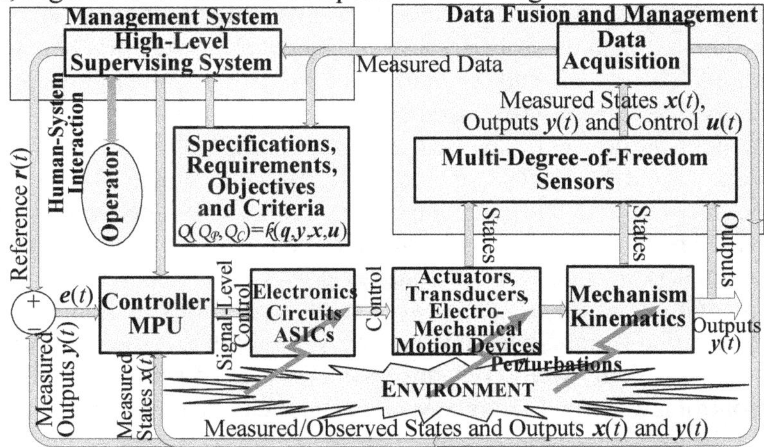

Figure 5.2. High-level functional diagram: For advanced-technology components and modules, designer finds system organization, procures hardware, solves control and data management problems, implements adequate sensing schemes, performs optimization, minimizes heterogeneities, and, implements algorithms and protocols.

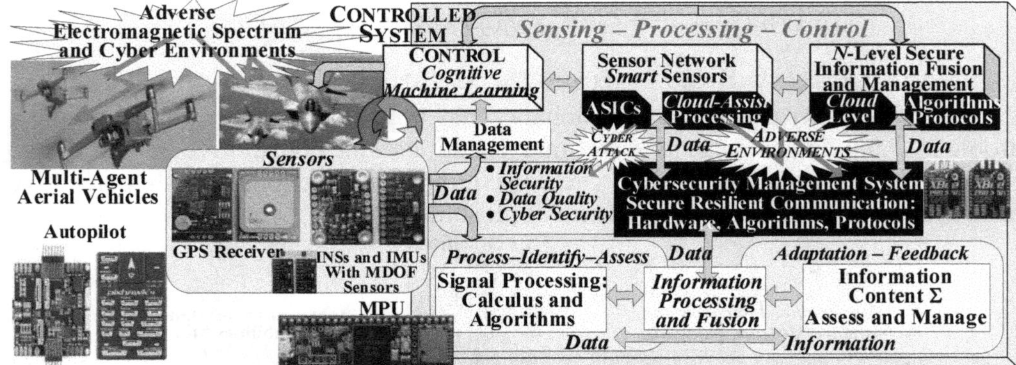

Figure 5.3. Physical domain, information domain and cyberspace: Aerial system functional diagram and images of key components.

System Dynamics – System analysis implies studies and evaluation using adequate techniques which support design and optimization tasks. Dynamic analysis implies:

1. Deterministic and statistical high-fidelity modeling, such as:
   - Development of governing equations of motion;
   - Identification and parameter estimation;
   - Fidelity analysis, consistent model reduction and model validation;
   - Probabilistic analysis;
2. Analytic and numeric analyses, solution of governing equations, and, data-intensive simulations;
3. Control laws design and dynamic optimization, nonlinear optimization, management systems design, and, implementation;
4. Performance and capabilities evaluations;
5. Evaluation of experimental data and validation.

Model Fidelity, Control Complexity and Hardware Solutions – The designer conducts multi-physics analysis, constructs exploratory high-fidelity models and investigate pragmatic control algorithms. Consistency, *implementability* and practicality of control laws are studied. There is no end for enhancing model fidelity and expanding control laws complexity. Data fusion, data management and governance schemes may result in considerable algorithmic and hardware (sensors, ASICs, MPU, etc.) impediments and limits. Models should be consistent and sufficiently accurate. *Minimal complexity* control laws, such as PID and other algorithms, can be synthesized using dynamic programming, Hamiltonian optimization and Lyapunov theory. These control laws may ensure suboptimal performance from analytic prospects, while optimality is ensured from implementation, integrity, physical system capabilities and functionality. The trade-off between algorithmic intricacy, hardware complexity, system performance and capabilities are studied. Many physical systems are open-loop stable. *Minimal complexity* control laws with directly measured states ensure near-optimal performance.

Proportional-Integral-Derivative Control Laws and State Feedback – System performance is improved and optimized by means of adequate hardware selection, as well as by design and implementation of consistent control laws and algorithms. For linear systems, one applies time- and frequency-domain methods, such as the Laplace and Fourier transforms, transfer functions, *modal* control, etc. To implement widely applied PID control laws, the measured tracking error $e(t)$ is used, $e(t)=r(t)-y(t)$. Physical states $\boldsymbol{x}(t)$ and outputs $\boldsymbol{y}(t)$ are measured, estimated or observed to implement feedback. For some systems, performance improvement can be achieved using advanced control laws. For many systems, PID control laws may guarantee near-optimal performance ensuring hardware and software simplicity. While physical systems are nonlinear with inherent limits, all systems are controllable. Advanced design methods are used if strengthen specifications are imposed in MIMO systems. High-precision sensors, multi-objective nonlinear optimization, data acquisition and nonlinear state feedback algorithms are considered addressing stability, disturbance rejection, vibration and noise minimization, etc. Nonlinear control laws are used in high-accuracy pointing systems, multi-degree-of-freedom robots and manipulators, high-precision servos, sound and audio systems, etc. Proportional integral derivative controllers are implemented in advanced robots, aerial and ground vehicles, and other platforms.

Optimality, Performance Measures and Nonlinear Programming – Optimality is procured by performance measures, objective functions, metrics, etc. For $\mathbf{M}(\boldsymbol{x},\boldsymbol{y},\boldsymbol{u})$, select decision variables $(\boldsymbol{q},\boldsymbol{x},\boldsymbol{y},\boldsymbol{u})$, and, minimizes-maximizes a real-valued objective function $\boldsymbol{\Omega}(\boldsymbol{q},\boldsymbol{x},\boldsymbol{y},\boldsymbol{u})$

$$\min_{q\in Q,x\in X,y\in Y,u\in U}\max \Omega(q,x,y,u),\ (\boldsymbol{q},\boldsymbol{x},\boldsymbol{y},\boldsymbol{u})\in Q\times X\times Y\times U,$$

subject to constraints, governing and constitutive equations

$$g(\boldsymbol{q},\boldsymbol{x},\boldsymbol{y},\boldsymbol{u})\leq 0,\ h(\boldsymbol{q},\boldsymbol{x},\boldsymbol{y},\boldsymbol{u})=0,\ \dot{\boldsymbol{x}}=F(\boldsymbol{x},\boldsymbol{u}),\ \boldsymbol{y}=H(\boldsymbol{x}),$$

where $g(\cdot)$, $h(\cdot)$, $F(\cdot)$ and $H(\cdot)$ are the real-valued functions.

Search and Heuristic Problems, Exploratory Optimization, and, Pragmatic Control – *Model-free*, model-following, *learning-enabled* and other design premises have being investigated. These concepts are usually restricted to open-loop stable systems. A *model-free iterative feedback tuning* implies homogeneous measurements and iterative optimization of $\boldsymbol{\Omega}(\boldsymbol{q},\boldsymbol{x},\boldsymbol{y},\boldsymbol{u})$ by reconfiguring a control scheme $\boldsymbol{u}=\boldsymbol{C}(\boldsymbol{k},\boldsymbol{q},\boldsymbol{x},\boldsymbol{y})$ and parameters $\boldsymbol{k}\in\mathbb{R}$, asserting system functionality. Minimizing objective functions $\boldsymbol{\Omega}(\cdot)$, performance functionals or indexes $J(\cdot)$, the gradient descent, search, stochastic gradient descent and other iterative optimization algorithms may yield a minimum of a continuously differentiable function $\boldsymbol{\Omega}(\cdot)$. Using the *learning* rate γ, one obtains

$$\boldsymbol{u}=\boldsymbol{C}(\boldsymbol{k},\boldsymbol{q},\boldsymbol{x},\boldsymbol{y}),\ \boldsymbol{k}_{n+1}=\boldsymbol{k}_n-\gamma\left(\nabla_{q,x,y,u}\Omega(q,x,y,\boldsymbol{u}_{u=C(k,q,x,y)})\right).$$

Various challenges arise considering in-line system reconfiguration. One may apply metaheuristic and genetic optimization for a postulated system models, evolutionary algorithms and machine learning to solve optimization problems in high- and infinite-dimensional search spaces on a class of admissible controls $\boldsymbol{u}=\varphi(t,\boldsymbol{x})$. Axiomatized formulations, description inadequacy, complexity, data heterogeneities, sensitivity, errors, instabilities and other obscurities of search and heuristic algorithms may not guarantee multi-physics consistency at device and system levels. These algorithms do not admit closed-form unique solutions and result in algorithmic complexities, hardware impediments and sensitivity.

In industrial control, process control and SCADA systems, practitioners design and implement pragmatic data-driven control schemes. For open-loop stable or marginally-stable systems $\mathbb{M}(\boldsymbol{x},\boldsymbol{y},\boldsymbol{u})$, designers implement fixed-structure or reconfigurable analog and digital PID control laws $\boldsymbol{u}=\varphi(t,\boldsymbol{e})$, $\boldsymbol{e}(t)=[\boldsymbol{r}(t)-\boldsymbol{y}(t)]$.

The fixed analog and digital PID control laws are given by (5.23) and (5.41) as

$$u(t)=k_p e(t)+k_i\int e(t)dt+k_d\frac{de(t)}{dt},$$

$$u_n=u_{n-1}+k_{e0}e_n+k_{e1}e_{n-1}+k_{e2}e_{n-2},\ k_{e0}=k_p+T_sk_i+\frac{1}{T_s}k_d,\ k_{e1}=-k_p-\frac{2}{T_s}k_d,\ k_{e2}=\frac{1}{T_s}k_d$$

with constant-coefficient or variable feedback gains k_{Fj}. These robust, low-latency *minimal complexity* controllers are implemented by low-power analog electronics and MPUs, ensuring near-optimal performance of broadly commercialized systems. Designers minimize the use of mathematic-centric optimization and design solutions, adjusting preconfigured and reconfigurable controllers on operational simulators, experimental platforms or functioning systems. Examples include MDOF robots, aerial and ground vehicles, fifth and sixth generations of fighter aircraft, etc. Reconfiguration of preset control structures and gain scheduling are accomplished.

Dynamic Optimization and Control Problems – Solving design problems, consider necessary and sufficient conditions for optimality, stationary points (minima, maxima and saddle points), search spaces and dimensionality, Lagrange multipliers optimality, as well as not constrained and constrained optimization.

Physical system governance and dynamics $\mathbb{M}:=\{\boldsymbol{x}: \boldsymbol{x}\in X(X_0,U,D), \boldsymbol{y}=H(\boldsymbol{x}), \boldsymbol{u}=\varphi(t,\boldsymbol{x})\}$ may be described by the governing and constitutive equations

$$\dot{x}=F(x,u,d),\ \mathcal{M}:=\{\dot{x}=F(x,u,d),\ y=H(x),\ u=\varphi(t,x)\},\ \mathcal{M}\subseteq\mathbb{M},$$

which yield a model. For a model $\mathcal{M}(\cdot)$, $x\in X\subset\mathbb{R}^n$ is the state vector with states evolving in X, $u\in U\subset\mathbb{R}^m$ is the control vector, $y\in Y\subset\mathbb{R}^b$ is the output vector, $y=H(x)$, $d\in D\subset\mathbb{R}^v$ is the disturbance vector, and, $F(\cdot):\mathbb{R}^n\times\mathbb{R}^m\times\mathbb{R}^v\to\mathbb{R}^n$ and $H(\cdot):\mathbb{R}^n\to\mathbb{R}^{b\times n}$ are the nonlinear maps.

Using state-dependent quantities $\boldsymbol{q}$, outputs $\boldsymbol{y}$, states $\boldsymbol{x}$ and controls $\boldsymbol{u}$ for a system $\mathbb{M}(t,\boldsymbol{x},\boldsymbol{y},\boldsymbol{u},\boldsymbol{d})$, using a model $\mathcal{M}(t,x,y,u,d)$, $\mathcal{M}\subseteq\mathbb{M}$, one may specify eigenvalues, minimize a performance functional which represent specifications and requirements mapped by positive definite integrands $\omega(\cdot)$, or, apply Lyapunov criteria, in order to design a control law $u=\varphi(t,x)$. Control function $u=\varphi(t,x)$ is found to ensure optimality. This control is implemented as $\boldsymbol{u}=\varphi(t,\boldsymbol{x})$. For example, minimizing a physics-consistent functional

$$J=\min_{t,x,u}\int_{t_0}^{t_f}\omega(t,x,u)dt,\ \omega(t,x,u)>0,\ \boldsymbol{q}=\kappa(t,x,y,u)$$

subject to a model $\mathcal{M}(\cdot)$, one finds a state-dependent control function u.

Performance functionals and indexes define control laws and feedback structure. The input-output and states correspondence should be ensured for a physical system $\mathbb{M}(\cdot)$ and its model $\mathcal{M}(\cdot)$. High fidelity assertion of physical systems $\mathbb{M}(t,\boldsymbol{x},\boldsymbol{y},\boldsymbol{u},\boldsymbol{d})$ assumes direct measurements of $(\boldsymbol{x},\boldsymbol{y},\boldsymbol{u},\boldsymbol{d})$. We focus on analytic design, closed-form solution of optimization problems, and proven control schemes. All-states feedback and *minimal complexity* control laws can be designed minimizing design-specific functionals and indexes. To guaranteed analytic solutions and uniqueness, apply the calculus of variation, Lagrange multipliers, Hamiltonian optimization, as well as Lyapunov stability and optimality criteria. Sufficient fidelity of descriptive models for physical systems are ensured. Mathematical models are not unique, assume applied postulates and laws of physics, observability of exhibited phenomena, etc. Explicit dynamic-probabilistic models are found assuming that system states are measured or observed.

Objective Function and Performance Functionals – System states and variables are used to describe physical quantities. Mathematical optimization (mathematical programming), numeric optimization and other concepts are applied. Linear programming, convex optimization and nonlinear programming problems are solved.
A mathematical optimization problem is formulated as:

Minimize $\min_{x\in X} f(x)$ subject to $g_i(x)\le b_i$, $i=1,2,\ldots$, $X=\{x: g_i(x)\le b_i, \forall i\}$,

with designed objective function $f(\cdot):X\to\mathbb{R}$, and, constraints $g_i(x)\le b_i$, $g(\cdot):X\to\mathbb{R}$.

The problem is to find the minimum (or maximum) value of an objective function $f(\cdot)$ subject to $x\in X$ and constraints $g_i(x)\le b_i$.

The *admissible* region for x is $X=\{x: g_i(x)\le b_i, \forall i\}$.

Solving the minimization problem, one finds $x_0\in X$, such that $f(x_0)\le f(x)$ for all $x\in X$.

For a local minimum x^*, $\|x-x^*\|\le\varepsilon, f(x^*)\le f(x)$.

For the maximization problem, one finds $x_0\in X$, such that $f(x_0)\ge f(x)$ for all $x\in X$.

Example 5.1. *Minimization Problems*
Minimize $\min(x^2+y^2)$ subject to $(x+1)^3+y^2=0$, $(x,y)\in\mathbb{R}$.

The curve $(x+1)^3+y^2=0$ is a degenerate semi-cubic elliptic curve, opening away from the y-axis with a cusp at $(-1, 0)$. For a function $f(x,y)=x^2+y^2$, using the Lagrange multiplier λ, the Hamiltonian function H is

$$H(x,y,\lambda)=x^2+y^2-\lambda\left[(x+1)^3+y^2\right].$$

The derivatives are

$$\frac{\partial H}{\partial x}=2x-3\lambda(x+1)^2=0,\ \frac{\partial H}{\partial y}=2y-2\lambda y=0,\ \frac{\partial H}{\partial \lambda}=-(x+1)^3-y^2=0.$$

From the second expression $\frac{\partial H}{\partial y}=2y-2\lambda y=0$, $(\lambda-1)y=0$.

To satisfy the third and first relationships, $y\neq 0$. Hence, $\lambda=1$.

The first expression gives a quadratic algebraic equation in x as $2x-3(x+1)^2=0$.

Therefore, a solution is sough using the third relationship $(x+1)^3+y^2=0$, which yields a unique singular point at $(-1, 0)$.

This $(x, y)=(-1, 0)$ satisfies $(x+1)^3+y^2=0$, with $x^2+y^2=1$.

Hence the minimum of $f(x,y)=x^2+y^2$ is $f(-1)=1$. ■

Dynamic Optimization – Consider the stabilization problem. Apply physics-consistent and mathematically-adequate quadratic functionals, which admit analytic solutions [4, 5]

$$J=\min_{x,u}\int_{t_0}^{t_f}\left(\sum_{i=1}^{n}q_{ii}x_i^2+\sum_{j=1}^{m}g_{jj}u_j^2\right)dt,\ J=\min_{x,u}\int_{0}^{\infty}\left(\sum_{i=1}^{n}q_{ii}x_i^2+\sum_{j=1}^{m}g_{jj}u_j^2\right)dt, x\in\mathbb{R}^n,\ u\in\mathbb{R}^m,$$

where q_{ii} and g_{jj} are the positive definite weighting coefficients, $(q_{ii},g_{jj})\in\mathbb{R}$.

Admissible optimal control laws $u=\varphi(t,x)$, $u\subseteq \mathbf{C}_u$ are designed by asserting optimality by minimizing

$$J=\min_{t,u}\int_{t_0}^{t_f}1dt,\ J=\min_{t,x,u}\int_{t_0}^{t_f}\left(\sum_{i=1}^{n}q_{ii}(t)|x_i|+\sum_{j=1}^{m}g_{jj}(t)|u_j|\right)dt,$$

and other functionals. Analytic closed-form solutions are found in [6-14].

Weighting Coefficients – *Arbitrary units* and *scaling* are applied to define the weighting coefficients (q_{ii},g_{jj}). In a functional

$$J=\min_{x,u}\int_0^\infty\left(\sum_{i=1}^n q_{ii}x_i^2+\sum_{j=1}^m g_{jj}u_j^2\right)dt,\quad q_{ii}=\frac{1}{\tilde{x}_{i\,\max}^2},\quad g_{jj}=\frac{1}{\tilde{u}_{j\,\max}^2},$$

the weighting coefficients may be chosen using scaling applying the estimates on the variations of states and controls $(\tilde{x}_{i\,\max},\tilde{u}_{j\,\max})$.

Dynamic Optimization, Analytic Design and Admissible Control – Using the tracking error $e(t)=r(t)-y(t)$, the functional may be given as

$$J=\min_{t,e,x,u}\int_0^\infty\left(\sum_{l=1}^b q_{ll}(t)e_l^2+\sum_{i=1}^n q_{ii}(t)x_i^2+\sum_{j=1}^m g_{jj}(t)u_j^2\right)dt,\ e\in\mathbb{R}^b,\ x\in\mathbb{R}^n,\ u\in\mathbb{R}^m,$$

$$J=\min_{t,e,x,u}\int_0^\infty\left(\sum_{l=1}^b q_{ll}(t)|e_l|+\sum_{i=1}^n q_{ii}(t)|x_i|+\sum_{j=1}^m g_{jj}(t)|u_j|\right)dt.$$

Control laws with state feedback and proportional-integral-derivative feedback are analytically designed on an admissible controller class $\mathbf{C}_{u(\cdot)\in U}$, implementable feedback and admissible structures $\varphi(\cdot)$, such as $u=\varphi(t,x,e)$, $u\subseteq\mathbf{C}_u$. As performance quantities $\boldsymbol{q}$, use the control efforts, transients, tracking error, etc. The control efforts are assessed using the positive definite integrands $|u|$ or u^{2n}, n=1,2,3,.... The control rate is evaluated as $|du/dt|$. One applies physics-consistent and device-specific specifications. For example, analysis of current and torque ripples yields quantitative assessment of efficiency, heating, vibration, noise, wearing, failures, etc.

Let the performance requirements are: (1) Tracking error $e(t)$; (2) Settling time; (3) Heat losses $P_{\text{loss}}\equiv i^2+\omega^2$; (4) Control energy $E_u\equiv u^2$. One has $\boldsymbol{q}=(t, e, P_{\text{loss}}, E_u)$.

Various performance integrands can be applied. For example,

$$J=\min_{e,u}\int_0^\infty(e^2+u^2)dt,\quad J=\min_{t,e}\int_0^\infty t|e|dt,\quad J=\min_{t,e}\int_0^\infty a^{-t}|e|dt,\ a>1,$$

$$J=\min_{t,e,u}\int_0^\infty\left(a^{-t}|e|+e^2+\int\tanh^{-1}u\,du\right)dt,\quad J=\min_{t,x,e,u}\int_0^\infty(a^{-t}|e|+te^2+|e|+e^2+i^2+\omega^2+u^2)dt.$$

Dynamic Evolutions and Transient Dynamics – Consider the output transient dynamics and two evolution envelopes I and II for output $y(t)$. The output response is illustrated in Figure 5.4.a for the step reference $r(t)$=const. System dynamics is stable because output is bounded and converges to the steady-state value $y_{\text{steady-state}}$, e.g.,

$$\lim_{t\to\infty}y(t)=y_{\text{steady-state}},\quad \lim_{t\to\infty}e(t)\to 0.$$

Let an optimal design yields the best performance for $y(t)$ as shown in Figure 5.4.a within an *achievable* evolution envelope II. If transient dynamics, tracking error, settling time and overshoot specifications are not achieved, and, an evolution envelope I is required, alternative hardware solutions should be considered. There are limits on maximum torque, force, acceleration, power, voltage, current, etc. The designer may guarantee the best *achievable* performance. For example, high-torque density actuators, light-weight kinematics, high-efficiency controller-drivers and other solutions should be used. Hardware physical limits may not be surpassed and overcome by software and algorithms. The designers are aware on physical limits, device specificity, technological constraints and other factors.

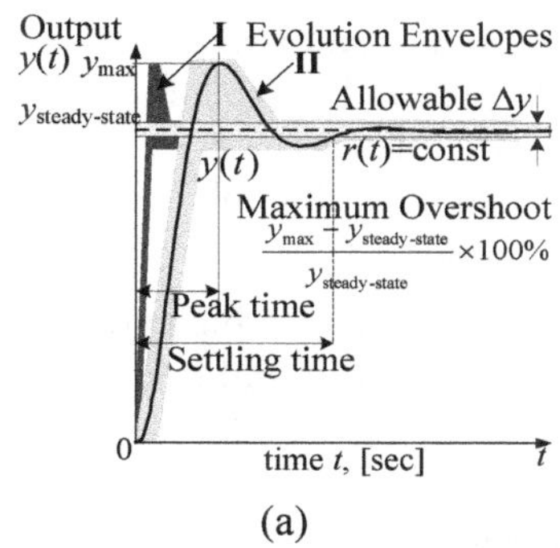

(a)

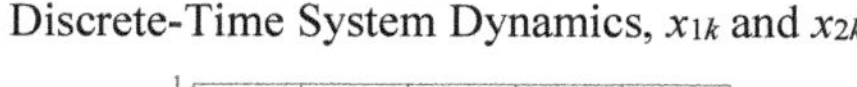

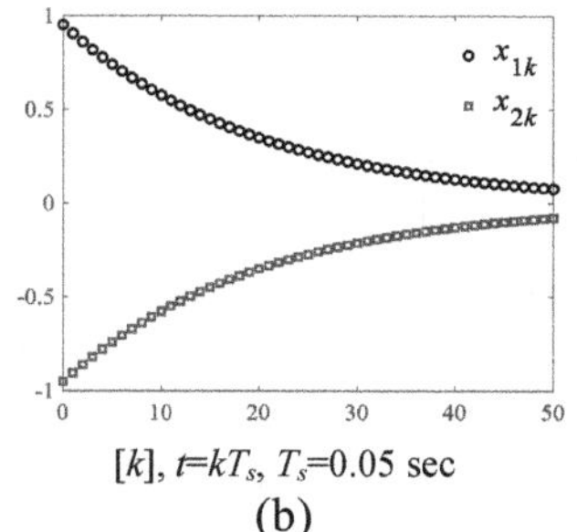

[k], $t=kT_s$, T_s=0.05 sec

(b)

Figure 5.4. (a) Output transient response and evolution envelopes;
(b) Second-order discrete-time system dynamics with initial conditions, Example 5.7.

Settling Time, Overshoot and Error – The settling time is time needed for the system output $y(t)$ to reach and stay within the steady-state value $y_{\text{steady-state}}$. The allowable difference between $y(t)$ and $y_{\text{steady-state}}$ is used to specify the settling time. Specifications on the difference Δy may vary from 5% to 0.1% or less. In pointing systems the required repositioning accuracy can be milliradians. The settling time is the minimum time after which the system response remains within the specified error, considering the steady-state value $y_{\text{steady-state}}$ and command $r(t)$. As depicted in Figure 5.4.a for $r(t)$=const, the peak time is the time required for the output $y(t)$ to reach the first peak of the overshoot. The overshoot is computed as $\text{overshoot} = \dfrac{y_{\max} - y_{\text{steady-state}}}{y_{\text{steady-state}}} \times 100\%$.

5.2. System Analysis: State-Space Equations and Transfer Functions

5.2.1. Continuous-Time Systems

Physical processes and systems are described by nonlinear governing and constitutive equations. One may linearize nonlinear equations yielding linear differential and difference equations, and, find transfer functions $G(s)$ and $G(z)$ in the s and z domains. The Laplace operator $s=d/dt$ and Laplace transforms are used for continuous-time systems. In state-space form, transient dynamics of linear systems with n states $x \in \mathbb{R}^n$ and m controls $u \in \mathbb{R}^m$ is described by a set of n linear first-order differential equations

$$\frac{dx_1}{dt} = a_{11}x_1 + a_{12}x_2 + \ldots + a_{1\,n-1}x_{n-1} + a_{1\,n}x_n + b_{11}u_1 + b_{12}u_2 + \ldots + b_{1\,m-1}u_{m-1} + b_{1\,m}u_m, x_1(t_0) = x_{10}, \quad (5.2)$$

$$\vdots$$

$$\frac{dx_n}{dt} = a_{n\,1}x_1 + a_{n\,2}x_2 + \ldots + a_{n\,n-1}x_{n-1} + a_{n\,n}x_n + b_{n\,1}u_1 + b_{n\,2}u_2 + \ldots + b_{n\,m-1}u_{m-1} + b_{n\,m}u_m,\ x_n(t_0) = x_{n0}.$$

In matrix form

$$\frac{dx}{dt} = Ax + Bu,\quad x(t_0) = x_0, \qquad (5.3)$$

$$\frac{dx}{dt} = \begin{bmatrix} \frac{dx_1}{dt} \\ \frac{dx_2}{dt} \\ \vdots \\ \frac{dx_{n-1}}{dt} \\ \frac{dx_n}{dt} \end{bmatrix} = \begin{bmatrix} a_{11} & a_{12} & \cdots & a_{1\,n-1} & a_{1\,n} \\ a_{21} & a_{22} & \cdots & a_{2\,n-1} & a_{2\,n} \\ \vdots & \vdots & \ddots & \vdots & \vdots \\ a_{n-1\,1} & a_{n-1\,2} & \cdots & a_{n-1\,n-1} & a_{n-1\,n} \\ a_{n\,1} & a_{n\,2} & \cdots & a_{n\,n-1} & a_{n\,n} \end{bmatrix} \begin{bmatrix} x_1 \\ x_2 \\ \vdots \\ x_{n-1} \\ x_n \end{bmatrix} + \begin{bmatrix} b_{11} & b_{12} & \cdots & b_{1\,m-1} & b_{1\,m} \\ b_{21} & b_{22} & \cdots & b_{2\,m-1} & b_{2\,m} \\ \vdots & \vdots & \ddots & \vdots & \vdots \\ b_{n-1\,1} & b_{n-1\,2} & \cdots & b_{n-1\,m-1} & b_{n-1\,m} \\ b_{n\,1} & b_{n\,2} & \cdots & b_{n\,m-1} & b_{n\,m} \end{bmatrix} \begin{bmatrix} u_1 \\ u_2 \\ \vdots \\ u_{m-1} \\ u_m \end{bmatrix}.$$

If system parameters are constant, matrices $A\in\mathbb{R}^{n\times n}$ and $B\in\mathbb{R}^{n\times m}$ are constant-coefficient. For an $n\times n$ matrix A, the characteristic polynomial $p_A(s)$ is found using the determinant, $p_A(s)=\det(sI-A)$, where $I\in\mathbb{R}^{n\times n}$ is the identity matrix.
The characteristic equation is found by equating the characteristic polynomial to zero

$$\det(sI-A)=0,\ |sI-A|=0,\ a_n s^n + a_{n-1}s^{n-1} + \ldots + a_1 s + a_0 = 0\,. \qquad (5.4)$$

The characteristic equation (5.4) yields the eigenvalues λ_i which are also called the characteristic roots or characteristic poles s_i, $\lambda_i\equiv s_i$. System is stable if real parts of all eigenvalues s_i are negative, $\forall\mathrm{Re}(s_i)<0$, $\lambda_i\equiv s_i$. Stability and dynamic analyses using eigenvalues are valid only for linear systems.

A transfer function can be found using the state-space equations applying the Laplace transform. Consider a linear time-invariant system (5.3) in the time and s domains

$$\frac{dx}{dt} = Ax + Bu\,,\ y=Hx,\ t\geq 0, \qquad (5.5)$$

$$sX(s) - x(t_0) = AX(s) + BU(s)\,,\ Y(s) = HX(s)\,,\ X(s) = (sI-A)^{-1}\left[x_0 + BU(s)\right],$$

where $y\in\mathbb{R}^b$ is the output vector, and, the output equation is $y=Hx$; $H\in\mathbb{R}^{b\times n}$ is the constant coefficients matrix; $(sI-A)\in\mathbb{R}^{n\times n}$ is the invertible matrix, $\det(sI-A)\neq 0$.

If the initial conditions are zero, $X(s) = (sI-A)^{-1}BU(s)$.

Using the system output $y(t)=Hx(t)$, $Y(s) = HX(s) = H(sI-A)^{-1}BU(s)$.

The transfer function is

$$G(s) = \frac{Y(s)}{U(s)} = H(sI-A)^{-1}B = H\Phi(s)B\,, \qquad (5.6)$$

where $\Phi(s)$ is the state transition matrix, $\Phi(s)=(sI-A)^{-1}$.

The state transition matrix in time domain $\Phi(t)$ is

$$\Phi(t) = e^{At},\ \Phi(t) = \mathcal{L}^{-1}\{\Phi(s)\} = \mathcal{L}^{-1}\left\{(sI-A)^{-1}\right\}. \qquad (5.7)$$

A unique solution of $\frac{dx}{dt} = Ax$ with initial conditions x_0 is $x(t) = \Phi(t)x_0 = e^{At}x_0$.

One has $\frac{dx}{dt} = \frac{d}{dt}(e^{At}x_0) = Ae^{At}x_0 = Ax$.

For the n-order differential equation, apply the Laplace transform to both sides assuming the initial conditions are zero, yielding

$$\sum_{i=0}^{n} a_i \frac{d^i y(t)}{dt^i} = \sum_{i=0}^{m} b_i \frac{d^i u(t)}{dt^i}\,,\ \left(\sum_{i=0}^{n} a_i s^i\right)Y(s) = \left(\sum_{i=0}^{m} b_i s^i\right)U(s)\,.$$

Hence, a transfer function is

$$G(s) = \frac{Y(s)}{U(s)} = \frac{b_m s^m + b_{m-1}s^{m-1} + \ldots + b_1 s + b_0}{a_n s^n + a_{n-1}s^{n-1} + \ldots + a_1 s + a_0}\,. \qquad (5.8)$$

Stability and Characteristic Equation – Stability of linear time-invariant systems is guaranteed if all characteristic eigenvalues, obtained by solving the characteristic equation (5.4) $p_A(s)=0$, $a_n s^n + a_{n-1}s^{n-1} + \ldots + a_1 s + a_0 = 0$, have negative real parts, $\forall\mathrm{Re}(s_i)<0$. One may perform partial fractioning of (5.8) to find time-domain system evolution.

Laplace and z Transforms – Table 5.1 documents the Laplace and z transforms.

The bilateral two-sided Laplace and z transforms are

$$X(s) = \mathcal{L}\{x(t)\} = \int_{-\infty}^{\infty} x(t)e^{-st}dt \text{ and } X(z) = \mathcal{Z}\{x[k]\} = \sum_{k=-\infty}^{\infty} x[k]e^{-k}\,.$$

The z and s operators are related as $z = e^{sT_s}$, where T_s is the sampling time.

Table 5.1. Laplace and z-transform Table

Laplace Transform $X(s)$	Time-Domain Signal $x(t)$	Discrete Time-Domain Signal $x(kT_s)$	z-transform $X(z)$
Bilateral Laplace transform $X(s) = \mathcal{L}\{x(t)\} = \int_{-\infty}^{\infty} x(t)e^{-st}dt$ Inverse Laplace transform $x(t) = \mathcal{L}^{-1}\{X(s)\}(t) = \frac{1}{2\pi j}\int_{\sigma-j\infty}^{\sigma+j\infty} X(s)e^{st}ds$		Bilateral z-transform $X(z) = \mathcal{Z}\{x[k]\} = \sum_{k=-\infty}^{\infty} x[k]e^{-k}$ Inverse z-transform $x[k] = \mathcal{Z}^{-1}\{X(z)\} = \frac{1}{2\pi j}\oint_C X(z)z^{k-1}dz$	
$\frac{1}{s}$	Unit step $1(t)$	$1(kT_s)$	$\frac{1}{1-z^{-1}} = \frac{z}{z-1}$
$\frac{1}{s^2}$	Ramp function $t1(t)$	$kT_s1(kT_s)$	$\frac{T_s z^{-1}}{(1-z^{-1})^2} = \frac{T_s z}{(z-1)^2}$
$\frac{2}{s^3}$	$t^2 1(t)$	$(kT_s)^2 1(kT_s)$	$\frac{T_s^2 z^{-1}(1+z^{-1})}{(1-z^{-1})^3}$
$\frac{6}{s^4}$	$t^3 1(t)$	$(kT_s)^3 1(kT_s)$	$\frac{T_s^3 z^{-1}(1+4z^{-1}+z^{-2})}{(1-z^{-1})^4}$
$\frac{24}{s^5}$	$t^4 1(t)$	$(kT_s)^4 1(kT_s)$	$\frac{T_s^4 z^{-1}(1+11z^{-1}+11z^{-2}+z^{-3})}{(1-z^{-1})^5}$
$\frac{1}{s+a}$	$e^{-at}1(t)$	$e^{-akT_s}1(kT_s)$	$\frac{1}{1-e^{-aT_s}z^{-1}}$
$\frac{a}{s(s+a)}$	$(1-e^{-at})1(t)$	$(1-e^{-akT_s})1(kT_s)$	$\frac{(1-e^{-aT_s})z^{-1}}{(1-z^{-1})(1-e^{-aT_s}z^{-1})}$
$\frac{b-a}{(s+a)(s+b)}$	$(e^{-at}-e^{-bt})1(t)$	$(e^{-akT_s}-e^{-bkT_s})1(kT_s)$	$\frac{(e^{-aT_s}-e^{-bT_s})z^{-1}}{(1-e^{-aT_s}z^{-1})(1-e^{-bT_s}z^{-1})}$
$\frac{1}{(s+a)^2}$	$te^{-at}1(t)$	$kT_s e^{-akT_s}1(kT_s)$	$\frac{T_s e^{-aT_s}z^{-1}}{(1-e^{-aT_s}z^{-1})^2}$
$\frac{s}{(s+a)^2}$	$(1-at)e^{-at}1(t)$	$(1-akT_s)e^{-akT_s}1(kT_s)$	$\frac{1-(1+aT_s)e^{-aT_s}z^{-1}}{(1-e^{-aT_s}z^{-1})^2}$
$\frac{\omega_0}{s^2+\omega_0^2}$	$\sin(\omega_0 t)1(t)$	$\sin(\omega_0 kT_s)1(kT_s)$	$\frac{z^{-1}\sin(\omega_0 T_s)}{1-2z^{-1}\cos(\omega_0 T_s)+z^{-2}}$
$\frac{s}{s^2+\omega_0^2}$	$\cos(\omega_0 t)1(t)$	$\cos(\omega_0 kT_s)1(kT_s)$	$\frac{1-z^{-1}\cos(\omega_0 T_s)}{1-2z^{-1}\cos(\omega_0 T_s)+z^{-2}}$
$\frac{\omega_0}{(s+a)^2+\omega_0^2}$	$e^{-at}\sin(\omega_0 t)1(t)$	$e^{-akT_s}\sin(\omega_0 kT_s)1(kT_s)$	$\frac{e^{-akT_s}z^{-1}\sin(\omega_0 T_s)}{1-2e^{-aT_s}z^{-1}\cos(\omega_0 T_s)+e^{-2aT_s}z^{-2}}$
$\frac{s+a}{(s+a)^2+\omega_0^2}$	$e^{-at}\cos(\omega_0 t)1(t)$	$e^{-akT_s}\cos(\omega_0 kT_s)1(kT_s)$	$\frac{1-e^{-aT_s}z^{-1}\cos(\omega_0 T_s)}{1-2e^{-aT_s}z^{-1}\cos(\omega_0 T_s)+e^{-2aT_s}z^{-2}}$

State Transition Matrix, Matrix Exponential and System Dynamics – Study dynamics of states $x(t)$ and outputs $y(t)$. For linear systems (5.5) $\frac{dx}{dt} = Ax + Bu$, $y=Hx$, $t\geq 0$, $t_0=0$, using the state transition matrix (5.7) $\Phi(t)=e^{At}$, a solution for the state vector $x(t)$ due to initial conditions x_0 and input $u(t)$ is

$$x(t) = \underbrace{\Phi(t)x_0}_{\text{Solution Due to Initial Conditions } x_0} + \underbrace{\int_0^t \Phi(t-\tau)Bu(\tau)d\tau}_{\text{Solution Due to Input } u(t)}, \; x(t) = e^{At}x_0 + \int_0^t e^{A(t-\tau)}Bu(\tau)d\tau, \; \Phi(t)=e^{At}, \quad (5.9)$$

$$y(t) = Hx = \underbrace{H\Phi(t)x_0}_{\text{Solution Due to Initial Conditions } x_0} + \underbrace{H\int_0^t \Phi(t-\tau)Bu(\tau)d\tau}_{\text{Solution Due to Input } u(t)}, \; y(t) = He^{At}x_0 + H\int_0^t e^{A(t-\tau)}Bu(\tau)d\tau, \; t\geq 0.$$

Analyzing states and outputs evolutions due to input $u(t)$, recall that for a continuous function f with an antiderivative F, $\int_a^b f(x)dx = F(b) - F(a)$.

Matrix Exponential – For a constant-coefficient square matrix $A\in\mathbb{R}^{n\times n}$, the state transition matrix and matrix exponential e^{At} satisfy

$$\frac{d}{dt}e^{At} = Ae^{At} = e^{At}A, \; \det(e^{At})\neq 0, \; e^{At}e^{-At}=e^{-At}e^{At}=e^{0}=I, \; [e^{At}]^{-1} = e^{-At}, \; \mathcal{L}\{e^{At}\} = (\lambda I - A)^{-1}.$$

A unique solution of $\frac{dx}{dt} = Ax$ with initial conditions x_0 is given as $x(t) = \Phi(t)x_0 = e^{At}x_0$.

We have $\frac{dx}{dt} = \frac{d}{dt}(e^{At}x_0) = Ae^{At}x_0 = Ax$, $x(t_0) = e^{A(t_0-t_0)}x_0 = Ix_0 = x_0$.

Deriving State Transition Matrix – The state transition matrix $\Phi(t)$ is computed by applying various approaches, such as:

1. Cayley-Hamilton theorem.
   Using eigenvalues $\boldsymbol{\lambda}$ and mappings $\beta(t)$ of $A\in\mathbb{R}^{n\times n}$, characterized by a characteristic polynomial $p_A(s)=\det(\lambda I-A)$ and characteristic equation $|\lambda I-A|=0$ which yields $\boldsymbol{\lambda}$,

$$\Phi(t) = e^{At} = \sum_{i=0}^{n-1}\beta_i(t)A^i, \quad \begin{bmatrix}\beta_0(t)\\ \beta_1(t)\\ \vdots\\ \beta_{n-1}(t)\end{bmatrix} = \begin{bmatrix}1 & \lambda_1 & \lambda_1^2 & \cdots & \lambda_1^{n-1}\\ 1 & \lambda_2 & \lambda_2^2 & \cdots & \lambda_2^{n-1}\\ \vdots & \vdots & \vdots & \ddots & \vdots\\ 1 & \lambda_n & \lambda_n^2 & \cdots & \lambda_n^{n-1}\end{bmatrix}^{-1} \begin{bmatrix}e^{\lambda_1 t}\\ e^{\lambda_2 t}\\ \vdots\\ e^{\lambda_n t}\end{bmatrix}; \quad (5.10)$$

2. Sylvester expansion (matrix theorem).
   Using eigenvalues $\boldsymbol{\lambda}$ and constituent matrices C_i

$$\Phi(t) = e^{At} = \sum_{i=1}^{n} e^{\lambda_i t}C_i, \; C_i = \frac{\prod_{j=1, j\neq i}^{n}\left[A-\lambda_j I\right]}{\prod_{j=1, j\neq i}^{n}\left(\lambda_i - \lambda_j\right)}, \; C_iC_j=0, \; i\neq j; \quad (5.11)$$

3. Eigenvalue-eigenvector pair $(\boldsymbol{\lambda},\boldsymbol{v})$ and *modal* matrix $\mathbf{M}\in\mathbb{R}^{n\times n}$,

$$\Phi(t) = e^{At} = \mathbf{M}e^{\Lambda t}\mathbf{M}^{-1}, \; e^{\Lambda t} = \begin{bmatrix}e^{\lambda_1 t} & \cdots & 0\\ \vdots & \ddots & \vdots\\ 0 & \cdots & e^{\lambda_n t}\end{bmatrix}, \; \boldsymbol{\Lambda} = \begin{bmatrix}\lambda_1 & \cdots & 0\\ \vdots & \ddots & \vdots\\ 0 & \cdots & \lambda_n\end{bmatrix}, \quad (5.12)$$

   $\mathbf{M}=[\boldsymbol{v}_1,\ldots,\boldsymbol{v}_n]$, $(\lambda_i I-A)\boldsymbol{v}_i=0$, $\mathbf{M}^{-1}A\mathbf{M}=\boldsymbol{\Lambda}$ and $\mathbf{M}\boldsymbol{\Lambda}\mathbf{M}^{-1}=A$;

4. Maclaurin series

$$\Phi(t)=e^{At}=\sum_{i=0}^{\infty}\frac{1}{i!}A^i t^i=I+At+\tfrac{1}{2!}A^2t^2+\tfrac{1}{3!}A^3t^3+\cdots; \tag{5.13}$$

5. Laplace transform (5.7) $\Phi(t)=\mathcal{L}^{-1}\{\Phi(s)\}=\mathcal{L}^{-1}\{(sI-A)^{-1}\}$, $\Phi(s)=(sI-A)^{-1}$.

Eigenvalues and Eigenvectors – The characteristic polynomial of a square $n\times n$ matrix A is an nth degree polynomial $p_A(\lambda)$, and, n eigenvalues are found from a characteristic equation $p_A(\lambda)=\det(\lambda I-A)=0$, $a_n\lambda^n+a_{n-1}\lambda^{n-1}+...+a_1\lambda+a_0=0$. A set of eigenvalues is called the *spectrum* of A. For each eigenvalue λ_i, there exists a subspace of vectors S_λ which satisfies an equation $A\boldsymbol{v}_i=\lambda_i\boldsymbol{v}_i$, $\boldsymbol{v}\in S_\lambda$, and, $(\lambda_i I-A)\boldsymbol{v}_i=0$.

Eigenspace, Eigenvectors and Modal Matrix – This *eigenspace* S_λ is unique, but the eigenvectors are not unique. *Eigenspaces* of different eigenvalues are independent, and, the eigenvectors of different eigenvalues are linearly independent. When a matrix is diagonalizable, the direct sum of *eigenspaces* is a vector space.

There exists a basis of eigenvectors. A *modal* matrix **M** is formed from a basis of eigenvectors as columns, $\mathbf{M}=[\boldsymbol{v}_1,\ldots,\boldsymbol{v}_n]$, where $\boldsymbol{v}_i$ are found solving $(\lambda_i I-A)\boldsymbol{v}_i=0$. For a given matrix $A\in\mathbb{R}^{n\times n}$, one finds n eigenvalues λ_i, $i=1,\ldots,n$. Solution $(\lambda_i I-A)\boldsymbol{v}_i=0$ yields n eigenvectors $\boldsymbol{v}_i$. Eigenvectors of an $n\times n$ matrix $A\in\mathbb{R}^{n\times n}$ are $\boldsymbol{v}_i$ in $\mathbb{R}^n$ such that $A\boldsymbol{v}_i=\lambda_i\boldsymbol{v}_i$.

Equation $A\boldsymbol{v}_i=\lambda_i\boldsymbol{v}_i$ has a nontrivial solution. If $A\boldsymbol{v}_i=\lambda_i\boldsymbol{v}_i$ for $\lambda_i\neq 0$, λ_i is the eigenvalue for $\boldsymbol{v}_i$, and, $\boldsymbol{v}_i$ is an eigenvector for λ_i. In Examples 5.2 to 5.5, the eigenvalues, matrix exponential e^{At}, eigenvectors, *modal* matrix **M** and other mathematical operators are found to analyze stability, solution of linear differential equations, transients, etc.

Cayley-Hamilton Theorem – For an $n\times n$ matrix $A\in\mathbb{R}^{n\times n}$, a characteristic polynomial (5.4) $p_A(\lambda)=\det(\lambda I-A)$ yields the characteristic equation $\det(\lambda I-A)=0$, $a_n\lambda^n+a_{n-1}\lambda^{n-1}+...+a_1\lambda+a_0=0$. A matrix exponential e^{At} is found as (5.10)

$$e^{At}=f(A)=\sum_{i=0}^{n-1}\beta_i(t)A^i,\quad \begin{bmatrix}\beta_0(t)\\ \beta_1(t)\\ \vdots\\ \beta_{n-1}(t)\end{bmatrix}=\begin{bmatrix}1 & \lambda_1 & \lambda_1^2 & \cdots & \lambda_1^{n-1}\\ 1 & \lambda_2 & \lambda_2^2 & \cdots & \lambda_2^{n-1}\\ \vdots & \vdots & \vdots & \ddots & \vdots\\ 1 & \lambda_n & \lambda_n^2 & \cdots & \lambda_n^{n-1}\end{bmatrix}^{-1}\begin{bmatrix}e^{\lambda_1 t}\\ e^{\lambda_2 t}\\ \vdots\\ e^{\lambda_n t}\end{bmatrix}.$$

The characteristic eigenvalues λ_i yield $\beta_i(t)$. ■

Inverse of a Square Matrix – Consider an $n\times n$ matrix $P\in\mathbb{R}^{n\times n}$, $P=\begin{bmatrix}p_{11} & \cdots & p_{1n}\\ \vdots & \ddots & \vdots\\ p_{n1} & \cdots & p_{nn}\end{bmatrix}$.

An inverse $n\times n$ matrix P^{-1} exists if P is nonsingular, $\det(P)\neq 0$. The Laplace expansion defines the determinant of matrix P in terms of determinants of minors M_{ij}. Using the scalar determinant $\det(P)$ and adjugate (adjoint) matrix $\operatorname{adj}(P)$, we have

$$P^{-1}=\frac{1}{\det(P)}\operatorname{adj}(P),\ \det(P)=\sum_{j=1}^{n}(-1)^{i+j}p_{ij}M_{ij},\ \operatorname{adj}(P)=\begin{bmatrix}c_{11} & \cdots & c_{1n}\\ \vdots & \ddots & \vdots\\ c_{n1} & \cdots & c_{nn}\end{bmatrix}^T,\ c_{ij}=(-1)^{i+j}M_{ij},$$

where M_{ij} is the minor, which corresponds to p_{ij}, and, defined as the determinant of the $(n-1)\times(n-1)$ matrix, obtained by eliminating the ith row and jth column; c_{ij} are the cofactors.

The adjugate matrix adj(P) is the transpose of the matrix of cofactors. One has $PP^{-1}=I$, $P^{-1}P=I$, $(P^{-1})^{-1}=P$, $(kP)^{-1}=k^{-1}P^{-1}$, $\det(P^{-1})=(\det P)^{-1}$, $\det(P^{-1})(\det P)=1$, $(P^T)^{-1}=(P^{-1})^T$, $(PQ)^{-1}=Q^{-1}P^{-1}$.

For $P=\begin{bmatrix} p_{11} & p_{12} \\ p_{21} & p_{22} \end{bmatrix}$, $P^{-1}=\begin{bmatrix} p_{11} & p_{12} \\ p_{21} & p_{22} \end{bmatrix}^{-1}=\dfrac{1}{p_{11}p_{22}-p_{12}p_{21}}\begin{bmatrix} p_{22} & -p_{12} \\ -p_{21} & p_{11} \end{bmatrix}$. ■

Example 5.2. *Second-Order System: State Transition Matrix and Matrix Exponential*

Consider a system

$$\frac{dx}{dt}=Ax,\quad \begin{bmatrix} \frac{dx_1}{dt} \\ \frac{dx_2}{dt} \end{bmatrix}=\begin{bmatrix} 1 & 2 \\ 5 & 4 \end{bmatrix}\begin{bmatrix} x_1 \\ x_2 \end{bmatrix},\ A=\begin{bmatrix} 1 & 2 \\ 5 & 4 \end{bmatrix},\ x_0=\begin{bmatrix} x_{10} \\ x_{20} \end{bmatrix}=\begin{bmatrix} 1 \\ -1 \end{bmatrix}.$$

The characteristic equation (5.4) $\det(\lambda I-A)=0$, $|\lambda I-A|=0$ yields

$$\left|\lambda\begin{bmatrix} 1 & 0 \\ 0 & 1 \end{bmatrix}-\begin{bmatrix} 1 & 2 \\ 5 & 4 \end{bmatrix}\right|=\left|\begin{bmatrix} \lambda-1 & -2 \\ -5 & \lambda-4 \end{bmatrix}\right|=\lambda^2-5\lambda-6=0\cdot$$

One computes two eigenvalues $\lambda_1=-1$ and $\lambda_2=6$.

Mappings $\beta_0(t)$ and $\beta_1(t)$ are found using (5.10) as

$$\begin{bmatrix} \beta_0(t) \\ \beta_1(t) \end{bmatrix}=\begin{bmatrix} 1 & \lambda_1 \\ 1 & \lambda_2 \end{bmatrix}^{-1}\begin{bmatrix} e^{\lambda_1 t} \\ e^{\lambda_2 t} \end{bmatrix}=\begin{bmatrix} 1 & -1 \\ 1 & 6 \end{bmatrix}^{-1}\begin{bmatrix} e^{-t} \\ e^{6t} \end{bmatrix}=\begin{bmatrix} \frac{6}{7}e^{-t}+\frac{1}{7}e^{6t} \\ -\frac{1}{7}e^{-t}+\frac{1}{7}e^{6t} \end{bmatrix}.$$

Using the Cayley-Hamilton theorem (5.10), the state transition matrix is

$$\Phi(t)=e^{At}=\sum_{i=0}^{n-1}\beta_i(t)\,A^i=\beta_0(t)I+\beta_1(t)A$$

$$=\left[\tfrac{6}{7}e^{-t}+\tfrac{1}{7}e^{6t}\right]\begin{bmatrix} 1 & 0 \\ 0 & 1 \end{bmatrix}+\left[-\tfrac{1}{7}e^{-t}+\tfrac{1}{7}e^{6t}\right]\begin{bmatrix} 1 & 2 \\ 5 & 4 \end{bmatrix}=\begin{bmatrix} \frac{5}{7}e^{-t}+\frac{2}{7}e^{6t} & -\frac{2}{7}e^{-t}+\frac{2}{7}e^{6t} \\ -\frac{5}{7}e^{-t}+\frac{5}{7}e^{6t} & \frac{2}{7}e^{-t}+\frac{5}{7}e^{6t} \end{bmatrix}.$$

Using the state transition matrix $\Phi(t)=e^{At}$, a system evolution due to initial conditions $x_0=[1,-1]$ is

$$x(t)=\Phi(t)x_0=e^{At}x_0=\begin{bmatrix} \frac{5}{7}e^{-t}+\frac{2}{7}e^{6t} & -\frac{2}{7}e^{-t}+\frac{2}{7}e^{6t} \\ -\frac{5}{7}e^{-t}+\frac{5}{7}e^{6t} & \frac{2}{7}e^{-t}+\frac{5}{7}e^{6t} \end{bmatrix}\begin{bmatrix} 1 \\ -1 \end{bmatrix}=\begin{bmatrix} e^{-t} \\ -e^{-t} \end{bmatrix}.$$

The MATLAB code to solve the aforementioned problems is

```
A=[1 2; 5 4]; E=eig(A); t=sym('t'); eAt=expm(A*t)
x0=[1; -1]; xt=eAt*x0
```

To use the Sylvester expansion (5.11), $\Phi(t)=e^{At}=\sum_{i=1}^{n}e^{\lambda_i t}C_i$, $C_i=\dfrac{\prod_{j=1,j\neq i}^{n}[A-\lambda_j I]}{\prod_{j=1,j\neq i}^{n}[\lambda_i-\lambda_j]}$,

compute eigenvalues. For A and computed eigenvalues $\lambda_1=-1$, $\lambda_2=6$, we have

$$C_1=\frac{\prod_{j=1,j\neq i}^{2}[A-\lambda_j I]}{\prod_{j=1,j\neq i}^{2}[\lambda_1-\lambda_j]}=\frac{A-\lambda_2 I}{\lambda_1-\lambda_2}=\frac{\begin{bmatrix} 1 & 2 \\ 5 & 4 \end{bmatrix}-6\begin{bmatrix} 1 & 0 \\ 0 & 1 \end{bmatrix}}{-7}=\begin{bmatrix} \frac{5}{7} & -\frac{2}{7} \\ -\frac{5}{7} & \frac{2}{7} \end{bmatrix},$$

$$C_2 = \frac{\prod_{j=1,j\neq i}^{2}[A-\lambda_j I]}{\prod_{j=1,j\neq i}^{2}[\lambda_2-\lambda_j]} = \frac{A-\lambda_1 I}{\lambda_2-\lambda_1} = \frac{\begin{bmatrix}1 & 2\\ 5 & 4\end{bmatrix}+\begin{bmatrix}1 & 0\\ 0 & 1\end{bmatrix}}{7} = \begin{bmatrix}\frac{2}{7} & \frac{2}{7}\\ \frac{5}{7} & \frac{5}{7}\end{bmatrix}.$$

Hence, one finds

$$\Phi(t) = e^{At} = \sum_{i=1}^{n} e^{\lambda_i t}C_i = e^{\lambda_1 t}C_1 + e^{\lambda_2 t}C_2 = e^{-t}\begin{bmatrix}\frac{5}{7} & -\frac{2}{7}\\ -\frac{5}{7} & \frac{2}{7}\end{bmatrix} + e^{6t}\begin{bmatrix}\frac{2}{7} & \frac{2}{7}\\ \frac{5}{7} & \frac{5}{7}\end{bmatrix} = \begin{bmatrix}\frac{5}{7}e^{-t}+\frac{2}{7}e^{6t} & -\frac{2}{7}e^{-t}+\frac{2}{7}e^{6t}\\ -\frac{5}{7}e^{-t}+\frac{5}{7}e^{6t} & \frac{2}{7}e^{-t}+\frac{5}{7}e^{6t}\end{bmatrix}.$$

■

Example 5.3. *Linear System With Complex Eigenvalues*

For differential equations

$$\frac{dx}{dt} = Ax, \quad \begin{bmatrix}\frac{dx_1}{dt}\\ \frac{dx_2}{dt}\end{bmatrix} = \begin{bmatrix}-1 & -1\\ 1 & -1\end{bmatrix}\begin{bmatrix}x_1\\ x_2\end{bmatrix}, \; A = \begin{bmatrix}-1 & -1\\ 1 & -1\end{bmatrix},$$

the solution is $x(t) = e^{At}x_0$.

For $A = \begin{bmatrix}-1 & -1\\ 1 & -1\end{bmatrix}$, the characteristic equation (5.4) is $\det(\lambda I - A)=0$, $(\lambda+1)^2+1=0$.

Hence, $\lambda_1 = -1+i$, $\lambda_2 = -1-i$.

From (5.10), using the Euler identity $\frac{1}{2}(e^{it}+e^{-it}) = \cos t$, $\frac{1}{2}i(e^{-it}-e^{it}) = \sin t$,

$$\begin{bmatrix}\beta_0(t)\\ \beta_1(t)\end{bmatrix} = \begin{bmatrix}1 & -1+i\\ 1 & -1-i\end{bmatrix}^{-1}\begin{bmatrix}e^{(-1+i)t}\\ e^{(-1-i)t}\end{bmatrix} = \begin{bmatrix}\frac{1}{2}(1-i) & \frac{1}{2}(1+i)\\ -\frac{1}{2}i & \frac{1}{2}i\end{bmatrix}\begin{bmatrix}e^{(-1+i)t}\\ e^{(-1-i)t}\end{bmatrix}$$

$$= \begin{bmatrix}\frac{1}{2}(e^{(-1+i)t} - ie^{(-1+i)t}) + \frac{1}{2}(e^{(-1-i)t} + ie^{(-1-i)t})\\ \frac{1}{2}i(-e^{(-1+i)t} + e^{(-1-i)t})\end{bmatrix} = \begin{bmatrix}e^{-t}\cos t + e^{-t}\sin t\\ e^{-t}\sin t\end{bmatrix}.$$

With derived $\beta_0(t) = e^{-t}\cos t + e^{-t}\sin t$ and $\beta_1(t) = e^{-t}\sin t$, one has

$$\Phi(t) = e^{At} = \beta_0(t)I + \beta_1(t)A = \left[e^{-t}\cos t + e^{-t}\sin t\right]\begin{bmatrix}1 & 0\\ 0 & 1\end{bmatrix} + e^{-t}\sin t\begin{bmatrix}-1 & -1\\ 1 & -1\end{bmatrix} = \begin{bmatrix}e^{-t}\cos t & -e^{-t}\sin t\\ e^{-t}\sin t & e^{-t}\cos t\end{bmatrix}.$$

Evolutions of $x_1(t)$ and $x_2(t)$ due to initial conditions (x_{10}, x_{20}) are found as

$$x(t) = e^{At}x_0, \quad \begin{bmatrix}x_1(t)\\ x_2(t)\end{bmatrix} = \begin{bmatrix}e^{-t}\cos t & -e^{-t}\sin t\\ e^{-t}\sin t & e^{-t}\cos t\end{bmatrix}\begin{bmatrix}x_{10}\\ x_{20}\end{bmatrix}.$$

■

Example 5.4. *Second-Order System: State Transition Matrix and System Dynamics*

For a system

$$\frac{dx}{dt} = Ax + Bu, \quad \begin{bmatrix}\frac{dx_1}{dt}\\ \frac{dx_2}{dt}\end{bmatrix} = \begin{bmatrix}0 & 1\\ -2 & -3\end{bmatrix}\begin{bmatrix}x_1\\ x_2\end{bmatrix} + \begin{bmatrix}0\\ 1\end{bmatrix}u, \; x_0 = \begin{bmatrix}x_{10}\\ x_{20}\end{bmatrix} = \begin{bmatrix}1\\ -1\end{bmatrix},$$

$y=Hx+Du$, $y=[1 \;\; 0]x+[0]u$, $y=x_1$,

find the state transition matrix, and, derive an analytic solution for state evolutions due to initial conditions $x_0=[x_{10}, x_{20}]$ if the input is a unit step, $u=1(t)$.

Using the Sylvester expansion (5.11), $\Phi(t)=e^{At}=\sum_{i=1}^{n} e^{\lambda_i t}C_i$.

The characteristic equation (5.4) $|\lambda I-A|=0$, $\lambda^2+3\lambda+2=0$ yields $\lambda_1=-1$, $\lambda_2=-2$.

Applying $C_i=\dfrac{\prod_{j=1,j\neq i}^{n}[A-\lambda_j I]}{\prod_{j=1,j\neq i}^{n}[\lambda_i-\lambda_j]}$, we find

$$C_1=\frac{\prod_{j=1,j\neq i}^{2}[A-\lambda_j I]}{\prod_{j=1,j\neq i}^{2}[\lambda_1-\lambda_j]}=\frac{A-\lambda_2 I}{\lambda_1-\lambda_2}=\frac{\begin{bmatrix}0 & 1\\ -2 & -3\end{bmatrix}+2\begin{bmatrix}1 & 0\\ 0 & 1\end{bmatrix}}{1}=\begin{bmatrix}2 & 1\\ -2 & -1\end{bmatrix},$$

$$C_2=\frac{\prod_{j=1,j\neq i}^{2}[A-\lambda_j I]}{\prod_{j=1,j\neq i}^{2}[\lambda_2-\lambda_j]}=\frac{A-\lambda_1 I}{\lambda_2-\lambda_1}=\frac{\begin{bmatrix}0 & 1\\ -2 & -3\end{bmatrix}+\begin{bmatrix}1 & 0\\ 0 & 1\end{bmatrix}}{-1}=\begin{bmatrix}-1 & -1\\ 2 & 2\end{bmatrix}.$$

Hence, $\Phi(t)$ is

$$\Phi(t)=e^{At}=\sum_{i=1}^{n} e^{\lambda_i t}C_i=e^{\lambda_1 t}C_1+e^{\lambda_2 t}C_2=e^{-t}\begin{bmatrix}2 & 1\\ -2 & -1\end{bmatrix}+e^{-2t}\begin{bmatrix}-1 & -1\\ 2 & 2\end{bmatrix}=\begin{bmatrix}2e^{-t}-e^{-2t} & e^{-t}-e^{-2t}\\ -2e^{-t}+2e^{-2t} & -e^{-t}+2e^{-2t}\end{bmatrix}.$$

The state transition matrix in the s-domain $\Phi(s)$ is

$$\Phi(s)=(sI-A)^{-1}=\begin{bmatrix}s & -1\\ 2 & s+3\end{bmatrix}^{-1}=\frac{1}{s^2+3s+2}\begin{bmatrix}s+3 & 1\\ -2 & s\end{bmatrix}=\begin{bmatrix}\dfrac{s+3}{s^2+3s+2} & \dfrac{1}{s^2+3s+2}\\ -\dfrac{2}{s^2+3s+2} & \dfrac{s}{s^2+3s+2}\end{bmatrix}.$$

The characteristic equation (5.4) $|sI-A|=0$, $s^2+3s+2=0$ yields $s_1=-1$, $s_2=-2$. Partial fractioning and inverse Laplace transforms, reported in Table 5.1, yield

$$\Phi(t)=e^{At}=\mathcal{L}^{-1}\left[(sI-A)^{-1}\right]=\begin{bmatrix}2e^{-t}-e^{-2t} & e^{-t}-e^{-2t}\\ -2e^{-t}+2e^{-2t} & -e^{-t}+2e^{-2t}\end{bmatrix}.$$

Using MATLAB, compute $\Phi(t)=e^{At}$ and $\Phi(t)=\mathcal{L}^{-1}\{\Phi(s)\}=\mathcal{L}^{-1}\{(sI-A)^{-1}\}$.

The codes and results are

```
>> A=[0 1; -2 -3]; t=sym('t'); eAt=expm(A*t)
eAt =  [ 2*exp(-t) - exp(-2*t),     exp(-t) - exp(-2*t)]
       [ 2*exp(-2*t) - 2*exp(-t), 2*exp(-2*t) - exp(-t)]
```

and

```
>> A=[0 1; -2 -3]; s=sym('s'); [n,n]=size(A); Fs=inv(s*eye(n)-A), Ft=ilaplace(Fs)
Fs =   [ (s + 3)/(s^2 + 3*s + 2), 1/(s^2 + 3*s + 2)]
       [      -2/(s^2 + 3*s + 2), s/(s^2 + 3*s + 2)]
Ft =  [ 2*exp(-t) - exp(-2*t),      exp(-t) - exp(-2*t)]
      [ 2*exp(-2*t) - 2*exp(-t), 2*exp(-2*t) - exp(-t)]
```

State dynamics for initial conditions x_0 and unit-step input $u=1(t)$ is (5.9)

$$x(t)=\begin{bmatrix}x_1(t)\\ x_2(t)\end{bmatrix}=e^{At}x_0+\int_0^t e^{A(t-\tau)}Bu(\tau)d\tau$$

$$=\begin{bmatrix}2e^{-t}-e^{-2t} & e^{-t}-e^{-2t}\\ -2e^{-t}+2e^{-2t} & -e^{-t}+2e^{-2t}\end{bmatrix}\begin{bmatrix}x_{10}\\ x_{20}\end{bmatrix}+\int_0^t\begin{bmatrix}2e^{-(t-\tau)}-e^{-2(t-\tau)} & e^{-(t-\tau)}-e^{-2(t-\tau)}\\ -2e^{-(t-\tau)}+2e^{-2(t-\tau)} & -e^{-(t-\tau)}+2e^{-2(t-\tau)}\end{bmatrix}\begin{bmatrix}0\\ 1\end{bmatrix}d\tau$$

$$=\begin{bmatrix}2e^{-t}-e^{-2t} & e^{-t}-e^{-2t}\\ -2e^{-t}+2e^{-2t} & -e^{-t}+2e^{-2t}\end{bmatrix}\begin{bmatrix}x_{10}\\ x_{20}\end{bmatrix}+\begin{bmatrix}0.5-e^{-t}+0.5^{-2t}\\ e^{-t}-e^{-2t}\end{bmatrix},\ t\geq 0.$$

For given, $x_0=\begin{bmatrix}x_{10}\\x_{20}\end{bmatrix}=\begin{bmatrix}1\\-1\end{bmatrix}$, we have $x(t)=\begin{bmatrix}x_1(t)\\x_2(t)\end{bmatrix}=\begin{bmatrix}\frac{1}{2}(1+e^{-2t})\\-e^{-2t}\end{bmatrix}, t\geq 0$.

Compute system transients due to initial conditions x_0=[1, −1] and input u=1(t) using MATLAB. The **int** command is used. To find a definite integral $\int_a^b f(x)dx$, use **int(f,x,a,b)**.

```
A=[0 1; -2 -3]; B=[0; 1]; t=sym('t'); tau=sym('tau'); u=sym('u'); eAt=expm(A*t); x0=[1; -1]; u=1;
xt=eAt*x0+int(expm(A*(t-tau))*B*u,tau,0,t), pretty(xt); x_t=simplify(xt)
```

We have $x(t)=\begin{bmatrix}x_1(t)\\x_2(t)\end{bmatrix}=\begin{bmatrix}\frac{1}{2}(1+e^{-2t})\\-e^{-2t}\end{bmatrix}$.

Consider other initial conditions and inputs.

Let x_0=[10, −10], $u=10e^{-t}\sin^2(10t)$. Hence,

```
x0=[10; -10]; u=10*exp(-t)*(sin(10*t))^2; xt=eAt*x0+int(expm(A*(t-tau))*B*u,tau,0,t), pretty(xt);
x_t=simplify(xt)
```

For output $y(t)=x_1(t)$, the output equation is $y=Hx+Du=[1\ \ 0]x+[0]u$.

The transfer function (5.6) is

$$G(s)=\frac{Y(s)}{U(s)}=H(sI-A)^{-1}B=H\Phi(s)B=\begin{bmatrix}1 & 0\end{bmatrix}\begin{bmatrix}\frac{s+3}{s^2+3s+2} & \frac{1}{s^2+3s+2}\\ -\frac{2}{s^2+3s+2} & \frac{s}{s^2+3s+2}\end{bmatrix}\begin{bmatrix}0\\1\end{bmatrix}=\frac{1}{s^2+3s+2}.$$

The MATLAB code to find $G(s)$ is

```
>> A=[0 1; -2 -3]; B=[0; 1]; H=[1 0]; D=0; [num, den]=ss2tf(A,B,H,D)
num =
   0   0   1
den =
   1   3   2
```

■

Example 5.5. Second-Order System: State Transition Matrix and Modal Matrix

Consider a system $\frac{dx}{dt}=Ax$, $\begin{bmatrix}\frac{dx_1}{dt}\\\frac{dx_2}{dt}\end{bmatrix}=\begin{bmatrix}0 & 1\\-6 & -5\end{bmatrix}\begin{bmatrix}x_1\\x_2\end{bmatrix}$.

Hence, $\Phi(s)=(sI-A)^{-1}=\begin{bmatrix}s & -1\\6 & s+5\end{bmatrix}^{-1}=\frac{1}{s^2+5s+6}\begin{bmatrix}s+5 & 1\\-6 & s\end{bmatrix}=\begin{bmatrix}\frac{s+5}{s^2+5s+6} & \frac{1}{s^2+5s+6}\\\frac{-6}{s^2+5s+6} & \frac{s}{s^2+5s+6}\end{bmatrix}$.

One has $(sI-A)=\begin{bmatrix}s & 0\\0 & s\end{bmatrix}-\begin{bmatrix}0 & 1\\-6 & -5\end{bmatrix}=\begin{bmatrix}s & -1\\6 & s+5\end{bmatrix}$, $|sI-A|=\begin{vmatrix}s & -1\\6 & s+5\end{vmatrix}=s^2+5s+6=0$.

The eigenvalues are $s_1=-2$ and $s_2=-3$.

The partial fractioning expansion yields $\Phi(t)=\mathcal{L}^{-1}\{\Phi(s)\}=\begin{bmatrix}3e^{-2t}-2e^{-3t} & e^{-2t}-e^{-3t}\\-6e^{-2t}+6e^{-3t} & -2e^{-2t}+3e^{-3t}\end{bmatrix}$.

Apply the Cayley-Hamilton theorem to find e^{At}. The eigenvalues are found using the characteristic equation $|\lambda I-A|=\begin{vmatrix}\lambda & -1\\6 & \lambda+5\end{vmatrix}=\lambda^2+5\lambda+6=0$, $\lambda_1=-2$ and $\lambda_2=-3$.

From (5.10), we have $\begin{bmatrix}\beta_0(t)\\\beta_1(t)\end{bmatrix}=\begin{bmatrix}1 & \lambda_1\\1 & \lambda_2\end{bmatrix}^{-1}\begin{bmatrix}e^{\lambda_1 t}\\e^{\lambda_2 t}\end{bmatrix}=\begin{bmatrix}1 & -2\\1 & -3\end{bmatrix}^{-1}\begin{bmatrix}e^{-2t}\\e^{-3t}\end{bmatrix}=\begin{bmatrix}3e^{-2t}-2e^{-3t}\\e^{-2t}-e^{-3t}\end{bmatrix}$.

Hence, $\beta_0(t)=3e^{-2t}-2e^{-3t}$ and $\beta_1(t)=e^{-2t}-e^{-3t}$.

The state transition matrix is

$$\Phi(t)=e^{At}=\beta_0(t)I+\beta_1(t)A$$

$$=\begin{bmatrix}3e^{-2t}-2e^{-3t} & 0\\ 0 & 3e^{-2t}-2e^{-3t}\end{bmatrix}+\begin{bmatrix}0 & e^{-2t}-e^{-3t}\\ -6e^{-2t}+6e^{-3t} & -5e^{-2t}+5e^{-3t}\end{bmatrix}=\begin{bmatrix}3e^{-2t}-2e^{-3t} & e^{-2t}-e^{-3t}\\ -6e^{-2t}+6e^{-3t} & -2e^{-2t}+3e^{-3t}\end{bmatrix}.$$

Investigate a tuple of the eigenvalues, *modal* matrix and state transition matrix $(\mathbf{\Lambda},\mathbf{M},\Phi)$. The computed eigenvalues and eigenvectors pair $(\boldsymbol{\lambda},\boldsymbol{v})$ for A, yield the resulting diagonal matrix of eigenvalues $\mathbf{\Lambda}$, diagonal matrix $e^{\Lambda t}$, and, *modal* matrix $\mathbf{M}$.

We found $\mathbf{\Lambda}=\begin{bmatrix}\lambda_1 & 0\\ 0 & \lambda_2\end{bmatrix}=\begin{bmatrix}-2 & 0\\ 0 & -3\end{bmatrix}$.

The eigenvectors are not unique. Solving $(\lambda_i I-A)\boldsymbol{v}_i=0$, one has

$$\mathbf{M}=\begin{bmatrix}1 & 1\\ -2 & -3\end{bmatrix},\ \mathbf{M}=\begin{bmatrix}\frac{1}{\sqrt{5}} & \frac{1}{\sqrt{10}}\\ \frac{-2}{\sqrt{5}} & \frac{-3}{\sqrt{10}}\end{bmatrix},\ \mathbf{M}=\begin{bmatrix}1 & -1\\ -2 & 3\end{bmatrix},\ \mathbf{M}=\begin{bmatrix}\frac{1}{\sqrt{5}} & \frac{-1}{\sqrt{10}}\\ \frac{-2}{\sqrt{5}} & \frac{3}{\sqrt{10}}\end{bmatrix},\ \text{etc.}$$

We have, $\mathbf{M}^{-1}A\mathbf{M}=\mathbf{\Lambda}$ and $\mathbf{M\Lambda M}^{-1}=A$. For $\mathbf{M}=\begin{bmatrix}1 & 1\\ -2 & -3\end{bmatrix}$,

$$\underbrace{\begin{bmatrix}3 & 1\\ -2 & -1\end{bmatrix}}_{\mathbf{M}^{-1}}\underbrace{\begin{bmatrix}0 & 1\\ -6 & -5\end{bmatrix}}_{A}\underbrace{\begin{bmatrix}1 & 1\\ -2 & -3\end{bmatrix}}_{\mathbf{M}}=\underbrace{\begin{bmatrix}-2 & 0\\ 0 & -3\end{bmatrix}}_{\mathbf{\Lambda}}\ \text{and}\ \underbrace{\begin{bmatrix}1 & 1\\ -2 & -3\end{bmatrix}}_{\mathbf{M}}\underbrace{\begin{bmatrix}-2 & 0\\ 0 & -3\end{bmatrix}}_{\mathbf{\Lambda}}\underbrace{\begin{bmatrix}3 & 1\\ -2 & -1\end{bmatrix}}_{\mathbf{M}^{-1}}=\underbrace{\begin{bmatrix}0 & 1\\ -6 & -5\end{bmatrix}}_{A}.$$

Using $(e^{\Lambda t},\mathbf{M})$, $e^{\Lambda t}=\begin{bmatrix}e^{-2t} & 0\\ 0 & e^{-3t}\end{bmatrix}$, the state transition matrix is found as (5.12)

$$\Phi(t)=e^{At}=\mathbf{M}e^{\Lambda t}\mathbf{M}^{-1},\ \Phi(t)=\begin{bmatrix}3e^{-2t}-2e^{-3t} & e^{-2t}-e^{-3t}\\ -6e^{-2t}+6e^{-3t} & -2e^{-2t}+3e^{-3t}\end{bmatrix}.$$ ■

Differential Equations and State-Space Models–The n-order linear system is described as

$$\frac{d^n y}{dt^n}+a_{n-1}\frac{d^{n-1}y}{dt^{n-1}}+\ldots+a_1\frac{dy}{dt}+a_0 y=u\cdot$$

Define the state variables as $x_1=y$, $x_2=\frac{dy}{dt}$, …., $x_n=\frac{d^{n-1}y}{dt^{n-1}}$.

Hence, $\frac{dx_1}{dt}=x_2$, $\frac{dx_2}{dt}=x_3$, …, $\frac{dx_n}{dt}=-a_0x_1-a_1x_2-\ldots-a_{n-1}x_n+u$.

One finds (5.5) as
$$\frac{dx}{dt}=\begin{bmatrix}\frac{dx_1}{dt}\\ \frac{dx_2}{dt}\\ \vdots\\ \frac{dx_{n-1}}{dt}\\ \frac{dx_n}{dt}\end{bmatrix}=\begin{bmatrix}0 & 1 & 0 & \cdots & 0 & 0\\ 0 & 0 & 1 & \cdots & 0 & 0\\ \vdots & \vdots & \vdots & \ddots & \vdots & \vdots\\ 0 & 0 & 0 & \cdots & 0 & 1\\ -a_0 & -a_1 & -a_2 & \cdots & -a_{n-2} & -a_{n-1}\end{bmatrix}\begin{bmatrix}x_1\\ x_2\\ \vdots\\ x_{n-1}\\ x_n\end{bmatrix}+\begin{bmatrix}0\\ 0\\ 0\\ 0\\ 1\end{bmatrix}u=Ax+Bu.$$

From $x_1=y$, the output equation is $y=[1\ 0\ \ldots\ 0\ 0]x=Hx$, $H=[1\ 0\ \ldots\ 0\ 0]$.

Nonlinear Systems and Physical Limits – Laws of physics yield nonlinear differential equations. There are hardware-defined control limits $\mathcal{U}_{\min}\le\mathcal{U}\le\mathcal{U}_{\max}$. The designed control law and a control function are bounded, $u_{\min}\le u\le u_{\max}$. The system output $y(t)$ may be a nonlinear function of states $x(t)$, $y=H(x)$. One has

$\dot{x}(t)=F(x,r,d)+B(x)u$, $y=H(x)$, $u_{\min}\le u\le u_{\max}$, $x(t_0)=x_0$,

where $F(\cdot):\mathbb{R}^n\times\mathbb{R}^b\times\mathbb{R}^v\to\mathbb{R}^n$, $B(\cdot):\mathbb{R}^n\to\mathbb{R}^{n\times m}$ and $H(\cdot):\mathbb{R}^n\to\mathbb{R}^{b\times n}$ are the nonlinear maps.

5.2.2. Discrete-Time Systems

Laws of physics, rigid-body dynamics, continuum mechanics, hydrodynamics, thermodynamics, fluid mechanics and electromagnetics yield governing differential equations with time-invariant and time-varying parameters. Physical devices and components are continuous. In ASICs, high-frequency switching is accomplished by analog field-effect transistors. Combinational digital circuits perform binary and n-ary arithmetic, logic and operations on integers. In combinational circuits, picosecond propagation, contamination and other delays are due to analog signal switching and transductions. Quartz crystal oscillators, *LC* circuits, operational amplifier, resonant tunneling diode and other oscillators generate kilohertz to terahertz frequency periodic analog signals, and, the frequency dividers and multipliers circuits are implemented. The outputs are analog high-frequency continuous-time signals.

Continuous- and Discrete-Time Signals – Mathematically, continuous-time signals $x(t)$ can be sampled and quantized at discrete instants of time, yielding descriptive discrete-time mapping $x_k \equiv x[k]$. A discrete-time signal $x[k]$ is a real-valued function of quantized $x(t)$ at time index k. The sampled $x[k]$ has a finite set of values in its sequence. Hence, signals are mathematically described as sequences $x[k]$, $-\infty<k<\infty$, where k is an integer. For the sampling time (sampling period) T_s, $x(t)=x[kT_s]$. The sampling time T_s is reciprocal is the sampling frequency.

Data sharing, data fusion, data management and processing are performed on discrete logic and discrete calculus. Continuous and discrete data domains and spaces are applied. Consider linear functions from $\mathbb{R}$ to the resulting vector space structure. Continuous and discrete domains and codomains are characterized by given topologies, vector spaces and maps. For two topological vector spaces V and W, a linear map is $M: V \to W$. There are unique finite dimensional topological vector spaces (V,W) over $\mathbb{R}$. Analog and digital data processing, derived applying differential, discrete and propositional calculi, are physically implemented by means of discrete logic. Investigating control of analog systems, one considers pertained topics in digital signal processing, such as sampling, analog-to-digital and digital-to-analog conversions, data fusion rates, latencies, etc. Analog signal processing is performed on continuous signals. Digital systems support information governance in discrete domain. The ASICs, MPUs and FPGAs physically implement digital data sharing, interfacing, management, coordination and control.

Discrete-Time Systems – Linear and nonlinear differential equations are discretized to support analytic design. Design of analog controllers and filters are performed in the continuous time domain, continuous frequency domain and s domain. Analog filters and control laws can be discretized. Analysis, control and optimization tasks are performed in discrete time and z domains.

Linear time-invariant systems (5.3) and (5.5) are discretized with the sampling time T_s, which depends on system dynamics and hardware solutions. Digital controllers, filters, observers, estimators, interface and data fusion are physically implemented by MPUs and FPGAs. Analog and digital paradigms are applied to design control laws, filters, state *observers*, etc.

The state-space difference equation for n states, m controls and b outputs is

$$x_{k+1} = A_k x_k + B_k u_k, \tag{5.14}$$

$$x_{k+1} = \begin{bmatrix} x_{k+1,1} \\ x_{k+1,2} \\ \vdots \\ x_{k+1,n-1} \\ x_{k+1,n} \end{bmatrix} = \begin{bmatrix} a_{k11} & a_{k12} & \cdots & a_{k1n-1} & a_{k1n} \\ a_{k21} & a_{k22} & \cdots & a_{k2n-1} & a_{k2n} \\ \vdots & \vdots & \ddots & \vdots & \vdots \\ a_{kn-11} & a_{kn-12} & \cdots & a_{kn-1n-1} & a_{kn-1n} \\ a_{kn1} & a_{kn2} & \cdots & a_{knn-1} & a_{knn} \end{bmatrix} \begin{bmatrix} x_{k1} \\ x_{k2} \\ \vdots \\ x_{kn-1} \\ x_{kn} \end{bmatrix} + \begin{bmatrix} b_{k11} & b_{k12} & \cdots & b_{k1m-1} & b_{k1m} \\ b_{k21} & b_{k22} & \cdots & b_{k2m-1} & b_{k2m} \\ \vdots & \vdots & \ddots & \vdots & \vdots \\ b_{kn-11} & b_{kn-12} & \cdots & b_{kn-1m-1} & b_{kn-1m} \\ b_{kn1} & b_{kn2} & \cdots & b_{knm-1} & b_{knm} \end{bmatrix} \begin{bmatrix} u_{k1} \\ u_{k2} \\ \vdots \\ u_{km-1} \\ u_{km} \end{bmatrix},$$

$$y_k = H_k x_k,\ x_{k=k_0} = x_{k0},\ k \in \mathbb{N}^0,$$

where $A_k \in \mathbb{R}^{n\times n}$ and $B_k \in \mathbb{R}^{n\times m}$ are the constant-coefficient matrices of coefficients; $H_k \in \mathbb{R}^{b\times n}$ is the output matrix, pertained to the output equation $y_k = H_k x_k$.

Notations – Following classical definitions, notations (x_k, y_k, u_k) imply $x_k \equiv x[k]$, $y_k \equiv y[k]$, $u_k \equiv u[k]$. The discrete system states, outputs and control may be denoted as (x_n, y_n, u_n), $x_n \equiv x[n]$, $y_n \equiv y[n]$, $u_n \equiv u[n]$. Difference equation (5.14) $x_{k+1} = A_k x_k + B_k u_k$ may be written as $x_k = A_k x_{k-1} + B_k u_{k-1}$. Continuous and discrete time are related as $t = kT_s$ or $t = nT_s$. ■

For an n-order linear difference equation with time-invariant coefficients and zero initial conditions $\sum_{i=0}^{n} a_i y_{n-i} = \sum_{i=0}^{m} b_i u_{n-i}$. The z-transform and transfer function are

$$\left(\sum_{i=0}^{n} a_i z^i\right) Y(z) = \left(\sum_{i=0}^{m} b_i z^i\right) U(z),\ G(z) = \frac{Y(z)}{U(z)} = \frac{b_m z^m + b_{m-1} z^{m-1} + \ldots + b_1 z + b_0}{a_n z^n + a_{n-1} z^{n-1} + \ldots + a_1 z + a_0},\ n \geq m. \tag{5.15}$$

Solution of Difference Equations and State Transition Matrix – Solution of linear difference equation (5.14) $x_{k+1} = A_k x_k + B_k u_k$, $x[k+1] = A_k x[k] + B_k u[k]$, $k \in \mathbb{N}^0$, with initial conditions $x[0]$ and inputs $u[k]$, is found using the inverse z-transform of $X(z)$ or the state transition matrix $\Phi[k] = A^k$. Using notations $A \equiv A_k$ or $A \equiv A_n$ and $B \equiv B_k$ or $B \equiv B_n$,

$x[1] = Ax[0] + Bu[0]$, $x[2] = Ax[1] + Bu[1] = A^2 x[0] + ABu[0] + B_k u[1]$, … .

Hence,

$$x[k] = A^k x[0] + \sum_{i=0}^{k-1} A^{k-i-1} Bu[i] = \underbrace{\Phi[k]}_{\Phi[k]=A^k} x[0] + \sum_{i=0}^{k-1} \Phi[k-i-1] Bu[i],\ k = 0,1,2,\ldots,\Phi[k] = A^k. \tag{5.16}$$

$$x[k] = A^{k-k_0} x[0] + \sum_{i=k_0}^{k-1} A^{k-i-1} Bu[i] = \underbrace{\Phi[k,k_0]}_{\Phi[k,k_0]=A^{k-k_0}} x[0] + \sum_{i=k_0}^{k-1} \Phi[k-i-1] Bu[i],\ k \geq k_0.$$

From (5.14), (5.16), $X(z) = z(zI - A)^{-1} x[0] + (zI - A)^{-1} BU(z)$ and the inverse z-transform $x[k] = Z^{-1}\{X(z)\}$, $x[k] = Z^{-1}\{z(zI - A)^{-1}\} x[0] + Z^{-1}\{(zI - A)^{-1} BU(z)\}$, we obtain

$$\Phi[k] = A^k,\ \Phi(z) = z(zI - A)^{-1},\ \Phi[k] = Z^{-1}\{\Phi(z)\} = Z^{-1}\{z(zI - A)^{-1}\}. \tag{5.17}$$

For the state transition matrix $\Phi[k]$ one has

$\Phi[0] = A^0 = I$, $x[0] = \Phi[0] x[0]$, $\Phi^i[k] = \Phi[ik]$, $(A^k)^i = A^{ik}$, $\Phi[k+1] = A\Phi[k]$.

Eigenvalues, System Stability and Lyapunov Equation – For $A_k \in \mathbb{R}^{n\times n}$, one finds the characteristic polynomial $p_A(z) = \det(zI - A_k)$, and, computes characteristic eigenvalues z

using the characteristic equation det($zI–A_k$)=0. For a given A_k, there are n eigenvalues z_i, i=1,…,n. A linear discrete-time system is asymptotically stable if $|z_i|\le 1$, $\forall z_i$, and $\lim_{k\to\infty} x[k]\to 0$. System is unstable if there exists an eigenvalue $|z_i|>1$, and, marginally stable if $|z_i|=1$. That is, a system is asymptotically stable if all eigenvalues are located in the unit circle of the complex plane.

A system is bounded-input bounded-output stable if and only if the impulse response is absolutely summable. Stability analysis may be performed using the Lyapunov stability theory solving a discrete Lyapunov equation (6.128) $A_k^T K_k A_k - K_k = -Q_k$. *Conjecture* 6.6 states that a linear discrete-time system $x_{k+1}=A_k x_k$ is stable, if for a positive definite matrix $Q_k\in\mathbb{R}^{n\times n}$ there exists a positive definite symmetric matrix $K_k\in\mathbb{R}^{n\times n}$ which is a solution of the Lyapunov equation.

State Transition Matrix For Discrete-Time Systems – The state transition matrix in z- and time domains are (5.16)

$$\Phi(z)=z(zI-A)^{-1},\ \Phi[k]=Z^{-1}\{\Phi(z)\}=Z^{-1}\left\{z(zI-A)^{-1}\right\},\ \Phi[k]=A^k.$$

Cayley-Hamilton Algorithm – Using (5.10), the state transition matrix is

$$\Phi[k]=A^k,\ A^k=\sum_{i=0}^{n-1}\beta_i[k]A^i,\ \begin{bmatrix}\beta_0[k]\\ \beta_1[k]\\ \vdots\\ \beta_{n-1}[k]\end{bmatrix}=\begin{bmatrix}1 & z_1 & z_1^2 & \cdots & z_1^{n-1}\\ 1 & z_2 & z_2^2 & \cdots & z_2^{n-1}\\ \vdots & \vdots & \vdots & \ddots & \vdots\\ 1 & z_n & z_n^2 & \cdots & z_n^{n-1}\end{bmatrix}^{-1}\begin{bmatrix}(z_1)^k\\ (z_2)^k\\ \vdots\\ (z_n)^k\end{bmatrix}.$$

The eigenvalue-eigenvector pair $(\boldsymbol{z},\boldsymbol{v})$ and *modal* matrix $\mathbf{M}\in\mathbb{R}^{n\times n}$ for A are found. One has $\mathbf{M}^{-1}A\mathbf{M}=\boldsymbol{Z}$ and $\mathbf{M}\boldsymbol{Z}\mathbf{M}^{-1}=A$. From (5.12), we obtain

$$\Phi[k]=A^k=\mathbf{M}\boldsymbol{Z}^k\mathbf{M}^{-1},\ \boldsymbol{Z}^k=\begin{bmatrix}(z_1)^k & \cdots & 0\\ \vdots & \ddots & \vdots\\ 0 & \cdots & (z_n)^k\end{bmatrix},\ \mathbf{M}=[\boldsymbol{v}_1,\ldots,\boldsymbol{v}_n],\ (z_iI-A)\boldsymbol{v}_i=0.$$

Example 5.6. *State Transition Matrix, Lyapunov Equation and System Dynamics*

Consider a system $x[k+1]=Ax[k]+Bu[k]$, $\begin{bmatrix}x_1[k+1]\\ x_2[k+1]\end{bmatrix}=\begin{bmatrix}0 & 1\\ -\frac{1}{6} & -\frac{5}{6}\end{bmatrix}\begin{bmatrix}x_1[k]\\ x_2[k]\end{bmatrix}+\begin{bmatrix}0\\ 1\end{bmatrix}u[k]$.

Analyze stability, find the state transition matrix, and, derive an analytic solution for state evolutions due to initial conditions $x[0]=\begin{bmatrix}1\\ 0\end{bmatrix}$ if input is $u[k]=(2)^{-k}1[k]$.

The state transition matrix $\Phi(z)$ is

$$\Phi(z)=z\left(zI-A\right)^{-1}=z\begin{bmatrix}z & -1\\ \frac{1}{6} & z+\frac{5}{6}\end{bmatrix}^{-1}=\frac{z}{z^2+\frac{5}{6}z+\frac{1}{6}}\begin{bmatrix}z+\frac{5}{6} & 1\\ -\frac{1}{6} & z\end{bmatrix}.$$

The characteristic equation is $|zI–A|=0$, $z^2+\frac{5}{6}z+\frac{1}{6}=0$, $z_1=-\frac{1}{2}$, $z_2=-\frac{1}{3}$.

Hence, a system is stable. Perform partial fractioning and apply inverse z-transforms.

$$\Phi[k]=\mathcal{Z}^{-1}\{\Phi[z]\}=\mathcal{Z}^{-1}\left\{\begin{bmatrix}\frac{-2z}{z+\frac{1}{2}}+\frac{3z}{z+\frac{1}{3}} & \frac{-6z}{z+\frac{1}{2}}+\frac{6z}{z+\frac{1}{3}}\\ \frac{z}{z+\frac{1}{2}}-\frac{z}{z+\frac{1}{3}} & \frac{3z}{z+\frac{1}{2}}-\frac{2z}{z+\frac{1}{3}}\end{bmatrix}\right\}=\begin{bmatrix}-2(-\frac{1}{2})^k+3(-\frac{1}{3})^k & -6(-\frac{1}{2})^k+6(-\frac{1}{3})^k\\ (-\frac{1}{2})^k-(-\frac{1}{3})^k & 3(-\frac{1}{2})^k-2(-\frac{1}{3})^k\end{bmatrix}.$$

A discrete linear systems is asymptotically stable because $|z_i|\le 1$, i=1,2, and, $\lim_{k\to\infty} x[k]\to 0$. That is, the zero-input response evolves to zero as time increases.

Apply the Cayley-Hamilton theorem to find $\Phi[k]=A^k$.

The characteristic eigenvalues are $z_1=-1/2$ and $z_2=-1/3$. We have

$$\begin{bmatrix}\beta_0[k]\\ \beta_1[k]\end{bmatrix}=\begin{bmatrix}1 & z_1\\ 1 & z_2\end{bmatrix}^{-1}\begin{bmatrix}(z_1)^k\\ (z_2)^k\end{bmatrix}=\begin{bmatrix}1 & -\frac{1}{2}\\ 1 & -\frac{1}{3}\end{bmatrix}^{-1}\begin{bmatrix}(-\frac{1}{2})^k\\ (-\frac{1}{3})^k\end{bmatrix}=\begin{bmatrix}-2(-\frac{1}{2})^k+3(-\frac{1}{3})^k\\ -6(-\frac{1}{2})^k+6(-\frac{1}{3})^k\end{bmatrix}.$$

Hence, $\beta_0[k]=-2(-\frac{1}{2})^k+3(-\frac{1}{3})^k$, $\beta_1[k]=-6(-\frac{1}{2})^k+6(-\frac{1}{3})^k$.

The state transition matrix is

$$\Phi[k]=A^k=\beta_0[k]I+\beta_1[k]A=\begin{bmatrix}-2(-\frac{1}{2})^k+3(-\frac{1}{3})^k & 0\\ 0 & -2(-\frac{1}{2})^k+3(-\frac{1}{3})^k\end{bmatrix}+\begin{bmatrix}0 & -6(-\frac{1}{2})^k+6(-\frac{1}{3})^k\\ (-\frac{1}{2})^k-(-\frac{1}{3})^k & 5(-\frac{1}{2})^k-5(-\frac{1}{3})^k\end{bmatrix}$$

$$=\begin{bmatrix}-2(-\frac{1}{2})^k+3(-\frac{1}{3})^k & -6(-\frac{1}{2})^k+6(-\frac{1}{3})^k\\ (-\frac{1}{2})^k-(-\frac{1}{3})^k & 3(-\frac{1}{2})^k-2(-\frac{1}{3})^k\end{bmatrix}.$$

Using (5.16) and (5.17), the system evolution with $x[0]=[1, 0]$ and $u[k]=(2)^{-k}1[k]$ is

$$x[k]=\Phi[k]x[0]+\sum_{i=0}^{k-1}\Phi[k-i-1]Bu[i],$$

$$\begin{bmatrix}x_1[k]\\ x_2[k]\end{bmatrix}=\begin{bmatrix}-2(-\frac{1}{2})^k+3(-\frac{1}{3})^k\\ (-\frac{1}{2})^k-(-\frac{1}{3})^k\end{bmatrix}+\sum_{i=0}^{k-1}\begin{bmatrix}-6(-\frac{1}{2})^{k-i-1}+6(-\frac{1}{3})^{k-i-1}\\ 3(-\frac{1}{2})^{k-i-1}-2(-\frac{1}{3})^{k-i-1}\end{bmatrix}(2)^{-i}.$$

One may develop a custom MATLAB code to calculate $\Phi[k]=A^k$ and verify solutions. For example, `A=[0 1; -1/6 -5/6]; k=sym('k'); Ak=A^k` may not yield expected results.

Finding an eigenvalue-eigenvector pair $(\boldsymbol{z},\boldsymbol{v})$ and *modal* matrix $\mathbf{M}\in\mathbb{R}^{2\times 2}$ for $A\in\mathbb{R}^{2\times 2}$, the state transition matrix is

$$\Phi[k]=A^k=\mathbf{M}\boldsymbol{Z}^k\mathbf{M}^{-1},\ \boldsymbol{Z}^k=\begin{bmatrix}(z_1)^k & 0\\ 0 & (z_2)^k\end{bmatrix},\ \mathbf{M}=[\boldsymbol{v}_1, \boldsymbol{v}_2],\ (z_iI-A)\boldsymbol{v}_i=0.$$

With $z_1=-\frac{1}{2}$, $z_2=-\frac{1}{3}$, one has $\mathbf{M}=\begin{bmatrix}2 & 3\\ -1 & -1\end{bmatrix}$ and $\mathbf{M}^{-1}=\begin{bmatrix}-1 & -3\\ 1 & 2\end{bmatrix}$.

Recall that $\mathbf{M}^{-1}A\mathbf{M}=\boldsymbol{Z}$ and $\mathbf{M}\boldsymbol{Z}\mathbf{M}^{-1}=A$. We obtain $\mathbf{M}^{-1}A\mathbf{M}=\boldsymbol{Z}=\begin{bmatrix}-\frac{1}{2} & 0\\ 0 & -\frac{1}{3}\end{bmatrix}$.

The state transition matrix is

$$\Phi[k]=A^k=\mathbf{M}\boldsymbol{Z}^k\mathbf{M}^{-1}=\begin{bmatrix}2 & 3\\ -1 & -1\end{bmatrix}\begin{bmatrix}(-\frac{1}{2})^k & 0\\ 0 & (-\frac{1}{3})^k\end{bmatrix}\begin{bmatrix}-1 & -3\\ 1 & 2\end{bmatrix}=\begin{bmatrix}-2(-\frac{1}{2})^k+3(-\frac{1}{3})^k & -6(-\frac{1}{2})^k+6(-\frac{1}{3})^k\\ (-\frac{1}{2})^k-(-\frac{1}{3})^k & 3(-\frac{1}{2})^k-2(-\frac{1}{3})^k\end{bmatrix}.$$

Lyapunov Equation – Apply the Lyapunov equation to analyze system stability.

Solve a discrete Lyapunov equation (6.128) $A_k^TK_kA_k-K_k=-Q_k$, $Q_k>0$.

Using notations of this section, the Lyapunov equation is $A^TK[k]A-K[k]=-Q[k]$.

For $Q_k = \begin{bmatrix} 1 & 0 \\ 0 & 1 \end{bmatrix}$, using the **eig**, **dlyap** and **det** commands, compute z_i, K_k and $\det(K_k)$ as

```
Ak=[0 1; -1/6 -5/6]; Qk=[1 0; 0 1]; Eig=eig(Ak), Kk=dlyap(Ak',Qk), det=det(Kk)
```

We have $K_k = \begin{bmatrix} 1.1167 & 0.5 \\ 0.5 & 4.2 \end{bmatrix}$ and $\det(K_k)$=4.44.

Hence, the system is stable. Recall that the eigenvalues are $z_1=-1/2$ and $z_2=-1/3$.

The **dlyap** command solves a discrete Lyapunov equation by using the matrix block substitutions and Schur decomposition.

Solve $A_k^T K_k A_k - K_k = -Q_k$ as a system of equations

```
Ak=[0 1; -1/6 -5/6]; Qk=[1 0; 0 1];
n=length(Ak); X=ones(n^2);
for i=1:n
  for j=1:n
    X((1:n)+n*(i-1),(1:n)+n*(j-1))=Ak*Ak(i,j);
end
  end
Kk=reshape((eye(n^2)-X)'\Qk(:),n,n)
```

We obtain $K_k = \begin{bmatrix} 1.1167 & 0.5 \\ 0.5 & 4.2 \end{bmatrix}$. ■

Discretization of Continuous-Time Systems – Using laws of physics, one finds differential equations. For linear systems (5.3) and (5.5) $\dot{x}=Ax+Bu$, a transfer function (5.8) $G(s)$ is found. To derive the state-space difference equation (5.14) $x_{k+1}=A_k x_k+B_k u_k$ and transfer function (5.15) $G(z)$, different approaches are applied. For continuous-time systems (5.5), the solution due to initial conditions x_0 and input u is given by (5.9) $x(t_f) = e^{A(t_f-t_0)}x_0 + \int_{t_0}^{t} e^{A(t_f-\tau)}Bu(\tau)d\tau$, $t>t_0$. With the sampling period T_s, define the initial time $t_0=kT_s$, while the final time is $t_f=(k+1)T_s$.

For the piecewise-constant input over a sampling time T_s, $u(t)=u[k]$, $kT_s \le t<(k+1)T_s$

$$x[k+1] = e^{AT_s}x[k] + \left[\int_{kT_s}^{(k+1)T_s} e^{A((k+1)T_s-\tau)}Bd\tau\right]u[k], \tag{5.18}$$

$$x_{k+1} = A_k x_k + B_k u_k .$$

Integration yields $\int e^{At}dt=A^{-1}e^{At}=e^{At}A^{-1}$. Hence, $B_k = A^{-1}(e^{AT_s} - I)B$.

For a system (5.5) with matrices (A,B,H,D), using T_s, one obtains (A_k,B_k,H_k,D_k)

$$A_k = e^{AT_s},\ B_k = A^{-1}(e^{AT_s} - I)B,\ H_k=H,\ D_k=D. \tag{5.19}$$

State-Space Models – There are a number of methods to derive the matrix exponential and find $A_k \in \mathbb{R}^{n\times n}$ and $B_k \in \mathbb{R}^{n\times m}$. Analytic and numeric solutions are found applying the Jordan-Chevalley decomposition, Jordan canonical form, Cayley-Hamilton theorem, Maclaurin series $A_k = e^{AT_s} = \sum_{i=0}^{\infty} \frac{1}{i!}(AT_s)^i$, etc.

The Maclaurin series is $e^x = \sum_{i=0}^{\infty} \frac{1}{i!}x^i = 1 + x + \frac{1}{2!}x^2 + \frac{1}{3!}x^3 + \ldots$.

Hence, $A_k = e^{AT_s} = \sum_{i=0}^{\infty} \frac{1}{i!}(AT_s)^i = I + AT_s + \frac{1}{2!}(AT_s)^2 + \frac{1}{3!}(AT_s)^3 + \ldots$.

Discretization complexity and accuracy can be asserted and evaluated.

Illustrative Example 5.1. For a square matrix A, compute the matrix exponential as
$$A_k = e^{AT_s} = \sum_{i=0}^{\infty} \tfrac{1}{i!}(AT_s)^i = I + AT_s + \tfrac{1}{2!}(AT_s)^2 + \tfrac{1}{3!}(AT_s)^3 + \dots .$$

Consider $A = \begin{bmatrix} 0 & 1 \\ -2 & -3 \end{bmatrix}$. Let the sampling time is T_s=0.05 sec. Compute A_k using an analytic relationship and Maclaurin series. MATLAB supports discretization.

```
A=[0 1; -2 -3]; Ts=0.05;  % Matrix A
Ak_exp=expm(A*Ts)   % Compute Ak using matrix exponential using the MATLAB expm command
% Compute Ak using Maclaurin series
A=Ts*A; Ak_series=eye(size(A))+A  % First two terms
Ai=A;    for i=2:1:5
         Ai=Ai*A/i;    % Compute Ai/i! from previous term A(i−1)/(i-1)!
Ak_series=Ak_series+Ai
end; Ak_exp, Ak_series
```

For $A = \begin{bmatrix} 0 & 1 \\ -2 & -3 \end{bmatrix}$, T_s=0.05 sec, using $A_k = e^{AT_s}$ and $A_k = \sum_{i=0}^{5} \tfrac{1}{i!}(AT_s)^i$, we have
$$A_k = \begin{bmatrix} 0.99762 & 0.046392 \\ -0.092784 & 0.85845 \end{bmatrix}.$$

As an illustrative example, generate a square matrix $A \in \mathbb{R}^{n\times n}$ with normally distributed pseudorandom entries a_{ij}, and, find A_k. For n=10, the MATLAB code is

```
n=10; A=randn(n,n); Ts=0.05; % Square matrix A with random aij
Ak_exp=expm(A*Ts)           % Compute Ak using matrix exponential
```
■

s- and z-Domains – The unilateral z-transform is the Laplace transform of the sampled signal. One has $X(s) = X(z)\big|_{z=e^{sT_s}}$.

From $z = e^{sT_s}$, we have $s = \frac{1}{T_s}\ln(z)$.

The expressions and series expansion for ln(z) are
$$z = e^{sT_s},\ s = \frac{1}{T_s}\ln(z), \ln(z) = 2\left[\frac{z-1}{z+1} + \frac{1}{3}\left(\frac{z-1}{z+1}\right)^3 + \frac{1}{5}\left(\frac{z-1}{z+1}\right)^5 + \dots\right], z > 0. \quad (5.20)$$

Transfer Function and Tustin Approximation – By truncating the series for ln(z), one obtains the Tustin approximation $\ln(z) \approx 2\frac{z-1}{z+1} = 2\frac{1-z^{-1}}{1+z^{-1}}$. The Tustin approximation is applied to the s-domain transfer functions $G(s)$, and, finding $G(z)$ for the specified T_s
$$G(z) \equiv G(s)\big|_{s=\frac{1}{T_s}\ln(z)\approx\frac{2}{T_s}\frac{z-1}{z+1}=\frac{2}{T_s}\frac{1-z^{-1}}{1+z^{-1}}},\ s = \frac{1}{T_s}\ln(z) \approx \frac{2}{T_s}\frac{z-1}{z+1} = \frac{2}{T_s}\frac{1-z^{-1}}{1+z^{-1}}. \quad (5.21)$$

Discretization of Nonlinear Systems – For nonlinear differential equations $\dot{x}=F(x)+B(x)u$, $y=H(x)$, nonlinear difference equations are found applying the forward rectangular rule or other approximations for dx/dt.

The Euler approximation gives $\frac{dx}{dt} \approx \frac{x(t+T_s)-x(t)}{T_s}$.

The sampling time T_s should be consistent with data fusion, sensors and ASICs, etc. Discretize nonlinear differential equations to perform design, and find control laws.

Applying $\left.\frac{dx}{dt}\right|_{t=kT_s} = \frac{x(kT_s + T_s) - x(kT_s)}{T_s}$, obtain

$$x_{k+1} = F_k(x_k) + B_k(x_k)u_k,\ y_k = H(x_k). \tag{5.22}$$

Example 5.7. Continuous and Discrete Second-Order Systems
Study a system examined in Example 5.4 and Illustrative Example 5.1

$$\frac{dx}{dt} = Ax + Bu, \quad \begin{bmatrix} \frac{dx_1}{dt} \\ \frac{dx_2}{dt} \end{bmatrix} = \begin{bmatrix} 0 & 1 \\ -2 & -3 \end{bmatrix}\begin{bmatrix} x_1 \\ x_2 \end{bmatrix} + \begin{bmatrix} 0 \\ 1 \end{bmatrix} u,\ y = x_1.$$

The transfer function is $G(s) = \frac{1}{s^2 + 3s + 2}$.

For $G(s)$, using (5.21), one obtains a transfer function $G(z)$.

Discretize a system using the MATLAB command **c2d**, with syntax **sysd=c2d(sys,Ts)** or **sysd=c2d(sys,Ts,method)**.

The following discretization methods are supported: (1) Zero-order hold 'Zoh', which implies the piecewise constant input over the sampling time T_s; (2) Triangle approximation 'Foh', which is the modified first-order hold; (3) 'Impulse' invariant discretization; (4) Bilinear 'Tustin' approximation; (5) 'Matched', which is the zero-pole matching method; (6) 'Least-squares' estimation.

MATLAB commands below perform discretization using the Tustin approximation (5.21), find a transfer function $G(z)$, calculate zeros and poles, solve difference equation, and, plot system dynamics. The discrete-time system dynamics with T_s=0.05 sec for initial conditions $\begin{bmatrix} x_{1\,k} \\ x_{2\,k} \end{bmatrix}_0 = \begin{bmatrix} 1 \\ -1 \end{bmatrix}$ is documented in Figure 5.4.b.

```
A=[0 1; -2 -3]; B=[0; 1]; H=[1 0]; D=0; % Continuous-time system: State-space model
sys=ss(A,B,H,D)             % Continuous-time system
Ts=0.05;                    % Sampling time
sys_d=c2d(sys,Ts,'Tustin') % Finding discrete-time system using Tustin approximation
G_z=tf(sys_d)               % Transfer function G(z)
poles=zpk(G_z)              % Zeros, poles and gain
% Discrete-time system: Finding state-space model
Ak=sys_d.a, Bk=sys_d.b, Hk=sys_d.c, Dk=sys_d.d
% Dynamics of a discrete-time system: Solution of difference equation with initial conditions and input
x0=[1 -1]';     % Initial conditions
u=0;            % Input
   x(:,1)=Ak*x0+Bk*u;
for k=1:50
   x(:,k+1)=Ak*x(:,k)+Bk*u;
end
k=0:50; plot(k,x(1,:),'ko',k,x(2,:),'bs','LineWidth',1.5);
legend('{\itx}_1_{\itk}','{\itx}_2_{\itk}','FontSize',22); legend boxoff;
```

The following transfer function and matrices are computed

```
G_z =
 0.0005807 z^2 + 0.001161 z + 0.0005807
 ---------------------------------------
        z^2 - 1.856 z + 0.8606
Ak =  9.9768e-01  4.6458e-02
     -9.2915e-02  8.5830e-01
Bk = 1.1614e-03
     4.6458e-02
Hk = 9.9884e-01   2.3229e-02
Dk = 5.8072e-04
```

Hence, $G(z)=\dfrac{0.0005807z^2+0.001161z+0.0005807}{z^2-1.856z+0.8606}$.

The linear continuous-time system $\dot{x}=Ax+Bu$ is discretized, finding the difference equation

$$x_{k+1}=A_kx_k+B_ku_k,\ \begin{bmatrix}x_{1\ k+1}\\ x_{2\ k+1}\end{bmatrix}=\begin{bmatrix}0.99768 & 0.046458\\ -0.092915 & 0.8583\end{bmatrix}\begin{bmatrix}x_{1k}\\ x_{2k}\end{bmatrix}+\begin{bmatrix}0.0011614\\ 0.046458\end{bmatrix}u_k.$$

The computed output equation matrices are H_k=[0.9988 0.0232], D_k=[0.00058].

The Tustin truncation (5.21) yields the approximation for A_k, B_k, H_k and D_k. Use the matrix exponential to analytically or numerically find A_k and B_k, given by (5.19)

$$A_k=e^{AT_s},\ B_k=A^{-1}(e^{AT_s}-I)B.$$

The output equation matrices are $H_k=H$ and $D_k=D$.

Compute A_k and B_k using MATLAB statements applying (5.19)

```
A=[0 1; -2 -3]; B=[0; 1]; H=[1 0]; D=0; Ts=0.05;
Ak=expm(A*Ts), Bk=inv(A)*(Ak-eye(size(A)))*B
```

The numeric results are

```
Ak =  9.9762e-01  4.6392e-02
     -9.2784e-02  8.5845e-01
Bk =  1.1893e-03
      4.6392e-02
```

Hence, $x_{k+1}=\begin{bmatrix}x_{1k+1}\\ x_{2k+1}\end{bmatrix}=A_kx_k+B_ku_k=\begin{bmatrix}0.99762 & 0.046392\\ -0.092784 & 0.85845\end{bmatrix}\begin{bmatrix}x_{1k}\\ x_{2k}\end{bmatrix}+\begin{bmatrix}0.0011893\\ 0.046392\end{bmatrix}u_k$,

$H_k=H$=[1 0], $D_k=D$=[0]. ∎

Example 5.8. *Euler Approximation and Discretization*

For the first-order linear constant-coefficient differential equation

$$\frac{dx}{dt}=-ax(t)+bu(t),$$

find the difference equation using the forward difference.

Using $t=kT_s$, we have $\left.\dfrac{dx}{dt}\right|_{t=kT_s}=-ax(kT_s)+bu(kT_s)$.

The forward rectangular rule, which is the Euler approximation, is $\dfrac{dx}{dt}\approx\dfrac{x(t+T_s)-x(t)}{T_s}$.

Thus, $\left.\dfrac{dx}{dt}\right|_{t=kT_s}=\dfrac{x(kT_s+T_s)-x(kT_s)}{T_s}$.

Using the forward difference, $\dfrac{x(kT_s+T_s)-x(kT_s)}{T_s}=-ax(kT_s)+bu(kT_s)$.

Denote $x(t)$ and $u(t)$ at discrete instances t_k and t_{k+1} as

$$x_k=x(t)\big|_{t=kT_s},\ x_{k+1}=x(t)\big|_{t=(k+1)T_s},\ u_k=u(t)\big|_{t=kT_s}.$$

Hence, $\dfrac{x_{k+1}-x_k}{T_s}=-ax_k+bu_k$, $x_{k+1}=x\left[(k+1)T_s\right]$, $x_k=x(kT_s)$, $u_k=u(kT_s)$.

The resulting difference equation is

$x_{k+1}=(1-aT_s)x_k+bT_su_k$, $x_{k+1}=a_kx_k+b_ku_k$, $a_k=(1-aT_s)$, $b_k=bT_s$.

or $x_k=\left(1-aT_s\right)x_{k-1}+bT_su_{k-1}=a_kx_{k-1}+b_ku_{k-1}$.

The transfer function is $G(z)=\dfrac{X(z)}{U(z)}=\dfrac{bT_sz^{-1}}{1-(1-aT_s)z^{-1}}=\dfrac{b_kz^{-1}}{1-a_kz^{-1}}=\dfrac{b_k}{z-a_k}$. ∎

5.3. Analog and Digital Proportional-Integral-Derivative Control

5.3.1. Analog Proportional-Integral-Derivative Control Laws

Examine control structures with the state and tracking error feedback. Optimization, stabilization and tracking control problems are solved by minimizing functionals considering electronic, electromechanical, mechanical and other systems. Control structures with state feedback are found by minimizing performance functionals and indexes. Proportional-integral-derivative (PID) control laws are designed using the tracking error $e(t)$ feedback. These algorithms ensure stability, reference $r(t)$ tracking, disturbances rejection, etc.

The goal is to design control laws $u\in U$, such that the closed-loop system evolutions (x,y,e) are bounded and converge to equilibrium (x_e,e_e) for initial conditions (x_0,e_0), admissible references r and disturbances d, guaranteeing:

1. $\lim_{t\to\infty}\|x(t)-x_e\|\le\delta_x$ and $\lim_{t\to\infty}\|x(t)-x_e\|\to 0$, $x\in X(X_0,U,R,D)$;
2. $\lim_{t\to\infty}\|e(t)\|\le\delta_e$ and $\lim_{t\to\infty}e(t)\to 0$, $e\in E(X,U,R,D)$.

A classical analog PID control law is

$$u(t)=\underbrace{k_pe(t)}_{\text{proportional feedback}}+\underbrace{k_i\int e(t)dt}_{\text{integral feedback}}+\underbrace{k_d\frac{de(t)}{dt}}_{\text{derivative feedback}},\ e(t)=r(t)-y(t),\ (k_p,k_i,k_d)>0,\quad(5.23)$$

where $e(t)$ is the tracking error, defined as difference between the reference and output, $e(t)=r(t)-y(t)$; k_p, k_i and k_d are the proportional, integral and derivative feedback gains, $k_p>0$, $k_i>0$ and $k_d>0$.

The diagram of a system with control law (5.23) is shown in Figure 5.5.

Using the Laplace operator $s=d/dt$, from (5.23), $U(s)=\left(k_p+\frac{k_i}{s}+k_ds\right)E(s)$.

Hence, the transfer function of a PID control law (5.23) is

$$G_{PID}(s)=\frac{U(s)}{E(s)}=\frac{k_ds^2+k_ps+k_i}{s}.\quad(5.24)$$

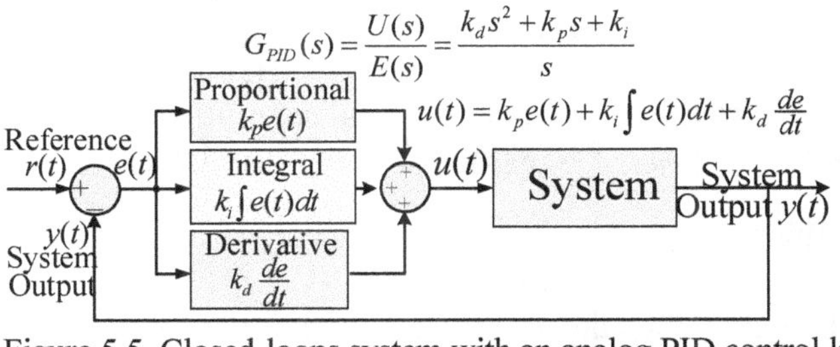

Figure 5.5. Closed-loops system with an analog PID control law

From (5.23), one has:

1. Proportional control law is $u(t)=k_pe(t)$;
2. Proportional-integral control is $u(t)=k_pe(t)+k_i\int e(t)dt$;
3. Proportional-derivative control law is $u(t)=k_pe(t)+k_d\frac{de(t)}{dt}$.

Linear and nonlinear control laws can be designed and implemented. Closed-loop systems with PID control laws in time- and s- domains are shown in Figures 5.6.a and b.

The inherent hardware-defined control limit is shown. If system is linear or can be linearized, as shown in Figure 5.6.b, the transfer function algebra, transfer function $G_{sys}(s)$, Laplace transforms, frequency-domain analysis, final-value theorem and other concepts of linear control theory are applied.

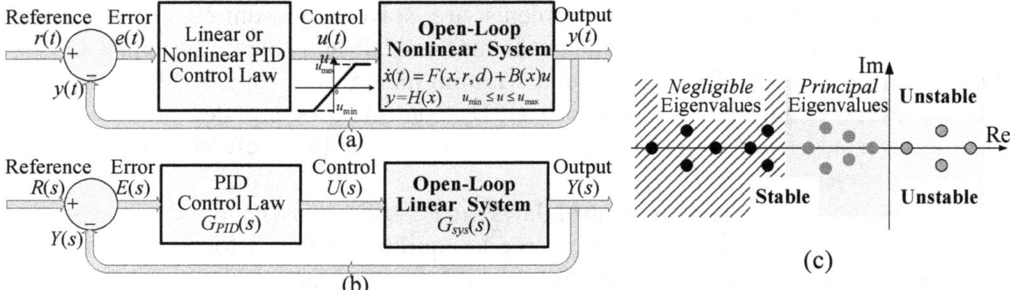

Figure 5.6. (a) Time-domain diagram of a nonlinear closed-loop system with a PID control law; (b) s-domain diagram of a linear closed-loop system with $G_{PID}(s)$ and $G_{sys}(s)$. Transfer function and tracking error are $G(s)=\dfrac{Y(s)}{R(s)}=\dfrac{G_{PID}(s)G_{sys}(s)}{1+G_{PID}(s)G_{sys}(s)}$ and $\lim\limits_{t\to\infty}e(t)=\lim\limits_{s\to 0}\dfrac{sR(s)}{1+G_{PID}G_{sys}}$;
(c) Linear system analysis: Characteristic eigenvalues in the complex plane.

Stability and tracking are achieved if:

1. Evolutions of system states $x(t)$ and output $y(t)$ are bounded for bounded inputs, and, $\lim\limits_{t\to\infty}\|x(t)-x_e\|\le\delta_x$;
2. System output $y(t)$ converges to the reference $r(t)$;
3. The tracking error $e(t)=r(t)-y(t)$ is bounded for bounded reference, and, $\lim\limits_{t\to\infty}\|e(t)\|\le\delta_e$, ideally $\lim\limits_{t\to\infty}e(t)\to 0$, $\forall r\in R$;
4. Specified tracking accuracy are achieved as $\lim\limits_{t\to\infty}|e(t)|\le\delta$, $\delta>0$, $\forall r\in R$.

Theoretically, a zero-error tracking can be ensured by assuming ideal measurements on $y(t)$, $r(t)$ and $e(t)$. Physical systems always exhibit a tracking error due to sensor errors, precision, digital-to-analog and analog-to-digital conversion errors, etc. In time domain and s-domain, the tracking error is $e(t)=r(t)-y(t)$, $E(s)=R(s)-Y(s)$.
For a closed-loop system, shown in Figure 5.6.b, the Laplace transform of the output is

$$Y(s)=G_{sys}(s)U(s)=G_{sys}(s)G_{PID}(s)E(s)=G_{sys}(s)G_{PID}(s)\left[R(s)-Y(s)\right]. \quad (5.25)$$

From (5.25), one finds $E(s)=\dfrac{R(s)}{1+G_{PID}(s)G_{sys}(s)}$.

The final value theorem yields the steady-state tracking error

$$\lim_{t\to\infty}e(t)=\lim_{s\to 0}\frac{sR(s)}{1+G_{PID}(s)G_{sys}(s)}. \quad (5.26)$$

The transfer function of the closed-loop linear systems is

$$G(s)=\frac{Y(s)}{R(s)}=\frac{G_{PID}(s)G_{sys}(s)}{1+G_{PID}(s)G_{sys}(s)},\ G(s)=k\frac{\prod_{i,j}\left(T_{n,i}s+1\right)\cdots\left(T_{n,j}^2s^2+2\xi_{n,j}T_{n,j}s+1\right)}{\prod_{i,j}\left(T_{d,i}s+1\right)\cdots\left(T_{d,j}^2s^2+2\xi_{d,j}T_{d,j}s+1\right)}, \quad (5.27)$$

where k is the constant factor; T_j and ξ_j are the time constants and damping coefficients.

The zeros and poles of a transfer function (5.27) can be found to evaluate stability margins, settling time, overshoot, etc. The proportional, integral and derivative feedback affect the eigenvalues, system dynamics, stability and other performance measures. The feedback gains k_p, k_i and k_d can be derived to ensure *principal* eigenvalues, while the negligible eigenvalues are located far left in the complex plane, see Figure 5.6.c. For linear systems, the *modal* control is accomplished assigning eigenvalues and tracking error.

Example 5.9. *One-Dimensional Motion of a Rigid Body: Force and Torque Control*
Study a *force* and *torque control* problems for a one-dimensional mechanical motion of frictionless mass. Disturbances and actuator dynamics are not considered. Translational and rotational equations of motion are found using the Newton second law.

One-dimensional motion of a lossless, not-perturbed rigid-body is

$$\frac{dx_1}{dt} = x_2,$$

$$\frac{dx_2}{dt} = u,\ y(t)=x_1(t).$$

States $x_1(t)$ and $x_2(t)$ denote the displacement and velocity, while u is the force or torque applied to control system motions. The output is the displacement, e.g., $y(t)=x_1(t)$.

The transfer function of the open-loop system is $G_{sys}(s)=\frac{1}{s^2}$.

The proportional-derivative control law is

$$u(t)=k_pe(t)+k_d\frac{de(t)}{dt},\ G_{PD}(s)=\frac{U(s)}{E(s)}=k_p+k_ds.$$

The transfer function of the closed-loop system is

$$G(s)=\frac{Y(s)}{R(s)}==\frac{G_{PID}(s)G_{sys}(s)}{1+G_{PID}(s)G_{sys}(s)}=\frac{(k_p+k_ds)\frac{1}{s^2}}{1+(k_p+k_ds)\frac{1}{s^2}}=\frac{k_ds+k_p}{s^2+k_ds+k_p}.$$

The characteristic equation is $s^2+k_ds+k_p=0$.

For any $k_p>0$ and $k_d>0$, the characteristic eigenvalues (s_1,s_2) have negative real parts. Hence, the closed-loop system is stable. One can specify the settling time and overshoot which yield characteristic eigenvalues, as well as (k_p,k_d).

Let the desired eigenvalues are $\lambda_1=-1$ and $\lambda_2=-1$. E.g., $(\lambda+1)(\lambda+1)=\lambda^2+2\lambda+1=0$.

The system and specified characteristic equations are

$s^2+k_ds+k_p=0$, $(\lambda+1)(\lambda+1)=\lambda^2+2\lambda+1=0$.

Equating the terms for s^1 and s^0, the feedback coefficients are $k_p=1$ and $k_d=2$.

The steady-state tracking error for the unit step reference $r(t)=1(t)$ with $R(s)=1/s$ is calculated using the final value theorem (5.26)

$$\lim_{t\to\infty}e(t)=\lim_{s\to 0}\frac{sR(s)}{1+G_{PID}(s)G_{sys}(s)}=\lim_{s\to 0}\frac{s\frac{1}{s}}{1+\left(k_p+k_ds\right)\frac{1}{s^2}}=\lim_{s\to 0}\frac{s^2}{s^2+k_ds+k_p}=0.$$

The steady-state error for the unit ramp function $r(t)=t$ with $R(s)=1/s^2$ is

$$\lim_{t\to\infty}e(t)=\lim_{s\to 0}\frac{sR(s)}{1+G_{PID}(s)G_{sys}(s)}=\lim_{s\to 0}\frac{s\frac{1}{s^2}}{1+\left(k_p+k_ds\right)\frac{1}{s^2}}=\lim_{s\to 0}\frac{s}{s^2+k_ds+k_p}=0.$$

For $r(t)=\cos(\omega_0t)$, one has $R(s)=\frac{s}{s^2+\omega_0^2}$. Hence,

$$\lim_{t\to\infty} e(t) = \lim_{s\to 0}\frac{sR(s)}{1+G_{PID}(s)G_{sys}(s)} = \lim_{s\to 0}\frac{s\dfrac{s}{s^2+\omega_0^2}}{1+\left(k_p+k_d s\right)\dfrac{1}{s^2}} = \lim_{s\to 0}\frac{s^4}{(s^2+\omega_0^2)(s^2+k_d s+k_p)} = 0\cdot$$

The system is stable, and, zero steady-state error is ensured. Evolutions of $x_1(t)$, $x_2(t)$ and $e(t)$ can be found by using the Laplace transform and state transition matrix $\Phi(t)$ for given $r(t)$ and initial conditions. The derivative feedback may result in sensitivity to noise and perturbations. Therefore, filters are used. ■

Example 5.10. *Control of a Permanent Magnet DC Motor*

Study an electric drive with a permanent magnet dc motor, covered in chapter 3. The output is the angular velocity ω_r. Using Kirchhoff's voltage law and Newton's second law of motion, a linear model is given by differential equations (3.19)

$$\frac{di_a}{dt} = \frac{1}{L_a}(-r_a i_a - k_a\omega_r + u_a),$$

$$\frac{d\omega_r}{dt} = \frac{1}{J}\left(k_a i_a - B_m\omega_r - T_L\right)\cdot$$

Denoting $x_1=i_a$, $x_2=\omega_r$, $u=u_a$, and the load torque T_L as disturbance $d=T_L$,

$$\frac{dx}{dt} = \begin{bmatrix}\dfrac{dx_1}{dt}\\ \dfrac{dx_2}{dt}\end{bmatrix} = \begin{bmatrix}-\dfrac{r_a}{L_a} & -\dfrac{k_a}{L_a}\\ \dfrac{k_a}{J} & -\dfrac{B_m}{J}\end{bmatrix}\begin{bmatrix}x_1\\ x_2\end{bmatrix} - \begin{bmatrix}0\\ \dfrac{1}{J}\end{bmatrix}T_L + \begin{bmatrix}\dfrac{1}{L_a}\\ 0\end{bmatrix}u = Ax + B_D d + Bu,\ y = x_2.$$

For an open-loop electric drive with output ω_r, we have a transfer function

$$G_{sys}(s) = \frac{Y(s)}{U(s)} = \frac{\Omega_r(s)}{U(s)} = \frac{k_a}{L_a Js^2 + (r_a J + L_a B_m)s + r_a B_m + k_a^2}\cdot$$

The characteristic equation is $L_a Js^2 + (r_a J + L_a B_m)s + r_a B_m + k_a^2 = 0$.

The solution of the quadratic equation $as^2+bs+c=0$ is $s_{1,2} = \dfrac{-b\pm\sqrt{b^2-4ac}}{2a}$.

Here, $a=L_aJ$, $b=(r_aJ+L_aB_m)$ and $c=(r_aB_m+k_a^2)$.

All motor parameters are positive. Hence, $(a,b,c)>0$. For any a, b and c, the real parts of eigenvalues are negative. Hence, the open-loop system is stable.

As the armature voltage u_a is applied, motor rotates at the particular angular velocity ω_r, which also depends on T_L. Study dynamics and stability using the linear control system theory. Any PID control laws will guarantee stability. Consider the proportional tracking control law

$u=k_pe$, $k_p>0$, $G_p(s)=k_p$.

With $y=\omega_r$, the tracking error is $e(t)=(r-y)=(r-\omega_r)$.

The transfer function of the closed-loop system is

$$G(s) = \frac{Y(s)}{R(s)} = \frac{\Omega_r(s)}{R(s)} = \frac{G_{PID}(s)G_{sys}(s)}{1+G_{PID}(s)G_{sys}(s)} = \frac{k_p k_a}{L_a Js^2 + (r_a J + L_a B_m)s + r_a B_m + k_a^2 + k_p k_a}\cdot$$

The characteristic equation is $L_a Js^2 + (r_a J + L_a B_m)s + r_a B_m + k_a^2 + k_p k_a = 0$.

For the closed-loop system, $a=L_aJ$, $b=(r_aJ+L_aB_m)$, $c=(r_aB_m+k_a^2+k_pk_a)$, $(a,b,c)>0$.

The stability is guaranteed because the real parts of all characteristic eigenvalues $s_{1,2}=\frac{-b\pm\sqrt{b^2-4ac}}{2a}$ are negative.

One may find the proportional feedback gain k_p of the control law $u=k_pe$ to ensure the specified tracking error. The steady-state error is found using (5.26).

For the unit step reference $r(t)=1$, $R(s)=1/s$.

Hence $$\lim_{t\to\infty} e(t)=\lim_{s\to 0}\frac{sR(s)}{1+k_pG_{sys}(s)}=\lim_{s\to 0}\frac{1}{1+k_p\dfrac{k_a}{L_aJs^2+(r_aJ+L_aB_m)s+r_aB_m+k_a^2}}$$

$$=\lim_{s\to 0}\frac{L_aJs^2+(r_aJ+L_aB_m)s+r_aB_m+k_a^2}{L_aJs^2+(r_aJ+L_aB_m)s+r_aB_m+k_a^2+k_pk_a}.$$

Let the sensor measures ω_r with accuracy 0.1%. Specify the control-dependent steady-state tracking error to be 0.1%. That is, $\lim_{t\to\infty} e(t)=e(\infty)=0.001$. Equation

$$\lim_{t\to\infty} e(t)=\lim_{s\to 0}\frac{L_aJs^2+(r_aJ+L_aB_m)s+r_aB_m+k_a^2}{L_aJs^2+(r_aJ+L_aB_m)s+r_aB_m+k_a^2+k_pk_a}=\frac{r_aB_m+k_a^2}{r_aB_m+k_a^2+k_pk_a}=0.001$$

yields k_p.

For illustrations, let the parameters are r_a=1 ohm, L_a=1 H, k_a=1 V-sec/rad, B_m=1 N-m-sec/rad and J=1 kg-m^2. One has k_p=1998.

The characteristic equation $L_aJs^2+(r_aJ+L_aB_m)s+r_aB_m+k_a^2+k_pk_a=s^2+2s+2000=0$ yields the eigenvalues $s_{1,2}=-1\pm 44.7i$. ■

Example 5.11. *Control of a Permanent Magnet DC Motor With a Current Feedback and Proportional-Integral Regulator*

For an electric drive with a permanent magnet dc motor, studied in *Example* 5.10, consider a PI control with a current feedback

$$u(t)=-k_Ii_a+k_pe+k_i\int e dt.$$

For an open-loop system, $G_{sys}(s)=\frac{\Omega(s)}{U(s)}=\frac{k_a}{L_aJs^2+(r_aJ+L_aB_m)s+r_aB_m+k_a^2}$.

The transfer function of an electric drive with a current feedback $-k_Ii_a$ is

$$G^*_{sys}(s)=\frac{k_a}{L_aJs^2+(r_aJ+k_IJ+L_aB_m)s+(r_a+k_I)B_m+k_a^2}.$$

For a closed-loop system, we have

$$G(s)=\frac{\Omega(s)}{R(s)}=\frac{G_{PID}(s)G^*_{sys}(s)}{1+G_{PID}(s)G^*_{sys}(s)}=\frac{\frac{k_ps+k_i}{s}G^*_{sys}(s)}{1+\frac{k_ps+k_i}{s}G^*_{sys}(s)}$$

$$=\frac{(k_ps+k_i)k_a}{L_aJs^3+(r_aJ+k_IJ+L_aB_m)s^2+(r_aB_m+k_IB_m+k_a^2+k_pk_a)s+k_ik_a}.$$

The characteristic equation is

$$L_aJs^3+(r_aJ+k_IJ+L_aB_m)s^2+(r_aB_m+k_IB_m+k_a^2+k_pk_a)s+k_ik_a=0.$$

Let the specified eigenvalues are $\lambda_1=-1$, $\lambda_2=-1$ and $\lambda_3=-1$.

The characteristic equation is $(\lambda+1)(\lambda+1)(\lambda+1)=\lambda^3+3\lambda^2+3\lambda+1=0$.

Comparison of these characteristic equations yields

$$\frac{r_aJ+k_1J+L_aB_m}{L_aJ}=3,\ \frac{r_aB_m+k_1B_m+k_a^2+k_pk_a}{L_aJ}=3,\ \frac{k_ik_a}{L_aJ}=1.$$

For illustrative purposes, assume all motor parameters are ones.
Hence, the feedback gains are k_1=1, k_p=0 and k_i=1.
The steady-state tracking error is found by using the final value theorem (5.26).
For $r(t)$=1, one has $R(s)$=1/s. We have

$$\lim_{t\to\infty}e(t)=\lim_{s\to0}\frac{sR(s)}{1+G_{PID}(s)G_{sys}^*(s)}=\lim_{s\to0}\frac{sR(s)}{1+\dfrac{k_ps+k_i}{s}G_{sys}^*(s)}$$

$$=\lim_{s\to0}\frac{s\left[L_aJs^2+(r_aJ+k_1J+L_aB_m)s+r_aB_m+k_1B_m+k_a^2\right]}{L_aJs^3+(r_aJ+k_1J+L_aB_m)s^2+(r_aB_m+k_1B_m+k_a^2+k_pk_a)s+k_ik_a}=0.$$

Hence, $\lim_{t\to\infty}e(t)=e(\infty)=0$, and, the steady-state tracking error is zero. ■

Consider a PID control structure with N_i integrals and N_d derivative terms

$$u(t)=\underbrace{k_pe(t)}_{\text{proportional}}+\underbrace{\sum_{j=1}^{N_i}\int\ldots\int k_{i_j}edt}_{\text{integral}}+\underbrace{\sum_{j=1}^{N_d}k_{d_j}\frac{d^je(t)}{dt^j}}_{\text{derivative}},U(s)=\left(k_p+\sum_{j=1}^{N_i}k_{i_j}\frac{1}{s^j}+\sum_{j=1}^{N_d}k_{d_j}s^j\right)E(s),\quad(5.28)$$

where N_i and N_d are the positive integers; k_{i_j} and k_{d_j} are the feedback coefficients.

From (5.28), one finds the corresponding transfer function $G_{PID}(s)$.
Nonlinear PID Control Laws – Nonlinear control laws have being designed and implemented. Use a nonlinear feedback mapping $\varphi(\cdot)$ as

$$u(t)=\varphi\left(e,\int edt,\tfrac{de}{dt}\right)=\underbrace{\sum_{z=1}^{Z_p}k_{p_z}e^{2z-1}(t)}_{\text{proportional}}+\underbrace{\sum_{j=1}^{N_i}\int\ldots\int\sum_{z=1}^{Z_i}k_{i_{j,z}}e^{2z-1}dt}_{\text{integral}}+\underbrace{\sum_{j=1}^{N_d}\sum_{z=1}^{Z_d}k_{d_{j,z}}\frac{d^je^{2z-1}(t)}{dt^j}}_{\text{derivative}},\quad(5.29)$$

where $\varphi(\cdot)$ is the tracking error mapping, defined by the PID control structure; Z_p, Z_i and Z_d are the positive integers; k_{p_z}, $k_{i_{j,z}}$ and $k_{d_{j,z}}$ are the proportional, integral and derivative feedback coefficients.

Example 5.12. *Linear and Nonlinear Proportional-Integral-Derivative Control Laws*
Consider (5.28). Let the integers (N, Z) are N_i=1, N_d=1, Z_p=1, Z_i=1 and Z_d=1.
We have a PID control law (5.23) $u(t)=k_pe(t)+k_i\int e(t)dt+k_d\frac{de}{dt}$.

For N_i=3 and N_d=1, $u(t)=k_pe(t)+k_{i1}\int e(t)dt+k_{i2}\iint e(t)dt+k_{i3}\iiint e(t)dt+k_d\frac{de(t)}{dt}$.

One finds a transfer function $G_{PID}(s)=\dfrac{U(s)}{E(s)}=\dfrac{k_ds^4+k_ps^3+k_{i1}s^2+k_{i2}s+k_{i3}}{s^3}$.

Let N_i=2, N_d=1, Z_p=3, Z_i=2 and Z_d=1. From (5.29), one obtains a nonlinear control law
$u(t)=k_{p1}e(t)+k_{p2}e^3(t)+k_{p3}e^5(t)+k_{i1,1}\int edt+k_{i2,1}\iint edt+k_{i1,2}\int e^3dt+k_{i2,2}\iint e^3dt+k_{d1,1}\frac{de}{dt}$. ■

Nonlinear Control Laws and Fractional Feedback Structures – Nonlinear control laws are aimed to improve system dynamics, enhance tracking accuracy, guarantee disturbance attenuation, etc. Sensitivity of control algorithms, stability and robustness impairments must be considered assessing sensor impediments, such as exhibited noise, errors, linearity, etc. Using the multi-index notations, the nonlinear PID-type control structure with fractional feedback terms, which lie in the first and third quadrants, is

$$u=\varphi\left(e,\int e dt,\frac{de}{dt}\right)$$
$$=\underbrace{\sum_{\substack{z=q_p\\ l=g_p}}^{Z_p,L_p} k_{p_{z,l}}\operatorname{sgn}(e)|e|^{\frac{2z-1}{2l-1}}}_{\text{proportional}}+\underbrace{\sum_{j=1}^{N_i}\int\ldots\int\sum_{\substack{z=q_i\\ l=g_i}}^{Z_i,L_i} k_{i_{j,z,l}}\operatorname{sgn}(e)|e|^{\frac{2z-1}{2l-1}}dt}_{\text{integral}}+\underbrace{\sum_{j=1}^{N_d}\sum_{\substack{z=q_d\\ l=g_d}}^{Z_d,L_d} k_{d_{j,z,l}}\operatorname{sgn}(e)\frac{d^j|e|^{\frac{2z-1}{2l-1}}}{dt^j}}_{\text{derivative}}, \quad (5.30)$$

where (Z_p,Z_i,Z_d), (L_p,L_i,L_d) and (N_i,N_d) are the non-negative integers; (q_p,q_i,q_d) and (g_p,g_i,g_d) are the integers, $(q_p,q_i,q_d)\geq 1$ and $(g_p,g_i,g_d)\geq 1$; $k_{p_{z,l}}$, $k_{i_{j,z,l}}$ and $k_{d_{j,z,l}}$ are the feedback gains.

Fractional Feedback and Notations – A fractional exponent is $a^{\frac{i}{j}}=(\sqrt[j]{a})^i=\sqrt[j]{a^i}$. The fractional exponent on e are real-valued, positive in the first quadrant, negative in the third quadrant, and, reduced notation is $|e|^{\frac{2z-1}{2l-1}}\in\mathbb{R}$. Fractional exponent feedback terms in (5.30) $\operatorname{sgn}(e)|e|^{\frac{2z-1}{2l-1}}$, $\operatorname{sgn}(e)|e|^{\frac{2z-1}{2l-1}}$ and $\operatorname{sgn}(e)|e|^{\frac{2z-1}{2l-1}}$, result in nonlinear feedback, such as $\operatorname{sgn}(e)|e(t)|^{1/3}$, $\operatorname{sgn}(e)|e(t)|^{1/7}$, $\operatorname{sgn}(e)|e(t)|^{3/7}$ and others. These feedback ensure large control effort $u(t)$ for small tracking error, improving accuracy and physical controllability.

Control Limits – Analytic design of bounded control laws and stability analysis are reported in chapter 6. The inherent hardware control limits $\mathcal{U}_{\min}\leq\mathcal{U}\leq\mathcal{U}_{\max}$ may be reached by control laws $u(t)$. One may design $u(t)$ which should not exceed the limits as $u_{\min}\leq u\leq u_{\max}$. For example, the duty ratio in PWM amplifiers and dc-dc regulators are constrained as $d_{D\min}\leq d_D\leq d_{D\max}$, $d_D\in[0\ \ 1]$ or $d_D\in[-1\ \ 1]$. For actuators and transducers, $\mathcal{U}_{\min}\leq\mathcal{U}\leq\mathcal{U}_{\max}$. If hardware limits are reached or there are nonlinear feedback, linear analysis may be inadequate. The eigenvalues-defined stability, *pole placement* design, *modal* control and other methods of linear control may yield inconsistencies. For physical systems, limits on voltage, current, charge, force, torque, power and other physical quantities are considered. There are mechanical limits on maximum angular and linear displacements, velocities, accelerations, etc. In addition to control limits $\mathcal{U}_{\min}\leq\mathcal{U}\leq\mathcal{U}_{\max}$, there are limits on physical variables $x_{\min}\leq x\leq x_{\max}$ as rated and peak (maximum allowed) voltages, currents, velocities and displacements are specified. The constrained PID control laws with linear and nonlinear feedback mappings and bound imposed are

$$u=\phi\left(\varphi(e,\int e dt,\tfrac{de}{dt})\right),\ u_{\min}\leq u\leq u_{\max},\ u_{\min}\leq\phi(\cdot)\leq u_{\max}, \quad (5.31)$$

$$u=\phi_{u_{\min}}^{u_{\max}}\left(\underbrace{\sum_{\substack{z=q_p\\ l=g_p}}^{Z_p,L_p} k_{p_{z,l}}\operatorname{sgn}(e)|e|^{\frac{2z-1}{2l-1}}}_{\text{proportional}}+\underbrace{\sum_{j=1}^{N_i}\int\ldots\int\sum_{\substack{z=q_i\\ l=g_i}}^{Z_i,L_i} k_{i_{j,z,l}}\operatorname{sgn}(e)|e|^{\frac{2z-1}{2l-1}}dt}_{\text{integral}}+\underbrace{\sum_{j=1}^{N_d}\sum_{\substack{z=q_d\\ l=g_d}}^{Z_d,L_d} k_{d_{j,z,l}}\operatorname{sgn}(e)\frac{d^j|e|^{\frac{2z-1}{2l-1}}}{dt^j}}_{\text{derivative}}\right).$$

Here, $\phi(\cdot)$ is the hardware-defined symmetric or asymmetric continuous, piecewise-continuous or discontinuous bounded function which represents the inherent hardware limits $\mathcal{U}_{\min}\leq\mathcal{U}\leq\mathcal{U}_{\max}$. Functions $\phi(\cdot)$ are reported in chapter 6 to describe $\mathcal{U}_{\min}\leq\mathcal{U}\leq\mathcal{U}_{\max}$, and, $u_{\min}\leq\phi(\cdot)\leq u_{\max}$. The range of $\phi(\cdot)$ is $[u_{\min},\ u_{\max}]$, while the domain of $\phi(\cdot)$ is $(-\infty,\ \infty)$.

The closed-loop system with a constrained control is shown in Figure 5.7.a. For the bounded PID control laws with linear proportional, integral and derivative terms,

$$u=\phi_{u_{\min}}^{u_{\max}}\left(k_p e+\sum_{j=1}^{N_i}\int\ldots\int k_{i_j}edt+\sum_{j=1}^{N_d}k_{d_j}\frac{d^j e}{dt^j}\right), u_{\min}\le u\le u_{\max},\ u_{\min}\le\phi(\cdot)\le u_{\max}. \quad (5.32)$$

Bounded functions $\phi(\cdot)$ are found for electronic, microelectronic, electromagnetic, electrostatic, hydraulic, mechanical and other devices. With symmetric or asymmetric $\phi(\cdot)$ and nonlinear feedback mapping $\varphi(\cdot)$, including fractional tracking error terms, one has

$$u=\phi_{u_{\min}}^{u_{\max}}\left(\sum_{m=1}^{M_p}k_{p_m}\operatorname{sgn}(e)|e|^{\sum_{p=1}^{L_p}\frac{2m-1}{2p-1}}+\sum_{j=1}^{N_i}\int\ldots\int\sum_{m=1}^{M_i}k_{i_{j,m}}\operatorname{sgn}(e)|e|^{\sum_{p=1}^{L_i}\frac{2m-1}{2p-1}}dt+\sum_{j=1}^{N_d}\sum_{m=1}^{M_d}k_{d_{j,m}}\operatorname{sgn}(e)\frac{d^j}{dt^j}|e|^{\sum_{p=1}^{L_d}\frac{2m-1}{2p-1}}\right),$$

$u_{\min}\le u\le u_{\max}$, $u_{\min}\le\phi(\cdot)\le u_{\max}$. (5.33)

As documented in Figure 5.7.a, control $u(t)$ varies between the minimum and maximum values, $u_{\min}\le u\le u_{\max}$. For a symmetric saturation $|u|\le u_{\max}$. If control function reaches bounds ($-u_{\max}$, $u_{\max}$), one has

$$u=\begin{cases}u_{\max}, & \varphi(\cdot)>u_{\max}, \text{ saturation on } u\\ \varphi(\cdot), & |\varphi(\cdot)|\le u_{\max}, \text{ linear control } u\\ -u_{\max}, & \varphi(\cdot)<u_{\max}, \text{ saturation on } u\end{cases}.$$

Due to control limits, nonlinear control theory is applied.

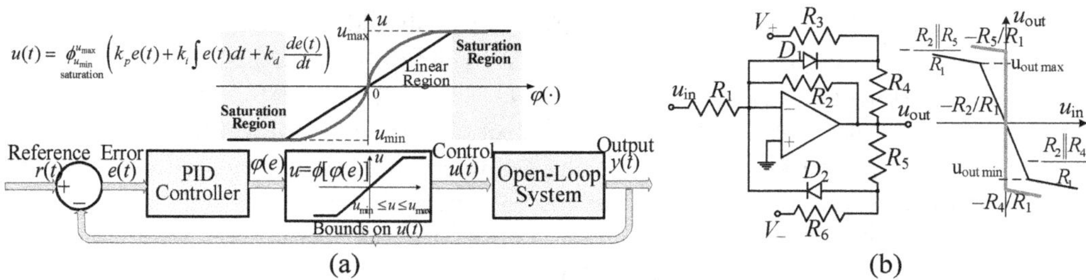

(a) (b)

Figure 5.7. (a) System with the constrained PID control law $u=\phi\left(\varphi(e,\int edt,\frac{de}{dt})\right)$, $u_{\min}\le u\le u_{\max}$; (b) Implementation of the variable-gain feedback using the operational amplifier with diodes.

Control laws are implemented by analog and digital controllers. The reference, output and error $e(t)$ are measured. Various hardware solutions exist to implement control laws. The inverted operational amplifier with a diode limiter circuit implements the nonlinear feedback as illustrated in Figure 5.7.b. The slops of the piecewise-continuous input-output characteristics (u_{in},u_{out}) are defined by the input and feedback resistors R_1 and R_2, as well by resistors connected to $\pm V$.

We have

$$u_{\text{out max}}=\frac{R_5}{R_6}V+\left(1+\frac{R_5}{R_6}\right)V_D,\quad u_{\text{out min}}=-\frac{R_4}{R_3}V-\left(1+\frac{R_4}{R_3}\right)V_D,$$

where V_D is the voltage drop at the diodes.

If R_2 is removed, one realized the comparator with the upper and lower values at $-R_5/R_1$ and $-R_4/R_1$. The implemented theoretical piecewise-continuous variable-gain feedback with input-output characteristic (u_{in},u_{out}) is continuous.

Example 5.13. *Control of a Servo With Permanent Magnet DC Motor*
Using high-torque density permanent magnet actuators, such as multi-pole permanent magnet synchronous motors and stepper motors, the direct-drive configuration can be due to low output mechanical angular velocity. For dc motors, planetary gearheads are used. Consider a servo which actuates a rotating stage as documented in Figure 5.8.a.

The platform angular displacement is a function of the rotor displacement. Using the gear ratio k_{gear}, the output equation is $y=Hx$, $y(t)=k_{gear}\theta_r(t)$. To change the angular velocity and displacement, one regulates the voltage applied to the armature winding u_a. The rated armature voltage for the motor is $\pm u_{\max}$. The rated current is $i_{a\max}$, and, the maximum angular velocity is $\omega_{r\max}$. We have $u_{\max}=30$ V, $-30\le u_a\le 30$ V, $i_{a\max}=0.15$ A, $\omega_{r\max}=150$ rad/sec, $r_a=200$ ohm, $L_a=0.002$ H, $k_a=0.2$ V-sec/rad, $J=2\times10^{-8}$ kg-m^2 and $B_m=5\times10^{-8}$ N-m-sec/rad. The reduction gear ratio is 100:1.

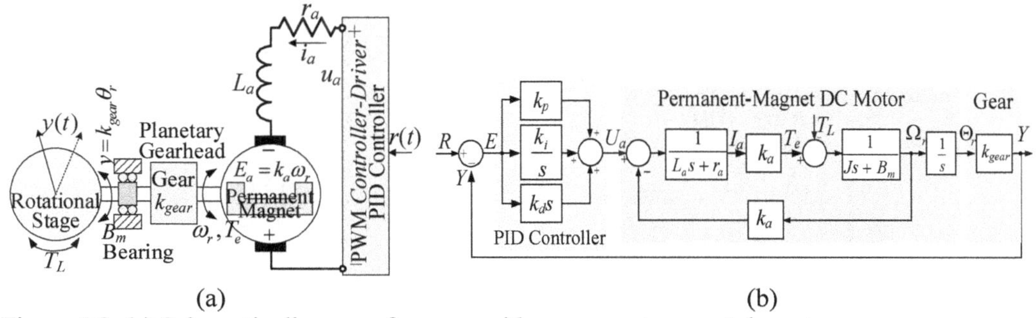

(a) (b)

Figure 5.8. (a) Schematic diagram of a servo with permanent magnet dc motor; (b) s-domain diagram of the closed-loop system with an analog PID control law.

The differential equations are

$$\frac{di_a}{dt}=\frac{1}{L_a}(-r_a i_a-k_a\omega_r+u_a),$$

$$\frac{d\omega_r}{dt}=\frac{1}{J}\left(T_e-T_{viscous}-T_L\right)=\frac{1}{J}\left(k_a i_a-B_m\omega_r-T_L\right),$$

$$\frac{d\theta_r}{dt}=\omega_r\,.$$

The output equation is

$y=k_{gear}\theta_r(t)$, $Y(s)=k_{gear}\Theta_r(s)$.

The s-domain diagram is documented in Figure 5.8.b.

The transfer function of an open-loop system is

$$G_{sys}(s)=\frac{Y(s)}{U_a(s)}=\frac{k_{gear}k_a}{s\left(L_aJs^2+\left(r_aJ+L_aB_m\right)s+r_aB_m+k_a^2\right)}.$$

For a PID control

$$u(t)=k_pe(t)+k_i\int e(t)dt+k_d\frac{de(t)}{dt},\; G_{PID}(s)=\frac{U(s)}{E(s)}=\frac{k_ds^2+k_ps+k_i}{s}.$$

The closed-loop transfer function is

$$G(s)=\frac{Y(s)}{R(s)}=\frac{G_{PID}(s)G_{sys}(s)}{1+G_{PID}(s)G_{sys}(s)}=\frac{\left(k_d s^2+k_p s+k_i\right)k_{gear}k_a}{s^2\left(L_a J s^2+\left(r_a J+L_a B_m\right)s+r_a B_m+k_a^2\right)+\left(k_d s^2+k_p s+k_i\right)k_{gear}k_a}$$

$$=\frac{\frac{k_d}{k_i}s^2+\frac{k_p}{k_i}s+1}{\frac{L_a J}{k_{gear}k_a k_i}s^4+\frac{r_a J+L_a B_m}{k_{gear}k_a k_i}s^3+\frac{r_a B_m+k_a^2+k_{gear}k_a k_d}{k_{gear}k_a k_i}s^2+\frac{k_p}{k_i}s+1}.$$

The numerical values of the numerator and denominator coefficients in the transfer function $G_{sys}(s)$ are found using MATLAB statements

```
format short e; ra=200; La=0.002; ka=0.2; J=0.00000002; Bm=0.00000005; kgear=0.01;
num_s=[ka*kgear]; den_s=[La*J (ra*J+La*Bm) (ra*Bm+ka^2) 0]; num_s, den_s
```

For the open-loop system, $G_{sys}(s)=\frac{Y(s)}{U(s)}=\frac{2\times10^{-3}}{s\left(4\times10^{-11}s^2+4\times10^{-6}s+4\times10^{-2}\right)}$.

The open-loop system is unstable because one of eigenvalues is at origin.

Using the **roots** command, the eigenvalues are computed

```
>> Eigenvalues=roots(den_s)
Eigenvalues =          0
               -8.8729e+04
               -1.1273e+04
```

The characteristic equation of the closed-loop system is

$$\frac{L_a J}{k_{gear}k_a k_i}s^4+\frac{r_a J+L_a B_m}{k_{gear}k_a k_i}s^3+\frac{r_a B_m+k_a^2+k_{gear}k_a k_d}{k_{gear}k_a k_i}s^2+\frac{k_p}{k_i}s+1=0.$$

The positive definite proportional, integral and derivative feedback coefficients (k_p, k_i, k_d)>0 affect the location of eigenvalues, defining the overshoot, settling time, tracking error, etc. Let k_p=25000, k_i=250 and k_d=2.5. The MATLAB statements are

```
kp=25000; ki=250; kd=2.5;
% Denominator of the closed-loop transfer function
den_c=[(La*J)/(kgear*ka*ki) (ra*J+La*Bm)/(kgear*ka*ki) (ra*Bm+ka^2+kgear*ka*kd)/(kgear*ka*ki) kp/ki 1];
Eigenvalues_Closed_Loop=roots(den_c) %Eigenvalues of the closed-loop system
```

The eigenvalues of the closed-loop system are found.

```
Eigenvalues_Closed_Loop = -8.7273e+04
                          -1.1482e+04
                          -1.2474e+03
                          -1.0000e-02
```

The closed-loop system is stable because the real parts of eigenvalues are negative. Study three cases: (1) k_p=25000, k_i=250 and k_d=2.5; (2) k_p=25000, k_i=25000000 and k_d=0; (3) k_p=25000, k_i=250 and k_d=0. To simulate the closed-loop system for $r(t)$=1, use the **lsim** command, and, the MATLAB code is

```
ra=200; La=0.002; ka=0.2; J=0.00000002; Bm=0.00000005; kgear=0.01; ref=1; % Reference: r(t)=1 rad
kp=25000; ki=250; kd=2.5;          % 1. PID feedback gains
% kp=25000; ki=25000000; kd=0;  % 2. PID feedback gains
% kp=25000; ki=250; kd=0;          % 3. PID feedback gains
num_c=[kd/ki kp/ki 1];  % Numerator and denominator of the closed-loop system
den_c=[(La*J)/(kgear*ka*ki) (ra*J+La*Bm)/(kgear*ka*ki) (ra*Bm+ka^2+kgear*ka*kd)/(kgear*ka*ki) kp/ki 1];
t=0:0.0001:0.015; u=ref*ones(size(t)); y=lsim(num_c,den_c,u,t);
plot(t,y,'k',y,u,'b--','LineWidth',3);
title('Angular Displacement, {\ity}({\itt})=0.01{\it\theta_r}, {\itr}({\itt})=1 rad','FontSize',18);
xlabel('Time [seconds]','FontSize',16);
ylabel('Output {\ity}({\itt}) and Reference {\itr}({\itt})','FontSize',16); axis([0 0.015,0 1.05])
```

The servo output $y(t)$ and reference $r(t)$ are illustrated in Figures 5.9 if $r(t)$=1 rad.

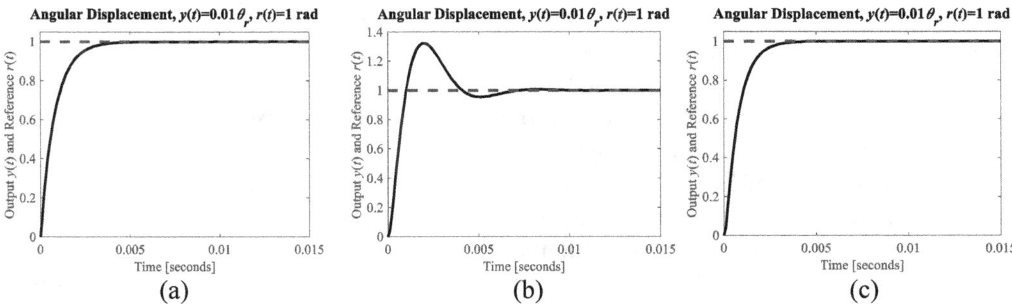

Figure 5.9. Dynamics of the closed-loop system with an analog PID control law: (a) k_p=25000, k_i=250 and k_d=2.5; (b) k_p=25000, k_i=25000000 and k_d=25; (c) k_p=25000, k_i=250 and k_d=0.

High-Performance, High-Bandwidth Direct-Drive Servos – Direct drives and servos are used to guarantee superior performance and capabilities. The adequate electromagnetic torque T_e must be developed to ensure the specified acceleration, loads rejection, etc. The actuator is directly connected to kinematics ensuring system simplicity, robustness, efficiency, moment of inertia minimization, size and weight reduction, etc. Furthermore, one eliminates backlash and dead zone, thereby ensuring repositioning accuracy, linearity, etc. With no gear, k_{gear}=1. We investigate the servo performance with motor and controller-driver limits $u_{\min}\le u\le u_{\max}$. Recall that u_a=±30 V, and, $-30\le u\le 30$ V. The SIMULINK diagram to perform simulations is similar to Figure 5.11.a. We use the PID, Saturation, nonlinear function f(u) and other blocks. The applied voltage is constrained. Not constrained and bounded PI control laws with k_p=1000 and k_i=1000 are

$$u = 1000e(t) + 1000\int e(t)dt,$$

$$u = \phi_{u_{\min}}^{u_{\max}}\left(1000e(t) + 1000\int e(t)dt\right), \; -30\le u\le 30 \text{ V}.$$

Transient dynamics of the system states, output evolutions and other quantitative performance variables are evaluated. Figure 5.10 documents the output $y(t)$ if the reference angular displacement is $r(t)=\pm1$ rad, $r(t)=\text{sgn}(\sin(\omega_0 t))$, $\omega_0=2\pi f$, f=15 Hz. The load $T_L(t)=\mp0.01$ N-m is applied at 0.03 sec, $T_L(t)\,|_{t=0.03\text{ sec}} = -0.01\text{sgn}(\sin(\omega_0 t))$, $\omega_0=2\pi f$, f=15 Hz. The reference magnitude and loads significantly affect control efforts, system dynamics, settling time, overshoot, etc. In case of the unconstrained control, the applied voltage reaches 2000 V. Physical limits and constraints result in many impediments, such as increased settling time, dynamic error, instability, sensitivity, etc. The control limits, nonlinear electromagnetics, friction and varying parameters must be studied.

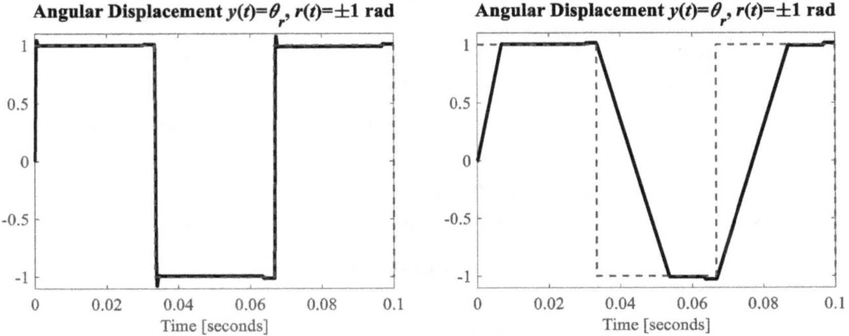

Figure 5.10. Closed-loop system dynamics with a linear control $u = 1000e(t) + 1000\int e(t)dt$, and, constrained control $u = \phi_{-30}^{+30}(1000e(t) + 1000\int e(t)dt)$, $r(t)=\pm1$ rad, $T_L(t)=\mp0.01$ N-m, f=15 Hz. ■

Example 5.14. *Proportional-Integral-Derivative Control for an Electric Drive*
For the industrial high-performance 34 NEMA-size permanent magnet dc motor, the parameters are: r_a=3.15 ohm, L_a=0.0066 H, k_a=0.16 V-sec/rad, J=0.0001 kg-m^2 and B_m=0.0001 N-m-sec/rad. The four-quadrant PWM amplifier supports high-frequency MOSFET switching, output voltage filtering, as well as the proportional-integral-derivative, current and angular velocity feedback. The rated and peak power, torque, angular velocity, voltage, current, overloading, impedance and other characteristics must be guaranteed by a chosen kinematics, actuator, *torque limiter*, *flexible shaft coupling* and PWM amplifier. The output PWM amplifier voltage and maximum armature voltage are limited as $-60\le \mathcal{U}\le 60$ V.

The output equation is $y=\omega_r(t)$, $Y(s)=\Omega_r(s)$.

The transfer function of a closed-loop system with PID control law is

$$G(s)=\frac{Y(s)}{R(s)}=\frac{G_{PID}(s)G_{sys}(s)}{1+G_{PID}(s)G_{sys}(s)}=\frac{\left(k_d s^2+k_p s+k_i\right)k_a}{s\left(L_aJs^2+\left(r_aJ+L_aB_m\right)s+r_aB_m+k_a^2\right)+\left(k_d s^2+k_p s+k_i\right)k_a}$$

$$=\frac{\left(k_d s^2+k_p s+k_i\right)k_a}{\frac{L_aJ}{k_ak_i}s^3+\frac{r_aJ+L_aB_m+k_ak_d}{k_ak_i}s^2+\frac{r_aB_m+k_a^2+k_ak_p}{k_ak_i}s+1}.$$

The characteristic equation is

$$\frac{L_aJ}{k_ak_i}s^3+\frac{r_aJ+L_aB_m+k_ak_d}{k_ak_i}s^2+\frac{r_aB_m+k_a^2+k_ak_p}{k_ak_i}s+1=0\,.$$

Specify the desired eigenvalues to be $\lambda_1=-0.1$ and $\lambda_{2,3}=-1450\pm4710i$.

The resulting feedback coefficients are k_p=100, k_i=10 and k_d=0.01.

Figure 5.11.a documents the SIMULINK diagram to perform simulations with not bounded and constrained PID control law

$$u=\phi_{u_{\min}}^{u_{\max}}\left(-k_Ii_a-k_\omega\omega_r+k_pe+k_i\int e\,dt+k_d\tfrac{de}{dt}\right),\ -60\le u\le 60.$$

For k_I=0 and k_ω=0, one has $u=\phi_{u_{\min}}^{u_{\max}}\left(k_pe+k_i\int e\,dt+k_d\tfrac{de}{dt}\right)$, $-60\le u\le 60$.

For not constrained control, control $u(t)$ reaches 6054.2 V, while the angular velocity exceeds ±1000 rad/sec for $r(t)$=±300 rad/sec. The assumed system linearity yields inadequate assessment. The *high gain* control may destabilize the open-loop stable systems due to constraints, nonlinearities, noise, measurement errors, sensitivity, etc. With an inherent hardware limit $-60\le \mathcal{U}\le 60$, perform analysis, assess settling time, evaluate overshoot, etc. Figure 5.11.b documents the electric drive output $y(t)=\omega_r(t)$ for reference angular velocity $r(t)$=±300 rad/sec, $r(t)$=300sgn(sin($\omega_0 t$)), $\omega_0=2\pi f$, f=5 Hz, and, load $T_L(t)$=±0.5 N-m, $T_L(t)\,|_{\tau=0.075\text{ sec}}$=0.5sgn(sin($\omega_0 t$)) N-m, $\omega_0=2\pi f$, f=5 Hz. The reference $r(t)$ and loads $T_L(t)$ affect the control efforts, system dynamics, settling time, overshoot, tracking error, etc. Applied voltage to motor is reported in Figure 5.11.b.

The tracking error and settling time are defined using the performance metrics.

To assess optimality, one may evaluate $J=\int_{t_0}^{t_f}t\left|e\right|dt$, $t\ge 0$.

Alternatively, a quadratic functional may be used $J=\int_{t_0}^{t_f}(qe^2+gu^2)dt$, $q\ge 0$, $g\ge 0$.

Evolution of $J(t)$ is assessed by simulating in SIMULINK, using experimental data.

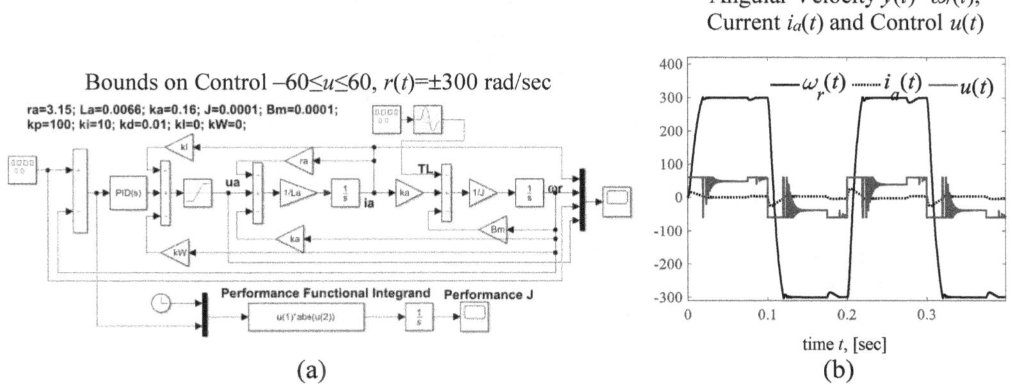

(a) (b)

Figure 5.11. (a) SIMULINK diagram of the closed-loop system with saturation $-60\le u\le 60$ V; (b) Closed-loop system dynamics with constrained control $u=\phi_{u_{\min}}^{u_{\max}}\left(k_p e+k_i\int edt+k_d\frac{de}{dt}\right)$, $-60\le u\le 60$, k_p=100, k_i=10, k_d=0.01. The reference angular velocity and loads are $r(t)=\pm 300$ rad/sec and $T_L(t)\,|_{\tau=0.075\text{ sec}}=\pm 0.5$ N-m. ■

MATLAB in Simulation, Analysis and Control of Dynamic Systems – MATLAB supports basic control law designs and analysis. Users learns and masters existing solutions, refines problems, and, develops new solutions to solve engineering problems. MATLAB demonstrates applications of simulation, analysis and control concepts to aerospace systems, robots, actuators and other systems. There are a number of various environments such as SIMULINK, as well as the Aerospace, Control System, Data Acquisition, Deep Learning, Instrument Control, MATLAB Coder, Model Predictive Control, Optimization, Robotics System, Robust Control, Signal Processing, Symbolic Math, System Identification and other toolboxes. Some solutions may be applied. Educational exercises not always suit industry- and technology-focused designs. One should verify simplifications, postulates and assumptions to solve engineering problems. Practitioners deal with complex systems hardware and software applying gained expertise and knowledge. MATLAB supports HDL code generation. The HDL Coder™ generates portable, synthesizable VHDL® and Verilog® code from MATLAB functions, SIMULINK models, and finite state machine Stateflow charts. The generated HDL code can be used with Xilinx FPGAs and Zynq SoC ARM processors. The FPGA IP cores are imported in Vivado IP Integrator. The HDL Coder empowers and automates code generation for development platforms with code verification.

Example 5.15. *Control of a Magnetic Levitation System*
In chapter 3, we studied translational variable reluctance electromechanical motion devices. Solenoids, electromagnets and magnetic levitation systems are used in various applications. Using laws of physics, nonlinear differential equations were derived. Various control approaches can be applied. Neural network and fuzzy logic concepts were studied as illustrative theoretical examples. Control methods and controllers must be implemented and substantiated for physical systems. Designed controllers must be system-complimentary, hardware-compliant, and, practical.

MATLAB offers various educational examples. Consider the NARMA-L2 control of a magnetic levitation system as reported in the Neural Network Toolbox, see Figure 5.12.a. The SIMULINK diagram is depicted in Figure 5.12.b. One validates and refines a "plant" model, derive and use adequate equations of motion for magnetic levitation systems, consider the applied voltage as a control variable, etc. We use the provided "plant" as a ready-to-use example. No changes are made to the "plant" model, coefficients and simulation settings. Compare a "neural controller", which may be impossible to implement, with a conventional PID control which can be implemented using a single operational amplifier.

The SIMULINK diagrams are reported in Figure 5.12 and 5.13.a. Simulation results are documented for a trained "neural controller". The saturation limits, which correspond to the "MARMA-L2 Controller", are $u_{\min}= -1$ and $u_{\max}=4$. The "neural controller" yields undesirable persistent high-frequency switching control activities, oscillations and overall inadequateness. The "plant" model is inconsistent. It assumes bidirectional electromagnetic force and other postulates. Furthermore, the axiomatized current-fed regulators, if chosen, imply pertained complexity and challenges.

A proportional control law

$$u = \phi_{u_{\min}}^{u_{\max}}(1000e), \quad u = \begin{cases} u_{\max}, & \varphi(\cdot) > u_{\max}, \text{ saturation on } u \\ \varphi(\cdot), & |\varphi(\cdot)| \le u_{\max}, \text{ linear control } u \\ u_{\min}, & \varphi(\cdot) < u_{\min}, \text{ saturation on } u \end{cases}$$

guarantees an adequate performance, and, ensures better performance than a hypothetical neural network solution. The "neural controller" yields the limit cycles and oscillations for $x_{\text{neural}}(t)$. PID control laws guarantee compliance and implementability.

Consider a proportional-derivative control law, given as

$$u = \text{saturation}_{u_{\min}}^{u_{\max}}\left(k_p e + k_d \frac{de}{dt}\right), \; e(t)=(r-y), \; k_p=1000, \; k_d=10.$$

The saturation block is used specifying $u_{\min}=-1$ and $u_{\max}=4$, see Figure 5.13.a. The simulated evolutions of control functions $i_{\text{neural}}(t)$ and $i(t)$, as well as the displacement dynamics $x_{\text{neural}}(t)$ and $x(t)$, are illustrated in Figures 5.13.b and c.

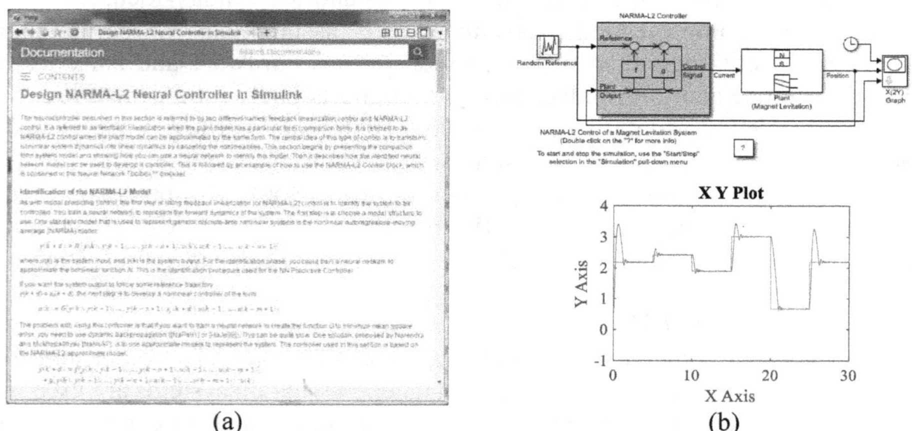

Figure 5.12. (a) NARMA-L2 "neural controller in Simulink", www.mathworks.com; (b) SIMULINK diagram: In the Command Window type **>> narmamaglev**. Simulation yields the system output for the reference input, *MATLAB R*2022, MathWorks, Inc., 2022.

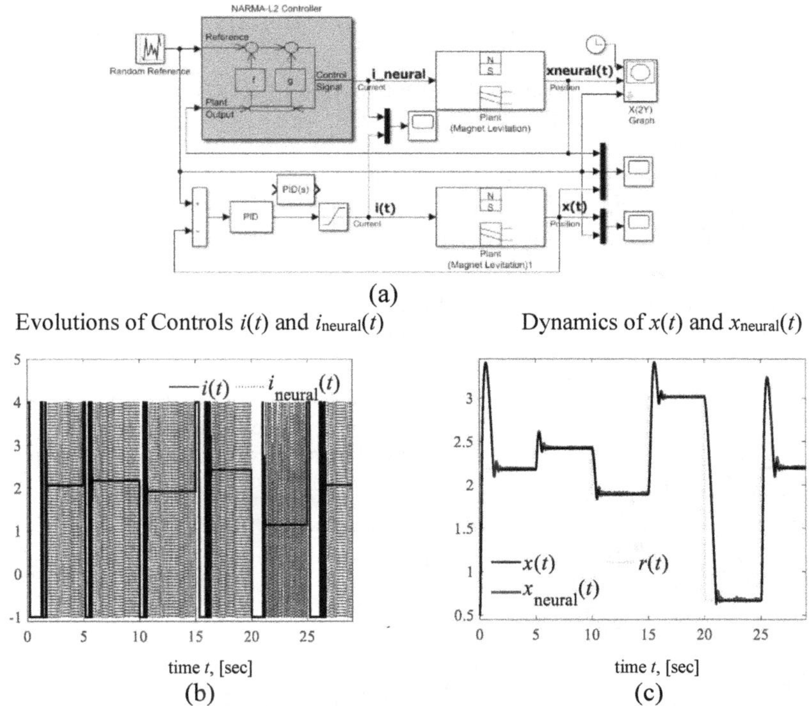

Figure 5.13. (a) SIMULINK diagram: NARMA-L2 "neural controller", and, a PID control law; (b) Evolutions of a proportional-derivative control $i(t)$ and "neural controller" $i_{\text{neural}}(t)$. The "trained neurocontroller" exhibits continuous switching $i_{\text{neural}}(t)$, while a proportional-derivative control law $u\equiv i$, $u=\phi_{u_{\min}}^{u_{\max}}(k_p e+k_d \frac{de}{dt})$, k_p=1000, k_d=10, $u_{\min}$= –1, $u_{\max}$=4 yields consistent tracking and continuous equilibrium for $i(t)$;
(c) Dynamics of displacement $x(t)$ and $x_{\text{neural}}(t)$ for reference $r(t)$.

For a magnetic levitation system, studied in section 3.3.1, a mathematical model is

$$\frac{di}{dt}=\frac{1}{L(x)+L_s}\left[-ri-iv\frac{\partial L(x)}{\partial x}+u\right],$$

$$\frac{dv}{dt}=\frac{1}{m}\left[\frac{1}{2}\frac{\partial L(x)}{\partial x}i^2-mg-F_\xi\right],$$

$$\frac{dx}{dt}=v,\ x_{\min}\le x\le x_{\max},$$

where $L(x)$ is the magnetizing inductance, which can be experimentally measured or derived using laws of physics; L_s is the self and leakage inductance; F_ξ is the disturbance force acting on the suspended mass.

For a magnetic levitation system, documented in Figure 5.14.a, the experimentally measured magnetizing inductance $L(x)$ and electromagnetic force $F_e(i,x)$ are found. As covered in Example 3.9, the physics-consistent $L(x)$ and $F_e(i,x)$ are

$$L(x)=\frac{ae^{-bx}}{(c+dx)},\ F_e(i,x)=\tfrac{1}{2}ae^{-bx}\frac{b(c+dx)+d}{(c+dx)^2}i^2,\ (a,b,c,d)>0.$$

The experimentally measured parameters are r_a=3.4 ohm, L_s=0.01 H and m=0.054 kg. For the magnetic levitation system, the one-directional electromagnetic force is

$$F_e(i,x)=\tfrac{1}{2}ae^{-bx}\frac{b(c+dx)+d}{(c+dx)^2}i^2,\ a=0.029,\ b=185,\ c=0.93,\ d=208.$$

The closed-loop system is designed with a proportional-integral control law $u(t)=\phi_{u_{\min}}^{u_{\max}}\left(k_pe+k_i\int edt\right)$, $0\le u\le 20$ V.

The variable-reluctance electromagnetic devices develop one-directional electromagnetic force F_e, and, the applied voltage is limited as $0\le u\le 20$ V. Design and stability analysis may be accomplished using the Lyapunov stability theory, covered in chapter 6. The feedback gains are found to be $k_p=3.51\times10^5$ and $k_i=4.49\times10^4$. The image of a magnetic levitation system is depicted in Figure 5.14.a. Figures 5.14.b report the experimental results in stabilization of the suspended mass, perturbation attenuation, as well as moving mass steering for $r=2$ mm.

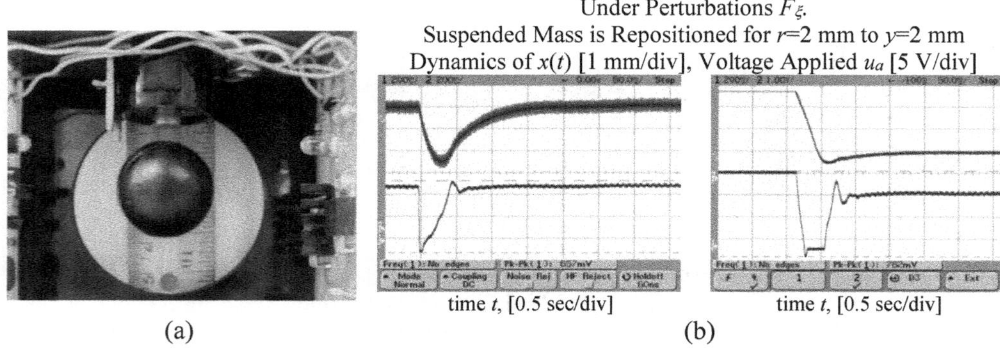

Figure 5.14. (a) Image of a magnetic levitation system with the suspended ferromagnetic ball; (b) Suspended mass stabilization under disturbances F_ξ, as well as steering if $r=2$ mm: Evolutions of the displacement x (top plot) and control voltage u (bottom plot). ■

5.3.2. Digital Proportional Integral Derivative Control

Microprocessors, DSPs and FPGAs are used to implement data fusion, data management, control, etc. Diagnostics, filtering, data acquisition and processing are performed using Boolean algebra, algebraic operations, discrete calculus and digital processing. Digital algorithms and protocols are implemented. For continuous-time systems, the measured analog $x(t)$, $y(t)$, $e(t)$, $r(t)$ and other quantities are sampled with the sampling time (also called sampling period and sampling interval) T_s. Analog and digital domains are related using transforms and operators, and, for time, $t=kT_s$, where k is the integer. In *hybrid* systems, analog physical modules are controlled using digital controllers. To design digital systems, differential equations are discretized yielding difference equations.

Discretization of linear state-space equations $\dot{x}=Ax+Bu$ as well as transfer functions $G(s)$ to $x_{k+1}=A_kx_k+B_ku_k$ and $G(z)$ were reported in section 5.2.2. One finds the difference equations for linear and nonlinear systems. Different methods are applied. There are analog and digital components in *hybrid* systems as shown in Figures 5.15 and 5.16. Linear and nonlinear continuous-time systems with digital controllers, *hybrid* circuits (A/D and D/A converters, data hold circuits, etc.), power electronics, analog actuators and kinematics are depicted.

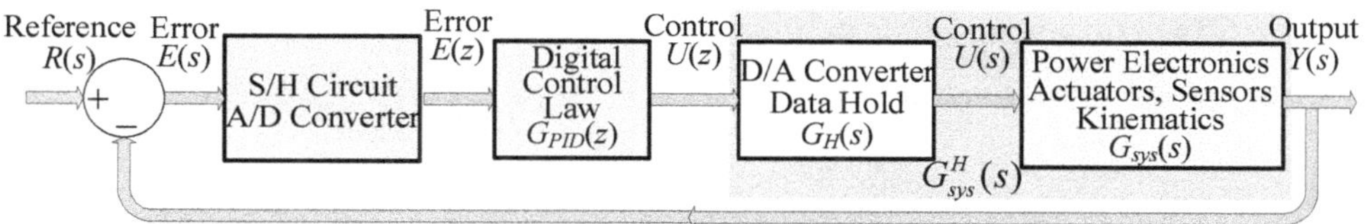

Figure 5.15. Linear *hybrid* systems with a digital PID control law

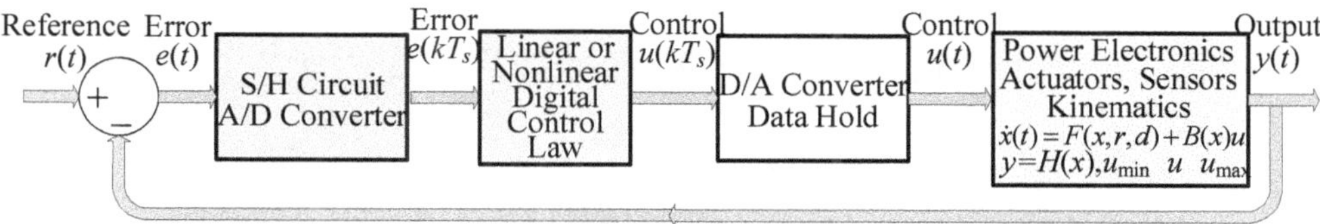

Figure 5.16. Nonlinear *hybrid* system with a digital control law

If system is described by linear constant-coefficient (time-invariant) differential equations, the closed-loop system is documented in Figure 5.15 using the transfer function for open-loop system $G_{sys}(s)$, data hold circuit $G_H(s)$, and digital control law $G_{PID}(z)$. To convert the discrete signals from a MPU to signals which drive transistors in PWM amplifiers, zero- and first-order data hold circuits are used to avoid complexity and time delay associated with high-order data holds. The N-order data hold circuit with the zero-order data hold is documented in Figure 5.17.

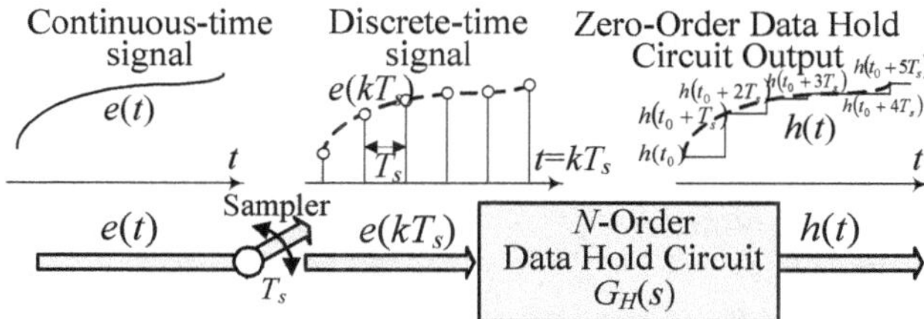

Figure 5.17. Sampler and N-order data hold circuit with zero-order data hold

For a zero-order data hold circuit, the piecewise continuous data hold output is

$$h(t)=\sum_{k=0}^{\infty}e(kT_s)[1(t-kT_s)-1(t-(k+1)T_s)]\cdot \tag{5.34}$$

The transfer function of the zero-order data hold is

$$G_H(s)=\frac{1-e^{-T_s s}}{s}\ . \tag{5.35}$$

The first-order data hold performs the direct linear extrapolation, expressed in time domain as

$$h(t)=1(t)+\frac{t}{T_s}1(t)-\frac{t-T_s}{T_s}1(t-T_s)-1(t-T_s)\,. \tag{5.36}$$

Hence, the transfer function of the first-order data hold is

$$G_H(s)=\frac{1}{s}+\frac{1}{T_s s^2}-\frac{1}{T_s s^2}e^{-T_s s}-\frac{1}{s}e^{-T_s s}=\left(1-e^{-T_s s}\right)\frac{T_s s+1}{T_s s^2}\,. \tag{5.37}$$

The system $G_{sys}(s)$ with the data hold circuit $G_H(s)$ yields

$$G_{sys}^H(s)=G_H(s)G_{sys}(s)\,. \tag{5.38}$$

Example 5.16. Discretization of Analog Control Laws
An analog proportional control law is $u(t)=k_p e(t)$, $G_P(s)=\dfrac{U(s)}{E(s)}=k_p$.

We have $G_P(z)=\dfrac{U(z)}{E(z)}=k_p$.

A proportional digital control law in common notations is
$u(kT_s)=k_p e(kT_s)$, $u[k]=k_p e[k]$, $u_k=k_p e_k$, or, $u[n]=k_p e[n]$, $u_n=k_p e_n$.

For an integral term $u(t)=k_i\int e(t)dt$, $G_I(s)=\dfrac{U(s)}{E(s)}=\dfrac{k_i}{s}$.

The Euler approximation yields $G_I(z)=\dfrac{U(z)}{E(z)}=k_i\dfrac{T_s}{1-z^{-1}}=k_i\dfrac{T_s z}{z-1}$.

Hence, $u_k=u_{k-1}+k_i T_s e_k$, $u_k=u[k]$.

The derivative term is $u(t)=k_d\dfrac{de(t)}{dt}$, $G_D(s)=k_d s$.

The trapezoidal approximation of the first difference yields

$$G_D(z)=\frac{U(z)}{E(z)}=k_d\frac{1-z^{-1}}{T_s}=k_d\frac{z-1}{T_s z}.$$

The algorithmic realization of the derivative controller is

$$u_k=\frac{k_d}{T_s}\left(e_k-e_{k-1}\right),\ u_k=u[k],\ e_k=e[k],\ e_{k-1}=e[k-1],$$

or

$$u_n=\frac{k_d}{T_s}\left(e_n-e_{n-1}\right),\ u_n=u[n],\ e_n=e[n],\ e_{n-1}=e[n-1].$$

■

Proportional-Integral-Derivative Control: Euler and Trapezoidal Approximations – Consider the PID control law, for which

$$u(t)=k_p e(t)+k_i\int e(t)dt+k_d\frac{de(t)}{dt},\ G_{PID}(s)=\frac{k_p s+k_i+k_d s^2}{s},\ U(s)=\frac{k_p s+k_i+k_d s^2}{s}E(s).$$

Using the Euler and trapezoidal approximations

$$G_{PID}(z)=\frac{U(z)}{E(z)}=k_p+k_i\frac{T_s}{1-z^{-1}}+\frac{k_d}{T_s}\left(1-z^{-1}\right)=\frac{\left(k_p+T_s k_i+\frac{1}{T_s}k_d\right)-\left(k_p+\frac{2}{T_s}k_d\right)z^{-1}+\frac{1}{T_s}k_d z^{-2}}{1-z^{-1}}. \tag{5.39}$$

From

$$U(z)\left(1-z^{-1}\right)=\left[\left(k_p+T_s k_i+\frac{1}{T_s}k_d\right)-\left(k_p+\frac{2}{T_s}k_d\right)z^{-1}+\frac{1}{T_s}k_d z^{-2}\right]E(z), \tag{5.40}$$

we have

$$u_n=u_{n-1}+k_{e0}e_n+k_{e1}e_{n-1}+k_{e2}e_{n-2},\ u_{n-1}=u[n-1],\ e_n=e[n],\ e_{n-1}=e[n-1],\ e_{n-2}=e[n-2], \tag{5.41}$$

where $k_{e0}=k_p+T_s k_i+\frac{1}{T_s}k_d$, $k_{e1}=-k_p-\frac{2}{T_s}k_d$ and $k_{e2}=\frac{1}{T_s}k_d$.

PID Control in the Reference-Output Form – The *reference-output* form of a digital PID control law is found by using $E(s)=R(s)-Y(s)$. In particular, one uses

$$G_{PID}(z)=\frac{U(z)}{E(z)}=\frac{U(z)}{R(z)-Y(z)}. \tag{5.42}$$

The feedback gains of the digital control laws are related to the proportional, integral and derivative coefficients of the analog PID control law $(k_p,k_i,k_d)>0$, as well as affected by the sampling time T_s. Different analytic and numerical approaches can be applied to derive feedback gains, such as approximating the integral term by the trapezoidal summation, expressing the derivative term by a two-point difference, etc. Approximations of integration (rectangular, trapezoidal, Tustin, bilinear, etc.) and differentiation (Euler, Taylor, backward difference, etc.) result in different transfer functions $G_{PID}(z)$ and expressions for $u(k)$. One may not add stand-alone proportional, integral and derivative terms to express u_n, which may yield inconsistent solutions.

PID Control and Controller Implementation: Tustin Approximation – Apply the Tustin approximation (5.21) $s=\frac{1}{T_s}\ln(z)\approx\frac{2}{T_s}\frac{1-z^{-1}}{1+z^{-1}}$.

The transfer function of the PID control law is $G_{PID}(s)=\frac{U(s)}{E(s)}=\frac{k_ds^2+k_ps+k_i}{s}$.

Hence, we have

$$G_{PID}(z)=\frac{U(z)}{E(z)}=\frac{k_d\left(\frac{2}{T_s}\frac{1-z^{-1}}{1+z^{-1}}\right)^2+k_p\frac{2}{T_s}\frac{1-z^{-1}}{1+z^{-1}}+k_i}{\frac{2}{T_s}\frac{1-z^{-1}}{1+z^{-1}}}$$
$$=\frac{\left(2T_sk_p+T_s^2k_i+4k_d\right)+\left(2T_s^2k_i-8k_d\right)z^{-1}+\left(-2T_sk_p+T_s^2k_i+4k_d\right)z^{-2}}{2T_s\left(1-z^{-2}\right)}. \tag{5.43}$$

For input $E(z)$ and output $U(z)$, one finds

$$\left[2T_s\left(1-z^{-2}\right)\right]U(z)=\left[\left(2T_sk_p+T_s^2k_i+4k_d\right)+\left(2T_s^2k_i-8k_d\right)z^{-1}+\left(-2T_sk_p+T_s^2k_i+4k_d\right)z^{-2}\right]E(z),$$
$$\left(1-z^{-2}\right)U(z)=\left(k_{e0}+k_{e1}z^{-1}+k_{e2}z^{-2}\right)E(z), \tag{5.44}$$
$$k_{e0}=k_p+\tfrac{1}{2}T_sk_i+\tfrac{2}{T_s}k_d,\quad k_{e1}=T_sk_i-\tfrac{4}{T_s}k_d,\quad k_{e2}=-k_p+\tfrac{1}{2}T_sk_i+\tfrac{2}{T_s}k_d.$$

From (5.44), one implements the digital controller using e_n, e_{n-1}, e_{n-2} and u_{n-2} as

$$u_n=u_{n-2}+k_{e0}e_n+k_{e1}e_{n-1}+k_{e2}e_{n-2},\ u_{n-2}=u[n-2],\ e_n=e[n],\ e_{n-1}=e[n-1],\ e_{n-2}=e[n-2]. \tag{5.45}$$

Expression (5.45) gives the control function u_n, computed using $(u_{n-2},e_n,e_{n-1},e_{n-2})$ at sampling instants. The gains (k_{e0},k_{e1},k_{e2}) are found using (k_p,k_i,k_d) and T_s. The sampling time T_s depends on the sensors bandwidth, data fusion rates, filters latencies, etc. Other control laws are discretized as documented in Practice Problem 5.5.

The closed-loop system with a digital control law $G_{PID}(z)$ is illustrated in Figure 5.15. The transfer function of the closed-loop systems is

$$G(z)=\frac{Y(z)}{R(z)}=\frac{G_{PID}(z)G_H(z)G_{sys}(z)}{1+G_{PID}(z)G_H(z)G_{sys}(z)}. \tag{5.46}$$

Analysis of linear discrete-time systems is conducted by computing the characteristic eigenvalues and steady-state error, performing simulations, etc.

Example 5.17. *Digital Servo With a Permanent Magnet DC Motor*
Consider a pointing servo, actuated by a permanent magnet dc motor. Analog control laws were examined in Example 5.13. Study the digital PID control laws. The objectives are to guarantee stability, attain the fast repositioning and minimize tracking error. The block diagram of the closed-loop system with a digital PID controller, A/D and D/A converters, and data hold circuit is documented in Figure 5.15. Using the transfer function of the open-loop system and zero-order data hold, we have

$$G_{sys}^{H}(s)=G_H(s)G_{sys}(s)=\frac{1-e^{-T_s s}}{s}\frac{k_{gear}k_a}{s\left(L_aJs^2+\left(r_aJ+L_aB_m\right)s+r_aB_m+k_a^2\right)}.$$

The parameters are r_a=200 ohm, L_a=0.002 H, k_a=0.2 V-sec/rad, J=2×10^{-8} kg-m^2, B_m=5×10^{-8} N-m-sec/rad and k_{gear}=100. The transfer function in z-domain is found using the **c2dm** command. The **filter** command supports simulations. We perform discretization, simulations and plotting using the following code.

```
ra=200; La=0.002; ka=0.2; J=0.00000002; Bm=0.00000005; kgear=0.01; % Motor parameters
% Numerator and denominator of the open-loop transfer function
num_s=[ka*kgear]; den_s=[La*J ra*J+La*Bm ra*Bm+ka^2 0]; num_s; den_s;
% Numerator and denominator of GD(z) with zero-order data hold
Ts=0.0001;          % Sampling time Ts
[num_dz,den_dz]=c2dm(num_s,den_s,Ts,'zoh'); num_dz; den_dz;
% Feedback gains of the digital control law
kp=1000; ki=1000; kd=1;
kd1=(2*Ts*kp+Ts^2*ki+4*kd)/(2*Ts); kd2=(2*Ts^2*ki-8*kd)/(2*Ts);kd3=(-2*Ts*kp+Ts^2*ki+4*kd)/(2*Ts);
% Numerator and denominator of the transfer function of the controller
num_pidz=[kd1 kd2 kd3]; den_pidz=[1 0 -1]; num_pidz; den_pidz;
% Numerator and denominator of the closed-loop transfer function G(z)
num_z=conv(num_pidz,num_dz); den_z=conv(den_pidz,den_dz)+conv(num_pidz,num_dz);
num_z; den_z;
k_final=1500; k=0:1:k_final;        % Samples, t=k*Ts
ref=1; r=ref*ones(1,k_final+1);  % Reference (command) input is r=1 rad
% Servo output y(k)
y=filter(num_z,den_z,r);
plot(k,y,'o','LineWidth',3); hold on; plot(k,y,'k','LineWidth',1.5); hold on;
plot(k,r,'b--','LineWidth',1); hold off; axis([0 1500, 0 1.05])
% title('Angular Displacement, {\ity}({\itk})=0.01{\it\theta_r}, {\itr}=1 rad','FontSize',16);
% xlabel('Discrete Time [{\itk}], Continuous Time {\itt=kT_s} [sec]','FontSize',16);
```

The open-loop system transfer function is $G_{sys}(s)=\dfrac{Y(s)}{U(s)}=\dfrac{2\times10^{-3}}{s\left(4\times10^{-11}s^2+4\times10^{-6}s+4\times10^{-2}\right)}$.

The sampling time T_s, defined by sensors, data fusion and protocols, significantly affects system dynamics. For T_s=0.0001 sec, and, a zero-order hold, one finds

$$G_{sys}^{H}(s)=G_H(z)G_{sys}(z)=\frac{1.65\times10^{-6}z^2+1.71\times10^{-6}z+2.63\times10^{-8}}{z^3-1.32z^2+0.324z-4.54\times10^{-5}}.$$

The transfer function of the digital PID control law is (5.43)

$$G_{PID}(z)=\frac{k_{d1}z^2+k_{d2}z+k_{d3}}{z^2-z}.$$

For k_p=1000, k_i=1000, k_d=1, T_s=0.0001 sec, using (5.43), we have k_{d1}=21000, k_{d2}=–40000 and k_{d3}=19000. The transfer function of the closed-loop system is

$$G(z)=\frac{Y(z)}{R(z)}=\frac{G_c(z)G_H(z)G_{sys}(z)}{1+G_c(z)G_H(z)G_{sys}(z)},\;G(z)=\frac{0.0346z^4-0.03z^3-0.0364z^2+0.0314z+0.0005}{z^5-1.29z^4-0.706z^3+1.29z^2-0.293z+0.000545}.$$

The output dynamics for the reference input $r(kT_s)$=1 rad, k≥0 is shown in Figure 5.18. The settling time is $t_{settling}$=1000×0.0001=0.1 sec, and, there is no overshoot.

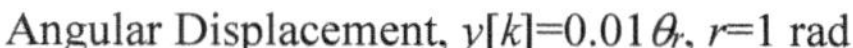

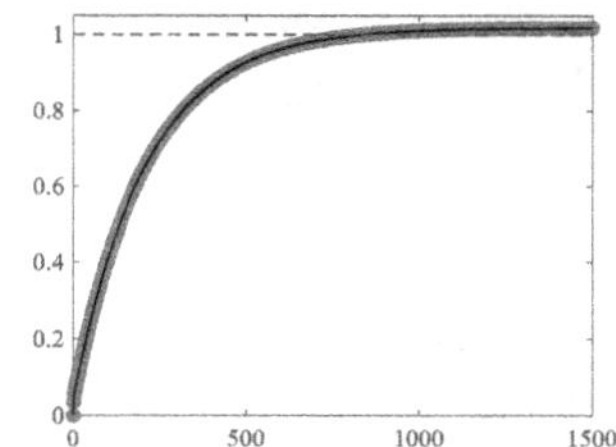

Discrete Time [k], Conterminous Time $t=kT_s$ [sec]

Figure 5.18. Output dynamics of the system with a digital PID control law

For different T_s, one recalculates the feedback gains. For sensor-consistent, data fusion dependent and MPU-supported T_s, one computes the feedback gains to ensure stability, adequate dynamics, etc. Furthermore, control limits must be examined. ■

5.4. *Modal* Control and State *Observer* Design

5.4.1. *Modal* Control

Adequately designed physical systems are controllable. These systems are described by governing and constitutive equations. Consider time-invariant MIMO linear systems (5.5) $\frac{dx}{dt}=Ax+Bu$, governance of which is given by (5.9). Specify the eigenvalues $\boldsymbol{\lambda}\in\boldsymbol{\Lambda}$, $\boldsymbol{\lambda}\in\mathbb{R}\times\mathbb{C}$ or $\forall\boldsymbol{\lambda}\in\mathbb{R}$, $\forall\text{Re}(\boldsymbol{\lambda}_F)<0$, and, design the state-dependent control law

$$u=-Kx, \tag{5.47}$$

to ensure the assigned eigenvalues $\boldsymbol{\lambda}$ and transient modes for the closed-loop system.

For a closed-loop MIMO system (5.5)-(5.47) under disturbances $d(t)$

$$\frac{dx}{dt}=Ax+Bu+B_Dd=\left(A-BK_F\right)x+B_Dd=\mathbf{A}x+B_Dd\,,\ \mathbf{A}=(A-BK_F),\ t\geq 0, \tag{5.48}$$

where $K_F\in\mathbb{R}^{m\times n}$ is the feedback gain matrix.

The *modal* matrix $\mathbf{M}$ was used to find the state transition matrix and examine the system dynamics. In particular,

$$\Phi(t)=e^{\mathbf{A}t}=\mathbf{M}e^{\Lambda t}\mathbf{M}^{-1},\ x(t)=e^{\mathbf{A}t}x_0=\mathbf{M}e^{\Lambda t}\mathbf{M}^{-1}x_0,\ e^{\Lambda t}=\begin{bmatrix} e^{\lambda_1 t} & \cdots & 0 \\ \vdots & \ddots & \vdots \\ 0 & \cdots & e^{\lambda_n t}\end{bmatrix}.$$

For a closed-loop system with $\mathbf{A}=(A-BK_F)$, $\mathbf{A}\in\mathbb{R}^{n\times n}$, the *modal* matrix of the eigenvectors is $\mathbf{M}=[\boldsymbol{v}_1,\ldots,\boldsymbol{v}_n]$, $\mathbf{M}\in\mathbb{R}^{n\times n}$, such that $\mathbf{M}^{-1}\mathbf{A}\mathbf{M}=\boldsymbol{\Lambda}$.

The diagonal matrix of eigenvalues is $\boldsymbol{\Lambda}=\begin{bmatrix} \lambda_1 & \cdots & 0 \\ \vdots & \ddots & \vdots \\ 0 & \cdots & \lambda_n\end{bmatrix}$, $\boldsymbol{\Lambda}\in\mathbb{R}^{n\times n}$.

The eigenvectors $\boldsymbol{v}_i$ are found by solving $(\lambda_i I-\mathbf{A})\boldsymbol{v}_i=0$.

The eigenvalues of the closed-loop system $\boldsymbol{\lambda}$ are found by using the characteristic polynomial $p_{\mathbf{A}}(\lambda)=\det(\boldsymbol{\lambda}I-\mathbf{A})$. The characteristic equation $\det(\boldsymbol{\lambda}I-\mathbf{A})=0$ yields eigenvalues $\boldsymbol{\lambda}$. For matrix $\mathbf{A}\in\mathbb{R}^{n\times n}$, one finds n eigenvalues λ_i, $i=1,\ldots,n$.

For a closed-loop system (5.48), one may ensure specified dynamic modes as $x(t)=e^{\mathbf{A}t}x_0=\mathbf{M}e^{\Lambda t}\mathbf{M}^{-1}x_0$ with $\boldsymbol{\lambda}\in\mathbb{R}\times\mathbb{C}$, $\forall\text{Re}(\boldsymbol{\lambda}_i)<0$.

A *modal* control law is a control law design scheme, while the modal matrix yields the state transition matrix to assess system governance.

Specifying eigenvalues $\boldsymbol{\lambda}\in\mathbb{R}\times\mathbb{C}$, $\forall\text{Re}(\boldsymbol{\lambda}_i)<0$, we find a *modal* control law (5.47).

Definition 5.1. *Controllability* – The linear system (5.5) $\frac{dx}{dt}=Ax+Bu$, $y=Hx$ is controllable if the controllability matrix

$$C=[B\ AB\ \ldots\ A^{n-1}B],\ C\in\mathbb{R}^{n\times n} \tag{5.49}$$

has rank n, rank(C)=n. If rank(C)=$r<n$, then, only r eigenvalues of $(A-BK_F)\in\mathbb{R}^{n\times n}$ can be assigned. ■

Conjecture 5.1. For a controllable system (5.5), the *modal* control law (5.47) $u=-K_Fx$ with a constant-coefficient feedback matrix K_F guarantees placement and correspondence of the closed-loop system eigenvalues

$$\alpha(\mathbf{s})=\det\left(sI-\left(A-BK_F\right)\right)=\det(sI-\mathbf{A})=0,\ \mathbf{A}=(A-BK_F),$$

to the specified eigenvalues

$$\beta(\boldsymbol{\lambda})=(\lambda_1,\lambda_2,\ldots,\lambda_n),\ \forall\text{Re}(\boldsymbol{\lambda}_i)<0,\ \alpha(\mathbf{s})=\beta(\boldsymbol{\lambda}),$$

and, the state transition matrix is $\Phi(t)=e^{\mathbf{A}t}=\mathbf{M}e^{\Lambda t}\mathbf{M}^{-1}$. ■

The assigned real and complex conjugate-pair eigenvalues are $\beta(\boldsymbol{\lambda})=(\lambda_1,\lambda_2,\ldots,\lambda_n)$, $\forall\text{Re}(\lambda_i)<0$. The feedback gain matrix K_F is found such that $\alpha(\mathbf{s})=\det[sI-(A-BK_F)]=0$ guarantees equality to the specified $\beta(\boldsymbol{\lambda})=(\lambda-\lambda_1)(\lambda-\lambda_2)\ldots(\lambda-\lambda_n)=0$. The Ackermann method is applied to find $K_F\in\mathbb{R}^{m\times n}$ for given $(A,B,\beta(\boldsymbol{\lambda}))$.

For servos, gimbals, industrial and mobile robots, aerial systems and other dynamic systems, tracking control problems may be solved using the extended system dynamics as documented in sections 5.7, 5.8 and 5.9.

Example 5.18. *Modal and State Transition Matrices, and, Modal Control*

Consider a system $\frac{dx}{dt}=Ax=\begin{bmatrix}0 & 1\\ -2 & -3\end{bmatrix}\begin{bmatrix}x_1\\ x_2\end{bmatrix}$, $A\in\mathbb{R}^{2\times 2}$.

In Example 5.5, the state transition matrix is found using the Caley-Hamilton theorem (5.10) as $\Phi(t)=e^{At}=\begin{bmatrix}2e^{-t}-e^{-2t} & e^{-t}-e^{-2t}\\ -2e^{-t}+2e^{-2t} & -e^{-t}+2e^{-2t}\end{bmatrix}$. Using a modal matrix $\mathbf{M}$,

$$\Phi(t)=e^{At}=\mathbf{M}e^{\Lambda t}\mathbf{M}^{-1},\ \mathbf{M}=\begin{bmatrix}v_{1_{\lambda_1}}, v_{2_{\lambda_2}}\end{bmatrix}=\begin{bmatrix}v_{1_1} & v_{2_1}\\ v_{1_2} & v_{2_2}\end{bmatrix},\ e^{\Lambda t}=\begin{bmatrix}e^{-\lambda_1 t} & 0\\ 0 & e^{-\lambda_2 t}\end{bmatrix}.$$

For $A=\begin{bmatrix}0 & 1\\ -2 & -3\end{bmatrix}$, the characteristic equations is $|\lambda I-A|=0$, $\lambda^2+3\lambda+2=0$.

The eigenvalues are $\lambda_1=-1$ and $\lambda_2=-2$. Therefore, $e^{\Lambda t}=\begin{bmatrix}e^{-t} & 0\\ 0 & e^{-2t}\end{bmatrix}$.

Find $\mathbf{M}=[\boldsymbol{v}_1, \boldsymbol{v}_2]$ by computing the eigenvectors $(\boldsymbol{v}_1,\boldsymbol{v}_2)$ solving $(\lambda_iI-A)\boldsymbol{v}_i=0$.

From $\begin{bmatrix} \lambda_i & -1 \\ 2 & \lambda_i+3 \end{bmatrix}\begin{bmatrix} v_{i_1} \\ v_{i_2} \end{bmatrix}=\begin{bmatrix} 0 \\ 0 \end{bmatrix}$, i=1,2, for $\lambda_1=-1, \lambda_2=-2$, the eigenvectors are

$\boldsymbol{v}_{1_{\lambda=-1}}=\begin{bmatrix} 1 \\ -1 \end{bmatrix}$, $\boldsymbol{v}_{2_{\lambda=-2}}=\begin{bmatrix} 1 \\ -2 \end{bmatrix}$.

Hence, $\mathbf{M}=\begin{bmatrix} 1 & 1 \\ -1 & -2 \end{bmatrix}$ and $\mathbf{M}^{-1}=\begin{bmatrix} 2 & 1 \\ -1 & -1 \end{bmatrix}$.

Therefore, $\Phi(t)=e^{At}=\mathbf{M}e^{\Lambda t}\mathbf{M}^{-1}=\begin{bmatrix} 1 & 1 \\ -1 & -2 \end{bmatrix}\begin{bmatrix} e^{-t} & 0 \\ 0 & e^{-2t} \end{bmatrix}\begin{bmatrix} 2 & 1 \\ -1 & -1 \end{bmatrix}=\begin{bmatrix} 2e^{-t}-e^{-2t} & e^{-t}-e^{-2t} \\ -2e^{-t}+2e^{-2t} & -e^{-t}+2e^{-2t} \end{bmatrix}$.

For a system $\frac{dx}{dt}=Ax=\begin{bmatrix} 0 & 1 \\ -2 & -3 \end{bmatrix}\begin{bmatrix} x_1 \\ x_2 \end{bmatrix}+\begin{bmatrix} 0 \\ 1 \end{bmatrix}u$,

find a control law $u=-K_Fx$ to guarantee the specified eigenvalues $\lambda_1=-10$ and $\lambda_2=-20$.

Find and compare two characteristic equations.

The characteristic equation for a closed-loop system with unknown (k_{F1},k_{F2}) is

$$\left|sI-(A-BK_F)\right|=\left|\begin{bmatrix} s & 0 \\ 0 & s \end{bmatrix}-\begin{bmatrix} 0 & 1 \\ -2 & -3 \end{bmatrix}-\begin{bmatrix} 0 \\ 1 \end{bmatrix}\begin{bmatrix} k_{F1} & k_{F2} \end{bmatrix}\right|=\left|\begin{bmatrix} s & -1 \\ 2+k_{F1} & s+3+k_{F2} \end{bmatrix}\right|=s^2+(3+k_{F2})s^2+2+k_{F1}=0.$$

A characteristic equation for the assigned eigenvalues $\lambda_1=-10$ and $\lambda_2=-20$ is $(\lambda+10)(\lambda+20)=\lambda^2+30\lambda+200=0$.

From $\alpha(\mathbf{s})=\beta(\boldsymbol{\lambda})$, we have k_{F1}=198 and k_{F2}=27.

The stabilizing control law (5.47), which guarantees $\lambda_1=-10$ and $\lambda_2=-20$, is $u=-K_Fx=-[198\ \ 27]x=-198x_1-27x_2$.

The MATLAB code below uses the **place**, **eig** and **acker** commands to solve this problem.

```
A=[0 1; -2 -3]; B=[0; 1];   % Matrices A and B
p=[-10 -20];                % Desired eigenvalues
KF=place(A,B,p); KF, KF=acker(A,B,p); KF  % Compute feedback matrix KF
Aclosed=A-B*KF, Eigenvalues=eig(Aclosed) % Compute closed-loop system matrix and eigenvalues
```

Analytic results are substantiated by finding

```
KF = 198.0000  27.0000
Eigenvalues =  -10.0000
               -20.0000
```

■

Example 5.19. *Control of a Servo*

Design a stabilizing control law (5.47), and, find K_F for a servo with a permanent magnet dc motor. The governing equations are reported in Example 5.13

$$\frac{di_a}{dt}=\frac{1}{L_a}(-r_ai_a-k_a\omega_r+u_a),\quad \frac{d\omega_r}{dt}=\frac{1}{J}(k_ai_a-B_m\omega_r-T_L),\quad \frac{d\theta_r}{dt}=\omega_r,\ y=\theta_r.$$

Let all parameters are equal to one. Hence,

$$\frac{dx}{dt}=Ax+Bu=\begin{bmatrix} -1 & -1 & 0 \\ 1 & -1 & 0 \\ 0 & 1 & 0 \end{bmatrix}x+\begin{bmatrix} 1 \\ 0 \\ 0 \end{bmatrix}u,\ x=\begin{bmatrix} x_1 \\ x_2 \\ x_3 \end{bmatrix}=\begin{bmatrix} i_a \\ \omega_r \\ \theta_r \end{bmatrix},\ y=Hx,\ H=[0\ 0\ 1],\ y=x_3,\ u=u_a.$$

The control law $u=-K_Fx$ should guarantee the specified eigenvalues.

Let $\lambda_1=-1$, $\lambda_2=-2$, $\lambda_3=-3$.

Find and compare two characteristic equations.

The characteristic equation for a closed-loop system is

$$|sI-(A-BK_F)| = \left|\begin{bmatrix} s & 0 & 0 \\ 0 & s & 0 \\ 0 & 0 & s \end{bmatrix} - \begin{bmatrix} -1 & -1 & 0 \\ 1 & -1 & 0 \\ 0 & 1 & 0 \end{bmatrix} + \begin{bmatrix} 1 \\ 0 \\ 0 \end{bmatrix} \begin{bmatrix} k_{F1} & k_{F2} & k_{F3} \end{bmatrix}\right| = \left|\begin{bmatrix} s+1+k_{F1} & 1+k_{F2} & k_{F3} \\ -1 & s+1 & 0 \\ 0 & -1 & s \end{bmatrix}\right|$$

$$= s^3 + (2+k_{F1})s^2 + (2+k_{F1}+k_{F2})s + k_{F3} = 0.$$

A characteristic equation for the assigned eigenvalues is
$(\lambda+1)(\lambda+2)(\lambda+3) = \lambda^3 + 6\lambda^2 + 11\lambda + 6 = 0$.

From $\alpha(\mathbf{s})=\beta(\boldsymbol{\lambda})$, we have $k_{F1}=4$, $k_{F2}=5$ and $k_{F3}=6$.

The stabilizing control law (5.47), which guarantees $\lambda_1=-1$, $\lambda_2=-2$, $\lambda_3=-3$, is
$u=-K_Fx=-[4\;\;5\;\;6]x=-4x_1-5x_2-6x_3$.

The MATLAB code solves this problem.

```
A=[-1 -1 0; 1 -1 0; 0 1 0]; B=[1; 0; 0]; H=[0 0 1]; D=[0]; % Matrices A, B, H and D
p=[-1 -2 -3];                                  % Desired eigenvalues
KF=place(A,B,p); KF, KF=acker(A,B,p); KF   % Compute feedback matrix KF
Aclosed=A-B*KF, Eigenvalues=eig(Aclosed) % Compute closed-loop system matrix and eigenvalues
```

The controllability and observability matrices, and, their ranks, are found as

```
C=ctrb(A,B); C, Crank=rank(C)   % Controllability matrix C, and, computing rank of C
O=obsv(A,H); O, Orank =rank(O) % Observability matrix O, and, computing rank of O
```

We have $C=\begin{bmatrix} 1 & -1 & 0 \\ 0 & 1 & -2 \\ 0 & 0 & 1 \end{bmatrix}$, rank($C$)=3, and, $O=\begin{bmatrix} 0 & 0 & 1 \\ 0 & 1 & 0 \\ 1 & -1 & 0 \end{bmatrix}$, rank($O$)=3. ■

Example 5.20. *Modal Control of a Servo*

For a servo, considered in Examples 5.13 and 5.19, design a stabilizing control law (5.47) $u=-K_Fx$. The parameters are r_a=3.15 ohm, L_a=0.0066 H, k_a=0.16 V-sec/rad, J=0.0001 kg-m^2, B_m=0.0001 N-m-sec/rad and k_{gear}=0.1.

The control is bounded as $-30 \le u_a \le 30$ V. The eigenvalues should be consistent with admissible dynamic modalities, system limits, etc. Inadequate eigenvalues result in high-gain control, sensitivity, overshoot, instabilities, etc. One has

$$\frac{dx}{dt} = Ax + Bu = \begin{bmatrix} -r_a/L_a & -k_a/L_a & 0 \\ k_a/J & -B_m/J & 0 \\ 0 & 1 & 0 \end{bmatrix} x + \begin{bmatrix} 1/L_a \\ 0 \\ 0 \end{bmatrix} u,\; y=Hx,\; H=[0\;\;0\;\;k_{gear}],\; y=k_{gear}x_3,\; u=u_a.$$

Design a stabilizing control law $u=-K_Fx$ to guarantee $\lambda_1=-100$, $\lambda_2=-500$, $\lambda_3=-1000$.

The MATLAB code to analyze controllability and observability by evaluating ranks of controllability and observability matrices (C,O), as well as to compute feedback matrix is

```
ra=3.15; La=0.0066; ka=0.16; J=0.0001; Bm=0.0001; kgear=0.1;  % Parameters
A=[-ra/La -ka/La 0; ka/J -Bm/J 0; 0 1 0]; B=[1/La; 0; 0]; H=[0 0 kgear]; D=[0]; % A, B, H and D
C=ctrb(A,B); Crank=rank(C), C, Crank  % Controllability matrix
O=obsv(A,H); Orank=rank(O), O, Orank % Observability matrix
p=[-100 -500 -1000]; KF=place(A,B,p); KF % Desired eigenvalues, and, computing feedback matrix KF
Aclosed=A-B*KF, Eigenvalues=eig(Aclosed)  % Closed-loop system matrix and eigenvalues
```

We have rank(C)=3, rank(O)=3 and K_F=[7.4 2.51 206.3].

The stabilizing control law is
$u=-K_Fx=-[7.4\;\;2.51\;\;206.3]x=-7.4x_1-2.51x_2-206.3x_3$.

Hence, $u_a=-7.4i_a-2.51\omega_r-206.3\theta_r$.

With no constraints on applied voltage u_a, transient dynamics is documented in Figure 5.19.a for initial conditions $\begin{bmatrix} x_{10} \\ x_{20} \\ x_{30} \end{bmatrix} = \begin{bmatrix} 10 \\ 25 \\ 1 \end{bmatrix}$. The stabilization problem is solved, and, the eigenvalues are $\boldsymbol{\lambda}=(-100, -500, -1000)$.

Even with an adequate choice of the eigenvalues, limits on control u are reached. Control limits significantly affect system dynamics as shown in Figure 5.19.b. One cannot use eigenvalues to characterize nonlinear systems.

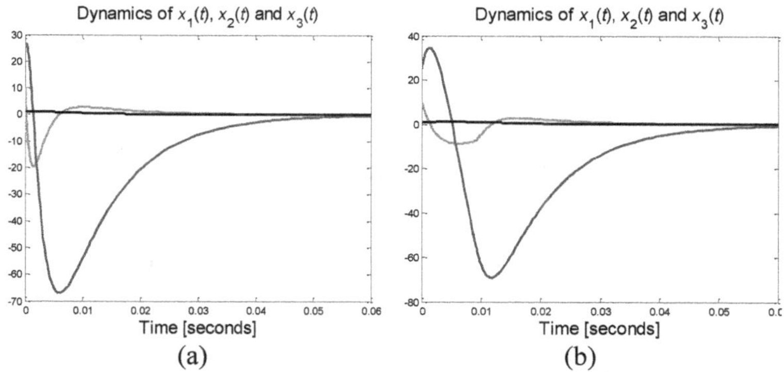

(a) (b)

Figure 5.19. (a) Dynamics of a linear closed-loop system (x_1,x_2,x_3), $\lambda_1=-100,\lambda_2=-500,\lambda_3=-1000$; (b) Dynamics of the closed-loop system if control is bounded as $-30\le u\le 30$. ■

5.4.2. State *Observer*

A state *observer* is designed using the measured inputs $u(t)$ and output $y(t)$ to observe states $x(t)$ as $\hat{x}(t)$. The equation of the state *observer* is

$$\frac{d\hat{x}}{dt} = A\hat{x} + Bu + K_E(y-\hat{y}) = \underbrace{(A-K_EH)}_{\bar{\mathbf{A}}=(A-K_EH)}\hat{x} + Bu + K_Ey = \bar{\mathbf{A}}\hat{x} + Bu + K_Ey, \quad (5.50)$$

where $\hat{x}(t)$ is the observed state vector which should yield the observed $x(t)$; $K_E\in\mathbb{R}^{n\times b}$ is the *observer* gain matrix.

Input and Output Pair – To design the state *observer*, one assumes the measured outputs and inputs pair (y,u). The input vector may comprise of controls $u(t)$, disturbances $d(t)$ and perturbations $w(t)$. That is, in (5.50), one has $u \equiv \begin{bmatrix} u_{\text{control}}(t) \\ d_{\text{disturbance}}(t) \\ w_{\text{perturbation}}(t) \end{bmatrix}$. Measurements of disturbances and perturbations with pertained noise cause challenges.

Definition 5.2. *Observability* – The linear system (5.5)

$$\frac{dx}{dt} = Ax + Bu,\; y=Hx$$

is observable if the observability matrix

$$O=[H \;\; HA \;\ldots\; HA^{n-1}]^T,\; O\in\mathbb{R}^{n\times n} \quad (5.51)$$

has rank n, rank$(O)=n$. If rank$(O)=r<n$, then, only r eigenvalues of $(A-K_EH)$, $K_E\in\mathbb{R}^{n\times b}$ can be assigned. If rank$(O)=n$, all states can be observed using a pair of vectors (u,y).■

The *observer* error is $\bar{e}(t)=x(t)-\hat{x}(t)$, $\bar{e}\in\mathbb{R}^n$.

The error evolution is described by differential equations, and, the eigenvalues are found using the characteristic equation

$$\frac{d\bar{e}}{dt}=\left(A-K_EH\right)\bar{e}=\bar{\mathbf{A}}\bar{e},\ \bar{e}(t_0)=\bar{e}_0,\ t\geq 0,\ |\lambda I-(A-K_EH)|=0,\ \bar{e}(t)=x(t)-\hat{x}(t),\quad (5.52)$$

$$\bar{e}(t)=e^{\bar{\mathbf{A}}t}\bar{e}_0=\bar{\mathbf{M}}e^{\mathbf{\Lambda}_E t}\bar{\mathbf{M}}^{-1}\bar{e}_0,\ e^{\mathbf{\Lambda}_E t}=\begin{bmatrix} e^{\lambda_{E1}t} & \cdots & 0\\ \vdots & \ddots & \vdots\\ 0 & \cdots & e^{\lambda_{En}t}\end{bmatrix},\ \mathbf{\Lambda}_E=\begin{bmatrix}\lambda_{E1} & \cdots & 0\\ \vdots & \ddots & \vdots\\ 0 & \cdots & \lambda_{En}\end{bmatrix}.$$

where $\bar{\mathbf{M}}$ is the *modal* matrix of the eigenvectors of $\bar{\mathbf{A}}\in\mathbb{R}^{n\times n}$, $\bar{\mathbf{M}}=[\bar{v}_1,\ldots,\bar{v}_n]$, $\bar{\mathbf{M}}^{-1}\bar{\mathbf{A}}\bar{\mathbf{M}}=\mathbf{\Lambda}_E$; $\mathbf{\Lambda}_E$ is the diagonal matrix of eigenvalues, $\mathbf{\Lambda}_E\in\mathbb{R}^{n\times n}$; $\bar{v}_i$ are the eigenvectors, found solving $(\lambda_i I-\bar{\mathbf{A}})\bar{v}_i=0$.

Observed states $\hat{x}(t)$ asymptotically converge to $x(t)$ if $\lim_{t\to\infty}\bar{e}(t)\to 0$.

The *observer* gain matrix K_E is found by specifying the eigenvalues. One solves the problem $\alpha(\mathbf{s})=\beta(\boldsymbol{\lambda}_E)$, finding the unknown matrix K_E.

Conjecture 5.2. If system (5.5)

$$\frac{dx}{dt}=Ax+Bu,\ y=Hx$$

is observable, the states can be observed using the measured input and output pair (u,y). There exists a constant-coefficient *observer* gain matrix $K_E\in\mathbb{R}^{n\times b}$ such that the assigned eigenvalues $\boldsymbol{\lambda}_E=(\lambda_{E1},\lambda_{E2},\ldots,\lambda_{En})$ for $(A-K_EH)\in\mathbb{R}^{n\times n}$,

$$\det\left(sI-\left(A-K_EH\right)\right)=\det(sI-\bar{\mathbf{A}})=0$$

are guaranteed by the state *observer* (5.50). The asymptotic convergence of the observed states $\hat{x}(t)$ to $x(t)$ is guaranteed, and, the state transition matrix $\bar{\Phi}(t)=e^{\bar{\mathbf{A}}t}=\bar{\mathbf{M}}e^{\mathbf{\Lambda}_E t}\bar{\mathbf{M}}^{-1}$ defines the governance of the *observer* error $\bar{e}(t)$. ■

5.4.3. *Modal* Control Law and State *Observer* Design

The closed-loop system (5.48) $\frac{dx}{dt}=Ax+Bu=\left(A-BK_F\right)x$, $y=Hx$ with the *modal* control law (5.47), and, state *observer* dynamics (5.50), are governed as

$$\begin{cases}\dfrac{dx}{dt}=Ax+Bu=Ax-BK_F\hat{x},\ y=Hx,\\[4pt] \dfrac{d\hat{x}}{dt}=\left(A-BK_F-K_EH\right)\hat{x}+K_Ey,\end{cases}\quad \begin{bmatrix}\dfrac{dx}{dt}\\[4pt] \dfrac{d\hat{x}}{dt}\end{bmatrix}=\underbrace{\begin{bmatrix}A & -BK_F\\ K_EH & A-BK_F-K_EH\end{bmatrix}}_{\boldsymbol{A}=\begin{bmatrix}A & -BK_F\\ K_EH & A-BK_F-K_EH\end{bmatrix}}\begin{bmatrix}x\\ \hat{x}\end{bmatrix}.\quad (5.53)$$

For the closed-loop system and state *observer* (5.53) with $\boldsymbol{A}=\begin{bmatrix}A & -BK_F\\ K_EH & A-BK_F-K_EH\end{bmatrix}$, there is a union of eigenvalues in (5.53)

$$\{|sI-(A-BK_F)|=0,\ |sI-(A-K_EH)|=0,\ |sI-(A-BK_F-K_EH)|=0\}.$$

The derived (5.53) yields a representative controller with *observer* diagrams, reported in Figures 5.20. For a system under disturbances, $\dot{x}=Ax+Bu+B_Dd$, vectors (u,d) are the system inputs to observe $\hat{x}$.
The observed states $\hat{x}$ are used to implement the controller, and, control law (5.47) is

$$u=-K_F\hat{x}. \qquad (5.54)$$

Illustrative examples for single-input single-output and MIMO systems are solved in Example 5.21 and Practice Problem 5.6.

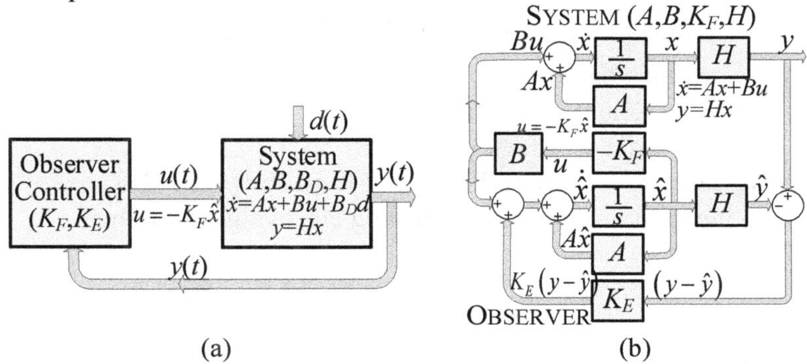

Figure 5.20. (a) State *observer* and controller diagram for
$\frac{dx}{dt}=Ax+Bu=Ax-BK_F\hat{x},\ y=Hx,\ \frac{d\hat{x}}{dt}=(A-BK_F-K_EH)\hat{x}+K_Ey=(A-BK_F)\hat{x}+K_E(y-\hat{y})$;
(b) Descriptive schematics of the time-invariant closed-loop system with the observed states $\hat{x}(t)$. The control law is $u=-K_F\hat{x}$.

System and Observer Eigenvalues – To guarantee stability, the closed-loop system and state *observer* eigenvalues $(\boldsymbol{\lambda}_F,\boldsymbol{\lambda}_E)$ must have negative real parts, $\forall\text{Re}(\lambda_{Fi},\lambda_{Ej})<0$.

The eigenvalues are defined by characteristic equations $\begin{cases}|sI-(A-BK_F)|=0,\\ |sI-(A-K_EH)|=0,\\ |sI-(A-BK_F-K_EH)|=0\end{cases}$.

The *observer* error vector $\bar{e}(t)=[x(t)-\hat{x}(t)]$ should converge faster than system states.
Select the *observer* eigenvalues such as $|\lambda_{Fi}|\ll|\lambda_{Ej}|$, $\text{Re}(\lambda_{Fj})\ll\text{Re}(\lambda_{Ej})$, $\forall(\lambda_{Fi},\lambda_{Ej})$.

If $\text{Im}(\lambda_i)\neq 0$, one has $0\le\tan^{-1}\left|\frac{\text{Im}(\lambda_i)}{\text{Re}(\lambda_i)}\right|<1$.

For real characteristic eigenvalues $\tan^{-1}\left(\frac{\text{Im}(\lambda_i)}{\text{Re}(\lambda_i)}\right)=0$.

If $|\text{Re}(\lambda_i)|=|\text{Im}(\lambda_i)|$, we have $\tan^{-1}\left|\frac{\text{Im}(\lambda_i)}{\text{Re}(\lambda_i)}\right|=0.7854$.

In addition to example in this sections, illustrative examples for *modal* control and state *observer* design are also reported in Practice Problem 5.6 and section 6.15.4.

Example 5.21. *Modal Control and State Observer Design*
Consider a servo, studied in Examples 5.13, 5.19 and 5.20. The output is the angular displacement θ_r. For illustration, assume all parameters are ones. We have

$$\frac{dx}{dt}=Ax+Bu=\begin{bmatrix}-1&-1&0\\1&-1&0\\0&1&0\end{bmatrix}x+\begin{bmatrix}1\\0\\0\end{bmatrix}u,\ x=\begin{bmatrix}x_1\\x_2\\x_3\end{bmatrix}=\begin{bmatrix}i_a\\\omega_r\\\theta_r\end{bmatrix},\ u=u_a$$

with an output $y=Hx=[0\ \ 0\ \ 1]x$, $y=x_3$.

Design a *modal* controller and state *observer*. Find the gain matrices K_F and K_E. In *modal* control and *observer* designs, the characteristic equations are

$|sI-(A-BK_F)|=0$, $|sI-(A-K_EH)|=0$.

Derive the feedback gain matrix K_F in stabilizing control law (5.54) $u=-K_F\hat{x}$ to guarantee eigenvalues $\boldsymbol{\lambda}_F$ as:

1. $\lambda_{F1}=-1$, $\lambda_{F2}=-2$ and $\lambda_{F3}=-3$;
2. $\lambda_{F1,2}=-2\pm 2j$ and $\lambda_{F3}=-3$.

For two cases, we compute $K_F=[4\ \ 5\ \ 6]$ and $K_F=[5\ \ 13\ \ 24]$.

Design the state *observer* by specifying eigenvalues $\boldsymbol{\lambda}_E$ to be:

1. $\lambda_{E1}=-5$, $\lambda_{E2}=-10$ and $\lambda_{E3}=-15$;
2. $\lambda_{E1,2}=-10\pm 5j$ and $\lambda_{E3}=-15$.

For the specified eigenvalues, we have $K_E=\begin{bmatrix}477\\217\\28\end{bmatrix}$ and $K_E=\begin{bmatrix}1452\\357\\33\end{bmatrix}$.

System dynamics and evolutions can be computed by using the **lsim**, **step** and other commands. Evolutions of $(x(t),\hat{x}(t))$ are reported in Figures 5.21.a. The observed states $\hat{x}(t)$ converge to $x(t)$. The output is $y(t)=x_3(t)$. A MATLAB code to compute (K_F,K_E), and, to assess dynamics for $(x(t),\hat{x}(t))$ is

```
A=[-1 -1 0; 1 -1 0; 0 1 0]; B=[1; 0; 0]; H=[0  0  1]; D=[0];  % State-space model matrices A, B, H and D
p=[-1 -2 -3]; KF=place(A,B,p); KF     % Specified closed-loop system eigenvalues, and, compute KF
p=[-2+2j -2-2j -3]; KF=place(A,B,p); KF
p=[-5 -10 -15]; KE=place (A',H',p)'; KE  % Specified observer eigenvalues, and, compute KE
p=[-10+5j -10-5j -15];  KE=place (A',H',p)'; KE
A_o=A-KE*H; eig(A_o);     % Observer system matrix and eigenvalues
x0=[-1; 1; 0]; t=[0:0.01:5]'; % Initial conditions and time
u=0*t;                       % Input, u=0
M=ss(A,B,H,D);               % Linear time-invariant system model
[y,t,x]=lsim(M,u,t,x0);      % Simulate system to find x(t) and y(t)
M_o=ss(A_o,KE,H,D);  % Observer model
[yhat,t,xhat]=lsim(M_o,y,t); % Simulate observer with specified initial conditions
figure(1); plot(t,x,'b',t,xhat,'k--','LineWidth',2.5);  % axis([0 5 -1.05 1.25]); % Plot states x and xhat
legend('{\itx}_1({\itt}), {\itx}_{10}=-1','{\itx}_2({\itt}), {\itx}_{20}=1', '{\itx}_3({\itt}), {\itx}_{30}=0', ...
'Observed {\itx}_1, {\itx}_2, {\itx}_3','NumColumns',1,'FontSize',20); legend boxoff;
```

Figures 5.21.c report the SIMULINK diagrams to simulate the state-space models, as well as a diagram for a closed-loop system with an *observer* and controller. The control law (5.54) $u=-K_F\hat{x}$ is investigated. The initial conditions are $x_0=[-1, 1,0]$ and $\hat{x}_0=[-0.5, 0.5, 0]$. Convergence of the observed states $\hat{x}(t)$ to system states $x(t)$ is guaranteed, and, $\lim_{t\to\infty}\bar{e}(t)=\lim_{t\to\infty}[\hat{x}(t)-x(t)]\to 0$. Closed-loop system is stable for any initial conditions. The *observer* error dynamics is ~1 sec, which depends on specified λ_{Ei}. The system transients are ~4 sec for the assigned λ_{Fi}.

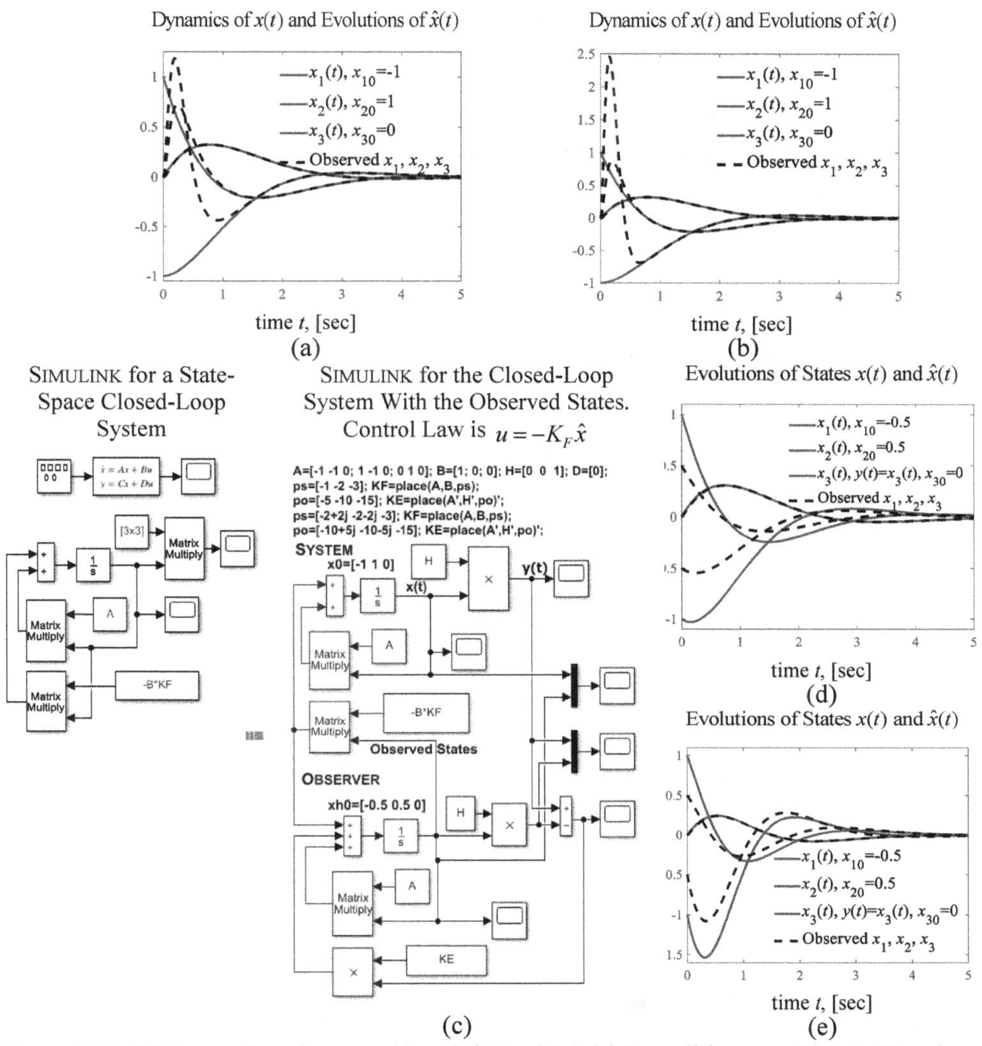

Figure 5.21. (a) Dynamics of states $x(t)$ and $\hat{x}(t)$ with initial conditions, x_0=[1, −1, 0] and $\hat{x}_0$=[0, 0, 0]. The *observer* eigenvalues are λ_{E1}= −5, λ_{E2}= −10, λ_{E3}= −15;

(b) Dynamics of states $x(t)$ and $\hat{x}(t)$. The *observer* eigenvalues are $\lambda_{1,2E}$= −10±5j, λ_{3E}= −15;

(c) Simulation of a system in SIMULINK applying the state-space notations. Closed-loop system with *observer* and control law $u=-K_F\hat{x}$. States $\hat{x}(t)$ are observed by using a measured input-output pair (u,y);

(d) Closed-loop system dynamics with controller and *observer* for λ_{F1}= −1, λ_{F2}= −2, λ_{F3}= −3, and, λ_{E1}= −5, λ_{E2}= −10, λ_{E3}= −15. Feedback and *observer* gain matrices are K_F=[4 5 6] and $K_E=\begin{bmatrix}477\\217\\28\end{bmatrix}$. The initial conditions are x_0=[−1, 1, 0] and $\hat{x}_0$=[−0.5, 0.5, 0];

(e) Closed-loop system dynamics with controller and *observer* for $\lambda_{F1,2}$= −2±2j, λ_{F3}= −3, and, $\lambda_{E1,2}$= −10±5j, λ_{E3}= −15. Feedback and *observer* matrices are K_F=[5 13 24], $K_E=\begin{bmatrix}1452\\357\\33\end{bmatrix}$. ■

State Observer and Modal Control Optimization and Search Algorithm 5.1: For the eigenvalues ($\boldsymbol{\lambda}_F,\boldsymbol{\lambda}_E$) within the specified range $(\boldsymbol{\Lambda}_F^{\Lambda},\boldsymbol{\Lambda}_E^{\Lambda})$

$$\boldsymbol{\lambda}_F \in \boldsymbol{\Lambda}_F^{\Lambda},\ \boldsymbol{\lambda}_E \in \boldsymbol{\Lambda}_E^{\Lambda},\ (\boldsymbol{\lambda}_F,\boldsymbol{\lambda}_E)\in\mathbb{R}\times\mathbb{C} \text{ or } \forall(\boldsymbol{\lambda}_F,\boldsymbol{\lambda}_E)\in\mathbb{R},\ \forall\boldsymbol{\lambda}_F<\forall\boldsymbol{\lambda}_E,\ \forall\mathrm{Re}(\boldsymbol{\lambda}_F,\boldsymbol{\lambda}_E)<0,$$

which define evolutions of the system states, *observer* and error vectors $\left(x(t),\hat{x}(t),\bar{e}(t)\right)$, find the controller and *observer* gain matrices (K_F,K_E) as

$$\text{minimize} \min_{\boldsymbol{\lambda}_F\in\boldsymbol{\Lambda}_F^{\Lambda},\boldsymbol{\lambda}_E\in\boldsymbol{\Lambda}_E^{\Lambda}} \left\{\left\|K_F\right\|_p,\ \left\|K_E\right\|_p\right\},$$

$$\text{subject to } (k_{Fij\,\min}\le k_{Fij}\le k_{Fij\,\max},\ k_{Eij\,\min}\le k_{Eij}\le k_{Eij\,\max}).$$

An eigenvalue pair ($\boldsymbol{\lambda}_F,\boldsymbol{\lambda}_E$) should guarantee asymptotic converge for $(x(t),\hat{x}(t),\bar{e}(t))$ as $\begin{cases}\lim_{t\to\infty} x(t)\to x_e \\ \lim_{t\to\infty}\hat{x}(t)\to x,\ \lim_{t\to\infty}\bar{e}(t)\to 0\end{cases}$, and, ensure criteria imposed on (K_F,K_E). ■

Design a *modal* control law (5.47) $u=-K_Fx$ and state *observer* (5.50) $\frac{d\hat{x}}{dt}=\left(A-K_EH\right)\hat{x}+Bu+K_Ey$, specifying a range ($\boldsymbol{\lambda}_F \in \boldsymbol{\Lambda}_F^{\Lambda},\ \boldsymbol{\lambda}_E \in \boldsymbol{\Lambda}_E^{\Lambda}$) to control governance of system (5.48) $\frac{dx}{dt}=Ax+Bu=\left(A-BK_F\right)x=\mathbf{A}x$, $u=-K_F\hat{x}$ and guarantee asymptotic convergence of the *observer* error, described by (5.52) $\frac{d\bar{e}}{dt}=\left(A-K_EH\right)\bar{e}=\bar{\mathbf{A}}\bar{e}$, $\bar{e}(t)=x(t)-\hat{x}(t)$, $\lim_{t\to\infty}\bar{e}(t)\to 0$. Design tasks focus on:

1. Define eigenvalues ($\boldsymbol{\lambda}_F,\boldsymbol{\lambda}_E$) within $(\boldsymbol{\Lambda}_F^{\Lambda},\boldsymbol{\Lambda}_E^{\Lambda})$ for eigenvalue-dependent system and state *observer* stable transient dynamics, which yield (K_F,K_E);
2. Consider a complex $\boldsymbol{\lambda}$-plane for $(\boldsymbol{\lambda}_F,\boldsymbol{\lambda}_E)\in\mathbb{R}\times\mathbb{C}$ with the angle and magnitude conditions or $\forall(\boldsymbol{\lambda}_F,\boldsymbol{\lambda}_E)\in\mathbb{R}$, asymptotes, natural frequency, damping ratio and stability margins. For ($\boldsymbol{\lambda}_F,\boldsymbol{\lambda}_E$), $\forall\mathrm{Re}(\boldsymbol{\lambda}_F,\boldsymbol{\lambda}_E)<0$, one has $0<\tan^{-1}\left|\frac{\mathrm{Im}(\lambda_i)}{\mathrm{Re}(\lambda_i)}\right|<1,\ \forall\lambda_i$;
3. Impose definiteness on controller feedback gains and *observer* gains (k_{Fij},k_{Eij}), specify matrix norms ($\|K_F\|_p,\|K_E\|_p$), as well as *admissible* ($k_{Fij\min}\le k_{Fij}\le k_{Fij\max}$, $k_{Eij\min}\le k_{Eij}\le k_{Eij\max}$), preventing high-gain feedback and *observer* gains which may yield limits on control efforts, sensitivity, instabilities, limit cycles, etc.;
4. Hardware implementation of the *observer* and controller (5.54) $u=-K_F\hat{x}$ for a physical system $\mathbf{M}(\cdot)$;
5. Characterization and evaluation of system performance.

Examples are solved in Practice Problem 5.6 and section 6.15.4.

5.5. Optimal Control and Dynamic Optimization

Solve dynamic optimization problems and design control laws by minimizing performance functionals. As reported in section 2.5, apply the Hamiltonian optimization, as well as a Lyapunov stability and optimality paradigm. Using the state-dependent quantities q, outputs y, states x and control u, one may derive real-valued integrands $\omega(\cdot)$ using a quintuple (t,q,x,y,u). Tracking error, settling time, bandwidth and other performance measures are specified and evaluated. Synthesizes scalar performance integrands $\omega(t,x,u)>0$, and, minimize a functional

$$J=\min_{t,x,u}\int_{t_0}^{t_f}\omega(t,x,u)dt,\ \omega(t,x,u)>0, \tag{5.55}$$

where $\omega(\cdot)$ is the positive definite performance integrand; t_0 and t_f are the initial and final time which define the time horizon

As emphasized in chapter 1 and section 5.1, the performance Q_P and capabilities Q_C measures $Q(Q_P,Q_C)$ are characterized by $q(x,y,u)$, mapped by scalar $\omega(t,x,u)$. Physics-consistent positive definite, continuous and differentiable integrand functions $\omega(\cdot)$ are synthesized to ensure analytic design, closed-form solution of optimization problem, etc.

Tracking Control – Minimize designed-by-specifications performance functional

$$J=\min_{x,e,u}\int_{t_0}^{t_f}\omega(x,e,u)dt,\ \omega(x,e,u)>0, \tag{5.56}$$

subject to system dynamics $\dot{x}(t)=f(t,x,u,r,d)$, $y=H(x)$, $e(t)=r(t)-y(t)$, $x(t_0)=x_0$, and, design a control law

$$u=\varphi(t,x,e),\ u\in U, \tag{5.57}$$

such that the closed-loop system evolutions $x\in X(X_0,U,R,D)$, $y\in Y(X,U,R,D)$ and $e\in E(X,U,R,D)$ are bounded and converge to equilibrium $(x_e,e_e)\in X\times E$ for initial conditions $x_0\in X$, references $r\in R$ and disturbances $d\in D$,

$$\lim_{t\to\infty}\|x(t)-x_e\|\le\delta_x,\ 0<\delta_x\ll 1,\text{ or, }\lim_{t\to\infty}\|x(t)-x_e\|\to 0,$$

and $$\lim_{t\to\infty}\|e(t)\|\le\delta_e,\ 0<\delta_e\ll 1,\text{ or, }\lim_{t\to\infty}e(t)\to 0.$$

In (5.57), linear and nonlinear time-, state- and tracking error mappings and structures are given by $\varphi(t,x,e)$, $\varphi(\cdot):\mathbb{R}_{\ge 0}\times\mathbb{R}^n\times\mathbb{R}^b\to\mathbb{R}^m$. For the inherent symmetric or asymmetric hardware-defined limits, $u_{\min}\le u\le u_{\max}$, one need to designs a control law $u=\phi(\varphi(t,x,e))$, $u\in U$, where $\phi(\cdot)$ is the continuous, piecewise-continuous of discontinuous bounded function which represents limits $u_{\min}\le u\le u_{\max}$. ■

Stabilization Problem – Minimize designed-by-specifications performance functional

$$J=\min_{x,u}\int_{t_0}^{t_f}\omega(x,u)dt,\ \omega(x,u)>0, \tag{5.58}$$

subject to system dynamics $\dot{x}(t)=f(t,x,u,r,d)$, $y=H(x)$, $x(t_0)=x_0$, and, design an optimal control law

$$u=\varphi(t,x),\ u\in U, \tag{5.59}$$

such that the closed-loop system evolutions $x\in X(X_0,U,D)$ and $y\in Y(X,U,D)$ are bounded and converge to equilibrium $x_e\in X$ for initial conditions $x_0\in X$ and disturbances $d\in D$,

$$\lim_{t\to\infty}\|x(t)-x_e\|\le\delta_x,\ 0<\delta_x\ll 1,\text{ or, }\lim_{t\to\infty}\|x(t)-x_e\|\to 0.$$ ■

Example 5.22. *Performance Integrands*
One designs constrained and not constrained control laws, as well as the *minimal complexity* algorithms using measured variables (x_m,e) as $u=\varphi(t,x_m)$ and $u=\varphi(t,x_m,e)$.
A linear quadratic regulator design is covered in section 5.6. In quadratic functional

$$J=\min_{x,u}\int_{t_0}^{t_f}(x^2+u^2)dt,\ \omega(x,u)=(x^2+u^2)>0.$$

The differentiable positive definite performance integrand $\omega(x,u)$ assesses system dynamics. Quadratic functionals with quadratic $\omega(x,u)$ yield a unique and analytic closed-form solution. One finds control laws and computes feedback gains. Positive definite continuous and discontinuous nonquadratic integrands $\omega(x,u)$ may be considered researching parametric, nonparametric, numeric and other optimizations. For example,

$J=\min_{t,u}\int_{t_0}^{t_f}1dt$, $J=\min_{t,x,u}\int_{t_0}^{t_f}\left(1+|x|\right)dt$, $J=\min_{x,u}\int_{t_0}^{t_f}\left(|x|+|u|\right)dt$, $J=\min_{x,u}\int_{t_0}^{t_f}\left(|x|x^2+|u|u^4\right)dt$.

Minimization of functionals should yield physical consistency and closed-form solutions. ■

Hamiltonian Function – Minimization and maximization problems are solved using the Hamiltonian dynamic optimization, calculus of variations, dynamic programming, maximum principle, nonlinear programming, etc.Consider the system, described by linear differential equations (5.3)

$\dot{x}(t)=Ax+Bu$, $x(t_0)=x_0$, $t\geq 0$.

For performance functional (5.58) $J=\min_{x,u}\int_{t_0}^{t_f}\omega(x,u)dt$, a scalar, continuous, and differentiable real-valued Hamiltonian function $H\in C^p$, $p\geq 2$, is

$$H(t,x,u,\boldsymbol{\lambda})=\omega(x,u)+\boldsymbol{\lambda}^T(t,x)(Ax+Bu), \tag{5.60}$$

$$H(t,x,u,\tfrac{\partial V}{\partial x})=\omega(x,u)+\left(\frac{\partial V}{\partial x}\right)^T(Ax+Bu),$$

where $\boldsymbol{\lambda}$ is the *costate* vector; $V(t,x)$ is the return function.

A general solution of optimization problem is reported in section 2.5.

Hamiltonian, Optimality and Hamilton-Jacobi Equation – A control function (5.59) $u=\varphi(t,x)$ is found by minimizing a scalar continuous Hamiltonian (5.60). Apply the stationarity condition $\frac{\partial H(x,u,\frac{\partial V}{\partial x})}{\partial u}=0$, known as the first-order necessary condition for optimality, to find an optimal control

$$\frac{\partial H(x,u,\frac{\partial V}{\partial x})}{\partial u}=\frac{\partial}{\partial u}\left[\omega(x,u)+\left(\frac{\partial V}{\partial x}\right)^T(Ax+Bu)\right]=0. \tag{5.61}$$

Second Derivative Test for Local Extrema – For a Hamiltonian $H(\cdot)$, global minimum and maximum correspond to $\frac{\partial^2 H(x,u,\frac{\partial V}{\partial x})}{\partial u\times\partial u^T}>0$ and $\frac{\partial^2 H(x,u,\frac{\partial V}{\partial x})}{\partial u\times\partial u^T}<0$. If $\frac{\partial^2 H(x,u,\frac{\partial V}{\partial x})}{\partial u\times\partial u^T}=0$, one has a saddle solution, the second derivative test is inconclusive, and, higher-order derivatives should be used. Hence, the second-order necessary conditions for optimality is $\frac{\partial^2 H(x,u,\frac{\partial V}{\partial x})}{\partial u\times\partial u^T}>0$.

As documented in section 2.5, a scalar real-valued continuous and differentiable return function $V(t,x)$ should satisfy the functional differential equation. That is, the closed-form solution must be found solving the Hamilton-Jacobi equation

$$-\frac{\partial V^*}{\partial t}=\omega(x,u^*)+\left(\frac{\partial V^*}{\partial x}\right)^T\left(F(x)+B(x)u^*\right),\ u^*=\varphi\left(t,\frac{\partial V^*}{\partial x}\right). \tag{5.62}$$

Optimality Conditions – We minimize the scalar-valued Hamiltonian to find an optimal control function $u^*(t,x)$ which ensures an optimal trajectory $x^*(t)$. The first-order partial derivative of the Hamiltonian $\frac{\partial}{\partial u}H(x,u,\frac{\partial V}{\partial x})$ (5.61) exists, and, a critical point yields a control function $u=\varphi(t,x)$. If the second-order partial derivative of a Hamiltonian $H(\cdot)$ at a critical point is positive, minimum is guaranteed. Integrands $\omega(x,u)$ with degree $\deg(\omega)\geq 2$, real-valued $(\lambda,\frac{\partial V}{\partial x})$, and, governing equations of motion $\dot{x}(t)=Ax+Bu$ or $\dot{x}(t)=F(x,u)$, yield a multivariate scalar differentiable Hamiltonian $H(\cdot)$ as well as solution of the functional equation (5.62) for a scalar function $V(t,x)$. One solves the Hamilton-Jacobi equation (5.62) by finding a return function $V(t,x)$ for given performance functional and equations of motion. We obtain an optimal control (2.61) $u^*=\varphi\left(t,x,\frac{\partial V^*}{\partial x}\right)$, where the return function $V^*(t,x)$ is a solution of (5.62).

Continuous Dynamic Optimization – To maximize or minimize a functional with performance integrand $\omega(x,u)$, subject to the governing dynamics, use the scalar Hamiltonian $H(\cdot)$ or Lagrangian $\Lambda(\cdot)$. For linear and nonlinear dynamic systems, define a scalar Lagrangian as

$$\Lambda=\omega(x,u)+\lambda^T(Ax+Bu),\ \Lambda=\omega(x,u)+\lambda^T\left[F(t,x)+B(x)u\right],$$

where λ is the Lagrange multiplier, $\lambda\geq 0$.

Using the Hamiltonian (5.60), for given boundary conditions, apply:

State equation $\dot{x}(t)=\frac{\partial H}{\partial \lambda}^T$; *Costate* equation $\dot{\lambda}=-\frac{\partial H}{\partial x}^T$; Stationarity condition $\frac{\partial H}{\partial u}=0$.

Minimax Problem – Dynamic optimization implies a solution of the minimax problems using a system of equations

$$\left\{\dot{x}(t)=\frac{\partial H}{\partial \lambda}^T,\ \dot{\lambda}=-\frac{\partial H}{\partial x}^T,\ \frac{\partial H}{\partial u}=0\right\}\text{ or }\left\{\dot{x}(t)=\frac{\partial \Lambda}{\partial \lambda}^T,\ \dot{\lambda}=-\frac{\partial \Lambda}{\partial x}^T,\ \frac{\partial \Lambda}{\partial u}=0\right\}.\quad(5.63)$$

5.6. Optimization of Linear Continuous-Time Systems: Linear Quadratic Regulator Problem

Consider a linear time-invariant system

$$\dot{x}(t)=Ax+Bu,\ x(t_0)=x_0,\quad(5.64)$$

where $A\in\mathbb{R}^{n\times n}$ and $B\in\mathbb{R}^{n\times m}$ are the constant-coefficient matrices.

Apply an optimization calculus, and, solve the linear quadratic regulator (LQR) problem. Consider a positive definite quadratic performance integrand $\omega(x,u)=\frac{1}{2}\left(x^TQx+u^TGu\right)$ with diagonal weighting matrices $(Q,G)>0$. A quadratic function is a quadratic polynomial of degree 2. The quadratic performance functional is

$$J=\min_{x,u}\int_{t_0}^{t_f}\tfrac{1}{2}\left(x^TQx+u^TGu\right)dt,\ Q\geq 0,\ G>0,\quad(5.65)$$

where $Q\in\mathbb{R}^{n\times n}$ is the positive semi-definite constant-coefficient diagonal weighting matrix; $G\in\mathbb{R}^{m\times m}$ is the positive definite constant-coefficient diagonal weighting matrix.

Weighting Coefficients – The exploratory choice of weighting coefficients of diagonal matrices (Q,G) is conducted using the estimates on variations of $(\tilde{x}_{i\max},\tilde{u}_{j\max})$, $q_{ii}=\frac{1}{\tilde{x}^2_{i\max}}$ and $g_{jj}=\frac{1}{\tilde{u}^2_{j\max}}$. Values q_{ii} are defined by *scaling* of specified variations on $\tilde{x}_{i\max}$, while g_{jj} are computed by using variations of control efforts $\tilde{u}_{j\max}$.

Problem Formulations – Minimize the quadratic performance functional (5.65)

$$J=\min_{x,u}\int_{t_0}^{t_f}\tfrac{1}{2}\left(x^TQx+u^TGu\right)dt$$

subject to the linear system (5.64)

$$\dot{x}(t)=Ax+Bu\,,$$

and, analytically design a control law by means of Hamiltonian optimization.

We find a closed-form solution to dynamic optimization problems.

Using (5.64) and (5.65), the real-valued *p*-differentiable Hamiltonian function is

$$H(x,u,\tfrac{\partial V}{\partial x})=\tfrac{1}{2}\left(x^TQx+u^TGu\right)+\left(\frac{\partial V}{\partial x}\right)^T(Ax+Bu)\,. \tag{5.66}$$

The Hamilton-Jacobi equation (5.62) is

$$-\frac{\partial V}{\partial t}=\tfrac{1}{2}\left(x^TQx+u^TGu\right)+\left(\frac{\partial V}{\partial x}\right)^T(Ax+Bu)\,. \tag{5.67}$$

Quadratic Function – A quadratic performance integrand is $\omega(x,u)=\tfrac{1}{2}\left(x^TQx+u^TGu\right)$, where weighting matrices $(Q,G)>0$ are diagonal. A quadratic function is a quadratic polynomial of degree 2. A quadratic function $f(x)=x^TQx+c^Tx+d$ is:

1. Convex if and only if $Q\geq 0$;
2. Strictly convex if and only if $Q>0$;
3. Concave if and only if $Q\leq 0$;
4. Strictly concave if and only if $Q<0$.

The derivative of the Hamiltonian $H(\cdot)$ exists. A control function $u(\cdot):[t_0,t_f)\rightarrow\mathbb{R}^m$ is found by using the first-order necessary condition for optimality (5.61).

From $\dfrac{\partial H(x,u,\frac{\partial V}{\partial x})}{\partial x}=u^TG+\left(\dfrac{\partial V}{\partial x}\right)^TB=0$, we have

$$u=-G^{-1}B^T\frac{\partial V}{\partial x}\,. \tag{5.68}$$

The second-order necessary condition for optimality is guaranteed because $G>0$, and, $\frac{\partial^2 H}{\partial u\times\partial u^T}=G>0$. Substitute (5.68) in (5.67). The Hamilton-Jacobi functional equation

$$\begin{aligned}-\frac{\partial V}{\partial t}&=\tfrac{1}{2}x^TQx+\tfrac{1}{2}\left(\frac{\partial V}{\partial x}\right)^TBG^{-1}B^T\frac{\partial V}{\partial x}+\left(\frac{\partial V}{\partial x}\right)^TAx-\left(\frac{\partial V}{\partial x}\right)^TBG^{-1}B^T\frac{\partial V}{\partial x}\\&=\tfrac{1}{2}x^TQx+\left(\frac{\partial V}{\partial x}\right)^TAx-\tfrac{1}{2}\left(\frac{\partial V}{\partial x}\right)^TBG^{-1}B^T\frac{\partial V}{\partial x}\end{aligned} \tag{5.69}$$

admits a unique solution. Equation (5.69) is satisfied by a quadratic return function

$$V(x)=\tfrac{1}{2}x^TKx\,, \tag{5.70}$$

where $K\in\mathbb{R}^{n\times n}$ is a symmetric matrix, $K=K^T$.

From (5.69) with (5.70), using the matrix identity $x^T KAx = \frac{1}{2}x^T(A^T K + KA)x$ and $\frac{\partial V}{\partial t} = \frac{\partial}{\partial t}(\frac{1}{2}x^T Kx) = \frac{1}{2}x^T \dot{K}x$, we have

$$-\tfrac{1}{2}x^T \dot{K}x = \tfrac{1}{2}x^T Qx + \tfrac{1}{2}x^T A^T Kx + \tfrac{1}{2}x^T KAx - \tfrac{1}{2}x^T KBG^{-1}B^T Kx, \qquad (5.71)$$

$$V(t_f, x) = \tfrac{1}{2}x^T K(t_f)x = \tfrac{1}{2}x^T K_f x,\ K(t_f) = K_f,$$

where the boundary condition $V(t_f, x) = \frac{1}{2}x^T K(t_f)x = \frac{1}{2}x^T K_f x$, $K(t_f)=K_f$ is considered.

The quadratic differential equation to find the unknown symmetric matrix K is

$$-\dot{K}(t) = Q + A^T K + KA - KBG^{-1}B^T K,\ K(t_f) = K_f. \qquad (5.72)$$

From (5.68) and (5.70), the control law with time-varying $K(t)$ and feedback gain matrix $K_F(t)$ are

$$u = -G^{-1}B^T K(t)x = -K_F(t)x,\ K_F(t) = G^{-1}B^T K(t). \qquad (5.73)$$

Infinite Time Horizon Optimization, Control Law and Algebraic Riccati Equation – If in functional (5.65) $t_f=\infty$, minimize

$$J = \min_{x,u} \int_{t_0}^{\infty} \tfrac{1}{2}\left(x^T Qx + u^T Gu\right)dt, \qquad (5.74)$$

subject to linear system (5.64).

The stationarity condition (5.61) for a Hamiltonian (5.66) yields (5.68) $u = -G^{-1}B^T \frac{\partial V}{\partial x}$.

The functional equation (5.69) with $\frac{\partial V}{\partial t} = 0$ becomes

$$0 = \tfrac{1}{2}x^T Qx + \left(\frac{\partial V}{\partial x}\right)^T Ax - \tfrac{1}{2}\left(\frac{\partial V}{\partial x}\right)^T BG^{-1}B^T \frac{\partial V}{\partial x}. \qquad (5.75)$$

This equation is satisfied by a positive definite quadratic return function (5.70) $V(x) = \frac{1}{2}x^T Kx$, which implies a positive-definite constant-coefficient matrix $K>0$.

The constant-coefficient matrix K is a solution of an algebraic equation

$$Q + A^T K + KA - KBG^{-1}B^T K = 0,\ K>0,\ K^T=K. \qquad (5.76)$$

For $(Q,G)>0$ and $J>0$, a symmetric matrix $K = \begin{bmatrix} k_{11} & \cdots & k_{1n} \\ \vdots & \ddots & \vdots \\ k_{n1} & \cdots & k_{nn} \end{bmatrix}$, $k_{ij}=k_{ji}$, $K=K^T$ must be positive definite. Positive definiteness of K can be verified using the Sylvester criterion.

Definition 5.3. A symmetric matrix K is positive definite if all its eigenvalues are positive, $\forall\lambda_i>0$. ■

A constant feedback gain control law and closed-loop system are

$$u = -G^{-1}B^T Kx = -K_F x, \qquad (5.77)$$

$$\dot{x}(t) = Ax + \underset{u=-G^{-1}B^T Kx}{Bu} = Ax - BG^{-1}B^T Kx = (A - BG^{-1}B^T K)x = (A - BK_F)x.$$

The eigenvalues of the matrix $(A - BG^{-1}B^T K) = (A - BK_F) \in \mathbb{R}^{n\times n}$ have negative real parts, ensuring stability. In (5.76), $BG^{-1}B^T>0$ and $KBG^{-1}B^T K>0$. Quadratic differential equation (5.72) and algebraic equation (5.76) admit solutions for $K(t)$ and K. These equations are solved using different approaches. For example, $K(t)$ is found by using a backward solution of differential equations (5.72).

Quadratic Equations and Hamilton-Jacobi Equation – One investigates error, precision, relative and absolute accuracy, stability and convergence of solutions of equations. Solutions are validated by substitution. A quadratic equation (5.76) $Q+A^TK+KA-KBG^{-1}B^TK=0$, $K>0$ admits a solution for a symmetric matrix $K\in\mathbb{R}^{n\times n}$, which is straightforwardly verifiable. The $\frac{1}{2}n(n+1)$ unknown k_{ij} are found.

Verify a solution for a functional equation (5.75), and, evaluate a $\frac{1}{2}n(n+1)$-dimensional manifold $\Delta(x)$ with $\frac{1}{2}n(n+1)$ quadratic terms of n-variables

$$\Delta(x)=\tfrac{1}{2}x^TQx+\left(\frac{\partial V}{\partial x}\right)^T Ax-\tfrac{1}{2}\left(\frac{\partial V}{\partial x}\right)^T BG^{-1}B^T\frac{\partial V}{\partial x},\ \Delta_{ij}=\Delta_{ji},\ x\in X,$$

where $V(x)=\frac{1}{2}x^TKx$ with a positive definite computed K.

For $\Delta(\cdot):\mathbb{R}^n\to\mathbb{R}$, an absolute manifold-valued errors $|\Delta_{ij}|_{\max}$ and $\|\Delta_{ij}\|_p$ are evaluated for $\forall x\in X$. These errors may associate with analytic and numeric errors, algorithmic stability, convergence, etc. For linear systems, minimization of a quadratic functional (5.74) yields a functional equation (5.75), satisfied by the quadratic return function (5.70). With computed K, one should obtain $\Delta(x)=0$, $\forall\Delta_{ij}=0$, $\forall x\in X$. Minimization of nonlinear functionals yields a Hamilton-Jacobi equation, approximated by nonquadratic return functions. One evaluates $|\Delta_{ij}|_{\max}$ and $\|\Delta_{ij}\|_p$, such as $|\Delta_{ij}|_{\max}\le\delta$ and $\|\Delta_{ij}\|_p\le\delta_p$, $\forall x\in X$. The volume, Hausdorff and other measures can be applied to evaluate a solution of the Hamilton-Jacobi equation.

Schur Decomposition and Solution of an Algebraic Equation – For equation (5.76), matrix algebra yields

$$Q+A^TK+KA-KBG^{-1}B^TK=0,\ K=K^T,\ K\in\mathbb{R}^{n\times n},$$

$$\begin{bmatrix}-K & I\end{bmatrix}\begin{bmatrix}A & -BG^{-1}B^T\\ -Q & -A^T\end{bmatrix}\begin{bmatrix}I\\ K\end{bmatrix}=\begin{bmatrix}-K & I\end{bmatrix}P\begin{bmatrix}I\\ K\end{bmatrix}=0,\ P=\begin{bmatrix}A & -BG^{-1}B^T\\ -Q & -A^T\end{bmatrix},\ P\in\mathbb{R}^{2n\times 2n}. \quad (5.78)$$

The degree $2n$ characteristic polynomial of matrix $P\in\mathbb{R}^{2n\times 2n}$ in a variable λ is $p_P(\lambda)=\det(\lambda I-P)$. The eigenvalues and eigenvectors of P are found. There exists the Schur decomposition for the square matrix $P\in\mathbb{R}^{2n\times 2n}$, which is not unique, such that

$$P=\mathbf{USU}^T,\ \mathbf{U}\in\mathbb{R}^{2n\times 2n},\ \mathbf{S}\in\mathbb{R}^{2n\times 2n},$$

$$\mathbf{U}=\begin{bmatrix}\mathbf{U}_{11} & \mathbf{U}_{12}\\ \mathbf{U}_{21} & \mathbf{U}_{22}\end{bmatrix},\ (\mathbf{U}_{11},\mathbf{U}_{12},\mathbf{U}_{21},\mathbf{U}_{22})\in\mathbb{R}^{n\times n},\ \mathbf{S}=\begin{bmatrix}\mathbf{S}_{11} & \mathbf{S}_{12}\\ 0 & \mathbf{S}_{22}\end{bmatrix},\ (\mathbf{S}_{11},\mathbf{S}_{12},\mathbf{S}_{22})\in\mathbb{R}^{n\times n},$$

where $\mathbf{U}$ is the orthonormal unitary matrix; $\mathbf{S}$ is the upper-triangular Schur matrix.

The inverse matrix $\mathbf{U}^{-1}$ is the conjugate transpose of $\mathbf{U}$, $\mathbf{U}^{-1}=\mathbf{U}^*$, and, $\mathbf{UU}^*=\mathbf{U}^*\mathbf{U}=I$. The upper-triangular Schur matrix $\mathbf{S}$ has the eigenvalues $\mathrm{Re}(\lambda_i)$ of P on its diagonal. One may order the eigenvalues as $\mathrm{Re}(\lambda_1)\ge\mathrm{Re}(\lambda_2)\ge\ldots$ or $\mathrm{Re}(\lambda_1)\le\mathrm{Re}(\lambda_2)\le\ldots$. The $O(n^3)$-complexity Gram-Schmidt orthonormalization yields a matrix pair $(\mathbf{U},\mathbf{S})$, partitioned as the $(n\times n)$ blocks.

Apply the Schur decomposition $P=\mathbf{USU}^T$ with eigenvalues, ordered as $\mathrm{Re}(\lambda_1)\le\mathrm{Re}(\lambda_2)\le\ldots$, and, solve of the matrix equation (5.78) as

$$\begin{bmatrix}-K & I\end{bmatrix}\mathbf{USU}^T\begin{bmatrix}I\\ K\end{bmatrix}=0,\ K=\mathbf{U}_{21}\mathbf{U}_{11}^{-1},\ K=K^T,\ K>0, \quad (5.79)$$

where $\mathbf{U}_{11}$ is invertable, $\det(\mathbf{U}_{11})\neq 0$.

MATLAB is used to solve (5.76). The **lqr** and **care** commands yield a solution for K. Custom MATLAB codes can be developed using matrix algebra, tools and algorithms. Examples 5.28 and 5.30, as well as section 5.11.1, report standard solvers and Schur decomposition with MATLAB codes to solve a quadratic matrix algebraic equation. ■

Example 5.23. *Quadratic Differential Equations: Riccati Equations*
A quadratic differential equation should be solved. Jacopo Francesco Riccati (1676-1754), a mathematician from Venice, found an analytic solutions for a class of differential equations

$$z'=a(t)z^2+b(t)z+c(t),$$

where $a(t)$, $b(t)$ and $c(t)$ are continuous functions of t.

If the coefficients (a,b,c) are constants, an equation

$$z'=az^2+bz+cz$$

is reduced to a separable differential equation.

The solution is found by integrating $\int \frac{dz}{az^2+bz+c}=\int dt$.

Integration of rational functions is performed. ■

Example 5.24. *Solution of Riccati Equations*
Solve an equation $z'=az^2+bz+cz$, $a=1$, $b=4$ and $c=8$.

Apply the substitution $w=\frac{1}{2}(z+2)$, and, use the integral $\int \frac{dw}{w^2+1}=\tan^{-1} w$.

We find $\int \frac{dz}{z^2+4z+8}=\int \frac{dz}{(z+2)^2+4}=\int \frac{d(z+2)}{(z+2)^2+2^2}=\frac{1}{2}\tan^{-1}\left(\frac{z+2}{2}\right)$.

Hence, $\frac{1}{2}\tan^{-1}\left(\frac{z+2}{2}\right)=t+C$

Consider $z'=az^2+bz+cz$, $a=1$, $b=1$ and $c=1$.

We have $\int \frac{dz}{z^2+z+1}=\int dt$.

From $\int \frac{dz}{z^2+z+1}=\int \frac{dz}{(z+\frac{1}{2})^2+(\frac{\sqrt{3}}{2})^2}=\frac{1}{\frac{1}{2}\sqrt{3}}\tan^{-1}\frac{z+\frac{1}{2}}{\frac{1}{2}\sqrt{3}}$,

one has $\frac{2}{\sqrt{3}}\tan^{-1}\frac{2z+1}{\sqrt{3}}=t+C$. ■

Example 5.25.
Consider the *force control* problem for a moving mass with viscous friction. The velocity v is the state variable. The applied force F_{applied} is control. The system dynamics is

$$\frac{dv}{dt}=\frac{1}{m}\sum F=\frac{1}{m}\left(F_{\text{applied}}-B_m v\right),\ x\equiv v,\ u\equiv F_{\text{applied}},$$

which yields $\dot{x}(t)=ax+bu$, $a=-B_m/m$, $b=1/m$.

Minimize the quadratic functional $J=\min_{x,u}\int_{t_0}^{t_f}\frac{1}{2}\left(qx^2+gu^2\right)dt$, $q\geq 0$, $g>0$.

The Hamiltonian (5.66) is

$$H(x,u,\tfrac{\partial V}{\partial x})=\underbrace{\tfrac{1}{2}\left(qx^2+gu^2\right)}_{\substack{\text{Performance Integrand } \omega(x,u)\\ J(x,u)=\min_{x,u}\int_{t_0}^{t_f}\frac{1}{2}(qx^2+gu^2)dt}}+\frac{\partial V}{\partial x}\underbrace{\frac{dx}{dt}}_{\substack{\text{System Dynamics}\\ \dot{x}(t)=ax+bu}}=\tfrac{1}{2}\left(qx^2+gu^2\right)+\frac{\partial V}{\partial x}\left(ax+bu\right).$$

From (5.61) $\frac{\partial H}{\partial u}=0$, one obtains $gu+\frac{\partial V}{\partial x}b=0$.

Hence, the control law is $u=-\frac{b}{g}\frac{\partial V}{\partial x}=-g^{-1}b\frac{\partial V}{\partial x}$.

A continuous and differentiable quadratic return function $V(x)=\frac{1}{2}kx^2$ satisfies the Hamilton-Jacobi equation (5.69). Solving

$$-\frac{\partial V}{\partial t}=\frac{1}{2}qx^2+\frac{\partial V}{\partial x}ax-\frac{1}{2}q\frac{b^2}{g}\frac{\partial V}{\partial x}^2, V(x)=\frac{1}{2}kx^2,$$

one obtains the Riccati equation (6.73) to solve for unknown $k(t)$

$$-\frac{dk}{dt}=q+2ak-\frac{b^2}{g}k^2.$$

An optimal control law is $u=-\frac{b}{g}\frac{\partial V}{\partial x}=-\frac{b}{g}k(t)x$.

If $t_f=\infty$, one solves a quadratic equation (5.76) $q+2ak-\frac{b^2}{g}k^2=0$, $k>0$.

The constant feedback gain control law is $u=-\frac{b}{g}\frac{\partial V}{\partial x}=-\frac{b}{g}kx$, $k>0$.

The closed-loop system

$$\dot{x}(t)=\left(a-\frac{b^2}{g}k\right)x$$

is stable if $(a-g^{-1}b^2k)<0$. This condition for stability is guaranteed.

The second-order necessary conditions for optimality $\frac{\partial^2 H}{\partial u^2}=g>0$ is guaranteed.

Consider a nonlinear system with viscous friction

$$\dot{x}(t)=\frac{1}{m}\sum F=\frac{1}{m}\left(u-B_m x-B_{m1}\,\mathrm{sgn}(x)|x|^{1/3}\right).$$

With the linear control law designed, the closed-loop system is

$$\dot{x}(t)=\frac{1}{m}\left(-\frac{b}{g}kx-B_m x-B_{m1}\,\mathrm{sgn}(x)|x|^{1/3}\right), k>0.$$

This system is stable which can be proven by applying the Lyapunov stability theory covered in chapter 6. ■

Example 5.26. *Optimal Control of the First-Order System: Riccati Equations*
Consider a system studied in Example 5.25. Assume $m=1$ and neglect viscous friction, $B_m=0$. The differential equation is

$$\dot{x}=u.$$

The state-space model (5.64) is $\dot{x}(t)=Ax+Bu$, $A=[0]$, $B=[1]$.

For $t_f=\infty$, the functional (5.65) becomes $J=\min_{x,u}\int_0^\infty \frac{1}{2}(x^2+u^2)dt$, $q=1$, $g=1$.

The algebraic equation (5.76) $1-k^2=0$, $k>0$, yields $k=1$.

A time-invariant control law is $u=-kx=-K_F x=-x$.

Applying the **lqr** command, one finds the return function coefficient k, feedback gain K_F, as well as the eigenvalue λ for the closed-loop system. One has

```
[KF,K,Eigenvalues]=lqr(0,1,1,1,0)
```

We obtain $k=1$, $K_F=1$, $\lambda=-1$.

The control law is $u=-kx=-K_F x=-x$.

The closed-loop system is stable.

Consider a functional $J=\min_{x,u}\int_0^{t_f}\tfrac{1}{2}(qx^2+gu^2)dt$, q=1, g=1, t_f=5 sec.

From the Hamiltonian $H=\tfrac{1}{2}(qx^2+gu^2)+\frac{\partial V}{\partial x}u$ and using $\frac{\partial H}{\partial u}=0$, the control law is $u=-\frac{\partial V}{\partial x}$.

The Hamilton-Jacobi equation (5.69) $-\frac{\partial V}{\partial t}=\tfrac{1}{2}x^2-\tfrac{1}{2}\left(\frac{\partial V}{\partial x}\right)^2$ is satisfied by the quadratic return function (5.70) $V(x)=\tfrac{1}{2}k(t)x^2$.

The unknown $k(t)$ of the return function $V(x)$ is found by solving (5.72) $-\dot{k}(t)=1-k^2(t),\ k(t_f)=0$.

From $\frac{dk}{dt}=k^2-1$, one has $\frac{dk}{k^2-1}=dt$. The integral is $\int\frac{dk}{k^2-a^2}=\frac{1}{2a}\ln\left|\frac{k-a}{k+a}\right|$.

Note that $\int\frac{dk}{a^2-c^2k^2}=\frac{1}{2ac}\ln\left|\frac{a+ck}{a-ck}\right|$, and, $\log_b y=x$ means $b^x=y$.

With the specified limits, a time-varying feedback coefficient is $k(t)=\frac{1-e^{-2(t_f-t)}}{1+e^{-2(t_f-t)}}$.

An optimal control law (5.73), denoted as u^*, $u^*(t)=-k(t)x=-\frac{1-e^{-2(t_f-t)}}{1+e^{-2(t_f-t)}}x$, guarantees the minimum of the quadratic functional, subject to the system dynamics.

A time-varying gain $k(t)$ if t_f=5 sec is illustrated in Figure 5.22.a. Simulations of the closed-loop system with $k(t)$ are performed. The SIMULINK diagram is documented in Figure 5.22.b. For $t_f=\infty$, we computed k=1. Simulate closed-loop systems with time-varying $k(t)$, as well as the time-invariant closed-loop system with k=1.

Figure 5.22.c reports system dynamics x^* with u^*, and, evolutions (x,u) for time-invariant control $u=-kx=-x$. Dynamics are nearly identical.

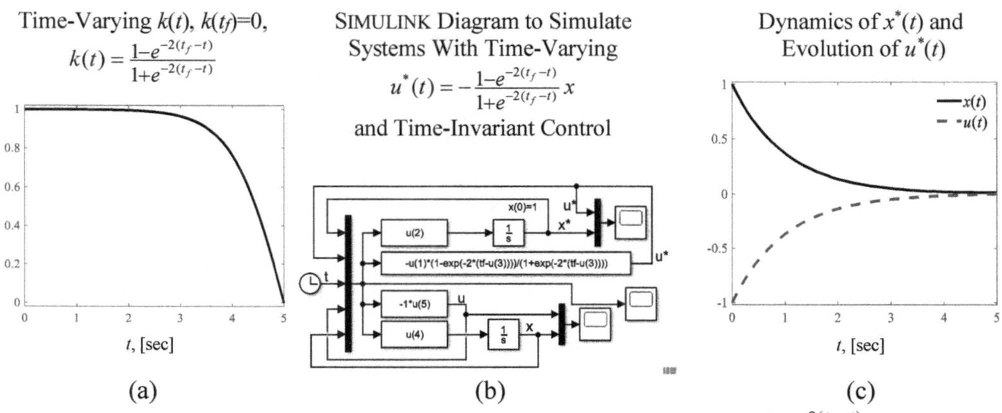

(a) (b) (c)

Figure 5.22. (a) Evolution of a time-varying feedback coefficient $k(t)=\frac{1-e^{-2(t_f-t)}}{1+e^{-2(t_f-t)}}$, t_f=5 sec;
(b) SIMULINK diagram to simulate closed-loop systems with control functions $u^*(t)$ and u. The governing equations with time-varying $k(t)$ and time-invariant feedback coefficient k are
$\frac{dx}{dt}=u,\ u^*=-k(t)x=-\frac{1-e^{-2(t_f-t)}}{1+e^{-2(t_f-t)}}x$, and, $\frac{dx}{dt}=u,\ u=-kx=-x$, k=1;
(c) System dynamics $x^*(t)$ and $u^*(t)$ in the closed-loop system. The initial condition is x_0=1. ■

Example 5.27. *Optimal Control of the Open-Loop Stable System*
Consider a system

$\dot{x} = -x + u$, x_0=1.

In the state-space, $\dot{x}(t) = Ax + Bu$, A=[–1] and B=[1].

Minimize a quadratic functional

$J = \min_{x,u} \int_0^{t_f} \frac{1}{2}(qx^2 + gu^2)dt$, t_f=5 sec

subject to $\dot{x} = -x + u$.

Let the estimates on variations of state and control are $\tilde{x}_{\max} = 1$ and $\tilde{u}_{\max} = 1$.

Hence, $q = \frac{1}{\tilde{x}^2_{\max}}$ and $g = \frac{1}{\tilde{u}^2_{\max}}$.

A scalar Hamiltonian function is $H = \frac{1}{2}(qx^2 + gu^2) + \frac{\partial V}{\partial x}(-x + u)$.

Applying $\frac{\partial H}{\partial u} = 0$, an optimal control law is $u = -\frac{\partial V}{\partial x}$.

Solution of the Hamilton-Jacobi equation (5.69) $-\frac{\partial V}{\partial t} = \frac{1}{2}x^2 - \frac{\partial V}{\partial x}x - \frac{1}{2}\left(\frac{\partial V}{\partial x}\right)^2$ is satisfied by the quadratic return function $V(x) = \frac{1}{2}k(t)x^2$.

A control law (5.73) is $u = -k(t)x$.

The unknown $k(t)$ of the return function (5.70) $V(x) = \frac{1}{2}k(t)x^2$ is found by solving (5.72)

$-\dot{k}(t) = 1 - 2k(t) - k^2(t),\ k(t_f) = 0$.

From $\dot{k}(t) = k^2 + 2k - 1$, we have $\frac{dk}{k^2 + 2k - 1} = dt$, yielding an integral $\int \frac{dk}{ak^2 + bk + c}$.

One has $\int \frac{dk}{ak^2 + bk + c} = \begin{cases} \frac{2}{\sqrt{4ac-b^2}} \tan^{-1} \frac{2ak+b}{\sqrt{4ac-b^2}}, & 4ac > b^2 \\ \frac{2}{\sqrt{b^2-4ac}} \ln \frac{2ak+b-\sqrt{b^2-4ac}}{2ak+b+\sqrt{b^2-4ac}}, & b^2 > 4ac. \\ -\frac{2}{2ak+b}, & b^2 = 4ac \end{cases}$

The characteristic roots (r_1,r_2) for $k^2+2k-1=0$ are $r_{1,2} = -1 \pm \sqrt{2}$.

Hence $\int \frac{dk}{ak^2 + bk + c} = \frac{1}{a(r_1 - r_2)} \ln \left| \frac{k - r_1}{k - r_2} \right|$, $b^2 > 4ac$.

For $\frac{dk}{(k - 1 + \sqrt{2})(k - 1 - \sqrt{2})} = dt$, fractioning yields $\frac{1}{2\sqrt{2}} \int_0^{t_f} \left(\frac{1}{k+1-\sqrt{2}} - \frac{1}{k+1+\sqrt{2}} \right) dk = \int_0^{t_f} dt$.

From $\ln\left(\frac{1-\sqrt{2}}{1+\sqrt{2}}\right) - \ln\left(\frac{k(t)+1-\sqrt{2}}{k(t)+1+\sqrt{2}}\right) = 2\sqrt{2}(t_f - t)$, using the product rule ln(xy)=ln(x)+ln(y),

we have $k(t) = \frac{\sqrt{2} - 1 + (1 - \sqrt{2})e^{-2\sqrt{2}(t_f - t)}}{1 + e^{-2\sqrt{2}(t_f - t)}}$.

Evolution of $k(t)$ is illustrated in Figure 5.23.a.

An optimal control law is $u^*(t) = -k(t)x = -\frac{\sqrt{2} - 1 + (1 - \sqrt{2})e^{-2\sqrt{2}(t_f - t)}}{1 + e^{-2\sqrt{2}(t_f - t)}}x$.

Transients for (x^*,u^*) are depicted in Figure 5.23.c. Example 6.8 reports solutions using the *costate* equation and stationarity condition.

Minimize $J=\min\limits_{x,u}\int_0^{\infty}\frac{1}{2}(x^2+u^2)dt$, $t_f=\infty$.

One finds $u=-kx$.

The quadratic Riccati equation to solve is $1-2k-k^2=0$, $k>0$.

Having computed $k=0.414$, a control law is $u=-0.414x$.

The SIMULINK diagram to simulate closed-loop systems is reported in Figure 5.23.b. Evolutions of closed-loop systems for (x^*,u^*) and (x,u) are reported in Figure 5.23.c.

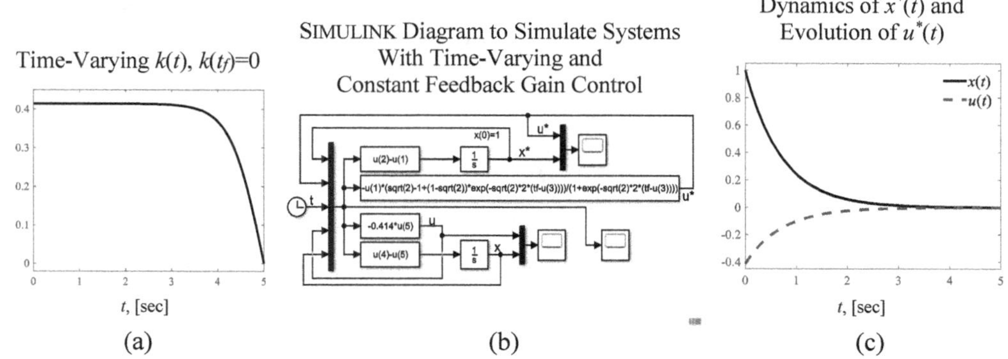

Figure 5.23. (a) Evolution of the time-varying coefficient $k(t)$, $k(t)=\dfrac{\sqrt{2}-1+(1-\sqrt{2})e^{-2\sqrt{2}(t_f-t)}}{1+e^{-2\sqrt{2}(t_f-t)}}$, $t_f=5$ sec;

(b) SIMULINK diagram to simulate closed-loop systems with optimal control $u^*(t)=-k(t)x$ and constant feedback gain control $u=-kx$. The closed-loop systems governing equations are

$$\frac{dx}{dt}=-x-k(t)x=-\left(1+\frac{\sqrt{2}-1+(1-\sqrt{2})e^{-2\sqrt{2}(t_f-t)}}{1+e^{-2\sqrt{2}(t_f-t)}}\right)x,\ \text{and,}\ \frac{dx}{dt}=-x-kx=-1.414x;$$

(c) Optimal system dynamics $x^*(t)$ and control evolution $u^*(t)$, $x_0=1$. ■

Example 5.28. *One-Dimensional Motion of a Rigid-Body: Infinite Time Horizon, Closed-Form Solution, and, Error Analysis in Solution of Hamilton-Jacobi Equation*

Consider a set of two first-order differential equations

$$\dot{x}_1=x_2,$$

$$\dot{x}_2=u.$$

The state variables are the displacement $x_1(t)$ and velocity $x_2(t)$, while u is the force or torque. Using the state-space notations (5.64), we have

$$\dot{x}(t)=Ax+Bu,\ \begin{bmatrix}\dot{x}_1\\ \dot{x}_2\end{bmatrix}=\begin{bmatrix}0&1\\0&0\end{bmatrix}\begin{bmatrix}x_1\\x_2\end{bmatrix}+\begin{bmatrix}0\\1\end{bmatrix}u,\ A=\begin{bmatrix}0&1\\0&0\end{bmatrix},\ B=\begin{bmatrix}0\\1\end{bmatrix}.$$

Infinite Time Horizon Optimization – Minimize the quadratic functional (5.65)

$$J=\min_{x,u}\int_0^{\infty}\frac{1}{2}\left(\begin{bmatrix}x_1&x_2\end{bmatrix}\begin{bmatrix}q_{11}&0\\0&q_{22}\end{bmatrix}\begin{bmatrix}x_1\\x_2\end{bmatrix}+uGu\right)dt,\ Q=\begin{bmatrix}q_{11}&0\\0&q_{22}\end{bmatrix},\ G=g.$$

That is, $J=\min\limits_{x,u}\int_0^{\infty}\frac{1}{2}\left(q_{11}x_1^2+q_{22}x_2^2+gu^2\right)dt$, $q_{11}\geq 0$, $q_{22}\geq 0$, $g>0$.

In explicit form, a scalar Hamiltonian function (5.66) is

$$H(x,u,\tfrac{\partial V}{\partial x}) = \tfrac{1}{2}q_{11}x_1^2 + \tfrac{1}{2}q_{22}x_2^2 + \tfrac{1}{2}gu^2 + \frac{\partial V}{\partial x_1}x_2 + \frac{\partial V}{\partial x_2}u \,.$$

The stationarity condition (5.61) yields $\frac{\partial}{\partial u}H(x,u,\tfrac{\partial V}{\partial x}) = gu + \frac{\partial V}{\partial x_2} = 0$.

Hence, $u = -\frac{1}{g}\frac{\partial V}{\partial x_2}$.

The Hamilton-Jacobi equation (5.69)

$$-\frac{\partial V}{\partial t} = \tfrac{1}{2}q_{11}x_1^2 + \tfrac{1}{2}q_{22}x_2^2 - \frac{\partial V}{\partial x_1}x_2 - \frac{1}{2g}\left(\frac{\partial V}{\partial x_2}\right)^2$$

for infinite time horizon is $\tfrac{1}{2}q_{11}x_1^2 + \tfrac{1}{2}q_{22}x_2^2 - \frac{\partial V}{\partial x_1}x_2 - \frac{1}{2g}\left(\frac{\partial V}{\partial x_2}\right)^2 = 0$.

The closed-form solution is found by a quadratic return function

$$V(x_1,x_2) = \tfrac{1}{2}x^TKx = \tfrac{1}{2}\begin{bmatrix} x_1 & x_2 \end{bmatrix}\begin{bmatrix} k_{11} & k_{12} \\ k_{21} & k_{22} \end{bmatrix}\begin{bmatrix} x_1 \\ x_2 \end{bmatrix} = \tfrac{1}{2}k_{11}x_1^2 + k_{12}x_1x_2 + \tfrac{1}{2}k_{22}x_2^2,\ k_{12} = k_{21}.$$

Hence, $u = -\frac{1}{g}\left(k_{12}x_1 + k_{22}x_2\right)$.

The algebraic Riccati equation are found by grouping terms for x_1^2, x_1x_2 and x_2^2.

Three quadratic equations for unknown (k_{11},k_{12},k_{22}) are

$q_{11} - \frac{1}{g}k_{12}^2 = 0$, $k_{11} - \frac{1}{g}k_{12}k_{22} = 0$, $q_{22} + 2k_{12} - \frac{1}{g}k_{22}^2 = 0$.

For q_{11}=100, q_{22}=10 and g=1, we have k_{11}=54.77, k_{12}=10, k_{21}=10 and k_{22}=5.48.

Using matrix notations, the Hamilton-Jacobi equation (5.69) is

$$-\frac{\partial V}{\partial t} = \tfrac{1}{2}x^TQx + \left(\frac{\partial V}{\partial x}\right)^T Ax - \tfrac{1}{2}\left(\frac{\partial V}{\partial x}\right)^T BG^{-1}B^T\frac{\partial V}{\partial x}\,.$$

For infinite time horizon, solve $\tfrac{1}{2}x^TQx + \left(\frac{\partial V}{\partial x}\right)^T Ax - \tfrac{1}{2}\left(\frac{\partial V}{\partial x}\right)^T BG^{-1}B^T\frac{\partial V}{\partial x} = 0$.

This equation admits a closed-form unique solution, given by a positive definite quadratic return function (5.70) $V(x) = \tfrac{1}{2}x^TKx$, K>0.

Hence, $V(x) = \tfrac{1}{2}\begin{bmatrix} x_1 & x_2 \end{bmatrix}\begin{bmatrix} k_{11} & k_{12} \\ k_{21} & k_{22} \end{bmatrix}\begin{bmatrix} x_1 \\ x_2 \end{bmatrix}$, $K = \begin{bmatrix} k_{11} & k_{12} \\ k_{21} & k_{22} \end{bmatrix}$, $k_{12} = k_{21}$.

The control law (5.77) is

$$u = -G^{-1}B^TKx = -G^{-1}\begin{bmatrix} 0 & 1 \end{bmatrix}\begin{bmatrix} k_{11} & k_{12} \\ k_{21} & k_{22} \end{bmatrix}\begin{bmatrix} x_1 \\ x_2 \end{bmatrix} = -\tfrac{1}{g}\left(k_{12}x_1 + k_{22}x_2\right).$$

The unknown matrix K is found solving an algebraic equation (5.76)

$Q + A^TK + KA - KBG^{-1}B^TK = 0$,

$$\begin{bmatrix} q_{11} & 0 \\ 0 & q_{22} \end{bmatrix} + \begin{bmatrix} 0 & 0 \\ 1 & 0 \end{bmatrix}\begin{bmatrix} k_{11} & k_{12} \\ k_{21} & k_{22} \end{bmatrix} + \begin{bmatrix} k_{11} & k_{12} \\ k_{21} & k_{22} \end{bmatrix}\begin{bmatrix} 0 & 1 \\ 0 & 0 \end{bmatrix} - \begin{bmatrix} k_{11} & k_{12} \\ k_{21} & k_{22} \end{bmatrix}\begin{bmatrix} 0 \\ 1 \end{bmatrix}g^{-1}\begin{bmatrix} 0 & 1 \end{bmatrix}\begin{bmatrix} k_{11} & k_{12} \\ k_{21} & k_{22} \end{bmatrix} = \begin{bmatrix} 0 & 0 \\ 0 & 0 \end{bmatrix}.$$

One finds three quadratic equations

$q_{11} - \frac{1}{g}k_{12}^2 = 0$, $k_{11} - \frac{1}{g}k_{12}k_{22} = 0$, $q_{22} + 2k_{12} - \frac{1}{g}k_{22}^2 = 0$.

Hence, $k_{11} = \frac{1}{g}k_{12}k_{22}$, $k_{12} = k_{21} = \pm\sqrt{q_{11}g}$, $k_{22} = \pm\sqrt{g(q_{22} + 2k_{12})}$.

The performance functional J is positive definite because the quadratic terms are used, and, q_{11}≥0, q_{22}≥0, g>0. Hence,

$k_{11}=\sqrt{q_{11}\left(q_{22}+2\sqrt{q_{11}g}\right)}$, $k_{12}=k_{21}=\sqrt{q_{11}g}$, $k_{22}=\sqrt{g\left(q_{22}+2\sqrt{q_{11}g}\right)}$.

The control law is

$$u=-\tfrac{1}{g}\left(\sqrt{q_{11}g}x_1+\sqrt{g\left(q_{22}+2\sqrt{q_{11}g}\right)}x_2\right)=-\sqrt{\tfrac{q_{11}}{g}}x_1-\sqrt{\tfrac{q_{22}+2\sqrt{q_{11}g}}{g}}x_2 .$$

We found an analytic solution using matrix calculus.

For q_{11}=100, q_{22}=10 and g=1, $K=\begin{bmatrix}k_{11} & k_{12}\\ k_{21} & k_{22}\end{bmatrix}=\begin{bmatrix}54.77 & 10\\ 10 & 5.48\end{bmatrix}$.

A control law is $u(t)=-10x_1-5.48x_2$.

Compute the feedback gains, matrix K and eigenvalues by applying the **lqr** and **care** commands

```
A=[0 1; 0 0]; B=[0;1]; Q=[100 0; 0 10]; G=[1];
[KF,K,Eigenvalues]=lqr(A,B,Q,G)   % lqr solver
[K,Eigenvalues,KF]=care(A,B,Q,G) % care solver
```

One finds

```
KF  = 10.0000  5.4772
K =    54.7723 10.0000
       10.0000  5.4772
Eigenvalues = -2.7386 + 1.5811i
              -2.7386 - 1.5811i
```

Schur Decomposition Algorithm to Solve Riccati Equation – Solve an algebraic Riccati equation (5.76) by using the Schur decomposition algorithm (5.78)-(5.79). For

$$P=\begin{bmatrix}A & -BG^{-1}B^T\\ -Q & -A^T\end{bmatrix},$$

the Schur decomposition is P=USU^T, $\mathbf{U}=\begin{bmatrix}\mathbf{U}_{11} & \mathbf{U}_{12}\\ \mathbf{U}_{21} & \mathbf{U}_{22}\end{bmatrix}$, $\mathbf{S}=\begin{bmatrix}\mathbf{S}_{11} & \mathbf{S}_{12}\\ 0 & \mathbf{S}_{22}\end{bmatrix}$.

The unknown matrix K is found as (5.79) $K=\mathbf{U}_{21}\mathbf{U}_{11}^{-1}$. The MATLAB code is

```
A=[0 1; 0 0]; B=[0;1]; Q=[100 0; 0 10]; G=[1]; n=length(A); P=[A -B*inv(G)*B'; -Q  -A'];
% Compute Schur decomposition matrices (s,u), P=u*s*u', u*u'=eye(size(P)), u'*u=eye(size(P))
[u,s]=schur(P);
[U,S]=ordschur(u,s,'lhp'); % Reorder eigenvalues of Schur decomposition
U11=U(1:n,1:n); U12=U(1:n,n+1:2*n); U21=U(n+1:2*n,1:n); U22=U(n+1:2*n,n+1:2*n);
K=U21*inv(U11), KF=inv(G)*B'*K
```

We have $K=\begin{bmatrix}k_{11} & k_{12}\\ k_{21} & k_{22}\end{bmatrix}=\begin{bmatrix}54.77 & 10\\ 10 & 5.48\end{bmatrix}$.

Hence, $u=-k_{21}x_1-k_{22}x_2=-K_{F1}x_1-K_{F2}x_2=-10x_1-5.48x_2$.

The closed-loop system

$$\dot{x}_1=x_2,$$

$$\dot{x}_2=-10x_1-5.48x_2$$

is stable because the eigenvalues $\lambda_{1,2}=-2.74\pm1.58j$ have negative real parts.

Figure 5.24.a documents a SIMULINK diagram for a closed-loop system. The system dynamics ($x_1(t)$,$x_2(t)$) is illustrated in Figure 5.24.b for initial conditions x_{10}=1, x_{20}= –1.

Hamilton-Jacobi Equation, Quadratic Return Function and Error Evaluation – A quadratic function $V(x_1,x_2)=\tfrac{1}{2}k_{11}x_1^2+k_{12}x_1x_2+\tfrac{1}{2}k_{22}x_2^2$ is a solution of the Hamilton-Jacobi equation (5.75)

$$\tfrac{1}{2}q_{11}x_1^2+\tfrac{1}{2}q_{22}x_2^2+\frac{\partial V}{\partial x_1}x_2-\frac{1}{2g}\left(\frac{\partial V}{\partial x_2}\right)^2=0.$$

An algebraic equation (5.76) $Q+A^TK+KA-KBG^{-1}B^TK=0$ yields an unknown positive definite matrix K with (k_{11},k_{12},k_{22}).

Investigate error, precision, relative and absolute accuracy of solved Riccati and Hamilton-Jacobi equations. One verifies $Q+A^TK+KA-KBG^{-1}B^TK=0$, $K>0$ by substitution. Evaluate a manifold $\Delta(x)$ for a functional equation (5.75)

$$\Delta(x_1^2,x_1x_2,x_2^2)=\tfrac{1}{2}q_{11}x_1^2+\tfrac{1}{2}q_{22}x_2^2+\frac{\partial V}{\partial x_1}x_2-\frac{1}{2g}\left(\frac{\partial V}{\partial x_2}\right)^2,$$

$V(x_1,x_2)=\tfrac{1}{2}k_{11}x_1^2+k_{12}x_1x_2+\tfrac{1}{2}k_{22}x_2^2$, k_{11}=54.77, $k_{12}=k_{21}$=10, k_{22}=5.48.

A surface for $\Delta(x_1^2,x_1x_2,x_2^2)$ is depicted in Figure 5.24.c. We have $\Delta(x)=0$, $\forall x\in X$. Our results imply that accurate solutions are found. The errors associate with analytic and numeric solutions. Our analysis is important in nonquadratic dynamic optimization. An illustrative example is documented in Example 6.39.

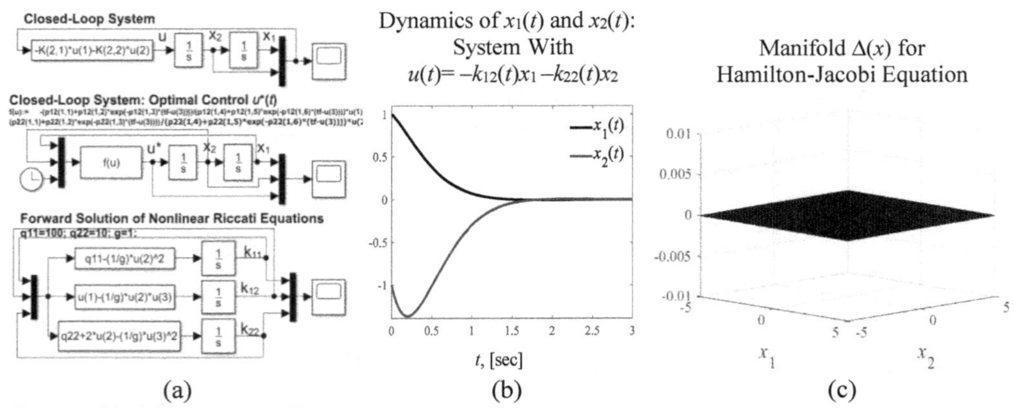

Figure 5.24. (a) SIMULINK diagrams to:
(i) Simulate a closed-loop system $\dot{x}_1=x_2$, $\dot{x}_2=u$, $u(t)=-10x_1-5.48x_2$;
(ii) Simulate a closed-loop system with an optimal control
$\dot{x}_1=x_2$, $\dot{x}_2=u$, $u^*(t)=-k_{12p}(t)x_1-k_{22p}(t)x_2$, Example 5.29;
(iii) Solve a set of nonlinear differential equations using the forward solution of Riccati equations $\dot{k}_{11}=q_{11}-\frac{1}{g}k_{12}^2$, $\dot{k}_{12}=k_{11}-\frac{1}{g}k_{12}k_{22}$, $\dot{k}_{22}=q_{22}+2k_{12}-\frac{1}{g}k_{22}^2$, $k_{ij}(t_f)=0$, t_f=3 sec, Example 5.29;
(b) Closed-loop system dynamics $\dot{x}_1=x_2$, $\dot{x}_2=u$, $u(t)=-10x_1-5.48x_2$, $x_{10}=1$, $x_{20}=-1$;
(c) Surface for the manifold $\Delta(x)$, $\Delta(x_1,x_2)=\frac{1}{2}q_{11}x_1^2+\frac{1}{2}q_{22}x_2^2+\frac{\partial V}{\partial x_1}x_2-\frac{1}{2g}(\frac{\partial V}{\partial x_2})^2$, $V(x_1,x_2)=\frac{1}{2}k_{11}x_1^2+k_{12}x_1x_2+\frac{1}{2}k_{22}x_2^2$, k_{11}=54.77, $k_{12}=k_{21}$=10, k_{22}=5.48, yielding $\Delta(x)=0$, $\forall(x_1,x_2)\in X$.■

Example 5.29. One-Dimensional Motion of a Rigid-Body: Finite Time Horizon Optimization and Numeric Solution
Consider a set of two first-order differential equations, studied in Example 5.28

$$\dot{x}_1=x_2,$$
$$\dot{x}_2=u.$$

In Example 5.28, an infinite horizon optimization problem was solved for $t_f=\infty$. A closed-form solution was found. Solve an optimal control problem by minimizing the quadratic functional (5.65)

$$J=\min_{x,u}\int_0^{t_f}\tfrac{1}{2}\left(q_{11}x_1^2+q_{22}x_2^2+gu^2\right)dt,\ q_{11}=100,\ q_{22}=10,\ g=1,\ t_f=3 \text{ sec.}$$

The control law (5.73) is

$$u=-G^{-1}B^TK(t)x=-\tfrac{1}{g}\left(k_{12}(t)x_1+k_{22}(t)x_2\right).$$

The differential Riccati equation (5.72) $-\dot{K}=Q+A^TK+KA-KBG^{-1}B^TK$ should be solved, finding $k_{ij}(t)$. The time horizon $t\in T$ and $K(t_f)$ are defined.

Let $K(t_f)=\begin{bmatrix}k_{11}(t_f) & k_{12}(t_f)\\ k_{21}(t_f) & k_{22}(t_f)\end{bmatrix}=\begin{bmatrix}0 & 0\\ 0 & 0\end{bmatrix}$ and $t_f=3$ sec.

The quadratic differential equations are

$$\dot{k}_{11}=-q_{11}+\tfrac{1}{g}k_{12}^2,\ k_{ij}(t_f)=k_{ij_f},$$

$$\dot{k}_{12}=-k_{11}+\tfrac{1}{g}k_{12}k_{22},$$

$$\dot{k}_{22}=-q_{22}-2k_{12}+\tfrac{1}{g}k_{22}^2.$$

For a system with $n\geq 2$, an analytic solution of a system of quadratic differential equations is difficult to find. One has a system of $\tfrac{1}{2}n(n+1)$ coupled quadratic differential equations, for with the conjectural solutions is sought as

$$k_{ij}(t)=\frac{p_{1_{ij}}+p_{2_{ij}}e^{-p_{3_{ij}}(t_f-t)}}{p_{4_{ij}}+p_{5_{ij}}e^{-p_{6_{ij}}(t_f-t)}},$$

where p_{ij} are the real-valued parameters to find.

The numeric solution is obtained using the MATLAB command **ode45** with a fixed step size differentiation. The differential equations are solved backward with $K(t_f)=0$, $t_f=3$ sec. The weighting coefficients are $q_{11}=100$, $q_{22}=10$ and $g=1$.

The evolutions of $k_{11}(t)$, $k_{12}(t)$ and $k_{22}(t)$ are documented in Figure 5.25.a by circles. With an adequate t_f, $K(t)$ converges to the equilibrium solutions, at which $k_{11}=54.7723$, $k_{12}=10$ and $k_{22}=5.4772$ at $t=0$ sec. Two m-files **ch51.m** and **ch52.m** are used.

```
% ch51.m
clear all; t0=0; tfinal=3; Nsteps=30; % Initial and final time, fixed step size solution
tspan=[tfinal t0]; tspan=linspace(tfinal,t0,Nsteps);
k0=[0 0 0]';                    % Initial conditions for k(t)
[t,k]=ode45('ch52',tspan,k0); % ode45 MATLAB solver
figure(1); plot(t,k(:,1),'k-',t,k(:,2),'b-',t,k(:,3),'m-','Linewidth',2); axis([0 3 0 57]);
legend('{\itk}_{11}({\itt})','{\itk}_{12}({\itt})','{\itk}_{22}({\itt})','Location','northwest','linewidth',2,'FontSize',20);
legend boxoff;
```

and

```
% ch52.m Numeric solution of nonlinear differential equations
function dkdt=difer(t,k);
q11=100; q22=10; g=1;  % Weighting coefficients
% System of three differential equations
dkdt=[-q11+(k(2,:)^2)/g;...        % first differential equation
      -k(1,:)+(k(2,:)*k(3,:))/g;...  % second differential equation
      -q22-2*k(2,:)+(k(3,:)^2)/g]; % third differential equation
```

Having found numeric solutions for $k_{11}(t)$, $k_{12}(t)$ and $k_{22}(t)$, perform least-squires estimation and interpolation to parameterize $k_{11p}(t)$, $k_{12p}(t)$ and $k_{22p}(t)$. Analytic solutions $k_{ij}(t)$ are unknown. Approximate solutions of high-order coupled Riccati equations by $k_{ij_p}(t)=\dfrac{p_{1_{ij}}+p_{2_{ij}}e^{-p_{3_{ij}}(t_f-t)}}{p_{4_{ij}}+p_{5_{ij}}e^{-p_{6_{ij}}(t_f-t)}}$. Time-varying $k_{ij_p}(t)$ are parameterized minimizing errors as

$$k_{11p}(t)=\frac{55.0937-55.1007e^{-8.8946(t_f-t)}}{1.0059+3.8566e^{-5.4347(t_f-t)}},\; k_{12p}(t)=\frac{10.0338-10.7166e^{-3.8534(t_f-t)}}{1.0024+7.3444e^{-7.0581(t_f-t)}},$$

$$k_{22p}(t)=\frac{5.5093-5.5101e^{-8.9138(t_f-t)}}{1.0058+3.8656e^{-5.4363(t_f-t)}}.$$

Figure 5.25.a depicts computed $(k_{11}(t),k_{12}(t),k_{22}(t))$ as well as evolutions of parameterized $(k_{11p}(t),k_{12p}(t),k_{22p}(t))$. The numeric error is less than 1%.

An optimal control law is

$u^*(t)=-k_{12p}(t)x_1-k_{22p}(t)x_2$.

Figure 5.24.a reports SIMULINK diagrams to simulate closed-loop systems with:

1. Control law with constant feedback gains, $u(t)=-k_{12}x_1-k_{22}x_2$;
2. Optimal control law with time-varying feedback gains, $u^*(t)=-k_{12p}(t)x_1-k_{22p}(t)x_2$;
3. Forward solution of a set of nonlinear Riccati equations.

An optimal transient dynamics $x_1^*(t)$, $x_2^*(t)$ and evolution of $u^*(t)$ are reported in Figure 5.25.b. These $k_{ij_p}(t)$ can be used to implement an optimal control.

For $K(t_f)\neq 0$, consider $k_{11}(t_f)=25$, $k_{12}(t_f)=2.5$ and $k_{22}(t_f)=2.5$, $t_f=3$ sec.

Computed $k_{11}(t)$, $k_{12}(t)$ and $k_{22}(t)$ are parametrized. For $0\leq t\leq 3$ sec, we have

$$k_{11p}(t)=\frac{56.2407-5.7886e^{-8.3108(t_f-t)}}{1.0269+0.9911e^{-5.6358(t_f-t)}},\; k_{12p}(t)=\frac{10.238-4.5853e^{-5.2471(t_f-t)}}{1.0238+1.2372e^{-6.03(t_f-t)}},$$

$$k_{22p}(t)=\frac{5.6208-0.1714e^{-16.1817(t_f-t)}}{1.0262+1.1536e^{-5.7206(t_f-t)}}.$$

Figure 5.25.c illustrates evolutions of $(k_{11}(t),k_{12}(t),k_{22}(t))$ as well as $(k_{11p}(t),k_{12p}(t),k_{22p}(t))$.

Closed-loop stability of systems, which evolve in $t\geq 0$ sec, $t\in T$ is guarantee with:

1. Constant feedback gains (k_{12},k_{22}), with which $u(t)=-k_{12}x_1-k_{22}x_2$, $0\leq t<\infty$ sec;
2. Time-varying feedback gains $k_{12p}(t)$ and $k_{22p}(t)$, and, optimal control law $u^*(t)=-k_{12p}(t)x_1-k_{22p}(t)x_2$ for $0\leq t\leq 3$ sec.

SIMULINK Solution of the Riccati Equation – Recall that $K>0$. SIMULINK may be used to solve the Riccati equation. The forward solutions of differential equations are found for

$$\dot{k}_{11}=q_{11}-\tfrac{1}{g}k_{12}^2,\; \dot{k}_{12}=k_{11}-\tfrac{1}{g}k_{12}k_{22},\; \dot{k}_{22}=q_{22}+2k_{12}-\tfrac{1}{g}k_{22}^2,\; q_{11}=100,\; q_{22}=10,\; g=1,\; t\geq 0.$$

SIMULINK diagram is reported in Figure 5.24.a. For $t\to\infty$, the resulting $k_{ij\,\text{forward}}(t)$ yield $k_{11}(t)|_{t\to\infty}=54.7723$, $k_{12}(t)|_{t\to\infty}=10$ and $k_{22}(t)|_{t\to\infty}=5.4772$. Evolutions of $k_{11\,\text{forward}}(t)$, $k_{12\,\text{forward}}(t)$ and $k_{22\,\text{forward}}(t)$, shown in Figure 5.25.d, may be refined to the backward solutions $k_{ij}(t)$ and parametrized.

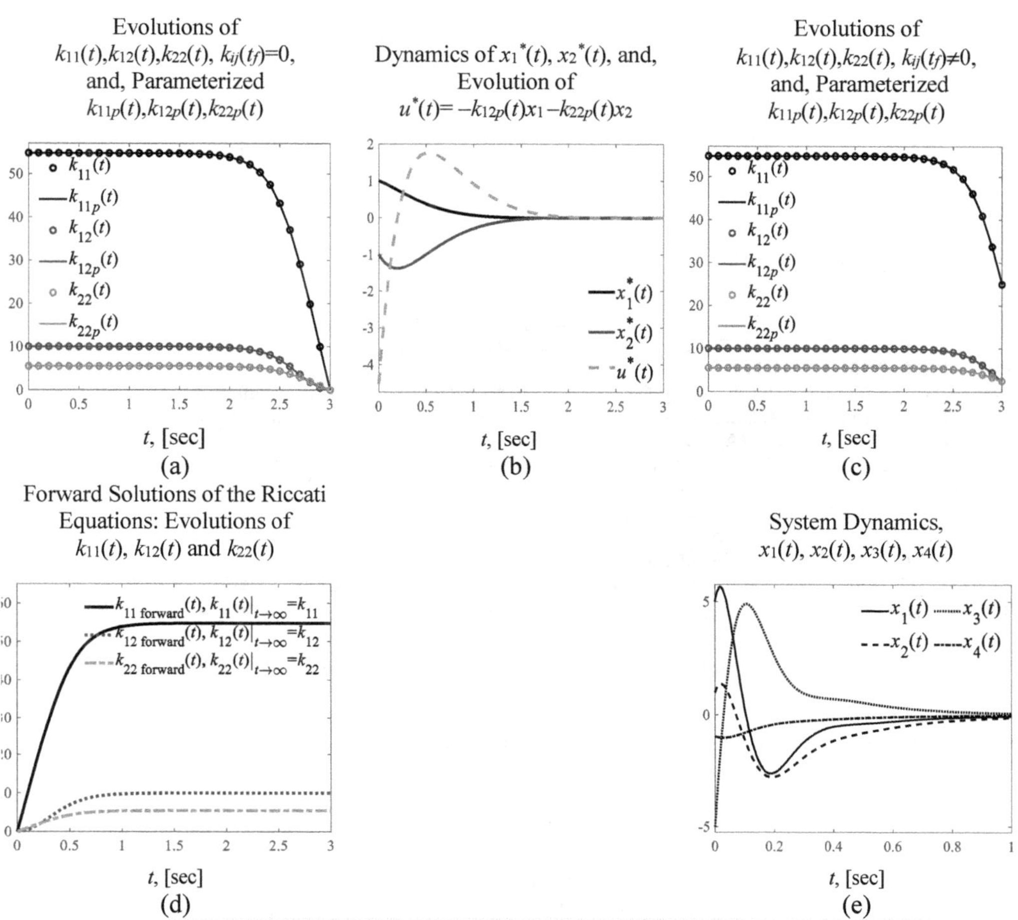

Figure 5.25. (a) Numeric solution of the Riccati equations if $k_{ij}(t_f)=0$, $t_f=3$ sec: Numeric solutions for $k_{11}(t)$, $k_{12}(t)$, $k_{22}(t)$ (circles), and, evolutions of parameterized $k_{11p}(t)$,$k_{12p}(t)$,$k_{22p}(t)$ (solid lines); (b) Closed-loop system dynamics with an optimal control law $u^*(t)=-k_{12p}(t)x_1-k_{22p}(t)x_2$. Evolutions of $x_1^*(t)$, $x_2^*(t)$ and $u^*(t)$. Initial conditions are $x_{10}=1$, $x_{20}=-1$;
(c) Solution of the Riccati equations if $k_{11}(t_f)=25$, $k_{12}(t_f)=2.5$, $k_{22}(t_f)=2.5$, $t_f=3$ sec: Numeric solutions for $k_{11}(t)$, $k_{12}(t)$, $k_{22}(t)$ (circles), and, evolutions of $k_{11p}(t)$, $k_{12p}(t)$, $k_{22p}(t)$ (solid lins);
(d) Forward solution of nonlinear differential equations, yielding $k_{11\text{forward}}(t)$, $k_{12\text{forward}}(t)$, $k_{22\text{forward}}(t)$;
(e) Dynamics of the states $x_1(t)$, $x_2(t)$, $x_3(t)$ and $x_4(t)$, Example 5.30. ■

Example 5.30. *Multi-Input Multi-Output System*
Consider a system

$$\dot{x}=Ax+Bu=\begin{bmatrix}-10 & 0 & -20 & 0\\ 0 & -10 & -10 & 0\\ 10 & 5 & -1 & 0\\ 0 & 0 & 1 & 0\end{bmatrix}\begin{bmatrix}x_1\\ x_2\\ x_3\\ x_4\end{bmatrix}+\begin{bmatrix}10 & 0\\ 0 & 0\\ 0 & 10\\ 0 & 0\end{bmatrix}\begin{bmatrix}u_1\\ u_2\end{bmatrix}.$$

The outputs are x_1 and x_4.

The output equation is $y=Hx+Du=\begin{bmatrix}1 & 0 & 0 & 0\\ 0 & 0 & 0 & 1\end{bmatrix}\begin{bmatrix}x_1\\ x_2\\ x_3\\ x_4\end{bmatrix}+\begin{bmatrix}0 & 0\\ 0 & 0\end{bmatrix}\begin{bmatrix}u_1\\ u_2\end{bmatrix}.$

Minimize the quadratic performance functional (5.65) with $t_f=\infty$

$$J=\min_{x,u}\int_0^\infty \tfrac{1}{2}\left(x^TQx+u^TGu\right)dt=\min_{x,u}\int_0^\infty \tfrac{1}{2}\left(\begin{bmatrix}x_1 & x_2 & x_3 & x_4\end{bmatrix}\begin{bmatrix}1&0&0&0\\0&1&0&0\\0&0&1&0\\0&0&0&100\end{bmatrix}\begin{bmatrix}x_1\\x_2\\x_3\\x_4\end{bmatrix}+\begin{bmatrix}u_1 & u_2\end{bmatrix}\begin{bmatrix}1&0\\0&1\end{bmatrix}\begin{bmatrix}u_1\\u_2\end{bmatrix}\right)dt$$

$$=\min_{x,u}\int_0^\infty \tfrac{1}{2}\left(x_1^2+x_2^2+x_3^2+100x_4^2+u_1^2+u_2^2\right)dt.$$

MATLAB supports control design and system simulation.

```
A=[-10 0 -20 0; 0 -10 -10 0; 10 5 -1 0; 0 0 1 0]; disp('eigenvalues_A'); disp(eig(A)); % Eigenvalues of A
B=[10 0; 0 0; 0 10; 0 0]; H=[1 0 0 0; 0 0 0 1]; D=[0];
Q=[1 0 0 0; 0 1 0 0; 0 0 1 0; 0 0 0 100]; G=[1 0; 0 1]; % Matrices of weighting coefficients
 [KF,K,Eigenvalues]=lqr(A,B,Q,G); % lqr solver: Feedback, return function coefficients and eigenvalues
% [K,Eigenvalues,KF]=care(A,B,Q,G) % care solver
disp('KF'); disp(KF); disp('K'); disp(K); disp('eigenvalues A-BKF'); disp(Eigenvalues);
A_closed_loop=A-B*KF; % Closed-loop system
t=0:0.002:1;  uu=[0*ones(max(size(t)),4)];
x0=[5 1 -5 -1];  % Initial conditions
[y,x]=lsim(A_closed_loop,B*KF,H,D,uu,t,x0); plot(t,x,'LineWidth',3);
plot(t,x(:,1),'k',t,x(:,2),'k--',t,x(:,3),'k:',t,x(:,4),'k-.','LineWidth',2); legend boxoff; axis([0 1 -5.25 5.75])
legend('{\itx}_1({\itt})','{\itx}_2({\itt})','{\itx}_3({\itt})','{\itx}_4({\itt})','NumColumns',2,'FontSize',22);
```

Matrix K, feedback gain matrix K_F, and eigenvalues of the closed-loop system $\dot{x}=\left(A-BG^{-1}B^TK\right)=\left(A-BK_F\right)$ are found. The algebraic equation (5.76) is solved using the **lqr** and **care** commands, yielding a symmetric matrix K as

$$K=\begin{bmatrix}0.0533&0.0071&0.0194&0.465\\0.0071&0.0543&0.0101&0.32\\0.0194&0.0101&0.1228&0.885\\0.465&0.32&0.885&25.16\end{bmatrix}.$$

The stabilizing control law is

$$u=\begin{bmatrix}u_1\\u_2\end{bmatrix}=-G^{-1}B^TKx=-K_Fx=-\begin{bmatrix}0.533&0.0706&0.194&4.652\\0.194&0.1014&1.228&8.852\end{bmatrix}\begin{bmatrix}x_1\\x_2\\x_3\\x_4\end{bmatrix}.$$

The closed-loop system eigenvalues are –3.965, –10.989 and –11.828±16.051j.

Dynamics of the closed-loop system states $x(t)$ are plotted in Figure 5.25.e for initial conditions x_{10}=5, x_{20}=1, x_{30}= –5 and x_{40}= –1. System is stable, and, $\lim_{t\to\infty} x(t)\to 0$.

Solution of Algebraic Riccati Equation Using the Schur Decomposition – An algebraic Riccati equation (5.76) is solved using the Schur decomposition (5.78)-(5.79). The algebraic Riccati equation was solved using the **lqr** and **care** commands. The custom MATLAB code solves an algebraic equation (5.76) $0=Q+A^TK+KA-KBG^{-1}B^TK$, K>0, K^T=K by using the Schur decomposition.

The feedback gain matrix K_F is computed.

```
n=length(A); P=[A -B*inv(G)*B'; -Q  -A'];
[u,s]=schur(P); % Compute Schur decomposition matrices (s,u), P=u*s*u', u*u'=eye(size(P)),
u'*u=eye(size(P))
[U,S]=ordschur(u,s,'lhp'); % Reorder eigenvalues of Schur decomposition
U11=U(1:n,1:n); U12=U(1:n,n+1:2*n); U21=U(n+1:2*n,1:n); U22=U(n+1:2*n,n+1:2*n);
K=U21*inv(U11), KF=inv(G)*B'*K
```

■

5.7. Tracking Control of Linear Continuous-Time Systems

Tracking Control – Optimal control and managements of MIMO systems, comprised of interconnected compliant components, should guarantee bounded reference tracking, stability and disturbance attenuation. Analog and digital tracking controllers are implemented in high-performance gimbals, pointing and steering systems, industrial and mobile robots, smart manufacturing, aerial systems and various processes.

Control laws are designed using the tracking error $e(t)=r(t)-y(t)$ and state feedback. For dynamic systems (5.64) $\dot{x}(t)=Ax+Bu$ with output $y(t)=Hx(t)$, synthesize the tracking optimal control law by minimizing the performance functional.

The tracking error is $e(t)=r(t)-y(t)=r(t)-Hx(t)$.

Denoting $e(t)=\dot{x}_{\text{ref}}(t)$, consider the system dynamics and error evolutions

$$\begin{cases}\dot{x}(t)=Ax+Bu,\ y=Hx\\ \dot{x}_{\text{ref}}(t)=r-y=r-Hx\end{cases},\ \dot{\mathbf{x}}(t)=\mathbf{A}\mathbf{x}+A_r r+\mathbf{B}u,\ \mathbf{x}=\begin{bmatrix}x\\ x_{\text{ref}}\end{bmatrix},\ e(t)=r(t)-y(t),\ \mathbf{x}_0(t_0)=\mathbf{x}_0,\quad(5.80)$$

where $\mathbf{A}$, $\mathbf{B}$ and A_r are the matrices, $\mathbf{A}=\begin{bmatrix}A & 0\\ -H & 0\end{bmatrix}$, $\mathbf{B}=\begin{bmatrix}B\\ 0\end{bmatrix}$ and $A_r=\begin{bmatrix}0\\ I\end{bmatrix}$.

Minimize the quadratic performance functional

$$J=\min_{\mathbf{x},u}\int_{t_0}^{t_f}\tfrac{1}{2}\left(\mathbf{x}^T Q\mathbf{x}+u^T Gu\right)dt \quad(5.81)$$

subject to (5.80). For (5.81) and (5.80), the Hamiltonian is

$$H\left(\mathbf{x},u,\tfrac{\partial V}{\partial \mathbf{x}}\right)=\tfrac{1}{2}\left(\mathbf{x}^T Q\mathbf{x}+u^T Gu\right)+\left(\frac{\partial V}{\partial \mathbf{x}}\right)^T(\mathbf{A}\mathbf{x}+\mathbf{B}u). \quad(5.82)$$

The first-order necessary condition for optimality (5.61) for (5.82) yields $\frac{\partial H}{\partial u}=u^T G+\left(\frac{\partial V}{\partial \mathbf{x}}\right)^T\mathbf{B}=0$. The control law is

$$u=-G^{-1}\mathbf{B}^T\frac{\partial V}{\partial \mathbf{x}}=-G^{-1}\begin{bmatrix}B\\ 0\end{bmatrix}^T\frac{\partial V}{\partial \mathbf{x}}. \quad(5.83)$$

Solution of the Hamilton-Jacobi functional equation

$$-\frac{\partial V}{\partial t}=\tfrac{1}{2}\mathbf{x}^T Q\mathbf{x}+\left(\frac{\partial V}{\partial \mathbf{x}}\right)^T\mathbf{A}\mathbf{x}-\tfrac{1}{2}\left(\frac{\partial V}{\partial \mathbf{x}}\right)^T\mathbf{B}G^{-1}\mathbf{B}^T\frac{\partial V}{\partial \mathbf{x}} \quad(5.84)$$

is satisfied by the quadratic return function

$$V(\mathbf{x})=\tfrac{1}{2}\mathbf{x}^T K\mathbf{x}. \quad(5.85)$$

From (5.84) and (5.85), the Riccati equation for an unknown K is

$$-\dot{K}=Q+\mathbf{A}^T K+K\mathbf{A}-K\mathbf{B}G^{-1}\mathbf{B}^T K,\ K(t_f)=K_f. \quad(5.86)$$

The control law with an integral state feedback $x_{\text{ref}}(t)=\int e(t)dt$ is

$$u(t)=-G^{-1}\mathbf{B}^T K\mathbf{x}(t)=-G^{-1}\begin{bmatrix}B\\ 0\end{bmatrix}^T K\begin{bmatrix}x\\ x_{\text{ref}}\end{bmatrix}=-G^{-1}\begin{bmatrix}B\\ 0\end{bmatrix}^T K\begin{bmatrix}x(t)\\ \int e(t)dt\end{bmatrix}. \quad(5.87)$$

For the extended state-space model (5.80), one may apply the *modal* control as well as the reported LQR tracking control law which yield the integral feedback term $\int e(t)dt$.

5.8. State Augmentation and Tracking Control

Design proportional-integral control laws with state feedback by applying the *state augmentation* method. Consider linear MIMO systems with the tracking error vector

$$\dot{x}(t)=Ax+Bu,\ x(t_0)=x_0,\ t\geq 0,\ y(t)=Hx(t), \tag{5.88}$$
$$e(t)=r(t)-y(t)=r(t)-Hx(t).$$

where $A\in\mathbb{R}^{n\times n}$, $B\in\mathbb{R}^{n\times m}$ and $H\in\mathbb{R}^{b\times n}$ are the matrices of the state-space and output equations.

Specify a stable evolution of the tracking error dynamics as

$$\dot{e}(t)=-I_E e+I\dot{r}-HAx-HBu,\ I_E\in\mathbb{R}^{b\times b}, \tag{5.89}$$

where I_E is the positive definite diagonal matrix, $I_E>0$; $I\in\mathbb{R}^{b\times b}$ is the identity matrix.

The error dynamics is described by (5.89). For example, $I_E=I$. From (5.88) and (5.89), the extended state vector is $\mathbf{x}(t)=\begin{bmatrix} x(t)\\ e(t)\end{bmatrix}$. One finds

$$\dot{\mathbf{x}}=\begin{bmatrix}\dot{x}\\ \dot{e}\end{bmatrix}=\begin{bmatrix}A & 0\\ -HA & -I_E\end{bmatrix}\begin{bmatrix}x\\ e\end{bmatrix}+\begin{bmatrix}0\\ I\end{bmatrix}\dot{r}+\begin{bmatrix}B\\ -HB\end{bmatrix}u=\mathbf{A}\mathbf{x}+A_r\dot{r}+\mathbf{B}u,\mathbf{x}=\begin{bmatrix}x(t)\\ e(t)\end{bmatrix},\mathbf{A}=\begin{bmatrix}A & 0\\ -HA & -I_E\end{bmatrix},\mathbf{B}=\begin{bmatrix}B\\ -HB\end{bmatrix}. \tag{5.90}$$

Define a vector $z=\begin{bmatrix}\mathbf{x}\\ u\end{bmatrix}$, and, consider the evolution of u, governed as

$$\dot{u}=-I_U u+I_V v, \tag{5.91}$$

where $v\in\mathbb{R}^m$ is the control action vector; $I_U\in\mathbb{R}^{m\times m}$ and $I_V\in\mathbb{R}^{m\times m}$ are the positive definite diagonal matrices, $I_U>0$, $I_V>0$. For example, $I_U=I$ and $I_V=I$.

Consider $z=\begin{bmatrix}\mathbf{x}\\ u\end{bmatrix}=\begin{bmatrix}x\\ e\\ u\end{bmatrix}$, $z\in\mathbb{R}^{n+b+m}$ and $v\in\mathbb{R}^m$, which evolve as

$$\dot{x}(t)=Ax+Bu, \tag{5.92}$$
$$\dot{e}(t)=-I_E e+\dot{r}-HAx-HBu,$$
$$\dot{u}=-I_U u+I_V v.$$

From (5.90) and (5.92),

$$\dot{z}(t)=\begin{bmatrix}\mathbf{A} & \mathbf{B}\\ 0 & -I_U\end{bmatrix}z+\begin{bmatrix}A_r\\ 0\end{bmatrix}\dot{r}+\begin{bmatrix}0\\ I_V\end{bmatrix}v=A_z z+A_{rz}\dot{r}+B_z v,\ z(t_0)=z_0, \tag{5.93}$$

where $A_z\in\mathbb{R}^{(n+b+m)\times(n+b+m)}$ and $B_z\in\mathbb{R}^{(n+b+m)\times m}$ are the constant-coefficient matrices.

Minimize the quadratic functional

$$J=\min_{z,v}\int_{t_0}^{t_f}\tfrac{1}{2}\left(z^T Q_z z+v^T G_z v\right)dt,\ Q_z\in\mathbb{R}^{(n+b+m)\times(n+b+m)},\ Q_z\geq 0,\ G_z\in\mathbb{R}^{m\times m},\ G_z>0, \tag{5.94}$$

subject to (5.93). A scalar Hamiltonian function is

$$H=\tfrac{1}{2}z^T Q_z z+\tfrac{1}{2}v^T G_z v+\left(\frac{\partial V}{\partial z}\right)^T\left(A_z z+B_z v\right). \tag{5.95}$$

Applying the first-order necessary condition for optimality (5.61) $\frac{\partial H}{\partial v}=0$,

$$v=-G_z^{-1}B_z^T\frac{\partial V}{\partial z}. \tag{5.96}$$

Solution of the Hamilton-Jacobi differential equation

$$-\frac{\partial V}{\partial t}=\tfrac{1}{2}z^TQ_zz+\left(\frac{\partial V}{\partial z}\right)^T A_zz-\tfrac{1}{2}\left(\frac{\partial V}{\partial z}\right)^T B_zG_z^{-1}B_z^T\frac{\partial V}{\partial z} \quad (5.97)$$

is satisfied by a continuous and differentiable quadratic return function

$$V(z)=\tfrac{1}{2}z^TKz. \quad (5.98)$$

Using (5.96) and (5.98), one obtains

$$v=-G_z^{-1}B_z^TKz. \quad (5.99)$$

Substituting (5.98) in (5.97), the Riccati equation is

$$-\dot{K}=Q_z+A_z^TK+KA_z-KB_zG_z^{-1}B_z^TK,\ K(t_f)=K_f. \quad (5.100)$$

Using (5.90) $\dot{\mathbf{x}}(t)=\mathbf{A}\mathbf{x}+\mathbf{B}u$, we have $u=\mathbf{B}^{-1}\left(\dot{\mathbf{x}}(t)-\mathbf{A}\mathbf{x}\right)$. Thus,

$$u=\mathbf{B}^{-1}(\dot{\mathbf{x}}(t)-\mathbf{A}\mathbf{x})=(\mathbf{B}^T\mathbf{B})^{-1}\mathbf{B}^T(\dot{\mathbf{x}}(t)-\mathbf{A}\mathbf{x}). \quad (5.101)$$

Applying (5.93), (5.100) and (5.101), one obtains

$$\dot{u}(t)=-G_z^{-1}B_z^TKz-I_Uu=-G_z^{-1}\begin{bmatrix}0\\I_V\end{bmatrix}^TKz-I_Uu=-G_z^{-1}\begin{bmatrix}0\\I_V\end{bmatrix}^T\begin{bmatrix}K_{11}&K_{21}^T\\K_{21}&K_{22}\end{bmatrix}\begin{bmatrix}\mathbf{x}\\u\end{bmatrix}-I_Uu \quad (5.102)$$

$$=-G_z^{-1}I_VK_{21}\mathbf{x}-\left(G_z^{-1}I_VK_{22}+I_U\right)u$$

$$=\left[-G_z^{-1}I_VK_{21}+\left(G_z^{-1}I_VK_{22}+I_U\right)\mathbf{B}^T\mathbf{A}\right]\mathbf{x}-\left(G_z^{-1}I_VK_{22}+I_U\right)\mathbf{B}^T\dot{\mathbf{x}}=K_{F2}\mathbf{x}+K_{F1}\dot{\mathbf{x}}.$$

From (5.102), a proportional-integral control law with state and tracking error feedback is

$$u=-\left(G_z^{-1}I_VK_{22}+I_U\right)\mathbf{B}^T\mathbf{x}+\int\left[-G_z^{-1}I_VK_{21}+\left(G_z^{-1}I_VK_{22}+I_U\right)\mathbf{B}^T\mathbf{A}\right]\mathbf{x}dt \quad (5.103)$$

$$=-\left(G_z^{-1}I_VK_{22}+I_U\right)\mathbf{B}^T\begin{bmatrix}x\\e\end{bmatrix}+\left[-G_z^{-1}I_VK_{21}+\left(G_z^{-1}I_VK_{22}+I_U\right)\mathbf{B}^T\mathbf{A}\right]\begin{bmatrix}\int xdt\\\int edt\end{bmatrix}$$

$$=K_{F1}\mathbf{x}+\int K_{F2}\mathbf{x}dt,\ \mathbf{x}=\begin{bmatrix}x\\e\end{bmatrix}.$$

Example 5.31. *Controllability and Observability.*
Assume system (5.88) is controllable and observable. That is, the controllability and observability matrices have full rank, and, rank([B AB ...$A^{(n-1)}B$])=n. System (5.93) is controllable and observable because for

C=[B_z A_zB_z ... $A_z^{(n-1)}B_z$], O=[C CA_z...$CA_z^{(n-1)}$]^T,

the ranks are

rank(C)=(n+b+m), rank(O)=(n+b+m). ■

Nonlinear Systems – For nonlinear systems, the proposed procedure is applied.
Consider a system

$$\dot{x}(t)=F(x)+B(x)u,\ y=Hx,\ e(t)=Nr(t)-y(t)=Nr(t)-Hx(t), \quad (5.104)$$

where $F(\cdot)$:$\mathbb{R}^n\to\mathbb{R}^n$ and $B(\cdot)$:$\mathbb{R}^n\to\mathbb{R}^{n\times m}$ are the continuous and Lipschitz nonlinear maps.

To guarantee stability, the tracking error and state evolutions must be bounded for bounded inputs $r(t)$. Specify the dynamic evolution of $e(e)$ as

$$\dot{e}(t)=-I_Ee+I\dot{r}-HF(x)-HB(x)u,\ I_E>0. \quad (5.105)$$

Dynamics of state and error vectors are

$$\dot{\mathbf{x}}(t)=\begin{bmatrix}\dot{x}\\ \dot{e}\end{bmatrix}=\begin{bmatrix}F(x)\\ -HF(x)\end{bmatrix}+\begin{bmatrix}0\\ -I_E\end{bmatrix}\begin{bmatrix}x\\ e\end{bmatrix}+\begin{bmatrix}0\\ I\end{bmatrix}\dot{r}+\begin{bmatrix}B(x)\\ -HB(x)\end{bmatrix}u=\mathbf{F}(\mathbf{x})+A_r\dot{r}+\mathbf{B}(x)u\,. \quad (5.106)$$

Denote $\mathbf{x}=\begin{bmatrix}x\\ e\end{bmatrix}$ and $z=\begin{bmatrix}\mathbf{x}\\ u\end{bmatrix}$, where the evolution of u is governed by (5.91) $\dot{u}=-I_U u+I_V v$. The augmented vector is

$$\dot{z}(t)=\begin{bmatrix}\dot{\mathbf{x}}\\ \dot{u}\end{bmatrix}=\begin{bmatrix}\mathbf{F}(\mathbf{x})\\ 0\end{bmatrix}+\begin{bmatrix}A_r\\ 0\end{bmatrix}\dot{r}+\begin{bmatrix}\mathbf{B}(x)u\\ -I_U u\end{bmatrix}+\begin{bmatrix}0\\ I_V\end{bmatrix}v=F_z(z)+A_{rz}\dot{r}+B_z v\,. \quad (5.107)$$

The controllability of (5.107) is guaranteed. Using the Lie bracket operator $\left[\mathrm{ad}_{F_z}^k B_z\right]=\left[F_x \ldots [F_z,B_z]\right]$ one finds that $C=\left[B_z \ldots [\mathrm{ad}_{F_z}^k B_z]\right]$ spans in $(n+b+m)$-space with the rank$(C)=(n+b+m)$.

Minimize the quadratic functional (5.94) $J=\min\limits_{z,v}\int_{t_0}^{t_f}\frac{1}{2}\left(z^T Q_z z+v^T G_z v\right)dt$, $Q\geq 0$, $G_z>0$ subject to (5.107). The resulting Hamiltonian is

$$H=\tfrac{1}{2}z^T Q_z z+\tfrac{1}{2}v^T G_z v+\left(\frac{\partial V}{\partial z}\right)^T\left[F_z(z)+B_z v\right]. \quad (5.108)$$

The first-order necessary condition for optimality (5.61) yields

$$v=-G_z^{-1}B_z^T\frac{\partial V}{\partial z}, \quad (5.109)$$

As reported in chapter 6, the solution of the Hamilton-Jacobi equation

$$-\frac{\partial V}{\partial t}=\tfrac{1}{2}z^T Q_z z+\frac{\partial V}{\partial z}^T F_z(z)-\tfrac{1}{2}\left(\frac{\partial V}{\partial z}\right)^T B_z G_z^{-1}B_z^T\frac{\partial V}{\partial z} \quad (5.110)$$

is approximated by the nonquadratic return function (6.24)

$$V(t,z)=\sum_{i,j}\kappa_{1_i}^T(t,z^{<d_i>})K_{ij}(t)\kappa_{2_j}(t,z^{<d_j>}),\ K_{ij}>0. \quad (5.111)$$

Symmetric matrices K_{ij} are found by solving (5.110). For quadratic return function (5.98) $V(z)=\frac{1}{2}z^T Kz$, a control function (5.109) is $v=-G_z^{-1}B_z^T Kz$. From (5.101)

$$\dot{u}(t)=-G_z^{-1}B_z^T\frac{\partial V}{\partial z}-I_U u=-G_z^{-1}\begin{bmatrix}0\\ I_V\end{bmatrix}^T\frac{\partial V(\mathbf{x},u)}{\partial[\mathbf{x}\,u]^T}-I_U u=-G_z^{-1}\begin{bmatrix}0\\ I_V\end{bmatrix}^T Kz-I_U u$$

$$=-G_z^{-1}\begin{bmatrix}0\\ I_V\end{bmatrix}^T\begin{bmatrix}K_{11} & K_{21}^T\\ K_{21} & K_{22}\end{bmatrix}\begin{bmatrix}\mathbf{x}\\ u\end{bmatrix}-I_U u=-G_z^{-1}I_V K_{21}\mathbf{x}-\left(G_z^{-1}I_V K_{22}+I_U\right)u \quad (5.112)$$

$$=\left[-G_z^{-1}I_V K_{21}\mathbf{x}+\left(G_z^{-1}I_V K_{22}+I_U\right)\mathbf{B}^T(x)\mathbf{F}(\mathbf{x})\right]-\left(G_z^{-1}I_V K_{22}+I_U\right)\mathbf{B}^T(x)\dot{\mathbf{x}}.$$

A proportional-integral control law with the state feedback and tracking error feedback is

$$u=-\left(G_z^{-1}I_V K_{22}+I_U\right)\mathbf{B}^T(x)\mathbf{x}+\int\left[-G_z^{-1}I_V K_{21}\mathbf{x}+\left(G_z^{-1}I_V K_{22}+I_U\right)\mathbf{B}^T(x)\mathbf{F}(\mathbf{x})\right]dt,\ \mathbf{x}=\begin{bmatrix}x\\ e\end{bmatrix}. \quad (5.113)$$

Example 5.32. *Proportional-Derivative and Tracking Control With State Feedback*
A one-dimensional motion of a moving frictionless mass is described by

$\dot{x}_1=u$,

$\dot{x}_2=x_1$, $y=x_2$.

Considering the *force control* problem, covered in Example 5.9.

For the PID control law, $u = k_p e + k_i \int e dt + k_d \frac{de}{dt}$, $e(t)=(r-y)=(r-x_2)$.

The transfer function and characteristic equation of the closed-loop system are

$$G(s) = \frac{k_d s^2 + k_p s + k_i}{s^3 + k_d s^2 + k_p s + k_i}, \quad s^3+k_d s^2+k_p s+k_i=0.$$

For the proportional-derivative control law $u = k_p e + k_d \frac{de}{dt}$, the transfer function and characteristic equation are $G(s) = \frac{k_d s + k_p}{s^2 + k_d s + k_p}$, $s^2+k_d s+k_p=0$.

Applying the *pole placement* approach, the feedback gains can be found to ensure the specified time constant T, damping coefficient ξ, overshoot, etc.

We have $T = \sqrt{\frac{1}{k_p}}$, $\xi = \frac{k_d}{2\sqrt{k_p}}$.

Let $T = \sqrt{\frac{1}{k_p}}$ =0.2 sec. Two damping coefficients are specified to be ξ=0.707 and ξ=1.

For T=0.2 sec and ξ=0.707, the feedback gains are k_p=25 and k_d=7.07.

For T=0.2 sec and ξ=1, we compute k_p=25 and k_d=10.

The resulting closed-loop output dynamics $y=x_2$ for two sets of feedback gains in $u = k_p e + k_d \frac{de}{dt}$ are documented in Figures 5.26 by dashed lines if $r(t)=\pm 1$.

Design a tracking control law using the augmented vector. The evolutions of dynamic tracking error and control efforts are

$$\dot{e}(t) = -e - x_1,$$

$$\dot{u}(t) = -u + v.$$

We obtain the governing equations (5.93)

$$\dot{z}(t) = A_z z + A_{rz}\dot{r} + B_z v, \; z = \begin{bmatrix} x_1 \\ x_2 \\ e \\ u \end{bmatrix}, \; y=Hx, \; A_z = \begin{bmatrix} 0 & 0 & 0 & 1 \\ 1 & 0 & 0 & 0 \\ -1 & 0 & -1 & 0 \\ 0 & 0 & 0 & -1 \end{bmatrix}, \; B_z = \begin{bmatrix} 0 \\ 0 \\ 0 \\ 1 \end{bmatrix}, \; H=[0 \;\; 1].$$

The controllability matrix $C=[B_z \; A_zB \; A_z^2B_z \; A_z^3B_z]$ is $C = \begin{bmatrix} 0 & 1 & -1 & 1 \\ 0 & 0 & 1 & -1 \\ 0 & 0 & -1 & 2 \\ 1 & -1 & 1 & -1 \end{bmatrix}$.

The rank of controllability matrix C and observability matrix $O=[C \; CA_z \; CA_z^2 \; CA_z^3]^T$ is rank(C)=4 and rank(O)=4. The number of uncontrollable and unobservable states is zero. The system with the extended vector z is controllable and observable.

Minimize $J = \min_{z,v} \int_0^\infty \frac{1}{2}\left(z^T Q_z z + v^T G_z v\right) dt, \; Q_z = \begin{bmatrix} 1 & 0 & 0 & 0 \\ 0 & 1 & 0 & 0 \\ 0 & 0 & 1\times 10^7 & 0 \\ 0 & 0 & 0 & 1 \end{bmatrix}, \; G_z = 1$.

Solving (5.100) $Q_z + KA_z + A_z^T K - KB_z G_z^{-1} B_z^T K = 0$, we obtain

$$K = \begin{bmatrix} 8.69\times 10^3 & 2.84\times 10^1 & -8.12\times 10^4 & 4.03\times 10^2 \\ 2.84\times 10^1 & 3.16\times 10^3 & 2.76\times 10^3 & 1 \\ -8.12\times 10^4 & 2.76\times 10^3 & 1.2\times 10^6 & -2.76\times 10^3 \\ 4.03\times 10^2 & 1 & -2.76\times 10^3 & 2.74\times 10^1 \end{bmatrix}, \; K>0, \; K=K^T.$$

The proportional-integral tracking control law (5.103) is

$$u = K_{F1}\begin{bmatrix} x_1 \\ x_2 \\ e \end{bmatrix} + K_{F2}\int \begin{bmatrix} x_1 \\ x_2 \\ e \end{bmatrix} dt, \; K_{F1}=[-27.4 \quad 0 \quad 0],\; K_{F2}=[-403 \quad -1 \quad 2758.4].$$

Figures 5.26 report simulations for the closed-loop system if $r(t)=\pm 1$. Designed proportional-derivative control law does not have a derivative feedback, and, ensures better performance.

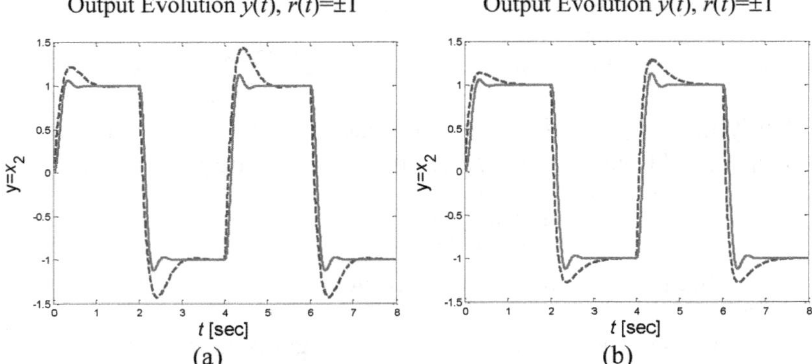

(a) (b)

Figure 5.26. (a) Output dynamics $y=x_2(t)$ of a system with proportional-derivative control $u = k_p e + k_d \frac{de}{dt}$, k_p=25, k_d=7.07 (dashed line), and, tracking control law $u = K_{F1}\begin{bmatrix} x_1 \\ x_2 \\ e \end{bmatrix} + K_{F2}\int \begin{bmatrix} x_1 \\ x_2 \\ e \end{bmatrix} dt$ (solid line);
(b) Output dynamics if $u = k_p e + k_d \frac{de}{dt}$, k_p=25, k_d=10 (dashed line) and tracking control law (solid line). ■

5.9. Optimal Tracking Control of Continuous-Time Systems

Consider a linear system

$$\dot{x}(t)=Ax+Bu,\; x(t_0)=x_0,\; t\geq 0, \qquad (5.114)$$
$$y(t)=Hx(t),\; e(t)=(r-y)=(r-Hx),$$

where $A\in\mathbb{R}^{n\times n}$, $B\in\mathbb{R}^{n\times m}$ and $H\in\mathbb{R}^{b\times n}$.

Investigate the *error state* $x_e(t)$ and tracking error $e(t)$ dynamics. Define the *error state* $x_e(t)$ and tracking error $e(t)$ governance as

$$\dot{x}_e(t) = I_X(r-y) = I_X(r-Hx),\; e(t)\equiv \dot{x}_e(t),\; I_X\in\mathbb{R}^{b\times b}, \qquad (5.115)$$
$$\dot{e}(t) = -I_E e + I_E^X x_e + \dot{r} - \dot{y} = -I_E e + I_E^X x_e + \dot{r} - HAx - HBu,\; I_E\in\mathbb{R}^{b\times b},\; I_E^X \in\mathbb{R}^{b\times b},$$

where I_X and I_E are the positive definite diagonal matrices, such as $I_X=I$ and $I_E=I$.

The *error state* and tracking error dynamics are stable. The eigenvalues $(\lambda_{x_e}, \lambda_e)$ for (x_e,e) are affected by (I_X, I_E). Evolutions of $x_e(t)$ and $e(t)$ are bounded for the bounded inputs, and, the tracking error depends on inputs.

For the state vector $\mathbf{x}(t) = \begin{bmatrix} x(t) \\ x_e(t) \\ e(t) \end{bmatrix}$, $\mathbf{x}\in\boldsymbol{X}\subset\mathbb{R}^{n+2b}$, using (5.114) and (5.115), the extended governing equations which describe system and error dynamics are

$$\dot{\mathbf{x}}(t)=\begin{bmatrix}\dot{x}(t)\\ \dot{x}_e(t)\\ \dot{e}(t)\end{bmatrix}=\mathbf{A}\mathbf{x}+A_r r+\mathbf{B}u=\begin{bmatrix}A & 0 & 0\\ -I_X H & 0 & 0\\ -HA & I_E^X & -I_E\end{bmatrix}\mathbf{x}+\begin{bmatrix}0 & 0\\ I_X & 0\\ 0 & I\end{bmatrix}\begin{bmatrix}r\\ \dot{r}\end{bmatrix}+\begin{bmatrix}B\\ 0\\ -HB\end{bmatrix}u,\ \mathbf{x}(t)=\begin{bmatrix}x(t)\\ x_e(t)\\ e(t)\end{bmatrix},\quad(5.116)$$

where $\mathbf{A}\in\mathbb{R}^{(n+2b)\times(n+2b)}$, $A_r\in\mathbb{R}^{(n+2b)\times 2b}$ and $\mathbf{B}\in\mathbb{R}^{(n+2b)\times m}$ are the matrices.

Stochastic Optimization: System States, Tracking Error and References – Evolutions of states $\mathbf{x}(t)$ are affected by references $r(t)$ and disturbances $d(t)$, which could be deterministic, random, continuous, piecewise continuous, etc. For random $r(t)$, stochastic optimization implies minimization of $J=\min_{t,x,e,u}\mathbf{E}\int_{t_0}^{t_f}\omega(t,x,e,u)dt$. Dynamic stochastic optimization complicates design and may yield system sensitivity with a minor overall gain. Furthermore, references $r(t)$ are not stationary, and, may not be statistically characterized *a priori* and *posteriori*. Stochastic optimization of dynamic systems, which is an open problem, may ensure optimality in probability under many assumptions. In physical systems, processes and applications, bounded references $r(t)$, perturbations and disturbances $d(t)$ are deterministic, superposition of deterministic and stochastic processes, or, stationary and non-stationary processes. Deterministic optimization ensures ε-global optimality, and, a local optimum with an objective functional differs by at most ε from the global for a given $\varepsilon>0$.

Non-stationary processes take finite values at any time, and, may not admit probabilistic models. Consider references

$|r(t)|<M,\ \forall t,\ t\geq 0,$

where M is the finite amplitude, $|M|<\infty$.

Typical references are periodic, aperiodic, symmetric, asymmetric continuous and piecewise continuous with $|r(t)|<\infty$. For example, periodic references with amplitude M, period T and frequency $f=1/T$ are

$r(t)=M\sin(\omega t)$, $r(t)=M\mathrm{sgn}[\sin(\omega t)]$, $\omega=2\pi f$.

Other illustrative patterns with $(a,b)>0$ are $r(t)=Me^{-at}\sin(\omega t-\phi)$,

$r(t)=Me^{-at}\mathrm{sgn}[\sin(\omega t)]$, $r(t)=Me^{-\sin(bt)}\sin(\omega t-\phi)$, $r(t)=Me^{-at}e^{\sin^2 bt}\sin(\omega t-\phi)$, etc.

Dynamic tracking of $r(t)$ should be ensured with minimal tracking error, allowed overshoot, minimal settling time, etc. Deterministic bounded references $r(t)$ affect evolutions of (x,y,e,u). References $r(t)$ are not system states, and, may not affect dynamic optimization schemes. For MIMO system with states $\mathbf{x}(t)=\begin{bmatrix}x(t)\\ x_e(t)\\ e(t)\end{bmatrix}$, $\begin{bmatrix}x_e(t)\\ e(t)\end{bmatrix}\equiv\begin{bmatrix}\int edt\\ e(t)\end{bmatrix}$ and $|r(t)|<M$, stability, boundedness and optimal governance of $\mathbf{x}(t)$ are ensured by solving stability and optimization problems in the Hamiltonian and Lyapunov sense.

Problem Formulation – Minimize the quadratic functional

$$J=\min_{\mathbf{x},u}\int_{t_0}^{t_f}\tfrac{1}{2}\left(\mathbf{x}^T Q\mathbf{x}+u^T Gu\right)dt,\ Q\geq 0,\ G>0,\quad(5.117)$$

subject to (5.116), and, design a control law.

The bounded references $r(t)$ are not the states. The Hamiltonian is

$$H(\mathbf{x},u,\tfrac{\partial V}{\partial \mathbf{x}})=\tfrac{1}{2}\mathbf{x}^T Q\mathbf{x}+\tfrac{1}{2}u^T Gu+\left(\frac{\partial V}{\partial \mathbf{x}}\right)^T(\mathbf{A}\mathbf{x}+\mathbf{B}u). \tag{5.118}$$

Applying the first-order necessary condition for optimality (5.22)

$$u=-G^{-1}\mathbf{B}^T\frac{\partial V}{\partial \mathbf{x}}. \tag{5.119}$$

Solution of the resulting Hamilton-Jacobi equation

$$-\frac{\partial V}{\partial t}=\tfrac{1}{2}\mathbf{x}^T Q\mathbf{x}+\left(\frac{\partial V}{\partial \mathbf{x}}\right)^T\mathbf{A}\mathbf{x}-\tfrac{1}{2}\left(\frac{\partial V}{\partial \mathbf{x}}\right)^T\mathbf{B}G^{-1}\mathbf{B}^T\frac{\partial V}{\partial \mathbf{x}} \tag{5.120}$$

is satisfied by a continuous and differentiable quadratic return function

$$V(\mathbf{x})=\tfrac{1}{2}\mathbf{x}^T K\mathbf{x},\ K=K^T,\ K>0. \tag{5.121}$$

Proportional-Integral Tracking Control With State Feedback – From (5.119) and (5.121), one obtains a control law

$$u=-G^{-1}\mathbf{B}^T K\mathbf{x},\ \mathbf{x}(t)=\begin{bmatrix}x(t)\\ x_e(t)\\ e(t)\end{bmatrix}. \tag{5.122}$$

Solution of the Riccati equation

$$-\dot{K}=Q+\mathbf{A}^T K+K\mathbf{A}-K\mathbf{B}G^{-1}\mathbf{B}^T K,\ K(t_f)=K_f \tag{5.123}$$

yields the unknown $K(t)$.

Proportional-Integral-Derivative Control – For system and error governance (5.116), consider the evolution of the second derivative of tracking error $\frac{d^2e}{dt^2}$ using a derivative state $\boldsymbol{e}(t)\equiv\frac{de}{dt}$, which evolves as

$$\dot{\boldsymbol{e}}(t)=-I_{\boldsymbol{e}}^E e-I_{\boldsymbol{e}}\boldsymbol{e}+\ddot{r}-\ddot{y}=-I_{\boldsymbol{e}}^E e-I_{\boldsymbol{e}}\boldsymbol{e}+\ddot{r}-HA\dot{x}-HB\dot{u},\ I_{\boldsymbol{e}}^E\in\mathbb{R}^{b\times b},\ I_{\boldsymbol{e}}\in\mathbb{R}^{b\times b},$$

with positive definite $I_{\boldsymbol{e}}^E$ and $I_{\boldsymbol{e}}$. Evolutions of $\boldsymbol{e}(t)$ are stable with the bounded input.

Using (5.114)-(5.116) and governance for $\boldsymbol{e}(t)$, one finds

$$\dot{\mathbf{x}}(t)=\begin{bmatrix}\dot{x}(t)\\ \dot{x}_e(t)\\ \dot{e}(t)\\ \dot{\boldsymbol{e}}(t)\end{bmatrix}=\mathbf{A}\mathbf{x}+A_r r+\mathbf{B}\mathbf{u},\ \mathbf{x}=\begin{bmatrix}x(t)\\ x_e(t)\\ e(t)\\ \boldsymbol{e}(t)\end{bmatrix},\ \mathbf{u}=\begin{bmatrix}u\\ \dot{u}\end{bmatrix},\ r\equiv\begin{bmatrix}r\\ \dot{r}\\ \ddot{r}\end{bmatrix},$$

$\mathbf{A}\in\mathbb{R}^{(n+3b)\times(n+3b)}$, $A_r\in\mathbb{R}^{(n+3b)\times 3b}$, $\mathbf{B}\in\mathbb{R}^{(n+3b)\times 2m}$.

Minimizing the quadratic functional

$$J=\min_{\mathbf{x},u}\int_{t_0}^{t_f}\tfrac{1}{2}\left(\mathbf{x}^T Q\mathbf{x}+u^T Gu\right)dt,\ Q\in\mathbb{R}^{(n+3b)\times(n+3b)},$$

the proportional-integral-derivative tracking control law with state feedback is

$$\mathbf{u}=\begin{bmatrix}u\\ \dot{u}\end{bmatrix}=-G^{-1}\mathbf{B}^T K\mathbf{x},\ \mathbf{x}=\begin{bmatrix}x\\ \int e\,dt\\ e\\ \frac{de}{dt}\end{bmatrix}.$$

From $\dot{x}(t)=Ax+Bu$, we have $u=B^{-1}(\dot{x}(t)-Ax)=(B^T B)^{-1}B^T(\dot{x}(t)-Ax)$.

If $HB\equiv 0$, the extended system governance and control law are

$$\dot{\mathbf{x}}(t) = \mathbf{A}\mathbf{x} + A_r r + \mathbf{B}u,\ u = -G^{-1}\mathbf{B}^T K\mathbf{x},\ \mathbf{x} = \begin{bmatrix} x \\ \int e dt \\ e \\ \frac{de}{dt} \end{bmatrix}.$$

Example 5.33. *Rotational Motion of a Satellite: Proportional-Integral-Derivative Control With State Feedback, and, Observer Design*

Consider the rotational motion of a satellite around the center of mass, as schematically depicted in Figure 5.27.a. Two or more liquid- or solid propellant thrusters, placed on opposite sides, are synchronously ignited. The thrust F_{thrust} is limited. For two thrusters on opposite sides, the total torque is

$T = R_{\perp}(F_{1\text{thrust}} + F_{2\text{thrust}})$,

where $R_{\perp}$ is the perpendicular radius.

Denote $u_1 \equiv T_1$ and $u_2 \equiv T_2$.

Consider synchronized controlled ignition and homogeneous thrust of identical opposing thrusters, such that $|u_1| = |u_2|$. Hence, the control input is $u \equiv T$, $T = T_1 + T_2$.

Control Limits – Due to inherent limits on the thrust $F_{i\ \text{thrust}}$, in the *arbitrary units*, we have $0 \leq (u_1, u_2) \leq ½$ and $0 \leq u \leq 1$. The opposing thrusters develop control efforts, $|u| \leq 1$. The constrained optimization for the studied problem will be examined in Examples 6.32 and 6.48. Assume thrusters develop continuous control function $u(t) \equiv T(t)$ which can be regulated. Solid propellants produce a short-duration *pulse* thrust and near-constant thrust. Many rocket engines do not ensure a uniform and continuous thrust. The engine-specific thrust production, controlled burn rate and other problems may be considered. Monopropellant, bipropellant and cryogenic liquid propellant thrusters may ensure controlled thrust.

The Newtonian dynamics yields the governing equations for rotational motions. In the *arbitrary units*, assuming continuous thrust for $J = 1$ kg-m^2, one has

$$\frac{d^2\theta}{dt^2} = u,\ u(t) \equiv T(t),\ y = \theta.$$

Using $x = \begin{bmatrix} x_1 \\ x_2 \end{bmatrix} = \begin{bmatrix} \omega \\ \theta \end{bmatrix}$, we obtain

$$\dot{x}_1 = u,$$

$$\dot{x}_2 = x_1,\ y = x_2,\ e(t) = (r - y) = (r - x_2),$$

where x_1 and x_2 are the angular velocity ω and displacement θ; u is the total reaction torque, which is assumed to be continuous.

Proportional-Integral Control – Find the extended system dynamics. The governing equations (5.116) with $I_X = 1$, $I_E = 1$ and $I_E^X = 0$ are

$$\dot{x}_1 = u,$$

$$\dot{x}_2 = x_1,\ y = x_2,$$

$$\dot{x}_e = r - y = r - x_2,$$

$$\dot{e} = -e + \dot{r} - \dot{y} = -e + \dot{r} - \dot{x}_2 = -e + \dot{r} - x_1.$$

For the state vector **x**, (5.116) yields

$$\dot{\mathbf{x}}=\mathbf{A}\mathbf{x}+A_r r+\mathbf{B}u,\ \begin{bmatrix}\dot{x}_1\\ \dot{x}_2\\ \dot{x}_e\\ \dot{e}\end{bmatrix}=\begin{bmatrix}0&0&0&0\\ 1&0&0&0\\ 0&-1&0&0\\ -1&0&0&-1\end{bmatrix}\begin{bmatrix}x_1\\ x_2\\ x_e\\ e\end{bmatrix}+\begin{bmatrix}0&0\\ 0&0\\ 1&0\\ 0&1\end{bmatrix}\begin{bmatrix}r\\ \dot{r}\end{bmatrix}+\begin{bmatrix}1\\ 0\\ 0\\ 0\end{bmatrix}u,\ \mathbf{x}=\begin{bmatrix}x_1\\ x_2\\ x_e\\ e\end{bmatrix}.$$

Minimize a quadratic infinite horizon functional (5.117)

$$J=\min_{\mathbf{x},u}\int_{t_0}^{\infty}\tfrac{1}{2}\left(\mathbf{x}^T Q\mathbf{x}+Gu^2\right)dt,\ t_f=\infty$$

subject to (5.116) $\dot{\mathbf{x}}=\mathbf{A}\mathbf{x}+A_r r+\mathbf{B}u$.

Equation (5.120) $0=\tfrac{1}{2}\mathbf{x}^T Q\mathbf{x}+\left(\frac{\partial V}{\partial \mathbf{x}}\right)^T\mathbf{A}\mathbf{x}-\tfrac{1}{2}\left(\frac{\partial V}{\partial \mathbf{x}}\right)^T\mathbf{B}G^{-1}\mathbf{B}^T\frac{\partial V}{\partial \mathbf{x}}$ is satisfied by the quadratic return function $V=\tfrac{1}{2}\mathbf{x}^T K\mathbf{x}$.

To compute the feedback gains of the tracking control law (5.122)

$$u=-G^{-1}\mathbf{B}^T K\mathbf{x}=-K_F\mathbf{x},\ \mathbf{x}(t)=\begin{bmatrix}x_1\\ x_2\\ \int edt\\ e\end{bmatrix},$$

solve an algebraic equation $Q+\mathbf{A}^T K+K\mathbf{A}-K\mathbf{B}G^{-1}\mathbf{B}^T K=0$, $K\in\mathbb{R}^{4\times 4}$, $K>0$.

The weighting coefficients matrices are $Q=\begin{bmatrix}1&0&0&0\\ 0&1&0&0\\ 0&0&q_{33}&0\\ 0&0&0&1\end{bmatrix}$ and $G=1$.

Transient dynamics and stability are affected by (Q,G).

For $q_{33}=2.5$ and $q_{33}=5$, we compute

$$K=\begin{bmatrix}2.71&3.13&-1.58&-0.059\\ 3.13&7.18&-4.38&0.28\\ -1.58&-4.38&4.94&-0.094\\ -0.059&0.28&-0.094&0.498\end{bmatrix}\ \text{and}\ K=\begin{bmatrix}2.95&3.8&-2.24&-0.05\\ 3.8&9.26&-6.7&0.3\\ -2.24&-6.7&8.49&-0.11\\ -0.05&0.3&-0.11&0.5\end{bmatrix}.$$

The feedback gain matrices are

$K_F=[2.71\ 3.13\ -1.58\ -0.059]$ and $K_F=[2.95\ 3.8\ -2.24\ -0.05]$.

Hence, for $q_{33}=2.5$ and $q_{33}=5$,

$$u=-2.71x_1-3.13x_2+1.58\int edt+0.059e$$

and

$$u=-2.95x_1-3.8x_2+2.24\int edt+0.05e.$$

The SIMULINK diagram is reported in Figure 5.27.b. The closed-loop system evolutions and control efforts for $r(t)=\pm 1$ are reported in Figures 5.27.c if $u=-2.95x_1-3.8x_2+2.24\int edt+0.05e$. Control laws with a high gain proportional feedback can be design specifying $q_{44}\gg 1$. Large feedback gains yield high control efforts which may not be ensured due to control limits. Computing of K and K_F takes ~0.01 sec. One may reconfigure control laws in near real-time achieving adaptive capabilities.

Observer Design – The measured variables are the angular displacement $\theta(t)$ and the tracking error $e(t)$. Wile the gyroscopes measure the angular rates, assume the angular velocity $\omega(t)$ is not measured. An *observer* is designed to observe $\hat{x}_1(t)$.

For a system

$$\dot{x}=\begin{bmatrix}\dot{x}_1\\ \dot{x}_2\end{bmatrix}=\begin{bmatrix}0 & 0\\ 1 & 0\end{bmatrix}\begin{bmatrix}x_1\\ x_2\end{bmatrix}+\begin{bmatrix}1\\ 0\end{bmatrix}u,\ y=x_2,$$

the equation of the state *observer* (5.50) is $\frac{d\hat{x}}{dt}=\left(A-K_E H\right)\hat{x}+Bu+K_E y$.

Let the *observer* eigenvalues are –5 and –10. One finds $K_E=[50\ \ 15]^T$.

The SIMULINK diagram for the closed-loop system with an observed is reported in Figure 5.27.b. System dynamics with a control law

$u=-2.95\hat{x}_1-3.8x_2+2.24\int edt+0.05e$

is illustrated in Figure 5.27.d. The initial conditions are $x_0=[-0.5, 0]$ and $\hat{x}_0=[0,\ 0]$.

Proportional-Integral-Derivative Control – Consider the extended dynamics with $\boldsymbol{e}(t)\equiv\frac{de}{dt}$. We have

$$\dot{x}_1=u,\ y=x_2,$$
$$\dot{x}_2=x_1,$$
$$\dot{x}_e=r-y=r-x_2,$$
$$\dot{e}=-e+\dot{r}-\dot{y}=-e+\dot{r}-\dot{x}_2=-e+\dot{r}-x_1,$$
$$\dot{\boldsymbol{e}}=-e-\boldsymbol{e}+\ddot{r}-\ddot{y}=-e-\boldsymbol{e}+\ddot{r}-\dot{x}_1=-e-\boldsymbol{e}+\ddot{r}-u.$$

Minimize a functional

$$J=\min_{\mathbf{x},u}\int_{t_0}^{\infty}\tfrac{1}{2}\left(\mathbf{x}^T Q\mathbf{x}+Gu^2\right)dt,\ Q=\begin{bmatrix}1&0&0&0&0\\0&1&0&0&0\\0&0&5&0&0\\0&0&0&1&0\\0&0&0&0&0.001\end{bmatrix},\ G=1$$

subject to

$$\dot{\mathbf{x}}(t)=\mathbf{A}\mathbf{x}+A_r r+\mathbf{B}u,\ \mathbf{x}=\begin{bmatrix}x_1\\ x_2\\ \int edt\\ e\\ \boldsymbol{e}\end{bmatrix},\ \boldsymbol{e}(t)\equiv\frac{de}{dt}.$$

Solving an algebraic equation $Q+K\mathbf{A}+\mathbf{A}^T K-K\mathbf{B}G^{-1}\mathbf{B}^T K=0$, we have

$$K=\begin{bmatrix}2.95&3.8&-2.24&-0.05&0.00048\\3.8&9.26&-6.7&0.3&0.00015\\-2.24&-6.7&8.49&-0.11&-0.00006\\-0.05&0.3&-0.11&0.5&-0.00025\\0.00048&0.00015&-0.00006&-0.00025&0.0005\end{bmatrix},\ K\in\mathbb{R}^{5\times5},\ K>0.$$

One finds the feedback gain matrix K_F.

The proportional-integral-derivative control law with state feedback is

$u=-2.95x_1-3.8x_2+2.24\int edt+0.05e+0.000025\frac{de}{dt}$.

System dynamics and evolution of $u(t)$ are illustrated in Figure 5.27.e.

Rotational motion of a satellite, controlled by two thrusters on opposite sides. Clockwise and counterclockwise rotation is ensured by thrusters with one-directional thrusts F_{thrust}.
Rotational torques is $T=R_{\perp}F_{\text{thrust}}$

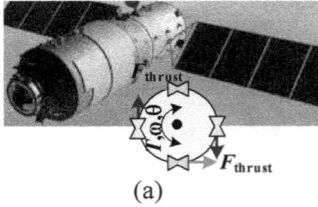

(a)

SIMULINK Diagrams:
Closed-loop system with proportional-integral and state feedback control law

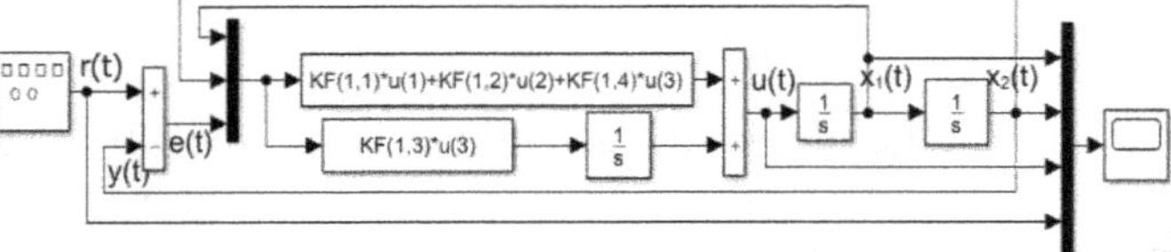

Closed-loop system with proportional-integral and state feedback control law, and, with an *observer* to compute $\hat{x}_1$

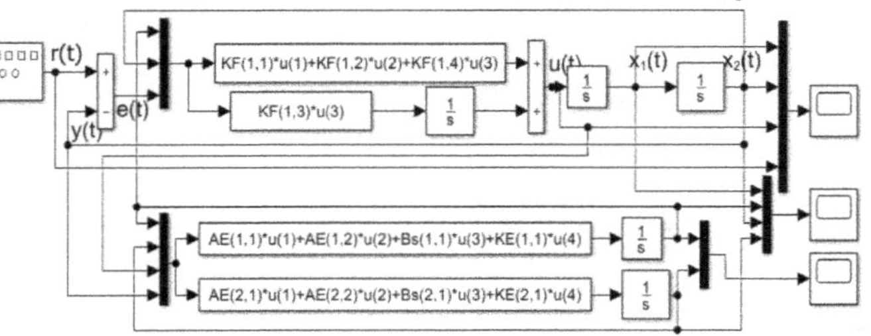

Closed-loop system with proportional-integral-derivative and state feedback control law

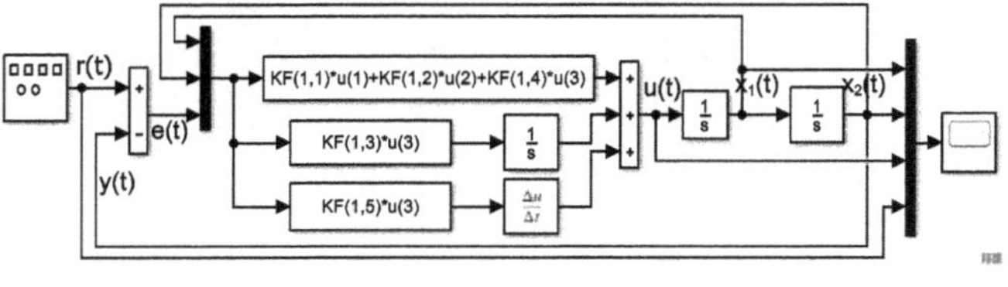

(b)

System Dynamics (x_1,x_2), Output $y=x_2$, Control Evolution $u(t)$, $u=-2.95x_1-3.8x_2+2.24\int edt+0.05e$

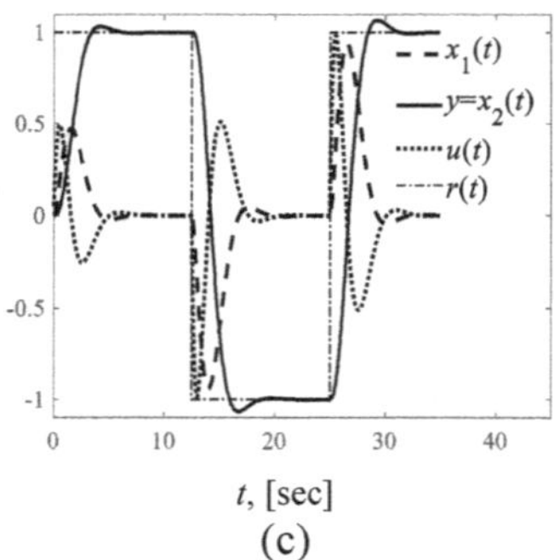

(c)

System Dynamics (x_1,x_2), Output $y=x_2$, Control Evolution $u(t)$, $u=-2.95\hat{x}_1-3.8x_2+2.24\int edt+0.05e$

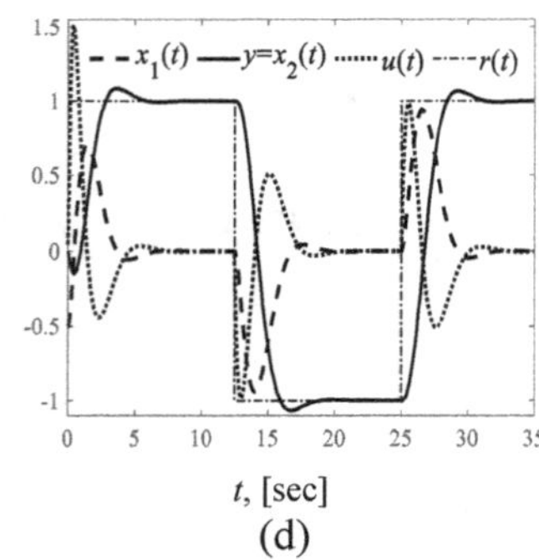

(d)

System Dynamics (x_1,x_2), Output $y=x_2$, Control Evolution $u(t)$, $u=-2.95x_1-3.8x_2$ $+2.24\int edt+0.05e+0.000025\frac{de}{dt}$

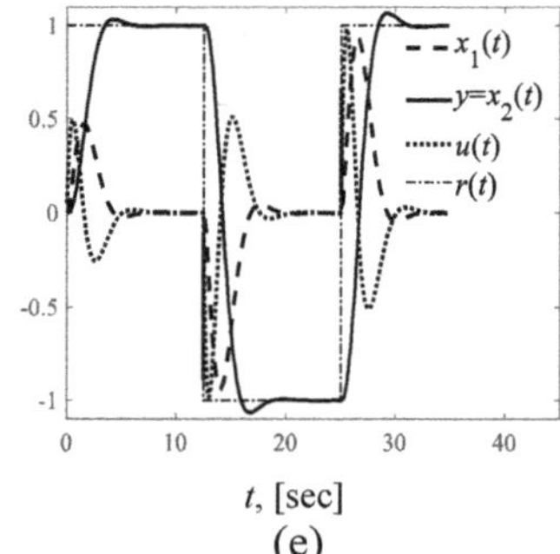

(e)

Figure 5.27. (a) Rotational motion of a satellite, F_{thrust}, $T=R_{\perp}F_{\text{thrust}}$;
(b) SIMULINK diagrams: Closed-loop systems with tracking control laws and *observer*;
(c) Closed-loop system dynamics with a control law $u=-2.95x_1-3.8x_2+2.24\int edt+0.05e$;
(d) Closed-loop systems with a control law $u=-2.95\hat{x}_1-3.8x_2+2.24\int edt+0.05e$ and *observer*;
(e) Closed-loop systems with a proportional-integral-derivative and state feedback control, $u=-2.95x_1-3.8x_2+2.24\int edt+0.05e+0.000025\frac{de}{dt}$.

Formal Description and Physical Implementation – The piecewise continuous reference $r(t)=\pm M$ may be represented as $r(t)=M\text{sgn}(\sin(\omega t))$, $r(t)=M\text{sgn}(\cos(\omega t))$ or other functions. Recall that

$$\frac{d}{dt}\text{sgn}(M\sin(\omega t)) = 2M\omega\cos(\omega t)\delta(\sin(\omega t)),\ \frac{d}{dt}\text{sgn}(M\cos(\omega t)) = 2M\omega\sin(\omega t)\delta(\cos(\omega t)),$$

where δ is the Dirac delta function.

It is possible to perform analysis specifying $r(t)=\pm M$. In numeric solutions and simulations, including analog processing, $r(t)\approx\pm M$. One may not physically implement the references $r(t)=\pm M$ and $r[n]=\pm M$.

Due to continuum electrostatics, energy transductions, transients and latencies, the switching-frequency in ASICs, operational amplifiers (*slew rate*) and MOSFETs may reach ~10 GHz. Switching and evolutions are continuous. To attain consistent numeric simulations, one accounts for the aforementioned transients, and, not alarmed by potential numeric inconsistencies which may be due to inadequateness and simplification of underlying physics. In physical systems, states $x(t)$, outputs $y(t)$, reference $r(t)$, tracking error $e(t)$, derivative de/d and controls $u(t)$ are continuous, and, analog signals are mapped as discrete processes. ■

5.10. Control of Linear Discrete-Time Systems: Design of Control Laws

5.10.1. Stabilizing Control: Linear Quadratic Regulators

Consider a discretized system, described by a difference equation

$$x_{n+1} = A_n x_n + B_n u_n,\ n \geq 0. \tag{5.124}$$

With a sampling time T_s and uniform sampling, denote $x_n=x[n]=x(nT_s)$ instead of $x_k=x[k]=x(kT_s)$ used in sections 5.2 and 5.3. Minimize the quadratic performance index

$$J = \min_{x_n,u_n} \sum_{n=0}^{N-1}\left[x_n^T Q_n x_n + u_n^T G_n u_n\right],\ Q_n \geq 0,\ G_n > 0 \tag{5.125}$$

subject to system (5.124).

Weighting Coefficient – The entries of diagonal matrices (Q_n,G_n) may be selected as

$$q_{ii_n} = \frac{1}{\tilde{x}_{i\,n\,\max}^2},\ g_{jj_n} = \frac{1}{\tilde{u}_{j\,n\,\max}^2},$$

where $\tilde{x}_{i\,n\,\max}$ and $\tilde{u}_{j\,n\,\max}$ are the expected variations of states and controls x_n and u_n.

For a system (5.124) and performance index (5.125), the Hamiltonian function is

$$H\left(x_n,u_n,V(x_{n+1})\right) = x_n^T Q_n x_n + u_n^T G_n u_n + V(x_{n+1}). \tag{5.126}$$

For a positive definite quadratic return function

$$V(x_n) = x_n^T K_n x_n,\ K_n>0, \tag{5.127}$$

we have

$$\begin{aligned} &H(x_n,u_n) = x_n^T Q_n x_n + u_n^T G_n u_n + V(x_{n+1}) = x_n^T Q_n x_n + u_n^T G_n u_n + x_{n+1}^T K_{n+1} x_{n+1} = \\ &x_n^T Q_n x_n + u_n^T G_n u_n + x_n^T A_n^T K_{n+1} A_n x_n + x_n^T A_n^T K_{n+1} B_n u_n + u_n^T B_n^T K_{n+1} A_n x_n + u_n^T B_n^T K_{n+1} B_n u_n. \end{aligned} \tag{5.128}$$

The first and second differences of the Hamiltonian $H(x_n,u_n)$ (5.128) exist. The finite forward difference is $\Delta H=H_{n+1}-H_n$, where Δ is the difference operator.

Using the first difference, one has

$$u_n^T G_n + x_n^T A_n^T K_{n+1} B_n + u_n^T B_n^T K_{n+1} B_n = 0. \tag{5.129}$$

The control law is

$$u_n = -\left(G_n + B_n^T K_{n+1} B_n\right)^{-1} B_n^T K_{n+1} A_n x_n. \tag{5.130}$$

Applying (5.126)-(5.128), (5.130) and

$$V(x_n) = x_n^T Q_n x_n + u_n^T G_n u_n + V(x_{n+1}), \tag{5.131}$$

we obtain

$$x_n^T K_n x_n = x_n^T Q_n x_n + x_n^T A_n^T K_{n+1} A_n x_n - x_n^T A_n^T K_{n+1} B_n \left(G_n + B_n^T K_{n+1} B_n\right)^{-1} B_n^T K_{n+1} A_n x. \tag{5.132}$$

The unknown symmetric matrix K_{n+1} is found by solving a recurrence equation

$$-K_n + Q_n + A_n^T K_{n+1} A_n - A_n^T K_{n+1} B_n \left(G_n + B_n^T K_{n+1} B_n\right)^{-1} B_n^T K_{n+1} A_n = 0, K_n = K_n^T. \tag{5.133}$$

For infinite time horizon, minimize the performance index (5.125) with $N \rightarrow \infty$

$$J = \min_{x_n, u_n} \sum_{n=0}^{\infty} \left[x_n^T Q_n x_n + u_n^T G_n u_n \right], Q_n \geq 0, G_n > 0 \tag{5.134}$$

subject to a discrete-time system (5.124). Using a scalar Hamiltonian (5.126) and quadratic return function (5.127), an optimal control law is

$$u_n = -\left(G_n + B_n^T K_n B_n\right)^{-1} B_n^T K_n A_n x_n. \tag{5.135}$$

The symmetric matrix K_n is found by solving an equation

$$-K_n + Q_n + A_n^T K_n A_n - A_n^T K_n B_n \left(G_n + B_n^T K_n B_n\right)^{-1} B_n^T K_n A_n = 0, \; K_n = K_n^T. \tag{5.136}$$

Matrix K_n is positive definite. The closed-loop systems (5.124) with (5.135) is

$$x_{n+1} = \left[A_n - B_n \left(G_n + B_n^T K_n B_n\right)^{-1} B_n^T K_n A_n \right] x_n. \tag{5.137}$$

MATLAB Solvers – MATLAB **dlqr** and **idare** commands solve equation (5.136) applying matrix algebra, partitioning, diagonalization and eigendecomposition. Eigenvalue-eigenvector factorizations, Newton methods, as well as recursive algorithms are developed, supported by existing and custom MATLAB codes. Pertained MATLAB codes are reported in Example 6.52.

Example 5.34. *Discretization and Control: Second-Order System*

For the second-order system

$$\dot{x} = Ax + Bu, \; \dot{x} = \begin{bmatrix} 0 & 1 \\ -2 & -3 \end{bmatrix} \begin{bmatrix} x_1 \\ x_2 \end{bmatrix} + \begin{bmatrix} 1 & 0 \\ 0 & 1 \end{bmatrix} \begin{bmatrix} u_1 \\ u_2 \end{bmatrix},$$

find the discrete-time model if T_s=0.05 sec.

Using (5.19) $A_n = e^{AT_s}$ and $B_n = A^{-1}(e^{AT_s} - I)B$, compute matrices (A_n, B_n) as

```
A=[0 1; -2 -3]; B=[1 0; 0 1]; Ts=0.05;
An=expm(A*Ts), Bn=inv(A)*(An-eye(size(A)))*B
```

We have

$$x_{n+1} = A_n x_n + B_n u_n, \; n \geq 0,$$

$$x_{n+1} = \begin{bmatrix} x_{1\,n+1} \\ x_{2\,n+1} \end{bmatrix} = \begin{bmatrix} 0.9976 & 0.0464 \\ -0.0928 & 0.8585 \end{bmatrix} \begin{bmatrix} x_{1n} \\ x_{2n} \end{bmatrix} + \begin{bmatrix} 0.05 & 0.00119 \\ -0.00238 & 0.04639 \end{bmatrix} \begin{bmatrix} u_{1n} \\ u_{2n} \end{bmatrix}.$$

Minimize the quadratic performance index (5.134)

$$J=\min_{x_n,u_n}\sum_{n=0}^{\infty}\left[\begin{bmatrix}x_{1n} & x_{2n}\end{bmatrix}\begin{bmatrix}1 & 0\\ 0 & 1\end{bmatrix}\begin{bmatrix}x_{1n}\\ x_{2n}\end{bmatrix}+\begin{bmatrix}u_{1n} & u_{2n}\end{bmatrix}\begin{bmatrix}10 & 0\\ 0 & 10\end{bmatrix}\begin{bmatrix}u_{1n}\\ u_{2n}\end{bmatrix}\right]=\min_{x_n,u_n}\sum_{n=0}^{\infty}\left(x_{1n}^2+x_{2n}^2+10u_{1n}^2+10u_{2n}^2\right).$$

The control law (5.135) is $u_n=-\left(G_n+B_n^TK_nB_n\right)^{-1}B_n^TK_nA_nx_n$.

Different MATLAB solvers are available to solve (5.136).

Using the **dlqr** command, compute K_n by solving (5.136), find feedback gain matrix, and, compute eigenvalues. For $Q_n=\begin{bmatrix}1 & 0\\ 0 & 1\end{bmatrix}$ and $G_n=\begin{bmatrix}10 & 0\\ 0 & 10\end{bmatrix}$, one has

```
Qn=1*eye(size(An)); Gn=10*eye(size(Bn));
[KF,Kn,Eigenvalues]=dlqr(An,Bn,Qn,Gn)
```

We found $K_n=\begin{bmatrix}23.52 & 4.319\\ 4.319 & 5.262\end{bmatrix}$, $K_{Fn}=\left(G_n+B_n^TK_{n+1}B_n\right)^{-1}B_n^TK_nA_n=\begin{bmatrix}0.1136 & 0.0227\\ 0.0203 & 0.0224\end{bmatrix}$.

A command **idare** computes K_n and feedback gain matrix K_{Fn}. In particular,

```
[Kn,KFn]=idare(An,Bn,Qn,Gn)
```

yields $K_n=\begin{bmatrix}23.52 & 4.319\\ 4.319 & 5.262\end{bmatrix}$, $K_{Fn}=\begin{bmatrix}0.1136 & 0.0227\\ 0.0203 & 0.0224\end{bmatrix}$.

The control law is

$$u_n=-\left(G_n+B_n^TK_nB_n\right)^{-1}B_n^TK_nA_nx_n=-\begin{bmatrix}0.1136 & 0.0227\\ 0.0203 & 0.0224\end{bmatrix}\begin{bmatrix}x_{1n}\\ x_{2n}\end{bmatrix},$$

$$u_{1n}=-0.1136x_{1n}-0.0227x_{2n},$$

$$u_{2n}=-0.0203x_{1n}-0.0224x_{2n}.$$

The eigenvalues of the closed-loop system are 0.9418 and 0.90758. System is stable because the eigenvalues are within the unit circle. ■

Example 5.35. *Discrete-Time System With Disturbances*

For the third-order system with states x_n, control u_n and disturbances d_n, consider

$$x_{n+1}=\begin{bmatrix}x_{1n+1}\\ x_{2n+1}\\ x_{3n+1}\end{bmatrix}=A_nx_n+B_{Dn}d_n+B_nu_n=\begin{bmatrix}1 & 1 & 2\\ 3 & 3 & 4\\ 5 & 5 & 6\end{bmatrix}\begin{bmatrix}x_{1n}\\ x_{2n}\\ x_{3n}\end{bmatrix}+\begin{bmatrix}10\\ 0\\ 0\end{bmatrix}d_n+\begin{bmatrix}10\\ 20\\ 30\end{bmatrix}u_n,\ n\geq 0.$$

Let the output equation is $y_n=H_nx_n+D_nu_n$, $H_n=[1\ \ 0\ \ 0]$, $D_n=[0]$.

Minimize the quadratic performance index (5.134)

$$J=\min_{x_n,u_n}\sum_{n=0}^{\infty}\left[\begin{bmatrix}x_{1n} & x_{2n} & x_{3n}\end{bmatrix}\begin{bmatrix}1 & 0 & 0\\ 0 & 10 & 0\\ 0 & 0 & 100\end{bmatrix}\begin{bmatrix}x_{1n}\\ x_{2n}\\ x_{3n}\end{bmatrix}+1000u_n^2\right]=\min_{x_n,u_n}\sum_{n=0}^{\infty}\left(x_{1n}^2+10x_{2n}^2+100x_{3n}^2+1000u_n^2\right).$$

The MATLAB code to find matrices K_n and K_{Fn} are

```
An=[1 1 2; 3 3 4; 5 5 6]; Bn=[10; 20; 30];
Qn=eye(size(An)); Qn(2,2)=10; Qn(3,3)=100; Gn=1000;
[KFn,Kn,Eigenvalies]=dlqr(An,Bn,Qn,Gn)
[Kn,KFn]=idare(An,Bn,Qn,Gn)
```

We find $K_n=\begin{bmatrix}48.4 & 47.4 & 31.1\\ 47.4 & 57.4 & 31.1\\ 31.1 & 31.1 & 139.8\end{bmatrix}$ and $K_{Fn}=\begin{bmatrix}0.155 & 0.155 & 0.2\end{bmatrix}$.

Commands **idare** and **dlqr** yield same K_n and K_{Fn}. The control law is

$$u_n=-0.155x_{1n}-0.155x_{2n}-0.2x_{3n}.$$

System is stable because the computed eigenvalues (–0.6736, 0.0388, 0) are within the unit circle. A closed-loop system is simulated using the **filter** command.

For a closed-loop system

$$x_{n+1} = \left[A_n - B_n \left(G_n + B_n^T K_n B_n \right)^{-1} B_n^T K_n A_n \right] x_n + B_{Dn} d_n ,$$

one may find the transfer function in the z-domain. Simulation results for the system output $y_n = x_{1n}$ are illustrated in Figures 5.28.a if the initial conditions are $x_{n0}=[0\ \ 5\ \ 10]$.

The disturbances are $d_n=0$, and, a unit sequence $d_n=1_n$. The MATLAB code is

```
An_closed=An-Bn*KFn; Bdn=[10; 0; 0]; Hn=[1 0 0]; Dn=[0];
[num,den]=ss2tf(An_closed,Bdn,Hn,Dn);
x0=[0 5 10]; n=0:1:20;
figure(1); dn=0*[ones(1,21)]; y=filter(num,den,dn,x0); stem(n,y,'ko','LineWidth',3); % dn=0;
figure(2); dn=1*[ones(1,21)]; y=filter(num,den,dn,x0); stem(n,y,'ko','LineWidth',3); % dn=1;
```

For the state-space model, the **dlsim** command is used to solve difference equations. States dynamics $x_n=[x_{1n}, x_{2n}, x_{3n}]$ and output evolutions $y_n=x_{1n}$ are documented in Figures 5.28.b for initial conditions $x_{n0}=[0\ \ 5\ \ 10]$ and $d_n=1_n$. We have

```
An_closed=An-Bn*KFn; Bdn=[10; 0; 0]; Hn=[1 0 0]; Dn=[0];
x0=[0 5 10]; n=0:1:20; dn=ones(1,length(n)); % d=zeros(1,length(n)); % dn=1 or dn=0
[y,x]=dlsim(An_closed,Bdn,Hn,Dn,dn,x0);
figure(1); stem(n,y,'ko','LineWidth',3);
figure(2); plot(n,x,'ko','LineWidth',3); xlabel('Discrete Time, [{\itn}]','FontSize',18);
plot(n,x(:,1),'ko', n,x(:,2),'bs', n,x(:,3),'gd', 'LineWidth',2); axis([0 20 -1 10.2])
legend('{\itx}_1_{\itn}','{\itx}_2_{\itn}','{\itx}_3_{\itn}','NumColumns',3,'FontSize',26); legend boxoff;
```

The closed-loop system is stable, and, the stabilization problem is solved. While simulations using the **filter** and **dlsim** commands yield same equilibriums, there are some discrepancies. In Example 5.36, as well as in chapter 6, the MATLAB codes are reported to solve linear and nonlinear difference equations.

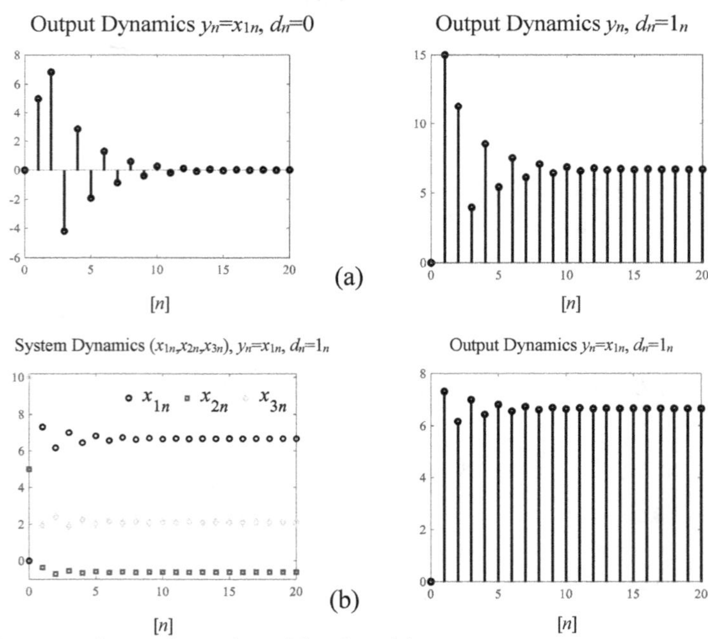

Figure 5.28. Output and state dynamics of the closed-loop system:
(a) Output dynamics of $y_n=x_{1n}$ due to initial conditions if $d_n=0$ and $d_n=1_n$, $\forall x_{n0}=0$;
(b) States (x_{1n}, x_{2n}, x_{3n}) and output $y_n=x_{1n}$ dynamics for a unit input sequence $d_n=1_n$, $x_{n0}=[0\ 5\ 10]$. ■

5.10.2. Tracking Control Laws

In section 5.10.1, stabilizing control laws were designed by minimizing (5.125) subject to (5.124). Investigate problems in design of tracking control laws. For linear systems (5.114) $\dot{x}(t)=Ax+Bu$, $y(t)=Hx(t)$, $e(t)=(r-y)=(r-Hx)$, the *error state* $x_e(t)$ and tracking error $e(t)$ dynamics are governed by (5.115). Discretize an extended model (5.116)

$$\mathbf{x}(t)=\begin{bmatrix}x(t)\\x_e(t)\\e(t)\end{bmatrix},\ \dot{\mathbf{x}}(t)=\begin{bmatrix}\dot{x}(t)\\\dot{x}_e(t)\\\dot{e}(t)\end{bmatrix}=\mathbf{Ax}+A_r r+\mathbf{B}u=\begin{bmatrix}A&0&0\\-I_X H&0&0\\-HA&I_E^X&-I_E\end{bmatrix}\mathbf{x}+\begin{bmatrix}0&0\\I_X&0\\0&I\end{bmatrix}\begin{bmatrix}r\\\dot{r}\end{bmatrix}+\begin{bmatrix}B\\0\\-HB\end{bmatrix}u$$

for a given T_s using (5.19) $\mathbf{A}_n=e^{\mathbf{A}T_s}$, $\mathbf{B}_n=\mathbf{A}^{-1}(e^{\mathbf{A}T_s}-I)\mathbf{B}$, $H_n=H$ and $D_n=D$.

We have

$$\mathbf{x}_{n+1}=\mathbf{A}_n\mathbf{x}_n+A_{rn}r_n+\mathbf{B}_n u_n,\ \mathbf{x}_n=\begin{bmatrix}x_n\\x_{en}\\e_n\end{bmatrix},\ n\ge 0. \tag{5.138}$$

Minimize a performance index

$$J=\min_{\mathbf{x}_n,u_n}\sum_{n=0}^{\infty}\left[\mathbf{x}_n^T Q_n\mathbf{x}_n+u_n^T G_n u_n\right],\ Q_n\ge 0,\ G_n>0, \tag{5.139}$$

subject to (5.138). Using a scalar Hamiltonian $H=\mathbf{x}_n^T Q_n\mathbf{x}_n+u_n^T G_n u_n+V(\mathbf{x}_{n+1})$ and the quadratic return function (5.127) $V(\mathbf{x}_n)=\mathbf{x}_n^T K_n\mathbf{x}_n$, $K_n>0$, the tracking control law is

$$u_n=-\left(G_n+\mathbf{B}_n^T K_n\mathbf{B}_n\right)^{-1}\mathbf{B}_n^T K_n\mathbf{A}_n\mathbf{x}_n. \tag{5.140}$$

The symmetric matrix K_n is computed by solving

$$-K_n+Q_n+\mathbf{A}_n^T K_n\mathbf{A}_n-\mathbf{A}_n^T K_n\mathbf{B}_n\left(G_n+\mathbf{B}_n^T K_n\mathbf{B}_n\right)^{-1}\mathbf{B}_n^T K_n\mathbf{A}_n=0,\ K_n=K_n^T,\ K_n>0. \tag{5.141}$$

Example 5.36. *Tracking Control of a Rotational Motion of Satellite*

The controlled rotational motion of a satellite was examined in Example 5.33. The governing equations for a rigid-body states and error dynamics is

$$\dot{\mathbf{x}}=\mathbf{Ax}+A_r r+\mathbf{B}u,\ \begin{bmatrix}\dot{x}_1\\\dot{x}_2\\\dot{x}_e\\\dot{e}\end{bmatrix}=\begin{bmatrix}0&0&0&0\\1&0&0&0\\0&-1&0&0\\-1&0&0&-1\end{bmatrix}\begin{bmatrix}x_1\\x_2\\x_e\\e\end{bmatrix}+\begin{bmatrix}0&0\\0&0\\1&0\\0&1\end{bmatrix}\begin{bmatrix}r\\\dot{r}\end{bmatrix}+\begin{bmatrix}1\\0\\0\\0\end{bmatrix}u,\ \mathbf{x}=\begin{bmatrix}x\\\int edt\\e\end{bmatrix}=\begin{bmatrix}x_1\\x_2\\\int edt\\e\end{bmatrix}=\begin{bmatrix}\omega\\\theta\\\int edt\\e\end{bmatrix},\ y=x_2.$$

Apply (5.19) to discretize a continuous-time system for T_s=0.1 sec.

From $\mathbf{A}_n=e^{\mathbf{A}T_s}$ and $\mathbf{B}_n=\mathbf{A}^{-1}(e^{\mathbf{A}T_s}-I)\mathbf{B}$, we have

$$\mathbf{x}_{n+1}=\mathbf{A}_n\mathbf{x}_n+A_{rn}r_n+\mathbf{B}_n u_n,\ \begin{bmatrix}x_{1n+1}\\x_{2n+1}\\x_{en+1}\\e_{n+1}\end{bmatrix}=\begin{bmatrix}1&0&0&0\\0.1&1&0&0\\-0.005&-0.1&1&0\\-0.095&0&0&0.905\end{bmatrix}\begin{bmatrix}x_{1n}\\x_{2n}\\x_{en}\\e_n\end{bmatrix}+\begin{bmatrix}0\\0\\0.1\\0\end{bmatrix}r_n+\begin{bmatrix}0.1\\0.005\\0\\-0.005\end{bmatrix}u_n,\ y_n=x_{2n}.$$

Minimize the quadratic performance index (5.139) $J=\min_{\mathbf{x}_n,u_n}\sum_{n=0}^{\infty}\left[\mathbf{x}_n^T Q_n\mathbf{x}_n+u_n^T G_n u_n\right]$.

The tracking control law (5.140) is

$$u_n = -\left(G_n + \mathbf{B}_n^T K_n \mathbf{B}_n\right)^{-1} \mathbf{B}_n^T K_n \mathbf{A}_n \mathbf{x}_n = -K_{Fn}\mathbf{x}_n = -K_{Fn}\begin{bmatrix} x_{1n} \\ x_{2n} \\ x_{en} \\ e_n \end{bmatrix}.$$

For $Q_n = \begin{bmatrix} 1 & 0 & 0 & 0 \\ 0 & 1 & 0 & 0 \\ 0 & 0 & 1 & 0 \\ 0 & 0 & 0 & 1 \end{bmatrix}$ and three G_n, a positive definite matrix K_n and feedback gain matrix K_{Fn} are computed. We have the following feedback gain matrices K_{Fn}:

1. For G_n=1, $K_n = \begin{bmatrix} 25.3 & 24.8 & -10.1 & -0.71 \\ 24.8 & 53.8 & -25.4 & 2.48 \\ -10.1 & -25.4 & 25.2 & -0.72 \\ -0.71 & 2.48 & -0.72 & 5.49 \end{bmatrix}$, K_{Fn}=[2.3 2.22 –0.884 –0.06];
2. For G_n=0.1, $K_n = \begin{bmatrix} 5.47 & 6.69 & -3.25 & -0.31 \\ 6.69 & 32.5 & -16.5 & 3.08 \\ -3.25 & -16.5 & 21.2 & -1 \\ -0.31 & 3.08 & -1 & 5.47 \end{bmatrix}$, K_{Fn}=[4.12 5.27 –2.48 –0.236];
3. For G_n=0.01, $K_n = \begin{bmatrix} 1.86 & 2.17 & -1.17 & -0.12 \\ 2.17 & 25.9 & -13.3 & 3.39 \\ -1.17 & -13.3 & 19.6 & -1.16 \\ -0.12 & 3.39 & -1.16 & 5.45 \end{bmatrix}$, K_{Fn}=[7.41 11.1 –5.64 –0.62].

MATLAB is used to perform discretization, design and solve difference equations.

```
% Continuous-time system: State-space model
A=[0 0 0 0; 1 0 0 0; 0 -1 0 0; -1 0 0 -1]; B=[1; 0; 0; 0]; H=[0 1 0 0]; D=0;
sys=ss(A,B,H,D)  % Continuous-time system
Ts=0.1;                 % Sampling time
% Discrete-time system: State-space model matrices An, Bn, Hn, Dn and Arn
An=expm(A*Ts), Bn=pinv(A)*(An-eye(size(A)))*B, Hn=H; Dn=D; Arn=Ts*[0; 0; 1; 0];
Qn=1*eye(size(An)); Gn=1; Gn=0.1; Gn=0.01;  % Weighting matrices, Gn=1, Gn=0.1 and Gn=0.01
[KFn,Kn,Eigenvalues]=dlqr(An,Bn,Qn,Gn)     % Discrete LQR design
% Transient dynamics of a discrete-time system: Solution of difference equation
x0=[0.1 -0.1 0 0]';       % Initial conditions
N1=100; N=3*N1;
% Computing x[n], u[n] and r[n]
for r=1:1:N
   r(1:N1)=1; r(N1+1:1:2*N1)=-1; r(2*N1+1:1:3*N1)=1;
end
   x(:,1)=(An-Bn*KFn)*x0+Arn*r(:,1);
for n=1:1:N
   x(:,n+1)=(An-Bn*KFn)*x(:,n)+Arn*r(:,n);
   u=-KFn*x;
end
n=0:1:N; plot(n,x(1,:),'bs',n,x(2,:),'ko',n,u(1,:),'rd','LineWidth',1); hold on;
plot(1:N,r(1, :),'k--','LineWidth',0.5);  hold off; axis([0 N -1.1 1.1]); axis([0 N -1.5 1.5]);
legend('{\itx}_1_{\itn}','{\ity_n}={\itx}_2_{\itn}', '{\itu}_{\itn}','Location','southeast','NumColumns',1,'FontSize',20);
legend boxoff;
```

Figures 5.29 documents dynamics of a closed-loop system with or three computed K_{Fn} for reference sequences r_n=±1$_n$ and initial conditions $\mathbf{x}_{n0}$=[0.1 –0.1 0 0]. Tracking is guaranteed. The settling time is reduced with feedback gains, computed for G_n=0.01, for which control efforts u_n increase because K_{Fn}=[7.41 11.1 –5.64 –0.62]. Selecting the weighting coefficients $q_{ii_n} = \frac{1}{\tilde{x}_{i\,n\,\max}^2}$, $g_{jj_n} = \frac{1}{\tilde{u}_{j\,n\,\max}^2}$, one deduces reasonable expected variations of states and controls (x_n,u_n) as $(\tilde{x}_{i\,n\,\max}, \tilde{u}_{j\,n\,\max})$.

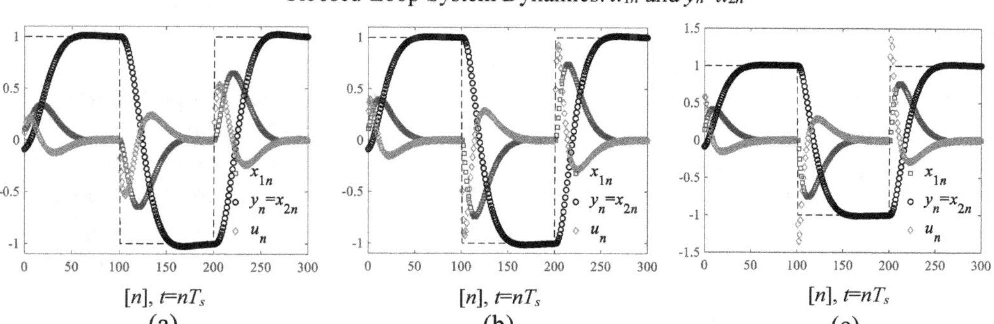

(a) (b) (c)

Figure 5.29. Closed-loop system dynamics and tracking of reference $r_n=\pm 1_n$. The control law is $u_n= -K_{Fn}\mathbf{x}_n$ with three feedback gain matrices K_{Fn}:
(a) K_{Fn}=[2.3 2.22 –0.884 –0.06]; (b) K_{Fn}=[4.12 5.27 –2.48 –0.236]; (c) K_{Fn}=[7.41 11.1 –5.64 –0.62].■

5.11. Analog and Digital Tracking Control of Aerial Vehicles

Study tracking control problems for aerial vehicles. Tracking control laws are designed for various fixed-wing and rotary-wing aircraft. Example are unmanned aerial vehicles, multi-rotor helicopters, passenger and cargo aircraft, military aircraft, etc. For aerial vehicles, the coupled translational and rotational dynamics are reported in section 2.4. Applying (2.43) and (2.47), the linearized models are derived in the considered flight envelope by using the vehicle velocity V, angle of attack α and sideslip angle β, $\begin{bmatrix} V \\ \alpha \\ \beta \end{bmatrix} = \begin{bmatrix} \sqrt{u^2+v^2+w^2} \\ \tan^{-1}\frac{w}{u} \\ \sin^{-1}\frac{v}{V} \end{bmatrix}$.

For the angular rates (p,q,r) and Euler angles (ϕ,θ,ψ), $\begin{bmatrix} \dot\phi \\ \dot\theta \\ \dot\psi \end{bmatrix} = \begin{bmatrix} 1 & \sin\phi\tan\theta & \cos\phi\tan\theta \\ 0 & \cos\phi & -\sin\phi \\ 0 & \sin\phi\sec\theta & \cos\phi\sec\theta \end{bmatrix}\begin{bmatrix} p \\ q \\ r \end{bmatrix}$.

Specifications on aerial vehicles are reported in Section 1.2. The Federal Aviation Administration specifications and military standards define and assert flying qualities for the maneuvering flight, operational and mission flight phases, basic and aerobatic maneuvering, etc. One evaluates stability, residual oscillations, pitch-roll-yaw responses, longitudinal and lateral pitch-roll-yaw control, maneuvering characteristics, controlled motions in maneuvering flight, operational and permissible flight envelopes, etc. Quantifiable specifications, performance indicators and capability metrics are asserted and evaluated using a quadruple $(\boldsymbol{q},\boldsymbol{x},\boldsymbol{y},\boldsymbol{u})$, stipulated by performance measures.

Using consistent models, minimizing functionals and indexes, as well as applying the stability and optimality criteria given by a Lyapunov pair $\left\{V(x)>0, \frac{dV(x)}{dt}<0\right\}$, $\{V(x_n)>0, \Delta V(x_n)<0\}$, solve tracking control problems. Decoupled longitudinal and lateral models, as well as the coupled dynamics are investigated. Nonlinear control problems are formulated and solved in chapter 6.

5.11.1. Analog and Digital Tracking Control of a Fixed-Wing Aerial Vehicle

Consider the longitudinal and lateral linear models for a fixed-wing unmanned aerial vehicle. Depending on flight phases, regimes and maneuvers, one uses pertinent states, outputs, descriptive models and constitutive relationships. For example, the outputs are the Euler angles, angle of attack, sideslip angle, altitude h, etc. Linear models are examined in this section, and, the constrained control problem is solved in section 6.8.2. The model parameters and coefficients are identified [15-17] for a near-straight and level flight at elevation 500 m, 25 m/s, 2° pitch angle, 1° angle of attack, zero sideslip angle and zero roll angle.

Longitudinal Dynamics – Study the longitudinal dynamics using the forward velocity $x_1=V$, angle of attack $x_2=\alpha$, pitch rate $x_3=q$ and pitch angle $x_4=\theta$. The outputs $y(t)$ are V and θ. The considered aerial vehicle is controlled by regulating thrust u_t and displacing elevators u_E. That is, $u_1=u_t$ and $u_2=u_E$. The engine thrust is positive. Linearized at equilibrium, the thrust control admits negative and positive values, regulated by changing the variable-pitch propeller angular velocity of rotation and propeller's blade pitch. The elevators are the pitch-control aerodynamic surfaces. The longitudinal model is

$$\begin{matrix} \dot{x}=Ax+Bu \\ y=Hx \end{matrix},\quad \begin{bmatrix} \dot{x}_1 \\ \dot{x}_2 \\ \dot{x}_3 \\ \dot{x}_4 \end{bmatrix}=\begin{bmatrix} -0.15 & -1.3 & 0 & -9.81 \\ 0 & -7.4 & 0.96 & 0 \\ 0.08 & -5.6 & -2.4 & 0 \\ 0 & 0 & 1 & 0 \end{bmatrix}\begin{bmatrix} x_1 \\ x_2 \\ x_3 \\ x_4 \end{bmatrix}+\begin{bmatrix} 5.3 & -0.29 \\ 0 & 0.074 \\ 0 & 1.6 \\ 0 & 0 \end{bmatrix}\begin{bmatrix} u_1 \\ u_2 \end{bmatrix},\ y=\begin{bmatrix} 1 & 0 & 0 & 0 \\ 0 & 0 & 0 & 1 \end{bmatrix}\begin{bmatrix} x_1 \\ x_2 \\ x_3 \\ x_4 \end{bmatrix},$$

$$x=\begin{bmatrix} x_1 \\ x_2 \\ x_3 \\ x_4 \end{bmatrix}=\begin{bmatrix} V \\ \alpha \\ q \\ \theta \end{bmatrix},\ y=\begin{bmatrix} x_1 \\ x_4 \end{bmatrix}=\begin{bmatrix} V \\ \theta \end{bmatrix},\ u=\begin{bmatrix} u_1 \\ u_2 \end{bmatrix}=\begin{bmatrix} u_t \\ u_E \end{bmatrix},\ x\in\mathbb{R}^4,\ y\in\mathbb{R}^2,\ u\in\mathbb{R}^2.$$

Analog Tracking Control – The tracking error $e(t)$ is

$$e=\begin{bmatrix} e_1 \\ e_2 \end{bmatrix}=\begin{bmatrix} r_1-y_1 \\ r_2-y_2 \end{bmatrix}=\begin{bmatrix} r_1-x_1 \\ r_2-x_4 \end{bmatrix}=\begin{bmatrix} r_1-V \\ r_2-\theta \end{bmatrix},\ e\in\mathbb{R}^2.$$

With the output vector $y=\begin{bmatrix} x_1 \\ x_4 \end{bmatrix}=\begin{bmatrix} V \\ \theta \end{bmatrix}$, the *error states* $x_e(t)$ dynamics and tracking error $e(t)$ evolutions are governed by (5.115)

$$\dot{x}_e(t)=I_X(r-y)=I_X(r-Hx),\ e(t)\equiv\dot{x}_e(t),\ x_e\in\mathbb{R}^2,\ I_X=I,\ I_X\in\mathbb{R}^{2\times 2},$$

$$\dot{e}(t)=-I_Ee+\dot{r}-\dot{y}=-I_Ee+\dot{r}-HAx-HBu,\ e\in\mathbb{R}^2,\ I_E=I,\ I_E\in\mathbb{R}^{2\times 2}.$$

We have $\begin{bmatrix} \dot{x}_{e1} \\ \dot{x}_{e2} \end{bmatrix}=\begin{bmatrix} 1 & 0 \\ 0 & 1 \end{bmatrix}\begin{bmatrix} e_1 \\ e_2 \end{bmatrix}=\begin{bmatrix} 1 & 0 \\ 0 & 1 \end{bmatrix}\left(\begin{bmatrix} r_1 \\ r_2 \end{bmatrix}-Hx\right),\ \begin{bmatrix} e_1 \\ e_2 \end{bmatrix}\equiv\begin{bmatrix} \dot{x}_{e1} \\ \dot{x}_{e2} \end{bmatrix}.$

The extended model dynamics (5.116) is

$$\dot{\mathbf{x}}(t)=\begin{bmatrix} \dot{x}(t) \\ \dot{x}_e(t) \\ \dot{e}(t) \end{bmatrix}=\mathbf{A}\mathbf{x}+A_rr+\mathbf{B}u=\begin{bmatrix} A & 0 & 0 \\ -I_XH & 0 & 0 \\ -HA & 0 & -I_E \end{bmatrix}\mathbf{x}+\begin{bmatrix} 0 & 0 \\ I_X & 0 \\ 0 & I \end{bmatrix}\begin{bmatrix} r \\ \dot{r} \end{bmatrix}+\begin{bmatrix} B \\ 0 \\ -HB \end{bmatrix}u,\ \mathbf{x}\in\mathbb{R}^8.$$

Minimize the performance functional (5.117)

$$J=\min_{\mathbf{x},u}\int_0^\infty \tfrac{1}{2}\left(\mathbf{x}^TQ\mathbf{x}+u^TGu\right)dt,$$

subject to the expanded system. The tracking control law (5.122) is

$$u=-G^{-1}\mathbf{B}^T K\mathbf{x}=-K_F\mathbf{x}=-K_F\begin{bmatrix} x \\ \int e_1 dt \\ \int e_2 dt \\ e_1 \\ e_2 \end{bmatrix},\; x=\begin{bmatrix} x_1 \\ x_2 \\ x_3 \\ x_4 \end{bmatrix}=\begin{bmatrix} V \\ \alpha \\ q \\ \theta \end{bmatrix},\; e=\begin{bmatrix} e_1 \\ e_2 \end{bmatrix}=\begin{bmatrix} r_1-V \\ r_2-\theta \end{bmatrix}.$$

Let $Q=I$, $Q\in\mathbb{R}^{8\times 8}$ and $G=I$, $G\in\mathbb{R}^{2\times 2}$, $G=\begin{bmatrix} 1 & 0 \\ 0 & 1 \end{bmatrix}$.

Solve the algebraic equation (5.123) $Q+\mathbf{A}^T K+K\mathbf{A}-K\mathbf{B}G^{-1}\mathbf{B}^T K=0$.

The MATLAB code with the **lqr** solver is

```
% Longitudinal Dynamics: State-space model, matrices of A and B of the model coefficients
A=[-0.15 -1.3  0 -9.81; 0  -7.4  0.96 0; 0.08 -5.6 -2.4  0;  0   0  1   0];
B=[5.3 -0.29; 0 0.074; 0  1.6; 0 0];
H=[1 0 0 0; 0 0 0 1]; D=[0 0; 0 0]; % H matrix: Outputs are the forward velocity and pitch Euler angle
IX=1*eye(2,2); IE=1*eye(2,2);     % Error dynamics matrices IX and IE
% Extended System and LQR Design
As=[A zeros(4,4); -IX*H zeros(2,4); -H*A zeros(2,2) -IE]; Bs=[B; zeros(2,2); -H*B];
Q=1*eye(8,8);Q(5,5)=1;Q(6,6)=1;Q(7,7)=1;Q(8,8)=1;G=1*eye(2,2);
[K,K,Eig]=lqr(As,Bs,Q,G); KF=inv(G)*Bs'*K
```

We find $K\in\mathbb{R}^{8\times 8}$, $K=\begin{bmatrix} 0.69 & -0.0207 & -0.0172 & -0.279 & -0.379 & -0.0236 & 0.446 & -0.0014 \\ -0.0207 & 0.356 & -0.412 & -1.255 & 0.0072 & 0.405 & 0.0037 & 0.0227 \\ -0.0172 & -0.412 & 0.672 & 2.02 & 0.021 & -0.644 & -0.00038 & -0.041 \\ -0.279 & -1.255 & 2.02 & 9.05 & 0.225 & -3.04 & 0.0227 & 0.271 \\ -0.379 & 0.0072 & 0.021 & 0.225 & 1.33 & -0.0339 & -0.192 & -0.0003 \\ -0.0236 & 0.405 & -0.644 & -3.04 & -0.0339 & 3.24 & -0.0069 & -0.0638 \\ 0.446 & 0.0037 & -0.00038 & 0.0227 & -0.192 & -0.0069 & 0.482 & -0.0003 \\ -0.0014 & 0.0227 & -0.041 & 0.271 & -0.0003 & -0.0638 & -0.0003 & 0.498 \end{bmatrix}$.

The longitudinal tracking control law is

$$u=-K_F\begin{bmatrix} x \\ \int e_1 dt \\ \int e_2 dt \\ e_1 \\ e_2 \end{bmatrix},\; K_F=\begin{bmatrix} 1.294 & -0.129 & -0.089 & -1.597 & -0.996 & -0.088 & -0.191 & -0.006 \\ -0.1 & -0.625 & 1.05 & 3.23 & 0.089 & -0.996 & 0.0101 & -0.064 \end{bmatrix}.$$

Use the Schur decomposition to solve an algebraic Riccati equation

$$Q+\mathbf{A}^T K+K\mathbf{A}-K\mathbf{B}G^{-1}\mathbf{B}^T K=0$$

and find $K\in\mathbb{R}^{8\times 8}$ as given by (5.78)-(5.79)

$$\begin{bmatrix} -K & I \end{bmatrix}P\begin{bmatrix} I \\ K \end{bmatrix}=0,\; P=\begin{bmatrix} \mathbf{A} & -\mathbf{B}G^{-1}\mathbf{B}^T \\ -Q & -\mathbf{A}^T \end{bmatrix},\; P\in\mathbb{R}^{16\times 16},$$

$$P=\mathbf{USU}^T,\; \mathbf{U}=\begin{bmatrix} \mathbf{U}_{11} & \mathbf{U}_{12} \\ \mathbf{U}_{21} & \mathbf{U}_{22} \end{bmatrix},\; \mathbf{S}=\begin{bmatrix} \mathbf{S}_{11} & \mathbf{S}_{12} \\ 0 & \mathbf{S}_{22} \end{bmatrix},\; K=\mathbf{U}_{21}\mathbf{U}_{11}^{-1}.$$

One yields the same K by using the MATLAB code

```
n=length(As); P=[As -Bs*inv(G)*Bs'; -Q  -As'];
[u,s]=schur(P); % Compute the Schur decomposition matrices (s,u), P=u*s*u' and u'*u=eye(size(P))
[U,S]=ordschur(u,s,'lhp'); % Reorder eigenvalues of Schur decomposition
U11=U(1:n,1:n); U12=U(1:n,n+1:2*n); U21=U(n+1:2*n,1:n); U22=U(n+1:2*n,n+1:2*n);  K=U21*inv(U11)
```

The SIMULINK diagram is depicted in Figure 5.30.a. Dynamics of the closed-loop systems are reported in Figures 5.30.b for states (V,α,q,θ) if the references are $r(t)=\begin{bmatrix} r_1(t) \\ r_2(t) \end{bmatrix}=\begin{bmatrix} V_{\text{ref}} \\ \theta_{\text{ref}} \end{bmatrix}=\begin{bmatrix} \mp 10\frac{\text{m}}{\text{sec}} \\ \pm 12.5^\circ \end{bmatrix}$.

Digital Tracking Control – Design a digital tracking control law. For T_s=0.1 sec, the Tustin approximation yields the difference equation (5.138). The discretized model is

$$\mathbf{x}_{n+1} = \mathbf{A}_n\mathbf{x}_n + A_{rn}r_n + \mathbf{B}_n u_n,\ \mathrm{y}_n = \mathbf{H}_n\mathbf{x}_n + D_n u_n,\ \begin{bmatrix} y_{1n} \\ y_{2n} \end{bmatrix} = \begin{bmatrix} x_{1n} \\ x_{4n} \end{bmatrix},$$

$$\begin{bmatrix} x_{1n+1} \\ x_{2n+1} \\ x_{3n+1} \\ x_{4n+1} \\ x_{e_1 n+1} \\ x_{e_2 n+1} \\ e_{1n+1} \\ e_{2n+1} \end{bmatrix} = \begin{bmatrix} 0.985 & -0.0831 & -0.0495 & -0.974 & 0 & 0 & 0 & 0 \\ 0.0003 & 0.462 & 0.0589 & -0.0001 & 0 & 0 & 0 & 0 \\ 0.007 & -0.344 & 0.769 & -0.0036 & 0 & 0 & 0 & 0 \\ 0.0004 & -0.0204 & 0.0882 & 1 & 0 & 0 & 0 & 0 \\ -0.0993 & 0.0049 & 0.0017 & 0.0488 & 1 & 0 & 0 & 0 \\ 0 & 0.0007 & -0.0046 & -0.1 & 0 & 1 & 0 & 0 \\ 0.0143 & 0.0784 & 0.0478 & 0.927 & 0 & 0 & 0.905 & 0 \\ -0.0004 & 0.0196 & -0.0838 & 0.0001 & 0 & 0 & 0 & 0.905 \end{bmatrix} \begin{bmatrix} x_{1n} \\ x_{2n} \\ x_{3n} \\ x_{4n} \\ x_{e_1 n} \\ x_{e_2 n} \\ e_{1n} \\ e_{2n} \end{bmatrix} + \begin{bmatrix} 0 & 0 \\ 0 & 0 \\ 0 & 0 \\ 0 & 0 \\ 0.1 & 0 \\ 0 & 0.1 \\ 0 & 0 \\ 0 & 0 \end{bmatrix} \begin{bmatrix} r_{1n} \\ r_{2n} \end{bmatrix} + \begin{bmatrix} 0.526 & -0.0319 \\ 0.0001 & 0.0108 \\ 0.0019 & 0.14 \\ 0.0001 & 0.0073 \\ 0 & 0 \\ 0 & 0 \\ -0.501 & 0.0304 \\ -0.0001 & -0.0071 \end{bmatrix} \begin{bmatrix} u_{1n} \\ u_{2n} \end{bmatrix}.$$

Minimize the quadratic performance index (5.139)

$$J = \min_{\mathbf{x}_n, u_n} \sum_{n=0}^{\infty} \left[\mathbf{x}_n^T Q_n \mathbf{x}_n + u_n^T G_n u_n \right],\ Q_n=I,\ Q_n \in \mathbb{R}^{8\times 8},\ G_n=I,\ G_n \in \mathbb{R}^{2\times 2},\ G_n = \begin{bmatrix} 1 & 0 \\ 0 & 1 \end{bmatrix}.$$

Equation (5.141) yields $K_n \in \mathbb{R}^{8\times 8}$, computed using MATLAB.
The tracking control law (5.140) is

$$u_n = -\left(G_n + \mathbf{B}_n^T K_n \mathbf{B}_n\right)^{-1} \mathbf{B}_n^T K_n \mathbf{A}_n \mathbf{x}_n = -K_{Fn}\mathbf{x}_n,$$

$$K_{Fn} = \begin{bmatrix} 0.908 & -0.115 & -0.103 & -1.588 & -0.676 & -0.087 & -0.123 & -0.0055 \\ -0.065 & -0.63 & 0.972 & 3.026 & 0.0618 & -0.919 & 0.0052 & -0.0556 \end{bmatrix}.$$

```
Ts=0.1; % Sampling time
Asd=expm(As*Ts), Bsd=pinv(As)*(Asd-eye(size(As)))*Bs % Discretization
Hs=[1 0 0 0 0 0 0 0; 0 0 0 1 0 0 0 0]; Ds=[0 0; 0 0];
% Discrete-time system
sys_C=ss(As,Bs,Hs,Ds);          % Continuous systems
sys_D=c2d(sys_C,Ts,'tustin'); % Discrete systems
Ad=sys_D.A; Bd=sys_D.B; Asd=Ad; Bsd=Bd;
Ark=Ts*[zeros(4,2); IX; zeros(2,2)]; % Reference matrix
Qd=1*eye(8,8); Gd=1*eye(2,2);
[KFd,Sd,Eigd]=dlqr(Asd,Bsd,Qd,Gd) % Discrete LQR design
% Solution of Difference Equation
x0=[0 0 0 0 0 0 0 0]'; % Initial conditions
% Square references r1 and r2
N=1000; k=0:1:N; r1=-10*square(((pi/250)*k),50); r2=12.5*square(((pi/250)*k),50); r=[r1; r2];
% Solving difference equation
x(:,1)=(Asd-Bsd*KFd)*x0+Ark*r(:,1);
for k=1:N
   x(:,k+1)=(Asd-Bsd*KFd)*x(:,k)+Ark*r(:,k);
   u=-KFd*x; end
% Plotting
k=0:1:N; figure(1); plot(k,x(1,:),'ko',k,x(4,:),'bs','LineWidth',1.5); axis([0 N -13.5 18]);
legend('{\ity}_1_{\itn}={\itx}_1_{\itn}={\itV}_{\itn}','{\ity}_2_{\itn}={\itx}_4_{\itn}={\it\theta}_{\itn}', ...
'Location','northwest','NumColumns',2,'FontSize',22); legend boxoff;
figure(2); plot(k,u,'ko','LineWidth',1.5); % Control evolutions
```

For the closed-loop system, the eigenvalues are computed as
0.942±0.0139i, 0.924±0.03i, 0.885, 0.651, 0.555 and 0.478.

Dynamics of the outputs (y_{1n},y_{2n}) for references r_{1n}=∓ 10 m/s and r_{2n}=±12.5° are reported in Figure 5.30.d.

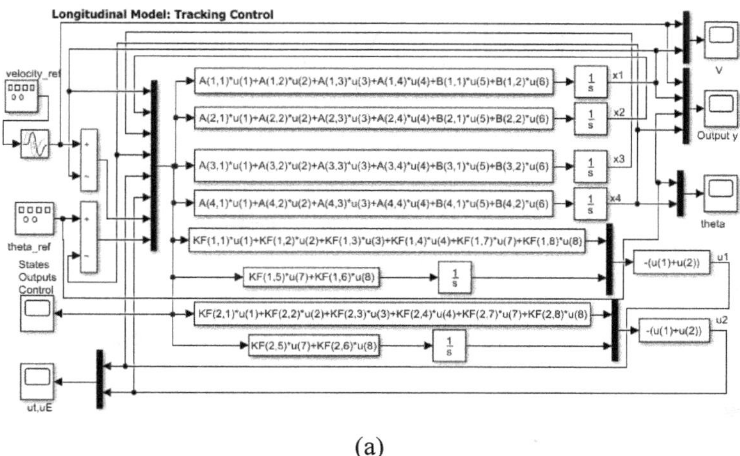

(a)

Dynamics of States x_1, x_2, x_3, x_4 Dynamics of Outputs $y_1=x_1$ and $y_2=x_4$ Outputs $y_{1n}=x_{1n}$ and $y_{2n}=x_{4n}$

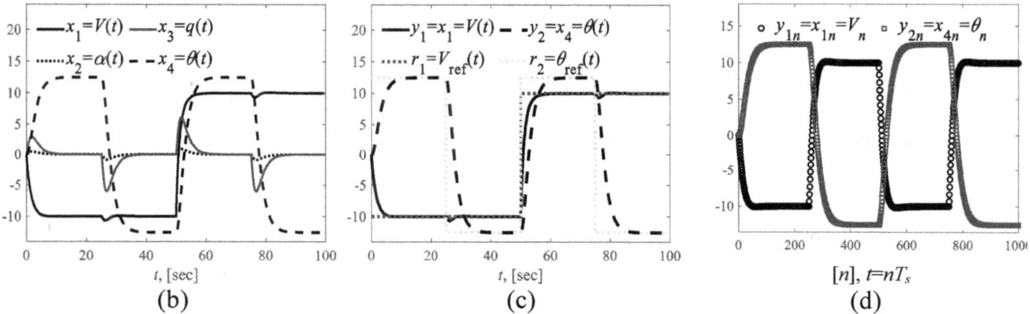

Figure 5.30. (a) SIMULINK diagram to simulate a closed-loop system;
(b) Dynamics of system states (x_1,x_2,x_3,x_4) for $r_1(t)=\mp 10$ m/s and $r_2(t)=\pm 12.5°$;
(c) Evolution of outputs (y_1,y_2) for $r_1(t)=\mp 10$ m/s and $r_2(t)=\pm 12.5°$;
(d) Discrete-time system: Evolution of outputs (y_{1n},y_{2n}) for $r_{1n}=\mp 10$ m/s and $r_{2n}=\pm 12.5°$.

Lateral Dynamics – The sideslip angle β, roll and yaw rates (p,r), and, roll and yaw angles (ϕ,ψ) are the states, $x\in\mathbb{R}^5$. Outputs $y(t)$ are the Euler angles ϕ and ψ, $y\in\mathbb{R}^2$. Displacements of ailerons (right and left ailerons are deflected differentially and synchronously) and rudder aerodynamic control surfaces are controls. Hence, $u_1=u_A$ and $u_2=u_R$, $u\in\mathbb{R}^2$. While physical limits are imposed on deflections of control surfaces, bounds on (u_A,u_R) are not considered in this section. Section 6.8.2 examines and solves the constrained control.

Hence, the states, outputs and control vectors are

$$x=\begin{bmatrix}x_1\\x_2\\x_3\\x_4\\x_5\end{bmatrix}=\begin{bmatrix}\beta\\p\\r\\\phi\\\psi\end{bmatrix},\ y=\begin{bmatrix}x_4\\x_5\end{bmatrix}=\begin{bmatrix}\phi\\\psi\end{bmatrix},\ u=\begin{bmatrix}u_1\\u_2\end{bmatrix}=\begin{bmatrix}u_A\\u_R\end{bmatrix},\quad x\in\mathbb{R}^5,\ y\in\mathbb{R}^2,\ u\in\mathbb{R}^2.$$

The differential equations for states (β,p,r,ϕ,ψ) and constitutive relationships for outputs (ϕ,ψ), controlled by (u_A,u_R) are

$$\dot{x} = Ax + Bu,\quad y = Hx,\quad \begin{bmatrix} \dot{x}_1 \\ \dot{x}_2 \\ \dot{x}_3 \\ \dot{x}_4 \\ \dot{x}_5 \end{bmatrix} = \begin{bmatrix} 0.08 & -0.07 & -0.94 & 0 & 0 \\ -0.51 & -3.8 & -0.45 & 0 & 0 \\ 5.9 & 1.3 & -2.2 & 0 & 0 \\ 0 & 1 & 0 & 0 & 0 \\ 0 & 0 & 1 & 0 & 0 \end{bmatrix} \begin{bmatrix} x_1 \\ x_2 \\ x_3 \\ x_4 \\ x_5 \end{bmatrix} + \begin{bmatrix} 0 & 0.13 \\ 17 & 0.28 \\ 0.07 & 0.95 \\ 0 & 0 \\ 0 & 0 \end{bmatrix} \begin{bmatrix} u_1 \\ u_2 \end{bmatrix},\ y = \begin{bmatrix} 0 & 0 & 0 & 1 & 0 \\ 0 & 0 & 0 & 0 & 1 \end{bmatrix} \begin{bmatrix} x_1 \\ x_2 \\ x_3 \\ x_4 \\ x_5 \end{bmatrix},$$

$$x = \begin{bmatrix} x_1 \\ x_2 \\ x_3 \\ x_4 \\ x_5 \end{bmatrix} = \begin{bmatrix} \beta \\ p \\ r \\ \phi \\ \psi \end{bmatrix},\ y = \begin{bmatrix} x_4 \\ x_5 \end{bmatrix} = \begin{bmatrix} \phi \\ \psi \end{bmatrix},\ u = \begin{bmatrix} u_1 \\ u_2 \end{bmatrix} = \begin{bmatrix} u_A \\ u_R \end{bmatrix},\quad x\in\mathbb{R}^5,\ y\in\mathbb{R}^2,\ u\in\mathbb{R}^2.$$

Dynamics of the *error states* $x_e(t)$ and tracking error vector $e(t)$ are

$\dot{x}_e(t) = I_X(r-y) = I_X(r-Hx)$, $e(t) \equiv \dot{x}_e(t)$, $x_e\in\mathbb{R}^2$, $I_X=I$, $I_X\in\mathbb{R}^{2\times 2}$,

$\dot{e}(t) = -I_E e + \dot{r} - \dot{y} = -I_E e + \dot{r} - HAx - HBu$, $e\in\mathbb{R}^2$, $I_E=I$, $I_E\in\mathbb{R}^{2\times 2}$.

The state-space governing equation is

$$\dot{\mathbf{x}}(t) = \begin{bmatrix} \dot{x}(t) \\ \dot{x}_e(t) \\ \dot{e}(t) \end{bmatrix} = \mathbf{A}\mathbf{x} + A_r r + \mathbf{B}u = \begin{bmatrix} A & 0 & 0 \\ -I_X H & 0 & 0 \\ -HA & 0 & -I_E \end{bmatrix} \mathbf{x} + \begin{bmatrix} 0 & 0 \\ I_X & 0 \\ 0 & I \end{bmatrix} \begin{bmatrix} r \\ \dot{r} \end{bmatrix} + \begin{bmatrix} B \\ 0 \\ -HB \end{bmatrix} u\,.$$

Minimizing the performance functional (5.117) $J = \min_{\mathbf{x},u} \int_0^\infty \frac{1}{2}\left(\mathbf{x}^T Q \mathbf{x} + u^T G u\right) dt$, a tracking control law (5.122) is found to be

$$u = -G^{-1}\mathbf{B}^T K \mathbf{x} = -K_F \mathbf{x} = -K_F \begin{bmatrix} x \\ \int e_1 dt \\ \int e_2 dt \\ e_1 \\ e_2 \end{bmatrix},\ x = \begin{bmatrix} x_1 \\ x_2 \\ x_3 \\ x_4 \\ x_5 \end{bmatrix} = \begin{bmatrix} \beta \\ p \\ r \\ \phi \\ \psi \end{bmatrix},\ e = \begin{bmatrix} e_1 \\ e_2 \end{bmatrix} = \begin{bmatrix} r_1 - \phi \\ r_2 - \psi \end{bmatrix}.$$

Let $Q=I$, $Q\in\mathbb{R}^{9\times 9}$ and $G=0.1I$, $G\in\mathbb{R}^{2\times 2}$, $G = \begin{bmatrix} 0.1 & 0 \\ 0 & 0.1 \end{bmatrix}$.

A control law with the feedback gain matrix K_F is

$$u = -K_F \begin{bmatrix} x \\ \int e_1 dt \\ \int e_2 dt \\ e_1 \\ e_2 \end{bmatrix},\ K_F = \begin{bmatrix} -0.4 & 3.07 & 0.574 & 5.81 & 0.876 & -3.13 & -0.418 & -0.39 & -0.055 \\ 4.56 & 0.07 & 2.44 & -0.847 & 7.12 & 0.418 & -3.13 & 0.045 & -0.293 \end{bmatrix}.$$

The MATLAB statements are

```
% Lateral Dynamics: State-Space Model. Matrices of A and B of the model coefficients
A=[0.08 -0.07 -0.94 0 0; -0.51 -3.8 -0.45 0 0; 5.9 1.3 -2.2 0 0; 0 1 0 0 0; 0 0 1 0 0];
B=[0 0.13; 17 0.28; 0.07 0.95; 0 0; 0 0];
H=[0 0 0 1 0; 0 0 0 0 1]; D=0; % H matrix: Outputs are the roll and yaw Euler angles
% Extended State-Space Model: States and error dynamics
IX=1*eye(2,2); IE=1*eye(2,2);
As=[A zeros(5,4); -IX*H zeros(2,4); -H*A zeros(2,2) -IE];
Bs=[B; zeros(2,2); -H*B];
Q=1*eye(9,9); Q(6,6)=1; Q(8,8)=1; G=0.1*eye(2,2);
[K,S,Eig]=lqr(As,Bs,Q,G); KF=inv(G)*Bs'*S
```

Figure 5.31.a illustrates dynamics of states $(\beta, r, p, \phi, \psi)$ for references $r(t) = \begin{bmatrix} r_1(t) \\ r_2(t) \end{bmatrix} = \begin{bmatrix} \phi_{\text{ref}} \\ \psi_{\text{ref}} \end{bmatrix} = \begin{bmatrix} \mp 25^\circ \\ \pm 45^\circ \end{bmatrix}$. Outputs (ϕ, ψ) are depicted in Figure 5.31.b.

Displacements of control surfaces are examined, and, controls efforts exceed the limits on (u_A,u_R) for $r(t)$ within flight envelope. Stability under the bounded control is ensured in an admissible flight envelope. Design of control laws for the considered aerial vehicle, with physical limits on controls $u_{i\min}\le u_i\le u_{i\max}$, are studied in section 6.8.2.

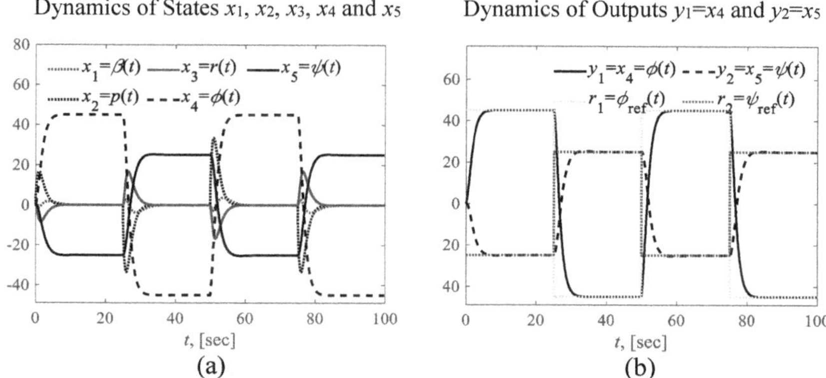

Figure 5.31. Dynamics of the closed-loop system:
(a) System states (x_1,x_2,x_3,x_4,x_5) for references $r_1(t)=\mp 25^\circ$ and $r_2(t)=\pm 45^\circ$.
(b) Evolution of outputs $y_1=\phi$ and $y_2=\psi$ for $r_1(t)=\mp 25^\circ$ and $r_2(t)=\pm 45^\circ$.

5.11.2. Analog and Digital Tracking Control of a Supersonic Fighter Aircraft

To enable control, agility, maneuverability and aerobatic capabilities, advanced airframes, aerodynamically and structurally optimized control surfaces, wing sweep configurations, thrust vectoring and other technologies have being researched and deployed. Consider a supersonic multirole fighter aircraft, controlled by steering right and left horizontal stabilizers $(\delta_{sR},\delta_{sL})$, flaperons $(\delta_{fR},\delta_{fL})$, canard δ_c and rudder δ_r. Symmetry of $(\delta_{sR},\delta_{sL})$ and $(\delta_{fR},\delta_{fL})$ yields δ_s and δ_f. Hydraulic and electromagnetic actuators displace aerodynamic control surfaces. Neglect fast electronics, sensors and actuators dynamics. Assume the input-output actuators and sensors linearity, no processing and interfacing latencies, neglect elasticity and unsteady aerodynamics, etc. To ensure controlled motions in maneuvering flight, tracking control laws are designed.

Consider nonlinear equations of motions, developed in section 2.4. For a straight and level flight,

$$\bar{u}\neq 0,\ \bar{v}=0,\ \bar{w}=0,\ \bar{p}=0,\ \bar{q}=0,\ \bar{r}=0,\ \bar{F}_X\neq 0,\ \bar{F}_Y=0,\ \bar{F}_Z\neq 0,\ \bar{L}=0,\ \bar{M}=0,\ \bar{N}=0.$$

One has $\bar{F}_X=mg\sin\bar{\theta}$ and $\bar{F}_Z=mg\cos\bar{\theta}$. Linearization is not always yields consistent results due to extended flight, service and performance envelopes, as well as longitudinal and lateral nonlinear coupling. Practice Problems 6.14.6 and 6.14.7 study nonlinear aerodynamics. Identification and parameter estimation can be performed during maneuvering flight in near real time. Flight regimes and maneuvers affect the longitudinal-lateral flight envelopes. Linearization of highly-coupled aerodynamics, asymmetry, inertial coupling and other factors are considered deriving models and designing adequate control schemes.

For states $x=[V,\alpha,q,\theta,\beta,p,r,\phi,\psi]$ at 5000 m elevation flight with V=270 m/s, a nonlinear model is (2.52)

$$\dot{x}(t) = F(x) + B(x)u,$$

$$\begin{bmatrix} \dot{V}(t) \\ \dot{\alpha}(t) \\ \dot{q}(t) \\ \dot{\theta}(t) \\ \dot{\beta}(t) \\ \dot{p}(t) \\ \dot{r}(t) \\ \dot{\phi}(t) \\ \dot{\psi}(t) \end{bmatrix} = \begin{bmatrix} \frac{1}{m}[-D\cos\beta + S\sin\beta - mg(\sin\alpha\cos\theta\cos\beta\cos\phi + \cos\theta\sin\beta\sin\phi - \cos\alpha\sin\theta\cos\beta) + T\cos\alpha\cos\beta] \\ q - (p\cos\alpha + r\sin\alpha)\tan\beta + \frac{g}{V\cos\beta}(\cos\alpha\cos\theta\cos\phi + \sin\alpha\sin\theta) \\ \frac{1}{I_{yy}}\left[(-I_{xx}+I_{zz})pr - I_{xz}(p^2 - r^2)\right] \\ q\cos\phi - r\sin\phi \\ p\sin\alpha - r\cos\alpha + \frac{g}{V}(\cos\theta\cos\beta\sin\phi + \cos\alpha\sin\theta\sin\beta - \sin\alpha\cos\theta\sin\beta\cos\phi) \\ \frac{1}{I_{xx}I_{zz} - I_{xz}^2}\left[I_{xz}(I_{xx} - I_{yy} + I_{zz})qp - (I_{xz}^2 - I_{yy}I_{zz} + I_{zz}^2)qr\right] \\ \frac{1}{I_{xx}I_{zz} - I_{xz}^2}\left[-(I_{xx}^2 - I_{xx}I_{yy} + I_{xz}^2)qp + I_{xz}(I_{xx} - I_{yy} + I_{zz})qr\right] \\ p + q\tan\theta\sin\phi + r\tan\theta\cos\phi \\ q\sec\theta\sin\phi + r\sec\theta\cos\phi \end{bmatrix} + B(x)\begin{bmatrix} \delta_s \\ \delta_f \\ \delta_c \\ \delta_r \end{bmatrix}.$$

For the twin-engine fighter aircraft, m=28000 kg, I_{xx}=2.75×10^4 kg-m^2, I_{xz}=–1280 kg-m^2, I_{yy}=2.51×10^5 kg-m^2, I_{zz}=2.38×10^5 kg-m^2, other I_{ij}=0, ρ=0.7364 kg/m^3, c_d=0.285 (at V=270 m/s) and A=4.9 m^2.

There is the longitudinal and lateral axes coupling in mappings $F(x)$ and $B(x)$.

The model is linearized using the MATLAB code, finding $\dot{x}=Ax+Bu$.

Matrix $A\in\mathbb{R}^{9\times 9}$ is found using the Jacobian for $F(x)$. Matrix $B\in\mathbb{R}^{9\times 4}$ is derived using equations for aerodynamic forces and moments $[(F_X,F_Y,F_Z),(L,M,N)]$.

The equilibrium for states $(V,\alpha,q,\theta,\beta,p,r,\phi,\psi)$ are

$\bar{x}=[270,\ 0.01,\ 0.025,\ 0.025,\ 0,\ 0.025,\ 0.025,\ 0,\ 0]$.

The forward velocity V is in m/s, the rates (q,p,r) are in rad/s, and, the Euler angles (θ,ϕ,ψ) and control surfaces displacement are in radians.

The MATLAB code is

```
syms m r cd A g Ixx Ixy Ixz Iyy Iyx Izz;
syms xd1 xd2 xd3 xd4 xd5 xd6 xd7 xd8 xd9;
syms x1 x2 x3 x4 x5 x6 x7 x8 x9;
x=[x1 x2 x3 x4 x5 x6 x7 x8 x9]; % States
xd1=(-0.5*r*cd*A*x1*x1*cos(x5)-m*g*(sin(x2)*cos(x4)*cos(x5)*cos(x8)+cos(x4)*sin(x5)*sin(x8)-
cos(x2)*sin(x4)*cos(x5)))/m;
xd2=x3-x6*cos(x2)*tan(x5)-
x7*sin(x2)*tan(x5)+(g/(x(1)*cos(x5)))*(cos(x2)*cos(x4)*cos(x8)+sin(x2)*sin(x4));
xd3=(1/Iyy)*((-Ixx+Izz)*x6*x7-Ixz*(x6^2-x7^2));
xd4=x3*cos(x8)-x7*sin(x8);
xd5=x6*sin(x2)-x7*cos(x2)+(g/x(1))*(cos(x4)*sin(x5)*sin(x8)+cos(x2)*sin(x4)*sin(x5)-
sin(x2)*cos(x4)*sin(x5)*cos(x8));
xd6=(1/(Ixx*Izz-Ixz^2))*(Ixz*(Ixx-Iyy+Izz)*x3*x6-(Ixz^2-Iyy*Izz+Izz^2)*x3*x7);
xd7=(1/(Ixx*Izz-Ixz^2))*(-(Ixx^2-Ixx*Iyy+Ixz^2)*x3*x6+Ixz*(Ixx-Iyy+Izz)*x3*x7);
xd8=x6+x3*tan(x4)*sin(x8)+x7*tan(x4)*cos(x8);
xd9=x3*sec(x4)*sin(x8)+x7*sec(x4)*cos(x8);
equations=[xd1 xd2 xd3 xd4 xd5 xd6 xd7 xd8 xd9];
V=270; g=9.81; m=28000; r=0.7364; cd=0.285; A=4.9; % Constants and parameters
Ixx=2.75e4; Ixy=0; Ixz=-1280; Iyy=2.51e5; Iyz=0; Izz=2.38e5;
A_jacobian=jacobian(equations,x);
x1=270; x2=0.01; x3=0.025; x4=0.025; x5=0; x6=0.025; x7=0.025; x8=0; x9=0; % Equilibria
A=double(subs(A_jacobian)) % Matrix A
```

The linearization yields

$\dot{x} = Ax + Bu,$

$$\begin{bmatrix} \dot{x}_1 \\ \dot{x}_2 \\ \dot{x}_3 \\ \dot{x}_4 \\ \dot{x}_5 \\ \dot{x}_6 \\ \dot{x}_7 \\ \dot{x}_8 \\ \dot{x}_9 \end{bmatrix} = \begin{bmatrix} -0.0099 & -9.81 & 0 & 9.81 & 0 & 0 & 0 & 0 & 0 \\ -0.00013 & 0.00054 & 1 & -0.00054 & -0.0252 & 0 & 0 & 0 & 0 \\ 0 & 0 & 0 & 0 & 0 & 0.0212 & 0.0207 & 0 & 0 \\ 0 & 0 & 1 & 0 & 0 & 0 & 0 & -0.025 & 0 \\ 0 & 0.0252 & 0 & 0 & 0.00054 & 0.01 & -1 & 0 & 0 \\ 0 & 0 & 0.0117 & 0 & 0 & -0.00007 & 0.0118 & 0 & 0 \\ 0 & 0 & 0.0234 & 0 & 0 & 0.0235 & -0.00007 & 0 & 0 \\ 0 & 0 & 0 & 0.025 & 0 & 1 & 0.025 & 0.00063 & 0 \\ 0 & 0 & 0 & 0.00063 & 0 & 0 & 1 & 0.025 & 0 \end{bmatrix} \begin{bmatrix} x_1 \\ x_2 \\ x_3 \\ x_4 \\ x_5 \\ x_6 \\ x_7 \\ x_8 \\ x_9 \end{bmatrix}$$

$$+ \begin{bmatrix} -0.0008 & -0.0005 & -0.0001 & -0.0007 \\ 0.093 & 0 & -0.024 & 0 \\ 5.3 & 0 & 1.9 & 0 \\ 0 & 0 & 0 & 0 \\ 0 & 0 & 0 & 0 \\ 0.053 & 6.8 & 0 & 0.07 \\ 0.081 & 0.07 & 0 & 8.2 \\ 0 & 0 & 0 & 0 \\ 0 & 0 & 0 & 0 \end{bmatrix} \begin{bmatrix} u_1 \\ u_2 \\ u_3 \\ u_4 \end{bmatrix}.$$

The states, controls and outputs are

$$x = \begin{bmatrix} x_1 \\ x_2 \\ x_3 \\ x_4 \\ x_5 \\ x_6 \\ x_7 \\ x_8 \\ x_9 \end{bmatrix} = \begin{bmatrix} V \\ \alpha \\ q \\ \theta \\ \beta \\ p \\ r \\ \phi \\ \psi \end{bmatrix},\ u = \begin{bmatrix} u_1 \\ u_2 \\ u_3 \\ u_4 \end{bmatrix} = \begin{bmatrix} \delta_s \\ \delta_f \\ \delta_c \\ \delta_r \end{bmatrix},\ y = \begin{bmatrix} y_1 \\ y_2 \\ y_3 \end{bmatrix} = Hx = \begin{bmatrix} 0 & 0 & 0 & 1 & 0 & 0 & 0 & 0 & 0 \\ 0 & 0 & 0 & 0 & 0 & 0 & 0 & 1 & 0 \\ 0 & 0 & 0 & 0 & 0 & 0 & 0 & 0 & 1 \end{bmatrix} x,\ y = \begin{bmatrix} \theta \\ \phi \\ \psi \end{bmatrix}.$$

The tracking error vector is

$$e(t) = \begin{bmatrix} e_1 \\ e_2 \\ e_3 \end{bmatrix} = \begin{bmatrix} r_1 - y_1 \\ r_2 - y_2 \\ r_3 - y_3 \end{bmatrix} = \begin{bmatrix} r_1 - x_4 \\ r_2 - x_8 \\ r_3 - x_9 \end{bmatrix} = \begin{bmatrix} r_1 - \theta \\ r_2 - \phi \\ r_3 - \psi \end{bmatrix}.$$

Evolutions of the *error states* $x_e(t)$ and tracking error dynamics are governed as

$\dot{x}_e(t) = I_X(r - y) = I_X(r - Hx)$, $e(t) \equiv \dot{x}_e(t)$, $x_e \in \mathbb{R}^3$,

$\dot{e}(t) = -I_E e + \dot{r} - \dot{y} = -I_E e + \dot{r} - HAx - HBu$, $e \in \mathbb{R}^3$.

The extended system dynamics (5.116) is

$$\dot{\mathbf{x}}(t) = \begin{bmatrix} \dot{x}(t) \\ \dot{x}_e(t) \\ \dot{e}(t) \end{bmatrix} = \mathbf{A}\mathbf{x} + A_r r + \mathbf{B}u = \begin{bmatrix} A & 0 & 0 \\ -I_X H & 0 & 0 \\ -HA & 0 & -I_E \end{bmatrix} \mathbf{x} + \begin{bmatrix} 0 & 0 \\ I_X & 0 \\ 0 & I \end{bmatrix} \begin{bmatrix} r \\ \dot{r} \end{bmatrix} + \begin{bmatrix} B \\ 0 \\ -HB \end{bmatrix} u \cdot$$

Minimize a quadratic functional (5.117)

$$J = \min_{\mathbf{x},u} \int_0^\infty \tfrac{1}{2}\left(\mathbf{x}^T Q \mathbf{x} + u^T G u\right) dt \cdot$$

A tracking control law (5.122) with the state x and error feedback is

$$u=-G^{-1}\mathbf{B}^T K\mathbf{x}=-K_F\mathbf{x}=-K_F\begin{bmatrix} x \\ \int e_1 dt \\ \int e_2 dt \\ \int e_3 dt \\ e_1 \\ e_2 \\ e_3 \end{bmatrix},\ e=\begin{bmatrix} e_1 \\ e_2 \\ e_3 \end{bmatrix}=\begin{bmatrix} r_1-\theta \\ r_2-\phi \\ r_3-\psi \end{bmatrix}.$$

Coefficients in $Q\in\mathbb{R}^{15\times 15}$ are $q_{1,1}=q_{2,2}=q_{3,3}=q_{4,4}=q_{5,5}=q_{6,6}=q_{7,7}=q_{8,8}=q_{9,9}=0.01$, $q_{10,10}=q_{11,11}=q_{12,12}=10$ and $q_{13,13}=q_{14,14}=q_{15,15}=1$, while, $G=I$, $G\in\mathbb{R}^{4\times 4}$.

The control law with computed feedback gain matrix K_F is

$$K_F=\begin{bmatrix} -0.0288 & 1.975 & 0.89 & 0.446 & 1.284 & 0.0043 & 0.0053 & -0.0148 & 1.295 & -2.943 & 0.0491 & 0.0276 & -0.0679 & -0.00052 & -0.00028 \\ -0.0028 & 0.0997 & -0.0048 & -0.101 & 0.212 & 0.866 & 0.0021 & 2.44 & 0.171 & 0.0308 & -3.155 & 0.0361 & 0.00051 & -0.0751 & 0.00051 \\ 0.0816 & -5.571 & 0.408 & 6.455 & -3.412 & -0.0017 & -0.0016 & 0.0185 & -3.424 & -1.156 & -0.21 & -0.117 & -0.0243 & -0.00046 & -0.00041 \\ -0.0034 & 0.12 & -0.0057 & -0.134 & 0.16 & 0.0035 & 0.772 & 0.0286 & 2.482 & 0.0173 & -0.0279 & -3.16 & 0.00023 & -0.00118 & -0.0779 \end{bmatrix}.$$

Figure 5.32.a depicts outputs (θ,ϕ,ψ) for $r(t)=\begin{bmatrix} r_1(t) \\ r_2(t) \\ r_3(t) \end{bmatrix}=\begin{bmatrix} \theta_{\text{ref}}(t) \\ \phi_{\text{ref}}(t) \\ \psi_{\text{ref}}(t) \end{bmatrix}=\begin{bmatrix} \pm 0.5\text{ rad} \\ \mp 0.75\text{ rad} \\ \mp 1\text{ rad} \end{bmatrix}$.

Discretize a system $\dot{\mathbf{x}}(t)=\mathbf{A}\mathbf{x}+A_r r+\mathbf{B}u$, and, design a tracking control law. For T_s=0.01 sec, use the Tustin approximation.

A discrete-time model (5.138) is found to be

$$\mathbf{x}_{n+1}=\mathbf{A}_n\mathbf{x}_n+A_{r\,n}r_n+\mathbf{B}_n u_n,\ y_n=\begin{bmatrix} y_{1\,n} \\ y_{2\,n} \\ y_{3\,n} \end{bmatrix}=\begin{bmatrix} x_{4\,n} \\ x_{8\,n} \\ x_{9\,n} \end{bmatrix},$$

$$\begin{bmatrix} x_{1\,n+1} \\ x_{2\,n+1} \\ x_{3\,n+1} \\ x_{4\,n+1} \\ x_{5\,n+1} \\ x_{6\,n+1} \\ x_{7\,n+1} \\ x_{8\,n+1} \\ x_{9\,n+1} \\ x_{e_1\,n+1} \\ x_{e_2\,n+1} \\ x_{e_3\,n+1} \\ e_{1\,n+1} \\ e_{2\,n+1} \\ e_{3\,n+1} \end{bmatrix}=\begin{bmatrix} 1 & -0.0981 & 0 & -0.0981 & 0 & 0 & 0 & 0 & 0 & 0 & 0 & 0 & 0 & 0 & 0 \\ 0 & 1 & 0.01 & 0 & -0.0003 & 0 & 0 & 0 & 0 & 0 & 0 & 0 & 0 & 0 & 0 \\ 0 & 0 & 1 & 0 & 0 & 0.0002 & 0.0002 & 0 & 0 & 0 & 0 & 0 & 0 & 0 & 0 \\ 0 & 0 & 0.01 & 1 & 0 & 0 & 0 & -0.0003 & 0 & 0 & 0 & 0 & 0 & 0 & 0 \\ 0 & 0.0003 & 0. & 0 & 1 & 0.00001 & -0.01 & 0 & 0 & 0 & 0 & 0 & 0 & 0 & 0 \\ 0 & 0 & 0.0001 & 0 & 0 & 1 & 0.0001 & 0 & 0 & 0 & 0 & 0 & 0 & 0 & 0 \\ 0 & 0 & 0.0002 & 0 & 0 & 0.0002 & 1 & 0 & 0 & 0 & 0 & 0 & 0 & 0 & 0 \\ 0 & 0 & 0 & 0.0003 & 0 & 0.01 & 0.0003 & 1 & 0 & 0 & 0 & 0 & 0 & 0 & 0 \\ 0 & 0 & 0 & 0 & 0 & 0 & 0.01 & 0.0003 & 1 & 0 & 0 & 0 & 0 & 0 & 0 \\ 0 & 0 & -0.0001 & -0.01 & 0 & 0 & 0 & 0 & 0 & 1 & 0 & 0 & 0 & 0 & 0 \\ 0 & 0 & 0 & 0 & 0 & -0.0001 & 0 & -0.01 & 0 & 0 & 1 & 0 & 0 & 0 & 0 \\ 0 & 0 & 0 & 0 & 0 & 0 & -0.0001 & 0 & -0.01 & 0 & 0 & 1 & 0 & 0 & 0 \\ 0 & 0 & -0.01 & 0 & 0 & 0 & 0 & 0.0002 & 0 & 0 & 0 & 0 & 0.99 & 0 & 0 \\ 0 & 0 & 0 & -0.0003 & 0 & -0.01 & -0.0002 & 0 & 0 & 0 & 0 & 0 & 0 & 0.99 & 0 \\ 0 & 0 & 0 & 0 & 0 & 0 & -0.01 & -0.0002 & 0 & 0 & 0 & 0 & 0 & 0 & 0.99 \end{bmatrix}\begin{bmatrix} x_{1\,n} \\ x_{2\,n} \\ x_{3\,n} \\ x_{4\,n} \\ x_{5\,n} \\ x_{6\,n} \\ x_{7\,n} \\ x_{8\,n} \\ x_{9\,n} \\ x_{e_1\,n} \\ x_{e_2\,n} \\ x_{e_3\,n} \\ e_{1\,n} \\ e_{2\,n} \\ e_{3\,n} \end{bmatrix}$$

$$+\begin{bmatrix} 0 & 0 & 0 \\ 0 & 0 & 0 \\ 0 & 0 & 0 \\ 0 & 0 & 0 \\ 0 & 0 & 0 \\ 0 & 0 & 0 \\ 0 & 0 & 0 \\ 0 & 0 & 0 \\ 0 & 0 & 0 \\ 0.01 & 0 & 0 \\ 0 & 0.01 & 0 \\ 0 & 0 & 0.01 \\ 0 & 0 & 0 \\ 0 & 0 & 0 \\ 0 & 0 & 0 \end{bmatrix}\begin{bmatrix} r_{1\,n} \\ r_{2\,n} \\ r_{3\,n} \end{bmatrix}+\begin{bmatrix} -0.0001 & 0 & 0.0001 & 0 \\ 0.0012 & 0 & -0.00015 & 0 \\ 0.053 & 0 & 0.019 & 0 \\ 0.00027 & 0 & 0.0001 & 0 \\ 0 & 0 & 0 & -0.00041 \\ 0.0005 & 0.068 & 0 & 0.0007 \\ 0.0008 & 0.00071 & 0 & 0.082 \\ 0 & 0.00034 & 0 & 0.0001 \\ 0 & 0 & 0 & 0.00041 \\ 0 & 0 & 0 & 0 \\ 0 & 0 & 0 & 0 \\ 0 & 0 & 0 & 0 \\ -0.00026 & 0 & -0.0001 & 0 \\ 0 & -0.00034 & 0 & 0 \\ 0 & 0 & 0 & -0.00041 \end{bmatrix}\begin{bmatrix} u_{1\,n} \\ u_{2\,n} \\ u_{3\,n} \\ u_{4\,n} \end{bmatrix}.$$

Solve a discrete LQR problem minimizing the quadratic performance index (5.139)

$$J = \min_{\mathbf{x}_n, u_n} \sum_{n=0}^{\infty} \left[\mathbf{x}_n^T Q_n \mathbf{x}_n + u_n^T G_n u_n \right].$$

The weighting matrices $Q_n \in \mathbb{R}^{15\times15}$ and $G_n \in \mathbb{R}^{4\times4}$ are same as in the analog control law design. The tracking control law (5.140) is

$$u_n = -\left(G_n + \mathbf{B}_n^T K_n \mathbf{B}_n\right)^{-1} \mathbf{B}_n^T K_n \mathbf{A}_n \mathbf{x}_n = -K_{Fn} \mathbf{x}_n.$$

Solving (5.141), compute K_n. The feedback gain matrix K_{Fn} is

$$K_{Fn} = \begin{bmatrix} -0.0288 & 1.975 & 0.878 & 0.396 & 1.284 & 0.0042 & 0.0052 & -0.0151 & 1.294 & -2.862 & 0.05 & 0.0285 & -0.0658 & -0.00049 & -0.00025 \\ -0.0028 & 0.0963 & -0.0047 & -0.0981 & 0.205 & 0.853 & 0.0019 & 2.384 & 0.164 & 0.0303 & -3.063 & 0.0362 & 0.0005 & -0.0726 & 0.00052 \\ 0.0814 & -5.567 & 0.403 & 6.432 & -3.406 & -0.0017 & -0.0016 & 0.0185 & -3.418 & -1.126 & -0.209 & -0.117 & -0.0235 & -0.00045 & -0.00041 \\ -0.0032 & 0.116 & -0.0056 & -0.13 & 0.154 & 0.0033 & 0.76 & 0.0272 & 2.42 & 0.0173 & -0.0258 & -3.061 & 0.00024 & -0.0011 & -0.0751 \end{bmatrix}.$$

The MATLAB code to discretize a system, design a control law, solve difference equations, and, plot dynamics is reported below.

```
As=[A zeros(9,6); -IX*H zeros(3,6); -H*A zeros(3,3) -IE]; % Extended State-Space Model
Bs=[B; zeros(3,4); -H*B]; Hs=[H zeros(3,6)]; D=[0];
sys=ss(As,Bs,Hs,D);       Ts=0.01;   % Sampling time
sys_d=c2d(sys,Ts,'tustin'); G_z=tf(sys_d); poles=zpk(G_z);
% Discrete time system: State space model
An=sys_d.a, Bn=sys_d.b, Hn=sys_d.c, Dn=sys_d.d, Arn=Ts*[zeros(9,3); IX; zeros(3,3)];
Qn=0.01*eye(15,15);  Gn=1*eye(4,4);
Qn(10,10)=10; Qn(11,11)=10; Qn(12,12)=10; Qn(13,13)=1; Qn(14,14)=1; Qn(15,15)=1;
[KFn,Kn,Eigenvalues]=dlqr(An,Bn,Qn,Gn) % Discrete LQR design
% Transient dynamics of a discrete-time system: Solution of difference equation
x0=[0 0 0 0 0 0 0 0 0 0 0 0 0 0 0]'; % Initial conditions
% Square references r(n)
clear x
N=3000; n=0:1:N;
r1=0.5*square(((pi/1000)*n),50); r2=-0.75*square(((pi/1000)*n),50); r3=-1*square(((pi/1000)*n),50);
r=[r1;r2;r3];
% Solve difference equations
x(:,1)=(An-Bn*KFn)*x0+Arn*r(:,1);
for n=1:1:N
   x(:,n+1)=(An-Bn*KFn)*x(:,n)+Arn*r(:,n);
   u=-KFn*x;
end
% Plotting
n=0:1:N;
figure(1);
plot(n,x(4,:),'ko',n,x(8,:),'bs',n,x(9,:),'rd','LineWidth',1.5); axis([0 N -1.15 1.45]); % Evolution of outputs
figure(2);
plot(1:10:3001,x(4,(1:10:end)),'ko',1:10:3001,x(8,(1:10:end)),'bs',1:10:3001,x(9,(1:10:end)),'rd','LineWidth',1.5);
axis([0 N -1.15 1.45]);
legend('{\ity}_1_{\itn}={\itx}_4_{\itn}={\it\theta}_{\itn}','{\ity}_2_{\itn}={\itx}_8_{\itn}={\it\phi}_{\itn}', ...
'{\ity}_3_{\itn}={\itx}_9_{\itn}={\it\psi}_{\itn}','Location','Northwest','NumColumns',2,'FontSize',18);
legend boxoff;
```

Dynamics of the closed-loop MIMO system are reported in Figure 5.32.b. Consistency of analog and digital control laws are demonstrated. Simulations support and substantiate modeling, analysis and design. Derived dynamics complies with high-g maneuvering capabilities, agility, controllability and other flying qualities of the aircraft. Practice Problems 6.14.6 and 6.14.7 document results in constrained control and dynamic optimization considering limits on displacements of aerodynamic control surfaces.

Displacements of aerodynamic control surfaces may exceed physical limits. Control laws should be designed for the expanded flight envelopes by investigating control limits in Practice Problems 6.14.6 and 6.14.7 [18].

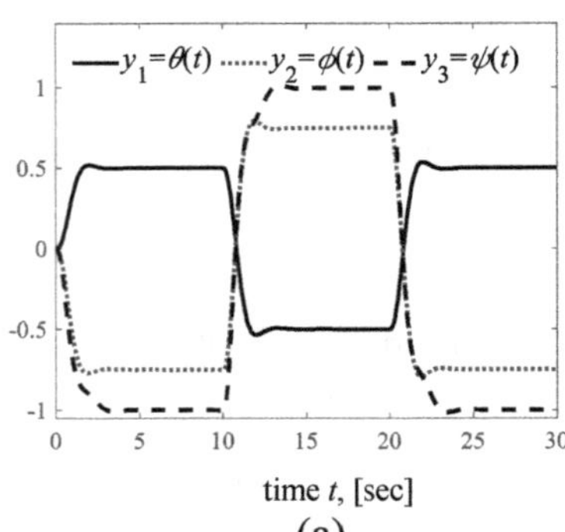

Dynamics of Outputs θ_n, ϕ_n and ψ_n, [rad]

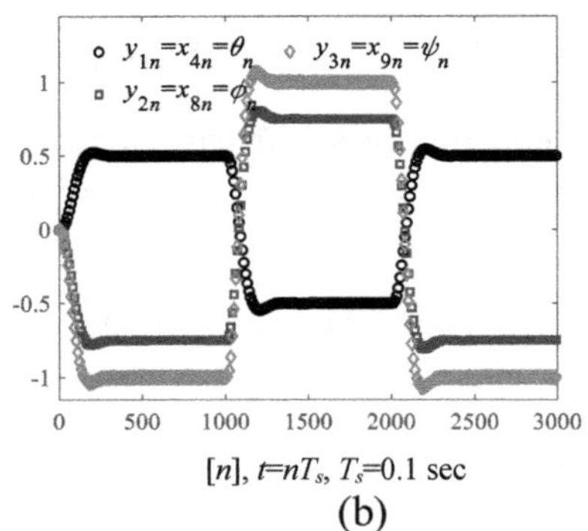

Figure 5.32. Output dynamics of the closed-loop systems with analog and digital tracking control laws:

(a) Continuous-time system: Dynamics of outputs $y=\begin{bmatrix} x_4(t) \\ x_8(t) \\ x_9(t) \end{bmatrix}=\begin{bmatrix} \theta(t) \\ \phi(t) \\ \psi(t) \end{bmatrix}$ if $r=\begin{bmatrix} r_1(t) \\ r_2(t) \\ r_3(t) \end{bmatrix}=\begin{bmatrix} \pm 0.5 \text{ rad} \\ \mp 0.75 \text{ rad} \\ \mp 1 \text{ rad} \end{bmatrix}$;

(b) Discrete-time system: Evolution of outputs $y_n=\begin{bmatrix} x_{4n} \\ x_{8n} \\ x_{9n} \end{bmatrix}=\begin{bmatrix} \theta_n \\ \phi_n \\ \psi_n \end{bmatrix}$ if $r_n=\begin{bmatrix} r_{1n} \\ r_{2n} \\ r_{3n} \end{bmatrix}=\begin{bmatrix} \pm 0.5 \text{ rad} \\ \mp 0.75 \text{ rad} \\ \mp 1 \text{ rad} \end{bmatrix}$.■

5.12. Multi-Degree-of-Freedom Sensors

Distributed Sensing, Data Management and Control – States and variables must be measured, observed or estimated with post-measurement filtering, data sharing, information fusion, data management and processing. Distributed sensing and processing support autonomy, command, control, coordination, etc. Sensing and data acquisition of spatiotemporal physical quantities imply:

1. Spatially-distributed sensing, processing and sharing;
2. On-sensor data governance and data analytics;
3. Information management, sharing and data fusion.

These consideration and topics pertain to the microsystem-technology integrated system-in-package MDOF sensors with ASICs and MPUs, which independently process measurements from each sensor, support data acquisition solutions. With multiple sensing and processing nodes, scalability and interoperability should be accomplished within multi-node multiple-level asynchronous data fusion and data management schemes. Integrated MDOF sensors are enabled by low-power MPUs and I^2C which process, interface and fuse data. On-node data-intensive computing supports distributed processing, ensuring data aggregation and data management with robust operation on sub-samples or hashed sets of data. Cooperative distributed processing is essential in *centralized*, *decentralized* and *behavioral* governance and coordination. To solve specific tasks and problems, an interacting network of semi-autonomous on-device processing is investigated. Tasks allocation, partitioning and coordination are implemented using software- and hardware-supported algorithms and protocols. Some analysis schemes are reported in chapters 6 and 7 by investigating hierarchical topology dynamic networks for spatiotemporal sensing and processing. Considered solutions support legacy systems as well as enable design of next generation of systems with distributed multi-node processing of spatiotemporal sensor data. Various integrated MDOF sensors have been widely commercialized, and, the problem is to integrate the existing solutions within systems under consideration.

Physical Data Quality – Physical data quality is critical in aerial, automotive, electromechanical, electronic, energy, mechanical, robotic and other systems. Signals are perturbed by distortions, electromagnetic spectrum, interference, noise, etc. Measurement transductions and energy conversion in acoustic, magnetic, piezoelectric, semiconductor and other sensors are affected by:

1. Device physics and physical phenomena;
2. Heterogeneity and properties of ferromagnetic, piezoelectric, semiconductor and other materials;
3. Thermal and shot noise due to magnetic and electrostatic field nonuniformity, spatiotemporal heat transfer, diffusion heterogeneity,crystal nonstoichiometry, lattice defects and misalignments, voids, etc.;
4. Acoustic, electromagnetic, flicker, thermal and other distortions;
5. Electromagnetic interference and electromagnetic spectrum.

Physical data quality is defined and depends on device physics, fabrication technology, sensor organization, ASICs, interface, algorithms and protocols, etc. Data quality is characterized by:

1. Accuracy, precision, linearity, errors, cross-axis-dependencies and couplings;
2. Dynamic response and bandwidth;
3. Signal to noise ratio and noise spectral density;
4. Statistical correlations and coherence;
5. Environmental sensitivity, immunity, robustness and range.

Measurements by inertial, magnetic, tactile and other sensors exhibit heterogeneities. Low, medium and high frequency noise and distortions, which are the superposition of stochastic and deterministic processes, should be attenuated by filters. Filters are designed applying probabilistic analysis, statistical models of signals and noise, stochastic optimization, deterministic methods, etc. Digital signal processing and signal conditioning are accomplished by MPUs and FPGAs.

Physical Measurements and Processing – Consider system-in-package microsystem-technology integrated sensors. Physical quantities $\mathbf{y}$ are measured by sensing elements which perform energy conversion and transductions. Sensing elements directly sense (measure) $\mathbf{y}(t)$, and, sensing elements output analog measurements as a vector $\bar{\mathbf{y}}$. These $\bar{\mathbf{y}}(t)$ are processed by built-in ASICs and MPUs, yielding the outputs $\bar{\mathbf{y}}[n]$ and $\hat{\mathbf{y}}[n]$.

One has, $\mathbf{y}(t)\Rightarrow\bar{\mathbf{y}}(t)\Rightarrow\bar{\mathbf{y}}[n]\Rightarrow\hat{\mathbf{y}}[n]$.

While the measured real-valued physical quantities $\mathbf{y}(t)$ are continuous $\bar{\mathbf{y}}(t)$, the system-in-package integrated sensors with MPUs perform:

1. Analog-to-analog, analog-to-digital, digital-to-digital and digital-to-analog conversions;
2. Nonlinearities and perturbation compensations;
3. Signal conditioning, filtering and estimations;
4. Peripheral interfacing and seamless communication.

Measured $\bar{\mathbf{y}}(t)$ and processed $(\bar{\mathbf{y}}[n],\hat{\mathbf{y}}[n])$ are perturbed and superimposed with noise $\mathbf{n}$, errors $\boldsymbol{\varepsilon}$ and distortions $\boldsymbol{\Delta}$ of different origins. Signal conditioning of measured $\mathbf{y}(t)$ implies amplification, buffering, isolation, pre-filtering, analog-to-analog and analog-to-digital conversion, range matching, interfacing and other processes, yielding $\bar{\mathbf{y}}(t)$. For $\mathbf{y}(t)\Rightarrow\bar{\mathbf{y}}(t)\Rightarrow\bar{\mathbf{y}}[n]\Rightarrow\hat{\mathbf{y}}[n]$, one has

$$\underset{\substack{\text{Physical}\\ \text{Quantities}}}{\mathbf{y}(t)}\ \underset{\substack{\text{Sensing}\\ \text{Elements}}}{\overset{\substack{\text{Transductions in}\\ \text{Semiconductors}\\ \text{and Motional}\\ \text{Structures}}}{\Rightarrow}}\ \underset{\substack{\text{Direct}\\ \text{Measurements}}}{\overset{\substack{\text{Sensing}\\ \text{Element}\\ \text{Outputs}}}{\bar{\mathbf{y}}(t)}}\ \underset{\substack{\text{Signal Conditioning}\\ \text{ASICs Processing}}}{\overset{\substack{\text{Processing}\\ \text{Calculus and}\\ \text{Algorithms}}}{\Rightarrow}}\ \underset{\substack{\text{Data}\\ \text{Fused}}}{\overset{\substack{\text{Digital}\\ \text{Outputs}}}{\bar{\mathbf{y}}[n]}}\ \underset{\text{ASICs Processing}}{\overset{\substack{\text{Filtering and}\\ \text{Estimation:}\\ \text{Processing}\\ \text{Calculus and}\\ \text{Algorithms}}}{\Rightarrow}}\ \underset{\substack{\text{Data}\\ \text{Management}}}{\overset{\substack{\text{Filtered}\\ \text{Outputs}}}{\hat{\mathbf{y}}[n]}}\ , \tag{5.142}$$

$$\bar{\mathbf{y}}=\mathbf{y}+f_{\bar{\mathbf{y}}}(\bar{\mathbf{n}}_{\bar{\mathbf{y}}},\bar{\boldsymbol{\varepsilon}}_{\bar{\mathbf{y}}},\bar{\boldsymbol{\Delta}}_{\bar{\mathbf{y}}}),\ \hat{\mathbf{y}}=\bar{\mathbf{y}}+f_{\hat{\mathbf{y}}}(\bar{\mathbf{n}}_{\bar{\mathbf{y}}},\bar{\boldsymbol{\varepsilon}}_{\bar{\mathbf{y}}},\bar{\boldsymbol{\Delta}}_{\bar{\mathbf{y}}},\hat{\mathbf{n}}_{\hat{\mathbf{y}}},\hat{\boldsymbol{\varepsilon}}_{\hat{\mathbf{y}}},\hat{\boldsymbol{\Delta}}_{\hat{\mathbf{y}}}).$$

Measured and processed $(\bar{\mathbf{y}}(t),\bar{\mathbf{y}}[n],\hat{\mathbf{y}}[n])$ are distorted by noise, errors and distortions $(\mathbf{n},\boldsymbol{\varepsilon},\boldsymbol{\Delta})$. These physics-, device-, circuits-, algorithms- and interference dependent $(\mathbf{n},\boldsymbol{\varepsilon},\boldsymbol{\Delta})$ are due to sensor heterogeneity, nonlinearities, temperature and environmental sensitivity, electromagnetic interference and spectrum, axes and transduction couplings, bias, drift, misalignment, sampling, quantization error, perturbations and other phenomena in sensing elements, microelectronic devices, ASICs, interconnect, etc. Sensor measurements and processed outputs $(\bar{\mathbf{y}}(t),\bar{\mathbf{y}}[n],\hat{\mathbf{y}}[n])$ are signals which are superposition of stochastic and deterministic processes, $(\boldsymbol{n}_{\bar{\mathbf{y}}(t)},\boldsymbol{n}_{\bar{\mathbf{y}}[n]},\boldsymbol{n}_{\hat{\mathbf{y}}[n]})$. For statistically independent and dependent random and deterministic $(\mathbf{n},\boldsymbol{\varepsilon},\boldsymbol{\Delta})$, investigate $(\bar{\mathbf{y}},\hat{\mathbf{y}})$ with pertained $(\boldsymbol{n}_{\bar{\mathbf{y}}(t)},\boldsymbol{n}_{\bar{\mathbf{y}}[n]},\boldsymbol{n}_{\hat{\mathbf{y}}[n]})$ which impair physical data quality, resulting in information losses.

For $(\bar{\mathbf{y}}(t),\bar{\mathbf{y}}[n],\hat{\mathbf{y}}[n])$, one performs probabilistic and statistical analyses. Nonlinear dependences and statistical models of high-dimensional random and deterministic processes $(\boldsymbol{n}_{\bar{\mathbf{y}}(t)},\boldsymbol{n}_{\bar{\mathbf{y}}[n]},\boldsymbol{n}_{\hat{\mathbf{y}}[n]})$ are very complex, and, the superposition principle is not applicable.

Microsystem-Technology Inertial Sensors in Guidance, Navigation and Control – Consider MDOF inertial measurement units (IMUs) and inertial navigation systems (INSs). These IMUs and INSs have being designed, fabricated and commercialized by Analog Devices, Bosch Sensortec, Epson Electronics, Fairchild Semiconductor, Honeywell, TDK InvenSense, Northrop Grumman, Panasonic, STMicroelectronics, Silicon Sensing, Systron Donner, etc. Three-degree-of-freedom accelerometers and gyroscopes, triaxial microelectronic magnetometer and barometric pressure sensor measure linear accelerations $\boldsymbol{a}$, angular rates $\boldsymbol{\omega}$, magnetic field $\boldsymbol{B}$ and pressure p. Physical quantities $\mathbf{y}(t)=[\boldsymbol{a},\boldsymbol{\omega},\boldsymbol{B},p]$, $\boldsymbol{a}=(a_x,a_y,a_z)$, $\boldsymbol{\omega}=(\omega_\phi,\omega_\theta,\omega_\psi)$, $\boldsymbol{B}=(B_x,B_y,B_z)$ are measured with noise, errors and distortions $(\mathbf{n},\boldsymbol{\varepsilon},\boldsymbol{\Delta})$. Analog to digital conversion $\bar{\mathbf{y}}(t)$ by ASICs yield $\bar{\mathbf{y}}[n]$ with $\boldsymbol{n}_{\bar{\mathbf{y}}[n]}$.

Physical quantities $\mathbf{y}=(\boldsymbol{a},\boldsymbol{\omega},\boldsymbol{B},p)$ are measured by electrostatic, optical, optoelectronic, motion and semiconductor sensing elements which yield sensed $\bar{\mathbf{y}}(t)=\left(\bar{\boldsymbol{a}}(t),\bar{\boldsymbol{\omega}}(t),\bar{\boldsymbol{B}}(t),\bar{p}(t)\right)$. These $\bar{\mathbf{y}}(t)$ may be characterized.

Many of IMUs and INSs output measured and processed data quadruples as

$$\bar{\mathbf{y}}[n]=\left(\bar{\boldsymbol{a}}[n],\bar{\boldsymbol{\omega}}[n],\bar{\boldsymbol{B}}[n],\bar{\boldsymbol{p}}[n]\right),\ \hat{\mathbf{y}}[n]=(\hat{\boldsymbol{a}}[n],\hat{\boldsymbol{\omega}}[n],\hat{\boldsymbol{B}}[n],\hat{\boldsymbol{p}}[n]).$$

These $(\bar{\mathbf{y}}[n],\hat{\mathbf{y}}[n])$ are asynchronously fused at different rates. The TDK InvenSense IMU MPU-9250/6500 and Bosch Sensortec INSs BNO055 and BNO080 are illustrated in Figures 5.33.a and b. These IMUs and INSs are installed on airframes, gimbals, mounts, robots, wearable electronics, etc. The NAVIO2 autopilot is shown in Figure 5.33.c. The objective is to accurately measure and estimate geolocation, position, orientation, motions and other quantities in pertained coordinate systems and body frame. Specifications are imposed on performance, capabilities and interoperability.

Processing algorithms should yield high-fidelity measurements and estimates on:

1. Multi-axis linear and angular accelerations $(\boldsymbol{a},\boldsymbol{\alpha})$, $\boldsymbol{a}=(a_x,a_y,a_z)$, $\boldsymbol{\alpha}=(\alpha_\theta,\alpha_\phi,\alpha_\psi)$;
2. Linear and angular velocities $(\boldsymbol{v},\boldsymbol{\omega})$, $\boldsymbol{v}=(v_x,v_y,v_z)$, $\boldsymbol{\omega}=(\omega_\theta,\omega_\phi,\omega_\psi)$;
3. Pitch θ, roll ϕ and yaw ψ Euler angles $\boldsymbol{\Theta}=(\theta,\phi,\psi)$;
4. Magnetic field $\boldsymbol{B}=(B_x,B_y,B_z)$;
5. Longitude and latitude $(\boldsymbol{\lambda},\boldsymbol{\varphi})$.

INSs BNO055 and 6 DOF IMU MPU-9250/6500 | Bosch Sensortec INSs BNO080 | NAVIO2 Autopilot

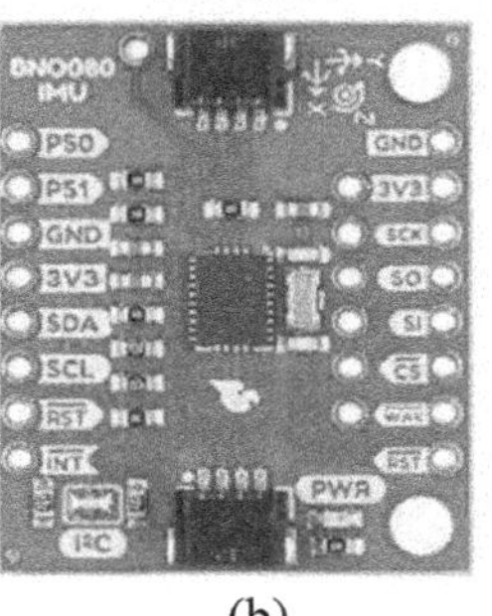

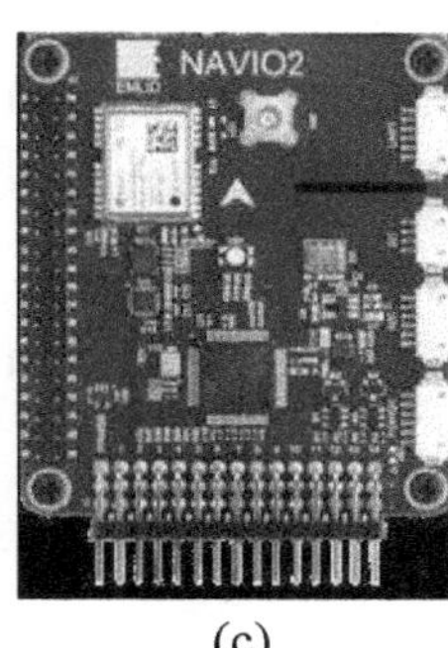

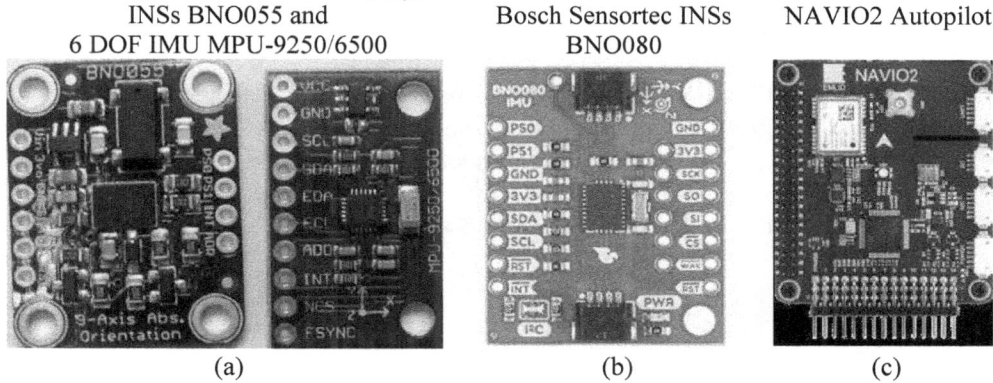

(a) (b) (c)

Figure 5.33. Integrated MDOF sensors measure linear accelerations, angular rates, magnetic field density and barometric pressure $\bar{\mathbf{y}}(t)=[\bar{\boldsymbol{a}}(t),\bar{\boldsymbol{\omega}}(t),\bar{\boldsymbol{B}}(t),\bar{\boldsymbol{p}}(t)]$. Measurements by triaxial accelerometer, gyroscope, magnetic field and barometer sensors are processed and fused.
(a) BNO055 and MPU-9250/6500: Six DOF IMU includes on-chip sensors, inter-integrated circuit (I^{2}C) interface ensuring 400 kHz communication with registers, SPI serial interface, and, 16-bit motion processor which performs signal conditioning and processing. The MPU-9250/6500 operates at 2.4 to 3.6 V, and, consumes ~10 mW. Measured $\bar{\mathbf{y}}(t)$ are digitized and processed. Measurements $\begin{bmatrix}\bar{\mathbf{y}}[n]\\ \hat{\mathbf{y}}[n]\end{bmatrix}=\begin{bmatrix}\bar{a}[n],\bar{\omega}[n]\\ \hat{a}[n],\hat{\omega}[n]\end{bmatrix},\ \begin{bmatrix}\bar{\boldsymbol{B}}[n]\\ \hat{\boldsymbol{B}}[n]\end{bmatrix}$ are fused at different data rates: 2000 Hz and 8000 Hz for $(\bar{\boldsymbol{a}}[n],\bar{\boldsymbol{\omega}}[n])$, up to 400 Hz for filtered $(\hat{\boldsymbol{a}}[n],\hat{\boldsymbol{\omega}}[n])$, and 8 Hz for $\hat{\boldsymbol{B}}[n]$;
(b) Absolute orientation INS, such as BNO055 and BNO080. The ARM Cortex MPUs implements filtering, estimation, data sharing algorithms and protocols. Relative and absolute orientation, Euler angles, directional cosines and motion estimates may be computed using orthogonal transformations, quaternion, rotation matrices and groups, tensor algebra and caculus, and curvilinear coordinates in three-dimensional coordinate systems and spaces;
(c) NAVIO2 autopilot with Raspberry PI processor (55×65 mm, 23 g shield and 54 g Raspberry Pi 2) has 1.2 GHz, 64-bit quad-core ARMv8 CPU, 1 GB RAM, 9 DOF MPU-9250, 9 DOF STMicroelectronics IMU LSM9DS1, MS5611 barometric pressure sensor, ultra-low power NEO-M8N Glonass/GPS/Beidou module, RC I/O co-processor, 14 PWM servo outputs, etc.

Rigid body dynamics, coordinate rotations and transformation matrices are documented in section 2.4. Coordinate systems and frames, used in aerial, ground and underwater vehicles, robots, motion platforms and navigation systems to support functionality, missions and applications are:

1. Geodetic coordinate system with the coordinate vector $\mathbf{r}_g = \begin{bmatrix} \text{Longitude} \\ \text{Latitude} \\ \text{Altitude} \end{bmatrix} = \begin{bmatrix} \boldsymbol{\lambda} \\ \boldsymbol{\varphi} \\ h \end{bmatrix}$;
2. Local north-east-down (NED) coordinate system, $\mathbf{r}_n = \begin{bmatrix} x_n \\ y_n \\ z_n \end{bmatrix}$, $\mathbf{v}_n = \begin{bmatrix} u_n \\ v_n \\ w_n \end{bmatrix}$, $\boldsymbol{a}_n = \begin{bmatrix} a_{x,n} \\ a_{y,n} \\ a_{z,n} \end{bmatrix}$;
3. Vehicle-carried NED system with $\mathbf{v}_{nv} = \begin{bmatrix} u_{nv} \\ v_{nv} \\ w_{nv} \end{bmatrix}$ and $\boldsymbol{a}_{nv} = \begin{bmatrix} a_{x,nv} \\ a_{y,nv} \\ a_{z,nv} \end{bmatrix}$;
4. Body frame system with $\mathbf{v}_B = \begin{bmatrix} u \\ v \\ w \end{bmatrix}$ and $\boldsymbol{a}_B = \begin{bmatrix} a_x \\ a_y \\ a_z \end{bmatrix}$.

For translational and rotational kinematics, using rotation matrices and transformation, one computes specified quantities. For example, one finds the north-east-down velocities (v_N,v_E,v_D), position, orientation, trajectory, heading and altitude h, etc.

Translational and rotational kinematics relationships, as well as the Euler angle dynamics, are applied

$$\dot{\mathbf{r}}_I(t) = \begin{bmatrix} \dot{x}_I(t) \\ \dot{y}_I(t) \\ \dot{z}_I(t) \end{bmatrix},\ \dot{\mathbf{r}}_I(t) = \mathbf{v}_I,\ \mathbf{v}_I = \mathcal{R}_B^I \mathbf{v}_B,\ \mathbf{v}_I = \begin{bmatrix} v_x \\ v_y \\ v_z \end{bmatrix},\ \mathbf{v}_B = \mathcal{R}_I^B \mathbf{v}_I,\ \mathbf{v}_B = \begin{bmatrix} u \\ v \\ w \end{bmatrix}, \quad (5.143)$$

$$\mathcal{R}_B^I = \begin{bmatrix} \cos\theta\cos\psi & \sin\phi\sin\theta\cos\psi - \cos\phi\sin\psi & \cos\phi\sin\theta\cos\psi + \sin\phi\sin\psi \\ \cos\theta\sin\psi & \sin\phi\sin\theta\sin\psi + \cos\phi\cos\psi & \cos\phi\sin\theta\sin\psi - \sin\phi\cos\psi \\ -\sin\theta & \sin\phi\cos\theta & \cos\phi\cos\theta \end{bmatrix},$$

$$\mathcal{R}_I^B = \begin{bmatrix} \cos\theta\cos\psi & \cos\theta\sin\psi & -\sin\theta \\ \sin\phi\sin\theta\cos\psi - \cos\phi\sin\psi & \sin\phi\sin\theta\sin\psi + \cos\phi\cos\psi & \sin\phi\cos\theta \\ \cos\phi\sin\theta\cos\psi + \sin\phi\sin\psi & \cos\phi\sin\theta\sin\psi - \sin\phi\cos\psi & \cos\phi\cos\theta \end{bmatrix},$$

$$\dot{\boldsymbol{\Theta}} = \begin{bmatrix} \dot{\phi} \\ \dot{\theta} \\ \dot{\psi} \end{bmatrix} = \begin{bmatrix} 1 & \sin\phi\tan\theta & \cos\phi\tan\theta \\ 0 & \cos\phi & -\sin\phi \\ 0 & \sin\phi\sec\theta & \cos\phi\sec\theta \end{bmatrix} \begin{bmatrix} p \\ q \\ r \end{bmatrix} = \mathcal{L}_B^I \boldsymbol{\omega}_B,\ \boldsymbol{\omega}_B = \begin{bmatrix} \omega_x \\ \omega_y \\ \omega_z \end{bmatrix}_B = \begin{bmatrix} p \\ q \\ r \end{bmatrix} = \begin{bmatrix} 1 & 0 & -\sin\theta \\ 0 & \cos\phi & \sin\phi\cos\theta \\ 0 & -\sin\phi & \cos\phi\cos\theta \end{bmatrix} \begin{bmatrix} \dot{\phi} \\ \dot{\theta} \\ \dot{\psi} \end{bmatrix} = \mathcal{L}_I^B \dot{\boldsymbol{\Theta}},$$

where u, v and w are the axial, lateral and normal velocities; $\mathcal{R}_B^I(\phi,\theta,\psi)$ is the rotation transformation matrix from body frame to inertial frame; $\mathcal{R}_I^B(\phi,\theta,\psi)$ is the rotation transformation matrix from inertial frame to body frame, $\mathcal{R}_I^B \mathcal{R}_B^I = I$; $\boldsymbol{\Theta}$ is the Euler angles vector; $\boldsymbol{\omega}_B$ is the body-axis angular rate vector, $\boldsymbol{\omega}_B = \begin{bmatrix} p \\ q \\ r \end{bmatrix}$; p, q and r are the body fixed roll, pitch and yaw rotation rates; $\mathcal{L}_B^I$ is the not orthonormal transformation matrix.

The velocity components along x-, y- and z-directions (u,v,w) are related to the forward velocity V, angle of attack α and sideslip angle β as

$$\begin{bmatrix} u \\ v \\ w \end{bmatrix} = \begin{bmatrix} V\cos\alpha\cos\beta \\ V\sin\beta \\ V\sin\alpha\cos\beta \end{bmatrix}, \quad \begin{bmatrix} V \\ \beta \\ \alpha \end{bmatrix} = \begin{bmatrix} \sqrt{u^2+v^2+w^2} \\ \sin^{-1}\frac{v}{V} \\ \tan^{-1}\frac{w}{u} \end{bmatrix}.$$

Consider measurements, typical physical data flow and fusion scheme. These tasks support processing algorithms on data analytics and control. For example, in aerial, automotive and robotic applications, under active electromagnetic spectrum, interferences, adversary activities, disruptive actions and jamming, we have

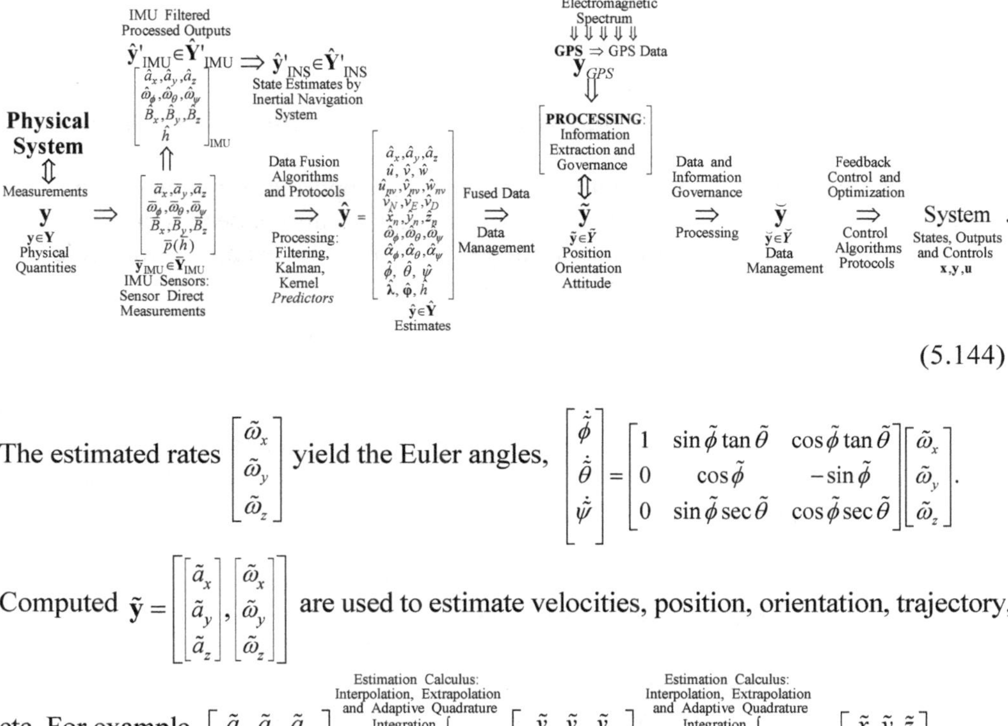

(5.144)

The estimated rates $\begin{bmatrix} \tilde{\omega}_x \\ \tilde{\omega}_y \\ \tilde{\omega}_z \end{bmatrix}$ yield the Euler angles, $\begin{bmatrix} \dot{\tilde{\phi}} \\ \dot{\tilde{\theta}} \\ \dot{\tilde{\psi}} \end{bmatrix} = \begin{bmatrix} 1 & \sin\tilde{\phi}\tan\tilde{\theta} & \cos\tilde{\phi}\tan\tilde{\theta} \\ 0 & \cos\tilde{\phi} & -\sin\tilde{\phi} \\ 0 & \sin\tilde{\phi}\sec\tilde{\theta} & \cos\tilde{\phi}\sec\tilde{\theta} \end{bmatrix} \begin{bmatrix} \tilde{\omega}_x \\ \tilde{\omega}_y \\ \tilde{\omega}_z \end{bmatrix}$.

Computed $\tilde{\mathbf{y}} = \left[\begin{bmatrix} \tilde{a}_x \\ \tilde{a}_y \\ \tilde{a}_z \end{bmatrix}, \begin{bmatrix} \tilde{\omega}_x \\ \tilde{\omega}_y \\ \tilde{\omega}_z \end{bmatrix} \right]$ are used to estimate velocities, position, orientation, trajectory,

etc. For example, $\underset{\text{Post-Processed IMU Outputs}}{\begin{bmatrix} \tilde{a}_x, \tilde{a}_y, \tilde{a}_z \\ \tilde{\omega}_\phi, \tilde{\omega}_\theta, \tilde{\omega}_\psi \end{bmatrix}} \overset{\text{Estimation Calculus: Interpolation, Extrapolation and Adaptive Quadrature Integration } \int}{\Longrightarrow} \underset{\text{Estimated Linear and Angular Velocities}}{\begin{bmatrix} \tilde{v}_x, \tilde{v}_y, \tilde{v}_z \\ \tilde{\omega}_\phi, \tilde{\omega}_\theta, \tilde{\omega}_\psi \end{bmatrix}} \overset{\text{Estimation Calculus: Interpolation, Extrapolation and Adaptive Quadrature Integration } \int}{\Longrightarrow} \underset{\text{Estimated Linear and Angular Positions}}{\begin{bmatrix} \tilde{x}, \tilde{y}, \tilde{z} \\ \tilde{\phi}, \tilde{\theta}, \tilde{\psi} \end{bmatrix}}$.

One designs integrated IMU-MPU/FPGA to ensure application-specific solutions featuring the component-to-module, module-to-system and system-on-module data sharing on measured $\bar{\mathbf{y}} \in \bar{\mathbf{Y}}$, yielding $\hat{\mathbf{y}} \in \hat{\mathbf{Y}}$ and $\tilde{\mathbf{y}} \in \tilde{\mathbf{Y}}$. Verification can be ensured by comparing data from redundant IMUs, INSs and GPS, as well as using *observers* and different data sources. As illustrated in data flow and fusion scheme (5.144), one may accomplish a *hybrid*, GPS-free and jamming-resistant inertial navigation, control and coordination. These tasks are important due to vulnerability of GPS to jamming, spoofing, denied environment, etc. To find $\tilde{\mathbf{y}} \in \tilde{\mathbf{Y}}$, the massively-parallelized kernel nonparametric regression and adaptive filtering of direct measurements $\bar{\mathbf{y}} \in \bar{\mathbf{Y}}$ are used. The matrix and Ricci calculi ensure robust solutions and near-real-time estimations. The IMU outputs $(\bar{\mathbf{y}},\hat{\mathbf{y}}')$ are superimposed with high-dimensional multi-source statistically-dependent $(\mathbf{n},\boldsymbol{\varepsilon},\boldsymbol{\Delta})$. Finding statistical models for $(\boldsymbol{n}_{\bar{\mathbf{y}}(t)}, \boldsymbol{n}_{\bar{\mathbf{y}}[n]}, \boldsymbol{n}_{\hat{\mathbf{y}}[n]})$, adaptive filter design and estimations are challenging problems. Consistent fusing algorithms and processing solutions are studied and implemented in a wide range of applications, platforms and operating environments.

Probabilistic analysis and statistical models are reported in the following sections. The probability distributions for the accelerometers and gyroscope channels, which output $(\hat{a}_x, \hat{a}_y, \hat{a}_z)$ and $(\hat{\omega}_\theta, \hat{\omega}_\phi, \hat{\omega}_\psi)$ with $(n_{\hat{a}_x}, n_{\hat{a}_y}, n_{\hat{a}_z})$ and $(n_{\hat{\omega}_\theta}, n_{\hat{\omega}_\phi}, n_{\hat{\omega}_\psi})$, are found in Example 5.47. Example 5.50 documents design of a digital filter $\mathcal{F}_{SC}(\mathcal{S},\mathcal{C})$, as well as in-flight studies for an aerial vehicle.

5.13. Descriptive Statistics: Statistical Models and Probabilistic Analysis

Descriptive, predictive and prescriptive data analytics support information sharing, data mining, control, failure and redundancy analyses, data quality assessment, situational awareness evaluation, etc. Probability theory is applied to examine stochastic and random processes, design filters, evaluate dependences, etc.

Definitions and Notations – Probability distributions $\mathcal{D}(\cdot)$ explicitly characterize continuous and discrete random variables and processes $X \equiv \{X(t)\}_{t\in T}$ or $X \equiv \{X[n]\}_{n\in \mathrm{N}}$. Probabilities $\mathbf{P}(X=x)$ of a random variable X taking a specific value $X=x$ are not directly measureable, and, one may calculates the probability for $x\in[a,b]$. For random continuous processes and variables X, probability distributions $\mathcal{D}(\cdot)$, under the hypotheses $\mathcal{h}\in\mathcal{H}$, are characterized by the probability density functions (pdf) $f_X(x)$ and cumulative distribution function (cdf) $F_X(x)$. The probability mass functions (pmf) $p_X(x)$ and cdf $F_X(x)$ are used to describe probability distributions $\mathcal{D}(\cdot)$ for a discrete random variable X. For example, if a random variable X represents the outcome of a single roll of a die, X takes values in $\{1, 2, 3, 4, 5, 6\}$.

Evaluating probabilities $\mathbf{P}(\cdot)$, for discrete X,

$$F_X(x) = \mathbf{P}(X \le x) = \sum_{x_i \le x} \mathbf{P}(X = x_i) = \sum_{x_i \le x} p_X(x_i),\ \mathbf{P}(X = b) = F_X(b) - \lim_{x\to b^-} F_X(x),\ p_X(x)=\mathbf{P}(X=x),$$

where X is a random variable, whereas x is a random number variable.

Notation $\mathbf{P}(X\le x)$ denotes the probability that the random variable X is less than or equal to the realization x, and, x is a fixed constant. The pmf and pdf satisfy

$p_X(x)\ge 0,\ f_X(x)\ge 0,\ \forall x\in\mathcal{S}.$

For a finite sample space, a statistical model for X is a pair $(\mathcal{S},\mathcal{P})$

$$\mathcal{R}_X=(\mathcal{S},\mathcal{P}),\ X\sim\mathcal{D}(\cdot), \tag{5.145}$$

where $\mathcal{S}$ is the set of observations; $\mathcal{P}$ is a set of probability distributions $\mathcal{D}(\cdot)$ on $\mathcal{S}$.

Parametric and Non-Parametric Statistical Models – Using the hypotheses $\mathcal{h}\in\mathcal{H}$, one identifies the probability distribution $X\sim\mathcal{D}(\cdot)$ within admissible probability distributions $\mathbf{D}$, $\mathcal{D}\in\mathbf{D}$, and, finds $\mathcal{R}_X=(\mathcal{S},\mathcal{D})$. For parameterized statistical models (5.145),

$\mathcal{D}=\{p_X(x;\phi):\phi\in\Phi\}$, $\mathcal{D}=\{f_X(x;\phi):\phi\in\Phi\}$ or $\mathcal{D}=\{F_X(x;\phi):\phi\in\Phi\}$, where the finite set $\Phi\in\mathbb{R}^d$ defines the model parameters.

For stationary random processes, assuming unique parameterization, $\{p_X(x;\phi_1)=p_X(x;\phi_2)|\phi_1=\phi_2\}$, $\{f_X(x;\phi_1)=f_X(x;\phi_2)|\phi_1=\phi_2\}$ and $\{F_X(x;\phi_1)=F_X(x;\phi_2)|\phi_1=\phi_2\}$.

The model is parametric if the descriptive functions $(p_X(x),F_X(x))$, $(f_X(x),F_X(x))$ and Φ exist, and, Φ has a finite dimension, $\Phi\subseteq\mathbb{R}^d$, where d is a positive integer, $d\in\mathbb{Z}^+$.

For non-paramedic statistical models

$\mathcal{D}=\{p_X(x;\phi):\phi\in\Phi\}$, $\mathcal{D}=\{f_X(x;\phi):\phi\in\Phi\}$ and $\mathcal{D}=\{F_X(x;\phi):\phi\in\Phi\}$,

the parameter set Φ is infinite.

A random process is an indexed collection of random variables defined on a probability space. For a continuous random variable, the probability of X belongs to an interval $A\in[a,b]$, $a<b$ is found using the Lebesgue-measurable pdf $f_X(x)$ as

$$\mathbf{P}(a\le X\le b)=\int_a^b f_X(x)dx,\ \mathbf{P}(X\in A)=\int_A f_X(x)dx,\ f_X(x)\ge 0,\ \forall x,\ \int_{-\infty}^{\infty} f_X(x)dx=1. \quad (5.146)$$

Example 5.37. *Gaussian Distribution*

If a random X is characterized by a single-variable Gaussian distribution $X\sim\mathcal{N}(\cdot)$,

$$\mathcal{D}=\{\ f_X(x;\mu,\sigma^2)=\frac{1}{\sqrt{2\pi}\sigma}e^{-\frac{(x-\mu)^2}{2\sigma^2}}:\mu\in\mathbb{R},\ \sigma>0\},$$

$$\int_{-\infty}^{\infty} f_X(x)dx=1,\ \mu=\int_{-\infty}^{\infty} xf_X(x)dx,\ \sigma^2=\int_{-\infty}^{\infty}(x-\mu)^2 f_X(x)dx.$$

The pdf and cdf are

$$f_X(x;\mu,\sigma^2)=\frac{1}{\sqrt{2\pi}\sigma}e^{-\frac{(x-\mu)^2}{2\sigma^2}},\ F_X(x;\mu,\sigma^2))=\tfrac{1}{2}\left[1+\operatorname{erf}\left(\tfrac{1}{\sqrt{2}\sigma}(x-\mu)\right)\right].$$

The parameters $\phi\in\Phi$ are the mean μ, standard deviation σ, $\sigma>0$, and, variance σ^2, $\sigma^2\in\mathbb{R}_{>0}$. The model dimension d is 2. ■

Example 5.38. *Statistically Independent Random Variables*

Consider a sum of N statistically independent random variables X_i, $Z=\sum_{i=1}^N X_i$.

The pdf $f_Z(z)$ of a sum $\sum_{i=1}^N X_i$ is the convolution of f_{X_i}

$$f_Z(z)=\left(f_{X_1}(x_1)*\ldots*f_{X_N}(x_N)\right)(z),$$

where $*$ is the convolution symbol.

Assuming a Gaussian distribution for X_i, $X_i\sim\mathcal{N}(\mu_i,\sigma_i^2)$, the pdfs are

$$f_{X_i}(x_i;\mu_i,\sigma_i^2)=\frac{1}{\sqrt{2\pi}\sigma_i}e^{-\frac{(x_i-\mu_i)^2}{2\sigma_i^2}},\ \mu_i\in\mathbb{R},\ \sigma_i>0.$$

For $Z=X_1+X_2$, find $Z\sim\mathcal{N}(\mu,\sigma^2)$. Consider statistically independent random X_1 and X_2 with

$$f(x_1)=\frac{1}{\sqrt{2\pi}\sigma_1}e^{-\frac{1}{2\sigma_1^2}(x_1-\mu_1)^2},\ f(x_2)=\frac{1}{\sqrt{2\pi}\sigma_2}e^{-\frac{1}{2\sigma_2^2}(x_2-\mu_2)^2},$$

We have $f_Z(z)=\left(f_{X_1}(x_1)*f_{X_2}(x_2)\right)(z)=\int_{-\infty}^{+\infty} f_{X_2}(z-x_1)f_{X_1}(x_1)dx_1=\int_{-\infty}^{+\infty} f_{X_1}(z-x_2)f_{X_2}(x_2)dx_2$.

Hence, $f_Z(z;\mu_z,\sigma_z^2)=\frac{1}{\sqrt{2\pi}\sqrt{\sigma_1^2+\sigma_2^2}}e^{-\frac{[z-(\mu_1+\mu_2)]^2}{2(\sigma_1^2+\sigma_2^2)}}$.

The mean value is the sum of individual means $\mu_i \in \mathbb{R}$, $\mu_z = \sum_{i=1}^{N} \mu_i$.

The variance is the sum of the individual variances $\sigma_z^2 = \sum_{i=1}^{N} \sigma_i^2$, $\forall \sigma_i > 0$.

Consider sensors' measurements. Assume that random noise and errors admit statistical model (5.145), characterized by probability distributions $\mathcal{D}_i \in \mathbf{D}$. Consider multi-sensor measurements using N sensors, letting that the random noise satisfies a normal distribution, $X_i \sim \mathcal{N}(\mu_i, \sigma_i^2)$.

For $X_i \sim \mathcal{N}(\mu_i, \sigma_i^2)$, averaging measurements, $\mu = \sum_{i=1}^{N} \mu_i$, $\mu \in \mathbb{R}$, $\sigma^2 = \sum_{i=1}^{N} \sigma_i^2$, $\sigma_i > 0$.

Minimization of the noise or error variance σ^2 cannot be ensured by aggregating and averaging measurements. The overall data homogeneity is ensured by: (1) High accuracy and high precision sensors with low noise, drift, misalignment error, etc.; (2) Electromagnetic shielding; (3) Low-noise, high-precision and high-resolution ASICs.■

Descriptive Functions – The cdf of a continuous random variable X is a real-valued continuous function $F_X(\cdot)$:$\mathbb{R} \to \mathbb{R}$, $F_X(x) = \mathbf{P}(X \le x)$, and, $F_X(x)$ is an integral of a pdf $f_X(x)$. That is, the right-continuous function $F_X(x)$ is an anti-derivative of $f_X(x)$.

The derivative of $F_X(x)$ yields $f_X(x)$, $f_X(x) = \frac{d}{dx} F_X(x)$.

For the Lebesgue-measurable $f_X(\cdot)$:$\mathbb{R} \to \mathbb{R}$ and $F_X(\cdot)$:$\mathbb{R} \to \mathbb{R}$ we have

$$F_X(x) = \mathbf{P}(X \le x) = \int_{-\infty}^{x} f_X(z)dz, \quad f_X(x) = \frac{d}{dx} F_X(x), \tag{5.147}$$

$$\lim_{x \to -\infty} F_X(x) = 0, \quad \lim_{x \to \infty} F_X(x) = 1, \quad \int_{-\infty}^{\infty} f_X(x)dx = 1.$$

The probability $\mathbf{P}(\cdot)$ and expectation $\mathbf{E}(\cdot)$ are

$$\mathbf{P}(a \le X \le b) = \int_{a}^{b} f_X(x)dx, \ a<b, \quad \mathbf{E}(X) = \int_{-\infty}^{\infty} x f_X(x)dx. \tag{5.148}$$

Histogram – A histogram $h_X(x)$ is a plot, surface or manifold the values of a variable or variables versus the count of observed numbers. For continuous variables, one defines an interval for variations of a variable value, defining a bin. A histogram maps an observed distribution $\mathcal{D}(\cdot)$. For a normalized histogram, such as $\sum_{x \in X} \bar{h}_X(x) = 1$, one computes the pdf $f_X(x)$ or pmf $p_X(x)$, for which $\int_{-\infty}^{\infty} f_X(x)dx = 1$ or $\sum_{x \in X} p(x) = 1$. With the number of observations N, for a given number of bins k, the histogram $h_X(x)$ is found, characterized, interpolated and parameterized. Number of bins k depends on N and variations of X. The Rice and Sturges relations are $k \cong \sqrt{N}$ ($k \cong 2\sqrt[3]{N}$), and, $k \cong \log_2 N + 1$. The Freedman-Diaconis equation, minimum cross-validation estimated error, and, other expressions are used to find k.

The cdf $F_X(x)$ defines the probability for a real-valued X. The cdf of a continuous X with a unimodal $f_X(x)$ is a non-decreasing right-continuous function such that $\lim_{x \to -\infty} F_X(x) \to 0$ and $\lim_{x \to \infty} F_X(x) \to 1$. These $F_X(x)$, $f_X(x)$ and $p_X(x)$ are obtained by accessing and characterizing processes and sequences. Using the tabulated frequencies within equally-spaced discrete intervals (*bins*), one finds a histogram $h_X(x)$, which yields the probability distribution under hypotheses $\mathcal{h}_i \in \mathcal{H}$. To find a statistical model:

- Examine a histogram $h_X(x)$, and, define a statistical model and probability distribution $\mathscr{R}_X=(\mathcal{S},\mathcal{P})$, $X\sim\mathcal{D}(\cdot)$, $\mathcal{D}\in\mathbf{D}$, $h\in\mathcal{H}$;
- Using the normalized histogram $\bar{h}_X(x)$, $\sum_{x\in X}\bar{h}_X(x)=1$, for $X\sim\mathcal{D}(\cdot)$, parameterize $(f_X(x),F_X(x))$ such as $\int_{-\infty}^{\infty}f_X(x)dx=1$, $\lim_{x\to-\infty}F_X(x)=0$, $\lim_{x\to\infty}F_X(x)=1$ solving the minimization problem

$$\min_{\substack{f_X\in(\mathcal{S},\mathcal{D})\\ \mathcal{D}=\{f_X(x;\phi):\phi\in\Phi\}}}\int_{-\infty}^{\infty}\left\|f_{X_{f_X\in(\mathcal{S},\mathcal{D})}}(x)-f_{X_{\bar{h}_X(x)}}(x)\right\|_l dx,\ \left\|f_{X_{f_X\in(\mathcal{S},\mathcal{D})}}(x)-f_{X_{\bar{h}_X(x)}}(x)\right\|_l\le\varepsilon_f,\ 0<\varepsilon_f\ll1,\quad(5.149)$$

$$\min_{\substack{F_X\in(\mathcal{S},\mathcal{D})\\ \mathcal{D}=\{F_X(x;\phi):\phi\in\Phi\}}}\int_{-\infty}^{\infty}\left\|F_{X_{F_X\in(\mathcal{S},\mathcal{D})}}(x)-F_{X_{\int_{-\infty}^{x}\bar{h}_X(y)dy}}(x)\right\|_l dx,\ \left\|F_{X_{F_X(x)\in(\mathcal{S},\mathcal{D})}}(x)-F_{X_{\int_{-\infty}^{x}\bar{h}_X(y)dy}}(x)\right\|_l\le\varepsilon_F,\ 0<\varepsilon_F\ll1;$$

- Investigate stationarity, test and validate hypotheses, and, validate models.

For a random variable X, one finds a continuous probability distribution $X\sim\mathcal{D}(\cdot)$. Interpolated and parameterized pdfs and cdfs are found as unimodal and multimodal $f_X(x)$ and $F_X(x)$ in $(\mathcal{S},\mathcal{D})$, $(f_X(x),F_X(x))\in(\mathcal{S},\mathcal{D})$. Analytic, parametric and nonparametric pdfs and cdfs are considered. For a pair $(F_X(x),f_X(x))$, one solves (5.149). The pdf $f_X(x)$ has an absolute maximum at x_0 such that $f_X(x_0)\ge f_X(x)$, $\forall x$.

Example 5.39. *Gaussian Distribution*
For a normal Gaussian distribution $X\sim\mathcal{N}(\mu,\sigma^2)$, the pdf and cdf are

$$f_X(x;\mu,\sigma^2)=\frac{1}{\sqrt{2\pi}\sigma}e^{-\frac{(x-\mu)^2}{2\sigma^2}},\quad F_X(x;\mu,\sigma^2))=\tfrac{1}{2}\left[1+\operatorname{erf}\left(\tfrac{1}{\sqrt{2}\sigma}(x-\mu)\right)\right].$$

These $F_X(x)$ and $f_X(x)$ satisfy (5.147) for any (μ,σ^2), $\mu\in\mathbb{R}$, $\sigma>0$, such that $\int_{-\infty}^{\infty}f_X(x)dx=1$, $\lim_{x\to-\infty}F_X(x)=0$, $\lim_{x\to\infty}F_X(x)=1$.

For $f_X(x)=\frac{1}{\sqrt{2\pi}\sigma}e^{-\frac{(x-\mu)^2}{2\sigma^2}}$, evaluate $\int_{-\infty}^{\infty}f_X(x)dx=\frac{1}{\sqrt{2\pi}\sigma}\int_{-\infty}^{\infty}e^{-\frac{(x-\mu)^2}{2\sigma^2}}dx$.

Let $u=\frac{x-\mu}{\sqrt{2}\sigma}$, which yields $x=\sqrt{2}\sigma u+\mu$. Hence, $dx=\sqrt{2}\sigma du$.

We have, $\int_{-\infty}^{\infty}f_X(x)dx=\frac{1}{\sqrt{2\pi}\sigma}\int_{-\infty}^{\infty}e^{-\frac{(x-\mu)^2}{2\sigma^2}}dx=\frac{1}{\sqrt{2\pi}\sigma}\int_{-\infty}^{\infty}e^{-u^2}\sqrt{2}\sigma du=\frac{1}{\sqrt{\pi}}\int_{-\infty}^{\infty}e^{-u^2}du=\frac{1}{\sqrt{\pi}}\sqrt{\pi}=1$.

The expectation is $\mathbf{E}[X]=\mu$. Using (5.148)

$$\mathbf{E}[X]=\int_{-\infty}^{\infty}xf_X(x)dx=\frac{1}{\sqrt{2\pi}\sigma}\int_{-\infty}^{\infty}xe^{-\frac{(x-\mu)^2}{2\sigma^2}}dx=\frac{1}{\sqrt{2\pi}\sigma}\int_{-\infty}^{\infty}(\sqrt{2}\sigma u+\mu)e^{-u^2}\sqrt{2}\sigma du$$
$$=\sqrt{\tfrac{2}{\pi}}\sigma\int_{-\infty}^{\infty}ue^{-u^2}du+\frac{\mu}{\sqrt{\pi}}\int_{-\infty}^{\infty}e^{-u^2}du=0+\frac{\mu}{\sqrt{\pi}}\sqrt{\pi}=\mu.$$

The variance is

$$\operatorname{var}(X)=\mathbf{E}\left[(X-\mathbf{E}[X])\cdot(X-\mathbf{E}[X])\right]=\mathbf{E}\left[(X-\mathbf{E}[X])^2\right]=\mathbf{E}\left[(X-\mu)^2\right]$$

$$=\frac{1}{\sqrt{2\pi}\sigma}\int_{-\infty}^{\infty}(x-\mu)^2e^{-\frac{(x-\mu)^2}{2\sigma^2}}dx=\frac{1}{\sqrt{2\pi}\sigma}\int_{-\infty}^{\infty}2\sigma^2u^2e^{-u^2}\sqrt{2}\sigma du=-\frac{\sigma^2}{\sqrt{\pi}}\int_{-\infty}^{\infty}u\frac{d}{du}(e^{-u^2})du=-0+\frac{\sigma^2}{\sqrt{\pi}}\sqrt{\pi}=\sigma^2.$$

Integration of the pdf within arbitrary limits is intractable.

For arbitrary a and b, one evaluates $\mathbf{P}(a\le X\le b)=\frac{1}{\sqrt{2\pi}\sigma}\int_a^b e^{-\frac{(x-\mu)^2}{2\sigma^2}}dx$. ■

Example 5.40. *Statistical Models of Noise, Errors and Failures*
Noise, interference, errors, failures and other random processes may be described by Gaussian, extreme value, Weibull and other distributions. Descriptive quantitative models $(\mathcal{S},\mathcal{D})$ are found and experimentally substantiated for many processes.

For normal distribution $X\sim\mathcal{N}(\mu,\sigma^2)$, the single variable pdf and cdf are

$$f_X(x;\mu,\sigma^2)=\frac{1}{\sqrt{2\pi}\sigma}e^{-\frac{(x-\mu)^2}{2\sigma^2}},\quad F_X(x;\mu,\sigma^2))=\tfrac{1}{2}\left[1+\operatorname{erf}\left(\tfrac{1}{\sqrt{2}\sigma}(x-\mu)\right)\right],\ \mu\in\mathbb{R},\ \sigma>0.$$

For extreme value distribution $\mathcal{EV}(\mu,\sigma)$ and generalized extreme value distribution $\mathcal{GEV}(\mu,\sigma,k)$,

$$f_X(x;\mu,\sigma)=\tfrac{1}{\sigma}e^{\frac{x-\mu}{\sigma}}e^{-e^{\frac{x-\mu}{\sigma}}},\quad f_X(x;\mu,\sigma,k)=\tfrac{1}{\sigma}e^{-\left(1+k\frac{x-\mu}{\sigma}\right)^{-\frac{1}{k}}}\left(1+\tfrac{k}{\sigma}(x-\mu)\right)^{-1-\frac{1}{k}},\mu\in\mathbb{R},\ \sigma>0,\ k\in\mathbb{R},$$

where μ, σ and k are the location, scale and shape parameters, $k\neq0$.

For continuous two- and three-parameter Weibull distributions $\mathcal{W}(\cdot)$,

$$f_X(x;a,b)=\begin{cases}\frac{b}{a}\left(\frac{x}{a}\right)^{b-1}e^{-\left(\frac{x}{a}\right)^b}, & x\geq0\\ 0, & x<0\end{cases},\quad f_X(x;a,b,c)=\begin{cases}\frac{b}{a}\left(\frac{x-c}{a}\right)^{b-1}e^{-\left(\frac{x-c}{a}\right)^b}, & x\geq c\\ 0, & x<c\end{cases},$$

where a, b and c are the scale, shape and location parameters, $(a,b)>0$, $(a,b,c)\in\mathbb{R}$.

For a two-parameter distribution $\mathcal{W}(a,b)$, $F_X(x)=\begin{cases}1-e^{-\left(\frac{x}{a}\right)^b}, & x\geq0\\ 0, & x<0\end{cases}$. ■

Example 5.41. *Normal and Extreme Value Distributions*
Consider pdfs for normal $\mathcal{N}(\mu,\sigma^2)$, extreme value $\mathcal{EV}(\mu,\sigma)$ and generalized extreme value $\mathcal{GEV}(\mu,\sigma,k)$ distributions

$$f_X(x;\mu,\sigma^2)=\frac{1}{\sqrt{2\pi}\sigma}e^{-\frac{(x-\mu)^2}{2\sigma^2}},\quad f_X(x;\mu,\sigma)=\tfrac{1}{\sigma}e^{\frac{x-\mu}{\sigma}}e^{-e^{\frac{x-\mu}{\sigma}}},\quad f_X(x;\mu,\sigma,k)=\tfrac{1}{\sigma}e^{-\left(1+k\frac{x-\mu}{\sigma}\right)^{-\frac{1}{k}}}\left(1+\tfrac{k}{\sigma}(x-\mu)\right)^{-1-\frac{1}{k}}.$$

Let μ=1, σ=0.5 and k=0.1. Plots for $f_X(x)$ are reported in Figure 5.34.a.

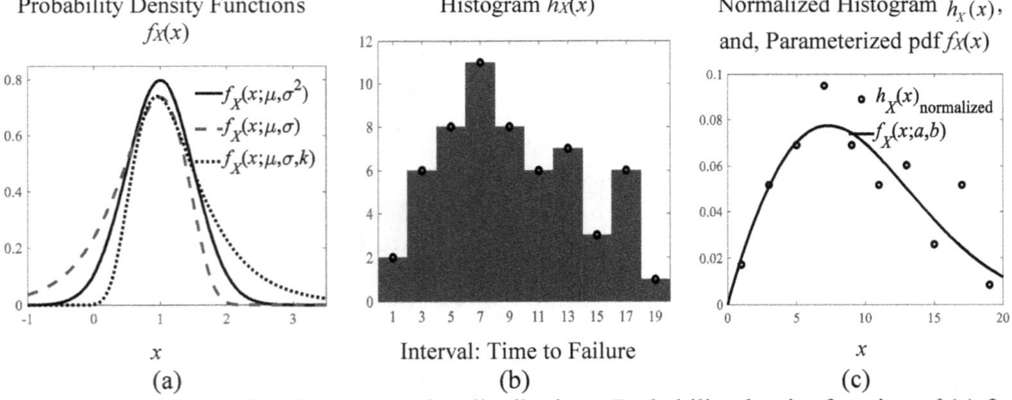

Figure 5.34. (a) Normal and extreme value distributions: Probability density functions $f_X(x)$ for $\mathcal{N}(\mu,\sigma^2)$, $\mathcal{EV}(\mu,\sigma)$ and $\mathcal{GEV}(\mu,\sigma,k)$, μ=1, σ=0.5, k=0.1;
(b) Histogram $h_X(x)$ for time-to-failure,which yields the Weibull distribution $\mathcal{W}(\cdot)$,Example 5.42;
(c) Normalized histogram $\bar{h}_X(x)$ (circles), and, parameterized pdf $f_X(x;a,b)=\frac{b}{a}\left(\frac{x}{a}\right)^{b-1}e^{-\left(\frac{x}{a}\right)^b}$, a=10.763, b=1.911. ■

Example 5.42. *Failure Analysis: Weibull Distribution*

Attenuation, fading, reliability and failure analyses are important in electronic, mechanical, power and other devices and systems. Maintenance and reliability analyses are performed using failures models to assess maintenance, services, replacement, safety, etc. One evaluates time to failure, as well as mean time between failures of devices, components, modules and systems. For identical components, the operating time to failure is measured. A failure model $(\mathcal{S},\mathcal{D})$ is found using the time-to-failure data set. Consider

time_to_failure=[0.9 1.2 2.3 2.5 2.9 3 3.4 3.9 4.4 4.7 5.1 5.2 5.4 5.7 5.7 5.9 6.2 6.4 6.5 6.5 6.7 7.1 7.3 7.4 7.9 7.9 8 8.2 8.3 8.7 8.8 8.8 8.9 9.3 9.8 10.4 11 11 11.4 11.5 11.9 12.1 12.2 12.4 12.7 13 13.2 13.5 14.1 14.8 15.9 16.1 16.2 16.7 16.9 17.2 18 18.7].

A histogram $h_X(x)$ is depicted in Figure 5.34.b, yielding the Weibull distribution, examined in Example 5.40. The pdf $f_X(x;a,b)=\frac{b}{a}\left(\frac{x}{a}\right)^{b-1}e^{-\left(\frac{x}{a}\right)^b}$, $x\geq 0$ is parametrized such that $\int_{-\infty}^{\infty} f_X(x)dx=1$, $(a,b)>0$. MATLAB is used to interpolate and parameterize $f_X(x)$, as well as to validate findings. Figure 5.34.c documents the normalized histogram $\bar{h}_X(x)$ and $f_X(x)$ with computed $a=10.763$ and $b=1.911$. ■

One considers various pertained distributions. For example:

1. The lognormal distrubution with
$$f_X(x;\mu,\sigma^2)=\frac{1}{\sqrt{2\pi}\sigma x}e^{-\frac{(\ln x-\mu)^2}{2\sigma^2}},\ F_X(x)=\frac{1}{2}+\frac{1}{2}\mathrm{erf}\left(\frac{\ln x-\mu}{\sqrt{2}\sigma}\right),\ \mu\in\mathbb{R},\ \sigma>0;$$
2. Rayleigh distribution, $f_X(x;\sigma)=\frac{1}{\sigma^2}xe^{-\frac{x^2}{2\sigma^2}}$, $F_X(x)=1-e^{-\frac{x^2}{2\sigma^2}}$, $\sigma>0$;
3. Inverse Gaussian distribution, $f_X(x;\mu,\lambda)=\sqrt{\frac{\lambda}{2\pi x^3}}e^{-\frac{\lambda(x-\mu)^2}{2\mu^2 x}}$, $(\mu,\lambda)>0$, $x\geq 0$.

Multimodal Distributions and Machine Learning Algorithms – Multimodal univariate distribution $X\sim\mathcal{D}(\cdot)$ can be found and parameterized as the mixture

$$f_X(x)=\sum_{i=1}^{N}a_i\frac{1}{\sqrt{2\pi}\sigma_i}e^{-\frac{(x-\mu_i)^2}{2\sigma_i^2}},\ f_X(x)=\sum_{i=1}^{N}a_i\frac{1}{\sigma_i}e^{\frac{x-\mu_i}{\sigma_i}}e^{-e^{\frac{x-\mu_i}{\sigma_i}}},\ \int_{-\infty}^{\infty}f_X(x)dx=1,\ \sum_i a_i=1,\ a_i\in\mathbb{R}.$$

The Gaussian mixture is used in clustering, *machine learning* and other algorithms. For example, a multimodal univariate distribution $X\sim\mathcal{D}(\cdot)$ may be characterized as a combination of unimodal functions

$$f_X(x)=\sum_{i=1}^{N}b_i\frac{1}{\sqrt{2\pi}\sigma_i}e^{-\frac{(x-\mu_i)^2}{2\sigma_i^2}}+\sum_{j=1}^{M}c_j\frac{1}{\sigma_j}e^{\frac{x-\mu_j}{\sigma_j}}e^{-e^{\frac{x-\mu_j}{\sigma_j}}},\ \sum_{i,j}(b_i+c_j)=1,\ \int_{-\infty}^{\infty}f_X(x)dx=1,\ (b_i,c_j)\in\mathbb{R}.$$

Conventional distributions and their *mixtures* may guarantee adequateness. Alternatively, distributions with multimodal pdfs $f_X(x)$, $f_X:\mathbb{R}\to\mathbb{R}$, $f_X(x)\geq 0$, $\int_{-\infty}^{\infty}f_X(x)dx=1$, $\forall x\in\mathbb{R}$ for continuous random X are documented in Table 5.2. High degree and fractional polynomials $P(x)$ are used to parameterize the exhibited $h_X(x)$, finding exclusive $f_X(x)$ and asserting $X\sim\mathcal{D}(\cdot)$, $\mathcal{D}\in\mathbb{D}$. The pdfs and pmfs are parameterized guaranteeing (5.147), solving and validating (5.149).

Table 5.2. Conventional and Multimodal Distributions: Continuous Random Variables

Probability Distributions $X\sim\mathcal{D}(\cdot)$	Conventional pdfs $f_X(x)$, $f_X(x)\geq 0$, $f_X:\mathbb{R}\to\mathbb{R}$, $\int_{-\infty}^{\infty} f_X(x)dx=1$	Multimodal pdfs $f_X(x)$, $f_X:\mathbb{R}\to\mathbb{R}$, $f_X(x)\geq 0$, $\exists f_X(x_0)_{\max}\geq f_X(x)$, $\int_{-\infty}^{\infty} f_X(x)dx=1$, $\forall x\in\mathbb{R}$
Normal $\mathcal{N}(\cdot)$ and Multimodal ${}^{\mathcal{M}}\mathcal{N}(\cdot)$	$f_X(x;\mu,\sigma^2)=\frac{1}{\sqrt{2\pi}\sigma}e^{-\frac{(x-\mu)^2}{2\sigma^2}}$, $x\in[-\infty,\infty]$, $\mu\in\mathbb{R}$, $\sigma>0$	Symmetric $f_X(x;\mu_n,\sigma^2,a_n,b)=\frac{1}{\sqrt{2\pi}b\sigma}e^{-\frac{1}{2\sigma^2}\sum_n a_n\lvert x-\mu_n\rvert^n}$, Asymmetric $f_X(x;\mu_n,\sigma^2,a_n,b)=\frac{1}{\sqrt{2\pi}b\sigma}e^{-\frac{1}{2\sigma^2}\sum_n a_n(x-\mu_n)^n}$, $f_X(x;\mu_n,\sigma^2,a_n,b)=\frac{1}{\sqrt{2\pi}b\sigma}e^{-\frac{1}{2\sigma^2}P(a_n,\mu_n,x)}$, $\mu_n\in\mathbb{R}$, $\sigma>0$, $a_n\in\mathbb{R}$, $b>0$, $\forall n$. Gaussian mixture $f_X(x)=\sum_{i=1}^{N}a_i\frac{1}{\sqrt{2\pi}\sigma_i}e^{-\frac{(x-\mu_i)^2}{2\sigma_i^2}}$, $\Sigma_i a_i=1$, $\mu_i\in\mathbb{R}$, $\sigma_i>0$, $a_i\in\mathbb{R}$
Extreme Value $\mathcal{EV}(\cdot)$ and Multimodal ${}^{\mathcal{M}}\mathcal{EV}(\cdot)$	$f_X(x;\mu,\sigma)=\frac{1}{\sigma}e^{\frac{x-\mu}{\sigma}}e^{-e^{-\frac{x-\mu}{\sigma}}}$, $\mu\in\mathbb{R}$, $\sigma>0$	Symmetric $f_X(x;\mu_n,\sigma,a_n,b_n,b)=\frac{1}{b\sigma}e^{\frac{1}{\sigma}\sum_n a_n\lvert x-\mu_n\rvert^n}e^{-e^{-\frac{1}{\sigma}\sum_n b_n\lvert x-\mu_n\rvert^n}}$, Asymmetric $f_X(x;\mu_n,\sigma,a_n,b_n,b)=\frac{1}{b\sigma}e^{\frac{1}{\sigma}\sum_n a_n(x-\mu_n)^n}e^{-e^{-\frac{1}{\sigma}\sum_n b_n(x-\mu_n)^n}}$, $f_X(x;\mu_n,\sigma,a_n,b_n,b)=\frac{1}{b\sigma}e^{\frac{1}{\sigma}P_1(a_n,\mu_n,x)}e^{-e^{-\frac{1}{\sigma}P_2(a_n,\mu_n,x)}}$, $\mu_n\in\mathbb{R}$, $\sigma>0$, $a_n\in\mathbb{R}$, $b_n\in\mathbb{R}$, $b>0$, $\forall n$. Mixture $f_X(x)=\sum_{i=1}^{N}a_i\frac{1}{\sigma_i}e^{\frac{x-\mu_i}{\sigma_i}}e^{-e^{-\frac{x-\mu_i}{\sigma_i}}}$, $\Sigma_i a_i=1$, $a_i\in\mathbb{R}$
Lognormal $\ln\mathcal{N}(\cdot)$ and Multimodal ${}^{\mathcal{M}}\ln\mathcal{GN}(\cdot)$	$f_X(x;\mu,\sigma^2)=\frac{1}{\sqrt{2\pi}\sigma x}e^{-\frac{(\ln x-\mu)^2}{2\sigma^2}}$, $x>0$, $\mu\in\mathbb{R}$, $\sigma>0$	$f_X(x;\mu_n,\sigma^2,a_n,b)=\frac{1}{\sqrt{2\pi}b\sigma x}e^{-\frac{1}{2\sigma^2}\sum_n a_n\lvert\ln x-\mu_n\rvert^n}$, $f_X(x;\mu_n,\sigma^2,a_n,b)=\frac{1}{\sqrt{2\pi}b\sigma x}e^{-\frac{1}{2\sigma^2}\sum_n a_n(\ln x-\mu_n)^n}$, $x>0$, $\mu_n\in\mathbb{R}$, $\sigma>0$, $a_n\in\mathbb{R}$, $b>0$, $\forall n$
Rayleigh and Multimodal Rayleigh	$f_X(x;\sigma)=\frac{1}{\sigma^2}xe^{-\frac{x^2}{2\sigma^2}}$, $x\geq 0$, $\sigma>0$	$f_X(x;\sigma,a_n,b)=\frac{1}{b\sigma^2}xe^{-\frac{1}{2\sigma^2}\sum_n a_n x^{2n}}$, $x\geq 0,\sigma>0,a_n\in\mathbb{R},b>0,\forall n$
Inverse Gaussian and Multimodal Inverse Gaussian	$f_X(x;\mu,\lambda)=\sqrt{\frac{\lambda}{2\pi x^3}}e^{-\frac{\lambda(x-\mu)^2}{2\mu^2 x}}$, $x>0$, $\mu>0$, $\lambda>0$	$f_X(x;\mu_n,\lambda,a_n)=\sqrt{\frac{\lambda}{2\pi x^3}}e^{-\frac{1}{2x}\sum_n\frac{1}{\mu_n^2}a_n\lvert x-\mu_n\rvert^n}$, $f_X(x;\mu_n,\lambda,a_n)=\sqrt{\frac{\lambda}{2\pi x^3}}e^{-\frac{1}{2x}\sum_n\frac{1}{\mu_n^2}a_n(x-\mu_n)^n}$, $x>0$, $\mu_n>0$, $\lambda>0$, $a_n\in\mathbb{R}$, $\forall n$

Example 5.43. *Maxwell-Boltzmann Unimodal and Multimodal Distributions*
For the Maxwell-Boltzmann distribution, the pdf is

$$f_X(x;a)=\sqrt{\frac{2}{\pi}}\frac{1}{a^3}x^2e^{-\frac{1}{2a^2}x^2},\ x\in(0,\infty),\ a>0,\ \mu=\sqrt{\frac{8}{\pi}}a,\ \sigma^2=\frac{3\pi-8}{\pi}a^2.$$

For a multimodal Maxwell-Boltzmann distribution,

$$f_X(x;a_m,a_n)=\sum_m a_m x^m\circ e^{-\sum_n a_n\lvert x\rvert^n},\ \int_{-\infty}^{\infty}f_X(x)dx=1,\ x\in(0,\infty),\ a_m\in\mathbb{R},\ a_n\in\mathbb{R},\ \forall(m,n),$$

$$f_X(x;a_m,a_n)=\sum_m a_m x^m\circ e^{-\sum_n a_n x^n},\ f_X(x;a_m,a_n)=P_1(a_m,x)\circ e^{-P_2(a_n,x)}.$$ ■

Example 5.44. *Uniqueness of Solutions*
Analytic solutions are found for multimodal distributions $X\sim\mathcal{D}(\cdot)$.
For a normal distribution $\mathcal{N}(\mu,\sigma^2)$ with domain $-\infty\le x\le\infty$, the single variable pdf is

$$f_X(x;\mu,\sigma^2)=\frac{1}{\sqrt{2\pi}\sigma}e^{-\frac{(x-\mu)^2}{2\sigma^2}},\ \mu\in\mathbb{R},\ \sigma>0.$$

For $\phi=e^{-x^2}$, one has $\frac{\partial}{\partial x}e^{-x^2}=-2xe^{-x^2}$, $\int e^{-x^2}dx=\frac{1}{2}\sqrt{\pi}\operatorname{erf}(x)$.

Considering $\phi=e^{-x-x^2}$, $\frac{\partial}{\partial x}e^{-x-x^2}=-(1+2x)e^{-x-x^2}$, obtain $\int e^{-x-x^2}dx=\frac{1}{2}\sqrt[4]{e}\sqrt{\pi}\operatorname{erf}(\frac{1}{2}+x)$.

For $\phi=e^{-\sum_{n=1}^{N}x^n}=e^{-x-x^2-x^3-x^4-\ldots}$, the global maximum, derivatives, indefinite and definite integrals exist. These pdfs are parameterized using (5.149).

Unique solutions for $(f_X(x), F_X(x))$ exist. ■

Example 5.45. *Multivariate Normal Distribution and* ***Multimodal Distribution***
Consider $\mathbf{X}=(X_1,\ldots,X_m)$ with independent X_i, $\mathbf{X}\sim\mathcal{N}(\boldsymbol{\mu},\Sigma)$, $\boldsymbol{\mu}\in\mathbb{R}^m$, $\Sigma\in\mathbb{R}^{m\times m}$.
A multivariate normal distribution of m-dimensional $\mathbf{X}$ is characterized by a pdf

$$f_{\mathbf{X}}(x_1,\ldots,x_m;\boldsymbol{\mu},\Sigma)=\left(\tfrac{1}{2\pi}\right)^{m/2}|\Sigma|^{-1/2}e^{-\frac{1}{2}[\mathbf{x}-\boldsymbol{\mu}]^T\Sigma^{-1}[\mathbf{x}-\boldsymbol{\mu}]},\ \left(\tfrac{1}{2\pi}\right)^{m/2}|\Sigma|^{-1/2}>0,$$

$$\boldsymbol{\mu}=\mathbf{E}[\mathbf{X}],\ \Sigma=\operatorname{cov}(\mathbf{X})=\mathbf{E}[(\mathbf{X}-\mathbf{E}[\mathbf{X}])(\mathbf{X}-\mathbf{E}[\mathbf{X}])^T],\ |\Sigma|>0,$$

where $\boldsymbol{\mu}$ is the mean vector, $\boldsymbol{\mu}\in\mathbb{R}^m$; Σ is the covariance matrix, $\Sigma\in\mathbb{R}^{m\times m}$, $|\Sigma|>0$.

There exists a matrix A, such that $\Sigma=A^TA$, $A\in\mathbb{R}^{m\times m}$.

For the m-variate multimodal distribution $\mathbf{X}\sim\mathcal{D}(\cdot)$, the pdf with $\int_{-\infty}^{\infty}f_{\mathbf{X}}(\mathbf{x})d\mathbf{x}=1$, is

$$f_{\mathbf{X}}(x_1,\ldots,x_m;\boldsymbol{\mu},b,A_n)=\frac{1}{\sqrt{(2\pi)^m b}}e^{-\sum_n[\mathbf{x}-\boldsymbol{\mu}]^{n/2\,T}A_n^{-1}A_n[\mathbf{x}-\boldsymbol{\mu}]^{n/2}},\ \boldsymbol{\mu}\in\mathbb{R},\ A_n\in\mathbb{R}^{m\times m},\ b\in\mathbb{R},\ b>0,\ |A_n|\neq0.$$

Consider a bivariate process $\mathbf{X}=(X_1,X_2)$, $\mathbf{X}\sim\mathcal{N}(\boldsymbol{\mu},\Sigma)$, $\boldsymbol{\mu}=\mathbf{E}[\mathbf{X}]=(\mathbf{E}[X_1],\mathbf{E}[X_2])$.

The joint pdf with $\int_{-\infty}^{\infty}\int_{-\infty}^{\infty}f_{X_1,X_2}(x_1,x_2)dx_1dx_2=1$, $(\mu_1,\mu_2)\in\mathbb{R}$, $(\sigma_{11},\sigma_{22})>0$, $\rho\in\mathbb{R}$, is

$$f_{X_1,X_2}(x_1,x_2;\boldsymbol{\mu},\Sigma)=\frac{1}{2\pi}|\Sigma|^{-1/2}e^{-\frac{1}{2}[\mathbf{x}-\boldsymbol{\mu}]^T\begin{bmatrix}\sigma_{11}^2 & \rho\sigma_{11}\sigma_{22}\\ \rho\sigma_{11}\sigma_{22} & \sigma_{22}^2\end{bmatrix}^{-1}[\mathbf{x}-\boldsymbol{\mu}]},\ \Sigma=\begin{bmatrix}\sigma_{11}^2 & \rho\sigma_{11}\sigma_{22}\\ \rho\sigma_{11}\sigma_{22} & \sigma_{22}^2\end{bmatrix},$$

$$f_{X_1,X_2}(x_1,x_2;\boldsymbol{\mu},\Sigma)=\frac{1}{2\pi\sigma_{11}\sigma_{22}\sqrt{1-\rho^2}}e^{-\frac{1}{2(1-\rho^2)}\left(\frac{(x_1-\mu_1)^2}{\sigma_{11}^2}-2\rho\frac{(x_1-\mu_1)(x_2-\mu_2)}{\sigma_{11}\sigma_{22}}+\frac{(x_2-\mu_2)^2}{\sigma_{22}^2}\right)},$$

where Σ is a nonsingular covariance matrix; ρ is the correlation between X_1 and X_2.
For a bivariate normal distribution $(X_1,X_2)\sim\mathcal{N}(\boldsymbol{\mu},\Sigma)$, let $\mu_1=0$, $\mu_2=0$, $\sigma_{11}=0.25$, $\sigma_{22}=0.5$ and $\rho=0.25$. The surface for a joint pdf is depicted in Figure 5.35.a. ■

Example 5.46. *Bimodal Distributions*
Consider a normal distribution $X\sim\mathcal{N}(\mu,\sigma^2)$ with the pdf

$$f_X(x;\mu,\sigma^2)=\frac{1}{\sqrt{2\pi}\sigma}e^{-\frac{(x-\mu)^2}{2\sigma^2}},\ \mu=0,\ \sigma=1.$$

A plot for $f_X(x)$ is depicted in Figure 5.35.b by solid line.

Let the resulting histograms $h_X(x)$ and $\bar{h}_X(x)$ yield a bimodal distribution, for which a statistical model $\mathcal{R}_X=(\mathcal{S},\mathcal{P})$, $X\sim\mathcal{D}(\cdot)$ should be found.

With $\sum_{x\in X}\bar{h}_X(x)=1$, pdfs $f_X(x;\phi)$ are interpolated and parameterized by $\{f_X(x;\phi):\phi\in\Phi\}$ with model parameters ϕ. Figure 5.35.b illustrates plots for $f_X(x)$ for two hypothetical $\bar{h}_X(x)$

$$f_X(x)=\frac{1}{\sqrt{2\pi}b\sigma}e^{-\frac{a_0+a_1x+a_2x^2+a_3x^3+a_4x^4}{2\sigma^2}}, \text{ such that } \int_{-\infty}^{\infty}f_X(x)dx=1.$$

Parameterize $f_X(x)$, finding for two $\bar{h}_X(x)$:

1. $\sigma=1$, $b=1.094$, $a_0=0$, $a_1=3$, $a_2=1$, $a_3=-3$, $a_4=1$;
2. $\sigma=1$, $b=1.82$, $a_0=0$, $a_1=1$, $a_2=1$, $a_3=-3$, $a_4=1$.

For a unimodal extreme value distribution $\mathcal{EV}(\mu,\sigma)$, $f_X(x)=\frac{1}{\sigma}e^{-\frac{x-\mu}{\sigma}}e^{-e^{-\frac{x-\mu}{\sigma}}}$, $\mu=0$, $\sigma=1$.

For two histograms $h_X(x)$ and $\bar{h}_X(x)$, consider a multimodal distribution with

$$f_X(x;\mu,\sigma,a_n,b_n,b)=\frac{1}{b\sigma}e^{\frac{1}{\sigma}\sum_n a_n(x-\mu)^n}e^{-e^{-\frac{1}{\sigma}\sum_n b_n(x-\mu)^n}}, \int_{-\infty}^{\infty}f_X(x)dx=1.$$

For the truncated bimodal pdf $f_X(x)=\frac{1}{b\sigma}e^{-\frac{a_0+a_1x+a_2x^2+a_3x^3+a_4x^4}{\sigma}}e^{-e^{\frac{a_0+a_1x+a_2x^2+a_3x^3+a_4x^4}{\sigma}}}$, the pdf is parameterized, and, parameters are found using (5.149) as:

1. $\mu=0$, $\sigma=1$, $b=0.883$, $a_0=0$, $a_1=1$, $a_2=-0.1$, $a_3=0.1$, $a_4=-0.1$;
2. $\mu=0$, $\sigma=1$, $b=0.929$, $a_0=0$, $a_1=1$, $a_2=-0.1$, $a_3=0.1$, $a_4=0.1$.

Figure 5.35.c documents the resulting plots.

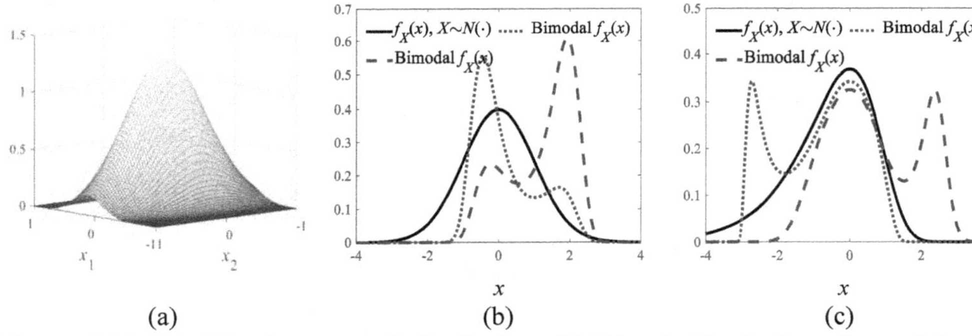

Figure 5.35. (a) Bivariate normal distribution $(X_1,X_2)\sim\mathcal{N}(\mu,\Sigma)$: Surface for a joint pdf $f_{X_1,X_2}(x_1,x_2)=\frac{1}{2\pi\sigma_{11}\sigma_{22}\sqrt{1-\rho^2}}e^{-\frac{1}{2(1-\rho^2)}\left(\frac{(x_1-\mu_1)^2}{\sigma_{11}^2}-2\rho\frac{(x_1-\mu_1)(x_2-\mu_2)}{\sigma_{11}\sigma_{22}}+\frac{(x_2-\mu_2)^2}{\sigma_{22}^2}\right)}$, $\mu_1=0$, $\mu_2=0$, $\sigma_{11}=0.25$, $\sigma_{22}=0.5$, $\rho=0.25$;

(b) Normal distribution $X\sim\mathcal{N}(\mu,\sigma^2)$, $f_X(x;\mu,\sigma^2)=\frac{1}{\sqrt{2\pi}\sigma}e^{-\frac{(x-\mu)^2}{2\sigma^2}}$, $\mu=0$, $\sigma=1$. Bimodal distribution, $f_X(x)=\frac{1}{\sqrt{2\pi}b\sigma}e^{-\frac{a_0+a_1x+a_2x^2+a_3x^3+a_4x^4}{2\sigma^2}}$, $\int_{-\infty}^{\infty}f_X(x)dx=1$, $\sigma=1$, $b=1.094$, $a_0=0$, $a_1=3$, $a_2=1$, $a_3=-3$, $a_4=1$ (dotted line), and, $\sigma=1$, $b=1.82$, $a_0=0$, $a_1=1$, $a_2=1$, $a_3=-3$, $a_4=1$ (dashed line);

(c) Extreme value unimodal distribution $\mathcal{EV}(\mu,\sigma)$, $f_X(x;\mu,\sigma)=\frac{1}{\sigma}e^{-\frac{x-\mu}{\sigma}}e^{-e^{-\frac{x-\mu}{\sigma}}}$, $\mu=0$, $\sigma=1$. Representative bimodal distributions, $f_X(x)=\frac{1}{b\sigma}e^{-\frac{a_0+a_1x+a_2x^2+a_3x^3+a_4x^4}{\sigma}}e^{-e^{\frac{a_0+a_1x+a_2x^2+a_3x^3+a_4x^4}{\sigma}}}$. ■

Example 5.47. *Statistical Model of Noise in MDOF Inertial Sensors*
Study a multi-axis accelerometer and gyroscope. At rest, the measured noise in the accelerometer and gyroscope channels $(n_{\bar{a}_x}, n_{\bar{a}_y}, n_{\bar{a}_z})$ and $(n_{\bar{\omega}_\theta}, n_{\bar{\omega}_\phi}, n_{\bar{\omega}_\psi})$ are depicted in Figures 5.36.a and b.

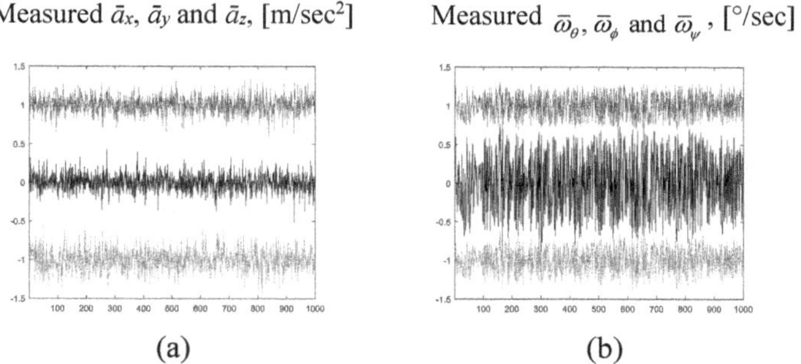

Figure 5.36. (a) Measured noise $(n_{\bar{a}_x}, n_{\bar{a}_y}, n_{\bar{a}_z})$ in the x, y and z linear acceleration axes $(\bar{a}_x, \bar{a}_y, \bar{a}_z)$ at rest, depicted by top, center and bottom plots with off-sets 1, 0 and −1;
(b) Noise $(n_{\bar{\omega}_\theta}, n_{\bar{\omega}_\phi}, n_{\bar{\omega}_\psi})$ in the pitch, roll and yaw angular rates channels $(\bar{\omega}_\theta, \bar{\omega}_\phi, \bar{\omega}_\psi)$ at rest.

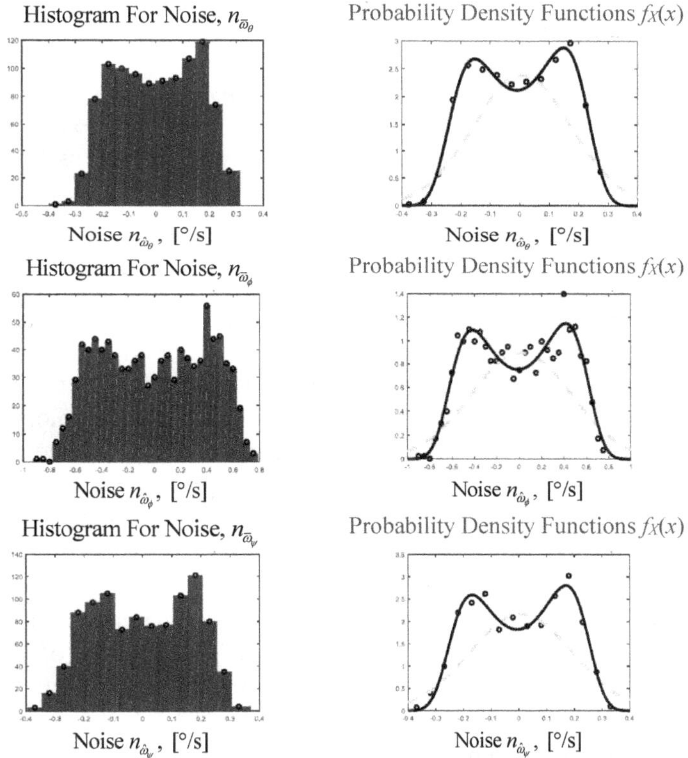

Figure 5.37. Histograms for noise in the gyroscope channels, and, parameterized pdfs for normal distribution $\mathcal{N}(\mu,\sigma^2)$ and bimodal distribution, depicted by the dashed and solid lines. For $(n_{\bar{\omega}_\theta}, n_{\bar{\omega}_\phi}, n_{\bar{\omega}_\psi})$, assuming $X \sim \mathcal{N}(\mu,\sigma)$, one has $\mathcal{N}(\mu=0.008, \sigma=0.167)$, $\mathcal{N}(\mu=0.0192, \sigma=0.447)$, $\mathcal{N}(\mu=0.0076, \sigma=0.184)$. For bimodal distributions,

$f_X(x;\ \sigma=0.0878,\ b=2.56,\ a_1=0.228,\ a_2=2.23,\ a_3=7.44,\ a_4=8)$,

$f_X(x;\ \sigma=0.29,\ b=1.24,\ a_1=-0.158,\ a_2=0.924,\ a_3=3.17,\ a_4=2.22)$,

and $f_X(x;\ \sigma=0.182,\ b=32.4,\ a_1=3.82,\ a_2=20.3,\ a_3=43,\ a_4=31.3)$.

Histograms for noise in the gyroscope channels are documented in Figures 5.37. The statistical models $(\mathcal{S},\mathcal{D})$, $\mathcal{D}=\{f_X(x;\phi):\phi\in\Phi\}$ are derived. Histograms $h_X(x)$ imply bimodal distributions. Consider

$$\mathcal{D}=\{\ f_X(x;\mu,\sigma^2)=\frac{1}{\sqrt{2\pi}\sigma}e^{-\frac{(x-\mu)^2}{2\sigma^2}}:\mu\in\mathbb{R},\ \sigma>0\},\ \int_{-\infty}^{\infty}f_X(x)dx=1,$$

and $$\mathcal{D}=\{\ f_X(x;\mu,\sigma^2,a_n,b)=\frac{1}{\sqrt{2\pi}b\sigma}e^{-\frac{1}{2\sigma^2}\sum_{n=0}^{\infty}a_n(x-\mu)^n}:\mu\in\mathbb{R},\ \sigma>0,\ b>0,\ a_n\in\mathbb{R},\ \forall n\}.$$

The normalized histograms $\bar{h}_X(x)$ are parametrized by unimodal and bimodal pdfs $f_X(x)$, reported in Figures 5.37 by dashed and solid lines. Asymmetric pdfs $f_X(x)=\frac{1}{\sqrt{2\pi}b\sigma}e^{-\frac{1}{2\sigma^2}\sum_n a_n(x-\mu)^n}$ are parameterized using (5.147) and (5.149). A bimodal distribution $\mathcal{D}=\{f_X(x;\phi):\phi\in\Phi\}$ ensures consistency and conformity, enables data fidelity, and empowers data fusion. ■

5.14. Digital Filters and Data Fusion

Design and implementation of analog filters are reported in section 3.7.2. Spatially distributed system-in-package MDOF sensors with integrated MPUs output the measured with pertained signal conditioning signals, as well as filtered and estimated variables. Application- and system specific off-the-shelf advanced sensors are widely used. Filters are designed by evaluating system bandwidth, dynamic modes, noise characteristics, immunity, sensitivity, compliance, etc.

To empower data analytics on large data sets and adaptive data management, consider topics on design of digital system and physical implementation. Dynamic data fusion, reconfigurable filtering and adaptive quadrature integration advance autonomy, modularity, interoperability, scalability, robustness and redundancy. Our goal is to leverage the latest microsystems, MEMS and processing technologies by designs, data fidelity enhancement, as well as ensuring compatibility and compliance with legacy solutions. High-performance processing, robust algorithms and protocols are applied. Various problems arise due to asynchronous fusion, data heterogeneity, uncertainties, multi-sensory multiple-layer fusion complexity, etc.

Using the system bandwidth, noise spectral density, as well as sensitivity of control and observer algorithms, one assesses system dynamic modes $\mathbf{M}(t)$, frequencies to preserve and noise frequencies to attenuate [19-24]. Design of low-pass, high-pass, band-pass and band-stop analog filters was performed in frequency and time domains. Frequency analysis is consistent in filters design. For the system frequencies $\Omega\in\mathbf{M}$, problem formulation, design consideration and specifications are reported in section 3.7.2. Sections 5.12 and 5.13 cover topics pertained to deterministic and random signals. Heterogeneity, distortions, electromagnetic interference and electromagnetic spectrum yield multisource noise $\boldsymbol{n}(t)=f(\boldsymbol{n}_i)$ in sensing elements, semiconductor devices and ASICs.

Design filters by evaluating the cut-off, pass-band and stop-band edge frequencies $(\omega_c,\omega_p,\omega_s)$ in $\Omega_{\boldsymbol{n}}\times\Omega$ considering:

1. System dynamic and frequency modes ($\mathbf{M},\Omega$);
2. Dynamic and frequency modalities ($\mathbf{M}_n,\Omega_n$) of a multispectral noise $n(t)$, $n[n]$.

For specified passband f_p, stopband f_s and sampling f_T frequencies, in digital filters, the passband and stopband frequencies (ω_p,ω_s) are $\omega_p=2\pi f_p/f_T$ and $\omega_s=2\pi f_s/f_T$.

There are analog and digital filter classes $\mathbf{C}_\mathcal{F}$, such as Bessel, biquad, Butterworth, Chebyshev, comb, elliptical, notch and others. Filters $\mathcal{F}_{SC}(\boldsymbol{S},\boldsymbol{C})\in\mathbf{C}_\mathcal{F}$ are characterized by structure $\boldsymbol{S}$ (order, degrees of numerator and denominator, fixed and dynamically reconfigurable, adaptive feedforward and feedback compensation schemes, etc.) and coefficients $\boldsymbol{C}$ (constant or time varying). In linear analog filters, a structure $\boldsymbol{S}$ is defined by the transfer function $G(j\omega)=\frac{N(j\omega)}{D(j\omega)}$ with particular degrees of numerator $N(j\omega)$ and denominator $D(j\omega)$. The filter coefficients $\boldsymbol{C}$ are computed finding the coefficients (a_i,b_i) of $N(j\omega)$ and $D(j\omega)$. One finds the recurrence relation of digital filters. Designs of analog filters $\mathcal{F}_{SC}(\boldsymbol{S},\boldsymbol{C})\in\mathbf{C}_\mathcal{F}$ are documented in section 3.7.2. Consider finite impulse response and infinite impulse response digital filers.

Recurrence Relation – A recurrence relation is an equation which recursively defines a sequence where the current values is a function of previous values. There are various digital filter classes, such as finite impulse response, infinite impulse response, autoregressive and others.

Finite Impulse Response Discrete Filters – The N-order constant-coefficient finite impulse response (FIR) filters, given as

$$y[n]=\sum_{k=0}^{N}b_kx[n-k]=b_0x[n]+b_1x[n-1]+\ldots+b_Nx[n-N]$$

are designed using frequency response specifications. The sampled impulse response $g[n]$, which defines $G(z)$ and coefficients of the nonrecursive FIR filter, is derived using an inverse discrete Fourier transform.

Finite-Order Linear Time-Invariant Impulse Response Filters – Design recursive infinite impulse response (IIR) filters

$$\mathcal{F}_{SC}(\boldsymbol{S},\boldsymbol{C})\equiv\left\{\begin{matrix}G_{M,N,a_l,b_k}(z)\\ y[n]=\sum_{k=0}^{N}b_kx[n-k]-\sum_{l=1}^{M}a_ly[n-l]\end{matrix}\right\}. \tag{5.150}$$

The transfer functions in the z- and frequency domains, as well as linear recurrence (difference) equation with M-feedback terms with feedback coefficients a_l and N-feedforward terms with feedforward coefficient b_k, are

$$G(z)=\frac{Y(z)}{X(z)}=\frac{N(z)}{D(z)}=\frac{\sum_{k=0}^{N}b_kz^{-k}}{1+\sum_{l=1}^{M}a_lz^{-l}},\quad G(e^{j\omega})=\frac{Y(e^{j\omega})}{X(e^{j\omega})}=\frac{\sum_{k=0}^{N}b_ke^{-j\omega k}}{1+\sum_{l=0}^{M}a_le^{-j\omega l}}, \tag{5.151}$$

$$y[n]=\sum_{k=0}^{N}b_kx[n-k]-\sum_{l=1}^{M}a_ly[n-l]=b_0x[n]+b_1x[n-1]+\ldots+b_Nx[n-N]-a_1y[n-1]-\ldots-a_My[n-M],$$

where $x[n]$ and $y[n]$ are the filter input and output; M and N are the feedback and feedforward orders; a_l and b_k are the feedback and feedforward coefficients.

As $\deg(D(z))\geq\deg(N(z))$, the recursive filter order is M. Coefficients b_k of numerator $N(z)$ correspond to the gains of the feedforward terms of recurrence equation (5.151). Coefficients a_l of denominator $D(z)$ correspond to the gains of the feedback terms of difference equation. Causal linear time invariant IIR filters are stable, and, all poles are inside of the unit circle, $|z_i|<1$. The sequence, generated by a recurrence relation, is called a recurrence sequence. The homogeneous recurrence equation (5.151) $y[n]=\sum_{k=0}^{N} b_k x[n-k]-\sum_{l=1}^{M} a_l y[n-l]$ is implemented by MPUs and FPGAs.

Digital Filters Design: Bilinear Transformation and Digital Frequency – Digital filters (5.150) can be designed using analog filters applying the bilinear z-transforms, reported in section 5.2.2. Using $z=e^{j\omega_n T_s}$, nonlinear relationships between analog frequency ω and normalized digital frequency ω_n are

$$\omega=\frac{2}{T_s}\tan(\tfrac{1}{2}\omega_n T_s),\ \omega_n=\frac{2}{T_s}\tan^{-1}(\tfrac{1}{2}\omega T_s),\ -\frac{\pi}{T_s}<\omega_n<\frac{\pi}{T_s},$$

where T_s is the sampling time.

Analog Filters, Discretization and Digital Filters – For signals, measured with noise and distortions, the specifications and analog filters design were covered in section 3.7.2. The filters class, structure and coefficients $\mathcal{F}_{SC}(\boldsymbol{S},\boldsymbol{C})\in\mathbf{C}_{\mathcal{F}}$ yield transfer functions $G(s)$. This $G(s)$ are found studying:

1. System frequencies and dynamic modes $(\Omega,\mathbf{M})$, $\Omega\in\mathbf{M}$;
2. Noise frequencies and dynamic modalities $(\Omega_n,\mathbf{M}_n)$, $\Omega_n\in\mathbf{M}_n$, as well as, noise power spectral density, signal-to-noise ratio, statistical characteristics of multispectral noise $\boldsymbol{n}$, etc.;
3. Systems design considerations;
4. Sensors, interfacing and processing hardware.

Consider the M-order filters $G(s)=\dfrac{N(s)}{D(s)}$. Apply a bilinear relationship (5.20)

$$s=\frac{1}{T_s}\ln(z)=\frac{2}{T_s}\sum_{i=0}^{\infty}\frac{1}{1+2i}\left(\frac{1-z^{-1}}{1+z^{-1}}\right)^{1+2i}=\frac{2}{T_s}\left[\frac{1-z^{-1}}{1+z^{-1}}+\frac{1}{3}\left(\frac{1-z^{-1}}{1+z^{-1}}\right)^3+\frac{1}{5}\left(\frac{1-z^{-1}}{1+z^{-1}}\right)^5+\ldots\right]$$ to find $G(z)$.

For the first-order Tustin approximation, we have

$$s=\frac{1}{T_s}\ln(z),\ \ln(z)\approx 2\frac{z-1}{z+1}=2\frac{1-z^{-1}}{1+z^{-1}}.$$

Considering the sensor data rate, processing capabilities, communication and interfacing, one finds T_s and yields $G(z)$

$$G(z)=G(s)\Big|_{s=\frac{2}{T_s}\frac{1-z^{-1}}{1+z^{-1}}}=\frac{N(s)}{D(s)}\bigg|_{s=\frac{2}{T_s}\frac{1-z^{-1}}{1+z^{-1}}},\ G(z)=\frac{Y(z)}{X(z)}=\frac{\sum_{k=0}^{N} b_k z^{-k}}{1+\sum_{l=1}^{M} a_l z^{-l}}=\frac{N(z)}{D(z)}. \quad (5.152)$$

The derived $G(z)$ yields the homogeneous linear recurrence equation (5.151) $y[n]=\sum_{k=0}^{N} b_k x[n-k]-\sum_{l=1}^{M} a_l y[n-l]$ with the M-feedback and N-feedforward orders. The feedback and feedforward coefficients are (a_l,b_k). Analog filter design yields the digital filer structure and coefficients (5.150) $\mathcal{F}_{SC}(\boldsymbol{S},\boldsymbol{C})$.

MATLAB Commands – MATLAB supports filters design and discretization using the **lp2lp**, **lp2hp**, **lp2bp**, **lp2bs**, **bilinear** and other commands. Practice Problem 5.11 illustrates the digital filter design and implementation using the second-order analog Butterworth filter, found for given system dynamic modes, specifications, and, sampling time T_s.

The filter transfer function $G(s)$ yields state-space differential equations $\dot{x}=Ax+Bu$, $y=Hx+Du$ with (A,B,H,D). For the hardware- and algorithm-defined T_s, find (A_k,B_k,H_k,D_k) using (5.19). The resulting recurrence equation (5.151) is found.

Example 5.48. *Second-Order IIR Notch Filter*
An ideal low-pass filter with the center notch frequency ω_0, $0<\omega_0\leq\pi$ has a frequency response $G(e^{j\omega})=\begin{cases}1, & |\omega|\leq\omega_0\\ 0, & \omega_0<|\omega|\leq\pi\end{cases}$. For the conjugate zeros $e^{j\omega_0}$ and conjugate poles $re^{j\omega_0}$, $0\leq r<1$, the second-order transfer function is

$$G(z)=\frac{(z-e^{j\omega_0})(z-e^{j\omega_0})}{(z-re^{j\omega_0})(z-re^{j\omega_0})}.$$

Using the Euler formula $e^{j\omega}=\cos(j\omega)+j\sin(j\omega)$,

$$G(z)=\frac{b_0+b_1z^{-1}+b_2z^{-2}}{1+a_1z^{-1}+a_2z^{-2}}=\frac{1-2\cos(\omega_0)z^{-1}+z^{-2}}{1-2r\cos(\omega_0)z^{-1}+r^2z^{-2}}.$$

Let $\omega_0=\pi/4$ and $r=0.95$. Figure 5.38 documents the magnitude and phase plots, computed using MATLAB. The specifications are ensured. In particular, $|G(e^{j\omega})|=1$ for $|\omega|<\omega_0$, and, $\phi\to 0$ for $\omega\to\pm\pi$.

```
w0=pi/4; r=0.95; w=-pi:pi/12800:pi; G=freqz([1 -2*cos(w0) 1],[1 -2*r*cos(w0) r^2],w);
figure(1); plot(w/pi,20*log10(abs(G)),'k-','LineWidth',2);
xlabel('Normalized Frequency, [\times\pi rad/sample]'); ylabel('Magnitude {\mid}{\itG}{\mid}, [dB]');
figure(2); plot(w/pi,phase(G),'k-','LineWidth',2);
xlabel('Normalized Frequency, [\times\pi rad/sample]'); ylabel('Phase {\phi}, [rad]');
```

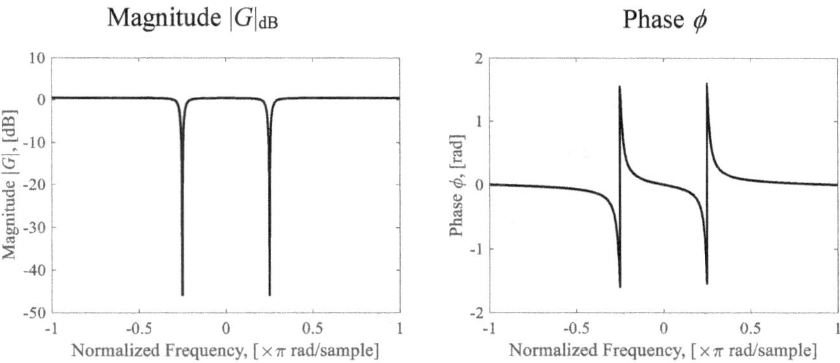

Figure 5.38. Magnitude $|G|_{dB}$ and phase responses for the second-order IIR notch filter. ■

Example 5.49. *Design of a Low-Pass IIR Filters*
Considering system evolution and bandwidth, define: (1) Pass-band frequency ω_p, $|\omega|<\omega_p$; (2) Stop-band frequency, $\omega_s<|\omega|\leq\pi$. Hence, $\omega_p<\omega_s$, and, the transition band is $\omega_p\leq\omega\leq\omega_s$.

The magnitudes of the filter response $G(e^{j\omega})$ in the pass-band and stop-band are

$$|G(e^{j\omega})-1|\leq\delta_p,\ |\omega|<\omega_p,$$

$$|G(e^{j\omega})|\leq\delta_s,\ \omega_s\leq|\omega|<\pi,$$

where δ_p and δ_s are the pass-band and stop-band ripples.

To design a low-pass filter, specify the filter order (N,M) and $(\omega_p,\omega_s,\delta_p,\delta_s)$. An ideal low-pass filter with cut-off frequency ω_c has a frequency response

$$G(e^{j\omega}) = \begin{cases} 1, & |\omega| \le \omega_c \\ 0, & \omega_c < |\omega| \le \pi \end{cases}.$$

The infinite impulse response is $g[n] = \frac{\omega_c}{\pi}\,\mathrm{sinc}\left(\frac{\omega_c n}{\pi}\right), -\infty < n < \infty$.

Specifying $n \in [-M, M]$, $g[n] = \begin{cases} \frac{\omega_c}{\pi}\,\mathrm{sinc}\left(\frac{\omega_c n}{\pi}\right), & n = -M, \ldots, 0, 1, \ldots, M \\ 0, & \text{otherwise} \end{cases}$.

MATLAB supports computing and plotting the truncated $g[n]$. For the specified cut-off frequency f_c=0.25, the window analysis is performed. Consider a signal, perturbed by a high power pseudorandom noise $\xi[n]$ which satisfies the normal distribution. Noise $\xi[n]$ should be filtered. The convolution in the time-domain is

$$y[n] = x[n] * g[n] \overset{z}{\leftrightarrow} X(z)G(z) = Y(z).$$

Figures 5.39 document $g[n]$, magnitudes, and $x[n]$. For noise $\xi[n]$, $\mu\approx 0$ and $\sigma^2 \approx 1$.

```
M=20; % Kernel length M
nM=0:1:M-1; p=nM-(M-1)/2;  fc=0.25; % Cut-off frequency;
g=2*fc*sinc(2*fc*p);
figure(1); stem(nM,g,'ko','LineWidth',2); grid; xlabel('{\itM}','FontSize',18); % Impulse response
[gw,w]=freqz(g); % Frequency Response
figure(2); plot(w/pi,abs(gw)); grid; xlabel('Frequency per sample, {\omega}','FontSize',18);
s=2*pi*(nM/(M-1)); W=1-1.25*cos(s)+0.25*cos(2*s); % Window
figure(3); stem(nM,W); grid; xlabel('{\itM}','FontSize',18);
c=g.*W; % Truncated sinc by window
figure(4); stem(nM,c); grid; xlabel('{\itM}','FontSize',18);
% Magnitude, Frequency Response of the Windowed sinc
figure(5); [gW,w]=freqz(c); plot(w/pi,abs(gW)); grid;
xlabel('Frequency per sample, {\omega}','FontSize',18);
figure(6); freqz(c); xlabel('Frequency'); % Frequency Response of the Windowed sinc Function, dB
% Signal x with a pseudorandom noise which satisfies the normal distribution
N=10000; NN=linspace(0,N,N+1);
Noise=1*randn(1,N+1); mean(Noise), var(Noise)
xn=sin(2*pi*NN/N)+2*sin(4*pi*NN/N)+cos(6*pi*NN/N)+Noise;
figure(7); plot(NN,xn); xlabel('[{\itn}]','FontSize',18);
xnc=conv(c,xn); plot(NN,xnc(M/2:N+M/2),'o'); % Convolution
w1=sinc(2*fc*p); [Px1,w1]=periodogram(xn); [Px2,w2]=periodogram(xnc(M/2:N+M/2));
figure(8); plot(w1,10*log10(Px1),w2,10*log10(Px2));
xlabel('{\omega}','FontSize',18); ylabel('10log_{10}P','FontSize',18); % Power
WN=M/2+1; b=(1/WN)*ones(1,WN); WN1=(WN-1)/2; % Window size
x1=conv(b,xn); Nx1=length(x1); % Compute convolution
figure(9); plot(NN,xn,'co',NN,x1(WN1:N+WN1),'ko'); xlabel('[{\itn}]','FontSize',18); % Plotting signals
[Px4,w4]=periodogram(x1(WN1:N+WN1));
figure(10); plot(w4,10*log(Px4)); xlabel('{\omega}','FontSize',18); ylabel('10log_{10}P','FontSize',18)
```

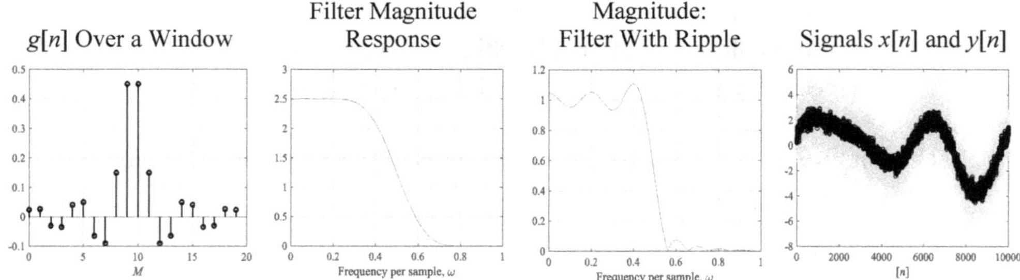

Figure 5.39. Filter design using impulse response. ■

The IIR, Bayesian, extended Kalman, Volterra and kernel adaptive filters, as well as *estimators* and observers are investigated in [19-24]. The IIR filters ensure:

1. Near real-time filter design and reconfiguration;
2. Algorithmic simplicity and adequate performance;
3. Data-centric processing, data fusion and retrieval schemes.

For multisource noise and distortions $\boldsymbol{n}=f(\boldsymbol{n}_i)$, analyze frequency modalities, spectral power density and probability distributions which affect physical data quality $\mathbf{Q}(\cdot)$. Using the measured $\bar{\mathbf{y}}$, filtered $\hat{\mathbf{y}}$ or fused $\tilde{\mathbf{y}}$ data $(\bar{\mathbf{y}},\hat{\mathbf{y}},\tilde{\mathbf{y}})$, one:

1. Characterizes dynamic modes and modalities for system $(\Omega,\mathbf{M})$, $\Omega\in\mathbf{M}$, $\Omega\in[\Omega_{\min}\ \Omega_{\max}]$ and noise $(\Omega_{\boldsymbol{n}},\mathbf{M}_{\boldsymbol{n}})$, $\Omega_{\boldsymbol{n}}\in\mathbf{M}_{\boldsymbol{n}}$, $\Omega_{\boldsymbol{n}}\in[\Omega_{\boldsymbol{n}\ \min}\ \Omega_{\boldsymbol{n}\ \max}]$ in time and frequency domains;
2. Asserts physical data quality $\mathbf{Q}(\cdot)$;
3. Designs and physically implements filters $\mathcal{F}_{SC}$ within an admissible filter class $\mathbf{C}_{\mathcal{F}}$, $\mathcal{F}_{SC}\in\mathbf{C}_{\mathcal{F}}$.

For time-invariant IIR filters, the structure and coefficients are $\boldsymbol{S}(M,N)$ and $\boldsymbol{C}(a_l,b_k)$. Filters are explicitly characterized by a structure $\boldsymbol{S}$ and coefficients $\boldsymbol{C}$, which may be computed and reconfigured in near real-time. The goal is to minimize information losses which are due to $(\mathbf{n},\boldsymbol{\varepsilon},\boldsymbol{\Delta})$, nonlinearities, delays and other factors. Using quantitative data homogeneity $\mathcal{M}_{\mathcal{H}}$ and data heterogeneity $\mathcal{M}_{\mathcal{L}}$ measures, filters $\mathcal{F}_{SC}(\boldsymbol{S},\boldsymbol{C})\in\mathbf{C}_{\mathcal{F}}$ should maximize data homogeneity and minimize data heterogeneity. Filter order and governing equations for $\mathcal{F}_{SC}(\boldsymbol{S},\boldsymbol{C})$ are found by solving the minimax problem, subject to constraints $g(\mathbf{Q},\boldsymbol{S},\boldsymbol{C})$.

For $\mathbf{y}\in\boldsymbol{Y}$, measured and processed tuple $(\bar{\mathbf{y}},\hat{\mathbf{y}},\tilde{\mathbf{y}})$ yields the measured and fused data sets

$$\mathcal{M}_{\text{data}}=\{(\bar{\mathbf{y}},\hat{\mathbf{y}},\tilde{\mathbf{y}})\in\bar{\boldsymbol{Y}}\times\hat{\boldsymbol{Y}}\times\tilde{\boldsymbol{Y}}\}. \qquad (5.153)$$

The quantitative deterministic and probabilistic measures for homogeneity and heterogeneity are given in spectral, statistical and temporal domains. One examines system amplitude preservation, noise attenuation, signal to noise ratio, ripple, error, latencies, linearity, variance, correlation, etc.

For example, the quantitative measures are

$$\mathcal{M}_{\mathcal{H}}=f_{\mathcal{H}}(\omega_c,\omega_p,\omega_s,\delta_p,\delta_s),\ \mathcal{M}_{\mathcal{L}}=f_{\mathcal{L}}(\omega_c,\omega_p,\omega_s,\delta_p,\delta_s), \qquad (5.154)$$

$$\mathcal{M}_{t,|G|}=\int_T t\frac{1}{\|\mathbf{y}\|_p}\|\bar{\mathbf{y}}-\tilde{\mathbf{y}}\|_p\,dt,\ \mathcal{M}_{\mathbf{n}}=\sqrt{\tfrac{1}{K}\sum_{i=1}^K\|\mathbf{n}\|_p},\ \mathcal{M}_{\mathbf{n}}=\tfrac{1}{K}\sum_{i=1}^K\|\mathbf{n}\|_p,\ \mathcal{M}_{\mathbf{n}}=\sqrt{\tfrac{1}{T}\int_T\|\mathbf{n}(t)\|_p\,dt},\ \mathcal{M}_{\mathbf{n}}=\tfrac{1}{T}\int_T\|\mathbf{n}(t)\|_p\,dt\,.$$

The Fourier transforms for $x(t)$ and $X(\omega)$ in time and frequency domains are

$$X(\omega)=\int_{-\infty}^{\infty}x(t)e^{-j\omega t}dt,\ x(t)=\frac{1}{2\pi}\int_{-\infty}^{\infty}X(\omega)e^{j\omega t}d\omega\,.$$

For discrete signals,

$$X[k]=\sum_{n=0}^{N-1}x[n]e^{-j\frac{2\pi}{N}kn},\ x[n]=\frac{1}{N}\sum_{k=0}^{N-1}X[k]e^{j\frac{2\pi}{N}kn}\,.$$

Adaptive and Reconfigurable Filters – Filter structure $\boldsymbol{S}$ yields linear, nonlinear or rational difference and recurrence equations with constant or varying coefficients $\boldsymbol{C}$. Design of adaptive and reconfigurable filters $\mathcal{F}_{SC}(\boldsymbol{S},\boldsymbol{C})$ yields a varying structure $\boldsymbol{S}(\bar{M},\bar{N})$ with order $\bar{M}$, degrees $(\bar{M},\bar{N})$, and, coefficients $\boldsymbol{C}(\bar{a}_l,\bar{b}_k)$. These $\boldsymbol{S}(\bar{M},\bar{N})$ and $\boldsymbol{C}(\bar{a}_l,\bar{b}_k)$ should be computed and reconfigured in real-time using specifications and a knowledge-based solutions library. The constraints are complexity of $\mathbf{C}_{\mathcal{F}}, (\bar{M},\bar{N})_{\max}$, robustness, sensitivity, etc.

Conjecture 5.3. Adaptive reconfigurable infinite impulse response filter

$$\mathcal{F}_{SC}(\boldsymbol{S},\boldsymbol{C})\in\mathbf{C}_{\mathcal{F}},\ \boldsymbol{S}=(\bar{M},\bar{N}),\ \boldsymbol{C}=(\bar{a}_l[n],\bar{b}_k[n]),\ \mathcal{F}_{SC}(\boldsymbol{S},\boldsymbol{C})\equiv\left\{\begin{matrix}G_{\bar{M},\bar{N},\bar{a}_l,\bar{b}_k}(z)\\ y[n]=\sum_{k=0}^{\bar{N}}\bar{b}_kx[n-k]-\sum_{l=1}^{\bar{M}}\bar{a}_ly[n-l]\end{matrix}\right\},$$

$$y[n]=\sum_{k=0}^{\bar{N}}\bar{b}_kx[n-k]-\sum_{l=1}^{\bar{M}}\bar{a}_ly[n-l]=\bar{b}_0x[n]+\bar{b}_1x[n-1]+\ldots+\bar{b}_{\bar{N}}x[n-\bar{N}]-\bar{a}_1y[n-1]-\ldots-\bar{a}_{\bar{M}}y[n-\bar{M}],$$

$$G(z)=\frac{Y(z)}{X(z)}=\frac{\sum_{k=0}^{\bar{N}}\bar{b}_kz^{-k}}{1+\sum_{l=0}^{\bar{M}}\bar{a}_lz^{-l}},\quad \sum_{n=0}^{\infty}|g[n]|<\infty,\ \forall|z_i|<1, \tag{5.155}$$

is designed by applying the measures $(\mathcal{M}_{\mathcal{H}},\mathcal{M}_{\mathcal{L}})$, and, solving a constrained minimax problem

$$\max_{\mathcal{M}_{\mathcal{H}}}\min_{\mathcal{M}_{\mathcal{L}}}\ _{\mathbf{M}\cup\Omega\cup\mathcal{F}_{SC}\cup\mathcal{M}_{\text{data}}\neq\varnothing}\left[\mathcal{M}_{\mathcal{H}}(\bar{\mathbf{y}},\hat{\mathbf{y}},\tilde{\mathbf{y}},\mathbf{Q},\boldsymbol{S},\boldsymbol{C}),\mathcal{M}_{\mathcal{L}}(\bar{\mathbf{y}},\hat{\mathbf{y}},\tilde{\mathbf{y}},\mathbf{Q},\boldsymbol{S},\boldsymbol{C}),g(\mathbf{Q},\boldsymbol{S},\boldsymbol{C})\right], \tag{5.156}$$

$$\mathcal{F}_{SC}(\boldsymbol{S},\boldsymbol{C})\in\mathbf{C}_{\mathcal{F}},\ \forall(\mathbf{M},\Omega,\mathcal{M}_{\text{data}},\mathbf{M}_{\boldsymbol{n}},\Omega_{\boldsymbol{n}})$$

using a finite set of *design variables* $(\mathbf{C}_{\mathcal{F}},\boldsymbol{S},\boldsymbol{C})$, *objective measures* $(\mathcal{M}_{\mathcal{H}\max}, \mathcal{M}_{\mathcal{L}\min})$ and constraints $g(\mathbf{Q},\boldsymbol{S},\boldsymbol{C})$. ■

Example 5.50. *Digital Filters in Inertial Sensing and Estimations: In-Flight Studies*
Digital filter design is experimentally verified on unmanned aerial vehicles. The IMUs measure $(\boldsymbol{a},\boldsymbol{\omega},\boldsymbol{B})$. The autopilot operates on the data fused from integrated IMUs and other sensors. Applied and experimental studies in asynchronous data fusion using measurements from MDOF sensors are reported in [25-30]. Key quantities of interest are estimated and affirmed within the data flow scheme (5.144).

Study solutions to support guidance, navigation and control, redundancy, robustness, as well as *hybrid* navigation schemes [25-31]. Adaptive filters are designed and implemented for IMUs, installed on an airframe. Using IMU measurements, filtering and data fusion are performed by the Teensy 4.1 development platform with ARM Cortex-M7 32-bit 600 MHz microprocessor. The fourth-order Butterworth filters were designed priory flights. Notch and Butterworth filters can be reconfirmed in real time. By estimating dynamic modes $\mathbf{M}(t,x)$ in time domain, and frequency modes $\Omega\in[\Omega_{\min}\ \Omega_{\max}]$ in frequency domain, the second-order adaptive filters are verified.

Using the measured $\boldsymbol{a}(t)$ as $\bar{\boldsymbol{a}}(t)$, the signal conditioning ASICs and ADCs on an IMU output $\bar{\boldsymbol{a}}[n]$. An adaptive filter processes $\bar{\boldsymbol{a}}[n]$ yielding $\tilde{\boldsymbol{a}}[n]$. Adaptive filter design assumes identification of $(\mathbf{M},\Omega)$, as well as characterization of multisource noise and distortions $\boldsymbol{n}=f(\boldsymbol{n}_i)$, finding $(\mathbf{M}_{\boldsymbol{n}},\Omega_{\boldsymbol{n}})$. Identification and parameter estimation problems are investigated for $\Omega\in[\Omega_{\min}\ \Omega_{\max}]$, and, a filter is designed for

$(\omega_p,\omega_c,\omega_s)\in\Omega_n\times\Omega$. Frequency modes are specified and can be estimated for different flight regimes and maneuvers. The quadrotor aerial vehicle exhibits maneuvering capabilities bandwidth ~10 Hz. Parameterization of the exhibited flight transients takes 0.01 sec, while, the design of the second- and fourth-order reconfigurable filters $\mathcal{F}_{SC}(\boldsymbol{S},\boldsymbol{C})$ takes from 0.01 to 0.04 sec. The computed center notch frequency $0<\omega_0\le\pi$, and, conjugate zeros and poles $(e^{j\omega_0}, re^{j\omega_0})$, $0\le r<1$, yield the unknown coefficients (a_i,b_i). Figures 5.40.a document the measured and filtered accelerations by an adaptive Butterworth filter during 225 sec flight. Measured and filtered accelerations are reported in Figures 5.40.b with the adaptively reconfigured second-order notch filter. The data rate is 200 Hz, e.g., data is fused with T_d=0.005 sec. Overall objectives are achieved. Adaptive filtering, redundant decentralized estimates, data processing and fusion from MDOF sensors are ensured.

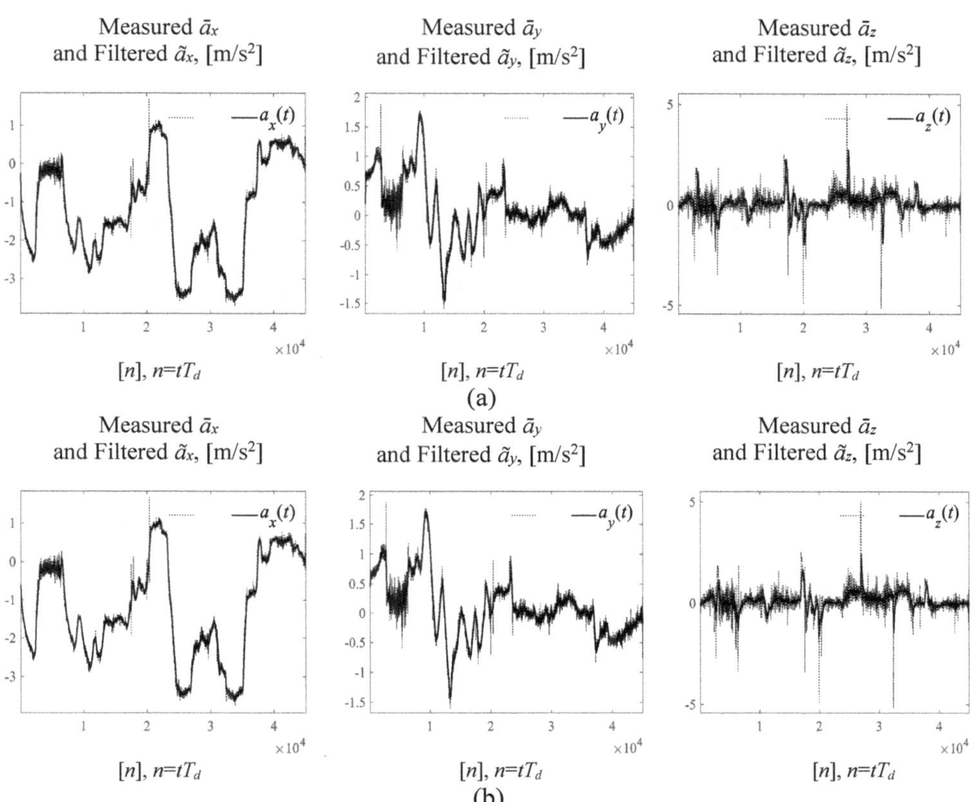

Figure 5.40. Maneuvering quadrotor helicopter flight: Measured accelerations $\bar{\boldsymbol{a}}=[\bar{a}_x,\bar{a}_y,\bar{a}_z]$ and filtered accelerations $\tilde{\boldsymbol{a}}=[\tilde{a}_x,\tilde{a}_y,\tilde{a}_z]$ are documented by dotted and solid lines:
(a) The fourth-order Butterworth filter in the (a_x,a_y) channels. Coefficients $(\bar{a}_l,\bar{b}_k)\big|_{t=100\text{ sec}}$ are $a_1=1$, $a_2=-3.94$, $a_3=5.82$, $a_4=-3.82$, $a_5=0.94$, $b_1=1.87\times10^{-8}$, $b_2=7.47\times10^{-8}$, $b_3=1.12\times10^{-7}$, $b_4=7.47\times10^{-8}$, $b_5=1.87\times10^{-8}$. For the a_z channel the coefficients are $a_1=1$, $a_2=-3.96$, $a_3=5.91$, $a_4=-3.91$, $a_5=0.97$, $b_1=1.19\times10^{-9}$, $b_2=4.74\times10^{-9}$, $b_3=7.11\times10^{-9}$, $b_4=4.74\times10^{-9}$, $b_5=1.19\times10^{-9}$;
(b) The second-order notch filter in the (a_x,a_y) channels. Coefficients $(\bar{a}_l,\bar{b}_k)\big|_{t=100\text{ sec}}$ are $a_1=1$, $a_2=-0.57$, $a_3=-0.414$, $b_1=0.293$, $b_2=-0.57$, $b_3=0.293$. For the a_z channel the coefficients are $a_1=1$, $a_2=-0.579$, $a_3=-0.414$, $b_1=0.293$, $b_2=-0.579$, $b_3=0.293$.

5.15. Practice Problems

Practice Problem 5.1. *Control of an Electric Drive With Permanent Magnet DC Motor*
Consider an electric drive with a permanent magnet dc motor studied in chapter 3 and in Examples 5.10, 5.11, 5.13, 5.14, 5.17, 5.19 and 5.20. The actuator is modeled by differential equations

$$\frac{di_a}{dt}=\frac{1}{L_a}(-r_a i_a-k_a\omega_r+u_a),$$

$$\frac{d\omega_r}{dt}=\frac{1}{J}\left(k_a i_a-B_m\omega_r-T_L\right),\ y=\omega_r.$$

The output is the angular velocity ω_r.

The tracking error is $e(t)=r-y=r-\omega_r$.

Consider tracking control problems with different control laws.

Control Law With the Proportional Tracking Error, Current and Angular Velocity Feedback – Consider a control law

$$u(t)=-k_I i_a-k_\omega\omega_r+k_p e,\ k_\omega=1.$$

Find the feedback gains to ensure the desired eigenvalues $\lambda_1=-10$ and $\lambda_2=-100$.

The transfer function of a motor is $G_{sys}(s)=\dfrac{\Omega(s)}{U(s)}=\dfrac{k_a}{L_aJs^2+\left(r_aJ+L_aB_m\right)s+r_aB_m+k_a^2}$.

With current and angular velocity feedback,

$$G^*_{sys}(s)=\frac{k_a}{L_aJs^2+\left(r_aJ+k_IJ+L_aB_m\right)s+r_aB_m+k_IB_m+k_a^2+k_ak_\omega}.$$

The transfer function for the closed-loop system is

$$G_\Sigma(s)=\frac{\Omega(s)}{R(s)}=\frac{G_{PID}(s)G^*_{sys}(s)}{1+G_{PID}(s)G^*_{sys}(s)}=\frac{k_pG^*_{sys}(s)}{1+k_pG^*_{sys}(s)}$$

$$=\frac{k_pk_a}{L_aJs^2+\left(r_aJ+k_IJ+L_aB_m\right)s+r_aB_m+k_IB_m+k_a^2+k_ak_\omega+k_pk_a}.$$

The characteristic equation is

$$s^2+\frac{r_aJ+k_IJ+L_aB_m}{L_aJ}s+\frac{r_aB_m+k_IB_m+k_a^2+k_ak_\omega+k_pk_a}{L_aJ}=0.$$

For the assigned $\lambda_1=-10$ and $\lambda_2=-100$, the characteristic equation is

$$(\lambda+10)(\lambda+100)=\lambda^2+110\lambda+1000=0.$$

Comparison of characteristic equations yields

$$\frac{r_aJ+k_IJ+L_aB_m}{L_aJ}=110,\quad \frac{r_aB_m+k_IB_m+k_a^2+k_ak_\omega+k_pk_a}{L_aJ}=1000.$$

For an illustrative example, let all motor parameters are equal to one. That is, r_a=1 ohm, L_a=1 H, k_a=1 V-sec/rad, B_m=1 N-m-sec/rad and J=1 kg-m^2. Letting $k_\omega=1$, we find the feedback gains $k_I=108$ and $k_p=889$.

The steady-state error is $\lim_{t\to\infty}e(t)=\lim_{s\to 0}\dfrac{sR(s)}{1+G_{PID}(s)G^*_{sys}(s)}$.

For the unit step reference $r(t)=1(t)$, $R(s)=1/s$. Using the final value theorem,

$$\lim_{t\to\infty} e(t) = \lim_{s\to 0} \frac{sR(s)}{1+G_{PID}(s)G^*_{sys}(s)} = \lim_{s\to 0} \frac{sR(s)}{1+k_p G^*_{sys}(s)}$$

$$= \lim_{s\to 0} \frac{1}{1+k_p \dfrac{k_a}{L_a J s^2 + (r_a J + k_I J + L_a B_m)s + r_a B_m + k_I B_m + k_a^2 + k_a k_\omega}}$$

$$= \lim_{s\to 0} \frac{L_a J s^2 + (r_a J + k_I J + L_a B_m)s + r_a B_m + k_I B_m + k_a^2 + k_a k_\omega}{L_a J s^2 + (r_a J + k_I J + L_a B_m)s + r_a B_m + k_I B_m + k_a^2 + k_a k_\omega + k_p k_a} = \frac{111}{1000} = 0.111.$$

Therefore, $\lim_{t\to\infty} e(t) = e(\infty) = 0.111$. To reduce the tracking error, reduce k_ω and change eigenvalues (λ_1,λ_2) to increase k_p. Consistent motor parameters are used.

Let $\lambda_1 = -10$, $\lambda_2 = -100$.

Specify the steady-state tracking error to be $\lim_{t\to\infty} e(t) = e(\infty) = 0.001$.

Find (k_I,k_ω,k_p) in $u(t) = -k_I i_a - k_\omega \omega_r + k_p e$.

From two characteristic equations, we have

$$\frac{r_a J + k_I J + L_a B_m}{L_a J} = -(\lambda_1+\lambda_2) = 110, \quad \frac{r_a B_m + k_I B_m + k_a^2 + k_a k_\omega + k_p k_a}{L_a J} = \lambda_1\lambda_2 = 1000.$$

The final value theorem

$$\lim_{t\to\infty} e(t) = \lim_{s\to 0} \frac{L_a J s^2 + (r_a J + k_I J + L_a B_m)s + r_a B_m + k_I B_m + k_a^2 + k_a k_\omega}{L_a J s^2 + (r_a J + k_I J + L_a B_m)s + r_a B_m + k_I B_m + k_a^2 + k_a k_\omega + k_p k_a}$$

with the specified steady-state tracking error $\lim_{t\to\infty} e(t) = e(\infty) = 0.001$ yields

$$\frac{r_a B_m + k_I B_m + k_a^2 + k_a k_\omega}{r_a B_m + k_I B_m + k_a^2 + k_a k_\omega + k_p k_a} = 0.001.$$

Using the motor parameters, solve three equations with three unknowns (k_I,k_ω,k_p)

$$\begin{bmatrix} J & 0 & 0 \\ B_m & k_a & k_a \\ (1-e(\infty))B_m & (1-e(\infty))k_a & -e(\infty)k_a \end{bmatrix} \begin{bmatrix} k_I \\ k_\omega \\ k_p \end{bmatrix} = \begin{bmatrix} -(\lambda_1+\lambda_2)L_a J - r_a J - L_a B_m \\ \lambda_1\lambda_2 L_a J - r_a B_m - k_a^2 \\ (e(\infty)-1)(r_a B_m + k_a^2) \end{bmatrix}.$$

Solve a hypothetical example when all parameters are ones. The solution is found as

```
ra=1; La=1; ka=1; J=1; Bm=1; s1=-10; s2=-100; steady_state_error=0.001;
A=[J 0 0; Bm ka ka; (1-steady_state_error)*Bm (1-steady_state_error)*ka -steady_state_error];
B=[-(s1+s2)*La*J-ra*J-La*Bm; s1*s2*La*J-ra*Bm-ka^2;
(steady_state_error-1)*(ra*Bm+ka^2)]; klkwkp=inv(A)*B
```

One has k_I=108, k_ω= −109 and k_p=999.

Using the motor parameters, assign eigenvalues (λ_1,λ_2) and tracking error $e(\infty)$. In $u(t) = -k_I i_a - k_\omega \omega_r + k_p e$, the feedback gains ($k_I,k_\omega,k_p$) must be positive, ($k_I,k_\omega,k_p$)>0. For a permanent magnet dc motor, r_a=3.15 ohm, L_a=0.0066 H, k_a=0.16 V-sec/rad, B_m=0.0001 N-m-sec/rad and J=0.00025 kg-m^2.

Let the desired eigenvalues are λ_1= −10000 and λ_2= −10000.

The specified tracking error is $\lim_{t\to\infty} e(t) = e(\infty) = 0.001$. Compute ($k_I,k_\omega,k_p$) as

```
ra=3.15; La=0.0066; ka=0.16; J=0.00025; Bm=0.0001;
s1=-10000; s2=-10000; steady_state_error=0.001;
A=[J 0 0; Bm ka ka; (1-steady_state_error)*Bm (1-steady_state_error)*ka -steady_state_error];
B=[-(s1+s2)*La*J-ra*J-La*Bm; s1*s2*La*J-ra*Bm-ka^2; (steady_state_error-1)*(ra*Bm+ka^2)];
klkwkp=inv(A)*B
```

We found k_I=128.85, k_ω=0.789 and k_p=1030.2. That is, $(k_I,k_\omega,k_p)>0$. High feedback gains yield high armature voltage. Limits on control are considered.

Control of Permanent Magnet DC Motor – Let a control law is $u(t)=-k_\omega \dfrac{d\omega_r}{dt}+k_p e$.

Assume all motor parameters are equal to one. Find feedback coefficients k_ω and k_p for $\lambda_1=-1$ and $\lambda_2=-3$. Using a transfer function of the permanent magnet dc motor $G_{sys}(s)$, a transfer function for a closed-loop system is

$$G_\Sigma(s)=\frac{\Omega(s)}{R(s)}=\frac{G_{PID}(s)\dfrac{G_{sys}(s)}{1+k_\omega sG_{sys}(s)}}{1+G_{PID}(s)\dfrac{G_{sys}(s)}{1+k_\omega sG_{sys}(s)}}=\frac{\dfrac{k_pk_a}{L_aJs^2+(r_aJ+L_aB_m)s+r_aB_m+k_a^2}}{1+\dfrac{(k_p+k_\omega s)k_a}{L_aJs^2+(r_aJ+L_aB_m)s+r_aB_m+k_a^2}}$$

$$=\frac{k_pk_a}{L_aJs^2+(r_aJ+L_aB_m+k_\omega k_a)s+r_aB_m+k_a^2+k_pk_a}.$$

The resulting characteristic equation $s^2+\dfrac{r_aJ+L_aB_m+k_\omega k_a}{L_aJ}s+\dfrac{r_aB_m+k_a^2+k_pk_a}{L_aJ}=0$ is compared with the assigned $(\lambda+1)(\lambda+3)=\lambda^2+4\lambda+3=0$.

One finds the feedback coefficients k_ω=2 and k_p=1.

Control of an Electric Drive With State Feedback – Investigate a control law with the current, angular velocity and integral feedback

$$u(t)=-k_Ii_a-k_\omega\omega_r+k_i\int e dt.$$

Let the desired eigenvalues are $\lambda_1=-1$, $\lambda_2=-1$ and $\lambda_3=-10$. With the current and angular velocity feedback k_Ii_a and $k_\omega\omega_r$, using $G_{sys}(s)$, we have

$$G^*_{sys}(s)=\frac{k_a}{L_aJs^2+(r_aJ+k_IJ+L_aB_m)s+(r_a+k_I)B_m+k_a^2+k_ak_\omega}.$$

The transfer function for a closed-loop system is

$$G_\Sigma(s)=\frac{\Omega(s)}{R(s)}=\frac{G_{PID}(s)G^*_{sys}(s)}{1+G_{PID}(s)G^*_{sys}(s)}=\frac{\dfrac{k_i}{s}G^*_{sys}(s)}{1+\dfrac{k_i}{s}G^*_{sys}(s)}$$

$$=\frac{k_ik_a}{L_aJs^3+(r_aJ+k_IJ+L_aB_m)s^2+(r_aB_m+k_IB_m+k_a^2+k_ak_\omega)s+k_ik_a}.$$

The characteristic equation is

$$L_aJs^3+(r_aJ+k_IJ+L_aB_m)s^2+(r_aB_m+k_IB_m+k_a^2+k_ak_\omega)s+k_ik_a=0.$$

The desired characteristic equation is $(\lambda+1)(\lambda+1)(\lambda+10)=\lambda^3+12\lambda^2+21\lambda+10=0$.

Comparison of

$$L_aJs^3+(r_aJ+k_IJ+L_aB_m)s^2+(r_aB_m+k_IB_m+k_a^2+k_ak_\omega)s+k_ik_a=0$$

and $\lambda^3+12\lambda^2+21\lambda+10=0$,

yields $\dfrac{r_aJ+k_IJ+L_aB_m}{L_aJ}=12$, $\dfrac{r_aB_m+k_IB_m+k_a^2+k_ak_\omega}{L_aJ}=21$, $\dfrac{k_ik_a}{L_aJ}=10$.

For any motor parameters, one finds (k_I, k_ω, k_i).

The steady-state error is $\lim_{t\to\infty} e(t) = \lim_{s\to 0}\dfrac{sR(s)}{1+G_{PID}(s)G^*_{sys}(s)}$.

For the unit step reference $r(t)=1(t)$, one has $R(s)=1/s$. Hence

$$\lim_{t\to\infty} e(t) = \lim_{s\to 0}\frac{sR(s)}{1+G_{PID}(s)G^*_{sys}(s)} = \lim_{s\to 0}\frac{sR(s)}{1+\frac{k_i}{s}G^*_{sys}(s)}$$

$$= \lim_{s\to 0}\frac{s\left[L_aJs^2+\left(r_aJ+k_IJ+L_aB_m\right)s+\left(r_a+k_I\right)B_m+k_a^2+k_ak_\omega\right]}{s\left[L_aJs^2+\left(r_aJ+k_IJ+L_aB_m\right)s+\left(r_a+k_I\right)B_m+k_a^2+k_ak_\omega\right]+k_ik_a} = 0.$$

Therefore, $\lim_{t\to\infty} e(t) = e(\infty) = 0$. ■

Practice Problem 5.2. Proportional-Integral-Derivative Control of an Electric Drive
Consider an electric drive with a permanent magnet dc motor, see Practice Problem 5.1.

A PID control law is $u(t) = k_pe + k_{i1}\int edt + k_{i2}\iint edt + k_d\frac{de}{dt}$.

The goal is to find the feedback gains, guaranteeing the characteristic eigenvalues $\lambda_1=-1$, $\lambda_2=-1$, $\lambda_3=-1$, $\lambda_4=-1$.

The transfer function for an electric drive is

$$G_{sys}(s) = \frac{\Omega(s)}{U(s)} = \frac{k_a}{L_aJs^2+\left(r_aJ+L_aB_m\right)s+r_aB_m+k_a^2} = \frac{k_a}{as^2+bs+c}.$$

Assume the motor parameters are one. Hence, $a=1$, $b=2$ and $c=2$.

The transfer function for the closed-loop system is

$$G(s) = \frac{\Omega(s)}{R(s)} = \frac{G_{PID}(s)G_{sys}(s)}{1+G_{PID}(s)G_{sys}(s)} = \frac{\dfrac{k_ps^2+k_{i1}s+k_{i2}+k_ds^3}{s^2}G_{sys}(s)}{1+\dfrac{k_ps^2+k_{i1}s+k_{i2}+k_ds^3}{s^2}G_{sys}(s)}$$

$$= \frac{k_a\left(k_ps^2+k_{i1}s+k_{i2}+k_ds^3\right)}{as^4+\left(b+k_ak_d\right)s^3+\left(c+k_ak_p\right)s^2+k_ak_{i1}s+k_ak_{i2}}.$$

The characteristic equation of the closed-loop system

$$s^4+\frac{b+k_ak_d}{a}s^3+\frac{c+k_ak_p}{a}s^2+\frac{k_ak_{i1}}{a}s+\frac{k_ak_{i2}}{a}=0$$

is compared with the desired characteristic equation

$$(\lambda+1)(\lambda+1)(\lambda+1)(\lambda+1)=\lambda^4+4\lambda^3+6\lambda^2+4\lambda+1=0.$$

The feedback gains are $k_p=4$, $k_{i1}=4$, $k_{i2}=2$ and $k_d=2$.

For the unit step input $r(t)=1(t)$ with $R(s)=1/s$, the steady-state error is

$$\lim_{t\to\infty} e(t) = \lim_{s\to 0}\frac{sR(s)}{1+G_{PID}(s)G_{sys}(s)} = \lim_{s\to 0}\frac{s^2\left(as^2+bs+c\right)}{s^2\left(as^2+bs+c\right)+\left(k_ps^2+k_{i1}s+k_{i2}+k_ds^3\right)k_a} = 0.$$

For the sinusoidal reference $r(t)=\sin(\omega_0t)1(t)$, $R(s)=\frac{\omega_0}{s^2+\omega_0^2}$, the steady-state error is

$$\lim_{t\to\infty} e(t) = \lim_{s\to 0}\frac{sR(s)}{1+G_{PID}(s)G_{sys}(s)} = \lim_{s\to 0}\frac{s\frac{\omega_0}{s^2+\omega_0^2}\left(as^2+bs+c\right)}{s^2\left(as^2+bs+c\right)+\left(k_ps^2+k_{i1}s+k_{i2}+k_ds^3\right)k_a} = 0.$$

Hence, $\lim_{t\to\infty} e(t) = e(\infty) = 0$. ■

Practice Problem 5.3. *State Transition Matrix and System Dynamics*

Study a system, consistent with the *force* and *torque* analysis and control problems

$$\frac{dx}{dt} = Ax + Bu,\quad \begin{bmatrix} \frac{dx_1}{dt} \\ \frac{dx_2}{dt} \end{bmatrix} = \begin{bmatrix} 0 & 1 \\ -3 & -4 \end{bmatrix}\begin{bmatrix} x_1 \\ x_2 \end{bmatrix} + \begin{bmatrix} 0 \\ 1 \end{bmatrix} u,\ y=x_1,\ y=Hx+Du,\ y=[1\quad 0]x+[0]u.$$

Let the initial conditions are $x_0 = \begin{bmatrix} x_{10} \\ x_{20} \end{bmatrix} = \begin{bmatrix} 1 \\ -1 \end{bmatrix}$. The input is a unit step, $u=1(t)$.

The characteristic equation (5.4) $|\lambda I - A|=0$ is

$$\left| \lambda \begin{bmatrix} 1 & 0 \\ 0 & 1 \end{bmatrix} - \begin{bmatrix} 0 & 1 \\ -3 & -4 \end{bmatrix} \right| = \left| \begin{bmatrix} \lambda & -1 \\ 3 & \lambda+4 \end{bmatrix} \right| = \lambda^2 + 4\lambda + 3 = 0 .$$

The eigenvalues are $\lambda_1 = -1$ and $\lambda_2 = -3$.

Apply the Cayley-Hamilton theorem (5.10). Mappings $\beta_0(t)$ and $\beta_1(t)$ are

$$\begin{bmatrix} \beta_0(t) \\ \beta_1(t) \end{bmatrix} = \begin{bmatrix} 1 & \lambda_1 \\ 1 & \lambda_2 \end{bmatrix}^{-1} \begin{bmatrix} e^{\lambda_1 t} \\ e^{\lambda_2 t} \end{bmatrix} = \begin{bmatrix} 1 & -1 \\ 1 & -3 \end{bmatrix}^{-1} \begin{bmatrix} e^{-t} \\ e^{-3t} \end{bmatrix} = \begin{bmatrix} \frac{3}{2}e^{-t} - \frac{1}{2}e^{-3t} \\ \frac{1}{2}e^{-t} - \frac{1}{2}e^{-3t} \end{bmatrix}.$$

The state transition matrix is

$$\Phi(t) = e^{At} = \sum_{i=0}^{n-1} \beta_i(t)\, A^i = \beta_0(t) I + \beta_1(t) A$$

$$= \left[\tfrac{3}{2}e^{-t} - \tfrac{1}{2}e^{-3t}\right]\begin{bmatrix} 1 & 0 \\ 0 & 1 \end{bmatrix} + \left[\tfrac{1}{2}e^{-t} - \tfrac{1}{2}e^{-3t}\right]\begin{bmatrix} 0 & 1 \\ -3 & -4 \end{bmatrix} = \begin{bmatrix} \frac{3}{2}e^{-t} - \frac{1}{2}e^{-3t} & \frac{1}{2}e^{-t} - \frac{1}{2}e^{-3t} \\ -\frac{3}{2}e^{-t} + \frac{3}{2}e^{-3t} & -\frac{1}{2}e^{-t} + \frac{3}{2}e^{-3t} \end{bmatrix}.$$

Apply equation (5.9) to find the state dynamics for initial conditions x_0 and unit-step input $u=1(t)$

$$x(t) = \begin{bmatrix} x_1(t) \\ x_2(t) \end{bmatrix} = e^{At}\begin{bmatrix} x_{10} \\ x_{20} \end{bmatrix} + \int_0^t e^{A(t-\tau)} Bu(\tau)d\tau = \begin{bmatrix} \frac{3}{2}e^{-t} - \frac{1}{2}e^{-3t} & \frac{1}{2}e^{-t} - \frac{1}{2}e^{-3t} \\ -\frac{3}{2}e^{-t} + \frac{3}{2}e^{-3t} & -\frac{1}{2}e^{-t} + \frac{3}{2}e^{-3t} \end{bmatrix}\begin{bmatrix} 1 \\ -1 \end{bmatrix}$$

$$+\int_0^t \begin{bmatrix} \frac{3}{2}e^{-(t-\tau)} - \frac{1}{2}e^{-3(t-\tau)} & \frac{1}{2}e^{-(t-\tau)} - \frac{1}{2}e^{-3(t-\tau)} \\ -\frac{3}{2}e^{-(t-\tau)} + \frac{3}{2}e^{-3(t-\tau)} & -\frac{1}{2}e^{-(t-\tau)} + \frac{3}{2}e^{-3(t-\tau)} \end{bmatrix}\begin{bmatrix} 0 \\ 1 \end{bmatrix} d\tau$$

$$= \begin{bmatrix} e^{-t} \\ -e^{-t} \end{bmatrix} + \begin{bmatrix} \frac{1}{2}e^{-(t-\tau)} - \frac{1}{6}e^{-3(t-\tau)}\Big|_0^t \\ -\frac{1}{2}e^{-(t-\tau)} + \frac{1}{2}e^{-3(t-\tau)}\Big|_0^t \end{bmatrix} = \begin{bmatrix} e^{-t} \\ -e^{-t} \end{bmatrix} + \begin{bmatrix} \frac{1}{2} - \frac{1}{2}e^{-t} - \frac{1}{6} + \frac{1}{6}e^{-3t} \\ -\frac{1}{2} + \frac{1}{2}e^{-t} + \frac{1}{2} - \frac{1}{2}e^{-3t} \end{bmatrix} = \begin{bmatrix} \frac{1}{3} + \frac{1}{2}e^{-t} + \frac{1}{6}e^{-3t} \\ -\frac{1}{2}e^{-t} - \frac{1}{2}e^{-3t} \end{bmatrix}, t \ge 0.$$

Compute system transients due to initial conditions $x_0=[1, -1]$ and input $u=1(t)$.

To find a definite integral $\int_a^b f(x)dx$, use **int(f,x,a,b)**. We have

```
A=[0 1; -3 -4]; B=[0; 1];
t=sym('t'); tau=sym('tau'); u=sym('u'); eAt=expm(A*t); x0=[1; -1]; u=1;
xt=eAt*x0+int(expm(A*(t-tau))*B*u,tau,0,t)
```

One has

```
xt =
     exp(-t)/2 + exp(-3*t)/6 + 1/3
   - exp(-t)/2 - exp(-3*t)/2
```

For output $y(t)=x_1(t)$, the output equation is $y=Hx+Du=[1\quad 0]x+[0]u$.

The transfer function (5.6) is

$$G(s)=\frac{Y(s)}{U(s)}=H\underbrace{(sI-A)^{-1}}_{\Phi(s)=(sI-A)^{-1}}B=H\Phi(s)B=\begin{bmatrix}1 & 0\end{bmatrix}\begin{bmatrix}\frac{s+4}{s^2+4s+3} & \frac{1}{s^2+4s+3}\\ -\frac{3}{s^2+4s+3} & \frac{s}{s^2+4s+3}\end{bmatrix}\begin{bmatrix}0\\1\end{bmatrix}=\frac{1}{s^2+4s+3}.$$

The MATLAB code

```
A=[0 1; -3 -4]; B=[0; 1]; H=[1 0]; D=0; [num, den]=ss2tf(A,B,H,D)
```

yields $G(s)$ as

```
num = 0   0   1
den =  1   4   3
```

■

Practice Problem 5.4. *Discrete-Time System: State Transition Matrix, System Dynamics and Stability*

Consider a system $x[k+1]=Ax[k]$, $\begin{bmatrix}x_1[k+1]\\x_2[k+1]\end{bmatrix}=\begin{bmatrix}0 & 1\\ \frac{1}{6} & -\frac{1}{6}\end{bmatrix}\begin{bmatrix}x_1[k]\\x_2[k]\end{bmatrix}$.

Solved problems related to Example 5.6. The state transition matrix $\Phi(z)$ is

$$\Phi(z)=z\left(zI-A\right)^{-1}=z\begin{bmatrix}z & -1\\ -\frac{1}{6} & z+\frac{1}{6}\end{bmatrix}^{-1}=\frac{z}{z^2+\frac{1}{6}z-\frac{1}{6}}\begin{bmatrix}z+\frac{1}{6} & 1\\ \frac{1}{6} & z\end{bmatrix}.$$

The characteristic equation is $|zI-A|=0$, $z^2+\frac{1}{6}z-\frac{1}{6}=0$.

Hence, the eigenvalues are $z_1=1/3$ and $z_2=-1/2$.

A discrete linear systems is asymptotically stable because $|z_i|\le 1, i=1,2$, and, $\lim_{k\to\infty} x[k]\to 0$.

The zero-input response evolves to zero as time increases as $k\to\infty$.

Performing partial fractioning and applying inverse z-transforms, one may find $\Phi[k]$ as $\Phi[k]=\mathcal{Z}^{-1}\{\Phi[z]\}$.

Finding an eigenvalue-eigenvector pair $(z,\boldsymbol{v})$ and *modal* matrix $\mathbf{M}\in\mathbb{R}^{2\times 2}$, the state transition matrix is

$$\Phi[k]=A^k=\mathbf{MZM}^{-1},\ \boldsymbol{Z}=\begin{bmatrix}(z_1)^k & 0\\ 0 & (z_2)^k\end{bmatrix},\ \mathbf{M}=[\boldsymbol{v}_1,\boldsymbol{v}_2],\ (z_iI-A)\boldsymbol{v}_i=0.$$

Apply the Cayley-Hamilton theorem (5.10), and, find $\Phi[k]=A^k$.

We find $\beta_0[k]$ and $\beta_1[k]$ as

$$\begin{bmatrix}\beta_0[k]\\ \beta_1[k]\end{bmatrix}=\begin{bmatrix}1 & z_1\\ 1 & z_2\end{bmatrix}^{-1}\begin{bmatrix}(z_1)^k\\ (z_2)^k\end{bmatrix}=\begin{bmatrix}1 & \frac{1}{3}\\ 1 & -\frac{1}{2}\end{bmatrix}^{-1}\begin{bmatrix}(\frac{1}{3})^k\\ (-\frac{1}{2})^k\end{bmatrix}=\begin{bmatrix}\frac{3}{5}(\frac{1}{3})^k+\frac{2}{5}(-\frac{1}{2})^k\\ \frac{6}{5}(\frac{1}{3})^k-\frac{6}{5}(-\frac{1}{2})^k\end{bmatrix}.$$

The state transition matrix is

$$\Phi[k]=A^k=\beta_0[k]I+\beta_1[k]A=\begin{bmatrix}\frac{3}{5}(\frac{1}{3})^k+\frac{2}{5}(-\frac{1}{2})^k & \frac{6}{5}(\frac{1}{3})^k-\frac{6}{5}(-\frac{1}{2})^k\\ \frac{1}{5}(\frac{1}{3})^k-\frac{1}{5}(-\frac{1}{2})^k & \frac{2}{5}(\frac{1}{3})^k+\frac{3}{5}(-\frac{1}{2})^k\end{bmatrix}.$$

Using (5.17), the system evolution for $x[0]=\begin{bmatrix}1\\-1\end{bmatrix}$ is

$$x[k]=\Phi[k]x[0],\ \begin{bmatrix}x_1[k]\\x_2[k]\end{bmatrix}=\begin{bmatrix}\frac{3}{5}(\frac{1}{3})^k+\frac{2}{5}(-\frac{1}{2})^k & \frac{6}{5}(\frac{1}{3})^k-\frac{6}{5}(-\frac{1}{2})^k\\ \frac{1}{5}(\frac{1}{3})^k-\frac{1}{5}(-\frac{1}{2})^k & \frac{2}{5}(\frac{1}{3})^k+\frac{3}{5}(-\frac{1}{2})^k\end{bmatrix}\begin{bmatrix}1\\-1\end{bmatrix}=\begin{bmatrix}-\frac{3}{5}(\frac{1}{3})^k+\frac{8}{5}(-\frac{1}{2})^k\\ -\frac{1}{5}(\frac{1}{3})^k-\frac{4}{5}(-\frac{1}{2})^k\end{bmatrix}.$$

Apply the Lyapunov equation to analyze system stability. This implies solution of a discrete Lyapunov equation (6.128) $A_k^T K_k A_k - K_k = -Q_k$, $Q_k>0$.

Let $Q_k = \begin{bmatrix} 1 & 0 \\ 0 & 1 \end{bmatrix}$. Using the **eig**, **dlyap** and **det** commands, find z_i, K_k and $\det(K_k)$ as

```
Ak=[0 1; 1/6 -1/6]; Qk=[1 0; 0 1]; Eig=eig(Ak), Kk=dlyap(Ak',Qk), Det=det(Kk)
```

We have $z_1=1/3$, $z_2=-1/2$, $K_k = \begin{bmatrix} 1.0595 & -0.0714 \\ -0.0714 & 2.1429 \end{bmatrix}$, $\det(K_k)=2.265$.

The **dlyap** command solves a discrete Lyapunov equation by using the matrix block substitutions and Schur decomposition.

Solve an equation $A_k^T K_k A_k - K_k = -Q_k$ using a custom MATLAB code

```
Ak=[0 1; 1/6 -1/6]; Qk=[1 0; 0 1];
n=length(Ak); X=ones(n^2);
for i=1:n
  for j=1:n
    X((1:n)+n*(i-1),(1:n)+n*(j-1))=Ak*Ak(i,j);
end
  end
Kk=reshape((eye(n^2)-X)'\Qk(:),n,n)
```

We obtain $K_k = \begin{bmatrix} 1.0595 & -0.0714 \\ -0.0714 & 2.1429 \end{bmatrix}$. The system is stable. ■

Practice Problem 5.5. Digital Control

The PID control law is $u = k_p e + k_i \int e dt + k_d \frac{de}{dt}$.

A transfer function is $G_{PID}(s) = \dfrac{U(s)}{E(s)} = \dfrac{k_d s^2 + k_p s + k_i}{s}$.

In section 5.3.2, for $G_{PID}(s)$, using Tustin approximation (5.21) $s = \frac{1}{T_s}\ln(z) \approx \frac{2}{T_s}\frac{1-z^{-1}}{1+z^{-1}}$,

we found (5.43) $G_{PID}(z) = \dfrac{U(z)}{E(z)} = \dfrac{\left(2T_s k_p + T_s^2 k_i + 4k_d\right) + \left(2T_s^2 k_i - 8k_d\right)z^{-1} + \left(-2T_s k_p + T_s^2 k_i + 4k_d\right)z^{-2}}{2T_s\left(1-z^{-2}\right)}$.

A digital controller to implement, uses e_n, e_{n-1}, e_{n-2} and u_{n-2}, and

$u_n = u_{n-2} + k_{e0}e_n + k_{e1}e_{n-1} + k_{e2}e_{n-2}$, $u_{n-2}=u[n-2]$, $e_n=e[n]$, $e_{n-1}=e[n-1]$, $e_{n-2}=e[n-2]$,

where $k_{e0} = k_p + \frac{1}{2}T_s k_i + \frac{2}{T_s}k_d$, $k_{e1} = T_s k_i - \frac{4}{T_s}k_d$ and $k_{e2} = -k_p + \frac{1}{2}T_s k_i + \frac{2}{T_s}k_d$.

Proportional-Integral Control – For $u = k_p e + k_i \int e dt$, the transfer function is

$$G_{PI}(s) = \frac{U(s)}{E(s)} = \frac{k_p s + k_i}{s}.$$

Using Tustin approximation, $G_{PI}(z) = \dfrac{U(z)}{E(z)} = \dfrac{k_p \frac{2}{T_s}\frac{1-z^{-1}}{1+z^{-1}} + k_i}{\frac{2}{T_s}\frac{1-z^{-1}}{1+z^{-1}}} = \dfrac{\left(2k_p + T_s k_i\right) + \left(-2k_p + T_s k_i\right)z^{-1}}{2\left(1-z^{-1}\right)}$.

Hence, a digital controller is $u_n = u_{n-1} + \left(k_p + \frac{1}{2}T_s k_i\right)e_n + \left(-k_p + \frac{1}{2}T_s k_i\right)e_{n-1}$.

Proportional-Derivative Control Law – For a proportional-derivative control law

$u = k_p e + k_d \frac{de}{dt}$, $G_{PD}(s) = \dfrac{U(s)}{E(s)} = k_p + k_d s$.

In the z-domain, $G_{PD}(z)=\dfrac{U(z)}{E(z)}=k_p+k_d\dfrac{2}{T_s}\dfrac{1-z^{-1}}{1+z^{-1}}=\dfrac{k_pT_s\left(1+z^{-1}\right)+k_d2\left(1-z^{-1}\right)}{T_s\left(1+z^{-1}\right)}$.

From $T_s\left(1+z^{-1}\right)U(z)=\left[T_s\left(1+z^{-1}\right)k_p+2\left(1-z^{-1}\right)k_d\right]E(z)$, we have

$u_n=-u_{n-1}+\left(k_p+\tfrac{2}{T_s}k_d\right)e_n+\left(k_p-\tfrac{2}{T_s}k_d\right)e_{n-1}$.

Integral Control Law – Consider an integral control $u=k_i\int edt$, $G_I(s)=\dfrac{U(s)}{E(s)}=\dfrac{k_i}{s}$.

We have $G_I(z)=\dfrac{U(z)}{E(z)}=\dfrac{k_iT_s\left(1+z^{-1}\right)}{2\left(1-z^{-1}\right)}$.

From $2\left(1-z^{-1}\right)U(z)=T_sk_i\left(1+z^{-1}\right)E(z)$, the digital controller to implement is

$u_n=u_{n-1}+\tfrac{1}{2}k_iT_s\left(e_n+e_{n-1}\right)$. ■

Practice Problem 5.6. *Modal Control and Observer Design for a Two-Link Robot*
Consider a linearized model for two-degree-of-freedom manipulator, examined in Example 2.8. Study the *torque control* problem. The linearized model is found in section 6.15.3 with the system states $x=\begin{bmatrix}x_1\\x_2\\x_3\\x_4\end{bmatrix}=\begin{bmatrix}\omega_1\\\theta_1\\\omega_2\\\theta_2\end{bmatrix}$ and controls $u=\begin{bmatrix}u_1\\u_2\end{bmatrix}=\begin{bmatrix}T_1\\T_2\end{bmatrix}$.

The outputs are angular displacements, $y=\begin{bmatrix}y_1\\y_2\end{bmatrix}=\begin{bmatrix}x_2\\x_4\end{bmatrix}=\begin{bmatrix}\theta_1\\\theta_2\end{bmatrix}$, $y=Hx=\begin{bmatrix}0&1&0&0\\0&0&0&1\end{bmatrix}x$.

For the linearized system, despite inconsistencies, denote $x\equiv\delta_x$ and $u\equiv\delta_x$.

The system states and controls are $\begin{bmatrix}x_1\\x_2\\x_3\\x_4\end{bmatrix}=\begin{bmatrix}\delta_{\omega_1}\\\delta_{\theta_1}\\\delta_{\omega_2}\\\delta_{\theta_2}\end{bmatrix}$, $\begin{bmatrix}u_1\\u_2\end{bmatrix}=\begin{bmatrix}\delta_{T_1}\\\delta_{T_2}\end{bmatrix}$.

We have $\dot{x}=\begin{bmatrix}\dot{x}_1\\\dot{x}_2\\\dot{x}_3\\\dot{x}_4\end{bmatrix}=Ax+Bu=\begin{bmatrix}-0.283&-11.25&-0.2355&-1.56\\1&0&0&0\\-0.2355&-9.364&-3.396&-22.5\\0&0&1&0\end{bmatrix}\begin{bmatrix}x_1\\x_2\\x_3\\x_4\end{bmatrix}+\begin{bmatrix}2.83&2.355\\0&0\\2.355&33.96\\0&0\end{bmatrix}\begin{bmatrix}u_1\\u_2\end{bmatrix}$.

As covered in section 5.4, design a *modal* control law and synthesize an *observer*. The characteristic equations are

$|sI-(A-BK_F)|=0$, $|sI-(A-K_EH)|=0$, $|sI-(A-BK_F-K_EH)|=0$.

The *modal* control law and feedback gain matrix K_F are found by specifying the closed-loop system eigenvalues.

Let $\lambda_{F1}=-5$, $\lambda_{F2}=-5$, $\lambda_{F3}=-10$, $\lambda_{F4}=-10$.

Using the measured outputs $y=\begin{bmatrix}y_1\\y_2\end{bmatrix}=\begin{bmatrix}x_2\\x_4\end{bmatrix}$ and controls $u=\begin{bmatrix}u_1\\u_2\end{bmatrix}$, the state *observer* is designed to observe not measured states $x_1(t)$ and $x_3(t)$.

The state *observer* gain matrix K_E is obtained by specifying the eigenvalues $\lambda_{E1}=-50$, $\lambda_{E2}=-50$, $\lambda_{E3}=-200$, $\lambda_{E4}=-200$.

The **place** command is used to find K_F and K_E.

```
A=[-0.283 -11.25 -0.2355 -1.56; 1 0 0 0; -0.2355 -9.364 -3.396 -22.5; 0 0 1 0];
B=[2.83 2.35; 0 0; 2.355 33.96; 0 0]; H=[0 1 0 0;0 0 0 1]; D=[0 0; 0 0];
ps=[-5 -5 -10 -10]; KF=place(A,B,ps)          % Modal control
po=[-50 -50 -200 -200]; KE=place(A',H',po)'  % State observer
```

We have $K_F = \begin{bmatrix} 5.524 & 14.772 & -0.389 & -1.289 \\ -0.39 & -1.3 & 0.369 & 0.9 \end{bmatrix}$ and $K_E = \begin{bmatrix} 9918.1 & -59.57 \\ 249.72 & -0.2355 \\ -67.37 & 9140.1 \\ -0.2355 & 246.6 \end{bmatrix}$.

With a feedback gain matrix K_F, a control law (5.54) with observed $\hat{x}(t)$ is $u = -K_F\hat{x}$.

Analyze the stabilizing control law with feedback gain matrix K_F, while an *observer* gain matrix is K_E. Figure 5.41.a reports the SIMULINK diagram to simulate a MIMO system in state-space notations. The outputs and controls pair (y,u) is measured. Dynamics of $(x(t), \hat{x}(t))$ and evolutions of $(y(t), \hat{y}(t))$ are reported in Figures 5.41.b. The initial conditions are x_0=[2, 1, –2, –1] and $\hat{x}_0$=[0, 0, 0, 0]. Tracking errors

$\bar{e}_{y_1}(t) = \hat{y}_1(t) - y_1(t)$, $\bar{e}_{y_2}(t) = \hat{y}_2(t) - y_2(t)$

converge to zero, see Figures 5.41.b. The observed states $\hat{x}(t)$ converge to $x(t)$, and, stability is guaranteed.The settling time for the *observer* and system are 0.1 and 0.3 sec.

Modal Control and State Observer Designs With Specifications – Sections 5.4 and 6.15.4 report dynamic optimality and search algorithms to design *modal* control laws (5.47) $u=-K_F x$ and (5.54) $u = -K_F\hat{x}$ with $\boldsymbol{\lambda}_F \in \mathbb{R}\times\mathbb{C}$ and $\forall\boldsymbol{\lambda}_F \in \mathbb{R}$, $\forall \text{Re}(\boldsymbol{\lambda}_F)<0$.

To ensure specified system governance and stability margins, specify positive definite feedback and *observer* gains $(k_{Fij},k_{Eij})>0$, and, admissible p-norms for (K_F,K_E).

Specify range$(\boldsymbol{\lambda}_F \in \Lambda_F^A,\ \boldsymbol{\lambda}_E \in \Lambda_E^A)$, and, find the eigenvalues $\forall(\boldsymbol{\lambda}_F,\boldsymbol{\lambda}_F)\in\mathbb{R}^4$, $\boldsymbol{\lambda}_i=(\lambda_1,\lambda_2,\lambda_3,\lambda_4)$ in the $\boldsymbol{\lambda}$-plane, to ensure settling time ~0.5 sec under the limits on the feedback and *observer* gains, $\forall(k_{Fij},k_{Eij})>0$.

Our objective is to find the controller and *observer* gains (K_F,K_E). System and *observer* governance with feedback gains $\forall(k_{Fij},k_{Eij})>0$ should be ensured within the range of real-valued eigenvalues $(\boldsymbol{\lambda}_F,\boldsymbol{\lambda}_E)$ as

$\Lambda_F^A \in [-5\ \ -100]$, $\forall\boldsymbol{\lambda}_F\in\mathbb{R}$, $\Lambda_E^A \in [-100\ \ -200]$, $\forall\boldsymbol{\lambda}_E\in\mathbb{R}$.

To avoid high gains, let $\|K_F\|_1\le 100$, $\|K_F\|_\infty\le 175$, $k_{F\max}\le 75$,
$\|K_E\|_1\le 22500$, $\|K_E\|_\infty\le 22500$, $k_{E\max}\le 22500$.

The *State Observer and Modal Control Algorithm* 5.1, reported in section 5.4. Let ($\Lambda_F^A \in [-5\ \ -100]$, $\forall\boldsymbol{\lambda}_F\in\mathbb{R}$, $\Lambda_E^A \in [-100\ \ -200]$, $\forall\boldsymbol{\lambda}_E\in\mathbb{R}$).

We obtain $\boldsymbol{\lambda}_F$=[–5, –7.5, –12.5, –75], $K_F = \begin{bmatrix} 11.9 & 61.3 & 8.887 & 70.21 \\ 0.0001 & 1.877 & 1.228 & 7.3 \end{bmatrix}$,

$\forall k_{Fij}>0$, $\|K_F\|_1=77.51$, $\|K_F\|_\infty=152.3$, $k_{F\max}=70.21$.

For a state *observer* (5.50) $\frac{d\hat{x}}{dt} = (A - K_E H)\hat{x} + Bu + K_E y$, one finds

$\boldsymbol{\lambda}_E$=[–125, –125, –175, –175], $K_E = \begin{bmatrix} 19597.6 & 3.145 \\ 287.23 & 0.191 \\ 1.324 & 20842.5 \\ 0.216 & 296.59 \end{bmatrix}$,

$\forall k_{Eij}>0$, $\|K_E\|_1=21142$, $\|K_E\|_\infty=20844$, $k_{E\max}=20842.5$.

The design objectives are guaranteed. The closed-loop system dynamics is similar to evolutions documented in Figures 5.41.b.

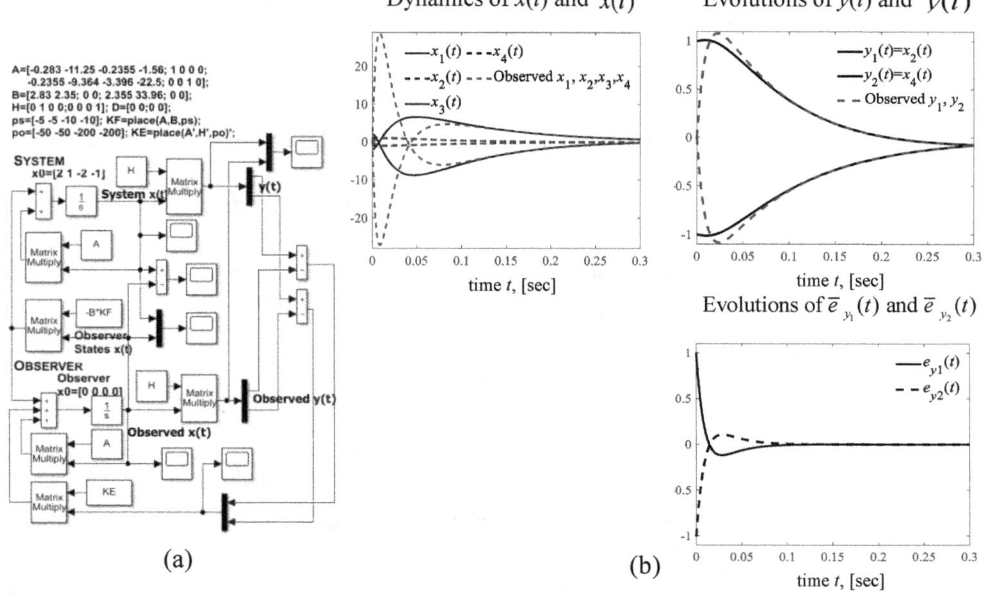

(a) (b)

Figure 5.41. Controller-*observer* structure for a MIMO system:
(a) SIMULINK diagram for a closed-loop system with the state *observer* and control law as $u=-K_F\hat{x}$, where $\hat{x}(t)$ is observed using $(u(t), y(t))$;
(b) Dynamics of states $\begin{bmatrix} x(t) \\ \hat{x}(t) \end{bmatrix}$, outputs $\begin{bmatrix} y(t) \\ \hat{y}(t) \end{bmatrix}$ and errors $\begin{bmatrix} \bar{e}_{y_1}(t)=\hat{y}_1(t)-y_1(t) \\ \bar{e}_{y_2}(t)=\hat{y}_2(t)-y_2(t) \end{bmatrix}$. Initial conditions are x_0=[2, 1, –2, –1] and $\hat{x}_0$=[0, 0, 0, 0]. ■

Practice Problem 5.7. Optimal Control of an Unstable System

Examples 5.26 and 5.27 investigate optimal control problems for the first-order systems

$$\dot{x}=u$$

and $$\dot{x}=-x+u.$$

Consider an open-loop unstable system $\dot{x}=x+u$.

Minimize a functional $J=\min_{x,u}\int_0^\infty \frac{1}{2}(x^2+u^2)dt$ subject to $\dot{x}=x+u$.

A scalar Hamiltonian function is $H(x,u,\frac{\partial V}{\partial x})=\frac{1}{2}x^2+\frac{1}{2}u^2+\frac{\partial V}{\partial x}(x+u)$.

Applying $\frac{\partial H}{\partial u}=0$, the control law is $u=-\frac{\partial V}{\partial x}$.

The Hamilton-Jacobi equation is $-\frac{\partial V}{\partial t}=\frac{1}{2}x^2+\frac{\partial V}{\partial x}x-\frac{1}{2}\left(\frac{\partial V}{\partial x}\right)^2$.

For t_f=∞, solve $0=\frac{1}{2}x^2+\frac{\partial V}{\partial x}x-\frac{1}{2}\left(\frac{\partial V}{\partial x}\right)^2$, satisfied by $V(x)=\frac{1}{2}kx^2$.

Grouping terms with x^2 one obtains an algebraic equation $1+2k-k^2=0$.

The control law is $u=-kx=-2.414x$, k=2.414.

The closed-loop system $\dot{x}=-1.414x$ is stable.

Consider minimization of a functional $J=\min_{x,u}\int_0^{t_f}\frac{1}{2}(x^2+u^2)dt$, t_f=5 sec.

The unknown $k(t)$ of the return function $V(x)$ is found by solving

$-\dot{k}=1+2k-k^2$, $k(t_f)$=0.

To analytically solve the Riccati equation, consider $\int_0^{t_f}\frac{dk}{k^2-2k-1}=\int_0^{t_f}dt$.

The real roots (r_1,r_2) for the quadratic equation k^2–2k–1=0 are $r_{1,2}=1\pm\sqrt{2}$.

The fractioning yields $\frac{1}{2\sqrt{2}}\int_0^{t_f}\left(\frac{1}{k-1-\sqrt{2}}-\frac{1}{k-1+\sqrt{2}}\right)dk=\int_0^{t_f}dt$.

From $\ln\left(\frac{k(t)-1-\sqrt{2}}{k(t)-1+\sqrt{2}}\right)-\ln\left(\frac{-1-\sqrt{2}}{-1+\sqrt{2}}\right)=-2\sqrt{2}(t_f-t)$, $\left(\frac{-1+\sqrt{2}}{-1-\sqrt{2}}\right)\left(\frac{k(t)-1-\sqrt{2}}{k(t)-1+\sqrt{2}}\right)=e^{-2\sqrt{2}(t_f-t)}$.

Hence, $k(t)=\dfrac{1-e^{-2\sqrt{2}(t_f-t)}}{\sqrt{2}-1+(\sqrt{2}+1)e^{-2\sqrt{2}(t_f-t)}}$.

An optimal control law is $u^*(t)=-k(t)x=-\dfrac{1-e^{-2\sqrt{2}(t_f-t)}}{\sqrt{2}-1+(\sqrt{2}+1)e^{-2\sqrt{2}(t_f-t)}}x$.

The SIMULINK diagram to simulate closed-loop systems with suboptimal control u and optimal control u^* are reported in Figures 5.42.a. Optimal dynamics $x^*(t)$ for an initial condition x_0= –1, as well as evolutions of $u^*(t)$ and $k(t)$, are illustrated in Figure 5.42.b.

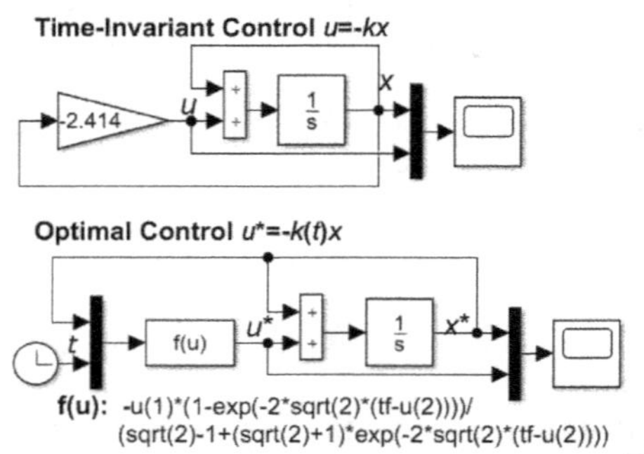

(a)

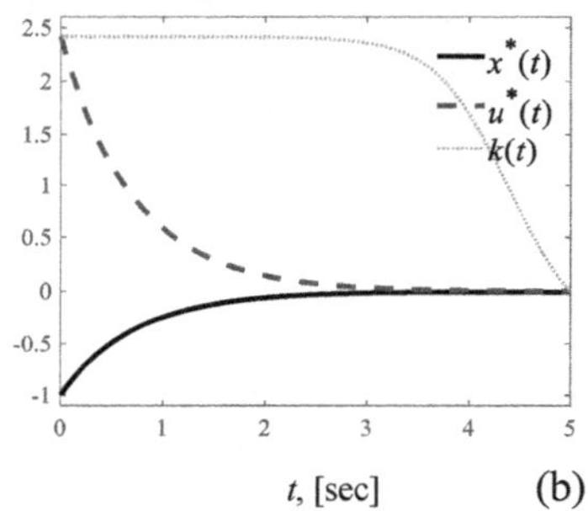

(b)

Figure 5.42. (a) SIMULINK diagrams to simulate closed-loop systems with control laws u= –kx and $u^*(t)$= –$k(t)x$. For time-invariant feedback gain k and time-varying $k(t)$, the governing equations are

$$\dot{x}=x+u=x-kx=(1-k)x,\ k=2.414,$$

and
$$\dot{x}=x+u^*=x-k(t)x=\left(1-\frac{1-e^{-2\sqrt{2}(t_f-t)}}{\sqrt{2}-1+(\sqrt{2}+1)e^{-2\sqrt{2}(t_f-t)}}\right)x;$$

(b) System dynamics $x^*(t)$ if x_0= –1, control evolution $u^*(t)$, and, evolution of time-varying $k(t)$. ■

Practice Problem 5.8. Discretization of a Nonlinear System and Tracking Control: Design for Continuous-Time and Discrete-Time Systems

Section 5.9 covers topics on optimal tracking control of continuous-time systems. Examples 5.33 and 5.36 study controlled rotational motion of a satellite in free space. Nonlinear governing equations for the rotational dynamics with aerodynamic drag are

$$\dot{x}_1=-x_1-x_1^3+u,$$

$$\dot{x}_2=x_1,\ y=x_2,\ e(t)=r(t)-y(t)=r(t)-x_2(t),$$

where x_1 and x_2 are the angular velocity ω and displacement θ, u is the controlled reaction torque.

The error dynamics is

$$\dot{x}_e = r - y = r - x_2,$$

$$\dot{e} = -e + \dot{r} - \dot{y} = -e + \dot{r} - \dot{x}_2 = -e + \dot{r} - x_1 .$$

Using the state vector $\mathbf{x} = \begin{bmatrix} x \\ \int edt \\ e \end{bmatrix} = \begin{bmatrix} x_1 \\ x_2 \\ \int edt \\ e \end{bmatrix}$, we have (5.116) with a nonlinear term $\mathbf{F}(\mathbf{x})$

$$\dot{\mathbf{x}} = \mathbf{A}\mathbf{x} + \mathbf{F}(\mathbf{x}) + A_r r + \mathbf{B}u, \begin{bmatrix} \dot{x}_1 \\ \dot{x}_2 \\ \dot{x}_e \\ \dot{e} \end{bmatrix} = \begin{bmatrix} -1 & 0 & 0 & 0 \\ 1 & 0 & 0 & 0 \\ 0 & -1 & 0 & 0 \\ -1 & 0 & 0 & -1 \end{bmatrix} \begin{bmatrix} x_1 \\ x_2 \\ x_e \\ e \end{bmatrix} + \begin{bmatrix} -x_1^3 \\ 0 \\ 0 \\ 0 \end{bmatrix} + \begin{bmatrix} 0 & 0 \\ 0 & 0 \\ 1 & 0 \\ 0 & 1 \end{bmatrix} \begin{bmatrix} r \\ \dot{r} \end{bmatrix} + \begin{bmatrix} 1 \\ 0 \\ 0 \\ 0 \end{bmatrix} u, \; y = x_2.$$

Design schemes for nonlinear systems are covered in chapter 6. While one may linearize the governing equations, design control laws for a linear model because the system is stable with any feedback on x_2. For example, a system $\begin{cases} \dot{x}_1 = -x_1 - x_1^3 - x_2 \\ \dot{x}_2 = x_1 \end{cases}$ is stable. Neglect a nonlinear term $\mathbf{F}(\mathbf{x})$ because a physics-consistent inherent nonlinearity $-x_1^3$ is a stabilizing term, which expands the stability margins.

To design the control law, minimize a quadratic functional (5.117)

$$J = \min_{\mathbf{x},u} \int_0^\infty \tfrac{1}{2}\left(\mathbf{x}^T Q \mathbf{x} + G u^2\right) dt$$

subject to $\dot{\mathbf{x}} = \mathbf{A}\mathbf{x} + A_r r + \mathbf{B}u, \; y = x_2$.

The Hamiltonian $H(\mathbf{x}, u, \frac{\partial V}{\partial \mathbf{x}})$ (5.118) yields a control law (5.122)

$$u = -G^{-1}\mathbf{B}^T \frac{\partial V(\mathbf{x})}{\partial \mathbf{x}} = -G^{-1}\mathbf{B}^T K\mathbf{x} = -K_F \mathbf{x}, \; \mathbf{x} = \begin{bmatrix} x \\ \int edt \\ e \end{bmatrix} = \begin{bmatrix} x_1 \\ x_2 \\ \int edt \\ e \end{bmatrix}.$$

Solution of the Hamilton-Jacobi equation (5.120) is satisfied by the quadratic return function (5.121) $V(\mathbf{x}) = \frac{1}{2}\mathbf{x}^T K \mathbf{x}$, $K \in \mathbb{R}^{4\times 4}$. Considering the quadratic terms, solve (5.123)

$$Q + \mathbf{A}^T K + K\mathbf{A} - K\mathbf{B}G^{-1}\mathbf{B}^T K = 0 .$$

For the weighting coefficients matrix $Q = \begin{bmatrix} 1 & 0 & 0 & 0 \\ 0 & 1 & 0 & 0 \\ 0 & 0 & 1 & 0 \\ 0 & 0 & 0 & 10 \end{bmatrix}$ and G=1, computing $K \in \mathbb{R}^{4\times 4}$ and $K_F \in \mathbb{R}^{1\times 4}$, find a control law

$$u = \left(-2x_1 - 2.87x_2 + \int edt + 0.612e\right).$$

The simulations can be performed similar to conducted in Example 5.33. For a linear model, discretization and discrete tracking control were investigated in Example 5.36. For a linear model, using (5.19) $\mathbf{A}_n = e^{\mathbf{A}T_s}$, $\mathbf{B}_n = \mathbf{A}^{-1}(e^{\mathbf{A}T_s} - I)\mathbf{B}$.

For a nonlinear system, apply the Euler approximation $\left.\frac{dx}{dt}\right|_{t=kT_s} = \frac{x(kT_s + T_s) - x(kT_s)}{T_s}$.

A nonlinear continuous-time system $\dot{\mathbf{x}} = \mathbf{A}\mathbf{x} + \mathbf{F}(\mathbf{x}) + A_r r + \mathbf{B}u$ is discretized as

$$\mathbf{x}_{n+1} = \mathbf{A}_n\mathbf{x}_n + \mathbf{F}_n(\mathbf{x}_n) + A_{rn}r_n + \mathbf{B}_n u_n, \quad \begin{bmatrix} x_{1n+1} \\ x_{2n+1} \\ x_{en+1} \\ e_{n+1} \end{bmatrix} = \begin{bmatrix} 1-T_s & 0 & 0 & 0 \\ T_s & 1 & 0 & 0 \\ 0 & -T_s & 1 & 0 \\ -T_s & 0 & 0 & 1-T_s \end{bmatrix} \begin{bmatrix} x_{1n} \\ x_{2n} \\ x_{en} \\ e_n \end{bmatrix} + \begin{bmatrix} -T_s x_{1n}^3 \\ 0 \\ 0 \\ 0 \end{bmatrix} + \begin{bmatrix} 0 \\ 0 \\ T_s \\ 0 \end{bmatrix} r_n + \begin{bmatrix} T_s \\ 0 \\ 0 \\ 0 \end{bmatrix} u_n.$$
$$y_n = x_{2n},$$

As in Example 5.36, the sampling period is T_s=0.1 sec. One finds discretized model matrices ($\mathbf{A}_n$,A_{rn},$\mathbf{B}_n$) and mapping $\mathbf{F}_n(\mathbf{x}_n)$. An aerodynamic drag enables stability, and, $-x_1^3$ and $-T_s x_{1n}^3$ are the stabilizing terms in continuous-time and discrete-time systems. Therefore, one may consider linear models, minimizing quadratic functionals and indexes. The resulting control laws with linear feedback will stabilize a nonlinear system, and, guarantee tracking.

To design a tracking control law, minimize the quadratic performance index (5.139)

$$J = \min_{\mathbf{x}_n, u_n} \sum_{n=0}^{\infty} \left[\mathbf{x}_n^T Q_n \mathbf{x}_n + u_n^T G_n u_n \right], \quad Q_n = \begin{bmatrix} 1 & 0 & 0 & 0 \\ 0 & 1 & 0 & 0 \\ 0 & 0 & 1 & 0 \\ 0 & 0 & 0 & 1 \end{bmatrix}.$$

Consider two different G_n are G_n=0.1 and G_n=0.01.

The tracking control law (5.140) is

$$u_n = -\left(G_n + \mathbf{B}_n^T K_n \mathbf{B}_n\right)^{-1} \mathbf{B}_n^T K_n \mathbf{A}_n \mathbf{x}_n = -K_{Fn}\mathbf{x}_n\,.$$

For G_n=0.1, $K_n = \begin{bmatrix} 5.08 & 8.29 & -3.88 & -0.4 \\ 8.29 & 37.2 & -18.8 & 2.92 \\ -3.88 & -18.8 & 22.3 & -0.93 \\ -0.4 & 2.92 & -0.93 & 5.22 \end{bmatrix}$ and K_{Fn}=[3.6 5.75 –2.57 –0.24].

For G_n=0.01, $K_n = \begin{bmatrix} 1.99 & 3.4 & -1.73 & -0.21 \\ 3.4 & 29.4 & -15.3 & 3.24 \\ -1.73 & -15.3 & 20.6 & -1.09 \\ -0.21 & 3.24 & -1.09 & 5.2 \end{bmatrix}$ and K_{Fn}=[7.2 11.9 –5.78 –0.63].

For a nonlinear discrete system, study tracking if reference is r_n=±1$_n$. Let the initial conditions are $\mathbf{x}_{n0}$=[0.1 –0.1 0 0]. Closed-loop system dynamics are documented in Figures 5.43 for computed feedback matrices K_{Fn}. Tracking is guaranteed with near-zero steady-state error.

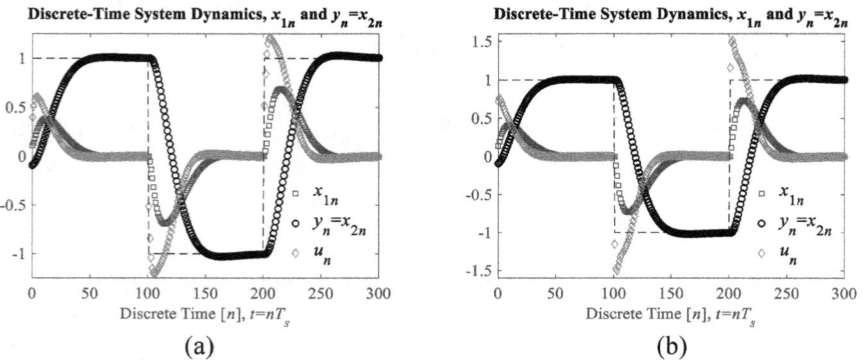

Figure 5.43. Closed-loop nonlinear system $\mathbf{x}_{n+1} = \mathbf{A}_n\mathbf{x}_n + \mathbf{F}_n(\mathbf{x}_n) + A_{rn}r_n + \mathbf{B}_n u_n$ with a tracking control law u_n= –$K_{Fn}\mathbf{x}_n$. System dynamics and tracking, r_n=±1$_n$: Evolutions of (x_{1n},x_{2n}), u_n and y_n=x_{2n}:
(a) Feedback gain matrix is K_{Fn}=[3.6 5.75 –2.57 –0.24];
(b) Feedback gain matrix is K_{Fn}=[7.2 11.9 –5.78 –0.63]. ■

Practice Problem 5.9. Control of Two-Phase Induction Motors

Practice Problems 3.8 to 3.11 report modeling and control principles, applied to induction motors. The Lagrange equations, Kirchhoff voltage law, Faraday law of induction, and *torsional-mechanical* equations of motion yield a not Cauchy form of governing equations

$$\frac{di_{as}}{dt}=\frac{1}{L_{ss}}\left[-r_s i_{as}-L_{ms}\left(\frac{di'_{ar}}{dt}\cos\theta_r-i'_{ar}\sin\theta_r\omega_r\right)+L_{ms}\left(\frac{di'_{br}}{dt}\sin\theta_r+i'_{br}\cos\theta_r\omega_r\right)+u_{as}\right],$$

$$\frac{di_{bs}}{dt}=\frac{1}{L_{ss}}\left[-r_s i_{bs}-L_{ms}\left(\frac{di'_{ar}}{dt}\sin\theta_r+i'_{ar}\cos\theta_r\omega_r\right)-L_{ms}\left(\frac{di'_{br}}{dt}\cos\theta_r-i'_{br}\sin\theta_r\omega_r\right)+u_{bs}\right],$$

$$\frac{di'_{ar}}{dt}=\frac{1}{L'_{rr}}\left[-r'_r i'_{ar}-L_{ms}\left(\frac{di_{as}}{dt}\cos\theta_r-i_{as}\sin\theta_r\omega_r\right)-L_{ms}\left(\frac{di_{bs}}{dt}\sin\theta_r+i_{bs}\cos\theta_r\omega_r\right)+u'_{ar}\right],$$

$$\frac{di'_{br}}{dt}=\frac{1}{L'_{rr}}\left[-r'_r i'_{br}+L_{ms}\left(\frac{di_{as}}{dt}\sin\theta_r+i_{as}\cos\theta_r\omega_r\right)-L_{ms}\left(\frac{di_{bs}}{dt}\cos\theta_r-i_{bs}\sin\theta_r\omega_r\right)+u'_{br}\right],$$

$$\frac{d\omega_r}{dt}=-\frac{P^2}{4J}L_{ms}\left[\left(i_{as}i'_{ar}+i_{bs}i'_{br}\right)\sin\theta_r+\left(i_{as}i'_{br}-i_{bs}i'_{ar}\right)\cos\theta_r\right]-\frac{B_m}{J}\omega_r-\frac{P}{2J}T_L,$$

$$\frac{d\theta_r}{dt}=\omega_r.$$

Consider the frequency control of induction motors. Cauchy's form of nonlinear governing equations were found in chapter 3. Open-loop induction motors, controlled by changing frequency of the applied phase voltages, or, by means of voltage-frequency control, are stable. The electromagnetic-consistent proportional or proportional-integral control will guarantee stability. Electromagnetic compliant frequency control must be implemented. The constant volts-per-hertz $\frac{V}{f}=\text{const}$ and varying $\frac{V}{f}\neq\text{const}$ control principles are implemented regulating the open-loop motors due to low *slip* electromagnetics.

For squirrel-cage induction motors, $u'_{ar}=0$ and $u'_{br}=0$. One changes the frequency and phase of the applied ac phase voltages (u_{as},u_{bs}) to guarantee the clockwise and counterclockwise rotation. Velocity regulation, stabilizing and tracking control, as well as load rejection must be accomplished.

Consider a 7.5 kW, 380 V (rms), 60 Hz, four-pole (P=4) A-class induction motor with the following parameters r_s=0.742 ohm, r_r=0.495 ohm, L_{ms}=0.128 H, L_{ls}=0.004 H, $L_{ss}=L_{ls}+L_{ms}$, L_{lr}=0.0065 H, $L_{rr}=L_{lr}+L_{ms}$, B_m=0.038 N-m-sec/rad and J=0.19 kg-m^2. Assume equal number of turns of the stator and rotor windings, $N_s=N_r$.

The balanced operation is guaranteed by applying the balanced phase voltages (u_{as},u_{bs}). Motor operates in the envelope $\omega_e\le\omega_r\le\omega_{r\,\text{critical}}$, where the synchronous angular velocity is $\omega_e=\frac{4\pi f}{P}$. For industrial A/B-class induction motors 1 to 1000 kW, the slip varies between 0.02 to 0.05. The *slip* is defined as $slip=\frac{\omega_e-\omega_r}{\omega_e}$.

Phase voltages (u_{as},u_{bs}) must form the balanced voltage set, and, the electrical angular velocity ω_r is defined by the frequency of phase voltages $\omega_f=2\pi f$.

The mechanical angular velocity is $\omega_{rm}=\frac{2}{P}\omega_r$.

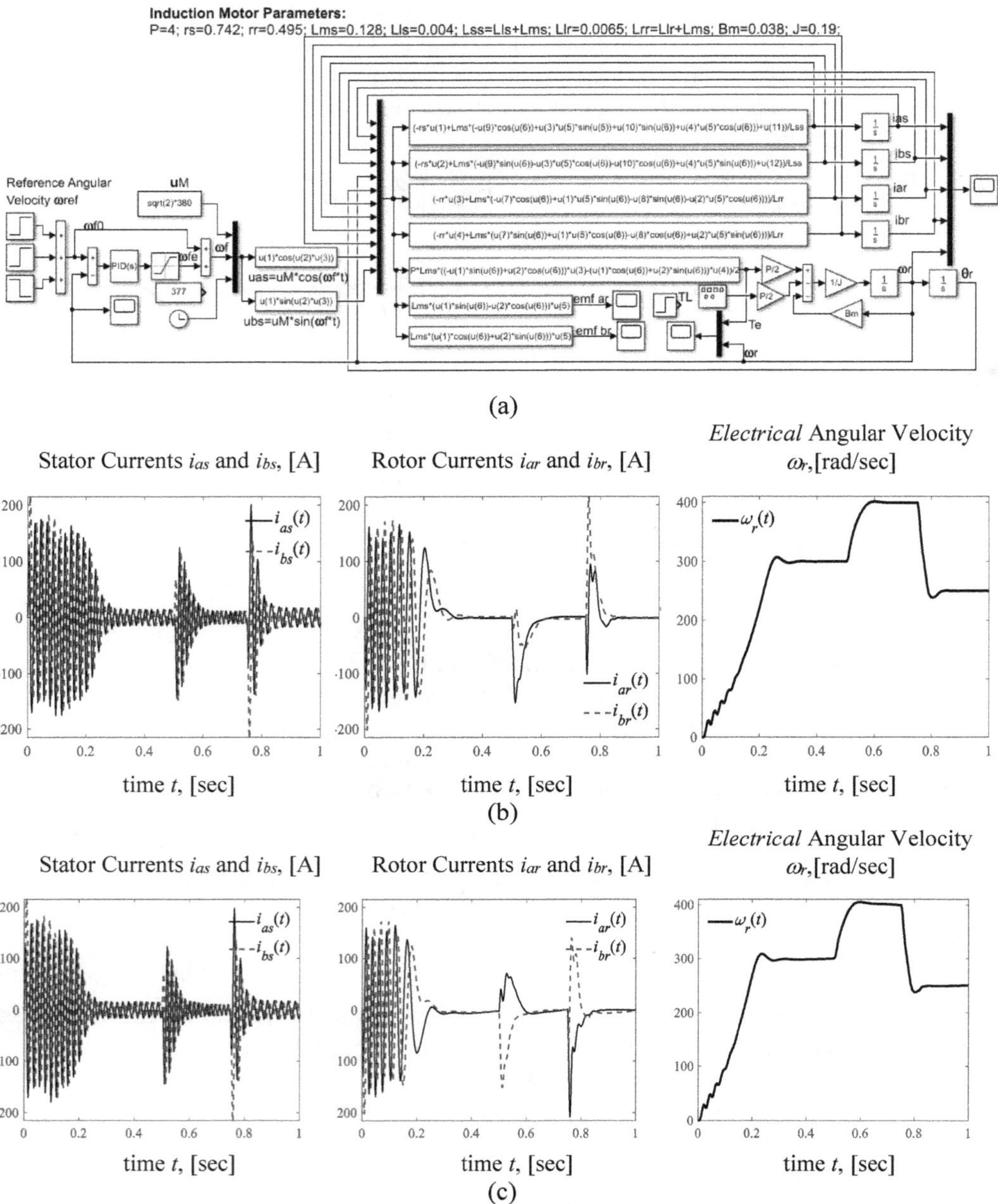

Figure 5.44. (a) SIMULINK diagram to model a closed-loop induction motor. The bounded balanced phase voltages are $u_{as}(t)=\sqrt{2}u_M\cos\left((\omega_{f_0}+\omega_{f_e})t\right)$, $u_{bs}(t)=\sqrt{2}u_M\sin\left((\omega_{f_0}+\omega_{f_e})t\right)$, u_M=380 V, $\omega_{f_0}\equiv\omega_{e_{\text{reference}}}$,

$$u\equiv\omega_{f_e},\ u=\phi_{u_{\min}}^{u_{\max}}\left(k_p e+k_i\int e dt\right),\ e(t)=[\omega_{\text{ref}}(t)-\omega_r(t)],\ u_{\min}\le u\le u_{\max},\ |u|\le 10\text{ Hz};$$

(b) Tracking of reference angular velocity: Dynamics of an induction motor, u_M=380 V, T_L=0;

(c) Tracking of reference angular velocity $\omega_{\text{ref}}=\begin{cases}300\ \frac{\text{rad}}{\text{sec}},\ 0\le t<0.5\text{ sec}\\ 400\ \frac{\text{rad}}{\text{sec}},\ 0.5\le t<0.75\text{ sec}\\ 250\ \frac{\text{rad}}{\text{sec}},\ 0.75\le t\le 1\text{ sec}\end{cases}$ under applied loads $T_L(t)$: Transient dynamics of a motor with loads $T_L(t)$=±20 N-m applied at 2 Hz.

Frequency $(\omega_{f_0}+\omega_{f_e})$ of balanced phase voltages (u_{as},u_{bs}) is controlled to regulate ω_r. We have

$u_{as}(t)=\sqrt{2}u_M\cos\left((\omega_{f_0}+\omega_{f_e})t\right)$, u_M=const, $\omega_{f_0}\equiv\omega_{e_{\text{reference}}}$, $u\equiv\omega_{f_e}$,

$u_{bs}(t)=\sqrt{2}u_M\sin\left((\omega_{f_0}+\omega_{f_e})t\right)$.

Notations: The angular and synchronous angular velocities are denotes as (ω_r,ω_e), while the angular frequencies of the applied phase voltages (u_{as},u_{bs}) are $(\omega_{f_0},\omega_{f_e})$, $\omega_f=2\pi f$.

Controlled voltage frequency is bounded as $\omega_{f_e\min}\le\omega_{f_e}\le\omega_{f_e\max}$.

For a proportional-integral frequency control,

$u\equiv\omega_{f_e}$, $u=\phi_{u_{\min}}^{u_{\max}}\left(k_pe+k_i\int edt\right)$, $e(t)=\omega_{\text{ref}}(t)-\omega_r(t)$, $u_{\min}\le u\le u_{\max}$.

For given $\omega_{f_0}\equiv\omega_{e_{\text{reference}}}$, motors operate in the angular velocity envelope $\omega_r\in[\omega_{r\ \text{critical}}\ \omega_e]$, $\omega_e=\frac{4\pi}{P}f$. The voltage frequency $\omega_{f_0}\equiv\omega_{e_{\text{reference}}}$ is found using the reference angular velocity ω_{ref}, which yields the corresponding $\omega_f=2\pi f$ for specified synchronous angular velocity ω_e. Induction motors rotate with the electrical angular velocity ω_r, and, $\omega_r<\omega_e$ at any load $T_L>0$ due to friction. Neglecting friction, for $T_L=0$, $\omega_r=\omega_e$. The regulated frequency of phase voltages (u_{as},u_{bs}) $u\equiv\omega_{f_e}$, is found using the tracking error $e(t)=\omega_{\text{ref}}(t)-\omega_r(t)$.

The limits on control are $|u|\le 10$ Hz, and, the feedback gains are $k_p=0.01$ and $k_i=0.01$.

The SIMULINK diagram is illustrated in Figure 5.44.a. The reference angular velocity is $\omega_{\text{ref}}=\begin{cases}300\ \frac{\text{rad}}{\text{sec}},\ 0\le t<0.5\text{ sec}\\400\ \frac{\text{rad}}{\text{sec}},\ 0.5\le t<0.75\text{ sec}\\250\ \frac{\text{rad}}{\text{sec}},\ 0.75\le t\le 1\text{ sec}\end{cases}$. Dynamics of the stator and rotor currents (i_{as},i_{bs},i_{ar},i_{br}) and angular velocity $\omega_r(t)$ are plotted in Figures 5.44.b and c if the load torque is $T_L=0$, as well as $T_L(t)=\pm 20$ N-m, applied at 2 Hz. Motor accelerates from stall, $\omega_{r0}=0$ rad/sec. Tracking of $\omega_{\text{ref}}(t)$, under loads applied $T_L(t)$, are guaranteed. Adequate acceleration capabilities and transient dynamics are achieved. The physics-consistent and hardware-compliant control law is designed and demonstrated. ■

Practice Problem 5.10. *Tracking Control of an Electric Drive With a Passive Filter and Compensator*

Investigate control solutions for an electromechanical system with a permanent magnet dc motor. Passive circuit schemes predate active analog filters, PWM power electronic schemes, analog controllers and digital electronic solutions. Many power electronics and communication systems implement passive circuits.

Consider an electric drive with a permanent magnet dc motor. Parameters are r_a=3.15 ohm, L_a=0.0066 H, k_a=0.16 V-sec/rad, B_m=0.0001 N-m-sec/rad and J=0.00025 kg-m^2. The motor bandwidth is 10 Hz, which implies that systemcs and transients are within ~0.1 sec. There is a noise in the measured angular velocity by a tachometer, with the noise frequency from 1000 to 10000 Hz.

Practice Problems 3.15 and 3.16 cover passive filters and regulators. Design the second-order notch filter by specifying the unit dc gain with attenuation up to –3 dB for the system frequency up to 100 Hz.

Noise, is attenuated by a notch filter, shown in Figure 5.45.a.

The transfer function is $G_{\text{filter}}(s)=\dfrac{U_0(s)}{U_1(s)}=\dfrac{R_3\left[LCs^2+(R_1+R_2)Cs+1\right]}{(R_2+R_3)LCs^2+\left[(R_1+R_2)R_3C+R_1R_2C+L\right]s+R_1+R_3}$.

In Practice Problem 3.15, we found R_1=1 ohm, R_2=100 ohm, R_3=1 ohm, L=0.01 H and C=0.01 F. The Bode plots are shown in Figure 3.72.b.

For the lag-lead regulator circuit schematics, shown in Figure 5.45.a.

The transfer function is $G_{\text{regulator}}(s)=\dfrac{R_{1R}R_{2R}C_{1R}C_{2R}s^2+(R_{1R}C_{1R}+R_{2R}C_{2R})s+1}{R_{1R}R_{2R}C_{1R}C_{2R}s^2+(R_{1R}C_{1R}+R_{1R}C_{2R}+R_{2R}C_{2R})s+1}$,

where R_{1R}=1 ohm, R_{2R}=1 ohm, C_{1R}=0.1 F and $C_{2R}=1\times10^{-5}$ F.

Passive regulators are characterized by a unit gain, which may not be sufficient. The steady-state error in the *unity feedback systems* is $\lim\limits_{t\to\infty}e(t)=\lim\limits_{s\to0}\dfrac{sR(s)}{1+G_{\text{regulator}}(s)G_{sys}(s)}$.

The notch filter gain is $\dfrac{R_3}{R_1+R_3}$. From $u_{th}=k_{th}\omega_r$, the tachometer coefficient is $k_{th}=\dfrac{u_{th}}{\omega_r}$, k_{th}=0.14 V-sec/rad. The sensor gain affects design. A *nonunity feedback system* is considered. Due to the sensor gain, filter attenuation and regulator gain, the balancing and compensating schemes are used to ensure tracking and minimize steady-state error. The scaling factor is calculated using the aforementioned gains. For the *nonunity feedback system*, schematically depicted by the SIMULINK diagram in Figure 5.46.a, the forward path transfer function is

$$G_F(s)=\frac{G_{\text{regulator}}(s)G_{sys}(s)}{1+G_{\text{regulator}}(s)G_{sys}(s)k_{th}G_{\text{filter}}(s)-G_{\text{regulator}}(s)G_{sys}(s)}.$$

The transfer function for a closed-loop system is $G_\Sigma(s)=\dfrac{\Omega(s)}{R(s)}=\dfrac{G_F(s)}{1+G_F(s)}$.

One computes the tracking error $\lim\limits_{t\to\infty}e(t)=\lim\limits_{s\to0}\dfrac{sR(s)}{1+G_F(s)}$, and, finds the scaling factor, improves system dynamics, minimizes hardware-dependent tracking error, etc.

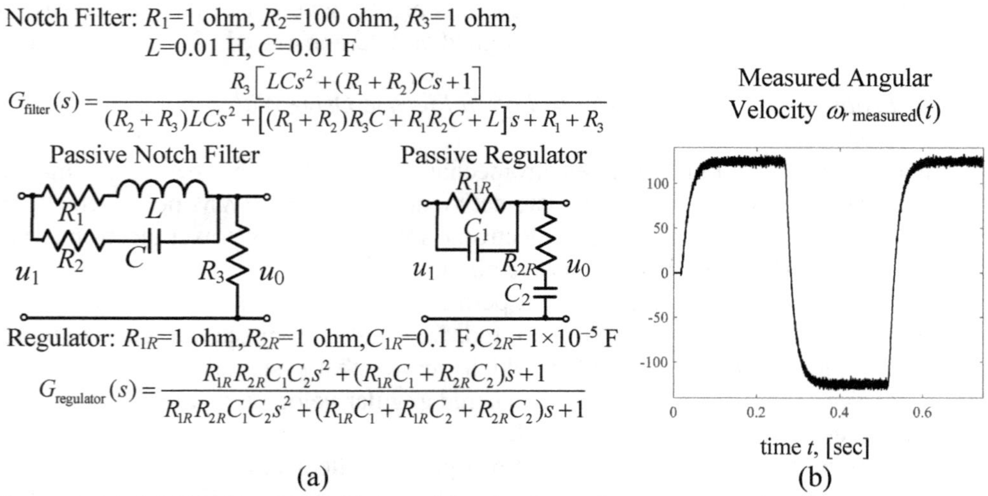

Notch Filter: R_1=1 ohm, R_2=100 ohm, R_3=1 ohm, L=0.01 H, C=0.01 F

$$G_{\text{filter}}(s)=\frac{R_3\left[LCs^2+(R_1+R_2)Cs+1\right]}{(R_2+R_3)LCs^2+\left[(R_1+R_2)R_3C+R_1R_2C+L\right]s+R_1+R_3}$$

Passive Notch Filter

Passive Regulator

Regulator: R_{1R}=1 ohm, R_{2R}=1 ohm, C_{1R}=0.1 F, $C_{2R}=1\times10^{-5}$ F

$$G_{\text{regulator}}(s)=\frac{R_{1R}R_{2R}C_1C_2s^2+(R_{1R}C_1+R_{2R}C_2)s+1}{R_{1R}R_{2R}C_1C_2s^2+(R_{1R}C_1+R_{1R}C_2+R_{2R}C_2)s+1}$$

(a)

Measured Angular Velocity $\omega_{r\,\text{measured}}(t)$

time t, [sec]

(b)

Figure 5.45. (a) Passive notch filter and lag-lead regulator to control a dc motor; (b) Measured angular velocity $\omega_{r\,\text{measured}}(t)$ tracks a reference $r(t)=\pm125$ rad/sec.

Experimental Studies – Experiments are performed for $r(t)=\pm125$ rad/sec and $T_L=0$ N-m. To ensure clockwise and counterclockwise rotation, the polarity of the armature voltage u_a is changed by using the voltage switching relay. Figure 5.45.b documents the measured angular velocity with high-power multisource multi-frequency noise. Filtering and tracking of $\omega_{\text{ref}}(t)$ are ensured with the steady-state error. At steady-state, the angular velocity may reach ±125 rad/sec with the compensating schemes.

Simulation Studies – A permanent magnet dc motor is described by equations (3.19)

$$\frac{di_a}{dt}=\frac{1}{L_a}\left(-r_a i_a-k_a\omega_r+u_a\right),\ -60\le u_a\le 60\text{ V},$$

$$\frac{d\omega_r}{dt}=\frac{1}{J}\left(k_a i_a-B_m\omega_r-T_L\right).$$

A SIMULINK diagram is illustrated in Figure 5.46.a. Dynamics of the armature current $i_a(t)$ and angular velocity $\omega_r(t)$ are depicted in Figure 5.46.b. Simulations are performed to analyze a closed-loop system with filter and regulator. In simulations, a Band-Limited White Noise block is used as a noise source.

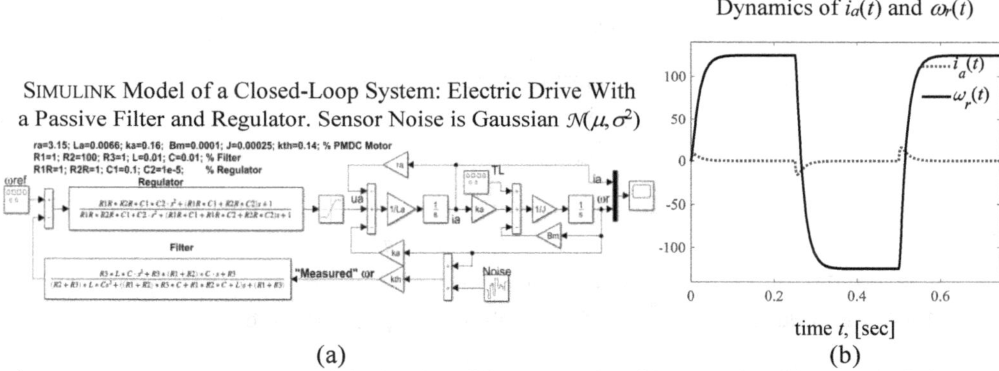

(a) (b)

Figure 5.46. (a) SIMULINK model of a closed-loop system with a passive filter and regulator; (b) Simulated dynamics of an electric drive, $r(t)=\pm125$ rad/sec. ■

Practice Problem 5.11. *Passive RLC Controlling Circuits: Lag-, Lead- and Lag-Lead Regulators*

As reported in chapter 3, passive filters and compensators (regulators) have been used. The passive *RLC* circuits implement analog regulators for closed-loop systems. These analog regulators are also referenced as compensators.

Regulators are illustrated in Figures 5.47. For linear systems, described by linear differential equations and transfer functions, the closed-loop system stability margins, characteristic eigenvalues, time constants, zeros, the steady-state error and other quantities can be examined.

The transfer functions of the lag-, lead- and lag-lead regulators are

$$G(s)=\frac{R_2}{R_1+R_2}\frac{s+\frac{1}{R_2C}}{s+\frac{1}{(R_1+R_2)C}},\quad G(s)=\frac{s+\frac{1}{R_1C}}{s+\frac{1}{R_1C}+\frac{1}{R_2C}},G(s)=\frac{(s+\frac{1}{R_1C_1})(s+\frac{1}{R_2C_2})}{s^2+(\frac{1}{R_1C_1}+\frac{1}{R_2C_1}+\frac{1}{R_2C_2})s+\frac{1}{R_1R_2C_1C_2}}.$$

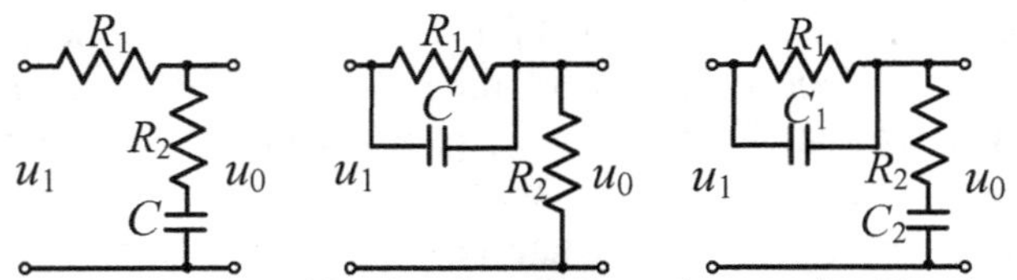

Figure 5.47. Lag-, lead- and lag-lead regulators, characterized by the transfer functions

$$G(s)=\frac{R_2}{R_1+R_2}\frac{s+\frac{1}{R_2C}}{s+\frac{1}{(R_1+R_2)C}},\ G(s)=\frac{s+\frac{1}{R_1C}}{s+\frac{1}{R_1C}+\frac{1}{R_2C}},\ G(s)=\frac{(s+\frac{1}{R_1C_1})(s+\frac{1}{R_2C_2})}{s^2+(\frac{1}{R_1C_1}+\frac{1}{R_2C_1}+\frac{1}{R_2C_2})s+\frac{1}{R_1R_2C_1C_2}}.$$

■

Practice Problem 5.12. *Digital Filters*

Design of analog filters was covered in section 3.7.2. As reported in section 5.14, using a bilinear transform one finds the transfer function in the z-domain using $z=e^{sT_s}$.

Having found a transfer function $G(s)$ for an analog filters $\mathcal{F}_{SC}(\mathbf{S},\mathbf{C})\in\mathbf{C}_{\mathcal{F}}$, apply the Tustin relationship (5.21). We have (5.152)

$$G(z)=G(s)\big|_{s=\frac{2}{T_s}\frac{1-z^{-1}}{1+z^{-1}}}=\frac{N(s)}{D(s)}\bigg|_{s=\frac{2}{T_s}\frac{1-z^{-1}}{1+z^{-1}}},\quad G(z)=\frac{Y(z)}{X(z)}=\frac{\sum_{k=0}^{N}b_kz^{-k}}{1+\sum_{l=1}^{M}a_lz^{-l}}=\frac{N(z)}{D(z)}.$$

In the frequency-domain, relationship $z=e^{j\omega}$ yields

$$G(e^{j\omega})=\frac{Y(e^{j\omega})}{X(e^{j\omega})}=\frac{\sum_{k=0}^{N}b_ke^{-j\omega k}}{1+\sum_{l=0}^{M}a_le^{-j\omega l}}.$$

Example 3.29 documents the Butterworth filters. Consider the second-order filter, and, the cut-off frequency is f_c=1000 Hz, $\omega_c=2\pi f_c$.

We find a digital filer structure and coefficients, $\mathcal{F}_{SC}\equiv G_{M,N,a_l,b_k}(z)$. From

```
n=2; f=1000; [num,den]=butter(n,2*pi*f,'low','s'); filter=tf(num,den), bode(num,den,'k')
```

the filter transfer function is $G(s)=\dfrac{3.95\times10^7}{s^2+8886s+3.95\times10^7}$.

Apply the Tustin relationship (5.21) $s=\dfrac{1}{T_s}\ln(z)\approx\dfrac{2}{T_s}\dfrac{z-1}{z+1}=\dfrac{2}{T_s}\dfrac{1-z^{-1}}{1+z^{-1}}$.

Let T_s=0.00001 sec. The MATLAB code is

```
Ts=0.00001; G_z=c2d(filter,Ts,'Tustin')
```

We compute

$$G(z)=\frac{Y(z)}{X(z)}=\frac{0.000944z^2+0.00189z+0.000944}{z^2-1.911z+0.915}=\frac{0.000944+0.00189z^{-1}+0.000944z^{-2}}{1-1.911z^{-1}+0.915z^{-2}}.$$

Hence,

$$\left(1-1.911z^{-1}+0.915z^{-2}\right)Y(z)=\left(0.000944+0.00189z^{-1}+0.000944z^{-2}\right)X(z).$$

A linear recurrence equation is

$$y_n=-0.915y_{n-2}+1.911y_{n-1}+0.000944x_n+0.00189x_{n-1}+0.000944x_{n-2}.$$ ■

Homework Problems

Homework Problem 5.1.
List requirements and specifications imposed on systems for specific applications, such as aerial, automotive, consumer electronics, manufacturing, robotic and transportation. Explain how control laws affect the system functionality and capabilities.

Homework Problem 5.2.
Describe challenges in design of control law. Report how to approach and solve control problems for physical systems using covered algorithms and control schemes.

Homework Problem 5.3.
For an open-loop stable system, relevant to the *force* and *torque control*

$$\dot{x}_1 = -x_1 + u,$$

$$\dot{x}_2 = x_1, \; y=x_2, \; e(t)=r(t)-y(t),$$

the PID control law is $u = k_{p1}e + k_{i1}\int edt + k_{i2}\iint edt + k_{i3}\iiint edt + k_d \frac{de}{dt}$, $(k_{p1},k_{i1},k_{i2},k_{i3},k_d)>0$.

Problems to solve:

1. Explain the role of feedback terms in the control law;
2. For a closed-loop system, investigate a given tracking control law studying the closed-loop system in SIMULINK. By trial and error, find the feedback gains $(k_{p1},k_{i1},k_{i2},k_{i3},k_d)>0$. Investigate the unit step, ramp and sinusoidal references $r(t)$ with amplitude $A=1$ with different frequency. For example, $r(t)=t$, $r(t)=t^2$, $r(t)=A\sin(\omega t)$, etc. Investigate how the feedback terms and feedback gains affect dynamics and tracking error;
3. Investigate the steady-state error for different references $r(t)$;
4. Propose additional feedback which may improve systems performance;
5. Using transfer function algebra, find a transfer function, specify desired eigenvalues, and find the feedback gains;
6. For a performance functional $J = \int_0^\infty \left(t\left|e\right| + \left|e\right| + x_1^2 + \left|x_2\right| + u^2\right)dt$, perform simulations in SIMULINK, and, quantitatively assert system performance for different feedback coefficients. Explain the role of integrands used in J.

Homework Problem 5.4.

Consider two systems $\dot{x} = Ax + Bu$, $\begin{bmatrix} \dot{x}_1 \\ \dot{x}_2 \end{bmatrix} = \begin{bmatrix} -2 & 1 \\ 2 & -3 \end{bmatrix} x + \begin{bmatrix} 1 \\ 0 \end{bmatrix} u$,

and $\dot{x} = Ax + Bu$, $\begin{bmatrix} \dot{x}_1 \\ \dot{x}_2 \end{bmatrix} = \begin{bmatrix} 0 & 1 \\ -3 & -4 \end{bmatrix} \begin{bmatrix} x_1 \\ x_2 \end{bmatrix} + \begin{bmatrix} 0 \\ 1 \end{bmatrix} u$.

Problems to solve:

1. Find the characteristic polynomials and eigenvalues;
2. Find the eigenvalues, *modal* and state transition matrices ($\mathbf{\Lambda}$, $\mathbf{M}$, $\Phi(s)$);
3. Using the Cayley-Hamilton theorem, compute e^{At};
4. Finding a *modal* matrix $\mathbf{M}$, and, derive $\Phi(t)$;
5. Find $x_1(t)$ and $x_2(t)$ for initial conditions $x_0=[1, -1]$ and unit step input $u(t)=1(t)$ and $u(t)=\sin(t)$. Use $x(t) = e^{At}x_0 + \int_0^t e^{A(t-\tau)}Bu(\tau)d\tau$;

6. Develop MATLAB codes to solve all above problems, see Examples 5.2 to 5.5. For example,

```
A=[-2 1; 2 -3]; B=[1; 0]; t=sym('t'); tau=sym('tau'); u=sym('u'); eAt=expm(A*t)
x0=[1; -1]; u=1; xt=eAt*x0+int(expm(A*(t-tau))*B*u,tau,0,t), pretty(xt); x_t=simplify(xt)
```

Homework Problem 5.5.

Consider a system $\dot{x} = Ax + Bu$, $\begin{bmatrix} \dot{x}_1 \\ \dot{x}_2 \end{bmatrix} = \begin{bmatrix} -2 & 0 \\ 1 & -1 \end{bmatrix} x + \begin{bmatrix} 1 \\ 0 \end{bmatrix} u$, $y=x_1$.

Problems to solve:

1. Find the transfer function and state transition matrix $\Phi(s)$;
2. Derive the state transition matrix $\Phi(t)$ by using different methods, such as Cayley-Hamilton theorem, Sylvester matrix theorem, modal matrix and inverse Laplace transform for $\Phi(s)$;
3. Applying the state transition matrix $\Phi(t)$, find $x_1(t)$ and $x_2(t)$ for initial conditions $x_0=[1, -1]$ if inputs are: step input $u(t)=1(t)$, $u(t)=e^{-t}$ and $u(t)=\sin(t)$;
4. Develop MATLAB codes to solve all problems assigned above.

Homework Problem 5.6.

Consider a system $x[k+1] = A_k x[k] + B_k u[k]$, $\begin{bmatrix} x_1[k+1] \\ x_2[k+1] \end{bmatrix} = \begin{bmatrix} 0 & 1 \\ \frac{3}{16} & -\frac{1}{2} \end{bmatrix} \begin{bmatrix} x_1[k] \\ x_2[k] \end{bmatrix} + \begin{bmatrix} 0 \\ 1 \end{bmatrix} u[k]$.

Using the results reported in Section 5.2.2 and Example 5.6, solve:

1. Find the state transition matrices $\Phi(z)$ and $\Phi[k]$;
2. For initial conditions $\begin{bmatrix} x_1[0] \\ x_2[0] \end{bmatrix} = \begin{bmatrix} 1 \\ -1 \end{bmatrix}$ and input $u[k] = (\frac{1}{2})^k$, find the system response using $\Phi[k]$;
3. Compute and plot dynamics for $x_1[k]$ and $x_2[k]$, $k=0,1,\ldots,9,10$;
4. Solve a discrete Lyapunov equation (6.128) $A_k^T K_k A_k - K_k = -Q_k$. Find K_k if $Q_k=I$.

Homework Problem 5.7.

Consider a second-order discrete-time system

$$x[k+1] = A_k x[k] + B_k u[k]$$

with a matrices (A_k, B_k) to be found using

```
n=2; Ak=randi([-2, 3], [n,n])/4, rank(Ak), Bk=round(2*rand(n,1))
```

Verify that rank A_k is 2, and, a system is controllable. For a controllable system, solve all problems assigned in Homework Problem 5.6.

Homework Problem 5.8.

Consider the *toque control* problem for a system, which describes rotational motion

$$\frac{d\omega}{dt} = -\omega - \theta + u, \quad \frac{d\theta}{dt} = \omega, \quad y=\theta.$$

Denoting $x = \begin{bmatrix} x_1 \\ x_2 \end{bmatrix} = \begin{bmatrix} \omega \\ \theta \end{bmatrix}$, a state-space model is $\dot{x} = \begin{bmatrix} \dot{x}_1 \\ \dot{x}_2 \end{bmatrix} = \begin{bmatrix} -1 & -1 \\ 1 & 0 \end{bmatrix} \begin{bmatrix} x_1 \\ x_2 \end{bmatrix} + \begin{bmatrix} 1 \\ 0 \end{bmatrix} u$, $y=[0 \;\; 1]x$.

Problems to solve:

1. Study controllability and observability;

2. Solve the *modal* control problem, and, find the feedback matrix K_F specifying the eigenvalues to be −10 and −100;
3. Report analytic solutions for closed-loop system evolutions $x_1(t)$ and $x_2(t)$ using $\Phi(t)$. The initial conditions are $x_0=[1, -1]$;
4. Design the *observer* if $x_1=\omega$ is not measured, while $y=\theta(t)$ is measured. Specify the *observer* eigenvalues, and, find the *observer* gain matrix K_E;
5. Solve *modal* control and *observer* design problems using the search algorithm specifying $\boldsymbol{\lambda}_F$ and $\boldsymbol{\lambda}_E$, and, positive definiteness of all feedback and *observer* gains. Perform simulations in SIMULINK;
6. Analytically design a stabilizing linear control law by minimizing $J=\min_{x,u}\int_0^{\infty}\frac{1}{2}\left(100x_1^2+10x_2^2+u^2\right)dt$. Simulate the closed-loop system in SIMULINK;
7. Analytically solve an optimization problem by minimizing $J=\min_{x,u}\int_0^{t_f}\frac{1}{2}\left(100x_1^2+10x_2^2+u^2\right)dt$, t_f=5 sec. Numerically solve the differential Riccati equations. Examine the evolutions of $k_{11}(t)$, $k_{12}(t)$ and $k_{22}(t)$. Interpolate, parameterize and find $k_{11p}(t)$, $k_{12p}(t)$ and $k_{22p}(t)$. Find $u^*(t)$ and evaluate $x^*(t)$;
8. Simulate the closed-loop system with an optimal control law in SIMULINK.

Homework Problem 5.9.

Study a tracking control for a system $\dot{x}=\begin{bmatrix}\dot{x}_1\\ \dot{x}_2\end{bmatrix}=\begin{bmatrix}-1 & -1\\ 1 & 0\end{bmatrix}\begin{bmatrix}x_1\\ x_2\end{bmatrix}+\begin{bmatrix}1\\ 0\end{bmatrix}u$, $y=[0\ \ 1]x$.

Problems to solve:

1. Propose a tracking PID control law using the tracking error $e(t)=r(t)-\theta(t)$. Find the feedback gains assigning adequate eigenvalues. Study system dynamics;
2. Analytically solve the tracking control problem minimizing $J=\min_{\mathbf{x},u}\int_0^{\infty}\frac{1}{2}\left(\mathbf{x}^TQ\mathbf{x}+Gu^2\right)dt$ subject to $\dot{\mathbf{x}}=\mathbf{A}\mathbf{x}+A_r r+\mathbf{B}u,\ y=x_2$. A tracking control law is $u=-G^{-1}\mathbf{B}^TK\mathbf{x}=-K_F\mathbf{x},\ \mathbf{x}=\begin{bmatrix}x_1\\ x_2\\ \int edt\\ e\end{bmatrix}$. Compute K_F for different (Q,G).

   Simulate closed-loop systems;
3. Design an *observer*, and, simulate the closed-loop systems with observed $\hat{x}_1(t)$. Analyze control functions $u(t)$ evolutions. Explain why one should study physical limits on control $u_{\min}\le u\le u_{\max}$.

Homework Problem 5.10.

Study a permanent magnet dc motor in an electric drive application. A dc motor is described by equations (3.19). In Examples 5.10 and 5.11, we studied

$$\frac{di_a}{dt}=\frac{1}{L_a}\left(-r_ai_a-k_a\omega_r+u_a\right),\ r_a=1 \text{ ohm},\ L_a=0.005 \text{ H},\ k_a=0.1 \text{ V-sec/rad},$$

$$\frac{d\omega_r}{dt}=\frac{1}{J}\left(k_ai_a-B_m\omega_r-T_L\right),\ B_m=0.005 \text{ N-m-sec/rad},\ J=0.1 \text{ kg-m}^2.$$

Problems to solve:

1. Derive a transfer function of the closed-loop electric drive system with a PID control law $u=k_pe+k_i\int edt+k_d\frac{de}{dt}$. Report the characteristic equation;

2. Using the characteristic equations, derive relationships and calculate (k_p, k_i, k_d) to guarantee the specified eigenvalues –10, –20 and –30;
3. Analyze electric derive dynamics with a PID control law in SIMULINK. By trial and error find alternative (k_p, k_i, k_d) and corresponding eigenvalues;
4. Solve the tracking control problem using the error dynamics by minimizing a performance functional $J = \min_{\mathbf{x},u} \int_0^\infty \frac{1}{2}\left(\mathbf{x}^T Q \mathbf{x} + G u^2\right) dt$ with different (Q,G). Find feedback gains, and, analyze system dynamics. Compare control laws designed.

Homework Problem 5.11.

Consider a pointing system actuated by a geared permanent magnet dc motor, see Example 5.13. The angular displacement of the pointing stage is $y(t)=k_{\text{gear}}\theta_r$. The system parameters are r_a=1 ohm, L_a=0.005 H, k_a=0.1 V-sec/rad, B_m=0.005 N-m-sec/rad, J=0.005 kg-m^2 and k_{gear}=0.1.

Problems to solve:

1. Design a PID control law specifying adequate eigenvalues. Study transient dynamics for $r(t)=\pm 1$ rad. Perform simulations in SIMULINK;
2. Design tracking control law by minimizing the quadratic functional $J = \min_{\mathbf{x},u} \int_0^\infty \frac{1}{2}\left(\mathbf{x}^T Q \mathbf{x} + G u^2\right) dt$ for different (Q,G). Simulate the closed-loop system;
3. Hardware limits are due to maximum voltage. For $-50 \le u \le 50$ V, assess system dynamics, and, examine how control limits affect system dynamics.

Homework Problem 5.12.

Consider a system $\dot{x} = \begin{bmatrix} \dot{x}_1 \\ \dot{x}_2 \end{bmatrix} = \begin{bmatrix} -1 & -1 \\ 1 & 0 \end{bmatrix} \begin{bmatrix} x_1 \\ x_2 \end{bmatrix} + \begin{bmatrix} 1 \\ 0 \end{bmatrix} u$, $y=[0 \;\; 1]x$.

Problems to solve:

1. Minimize the quadratic functional $J = \min_{x,u} \int_0^\infty \frac{1}{2}\left(q_{11}x_1^2 + q_{22}x_2^2 + g u^2\right) dt$, and, analytically design an analog control law by solving the LQR problem. Examine how the weighting coefficients (q_{11}, q_{22}, g) affect feedback gains, characteristic eigenvalues and system dynamics;
2. Solve optimal control problem minimizing $J = \min_{x,u} \int_0^{t_f} \frac{1}{2}\left(x_1^2 + x_2^2 + u^2\right) dt$, t_f=5 sec. Numerically solve the Riccati equation, interpolate numeric solutions, and, parameterize $k_{ij}(t)$ to define an optimal control law;
3. Discretize a system if T_s=0.01 sec. Design a stabilizing digital control law minimizing $J = \min_{\mathbf{x}_n, u_n} \sum_{n=0}^{\infty}\left(x_n^T Q_n x_n + G_n u_n^2\right)$. Examine different (Q_n, G_n) by computing feedback gains, and, simulating closed-loop systems;
4. Report advantages and disadvantages of analog and digital systems from a viewpoint of control laws design, as well as hardware implementation.

Homework Problem 5.13.

For a servo with axial topology actuator, assume that the self-inductance is negligibly small, and, the wedge-shaped NdFeB magnets are magnetized through thickness ensuring near-sigmoid magnetization. As reported in Example 3.20, the governing equations are

$$\frac{d\omega_r}{dt}=\frac{1}{J}\left[-B_m\omega_r+\frac{2}{r_a}R_\perp Nl_{eq}B_{\max}u_a\right],\ \frac{d\theta_r}{dt}=\omega_r,\ y=\theta_r,$$

where $J=5\times10^{-6}$ kg-m^2, $B_m=2\times10^{-6}$ N-m-s/rad, $r_a=30$ ohm, $R_\perp=0.02$ m, $N=100$, $l_{eq}=0.0125$ m and $B_{\max}=1$ T.

Problems to solve:

1. Design a PID control law specifying the characteristic equation for a closed-loop system $(T_1s+1)(T_2^2s^2+2\xi T_2s+1)=0$ with $T_1=0.01$ sec, $T_2=0.005$ sec and damping coefficient $\xi=0.707$. Report k_p, k_i and k_d;
2. Perform simulations of the closed-loop system for a reference $r(t)=\pm0.1$ rad;
3. Refine $(k_p,k_i,k_d)>0$ to ensure adequate performance guaranteeing the settling time less than 0.05 sec with overshoot less than 5%. Report the resulting (T_1,T_2,ξ). By trial and error, select (k_p,k_i,k_d), and find eigenvalues;
4. Investigate closed-loop system performance with a control limit $|u|\le10$ V;
5. Design a tracking control law minimizing $J=\min_{\mathbf{x},u}\int_0^\infty\frac{1}{2}\left(\mathbf{x}^TQ\mathbf{x}+Gu^2\right)dt$ for different Q and G. Perform simulations, and, investigate a closed-loop system dynamics considering limits $|u|\le10$ V;
6. Discretize a system for $T_s=0.0001$ sec;
7. Design a digital control law by solving the tracking LQR problem minimizing $J=\min_{\mathbf{x}_n,u_n}\sum_{n=0}^{\infty}\left(\mathbf{x}_n^TQ_n\mathbf{x}_n+G_nu_n^2\right)$. Perform simulations of a closed-loop system. Examine how (Q_n,G_n) affect the feedback matrix K_{Fn} and system dynamics.

References

1. A. M. Lyapunov, "Stability of motions," *Trans. Mathematic Society*, vol. 17, pp. 253-333, 1893.
2. N. Minorsky, "Directional stability of automatically steered bodies," *J. American Society of Naval Engineers*, vol. 34, no. 2, pp. 280-309, 1922.
3. R. Bellman, *Dynamic Programming*, Princeton University Press, NJ, 1957.
4. R. E. Kalman and J. E. Bertram, "Control system analysis and design via the "second method" of Lyapunov. I. Continuous-time systems," *Trans. ASME J. Basic Eng.*, vol. 82, no. 2, pp. 371-393, 1960.
5. R. E. Kalman, "Contributions to the theory of optimal control," *Boletin de la Sociedad Matematica Mexicana*, vol. 5, pp. 102-119, 1960.
6. B. C. Kuo, *Automatic Control Systems*, Prentice Hall, Englewood Cliffs, NJ, 1995.
7. S. E. Lyshevski, *Control Systems Theory with Engineering Applications*, Birkhauser, Boston, MA, 2000.
8. N. S. Nise, *Control Systems Engineering*, Wiley; NY, 2014.
9. A. Poznyak, *Advanced Mathematical Tools for Automatic Control Engineers: Volume* 1: *Deterministic Systems*, *Volume* 2: *Stochastic Systems*, Elsevier Science, NY, 2010.
10. K. Ogata, *Discrete-Time Control Systems*, Prentice-Hall, Upper Saddle River, NJ, 1995.
11. K. Ogata, *Modern Control Engineering*, Prentice-Hall, Upper Saddle River, NJ, 1997.
12. J. J. E. Slotine and W. Li, *Applied Nonlinear Control*, Pearson, NJ, 1991.
13. R. F. Stengel, *Optimal Control and Estimation*, Dover Pub., NY, 2012.

14. S. H. Zak, *Systems and Control*, Oxford University Press, NY, 2002.
15. S. E. Lyshevski and K. Dunipace, "Identification and tracking control of aircraft from real-time perspectives," *Proc. IEEE Conf. Control Applications,* Hartford, CT, pp. 499-504, 1997.
16. S. E. Lyshevski, K. Dunipace and R. Colgren, "Identification and reconfigurable control of multivariable aircraft," *Proc. American Control Conf.,* San Diego, CA, vol. 4, pp. 2732-2726, 1999.
17. S. E. Lyshevski, "Identification of nonlinear flight dynamics: Theory and practice," *IEEE Trans. on Aerospace and Electronic Systems,* vol. 36, no. 2, pp. 383-392, 2000.
18. S. E. Lyshevski, "Aircraft flight control system design under state and control bounds," *IEEE Trans. on Aerospace and Electronic Systems,* vol. 34, no. 1, pp. 257-263, 1998.
19. A. Berman, P. Zarchan and B. Lewis, "Comparisons between the extended Kalman filter and the state-dependent Riccati estimator," *J. Guidance, Control, and Dynamics*, vol. 37, no. 5, pp. 1556- 1567, 2014.
20. A. G. Deczky, "Synthesis of recursive digital filters using the minimum *p*-error criterion," *IEEE Trans. Audio and Electroacoustics*, vol. AU-20, no. 4, pp. 257-263, 1972.
21. S. Haykin, *Adaptive Filter Theory*, Pearson, 2013.
22. R. E. Kalman and R. S. Bucy, "New results in linear filtering and prediction theory," *ASME Trans., Series D, J. Basic Engineering*, vol. 83, no. 1, pp. 95-108, 1961.
23. H. J. Kushner, "Dynamic equations for optimal nonlinear filtering," *J. Differential Equations*, vol. 3, issue 2, pp. 179-190, 1967.
24. K. Reif and R. Unbehauen, "The extended Kalman filter as an exponential observer for nonlinear systems," *IEEE Trans. Signal Processing*, vol. 47, no. 8, pp. 2324-2328, 1999.
25. B. Abbott and S. E. Lyshevski, "Signal processing in MEMS inertial measurement units for dynamic motional control," *Proc. Conf. Electronics and Nanotechnology*, pp. 309-314, 2016.
26. J. A. Allison and S. E. Lyshevski, "Aerial systems with micro sensors: Data acquisition and descriptive reality in physical and virtual environments," *Proc. Conf. Electronics and Nanotechnology*, pp. 669-674, 2018.
27. L. Herlihy and S. E. Lyshevski, "Real-time dynamic data processing and fusion: Multi-degree-of-freedom MEMS inertial sensors," *Proc. Conf. Electronics and Nanotechnology*, pp. 631-637, 2018.
28. S. E. Lyshevski, "Distributed processing and data fusion in dynamic systems with microsystem sensor networks," *Proc. Conf. Electronics and Nanotechnology*, pp. 648-653, 2018.
29. S. E. Lyshevski, "Signal processing in cyber-physical MEMS sensors: Inertial measurement and navigation systems," *IEEE Trans. Industrial Electronics*, vol. 64, no. 12, pp. 9618-9625, 2017.
30. M. Rantanen, N. Mastronarde, J. Hudack and K. Dantu, "Decentralized tasks allocation in lossy networks: A simulation study," *Proc. Conf. Sensing, Communication, and Networks*, 2019.
31. H. Carvalho, P. Del Moral, A. Monin and G. Salut, "Optimal nonlinear filtering in GPS/INS integration," *IEEE Trans. Aerospace and Electronic Systems*, vol. 33, issue 3, pp. 835-850, 1997.

6 Design, Control and Optimization of Nonlinear Dynamic Systems

6.1. Introduction to Nonlinear Control and Optimization

All physical systems are nonlinear, and, there are inherent physical limits on states x, outputs y and controls u. All consistently designed physical systems are controllable, and, many systems are open-loop stable. Practical solutions and control schemes are found to stabilize systems if marginal stability, metastability and instabilities in small are observed. Proportional-integral-derivative and state feedback controllers can be designed and reconfigured to ensure near-optimal governance physical systems $\mathbf{M}(x,y,u)$. Challenges and impediments arise in MIMO systems which exhibit multi-axis-coupling, instabilities, resonance, flexible modes, unsteady flows, sensitivity, etc. A physical controllability is always guaranteed in adequately designed systems.

System functionality is accomplished by means of optimizations and control. Specifications are imposed on system performance $\mathcal{Q}_{\mathcal{P}}$ and capabilities $\mathcal{Q}_{C}$. Measures $\mathcal{Q}(\mathcal{Q}_{\mathcal{P}},\mathcal{Q}_{C})$ are evaluated using the variable-dependent quantities $\boldsymbol{q}(\boldsymbol{x},\boldsymbol{y},\boldsymbol{u})$. In design of control laws, apply performance measures, and, minimize performance functionals and indexes.

Nonlinear differential equations were derived in chapters 2 and 3. Chapters 4 and 5 cover system identification, parameter estimation and observers design. Many systems are not linearizable. The system output $y(t)$ is a function of states $x(t)$, $y=H(x)$. In aerial, robotic and other MIMO systems, digital control algorithms are implemented by MPUs and FPGAs using the measurements by integrated MDOF sensors. Sensors asynchronously output processed measurements by means of data fusion algorithms and protocols. One designs analog control laws $u(t)$. Analog-to-digital and digital-to-analog signal processing and pulse width modulation (PWM) are implemented by specialized controllers-drivers with ASICs. Analog filters and controllers can be discretized. Digital control laws u_k are implemented by MPUs and FPGAs which output high switching frequency signals which control MOSFETs drivers, thereby regulating MOSFET switching in four-quadrant power stages.

Continuity of physical transductions and energy conversion implies continuity of devices and processes, which are governed by regulating analog transitions controlled by digital signaling, logic, processing and computing. Recall that switching of MOSFETs is continuous with nanoseconds transients. Continuum electromagnetics, mechanics and thermodynamics yield nonlinear differential equations. For controllable spatiotemporal processes, system-specific optimization is performed.

Control laws are designed by minimizing performance functionals and indexes subject to descriptive models, described by governing and constitutive deterministic and stochastic differential, difference, algebraic, transcendental and other equations. Nonlinear optimization for continuous-time systems is reported in [1-6]. Design of discrete-time systems are documented in [2-4]. Linear and nonlinear control laws $u(t)$ are designed, and, implemented by analog and digital controllers. Analog control laws and filters are discretized for hardware-supported sampling period T_s considering processing schemes, algorithms, data fusion rate, communication protocols, etc. High-performance MPUs and FPGAs ensure unprecedented capabilities implementing adaptive processing, data management, control and communication schemes. Bounded control laws $u_{\min}\le u(\cdot)\le u_{\max}$ should be designed within admissible hardware-determined physical limits on control $\mathcal{U}_{\min}\le\mathcal{U}\le\mathcal{U}_{\max}$.

Continuous- and Discrete-Time Equations of Motion – Complying with system modular organization, underlying device physics and fidelity, sensing and control schemes, as well as hardware implementation, the governing and constitutive equations for physical systems $\mathbb{M}(x,y,u,r,d)$ are

$$\text{System Governing Model: } \dot{x}(t)=F(x,r,d)+B(x)u,\ y=H(x),\ t\ge 0, \qquad (6.1)$$
$$\text{Discretized Model: } x_{k+1}=F(x_k,r_k,d_k)+B(x_k)u_k,\ y_k=H(x_k),\ k\ge 0,$$

where $x\in X\subset\mathbb{R}^n$ is the state vector with states evolving in X; $u\in U\subset\mathbb{R}^m$ is the control vector, and, u can be bounded such as $u_{\min}\le u\le u_{\max}$; $r\in R\subset\mathbb{R}^b$ and $y\in Y\subset\mathbb{R}^b$ are the reference and output vectors; $d\in D\subset\mathbb{R}^v$ is the disturbance vector; $F(\cdot):\mathbb{R}^n\times\mathbb{R}^b\times\mathbb{R}^v\to\mathbb{R}^n$, $B(\cdot):\mathbb{R}^n\to\mathbb{R}^{n\times m}$ and $H(\cdot):\mathbb{R}^n\to\mathbb{R}^{b\times n}$ are the nonlinear maps.

Nonlinear physical systems $\mathbb{M}(t,x,y,u,d,r)$, modelled by governing and constitutive equations (6.1), are depicted in Figure 6.1.a. Governance, analysis, control and optimization at the devices, modules and system levels are considered. Physical systems are comprised of different components and subsystems. Laws of physics yield describing equations for states, controls, outputs and disturbances. For z subsystems $\mathrm{M}_i(t,\boldsymbol{x}_i,\boldsymbol{y}_i,\boldsymbol{u}_i,\boldsymbol{r}_i,\boldsymbol{d}_i)$, $\mathrm{M}_i\in\mathbb{M}$, $i=1,\ldots z$, a high-fidelity model $\mathcal{M}(\cdot)$ with aggregation mappings $(F_{ij},R_{ij},D_{ij},H_{ij})$ is

$$\mathcal{M}\equiv\left\{\begin{bmatrix}\dot{x}_1\\ \vdots\\ \dot{x}_z\end{bmatrix}=\begin{bmatrix}\sum_{i,j}F_{1_{ij}}(\boldsymbol{x}_1,\boldsymbol{x}_{ij},\boldsymbol{u}_1,\boldsymbol{u}_{ij})\\ \vdots\\ \sum_{i,j}F_{z_{ij}}(\boldsymbol{x}_z,\boldsymbol{x}_{ij},\boldsymbol{u}_z,\boldsymbol{u}_{ij})\end{bmatrix}+\begin{bmatrix}\sum_{i,j}R_{1_{ij}}(\boldsymbol{r}_1,\boldsymbol{r}_{ij})\\ \vdots\\ \sum_{i,j}R_{z_{ij}}(\boldsymbol{r}_z,\boldsymbol{r}_{ij})\end{bmatrix}+\begin{bmatrix}\sum_{i,j}D_{1_{ij}}(\boldsymbol{d}_1,\boldsymbol{d}_{ij})\\ \vdots\\ \sum_{i,j}D_{z_{ij}}(\boldsymbol{d}_z,\boldsymbol{d}_{ij})\end{bmatrix},\ \begin{bmatrix}y_1\\ \vdots\\ y_z\end{bmatrix}=\begin{bmatrix}\sum_{i,j}H_{1_{ij}}(\boldsymbol{x}_1,\boldsymbol{x}_{ij})\\ \vdots\\ \sum_{i,j}H_{z_{ij}}(\boldsymbol{x}_z,\boldsymbol{x}_{ij})\end{bmatrix}\right\}. \qquad (6.2)$$

Decoupling and aggregation are performed in multi-physics and cyber spaces. State-space modeling, design and implementation tasks are carried out by integrating analysis at device, component, module and system levels.

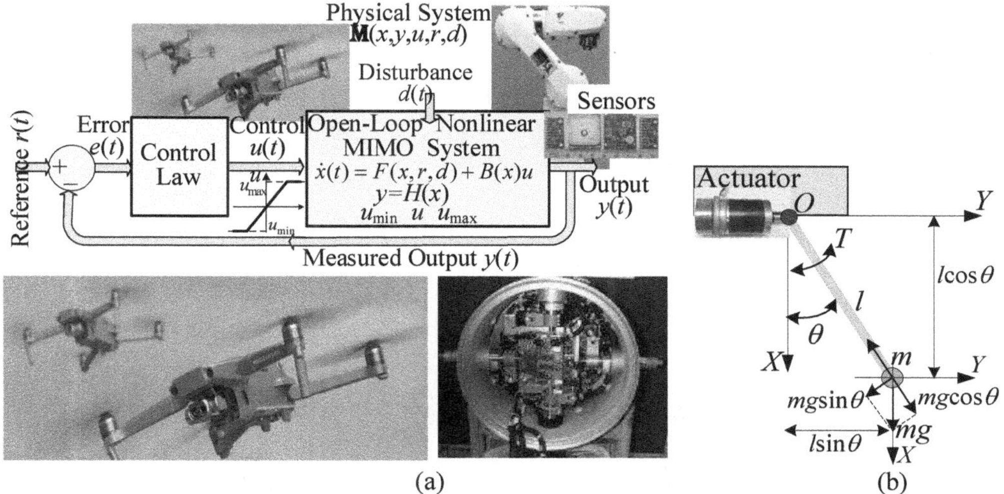

Figure 6.1. (a) High-level diagram of a nonlinear closed-loop system, and, controlled physical systems with various components and modules which guarantee compliance and complementarity; (b) Representative robotic arm with an actuator, Example 6.1.

6.2. Analytic and Numeric Methods in Nonlinear Analyses

6.2.1. Linearization of Nonlinear Systems

Some nonlinear systems can be linearized while ensuring descriptive fidelity. Consider nonlinear ordinary differential equations

$$\dot{x}(t) = F(x,u),\ x(t_0)=x_0,\ x\in X\subset\mathbb{R}^n,\ u\in U\subset\mathbb{R}^m,\ F(\cdot):\mathbb{R}^n\times\mathbb{R}^m\to\mathbb{R}^n, \qquad (6.3)$$

$$\begin{cases} \dot{x}_1 = f_1(x_1,x_2,\ldots,x_n,u_1,u_2,\ldots,u_m), \\ \vdots \\ \dot{x}_n = f_n(x_1,x_2,\ldots,x_n,u_1,u_2,\ldots,u_m), \end{cases} \quad J = \begin{bmatrix} \dfrac{\partial f_1}{\partial x_1} & \cdots & \dfrac{\partial f_1}{\partial x_n} \\ \vdots & \ddots & \vdots \\ \dfrac{\partial f_n}{\partial x_1} & \cdots & \dfrac{\partial f_n}{\partial x_n} \end{bmatrix},\ J_u = \begin{bmatrix} \dfrac{\partial f_1}{\partial u_1} & \cdots & \dfrac{\partial f_1}{\partial u_m} \\ \vdots & \ddots & \vdots \\ \dfrac{\partial f_n}{\partial u_1} & \cdots & \dfrac{\partial f_n}{\partial u_m} \end{bmatrix},$$

where $f_i(\cdot)$ are the continuously differentiable real-valued functions, $f_i(\cdot):\mathbb{R}^n\times\mathbb{R}^m\to\mathbb{R}$.

The Jacobians (J,J_u) are mappings of the first-order partial derivatives of vector-valued functions. For an *equilibrium state* vector $x_e=\bar{x}$, $\bar{x}\in\mathbb{R}^n$, there is an *equilibrium input* vector $\bar{u}\in\mathbb{R}^m$, such that

$$\dot{x}_e = F(x_e,\bar{u}) = F(\bar{x},\bar{u}) = 0. \qquad (6.4)$$

At equilibrium $\bar{x}$ for the constant input $u\equiv\bar{u}$, one has $x=\bar{x}$, $\forall t$. For (6.3), define

$$x = \bar{x}+\delta_x,\ u = \bar{u}+\delta_u. \qquad (6.5)$$

From (6.3)-(6.5), $\delta_x = x-\bar{x}$, $\delta_u = u-\bar{u}$. We have

$$\dot{\delta}_x(t) = F(\bar{x}+\delta_x,\bar{u}+\delta_u). \qquad (6.6)$$

Neglecting high-order terms, the Taylor expansion yields

$$\dot{\delta}_x(t) \approx F(\bar{x},\bar{u}) + \left.\frac{\partial F}{\partial x}\right|_{\substack{x=\bar{x}\\u=\bar{u}}}\delta_x + \left.\frac{\partial F}{\partial u}\right|_{\substack{x=\bar{x}\\u=\bar{u}}}\delta_u, \qquad (6.7)$$

where $F(\bar{x},\bar{u})=0$.

Hence, from (6.7) and (6.4), for the state variations $\delta_x(t)$ in $X_\delta \subset X$ with inputs $\delta_u(t)$ in $U_\delta \subset U$, the governing equation is

$$\dot{\delta}_x(t)=\left.\frac{\partial F}{\partial x}\right|_{\substack{x=\bar{x}\\u=\bar{u}}}\delta_x+\left.\frac{\partial F}{\partial u}\right|_{\substack{x=\bar{x}\\u=\bar{u}}}\delta_u=J\big|_{\substack{x=\bar{x}\\u=\bar{u}}}\delta_x+J_u\big|_{\substack{x=\bar{x}\\u=\bar{u}}}\delta_u=\left.\begin{bmatrix}\frac{\partial f_1}{\partial x_1} & \cdots & \frac{\partial f_1}{\partial x_n}\\ \vdots & \ddots & \vdots\\ \frac{\partial f_n}{\partial x_1} & \cdots & \frac{\partial f_n}{\partial x_n}\end{bmatrix}\right|_{\substack{x=\bar{x}\\u=\bar{u}}}\delta_x+\left.\begin{bmatrix}\frac{\partial f_1}{\partial u_1} & \cdots & \frac{\partial f_1}{\partial u_m}\\ \vdots & \ddots & \vdots\\ \frac{\partial f_n}{\partial u_1} & \cdots & \frac{\partial f_n}{\partial u_m}\end{bmatrix}\right|_{\substack{x=\bar{x}\\u=\bar{u}}}\delta_u,$$

$$\dot{\delta}_x(t)=A\delta_x+B\delta_u, \tag{6.8}$$

where $A\in\mathbb{R}^{n\times n}$ and $B\in\mathbb{R}^{n\times m}$ are the matrices.

The existence of $J=\left.\frac{\partial F}{\partial x}\right|_{\substack{x=\bar{x}\\u=\bar{u}}}$ and $J_u=\left.\frac{\partial F}{\partial u}\right|_{\substack{x=\bar{x}\\u=\bar{u}}}$ at $(\bar{x},\bar{u})\subset X\times U$ is guaranteed because $F(x,u)$ is a continuously differentiable function which is locally Lipschitz.

Nomenclature – Linearization of nonlinear systems (6.3) at equilibria $(\bar{x},\bar{u})$ yields the linearized system evolutions for $\delta_x(t)$ and $\delta_u(t)$. The resulting equation (6.8) $\dot{\delta}_x(t)=A\delta_x+B\delta_u$ is usually written as $\dot{x}=Ax+Bu$ by denoting $x\equiv\delta_x$ and $u\equiv\delta_u$. In a small neighborhood (X_δ, U_δ), it is assumed that differential equations (6.3) assume linearizability and that the local homeomorphism exists, such that $A=\left.\frac{\partial F}{\partial x}\right|_{\substack{x=\bar{x}\\u=\bar{u}}}$, $B=\left.\frac{\partial F}{\partial u}\right|_{\substack{x=\bar{x}\\u=\bar{u}}}$ with $x_\delta\in X_\delta\subset X$, $u_\delta\in U_\delta\subset U$. The Hartman-Grobman, Hadamard-Perron and other approaches depend on $F(x,u)$ and (x_e,u_e). Linearization may yield inconclusive results, and, inadequate for some classes of differential equations.

Example 6.1. *Lumped-Parameter Model for a Robotic Arm*

In Examples 2.5 and 2.7, the governing equation of motion for a pendulum is

$$\frac{d^2\theta}{dt^2}=\frac{1}{J}\left(-B_m\omega-mgl\sin\theta+T\right).$$

Figure 6.1.b illustrates an arm with a point mass m. The arm is actuated by a servomotor. The angular velocity and displacement are states ω and θ. The actuator develops a toque T to yield motions. Considering the *torque control* premise, $u\equiv T$. A set of two first-order differential equations is

$$\frac{d\omega}{dt}=\frac{1}{J}\left(-B_m\omega-mgl\sin\theta+u\right),\ u\equiv T,$$

$$\frac{d\theta}{dt}=\omega.$$

Using $\begin{bmatrix}x_1\\x_2\end{bmatrix}=\begin{bmatrix}\omega\\\theta\end{bmatrix}$, a nonlinear model in the state-space form (6.3) is

$$\dot{x}=\begin{bmatrix}\dot{x}_1\\\dot{x}_2\end{bmatrix}=\begin{bmatrix}-\frac{B_m}{J}x_1-\frac{mgl}{J}\sin x_2+\frac{1}{J}u\\x_1\end{bmatrix},\quad y=\begin{bmatrix}0 & 1\end{bmatrix}\begin{bmatrix}x_1\\x_2\end{bmatrix}=x_2.$$

From $\dot{x}=F(x,u)=0$ for $(x_1,x_2,u)=(0, 0, 0)$, one concludes that $x_1(t)$ and $x_2(t)$ are at equilibrium. If the arm is at the vertical downward position with $x_1=0$, $x_2=0$, $u=0$, the arm stays in that position for all t. An equilibrium point is $x=[0\ \ 0]^T$.

For $x_1=0$, $x_2=0$, $u=0$, from $-\frac{mgl}{J}\sin x_2 = 0$, we have $\bar{x} = \begin{bmatrix} \bar{x}_1 \\ \bar{x}_2 \end{bmatrix} = \begin{bmatrix} 0 \\ k\pi \end{bmatrix}$, where k is an integer.

To finds an equilibrium for $u\neq0$, apply (6.4) $\dot{x}_e = F(x_e,\bar{u}) = F(\bar{x},\bar{u}) = 0$.

We have $\begin{bmatrix} -\frac{B_m}{J}\bar{x}_1 - \frac{mgl}{J}\sin\bar{x}_2 + \frac{1}{J}\bar{u} \\ \bar{x}_1 \end{bmatrix} = \begin{bmatrix} 0 \\ 0 \end{bmatrix}$.

For $u\neq0$ and zero angular velocity $x_1=0$, find the angular displacement $x_2\neq0$ as $\sin\bar{x}_2 = \frac{1}{mgl}\bar{u}$. Hence, $\bar{x} = \begin{bmatrix} \bar{x}_1 \\ \bar{x}_2 \end{bmatrix} = \begin{bmatrix} 0 \\ \bar{x}_2 \end{bmatrix}$. That is, for specified $\bar{x}_2$, compute $\bar{u}$.

Apply (6.8) for the *equilibrium state* and *input* $(\bar{x},\bar{u})$. The governing state-space equation for evolutions of $\delta_x(t)$ with control $\delta_u(t)$ are

$$\dot{\delta}_x(t) = \begin{bmatrix} \dot{\delta}_{x_1} \\ \dot{\delta}_{x_2} \end{bmatrix} = \left.\frac{\partial F}{\partial x}\right|_{\substack{x=\bar{x}\\u=\bar{u}}} \delta_x + \left.\frac{\partial F}{\partial u}\right|_{\substack{x=\bar{x}\\u=\bar{u}}} \delta_u = A\delta_x + B\delta_u = \left.\begin{bmatrix} -\frac{B_m}{J} & -\frac{mgl}{J}\cos x_2 \\ 1 & 0 \end{bmatrix}\right|_{x=\bar{x}} \begin{bmatrix} \delta_{x_1} \\ \delta_{x_2} \end{bmatrix} + \begin{bmatrix} \frac{1}{J} \\ 0 \end{bmatrix}\delta_u,$$

$\delta_x = x - \bar{x}$, $\delta_u = u - \bar{u}$.

For an arm in downward position $(\bar{x},\bar{u}) = (\bar{x}_1,\bar{x}_2,\bar{u}) = (0,0,0)$, one has

$$\dot{\delta}_x(t) = \begin{bmatrix} \dot{\delta}_{x_1} \\ \dot{\delta}_{x_2} \end{bmatrix} = A\delta_x + B\delta_u = \begin{bmatrix} -\frac{B_m}{J} & -\frac{mgl}{J} \\ 1 & 0 \end{bmatrix} \begin{bmatrix} \delta_{x_1} \\ \delta_{x_2} \end{bmatrix} + \begin{bmatrix} \frac{1}{J} \\ 0 \end{bmatrix}\delta_u.$$

If arm in upward position $(\bar{x},\bar{u}) = (\bar{x}_1,\bar{x}_2,\bar{u}) = (0,\ \pi,\ 0)$

$$\dot{\delta}_x(t) = \begin{bmatrix} \dot{\delta}_{x_1} \\ \dot{\delta}_{x_2} \end{bmatrix} = A\delta_x + B\delta_u = \begin{bmatrix} -\frac{B_m}{J} & \frac{mgl}{J} \\ 1 & 0 \end{bmatrix} \begin{bmatrix} \delta_{x_1} \\ \delta_{x_2} \end{bmatrix} + \begin{bmatrix} \frac{1}{J} \\ 0 \end{bmatrix}\delta_u,$$

where δ_x and δ_u are the variations of x and u from equilibria.

The downward position is stable. The upward position (k=1,3,…) is unstable.

The linearized model is consistent for small variations of states and input. ■

Example 6.2. Analytic Solutions and Symbolic Notations

Analytic solutions can be found using symbolic computations.

For $A(x) = \begin{bmatrix} \cos ax & \sin bx \\ cx^3 & xe^{-dx} \end{bmatrix}$, find the inverse A^{-1}, verify results by validating $A^{-1}A=I$, and, differentiate $A(x)$ element-by-element. The MATLAB statements are

```
syms a b c d x; A=[cos(a*x), sin(b*x); c*x^3, x*exp(-d*x)]
inverseA=inv(A), I=simplify(inverseA*A), dAdx=simplify(diff(A))
```

The results are

```
A = [ cos(a*x),   sin(b*x)]
    [ c*x^3,   x*exp(-d*x)]
inverseA = [       1/(cos(a*x) - c*x^2*exp(d*x)*sin(b*x)), -(exp(d*x)*sin(b*x))/(x*(cos(a*x) - c*x^2*exp(d*x)*sin(b*x)))]
[ -(c*x^2*exp(d*x))/(cos(a*x) - c*x^2*exp(d*x)*sin(b*x)),  (exp(d*x)*cos(a*x))/(x*(cos(a*x) - c*x^2*exp(d*x)*sin(b*x)))]
I = [ 1, 0]
    [ 0, 1]
dAdx = [ -a*sin(a*x),          b*cos(b*x)]
       [ 3*c*x^2,   -exp(-d*x)*(d*x - 1)]
```

We have $A^{-1}=\begin{bmatrix}\dfrac{1}{\cos ax-cx^2e^{dx}\sin bx} & \dfrac{-e^{dx}\sin bx}{x\cos ax-cx^2e^{dx}\sin bx}\\ \dfrac{-cx^2e^{dx}}{\cos ax-cx^2e^{dx}\sin bx} & \dfrac{e^{dx}\cos ax}{x\cos ax-cx^2e^{dx}\sin bx}\end{bmatrix}$, $A^{-1}A=\begin{bmatrix}1 & 0\\ 0 & 1\end{bmatrix}$.

Differentiation yields $\dfrac{\partial A}{\partial x}=\begin{bmatrix}-a\sin ax & b\cos bx\\ 3cx^2 & -e^{-dx}(dx-1)\end{bmatrix}$. ■

Example 6.3. *Coordinate Transformations*
The transformation from Euclidean (x,y,z) to spherical (r,θ,ϕ) coordinates is

$x=r\cos(\theta)\sin\phi,\ y=r\sin(\theta)\sin\phi,\ z=r\cos\phi,$

where r is the radial distance; θ and ϕ are the polar and azimuthal angles.

Differentiate a column vector with respect to a row vector, and, find the Jacobian. The statements

```
syms r q f; x=r*cos(q)*sin(f); y=r*sin(q)*sin(f); z=r*cos(f); J=jacobian([x y z], [r q f])
```

yield the resulting Jacobian

```
J = [ cos(q)*sin(f), -r*sin(f)*sin(q), r*cos(f)*cos(q)]
    [ sin(f)*sin(q),  r*cos(q)*sin(f), r*cos(f)*sin(q)]
    [       cos(f),              0,           -r*sin(f)]
```

That is, $J=\begin{bmatrix}\cos\theta\sin\phi & -r\sin\theta\sin\phi & r\cos\theta\cos\phi\\ \sin\theta\sin\phi & r\cos\theta\sin\phi & r\sin\theta\cos\phi\\ \cos\phi & 0 & -r\sin\phi\end{bmatrix}$. ■

For nonlinear systems, perturbed by disturbances $d(t)$

$$\dot{x}(t)=F(x,u,d),\ x(t_0)=x_0,\ F(\cdot):\mathbb{R}^{n}\times\mathbb{R}^{m}\times\mathbb{R}^{v}\to\mathbb{R}^{n}. \quad (6.9)$$

Linearization at $(\bar{x},\bar{u},\bar{d})$ yields

$$\dot{\delta}_x(t)=\left.\frac{\partial F}{\partial x}\right|_{\substack{x=\bar{x}\\ u=\bar{u}\\ d=\bar{d}}}\delta_x+\left.\frac{\partial F}{\partial u}\right|_{\substack{x=\bar{x}\\ u=\bar{u}\\ d=\bar{d}}}\delta_u+\left.\frac{\partial F}{\partial d}\right|_{\substack{x=\bar{x}\\ u=\bar{u}\\ d=\bar{d}}}\delta_d=A\delta_x+B\delta_u+B_D\delta_d,\ A\in\mathbb{R}^{n\times n},\ B\in\mathbb{R}^{n\times m},\ B_D\in\mathbb{R}^{n\times v}. \quad (6.10)$$

Stochastic Nonlinear Systems – Many physical systems exhibit randomness or perturbed by random perturbations, such as noise, interference and other stationary and nonstationary stochastic processes. Independent, correlated, additive and multiplicative multivariate perturbations are studied. It is impossible to find analytic solutions, and, difficult to analyze high-order nonlinear stochastic differential equations.

Consider nonlinear systems with deterministic disturbances and stochastic perturbations (d,ξ)

$$\dot{x}(t)=F(x,u,d,\xi),\ x(t_0)=x_0,\ F(\cdot):\mathbb{R}^{n}\times\mathbb{R}^{m}\times\mathbb{R}^{v}\times\mathbb{R}^{p}\to\mathbb{R}^{n}, \quad (6.11)$$

where $\xi\in\Xi\subset\mathbb{R}^{p}$ are the continuous stochastic perturbations, characterized by distributions, spectral densities, etc.

The Lévy, Poisson, Markov, Wiener and other processes are considered. White noise $\mathbb{E}[\xi(t)]=0$, $\mathbb{E}[\xi(t)\xi(t')]=\delta(t-t')$ is the time derivative of the Wiener process $W\in\mathcal{W}$, $\xi(t)=dW/dt$. Linearization $\dot{x}(t)=F(x,u,d,\frac{dW}{dt})$ may be inadequate, and, a nonlinear stochastic analyses is obscured.

Nonlinear Systems and Linearization – Many systems exhibit not-linearizable electromagnetic, hydrodynamic, kinematic, plasma and other instabilities and nonlinearities. Linearization may be inadequate despite phenomena continuity and continuously differentiable real-valued functions $F(x,u,d)$. Examples are static and dynamic hysteresis, unsteady elasto-hydrodynamics, turbulent flow, unsteady vorticity, etc. One examines discontinuities, symmetric and asymmetric periodic solutions, limit cycles, local instabilities, oscillation, mixed-mode periodicity, etc.

Laws of physics, applied to describe phenomena, may yield dimensionality, class, order and structure of differential equations (6.1), (6.9) and (6.11). One should not oversimplify and overcomplicate models, ignore or linearize inherent stabilizing nonlinearities, etc. Focused studies are concentrated on:

1. Exhibited and observed phenomena;
2. Laws of physics, modeling fidelity, adequateness, and, analysis completeness;
3. Data regularity and fidelity;
4. Model complexity, adequacy and compliance in carrying out design, optimization and integration by means of characterization and verification.

Linearized models, given as

$$\dot{x}(t) = Ax + Bu,\ x(t_0)=x_0, \tag{6.12}$$

may yield adequate preliminary analysis. As documented in (6.8), differential equation $\dot{\delta}_x(t) = A\delta_x + B\delta_u$ describes evolutions of variations of $\delta_x(t)$ and $\delta_u(t)$ in a small neighborhood (X_δ, U_δ). Therefore, in (6.12) $\dot{x}=Ax+Bu$, $x\equiv\delta_x$ and $u\equiv\delta_u$.

6.2.2. Analysis of Nonlinear Systems

Using laws of physics, one develops and parameterizes models, investigates system evolutions, finds analytic and numeric solutions to differential equations, substantiates consistency, and, investigates fidelity. Physical systems are described by nonlinear differential equations (6.3) $\dot{x}(t) = F(x,u)$.

Second-Order Systems and Governing Equations – Consider the second-order nonlinear ordinary differential equation which admit analytic solutions

$$\begin{aligned}\dot{x}_1 &= f_1(x_1,x_2),\\ \dot{x}_2 &= f_2(x_1,x_2).\end{aligned} \tag{6.13}$$

A *vector field* is a vector-valued function $\mathbf{f}(\cdot):\mathbb{R}^n\rightarrow\mathbb{R}^n$ from a vector space to itself. For (6.13), consider the plane vector fields. The arrows at *critical points* in the plane represent the vectors at those points. A *critical point* corresponds to

$$\begin{aligned}\dot{x}_1&=f_1(x_1,x_2)=0,\\ \dot{x}_2&=f_2(x_1,x_2)=0.\end{aligned} \tag{6.14}$$

Equations $f_1(x_1,x_2)=0$ and $f_2(x_1,x_2)=0$ yield curves in the xy plane. The equilibria occur where two isoclines intersect. One can apply analytic methods, isocline method, or numerically solve differential equations. From (6.13) we have

$$\frac{\frac{dx_2}{dt}}{\frac{dx_1}{dt}} = \frac{dx_2}{dx_1} = \frac{f_2(x_1,x_2)}{f_1(x_1,x_2)}. \tag{6.15}$$

An isocline is a locus of points with a given tangent slope, and, $\frac{dx_2}{dx_1} = \frac{f_2(x_1,x_2)}{f_1(x_1,x_2)} = c$.

Example 6.4. *Analysis of a Nonlinear System*
Consider the second-order system which cannot be linearized

$$\dot{x}_1 = x_1(1 - x_1^2 - x_2^2) - ax_2\,,$$
$$\dot{x}_2 = ax_1 + x_2(1 - x_1^2 - x_2^2)\,.$$

Find analytic solutions pertained to system stability. For nonlinear systems, the Lyapunov stability theory yields consistent results as covered in section 6.10.
Consider a positive definite continuously differentiable function $V(x_1, x_2) = \frac{1}{2}x_1^2 + \frac{1}{2}x_2^2$.
The total derivative is $\frac{dV(x_1,x_2)}{dt} = \frac{dV}{dx_1}\frac{dx_1}{dt} + \frac{dV}{dx_2}\frac{dx_2}{dt} = x_1^2(1 - x_1^2 - x_2^2) + x_2^2(1 - x_1^2 - x_2^2)$.

For $V(x_1,x_2)>0$: (1) $\frac{dV}{dt} \ge 0$ if $(x_1^2 + x_2^2) \le 1$; (2) $\frac{dV}{dt} < 0$ if $(x_1^2 + x_2^2) > 1$.

The equation for a unit circle is $x_1^2 + x_2^2 = 1$. Per Conjecture 6.2:

(i) System is stable in large because $\frac{dV}{dt} < 0$ for $(x_1^2 + x_2^2) > 1$, $(x_1,x_2)_\Delta \in X_\Delta$, $X_\delta \subset X_\Delta \in X$;

(ii) System is unstable in small when $(x_1^2 + x_2^2) < 1$ because $\frac{dV}{dt} > 0$, and, states (x_1,x_2) exhibit periodic orbits in small $(x_1,x_2)_\delta \in X_\delta$.

Find an explicit analytic solution applying the polar coordinates
$x_1 = r\cos\theta$, $x_2 = r\sin\theta$.

Hence $x_1^2 + x_2^2 = r^2$, $\frac{x_2}{x_1} = \frac{r\sin\theta}{r\cos\theta} = \tan\theta$.

Using the chain and quotient rules, the differentiation yields

$$\frac{dr}{dt} = \frac{d}{dt}(x_1^2 + x_2^2)^{1/2} = \frac{1}{r}\left(x_1\frac{dx_1}{dt} + x_2\frac{dx_2}{dt}\right) = \frac{1}{r}\left(-ax_1x_2 + x_1^2(1 - r^2) + ax_1x_2 + x_2^2(1 - r^2)\right) = r(1 - r^2),$$

$$\frac{d\theta}{dt} = \frac{d}{dt}\tan^{-1}\frac{x_2}{x_1} = \frac{\frac{dx_2}{dt}x_1 - x_2\frac{dx_1}{dt}}{r^2} = a.$$

Solutions for the first-order Bernoulli ordinary differential equation $\frac{dr}{dt} = r(1 - r^2)$, as well as for an ordinary differential equation $\frac{d\theta}{dt} = a$ are

$$r(t) = \frac{1}{\sqrt{1 + (r_0^{-2} - 1)e^{-2t}}},\quad \theta(t) = \theta_0 + at\,.$$

A constant a defines the frequency of periodic evolutions $x_1(t)$ and $x_2(t)$. System is stable in large, and, solutions (x_1,x_2) are bounded for $a>0$, $a<0$ and $a=0$. In particular:

1. For $a>0$ and initial radius $r_0>1$, system trajectories (x_1,x_2) spiral towards periodic orbits. The bounded solutions cyclically evolve with the amplitude 1. If $a>0$, $r_0=1$, the states periodically evolve, and, the amplitude is 1. If $a>0$, $r_0<1$, the system trajectory spirals outwards, and, cyclically evolve within the orbits with a unit amplitude. The critical point (0,0) is an unstable spiral point. The unit circle is a stable closed trajectory;
2. If $a<0$, trajectories spirals clockwise;
3. For $a=0$ one has $\theta(t)=\theta_0$. The origin is a nodal source. All points are critical.

The studied second-order system exhibits periodic solutions for $a\neq 0$. For $a=0$, the system states $x_1(t)$ and $x_2(t)$ evolve to equilibrium. The solutions are bounded as $|x_1|\le 1$ and $|x_2|\le 1$.

Perform numeric studies. For a=5 and a= –5, the phase portraits are documented in Figures 6.2.a considering four initial conditions

$x_0=[x_{10}, x_{20}]=\{[-2, -2], [-0.1, -0.1], [0.1, 0.1], [2, 2]\}$.

Nonlinear ordinary differential equations are solved using the **ode45** solver. Dynamics of $x_1(t)$ and $x_2(t)$ are shown in Figures 6.2.b. The MATLAB code is

```
a=-5; f=@(t,x) [x(1)*(1-x(1)^2-x(2)^2)-a*x(2);a*x(1)+x(2)*(1-x(1)^2-x(2)^2)]; % a=5 and a=-5
% Calculate the phase trajectories for initial conditions
for x0=[-2 -0.1 0.1 2]
t0tfinal=[0,12]; InitialConditions=[x0, x0];  % Initial and final time, and, initial conditions x10 and x20
set=odeset('RelTol',1e-5,'AbsTol',1e-7);  % Specifying numeric accuracy in solving differential equations
[t,x]=ode45(f,t0tfinal,InitialConditions,set); % Solution of ordinary differential equations
plot(x(:,1),x(:,2),'LineWidth',2); axis([-2.1 2.1 -2.1 2.1]); hold on;   % Plot phase trajectories
figure(1); plot(x(1,1),x(1,2),'ko', 'LineWidth',2);        hold on;   % Initial conditions on plot
xlabel('{\itx}_1','FontSize',22); ylabel('{\itx}_2','FontSize',24);
end; hold off;
% Solution of differential equations with two sets of initial conditions
for x0=[0.1 2]
InitialConditions=[x0, -x0];
[t,x]=ode45(f,t0tfinal,InitialConditions,set); % Solution of ordinary differential equations
figure(2); plot(t,x(:,1),'k',t,x(:,2),'k--','LineWidth',2); axis([0 5 -2.1 2.1]); hold on; % Plot system dynamics
legend('{\itx}_1','{\itx}_2','Location','northeast','NumColumns',2,'FontSize',22); legend boxoff;
end; hold off
```

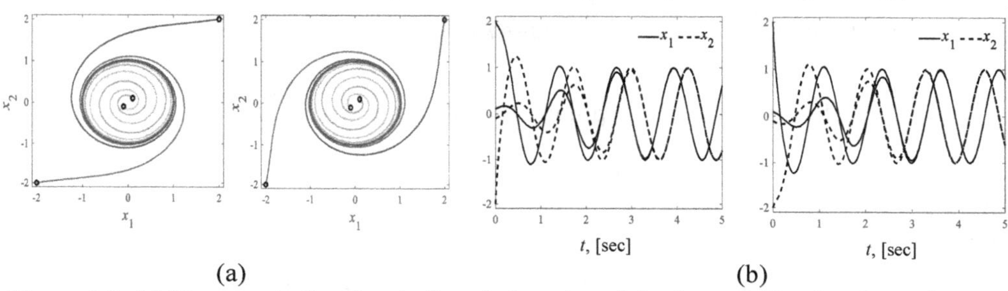

(a) (b)

Figure 6.2. (a) Phase portraits of periodic solutions (x_1,x_2) for (x_{10},x_{20}) if a=5 and a= –5; (b) Evolutions of $x_1(t)$ and $x_2(t)$, $[x_{10}, x_{20}]=\{[2, -2], [0.1, -0.1]\}$, a=5 and a= –5. ■

Example 6.5. *Reaction-Diffusion Model*

Consider the governing equations for a simplified model of reaction-diffusion process

$$\dot{x}_1 = -x_1 + 2x_2 - x_1^3 + u_1,$$

$$\dot{x}_2 = 2x_1 - x_2 - x_2^3 + u_2,$$

with Jacobians $J = \begin{bmatrix} \frac{\partial f_1}{\partial x_1} & \frac{\partial f_1}{\partial x_2} \\ \frac{\partial f_2}{\partial x_1} & \frac{\partial f_2}{\partial x_2} \end{bmatrix} = \begin{bmatrix} -1-3x_1^2 & 2 \\ 2 & -1-3x_2^2 \end{bmatrix}$, $J_u = \begin{bmatrix} \frac{\partial f_1}{\partial u_1} & \frac{\partial f_1}{\partial u_2} \\ \frac{\partial f_2}{\partial u_1} & \frac{\partial f_2}{\partial u_2} \end{bmatrix} = \begin{bmatrix} 1 & 0 \\ 0 & 1 \end{bmatrix}$.

The open-loop system with u_1=0, u_2=0 is asymptotically stable for all $x\in X$, $\forall t$.

Apply a positive definite Lyapunov function

$V(x_1, x_2) = \frac{1}{2}x_1^2 + \frac{1}{2}x_2^2$.

The total derivative is

$$\frac{dV(x_1,x_2)}{dt} = \frac{dV}{dx_1}\frac{dx_1}{dt} + \frac{dV}{dx_2}\frac{dx_2}{dt} = -x_1^2 + 4x_1x_2 - x_2^2 - x_1^4 - x_2^4.$$

For $V(x_1,x_2)>0$, $\frac{dV(x_1,x_2)}{dt} < 0$. Hence, the open-loop system is asymptotically stable, and, there are no limit cycles or bifurcations.

Examine the open-loop system by finding equilibrium state x_e by solving $F(x_e)=0$, $\forall x \in X$. With $u_1=0$ and $u_2=0$, we have

$$\dot{x}_1 = -x_1 + 2x_2 - x_1^3 = -2(x_1 - x_2) + x_1(1 - x_1^2) = 0,$$

$$\dot{x}_2 = 2x_1 - x_2 - x_2^3 = 2(x_1 - x_2) + x_2(1 - x_2^2) = 0.$$

Subtracting the second equation from the first, $-(x_1 - x_2)(3 + x_1^2 + x_1x_2 + x_2^2) = 0$.

Taking note that $(3 + x_1^2 + x_1x_2 + x_2^2) > 0$, one yields $x_1=x_2$.

From $\dot{x}_1 = -x_1 + 2x_2 - x_1^3 = 0$, $x_1=x_2$, we obtain $x_1(1 - x_1^2) = 0$.

The equilibria x_e are at (0,0), (1,1) and (−1,−1), at which $x(t)|_{t\to\infty} \to x_e$.

The Jacobian is $J = \begin{bmatrix} \dfrac{\partial f_1}{\partial x_1} & \dfrac{\partial f_1}{\partial x_2} \\ \dfrac{\partial f_2}{\partial x_1} & \dfrac{\partial f_2}{\partial x_2} \end{bmatrix} = \begin{bmatrix} -1-3x_1^2 & 2 \\ 2 & -1-3x_2^2 \end{bmatrix}$.

At (0,0), one has $J = \begin{bmatrix} -1 & 2 \\ 2 & -1 \end{bmatrix}$.

The characteristic equation $|J - \lambda I| = (\lambda+1)^2 - 4 = 0$ yields the eigenvalues $\lambda_1 = -3$ and $\lambda_2 = 1$.

The equilibrium is a saddle point. At (1,1) and (−1,−1), $J = \begin{bmatrix} -4 & 2 \\ 2 & -4 \end{bmatrix}$.

The characteristic equation $|J - \lambda I| = (\lambda+4)^2 - 4 = 0$ yields $\lambda_1 = -2$ and $\lambda_2 = -6$. The equilibria are the stable nodes.

Applying the Bendixson theorem, evaluate the divergence. The negative definiteness of divergence $\nabla \cdot \mathbf{f} = \dfrac{\partial f_1(x_1,x_2)}{\partial x_1} + \dfrac{\partial f_2(x_1,x_2)}{\partial x_2} = -1 - 3x_1^2 - 1 - 3x_2^2 = -2 - 3(x_1^2 + x_2^2) < 0$ indicates that there are no closed orbits, no limit cycles, and, no bifurcations.

For the considered system

$$\dot{x}_1 = -x_1 + 2x_2 - x_1^3,$$

$$\dot{x}_2 = 2x_1 - x_2 - x_2^3,$$

surfaces $f_1(x_1,x_2)$ and $f_2(x_1,x_2)$ are documented in Figure 6.3.a.

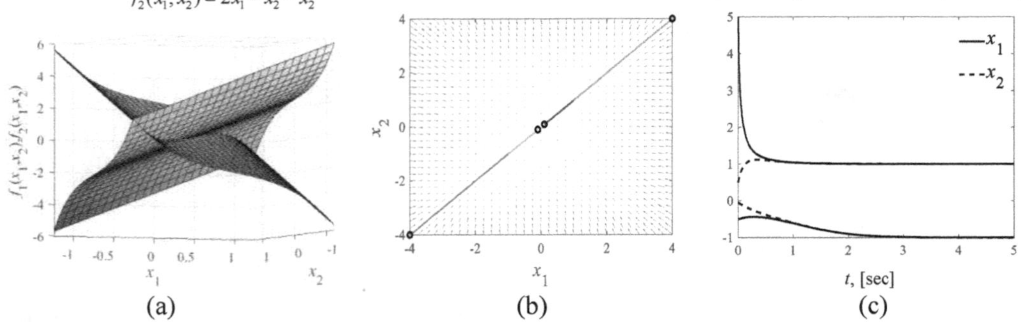

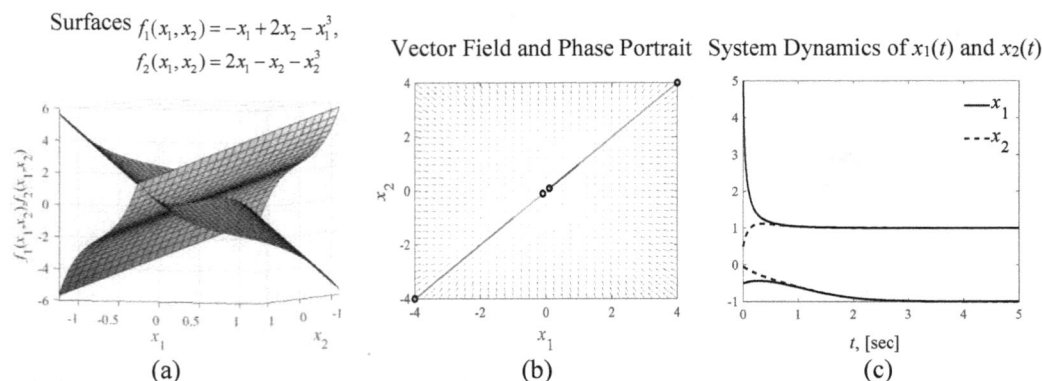

Figure 6.3. (a) Surfaces $f_1(x_1,x_2) = -x_1 + 2x_2 - x_1^3$ and $f_2(x_1,x_2) = 2x_1 - x_2 - x_2^3$;
(b) Vector field and phase portrait with different initial conditions;
(c) Dynamics of $x_1(t)$ and $x_2(t)$ for initial conditions $[x_{10}, x_{20}]=[5, 0.5]$, $[x_{10}, x_{20}]=[-0.5, -0.05]$.

Figure 6.3.b shows the vector field for different initial conditions

$[x_{10}, x_{20}]=\{[-4, -4], [-0.1, -0.1], [0.1, 0.1], [4, 4]\}$.

It was found that the equilibria are at (0,0), (1, 1) and (−1,−1). The transient dynamics of $x_1(t)$ and $x_2(t)$ for different initial conditions are depicted in Figure 6.3.c. Differential equations are solved using MATLAB.

```
f1=@(x1,x2) -x1+2*x2-x1.^3; f2=@(x1,x2) 2*x1-x2-x2.^3;
[X,Y]=meshgrid(-1.25:0.1:1.25,-1.25:0.1:1.25);
figure(1); surf(X,Y,f1(X,Y)); hold on; surf(X,Y,f2(X,Y)); hold off
xlabel('{\itx}_1','FontSize',20); ylabel('{\itx}_2','FontSize',20);
zlabel('{\itf}_1({\itx}_1,{\itx}_2),{\itf}_2({\itx}_1,{\itx}_2)','FontSize',18);
[x1,x2]=meshgrid(-4:0.25:4, -4:0.25:4);
x1dot=-x1+2*x2-x1.^3;   % dx1/dt=f1(x1,x2)
x2dot=2*x1-x2-x2.^3;    % dx2/dt=f2(x1,x2)
% Calculate the vector field using the quiver command
figure(2); quiver(x1,x2,x1dot,x2dot); axis([-4 4 -4 4]); hold on;
f=@(t,x) [-x(1)+2*x(2)-x(1)^3; 2*x(1)-x(2)-x(2)^3]; % Differential equations
% Calculate the phase trajectories for different initial conditions
for x0=[-4 -0.1 0.1 4]
t0tfinal=[0,5]; InitialConditions=[x0, x0]; set=odeset('RelTol',1e-5,'AbsTol',1e-7);
[t,x]=ode45(f,t0tfinal,InitialConditions,set); % Solution of ordinary differential equations
plot(x(:,1),x(:,2),'LineWidth',2);          hold on;    % Plot phase trajectories
plot(x(1,1),x(1,2),'ko', 'LineWidth',2);  hold on;     % Initial conditions
xlabel('{\itx}_1','FontSize',22); ylabel('{\itx}_2','FontSize',22);
end; hold off;
% Solution of differential equations with two initial conditions
for x0=[-0.5 5]
InitialConditions=[1*x0, 0.1*x0];
[t,x]=ode45(f,t0tfinal,InitialConditions,set);  % Solution of ordinary differential equations
figure(3); plot(t,x(:,1),'k',t,x(:,2),'k--','LineWidth',2); hold on;   % Plot system dynamics
legend('{\itx}_1','{\itx}_2','FontSize',24); legend boxoff;
end; hold off
```

■

Example 6.6. *Lotka-Volterra Differential Equation: Analytic and Numeric Solutions*

Numeric solutions of the Lotka-Volterra equations were reported in Examples 2.23 and 2.24. Equations

$$\dot{x}_1 = a_{11}x_1 - a_{12}x_1x_2,\ x_1(t_0)=x_{10},\ x_{10}>0,\ \forall(a_{ii},a_{ij})>0,$$

$$\dot{x}_2 = -a_{21}x_2 + a_{22}x_1x_2,\ x_2(t_0)=x_{20},\ x_{20}>0$$

admit solutions for $(x_1,x_2)>0$ with $(x_{10},x_{20})>0$. The linearization yields inadequate results. One finds isoclines and solves differential equations. Equations of motion yield

$$\frac{\frac{dx_2}{dt}}{\frac{dx_1}{dt}} = \frac{dx_2}{dx_1} = \frac{f_2(x_1,x_2)}{f_1(x_1,x_2)} = \frac{-a_{21}x_2 + a_{22}x_1x_2}{a_{11}x_1 - a_{12}x_1x_2} = \frac{x_2(-a_{21}+a_{22}x_1)}{x_1(a_{11}-a_{12}x_2)}.$$

Separate variables to solve the second-order nonlinear differential equations.

Integration $\int \frac{-a_{21}+a_{22}x_1}{x_1}dx_1 = \int \frac{a_{11}-a_{12}x_2}{x_2}dx_2$

yields $-a_{21}\ln x_1 + a_{22}x_1 - a_{11}\ln x_2 + a_{12}x_2 = c$,

where c is the constant which depends on the initial conditions.

For a considered class of differential equations, we conclude that $(x_1,x_2)>0$, $\forall(x_{10},x_{20})>0$.

Let $a_{11}=1$, $a_{12}=1$, $a_{21}=1$ and $a_{22}=1$.

The vector field and phase portraits for different initial conditions

$[x_{10}, x_{20}]=\{[0.5, 0.5], [1, 1], [1.5, 1.5], [2, 2], [2.5, 2.5], [3, 3]\}$

are illustrated in Figure 6.4.a.

Nonlinear ordinary differential equations are numerically solved. Transient dynamics are depicted in Figure 6.4.b. The MATLAB code is

```
[x1,x2]=meshgrid(-0.1:0.25:4.5, -0.1:0.25:4.5); a11=1; a12=1; a21=1; a22=1;
dx1=a11*x1-a12*x1.*x2;   % Lotka-Volterra equation dx1/dt=f1(x1,x2)
dx2=-a21*x2+a22*x1.*x2;  %                          dx2/dt=f2(x1,x2)
% Calculation of the vector field using the quiver command
figure(1);  quiver(x1,x2,dx1,dx2,'b-','LineWidth',1.5); axis([0 4.5 0 4.5]); hold on;
f=@(t,x) [a11*x(1)-a12*x(1)*x(2); -a21*x(2)+a22*x(1)*x(2)]; % Differential equations
% Calculate phase trajectories for different initial conditions
for x0=0.5:0.5:3
t0tfinal=[0,25]; InitialConditions=[x0, x0];  set=odeset('RelTol',1e-5,'AbsTol',1e-7);
[t,x]=ode45(f,t0tfinal, InitialConditions, set);    % Solution of ordinary differential equations
plot(x(:,1),x(:,2),'LineWidth',2);           hold on;  % Plot of phase trajectories: Periodic solutions
plot(x(1,1),x(1,2),'ko', 'LineWidth',2);  hold on;   % Initial conditions
xlabel('{\itx}_1','FontSize',22); ylabel('{\itx}_2','FontSize',22); end; hold off;
% Solution of differential equations with two initial conditions
for x0=[0.5 3]
InitialConditions=[x0, x0]; [t,x]=ode45(f,t0tfinal, InitialConditions, set);
figure(2); plot(t,x(:,1),'k',t,x(:,2),'k--','LineWidth',2); axis([0 23 0 4.5]); hold on;
legend('{\itx}_1','{\itx}_2','FontSize',24); legend boxoff;
end; hold off
```

Vector Field, Phase Portraits and Orbits of
$-a_{21}\ln x_1 + a_{22}x_1 - a_{11}\ln x_2 + a_{12}x_2 = c$

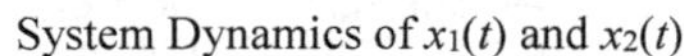

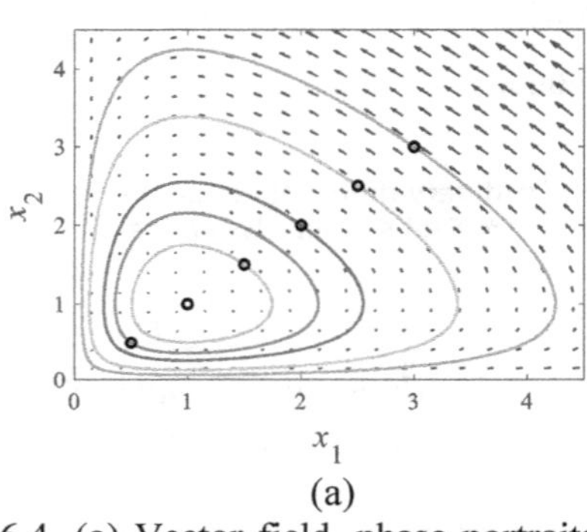

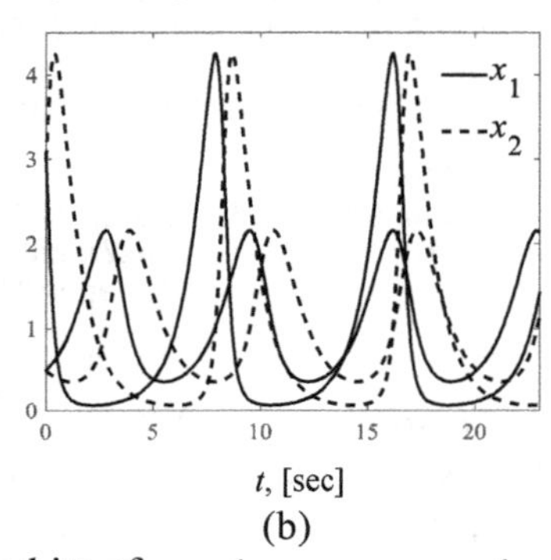

(a) (b)

Figure 6.4. (a) Vector field, phase portraits and orbits of $-a_{21}\ln x_1 + a_{22}x_1 - a_{11}\ln x_2 + a_{12}x_2 = c$ for different initial conditions (x_{10}, x_{20});
(b) Evolutions of $x_1(t)$ and $x_2(t)$ if initial conditions are $[x_{10}, x_{20}]=[0.5, 0.5]$ and $[x_{10}, x_{20}]=[3, 3]$.

Modified Lotka-Volterra Equation – Structures of Lotka-Volterra, Lorenz and other equations can be modified considering adequacy and consistency with the laws of physics, phenomena exhibited and processes observed.

A modified Lotka-Volterra nonlinear ordinary differential equation

$$\dot{x}_1 = a_{11}x_1 - a_{12}x_1x_2 - a_{13}x_1x_2^2,\ (x_{10},x_{20})>0,\ \forall(a_{ii},a_{ij})>0,$$

$$\dot{x}_2 = -a_{21}x_2 + a_{22}x_1x_2 + a_{23}x_1^2x_2,$$

admits evolutions with limit cycles, such as $(x_1,x_2)>0$, $\forall(x_{10},x_{20})>0$, $\forall t$.

From $\dfrac{dx_2}{dx_1} = \dfrac{f_2(x_1,x_2)}{f_1(x_1,x_2)} = \dfrac{x_2(-a_{21} + a_{22}x_1 + a_{23}x_1^2)}{x_1(a_{11} - a_{12}x_2 - a_{13}x_2^2)}$,

we have $\displaystyle\int \frac{-a_{21} + a_{22}x_1 + a_{23}x_1^2}{x_1}dx_1 = \int \frac{a_{11} - a_{12}x_2 - a_{13}x_2^2}{x_2}dx_2$.

Hence, $-a_{21}\ln x_1 + a_{22}x_1 + \frac{1}{2}a_{23}x_1^2 - a_{11}\ln x_2 + a_{12}x_2 + \frac{1}{2}a_{13}x_2^2 = c$.

Let a_{11}=1, a_{12}=1, a_{13}=1, a_{21}=1, a_{22}=1 and a_{23}=1. Figure 6.5.a illustrates the vector field and phase portraits for different initial conditions $(x_{10},x_{20})>0$. System dynamics with periodic limit cycles are reported in Figure 6.5.b.

Vector Field, Phase Portraits and Orbits of
$-a_{21}\ln x_1 + a_{22}x_1 + \frac{1}{2}a_{23}x_1^2 - a_{11}\ln x_2 + a_{12}x_2 + \frac{1}{2}a_{13}x_2^2 = c$

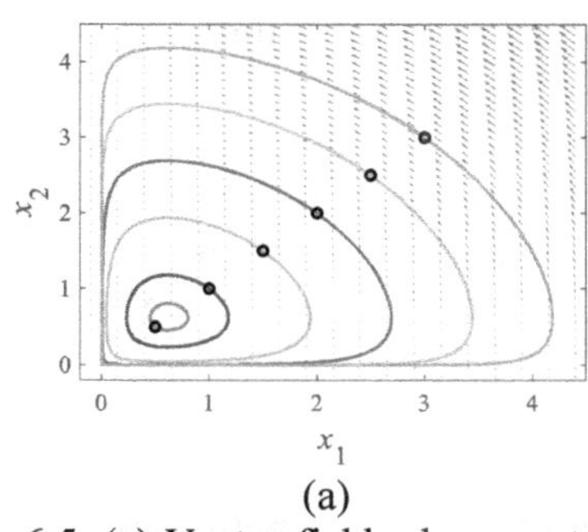

(a)

System Dynamics of $x_1(t)$ and $x_2(t)$

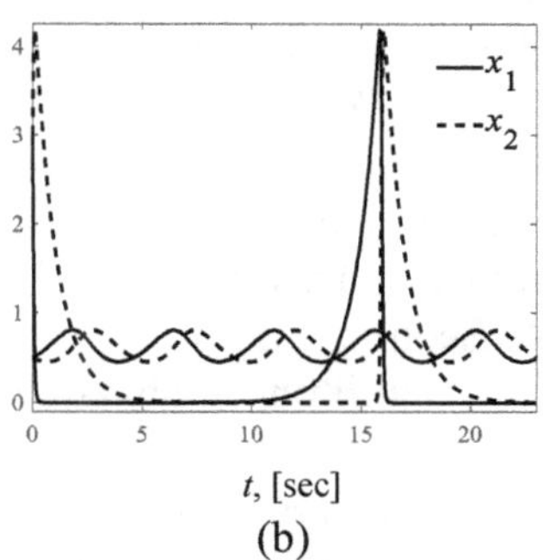

(b)

Figure 6.5. (a) Vector field, phase portraits and orbits for a modified Lotka-Volterra equation with different initial conditions (x_{10}, x_{20}), shown by circles;
(b) Evolutions of $x_1(t)$ and $x_2(t)$ for initial conditions $[x_{10}, x_{20}]=[0.5, 0.5]$ and $[x_{10}, x_{20}]=[3, 3]$. ■

6.3. Optimal Control of Nonlinear Systems

Dynamic optimization, Hamiltonian function and Hamilton-Jacobi equation were introduced in section 2.5 and chapter 5. For a nonlinear system $\dot{x}(t)=F(t,x)+B(t,x)u$, using positive definite continuous scalar functions $V(t,x)$, $V(\cdot):\mathbb{R}_{\geq 0}\times\mathbb{R}^n\to\mathbb{R}$, $V\in C^k$ $(k\geq 2)$ and $\omega(t,x,u)$, $\omega(\cdot):\mathbb{R}_{\geq 0}\times\mathbb{R}^n\times\mathbb{R}^m\to\mathbb{R}$, $\omega\in C^l$ $(l\geq 2)$, the Hamiltonian (2.61)

$$H(t,x,u,\tfrac{\partial V}{\partial x}) \equiv \omega(t,x,u) + \left(\frac{\partial V}{\partial x}\right)^T \left[F(t,x)+B(t,x)u\right]$$

is minimized finding an optimal control function (2.63) $u^* = \varphi\left(t,\frac{\partial V^*}{\partial x}\right)$.

A scalar function $V^*(t,x)$ satisfies the Hamilton-Jacobi equation (2.64)

$$-\frac{\partial V^*}{\partial t} = \omega(t,x,u^*) + \left(\frac{\partial V^*}{\partial x}\right)^T \left[F(t,x)+B(t,x)u^*\right],\ u^* = \varphi\left(t,\tfrac{\partial V^*}{\partial x}\right),\ V\in C^k.$$

Solutions of nonlinear optimization problems with descriptive examples are reported in this chapter. Control problems were solved in chapter 5 by minimizing quadratic performance functionals and indexes subject to linear system dynamics. Linear and nonlinear control problems can be solved applying calculus of variations, minimax Hamiltonian optimization, dynamic programming, nonlinear programming, Lyapunov theory, etc. To solve stabilization and tracking problems, consider

$$J = \min_{x,u}\int_{t_0}^{t_f}\omega(x,u)dt,\ \omega(x,u)>0,\quad J = \min_{x,u}\int_{t_0}^{t_f}\omega(x,u,\tfrac{\partial V}{\partial x})dt,\ \omega(x,u,\tfrac{\partial V}{\partial x})>0,$$

$$J = \min_{x,e,u}\int_{t_0}^{t_f}\omega(x,e,u)dt,\ \omega(x,e,u)>0, \tag{6.16}$$

where t_0 and t_f are the initial and final time which define the time horizon.

Physics-consistent positive definite, continuous and differentiable integrand functions $\omega(\cdot)$ are synthesized. One also may consider a functional $J = \min_{x,u} \int_{t_0}^{t_f} \omega(x,u,\frac{\partial V}{\partial x})dt$ to find control laws and solve the Hamilton-Jacobi equations.

Stabilizing control laws

$$u=\varphi(t,x),\ \varphi(\cdot):\mathbb{R}_{\geq 0}\times\mathbb{R}^n\to\mathbb{R}^m, \quad (6.17)$$

are designed by minimizing functionals (6.16) subject to system dynamics

$$\dot{x}(t)=F(x)+B(x)u,\ x(t_0)=x_0, \quad (6.18)$$

where $F(\cdot):\mathbb{R}^n\to\mathbb{R}^n$ and $B(\cdot):\mathbb{R}^n\to\mathbb{R}^{n\times m}$ are the continuous Lipschitz maps.

Dynamic optimization implies solution of a minimization, maximization or minimax problems. To find optimal control laws, necessary conditions for optimality are applied. Section 2.5 covers introduction to dynamic optimization, as well as solution of the Hamilton-Jacobi equation (2.65).

6.3.1. Optimality and Lagrange Multipliers

To minimize the performance functional (6.16) on $t\in[t_0\ t_f]$ with design-specific integrand $\omega(\cdot)$, subject to (6.18), consider the real-valued k-differentiable scalar Hamiltonian function

$$H(t,x,u,\boldsymbol{\lambda}) = \omega(x,u) + \boldsymbol{\lambda}^T(t,x)\left[F(x)+B(x)u\right], \quad (6.19)$$

$$\dot{x} = \left(\frac{\partial H}{\partial \boldsymbol{\lambda}}\right)^T = F(x)+B(x)u\,,$$

$$\dot{\boldsymbol{\lambda}}^T = -\frac{\partial H(t,x,u,\boldsymbol{\lambda})}{\partial x},$$

where $\boldsymbol{\lambda}$ is the *costate* vector with $\boldsymbol{\lambda}(t_f)$.

To minimize the Hamiltonian $\min_{u\in U}\left[H(t,x,u,\boldsymbol{\lambda})\right]$, consider a *costate equation* $\dot{\boldsymbol{\lambda}}^T = -\frac{\partial H}{\partial x}$ and stationarity condition $\frac{\partial H(t,x,u,\boldsymbol{\lambda})}{\partial u} = 0$, which is a necessary conditions for optimality.

That is, we have

$$\dot{\boldsymbol{\lambda}}^T = -\frac{\partial H(t,x,u,\boldsymbol{\lambda})}{\partial x}, \quad (6.20)$$

$$\frac{\partial H(t,x,u,\boldsymbol{\lambda})}{\partial u} = 0.$$

One finds an optimal control $u^*=\varphi(t,x)$ and state evolutions $x^*(t)$ solving a boundary-value problem with the boundary conditions $x(t_0)$ and $\boldsymbol{\lambda}(t_f)$. We use the Lagrange multiplier $\boldsymbol{\lambda}(t,x)$ and a continuous differentiable return function $V(t,x)$, as reported in section 2.5. These $\boldsymbol{\lambda}(t,x)$ or $V(t,x)$ should satisfy the Hamilton-Jacobi equation (2.62) and (5.63) $-\frac{\partial V^*}{\partial t} = \omega(x,u^*) + \left(\frac{\partial V^*}{\partial x}\right)^T\left(F(x)+B(x)u^*\right)$, $\omega\in C^l$, $V\in C^k$. An optimal trajectory $x^*(t)$ and control function $u^*(t)$ can be derived using (6.20).

Example 6.7. *Optimal Control Laws Design and State Feedback*

Optimal state and control evolutions, as well as optimal control law, are found by minimizing a quadratic functional

$J=\min_{x,u}\int_{t_0}^{t_f}\tfrac{1}{2}(x^2+0.25u^2)dt$, t_0=0, t_f=3 sec

subject to the first-order system $\dot{x}(t)=u$, $x_0=1$.

The boundary conditions are $x_0=1$ and $\lambda(t_f)=0$.

For the Hamiltonian (6.19)

$H(x,u,\lambda)=\tfrac{1}{2}(x^2+0.25u^2)+\lambda u$,

the *costate equation* (6.20) is $\dot{\lambda}=-\dfrac{\partial H}{\partial x}=-x$.

The stationarity condition (6.20) $\dfrac{\partial H}{\partial u}=0.25u+\lambda=0$ yields

$u=-4\lambda$.

Solve the two-point boundary value problem

$\dot{x}(t)=-4\lambda$, $x_0=1$,

$\dot{\lambda}(t)=-x$, $\lambda(t_f)=0$.

For resulting $\begin{bmatrix}\dot{x}\\ \dot{\lambda}\end{bmatrix}=A\begin{bmatrix}x\\ \lambda\end{bmatrix}=\begin{bmatrix}0 & -4\\ -1 & 0\end{bmatrix}\begin{bmatrix}x\\ \lambda\end{bmatrix}$,

the solution is $\begin{bmatrix}x(t)\\ \lambda(t)\end{bmatrix}=c_1\mathbf{v}_1e^{s_1t}+c_2\mathbf{v}_2e^{s_2t}$.

The characteristic equation $\det(sI-A)=s^2-4=0$ yields eigenvalues $s_1=2$, $s_2=-2$. In general, for a given $A\in\mathbb{R}^{n\times n}$, one finds n eigenvalues s_i and eigenvectors $\mathbf{v}_i$, solving

$A\mathbf{v}_i=s_i\mathbf{v}_i$, $(sI_i-A)\mathbf{v}_i=0$,

where the characteristic eigenvalues s_i may be complex even if A is real.

The eigenvectors are $\mathbf{v}_1=\begin{bmatrix}2\\ -1\end{bmatrix}$ and $\mathbf{v}_2=\begin{bmatrix}2\\ 1\end{bmatrix}$.

Using the boundary conditions $x_0=1$ and $\lambda(t_f)=0$, we have

$1=2c_1+2c_2$,

$0=-c_1e^{2t_f}+c_2e^{-2t_f}$.

Hence, $c_1=\dfrac{1}{2(1+e^{4t_f})}$, $c_2=\dfrac{1}{2(1+e^{-4t_f})}$.

From $\begin{bmatrix}x(t)\\ \lambda(t)\end{bmatrix}=c_1\mathbf{v}_1e^{s_1t}+c_2\mathbf{v}_2e^{s_2t}$,

one obtains $\begin{bmatrix}x(t)\\ \lambda(t)\end{bmatrix}=\dfrac{1}{2(1+e^{4t_f})}\begin{bmatrix}2\\ -1\end{bmatrix}e^{2t}+\dfrac{1}{2(1+e^{-4t_f})}\begin{bmatrix}2\\ 1\end{bmatrix}e^{-2t}$.

Hence, optimization yields optimal evolutions of (x^*,u^*) as

$x^*(t)=\dfrac{1}{1+e^{4t_f}}e^{2t}+\dfrac{1}{1+e^{-4t_f}}e^{-2t}$, $u^*(t)=-4\lambda(t)=\dfrac{2}{1+e^{4t_f}}e^{2t}-\dfrac{2}{1+e^{-4t_f}}e^{-2t}$.

Figure 6.6.a documents evolutions of $x^*(t)$ and $u^*(t)$.

Optimal Control and State Feedback – Derived results do not explicitly define a control law $u^*(t)$ as a function of state $x(t)$. To find an optimal control and closed-loop system with state feedback, consider

$\lambda(t)=k(t)x(t)$, $k(t_f)=0$.

Hence, $\dot{\lambda}(t)=\dot{k}(t)x(t)+k(t)\dot{x}(t)$.

Using the *costate equation* $\dot{\lambda} = -x$ and the stationarity condition $u = -4\lambda$,
$-x(t) = \dot{k}(t)x(t) - k(t)4\lambda(t) = \dot{k}(t)x(t) - 4k^2(t)x(t)$.
We obtain a quadratic Riccati differential equation
$\dot{k}(t) = 4k^2(t) - 1$, $k(t_f)=0$.
From $\dfrac{dk}{k^2 - 0.25} = 4dt$, one has
$2\tanh^{-1}(2k) = -4t + c$.
The boundary condition $k(t_f)=0$ gives $c=4t_f$.
Therefore, $k(t) = \frac{1}{2}\tanh\left(2(t_f - t)\right)$.
An optimal control law is
$u^*(t) = -4\lambda(t) = -4k(t)x(t) = -2\tanh\left(2(t_f - t)\right)x(t)$.
If $t_f \to \infty$, $k(t) \to 2$.
For suboptimal solution, minimize $J = \min\limits_{x,u} \int_{t_0}^{\infty} \frac{1}{2}(x^2 + 0.25u^2)dt$.
Using the LQR concept, we have $u = -2x$.

Simulations are performed for $t_f=3$ sec. Figure 6.6.b depicts diagrams to simulate closed-loop systems

$\dot{x}^*(t) = u^*$, $u^*(t) = -2\tanh\left(2(t_f - t)\right)x(t)$, $x_0=1$

and $\dot{x}(t) = u$, $u(t) = -2x(t)$, $x_0=1$.

Evolutions of $x^*(t)$, $u^*(t)$ and $k(t)$ are reported in Figure 6.7.c. The final value of J, calculated using SIMULINK, is 0.25. Dynamics of $x^*(t)$ and $x(t)$ are almost identical.

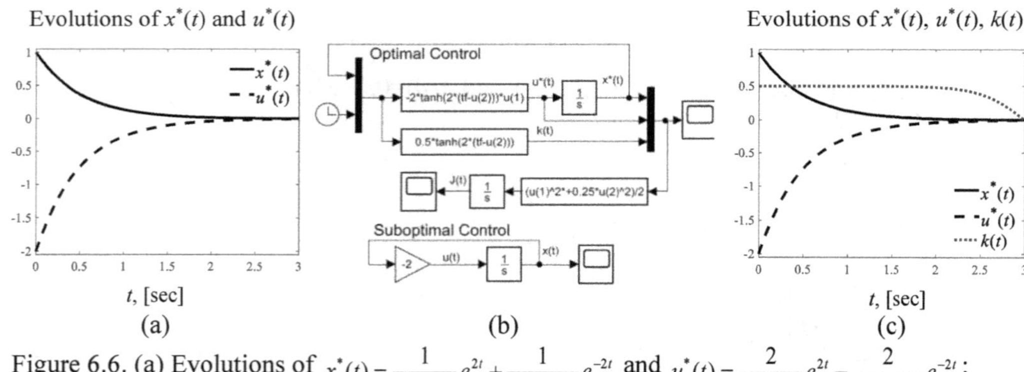

Figure 6.6. (a) Evolutions of $x^*(t) = \dfrac{1}{1+e^{4t_f}}e^{2t} + \dfrac{1}{1+e^{-4t_f}}e^{-2t}$ and $u^*(t) = \dfrac{2}{1+e^{4t_f}}e^{2t} - \dfrac{2}{1+e^{-4t_f}}e^{-2t}$;
(b) SIMULINK diagrams for optimal and suboptimal control laws;
(c) Closed-loop system: Evolutions of $x^*(t)$, $u^*(t)$ and $k(t)$. ■

Example 6.8. Control, Optimality and State Feedback

In Example 5.27, minimizing $J = \min\limits_{x,u} \int_0^{t_f} \frac{1}{2}(x^2 + u^2)dt$, $t_f=5$ sec subject to

$\dot{x} = -x + u$, $x_0=1$,

an optimal control law was found as $u^* = -k(t)x = -\dfrac{\sqrt{2} - 1 + (1-\sqrt{2})e^{-2\sqrt{2}(t_f - t)}}{1 + e^{-2\sqrt{2}(t_f - t)}}x$.

A system is simulated, and, evolutions for (x,u) are depicted in Figures 5.23.c and 6.7.a.

Apply an optimization concept, and, find optimal evolutions (x^*,u^*) by minimizing

$$J = \min_{x,u} \int_{t_0}^{t_f} \tfrac{1}{2}(x^2 + u^2)dt,\ t_0=0,\ t_f=5 \text{ sec}$$

subject to $\dot{x} = -x + u$, $x_0=1$.

The boundary conditions are $x_0=1$ and $\lambda(t_f)=0$.

For the Hamiltonian (6.19)

$$H(x,u,\lambda) = \tfrac{1}{2}(x^2 + u^2) + \lambda(-x+u),$$

the *costate equation* (6.20) is $\dot{\lambda} = -\dfrac{\partial H}{\partial x} = -x + \lambda$.

The stationarity condition (6.20) $\dfrac{\partial H}{\partial u} = u + \lambda = 0$ yields $u= -\lambda$.

One solves a two-point boundary value problem

$$\begin{bmatrix} \dot{x} \\ \dot{\lambda} \end{bmatrix} = A \begin{bmatrix} x \\ \lambda \end{bmatrix} = \begin{bmatrix} -1 & -1 \\ -1 & 1 \end{bmatrix} \begin{bmatrix} x \\ \lambda \end{bmatrix},\ x_0=1,\ \lambda(t_f)=0.$$

The solution is $\begin{bmatrix} x(t) \\ \lambda(t) \end{bmatrix} = c_1 \boldsymbol{v}_1 e^{s_1 t} + c_2 \boldsymbol{v}_2 e^{s_2 t}$.

The characteristic equation $\det(sI-A)=0$ yields $(s^2-2)=0$, and, $s_{1,2} = \pm\sqrt{2}$.

Solving $(sI_i-A)\boldsymbol{v}_i=0$ for $s_i = s_{1,2} = \pm\sqrt{2}$, the eigenvectors are

$$\boldsymbol{v}_1 = \begin{bmatrix} 1-\sqrt{2} \\ 1 \end{bmatrix},\ \boldsymbol{v}_2 = \begin{bmatrix} 1+\sqrt{2} \\ 1 \end{bmatrix}.$$

Hence, $\begin{bmatrix} x(t) \\ \lambda(t) \end{bmatrix} = c_1 \begin{bmatrix} 1-\sqrt{2} \\ 1 \end{bmatrix} e^{\sqrt{2}t} + c_2 \begin{bmatrix} 1+\sqrt{2} \\ 1 \end{bmatrix} e^{-\sqrt{2}t}$.

Using the boundary conditions $x_0=1$ and $\lambda(t_f)=0$, from

$$x(t) = c_1(1-\sqrt{2})e^{\sqrt{2}t} + c_2(1+\sqrt{2})e^{-\sqrt{2}t},$$

$$\lambda(t) = c_1 e^{\sqrt{2}t} + c_2 e^{-\sqrt{2}t},$$

one has $1 = c_1(1-\sqrt{2})e^0 + c_2(1+\sqrt{2})e^0$,

$$0 = c_1 e^{\sqrt{2}t_f} + c_2 e^{-\sqrt{2}t_f}.$$

Hence, $c_1 = \dfrac{1}{1-\sqrt{2}} + \left(1+\sqrt{2}\right)^3 \dfrac{e^{\sqrt{2}t_f}}{e^{-\sqrt{2}t_f} + (1+\sqrt{2})^2 e^{\sqrt{2}t_f}}$, $c_2 = \left(1+\sqrt{2}\right) \dfrac{e^{\sqrt{2}t_f}}{e^{-\sqrt{2}t_f} + (1+\sqrt{2})^2 e^{\sqrt{2}t_f}}$.

Solving the optimization problem, optimal evolutions (x^*,u^*) in time domain are

$$x^*(t) = \left(1-\sqrt{2}\right)\left[\frac{1}{1-\sqrt{2}} + \left(1+\sqrt{2}\right)^3 \frac{e^{\sqrt{2}t_f}}{e^{-\sqrt{2}t_f} + (1+\sqrt{2})^2 e^{\sqrt{2}t_f}}\right] e^{\sqrt{2}t} + \left(1+\sqrt{2}\right)^2 \frac{e^{\sqrt{2}t_f}}{e^{-\sqrt{2}t_f} + (1+\sqrt{2})^2 e^{\sqrt{2}t_f}} e^{-\sqrt{2}t},$$

$$u^*(t) = -\lambda(t) = -\left[\frac{1}{1-\sqrt{2}} + \left(1+\sqrt{2}\right)^3 \frac{e^{\sqrt{2}t_f}}{e^{-\sqrt{2}t_f} + (1+\sqrt{2})^2 e^{\sqrt{2}t_f}}\right] e^{\sqrt{2}t} - \left(1+\sqrt{2}\right) \frac{e^{\sqrt{2}t_f}}{e^{-\sqrt{2}t_f} + (1+\sqrt{2})^2 e^{\sqrt{2}t_f}} e^{-\sqrt{2}t}.$$

Figure 6.7.b documents evolutions of $x^*(t)$ and $u^*(t)$.

To explicitly define an optimal control law $u^*(t)$ as a function of state $x(t)$, apply

$\lambda(t)=k(t)x(t)$, $k(t_f)=0$.

Thus, $\dot{\lambda}(t) = \dot{k}(t)x(t) + k(t)\dot{x}(t)$.

Using the *costate equation* $\dot{\lambda} = -x + \lambda$ and the stationarity condition $u = -\lambda$, one obtains a differential equation for $k(t)$

$$-\dot{k}(t) = 1 - 2k(t) - k^2(t),\ k(t_f)=0, \text{ which yields } k(t) = \frac{\sqrt{2} - 1 + (1 - \sqrt{2})e^{-2\sqrt{2}(t_f - t)}}{1 + e^{-2\sqrt{2}(t_f - t)}}.$$

As a solution, reported in Example 5.27, we derived $(x^*(t), k(t), u^*(t))$. In particular, an optimal control law is $u^* = -\lambda = -k(t)x = -\dfrac{\sqrt{2} - 1 + (1 - \sqrt{2})e^{-2\sqrt{2}(t_f - t)}}{1 + e^{-2\sqrt{2}(t_f - t)}}x.$

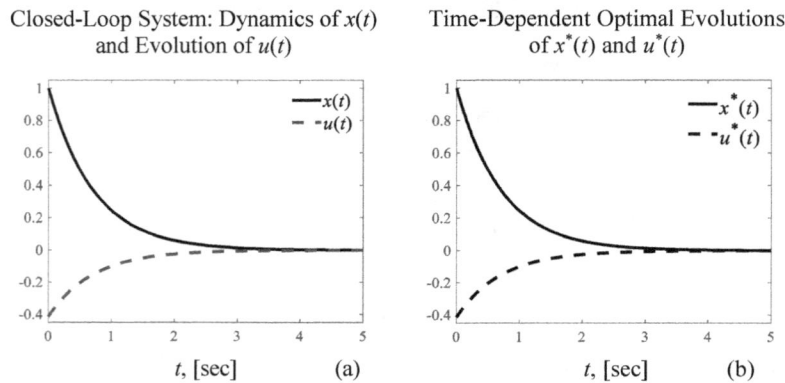

The closed-loop system $\dot{x} = -x + u^*$ evolves as illustrated in Figures 5.23.c and 6.7.a.

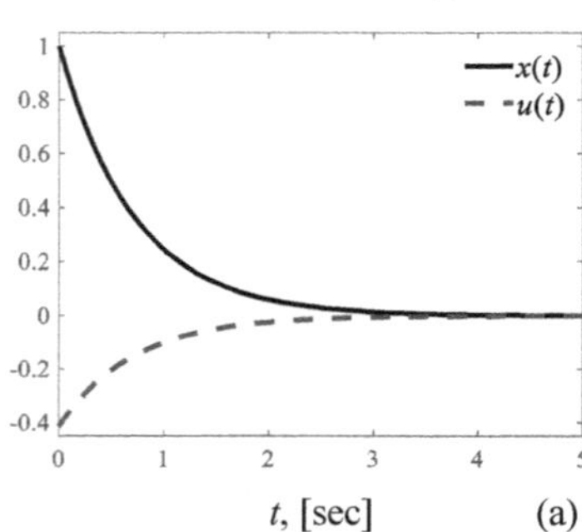

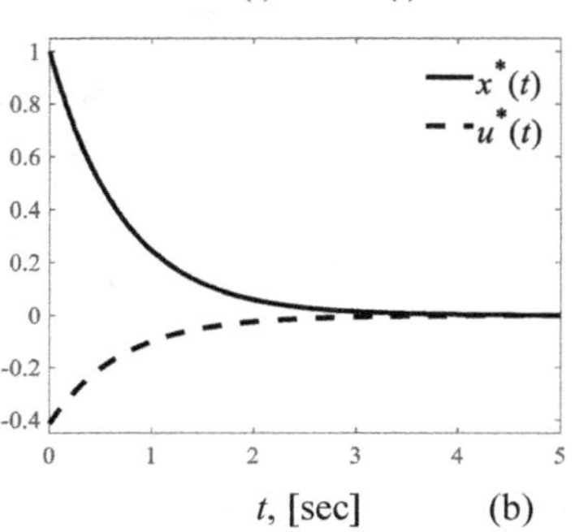

Figure 6.7. (a) State and control evolutions, $u^* = -k(t)x = -\dfrac{\sqrt{2} - 1 + (1 - \sqrt{2})e^{-2\sqrt{2}(t_f - t)}}{1 + e^{-2\sqrt{2}(t_f - t)}}x$;

(b) Evolutions of $x^*(t) = c_1(1 - \sqrt{2})e^{\sqrt{2}t} + c_2(1 + \sqrt{2})e^{-\sqrt{2}t}$ and $u^*(t) = -c_1 e^{\sqrt{2}t} - c_2 e^{-\sqrt{2}t}$. ■

6.3.2. Optimality and Hamilton-Jacobi Equation

To minimize the performance functional (6.16) $J = \min\limits_{x,u} \int_{t_0}^{t_f} \omega(x,u)dt$ with integrand $\omega(\cdot)$, subject to (6.18), consider the real-valued k-differentiable Hamiltonian function and continuous return function $V(t,x)$

$$H(t,x,u,\tfrac{\partial V}{\partial x}) = \omega(x,u) + \left(\frac{\partial V}{\partial x}\right)^T [F(x) + B(x)u] \cdot \tag{6.21}$$

The derivative $\dfrac{\partial H(t,x,u,\frac{\partial V}{\partial x})}{\partial u}$ exists. Control function $u(\cdot):\mathbb{R}_{\ge 0}\times\mathbb{R}^n \to \mathbb{R}^m$ is found using the first-order necessary condition for optimality

$$\frac{\partial H(t,x,u,\frac{\partial V}{\partial x})}{\partial u} = 0 \cdot \tag{6.22}$$

Minimize the Hamiltonian $\min\limits_{u\in U}\left[H(t,x,u,\tfrac{\partial V}{\partial x})\right]$, find a control function $u=\varphi(t,x)$, and, solve the Hamilton-Jacobi equation

$$-\frac{\partial V}{\partial t} = \omega(x,u) + \left(\frac{\partial V}{\partial x}\right)^T [F(x) + B(x)u],\ u = \varphi(t,x) \cdot \tag{6.23}$$

Solution of the Hamilton-Jacobi equation (6.23) is found applying a positive definite, continuous and differentiable return function

$$V(t,x)=\sum_{i=N_0,j=M_0}^{N,M}\tfrac{j}{2i}(x^{i/j})^T K_{ij}(t)x^{i/j},\ K_{ij}(\cdot):\mathbb{R}_{\geq 0}\rightarrow\mathbb{R}^{n\times n},\ N_0\geq 1,\ M_0\geq 1, \qquad (6.24)$$

$$V(t,x)=\sum_{i,j}(x^{<d_i>})^T K_{ij}(t)x^{<d_j>},\ V(t,x)=\sum_{i,j}\kappa_{1_i}^T(t,x^{<d_i>})K_{ij}(t)\kappa_{2_j}(t,x^{<d_j>}),\ (d_i,d_j)\in\mathbb{N},$$

where $x^{i/j}$ are real-valued, $(x^{i/j})^T Ix^{i/j}>0$, $I\in\mathbb{R}^{n\times n}$, $\forall(i,j)$; $\kappa_1(\cdot)$ and $\kappa_2(\cdot)$ are the real-valued functions, $\kappa_{1_i}^T(x^{<d_i>})I\kappa_{2_j}(x^{<d_j>})>0$, $\forall(i,j)$, $(d_i,d_j)\in\mathbb{N}$.

Hamilton-Jacobi Equation, Closed-Form Solution, and, Return Functions – Positive definite functions $V(t,x)$ (6.24) depend on $F(x)$, $B(x)$ and $\omega(x,u)$.

Solutions of an optimization problem

$$\begin{cases}\dot{x}=F(x)+B(x)u,\ u^*=\varphi\left(t,\frac{\partial V^*}{\partial x}\right),\ u^*\in U,\\ J=\min\limits_{x,u}\int_{t_0}^{t_f}\omega(x,u)dt,\ \omega(x,u)>0,\\ \min\limits_{u\in U}\left[H(t,x,u,\frac{\partial V}{\partial x})\right],\ H(t,x,u,\frac{\partial V}{\partial x})=\omega(x,u)+\frac{\partial V}{\partial x}^T\left[F(x)+B(x)u\right],\\ -\frac{\partial V^*}{\partial t}=\omega(x,u^*)+\left(\frac{\partial V^*}{\partial x}\right)^T\left[F(x)+B(x)u^*\right],\end{cases}$$

is sought on a particular class of $V(t,x)$. For Lebesgue integrable functions (H,V), $\left(\partial_u^1 H,\ \partial_u^2 H,\ \partial_x V\right)$ exist, and, $V(t,x)=\sum_{i,j}\kappa_{1_i}^T(x^{<d_i>})K_{ij}\kappa_{2_j}(x^{<d_j>})$ should satisfy a functional equation (6.23). One finds $K_{ij}(t)$ and optimal control law.

For $t_f=\infty$, an algebraic equation (6.23)

$$0=\omega(x,u)+\left(\frac{\partial V}{\partial x}\right)^T\left[F(x)+B(x)u\right],\ u=\varphi\left(\frac{\partial V}{\partial x}\right)$$

is satisfied by (6.24), solved, and, yields positive definite K_{ij}. Examples 6.9. 6.10, 6.37 and 6.39 report the use of real-valued functions $\kappa(\cdot)$ to define $V(x)$, solve (6.23) for $x\in X$, and, evaluate the accuracy of solution. Depending on $F(x)$ and $B(x)$, one may design positive definite performance integrands $\omega(x,u)$ or $\omega(x,u,\frac{\partial V}{\partial x})$ to ensure analytic closed-form solution of the Hamilton-Jacobi equation on a class of continuous scalar functions $V(x)$. One should not cancel inherent stabilizing nonlinearities, ignore states and control dependences, etc. For a particular classes of differential equations, such as Riccati-Abel and high-degree polynomial differential equations, a general solution can be found using the integrating factor, variation of parameters and other methods. Physics-consistent and design-specific explicit $\omega(x,u)$ or implicit $\omega(x,u,\frac{\partial V}{\partial x})$ are applied to find closed form analytic, Taylor series and numeric solutions of the optimization problem.

Example 6.9. *Return Functions*

One finds (6.24) to ensure physics and design consistencies.

For N_0=1, M_0=1, N=1 and M=1, the quadratic return function is $V(t,x)=\frac{1}{2}x^T K_{11}(t)x$.

If N=2 and M=1, $V(t,x)=\frac{1}{2}x^T K_{11}(t)x+\frac{1}{4}(x^2)^T K_{21}(t)x^2$.

For N=1 and M=2, using real-valued κ, $V(t,x)=\frac{1}{2}x^T K_{11}(t)x+(\sqrt{|x|})^T K_{12}(t)\sqrt{|x|}$.

One may consider

$$V(t,x)=\sum_{i=N_0,j=M_0}^{N,M}\frac{2j-1}{2i-1}(x^{\frac{2i-1}{2j-1}})^T K_{ij}(t)x^{\frac{2i-1}{2j-1}},\ V(t,x)=\tfrac{1}{2}x^T K(t)x+\sum_j\tfrac{1}{2j}(|x|^{1/j})^T K_{jj}(t)|x|^{1/j}.\ \blacksquare$$

Example 6.10. *Return Functions, Monomials and Polynomials*
For n variables $x\in\mathbb{R}^n$, consider a scalar function $V(x)$ given as a polynomial of degree $2d$. A vector $x^{\langle d\rangle}$ contains all monomials in x with degree less than or equal to d.
Monomial and Polynomial – A monomial is an expression having a product of powers of variables with nonnegative integer exponents. For (x_1,x_2,x_3,x_4), $kx_1^a x_2^b x_3^c x_4^e$ is a monomial with nonnegative integers $(a,b,c,e)\in\mathbb{N}$ and coefficient $k\neq 0$. The degree of a monomial is the sum of all exponents of variables. The set of monomials is a subset of multivariate polynomials. For a complete homogeneous symmetric polynomial having n variables with degrees $(1,\ldots,p)$, the number of terms is $\binom{n+p}{p}-1$.

The scalar function $V(x)$ can be expressed as the product of vector $x^{\langle d\rangle}$ using a square matrix K,

$$V(x)=(x^{\langle d\rangle})^T Kx^{\langle d\rangle}=\sum_{i,j} k_{ij}x_i^{\langle d\rangle}x_j^{\langle d\rangle},\ d\in\mathbb{N}.$$

For example, for a scalar $V(x)=k_{11}x_1^2+2k_{12}x_1x_2+k_{22}x_2^2+k_{33}x_1^4+2k_{34}x_1^2x_2^2+k_{44}x_2^4$

$$V(x)=\begin{bmatrix}x_1 & x_2 & x_1^2 & x_2^2\end{bmatrix}\begin{bmatrix}k_{11} & k_{12} & 0 & 0\\ k_{21} & k_{22} & 0 & 0\\ 0 & 0 & k_{33} & k_{34}\\ 0 & 0 & k_{43} & k_{44}\end{bmatrix}\begin{bmatrix}x_1\\ x_2\\ x_1^2\\ x_2^2\end{bmatrix},\ k_{ij}=k_{ji}.$$

Denote $x=\begin{bmatrix}x_1\\ x_2\end{bmatrix}$. In the vector-matrix notations, $V(x)=\begin{bmatrix}x & x^2\end{bmatrix}\begin{bmatrix}K_1 & 0\\ 0 & K_2\end{bmatrix}\begin{bmatrix}x\\ x^2\end{bmatrix}$, $(K_1,K_2)\in\mathbb{R}^{2\times 2}$.

One may consider $V(x)=(x^{\langle d_1\rangle})^T Kx^{\langle d_2\rangle}$, $(d_1,d_2)\in\mathbb{N}^0$.

Denote $x=\begin{bmatrix}x_1\\ x_2\end{bmatrix}$. We have $V(x)=\begin{bmatrix}1 & x^2\end{bmatrix}\begin{bmatrix}K_{11} & K_{12}\\ K_{21} & K_{22}\end{bmatrix}\begin{bmatrix}x^2\\ x^4\end{bmatrix}$, $(K_{11},K_{12},K_{21},K_{22})\in\mathbb{R}^{2\times 2}$.

High-degree and fractional multivariate polynomials, as well as matrix calculus are used to find $V(x)$ for $x\in\mathbb{R}^n$. The reported concept supports search algorithms in design of return functions, solution of functional equations, and, finding control laws.■

Minimization of Functionals With Explicit Nonquadratic Integrands – Minimize the performance functional (6.16) using the integrand $\omega(x,u)=\omega_x(x)+\omega_u(u)$, $\omega_u(u)=\frac{1}{2}u^T Gu$. That is, minimize

$$J=\min_{x,u}\int_{t_0}^{t_f}\left(\omega_x(x)+\tfrac{1}{2}u^T Gu\right)dt \tag{6.25}$$

subject to a time-invariant system (6.18) $\dot{x}(t)=F(x)+B(x)u$.

Find a control law (6.17) $u=\varphi(x)$. The Hamiltonian (6.21) for (6.25) is

$$H(x,u,\tfrac{\partial V}{\partial x})=\omega_x(x)+\tfrac{1}{2}u^T Gu+\left(\frac{\partial V}{\partial x}\right)^T\left[F(x)+B(x)u\right]. \tag{6.26}$$

Applying (6.22), one has $\frac{\partial H}{\partial u}=u^T G+\left(\frac{\partial V}{\partial x}\right)^T B(x)=0$.

Hence, a control law is

$$u = -G^{-1}B(x)^T \frac{\partial V}{\partial x}. \quad (6.27)$$

The second-order necessary condition for optimality is guaranteed, $\frac{\partial^2 H}{\partial u \times \partial u^T} = G > 0$, $G>0$. Substituting (6.27) in (6.26), find the Hamilton-Jacobi equation

$$-\frac{\partial V}{\partial t} = \omega_x(x) + \left(\frac{\partial V}{\partial x}\right)^T F(x) - \tfrac{1}{2}\left(\frac{\partial V}{\partial x}\right)^T B(x)G^{-1}B(x)^T \frac{\partial V}{\partial x}. \quad (6.28)$$

Applying $V(x)$ which satisfies (6.28), solve the Hamilton-Jacobi equation and obtain explicit control function (6.27). If $t_f=\infty$, the functional equation (6.28) becomes

$$\omega_x(x) + \left(\frac{\partial V}{\partial x}\right)^T F(x) - \tfrac{1}{2}\left(\frac{\partial V}{\partial x}\right)^T B(x)G^{-1}B(x)^T \frac{\partial V}{\partial x} = 0. \quad (6.29)$$

Solution of Hamilton-Jacobi Equation and Accuracy Analysis – One solves the Hamilton-Jacobi equation by approximating a solution by return functions $V(x)$. Study a N-dimensional manifold $\Delta(x)$

$$\Delta(x) = \omega_x(x) + \left(\frac{\partial V}{\partial x}\right)^T F(x) - \tfrac{1}{2}\left(\frac{\partial V}{\partial x}\right)^T B(x)G^{-1}B(x)^T \frac{\partial V}{\partial x}, \ x \in X,$$

which is as a multivariate polynomials with N terms of n-variables. For $\Delta(\cdot):\mathbb{R}^n \to \mathbb{R}^N$, an absolute manifold-valued errors ($|\Delta_{ij}|_{\max}, \|\Delta_{ij}\|_p$) are evaluated for $x \in X$. Solution of the functional equation should yield $|\Delta_{ij}|_{\max} \le \delta$ and $\|\Delta_{ij}\|_p \le \delta_p$ for $x \in X$. The volume, Hausdorff and other measures can be applied. Illustrative studies are reported in Examples 5.28 and 6.39.

Implicit Integrands and Dynamic Optimization – Physics-consistent integrands $\omega(x,u)$ are used, for example $\omega(x,u) \equiv \tilde{\omega}(\mathbf{E})$. One formulates and solve the optimization problem, which should admit a solution. One may apply implicit design-specific performance integrand $\omega(x,u,\frac{\partial V}{\partial x}) > 0$, $\forall x \in X$, $\forall u \in U$, $\forall t \in T$. That is, functional (6.16)

$$J = \min_{x,u} \int_{t_0}^{t_f} \omega(x,u,\tfrac{\partial V}{\partial x})dt, \ J = \min_{x,u} \int_{t_0}^{t_f} \left(\omega_x(x,\tfrac{\partial V}{\partial x}) + \tfrac{1}{2}u^T G u\right)dt, \ \omega_x(x,\tfrac{\partial V}{\partial x}) > 0$$

and return functions (6.24) $V(t,x)$ are designed to solve the Hamilton-Jacobi equation (6.28) which admit a closed-form solution. Example 6.11 reports the solution for the second-order system.

Example 6.11. *Nonlinear System and Control*

Consider a nonlinear time-invariant system $\begin{bmatrix}\dot{x}_1\\ \dot{x}_2\end{bmatrix} = \begin{bmatrix}-1 & -1\\ 1 & 0\end{bmatrix}\begin{bmatrix}x_1\\ x_2\end{bmatrix} + \begin{bmatrix}x_1^3 - x_1^5 - x_1x_2^2\\ 0\end{bmatrix} + \begin{bmatrix}1\\ 0\end{bmatrix}u.$

The open-loop system is stable, which can be proven using the Lyapunov stability theory. For a positive definite continuously differentiable Lyapunov function

$$V_L(x_1,x_2) = \tfrac{1}{2}\left(x_1^2 + x_2^2\right), \ V_L>0,$$

one has, $\frac{dV_L}{dt} = -x_1^2 + x_1^4 - x_1^6 - x_1^2x_2^2$, $\frac{dV_L}{dt} < 0$.

Solve an optimal control problem minimizing the performance functional (6.25) with $\omega_x(x)>0$

$$J=\min_{x,u}\int_0^{\infty}\left(\tfrac{1}{2}x_1^2+\tfrac{1}{2}x_2^2+x_1^4+x_1^2x_2^2+x_1^6+\tfrac{1}{2}u^2\right)dt\,.$$

The Hamiltonian function (6.26) is

$$H(x,u,\tfrac{\partial V}{\partial x})=\tfrac{1}{2}x_1^2+\tfrac{1}{2}x_2^2+x_1^4+x_1^2x_2^2+x_1^6+\tfrac{1}{2}u^2+\frac{dV}{dx_1}\left(-x_1-x_2+x_1^3-x_1^5-x_1x_2^2+u\right)+\frac{dV}{dx_2}x_1\,.$$

Applying (6.22) $\frac{\partial}{\partial u}H(x,u,\frac{\partial V}{\partial x})=0$, the control law (6.27) is $u=-\dfrac{dV}{dx_1}$.

Approximate solution of equation (6.29)

$$\tfrac{1}{2}x_1^2+\tfrac{1}{2}x_2^2+x_1^4+x_1^2x_2^2+x_1^6+\frac{dV}{dx_1}\left(-x_1-x_2+x_1^3-x_1^5-x_1x_2^2\right)+\frac{dV}{dx_2}x_1-\tfrac{1}{2}\left(\frac{dV}{dx_1}\right)^2=0$$

using the truncated return function (6.24)

$$V(x_1,x_2)=\tfrac{1}{2}k_{11}x_1^2+k_{12}x_1x_2+\tfrac{1}{2}k_{22}x_2^2+\tfrac{1}{4}k_{41}x_1^4+\tfrac{1}{2}k_{42}x_1^2x_2^2+\tfrac{1}{6}k_{61}x_1^6\,.$$

Grouping the terms, the quadratic, quartic and sixtic term equations are

x_1^2: $\tfrac{1}{2}-k_{11}+k_{12}-\tfrac{1}{2}k_{11}^2=0$, x_1x_2: $-k_{11}-k_{12}-k_{11}k_{12}+k_{22}=0$,

x_2^2: $\tfrac{1}{2}-k_{12}-\tfrac{1}{2}k_{12}^2=0$, x_1^4: $1+k_{11}-k_{11}k_{41}-k_{41}=0$,

$x_1^2x_2^2$: $1-k_{11}-k_{11}k_{42}-k_{42}=0$, x_1^6: $1-k_{11}+k_{41}-\tfrac{1}{2}k_{41}^2-k_{11}k_{61}-k_{61}=0$.

These equations yield the unknown coefficients k_{11}, k_{12}, k_{22}, k_{41}, k_{42} and k_{61}.

The first three equations result in k_{11}=0.682, k_{12}=0.414, k_{22}=1.378.

We compute k_{41}=1, k_{42}=0.189 and k_{61}=0.486.

The control law is

$$u=-\left(k_{11}x_1+k_{12}x_2+k_{41}x_1^3+k_{42}x_1x_2^2+k_{61}x_1^5\right)=-\left(0.682x_1+0.414x_2+x_1^3+0.189x_1x_2^2+0.486x_1^5\right).$$

The destabilizing term in the equations of motion x_1^3, is compensated by the term $-k_{41}x_1^3$, k_{41}=1 in the derived control law.

Nonlinear open-loop system is stable, and, $-x_1^5$ is a stabilizing term. Linearization of nonlinear system $\dot{x}=F(x)+Bu$ to $\dot{x}=Ax+Bu$ is considered. The linearization at the *equilibrium state* $\bar{x}=\begin{bmatrix}\bar{x}_1\\ \bar{x}_2\end{bmatrix}=\begin{bmatrix}0\\0\end{bmatrix}$ yields $\bar{u}=0$. The linearized $\dot{x}=Ax+Bu$ is stable. The LQR design for a linear system

$$\begin{bmatrix}\dot{x}_1\\ \dot{x}_2\end{bmatrix}=\begin{bmatrix}-1&-1\\1&0\end{bmatrix}\begin{bmatrix}x_1\\x_2\end{bmatrix}+\begin{bmatrix}1\\0\end{bmatrix}u$$

yields $u=-\left(k_{11}x_1+k_{12}x_2\right)=-0.682x_1-0.414x_2$.

The closed-loop nonlinear system is stable.

Minimize not implicitly defined functional

$$J=\min_{x,u}\int_0^{\infty}[\underbrace{\tfrac{1}{2}x_1^2+\tfrac{1}{2}x_2^2+\frac{\partial V}{\partial x_1}\left(-x_1^3+x_1^5+x_1x_2^2\right)+\tfrac{1}{2}u^2}_{\omega_x(x,\frac{\partial V}{\partial x})>0,\ V(x_1,x_2)=\frac{1}{2}k_{11}x_1^2+k_{12}x_1x_2+\frac{1}{2}k_{22}x_2^2}]dt$$

subject to a nonlinear system.

This yields the algebraic equation (6.29), satisfied by the quadratic return function

$$V(x_1,x_2)=\tfrac{1}{2}k_{11}x_1^2+k_{12}x_1x_2+\tfrac{1}{2}k_{22}x_2^2\,.$$

In an explicit functional, $\omega_x(x,\frac{\partial V}{\partial x})>0$ because $\frac{\partial V}{\partial x_1}\left(-x_1^3+x_1^5+x_1x_2^2\right)>0$.

We found the quadratic Lyapunov function, and, a stabilizing control law is $u=-\left(k_{11}x_1+k_{12}x_2\right)=-0.682x_1-0.414x_2$. ■

Example 6.12. *Controlled Motion of a Maneuvering Aerial Vehicle*

To control guided missiles, one considers rigid-body dynamics, studies translational and rotational motions, as well as investigates unsteady aerodynamics, guidance and control. Thrust, specific impulse, system-specific impulse, specific thrust, forces and moments are considered.

The aerodynamic drag, and, aerodynamic force on control surfaces, may be approximated as

$$F_{\text{aerodynamic drag}}=f(v)\approx\tfrac{1}{2}\rho A\left(c_1v+c_2v|v|\right),\ F_{\text{aerodynamic force}}=f(v,\delta)\approx\tfrac{1}{2}\rho A\left(c_3v+c_4v|v|\right)\delta,$$

where ρ is the air density; c_i are the drag coefficients which are functions of the Mach number, body shape, nose geometry, angle-of-attack, air compressibility, viscosity, etc.; A is the effective area; v is the velocity; δ is the displacement of the control surfaces.

Actuators reposition aerodynamic surfaces, such as canards, wings and tail fins. There are different aerodynamic configurations and control surface configurations, such as "+" and "x". Different combinations of four fins are used to typify the aileron δ_A, elevator δ_E and rudder δ_R to generate aerodynamic moments and forces. For the "+" and "x" configurations,

$$\delta_{A\text{"+"}}=\delta_1+\delta_2+\delta_3+\delta_4,\ \delta_{E\text{"+"}}=\delta_2-\delta_4,\ \delta_{R\text{"+"}}=-\delta_1+\delta_3,$$

and, $$\delta_{A\text{"x"}}=\delta_1+\delta_2+\delta_3+\delta_4,\ \delta_{E\text{"x"}}=\delta_1+\delta_2-\delta_3-\delta_4,\ \delta_{R\text{"x"}}=-\delta_1+\delta_2+\delta_3-\delta_4.$$

There are different missions and flight phases, such as autopilot, altitude hold, cruise control, etc. For example, the induced roll is attenuated by roll stabilization. A roll control is accomplished by regulating the aerodynamic surface displacement δ. Aerodynamic drag forces and unsteady aerodynamic flows are nonlinear functions of flight conditions, velocity, angle of attack, etc. Consider the governing equations which typify one-dimensional rotational motion. Study attitude dynamics and roll control.

Stabilization and tracking control problems are considered.

For roll stabilization, consider typified governing equations denoting $\begin{bmatrix}x_1\\x_2\end{bmatrix}=\begin{bmatrix}\omega_\phi\\\phi\end{bmatrix}$

$$\dot{x}_1=-a_1x_1-a_2x_1|x_1|+\underbrace{(b_1+b_2|x_1|+b_3x_1^2)}_{b(x_1)}u,\ b(x_1)>0,\ \forall x_1,$$

$$\dot{x}_2=x_1.$$

Assuming a_i=1 and b_i=1,

$$\dot{x}_1=-x_1-x_1|x_1|+(1+|x_1|+x_1^2)u,$$

$$\dot{x}_2=x_1.$$

One should not apply the feedback linearization or other technique to cancel the inherent stabilizing term $-x_1|x_1|$, or, simplify the state-dependent control and aerodynamic force. One has $b(x_1)>0$, $\forall x_1$. The performance functional is designed using aerodynamic, energy and design considerations. Minimize an implicit functional

$$J=\min_{x,u}\int_0^\infty \omega(x,u,\tfrac{\partial V}{\partial x})dt$$

$$=\min_{x,u}\int_0^\infty\left[\tfrac{1}{2}x_1^2+\tfrac{1}{2}x_2^2+\left[x_1|x_1|\tfrac{dV}{dx_1}+\tfrac{1}{2}\left(|x_1|+x_1^2\right)\left(\tfrac{dV}{dx_1}\right)^2\right]+\tfrac{1}{2}\left(1+|x_1|+x_1^2\right)u^2\right]dt,$$

for which $\omega(x,u,\tfrac{\partial V}{\partial x})>0$ should be guaranteed, $\left[x_1|x_1|\tfrac{dV}{dx_1}+\tfrac{1}{2}\left(|x_1|+x_1^2\right)\left(\tfrac{dV}{dx_1}\right)^2\right]>0$.

The resulting Hamiltonian is

$$H(x,u,\tfrac{\partial V}{\partial x})=\tfrac{1}{2}x_1^2+\tfrac{1}{2}x_2^2+\left[x_1|x_1|\tfrac{dV}{dx_1}+\tfrac{1}{2}\left(|x_1|+x_1^2\right)\left(\tfrac{dV}{dx_1}\right)^2\right]+\tfrac{1}{2}\left(1+|x_1|+x_1^2\right)u^2$$

$$+\tfrac{dV}{dx_1}\left[-x_1-x_1|x_1|+\left(1+|x_1|+x_1^2\right)u\right]+\tfrac{dV}{dx_2}x_1.$$

Apply (6.22), $\dfrac{\partial H}{\partial u}=\left(1+|x_1|+x_1^2\right)u+\dfrac{\partial V}{\partial x_1}\left(1+|x_1|+x_1^2\right)=0$.

An optimal control law is $u=-\dfrac{\partial V}{\partial x_1}$.

Solution of the resulting equation $\tfrac{1}{2}x_1^2+\tfrac{1}{2}x_2^2-\dfrac{dV}{dx_1}x_1+\dfrac{dV}{dx_2}x_1-\tfrac{1}{2}\left(\dfrac{dV}{dx_1}\right)^2=0$ is satisfied by the quadratic return function $V=\tfrac{1}{2}k_{11}x_1^2+k_{12}x_1x_2+\tfrac{1}{2}k_{22}x_2^2$.

Grouping terms for x_1^2, x_1x_2 and x_2^2, we have three equations with three unknowns

$\tfrac{1}{2}-k_{11}+k_{12}-\tfrac{1}{2}k_{11}^2=0$, $-k_{12}+k_{22}-k_{11}k_{12}=0$, $\tfrac{1}{2}-\tfrac{1}{2}k_{12}^2=0$.

Thus, $k_{11}=1$, $k_{12}=1$ and $k_{22}=2$. The control law is

$u=-x_1-x_2$.

We minimized a nonquadratic functional with a positive definite $\omega(x,u,\tfrac{\partial V}{\partial x})>0$, $\forall x\in X$, $\forall u\in U$, and, found $(k_{11},k_{12},k_{22})>0$. There is a trade-off in finding analytic solutions and solving nonlinear optimization problems using high-fidelity models, nonquadratic functionals, return functions, etc. Dynamics of the closed-loop system with initial conditions $\begin{bmatrix}x_{10}\\x_{20}\end{bmatrix}=\begin{bmatrix}0.1\\-0.1\end{bmatrix}$ and $\begin{bmatrix}x_{10}\\x_{20}\end{bmatrix}=\begin{bmatrix}1\\-1\end{bmatrix}$ are documented in Figures 6.8.

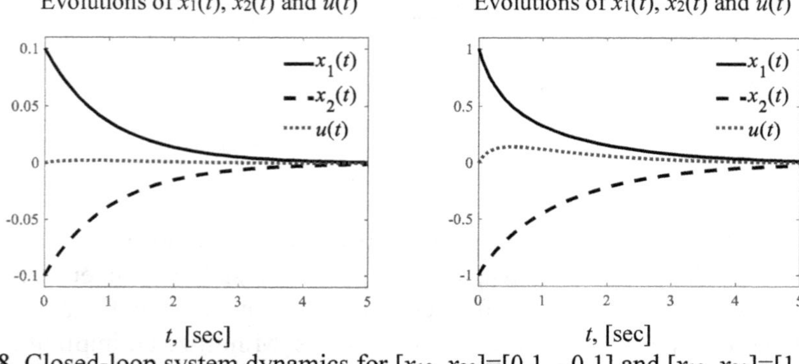

Figure 6.8. Closed-loop system dynamics for $[x_{10}, x_{20}]=[0.1, -0.1]$ and $[x_{10}, x_{20}]=[1, -1]$. ■

6.4. Tracking Control of Nonlinear Continuous-Time Systems

Design control laws with tracking error $e(t)$ and state $x(t)$ feedback. For linear systems, tracking control problems were studied and solved in sections 5.7, 5.8 and 5.9. For a nonlinear system (6.18), the output is $y(t)$, $y=Hx$, and, the tracking error is $e(t)=r(t)-y(t)$, where $r(t)$ is the reference.

Consider the system state dynamics with the outputs equation

$$\dot{x}(t)=F(x)+B(x)u,\ x(t_0)=x_0, \tag{6.30}$$
$$y=Hx,\ e(t)=r(t)-y(t)=r(t)-Hx(t).$$

The governing equations for *error states* $x_e(t)$ and tracking error $e(t)$ are

$$\dot{x}_e(t)=I_X(r-y)=I_X(r-Hx),\ e(t)\equiv\dot{x}_e(t),\ I_X\in\mathbb{R}^{b\times b}, \tag{6.31}$$
$$\dot{e}(t)=-I_E e+I_E^X x_e+\dot{r}-\dot{y}=-I_E e+I_E^X x_e+\dot{r}-HF(x)-HB(x)u,\ (I_X,I_E,I_E^X)\in\mathbb{R}^{b\times b},$$

where I_X and I_E are the positive definite diagonal matrices, such as $I_X=I$ and $I_E=I$.

Equations (6.31) explicitly define the error dynamics.

The extended vector is $\mathbf{x}(t)=\begin{bmatrix}x(t)\\x_e(t)\\e(t)\end{bmatrix}$, $\mathbf{x}\in\mathbf{X}\subset\mathbb{R}^{n+2b}$. From (6.30) and (6.31)

$$\dot{\mathbf{x}}(t)=\begin{bmatrix}\dot{x}(t)\\\dot{x}_e(t)\\\dot{e}(t)\end{bmatrix}=\mathbf{F}(\mathbf{x})+\mathbf{A}\mathbf{x}+A_r r+\mathbf{B}(\mathbf{x})u=\begin{bmatrix}F(x)\\0\\-HF(x)\end{bmatrix}+\begin{bmatrix}0&0&0\\-I_XH&0&0\\0&I_E^X&-I_E\end{bmatrix}\mathbf{x}+\begin{bmatrix}0&0\\I_X&0\\0&I\end{bmatrix}\begin{bmatrix}r\\\dot{r}\end{bmatrix}+\begin{bmatrix}B(x)\\0\\-HB(x)\end{bmatrix}u. \tag{6.32}$$

Minimize the performance functional

$$J=\min_{x,x_e,e,u}\int_{t_0}^{t_f}\omega(x,x_e,e,u)dt=\min_{\mathbf{x},u}\int_{t_0}^{t_f}\omega(\mathbf{x},u)dt,\ \omega(\mathbf{x},u)>0 \tag{6.33}$$

subject to system (6.32). Stability, boundedness and optimal evolutions of $\begin{bmatrix}x_e(t)\\e(t)\end{bmatrix}\equiv\begin{bmatrix}\int e(t)dt\\e(t)\end{bmatrix}$ should be ensured. In (6.33), $\omega(\cdot)$ is a positive definite performance integrand. Define functional (6.33) as

$$J=\min_{\mathbf{x},u}\int_{t_0}^{t_f}\left(\omega_x(\mathbf{x})+\tfrac{1}{2}u^TGu\right)dt,\ G\in\mathbb{R}^{m\times m},\ G>0. \tag{6.34}$$

As reported in section 5.9, continuous and piecewise continuous bounded references $r(t)$ assume finite values, such that $|r(t)|<M$, $\forall t$, where M is the finite magnitude, $|M|<\infty$. Deterministic $r(t)$ do not explicitly alter an optimization scheme. However, stability, boundedness and optimal governance of $(x,\int edt,e,y,u)$ are affected by $r(t)$, particularly due to physical limits on states and controls. The minimax optimization and Lyapunov stability principles are applied. For the resulting Hamiltonian function

$$H(\mathbf{x},u,\tfrac{\partial V}{\partial\mathbf{x}})=\omega_x(\mathbf{x})+\tfrac{1}{2}u^TGu+\left(\frac{\partial V}{\partial\mathbf{x}}\right)^T\left[\mathbf{F}(\mathbf{x})+\mathbf{A}\mathbf{x}+\mathbf{B}(\mathbf{x})u\right], \tag{6.35}$$

the first-order necessary condition for optimality (6.22) yields

$$u=-G^{-1}\mathbf{B}^T(\mathbf{x})\frac{\partial V}{\partial\mathbf{x}},\ \mathbf{x}=\begin{bmatrix}x\\\int edt\\e\end{bmatrix}. \tag{6.36}$$

Solution of the Hamilton-Jacobi equation

$$-\frac{\partial V}{\partial t}=\boldsymbol{\omega}_x(\mathbf{x})+\left(\frac{\partial V}{\partial \mathbf{x}}\right)^T\left[\mathbf{F}(\mathbf{x})+\mathbf{A}\mathbf{x}\right]-\tfrac{1}{2}\left(\frac{\partial V}{\partial \mathbf{x}}\right)^T\mathbf{B}(\mathbf{x})G^{-1}\mathbf{B}^T(\mathbf{x})\frac{\partial V}{\partial \mathbf{x}} \quad (6.37)$$

is satisfied or approximated by a continuous and differentiable return function

$$V(t,\mathbf{x})=\sum_{i=N_0,j=M_0}^{N,M}\tfrac{j}{2i}(\mathbf{x}^{i/j})^T K_{ij}(t)\mathbf{x}^{i/j},\; V(t,\mathbf{x})=\sum_{i,j}\kappa_{1_i}^T(t,\mathbf{x}^{<d_i>})K_{ij}(t)\kappa_{2_j}(t,\mathbf{x}^{<d_j>}),\; K_{ij}>0, \quad (6.38)$$

where $\kappa_1(\cdot)$ and $\kappa_2(\cdot)$ are the real-valued functions, $\kappa_{1_i}^T(t,\mathbf{x}^{<d_i>})I\kappa_{2_j}(t,\mathbf{x}^{<d_j>})>0, \forall(i,j),(d_i,d_j)\in\mathbb{N}$. The positive definite return function $V(t,\mathbf{x})$ depends on $\boldsymbol{\omega}_x(\mathbf{x})$ and $\mathbf{F}(\mathbf{x})$. Solving (6.37), compute K_{ij}, obtain control feedback structure $\frac{\partial V}{\partial \mathbf{x}}$ in (6.36), and, find feedback gains.

Fractional Exponent and Multiple-Integral Feedback – Section 6.10.4 focuses on nonlinear *minimal complexity* control with directly measured variables. Tracking control laws are designed with high degree and fractional exponent real-valued feedback on $e(t)$ and multiple-integral feedback $\sum_j k_j\int\ldots\int e dt$. One applies Lyapunov's optimization as documented by (6.119)-(6.121), or minimizes a functional. In particular,

$$\begin{cases}\text{Minimize } J=\min_{\mathbf{x},u}\int_{t_0}^{t_f}\left(\omega_x(\mathbf{x})+\tfrac{1}{2}u^TGu\right)dt\\[2ex] \text{subject to } \dot{\mathbf{x}}(t)=\begin{bmatrix}\dot{x}(t)\\ \begin{bmatrix}\dot{x}_{e_1}(t)\\ \dot{x}_{e_2}(t)\\ \vdots\\ \dot{x}_{e_j}(t)\end{bmatrix}\\ \dot{e}(t)\end{bmatrix}=\begin{bmatrix}F(x)+B(x)u\\ \begin{bmatrix}-I_XHx+I_Xr\\ x_{e_1}\\ \vdots\\ x_{e_{j-1}}\end{bmatrix}\\ -HF(x)-I_Ee-\sum_{z,l}I_Ee^{\frac{2z-1}{2l-1}}+I\dot{r}-HB(x)u\end{bmatrix},\; y=Hx.\end{cases}$$

6.5. Discontinuous and Continuous Control Limits

The realm of actual systems, as opposed to abstract and idealized studies, necessitate consideration of physical limits on control $u_{\min}\le u\le u_{\max}$. In electromagnetic, electrostatic, electronic, thermodynamic (gas turbine, internal combustion, rocket engines, etc.), fluidic, hydraulic and other devices and systems, there are continuous limits on control. Considering the *force* and *torque* control, the applied force and torque (F,T) in mechanisms are limited $u_{\min}\le u\le u_{\max}$.

Continuity and smoothness of control are ensured despite near-discontinuity in some systems, such as stepped gears, clutches and linkages, which may exhibit backlash, stiction, *dead zone*, etc. However, the force and torque *limiters*, *overload clutches* and *flexible shaft couplings* are used to prevent discontinuities and overloading yielding limits continuousness. Furthermore, force and torque are developed by hydraulic, electromagnetic, electrostatic, piezoelectric and other actuators, which are continuous. Nonlinearities and limits are considered to adequately formulate and solve design and optimization problems.

6.5.1. Description of Control Limits

Due to inherent physical limits $\mathcal{U}_{min}\le\mathcal{U}\le\mathcal{U}_{max}$, bounded control law should be analytically designed such that the control function is bounded as $u_{min}\le u(\cdot)\le u_{max}$, $u\in U$. Consider a broad class of memoryless physical limits on control, such as:

1. Symmetric or asymmetric saturation with constant or varying gain;
2. Saturation with *dead zone* and piecewise-continuous gains;
3. Switch, relay, and, relay with *dead zone*.

Physical Limits on Control Efforts, and, Bounds on Control Functions – Our objective is to design:

1. Bounded control laws $u=\phi[\varphi(t,x)]$, $u_{min}\le u(\cdot)\le u_{max}$ with discontinuous functions $\phi(\cdot)$, $u_{min}\le\phi(\cdot)\le u_{max}$;
2. Bounded control laws $u=\Phi[\varphi(t,x)]$, $u_{min}\le u(\cdot)\le u_{max}$ with continuous functions $\Phi(\cdot)$, $u_{min}\le\Phi(\cdot)\le u_{max}$.

We derive optimal control laws $u=\phi[\varphi(t,x)]$ and $u=\Phi[\varphi(t,x)]$ with the feedback structure $\varphi(\cdot)$, and, parametric and feedback terms continuity.

Bounded Functions – For given physical limits $\mathcal{U}$, find real-valued bounded functions $\phi(\cdot)$ and $\Phi(\cdot)$ with range $[u_{min}, u_{max}]$ or (u_{min}, u_{max}) and domain $(-\infty,\infty)$, which describe $\mathcal{U}_{min}\le\mathcal{U}\le\mathcal{U}_{max}$, such that

$$\min_{x\in X}\int_{-\infty}^{\infty}\left\|\mathcal{U}-\phi(x)\right\|_p dx,\ \left\|\mathcal{U}-\phi(\cdot)\right\|_p\le\varepsilon,\ |\phi(x)|\le M,\ \phi\subseteq\mathcal{U}_\varepsilon,\ p\ge 1,$$

$$\min_{x\in X}\int_{-\infty}^{\infty}\left\|\mathcal{U}-\Phi(x)\right\|_p dx,\ \left\|\mathcal{U}-\Phi(\cdot)\right\|_p\le\varepsilon,\ |\Phi(x)|\le M,\ \Phi\subseteq\mathcal{U}_\varepsilon,\ p\ge 1.$$

Discontinuous Function – For a discontinuous or piecewise continuous function ϕ, $\phi\in C^0$, $\phi\notin C^0$ or $\phi\notin C^k$, $k\ge 1$. The range and domain of $\phi(\cdot)$ are $[u_{min}, u_{max}]$ and $(-\infty,\infty)$.

Continuous Function – If $\Phi\in C^k$, $k\ge 1$, then, $\Phi(\cdot)$ is smooth and continuously differentiable, and, control function $u(\cdot)$ is continuous. Function $\Phi\in C^k$ is a C^k continuously differentiable function if first k derivatives $\Phi'(x)$, $\Phi''(x)$, ..., $\Phi^{(k)}(x)$ exist, and, the $\Phi^{(k)}$ derivative is continuous. The range of function $\Phi(\cdot)$ is (u_{min}, u_{max}), or, $[u_{min}, u_{max}]$, while the domain is $(-\infty,\infty)$. A function Φ is said to be analytic and of class C^ω, if Φ is smooth, and, if its Taylor series expansion around any point in its domain converges to the function in some neighborhood of the point.

Saturation and Relay Limits – Figure 6.8.a illustrates the saturation limit $-1\le\mathcal{U}\le 1$, as well as the infinitely-differentiable odd function $\Phi(x)=\tanh(x)$, $-1<\Phi<1$ with range $(-1,1)$, domain $(-\infty,\infty)$, and $\lim_{x\to\pm\infty}\tanh(x)\to\pm 1$.

The hyperbolic tangent is a strictly increasing function with a domain $\mathbb{R}$ and range $\{\Phi(x):\ -1<\Phi<1\}$, such that $\tanh(x):\mathbb{R}\to(-1, 1)$.

Assume control efforts admit ideal relay switching with limits $\mathcal{U}=\pm 1$.

A control law is $u=\phi[\varphi(t,x)]=\text{sgn}[\varphi(t,x)], u=\pm 1$. Here, ϕ is a sign (signum) function with:

1. A jump discontinuity at 0;
2. Undefined domain at $x=0$ because $\phi(0)$ and $\lim_{x\to 0}\text{sgn}(x)$ do not exist, while two-sided limits exist and not equal, $\lim_{x\to 0^+}\text{sgn}(x)=1$, $\lim_{x\to 0^-}\text{sgn}(x)=-1$.

Relay switching $\mathcal{U}$=±1 and a discontinuous control function $\phi(x)$=sgn(x) are reported in Figure 6.9.b. Approximate discontinuous $\mathcal{U}$=±1 by continuously differentiable functions $\Phi\in C^k$, $k\geq 1$ with range (–1, 1). For exhibited continuous switching $\mathcal{U}\cong\pm 1$, apply the infinitely-differentiable function $\Phi(x)=\tanh(ax)$ with $a\gg 1$, $-1<\Phi<1$. Figure 6.9.b illustrates $\Phi(x)=\tanh(ax)$, a=100.

For a relay with *dead zone*, to describe $\mathcal{U}$, consider a continuous function $\Phi(x)=\tanh^{2n+1}(ax)$, $n>1$, $a\gg 1$, $-1<\Phi<1$. The hyperbolic function $\Phi(x)=\tanh^{2n+1}(ax)$, n=10, a=50 is plotted in Figure 6.9.b. Descriptive findings are reported in section 6.7.

Saturation Limit $-1\leq\mathcal{U}\leq 1$ and $-1<\Phi<1$ Relay Limits $\mathcal{U}$=±1 and $\mathcal{U}\cong\pm 1$ with $\Phi\cong\pm 1$

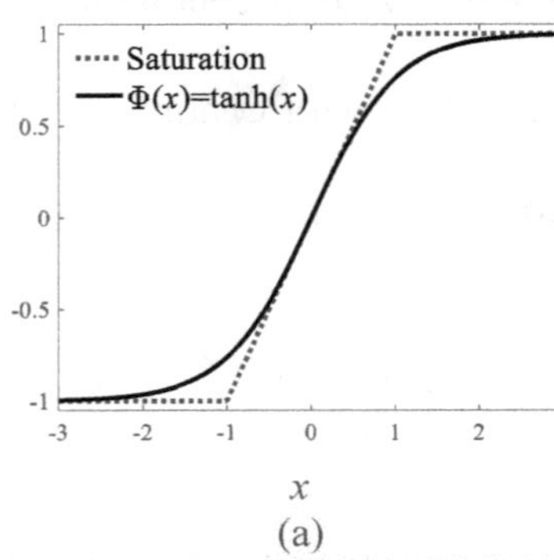

(a)

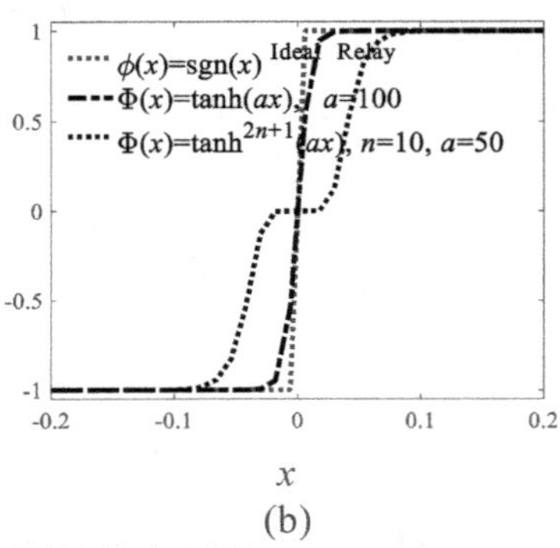

(b)

Figure 6.9. Symmetric limits on control $-1\leq\mathcal{U}\leq 1$, described by bounded discontinuous $\phi\notin C^k$ and continuous $\Phi\in C^k$ functions:

(a) Saturation $-1\leq\mathcal{U}\leq 1$ is described by $\Phi(x)=\tanh(x)$, $-1<\Phi<1$.

For p=1, $\int_X \left|\mathcal{U}-\Phi(x)\right| dx = 0.3814$, $-3\leq x\leq 3$;

(b) Relays $\mathcal{U}$=±1 and $\mathcal{U}\cong\pm 1$ are described by a discontinuous function $\phi(x)$=sgn(x) and an infinitely-differentiable function $\Phi(x)=\tanh^{2n+1}(ax)$, $n>1$, $a\gg 1$, $-1<\Phi<1$, $\Phi\in C^k$. Plots for $\Phi(x)=\tanh(ax)$, a=100 and $\Phi(x)=\tanh^{2n+1}(ax)$, n=10, a=50, $\Phi\cong\pm 1$.

Truncated series, Padé approximant and other approaches can be applied to describe Φ and solve pertained Hamilton-Jacobi equations. For $\Phi(x)=u_{\max}\tanh(ax)$, the hyperbolic tangent has an asymptotes $\pm u_{\max}$, while approximating polynomials may not have horizontal asymptotes. From $\tanh x = \mathrm{sgn}(x)\left(1-\dfrac{2}{e^{2|x|}+1}\right)$, we have $\tanh x \approx \mathrm{sgn}(x)\left(1-2e^{-2|x|}+2e^{-4|x|}-2e^{-6|x|}+\ldots\right)$ with the error $\sim R(e^{-2N|x|})$.

Geometric power series may not yield an inadequate approximation at origin. The Padé approximant for e^x is $e^x \approx \dfrac{p_i(x)}{p_i(-x)}$, $p_i(x)=\sum_{j=0}^{i}\dfrac{\binom{i}{j}}{j!\binom{2i}{j}}x^j$.

From $\tanh x = \dfrac{e^x-e^{-x}}{e^x+e^{-x}} = \dfrac{e^{2x}-1}{e^{2x}+1}$, a rational function is $\tanh x \approx \dfrac{x(10+x^2)(60+x^2)}{600+270x^2+11x^4+\frac{1}{24}x^6}$.

Truncations of tanh(x) yields rational functions as $\tanh x = \cfrac{x}{1+\cfrac{x^2}{3+\cfrac{x^2}{5+\ldots}}}$.

6.5.2. Optimal Control and Dynamic Optimization

For systems with $u_{min} \le u \le u_{max}$, using descriptive functions (ϕ, Φ), the constrained control laws $u \in U$ should be designed by minimizing functionals, finding the feedback structure $\varphi(\cdot)$ and parametric feedback gains $k_F(\cdot)$. For discontinuous $\phi \notin C^k$, a time-invariant switching manifold $\varphi(x)$ may be denoted as $\upsilon(x)$. Discontinuous control laws $u = \phi[\varphi(t,x)]$, such as $u = -\text{sgn}[\varphi(t,x)]$ with switching $u = \pm 1$, are designed by minimizing

$$J = \min_{t, u \in U} \int_{t_0}^{t_f} 1 dt \tag{6.39}$$

subject to a dynamic system

$$\dot{x}(t) = F(x) + B(x)u, \; u_{min} \le u \le u_{max}, \; u \in U. \tag{6.40}$$

Discontinuous Control and Switching Manifold – Minimum-time, sliding mode and maximum principle discontinuous control $u = -u_{max}\text{sgn}(\cdot)$ or $u = -\text{sgn}(\cdot)$ imply switching $u = \pm u_{max}$ or $u = \pm 1$.

Using the Hamiltonian (6.21), in section 6.3 we solved a minimization problem $\min_{u \in U} \left[H\left(t, x, u, \frac{\partial V}{\partial x}\right) \right]$ and found an optimal control u^* with optimal state evolution $x^*(t)$ by using (6.20) $\dot{\boldsymbol{\lambda}}^T = -\frac{\partial H(t,x,u,\boldsymbol{\lambda})}{\partial x}$ and $\frac{\partial H(t,x,u,\boldsymbol{\lambda})}{\partial u} = 0$. An optimal control is found by solving the boundary-value problem, demonstrated in Examples 6.7 and 6.8. This scheme implies deriving the Lagrange multiplier vector λ^* or function V^* such that

$$H(t, x^*, u^*, \boldsymbol{\lambda}^*) \le H(t, x^*, u, \boldsymbol{\lambda}^*), \; H\left(t, x^*, u^*, \frac{\partial V^*}{\partial x}\right) \le H\left(t, x^*, u, \frac{\partial V^*}{\partial x}\right).$$

Minimum-time discontinuous control

$$u = \phi[\varphi(x)] = -\text{sgn}\left(\frac{\partial V}{\partial x}\right), \; u = \pm 1$$

can be derived by applying calculus of variations, Hamiltonian optimization and Lyapunov concept. Calculus of variations, as demonstrated in Example 6.16, assumes that control $u = \pm u_{max}\text{sgn}[\varphi(x)]$ takes the value $\pm u_{max}$, differential equations (6.40) can admit solutions, and, the switching manifold can be deduced. This concept suits a limited class of low-order single-input/single-output systems. Therefore, the Hamiltonian optimization and Lyapunov concept are applied.

Solve the optimization problem by minimizing (6.39)

$$J = \min_{t, u \in U} \int_{t_0}^{t_f} 1 dt = (t_f - t_0), \; \min_{u \in U} \left[H\left(x, u, \frac{\partial V}{\partial x}\right) \right], \; u = \pm 1, \; u^2 \equiv 1 \tag{6.41}$$

subject to (6.40).

For a given functional (6.41) and $u \in U$, a Hamiltonian function is

$$H\left(x, u, \frac{\partial V}{\partial x}\right) = 1 + \left(\frac{\partial V}{\partial x}\right)^T \left[F(x) + B(x)u\right], \; u = \pm 1, \; u^2 \equiv 1. \tag{6.42}$$

The first-order necessary condition for optimality yields the discontinuous switching control

$$u = -\text{sgn}\left[\varphi(x)\right] = -\text{sgn}\left(B^T(x)\frac{\partial V}{\partial x}\right), \; u = \pm 1, \; u \in U. \tag{6.43}$$

The Hamilton-Jacobi equation is

$$-\frac{\partial V}{\partial t} = 1 + \left(\frac{\partial V}{\partial x}\right)^T F(x) - \left(\frac{\partial V}{\partial x}\right)^T B(x)\text{sgn}\left(B^T(x)\frac{\partial V}{\partial x}\right). \tag{6.44}$$

The Hamiltonian (6.42) admits the *costate equation* (6.20) $\dot{\boldsymbol{\lambda}}^T = -\frac{\partial H(t,x,u,\boldsymbol{\lambda})}{\partial x}$.

The Lagrange multiplier $\boldsymbol{\lambda}$ and return function $V(x)$ must ensure solution of (6.44), yielding a switching manifold $\varphi(x) = B^T(x)\frac{\partial V}{\partial x}$ and feedback gains.

Control laws with discontinuous switching cause impediments due to implementation obscurity, sensitivity, losses and other impairments.

Example 6.13. *Discontinuous Control*
With control limits $-\mathcal{U}_{\max} \leq \mathcal{U} \leq \mathcal{U}_{\max}$, such as $\mathcal{U} = \pm\mathcal{U}_{\max}$, to design a control function $u=\phi[\varphi(x)]$ with $u=\pm u_{\max}$, minimize the performance functional (6.39) is

$$J = \min_{t,u\in U} \frac{1}{u_{\max}} \int_{t_0}^{t_f} 1dt .$$

For a nonlinear system $\dot{x}(t) = F(x) + B(x)u$, the Hamiltonian is

$$H(x,u,\tfrac{\partial V}{\partial x}) = \frac{1}{u_{\max}} + \left(\frac{\partial V}{\partial x}\right)^T \left[F(x) + B(x)u\right].$$

The first-order necessary condition (6.22) yields

$$u = -u_{\max}\operatorname{sgn}\left[\varphi(x)\right] = -u_{\max}\operatorname{sgn}\left(B^T(x)\frac{\partial V}{\partial x}\right),\ u=\pm u_{\max},\ u\in U.$$ ■

Discontinuous Switching Manifold – In derived control law (6.43), the switching hypersurface is $\varphi(x) = B^T(x)\frac{\partial V}{\partial x}$, $\varphi(\cdot):\mathbb{R}^n\to\mathbb{R}^m$. Control laws are aimed to maintain a system within $\varphi(x)=0$. For $\varphi(x)\neq 0$, switching $u=\pm 1$ or $u=\pm u_{\max}$ is aimed to guarantee system equilibrium, at which $\varphi(x)=0$.

Due to state evolutions, perturbations, noise, interference, measurement impediments, data fusion, ASICs errors, electromagnetic interference and other factors, physical systems $\mathbb{M}(t,x,u,y,d,r)$ evolve with $x\neq 0$, $\forall t$. Correspondingly $\varphi(x)\neq 0$, $\forall(t,x)$. This implies persistent switching $u=\pm 1$ or $u=\pm u_{\max}$.

There are numerous impediments of discontinuous switching control laws. It is unlikely one may implement discontinuous switching even if desired. For example, the *pulse* thrust by solid-propellant thrusters, or, ultra-fast-switching of transistors and operational amplifiers, may typify relay characteristics being inherently continuous.

Continuous Control – Smooth functions $\Phi\in C^k$, $k\geq 1$ are used to support design of discontinuous control laws. Having found (6.43) $u= -\phi[\varphi(x)]$, investigate design using physics-consistent continuous control. For given control limits $\mathcal{U}_{\min} \leq \mathcal{U} \leq \mathcal{U}_{\max}$, map discontinuous $\phi\notin C^k$ by a continuous function $\Phi\in C^k, k\geq 1$, $\Phi\subseteq\mathcal{U}_\varepsilon$. A continuous control

$$u = -\Phi\left[\varphi(x)\right] = -\Phi\left(B^T(x)\frac{\partial V}{\partial x}\right),\ \varphi(x) = B^T(x)\frac{\partial V}{\partial x},\ \Phi\cong\pm 1,\ u_{\min}\leq u\leq u_{\max},\ u\in U \quad (6.45)$$

is consistent with (6.43), and, yields a solution of minimization problem.

One minimizes

$$J = \min_{t,x,u\in U} \int_{t_0}^{t_f} \left[\boldsymbol{\omega}_x(x) + \int \left(\Phi^{-1}(u)\right)^T du\right] dt = \min_{t,x,u\in U} \int_{t_0}^{t_f} \left[\boldsymbol{\omega}_x(x) + 1\right] dt,\ \int\left(\Phi^{-1}(u)\right)^T du \approx 1. \quad (6.46)$$

In (6.46), as reported in section 6.7, $\int\left(\Phi^{-1}(u)\right)^T du \approx 1$. Control law (6.43) assumes an ideal relay switching. A relay with *deadband* (neutral zone with $\delta \ll 1$), relay with *dead zone* $\pm\delta$, three-position relay and other limits $\mathcal{U}_{\min}\le\mathcal{U}\le\mathcal{U}_{\max}$ may be studied. For a relay with *dead zone* $\pm\delta$, a discontinuous $u=-\operatorname{sgn}\left(B^T(x)\frac{\partial V}{\partial x}\right)\Bigg|_{\begin{cases}u=\pm1,\left|B^T(x)\frac{\partial V}{\partial x}\right|\ge\delta\\ u=0,\left|B^T(x)\frac{\partial V}{\partial x}\right|<\delta\end{cases}}$, $\delta\ll1$, $-1\le u\le1$, is mapped by (6.45).

With a functional (6.46), solution of the Hamilton-Jacobi equation

$$-\frac{\partial V}{\partial t}=\omega_x(x)+\Phi^T\left(B^T(x)\frac{\partial V}{\partial x}\right)\Sigma\Phi\left(B^T(x)\frac{\partial V}{\partial x}\right)+\left(\frac{\partial V}{\partial x}\right)^T F(x)-\left(\frac{\partial V}{\partial x}\right)^T B(x)\Phi\left(B^T(x)\frac{\partial V}{\partial x}\right),$$
$$\Sigma\in\mathbb{R}^{m\times m},\ \Sigma>0 \qquad (6.47)$$

is approximated by a return function $V(t,x)$ (6.24). Boundedness of $\Phi(\cdot)$, positive definiteness of a diagonal matrix $\Sigma\in\mathbb{R}^{m\times m}$, $\left(\frac{\partial V}{\partial x}\right)^T B(x)\Phi(\cdot)>0$ and $\Phi^T(\cdot)\Sigma\Phi(\cdot)>0$, yields

$$\left(\frac{\partial V}{\partial x}\right)^T B(x)\Phi\left(B^T(x)\frac{\partial V}{\partial x}\right)>\Phi^T\left(B^T(x)\frac{\partial V}{\partial x}\right)\Sigma\Phi\left(B^T(x)\frac{\partial V}{\partial x}\right).$$

Solution of (6.47) exists and converges.

Example 6.14. *Discontinuous and Continuous Controls*

Discontinuous control limits $\mathcal{U}_{\min}\le\mathcal{U}\le\mathcal{U}_{\max}$ are described by a continuous function $\Phi\in C^k$, $k\ge1$. For a discontinuous relay $\mathcal{U}=\pm1$ and $u=\operatorname{sgn}[\varphi(x)]=-\operatorname{sgn}\left(\frac{\partial V}{\partial x}\right)$, $x\equiv\frac{\partial V}{\partial x}$, apply an infinitely-differentiable hyperbolic function Φ.

For $\mathcal{U}=\pm1$, consider $\mathcal{U}\cong\pm1$, and, $u=-\operatorname{sgn}\left(\frac{\partial V}{\partial x}\right)\approx-\tanh\left(a\frac{\partial V}{\partial x}\right)$.

For relays with *deadband* and relay with *dead zone* $\pm\delta$, limits $\mathcal{U}=\pm1|_{\pm\delta}$ imply finding a discontinuous or continuous control functions as

$$u=-\operatorname{sgn}\left(\frac{\partial V}{\partial x}\right)\bigg|_{\pm\delta},\quad u=-\tanh^{2n+1}\left(a\frac{\partial V}{\partial x}\right),\ n>1,\ a\gg1.$$

The range and domain of infinitely-differentiable functions $\tanh\left(a\frac{\partial V}{\partial x}\right)$ and $\tanh^{2n+1}\left(a\frac{\partial V}{\partial x}\right)$ are $(-1, 1)$ and $(-\infty,\infty)$. Furthermore, $\tanh\left(-a\frac{\partial V}{\partial x}\right)=-\tanh\left(a\frac{\partial V}{\partial x}\right)$.

For $\mathcal{U}\cong\pm\mathcal{U}_{\max}$, one has $u=-u_{\max}\tanh[\varphi(x)]=-u_{\max}\tanh\left(aB^T(x)\frac{\partial V}{\partial x}\right)$, $-u_{\max}<u<u_{\max}$,

$$u=-u_{\max}\tanh^{2n+1}[\varphi(x)]=-u_{\max}\tanh^{2n+1}\left(aB^T(x)\frac{\partial V}{\partial x}\right),\quad a\gg1.$$

Example 6.21 studies continuous functions $\Phi\in C^k$, $k\ge1$ which approximate $\mathcal{U}_{\min}\le\mathcal{U}\le\mathcal{U}_{\max}$, $|\mathcal{U}_{\max}|=|\mathcal{U}_{\min}|$, such as $\Phi\subseteq\mathcal{U}_\varepsilon$. For saturation, arctangent, Gudermannian $\operatorname{gd}(x)$ and other functions are used as $\Phi(x)=\operatorname{erf}(\frac{\sqrt{\pi}}{2}x)$, $\Phi(x)=\frac{2}{\pi}\arctan(\frac{\pi}{2}x)$, $\Phi(x)=\frac{2}{\pi}\operatorname{gd}(\frac{\pi}{2}x)$, $-1<\Phi<1$, $\Phi(x)=\frac{x}{\sqrt{1+x^2}}$, $\Phi(x)=\frac{x}{1+|x|}$. ■

Control Energy, Low-Power Design and Power Estimates – Discontinuous control $u=\phi[\varphi(x)]$, such as

$$u=-u_{\max}\operatorname{sgn}[\varphi(x)],\quad u=\begin{cases}u_{\max} & \text{if } \varphi(x)<0\\ -u_{\max} & \text{if } \varphi(x)>0\end{cases},$$

yields persistent switching $u=\pm u_{\max}$, including near-equilibrium. Control takes values $u=u_{\max}$ and $u=-u_{\max}$ for infinity small $\delta(x)$, perturbed states, systematic and random measurement errors, noise, etc. These yield adverse effects, including excessive power consumption, losses, state switching and other impediments.

Continuous control

$u=\Phi[\varphi(x)]$, $u\in U$, $\Phi\in C^k$, $k\geq 1$

minimizes system energy, which is of importance considering a problem $\min_{x\in X, u\in U} \|u\|_p$. This ensures efficiency and adequate transient dynamics. Minimization of control energy guarantees low-power design and low-power operation.

The energy is the integral of power, $E=\int_{t_0}^{t_f} P(t)dt$.

The electrical power is $P=VI=\frac{1}{R}V^2$.

The control energy E_u may be evaluated as $E_u=\int_{t_0}^{t_f} u^2 dt$ or $E_u=\int_{t_0}^{t_f} \|u\|_p\, dt$, $p\geq 1$.

Example 6.15. *Discontinuous and Continuous Control, and, Energy Estimates*

Consider a system with $\mathcal{U}_{\min}\leq\mathcal{U}\leq\mathcal{U}_{\max}$. Let the governing equations are

$\dot{x}_1(t)=-x_1+x_2+x_2^7+x_1^2u_1$, $-1\leq u_1\leq 1$,

$\dot{x}_2(t)=-x_2-x_2^3+x_1^3x_2^5u_2$, $-1\leq u_2\leq 1$.

Synthesize and examine different control laws for a system with control limits $\mathcal{U}=\pm 1$ and $\mathcal{U}\equiv\pm 1$.

The control bounds are:

1. $u_1=\pm 1$, $u_2=\pm 1$ for $\mathcal{U}=\pm 1$;
2. $u_1\equiv\pm 1$, $u_2\equiv\pm 1$ for $\mathcal{U}\equiv\pm 1$.

Minimizing (6.39) $J=\min_{t,u\in U}\int_{t_0}^{t_f} 1dt=(t_f-t_0)$, consider the Hamiltonian (6.42)

$$H(x,u,\tfrac{\partial V}{\partial x})=1+\left(\tfrac{\partial V}{\partial x}\right)^T[F(x)+B(x)u]$$

$$=1+\tfrac{\partial V}{\partial x_1}\left(-x_1+x_2+x_2^7+x_1^2u_1\right)+\tfrac{\partial V}{\partial x_2}\left(-x_2-x_2^3+x_1^3x_2^5u_2\right),\ u_1=\pm 1,\ u_1^2\equiv 1,\ u_2=\pm 1,\ u_2^2\equiv 1.$$

The necessary condition for optimality (6.22) yields a discontinuous control law (6.43)

$u_1=-\operatorname{sgn}\left(x_1^2\frac{\partial V}{\partial x_1}\right)$, $u_1=\pm 1$,

$u_2=-\operatorname{sgn}\left(x_1^3x_2^5\frac{\partial V}{\partial x_2}\right)$, $u_2=\pm 1$.

To solve the Hamilton-Jacobi equation (6.44), consider

$\operatorname{sgn}[\varphi(x)]\approx\tanh[a\varphi(x)]$, $a\gg 1$.

That is, $u=\Phi[\varphi(x)]$ with infinitely-differentiable hyperbolic tangent function $\Phi(x)=\tanh[\varphi(x)]$. In particular,

$u=\tanh[a\varphi(x)]$, $a=100$, $\Phi\equiv\pm 1$, $\Phi\subseteq\mathcal{U}_\varepsilon$.

To solve the Hamilton-Jacobi equation (6.44), the Taylor, Mercator and other series are applied to approximate $\tanh(z)$. The Taylor series is

$$\tanh(z)=\sum_{n=1}^{\infty}\tfrac{1}{(2n)!}2^{2n}(2^{2n}-1)B_{2n}z^{2n-1}=z-\tfrac{1}{3}z^3+\tfrac{2}{15}z^5-\tfrac{17}{315}z^7+\tfrac{68}{2835}z^9-\ldots,\ |z|<\tfrac{1}{2}\pi,$$

where B_m is the mth Bernoulli number, $B_m(n)=\sum_{k=0}^{m}\sum_{l=0}^{k}(-1)^l\begin{bmatrix}k\\l\end{bmatrix}\frac{(n+l)^m}{k+1}$.

Consider the return function is $V(x)=\frac{1}{2}k_{11}x_1^2+k_{12}x_1x_2+\frac{1}{2}k_{22}x_2^2$.

For a=100, we find k_{11}=0.328, k_{12}=0.486 and k_{22}=1.4.

One obtains

$u_1=-\tanh[100x_1^2(0.328x_1+0.486x_2)]$,

$u_2=-\tanh[100x_1^3x_2^5(0.486x_1+1.4x_2)]$.

For a discontinuous control

$u_1=-\text{sgn}[x_1^2(0.328x_1+0.486x_2)]$,

$u_2=-\text{sgn}[x_1^3x_2^5(0.486x_1+1.4x_2)]$.

To simulate differential equations with relay switching $u_1=\pm1$ and $u_2=\pm1$, robust numeric algorithms are used choosing an adequate solver. The transient dynamics for states (x_1,x_2) and control switching $(u_1=\pm1, u_2=\pm1)$ are documented in Figure 6.9.a for initial conditions $x_{10}=5$ and $x_{20}=-5$.

To reduce persistent switching $(u_1=\pm1, u_2=\pm1)$, use the *deadband* (neutral zone) with δ=0.001. One observes switching $u_1=\pm1$ and $u_2=\pm1$ during transients $x_1(t)$ and $x_2(t)$, which occur at near-equilibrium. This yields adverse effects.

Relay control may be inadequate. Figure 6.9.b illustrates dynamics for (x_1,x_2) and evolutions of (u_1,u_2) when $u=\tanh[a\varphi(x)]$, a=100. Adequate system transients and control evolutions are achieved.

Minimize

$$J=\min_{x,u}\int_0^\infty \tfrac{1}{2}(x_1^2+x_2^2+u_1^2+u_2^2)dt \ .$$

The Hamilton-Jacobi equation is approximated by

$V=\frac{1}{2}k_{11}x_1^2+k_{12}x_1x_2+\frac{1}{2}k_{22}x_2^2$.

A continuous control is

$u_1=-x_1^2(0.407x_1+0.14x_2)$,

$u_2=-x_1^3x_2^5(0.14x_1+0.503x_2)$.

System evolutions with $-1\le(u_1,u_2)\le1$ are documented in Figure 6.9.c. The closed-loop system is stable and exhibits adequate dynamics. Transient dynamics of states $x_1(t)$ and $x_2(t)$ are almost identical for different control laws.

Continuous control laws $u=\Phi[\varphi(x)]$, $\Phi\in C^k$ minimize the energy E_u and $\min_{x\in X, u\in U}\|u\|_p$, ensure adequate control activities, guarantee efficiency and optimality with minimal settling time and adequate transient dynamics, evaluated by a measure

$$\int_{t_0}^{t_f}\left(\|x_1\|_p+\|x_2\|_p\right)dt \ .$$

Minimization of control energy is important to guarantee low-power operation. We evaluate control energy as

$$E_u=\int_{t_0}^{t_f}\left(\|u_1\|_p+\|u_2\|_p\right)dt \ .$$

For p=1 and p=2, we have $E_u=\int_{t_0}^{t_f}\left(|u_1|+|u_2|\right)dt$ and $E_u=\int_{t_0}^{t_f}(u_1^2+u_2^2)dt$.

Computed E_u are reported in Figures 6.9 annotations.

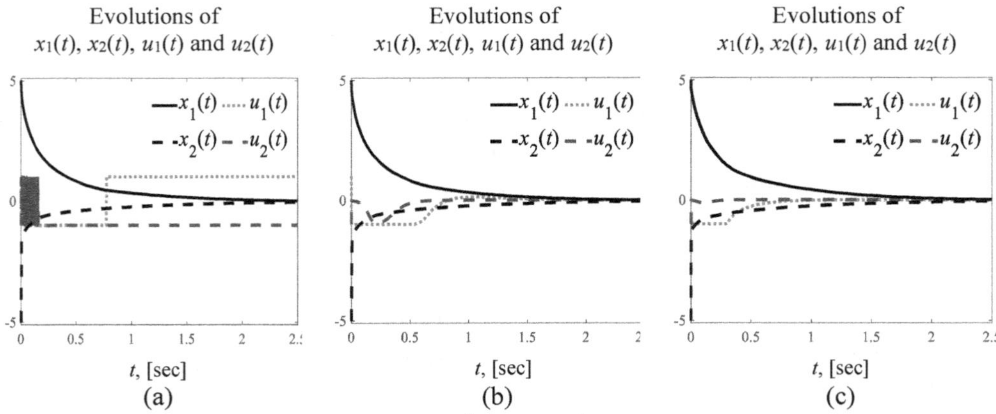

Figure 6.9. (a) Relay discontinuous control $\begin{cases} u_1 = -\text{sgn}[x_1^2(0.328x_1 + 0.486x_2)] \\ u_2 = -\text{sgn}[x_1^3 x_2^5(0.486x_1 + 1.4x_2)] \end{cases}$. Dynamics of (x_1, x_2) and switching activity (u_1, u_2) for $x_{10}=5$, $x_{20}=-5$. The energy estimates are $E_{u,p=1}=5$ and $E_{u,p=2}=5$ for $t\in[0\ \ 2.5]$ sec. Note that $E_u\to\infty$ as $t_f\to\infty$;
(b) Evolutions of (x_1, x_2) and (u_1, u_2) for $\begin{cases} u_1 = -\tanh[100x_1^2(0.328x_1 + 0.486x_2)] \\ u_2 = -\tanh[100x_1^3 x_2^5(0.486x_1 + 1.4x_2)] \end{cases}$. Finite E_u are $E_{u,p=1}=1$ and $E_{u,p=2}=0.813$;
(c) System dynamics (x_1, x_2) and evolutions of (u_1, u_2) in the closed-loop system with control $\begin{cases} u_1 = -x_1^2(0.407x_1 + 0.14x_2) \\ u_2 = -x_1^3 x_2^5(0.14x_1 + 0.503x_2) \end{cases}$ constrained to limits as $-1\le(u_1, u_2)\le 1$. Computed control energies are $E_{u,p=1}=0.48$ and $E_{u,p=2}=0.37$. ■

Example 6.16. *Calculus of Variations, System Dynamics and Control*
Consider a controlled system with $-1\le\mathcal{U}\le 1$, described by the second-order separable differential equations

$$\dot{x}_1(t) = x_2,$$
$$\dot{x}_2(t) = u.$$

Control may assume:

1. An ideal relay switching $\mathcal{U}=\pm1$ for which $\lim\limits_{x\to0^+, x\to0^-} \mathcal{U}$ exists;
2. Piecewise-continuous or continuous switching, such as $\mathcal{U}\cong\pm1$, for which $\lim\limits_{x\to0^+, x\to0, x\to0^-} \mathcal{U}$ exists.

Examine the design problem for a *torque-force* control for mechanical systems. Assume a relay switching and near-relay control function. Equations of motion are

$$\dot{x}_1(t) = x_2,$$
$$\dot{x}_2(t) = u,\ -1\le u\le 1,\ u=\pm1.$$

Discontinuous and continuous control laws $u=\phi[\varphi(x)]$ and $u=\Phi[\varphi(x)]$ are sought within:

1. Relay switching $u=\pm1$;
2. Smooth switching $u\cong\pm1$.

Calculus of variations is applied to design a discontinuous control function $u=\phi[\varphi(x)]$, $\phi\notin C^0, \phi\in C^0$. Consider a relay switching for which control takes values $u=\pm1$.

If u=1, from $\begin{cases}\dot{x}_1(t)=x_2\\ \dot{x}_2(t)=1\end{cases}$, one has $\dfrac{dx_2}{dx_1}=\dfrac{1}{x_2}$.

The integration $\int x_2dx_2=\int dx_1$ gives $\tfrac{1}{2}x_2^2=x_1+c_1$.

If u=−1, from $\begin{cases}\dot{x}_1(t)=x_2\\ \dot{x}_2(t)=-1\end{cases}$, we obtain $\dfrac{dx_2}{dx_1}=-\dfrac{1}{x_2}$.

Hence $\int x_2dx_2=-\int dx_1$, and, $\tfrac{1}{2}x_2^2=-x_1+c_2$.

For u=±1 and admissible evolutions of (x_1,x_2), comparison of $\begin{cases}\tfrac{1}{2}x_2^2=x_1+c_1 & \text{for } u=1\\ \tfrac{1}{2}x_2^2=-x_1+c_2 & \text{for } u=-1\end{cases}$ yields an equality $x_1+\tfrac{1}{2}x_2\,|\,x_2|=0$ at which $\lim_{x\to 0}\operatorname{sgn}(x)$ does not exist, while two-sided limits exist and not equal, $\lim_{x\to 0^+}\operatorname{sgn}(x)=1$, $\lim_{x\to 0^-}\operatorname{sgn}(x)=-1$. Control switches as u=±1 if $x_1+\tfrac{1}{2}x_2\,|\,x_2|\neq 0$.

With the derived switching curve $\varphi(x)=x_1+\tfrac{1}{2}x_2|x_2|$, we have

$u=-\operatorname{sgn}[\varphi(x)]=-\operatorname{sgn}\left(x_1+\tfrac{1}{2}x_2|x_2|\right)$, u=±1,

$$u=\begin{cases}1 & \text{if } \varphi(x_1,x_2)<0\\ -1 & \text{if } \varphi(x_1,x_2)>0\end{cases},\quad \varphi(x)=\left(x_1+\tfrac{1}{2}x_2|x_2|\right).$$

Using calculus of variations, we analytically derived a control law by analyzing solutions of differential equations for u=±1. The switching curve and phase-plane evolutions of (x_1,x_2) for different initial conditions (x_{10},x_{20}) are documented in Figure 6.10.a. The switching curve $\varphi(x)$ is in the second and fourth quadrants for [x_1,x_2]∈(−∞,∞). The system dynamics is reported in Figures 6.11 for different (x_{10},x_{20}).

The Hamiltonian optimization is applied. Minimize (6.39) $J=\min_{t,u\in U}\int_{t_0}^{t_f}1dt$.

From $H(x,u,\frac{\partial V}{\partial x})=1+\dfrac{\partial V}{\partial x_1}x_2+\dfrac{\partial V}{\partial x_2}u$, u=±1, u^2≡1,

one has $\dfrac{\partial H}{\partial u}=\dfrac{\partial}{\partial u}\left(1+\dfrac{\partial V}{\partial x_1}x_2+\dfrac{\partial V}{\partial x_2}u\right)=0$.

One yields a discontinuous control (6.43) $u=-\operatorname{sgn}\left(\dfrac{\partial V}{\partial x_2}\right)$.

Equation (6.44) $-\dfrac{\partial V}{\partial t}=1+\dfrac{\partial V}{\partial x_1}x_2-\dfrac{\partial V}{\partial x_2}\operatorname{sgn}\left(\dfrac{\partial V}{\partial x_2}\right)$

is solved by approximating limits $\mathcal{U}$≅±1 by $\Phi\subseteq\mathcal{U}_\varepsilon$, $\Phi\in C^k$, k≥1.

Consider u=Φ[$\varphi(x)$] with the infinitely-differentiable hyperbolic function Φ.

One obtains a continuous control $u=-\tanh\left(a\dfrac{\partial V}{\partial x_2}\right), a\gg 1$.

Solution of (6.47)

$$-\frac{\partial V}{\partial t}=\boldsymbol{\omega}_x(x)+\sigma\tanh^2\left(a\frac{\partial V}{\partial x_2}\right)+\frac{\partial V}{\partial x_1}x_2-\frac{\partial V}{\partial x_2}\tanh\left(a\frac{\partial V}{\partial x_2}\right),\ \boldsymbol{\omega}_x(x)\ll 1,\ 0<\sigma<1$$

is found using a nonquadratic, positive definite continuous and differentiable return functions (6.24).

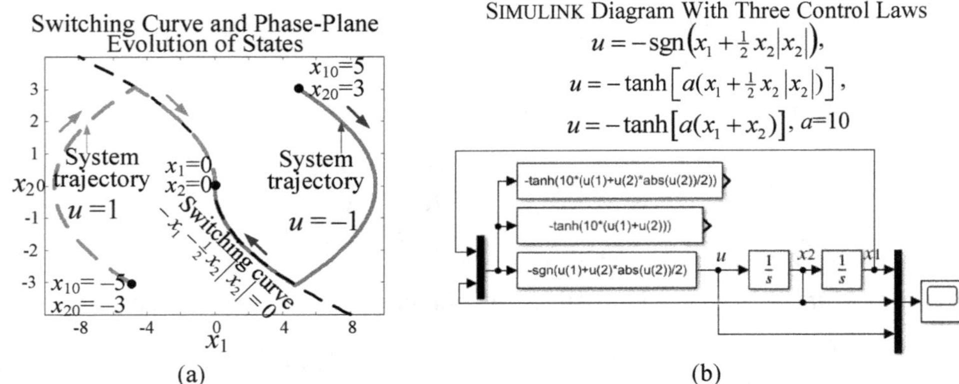

Figure 6.10. (a) Phase-plane evolution and switching curve $\varphi(x) = x_1 + \frac{1}{2}x_2|x_2|$ for a closed-loop system $\dot{x}_1(t) = x_2, \dot{x}_2(t) = u$, u=±1 with a control law $u = -\text{sgn}[\varphi(x)] = -\text{sgn}(x_1 + \frac{1}{2}x_2|x_2|)$; (b) SIMULINK diagram to simulate a closed-loop system with different control laws.

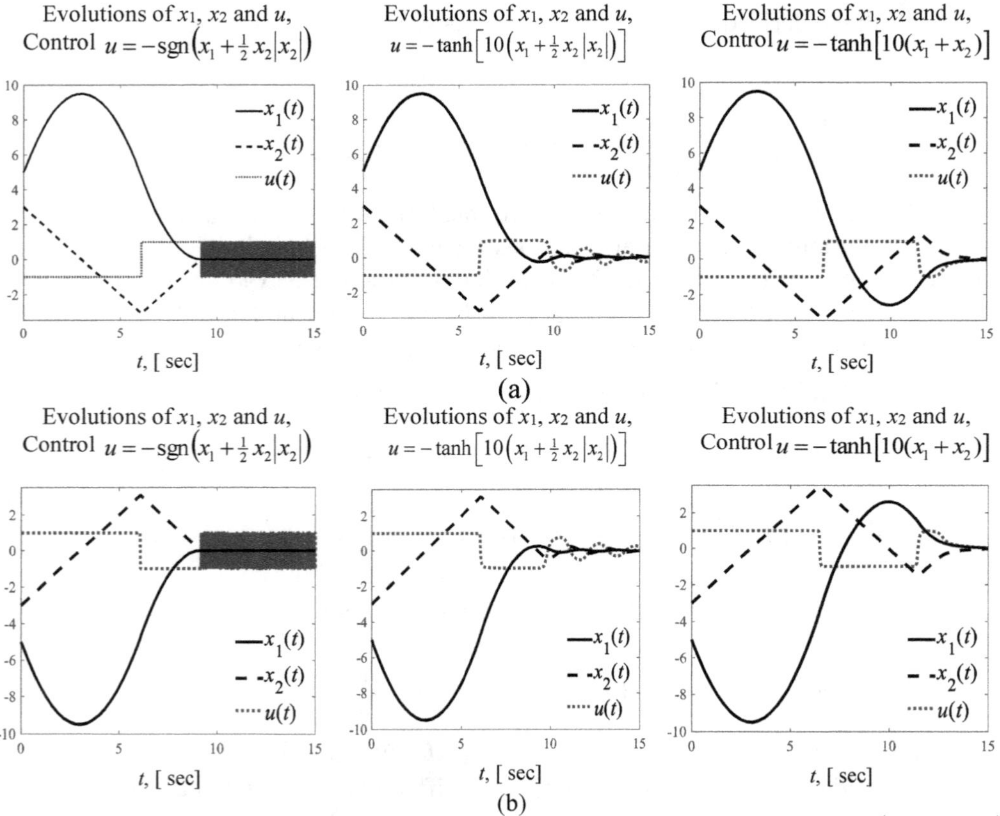

Figure 6.11. (a) Dynamics of (x_1,x_2) and evolutions of $u(t)$ for $u = -\text{sgn}(x_1 + \frac{1}{2}x_2|x_2|)$, $u = -\tanh[10(x_1 + \frac{1}{2}x_2|x_2|)]$ and $u = -\tanh[10(x_1 + x_2)]$. Initial conditions are x_{10}=5, x_{20}=3; (b) System dynamics (x_1,x_2) and evolutions of $u(t)$ for three control laws, x_0=[−5, −3].

Consider a return function $V(x_1, x_2) = \frac{1}{2}k_{11}x_1^2 + k_{12}x_1x_2 + \frac{1}{3}k_{22}x_2^2|x_2|$,

or, $V(x_1, x_2) = \frac{1}{2}k_{11}x_1^2 + k_{12}x_1x_2 + \frac{1}{3}k_{22}x_2^3\,\text{sgn}(x_2)$, $\text{sgn}(x) = \frac{x}{|x|} = \frac{|x|}{x}$.

For tanh(z), $z \equiv a\frac{\partial V}{\partial x}$, one has $\tanh(z) = z - \frac{1}{3}z^3 + \frac{2}{15}z^5 - \frac{17}{315}z^7 + \ldots, |z| < \frac{1}{2}\pi$.

For a=10, solve an algebraic equation

$$\boldsymbol{\omega}_x(x) + \sigma \tanh^2\left(a\frac{\partial V}{\partial x_2}\right) + \frac{\partial V}{\partial x_1}x_2 - \frac{\partial V}{\partial x_2}\tanh\left(a\frac{\partial V}{\partial x_2}\right) = 0, \ \boldsymbol{\omega}_x(x) \ll 1, \ 0<\sigma<1.$$

One finds unknown (k_{11},k_{12},k_{22}) of $V(x_1,x_2) = \frac{1}{2}k_{11}x_1^2 + k_{12}x_1x_2 + \frac{1}{3}k_{22}x_2^2|x_2|$.

Control law (6.45) is

$$u = -\tanh\left[a(x_1 + \frac{1}{2}x_2|x_2|)\right], \ a=10.$$

Approximating the Hamilton-Jacobi equation by

$$V(x_1,x_2) = \frac{1}{2}k_{11}x_1^2 + k_{12}x_1x_2 + \frac{1}{2}k_{22}x_2^2,$$

we have $u = -\tanh\left[a(x_1 + x_2)\right]$.

Figure 6.10.b illustrates a SIMULINK diagram with three control laws. Evolutions of $x_1(t)$, $x_2(t)$ and $u(t)$ are reported in Figures 6.11. There are advantages of continuous control (6.45) $u=\Phi[\varphi(x)]$, $\Phi\in C^k$, $k\geq 1$. ■

Example 6.17. *Discontinuous and Continuous Control*

For a system, considered in Example 6.16, switching control laws

$$u = -\text{sgn}\left(x_1 + \frac{1}{2}x_2|x_2|\right), \ u = -\text{sgn}\left(x_1 + x_2\right).$$

may be referenced as a discontinuous hard-switching sliding mode control.

Continuous control laws with nonlinear manifolds $\varphi(\cdot)$ are found using return functions (6.24)

$$V(x_1,x_2) = \frac{1}{2}k_{11}x_1^2 + k_{12}x_1x_2 + \frac{1}{2}k_{22}x_2^2, \ V(x_1,x_2) = \frac{1}{2}k_{11}x_1^2 + k_{12}x_1x_2 + \frac{1}{3}k_{22}x_2^2|x_2|.$$

In particular,

$$u = -\tanh\left[a(x_1 + \frac{1}{2}x_2|x_2|)\right],$$

$$u = -\tanh\left[a\left(x_1 + x_2\right)\right].$$

Simulations of closed-loop systems are reported in Figures 6.11. ■

Example 6.18. *Calculus of Variations and Hamiltonian Optimization*

For a system with control bounds $-1\leq \mathcal{U}\leq 1$, $\mathcal{U}=\pm 1$, consider the governing equation

$$\dot{x}_1(t) = x_2,$$

$$\dot{x}_2(t) = -x_1 - x_1^3 + u, \ -1\leq u\leq 1, \ u=\pm 1.$$

Calculus of Variations – Control function takes values $u=\pm 1$.

For u=1, one has $\frac{dx_2}{dx_1} = \frac{-x_1 - x_1^3 + 1}{x_2}$.

The integration $\int x_2 dx_2 = \int\left(-x_1 - x_1^3 + 1\right)dx_1$ yields

$$\frac{1}{2}x_2^2 = -\frac{1}{2}x_1^2 - \frac{1}{4}x_1^4 + x_1 + c_1.$$

For $u=-1$, from $\frac{dx_2}{dx_1} = \frac{-x_1 - x_1^3 - 1}{x_2}$, we have $\int x_2 dx_2 = \int\left(-x_1 - x_1^3 - 1\right)dx_1$.

Thus, $\frac{1}{2}x_2^2 = -\frac{1}{2}x_1^2 - \frac{1}{4}x_1^4 - x_1 + c_2$.

The switching $u=\pm 1$ occurs at $x_1 + \frac{1}{2}x_1|x_1| + \frac{1}{4}x_1^3|x_1| + \frac{1}{2}x_2|x_2| = 0$.

A switching curve $\varphi(x)=x_1+\frac{1}{2}x_1|x_1|+\frac{1}{4}x_1^3|x_1|+\frac{1}{2}x_2|x_2|$ is reported in Figure 6.12.a.

A discontinuous control is

$u=-\operatorname{sgn}[\varphi(x)]=-\operatorname{sgn}\left(x_1+\frac{1}{2}x_1|x_1|+\frac{1}{4}x_1^3|x_1|+\frac{1}{2}x_2|x_2|\right)$, $u=\pm 1$.

Hamiltonian Optimization – Minimize a functional (6.39) $J=\min_{t,u\in U}\int_{t_0}^{t_f}1dt$.

For a Hamiltonian (6.42) $H(x,u,\frac{\partial V}{\partial x})=1+\frac{\partial V}{\partial x_1}x_2+\frac{\partial V}{\partial x_2}(-x_1-x_1^3+u)$,

apply the first-order condition for optimality (6.22)

$$\frac{\partial H}{\partial u}=\frac{\partial}{\partial u}\left(1+\frac{\partial V}{\partial x_1}x_2+\frac{\partial V}{\partial x_2}(-x_1-x_1^3+u)\right)=0,\ u=\pm 1,\ u^2\equiv 1.$$

A discontinuous control law (6.43) is

$$u=-\operatorname{sgn}\left(\frac{\partial V}{\partial x_2}\right).$$

A continuous control (6.45) is

$$u=-\tanh\left(a\frac{\partial V}{\partial x_2}\right),\ a\gg 1.$$

Solve (6.47)

$$\boldsymbol{\omega}_x(x)+\sigma\tanh^2\left(a\frac{\partial V}{\partial x_2}\right)+\frac{\partial V}{\partial x_1}x_2+\frac{\partial V}{\partial x_2}(-x_1-x_1^3)-\frac{\partial V}{\partial x_2}\tanh\left(a\frac{\partial V}{\partial x_2}\right)=0,\ \boldsymbol{\omega}_x(x)\ll 1,\ 0<\sigma<1$$

approximating a solution by a return function

$$V(x_1,x_2)=\tfrac{1}{2}k_{11}x_1^2+k_{12}x_1x_2+k_{13}x_1|x_1|x_2+\tfrac{1}{3}k_{14}x_2^2|x_2|+k_{15}x_1^3|x_1|x_2+\tfrac{1}{6}k_{16}x_1^6.$$

One finds k_{ij}, and a control law is

$$u=-\tanh\left[a\left(x_1+\tfrac{1}{2}x_1|x_1|+\tfrac{1}{4}x_1^3|x_1|+\tfrac{1}{2}x_2|x_2|\right)\right],\ a=10,\ -1<u<1.$$

Evolutions of $x_1(t)$, $x_2(t)$ and $u(t)$ for designed discontinuous and continuous control laws are illustrated in Figures 6.12.b and c. The continuous control yields consistency.

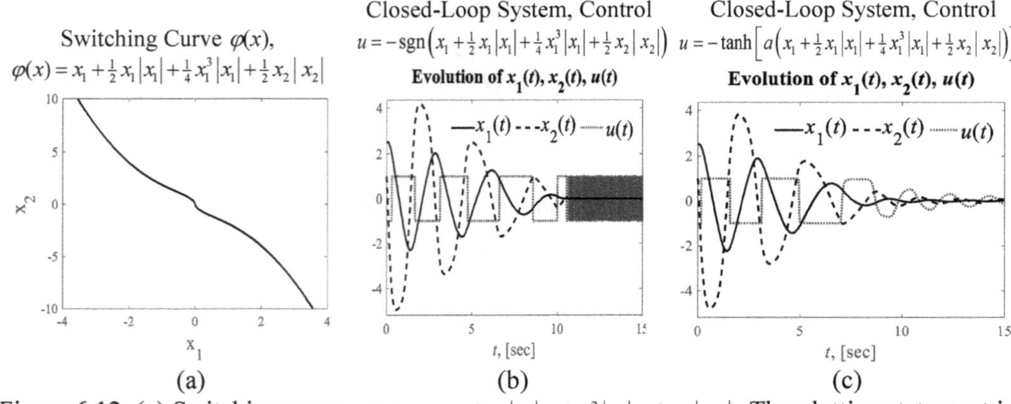

Figure 6.12. (a) Switching curve $\varphi(x)=x_1+\frac{1}{2}x_1|x_1|+\frac{1}{4}x_1^3|x_1|+\frac{1}{2}x_2|x_2|$. The plotting statement is

```
f=ezplot('x1+x1*abs(x1)/2+x1^3*abs(x1)/4+x2*abs(x2)/2',[-4 4 -10 10]); set(f,'Color','black','LineWidth',2)
```

(b) Transients of states (x_1,x_2) and evolutions of $u(t)$ if $u=-\operatorname{sgn}\left(x_1+\frac{1}{2}x_1|x_1|+\frac{1}{4}x_1^3|x_1|+\frac{1}{2}x_2|x_2|\right)$;

(c) Dynamics of (x_1,x_2) and evolution of $u(t)$ for $u=-\tanh\left[a\left(x_1+\frac{1}{2}x_1|x_1|+\frac{1}{4}x_1^3|x_1|+\frac{1}{2}x_2|x_2|\right)\right]$, $a=10$. Initial conditions are $x_{10}=2.5$ and $x_{20}=1$. ■

Example 6.19.

Consider a rigid-body system, governed by equations

$$\dot{x}_1 = -a(t)x_1 + b(t)u,$$

$$\dot{x}_2 = x_1, \; -1 \le u \le 1,$$

where x_1 is the velocity; x_2 is the displacement.

For translational and rotational dynamics, time-varying $a(t)$ and $b(t)$ are $a=B_v/m$ or $a=B_m/J$, and, $b=1/m$ or $b=1/J$. Friction and drag coefficients, mass, moment of inertia and other parameters vary. Varying $a(\cdot)\ge 0$ and $b(\cdot)>0$ are bounded as $a_{\min}\le a(\cdot)\le a_{\max}$ and $b_{\min}\le b(\cdot)\le b_{\max}$.

Design control laws, and, examine system robustness and *sensitivity* to parameter variations. Variations of $a(\cdot)\in[a_{\min}\ a_{\max}]$ and $b(\cdot)\in[b_{\min}\ b_{\max}]$ are continuous.

Let $a=a_0+a_v e^{-t}\sin(10\pi t)$, $b=b_0+b_v\cos(10\pi t)$, $a_0=2$, $a_v=2$, $b_0=1$, $b_v=0.75$.

Using the quadratic function

$$V(x_1, x_2) = \tfrac{1}{2}k_{11}x_1^2 + k_{12}x_1x_2 + \tfrac{1}{2}k_{22}x_2^2,$$

for $a_0=2$, $b_0=1$, (k_{11}, k_{12}, k_{22}) are computed. Control laws with a linear switching are

$$u = -\text{sgn}(x_1+x_2),$$

$$u = -\tanh[10(x_1+x_2)], \; \varphi(x)=10(x_1+x_2).$$

SIMULINK diagram to simulate the closed-loop systems with two control laws is reported in Figure 6.13.a. A discontinuous control $u=-\text{sgn}(x_1+x_2)$ results in persistent switching $u=\pm 1$ as documented in Figure 6.13.b. Due to time-varying $a(t)$ and $b(t)$, which is typical in physical systems, there are undesirable high frequency switching.

A control law $u=-\tanh[10(x_1+x_2)]$ yields to continuous control activity, guaranteeing an adequate performance, as depicted in Figure 6.13.c. Discontinuous switching of force and torque cannot be achieved due to actuators' transients. Furthermore, flexible actuator–mechanism *torque limiters*, *couplers* and *dampers* are used to prevent malfunctions and breakdowns. These *torque limiters* attenuate high instantaneous torques, preventing mechanical failures.

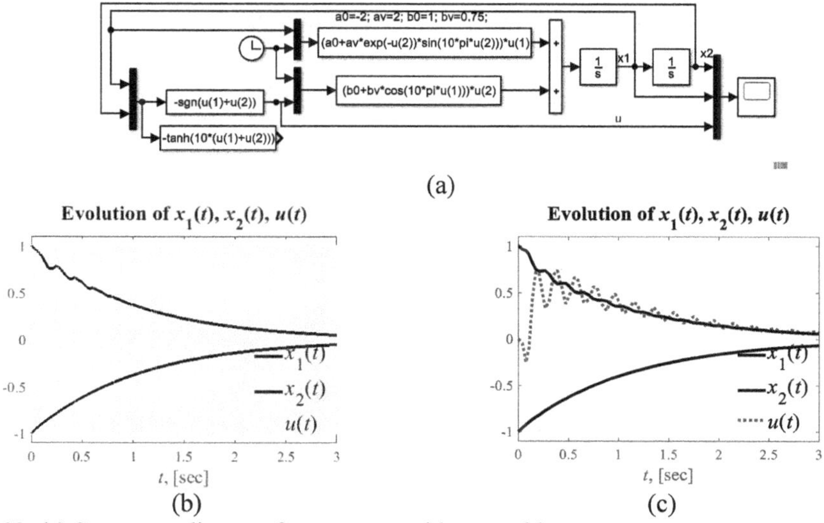

Figure 6.13. (a) SIMULINK diagram for a system with control laws $u=-\text{sgn}(x_1+x_2)$, $u=-\tanh[10(x_1+x_2)]$;
(b) Closed-loop system dynamics with discontinuous control law $u=-\text{sgn}(x_1+x_2)$,$[x_{10}, x_{20}]=[1,-1]$;
(c) System dynamics with a continuous control law $u=-\tanh[10(x_1+x_2)]$, $x_{10}=1$ and $x_{20}=-1$. ■

6.6. Discontinuous Tracking Control and Switching Modes

Control laws can be analytically designed using calculus of variation, Hamiltonian optimization and Lyapunov theory. Mathematically, solution of a minimum-time problem yields discontinuous control (6.43) $u=-\text{sgn}[\varphi(x)]$ with switching curves, surfaces or manifolds $\varphi(x)$.

To solve tracking control problems, states and tracking errors dynamics as reported in sections 5.9 and 6.4. Minimize functionals

$$J=\min_{t,u\in U}\int_{t_0}^{t_f}1dt,\ J=\min_{t,\mathbf{x},u\in U}\int_{t_0}^{t_f}[\omega_x(\mathbf{x})+1]dt, \tag{6.48}$$

subject to system (6.32) with states $\mathbf{x}(t)=\begin{bmatrix}x(t)\\x_e(t)\\e(t)\end{bmatrix}=\begin{bmatrix}x(t)\\\int edt\\e(t)\end{bmatrix}$, $\mathbf{x}\in\mathbf{X}\subset\mathbb{R}^{n+2b}$ and bounds on control $u\in U$

$$\dot{\mathbf{x}}(t)=\begin{bmatrix}\dot{x}(t)\\\dot{x}_e(t)\\\dot{e}(t)\end{bmatrix}=\mathbf{F}(\mathbf{x})+\mathbf{A}\mathbf{x}+A_r r+\mathbf{B}(\mathbf{x})u=\begin{bmatrix}F(x)\\0\\-HF(x)\end{bmatrix}+\begin{bmatrix}0&0&0\\-I_XH&0&0\\0&I_E^X&-I_E\end{bmatrix}\mathbf{x}+\begin{bmatrix}0&0\\I_X&0\\0&I\end{bmatrix}\begin{bmatrix}r\\\dot{r}\end{bmatrix}+\begin{bmatrix}B(x)\\0\\-HB(x)\end{bmatrix}u. \tag{6.49}$$

The Hamiltonian is

$$H(\mathbf{x},u,\tfrac{\partial V}{\partial \mathbf{x}})=1+\left(\frac{\partial V}{\partial \mathbf{x}}\right)^T[\mathbf{F}(\mathbf{x})+\mathbf{A}\mathbf{x}+\mathbf{B}(\mathbf{x})u],\ u^2=1,\ u\in U. \tag{6.50}$$

The necessary condition for optimality $\dfrac{\partial H(\mathbf{x},u,\frac{\partial V}{\partial \mathbf{x}})}{\partial u}=0$ yields

$$u=-\text{sgn}[\varphi(\mathbf{x})]=-\text{sgn}\left(\mathbf{B}^T(\mathbf{x})\frac{\partial V}{\partial \mathbf{x}}\right),\ \varphi(\mathbf{x})=\mathbf{B}^T(\mathbf{x})\frac{\partial V}{\partial \mathbf{x}},\ -1\le u\le 1,\ \mathbf{x}(t)=\begin{bmatrix}x(t)\\x_e(t)\\e(t)\end{bmatrix}=\begin{bmatrix}x(t)\\\int edt\\e(t)\end{bmatrix}, \tag{6.51}$$

with switching at $\varphi(\mathbf{x})=0$.

For $-u_{\max}\le u\le u_{\max}$ and control switching $u=\pm u_{\max}$,minimize $J=\min_{t,u\in U}\frac{1}{u_{\max}}\int_{t_0}^{t_f}1dt$,yielding

$$u=-u_{\max}\,\text{sgn}[\varphi(\mathbf{x})]=-u_{\max}\,\text{sgn}\left(\mathbf{B}^T(\mathbf{x})\frac{\partial V}{\partial \mathbf{x}}\right),\ u(\mathbf{x})=\begin{cases}u_{\max},\ \forall\varphi(\mathbf{x})<0\\-u_{\max},\ \forall\varphi(\mathbf{x})>0\end{cases},\ u\in U. \tag{6.52}$$

Discontinuous control laws (6.43) and (6.52) yield discontinuous switching $u=\pm1$ or $u=\pm u_{\max}$ for $\varphi(x)\neq0$, and referenced as sliding mode control [7-10].

The switching manifold is

$$\mathcal{M}=\{\mathbf{x}\in X\times E\mid\varphi(\mathbf{x})=0\}=\bigcap_{i=1}^{m}\{\mathbf{x}\in X\times E\mid\varphi_i(\mathbf{x})=0\}. \tag{6.53}$$

The Hamilton-Jacobi equation to solve is

$$-\frac{\partial V}{\partial t}=1+\left(\frac{\partial V}{\partial \mathbf{x}}\right)^T[\mathbf{F}(\mathbf{x})+\mathbf{A}\mathbf{x}]-\left(\frac{\partial V}{\partial \mathbf{x}}\right)^T\mathbf{B}(\mathbf{x})\text{sgn}\left(\mathbf{B}^T(\mathbf{x})\frac{\partial V}{\partial \mathbf{x}}\right). \tag{6.54}$$

Considere a continuous physics-consistent control function, such as $\mathcal{U}\cong\pm1$, and, $\Phi\subseteq\mathcal{U}_\varepsilon$, $\Phi\in C^k$, $k\ge1$. We have

$$u=-\Phi[\varphi(\mathbf{x})]=-\Phi\left(\mathbf{B}^T(\mathbf{x})\frac{\partial V}{\partial \mathbf{x}}\right),\ u\in U.$$

Solution of the Hamilton-Jacobi equation

$$-\frac{\partial V}{\partial t}=\Phi^T\left(\mathbf{B}^T(\mathbf{x})\frac{\partial V}{\partial \mathbf{x}}\right)\Sigma\Phi\left(\mathbf{B}^T(\mathbf{x})\frac{\partial V}{\partial \mathbf{x}}\right)+\left(\frac{\partial V}{\partial \mathbf{x}}\right)^T[\mathbf{F}(\mathbf{x})+\mathbf{A}\mathbf{x}]-\left(\frac{\partial V}{\partial \mathbf{x}}\right)^T\mathbf{B}(\mathbf{x})\Phi\left(\mathbf{B}^T(\mathbf{x})\frac{\partial V}{\partial \mathbf{x}}\right), \quad (6.56)$$

$$\left(\frac{\partial V}{\partial \mathbf{x}}\right)^T\mathbf{B}(\mathbf{x})\Phi\left(\mathbf{B}^T(\mathbf{x})\frac{\partial V}{\partial \mathbf{x}}\right)>\Phi^T\left(\mathbf{B}^T(\mathbf{x})\frac{\partial V}{\partial \mathbf{x}}\right)\Sigma\Phi\left(\mathbf{B}^T(\mathbf{x})\frac{\partial V}{\partial \mathbf{x}}\right),\ \Sigma>0,$$

exists, and, approximated by the continuous and differentiable return function

$$V(\mathbf{x})=\sum_{i=N_0,j=M_0}^{N,M}\tfrac{j}{2i}(\mathbf{x}^{i/j})^T K_{ij}(t)\mathbf{x}^{i/j}, \quad (6.57)$$

$$V(\mathbf{x})=\sum_{i,j}\kappa_{1_i}^T(\mathbf{x}^{<d_i>})K_{ij}(t)\kappa_{2_j}(\mathbf{x}^{<d_j>}),\ K_{ij}>0,\ \kappa_{1_i}^T(\mathbf{x}^{<d_i>})I\kappa_{2_j}(\mathbf{x}^{<d_j>})>0.$$

Example 6.20. *Tracking Control of Rotational Motion of a Missile*

Examples 5.33 and 5.36 investigated rotational motion of a satellite around the center of mass. Consider one-dimensional rotational dynamics of a maneuvering aerial vehicle. To control guided missiles, actuators deflect the aerodynamic surfaces, such as canards, wings and tail fins. The induced roll attenuated by roll stabilization. Roll control is accomplished by regulating the wing displacement δ, which is bounded, $\delta_{\min}\le\delta\le\delta_{\max}$, $u\equiv\delta$. A typifying governing equation is

$\dot{x}_1=-x_1-x_1|x_1|+u$, $-1\le u\le 1$,

$\dot{x}_2=x_1$, $y=x_2$, $e=(r-y)=(r-x_2)$.

Solve the tracking control problems. The extended system dynamics (6.49) is

$\dot{x}_1=-x_1-x_1|x_1|+u$,

$\dot{x}_2=x_1$,

$\dot{x}_e=r-y=r-x_2$,

$\dot{e}=-e+\dot{r}-\dot{x}_2=-e+\dot{r}-x_1$.

In the state-space form

$$\dot{\mathbf{x}}=\mathbf{F}(\mathbf{x})+\mathbf{A}\mathbf{x}+A_r r+\mathbf{B}u,\quad y=x_2,\ -1\le u\le 1 \qquad \begin{bmatrix}\dot{x}_1\\ \dot{x}_2\\ \dot{x}_e\\ \dot{e}\end{bmatrix}=\begin{bmatrix}-x_1|x_1|\\ 0\\ 0\\ 0\end{bmatrix}+\begin{bmatrix}-1&0&0&0\\ 1&0&0&0\\ 0&-1&0&0\\ -1&0&0&-1\end{bmatrix}\begin{bmatrix}x_1\\ x_2\\ x_e\\ e\end{bmatrix}+\begin{bmatrix}0&0\\ 0&0\\ 1&0\\ 0&1\end{bmatrix}\begin{bmatrix}r\\ \dot{r}\end{bmatrix}+\begin{bmatrix}1\\ 0\\ 0\\ 0\end{bmatrix}u.$$

Minimize $J=\min_{\mathbf{x},-1\le u\le 1}\int_0^\infty[\omega(\mathbf{x})+1]dt$ subject to the system dynamics.

Finding the resulting Hamiltonian (6.50), apply the first-order necessary condition for optimality (6.22). A discontinuous control law (6.51) is

$$u=-\operatorname{sgn}[\varphi(\mathbf{x})]=-\operatorname{sgn}\left(\mathbf{B}^T\frac{\partial V}{\partial \mathbf{x}}\right)=-\operatorname{sgn}\left(\frac{\partial V}{\partial x_1}\right).$$

For a continuous control $u=-\Phi[\varphi(\mathbf{x})]$, $u=-\Phi\left(\mathbf{B}^T\frac{\partial V}{\partial \mathbf{x}}\right)=-\tanh\left(a\frac{\partial V}{\partial x_1}\right)$, $a\gg 1$.

The Hamilton-Jacobi equation is

$$-\frac{\partial V}{\partial t}=\tfrac{1}{2}\mathbf{x}^T Q\mathbf{x}+\sigma\tanh^2\left(a\frac{\partial V}{\partial x_1}\right)-\frac{\partial V}{\partial x_1}(x_1+x_1|x_1|)+\frac{\partial V}{\partial x_2}x_1-\frac{\partial V}{\partial x_e}x_2-\frac{\partial V}{\partial e}(x_1+e)-\frac{\partial V}{\partial x_1}\tanh\left(a\frac{\partial V}{\partial x_1}\right).$$

Solution of the Hamilton-Jacobi equation is approximated by the quadratic return function $V(\mathbf{x})=\tfrac{1}{2}\mathbf{x}^T K\mathbf{x}$. Hence,

$$u=-\Phi[\varphi(\mathbf{x})]=-\Phi\left(\mathbf{B}^T\frac{\partial V}{\partial \mathbf{x}}\right)=-\tanh\left(a\frac{\partial V}{\partial x_1}\right)=-\tanh[a(k_{11}x_1+k_{12}x_2+k_{13}x_3+k_{14}x_4)].$$

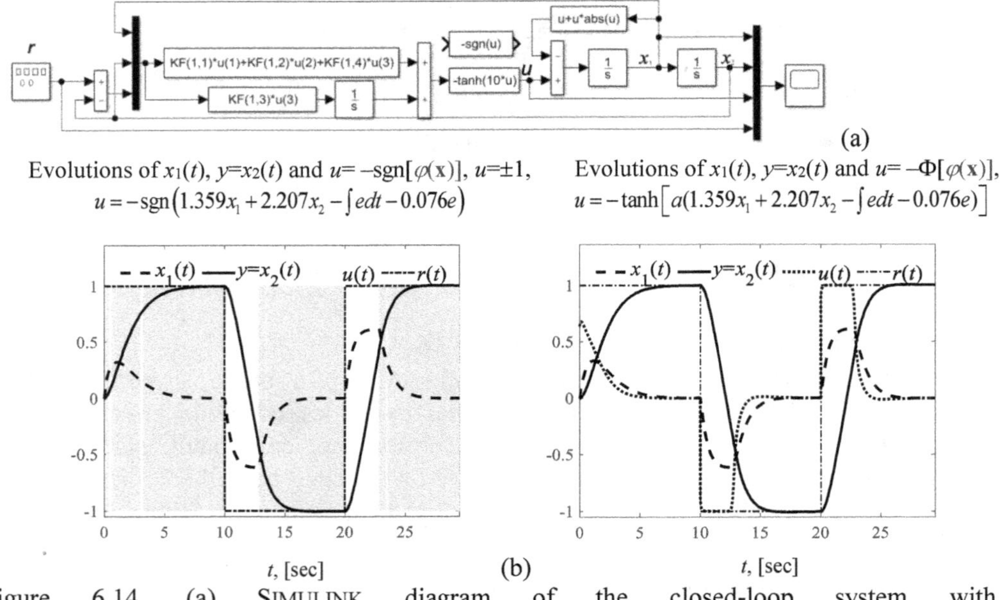

Figure 6.14. (a) SIMULINK diagram of the closed-loop system with $u=-\text{sgn}\left(1.359x_1+2.207x_2-\int edt-0.076e\right)$ and $u=-\tanh\left[a(1.359x_1+2.207x_2-\int edt-0.076e)\right]$, $a=10$; (b) Closed-loop system dynamics with discontinuous tracking control $u=-\phi[\varphi(\mathbf{x})]$, $u=\pm1$, and, continuous control $u=-\Phi[\varphi(\mathbf{x})]$. State $x_1(t)$, output $y=x_2(t)$ and control $u(t)$ evolve for $r(t)=\pm1$.

For $a=10$ and $Q=\begin{bmatrix}0&0&0&0\\0&0&0&0\\0&0&1&0\\0&0&0&1\end{bmatrix}$, one obtains $K=\begin{bmatrix}1.359&2.2067&-1&-0.0757\\2.2067&4.4483&-2.4347&0.2428\\-1&-2.4347&2.2067&-0.0757\\-0.0757&0.2428&-0.0757&0.497\end{bmatrix}$.

We find
$$u=-\text{sgn}\left(1.359x_1+2.207x_2-\int edt-0.076e\right),$$
and
$$u=-\tanh\left[a(1.359x_1+2.207x_2-\int edt-0.076e)\right],\ a=10.$$

The closed-loop system is simulated with discontinuous and continuous control laws. SIMULINK diagram is shown in Figure 6.14.a. Figure 6.14.b illustrates dynamics of $x_1(t)$ and $y(t)=x_2(t)$ if $r(t)=\pm1$. Evolutions of discontinuous $u=-\phi[\varphi(\mathbf{x})]$, $u=\pm1$ and continuous $u=-\Phi[\varphi(\mathbf{x})]$ control are reported. ■

6.7. Constrained Control of Systems

Solution of a constrained optimization problem using calculus of variations and maximum principle is complex, and, demonstrated for low-order systems. Solve a dynamic optimization problem by minimizing physics and design-consistent functionals [11-13]. Consider a system with symmetric $|\mathcal{U}_{\max}|=|\mathcal{U}_{\min}|$ and asymmetric $|\mathcal{U}_{\max}|\neq|\mathcal{U}_{\min}|$ limits $\mathcal{U}_{\min}\leq\mathcal{U}\leq\mathcal{U}_{\max}$. System governance is described as

$$\dot{x}(t)=F(x)+B(x)u,\ x(t_0)=x_0,\ u=\Phi[\varphi(x)],\ u_{\min}\leq u\leq u_{\max},\ u\in U,\ t\geq0, \tag{6.58}$$

where $\Phi(\cdot)$ is the bounded, real-valued, globally Lipschitz continuous function which describes physical limits, such as $\Phi\subseteq\mathcal{U}_\varepsilon$, $\Phi\in C^k$, $k\geq1$.

Symmetric and Asymmetric Control Limits, and, Bounds Description – Symmetric limits $\mathcal{U}_{\min}\leq\mathcal{U}\leq\mathcal{U}_{\max}$, $|\mathcal{U}_{\max}|=|\mathcal{U}_{\min}|$ have been studied. Asymmetric limits $\mathcal{U}_{\min}\leq\mathcal{U}\leq\mathcal{U}_{\max}$, $|\mathcal{U}_{\max}|\neq|\mathcal{U}_{\min}|$ are exhibited by electromagnetic, electrostatic, electronic, photonic, hydraulic, pneumatic, thermal and other devices.

For $\mathcal{U}$, find descriptive $\Phi(\cdot)$, and, design a continuous bounded control

$$u=-\underset{\substack{\text{Physical Limits } \mathcal{U}_{\min}\leq\mathcal{U}\leq\mathcal{U}_{\max} \\ \text{Limit Description } \Phi\subseteq\mathcal{U}_{\varepsilon} \\ u_{\min}\leq u\leq u_{\max}}}{\Phi}\underset{\substack{\text{Nonlinear Feedback} \\ \text{Structure} \\ \varphi(x)\equiv G^{-1}B^{T}(x)\frac{\partial V}{\partial x}}}{\left[\varphi(x)\right]},\quad u_{\min}<\Phi<u_{\max},\ u\in U\subset\mathbb{R}^{m}, \tag{6.59}$$

$$\min_{x\in X}\int_{-\infty}^{\infty}\left\|\mathcal{U}-\Phi(x)\right\|_{p}dx,\ \left\|\mathcal{U}-\Phi(x)\right\|_{p}\leq\varepsilon,\ \Phi\subseteq\mathcal{U}_{\varepsilon},\ p\geq1.$$

The Hausdorff distance, Fréchet measure and other estimates are applied to minimize $\min_{\Phi\subseteq\mathcal{U}}\left\|\mathcal{U}-\Phi[\varphi(x)]\right\|_{p}$ on a class of bounded, continuous and invertable functions $\Phi\in C^{k}, k\geq1$.

6.7.1. Physical Limits and Bounds on Control

Continuous, piecewise continuous or discontinuous symmetric and asymmetric limits $\mathcal{U}_{\min}\leq\mathcal{U}\leq\mathcal{U}_{\max}$ are described by bounded functions $\Phi(\cdot):\mathbb{R}^{n}\to\mathbb{R}^{m}$, $\Phi\subseteq\mathcal{U}_{\varepsilon}$, $\forall x\in X$ [3, 4, 11]. Continuous algebraic and transcendental (exponential, hyperbolic, logarithmic and trigonometric) functions $\Phi\in C^{k}$ are applied.

Limits and Continuous Bounded Control – Necessary and sufficient conditions that a control function $u(\cdot):\mathbb{R}^{n}\to\mathbb{R}^{m}$ is continuous is continuity of $\Phi(\cdot)$ for all $x\in X$ on $[t_0,t_f)$. We use a Lebesgue measurable, Lipschitz function Φ for which there exists an inverse, continuous and integrable function Φ^{-1}, $(\Phi^{-1})^{-1}=\Phi$. Continuous control $u=\Phi[\varphi(x)]$, $u(\cdot):\mathbb{R}^{n}\to\mathbb{R}^{m}$ is a differentiable function whose domain and range are sets X and U. Design a constrained control lawa within $U=\{u\in\mathbb{R}^{m}\,|\,u_{\min}\leq u\leq u_{\max}\}$, $u\in U\subset\mathbb{R}^{m}$.

Example 6.21. *Symmetric Limits*

For symmetric physical limits, $\mathcal{U}_{\min}\leq\mathcal{U}\leq\mathcal{U}_{\max}$, $|\mathcal{U}_{\max}|=|\mathcal{U}_{\min}|$. The sign, sigmoid, relay, saturation and other limits are described by continuous and invertable functions $\Phi\subseteq\mathcal{U}_{\varepsilon}$. To design control laws $u\in U$, $U=\{u\in\mathbb{R}^{m}\ |\ u_{i\min}\leq u_i\leq u_{i\max},\ |u_{i\max}|=|u_{i\min}|,\ i=1,\ldots,m\}$, consider an infinitely-differentiable hyperbolic tangent function

$$u=\Phi(x)=u_{\max}\tanh^{2n+1}(ax),\ x\equiv\varphi(\partial V/\partial x),\ u_{\min}<\Phi<u_{\max},\ n=0,1,\ldots$$

Function $\Phi(x)$ has a range $(u_{\min},u_{\max})$ and domain $(-\infty,\infty)$. For saturation,

$$u=\underset{u_{\min}\leq u\leq u_{\max}}{\Phi}[\varphi(x)],\ u=-\max\left[u_{\min},\min(u_{\max},\varphi(x))\right]=\begin{cases} u_{\max},\ \varphi(x)\geq u_{\max} \\ -\varphi(x),\ u_{\min}<\varphi(x)<u_{\max},\ u_{\min}<u<u_{\max}. \\ u_{\min},\ \varphi(x)\leq u_{\min}\end{cases}$$

Hyperbolic tangent, arctangent, Gudermannian $\text{gd}(x)=\arcsin(\tanh(x))$ and other algebraic functions approximate continuous, piecewise continuous and discontinuous limits $\mathcal{U}$, $\Phi\subseteq\mathcal{U}_{\varepsilon}$. For $-1\leq\mathcal{U}\leq1$, one may consider $-1<\Phi<1$ as

$$\Phi(x)=\text{erf}\left(\tfrac{\sqrt{\pi}}{2}x\right),\ \Phi(x)=\tfrac{2}{\pi}\arctan\left(\tfrac{\pi}{2}x\right),\ \Phi(x)=\tfrac{2}{\pi}\text{gd}\left(\tfrac{\pi}{2}x\right),$$

$$\Phi(x)=\frac{x}{\sqrt{1+x^{2}}},\ \Phi(x)=\frac{x}{1+|x|},\ \Phi(x)=\frac{x}{(1+|x|^{k})^{1/k}}.$$

Hyperbolic Tangent – For the hyperbolic tangent
$\Phi(x)=\tanh(x)$, $-1<\Phi<1$, $\Phi(0)=0$, $|\Phi(x)|<1$, $\Phi'(0)=1$, $\lim\limits_{|x|\to\infty} = \Phi'(x)=0$, $\int|\Phi'(x)|dx<\infty$.

Function $\Phi(x)=u_{\max}\tanh^{2n+1}(ax)$, $x\equiv\varphi(\partial V/\partial x)$, $a>0$ describes:

(1) Symmetric saturation limits, $n\geq 0$, $a>0$, $u_{\min}\leq u\leq u_{\max}$, $-u_{\max}<\Phi<u_{\max}$;

(2) Symmetric sign and relay limits, $n\geq 0$, $a\gg 1$, $u_{\min}\leq u\leq u_{\max}$, $-u_{\max}<\Phi<u_{\max}$.

Function $\Phi(x)=\tanh^{2n+1}ax$, $a>0$ has a range $(-1, 1)$, domain $(-\infty, \infty)$, and $\lim\limits_{x\to\pm\infty}\tanh^{2n+1}(ax)\to\pm 1$.

For $u=\Phi(x)=\tanh ax=\dfrac{\sinh ax}{\cosh ax}=\dfrac{e^{ax}-e^{-ax}}{e^{ax}+e^{-ax}}$, the inverse function is $\Phi^{-1}(u)=\frac{1}{a}\tanh^{-1}u=\frac{1}{2a}\ln\left(\frac{1+u}{1-u}\right)$.

One has $\dfrac{d}{dx}\tanh(ax)=a\mathrm{sech}^2(ax)$, $\dfrac{d}{du}\tanh^{-1}au=\dfrac{a}{1-a^2u^2}$,

$\int\tanh ax dx=\frac{1}{a}\ln(\cosh ax)$, $\int\tanh^{-1}(au)du=u\tanh^{-1}(au)+\frac{1}{2a}\ln(1-a^2u^2)$.

For the sign, relay and saturation limits $-1\leq\mathcal{U}\leq 1$, plots for $\Phi\subseteq\mathcal{U}_\varepsilon$ are depicted in Figures 6.15.a and b. For continuous $\Phi(x)=\tanh^{2n+1}ax$, an inverse and integrable function $\Phi^{-1}(u)$ exists.

If $\Phi(x)=\tanh^9(ax)$, $n=4$, one has

$$\Phi^{-1}(u)=\tfrac{1}{a}\tanh^{-1}\sqrt[9]{u},\ \tfrac{1}{a}\int\tanh^{-1}\sqrt[9]{u}du=\tfrac{1}{a}\left[u\tanh^{-1}\sqrt[9]{u}+\tfrac{1}{2}\ln(1-u^{\frac{2}{9}})+\tfrac{1}{24}u^{\frac{2}{9}}(4u^{\frac{4}{9}}+3u^{\frac{2}{3}}+6u^{\frac{2}{9}}+12)\right].$$

For the considered (Φ, Φ^{-1}), the domain and range are:

1. Domain $\tanh(x):\mathbb{R}\to(-\infty,\infty)$, $\tanh^{-1}(x):\mathbb{R}\to(-1, 1)$;
2. Range $\tanh(x):\mathbb{R}\to(-1, 1)$, $\tanh^{-1}(x):\mathbb{R}\to(-\infty,\infty)$.

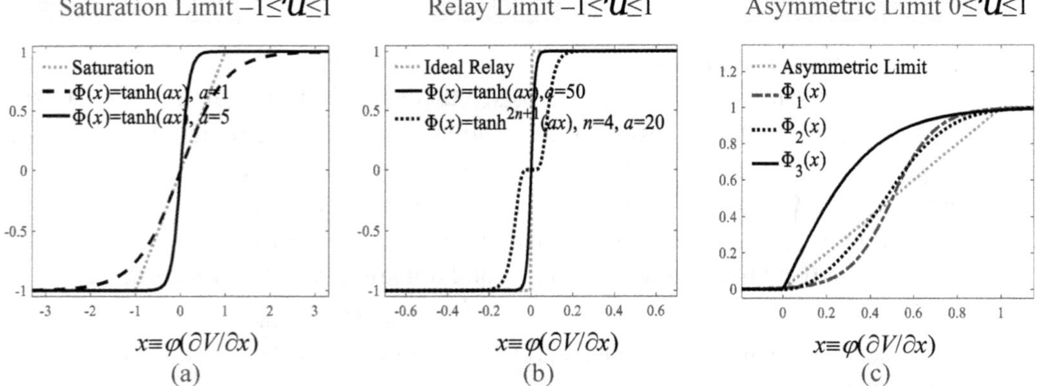

Figure 6.15. Symmetric $-1\leq\mathcal{U}\leq 1$ and asymmetric $0\leq\mathcal{U}\leq 1$ physical limits, described by continuous bounded functions $\Phi(x)$, $\Phi\subseteq\mathcal{U}_\varepsilon$:
(a) Saturation $-1\leq\mathcal{U}\leq 1$, described by range-restricted $\Phi(x)=\tanh(ax)$, $a=1$ and $a=5$, $-1<\Phi<1$;
(b) Ideal relay and relay with *dead zone* $-1\leq\mathcal{U}\leq 1$, described by $\Phi(x)=\tanh^{2n+1}(ax)$. Plots for $\Phi(x)=\tanh(ax)$, $a=50$, and, $\Phi(x)=\tanh^{2n+1}(ax)$, $n=4$, $a=20$, $-1<\Phi<1$;
(c) Asymmetric limits $0\leq\mathcal{U}\leq 1$, described by $0<\Phi<1$,
$\Phi_1(x)=\dfrac{1}{1+e^{-a(x-b)}}\Big|_{a=10,\ b=0.5}$, $\Phi_2(x)=\dfrac{e^{ax^2}-e^{-ax|x|}}{e^{ax^2}+e^{-ax|x|}}$, $a=2.5$, $\Phi_3(x)=\tanh\left[\frac{1}{2}a(|x|+x)\right]$, $a=2.5$. ■

Example 6.22. *Asymmetric Control Limits*

Typical asymmetric limits in physical systems are $0\le\mathcal{U}\le\mathcal{U}_{\max}$. One finds $0\le\Phi\le u_{\max}$, $\Phi\subseteq\mathcal{U}_\varepsilon$ to design control laws

$u\in U,\ 0\le u_i\le u_{i\max},\ i=1,\dots,m.$

Limits $0\le\mathcal{U}\le 1$ are described by continuous, differentiable and integrable functions $\Phi\subseteq\mathcal{U}_\varepsilon$ as

$$\Phi_1(x)=\frac{1}{1+e^{-a(x-b)}},\ \Phi_2(x)=\frac{e^{ax^2}-e^{-ax|x|}}{e^{ax^2}+e^{-ax|x|}},\ \Phi_3(x)=\tanh\left[\tfrac{1}{2}a\left(|x|+x\right)\right], 0<\Phi<1,\ (a,b)\in\mathbb{R}.$$

Plots for these $\Phi(x)$ are reported in Figure 6.15.c. ■

6.7.2. Design of Constrained Control Laws

To design control laws for physical systems with limits $\mathcal{U}_{\min}\le\mathcal{U}\le\mathcal{U}_{\max}$, derive physics-consistent functionals

$$J=\min_{x,u\in U}\int_{t_0}^{t_f}\left[\omega_x(x)+\omega_u(u)\right]dt$$

with positive definite integrands $(\omega_x(x),\omega_u(u))>0$. Minimize functionals subject to system models with $\Phi\subseteq\mathcal{U}_\varepsilon$, and, find a closed-form solution.

Example 6.23. *Quadratic Performance Integrands*

In chapter 5, the LQR problem is solved by minimizing the quadratic functional $J=\min_{x,u}\int_{t_0}^{t_f}\tfrac{1}{2}\left(x^TQx+u^TGu\right)dt$ using the energy-narrated integrands $\tfrac{1}{2}x^TQx$ and $\tfrac{1}{2}u^TGu$. Weighting matrices ($Q\ge0$, $G>0$) affect feedback gains, state evolutions $x(t)$, etc. The diagonal matrices $Q\in\mathbb{R}^{n\times n}$ and $G\in\mathbb{R}^{m\times m}$ are

$$Q=\begin{bmatrix}q_{11}&\cdots&0\\ \vdots&\ddots&\vdots\\ 0&\cdots&q_{nn}\end{bmatrix}=\begin{bmatrix}\frac{1}{\tilde{x}_{1\max}^2}&\cdots&0\\ \vdots&\ddots&\vdots\\ 0&\cdots&\frac{1}{\tilde{x}_{n\max}^2}\end{bmatrix},\ G=\begin{bmatrix}g_{11}&\cdots&0\\ \vdots&\ddots&\vdots\\ 0&\cdots&g_{mm}\end{bmatrix}=\begin{bmatrix}\frac{1}{\tilde{u}_{1\max}^2}&\cdots&0\\ \vdots&\ddots&\vdots\\ 0&\cdots&\frac{1}{\tilde{u}_{m\max}^2}\end{bmatrix}.$$

Integrands define and scale energy terms by applying the factoring coefficients which depend on the energy-associated coefficients (p_{ii},l_{ii}) such as m, J, k_s, r, L, C, etc. In the quadratic integrands $\tfrac{1}{2}x^TQx$ and $\tfrac{1}{2}u^TGu$, the parameterized weighting matrices are

$$Q=\begin{bmatrix}p_{11}q_{11}&\cdots&0\\ \vdots&\ddots&\vdots\\ 0&\cdots&p_{nn}q_{nn}\end{bmatrix},\ G=\begin{bmatrix}l_{11}g_{11}&\cdots&0\\ \vdots&\ddots&\vdots\\ 0&\cdots&l_{mm}g_{mm}\end{bmatrix},\ q_{ii}\ge0,\ p_{ii}>0,\ g_{ii}>0,\ l_{ii}>0.$$ ■

Performance Integrands, Energy Measures and Limits – Using kinetic, potential and dissipating energies (Γ,Π,D), positive definite integrands $\omega_x(\cdot)$ and $\omega_u(\cdot)$ assess the state- and control-dependent energies. These $\omega_x(\cdot)$ and $\omega_u(\cdot)$ are scaled using the expected evolutions and variations of $(\tilde{x}_{i\max},\tilde{u}_{i\max})$. The kinetic energy $\Gamma(x)$, stored in the moving mass and inductor, are $\tfrac{1}{2}mv^2$, $\tfrac{1}{2}J\omega^2$ and $\tfrac{1}{2}Li^2$. The potential energy of the spring and capacitor are $\Pi\sim f(x^2,x^4,\dots)$, $\Pi\sim f(\theta^2,\theta^4,\dots)$ and $\tfrac{1}{2}CV^2$. The dissipating energy of a moving mass with nonlinear thermodynamics, friction and damping is a nonlinear function of velocity $D\sim f(v^{1/3},v^{4/3},v^2,\dots)$, $D\sim f(\omega^{1/3},\omega^{4/3},\omega^2,\dots)$, etc. One finds the state- and control-depended positive definite and bounded total energies $0\le E_x(x)\le E_{x\max}$ and $0\le E_u(u)\le E_{u\max}$.

Example 6.24. *Nonquadratic Energy-Dependent Integrands*
Positive-definite energy-dependent integrand $\omega_x(x)>0$, $\forall x\in X$ can be explicitly expressed by high-degree and fractional multivariate polynomials. Monomials and polynomials are reported in Example 6.10.
Scalar $\omega_x(x)$ can be expressed as product of a vector $x^{\langle l\rangle}$. Using a square matrix Q_l,

$$\omega_x(x)=(x^{\langle l\rangle})^T Q_d x^{\langle l\rangle}=\sum_{i,j} q_{ij}x_i^{\langle l\rangle}x_j^{\langle l\rangle},\ \omega_x(x)>0,\ \forall x\in X,\ q_{ij}\geq 0,\ l\in\mathbb{N},$$

$$\omega_x(x)=(x^{\langle l_1\rangle})^T Q_l x^{\langle l_2\rangle},\ \omega_x(x)>0,\ \forall x\in X,\ (l_1,l_2)\in\mathbb{N}^0.$$

Integrand With Quadratic, Quartic and Sextic Terms – Consider positive definite integrands $\omega_x(x)$ on two states $x=[x_1\ \ x_2]^T$. As examples, the multivariate polynomials with quadratic, quartic and sextic terms are

$$\omega_x(x)=\begin{bmatrix} x_1 & x_2 & x_1^2 & x_2^2\end{bmatrix}\begin{bmatrix} q_{11} & 0 & 0 & 0\\ 0 & q_{22} & 0 & 0\\ 0 & 0 & q_{33} & q_{34}\\ 0 & 0 & q_{43} & q_{44}\end{bmatrix}\begin{bmatrix} x_1\\ x_2\\ x_1^2\\ x_2^2\end{bmatrix}=q_{11}x_1^2+q_{22}x_2^2+q_{33}x_1^4+2q_{34}x_1^2x_2^2+q_{44}x_2^4,\ Q_l\in\mathbb{R}^{4\times 4},\ q_{ij}=q_{ji},$$

$$\omega_x(x)=\begin{bmatrix} x_1 & x_2 & x_1^2 & x_2^2\end{bmatrix}\begin{bmatrix} q_{11} & 0 & 0 & 0\\ 0 & q_{22} & 0 & 0\\ 0 & 0 & q_{33} & q_{34}\\ 0 & 0 & q_{43} & q_{44}\end{bmatrix}\begin{bmatrix} x_1\\ x_2\\ x_1^4\\ x_2^4\end{bmatrix}=q_{11}x_1^2+q_{22}x_2^2+q_{33}x_1^6+q_{43}x_1^4x_2^2+q_{34}x_1^2x_2^4+q_{44}x_2^6.\qquad\blacksquare$$

Control Limits, Energy, and Control Integrand – Laws of physics and device physics yield energy and control limits $0\leq E_u\leq E_{u\max}$ and $\mathcal{U}_{\min}\leq\mathcal{U}\leq\mathcal{U}_{\max}$. For example, applied force and torque are found using the coenergy,

$$F=-\nabla W_c,\ T=-\nabla W_c.$$

Define a positive definite, limit-defined $(u_{\min},u_{\max})$ and domain-specific real-valued integrand $\omega_u(\cdot):\mathbb{R}^m\to\mathbb{R}$ as

$$\omega_u(u)=\int\left(\Phi^{-1}(u)\right)^T G du,\ \omega_u(u)>0,\ G>0,\ \forall u\in U.$$

Real-valued odd functions $\Phi(\cdot)$ and $\Phi^{-1}(\cdot)$ lie in the first and third quadrants. Integrand $\omega_u(u)=\int\left(\Phi^{-1}(u)\right)^T G du$ is explicitly defined for $u_{\min}\leq u\leq u_{\max}$, and, $\omega_u(u)>0$, $\forall u\in U$, see Examples 6.25, 6.26 and 6.27. Real-valued performance integrands are derived by applying the energy-dependent and specified-by-constrains positive definite $\omega_x(x)$ and $\omega_u(u)$. Minimize a nonquadratic functional

$$J=\min_{x,u\in U}\int_{t_0}^{t_f}\left[\omega_x(x)+\omega_u(u)\right]dt=\min_{x,u\in U}\int_{t_0}^{t_f}\left[\omega_x(x)+\int\left(\Phi^{-1}(u)\right)^T G du\right]dt,\ \omega_x(x)>0, \omega_u(u)>0, \quad (6.60)$$

subject to system dynamics (6.58) $\dot{x}(t)=F(x)+B(x)u$.

The Hamiltonian function is

$$H(x,u,\tfrac{\partial V}{\partial x})=\omega_x(x)+\int\left(\Phi^{-1}(u)\right)^T G du+\left(\frac{\partial V}{\partial x}\right)^T\left[F(x)+B(x)u\right]. \quad (6.61)$$

The necessary condition $\dfrac{\partial H(x,u,\frac{\partial V}{\partial x})}{\partial u}=\left(\Phi^{-1}(u)\right)^T G+\left(\dfrac{\partial V}{\partial x}\right)^T B(x)=0$ yields a constrained control

$$u=-\Phi\left(G^{-1}B^T(x)\frac{\partial V}{\partial x}\right),\ u=-\underset{\substack{\text{Physical Limits}\\ \mathcal{U}_{\min}\leq\mathcal{U}\leq\mathcal{U}_{\max}\\ \text{Limit Description }\Phi\subseteq\mathcal{U}_c\\ u_{\min}\leq u\leq u_{\max}}}{\Phi}\ \underset{\substack{\text{Nonlinear Feedback}\\ \text{Structure}\\ \varphi(x)=G^{-1}B^T(x)\frac{\partial V}{\partial x}}}{\left[\varphi(x)\right]}=-\Phi\left(G^{-1}B^T(x)\frac{\partial V}{\partial x}\right),\ u\in U. \quad (6.62)$$

An integrand $\omega_u(u)=G\int\tanh^{-1}u\,du$ yields analytically designed constrained control (6.62), guarantees $\frac{\partial^2 H}{\partial u^T\times\partial u}=\frac{\partial}{\partial u}\left(\Phi^{-1}(u)\right)^T G>0$, consistent with limits imposed, and, ensures a closed-form solution of the Hamilton-Jacobi equation

$$-\frac{\partial V}{\partial t}=\omega_x(x)+\int\left(\Phi^{-1}(u)\right)^T G\,du+\left(\frac{\partial V}{\partial x}\right)^T\left[F(x)+B(x)u\right]. \tag{6.63}$$

Integration by parts $\int v\,dw=vw-\int w\,dv$ yields

$$\int\left(\Phi^{-1}(u)\right)^T G\,du=\int\left[\Phi^{-1}\left(\Phi\left(G^{-1}B^T(x)\frac{\partial V}{\partial x}\right)\right)\right]^T G\,d\left[\Phi\left(G^{-1}B^T(x)\frac{\partial V}{\partial x}\right)\right]$$
$$=\left(\frac{\partial V}{\partial x}\right)^T B(x)\Phi\left(G^{-1}B^T(x)\frac{\partial V}{\partial x}\right)-\int\Phi^T\left(G^{-1}B^T(x)\frac{\partial V}{\partial x}\right)d\left(B^T(x)\frac{\partial V}{\partial x}\right). \tag{6.64}$$

From (6.63) and (6.64) one obtains the Hamilton-Jacobi equation

$$-\frac{\partial V}{\partial t}=\omega_x(x)+\left(\frac{\partial V}{\partial x}\right)^T F(x)-\int\Phi^T\left(G^{-1}B^T(x)\frac{\partial V}{\partial x}\right)d\left(B^T(x)\frac{\partial V}{\partial x}\right). \tag{6.65}$$

Positive definite $\int\Phi^T\left(G^{-1}B^T(x)\frac{\partial V}{\partial x}\right)d\left(B^T(x)\frac{\partial V}{\partial x}\right)$ is found for a given real-valued continuous $\Phi(\cdot)$, admits series approximations or approximating kernels to ensure a solution of (6.65). The positive definite real-valued, continuous and differentiable return functions $V(t,x)$ are (6.24)

$$V(t,x)=\sum_{i=N_0,j=M_0}^{N,M}\frac{j}{2i}(x^{i/j})^T K_{ij}(t)x^{i/j},\ V(t,x)=\sum_{i,j}\kappa_{1_i}^T(t,x^{<d_i>})K_{ij}(t)\kappa_{2_j}(t,x^{<d_j>}),\ K_{ij}(\cdot):\mathbb{R}_{\geq 0}\to\mathbb{R}^{n\times n}.$$

Return function $V(t,x)$, given by real-valued continuous and differentiable monomials in multivariate polynomials, depends on $F(x)$, $B(x)$, $\Phi\subseteq\mathcal{U}_\varepsilon$, $\omega_x(x)$, $\int\Phi^T\left(G^{-1}B^T(x)\frac{\partial V}{\partial x}\right)d\left(B^T(x)\frac{\partial V}{\partial x}\right)$, etc. A monomial is a product of variables x and $\kappa(x)$ with nonnegative integer exponents, including repetitions.

Example 6.25. *Performance Integrands*

Consider a symmetric saturation limit $-1\leq\mathcal{U}\leq 1$. Let $\Phi(x)=\tanh(x)$, $x\equiv\varphi(\partial V/\partial x)$, $\Phi\subseteq\mathcal{U}_\varepsilon$. A bounded control function

$$u=\Phi(x)=\tanh(x),\ x\equiv\varphi(\partial V/\partial x)\ \text{or}\ u=-\Phi\left(G^{-1}B^T(x)\frac{\partial V}{\partial x}\right)=-\tanh\left(G^{-1}B^T(x)\frac{\partial V}{\partial x}\right)$$

is found by minimizing (6.60) with $\omega_u(u)=G\int\tanh^{-1}u\,du$, $|u|<1$.

Plots for the defined-by-limits and specified-by-constrains integrand

$$\omega_u(u)=G\int\tanh^{-1}u\,du=G\left[u\tanh^{-1}u+\tfrac{1}{2}\ln(1-u^2)\right],\ |u|<1,\ G=1,$$

as well as for a quadratic integrand $\omega_u(u)=\frac{1}{2}u^2$, are reported in Figure 6.16.a.

Example 6.26. *Symmetric Limit*

For a saturation $-\mathcal{U}_{\max}\leq\mathcal{U}\leq\mathcal{U}_{\max}$, consider $\Phi(x)=u_{\max}\tanh(x)$ with range $(-u_{\max}, u_{\max})$ and domain $(-\infty,\infty)$, and, $\Phi^{-1}(u)=\tanh^{-1}\left(\frac{1}{u_{\max}}u\right)$ with range $(-\infty,\infty)$ and domain $(-u_{\max}, u_{\max})$.

A positive-definite integrand $\omega_u(u)>0$, $\forall u\in U$, $-u_{\max}<u<u_{\max}$, is

$$\omega_u(u)=G\int\tanh^{-1}\left(\frac{1}{u_{\max}}u\right)du=G\left[u\tanh^{-1}\left(\frac{1}{u_{\max}}u\right)+\frac{u_{\max}}{2}\ln\left(u_{\max}^2-u^2\right)\right],$$

Function $\Phi(x)=u_{\max}\tanh(ax)$, $a>0$ is studied in Example 6.33. ■

Performance Integrands $\omega_u(u)=\frac{1}{2}u^2$ and $\omega_u(u)=\int \tanh^{-1} u du = u\tanh^{-1} u + \frac{1}{2}\ln(1-u^2)$

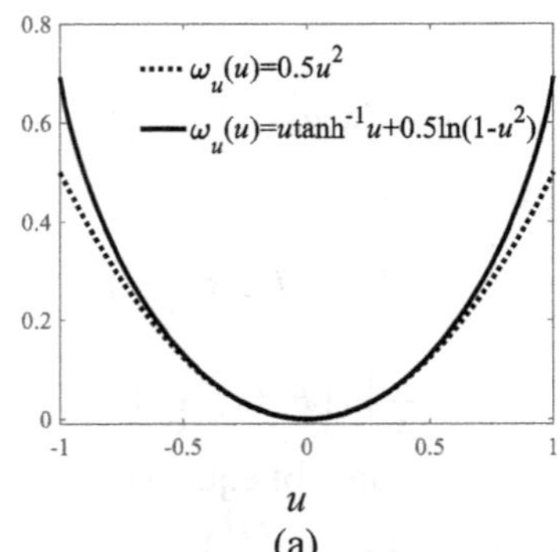

(a)

Asymmetric Bounds: Plots for $\Phi(x)=\dfrac{1}{1+e^{-a(x-b)}}$ and $\Phi^{-1}(u)=\frac{1}{a}\left[-\ln\left(\frac{1-u}{u}\right)+ab\right]$

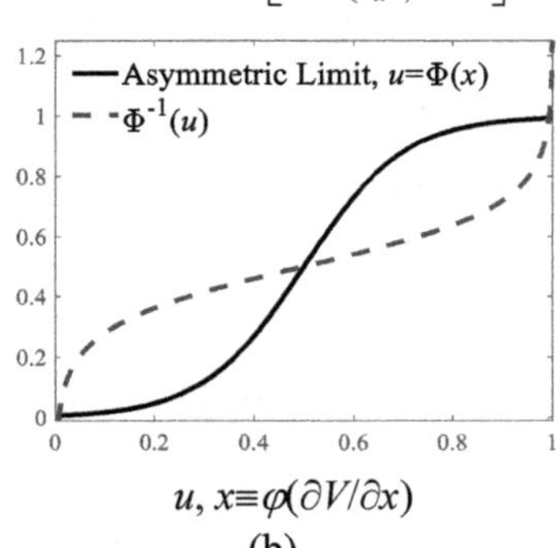

(b)

Asymmetric Bounds: Plots for $\omega_u(u)=\int \frac{1}{a}\left[-\ln\left(\frac{1-u}{u}\right)+ab\right]du = bu-\frac{1}{a}u\ln\left(\frac{1-u}{u}\right)+\frac{1}{a}\ln\left|1-u\right|$ and $\int \Phi(x)dx=\int \dfrac{1}{1+e^{-a(x-b)}}dx = x-b+\frac{1}{a}\ln\left|1+e^{-a(x-b)}\right|$

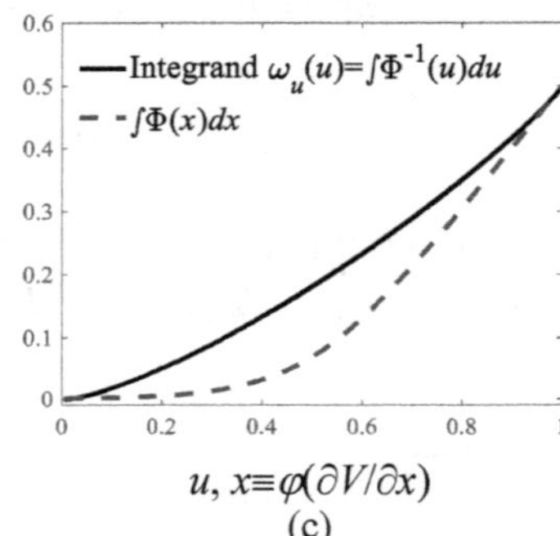

(c)

Functions ln(cosh(x)) and Approximation $\ln(\cosh(x))\approx\frac{1}{3}x^2$

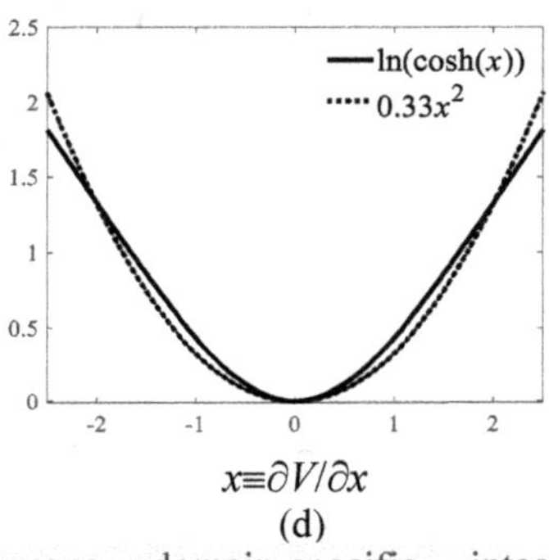

(d)

Figure 6.16. (a) Plots for defined-by-limits continuous domain-specific integrand $\omega_u(u)=G\int \tanh^{-1} u du = G\left[u\tanh^{-1} u+\frac{1}{2}\ln(1-u^2)\right]$, $|u|<1$, G=1, and, quadratic integrand $\omega_u(u)=\frac{1}{2}u^2$;

(b) Asymmetric limit $0\le \mathcal{U}\le 1$, described by $\Phi(x)=\dfrac{1}{1+e^{-a(x-b)}}$, a=10, b=0.5, $0<\Phi<1$, with $\Phi^{-1}(u)=\frac{1}{a}\left[-\ln\left(\frac{1-u}{u}\right)+ab\right]$, $0<u<1$, Example 6.27;

(c) Plot for a positive definite domain-specific continuous integrand $\omega_u(u)=\int \Phi^{-1}(u)du=\int \frac{1}{a}\left[-\ln\left(\frac{1-u}{u}\right)+ab\right]du = bu-\frac{1}{a}u\ln\left(\frac{1-u}{u}\right)+\frac{1}{a}\ln\left|1-u\right|$, and, plot for $\int \Phi(x)dx=\int \dfrac{1}{1+e^{-a(x-b)}}dx = x-b+\frac{1}{a}\ln\left|1+e^{-a(x-b)}\right|$, $x\equiv\varphi(B^T(x)\frac{\partial V}{\partial x})$, Example 6.27;

(d) Plots for ln(cosh(x)) and approximation $\ln(\cosh(x))\approx\frac{1}{3}x^2$, Example 6.28. ∎

Example 6.27. *Asymmetric Limit*
For an asymmetric limit $0\le\mathcal{U}\le 1$, see Example 6.22.

Investigate a continuous function $\Phi(x)=\dfrac{1}{1+e^{-a(x-b)}}$, $0<\Phi<1$, a=10, b=0.5.

We have $\Phi^{-1}(u)=\frac{1}{a}\left[-\ln\left(\frac{1-u}{u}\right)+ab\right]$, $0<u<1$.

Functions $\Phi(x)$ and $\Phi^{-1}(u)$ are plotted in Figure 6.16.b.

The positive definite integrand is

$\omega_u(u)=\int \Phi^{-1}(u)du=\int \frac{1}{a}\left[-\ln\left(\frac{1-u}{u}\right)+ab\right]du=bu-\frac{1}{a}u\ln\left(\frac{1-u}{u}\right)+\frac{1}{a}\ln|1-u|$, $0<u<1$.

In the Hamilton-Jacobi equation (6.65), the term is

$$\int\left[\Phi\left(G^{-1}B^T(x)\frac{\partial V}{\partial x}\right)\right]^T d\left(B^T(x)\frac{\partial V}{\partial x}\right)=\int \frac{1}{1+e^{-a\left((B^T(x)\frac{\partial V}{\partial x}-b)\right)}}d\left(B^T(x)\frac{\partial V}{\partial x}\right)$$

$$=B^T(x)\frac{\partial V}{\partial x}-b+\frac{1}{a}\ln\left|1+e^{-a\left((B^T(x)\frac{\partial V}{\partial x}-b)\right)}\right|,\ G=1.$$

Figure 6.16.c reports plots for $\omega_u(u)$ and $\int\Phi(x)dx=\int\frac{1}{1+e^{-a(x-b)}}dx=x-b+\frac{1}{a}\ln\left|1+e^{-a(x-b)}\right|$.■

Example 6.28. *System With a Saturation Limit on Control*

Consider a system with $-1\le \mathcal{U}\le 1$, described as

$\dot{x}(t)=ax+bu$, $b>0$, $-1<u<1$, $U=\{|u|<1\}$.

For $u=\Phi(x)=\tanh(x)$ with $\Phi^{-1}(u)=\tanh^{-1}(u)$, a functional (6.60) to minimize is

$$J=\min_{x,u\in U}\int_0^\infty(\underbrace{\tfrac{1}{2}qx^2}_{\omega_x(x)}+\underbrace{\int\tanh^{-1}udu}_{\omega_u(u)})dt\,.$$

A Hamiltonian function (6.61) is $H(x,u,\frac{\partial V}{\partial x})=\frac{1}{2}qx^2+\int\tanh^{-1}udu+\frac{\partial V}{\partial x}(ax+bu)$.

Apply the necessary condition for optimality (6.22). The constrained control (6.62) is

$u=-\tanh\left(b\frac{\partial V}{\partial x}\right)$.

The second-order necessary condition for optimality is satisfied because

$\frac{\partial^2 H}{\partial u^2}=\frac{\partial}{\partial u}\left(\tanh^{-1}(u)\right)=\frac{1}{1-u^2}>0$, $|u|<1$.

One finds (6.65)

$$-\frac{\partial V}{\partial t}=\frac{1}{2}qx^2+\frac{\partial V}{\partial x}ax-\int\tanh\left(b\frac{\partial V}{\partial x}\right)d\left(b\frac{\partial V}{\partial x}\right),\ \int\tanh(b\frac{\partial V}{\partial x})d(b\frac{\partial V}{\partial x})=\ln\left(\cosh(b\frac{\partial V}{\partial x})\right).$$

Hence, $-\frac{\partial V}{\partial t}=\frac{1}{2}qx^2+\frac{\partial V}{\partial x}ax-\ln\left(\cosh\left(b\frac{\partial V}{\partial x}\right)\right)$.

The Taylor, Mercator and other series are applied to approximate $\ln\left(\cosh(x)\right)$, $x\equiv b\frac{\partial V}{\partial x}$.

For small w, $|w-1|\le 1$, $w\ne 0$,

$$\ln w=(w-1)-\tfrac{1}{2}(w-1)^2+\tfrac{1}{3}(w-1)^3-...=\sum_{n=1}^{\infty}\tfrac{(-1)^{n-1}}{n}(w-1)^n,$$

$$\cosh x=1+\tfrac{1}{2!}x^2+\tfrac{1}{4!}x^4+...=\sum_{n=0}^{\infty}\tfrac{1}{2n!}x^{2n}.$$

We have $\ln\left(\cosh(x)\right)=\frac{1}{2}x^2-\frac{1}{12}x^4+\frac{1}{45}x^6-\frac{17}{2520}x^8+O(x)$,

$$\ln\left(\cosh(x)\right)=\ln\left(\tfrac{1}{2}(e^x+e^{-x})\right)=\ln\left(e^x+e^{-x}\right)-\ln(2)=x-\ln(2)+\ln\left(1+e^{-2x}\right)=x-\ln(2)+\sum_{k=1}^{\infty}(-1)^{k+1}\tfrac{1}{k}e^{-2kx}.$$

Depending on variations of $x\equiv G^{-1}B^T(x)\frac{\partial V}{\partial x}$, truncate a function $\ln\left(\cosh\left(G^{-1}B^T(x)\frac{\partial V}{\partial x}\right)\right)$.

For example, $\ln\left(\cosh\left(G^{-1}B^T(x)\frac{\partial V}{\partial x}\right)\right)=k\frac{\partial V}{\partial x}^T B(x)G^{-1}B^T(x)\frac{\partial V}{\partial x}$, $k>0$.

In (X,U), there are the lower and upper limits lim[inf(·)] and lim[sup(·)] for the range and domain of $\Phi(\cdot)$, $\Phi^{-1}(\cdot)$, $\varphi(\cdot)$ and $\int\left[\Phi\left(G^{-1}B^T(x)\frac{\partial V}{\partial x}\right)\right]^T d\left(B^T(x)\frac{\partial V}{\partial x}\right)$. The truncation error is estimated, and, solutions may be refined. For $\left|G^{-1}B^T(x)\frac{\partial V}{\partial x}\right| \le 2.5$, the least-squares approximation gives $\ln\left(\cosh\left(G^{-1}B^T(x)\frac{\partial V}{\partial x}\right)\right) = \frac{1}{3}\frac{\partial V}{\partial x}^T B(x)G^{-1}B^T(x)\frac{\partial V}{\partial x}$. Plots for $\ln(\cosh(x))$ and $\ln(\cosh(x)) \approx \frac{1}{3}x^2$, $|x|\le 2.5$ are reported in Figure 6.16.d.

Solve $\frac{1}{2}qx^2 + \frac{\partial V}{\partial x}ax = \frac{1}{3}\left(b\frac{\partial V}{\partial x}\right)^2$, $V(x) = \frac{1}{2}kx^2$.

From the quadratic equation $-0.33b^2k^2 + ak + \frac{1}{2}q = 0$, find $k>0$.

The constrained control law is

$$u = -\tanh(bkx),\ k>0,\ k = \frac{a + \sqrt{a^2 + 0.66b^2q}}{0.66b^2}.$$

The closed-loop system $\dot{x}(t) = ax - b\tanh(bkx)$ is stable for $\forall x \in X$ if $a<0$. If $a>0$, the sufficient condition for stability is $|b\tanh(bkx)|>ax$, $\forall x\in X$, $\forall t\in[t_0,t_f)$. ■

Example 6.29. *Second-Order System*

Consider a system with a saturation limit on control

$$\dot{x}_1(t) = ax_1 + bu,\ b>0,\ -1<u<1,\ U=\{|u|<1\},$$

$$\dot{x}_2(t) = x_1.$$

Use infinitely-differentiable hyperbolic functions $u=\Phi(x)=\tanh(x)$, $\Phi^{-1}(u)=\tanh^{-1}(u)$.

Minimize a positive definite functional

$$J = \min_{x,u\in U}\int_0^\infty \Big(\underbrace{\tfrac{1}{2}q_{11}x_1^2 + \tfrac{1}{2}q_{22}x_2^2}_{\omega_x(x)} + \underbrace{\int \tanh^{-1} u\,du}_{\omega_u(u)}\Big)dt.$$

The Hamiltonian (6.61) is

$$H(x,u,\tfrac{\partial V}{\partial x}) = \tfrac{1}{2}q_{11}x_1^2 + \tfrac{1}{2}q_{22}x_2^2 + \int \tanh^{-1} u\,du + \frac{\partial V}{\partial x_1}(ax_1 + bu) + \frac{\partial V}{\partial x_2}x_1.$$

From (6.22) $\partial H/\partial u=0$, we find a control law (6.62)

$$u = -\tanh\left(b\frac{\partial V}{\partial x_1}\right),\ u\in U.$$

Solve (6.65)

$$\tfrac{1}{2}q_{11}x_1^2 + \tfrac{1}{2}q_{22}x_2^2 + \frac{\partial V}{\partial x_1}ax_1 + \frac{\partial V}{\partial x_2}x_1 - \int \tanh\left(b\frac{\partial V}{\partial x_1}\right)d\left(b\frac{\partial V}{\partial x_1}\right) = 0.$$

From $\int \tanh\left(b\frac{\partial V}{\partial x_1}\right)d\left(b\frac{\partial V}{\partial x_1}\right) = \ln\left(\cosh\left(b\frac{\partial V}{\partial x_1}\right)\right)$, we have

$$-\frac{\partial V}{\partial t} = \tfrac{1}{2}q_{11}x_1^2 + \tfrac{1}{2}q_{22}x_2^2 + \frac{\partial V}{\partial x_1}ax_1 + \frac{\partial V}{\partial x_2}x_1 - \ln\left(\cosh\left(b\frac{\partial V}{\partial x_1}\right)\right).$$

Approximate a positive definite ln(cosh(x)) by using the least squares.

For $-1 \le b\frac{\partial V}{\partial x_1} \le 1$, $\ln\left(\cosh\left(b\frac{\partial V}{\partial x_1}\right)\right) \approx 0.45\left(b\frac{\partial V}{\partial x_1}\right)^2$.

Solution is found using $V = \frac{1}{2}k_{11}x_1^2 + k_{12}x_1x_2 + \frac{1}{2}k_{22}x_2^2$.

Assume $a=1$ and $b=1$. Let $q_{11}=1$ and $q_{22}=1$. One finds $k_{11}=3.28$, $k_{12}=1.05$, $k_{22}=2.05$.

The constrained control is

$$u = -\tanh(3.28x_1 + 1.05x_2).$$ ■

6.8. Constrained Tracking Control of Dynamic Systems

6.8.1. Design of Tracking Control Laws

Section 6.4 describes design of tracking control laws. Constrained optimization problem is formulated and solved in section 6.7. Physical limits on control $\mathcal{U}_{\min}\le\mathcal{U}\le\mathcal{U}_{\max}$ are described by a bounded continuous function $\Phi\subseteq\mathcal{U}_\varepsilon$, $u_{\min}\le\Phi\le u_{\max}$.

To analytically design tracking control laws, examine governance of system states $x(t)$, *error states* $x_e(t)$ and tracking errors $e(t)$. The extended system is

$$\dot{x}(t)=F(x)+B(x)u,\ u_{\min}\le u\le u_{\max},\ u\in U,\ y=Hx,\ e(t)=(r-y), \qquad (6.66)$$

$$\dot{x}_e(t)=I_X(r-y)=I_X(r-Hx),\ e(t)\equiv\dot{x}_e(t),$$

$$\dot{e}(t)=-I_Ee+I_E^Xx_e+\dot{r}-\dot{y}=-I_Ee+I_E^Xx_e+\dot{r}-HF(x)-HB(x)u,$$

$$\dot{\mathbf{x}}(t)=\begin{bmatrix}\dot{x}(t)\\ \dot{x}_e(t)\\ \dot{e}(t)\end{bmatrix}=\mathbf{F}(\mathbf{x})+\mathbf{A}\mathbf{x}+A_rr+\mathbf{B}(\mathbf{x})u=\begin{bmatrix}F(x)\\ 0\\ -HF(x)\end{bmatrix}+\begin{bmatrix}0&0&0\\ -I_XH&0&0\\ 0&I_E^X&-I_E\end{bmatrix}\mathbf{x}+\begin{bmatrix}0&0\\ I_X&0\\ 0&I\end{bmatrix}\begin{bmatrix}r\\ \dot{r}\end{bmatrix}+\begin{bmatrix}B(x)\\ 0\\ -HB(x)\end{bmatrix}u,$$

where $x\in X\subset\mathbb{R}^n$ is the state vector; $u\in U\subset\mathbb{R}^m$ is the control vector, and, the control function must be found within a set $U=\{u\in\mathbb{R}^m|\ u_{\min}\le u\le u_{\max}\}$; $y\in Y\subset\mathbb{R}^b$, $r\in R\subset\mathbb{R}^b$ and $e\in E\subset\mathbb{R}^b$ are the output, reference and tracking error vectors.

Example 6.30. *Physical and Mathematical Controllability*
Physical controllability of systems, states $x(t)$, outputs $y(t)$ and extended state vector $\mathbf{x}(t)=\begin{bmatrix}x(t)\\ x_e(t)\\ e(t)\end{bmatrix}$ are guaranteed because adequately designed physical systems are inherently controllable. Electromagnetic, mechanical, electromechanical, electronic and other systems are designed to guarantee controllability of processes and transductions. Model fidelity and descriptive complexity should not cause impediments. Mathematically, for (6.66), using the Lie bracket operator $\left[\mathrm{ad}_{\mathbf{F}}^k\mathbf{B}\right]=\left[\mathbf{F}\,...\,[\mathbf{F},\mathbf{B}]\right]$ one finds that $C=\left[\mathbf{B}\ldots[\mathrm{ad}_{\mathbf{F}}^k\mathbf{B}]\right]$ spans the $(n+2b)$ space with the rank $(n+2b)$. ■

Analytically design a control law. Minimize a performance functional

$$J=\min_{\mathbf{x},u\in U}\int_{t_0}^{t_f}\left[\omega_x(\mathbf{x})+\int\left(\Phi^{-1}(u)\right)^TGdu\right]dt,\ \omega_x(\mathbf{x})>0,\ G>0 \qquad (6.67)$$

subject to (6.66).

In (6.67), a positive definite $\omega_x(\mathbf{x})$ may be given as $\omega_x(\mathbf{x})=(\mathbf{x}^{<d_l>})^TQ\mathbf{x}^{<d_k>}$, $(l,k)\in\mathbb{N}$, where Q is a square matrix. The integrand $\omega_x(\mathbf{x})$ defines the specifications on the state dynamics $\mathbf{x}(t)$, as well as affects the solution of the Hamilton-Jacobi equation.

A scalar Hamiltonian function is

$$H(\mathbf{x},u,\tfrac{\partial V}{\partial\mathbf{x}})=\omega_x(\mathbf{x})+\int\left(\Phi^{-1}(u)\right)^TGdu+\left(\frac{\partial V}{\partial\mathbf{x}}\right)^T\left[\mathbf{F}(\mathbf{x})+\mathbf{A}\mathbf{x}+\mathbf{B}(\mathbf{x})u\right]. \qquad (6.68)$$

For (6.68), apply the necessary condition for optimality (6.22) $\dfrac{\partial H(\mathbf{x},u,\frac{\partial V}{\partial\mathbf{x}})}{\partial u}=0$.

A constrained tracking control law is

$$u=-\Phi\left(G^{-1}\mathbf{B}^T(\mathbf{x})\frac{\partial V}{\partial \mathbf{x}}\right),\ \mathbf{x}(t)=\begin{bmatrix}x(t)\\x_e(t)\\e(t)\end{bmatrix}=\begin{bmatrix}x(t)\\\int edt\\e(t)\end{bmatrix},\ u\in U. \tag{6.69}$$

Note that $\frac{\partial^2 H(\mathbf{x},u,\frac{\partial V}{\partial \mathbf{x}})}{\partial u\times\partial u^T}>0$. The Hamilton-Jacobi equation

$$-\frac{\partial V}{\partial t}=\omega_x(\mathbf{x})+\left(\frac{\partial V}{\partial \mathbf{x}}\right)^T[\mathbf{F}(\mathbf{x})+\mathbf{A}\mathbf{x}]-\int\Phi^T\left(G^{-1}\mathbf{B}^T(\mathbf{x})\frac{\partial V}{\partial \mathbf{x}}\right)d\left(\mathbf{B}^T(\mathbf{x})\frac{\partial V}{\partial \mathbf{x}}\right), \tag{6.70}$$

$$\int\Phi^T\left(G^{-1}\mathbf{B}^T(\mathbf{x})\frac{\partial V}{\partial \mathbf{x}}\right)d\left(\mathbf{B}^T(\mathbf{x})\frac{\partial V}{\partial \mathbf{x}}\right)>0$$

is solved using a positive definite return function $V(t,\mathbf{x})$

$$V(t,\mathbf{x})=\sum_{i=N_0,j=M_0}^{N,M}\tfrac{j}{2i}(\mathbf{x}^{i/j})^T K_{ij}(t)\mathbf{x}^{i/j},\ K_{ij}(\cdot):\mathbb{R}_{\ge 0}\to\mathbb{R}^{n\times n},\ K_{ij}>0,\ N_0\ge 1,\ M_0\ge 1,$$

$$V(t,\mathbf{x})=\sum_{i,j}\kappa_{1_i}^T(\mathbf{x}^{<d_i>})K_{ij}(t)\kappa_{2_j}(\mathbf{x}^{<d_j>}),\ (i,j)\in\mathbb{N}. \tag{6.71}$$

For a performance functional (6.67) $J=\min_{\mathbf{x},u\in U}\int_0^\infty\left[\omega_x(\mathbf{x})+\int\left(\Phi^{-1}(u)\right)^T Gdu\right]dt$, $t_f=\infty$, solve

$$0=\omega_x(\mathbf{x})+\left(\frac{\partial V}{\partial \mathbf{x}}\right)^T[\mathbf{F}(\mathbf{x})+\mathbf{A}\mathbf{x}]-\underbrace{\int\Phi^T\left(G^{-1}\mathbf{B}^T(\mathbf{x})\frac{\partial V}{\partial \mathbf{x}}\right)d\left(\mathbf{B}^T(\mathbf{x})\frac{\partial V}{\partial \mathbf{x}}\right)}_{\int\Phi^T\left(G^{-1}\mathbf{B}^T(\mathbf{x})\frac{\partial V}{\partial \mathbf{x}}\right)d\left(\mathbf{B}^T(\mathbf{x})\frac{\partial V}{\partial \mathbf{x}}\right)>0}, \tag{6.72}$$

using a scalar function $V(\mathbf{x})$

$$V(\mathbf{x})=\sum_{i=N_0,j=M_0}^{N,M}\tfrac{j}{2i}(\mathbf{x}^{i/j})^T K_{ij}\mathbf{x}^{i/j},\ V(\mathbf{x})=\sum_{i,j}\kappa_{1_i}^T(\mathbf{x}^{<d_i>})K_{ij}\kappa_{2_j}(\mathbf{x}^{<d_j>}), \tag{6.73}$$

$K_{ij}\in\mathbb{R}^{n\times n}$, $K_{ij}>0$, $N_0\ge 1$, $M_0\ge 1$.

Solution of the Hamilton-Jacobi Equations – Analytic and numeric solutions of equations (6.70) and (6.72) are studied. System nonlinearities $\mathbf{F}(\mathbf{x})$, $\mathbf{B}(\mathbf{x})$ and $\Phi\subseteq\mathcal{U}_\varepsilon$ affect solutions of Hamilton-Jacobi equations. Closed-form solutions to dynamic optimization problems were found. As discussed in section 6.3.2, one uses adequate return functions $V(t,\mathbf{x})$ and $V(\mathbf{x})$ to approximate solutions.

Symmetric Control Limits and Design of a Control Function

Examine symmetric control limits $-u_{\max}\le\mathcal{U}\le u_{\max}$. For relay, switch and saturation limits, an infinitely differentiable function $\Phi\subseteq\mathcal{U}_\varepsilon$ with range $(-u_{\max}, u_{\max})$ is

$\Phi(x)=u_{\max}\tanh^{2n+1}(ax)$, $-u_{\max}<\Phi<u_{\max}$, $a>0$, $n\ge 0$.

Consider $\Phi(x)=u_{\max}\tanh(ax)$, $-u_{\max}<\Phi<u_{\max}$, $a>0$.

The inverse function is $\Phi^{-1}(u)=\frac{1}{a}\tanh^{-1}(\frac{1}{u_{\max}}u)$, $-u_{\max}<u<u_{\max}$.

A positive definite integrand $\omega_u(u)$ is designed to analytically find a bounded control function as

$$\omega_u(u)=\int\tfrac{1}{a}\tanh^{-1}(\tfrac{1}{u_{\max}}u)du,\ -u_{\max}<u<u_{\max}.$$

A nonquadratic performance $\omega_x(\mathbf{x})$ is synthesized to impose specifications on the dynamics of $\mathbf{x}(t)$, as well as support an analytic solution of the Hamilton-Jacobi equation. For example, $\omega_x(\mathbf{x})=(\mathbf{x}^{<d_l>})^T Q\mathbf{x}^{<d_k>}$, $(l,k)\in\mathbb{N}$, where Q is a square matrix.

Minimize a functional

$$J=\min_{\mathbf{x},|u|\le u_{\max}}\int_0^\infty\left[\underbrace{\boldsymbol{\omega}_x(\mathbf{x})}_{\boldsymbol{\omega}_x(\mathbf{x})=(\mathbf{x}^{<d_l>})^T Q\mathbf{x}^{<d_k>}}+\int\tfrac{1}{a}\tanh^{-1}(\tfrac{1}{u_{\max}}u)G\,du\right]dt$$

subject to (6.66) $\dot{\mathbf{x}}(t)=\mathbf{F}(\mathbf{x})+\mathbf{A}\mathbf{x}+\mathbf{B}u$. We find a bounded control law as

$u=-u_{\max}\tanh(aG^{-1}\mathbf{B}^T\frac{\partial V}{\partial\mathbf{x}})$, $-u_{\max}<u<u_{\max}$.

Solution of the Hamilton-Jacobi equation

$$\underbrace{\boldsymbol{\omega}_x(\mathbf{x})}_{\boldsymbol{\omega}_x(\mathbf{x})=(\mathbf{x}^{<d_l>})^T Q\mathbf{x}^{<d_k>}}+\left(\frac{\partial V}{\partial\mathbf{x}}\right)^T(\mathbf{F}(\mathbf{x})+\mathbf{A}\mathbf{x})-\int u_{\max}\tanh(aG^{-1}\mathbf{B}^T\frac{\partial V}{\partial\mathbf{x}})d(\mathbf{B}^T\frac{\partial V}{\partial\mathbf{x}})=0,$$

$$\int u_{\max}\tanh(aG^{-1}\mathbf{B}^T\frac{\partial V}{\partial\mathbf{x}})d(\mathbf{B}^T\frac{\partial V}{\partial\mathbf{x}})=\frac{u_{\max}}{a}G\ln\left(\cosh(aG^{-1}\mathbf{B}^T\frac{\partial V}{\partial\mathbf{x}})\right),$$

is approximated by a return function (6.73).

MIMO Case-Study: Multivariate Control and Systems With Range-Explicit Limits

For multi-input systems with $u\in\mathbb{R}^m$, investigate symmetric limits on control

$-\mathcal{U}_{\max}\le\mathcal{U}\le\mathcal{U}_{\max}$, $\mathcal{U}=[\mathcal{U}_1,\ldots,\mathcal{U}_m]$, $\mathcal{U}\in\mathbb{R}^m$.

For relay, saturation and other limits, find $\Phi=[\Phi_1,\ldots,\Phi_m]$, $\Phi\in\mathbb{R}^m$, such that $\Phi\subseteq\mathcal{U}_\varepsilon$.

Let $\Phi(x)=\boldsymbol{u}_{\max}\tanh(\boldsymbol{a}x)$, $x\equiv\frac{\partial V}{\partial\mathbf{x}}$,

$$\boldsymbol{u}_{\max}=\begin{bmatrix}u_{1\max}&\cdots&0\\\vdots&\ddots&\vdots\\0&\cdots&u_{m\max}\end{bmatrix},\quad\boldsymbol{a}=\begin{bmatrix}a_1&\cdots&0\\\vdots&\ddots&\vdots\\0&\cdots&a_m\end{bmatrix},\quad\boldsymbol{u}_{\max}\in\mathbb{R}^{m\times m},\ \boldsymbol{a}\in\mathbb{R}^{m\times m}.$$

One has $\Phi^{-1}(u)=\boldsymbol{a}^{-1}\tanh^{-1}(\boldsymbol{u}_{\max}^{-1}u)$.

Minimize functional (6.67) with positive definite integrands $\omega_x(\mathbf{x})>0$ and $\boldsymbol{\omega}_u(u)=\int\left(\boldsymbol{a}^{-1}\tanh^{-1}(\boldsymbol{u}_{\max}^{-1}u)\right)^T G\,du>0$

$$J=\min_{\mathbf{x},u\in U}\int_{t_0}^{t_f}\left[\boldsymbol{\omega}_x(\mathbf{x})+\int\left(\Phi^{-1}(u)\right)^T G\,du\right]dt=\min_{\mathbf{x},u\in U}\int_{t_0}^{t_f}\left[\boldsymbol{\omega}_x(\mathbf{x})+\int\left(\boldsymbol{a}^{-1}\tanh^{-1}(\boldsymbol{u}_{\max}^{-1}u)\right)^T G\,du\right]dt$$

subject to (6.66) $\dot{\mathbf{x}}(t)=\mathbf{F}(\mathbf{x})+\mathbf{A}\mathbf{x}+\mathbf{B}u$.

Applying (6.22) $\frac{\partial}{\partial u}H(\mathbf{x},u,\frac{\partial V}{\partial\mathbf{x}})=0$ for Hamiltonian, the constrained control is

$$u=-\Phi\left(\frac{\partial V}{\partial\mathbf{x}}\right)=-\boldsymbol{u}_{\max}\tanh\left(\boldsymbol{a}G^{-1}\mathbf{B}^T\frac{\partial V}{\partial\mathbf{x}}\right),\quad\boldsymbol{u}_{\max}=\begin{bmatrix}u_{1\max}&\cdots&0\\\vdots&\ddots&\vdots\\0&\cdots&u_{m\max}\end{bmatrix},\quad\boldsymbol{a}=\begin{bmatrix}a_1&\cdots&0\\\vdots&\ddots&\vdots\\0&\cdots&a_m\end{bmatrix}.$$

The return function $V(t,\mathbf{x})$ should satisfy the Hamilton-Jacobi equation

$$-\frac{\partial V}{\partial t}=\boldsymbol{\omega}_x(\mathbf{x})+\left(\frac{\partial V}{\partial\mathbf{x}}\right)^T(\mathbf{F}(\mathbf{x})+\mathbf{A}\mathbf{x})-\int\left[\boldsymbol{u}_{\max}\tanh\left(\boldsymbol{a}G^{-1}\mathbf{B}^T\frac{\partial V}{\partial\mathbf{x}}\right)\right]^T d\left(\mathbf{B}^T\frac{\partial V}{\partial\mathbf{x}}\right),$$

$$\int\left[\boldsymbol{u}_{\max}\tanh(\boldsymbol{a}G^{-1}\mathbf{B}^T\frac{\partial V}{\partial\mathbf{x}})\right]^T d(\mathbf{B}^T\frac{\partial V}{\partial\mathbf{x}})=\left[\left(\ln\left(\cosh(\boldsymbol{a}G^{-1}\mathbf{B}^T\frac{\partial V}{\partial\mathbf{x}})\right)\right)^{\frac{1}{2}}\right]^T\boldsymbol{u}_{\max}\boldsymbol{a}^{-1}G\left(\ln\left(\cosh(\boldsymbol{a}G^{-1}\mathbf{B}^T\frac{\partial V}{\partial\mathbf{x}})\right)\right)^{\frac{1}{2}},$$

$\cosh(\boldsymbol{a}G^{-1}\mathbf{B}^T\frac{\partial V}{\partial\mathbf{x}})\ge1$, $\cosh(-x)=\cosh(x)$, $\lim_{x\to\infty}\cosh(x)=\infty$, $\cosh(x)\ge1$, $\forall x$.

A formulated problem and proposed scheme admits a closed-form solution.

One solves $\boldsymbol{\omega}_x(\mathbf{x})+\left(\frac{\partial V}{\partial\mathbf{x}}\right)^T(\mathbf{F}(\mathbf{x})+\mathbf{A}\mathbf{x})-\int\left[\boldsymbol{u}_{\max}\tanh\left(\boldsymbol{a}G^{-1}\mathbf{B}^T\frac{\partial V}{\partial\mathbf{x}}\right)\right]^T d\left(\mathbf{B}^T\frac{\partial V}{\partial\mathbf{x}}\right)=0$.

For given $-\mathcal{U}_{\max}\le\mathcal{U}\le\mathcal{U}_{\max}$, one may study $\Phi(x)=\boldsymbol{u}_{\max}\tanh^{2n+1}(\boldsymbol{a}x)$ and other $\Phi\subseteq\mathcal{U}_\varepsilon$.

Weighting Matrices – For linear systems, diagonal positive definite weighting matrices (Q,G) and weighting coefficients (q_{ii}, g_{jj}) can be scaled and defined by using the admissible variations of states and controls $(\tilde{x}_{i\ \max}, \tilde{u}_{j\ \max})$ as, $g_{jj}=\frac{1}{\tilde{u}_{j\ \max}^2}$. In the constrained control laws design, (q_{ii},g_{jj}) in (Q,G) should be selected taking into account additional factors considering ($\tilde{x}_{i\ \max}$, $u_{\min}\le u\le u_{\max}$, $u_{\min}<\Phi<u_{\max}$, a).

Example 6.31. *Second-Order System With Limits on Control*
Consider a system with saturation $-1\le u\le 1$. Design tracking control laws for

$$\dot{x}_1=u,$$

$$\dot{x}_2=x_1,\ y=x_2.$$

The tracking error is $e(t)=(r-y)=(r-x_2)$.

Linear System With No Control Limits – For a system with a transfer function $G_{sys}(s)=\frac{1}{s^2}$, the transfer function and characteristic equation for a closed-loop system with a proportional-derivative control law $u=k_pe+k_d\frac{de}{dt}$ are

$$G(s)=\frac{k_ds+k_p}{s^2+k_ds+k_p},\ s^2+k_ds+k_p=0.$$

While for systems with limits, linear control theory may not be applied, apply a *modal* control concept. We find the feedback gains (k_p,k_d) to ensure the specified time constant T and damping coefficient ξ. The relationships are $T=\sqrt{\frac{1}{k_p}}$ and $\xi=\frac{k_d}{2\sqrt{k_p}}$.

Let T=1 sec and ξ=0.707. We have $u=k_pe+k_d\frac{de}{dt}$, k_p=1, k_d=1.41.

For T=1 sec and ξ=1, one finds k_p=1 and k_d=2.

One defines adequate (T,ξ) because high-gain control with large (k_p,k_d) results in instabilities due to control limits and implementation impediments.

Tracking Control With Limits – Synthesize a constrained tracking control law. Dynamics without overshoot is desired. This can be achieved by minimizing consistent performance functionals, using adequate integrands, and, applying scaled weighting coefficients. Define the error governance as

$$\dot{x}_e=r-y=r-x_2,$$

$$\dot{e}=-e+\dot{r}-\dot{x}_2=-e+\dot{r}-x_1.$$

For the extended state vector $\mathbf{x}$, obtain (6.66) as

$$\dot{\mathbf{x}}=\mathbf{A}\mathbf{x}+A_rr+\mathbf{B}u,\ \begin{bmatrix}\dot{x}_1\\ \dot{x}_2\\ \dot{x}_e\\ \dot{e}\end{bmatrix}=\begin{bmatrix}0&0&0&0\\ 1&0&0&0\\ 0&-1&0&0\\ -1&0&0&-1\end{bmatrix}\begin{bmatrix}x_1\\ x_2\\ x_e\\ e\end{bmatrix}+\begin{bmatrix}0&0\\ 0&0\\ 1&0\\ 0&1\end{bmatrix}\begin{bmatrix}r\\ \dot{r}\end{bmatrix}+\begin{bmatrix}1\\ 0\\ 0\\ 0\end{bmatrix}u,\ \mathbf{x}(t)=\begin{bmatrix}x_1\\ x_2\\ x_e\\ e\end{bmatrix},-1\le u\le 1,\ y=x_2.$$

The controllability and observability matrices

C=[**B AB A**^{2}**B A**^{3}**B**], $C=\begin{bmatrix}1&0&0&0\\ 0&1&0&0\\ 0&0&-1&0\\ 0&-1&1&-1\end{bmatrix}$ and O=[C C**A** C**A**^{2} C**A**^{3}]^T

have full ranks, rank(C)=4, rank(O)=4. The number of uncontrollable and unobservable states is zero. The system with the extended state vector **x** is controllable and observable.

Control limits and high-gain control yield instabilities.

The saturation limit $-1\le\mathcal{U}\le 1$ is described by an infinitely-differentiable function $\Phi(x)=\tanh(x)$, $-1<\Phi<1$, $\Phi\subseteq\mathcal{U}_\varepsilon$. Minimize functional (6.67)

$$J=\min_{\mathbf{x},-1\le u\le 1}\int_0^\infty\left[\tfrac{1}{2}\mathbf{x}^TQ\mathbf{x}+G\int\tanh^{-1}du\right]dt,\ G>0,$$

subject to the fourth-order system. Using the Hamiltonian (6.68), find a control law (6.69)

$$u=-\tanh\left(G^{-1}\mathbf{B}^T\frac{\partial V}{\partial\mathbf{x}}\right),\ \mathbf{x}(t)=\begin{bmatrix}x_1(t)\\x_2(t)\\x_e(t)\\e(t)\end{bmatrix}=\begin{bmatrix}x_1(t)\\x_2(t)\\\int edt\\e(t)\end{bmatrix},\ u\in U.$$

Letting $Q=I$ and $G=1$, solve (6.72)

$$\tfrac{1}{2}\mathbf{x}^TQ\mathbf{x}+\left(\frac{\partial V}{\partial\mathbf{x}}\right)^T\mathbf{A}\mathbf{x}-\int\tanh\left(\mathbf{B}^T(\mathbf{x})\frac{\partial V}{\partial\mathbf{x}}\right)\left(\mathbf{B}^T(\mathbf{x})\frac{\partial V}{\partial\mathbf{x}}\right)=0,$$

$$\int\tanh\left(\mathbf{B}^T\tfrac{\partial V}{\partial\mathbf{x}}\right)d\left(\mathbf{B}^T\tfrac{\partial V}{\partial\mathbf{x}}\right)=\ln\cosh\left(\mathbf{B}^T\tfrac{\partial V}{\partial\mathbf{x}}\right)\approx\frac{1}{3}\left(\frac{\partial V}{\partial\mathbf{x}}\right)^T\mathbf{B}\mathbf{B}^T\frac{\partial V}{\partial\mathbf{x}}.$$

Approximate a solution by a quadratic return function (6.73) $V(\mathbf{x})=\frac{1}{2}\mathbf{x}^TK\mathbf{x}$.

Solving $Q+\mathbf{A}^TK+K\mathbf{A}-\frac{2}{3}K\mathbf{B}G^{-1}\mathbf{B}^TK=0$, we have $K=\begin{bmatrix}3.343 & 3.144 & -1.225 & -0.0811\\3.144 & 6.02 & -2.796 & 0.236\\-1.225 & -2.796 & 2.567 & -0.0662\\-0.0811 & 0.236 & -0.0662 & 0.498\end{bmatrix}$.

The constrained control law is $u=-\tanh\left(K_F\begin{bmatrix}x_1\\x_2\\\int edt\\e\end{bmatrix}\right)$, K_F=[3.34 3.14 –1.225 –0.081].

Figures 6.17 report simulation results for a closed-loop system if $r(t)=\pm1$ studying a proportional-derivative control $u=k_pe+k_d\frac{de}{dt}$ and constrained control $u=-\tanh\left(3.34x_1+3.14x_2-1.23\int edt-0.0811e\right)$.

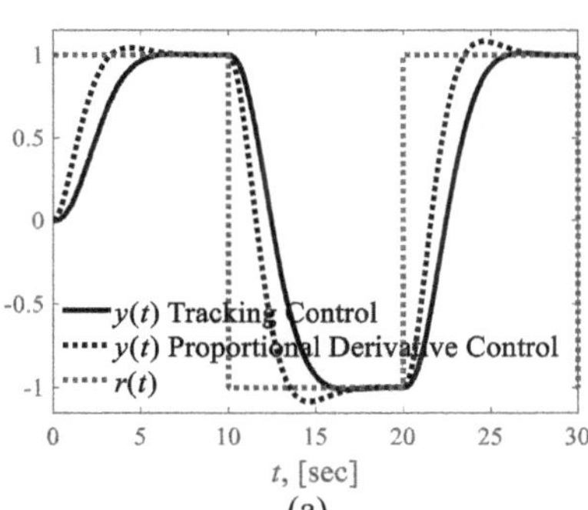

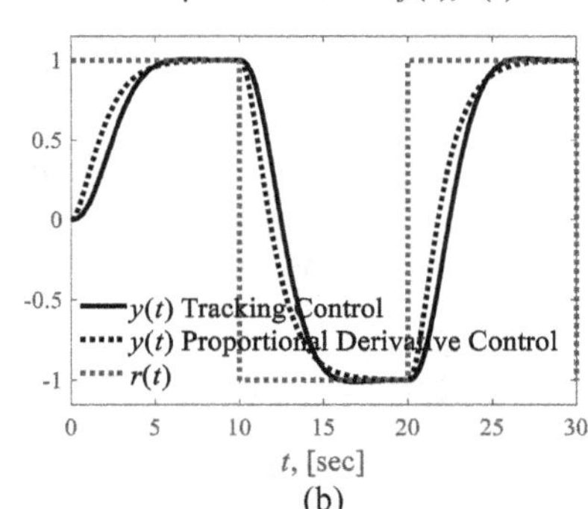

Figure 6.17. Output dynamics $y=x_2$ of closed-loop systems with the proportional-derivative control $u=k_pe+k_d\frac{de}{dt}$, and, control $u=-\tanh\left(3.34x_1+3.14x_2-1.23\int edt-0.0811e\right)$, $r(t)=\pm1$:
(a) Proportional-derivative control law (k_p=1, k_d=1.41) and constrained tracking control law $u=-\tanh\left(3.34x_1+3.14x_2-1.23\int edt-0.0811e\right)$: Evolutions are depicted by dotted and solid lines;
(b) Proportional-derivative control law (k_p=1, k_d=2) and tracking control law. ■

Example 6.32. *Tracking Control of a Satellite*
The rotational motion of a satellite around the center of mass is described by

$$\frac{d^2\theta}{dt^2}=u,\ y=\theta.$$

Tracking control was investigated in Example 5.33 not considering thrusters' nonlinearities. For synchronously ignited liquid-propellant opposing thrusters, one has $u_{\min}\leq u\leq u_{\max}$. Control nonlinearities are described by

$\Phi(x)=u_{\max}\tanh^{2n+1}(ax)$, $-u_{\max}<\Phi<u_{\max}$, $a>0$, $n\geq 0$.

For $-1\leq u\leq 1$, to ensure $\Phi\subseteq U_\varepsilon$, consider $\Phi(x)=\tanh^3(ax)$, $-1<\Phi<1$, $n=1$, $a=10$.

A model of rotational motion $\begin{bmatrix}\dot{x}_1\\ \dot{x}_2\end{bmatrix}=\begin{bmatrix}0&0\\1&0\end{bmatrix}\begin{bmatrix}x_1\\ x_2\end{bmatrix}+\begin{bmatrix}1\\0\end{bmatrix}u,\ -1\leq u\leq 1,\ y=x_2$ is combined with the error dynamics (6.66). An extended system is governed by

$$\dot{\mathbf{x}}=\mathbf{A}\mathbf{x}+A_r r+\mathbf{B}u,\quad \begin{bmatrix}\dot{x}_1\\ \dot{x}_2\\ \dot{x}_e\\ \dot{e}\end{bmatrix}=\begin{bmatrix}0&0&0&0\\1&0&0&0\\0&-1&0&0\\-1&0&0&-1\end{bmatrix}\begin{bmatrix}x_1\\ x_2\\ x_e\\ e\end{bmatrix}+\begin{bmatrix}0&0\\0&0\\1&0\\0&1\end{bmatrix}\begin{bmatrix}r\\ \dot{r}\end{bmatrix}+\begin{bmatrix}1\\0\\0\\0\end{bmatrix}u,\ -1\leq u\leq 1,\ y=x_2.$$

Minimize a physics-consistent functional (6.67)

$$J=\min_{\mathbf{x},|u|\leq 1}\int_0^\infty\left[\tfrac{1}{2}\mathbf{x}^TQ\mathbf{x}+G\int\tfrac{1}{a}\tanh^{-1}\sqrt[3]{u}du\right]dt\cdot$$

The necessary condition for optimality (6.22) yields

$$u=-\tanh^3\left(aG^{-1}\mathbf{B}^T\frac{\partial V}{\partial\mathbf{x}}\right),\ \mathbf{x}(t)=\begin{bmatrix}x_1(t)\\ x_2(t)\\ \int edt\\ e(t)\end{bmatrix},\ u\in U.$$

The Hamilton-Jacobi equation (6.70) is

$$-\frac{\partial V}{\partial t}=\tfrac{1}{2}\mathbf{x}^TQ\mathbf{x}+\left(\frac{\partial V}{\partial\mathbf{x}}\right)^T\mathbf{A}\mathbf{x}-\int\tanh^3\left(aG^{-1}\mathbf{B}^T\frac{\partial V}{\partial\mathbf{x}}\right)d\left(\mathbf{B}^T\frac{\partial V}{\partial\mathbf{x}}\right).$$

For the resulting integral, $\int\tanh^g xdx=-\frac{1}{g-1}\tanh^{g-1}x+\int\tanh^{g-2}xdx,\ g\neq 1$.

Hence, $\int\tanh^3(aG^{-1}x)dx=\frac{1}{a}G\left[\frac{1}{2}\mathrm{sech}^2(ax)+\ln\cosh(ax)+c\right]$,

$\mathrm{sech}\,ax=1-\frac{1}{2}ax^2+\frac{5}{24}ax^4-\frac{61}{720}ax^6+\frac{277}{8064}ax^8-\ldots,\ |ax|<\frac{1}{2}\pi,\ x\equiv\mathbf{B}^T\frac{\partial V}{\partial\mathbf{x}}$.

We have $\int\tanh^3\left(aG^{-1}\mathbf{B}^T\frac{\partial V}{\partial\mathbf{x}}\right)d\left(\mathbf{B}^T\frac{\partial V}{\partial\mathbf{x}}\right)\approx\frac{k}{a}\left(\frac{\partial V}{\partial\mathbf{x}}\right)^T\mathbf{B}G\mathbf{B}^T\frac{\partial V}{\partial\mathbf{x}}$, $k>0$,

where k depends on variations of argument, and, can be found using the least squares approximation.

Solve an algebraic equation (6.72) $\frac{1}{2}\mathbf{x}^TQ\mathbf{x}+\frac{\partial V}{\partial\mathbf{x}}^T\mathbf{A}\mathbf{x}-\frac{k}{a}\frac{\partial V}{\partial\mathbf{x}}^T\mathbf{B}G\mathbf{B}^T\frac{\partial V}{\partial\mathbf{x}}=0$, approximating a solution by a quadratic return function (6.73) $V(\mathbf{x})=\frac{1}{2}\mathbf{x}^TK\mathbf{x}$.

Specifying evolutions of $(\mathbf{x},u)$ and control limits, consider $Q=\begin{bmatrix}1&0&0&0\\0&1&0&0\\0&0&100&0\\0&0&0&100\end{bmatrix}$ and $G=1$.

The unknown K is computed by solving $Q+\mathbf{A}^TK+K\mathbf{A}-\frac{k}{a}K\mathbf{B}G\mathbf{B}^TK=0$, $a=10$, $k=1$.

We have $K=\begin{bmatrix} 18.71 & 31.36 & -22.36 & -3.16 \\ 31.36 & 129 & -97.84 & 33.99 \\ -22.36 & -97.84 & 140.25 & -14.15 \\ -3.16 & 33.99 & -14.15 & 49 \end{bmatrix}$, $K>0$.

The constrained control law (6.69) and feedback gains are

$$u=-\tanh^3\left(aG^{-1}\mathbf{B}^T\frac{\partial V}{\partial \mathbf{x}}\right)=-\tanh^3\left(aG^{-1}\mathbf{B}^T K\mathbf{x}\right)=-\tanh^3\left(K_F\mathbf{x}\right),\ -1<u<1,$$

$$u=-\tanh^3\left(187.15x_1+313.6x_2-223.61\int e dt-31.64e\right).$$

Computation of feedback coefficients takes less than 0.01 sec, suiting near-real-time reconfigurable and adaptive control. Figure 6.18.a depicts a SIMULINK diagram. Simulation of a system with the designed tracking control law is reported in Figure 6.18.b for $r(t)=\pm1$. The control structure $\varphi(\mathbf{x})$ exceeds limits, and, the control function is bounded $-1<u<1$. Angular velocity x_1, output angular displacement $y=x_2$ and control $u(t)$ adequately evolve for references $r(t)$ within an operating envelope.

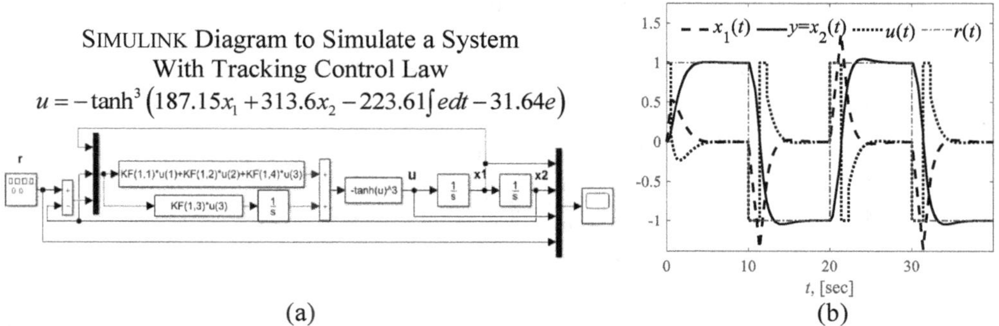

(a) (b)

Figure 6.18. Closed-loop system with a control law $u=-\tanh^3(K_F\mathbf{x})$:
(a) SIMULINK diagram to simulate a closed-loop system;
(b) Dynamics and tracking of the closed-loop system, references are $r(t)=\pm1$. ■

6.8.2. Constrained Tracking Control of a Fixed-Wing Aerial Vehicle

The decoupled longitudinal and lateral models of an aerial vehicle were studied in section 5.11.1.

Longitudinal Dynamics – The state variables are the forward velocity $x_1=V$, angle of attack $x_2=\alpha$, pitch rate $x_3=q$ and pitch angle $x_4=\theta$. With the state vector $x\in\mathbb{R}^4$, the outputs are V and θ, $y=[V, \theta]$, $y\in\mathbb{R}^2$.

For the considered airframe, there are the physical saturation limits on controls $u_{i\min}\le u_i\le u_{i\max}$. The thrust $T=f(c_T,\omega,D)\approx c_T\rho\omega^2D^4$ is a nonlinear function of the propeller angular velocity ω, propeller diameter D, number of blades, blade leaf, width and thickness, thread pitch, etc. The thrust coefficient c_T depends on the advance ratio, blade angle, propeller aerodynamics, velocity of propeller rotation ω, etc. One has $u_{t\min}\le u_t\le u_{t\max}$ with specific atmospheric pressure, air density, flight attitude, etc.

The displacement of elevators, which are the pitch-control aerodynamic surfaces, are limited, $-15^\circ\le u_E\le15^\circ$. For a model, linearized at equilibrium $(\bar{x},\bar{u}),\bar{u}=\begin{bmatrix}\bar{u}_t\\ \bar{u}_E\end{bmatrix}$, physical limits on controls are $u_{i\min}\le u_i\le u_{i\max}$. For the propeller velocity $\bar{\omega}$, one obtains $\bar{u}_t$, for which $-30\le\delta u_t\le30$ N, $u_1\equiv\delta u_t$. For $\bar{u}_E=0$, $u_2\equiv\delta u_E$.

Model is linearized and coefficients are found using the flight at elevation 500 m, 25 m/s, 1° pitch angle, zero angle of attack, zero sideslip angle and zero roll angle.

The longitudinal model is

$$\dot{x} = Ax + Bu,\quad \begin{bmatrix}\dot{x}_1\\ \dot{x}_2\\ \dot{x}_3\\ \dot{x}_4\end{bmatrix} = \begin{bmatrix}-0.15 & -1.3 & 0 & -9.81\\ 0 & -7.4 & 0.96 & 0\\ 0.1 & -5.6 & -2.4 & 0\\ 0 & 0 & 1 & 0\end{bmatrix}\begin{bmatrix}x_1\\ x_2\\ x_3\\ x_4\end{bmatrix} + \begin{bmatrix}5.3 & -0.29\\ 0 & 0.074\\ 0 & 1.6\\ 0 & 0\end{bmatrix}\begin{bmatrix}u_1\\ u_2\end{bmatrix},\ y = \begin{bmatrix}1 & 0 & 0 & 0\\ 0 & 0 & 0 & 1\end{bmatrix}\begin{bmatrix}x_1\\ x_2\\ x_3\\ x_4\end{bmatrix},$$

$$x = \begin{bmatrix}x_1\\ x_2\\ x_3\\ x_4\end{bmatrix} = \begin{bmatrix}V\\ \alpha\\ q\\ \theta\end{bmatrix},\ y = \begin{bmatrix}x_1\\ x_4\end{bmatrix} = \begin{bmatrix}V\\ \theta\end{bmatrix},\ u = \begin{bmatrix}u_1\\ u_2\end{bmatrix} = \begin{bmatrix}u_t\\ u_E\end{bmatrix},\ -30 \le u_1 \le 30,\ -15 \le u_2 \le 15.$$

Lateral Dynamics – The sideslip angle β, roll and yaw rates (p,r), roll and yaw angles (ϕ,ψ) are the lateral states, $x\in\mathbb{R}^5$. The outputs are ϕ and ψ, $y=[\phi, \psi]$, $y\in\mathbb{R}^2$. Physical limits $\mathcal{U}_{i\min}\le\mathcal{U}_i\le\mathcal{U}_{i\max}$ are imposed on displacements of ailerons (right and left ailerons are deflected differentially and simultaneously) and rudder. For equilibria $\bar{u}_A = 0$ and $\bar{u}_R = 0$, control limits are $-7.5°\le u_A\le 7.5°$, $u_1\equiv u_A$, $u_1\equiv\delta u_A$, $-20°\le u_R\le 20°$, $u_2\equiv u_R$, $u_2\equiv\delta u_R$, $u\in\mathbb{R}^2$. The governing and output equations are

$$\dot{x} = Ax + Bu,\ \begin{bmatrix}\dot{x}_1\\ \dot{x}_2\\ \dot{x}_3\\ \dot{x}_4\\ \dot{x}_5\end{bmatrix} = \begin{bmatrix}0.08 & -0.07 & -0.94 & 0 & 0\\ -0.51 & -3.8 & -0.45 & 0 & 0\\ 5.9 & 1.3 & -2.2 & 0 & 0\\ 0 & 1 & 0 & 0 & 0\\ 0 & 0 & 1 & 0 & 0\end{bmatrix}\begin{bmatrix}x_1\\ x_2\\ x_3\\ x_4\\ x_5\end{bmatrix} + \begin{bmatrix}0 & 0.13\\ 17 & 0.28\\ 0.07 & 0.95\\ 0 & 0\\ 0 & 0\end{bmatrix}\begin{bmatrix}u_1\\ u_2\end{bmatrix},\ y = \begin{bmatrix}0 & 0 & 0 & 1 & 0\\ 0 & 0 & 0 & 0 & 1\end{bmatrix}\begin{bmatrix}x_1\\ x_2\\ x_3\\ x_4\\ x_5\end{bmatrix}.$$

The states, controls and outputs are

$$x = \begin{bmatrix}x_1\\ x_2\\ x_3\\ x_4\\ x_5\end{bmatrix} = \begin{bmatrix}\beta\\ p\\ r\\ \phi\\ \psi\end{bmatrix},\ y = \begin{bmatrix}x_4\\ x_5\end{bmatrix} = \begin{bmatrix}\phi\\ \psi\end{bmatrix},\ u = \begin{bmatrix}u_1\\ u_2\end{bmatrix} = \begin{bmatrix}u_A\\ u_R\end{bmatrix},\ -7.5 \le u_1 \le 7.5,\ -20 \le u_2 \le 20.$$

For the vectors of *error states* $x_e(t)$ and tracking error $e(t)$, the dynamics are

$\dot{x}_e(t) = I_X(r-y) = I_X(r-Hx)$, $e(t)\equiv\dot{x}_e(t)$, $x_e\in\mathbb{R}^2$,

$\dot{e}(t) = -I_E e + \dot{r} - \dot{y} = -I_E e + \dot{r} - HAx - HBu$, $e\in\mathbb{R}^2$.

Solve longitudinal and lateral tracking control problems using the extended governing dynamics (6.66)

$$\dot{\mathbf{x}}(t) = \begin{bmatrix}\dot{x}(t)\\ \dot{x}_e(t)\\ \dot{e}(t)\end{bmatrix} = \mathbf{A}\mathbf{x} + A_r r + \mathbf{B}u = \begin{bmatrix}A & 0 & 0\\ -I_X H & 0 & 0\\ -HA & 0 & -I_E\end{bmatrix}\mathbf{x} + \begin{bmatrix}0 & 0\\ I_X & 0\\ 0 & I\end{bmatrix}\begin{bmatrix}r\\ \dot{r}\end{bmatrix} + \begin{bmatrix}B\\ 0\\ -HB\end{bmatrix}u\,.$$

The saturation limits $-\mathcal{U}_{i\max}\le\mathcal{U}_i\le\mathcal{U}_{i\max}$ are described by

$\Phi_i = u_{i\max}\tanh(a_i x)$, $\forall a_i=1$, $\Phi\subseteq\mathcal{U}_\varepsilon$, $-u_{i\max}<\Phi_i<u_{i\max}$, $u_{i\max}>0$, $i=1,2$.

For $\Phi=\boldsymbol{u}_{\max}\tanh(\boldsymbol{a}x)$, $\Phi=[\Phi_1,\Phi_2]$, $\Phi\subseteq\mathcal{U}_\varepsilon$, minimize a nonquadratic performance functional (6.67)

$$J=\min_{\substack{\mathbf{x}\\|u_1|\le u_{1\max}\\|u_2|\le u_{2\max}}}\int_0^\infty\left[\tfrac{1}{2}\mathbf{x}^TQ\mathbf{x}+\int\left(\boldsymbol{a}^{-1}\tanh^{-1}(\boldsymbol{u}_{\max}^{-1}u)\right)^TG\,du\right]dt,\ \boldsymbol{u}_{\max}=\begin{bmatrix}u_{1\max}&0\\0&u_{2\max}\end{bmatrix},\ \boldsymbol{a}=\begin{bmatrix}a_1&0\\0&a_2\end{bmatrix}.$$

The constrained tracking control law (6.69) is

$$u=-\boldsymbol{u}_{\max}\tanh(\boldsymbol{a}G^{-1}\mathbf{B}^TK\mathbf{x})=-\begin{bmatrix}u_{1\max}&0\\0&u_{2\max}\end{bmatrix}\tanh\left(K_F\begin{bmatrix}x\\\int e_1dt\\\int e_2dt\\e_1\\e_2\end{bmatrix}\right).$$

Solution of equation (6.72)

$$\tfrac{1}{2}\mathbf{x}^TQ\mathbf{x}+\left(\frac{\partial V}{d\mathbf{x}}\right)^T\mathbf{A}\mathbf{x}-\int\left[\boldsymbol{u}_{\max}\tanh\left(\boldsymbol{a}G^{-1}\mathbf{B}^T\frac{\partial V}{d\mathbf{x}}\right)\right]^Td\left(\mathbf{B}^T\frac{\partial V}{d\mathbf{x}}\right)=0,$$

$$\int\left[\boldsymbol{u}_{\max}\tanh(\boldsymbol{a}G^{-1}\mathbf{B}^T\tfrac{\partial V}{\partial\mathbf{x}})\right]^Td(\mathbf{B}^T\tfrac{\partial V}{\partial\mathbf{x}})$$

$$=\left[\left(\ln\left(\cosh(\boldsymbol{a}G^{-1}\mathbf{B}^T\tfrac{\partial V}{\partial\mathbf{x}})\right)\right)^{\frac{1}{2}}\right]^T\boldsymbol{u}_{\max}\boldsymbol{a}^{-1}G\left(\ln\left(\cosh(\boldsymbol{a}G^{-1}\mathbf{B}^T\tfrac{\partial V}{\partial\mathbf{x}})\right)\right)^{\frac{1}{2}}\approx\frac{9}{20}\frac{\partial V}{d\mathbf{x}}^T\mathbf{B}\boldsymbol{u}_{\max}\boldsymbol{a}G^{-1}\mathbf{B}^T\frac{\partial V}{d\mathbf{x}},$$

is found using the quadratic return function (6.73) $V(\mathbf{x})=\frac{1}{2}\mathbf{x}^TK\mathbf{x}$.

Longitudinal Control – In design of a longitudinal control law,

$$\boldsymbol{a}=\begin{bmatrix}a_1&0\\0&a_2\end{bmatrix}=\begin{bmatrix}1&0\\0&1\end{bmatrix},\ Q=I,\ Q\in\mathbb{R}^{8\times8},\ G=I,\ G\in\mathbb{R}^{2\times2},\ G=\begin{bmatrix}1&0\\0&1\end{bmatrix}.$$

A tracking control law is

$$u=-\begin{bmatrix}u_{1\max}&0\\0&u_{2\max}\end{bmatrix}\tanh\left(K_F\begin{bmatrix}x\\\int e_1dt\\\int e_2dt\\e_1\\e_2\end{bmatrix}\right),\ x=\begin{bmatrix}x_1\\x_2\\x_3\\x_4\end{bmatrix}=\begin{bmatrix}V\\\alpha\\q\\\theta\end{bmatrix},\ e=\begin{bmatrix}e_1\\e_2\end{bmatrix}=\begin{bmatrix}r_1-V\\r_2-\theta\end{bmatrix},\ u_{1\max}=30,\ u_{2\max}=15.$$

The feedback gain matrix K_F is found, and

$$K_F=\begin{bmatrix}0.232&-0.0236&0.0036&-0.06&-0.192&-0.0044&-0.0407&-0.0004\\0.0011&-0.272&0.355&0.583&0.0063&-0.272&0.001&-0.0263\end{bmatrix}.$$

Lateral Control – Solve a lateral control problem. One has

$$\boldsymbol{a}=\begin{bmatrix}a_1&0\\0&a_2\end{bmatrix}=\begin{bmatrix}1&0\\0&1\end{bmatrix},\ Q=I,\ Q\in\mathbb{R}^{9\times9},\ G=I,\ G\in\mathbb{R}^{2\times2},\ G=\begin{bmatrix}1&0\\0&1\end{bmatrix}.$$

Control law is

$$u=-\begin{bmatrix}u_{1\max}&0\\0&u_{2\max}\end{bmatrix}\tanh\left(K_F\begin{bmatrix}x\\\int e_1dt\\\int e_2dt\\e_1\\e_2\end{bmatrix}\right),\ x=\begin{bmatrix}x_1\\x_2\\x_3\\x_4\\x_5\end{bmatrix}=\begin{bmatrix}\beta\\p\\r\\\phi\\\psi\end{bmatrix},\ e=\begin{bmatrix}e_1\\e_2\end{bmatrix}=\begin{bmatrix}r_1-\phi\\r_2-\psi\end{bmatrix},u_{1\max}=7.5,\ u_{2\max}=20,$$

$$K_F=\begin{bmatrix}0.234&-0.011&0.0062&-0.0513&-0.192&-0.0087&-0.0407&-0.0009\\-0.0109&-0.097&0.251&0.556&0.0123&-0.272&0.0024&-0.029\end{bmatrix}.$$

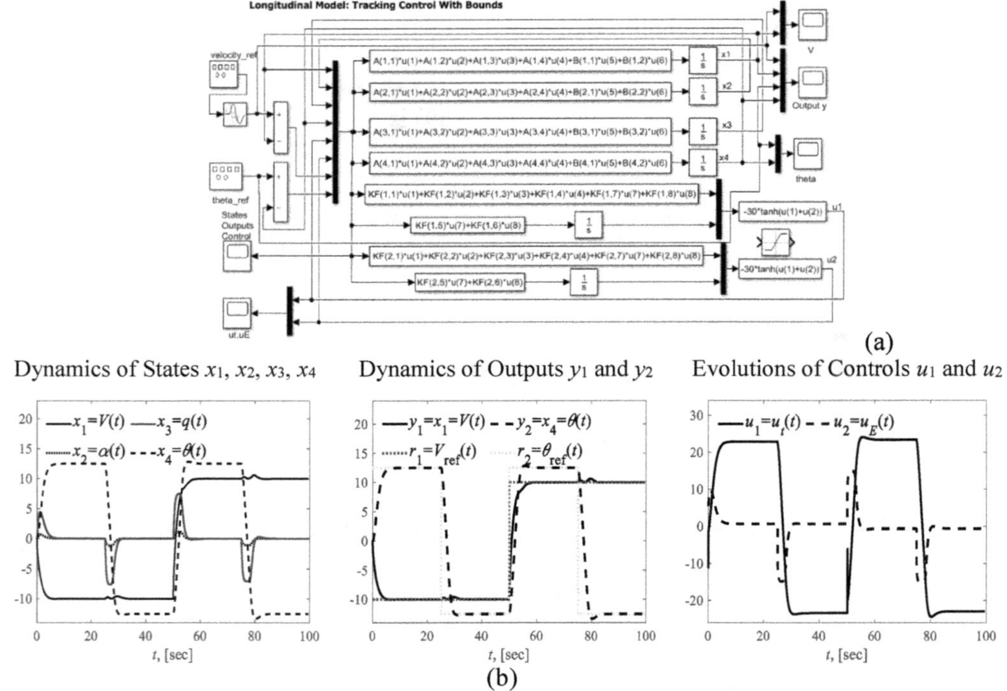

Figure 6.19. (a) Longitudinal Dynamics: SIMULINK diagram to simulate a closed-loop system; (b) States (x_1, x_2, x_3, x_4), outputs (y_1, y_2) and controls (u_1, u_2) for $r_1(t) = \mp 10$ m/s and $r_2(t) = \pm 12.5°$.

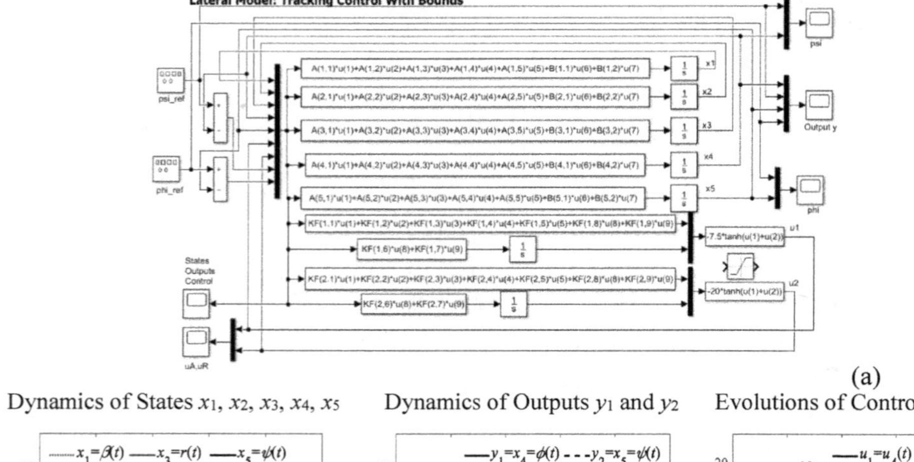

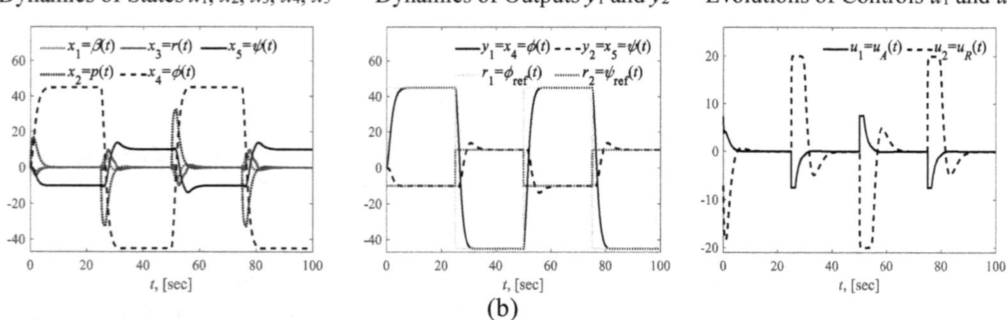

Figure 6.20. (a) Lateral Dynamics: SIMULINK diagram to simulate a closed-loop system; (b) States $(x_1, x_2, x_3, x_4, x_5)$, outputs (y_1, y_2) and controls (u_1, u_2) for $r_1(t) = \pm 10°$ and $r_2(t) = \pm 45°$.

Simulation Results – Figures 6.19.a and 6.20.a depict SIMULINK diagrams. Aerial vehicle is simulated considering maneuvers with outputs (V,θ) and (ϕ,ψ) at which deflections of aerodynamic control surfaces are reaching limits $\mathcal{U}_{i\min}\le\mathcal{U}_i\le\mathcal{U}_{i\max}$.

For longitudinal and lateral maneuvers, the references are

$$r=\begin{bmatrix} r_1\\ r_2\end{bmatrix}=\begin{bmatrix} V_{\text{ref}}\\ \theta_{\text{ref}}\end{bmatrix}=\begin{bmatrix} \mp 10\frac{\text{m}}{\text{sec}}\\ \pm 12.5^{\circ}\end{bmatrix},\ r=\begin{bmatrix} r_1\\ r_2\end{bmatrix}=\begin{bmatrix} \phi_{\text{ref}}\\ \psi_{\text{ref}}\end{bmatrix}=\begin{bmatrix} \pm 10^{\circ}\\ \pm 45^{\circ}\end{bmatrix}.$$

System outputs (V,θ) and (ϕ,ψ) track references $r(t)$. For a longitudinal model, evolutions of states (V,α,q,θ), outputs (V,θ) and controls (u_t,u_E) are reported in Figures 6.19.b. Figures 6.20.b depict the transients of lateral states (β,r,p,ϕ,ψ), outputs (ϕ,ψ), as well as controls (u_A,u_R). Adequate performance is ensured despite control limits. Dynamic optimization yields feedback gains k_{Fij}, and, the feedback matrices K_F are examined. For MIMO systems, one investigates states and outputs dependencies and coupling, feedback structure, as well as definiteness of feedback gains k_{Fij}. One considers: (i) Axis-coupled aerial vehicle statics and dynamics; (ii) Flight modes, phases and maneuvers; (iii) Flight envelope with augmentation, control and compensation schemes; (iv) Dependences of variables, forces, moments and displacement of aerodynamic control surfaces; (v) Maximum angle of attack, sideslip angle and rates, defined by airframes; etc. Due to linearization and simplifications, inconsistencies may arise. For aerial systems, aerodynamic coupling, gain scheduling, axes decoupling and other solutions are considered. The goal is to minimize contradicting factors for aerial vehicles maneuvers and flight phases. Control, axial dependences and coupled translational-rotational dynamics are investigated. Nonlinear aerodynamics is considered focusing on multi-objective decisions and prioritizations by minimizing consistent functionals, applying adequate integrands, choosing pertinent weighting matrices, etc.

6.9. Dynamic Optimization and Nonquadratic Performance Functionals

Design and Optimality – Control laws, robustness, stability and performance are predefined by model fidelity, design schemes, integrands and minimizing functionals, etc. Dynamic optimization and optimal design imply minimization of consistent quadratic and nonquadratic functionals subject to the system model $\mathcal{M}$ which evolves in $X\times U$ on $[t_0,t_f)$, and, finding control laws which must be compliant with hardware.

For physical systems $\mathbb{M}:=\{x: x\in X(X_0,U), y=H(x)\}$, using the measured states and outputs (x,y), optimality is asserted by performance functionals. Assume a seamless correspondence of the model and physical system variables.

Control law (6.62) $u=-\Phi\left(G^{-1}B^T(x)\frac{\partial V(x)}{\partial x}\right)$, $u\in U$ is designed by minimizing functional (6.60) $J=\min\limits_{x,u\in U}\int_{t_0}^{t_f}\Big[\omega_x(x)+\underbrace{\int\left(\Phi^{-1}(u)\right)^T G du}_{\omega_u(u)}\Big]dt$. Performance integrand $\omega_u(u)$ is synthesized using the bounded, integrable, real-analytic continuous function Φ, such that $\Phi\subseteq\mathcal{U}_\varepsilon$. Physics-consistent and design-specific high-degree multivariate integrands $\omega_x(\cdot):\mathbb{R}^n\to\mathbb{R}$ and $\omega_u(\cdot):\mathbb{R}^m\to\mathbb{R}$ are found, and, a nonlinear dynamic optimization problem is solved.

Linear and nonlinear models are described by differential equations (6.12) $\dot{x}(t) = Ax + Bu$ and (6.18) $\dot{x}(t) = F(x) + B(x)u$.

Minimize a performance functional $J = \min_{x,u} \int_{t_0}^{t_f} \omega(x,u)dt$, $\omega(\cdot,\cdot):\mathbb{R}^n\times\mathbb{R}^m\to\mathbb{R}$ as

$$J = \min_{x,u} \int_{t_0}^{t_f} \tfrac{1}{2}\left[\omega_1^T(x)Q\omega_1(x) + \dot{\omega}_2^T(x)P\dot{\omega}_2(x)\right]dt = \min_{x,u} \int_{t_0}^{t_f} \tfrac{1}{2}\left[\omega_1^T(x)Q\omega_1(x) + \dot{x}^T \frac{\partial\omega_2}{\partial x} P \frac{\partial\omega_2}{\partial x}\dot{x}\right]dt$$

$$= \min_{x,u} \int_{t_0}^{t_f} \tfrac{1}{2}\left[\omega_1^T(x)Q\omega_1(x) + \dot{x}^T P_\omega(\tfrac{\partial\omega_2}{\partial x})\dot{x}\right]dt = \min_{x,u} \int_{t_0}^{t_f} \tfrac{1}{2}\left[\omega_1^T(x)Q\omega_1(x) + \dot{x}^T \begin{bmatrix} p_{11}\left(\frac{\partial\omega_2}{\partial x_1}\right)^2 & \cdots & 0 \\ \vdots & \ddots & \vdots \\ 0 & \cdots & p_{nn}\left(\frac{\partial\omega_2}{\partial x_n}\right)^2 \end{bmatrix} \dot{x}\right] dt,$$

(6.74)

where $\omega_1(x)$ and $\omega_2(x)$ are the differentiable real-analytic continuous odd functions; $P_\omega(\cdot)$ is the diagonal mapping, $P_\omega(\cdot):\mathbb{R}^n\to\mathbb{R}^{n\times n}$, $P_\omega(\cdot)>0$,

$$P_\omega(\tfrac{\partial\omega_2}{\partial x}) = \frac{\partial\omega_2}{\partial x} P \frac{\partial\omega_2}{\partial x} = \begin{bmatrix} p_{11}\left(\frac{\partial\omega_2}{\partial x_1}\right)^2 & \cdots & 0 \\ \vdots & \ddots & \vdots \\ 0 & \cdots & p_{nn}\left(\frac{\partial\omega_2}{\partial x_n}\right)^2 \end{bmatrix},$$

$\forall p_{ii}>0$; $Q\in\mathbb{R}^{n\times n}$ and $P\in\mathbb{R}^{n\times n}$ are the positive definite diagonal matrices.

System performance and stability are specified by positive definite scalar integrands $\omega_1^T(x)Q\omega_1(x)$ and $\dot{\omega}_2^T(x)P\dot{\omega}_2(x)$. Odd functions $\omega_1(x)$ and $\omega_2(x)$ are symmetric within origin, and, $\omega_1(-x)=-\omega_1(x)$, $\omega_2(-x)=-\omega_2(x)$. The examples of odd functions are x^{2n+1}, $\sinh^{2n+1}(x)$, $\tanh^{2n+1}(x)$, $\operatorname{atanh}^{2n+1}(x)$ and others, $n\geq 0$.

Energy and Performance Integrands – Scalar positive definite integrands $\omega_1^T(x)Q\omega_1(x)$ and $\dot{\omega}_2^T(x)P\dot{\omega}_2(x)$ assess and assert the *admissible* energy and work changes in systems. There is a gradient vector field $\nabla\omega_2$ and total derivative $\dot{\omega}_2(x)$ duality. For a scalar function $\omega_2(x)$,

$$\dot{\omega}_2(x) = \frac{\partial\omega_2}{\partial x}\dot{x} = \begin{bmatrix} \frac{\partial\omega_2}{\partial x_1} & \cdots & 0 \\ \vdots & \ddots & \vdots \\ 0 & \cdots & \frac{\partial\omega_2}{\partial x_n} \end{bmatrix} \begin{bmatrix} \dot{x}_1 \\ \vdots \\ \dot{x}_n \end{bmatrix}, \dot{\omega}_2(\cdot,\cdot):\mathbb{R}^n\times\mathbb{R}^m\to\mathbb{R}, \frac{\partial\omega_2}{\partial x}:\mathbb{R}^n\to\mathbb{R}^{n\times n}. \quad (6.75)$$

Scalar $\omega_1^T(x)Q\omega_1(x)$ and $\dot{\omega}_2^T(x)P\dot{\omega}_2(x)$ are consistent with laws of physics. One selects and defines physics- and system-consistent, differentiable and integrable real-valued functions $\omega_1(x)$ and $\omega_2(x)$.

Example 6.33. *Integrands and Functions*
Functions $\omega_1(x)=x$, $\omega_2(x)=x$ yield a quadratic positive definite integrands. For a linear system $\dot{x}=Ax+Bu$, the functional is

$$J = \min_{x,u} \int_{t_0}^{t_f} \tfrac{1}{2}\left[x^TQx + (Ax+Bu)^T P(Ax+Bu)\right]dt, \; Q\in\mathbb{R}^{n\times n}, P\in\mathbb{R}^{n\times n}$$

Functions

$\omega_i(x)=x^{2n+1}$, $\omega_i(x)=\tanh^{2n+1}(x)$ and $\omega_i(x)=e^{-|x|}$, $i=1,2$, $n\geq 0$

result in nonquadratic functionals with positive definite integrands. ■

Example 6.34. *Controlled Systems Motion and Integrands*

Classical mechanics defines the force and torque as

$F=\nabla W$, $T=\nabla W$,

where W is the positive definite energy function.

The gradient vector field $\nabla\omega_2$ pertains to acting force, *generalized force*, actuation and motion. A positive definite energy- and control consistent integrand

$$\tfrac{1}{2}\dot{\omega}_2^T(x)P\dot{\omega}_2(x)=\tfrac{1}{2}\dot{x}^T\frac{\partial\omega_2}{\partial x}P\frac{\partial\omega_2}{\partial x}\dot{x}$$

is adequate solving control and governance optimization problems. ■

In general, $\omega_1(x)\neq\omega_2(x)$. Using $\omega(x)=\omega_1(x)=\omega_2(x)$, functional (6.74) becomes

$$J=\min_{x,u}\int_{t_0}^{t_f}\tfrac{1}{2}\left[\omega^T(x)Q\omega(x)+\dot{\omega}^T(x)P\dot{\omega}(x)\right]dt=\min_{x,u}\int_{t_0}^{t_f}\tfrac{1}{2}\left[\omega^T(x)Q\omega(x)+\dot{x}^T\frac{\partial\omega}{\partial x}P\frac{\partial\omega}{\partial x}\dot{x}\right]dt,(Q,P)>0, \quad (6.76)$$

Example 6.35. *Performance Integrands*

Consider a linear system

$\dot{x}=u$.

Functional (6.76) with $\omega(x)=x$, $Q=1$ and $P=1$, yields

$$J=\min_{x,u}\int_{t_0}^{t_f}\tfrac{1}{2}\left(x^2+\dot{x}^2\right)dt=\min_{x,u}\int_{t_0}^{t_f}\tfrac{1}{2}\left(x^2+u^2\right)dt.$$

For $\omega(x)=\tanh(x)$, from $\frac{d}{dx}\tanh x=\mathrm{sech}^2 x$, we have

$$J=\min_{x,u}\int_{t_0}^{t_f}\omega(x,u)dt=\min_{x,u}\int_{t_0}^{t_f}\tfrac{1}{2}\left[\tanh^2 x+\left(\tfrac{d}{dx}\tanh x\right)^2\dot{x}^2\right]dt=\min_{x,u}\int_{t_0}^{t_f}\tfrac{1}{2}\left(\tanh^2 x+\mathrm{sech}^4 x u^2\right)dt.$$

Surfaces for positive definite performance integrands $\omega(x,u)=\tfrac{1}{2}\left(x^2+u^2\right)$ and $\omega(x,u)=\tfrac{1}{2}\left(\tanh^2 x+\mathrm{sech}^4 x u^2\right)$ are depicted in Figures 6.21. The MATLAB code is

```
x=linspace(-2,2, 50); u=x; [X,U]=meshgrid(x,u);
W1=(X.*X+U.*U)./2; W2=(tanh(X).^2+(sech(X).^4).*U.^2)./2;
figure(1); surf(x,u,W1); xlabel('{\itx}','FontSize',22'); ylabel('{\itu}','FontSize',22');
figure(2); surf(x,u,W2); xlabel('{\itx}','FontSize',22'); ylabel('{\itu}','FontSize',22');
```

Integrand $\omega(x,u)=\tfrac{1}{2}\left(x^2+u^2\right)$

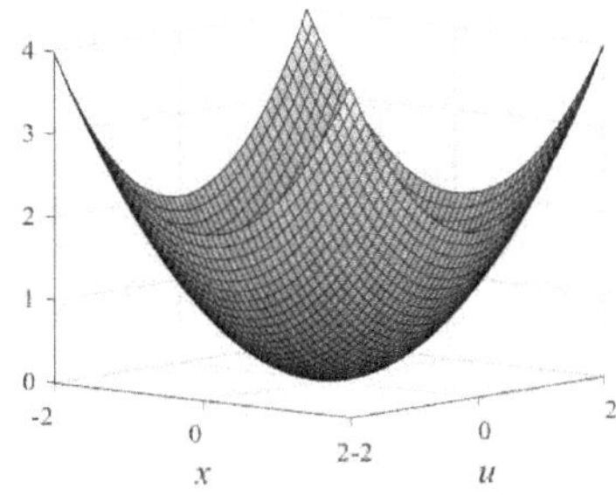

Integrand $\omega(x,u)=\tfrac{1}{2}\left(\tanh^2 x+\mathrm{sech}^4 x u^2\right)$

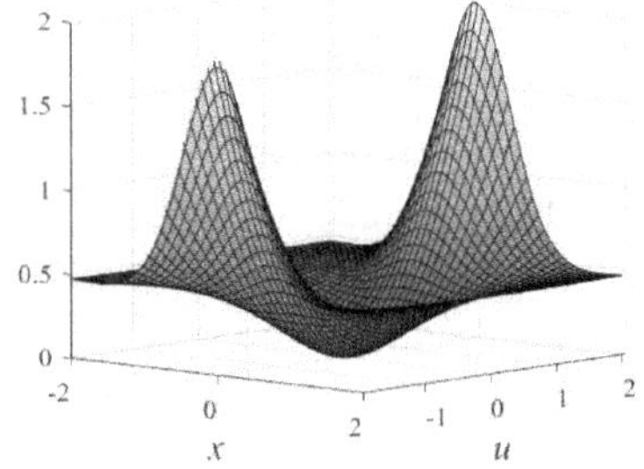

Figure 6.21. Surfaces for integrands $\omega(x,u)=\tfrac{1}{2}\left(x^2+u^2\right)$ and $\omega(x,u)=\tfrac{1}{2}\left(\tanh^2 x+\mathrm{sech}^4 x u^2\right)$. ■

For a performance functional (6.74), considering linear and nonlinear systems (6.12) and (6.18), the Hamiltonian functions are

$$H(x,u,\tfrac{\partial V}{\partial x})=\tfrac{1}{2}\boldsymbol{\omega}_1^T(x)Q\boldsymbol{\omega}_1(x)+\tfrac{1}{2}\dot{\boldsymbol{\omega}}_2^T(x)P\dot{\boldsymbol{\omega}}_2(x)+\left(\frac{\partial V}{\partial x}\right)^T(Ax+Bu)$$
$$=\tfrac{1}{2}\boldsymbol{\omega}_1^T(x)Q\boldsymbol{\omega}_1(x)+\tfrac{1}{2}(Ax+Bu)^T P_\omega(\tfrac{\partial\boldsymbol{\omega}_2}{\partial x})(Ax+Bu)+\left(\frac{\partial V}{\partial x}\right)^T(Ax+Bu), \quad (6.77)$$

$$H(x,u,\tfrac{\partial V}{\partial x})=\tfrac{1}{2}\boldsymbol{\omega}_1^T(x)Q\boldsymbol{\omega}_1(x)+\tfrac{1}{2}(F(x)+B(x)u)^T P_\omega(\tfrac{\partial\boldsymbol{\omega}_2}{\partial x})(F(x)+B(x)u)+\left(\frac{\partial V}{\partial x}\right)^T(F(x)+B(x)u).$$

The first-order necessary condition for optimality (6.22) yields optimal control laws for linear and nonlinear systems

$$u=-\left(B^TP_\omega(\tfrac{\partial\boldsymbol{\omega}_2}{\partial x})B\right)^{-1}B^T\left(P_\omega(\tfrac{\partial\boldsymbol{\omega}_2}{\partial x})Ax+\frac{\partial V}{\partial x}\right),\ P_\omega(\cdot):\mathbb{R}^n\rightarrow\mathbb{R}^{n\times n}, \quad (6.78)$$
$$u=-\left(B^T(x)P_\omega(\tfrac{\partial\boldsymbol{\omega}_2}{\partial x})B(x)\right)^{-1}B^T(x)\left(P_\omega(\tfrac{\partial\boldsymbol{\omega}_2}{\partial x})F(x)+\frac{\partial V}{\partial x}\right).$$

The second-order condition for optimality is guaranteed because $\dfrac{\partial^2 H}{\partial u\times\partial u^T}=B^TP_\omega(\tfrac{\partial\boldsymbol{\omega}_2}{\partial x})B>0$.

From (6.77) and (6.78), one obtains the Hamilton-Jacobi equation

$$-\frac{\partial V}{\partial t}=\boldsymbol{\omega}_1^T(x)Q\boldsymbol{\omega}_1(x)-\left(\frac{\partial V}{\partial x}\right)^TB\left(B^TP_\omega(\tfrac{\partial\boldsymbol{\omega}_2}{\partial x})B\right)^{-1}B^T\frac{\partial V}{\partial x}. \quad (6.79)$$

Solution of (6.79) is found by using (6.24)

$$V(t,x)=\sum_{i=N_0,j=M_0}^{N,M}\tfrac{j}{2i}(x^{i/j})^TK_{ij}(t)x^{i/j},\ V(t,x)=\sum_{i,j}\kappa_{1_i}^T(t,x^{<d_i>})K_{ij}(t)\kappa_{2_j}(t,x^{<d_j>}),\ N_0\geq 1, M_0\geq 1. \quad (6.80)$$

Optimality, dynamics and stability are affected by a functional, integrands $\boldsymbol{\omega}_1^T(x)Q\boldsymbol{\omega}_1(x)$ and $\dot{\boldsymbol{\omega}}_2^T(x)P\dot{\boldsymbol{\omega}}_2(x)$, return functions, and, control laws.

Quadratic Optimization – Performance integrands $\boldsymbol{\omega}_1^T(x)Q\boldsymbol{\omega}_1(x)$ and $\dot{\boldsymbol{\omega}}_2^T(x)P\dot{\boldsymbol{\omega}}_2(x)$ affect system dynamics, control law structure, optimality and stability. Integrable and differentiable functions $\boldsymbol{\omega}_1(x)$ and $\boldsymbol{\omega}_2(x)$ are consistent with laws of physics, device physics and control objectives. Minimize (6.76) with $\boldsymbol{\omega}(x)=x$

$$J=\min_{x,u}\int_{t_0}^{t_f}\tfrac{1}{2}\left[x^TQx+(Ax+Bu)^TP(Ax+Bu)\right]dt,\ Q\in\mathbb{R}^{n\times n},\ P\in\mathbb{R}^{n\times n} \quad (6.81)$$

subject to (6.12) $\dot{x}=Ax+Bu$. With $(Q,P)>0$, the positive definite integrands are $\tfrac{1}{2}x^TQx>0$ and $\tfrac{1}{2}(Ax+Bu)^TP(Ax+Bu)>0$.

The Hamiltonian function is

$$H(x,u,\tfrac{\partial V}{\partial x})=\tfrac{1}{2}\left[x^TQx+(Ax+Bu)^TP(Ax+Bu)\right]+\left(\frac{\partial V}{\partial x}\right)^T(Ax+Bu). \quad (6.82)$$

Applying (6.22), an optimal control is

$$u=-(B^TPB)^{-1}B^T\left(PAx+\frac{\partial V}{\partial x}\right). \quad (6.83)$$

The Hamilton-Jacobi equation (6.79)

$$-\frac{\partial V}{\partial t}=\tfrac{1}{2}x^TQx-\tfrac{1}{2}\left(\frac{\partial V}{\partial x}\right)^TB\left(B^TPB\right)^{-1}B^T\frac{\partial V}{\partial x}. \quad (6.84)$$

Equation (6.84) is satisfied by the quadratic return function $V=\frac{1}{2}x^TKx$.

From (6.83) with $V=\frac{1}{2}x^TKx$, we have

$$u=-(B^TPB)^{-1}B^T(PA+K)x, \tag{6.85}$$
$$-\dot{K}=Q-KB(B^TPB)^{-1}B^TK,\ K(t_f)=K_f.$$

For invertable L and D, the matrix product LD is invertible, and, $(LD)^{-1}=D^{-1}L^{-1}$. We derived an equation to compute a symmetric $K(t)$.

Dynamic Optimization – The reported concept ensures consistency to quadratic regulator and nonlinear optimization problems. Minimize a functional

$$J=\min_{x,u}\int_{t_0}^{t_f}\tfrac{1}{2}\left[\omega_1^T(x)Q\omega_1(x)+\dot{\omega}_2^T(x)P\dot{\omega}_2(x)\right]dt=\min_{x,u}\int_{t_0}^{t_f}\tfrac{1}{2}\left[\omega_1^T(x)Q\omega_1(x)+\dot{x}^T\circ P_\omega(\tfrac{\partial\omega_2}{\partial x})\circ\dot{x}\right]dt$$

$$=\min_{x,u}\int_{t_0}^{t_f}\tfrac{1}{2}\left[\omega_1^T(x)Q\omega_1(x)+\dot{x}^T\circ\begin{bmatrix}p_{11}\left(\frac{\partial\omega_2}{\partial x_1}\right)^2 & \cdots & 0\\ \vdots & \ddots & \vdots\\ 0 & \cdots & p_{nn}\left(\frac{\partial\omega_2}{\partial x_n}\right)^2\end{bmatrix}\circ\dot{x}\right]dt, \tag{6.86}$$

where $\circ$ is the element-wise Hadamard product.

Hamiltonian functions (6.77) for linear and nonlinear systems (6.12) and (6.18) are

$$H(x,u,\tfrac{\partial V}{\partial x})=\tfrac{1}{2}\omega_1^T(x)Q\omega_1(x)+\tfrac{1}{2}\dot{x}^T\circ\frac{\partial\omega_2}{\partial x}P\frac{\partial\omega_2}{\partial x}\circ\dot{x}+\left(\frac{\partial V}{\partial x}\right)^T(Ax+Bu)$$
$$=\tfrac{1}{2}\omega_1^T(x)Q\omega_1(x)+\tfrac{1}{2}x^TA^TP_\omega(\tfrac{\partial\omega_2}{\partial x})Ax+\tfrac{1}{2}u^TB^TP_\omega(\tfrac{\partial\omega_2}{\partial x})Bu+\left(\frac{\partial V}{\partial x}\right)^T(Ax+Bu), \tag{6.87}$$
$$H(x,u,\tfrac{\partial V}{\partial x})=\tfrac{1}{2}\omega_1^T(x)Q\omega_1(x)+\tfrac{1}{2}F^T(x)P_\omega(\tfrac{\partial\omega_2}{\partial x})F(x)+\tfrac{1}{2}u^TB^T(x)P_\omega(\tfrac{\partial\omega_2}{\partial x})B(x)u$$
$$+\left(\frac{\partial V}{\partial x}\right)^T(F(x)+B(x)u).$$

The necessary condition for optimality (6.22) yields control laws

$$u=-\left(B^TP_\omega(\tfrac{\partial\omega_2}{\partial x})B\right)^{-1}B^T\frac{\partial V}{\partial x}, \tag{6.88}$$
$$u=-\left(B^T(x)P_\omega(\tfrac{\partial\omega_2}{\partial x})B(x)\right)^{-1}B^T(x)\frac{\partial V}{\partial x}.$$

The Hamilton-Jacobi equations

$$-\frac{\partial V}{\partial t}=\tfrac{1}{2}\omega_1^T(x)Q\omega_1(x)+\tfrac{1}{2}x^TA^TP_\omega(\tfrac{\partial\omega_2}{\partial x})Ax+\left(\frac{\partial V}{\partial x}\right)^TAx-\tfrac{1}{2}\left(\frac{\partial V}{\partial x}\right)^TB\left(B^TP_\omega(\tfrac{\partial\omega_2}{\partial x})B\right)^{-1}B^T\frac{\partial V}{\partial x},$$
$$-\frac{\partial V}{\partial t}=\tfrac{1}{2}\omega_1^T(x)Q\omega_1(x)+\tfrac{1}{2}F^T(x)P_\omega(\tfrac{\partial\omega_2}{\partial x})F(x)+\left(\frac{\partial V}{\partial x}\right)^TF(x) \tag{6.89}$$
$$-\tfrac{1}{2}\left(\frac{\partial V}{\partial x}\right)^TB(x)\left(B^T(x)P_\omega(\tfrac{\partial\omega_2}{\partial x})B(x)\right)^{-1}B^T(x)\frac{\partial V}{\partial x}$$

are satisfied or approximated by the return function (6.80). Solving (6.89), one obtains control law structure and feedback gains in (6.88). Nonquadratic functionals with integrands $(\omega_1(x),\omega_2(x))$ are used and minimized, closed-form solution of the functional Hamilton-Jacobi equations is found, and, the nonlinear control laws are designed.

Illustrative Studies 6.1. *Equivalency of Regulators Design*
For linear systems $\dot{x}=Ax+Bu$, consider a quadratic functional, found using $\omega(x)=\omega_1(x)=\omega_2(x)$, $\omega(x)=x$. Minimize (6.86) with $\omega(x)=\omega_1(x)=\omega_2(x)$, $\omega(x)=x$

$$J=\min_{x,u}\int_{t_0}^{t_f}\tfrac{1}{2}\left[x^TQx+\dot{x}^T\circ P\circ\dot{x}\right]dt=\min_{x,u}\int_{t_0}^{t_f}\tfrac{1}{2}\left[x^TQx+(Ax+Bu)^T\circ P\circ(Ax+Bu)\right]dt$$

$$=\min_{x,u}\int_{t_0}^{t_f}\tfrac{1}{2}\left[x^TQx+x^TA^TPAx+u^TB^TPBu\right]dt,$$

where $\circ$ is the element-wise Hadamard product.

The Hamilton-Jacobi functional equation (6.89) is satisfied by a quadratic return function $V=\frac{1}{2}x^TKx$. From (6.88), a linear control law is

$u=-(B^TPB)^{-1}B^TKx$, $-\dot{K}=Q+A^TPA+A^TK+KA-KB(B^TPB)^{-1}B^TK$, $K(t_f)=K_f$.

Recall that a solution of the LQR problem yields

$u=-G^{-1}B^TKx$, $-\dot{K}=Q+A^TK+KA-KBG^{-1}B^TK$, $K(t_f)=K_f$. ■

Example 6.36. *Design of Optimal Control Laws*
Open-Loop Unstable System – Consider the first-order unstable system

$\dot{x}=ax+bu$, $(a,b)>0$.

Minimize the quadratic performance functional (6.81), where $\omega(x)=x$. Hence,

$$J=\min_{x,u}\int_{t_0}^{\infty}\tfrac{1}{2}\left[Q\omega^2(x)+P\left(\tfrac{\partial\omega}{\partial x}\right)^2(ax+bu)^2\right]dt=\min_{x,u}\int_{t_0}^{\infty}\tfrac{1}{2}\left(x^2+a^2x^2+2abxu+b^2u^2\right)dt,\ Q=1,\ P=1.$$

Using the quadratic return function $V=\frac{1}{2}kx^2$, a linear control law (6.85) is

$u=-\frac{1}{b}(a+k)x$.

A differential equation $-\dot{k}=1-k^2$ with $t_f=\infty$, yields $1-k^2=0$. We obtain $k=1$.

The control law is $u=-\frac{1}{b}(a+1)x$, $a>0$.

The closed-loop system is stable, and, evolves as $\dot{x}=-x$.

Open-Loop Stable System – Study a system

$\dot{x}=ax+bu$, $a<0$, $b>0$.

Minimize the quadratic performance functional (6.86) with $\omega(x)=x$, which yields

$$J=\min_{x,u}\int_{t_0}^{\infty}\tfrac{1}{2}\left[Q\omega^2(x)+(ax+bu)\circ P\left(\tfrac{\partial\omega}{\partial x}\right)^2\circ(ax+bu)\right]dt=\min_{x,u}\int_{t_0}^{\infty}\tfrac{1}{2}\left(Qx^2+Pa^2x^2+Pb^2u^2\right)dt\cdot$$

A solution of the Hamilton-Jacobi equation is satisfied by the quadratic return function $V=\frac{1}{2}kx^2$. Let $Q=1$ and $P=1$. The control law is

$u=-\frac{1}{b}kx$.

Let $a=-1$ and $b=1$.

The resulting differential equation is $-\dot{k}=2-2k-k^2$.

For $t_f=\infty$, solution of $-k^2-2k+2=0$ yields $k=0.732$.

The control law is $u=-\frac{1}{b}kx=-0.732x$.

The closed-loop system $\dot{x}=-1.732x$ is stable. ■

Example 6.37. *Control Law Design by Minimizing Nonquadratic Functionals*
Consider an open-loop unstable system

$\dot{x}=ax+bu$, $(a,b)>0$.

Minimize the performance functional (6.76)

$$J=\min_{x,u}\int_{t_0}^{\infty}\tfrac{1}{2}\left[Q\omega^2(x)+P\left(\frac{\partial\omega}{\partial x}\right)^2\left(a^2x^2+2abxu+b^2u^2\right)\right]dt\,.$$

Positive definite nonquadratic integrands are found for $\omega(x)=\tanh(x)$. For $Q=1$ and $P=1$,

$$J=\min_{x,u}\int_{t_0}^{\infty}\tfrac{1}{2}\left[\tanh^2 x+\operatorname{sech}^4 x\left(a^2x^2+2abxu+b^2u^2\right)\right]dt\,.$$

For $x\ll 1$, $\tanh^2 x\approx x^2$ and $\operatorname{sech}^4 x\approx 1$. We have a quadratic functional

$$J\approx\min_{x,u}\int_{t_0}^{\infty}\tfrac{1}{2}\left[x^2+a^2x^2+2abxu+b^2u^2\right]dt\,.$$

For $x>1$, $\tanh^2 x\approx 1$ and $\operatorname{sech}^4 x\approx 0$. The performance functional becomes

$$J\approx\min_{t,u}\int_{t_0}^{\infty}\tfrac{1}{2}\,dt\,.$$

Recall that $J=\min_{t,u\in U}\int_{t_0}^{t_f}1dt$ was used to solve the minimum-time control problem which yields a discontinuous control function.

Applying the necessary condition for optimality (6.22), from (6.78)

$$u=-\frac{a}{b}x-\frac{1}{b\operatorname{sech}^4 x}\frac{\partial V}{\partial x}\,.$$

The algebraic equation (6.79) is $\tfrac{1}{2}\tanh^2 x-\dfrac{1}{2\operatorname{sech}^4 x}\left(\dfrac{\partial V}{\partial x}\right)^2=0$.

This equation is solved approximating a solution by return functions $V(x)$, as well as expressing continuous functions $\tanh^2 x$ and $\operatorname{sech}^4 x$ for $x\in X$. Quadratic and nonquadratic return functions are applied.

Quadratic Return Function – For $x\ll 1$, solution of an equation

$$\tfrac{1}{2}\tanh^2 x-\frac{1}{2\operatorname{sech}^4 x}\left(\frac{\partial V}{\partial x}\right)^2=0$$

is approximated by the quadratic return function $V=\tfrac{1}{2}kx^2$.

We have $u=-\dfrac{a}{b}x-\dfrac{1}{b\operatorname{sech}^4 x}kx$.

Using $\tanh^2 x\approx x^2$ and $\operatorname{sech}^4 x\approx 1$, one finds $1-k^2=0$. Hence, $k=1$.

The closed-loop system dynamics is $\dot{x}=-\dfrac{k}{\operatorname{sech}^4 x}x$.

We have a nonlinear high-gain control law.

If $x\ll 1$, one has $\operatorname{sech}^4 x\approx 1$. Therefore, $u\approx-\dfrac{a+k}{b}x$.

Nonquadratic Return Function – To solve the algebraic equation

$$\tfrac{1}{2}\tanh^2 x-\frac{1}{2\operatorname{sech}^4 x}\left(\frac{\partial V}{\partial x}\right)^2=0,$$

apply a nonquadratic positive definite return function (6.80) $V(x)=k\left(x\tanh x-\ln\cosh x\right)$.

For $V(x)>0$, one has $\frac{\partial V}{\partial x}=kx\,\text{sech}^2 x$.

Solution of equation $\frac{1}{2}\tanh^2 x-\frac{1}{2}k^2x^2=0$ is found using $\tanh^2 x\approx x^2$. We have $k=1$.

A control law is $u=-\frac{a}{b}x-\frac{1}{b\text{sech}^4 x}\frac{\partial V}{\partial x}=-\frac{a}{b}x-\frac{1}{b\text{sech}^2 x}kx$, $k=1$.

Solution of the Hamilton-Jacobi equation and control law structure depend on return function (6.80). A consistent design implies finding a return function $V(x)$ which satisfies or approximates the Hamilton-Jacobi equation.

Study transient dynamics of the closed-loop system

$\dot{x}=ax+bu$, $a=1$, $b=1$.

We derived control laws $u=-\frac{a}{b}x-\frac{1}{b\text{sech}^4 x}kx$, $k=1$,

and $u=-\frac{a}{b}x-\frac{1}{b\text{sech}^2 x}kx$, $k=1$.

For three initial conditions $x_0=10$, $x_0=1$ and $x_0=-5$, dynamics of $x(t)$ are reported in Figures 6.22.a and b. The control bounds, if imposed, will increase the settling time.

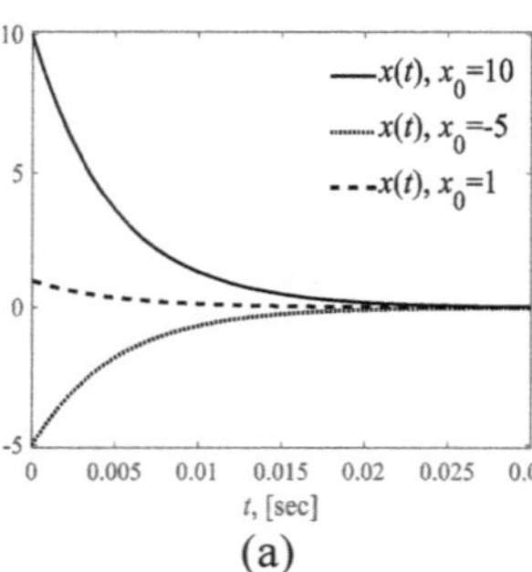

(a)

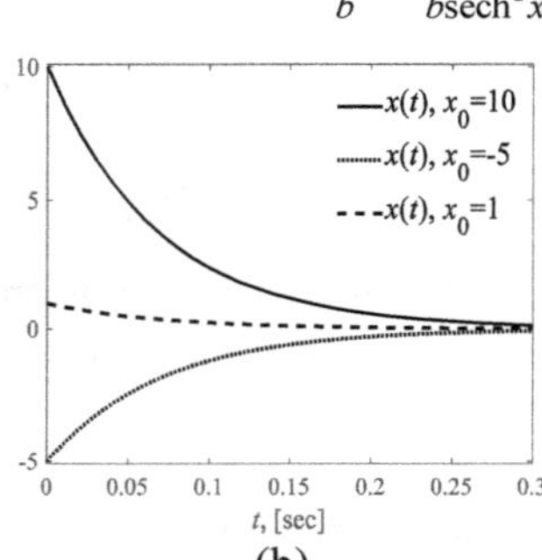

(b)

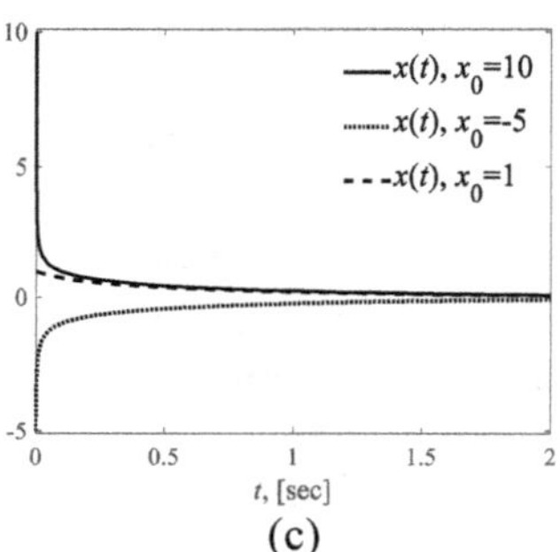

(c)

Figure 6.22. (a) Transients of the closed-loop system $\dot{x}=ax+bu$, $u=-\frac{a}{b}x-\frac{1}{b\text{sech}^4 x}kx$, $k=1$ for different initial conditions $x_0=10$, $x_0=1$ and $x_0=-5$;
(b) Dynamics of $x(t)$ for a system $\dot{x}=ax+bu$, $u=-\frac{a}{b}x-\frac{1}{b\text{sech}^2 x}kx$, $k=1$ with different x_0;
(c) Transients of the system $\dot{x}=ax+bu$, $u=-\frac{a}{b}x-\frac{1}{b}(x+2x^3+x^5)$ for different x_0, Example 6.38.■

Example 6.38. *Nonquadratic Performance Functional, and, Control Law Design*
Consider an unstable system

$\dot{x}=ax+bu$, $(a,b)>0$.

Find the control law by minimizing the performance functional (6.76) with $\omega(x)=\omega_1(x)=\omega_2(x)$, $\omega(x)=\frac{2}{\pi}\arctan x$, $Q=1$, $P=1$. We have

$$J=\min_{x,u}\int_{t_0}^{\infty}\frac{1}{2}\left[\left(\frac{2}{\pi}\arctan x\right)^2+\left(\frac{d}{dx}\left(\frac{2}{\pi}\arctan x\right)\right)^2(ax+bu)^2\right]dt,\ \frac{d}{dx}\left(\frac{2}{\pi}\arctan x\right)=\frac{2}{\pi}\frac{1}{x^2+1}.$$

Hence, $J=\min_{x,u}\int_0^\infty \frac{1}{2}\left[\left(\frac{2}{\pi}\arctan x\right)^2+\frac{4}{\pi^2}\left(\frac{1}{x^2+1}\right)^2(ax+bu)^2\right]dt$.

For resulting Hamiltonian (6.77)

$$H(x,u,\tfrac{\partial V}{\partial x})=\frac{1}{2}\left(\frac{4}{\pi^2}\arctan^2 x+\frac{4}{\pi^2}\frac{1}{(x^2+1)^2}(ax+bu)^2\right)+\frac{\partial V}{\partial x}(ax+bu),$$

apply the necessary condition for optimality (6.22). From

$$\frac{\partial H(x,u,\frac{\partial V}{\partial x})}{\partial u}=\frac{4}{\pi^2}\frac{1}{(x^2+1)^2}abx+\frac{4}{\pi^2}\frac{1}{(x^2+1)^2}b^2u+\frac{\partial V}{\partial x}b=0,$$

an optimal control law is

$$u=-\frac{a}{b}x-\frac{\pi^2(x^2+1)^2}{4b}\frac{\partial V}{\partial x}.$$

From $\frac{2}{\pi^2}\arctan^2 x+\frac{2}{\pi^2}\frac{1}{(x^2+1)^2}(ax+bu)^2+\frac{\partial V}{\partial x}(ax+bu)=0$, one has

$$\frac{2}{\pi^2}\arctan^2 x+\frac{2}{\pi^2}\frac{1}{(x^2+1)^2}\left(\frac{\pi^2(x^2+1)}{4}\frac{\partial V}{\partial x}\right)^2-\frac{\partial V}{\partial x}\frac{\pi^2(x^2+1)^2}{4}\frac{\partial V}{\partial x}=0.$$

The resulting equation

$$\frac{2}{\pi^2}\arctan^2 x-\frac{\pi^2(x^2+1)^2}{8}\left(\frac{\partial V}{\partial x}\right)^2=0$$

is approximated by the quadratic return function $V=\frac{1}{2}kx^2$.

From $\arctan x=x-\frac{1}{3}x^3+\frac{1}{5}x^5-\ldots\approx x$, we have

$$\frac{2}{\pi^2}x^2-\frac{\pi^2}{8}(1+x^2)^2k^2x^2=0.$$

Grouping the terms with x^2, obtain $\frac{2}{\pi^2}-\frac{\pi^2}{8}k^2=0$.

Hence, $k=\pm\sqrt{\frac{16}{\pi^4}}$, $k=\frac{4}{\pi^2}$.

From $u=-\frac{a}{b}x-\frac{\pi^2(x^2+1)^2}{4b}\frac{\partial V}{\partial x}$, $V=\frac{1}{2}kx^2$, we have

$$u=-\frac{a}{b}x-\frac{1}{b}(x+2x^3+x^5).$$

The closed-loop system is

$$\dot{x}=ax+bu=ax+b\left(-\frac{a}{b}x-\frac{1}{b}(x+2x^3+x^5)\right)=-x-2x^3-x^5.$$

Stability is examined by using a positive definite quadratic Lyapunov function $V_L=\frac{1}{2}x^2$.

The total derivative $\frac{dV_L}{dt}=\frac{dV_L}{dx}\frac{dx}{dt}=-x^2-2x^4-x^6<0$ is negative definite.

The closed-loop system is asymptotically stable.

Simulate a closed-loop system

$$\dot{x}=ax+bu,\ u=-\frac{a}{b}x-\frac{1}{b}(x+2x^3+x^5)$$

with different initial conditions $x_0=10$, $x_0=1$ and $x_0=-5$.

Dynamics of $x(t)$ are reported in Figure 6.22.c. ■

Example 6.39. *Nonlinear Optimization and Hamilton-Jacobi Equation*
Study the second-order unstable system

$$\dot{x}_1 = ax_1 + bu, \ (a,b)>0,$$
$$\dot{x}_2 = x_1.$$

Minimize a functional (6.76) if $\omega(x)=\omega_1(x)=\omega_2(x)=\tanh(x)$, $Q=I$ and $P=I$

$$J=\min_{x,u}\int_{t_0}^{\infty}\tfrac{1}{2}\left\{\begin{bmatrix}\tanh x_1 & \tanh x_2\end{bmatrix}\begin{bmatrix}1 & 0\\ 0 & 1\end{bmatrix}\begin{bmatrix}\tanh x_1\\ \tanh x_2\end{bmatrix}+\begin{bmatrix}\dot{x}_1\text{sech}^2x_1 & \dot{x}_2\text{sech}^2x_2\end{bmatrix}\begin{bmatrix}1 & 0\\ 0 & 1\end{bmatrix}\begin{bmatrix}\dot{x}_1\text{sech}^2x_1\\ \dot{x}_2\text{sech}^2x_2\end{bmatrix}\right\}dt$$

$$=\min_{x,u}\int_{t_0}^{\infty}\tfrac{1}{2}\left[\tanh^2 x_1+\tanh^2 x_2+\text{sech}^4x_1\left(a^2x_1^2+2abx_1u+b^2u^2\right)+x_1^2\text{sech}^4x_2\right]dt.$$

A scalar continuous Hamiltonian function is

$$H(x,u,\tfrac{\partial V}{\partial x})=\tfrac{1}{2}\left[\tanh^2 x_1+\tanh^2 x_2+\text{sech}^4x_1\left(a^2x_1^2+2abx_1u+b^2u^2\right)+x_1^2\text{sech}^4x_2\right]$$
$$+\frac{\partial V}{\partial x_1}\left(ax_1+bu\right)+\frac{\partial V}{\partial x_2}x_1.$$

Apply a necessary condition for optimality (6.22). A control law (6.78) is

$$u=-\frac{a}{b}x_1-\frac{1}{b\text{sech}^4x_1}\frac{\partial V}{\partial x_1}.$$

To solve (6.79) $\tfrac{1}{2}\tanh^2 x_1+\tfrac{1}{2}\tanh^2 x_2+\tfrac{1}{2}x_1^2\text{sech}^4x_2-\frac{1}{2\text{sech}^4x_1}\left(\frac{\partial V}{\partial x_1}\right)^2+\frac{\partial V}{\partial x_2}x_1=0$, apply different return functions $V(x)$, and, approximate the hyperbolic functions.

The Taylor series are

$$\tanh x = x-\tfrac{1}{3}x^3+\tfrac{2}{15}x^5-\tfrac{17}{315}x^7+\ldots=\sum_{n=1}^{\infty}\tfrac{1}{(2n)!}2^{2n}(2^{2n}-1)B_{2n}x^{2n-1}$$

and $$\text{sech}x = 1-\tfrac{1}{2}x^2+\tfrac{5}{24}x^4-\tfrac{61}{720}x^6+\ldots=\sum_{n=1}^{\infty}\tfrac{1}{(2n)!}E_{2n}x^{2n},$$

where B_n and E_n are the nth Bernoulli and Euler numbers.

If $|x_1|\le 1$ and $|x_2|\le 1$, $\tanh x\approx x$, and, if $x_1\ll 1$, $\text{sech}x\approx 1$.

Quadratic Return Function – Approximate solution of the algebraic equation by

$$V=\tfrac{1}{2}k_{11}x_1^2+k_{12}x_1x_2+\tfrac{1}{2}k_{22}x_2^2.$$

Equating the terms with x_1^2, x_1x_2 and x_2^2, for $a=1$ and $b=1$, one finds the coefficients of $V(x)$ to be $k_{11}=1.41$, $k_{12}=1$, $k_{22}=1.41$.

One has $u=-\frac{a}{b}x_1-\frac{1}{b\text{sech}^4x_1}(1.41x_1+x_2)$.

The closed-loop system is

$$\dot{x}_1=-\frac{1}{\text{sech}^4x_1}\frac{\partial V}{\partial x_1},$$
$$\dot{x}_2=x_1.$$

Figures 6.23.a document the transient dynamics of the closed-loop system with $u=-\frac{a}{b}x_1-\frac{1}{b\text{sech}^4x_1}(1.41x_1+x_2)$ if the initial conditions are $\begin{bmatrix}x_{10}\\ x_{20}\end{bmatrix}=\begin{bmatrix}0.5\\ -0.5\end{bmatrix}$, $\begin{bmatrix}x_{10}\\ x_{20}\end{bmatrix}=\begin{bmatrix}2\\ -2\end{bmatrix}$.
A nonlinear control law ensures suboptimal solutions with respect to the minimized functional because the Hamilton-Jacobi equation is approximated.

Stability Analysis – Examine stability of the closed-loop system. For a positive definite Lyapunov function

$$V_L = \tfrac{1}{2}x_1^2 + x_1x_2 + \tfrac{1}{2}x_2^2,$$

the total derivative is

$$\frac{dV_L}{dt} = -\frac{\partial V_L}{\partial x_1}\frac{1}{\operatorname{sech}^4 x_1}\frac{\partial V}{\partial x_1} + \frac{\partial V_L}{\partial x_2}x_1 = -(x_1+x_2)\frac{1}{\operatorname{sech}^4 x_1}(1.41x_1+x_2)+(x_1+x_2)x_1 < 0,$$

is negative definite. Recall that $\operatorname{sech}^4 x_1>0$. Hence, the system is asymptotically stable.

The surfaces for $V_L(x_1,x_2)>0$ and $dV_L/dt<0$ are depicted in Figure 6.23.b.

The MATLAB code is

```
x=linspace(-1,1,50); y=x; [x1,x2]=meshgrid(x,y); V=0.5*(x1.^2+x2.^2)+x1.*x2;
dV=(x1+x2).*x1- ((x1+x2).*(1.41*x1+x2))./(sech(x1).^4); hold off ;
surf(x,y,V) ; hold on ; surf(x,y,dV); xlabel('{\itx}_1','FontSize',22); ylabel('{\itx}_2','FontSize',22);
```

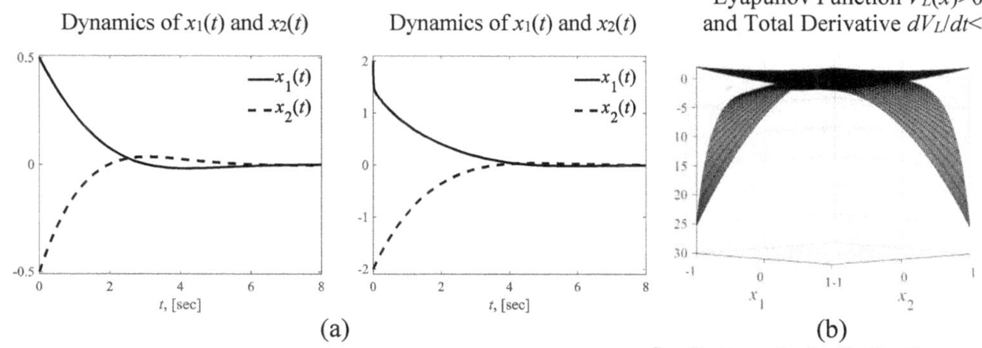

Figure 6.23. (a) Dynamics of the closed-loop system (x_1,x_2), $\begin{bmatrix} x_{10} \\ x_{20} \end{bmatrix} = \begin{bmatrix} 0.5 \\ -0.5 \end{bmatrix}$, $\begin{bmatrix} x_{10} \\ x_{20} \end{bmatrix} = \begin{bmatrix} 2 \\ -2 \end{bmatrix}$;
(b) Surfaces depict a positive definite Lyapunov function $V_L = \tfrac{1}{2}x_1^2 + x_1x_2 + \tfrac{1}{2}x_2^2$, $V_L(x)>0$, and, a negative definite total derivative $dV_L(x)/dt<0$.

Nonquadratic Return Function – In $(x_1,x_2)\in X$, solve a Hamilton-Jacobi equation

$$\tfrac{1}{2}\tanh^2 x_1 + \tfrac{1}{2}\tanh^2 x_2 + \tfrac{1}{2}x_1^2\operatorname{sech}^4 x_2 - \frac{1}{2\operatorname{sech}^4 x_1}\left(\frac{\partial V}{\partial x_1}\right)^2 + \frac{\partial V}{\partial x_2}x_1 = 0.$$

Apply a return function (6.80)

$$V(x_1,x_2) = k_{11}(x_1\tanh x_1 - \ln\cosh x_1) + k_{12}(\tanh x_1)x_2 + \tfrac{1}{2}k_{22}x_2^2, \; V(x_1,x_2)>0.$$

One has $\dfrac{\partial V}{\partial x_1} = k_{11}x_1\operatorname{sech}^2 x_1 + k_{12}(\operatorname{sech}^2 x_1)x_2$, $\dfrac{\partial V}{\partial x_2} = k_{12}\tanh x_1 + k_{22}x_2$.

Solve an equation

$$\tfrac{1}{2}\tanh^2 x_1 + \tfrac{1}{2}\tanh^2 x_2 + \tfrac{1}{2}x_1^2\operatorname{sech}^4 x_2 - \tfrac{1}{2}(k_{11}x_1 + k_{12}x_2)^2 + (k_{12}\tanh x_1 + k_{22}x_2)x_1 = 0,$$

where $\tanh x\approx x$, $\tanh^2 x\approx x^2$ and $\operatorname{sech}^4 x\approx 0$.

Grouping terms with x_1^2, x_1x_2 and x_2^2, solve a system of quadratic equations

$$\tfrac{1}{2} - \tfrac{1}{2}k_{11}^2 + k_{12} = 0,\; -k_{11}k_{12} + k_{22} = 0,\; \tfrac{1}{2} - \tfrac{1}{2}k_{12}^2 = 0,\; k_{11} = \sqrt{3},\; k_{12}=1,\; k_{22} = \sqrt{3}.$$

A nonlinear control law is $u = -\dfrac{a}{b}x_1 - \dfrac{1}{b\operatorname{sech}^2 x_1}(\sqrt{3}x_1 + x_2)$, $a=1$, $b=1$.

Figures 6.24.a report transient dynamics of the closed-loop system.

Control laws depend on return functions $V(x)$. A consistent $V(x_1,x_2)$ was applied to approximate solution of a Hamilton-Jacobi equation. We find a control law.

A surface for a positive definite return function

$V(x_1,x_2)=k_{11}(x_1\tanh x_1-\ln\cosh x_1)+k_{12}(\tanh x_1)x_2+\frac{1}{2}k_{22}x_2^2$, $k_{11}=\sqrt{3}$, $k_{12}=1$, $k_{22}=\sqrt{3}$ is documented in Figure 6.24.b.

Solution of the Hamilton-Jacobi Equation – The error for solved Hamilton-Jacobi equation is evaluated using the manifold $\Delta(x)$, given as

$$\Delta(x_1,x_2)=\tfrac{1}{2}\tanh^2 x_1+\tfrac{1}{2}\tanh^2 x_2+\tfrac{1}{2}x_1^2\text{sech}^4 x_2-\frac{1}{2\text{sech}^4 x_1}\left(\frac{\partial V}{\partial x_1}\right)^2+\frac{\partial V}{\partial x_2}x_1,$$

$$V(x_1,x_2)=k_{11}(x_1\tanh x_1-\ln\cosh x_1)+k_{12}(\tanh x_1)x_2+\tfrac{1}{2}k_{22}x_2^2,\ k_{11}=\sqrt{3},\ k_{12}=1,\ k_{22}=\sqrt{3}.$$

The manifold $\Delta(x)$ is evaluated for $(x_1,x_2)\in X$. Figure 6.24.c depicts the surface for $\Delta(x)$, and, errors $|\Delta_{ij}|_{\max}$ and $\|\Delta_{ij}\|_p$ depend on $(x_1,x_2)\in X$. The acceptable accuracy is ensured. Recall that for linear systems, solution of the Hamilton-Jacobi equation is approximated by a quadratic return function. Numeric analysis is reported in Example 5.28, where the error solving the Hamilton-Jacobi equation is $\forall\Delta_{ij}=0$.

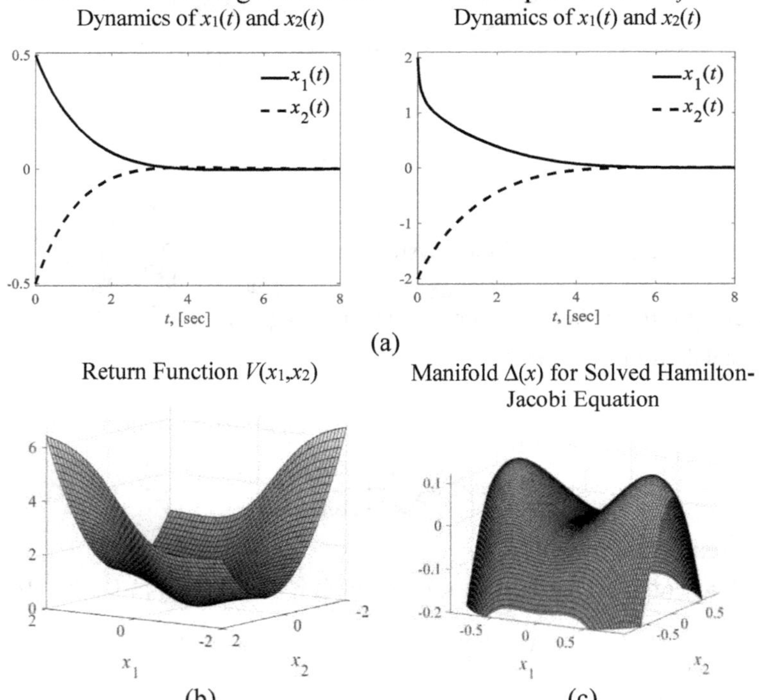

Figure 6.24. (a) Dynamics of the closed-loop system (x_1,x_2) with a nonlinear control law $u=-\frac{a}{b}x_1-\frac{1}{b\text{sech}^2 x_1}(\sqrt{3}x_1+x_2)$, $a=1$, $b=1$. The initial conditions are $\begin{bmatrix}x_{10}\\x_{20}\end{bmatrix}=\begin{bmatrix}0.5\\-0.5\end{bmatrix}$, $\begin{bmatrix}x_{10}\\x_{20}\end{bmatrix}=\begin{bmatrix}2\\-2\end{bmatrix}$;

(b) Surface for a continuous positive definite return function $V(x_1,x_2)=k_{11}(x_1\tanh x_1-\ln\cosh x_1)+k_{12}(\tanh x_1)x_2+\frac{1}{2}k_{22}x_2^2$, $k_{11}=\sqrt{3}$, $k_{12}=1$, $k_{22}=\sqrt{3}$;

(c) Surface for the manifold $\Delta(x)$, associated with the solution of the Hamilton-Jacobi equation, $\Delta(x_1,x_2)=\frac{1}{2}\tanh^2 x_1+\frac{1}{2}\tanh^2 x_2+\frac{1}{2}x_1^2\text{sech}^4 x_2-\frac{1}{2}(k_{11}x_1+k_{12}x_2)^2+(k_{12}\tanh x_1+k_{22}x_2)x_1$, $k_{11}=\sqrt{3}$, $k_{12}=1$, $k_{22}=\sqrt{3}$, $(x_1,x_2)\in X$. ■

6.10. Lyapunov Stability Theory in Analysis and Design

6.10.1. Stability Analysis of Dynamic Systems

Introductory studies in stability and optimality in a sense of system energy and energy governance have being discussed in previous chapters. In chapter 2, the system and energy governance were covered, and, a set of equations (2.59)

$$\begin{cases} \text{System Governance:} & \dot{x} = F(t,x) + B(t,x)u,\ u = \varphi(t,x),\ u \in U, \\ \text{System Stability:} & V(t,x) = \eta[\mathbb{E}(t,x)] > 0,\ \dfrac{dV(t,x)}{dt} < 0,\ \forall (t,x) \in T \times X, \\ \text{Energy Governance:} & \dfrac{dV}{dt} = \dfrac{\partial V}{\partial t} + \left(\dfrac{\partial V}{\partial x}\right)^T \left[F(t,x) + B(t,x)u\right],\ \dfrac{dV}{dt} \le -\boldsymbol{\omega}(t,x,u),\ \boldsymbol{\omega}(t,x,u) > 0 \end{cases}$$

was applied using a continuous and differentiable positive definite energy-dependent function $V(t,x) \equiv \eta[\mathbb{E}(t,x)]$. To guarantee stability, aapply and analyze (2.56) or (2.57)

$$\left\{\mathbb{E}(t,x) > 0,\ \frac{d\mathbb{E}(t,x)}{dt} < 0\right\},\ \left\{V(t,x) = \eta[\mathbb{E}(t,x)] > 0,\ \frac{dV(t,x)}{dt} < 0\right\},$$

$$\frac{dV}{dt} \le -\boldsymbol{\omega}(t,x,u),\ \boldsymbol{\omega}(t,x,u) > 0,\ \forall (t,x) \in T \times X \times U.$$

Examine stability of time-varying and time-invariant systems, described by nonlinear differential equations, using the Lyapunov stability theory [15].

Theorem 6.1. A nonlinear time-varying system

$$\dot{x}(t) = F(t,x),\ x(t_0) = x_0,\ t \ge 0. \tag{6.90}$$

is asymptotically stable if there exists a positive definite, continuously differentiable time-varying scalar function of all states $V(t,x) > 0$, $V(\cdot,\cdot): \mathbb{R}_{\ge 0} \times \mathbb{R}^n \to \mathbb{R}$ with continuous first-order partial derivatives with respect to t and x, and, the total derivative of $V(t,x)$ is negative definite

$$\frac{dV(t,x)}{dt} = \frac{\partial V}{\partial t} + \left(\frac{\partial V}{\partial x}\right)^T \frac{dx}{dt} = \frac{\partial V}{\partial t} + \left(\frac{\partial V}{\partial x}\right)^T F(t,x) < 0, \tag{6.91}$$

If $V(t,x) > 0$ and $\frac{dV(t,x)}{dt} < 0$, states evolve to a stable equilibrium x_e as $x(t)|_{t\to\infty} \to x_e$, $\lim_{t\to\infty} \|x(t) - x_e\| = 0$, $F(t,x_e) = 0$, $\forall (t,x) \in T \times X$. Sufficient conditions for asymptotic stability are defined by a Lyapunov pair $\left\{V(t,x), \frac{dV(t,x)}{dt}\right\}$ as

$$\left\{V(t,x) > 0,\ \frac{dV(t,x)}{dt} < 0\right\},\ \forall (t,x) \in T \times X. \tag{6.92}$$

■

Definition 6.1. A real-valued function $V(\cdot,\cdot): \mathbb{R}_{\ge 0} \times \mathbb{R}^n \to \mathbb{R}$ is positive definite if:

1. $V(t,x) > 0$, $t \ge 0$, $\forall (t,x) \ne 0$;
2. $V(t,x) \ge 0$, $t \ge 0$, $\forall (t,x)$;
3. $V(t,x) = 0$, $\forall (t,x) = 0$. ■

Definition 6.2. A multivariate function $V(\cdot): X \subset \mathbb{R}^n \to \mathbb{R}$, defined on an open set $X \subset \mathbb{R}^n$ is of class C^k on X, $V \in C^k$, $k \ge 1$ if all partial derivatives exist and continuous. For a smooth and continuously differentiable function $V \in C^k$, $k \ge 1$, there exist the first k derivatives $V'(x)$, $V''(x)$, ..., $V^{(k)}(x)$, such that the $V^{(k)}$ derivative is continuous. ■

Definition 6.3. Function $V(x)$ is smooth and infinitely differentiable of class C^∞ if $V(x)$ has derivatives of all orders. ■

Conjecture 6.1. Dynamic time-invariant system (6.90) $\dot{x}=F(x)$ is exponentially stable, and, $\|x(t)-x_e\| \le \alpha\|x_0 - x_e\|e^{-\beta t}$, $(\alpha,\beta)>0$, if there exists a continuously differentiable scalar function of all states $V(x)$, unbounded functions $\rho_1(\cdot)$ and $\rho_2(\cdot)$, and, continuous strictly increasing function $\rho_3(\cdot)$, $\rho_3(0)=0$ such that

$$\rho_1\left(\|x\|_p\right) \le V(t,x) \le \rho_2\left(\|x\|_p\right),\ \frac{dV}{dt} \le -\rho_3\left(\|x\|_p\right).$$ ■

Conjecture 6.2. Dynamic system (6.90) $\dot{x}=F(t,x)$ and equilibrium state x_e are unstable in small $x_\delta \in X_\delta$ and stable in large $x\in X$ if for a positive definite, continuously differentiable scalar function of all states $V(t,x)>0$, the total derivative is positive definite in small X_δ, and, negative definite in large X_Δ, such as $\left.\frac{dV}{dt}\right|_{x\in X_\delta} > 0$, $\left.\frac{dV}{dt}\right|_{x\in X_\Delta} < 0$. Furthermore, $X_\delta \subset X_\Delta \in X$, and, solutions $x(t)$ are bounded, such that $\|x(t)-x_e\| \le \varepsilon$, $\varepsilon>0$, $\forall(t,x)\in T\times X$. ■

Lyapunov Function and Energy – For given governing equations (6.90), one designs the Lyapunov function $V(t,x)$, and, finds evolution and definiteness of $dV(t,x)/dt$. As reported in sections 2.5 and 6.10.2, for a positive definite real-valued total energy $\mathbf{E}=(\Gamma+\Pi+D)>0$, $\forall(t,x)\in T\times X$, which is a continuously differentiable scalar function, the evolution and definiteness of $\frac{d\mathbf{E}(t,x)}{dt}$ are informative and descriptive. A system evolves to the equilibrium state if the energy decreases, which implies a negative definiteness of the total derivative $\frac{d\mathbf{E}(t,x)}{dt}$. Hence, system is stable if

$$\left\{\mathbf{E}(t,x)>0,\ \frac{d\mathbf{E}(t,x)}{dt}<0\right\},\ \text{where } \mathbf{E}(t,x)>0,\ \forall(t,x).$$

In macroscopic systems, one considers the equilibrium state, while microscopic quantum-mechanical systems are characterized by the ground state. Having found a time-invariant or time-varying positive definite energy function $\mathbf{E}(t,x)$, define physics-consistent

$$V(t,x)\equiv\mathbf{E}(t,x) \text{ or } V(t,x)\equiv\eta[\mathbf{E}(t,x)],$$

and, evaluate $dV(t,x)/dt$. For a nonlinear system, studied in Example 6.5, we found the equilibrium state x_e such that $x(t)|_{t\to\infty}\to x_e$. Results, reported in Conjecture 6.2, are illustrated in Example 6.4 and Practice Problem 6.14.8.

Example 6.40. *Stability of a Dynamic System*

Study a system, described as

$$\dot{x}_1(t) = x_1 + x_1^3 x_2^2 + u_1,\ t\ge 0,$$

$$\dot{x}_2(t) = x_2 + u_2.$$

A positive definite quadratic function of all states (x_1,x_2) is $V(x_1,x_2)=\frac{1}{2}\left(x_1^2+x_2^2\right)$.

The open-loop system with $u_1=0$ and $u_2=0$ is

$$\dot{x}_1(t) = x_1 + x_1^3 x_2^2,$$

$$\dot{x}_2(t) = x_2$$

This system is unstable because $\frac{dV}{dt}=\frac{\partial V}{\partial x_1}\frac{dx_1}{dt}+\frac{\partial V}{\partial x_2}\frac{dx_2}{dt}=x_1^2+x_1^4x_2^2+x_2^2>0$.

Consider control functions u_1 and u_2 as

$u_1=-2x_1-x_1^3-2x_1^3x_2^2$, $u_2=-2x_2-x_2^3$.

The resulting equations for a closed-loop system are

$\dot{x}_1(t)=-x_1-x_1^3-x_1^3x_2^2$,

$\dot{x}_2(t)=-x_2-x_2^3$.

For a scalar positive definite function $V(x_1,x_2)=\frac{1}{2}\left(x_1^2+x_2^2\right)$, we have

$$\frac{dV(x_1,x_2)}{dt}=\left(\frac{\partial V}{\partial x}\right)^T\frac{dx}{dt}=\left(\frac{\partial V}{\partial x}\right)^T F(x)=\frac{\partial V}{\partial x_1}\left(-x_1-x_1^3-x_1^3x_2^2\right)+\frac{\partial V}{\partial x_2}\left(-x_2-x_2^3\right)$$

$$=-x_1^2-x_1^4-x_1^4x_2^2-x_2^2-x_2^4.$$

The total derivative is negative definite $\frac{dV(x_1,x_2)}{dt}<0$, $\forall x\in X$.

The closed-loop system is asymptotically stable. ■

Example 6.41. *Stability of a Time-Invariant System*

Examine a time-invariant system

$\dot{x}_1(t)=-x_1+x_2$, $t\geq0$,

$\dot{x}_2(t)=-x_1-x_2-x_2\left|x_2\right|$.

Consider a positive definite scalar function of (x_1,x_2), $V(x_1,x_2)=\frac{1}{2}\left(x_1^2+x_2^2\right)$, $V(x_1,x_2)>0$.

The total derivative is

$$\frac{dV(x_1,x_2)}{dt}=\left(\frac{\partial V}{\partial x}\right)^T\frac{dx}{dt}=\left(\frac{\partial V}{\partial x}\right)^T F(x)=\frac{\partial V}{\partial x_1}\dot{x}_1+\frac{\partial V}{\partial x_2}\dot{x}_2=x_1\left(-x_1+x_2\right)+x_2\left(-x_1-x_2-x_2\left|x_2\right|\right)$$

$$=-x_1^2-x_2^2\left(1+\left|x_2\right|\right).$$

Therefore, $\frac{dV(x_1,x_2)}{dt}<0$.

The equilibrium state is asymptotically stable, the quadratic function $V(x_1,x_2)=\frac{1}{2}\left(x_1^2+x_2^2\right)$ is the Lyapunov function. Furthermore, $\lim_{t\to\infty}\left\|x(t)-x_e\right\|=0$, $\forall x\in X$. ■

Example 6.42. *Stability of Permanent-Magnet Synchronous Motors*

Consider a permanent-magnet synchronous motor in the rotor reference frame. Using the *quadrature* and *direct* currents and voltages operators $[(i_{qs},i_{ds}),(u_{qs},u_{ds})]$, the governing equations are found (3.86). Study control schemes using the voltages operators (u_{qs},u_{ds},u_{0s}), while ac motors are regulated by changing the phase voltages (u_{as},u_{bs},u_{cs}). The balanced voltage set is expressed by (3.88).

The *quadrature* voltage operator u_{qs}^r is regulated, while $u_{ds}^r=0$, $u_{0s}^r=0$.

The mathematical model, assuming T_L=0, is

$$\frac{di_{qs}^r}{dt}=-\frac{r_s}{L_{ss}}i_{qs}^r-\frac{\psi_m}{L_{ss}}\omega_r-i_{ds}^r\omega_r+\frac{1}{L_{ss}}u_{qs}^r,\ \frac{di_{ds}^r}{dt}=-\frac{r_s}{L_{ss}}i_{ds}^r+i_{qs}^r\omega_r+\frac{1}{L_{ss}}u_{ds}^r,\ \frac{d\omega_r}{dt}=\frac{3P^2\psi_m}{8J}i_{qs}^r-\frac{B_m}{J}\omega_r.$$

For a quadratic positive definite Lyapunov function

$$V(i_{qs}^r, i_{ds}^r, \omega_r) = \tfrac{1}{2}(i_{qs}^{r\,2} + i_{ds}^{r\,2} + \omega_r^2),$$

the total derivative for an open-loop system is

$$\frac{dV(i_{qs}^r, i_{ds}^r, \omega_r)}{dt} = -\frac{r_s}{L_{ss}} i_{qs}^{r\,2} - \frac{r_s}{L_{ss}} i_{ds}^{r\,2} - \frac{B_m}{J}\omega_r^{\,2} - \frac{8J\psi_m - 3P^2 L_{ss}\psi_m}{8JL_{ss}} i_{qs}^r \omega_r .$$

Thus, $\dfrac{dV(i_{qs}^r, i_{ds}^r, \omega_r)}{dt} < 0$.

Hence, an open-loop electric drive is asymptotically stable.

Let $u_{qs}^r = -k_I i_{qs}^r - k_\omega \omega_r$, $u_{qs}^r = 0$, $u_{ds}^r = 0$.

For a positive definite $V(i_{qs}^r, i_{ds}^r, \omega_r) = \tfrac{1}{2}(i_{qs}^{r\,2} + i_{ds}^{r\,2} + \omega_r^2)$, we have

$$\frac{dV(i_{qs}^r, i_{ds}^r, \omega_r)}{dt} = -\frac{r_s + k_I}{L_{ss}} i_{qs}^{r\,2} - \frac{r_s}{L_{ss}} i_{ds}^{r\,2} - \frac{B_m}{J}\omega_r^{\,2} - \frac{8J(\psi_m + k_\omega) - 3P^2 L_{ss}\psi_m}{8JL_{ss}} i_{qs}^r \omega_r .$$

Hence, $V(i_{qs}^r, i_{ds}^r, \omega_r) > 0$ and $\dfrac{dV(i_{qs}^r, i_{ds}^r, \omega_r)}{dt} < 0$.

The conditions for asymptotic stability are guaranteed.

The rate *dV/dt* defines system dynamics and convergence to equilibrium. ■

Example 6.43. *Analysis of a Time-Varying System*

Consider a time-varying system

$$\dot{x}_1(t) = -x_1 + x_2^3, \; t \ge 0,$$

$$\dot{x}_2(t) = -e^{-10t} x_1 x_2^2 - 5x_2 - x_2^3 .$$

A scalar time-dependent positive definite function $V(t,x_1,x_2)>0$ is chosen as $V(t, x_1, x_2) = \tfrac{1}{2}\left(x_1^2 + e^{10t} x_2^2\right)$.

The total derivative is

$$\frac{dV(t, x_1, x_2)}{dt} = \frac{\partial V}{\partial t} + \frac{\partial V}{\partial x_1}\left(-x_1 + x_2^3\right) + \frac{\partial V}{\partial x_2}\left(-e^{-10t} x_1 x_2^2 - 5x_2 - x_2^3\right) = -x_1^2 - e^{10t} x_2^4 .$$

The total derivative is time-dependent, and, $\dfrac{dV(t, x_1, x_2)}{dt} < 0$, $t \ge 0$.

Hence, $V(t, x_1, x_2) = \tfrac{1}{2}\left(x_1^2 + e^{10t} x_2^2\right)$ is a Lyapunov function, and, a system is asymptotically stable. ■

Example 6.44. *Stability of a Time-Varying System*

Study a time-varying second-order system

$$\dot{x}_1(t) = -x_1 - \frac{1}{1+t} x_1^3 + x_2, \; t \ge 0,$$

$$\dot{x}_2(t) = -x_1 - x_2 - \frac{1}{1+t} x_2^3 .$$

Consider a scalar positive definite function

$V(t, x_1, x_2) = \tfrac{1}{2}(1+t)\left(x_1^2 + x_2^2\right) > 0$, $t \ge 0$, $\forall (t,x) \in T \times X$.

The total derivative is

$$\frac{dV(t,x_1,x_2)}{dt}=\frac{\partial V}{\partial t}+\frac{\partial V}{\partial x_1}\left(-x_1-\frac{1}{1+t}x_1-x_1^3+x_2\right)+\frac{\partial V}{\partial x_2}\left(x_1-x_2-\frac{1}{1+t}x_2^3\right)$$
$$=-x_1^2\left(\tfrac{1}{2}+t+x_1^2\right)-x_2^2\left(\tfrac{1}{2}+t+x_2^2\right).$$

Hence, $V(t,x_1,x_2)>0$ and $\frac{dV(t,x_1,x_2)}{dt}<0$, $t\geq 0$, $\forall(t,x)\in T\times X$.

A system is asymptotically stable. ■

Example 6.45. *Analysis of Time-Varying Dynamic Systems*
Consider two time-varying models of one-dimensional motion, given as

$$\dot{x}_1(t)=\frac{1}{a+e^{-t}}(-2x_1-2x_2),\ a>0,\ t\geq 0,$$
$$\dot{x}_2(t)=x_1,$$

and

$$\dot{x}_1(t)=-2x_1-\frac{1}{a+e^{-t}}x_1-2x_2-\frac{1}{a+e^{-t}}x_2,\ a>0,\ t\geq 0,$$
$$\dot{x}_2(t)=x_1.$$

A positive definite time-dependent function is $V(t,x_1,x_2)=\frac{1}{2}e^t\left(x_1^2+2x_1x_2+x_2^2\right)$.

For dynamic systems with a continuous time-varying parameter $\frac{1}{a+e^{-t}}>0$, $t\geq 0$, characterized by a range $\{z\in\mathbb{R}: 0<z<1/a\}$, $a>0$, one has

$$\frac{dV(t,x_1,x_2)}{dt}=\frac{\partial V}{\partial t}+\frac{\partial V}{\partial x_1}\dot{x}_1+\frac{\partial V}{\partial x_2}\dot{x}_2<0,\ \forall(t,x_1,x_2)\neq 0,\ t\geq 0.$$

Systems are asymptotically stable. ■

6.10.2. Design of Lyapunov Functions and Control Laws

The Lyapunov paradigm is a foundation in stability analysis, control law design and dynamic optimization. Applying laws of physics, one performs concurrent analyses and derives differential equations. For example, the *generalized coordinates* q, *generalized momentum* p and *generalized forces* Q are used. Using (q,p) or system variables x, as well as time-invariant and time-varying parameters, the system energy is applied to derive equations of motion as reported in chapters 2 and 3. Scalar positive definite Lyapunov functions $V(x)$ can be designed using an energy tuple (Γ,Π,D) with positive definite kinetic, potential and dissipating energies. The quadratic measured may be used, finding $\Gamma(x)\propto x^2$, $\Pi(x)\propto x^2$ and $D(x)\propto x^2$. For multi-body systems, using the *generalized coordinates* $\boldsymbol{q}$, the kinetic energy is $\Gamma=\sum_i\Gamma_i=\frac{1}{2}\dot{\boldsymbol{q}}^TM(\boldsymbol{q})\dot{\boldsymbol{q}}$, where $M(\boldsymbol{q})$ is the mass matrix. The total energy is found as high degree and fractional multivariate polynomials, $\mathbb{E}(t,x)=[\Gamma(t,x)+\Pi(t,x)+D(t,x)]$, $\mathbb{E}(t,x)>0$, $\forall(t,x)\in T\times X$.

Design of Lyapunov Functions – A positive definite total energy $\mathbb{E}(t,x)$ is found. For continuous real-valued $\mathbb{E}(t,x)$, there are admissible variations and evolutions of $\frac{d\mathbb{E}(t,x)}{dt}$.

A system evolves towards an equilibrium if the energy rate of change $\frac{d\mathbb{E}(t,x)}{dt}$ decreases.

Controlled dynamic governance of an unperturbed system, characterized by a pair $\left\{\mathrm{E}(t,x)>0,\frac{d\mathrm{E}(t,x)}{dt}\right\}$, evolves to a stable equilibrium x_e as $x(t)\to x_e$ if $\frac{d\mathrm{E}(t,x)}{dt}<0$.

Furthermore, $\lim_{t\to\infty}\|x(t)-x_e\|=0$ if $\frac{d\mathrm{E}(t,x)}{dt}=0$.

Nonequilibrium of $\left\{\mathrm{E}(t,x),\frac{d\mathrm{E}(t,x)}{dt}\right\}$ implies dynamic evolutions of $x(t)$.

One finds time-invariant and time-varying continuously differentiable Lyapunov functions as:

1. $V(x)\equiv\mathrm{E}(x)$ or $V(x)\equiv\eta[\mathrm{E}(x)]$, $\mathrm{E}(x)=[\Gamma(x)+\Pi(x)+D(x)]$, $\mathrm{E}>0$, $\forall x\in X$;
2. $V(t,x)\equiv\mathrm{E}(t,x)$ or $V(t,x)\equiv\eta[\mathrm{E}(t,x)]$, $\mathrm{E}(t,x)=[\Gamma(t,x)+\Pi(t,x)+D(t,x)]$, $\mathrm{E}>0$, $\forall(t,x)\in T\times X$.

System energy, stability analysis and system optimization suggest the use of quadratic and nonquadratic continuously differentiable functions $V(x)$ and $V(t,x)$, expressed using real-valued functions (σ_1,σ_2) and parameter-dependent matrices (K,K_{ij}) as

$$V(x)=\tfrac{1}{2}x^TKx,\ V(t,x)=\tfrac{1}{2}x^TK(t)x,\ K>0,\ V(t,x)>0,\ \forall(t,x)\neq 0, \qquad (6.93)$$

$$V(t,x)=\sum_{i,j}\sigma_{1_i}^T(t,x^{<d_i>})K_{ij}(t)\sigma_{2_j}(t,x^{<d_j>}),\ V(t,x)>0,\ t\geq 0,\ \forall(t,x)\neq 0,$$

where $\sigma_1(\cdot)$ and $\sigma_2(\cdot)$ are the real-valued functions, $\sigma_{1_i}^T(t,x^{<d_i>})I\sigma_{2_j}(t,x^{<d_j>})>0, \forall(i,j),(d_i,d_j)\in\mathbb{N}$.

Example 6.46. *Energy, Nonlinear System Mapping F(x) and Lyapunov Function*
Consider a system, described by governing equations

$$\dot{x}_1=-x_1-x_1^3+x_2,\ t\geq 0,$$

$$\dot{x}_2=x_1-x_2-x_2^3.$$

For a quadratic positive definite function

$$V(x_1,x_2)=\tfrac{1}{2}\left(x_1^2+x_2^2\right),\ V(x)>0,$$

we obtain $\frac{dV}{dt}=-x_1^2-x_1^4+2x_1x_2-x_2^2-x_2^4$, $\frac{dV}{dt}<0$.

Hence, a system is asymptotically stable, and, $V(x_1,x_2)=\tfrac{1}{2}\left(x_1^2+x_2^2\right)$ is a Lyapunov function.

One may use the pertained positive definite $\mathrm{E}(x)$, applying $V(x)\equiv\eta[\mathrm{E}(x)]$.
For example, $V(x_1,x_2)=\tfrac{1}{4}\left(x_1^4+x_2^4\right), V(x_1,x_2)=\tfrac{1}{4}\left(x_1^4+2x_1^2x_2^2+x_2^4\right)$,

$$V(x_1,x_2)=\tfrac{1}{2}\left(x_1^2+x_2^2\right)+\tfrac{1}{4}\left(x_1^4+2x_1^2x_2^2+x_2^4\right),\ \text{etc.}$$

For (6.90) $\dot{x}=F(x)$, consider

$V(x)=F^T(x)F(x)$.

We have

$$V(x)=F^T(x)F(x)=\left[\left(-x_1-x_1^3+x_2\right),\left(x_1-x_2-x_2^3\right)\right]\begin{bmatrix}-x_1-x_1^3+x_2\\ x_1-x_2-x_2^3\end{bmatrix}=\left(-x_1-x_1^3+x_2\right)^2+\left(x_1-x_2-x_2^3\right)^2$$

$$=2x_1^2-4x_1x_2+2x_2^2+2x_1^4-2x_1x_2^3-2x_1^3x_2+2x_2^4+x_1^6+x_2^6.$$

For $V(x)\equiv\eta[\mathrm{E}(x)]$ and $V(x)=F^T(x)F(x)$, $V(x)>0$.

One finds $dV/dt<0$. System is asymptotically stable. ■

Lyapunov Function, Stability and Optimality – Consider a linear time-invariant open-loop system $\dot{x}(t)=Ax$, or, a closed-loop system

$$\dot{x}(t)=A_{\text{open loop}}x+Bu,\; u=-K_F x,\; A=(A_{\text{open loop}}-BK_F).$$

That is

$$\dot{x}(t)=A_{\text{open loop}}x+Bu=(A_{\text{open loop}}-BK_F)x=Ax,\; A=(A_{\text{open loop}}-BK_F),\; u=-K_F x. \quad (6.94)$$

Using an all states positive definite quadratic function $V=\frac{1}{2}x^TKx$ with a constant-coefficient matrix $K>0$, for (6.94) $\dot{x}(t)=Ax$, one finds

$$\frac{dV}{dt}=\frac{1}{2}(\dot{x}^TKx+x^TK\dot{x})=\frac{1}{2}x^T(A^TK+KA)x,\; K=K^T. \quad (6.95)$$

A Lyapunov pair (6.92) stipulates conditions for stability and optimality. Define negative definiteness of the total derivative dV/dt using a positive definite function $\omega_x(x)=\frac{1}{2}x^TQx$, $Q>0$. Specify a dynamic optimality stipulating evolution of dV/dt as

$$\frac{dV}{dt}=-\omega_x(x)=-\frac{1}{2}x^TQx,\; \omega_x(x)>0,\; Q>0. \quad (6.96)$$

Using (6.95) and (6.96), in the quadratic sense, from $\frac{1}{2}x^T(A^TK+KA)x=-\frac{1}{2}x^TQx$, the linear Lyapunov equation to find $K>0$ for given (A,Q) is

$$A^TK+KA=-Q,\; Q>0. \quad (6.97)$$

Conjecture 6.3. Asymptotic stability of a linear system (6.94) is guaranteed if for a positive definite matrix $Q>0$ with $\frac{dV}{dt}=-\frac{1}{2}x^TQx<0$, there exists a positive definite symmetric matrix $K>0$ which is a solution of the Lyapunov equation (6.97). Positive definite $K>0$ implies $V(x)=\frac{1}{2}x^TKx>0$ for a given rate of change (6.96) $\frac{dV}{dt}=-\frac{1}{2}x^TQx$, $Q>0$, and, $x(t)|_{t\to\infty}\to x_e$. ■

Conjecture 6.4. For a system (6.94), apply a positive definite function (6.93) $V(x)=\frac{1}{2}x^TKx>0$ with parameter-dependent matrix $K>0$, such that $V(x)=\frac{1}{2}x^TKx>0$. The sufficient conditions for asymptotic stability (6.91) are guaranteed if the Lyapunov equation (6.97) yields a positive definite matrix $Q>0$, which defines the rate of change (6.96) $\frac{dV}{dt}=-\frac{1}{2}x^TQx<0$, characterizing system optimality and evolutions of $x(t)$ to equilibrium x_e, $x(t)|_{t\to\infty}\to x_e$. ■

Stability, Lyapunov Function, State Transition Matrix and System Evolutions – For stable systems, the minimum and maximum eigenvalues are positive for positive definite K and Q, e.g., $\lambda_{\min}(K)>0$, $\lambda_{\max}(K)>0$, $\lambda_{\min}(Q)>0$ and $\lambda_{\max}(Q)>0$.

Inequalities

$$\lambda_{\min}(Q)x^Tx\le x^TQx\le\lambda_{\max}(Q)x^Tx,\; \lambda_{\min}(K)x^Tx\le x^TKx\le\lambda_{\max}(K)x^Tx$$

yield

$$-\frac{\frac{d}{dt}V(x)}{V(x)}\ge\frac{\lambda_{\min}(Q)}{\lambda_{\max}(K)}\frac{x^Tx}{x^Tx}=c,\; \frac{dV(x)}{dt}\le-cV,\; V(x)\le e^{-ct}V(0), \quad (6.98)$$

where $c=\frac{\lambda_{\min}(Q)}{\lambda_{\max}(K)}$, $c>0$.

For positive definite $(K,Q)>0$, the convergence of $V(t,x)$ is guaranteed, and,

$\lim_{t\to\infty} V(t,x)\to 0$, $\lim_{t\to\infty} \frac{dV}{dt}\to 0$.

For a linear system (6.94) $\dot{x}(t)=Ax$, as covered in section 5.2.1, a unique solution with initial conditions x_0 is found using the state transition matrix $\Phi(t)$, and,

$x(t)=\Phi(t)x_0=e^{At}x_0$.

Note that $\|e^{At}\|\le \beta e^{t[\max \mathrm{Re}(\lambda_i)]}$, β>0, $\forall t$.

System is asymptotically stable if $\lim_{t\to\infty} x(t)\to 0$, and, $\lim_{t\to\infty}\Phi(t)\to 0$.

A system is stable if evolutions of $(x(t),\Phi(t))$ are bounded,

$\|x(t)\|\le\varepsilon,\ \|\Phi(t)\|\le M, \forall x_0,\ t\ge 0$.

For a bounded $\Phi(t)$,

$\|x(t)\|=\|\Phi(t)x_0\|\le\|\Phi(t)\|\|x_0\|\le M\|x_0\|\le\varepsilon,\ \forall x_0,\ t\ge 0$.

For a positive definite Lyapunov function (6.93) $V(x)=\frac{1}{2}x^TKx>0$, K>0, and, negative definite rate of change (6.96) $\frac{dV}{dt}=-\frac{1}{2}x^TQx$, Q>0, we have

$\lim_{t\to\infty} V(t,x)\to 0$, $V(x)=\frac{1}{2}x^TKx=\frac{1}{2}x_0^T\Phi^T(t)K\Phi(t)x_0=\frac{1}{2}x_0^T(e^{At})^TKe^{At}x_0$,

and $\lim_{t\to\infty}\frac{dV}{dt}\to 0$, $\frac{dV}{dt}=-\frac{1}{2}x^TQx=-\frac{1}{2}x_0^T\Phi^T(t)Q\Phi(t)x_0=-\frac{1}{2}x_0^T(e^{At})^TQe^{At}x_0$,

where $|\lambda_{\max}(K)|\le\|K\|$ and $|\lambda_{\max}(Q)|\le\|Q\|$ for any matrix norm, including $\|\cdot\|_2$, and, there are the continuous bounds on the matrix exponential e^{At}.

Positive Definite Matrix, Eigenvalues and Eigenvectors – A matrix $K\in\mathbb{R}^{n\times n}$ is a positive definite matrix if and only if x^TKx>0 for all nonzero real-valued vector $x\in\mathbb{R}^n$, $\forall x\neq 0$. A positive definite matrix have strictly positive eigenvalues λ. Consider eigenvectors $v_i\neq 0$ with associated eigenvalues λ_i, $Kv_i=\lambda_i v_i$.

From $\mathbf{v}^T\mathbf{v}=\sum_{i=1}^n v_i^2$, $\mathbf{v}^T\mathbf{v}$ is strictly positive.

For K>0, we have $v_i^TKv_i>0$, $\lambda_i v_i^Tv_i>0$, $\forall(v_i,\lambda_i)$ and $\mathbf{v}^TK\mathbf{v}$>0.

Hence, for a positive definite K, all λ_i are strictly positive.

Uniqueness of Solution of Lyapunov Equation – If system is stable, for a positive definite Q, there exists a real-valued unique matrix K, such as

$$K=\int_0^\infty e^{A^Tt}Qe^{At}dt,\ K>0, \|K\|_2<\infty. \tag{6.99}$$

Using (6.99), one yields the Lyapunov equation (6.97)

$$A^TK+KA=\int_0^\infty\left(A^Te^{A^Tt}Qe^{At}+e^{A^Tt}Qe^{At}A\right)dt=\int_0^\infty\frac{d}{dt}\left(e^{A^Tt}Qe^{At}\right)dt=\lim_{t\to\infty}\left(e^{A^Tt}Qe^{At}\right)-Q=-Q$$

because $\lim_{t\to\infty}\|e^{At}\|_2=0$.

Matrix K is symmetric, $K=K^T$. Considering (6.97), one has

$A^TK+KA=-Q$ and $A^TK^T+K^TA=-Q$.

From $A^T(K-K^T)+(K-K^T)A=0$, we find $\frac{d}{dt}\left(e^{A^Tt}(K-K^T)e^{At}\right)=0$.

This implies $K=K^T$.

Lyapunov Equation and Control – Stability is defined by a Lyapunov pair (6.92) $\left\{V(x)>0, \frac{dV}{dt}<0\right\}$ [3, 15-17]. The Lyapunov equation (6.97) $A^TK+KA=-Q$ yields K or Q as formulated by Conjectures 6.3 and 6.4. One may apply optimization and search algorithms specifying positive definite (K,Q), eigenvalues, system dynamics, dV/dt evolution, etc. For linear systems, these concepts yield a control law $u=-K_Fx$ with the feedback gain matrix K_F in a closed-loop system (6.94) $\dot{x}(t)=Ax,\ A=(A_{\text{open loop}}-BK_F)$.

Quadratic Regulator Problem, Quadratic Measures and Control Laws Design – For a linear dynamic system (6.12) $\dot{x}(t)=Ax+Bu$, consider a Lyapunov pair (6.92) $\left\{V(x)>0, \frac{dV}{dt}<0\right\}$ with a quadratic function (6.93) $V=\frac{1}{2}x^TKx$, $K=K^T$.

We have

$$\frac{dV}{dt}=\tfrac{1}{2}\left(\dot{x}^TKx+x^TK\dot{x}\right)=\tfrac{1}{2}x^T\left(A^TK+KA\right)x+\tfrac{1}{2}\left(u^TB^TKx+x^TKBu\right). \qquad (6.100)$$

Using quadratic measures on the states and controls, specify the negative definite rate of change dV/dt as $\frac{dV}{dt}=-\frac{1}{2}x^TQx-\frac{1}{2}u^TGu$.

Hence, $\frac{dV}{dt}<0$, $\forall(x,u)$ because $(Q,G)>0$. One has

$$\begin{cases}\text{Dynamic System: } \dot{x}(t)=Ax+Bu,\\ \text{Stability Criteria: } \left\{V(x)>0, \frac{dV}{dt}<0\right\}, \forall x\in X, \forall u\in U, \forall t,\\ \text{Lyapunov Function and Derivative: } V(x)=\frac{1}{2}x^TKx,\ \frac{dV}{dt}=\left(\frac{\partial V}{\partial x}\right)^T\underset{\dot{x}(t)=Ax+Bu}{(Ax+Bu)},\\ \text{Optimality Criteria: } \frac{dV}{dt}=-\frac{1}{2}\left[x^TQx+u^TGu\right],\ (Q,G)>0,\ \frac{dV}{dt}<0,\ \forall(x,u),\end{cases} \qquad (6.101)$$

where $K>0$, $Q>0$ and $G>0$.

Problem Formulation and Solution – Minimize

$$\frac{dV}{dt}=-\tfrac{1}{2}\left[x^TQx+u^TGu\right],\ \frac{dV}{dt}<0,\ V=\tfrac{1}{2}x^TKx,\ K>0,\ V(x)>0,$$

subject to a system (6.12) $\dot{x}(t)=Ax+Bu$, and, design a control law.

For a scalar Hamiltonian

$$H(x,u,\tfrac{\partial V}{\partial x})=\tfrac{1}{2}x^TQx+\tfrac{1}{2}u^TGu+\left(\tfrac{\partial V}{\partial x}\right)^T(Ax+Bu)=\tfrac{1}{2}x^TQx+\tfrac{1}{2}u^TGu+\tfrac{1}{2}x^T\left(A^TK+KA\right)x+x^TKBu, \qquad (6.102)$$

the stationarity condition (6.22) $\frac{\partial}{\partial u}H(x,u,\frac{\partial V}{\partial x})=0$ yields a control law

$$u=-G^{-1}B^TKx. \qquad (6.103)$$

From (6.102) with (6.103), the positive definite symmetric matrix K is found by solving the algebraic equation

$$Q+A^TK+KA-KBG^{-1}B^TK=0,\ K>0,\ K=K^T. \qquad (6.104)$$

These results are equivalent to our findings in section 5.6 where the control design problem was solved by minimizing the quadratic performance functional $J=\min_{x,u}\int_{t_0}^{\infty}\frac{1}{2}\left(x^TQx+u^TGu\right)dt$. We studied a Lyapunov pair (6.92), given by (6.101) with a quadratic $V(x)>0$ and quadratic measures on $dV/dt<0$. The consistency between stability and optimality are ensured. Sensitivity and robustness are assessed using (6.101).

Conjecture 6.5. All-state feedback control law (6.103) guarantees stability of a system (6.12)-(6.103) $\dot{x}(t)=Ax+Bu$, $u=-G^{-1}B^TKx$. Control law structure and feedback gains are derived using stability and optimality criteria given by a Lyapunov pair (6.101) $\left\{\begin{array}{l} V(x)>0,\ \frac{dV(x)}{dt}<0, \\ V(x)=\frac{1}{2}x^TKx, \frac{dV}{dt}=-\frac{1}{2}[x^T\underset{Q>0}{Q}x+u^T\underset{G>0}{G}u] \end{array}\right\}$. Solution of equation (6.104) yields a positive definite K, and, $V(x)$ evolves as $V(x)\le e^{-ct}V(0)$, $c=\frac{\lambda_{\min}(Q+KBG^{-1}B^TK)}{\lambda_{\max}(K)}$, $c>0$. The closed-loop system evolves as $x(t)=e^{(A-BG^{-1}B^TK)t}x_0$, $\Phi(t)=e^{(A-BG^{-1}B^TK)t}$. ■

Lyapunov Function and System Evolutions – With positive definite (K,Q,G), one has
$\lambda_{\min}(K)>0$, $\lambda_{\max}(K)>0$, $\lambda_{\min}(Q+KBG^{-1}B^TK)>0$, $\lambda_{\max}(Q+KBG^{-1}B^TK)>0$.
Furthermore,

$$\lambda_{\min}(Q+KBG^{-1}B^TK)x^Tx\le x^T(Q+KBG^{-1}B^TK)x\le\lambda_{\max}(Q+KBG^{-1}B^TK)x^Tx,$$

$$\lambda_{\min}(K)x^Tx\le x^TKx\le\lambda_{\max}(K)x^Tx.$$

Therefore,

$$-\frac{\frac{d}{dt}V(x)}{V(x)}\ge\frac{\lambda_{\min}(Q+KBG^{-1}B^TK)}{\lambda_{\max}(K)}\frac{x^Tx}{x^Tx}=c,\quad \frac{dV(x)}{dt}\le -cV,\quad V(x)\le e^{-ct}V(0),$$

where $c=\frac{\lambda_{\min}(Q+KBG^{-1}B^TK)}{\lambda_{\max}(K)}$, $c>0$.

Example 6.47. *Lyapunov Equation and Control Law Design*
Consider a dynamic system

$$\dot{x}=\begin{bmatrix}\dot{x}_1\\ \dot{x}_2\end{bmatrix}=\begin{bmatrix}-1 & -2\\ 1 & 0\end{bmatrix}\begin{bmatrix}x_1\\ x_2\end{bmatrix}+\begin{bmatrix}1\\ 0\end{bmatrix}u.$$

The eigenvalues of A are found solving the characteristic equation $|\lambda I-A|=0$. We have

$$|\lambda I-A|=\left|\begin{bmatrix}\lambda & 0\\ 0 & \lambda\end{bmatrix}-\begin{bmatrix}-1 & -2\\ 1 & 0\end{bmatrix}\right|=\left|\begin{bmatrix}\lambda+1 & 2\\ -1 & \lambda\end{bmatrix}\right|=\lambda^2+\lambda+2=0,\ \lambda_{1,2}=-0.5\pm j\tfrac{\sqrt{7}}{2}.$$

Consider a positive definite function $V(x_1,x_2)=x_1^2+x_1x_2+\frac{1}{2}x_2^2$.

The total derivative is

$$\frac{dV(x_1,x_2)}{dt}=(2x_1+x_2)(-x_1-2x_2)+(x_1+x_2)x_1=-x_1^2-4x_1x_2-2x_2^2.$$

System is asymptotically stable because $\left\{V(x_1,x_2)>0,\ \frac{dV}{dt}<0\right\}$.

Using Conjecture 6.3, study stability by applying the Lyapunov equation (6.97).

Specify $\frac{dV}{dt}=-x^TQx$, $\frac{dV}{dt}=-\begin{bmatrix}x_1 & x_2\end{bmatrix}\begin{bmatrix}1 & 0\\ 0 & 1\end{bmatrix}\begin{bmatrix}x_1\\ x_2\end{bmatrix}=-x_1^2-x_2^2$.

For $Q=\begin{bmatrix}1 & 0\\ 0 & 1\end{bmatrix}$, $Q>0$, solve

$$A^TK+KA=-Q,\ \begin{bmatrix}-1 & 1\\ -2 & 0\end{bmatrix}\begin{bmatrix}k_{11} & k_{12}\\ k_{12} & k_{22}\end{bmatrix}+\begin{bmatrix}k_{11} & k_{12}\\ k_{12} & k_{22}\end{bmatrix}\begin{bmatrix}-1 & -2\\ 1 & 0\end{bmatrix}=-\begin{bmatrix}1 & 0\\ 0 & 1\end{bmatrix}.$$

The resulting equations

$-2k_{11}+2k_{12}=-1,\ -2k_{11}-k_{12}+k_{22}=0,\ -4k_{12}=-1$

yield three unknowns (k_{11},k_{12},k_{22}). We have $k_{11}=0.75$, $k_{12}=0.25$ and $k_{22}=1.75$.

For $K=\begin{bmatrix}k_{11} & k_{12}\\ k_{12} & k_{22}\end{bmatrix}=\begin{bmatrix}0.75 & 0.25\\ 0.25 & 1.75\end{bmatrix}$, $\det(K)>0$, $K>0$.

Hence, the open-loop system is stable.

Using the **eig**, **lyap** and **det** MATLAB commands, compute λ_i, K and $\det(K)$ as

```
A=[-1 -2; 1 0]; B=[1; 0]; Q=[1 0; 0 1]; E=eig(A), K=lyap(A',Q), det=det(K)
```

Design a control law applying results reported in Conjecture 6.5. Specify

$\frac{dV}{dt}=-\frac{1}{2}x^TQx-\frac{1}{2}u^TGu$, $Q=\begin{bmatrix}1 & 0\\ 0 & 1\end{bmatrix}$, $G=1$.

The control law (6.103) is $u=-G^{-1}B^TKx=-K_Fx$.

Matrix K of $V=\frac{1}{2}x^TKx$, $K>0$ is computed by solving equation (6.104).

The unknown matrices K and K_F are $K=\begin{bmatrix}k_{11} & k_{12}\\ k_{12} & k_{22}\end{bmatrix}=\begin{bmatrix}0.572 & 0.236\\ 0.236 & 1.516\end{bmatrix}$, $K_F=[0.572\ \ 0.236]$.

Furthermore, $\det(K)=0.812$, and, $V(x)>0$.

The control law is $u=-0.572x_1-0.236x_2$.

The eigenvalues of the closed-loop system are $\lambda_{1,2}=-0.786\pm j1.272$.

We have $V(x)>0$ and $dV/dt<0$.

Furthermore, $V(x)\le e^{-ct}V(0)$, $c=\frac{\lambda_{\min}(Q+KBG^{-1}B^TK)}{\lambda_{\max}(K)}$, $c=0.6363$. ■

6.10.3. Tracking Control Laws For Systems With Control Limits

Solve a tracking control problem for physical systems with control limits $u_{\min}\le u\le u_{\max}$. As reported in section 6.8, the governing equations (6.66) with $u_{\min}\le u\le u_{\max}$ are

$$\dot{\mathbf{x}}(t)=\begin{bmatrix}\dot{x}(t)\\ \dot{x}_e(t)\\ \dot{e}(t)\end{bmatrix}=\mathbf{F}(\mathbf{x})+\mathbf{A}\mathbf{x}+A_rr+\mathbf{B}(\mathbf{x})u=\begin{bmatrix}F(x)\\ 0\\ -HF(x)\end{bmatrix}+\begin{bmatrix}0 & 0 & 0\\ -I_XH & 0 & 0\\ 0 & I_E^X & -I_E\end{bmatrix}\mathbf{x}+\begin{bmatrix}0 & 0\\ I_X & 0\\ 0 & I\end{bmatrix}\begin{bmatrix}r\\ \dot{r}\end{bmatrix}+\begin{bmatrix}B(x)\\ 0\\ -HB(x)\end{bmatrix}u,$$

$y=Hx$, $u\in U$, $u_{\min}\le u\le u_{\max}$. (6.105)

Consider a Lyapunov pair (6.92) $\left\{V(\mathbf{x})>0,\frac{dV}{dt}<0\right\}$ with a Lyapunov function (6.93)

$$V(\mathbf{x})=\sum_{i,j}\sigma_{1_i}^T(\mathbf{x}^{<d_i>})K_{ij}\sigma_{2_j}(\mathbf{x}^{<d_j>})$$

and *admissible* $\frac{dV}{dt}\le-\boldsymbol{\omega}_x(\mathbf{x})-\boldsymbol{\omega}_u(u)$.

Apply positive definite state and control performance measures $\boldsymbol{\omega}_x(\mathbf{x})$ and $\boldsymbol{\omega}_u(u)$, which are physics-consistent and design-specific. The limits on the rate of energy changes are specified by positive definite $\boldsymbol{\omega}_x(\mathbf{x})$ and $\boldsymbol{\omega}_u(u)$. As reported in Example 6.24,

$\boldsymbol{\omega}_x(\mathbf{x})=(\mathbf{x}^{\langle q\rangle})^TQ_q\mathbf{x}^{\langle q\rangle}$, $\boldsymbol{\omega}_x(\mathbf{x})=\omega_1^T(\mathbf{x}^{<q_1>})Q_q\omega_2(\mathbf{x}^{<q_2>})$, $q\in\mathbb{N}$, $(q_1,q_2)\in\mathbb{N}$,

$\boldsymbol{\omega}_u(u)=\int\left(\Phi^{-1}(u)\right)^TGdu$, where $\omega_1(\cdot)$ and $\omega_2(\cdot)$ are the real-valued functions; Φ is a bounded, integrable, real-analytic continuous function, $\Phi \subseteq \mathcal{U}_\varepsilon$.

Physical Stability and Optimality – An asymptotically stable dynamic governance is characterized by an energy pair $\left\{\mathbb{E}(t,x)>0, \frac{d\mathbb{E}(t,x)}{dt}\right\}$, and, system evolves to a stable equilibrium x_e as $x(t)\to x_e$ with $\frac{d\mathbb{E}(t,x)}{dt}<0$, and, $\lim_{t\to\infty}\|x(t)-x_e\|=0$ for $\frac{d\mathbb{E}(t,x)}{dt}=0$. Evolutions $\left\{\mathbb{E}(t,x)>0, \frac{d\mathbb{E}(t,x)}{dt}\right\}$ affects transients in physical systems.

Apply a Lyapunov pair with $V(\mathbf{x})\equiv\mathbb{E}(\mathbf{x})$ or $V(\mathbf{x})\equiv\eta[\mathbb{E}(\mathbf{x})]$. Define stability margins and system governance measures, specified by $\left\{V(\mathbf{x})>0, \frac{dV}{dt}<0\right\}$. Then minimize dV/dt with nonquadratic measures $(\omega_x(\mathbf{x}),\omega_u(u))$ subject to system energy and control limits. Hence, we have

$$\begin{cases}\text{Dynamic System: } \dot{\mathbf{x}}(t)=\mathbf{F}(\mathbf{x})+\mathbf{A}\mathbf{x}+A_r r+\mathbf{B}(\mathbf{x})u,\ u\in U,\\ \text{Stability Criteria: } \left\{V(\mathbf{x})>0, \frac{dV(\mathbf{x})}{dt}<0\right\}, \forall\mathbf{x}\in X, \forall u\in U, \forall t,\\ \frac{dV}{dt}=\left(\frac{\partial V}{\partial\mathbf{x}}\right)^T\left[\mathbf{F}(\mathbf{x})+\mathbf{A}\mathbf{x}+A_r r+\mathbf{B}(\mathbf{x})u\right],\\ \text{Optimality Criteria: } \frac{dV}{dt}=-\underset{\omega_x(\mathbf{x})>0}{\omega_x(\mathbf{x})}-\underset{\omega_u(u)>0}{\omega_u(u)},\ \frac{dV}{dt}<0,\ \forall(\mathbf{x},u),\\ \text{States and Controls Measures: } \omega_x(\mathbf{x})=(\mathbf{x}^{\langle q\rangle})^T Q_q\mathbf{x}^{\langle q\rangle}, \omega_u(u)=\int\left(\Phi^{-1}(u)\right)^T G\,du.\end{cases}\tag{6.106}$$

Problem Formulation and Solution – Minimize the performance measure

$$\frac{dV}{dt}=-\left[\omega_x(\mathbf{x})+\omega_u(u)\right]=-\left[(\mathbf{x}^{\langle q\rangle})^T Q_q\mathbf{x}^{\langle q\rangle}+\int\left(\Phi^{-1}(u)\right)^T G\,du\right],\ \omega_x(\mathbf{x})>0,\ \omega_u(u)>0,$$

subject to (6.105) $\dot{\mathbf{x}}(t)=\mathbf{F}(\mathbf{x})+\mathbf{A}\mathbf{x}+A_r r+\mathbf{B}(\mathbf{x})u$, design $V(\mathbf{x})$, and, derive a control law.

From (6.106), the Hamiltonian is

$$H(\mathbf{x},u,\tfrac{dV}{d\mathbf{x}})=(\mathbf{x}^{\langle q\rangle})^T Q_q\mathbf{x}^{\langle q\rangle}+\int\left(\Phi^{-1}(u)\right)^T G\,du+\left(\frac{dV}{d\mathbf{x}}\right)^T\left[\mathbf{F}(\mathbf{x})+\mathbf{A}\mathbf{x}+\mathbf{B}(\mathbf{x})u\right].\tag{6.107}$$

Apply the stationarity conditions $\left\{\frac{\partial H}{\partial u}=0, \frac{\partial^2 H}{\partial^2 u}>0\right\}$.

For (6.107), using $\frac{\partial}{\partial u}H(\mathbf{x},u,\frac{dV}{d\mathbf{x}})=0$, a constrained control law $u_{\min}\le u\le u_{\max}$ is

$$u=-\Phi\left(G^{-1}\mathbf{B}^T(\mathbf{x})\frac{dV}{d\mathbf{x}}\right),\ \mathbf{x}(t)=\begin{bmatrix}x(t)\\ x_e(t)\\ e(t)\end{bmatrix}=\begin{bmatrix}x(t)\\ \int e\,dt\\ e(t)\end{bmatrix},\ u\in U.\tag{6.108}$$

Solution of

$$(\mathbf{x}^{\langle q\rangle})^T Q_q\mathbf{x}^{\langle q\rangle}+\left(\frac{dV}{d\mathbf{x}}\right)^T\left[\mathbf{F}(\mathbf{x})+\mathbf{A}\mathbf{x}\right]-\int\Phi^T\left(G^{-1}\mathbf{B}^T(\mathbf{x})\frac{dV}{d\mathbf{x}}\right)d\left(\mathbf{B}^T(\mathbf{x})\frac{dV}{d\mathbf{x}}\right)=0\tag{6.109}$$

is satisfied or approximated by (6.93) $V(\mathbf{x})=\sum_{i,j}\sigma_{1_i}^T(\mathbf{x}^{<d_i>})K_{ij}\sigma_{2_j}(\mathbf{x}^{<d_j>})$, yielding positive definite $K_{ij}>0$, such that $V(\mathbf{x})>0$.

Alternatively, similar to *Conjecture* 6.5, for a closed-loop system (6.105)-(6.108) and a Lyapunov pair (6.92) $\left\{V(\mathbf{x})=\sum_{i,j}\sigma_{1_i}^T(\mathbf{x}^{<d_i>})K_{ij}\sigma_{2_j}(\mathbf{x}^{<d_j>})>0, \frac{dV}{dt}<0\right\}$, there exist positive definite $\omega_x(\mathbf{x})$ and $\omega_u(u)$, such that

$$\frac{dV}{dt}=-\left[(\mathbf{x}^{\langle q\rangle})^T Q_q\mathbf{x}^{\langle q\rangle}+\int\left(\Phi^{-1}(u)\right)^T Gdu\right]<0\cdot \tag{6.110}$$

Recall that $\omega_u(u)=\int\left(\Phi^{-1}(u)\right)^T Gdu$ is positive definite, $\omega_u(u)>0$. Assessing optimality and stability, optimization yields control laws, designed by considering a Lyapunov pair and minimizing pertained measures. To implement these control laws, one needs all variables to be measured or observed. The designer may apply the Lyapunov concept if not all variables are measured. One synthesizes *minimal complexity* control laws with directly measured variables to reduce complexity, minimize number of sensors and ASICs, reduce power consumption, minimize size, ensure robustness, etc. Designed-by-specifications hardware solutions are found.

6.10.4. *Minimal Complexity* Control Laws

To reduce hardware complexity, using directly measurable variables, design *minimal complexity* control laws which will ensure stability and optimal performance. Control laws affect system dynamics, stability, robustness, etc. A control function u should guarantee the specified rate of change $dV/dt<0$ and evolutions of $V(t)$, which affect settling time, stability margins, robustness to parameter and structural variations, etc. Reconfigurable feedback and variable gains may be computed in near real time investigating an adaptive design. *Minimal complexity* control implies synthesis of control functions with directly measured system variables $x_{\mathbf{m}}$ and tracking error e.

Consider a system (6.105)

$$\dot{\mathbf{x}}(t)=\begin{bmatrix}\dot{x}(t)\\ \dot{x}_e(t)\\ \dot{e}(t)\end{bmatrix}=\mathbf{F}(\mathbf{x})+\mathbf{A}\mathbf{x}+A_r r+\mathbf{B}(\mathbf{x})u=\begin{bmatrix}F(x)\\ 0\\ -HF(x)\end{bmatrix}+\begin{bmatrix}0&0&0\\ -I_X H&0&0\\ 0&I_E^X&-I_E\end{bmatrix}\mathbf{x}+\begin{bmatrix}0&0\\ I_X&0\\ 0&I\end{bmatrix}\begin{bmatrix}r\\ \dot{r}\end{bmatrix}+\begin{bmatrix}B(x)\\ 0\\ -HB(x)\end{bmatrix}u$$

with the measured $(x_{\mathbf{m}},e)$. Stability and optimality are specified and studied using a Lyapunov pair $\left\{V_{\mathbf{m}}(\mathbf{x})>0, \frac{dV_{\mathbf{m}}}{dt}<0\right\}$. Consider governing equations, stability criterial and optimality measures as

$$\begin{cases}\text{Dynamic System: } \dot{\mathbf{x}}(t)=\mathbf{F}(\mathbf{x})+\mathbf{A}\mathbf{x}+A_r r+\mathbf{B}(\mathbf{x})u,\ u\in U, \qquad (6.111)\\[4pt] \text{Stability Criteria: } V_{\mathbf{m}}(\mathbf{x})\equiv V(\mathbf{x}), \left\{V_{\mathbf{m}}(\mathbf{x})>0, \frac{dV_{\mathbf{m}}(\mathbf{x})}{dt}<0\right\}, \forall\mathbf{x}\in X, \forall u\in U, \forall t,\\[4pt] \frac{dV_{\mathbf{m}}}{dt}=\left(\frac{\partial V_{\mathbf{m}}}{\partial\mathbf{x}}\right)^T\frac{d\mathbf{x}}{dt}, y=Hx,\\[4pt] \dot{\mathbf{x}}(t)=\begin{bmatrix}\dot{x}(t)\\ \dot{x}_e(t)\\ \dot{e}(t)\end{bmatrix}=\mathbf{F}(\mathbf{x})+\mathbf{A}\mathbf{x}+A_r r+\mathbf{B}(\mathbf{x})u=\begin{bmatrix}F(x)\\ 0\\ -HF(x)\end{bmatrix}+\begin{bmatrix}0&0&0\\ -I_X H&0&0\\ 0&I_E^X&-I_E\end{bmatrix}\mathbf{x}+\begin{bmatrix}0&0\\ I_X&0\\ 0&I\end{bmatrix}\begin{bmatrix}r\\ \dot{r}\end{bmatrix}+\begin{bmatrix}B(x)\\ 0\\ -HB(x)\end{bmatrix}u,\\[4pt] \text{Optimality Criteria: } \frac{dV_{\mathbf{m}}}{dt}=-\underset{\omega_x(\mathbf{x})>0}{\omega_x(\mathbf{x})}-\underset{\omega_u(u)>0}{\omega_u(u)},\ \frac{dV_{\mathbf{m}}}{dt}<0,\ \forall(\mathbf{x},u),\\[4pt] \text{States and Controls Measures: } \omega_x(\mathbf{x})=(\mathbf{x}^{\langle q\rangle})^T Q_q\mathbf{x}^{\langle q\rangle}, \omega_u(u)=\int\left(\Phi^{-1}(u)\right)^T Gdu.\end{cases}$$

One may truncate $V(\mathbf{x})$, given by (6.93). Apply a positive definite function of all states

$$V_{\mathbf{m}}(\mathbf{x})=\sum_{i,j}\sigma_{\mathbf{m}1_i}^T(\mathbf{x}^{<d_i>})K_{\mathbf{m}\,ij}\sigma_{\mathbf{m}2_j}(\mathbf{x}^{<d_j>}) \quad (6.112)$$

to yield directly measured ($x_{\mathbf{m}}$,e) in the feedback structure $u=\Phi\left[\varphi(x_{\mathbf{m}},\int edt,e)\right]$.

Minimize $\frac{dV_{\mathbf{m}}}{dt}=-\left[\omega_x(\mathbf{x})+\omega_u(u)\right]=-\left[(\mathbf{x}^{\langle q\rangle})^T Q_q\mathbf{x}^{\langle q\rangle}+\int\left(\Phi^{-1}(u)\right)^T G du\right]$,

subject to a system (6.105). A scalar Hamiltonian function as

$$H(\mathbf{x},u,\tfrac{\partial V_{\mathbf{m}}}{\partial\mathbf{x}})=(\mathbf{x}^{\langle q\rangle})^T Q_q\mathbf{x}^{\langle q\rangle}+\int\left(\Phi^{-1}(u)\right)^T G du+\left(\frac{\partial V_{\mathbf{m}}}{\partial\mathbf{x}}\right)^T\left[\mathbf{F}(\mathbf{x})+\mathbf{A}\mathbf{x}+\mathbf{B}(\mathbf{x})u\right]. \quad (6.113)$$

A stationarity condition (6.22) $\frac{\partial}{\partial u}H(\mathbf{x},u,\frac{\partial V_{\mathbf{m}}}{\partial\mathbf{x}})=0$ yields a bounded *minimal complexity* control law

$$u=\underset{\varphi(\cdot)=-G^{-1}\mathbf{B}^T(\mathbf{x})\frac{\partial V_{\mathbf{m}}}{\partial\mathbf{x}}}{\Phi\left[\varphi(x_{\mathbf{m}},\int edt,e)\right]}=-\Phi\left(G^{-1}\mathbf{B}^T(\mathbf{x})\frac{\partial V_{\mathbf{m}}}{\partial\mathbf{x}}\right),\ u\in U. \quad (6.114)$$

Function $V_{\mathbf{m}}(\mathbf{x})$ (6.112) should satisfy a nonquadratic equation

$$(\mathbf{x}^{\langle q\rangle})^T Q_q\mathbf{x}^{\langle q\rangle}+\left(\frac{\partial V_{\mathbf{m}}}{\partial\mathbf{x}}\right)^T\left[\mathbf{A}\mathbf{x}+\mathbf{F}(\mathbf{x})\right]-\int\Phi^T\left(G^{-1}\mathbf{B}^T(\mathbf{x})\frac{\partial V_{\mathbf{m}}}{\partial\mathbf{x}}\right)d\left(\mathbf{B}^T(\mathbf{x})\frac{\partial V_{\mathbf{m}}}{\partial\mathbf{x}}\right)=0. \quad (6.115)$$

Solving (6.115), one computes $K_{\mathbf{m}\,ij}$, and, finds a feedback structure $\varphi(x_{\mathbf{m}},\int edt,e)$ and feedback coefficients. Positive definite $V_{\mathbf{m}}(\mathbf{x})$, $\omega_x(\mathbf{x})$ and $\omega_u(u)$ affect a control law structure, closed-loop system governance, stability and robustness.

Tracking control laws with $e(t)$ and $x_e(t)\equiv\int edt$ can be designed on a class of proportional-integral control laws with measured state feedback

$$u=\underset{\varphi(\cdot)=-G^{-1}\mathbf{B}^T(\mathbf{x})\frac{\partial V_{\mathbf{m}}}{\partial\mathbf{x}}}{\Phi\left[\varphi(\int edt,e)\right]}=-\Phi\left(G^{-1}\mathbf{B}^T(\mathbf{x})\frac{\partial V_{\mathbf{m}}}{\partial\mathbf{x}}\right)=\Phi\left(k_p e+k_i\int edt\right),\ u\in U. \quad (6.116)$$

Unconstrained Control – A tracking control law

$$u=\underset{\varphi(\cdot)=-G^{-1}\mathbf{B}^T(\mathbf{x})\frac{\partial V_{\mathbf{m}}}{\partial\mathbf{x}}}{\varphi(x_{\mathbf{m}},\int edt,e)}=-G^{-1}\mathbf{B}^T(\mathbf{x})\frac{\partial V_{\mathbf{m}}}{\partial\mathbf{x}} \quad (6.117)$$

is designed by using the sufficient conditions for stability and optimality (6.92) and (6.111). In particular,

$$\left\{V_{\mathbf{m}}(\mathbf{x})>0,\ \frac{dV_{\mathbf{m}}}{dt}<0\right\},\ \frac{dV_{\mathbf{m}}}{dt}\le-\left[(\mathbf{x}^{\langle q\rangle})^T Q_q\mathbf{x}^{\langle q\rangle}+\tfrac{1}{2}u^T G u\right].$$

To find a closed-form solution of an optimization problem, a convex inequality problem is solved applying affine equality constraints and convex objective function. Minimize

$\frac{dV_{\mathbf{m}}}{dt}=-\left[(\mathbf{x}^{\langle q\rangle})^T Q_q\mathbf{x}^{\langle q\rangle}+\tfrac{1}{2}u^T G u\right]$ subject to (6.105) $\dot{\mathbf{x}}(t)=\mathbf{F}(\mathbf{x})+\mathbf{A}\mathbf{x}+A_r r+\mathbf{B}(\mathbf{x})u$,

and, analytically design a control law. Apply $\left\{\frac{\partial H}{\partial u}=0,\ \frac{\partial^2 H}{\partial u^2}>0\right\}$ for a scalar Hamiltonian

$$H(\mathbf{x},u,\tfrac{dV_{\mathbf{m}}}{d\mathbf{x}})=(\mathbf{x}^{\langle q\rangle})^T Q_q\mathbf{x}^{\langle q\rangle}+\tfrac{1}{2}u^T G u+\left(\frac{\partial V_{\mathbf{m}}}{\partial\mathbf{x}}\right)^T\left[\mathbf{F}(\mathbf{x})+\mathbf{A}\mathbf{x}+\mathbf{B}(\mathbf{x})u\right].$$

For Lyapunov pair with quadratic measures

$$\left\{V_{\mathbf{m}}(\mathbf{x})>0,\ \frac{dV_{\mathbf{m}}}{dt}\le-\left[\tfrac{1}{2}\mathbf{x}^T Q\mathbf{x}+\tfrac{1}{2}u^T G u\right]\right\},\ (Q,G)>0,$$

one finds and solves

$$\tfrac{1}{2}\mathbf{x}^T Q\mathbf{x}+\left(\frac{\partial V_{\mathbf{m}}}{\partial \mathbf{x}}\right)^T\left[\mathbf{F}(\mathbf{x})+\mathbf{A}\mathbf{x}\right]-\tfrac{1}{2}\left(\frac{\partial V_{\mathbf{m}}}{\partial \mathbf{x}}\right)^T\mathbf{B}(\mathbf{x})G^{-1}\mathbf{B}^T(\mathbf{x})\frac{\partial V_{\mathbf{m}}}{\partial \mathbf{x}}=0. \qquad (6.118)$$

Solution of a nonlinear optimization problem yields a *minimal complexity* control law.

Nonlinear Feedback, Fractional Exponent Feedback, and, Multiple-Integral Tracking Error – Consider design of control laws to derive high degree and fractional exponent real-valued feedback for the tracking error e, as well as multiple-integrals terms $\sum_{j=1}^{N_i}\int\underset{j}{\dots}\int k_{i_j}\,edt$.

Define the extended model

$$\dot{\mathbf{x}}(t)=\begin{bmatrix}\dot{x}(t)\\ \begin{bmatrix}\dot{x}_{e_1}(t)\\ \dot{x}_{e_2}(t)\\ \vdots\\ \dot{x}_{e_j}(t)\end{bmatrix}\\ \dot{e}(t)\end{bmatrix}=\begin{bmatrix}F(x)+B(x)u\\ \begin{bmatrix}-I_X Hx+I_X r\\ x_{e_1}\\ \vdots\\ x_{e_{j-1}}\end{bmatrix}\\ -HF(x)-I_E e-\sum_{z,l} I_E e^{\frac{2z-1}{2l-1}}+I\dot{r}-HB(x)u\end{bmatrix},\ y=Hx. \qquad (6.119)$$

Nonlinear Control Structure – The feedback structure is comprised of real-valued linear, high-degree and fractional exponent feedback on the tracking error $\sum_{z,l} e^{\frac{2z-1}{2l-1}}(t)$, as well as multiple integral feedback on $e(t)$ as $\sum_{j=1}^{N_i}\int\underset{j}{\dots}\int k_{i_j}\,edt$.

The tracking error feedback are in the second and fourth quadrants.

Illustrative Studies 6.2. *Linear and Fractional Feedback*

For Z=1 and L=4, we have $\sum_{z=1,l=1}^{Z,L} e^{\frac{2z-1}{2l-1}}(t)=e+\text{sgn}(e)|e|^{1/3}+\text{sgn}(e)|e|^{1/5}+\text{sgn}(e)|e|^{1/7}$.

One has $\text{sgn}(e)|e|^{1/3}=\text{sgn}(e)\sqrt[3]{|e|}=\begin{cases}\sqrt[3]{e},\ e>0\\ -\sqrt[3]{-e},\ e<0\end{cases}=\begin{cases}e^{1/3},\ e>0\\ -(-e)^{1/3},\ e<0\end{cases}$, $\text{sgn}(e)=\frac{e}{|e|}=\frac{|e|}{e}$.

As an illustration, a positive definite function

$V_{\mathbf{m}}(\mathbf{x})=c(x_u)+\ldots+x_u e+x_u\,\text{sgn}(e)|e|^{1/3}+x_u\,\text{sgn}(e)|e|^{1/5}+x_u\,\text{sgn}(e)|e|^{1/7}+\ldots+|e|^{2/3}+|e|^{2/5}+|e|^{2/7}$

yields linear and fractional feedback terms

$$\mathbf{B}^T(\mathbf{x})\frac{\partial V_{\mathbf{m}}(\mathbf{x})}{\partial \mathbf{x}}\equiv\frac{\partial V_{\mathbf{m}}(\mathbf{x})}{\partial x_u}=\ldots+e+\text{sgn}(e)|e|^{1/3}+\text{sgn}(e)|e|^{1/5}+\text{sgn}(e)|e|^{1/7}+\ldots$$

Here, $c(x_u)$ is the real-valued Cantor function, which is uniformly Hölder continuous and has zero derivative. ■

Bounded Control With Nonlinear Feedback Structure and Multiple-Integral Tracking Error Terms – Nonlinear control laws are designed using requirements imposed on a Lyapunov pair $\left\{V_{\mathbf{m}}(\mathbf{x})>0,\ \frac{dV_{\mathbf{m}}}{dt}<0\right\}$ subject to system dynamics. Applying performance measures $(\omega_x(\mathbf{x}),\omega_u(u))>0$, specify system optimality and stability as

$$\frac{dV_{\mathbf{m}}}{dt} \le -\boldsymbol{\omega}_x(\mathbf{x}) - \boldsymbol{\omega}_u(u), \quad \frac{dV_{\mathbf{m}}}{dt} = -\left[(\mathbf{x}^{\langle q\rangle})^T Q_q \mathbf{x}^{\langle q\rangle} + \int \left(\Phi^{-1}(u)\right)^T G du\right]. \tag{6.120}$$

Minimization yields a bounded control law $u = \Phi[\varphi(\mathbf{x})] = -\Phi\left[G^{-1}\mathbf{B}^T(\mathbf{x})\frac{\partial V_{\mathbf{m}}}{\partial \mathbf{x}}\right]$.

A control structure $\varphi(\cdot) = -G^{-1}\mathbf{B}^T(\mathbf{x})\frac{\partial V_{\mathbf{m}}}{\partial \mathbf{x}}$ yields high-degree multivariate feedback and fractional exponent Newton-Puiseux polynomial feedback.

A tracking control law is designed on a class of proportional-integral algorithms

$$u = \Phi\left(\varphi(e)\right) = \Phi\left(\underbrace{\sum_{z=q_p,l=g_p}^{Z_p,L_p} k_{p_{z,l}} e^{\frac{2z-1}{2l-1}}}_{\text{proportional}} + \underbrace{\sum_{j=1}^{N_i}\int\ldots\int \sum_{z=q_i,l=g_i}^{Z_i,L_i} k_{i_{j,z,l}} e^{\frac{2z-1}{2l-1}} dt}_{\text{integral}}\right), u_{\min} \le u \le u_{\max}, \tag{6.121}$$

where Z_p, Z_i, L_p, L_i and N_i are the integers, $(Z_p,Z_i,L_p,L_i,N_i)\ge 1$; q_p, q_i, g_p and g_i are the nonnegative integers, $(q_p,q_i)\ge 1$ and $(g_p,g_i)\ge 1$; $k_{p_{z,l}}$ and $k_{i_{j,z,l}}$ are the feedback coefficients.

Example 6.48. *Controlled Rotational Motion of an Aerial Vehicle*

In Examples 5.33, 5.36, 6.12, 6.20 and 6.32, control laws are designed for aerial vehicles. Consider design of *minimal complexity* control laws.

System states (x_1,x_2) and an tracking error $e(t)$ are governed by equations

$$\dot{x}_1 = -x_1 - x_1|x_1| + u,$$
$$\dot{x}_2 = x_1,$$
$$\dot{x}_e = r - y = r - x_2,$$
$$\dot{e} = -e + \dot{r} - \dot{x}_2 = -e + \dot{r} - x_1.$$

Hence,
$$\dot{\mathbf{x}}(t) = \begin{bmatrix}\dot{x}_1(t)\\ \dot{x}_2(t)\\ \dot{x}_e(t)\\ \dot{e}(t)\end{bmatrix} = \mathbf{F}(\mathbf{x}) + \mathbf{A}\mathbf{x} + A_r r + \mathbf{B}u = \begin{bmatrix}-x_1|x_1|\\ 0\\ 0\\ 0\end{bmatrix} + \begin{bmatrix}-1 & 0 & 0 & 0\\ 1 & 0 & 0 & 0\\ 0 & -1 & 0 & 0\\ -1 & 0 & 0 & -1\end{bmatrix}\begin{bmatrix}x_1\\ x_2\\ x_e\\ e\end{bmatrix} + \begin{bmatrix}0 & 0\\ 0 & 0\\ 1 & 0\\ 0 & 1\end{bmatrix}\begin{bmatrix}r\\ \dot{r}\end{bmatrix} + \begin{bmatrix}1\\ 0\\ 0\\ 0\end{bmatrix}u, y = x_2.$$

The directly measured variables are $x_{\mathbf{m}}=[x_2, e]$. Note that x_1 can be estimated. To design control laws, one may apply different positive definite functions.

For example, $V_{\mathbf{m}}(\mathbf{x}) = k_{11}c(x_1) + \frac{1}{2}\mathbf{x}^T\begin{bmatrix}0 & k_{12} & k_{13} & k_{14}\\ k_{21} & k_{22} & k_{23} & k_{24}\\ k_{31} & k_{32} & k_{33} & k_{34}\\ k_{41} & k_{42} & k_{43} & k_{44}\end{bmatrix}\mathbf{x}$.

Consider a truncated function

$$V_{\mathbf{m}}(\mathbf{x}) = k_{11}c(x_1) + k_{12}x_1x_2 + k_{13}x_1x_e + k_{14}x_1e + \tfrac{1}{2}(k_{22}x_2^2 + k_{33}x_e^2 + k_{44}e^2),$$

where $c(x_1)$ is the real-valued Cantor function which is uniformly Hölder continuous and has zero derivative, see section 2.5.

Unconstrained Control – Specify $\frac{dV_{\mathbf{m}}}{dt} \le -\frac{1}{2}\left(x_1^2 + x_2^2 + x_e^2 + 10e^2 + u^2\right)$.

Fot $\frac{dV_{\mathbf{m}}}{dt} = -\frac{1}{2}\left(x_1^2 + x_2^2 + x_e^2 + 10e^2 + u^2\right)$, a scalar Hamiltonian function is

$$H(\mathbf{x},u,\tfrac{\partial V_{\mathbf{m}}}{\partial \mathbf{x}}) = \tfrac{1}{2}\mathbf{x}^T\begin{bmatrix}1 & 0 & 0 & 0\\ 0 & 1 & 0 & 0\\ 0 & 0 & 1 & 0\\ 0 & 0 & 0 & 10\end{bmatrix}\mathbf{x} + \tfrac{1}{2}u^2 + \left(\frac{\partial V_{\mathbf{m}}}{\partial \mathbf{x}}\right)^T\left[\mathbf{F}(\mathbf{x}) + \mathbf{A}\mathbf{x} + \mathbf{B}u\right].$$

One finds a control law (6.117). Solution of the algebraic equation (6.118) yields the unknown k_{ii} and k_{ij}. Using a term for x_1e, $2k_{14}+k_{44}=0$.

Grouping terms for x_2^2, x_e^2 and e^2, solve equations

$-k_{12}^2+1=0$, $-k_{13}^2+1=0$, $-k_{14}^2+4k_{14}+10=0$.

Hence, $k_{12}=1$, $k_{13}=-1$ and $k_{14}=-1.74$.

One has $u=\varphi\left(x_{\mathbf{m}},\int edt,e\right)=-G^{-1}\mathbf{B}^T\frac{\partial V_{\mathbf{m}}}{\partial \mathbf{x}}=-G^{-1}\frac{\partial V_{\mathbf{m}}}{\partial x_1}=-x_2+\int edt+1.74e$, $G=1$.

System is stable, and, $\left\{V_{\mathbf{m}}(\mathbf{x})>0, \frac{dV_{\mathbf{m}}}{dt}<0\right\}$. Figure 6.25.a illustrates dynamics of the closed-loop system. Evolutions for (x_1,x_2) and $y=x_2$ are shown for reference $r(t)=\pm 1$.

Constrained Control – Consider the symmetric saturation limits on control, $-1\le u\le 1$.

The Lyapunov function is

$$V_{\mathbf{m}}(\mathbf{x})=k_{11}c(x_1)+k_{12}x_1x_2+k_{13}x_1x_e+k_{14}x_1e+\tfrac{1}{2}(k_{22}x_2^2+k_{33}x_e^2+k_{44}e^2).$$

Specify $\frac{dV_{\mathbf{m}}(x,e)}{dt}\le-\frac{1}{2}\left(x_1^2+x_2^2+0.25x_e^2+10e^2\right)-\int\tanh^{-1}udu$ to find a Hamiltonian. The resulting functional equation is found using the reported $V_{\mathbf{m}}(\mathbf{x})$, and, $\int\tanh\left(\mathbf{B}^T\frac{\partial V_{\mathbf{m}}}{\partial \mathbf{x}}\right)d\left(\mathbf{B}^T\frac{\partial V_{\mathbf{m}}}{\partial \mathbf{x}}\right)=\ln\cosh\left(\mathbf{B}^T\frac{\partial V_{\mathbf{m}}}{\partial \mathbf{x}}\right)\approx\frac{1}{3}\left(\frac{\partial V_{\mathbf{m}}}{\partial \mathbf{x}}\right)^T\mathbf{B}\mathbf{B}^T\frac{\partial V_{\mathbf{m}}}{\partial \mathbf{x}}$, A solution yields (k_{ii},k_{ij}).

We have $k_{12}=1.22$, $k_{13}=-0.612$, $k_{14}=-1.9$. The bounded control law (6.114) is

$$u=\Phi\left(\varphi(x_{\mathbf{m}},\textstyle\int edt,e)\right)=-\Phi\left(\mathbf{B}^T\frac{\partial V_{\mathbf{m}}}{\partial \mathbf{x}}\right)=-\Phi\left(\frac{\partial V_{\mathbf{m}}}{\partial x_1}\right)$$

$$=-\tanh\left(k_{12}x_2+k_{13}x_e+k_{14}e\right)=\tanh\left(-1.22x_2+0.612\textstyle\int edt+1.9e\right), -1<u<1.$$

The closed-loop system is simulated as reported in Figure 6.25.b. Dynamics of (x_1,x_2) and $y=x_2$ are illustrated for $r(t)=\pm 1$. For $(x_1,x_2)\in X$, approximation $\int\tanh\left(\frac{\partial V_{\mathbf{m}}}{\partial x_1}\right)d\left(\frac{\partial V_{\mathbf{m}}}{\partial x_1}\right)=\ln\cosh\left(\frac{\partial V_{\mathbf{m}}}{\partial x_1}\right)\approx\frac{1}{3}\left(\frac{\partial V_{\mathbf{m}}}{\partial x_1}\right)^2$ is valid. Practice Problem 6.14.5 solves the tracking control problem using only tracking error $e(t)$.

Control With the Fractional Exponent Error Feedback – Consider (6.119) as

$$\dot{\mathbf{x}}(t)=\begin{bmatrix}\dot{x}_1(t)\\ \dot{x}_2(t)\\ \dot{x}_e(t)\\ \dot{e}(t)\end{bmatrix}=\mathbf{F}(\mathbf{x})+\mathbf{A}\mathbf{x}+A_r r+\mathbf{B}u=\begin{bmatrix}-x_1|x_1|\\ 0\\ 0\\ -\operatorname{sgn}(e)|e|^{1/3}\end{bmatrix}+\begin{bmatrix}-1&0&0&0\\1&0&0&0\\0&-1&0&0\\-1&0&0&-1\end{bmatrix}\begin{bmatrix}x_1\\x_2\\x_e\\e\end{bmatrix}+\begin{bmatrix}0&0\\0&0\\1&0\\0&1\end{bmatrix}\begin{bmatrix}r\\ \dot{r}\end{bmatrix}+\begin{bmatrix}1\\0\\0\\0\end{bmatrix}u,\ y=x_2.$$

The differential equation for the tracking error has a term $\operatorname{sgn}(e)|e^{1/3}|$. A constrained control law is designed specifying system stability and optimality by a Lyapunov pair

$$V_{\mathbf{m}}(\mathbf{x})=k_{11}c(x_1)+k_{12}x_1x_2+k_{13}x_1x_e+k_{14}x_1e+k_{15}x_1\operatorname{sgn}(e)|e|^{1/3}+\tfrac{1}{2}(k_{22}x_2^2+k_{33}x_e^2+k_{44}e^2),$$

$$\frac{dV_{\mathbf{m}}(\mathbf{x})}{dt}\le-\tfrac{1}{2}\left(x_1^2+x_2^2+0.25x_e^2+10e^2+|e|^{2/3}\right)-\int\tanh^{-1}udu\,.$$

The unknown coefficients (k_{ii},k_{ij}) of $V_{\mathbf{m}}(\mathbf{x})$ are computed.

Unknown k_{15} is found by grouping the term for $|e|^{2/3}$. From $-\frac{1}{3}k_{15}^2+\frac{1}{2}=0$, $k_{15}=1.22$.

We analytically design a constrained control law

$$u=\Phi\left(\varphi(x_{\mathrm{m}},\textstyle\int edt,e)\right)=-\Phi\left(\mathbf{B}^T\frac{\partial V_{\mathrm{m}}}{\partial\mathbf{x}}\right)=-\Phi\left(\frac{\partial V_{\mathrm{m}}}{\partial x_1}\right)=-\tanh\left(k_{12}x_2+k_{13}x_e+k_{14}e+k_{15}\,\mathrm{sgn}(e)|e|^{1/3}\right)$$

$$=\tanh\left(-1.22x_2+0.612\textstyle\int edt+1.9e+1.22\,\mathrm{sgn}(e)|e|^{1/3}\right),\ -1<u<1.$$

System dynamics with a bounded control and feedback $\mathrm{sgn}(e)|e^{1/3}|$ is reported in Figure 6.25.c. The fractional exponent feedback affects settling time and error.

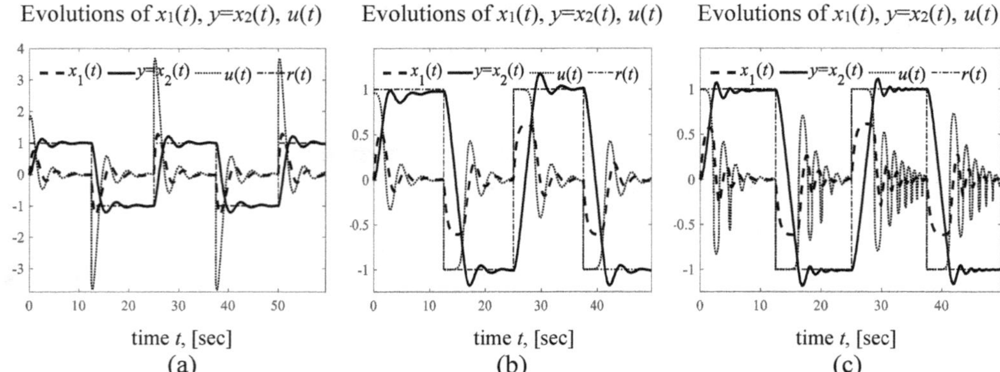

Figure 6.25. (a) System dynamics with a linear not constrained *minimal complexity* control;
(b) System dynamics with a bounded tracking control law, $r(t)=\pm1$;
(c) System dynamics with a constrained tracking control law with fractional feedback. ■

Example 6.49. *Minimal Complexity Control of a Lead Magnesium Niobate Actuator*
Open-loop stable solid state electrostrictive and piezoelectric actuators exhibit highly nonlinear characteristics. The displacement of ceramic actuators is a nonlinear function of the applied voltage or polarization P. For an isotropic polymer, the longitudinal and transverse strains are

$$s_l=q_{33}P^2,\ s_t=q_{13}P^2,$$

where q_{ij} are the charge-dependent electrostrictive coefficients, $q_{33}<0$, $q_{13}>0$.

Energy conversions and transductions are nonlinear, and, exhibit hysteresis. Under number of assumptions, the actuators with hysteresis are modeled as

$$m\frac{d^2y}{dt^2}+b\frac{dy}{dt}+ky+cz_h=du,\ y_{\min}\le y\le y_{\max},\ u_{\min}\le u\le u_{\max},$$

$$\dot{z}_h=-a_{31}z_h|\dot{u}|+a_{32}|z_h|\dot{u}+b_{31}\dot{u},$$

where y is the actuator displacement; m the *effective* mass; b, k, c and d are the parameters; a_{31}, a_{32} and b_{31} are the constants.

However, actuators are open-loop stable and controllable. Proportional and proportional-integral tracking control laws can be implemented. Feedback gains (k_p,k_i,k_d) are derived for particular ceramics, longitudinal and transverse strains as functions of the applied electric field (voltage), hysteresis loop, maximum displacement, noise, sensor gain, filters and PWM controller-driver topologies, etc. Actuators exhibit highly nonlinear transductions, and, the *hidden* hysteresis state z_h cannot be directly measured or observed. Not measured velocity may not be observed because the micrometer displacement $y(t)$ is measured with multi-frequency noise. Tracking control laws with the directly measured displacement and tracking error are used. Control laws should ensure high-precision repositioning, steering, stability, etc.

Study a high-strain, low-hysteresis lead magnesium niobate $Pb(Mg_{1/3}Nb_{2/3})O_3$ electrostrictive actuator. There are disks and plates configurations, and, multilayer stacks are separated with electrodes. These transducers are fabricated using ceramic materials. A relaxor perovskite ensures stability, high strain and low 5 to 10% hysteresis near a diffuse Currie transition. The eight-layer actuators and drivers are depicted in Figure 6.26.a. The applied voltage is limited, $u_{min} \le \mathcal{U} \le u_{max}$. For the considered transducer, $0 \le \mathcal{U} \le 100$ V.

Laws of physics yield descriptive equations and constitutive relationships. Actuator dynamics is characterized by nonlinear displacement-voltage curve $x=f(u)$ with saturation and hysteresis. The curves $x=f(u)$ are experimentally measured applying the sinusoidal voltage $u(t)$ and measuring the displacement $x(t)$. Catalogs for the lead magnesium niobate, lead zirconate titanate and other transducers report these nonlinear dependencies $x=f(u)$. Equations of motion are

$$\dot{x}_1 = -a_{11}x_1 - a_{12}x_1|x_1| - a_{13}x_2 - a_{14}x_2|x_2| - a_{15}x_3 + b_{11}u, \; x_{2\,min} \le x_2 \le x_{2\,max}, \; u_{min} \le u \le u_{max},$$

$$\dot{x}_2 = x_1,$$

$$\dot{x}_3 = -a_{31}x_3\left|\frac{du}{dt}\right| + a_{32}|x_3|\frac{du}{dt} + b_{31}\frac{du}{dt}.$$

States x_1 and x_2 are the velocity and displacement, $x_1=v$, $x_2=x$. The hysteresis is described by using x_3, which is not measurable and not observable. Actuator dynamics is studied evaluating the experimental data. Using actuator transients, parameters $[(a_{11},a_{12},a_{13},a_{14},a_{15},b_{11}),(a_{31},a_{32},b_{31})]$ are estimated. The exhibited hysteresis is investigated, and, described by $-a_{31}x_3\left|\frac{du}{dt}\right| + a_{32}|x_3|\frac{du}{dt} + b_{31}\frac{du}{dt} = 0$.

Using the experimental results, the parameters are found to be $a_{11}=179$, $a_{12}=2.4\times10^4$, $a_{13}=9.47\times10^3$, $a_{14}=5.1\times10^6$, $a_{15}=2.37\times10^3$, $a_{31}=0.047$, $a_{32}=0.013$, $b_{11}=0.001$ and $b_{32}=5.62\times10^{-8}$.

As illustrated in Examples 6.22 and 6.27, asymmetric limits are described by a continuous functions $\Phi \subseteq \mathcal{U}_\varepsilon$.

For $0 \le \mathcal{U} \le 100$, $\Phi\left(\mathbf{B}^T(\mathbf{x})\frac{\partial V_{\mathsf{m}}}{\partial \mathbf{x}}\right) = \dfrac{u_{max}}{1+e^{-a\left(\mathbf{B}^T(\mathbf{x})\frac{\partial V_{\mathsf{m}}}{\partial \mathbf{x}}-b\right)}}$, $u_{max}=100$, $a=10$, $b=0.5$, $0<u<100$.

For a chosen $\Phi(x)$, function $\Phi^{-1}(u)$ exists as studied in Example 6.27. Design a *minimal complexity* control law using a measured tracking error $e(t)$. Considering the state and error dynamics (6.105), a control law (6.121) is designed with linear proportional and integral feedback. The Lyapunov pair $\left\{V_{\mathsf{m}}(\mathbf{x})>0, \frac{dV_{\mathsf{m}}}{dt}<0\right\}$ is

$$V_{\mathsf{m}}(\mathbf{x}) = k_{11}c(x_1) + k_{14}x_1x_e + k_{15}x_1e + \tfrac{1}{2}(k_{22}x_2^2 + k_{33}x_3^2 + k_{44}x_e^2 + k_{55}e^2),$$

$$\frac{dV_{\mathsf{m}}}{dt} \le -x_1^2 - x_2^2 - x_3^2 - 1\times10^{12}x_e^2 - 1\times10^{12}e^2 - \int \Phi^{-1}(u)du.$$

Derived constrained control law

$$u = -\Phi\left(\mathbf{B}^T\frac{\partial V_{\mathsf{m}}}{\partial \mathbf{x}}\right) = \Phi\left(9.05\times10^7\int e dt + 1.48\times10^9 e\right), \; 0<u<100 \text{ V}$$

guarantees stability, ensures tracking, and assures the specified $dV_{\mathsf{m}}(\mathbf{x})/dt$. The control law is experimentally substantiated. Figures 6.26.b illustrate transient dynamics for the output $y=x_2(t)$ if the reference is a sequence of steps $r(t)=\begin{cases}10\\0\end{cases}$ μm.

Designed control laws can be implemented using commercialized PWM controllers-drivers. Adequate displacement sensors and filters are implemented.

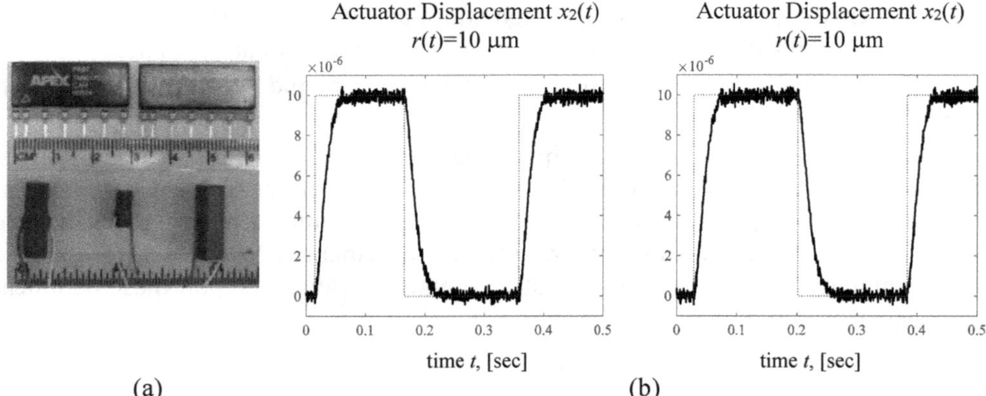

(a) (b)

Figure 6.26. (a) Eight-layer high-strain, low-hysteresis lead magnesium niobate actuators (4 and 8 mm diameter, 0.125 and 0.25 mm thickness separated multilayer stacks with electrodes) and PWM drivers;

(b) Actuator repositioning with displacement $y(t)=x_2(t)$ for references $r(t)=\begin{cases}10\\0\end{cases}$ μm . ■

Example 6.50. *Control of an Electrostatic Microactuator*

Electrostatic transducers were studied in section 3.2. The governing equations were developed. The states are the linear velocity v and displacement x, $\begin{bmatrix}x_1\\x_2\end{bmatrix}=\begin{bmatrix}v\\x\end{bmatrix}$. Example 3.2 documents experimental studies of an actuator. There are limits on the applied voltage, and, $0\le\mathcal{U}\le17.5$ V. As the voltage is applied, the movable plate repositions reducing the separation (x_0-x_2), $x_2\ge0$, and, $x_2<x_0$. That is, $x_2|_{\mathcal{V}=0}=0$ and $x_2|_{\mathcal{V}\ne0}\ne0$. The model (3.8) is

$$\dot{x}_1=\frac{1}{m}\left[\left(-c_1x_1-c_2x_1|x_1|-c_3x_1^3-k_1x_2-k_2x_2|x_2|-k_3x_2^3\right)+\frac{\varepsilon_0\varepsilon_rA}{2(x_0-x_2)^2}\mathcal{V}^2\right], x_2<x_0, \forall x_2, 0\le\mathcal{V}\le\mathcal{V}_{\max},$$

$$\dot{x}_2=x_1,\ y=x_2,$$

the coefficients vary. In particular, $c_i\in[c_{i\min}\ \ c_{i\max}]$, $c_i=c_{i0}\pm\Delta c_i$, $c_{i\min}>0$, $k_i\in[k_{i\min}\ \ k_{i\max}]$, $k_i=k_{i0}\pm\Delta k_i$, $k_{i\min}>0$.

To study open-loop system stability, consider a quadratic Lyapunov function

$V(x_1,x_2)=\frac{1}{2}x_1^2+x_1x_2+\frac{1}{2}x_2^2$, $V(x_1,x_2)>0$.

The total derivative is

$$\frac{dV(x_1,x_2)}{dt}=\frac{dV}{dx_1}\dot{x}_1+\frac{dV}{dx_2}\dot{x}_2=(x_1+x_2)\frac{1}{m}\left(-c_1x_1-c_2x_1|x_1|-c_3x_1^3-k_1x_2-k_2x_2|x_2|-k_3x_2^3\right)+(x_1+x_2)x_1$$

$$=\frac{1}{m}\left(-(c_1-m)x_1^2-c_2x_1^2|x_1|-c_3x_1^4-(k_1+c_1-m)x_1x_2-k_3x_1x_2^3-c_3x_1^3x_2-c_2x_1x_2|x_1|-k_2x_1x_2|x_2|-k_1x_2^2-k_2x_2^2|x_2|-k_3x_2^4\right).$$

Despite the fact that (c_i,k_i) vary, parameters are bounded and positive, $(c_i,k_i)>0$. Hence, $\frac{dV(x_1,x_2)}{dt}<0$, $\forall(c_i,k_i)>0, x\ne0$. The open-loop system is robustly asymptotically stable.

The applied voltage is limited as $0\le\mathcal{V}\le17.5$ V, and, the control u is bounded, $u=\mathcal{V}^2$, $0\le u\le u_{\max}$. Design a *minimal complexity* control law using the directly measured tracking error $e(t)$. For the extended system (6.119), consider a Lyapunov function

$V_{\mathbf{m}}(\mathbf{x}) = k_{11}c(x_1) + k_{13}x_1x_e + k_{14}x_1e + k_{15}x_1\,\mathrm{sgn}(e)|e|^{1/3} + \frac{1}{2}\left(k_{22}x_2^2 + k_{33}x_e^2 + k_{44}e^2\right)$, $V_{\mathbf{m}}(\mathbf{x})>0$,
where $c(x_1)$ is the real-valued Cantor function.

With the asymmetric limits, examined in Examples 6.27 and 6.48, let

$$\frac{dV_{\mathbf{m}}}{dt} \le -\left(x_1^2 + x_2^2 + 1\times10^{18}x_e^2 + 1\times10^{20}e^2 + 1\times10^{12}|e|^{2/3} + 1\times10^{-2}\int \Phi^{-1}(u)du\right).$$

The asymmetric limit is described by

$$\Phi\left(G^{-1}\mathbf{B}^T(\mathbf{x})\frac{\partial V_{\mathbf{m}}}{\partial \mathbf{x}}\right) = \frac{u_{\max}}{1+e^{-a\left(G^{-1}\mathbf{B}^T(\mathbf{x})\frac{\partial V_{\mathbf{m}}}{\partial \mathbf{x}}-b\right)}},\ u_{\max}=17.5\text{ V},\ a=100,\ b=0.05,\ 0<u<17.5.$$

Design the proportional-integral control law (6.121) with $Z_p=1$, $L_p=2$, $q_p=1$, $g_p=1$, $N_i=1$, $Z_i=1$, $L_i=1$, $q_i=1$, $g_i=1$. For $V_{\mathbf{m}}(\mathbf{x})$, the control law (6.121) is

$$u = \Phi\left(k_{p_{1,1}}e + k_{p_{1,2}}\,\mathrm{sgn}(e)|e|^{1/3} + k_{i_{1,1,1}}\int e\,dt\right),\ 0<u<17.5.$$

The microactuator parameters are $m=5.1\times10^{-9}$ kg, $\varepsilon_0=8.85\times10^{-12}$, $\varepsilon_r=1$, $A=6.19\times10^{-8}$ m^2, $x_0=7.5$ μm, $c_1=5.8\times10^{-5}$ N-s/m, $c_2=9.4\times10^{-4}$ N-s^2/m^2, $c_3=2.8\times10^{-5}$ N-s^3/m^3, $k_1=2.4\times10^{-2}$ N/m, $k_2=3.2\times10^5$ N/m^2 and $k_3=7.4\times10^4$ N/m^3.

A control law, designed using $\left\{V_{\mathbf{m}}(\mathbf{x})>0,\ \frac{dV_{\mathbf{m}}}{dt}<0\right\}$ with derived structure and computed feedback coefficients $(k_{p_{1,1}}, k_{p_{1,2}}, k_{i_{1,1,1}})$, is

$$u = -\Phi\left(G^{-1}\mathbf{B}^T\frac{\partial V_{\mathbf{m}}}{\partial \mathbf{x}}\right) = \Phi\left(9.5\times10^8\int e\,dt + 6.4\times10^8 e + 8.1\times10^5\,\mathrm{sgn}(e)|e|^{1/3}\right),\ 0<u<17.5.$$

The designed control law ensures adequate performance and compliance with sensing and controlling electronics. The experimental result are reported in Figure 6.27 for step references $r(t)=\begin{cases}2.5\\0\end{cases}$ μm and $r(t)=\begin{cases}5\\0\end{cases}$ μm, applied at f=200 Hz. The repositioning static error is ±0.35 μm due to sensing nonlinearities, dynamic error, calibration error, high noise and other heterogeneities. The maximum overshoot does not exceed 14%, and, the settling time is 0.00061 sec.

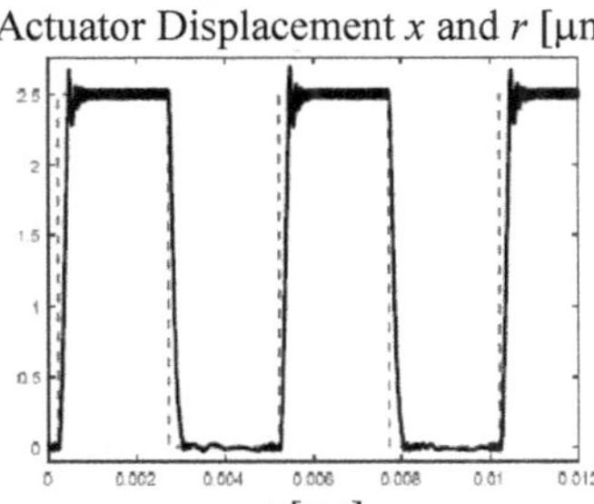

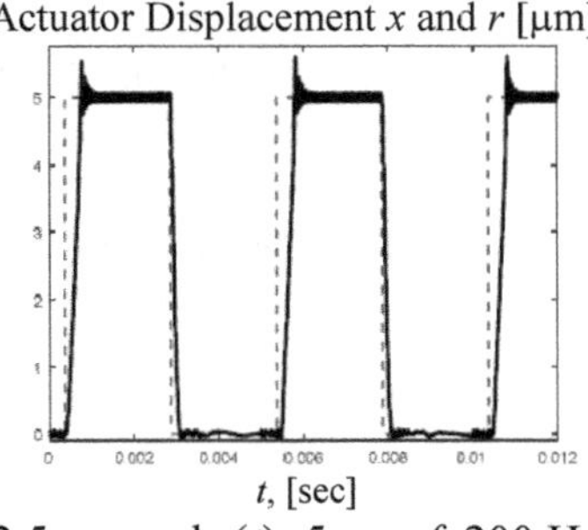

Figure 6.27. Actuator repositioning for $r(t)$=2.5 μm and $r(t)$=5 μm, f=200 Hz. ■

Example 6.51. *Control of a dc-dc PWM Voltage Regulator*

High switching frequency PWM dc-dc voltage regulators are nonlinear systems with inherent control limits on duty cycle d_D. For one-, two- and four-quadrant regulators, the physical limits are $0\le d_D\le 1$ and $-1\le d_D\le 1$. Designed and commercialized isolated and non-isolated dc-dc regulator topologies ensure open-loop stability. These regulators are controllable in full operating envelopes. That is, the output voltage is controlled in the topology- and power MOSFETs-defined range within the admissible currents and loads. Furthermore, the switching of the PWM output voltage may be adjusted. Output LC and

LCL filters ensure admissible switching and current ripple at the output terminal. Proportional and proportional-integral control laws, as well as active analog filters, guarantee adequate functionality. These filters, controllers and auxiliary circuits are integrated into a single package. The inductors and capacitors to implement *LC* filters.

For dc-dc voltage regulators, nonlinear differential equations are derived using the *averaging* concept considering the duty cycle as control, $u \equiv d_D$. A one-quadrant *step-down* dc-dc regulator with a switching MOSFET was studied in section 3.8.1. Controlled switching are regulated by a comparator. A duty cycle $d_D = t_{on}/(t_{on}+t_{off})$ is constrained, $0 \le d_D \le 1$. The *LC* filter ensures specified voltage waveforms and current ripples at the load with R_L and L_L in series. Figures 6.28 document a *step-down* regulator schematics, as well as images of the TPS5410D 8-pin small online integrated circuit (SOIC) package. The signal-level control voltage u_c is compared to the triangular voltage u_t. By varying duty cycle $0 \le d_D \le 1$, one regulates the output voltage u_{RL}. With an asymmetric limit $0 \le u \le 1$, denote $u \equiv d_D$, and, $0 \le u \le 1$. The goal is to stabilize the output voltage u_{RL} at the specified value if load varies, or, ensure reference tracking.

A mathematical model is given by differential equations, derived in section 3.8.1. The states are the voltage at the *LC* filter capacitor u_C, current through the filter inductor i_L, and, current in the load i_{RL}.

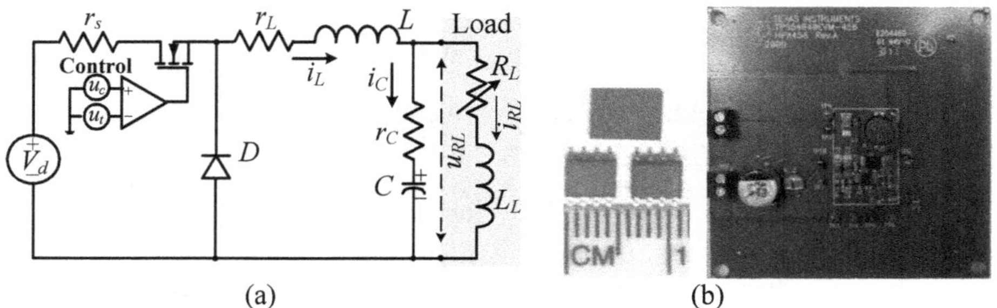

Figure 6.28. (a) Schematics of a *step-down* PWM regulator with the *LC* filter and $R_L L_L$ load; (b) Texas Instruments 8-pin SOIC package TPS5410D regulator with a high side n-channel MOSFET. An evaluation board TPS54040EVM with 5.5 to 36 V input voltage, ensures 1.23 to 31 V output voltage, 1 A output current, and, up to 500 kHz switching frequency.

Using the *averaging* concept, analysis of the governing equations when $d_D=0$ and $d_D=1$, yields the resulting model for $0 \le d_D \le 1$ as given by (3.107)

$$\frac{du_C}{dt} = \frac{1}{C}(i_L - i_{RL}),$$

$$\frac{di_L}{dt} = \frac{1}{L}\left[-u_C - (r_L + r_C)i_L + r_C i_{RL} - r_s i_L u + V_d u\right], \quad u \equiv d_D,\ 0 \le u \le 1,$$

$$\frac{di_{RL}}{dt} = \frac{1}{L_L}\left[u_C + r_C i_L - (R_L + r_C)i_{RL}\right].$$

The output voltage is $u_{RL}=(u_C+r_C i_L - r_C i_{RL})$. Simulations for open-loop and closed-loop PWM regulator were performed in Example 3.35. The dc-dc regulator is open-loop stable and controllable. Resistance r_s is negligibly small, and, nonlinear term $r_s i_L u$ can be neglected.

In the state-space form, a system is

$$\dot{x} = Ax + Bu = \begin{bmatrix} 0 & \frac{1}{C} & -\frac{1}{C} \\ -\frac{1}{L} & -\frac{r_L + r_C}{L} & \frac{r_C}{L} \\ \frac{1}{L_L} & \frac{r_C}{L_L} & -\frac{R_L + r_C}{L_L} \end{bmatrix} \begin{bmatrix} u_C \\ i_L \\ i_{RL} \end{bmatrix} + \begin{bmatrix} 0 \\ \frac{1}{L} V_d \\ 0 \end{bmatrix} u, \; x = \begin{bmatrix} x_1 \\ x_2 \\ x_3 \end{bmatrix} = \begin{bmatrix} u_C \\ i_L \\ i_{RL} \end{bmatrix}.$$

Experimental studies are performed for a PWM regulator with r_s=0.01 ohm, r_L=0.093 ohm, r_C=0.065 ohm, $C=1\times10^{-4}$ F, $L=5\times10^{-6}$ H and L_L=0.007 H. The dc input voltage is V_d=25 V. For r_s=0 and R_L=100 ohm, the characteristic eigenvalues of matrix A are $\lambda_{1,2}=-15803\pm41853i$, $\lambda_3=-14289$.

Linear Quadratic Regulator and Tracking Control – Tracking control laws should be designed to optimize dynamics, guarantee stability, ensure tracking, etc. Consider a linear model neglecting control limits.

With the regulator output $y=u_{RL}$, we have $y=(x_1+r_Cx_2-r_Cx_3)$.

Combine the governing equations for the dc-dc regulator states and dynamics of the tracking error states (x_e,e)

$\dot{x}_e(t) = r - y = r - Hx = r - x_1 - r_C x_2 + r_C x_3$, $e(t)=(r-y)$, $y=Hx$, $\underline{H}=[1 \;\; r_C \;\; -r_C]$,

$\dot{e}(t) = -e + \dot{r} - \dot{y} = -e + \dot{r} - H\dot{x}$.

We find (6.66)

$$\dot{\mathbf{x}}(t) = \begin{bmatrix} \dot{x}(t) \\ \dot{x}_e(t) \\ \dot{e}(t) \end{bmatrix} = \mathbf{A}\mathbf{x} + A_r r + \mathbf{B}u = \begin{bmatrix} A & 0 & 0 \\ -I_X H & 0 & 0 \\ -HA & 0 & -I_E \end{bmatrix} \mathbf{x} + \begin{bmatrix} 0 & 0 \\ I_X & 0 \\ 0 & I \end{bmatrix} \begin{bmatrix} r \\ \dot{r} \end{bmatrix} + \begin{bmatrix} B \\ 0 \\ -HB \end{bmatrix} u, \; \mathbf{x}(t) = \begin{bmatrix} x(t) \\ x_e(t) \\ e(t) \end{bmatrix}, \; I_X=I, \; I_E=I.$$

The tracking error is used in the extended state vector $\mathbf{x}(t)$ with $x_e(t) \equiv \int e(t)$ and $e(t)$.

For stable states and error dynamics, the eigenvalues for (x,x_e,e) are the negative.

Minimize a quadratic performance functional

$$J = \min_{\mathbf{x},u} \int_0^\infty \tfrac{1}{2}\left(\mathbf{x}^T Q \mathbf{x} + G u^2\right) dt, \; Q\in\mathbb{R}^{5\times5}, \; G\in\mathbb{R}$$

subject to the extended system dynamics.

The Hamiltonian and stationarity condition $\frac{\partial}{\partial u} H(\mathbf{x}, u, \frac{\partial V}{\partial \mathbf{x}}) = 0$ yield a control law $u = -G^{-1}\mathbf{B}^T \frac{\partial V}{\partial \mathbf{x}}$. The Hamilton-Jacobi functional equation is satisfied by the quadratic return function $V(\mathbf{x}) = \frac{1}{2}\mathbf{x}^T K \mathbf{x}$

A linear tracking control law is

$$u = -G^{-1}\mathbf{B}^T K \mathbf{x}.$$

Solution of the algebraic equation yields a positive-definite matrix $K\in\mathbb{R}^{5\times5}$

$$Q + \mathbf{A}^T K + K\mathbf{A} - K\mathbf{B}G^{-1}\mathbf{B}^T K = 0, \; K>0, \; K^T=K.$$

Let $q_{11}=0$, $q_{22}=0$, $q_{33}=1$, $q_{44}=10$, $q_{55}=10$ and $G=1$. The tracking control law is

$$u = -G^{-1}\mathbf{B}^T K\mathbf{x} = -0.00089u_C - 0.0131i_L - 0.0442i_{RL} + 3.16\int e dt + 1.38e.$$

Lyapunov Stability and Minimal Complexity Control – Apply a Lyapunov pair (6.92) $\left\{V(\mathbf{x}) > 0, \frac{dV(\mathbf{x})}{dt} < 0\right\}$, which guarantees system stability. Specify the rate of change dV/dt, which defines system optimality in a sense of system states and tracking error governance and Lyapunov's stability, e.g., stability margins, robustness, sensitivity to parameter variations, etc. Apply quadratic measures on states and control $(\mathbf{x},u)$, finding

$$\frac{dV}{dt} = -\tfrac{1}{2}\mathbf{x}^T Q\mathbf{x} - \tfrac{1}{2}u^T Gu,\ \frac{dV}{dt} < 0,\ (Q,G)>0.$$

Minimize $\frac{dV}{dt} = -\tfrac{1}{2}\mathbf{x}^T Q\mathbf{x} - \tfrac{1}{2}u^T Gu$ subject to the system dynamics, for which the total derivative of $V(\mathbf{x})$ is

$$\frac{dV}{dt} = \left(\frac{\partial V}{\partial \mathbf{x}}\right)^T (\mathbf{Ax} + A_r r + \mathbf{B}u).$$

A scalar Hamiltonian function is

$$H(\mathbf{x}, u, \tfrac{\partial V}{\partial \mathbf{x}}) = \tfrac{1}{2}\mathbf{x}^T Q\mathbf{x} + \tfrac{1}{2}u^T Gu + \left(\frac{\partial V}{\partial \mathbf{x}}\right)^T (\mathbf{Ax} + \mathbf{B}u).$$

Applying the stationarity condition, the control law is $u = -G^{-1}\mathbf{B}^T \frac{\partial V}{\partial \mathbf{x}}$.

Equation

$$\tfrac{1}{2}\mathbf{x}^T Q\mathbf{x} + \tfrac{1}{2}u^T Gu + \left(\frac{\partial V}{\partial \mathbf{x}}\right)^T (\mathbf{Ax} + \mathbf{B}u) = 0,\ u = -G^{-1}\mathbf{B}^T \frac{\partial V}{\partial \mathbf{x}}$$

admits a solution by $V(\mathbf{x}) = \tfrac{1}{2}\mathbf{x}^T K\mathbf{x}$. Hence,

$$u = -G^{-1}\mathbf{B}^T K\mathbf{x},\ Q + \mathbf{A}^T K + K\mathbf{A} - K\mathbf{B}G^{-1}\mathbf{B}^T K = 0,\ K>0,\ K^T=K.$$

There is an equivalency in the LQR and Lyapunov designs. Same control structure and algebraic equation to compute K are found. Minimization of the negative definite $\frac{dV(\mathbf{x})}{dt}$ yields a control structure and equation to find K. One has $V(\mathbf{x})>0$ if $K>0$.

Minimal Complexity Control – To ensure hardware and algorithmic implementability, maximize reliability and improve robustness, design control laws with the measured tracking error $e(t)$ not using the state feedback. This solution does not require use of sensors with ASICs, simplifies algorithmic simplicity, etc.

Design a *minimal complexity* control law selecting a Lyapunov function of all states $V_{\mathbf{m}}(\mathbf{x})$ and $\frac{dV_{\mathbf{m}}(\mathbf{x})}{dt}$ as

$$\left\{V_{\mathbf{m}}(\mathbf{x}) > 0,\ \frac{dV_{\mathbf{m}}(\mathbf{x})}{dt} < 0\right\},$$

such as control $u = -G^{-1}\mathbf{B}^T \frac{\partial V_{\mathbf{m}}}{\partial \mathbf{x}}$ is a function of the tracking error.

That is, our objective is to find $u = -G^{-1}\mathbf{B}^T \frac{\partial V_{\mathbf{m}}}{\partial \mathbf{x}} = \varphi\left(\int e dt, e\right)$

Consider a continuous scalar function of all states

$$V_{\mathbf{m}}(\mathbf{x}) = k_{22}c(x_2) + k_{24}x_2x_e + k_{25}x_2e + \tfrac{1}{2}(k_{11}x_1^2 + k_{33}x_3^2 + k_{44}x_e^2 + k_{55}e^2),$$

and, specify the total derivative to be

$$\frac{dV_{\mathbf{m}}}{dt} = -\tfrac{1}{2}\left(x_1^2 + x_2^2 + x_3^2 + 10x_e^2 + 10e^2 + u^2\right),$$

To define $V_{\mathbf{m}}(\mathbf{x})$, use the real-valued Cantor function $c(\cdot)$ which is uniformly Hölder continuous and has zero derivative.

Minimize $\frac{dV_{\mathbf{m}}}{dt}$ subject to a system, for which $\frac{dV_{\mathbf{m}}}{dt} = \left(\frac{\partial V_{\mathbf{m}}}{\partial \mathbf{x}}\right)^T (\mathbf{Ax} + A_r r + \mathbf{B}u)$.

A scalar Hamiltonian function is

$$H(x, u, \tfrac{\partial V_{\mathbf{m}}}{\partial \mathbf{x}}) = \tfrac{1}{2}\left(x_1^2 + x_2^2 + x_3^2 + 10x_e^2 + 10e^2 + u^2\right) + \left(\frac{\partial V_{\mathbf{m}}}{\partial \mathbf{x}}\right)^T (\mathbf{Ax} + \mathbf{B}u).$$

We obtain

$$u=-G^{-1}\mathbf{B}^T\frac{\partial V_{\mathbf{m}}}{\partial\mathbf{x}}=-G^{-1}\frac{V_d}{L}\frac{\partial V_{\mathbf{m}}}{\partial x_2}=-G^{-1}\frac{V_d}{L}\left(k_{24}x_e+k_{25}e\right)=-G^{-1}\frac{V_d}{L}\left(k_{24}\int edt+k_{25}e\right).$$

Having found k_{ij}, one concludes $V_{\mathbf{m}}(\mathbf{x})>0$. A proportional-integral control law is $u=-3.16x_e-3.16e=-3.16\int edt-3.16e$.

Experimental studies are conducted. The closed-loop dynamics are reported in Figures 6.29 for varying load R_L if the output voltage u_{RL} is specified to be 3.96 V and 12.1 V. The peak loads are: (1) R_L=4 ohm for $u_{RL\ \text{output}}$=3.96 V; (2) R_L=12 ohm for $u_{RL\ \text{output}}$ =12.1 V. The output voltage u_{RL} remains as specified if loads are applied. Tracking is achieved in an operating envelope. Experiments demonstrate that bounds and nonlinearities affect performance. Regulator governance and efficiency depend on an operating envelope, defined by input and output voltages (V_d, $u_{RL\ \text{output}}$), loads, load impedance, switching frequency, MOSFET switching schemes, etc.

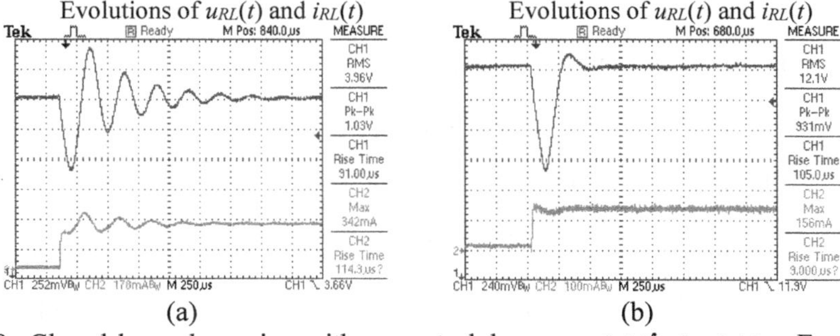

Figure 6.29. Closed-loop dynamics with a control law $u=-3.16\int edt-3.16e$: Evolutions of voltage $u_{RL}(t)$ and current $i_{RL}(t)$ (top and bottom lines) if the R_l load is applied:
(a) Output voltage is stabilized, $u_{RL\ \text{output}}$=3.96 V with steady-state current i_{RL}=0.342 A, R_L=11.6 ohm;
(b) Output voltage s stabilized, $u_{RL\ \text{output}}$=12.1 V. Steady-state current is i_{RL}=0.156 A, R_L=77.6 ohm.

Physical Limits and Constrained Control – Consider an asymmetric physical limit on the duty cycle $0\le\mathcal{U}\le 1$. Tracking bounded control laws are analytically designed minimizing

$$J=\min_{t,u\in U}\int_0^{t_f}1dt \text{ and } J=\min_{\mathbf{x},u\in U}\int_0^{\infty}\left(\tfrac{1}{2}\mathbf{x}^TQ\mathbf{x}+G\int\Phi^{-1}(u)du\right)dt.$$

Calculus of variations and maximum principle may not solve minimum time problem for high-order systems. A switching manifold $\varphi(\mathbf{x})$ cannot be derived because the maximum principle may not admit an analytic solution. A considered optimization problem cannot be solved by separation of variables.

Near-Discontinuous Control – A discontinuous control law (6.51) with symmetric relay switching $u=\phi[\varphi(\mathbf{x})]=\begin{cases}1, & \varphi(\mathbf{x})>0\\ -1, & \varphi(\mathbf{x})\le 0\end{cases}$ may be found by applying the maximum principle, as well as the stationarity condition on a scalar Hamiltonian function (6.50)

$$H(\mathbf{x},u,\tfrac{\partial V}{\partial\mathbf{x}})=1+\left(\frac{\partial V}{\partial\mathbf{x}}\right)^T\left(\mathbf{A}\mathbf{x}+\mathbf{B}u\right),\ u\in U.$$

However, the switching manifold $\varphi(\mathbf{x})$ cannot be found with high dimensional state manifold, asymmetric switching, unknown return function, not separable variables, etc.

The right-continuous Heaviside unit step function with a left finite limit is $\mathcal{H}(\mathbf{x})=\begin{cases}1, & \mathbf{x}>0\\ 0, & \mathbf{x}\le 0\end{cases}$.

A discontinuous asymmetric control limit is described by the Heaviside function $\mathcal{H}_{[0,\infty)}(\cdot)$, continuous on open semi-infinite interval, such that $\lim_{x\to 0, x\to 0^+} \mathcal{U}$ exists.

That is $u = \mathcal{H}[\varphi(\mathbf{x})] = \begin{cases} 1, & \varphi(\mathbf{x}) > 0 \\ 0, & \varphi(\mathbf{x}) \le 0 \end{cases}$.

A physical limit $0 \le d_D \le 1$ admits a description by a continuous function

$$\Phi(x) = \frac{1}{1+e^{-a(x-b)}},\ x \equiv \frac{\partial V}{\partial \mathbf{x}},\ 0<\Phi<1,\ a \gg 1,\ b \ll 1,\ \Phi \subseteq \mathcal{U}_\varepsilon.$$

This $\Phi(x)$ and $\Phi^{-1}(u)$ were examined in Examples 6.22 and 6.27.

To approximate a solution of the discontinuous control problem with asymmetric limits, minimize a functional

$$J = \min_{t,u\in U} \int_0^\infty \left(G \int \Phi^{-1}(u) du \right) dt,\ \Phi^{-1}(u) = \frac{1}{a}\left[-\ln\left(\frac{1-u}{u}\right) + ab \right],\ 0<u<1,$$

which results in the Hamiltonian $H(\mathbf{x}, u, \frac{\partial V}{\partial \mathbf{x}}) = G \int \Phi^{-1}(u) du + \left(\frac{\partial V}{\partial \mathbf{x}}\right)^T (\mathbf{A}\mathbf{x} + \mathbf{B}u)$.

One obtains $u = -\Phi\left(G^{-1}\mathbf{B}^T \frac{\partial V}{\partial \mathbf{x}}\right)$, $0<u<1$, $G=1$.

The Hamilton-Jacobi equation

$$-\frac{\partial V}{\partial t} = \left(\frac{\partial V}{\partial \mathbf{x}}\right)^T \mathbf{A}\mathbf{x} - \int \Phi\left(G^{-1}\mathbf{B}^T \frac{\partial V}{\partial \mathbf{x}}\right) d\left(\mathbf{B}^T \frac{\partial V}{\partial \mathbf{x}}\right),$$

$$\int \Phi\left(G^{-1}\mathbf{B}^T \frac{\partial V}{\partial \mathbf{x}}\right) d\left(\mathbf{B}^T \frac{\partial V}{\partial \mathbf{x}}\right) = \int \frac{1}{1+e^{-a\left((\mathbf{B}^T \frac{\partial V}{\partial \mathbf{x}} - b)\right)}} d\left(\mathbf{B}^T \frac{\partial V}{\partial \mathbf{x}}\right) = \mathbf{B}^T \frac{\partial V}{\partial \mathbf{x}} - b + \frac{1}{a} \ln\left| 1 + e^{-a\left((\mathbf{B}^T \frac{\partial V}{\partial \mathbf{x}} - b)\right)} \right|$$

admits a solution, given by a return function $V(\mathbf{x})$.

In an operating envelope $(x,u) \in X \times U$, find

$$\int \Phi\left(G^{-1}\mathbf{B}^T \frac{\partial V}{\partial \mathbf{x}}\right) d\left(\mathbf{B}^T \frac{\partial V}{\partial \mathbf{x}}\right) \approx k \left(\frac{\partial V}{\partial \mathbf{x}}\right)^T \mathbf{B} G^{-1} \mathbf{B}^T \frac{\partial V}{\partial \mathbf{x}},$$

The least-squares approximation yields k.

Discontinuous Control – A discontinuous control function is

$$u = \mathcal{H}_{[0,\infty)}\left(-G^{-1}\mathbf{B}^T \frac{\partial V}{\partial \mathbf{x}}\right) \approx \tilde{\mathcal{H}}_{[0,\infty)}\left(-G^{-1}\mathbf{B}^T \frac{\partial V}{\partial \mathbf{x}}\right) = -\Phi\left(G^{-1}\mathbf{B}^T \frac{\partial V}{\partial \mathbf{x}}\right).$$

The Hamilton-Jacobi equation is solved for $\Phi \subseteq \mathcal{U}_\varepsilon$ with $a=2500$ and $b=0.0025$. Using the quadratic return function $V(\mathbf{x}) = \frac{1}{2}\mathbf{x}^T K \mathbf{x}$, $G=1$, $k=1$, we have

$$u = -\Phi\left(\mathbf{B}^T \frac{\partial V}{\partial \mathbf{x}}\right) = \Phi\left(-0.0008u_C - 0.02i_L - 0.048i_{RL} + 4.47\int e dt + 1.53e\right),\ 0<\Phi<1.$$

With small state feedback gains, using the *near-discontinuous* $\Phi\left(G^{-1}\mathbf{B}^T \frac{\partial V}{\partial \mathbf{x}}\right) = \tilde{\mathcal{H}}_{[0,\infty)}\left(G^{-1}\mathbf{B}^T \frac{\partial V}{\partial \mathbf{x}}\right)$, implement a control law

$$u = -\Phi\left(\mathbf{B}^T \frac{\partial V}{\partial \mathbf{x}}\right) = \Phi\left(4.47\int e dt + 1.53e\right),\ 0<\Phi<1,\ 0<u<1,$$

$$u = \tilde{\mathcal{H}}_{[0,\infty)}\left(4.47\int e dt + 1.53e\right).$$

Continuum Device Physics and Hard Switching – A continuum electrostatic device physics of all semiconductor devices yields transient dynamics and latencies in MOSFETs, MOSFET drivers, comparators, etc. With continuum semiconductor devices, energy transductions in the *RLC* circuits, parasitic capacitances and other effects, one cannot implement a *hard-switching* signum control function. Transients in ultra-fast low-power switching analog and digital electronic devices may approach a near-sigmoid switching. Even in a high-speed comparators and ultra-fast switching low-power MOSFETs, transients are within nanoseconds. The regulator switching frequency is 500 kHz. To implement a near-discontinuous control, *hard-switching* of MOSFET must be applied. There numerous impediments in attempts to implement near-discontinuous control and switching schemes. Experimental results are reported in Figures 6.30. Implementation of a near-discontinuous schemes and pertained control yield switching at any loads and disturbances. In addition, there is a sensitivity to noise, electromagnetic interferance and measurement impediments. As a result, dc-dc regulators will exhibit limit cycles, persistent voltage chattering and current ripple, etc. The overall performance is inadequate due to losses, heating, low reliability, low efficiency, etc.

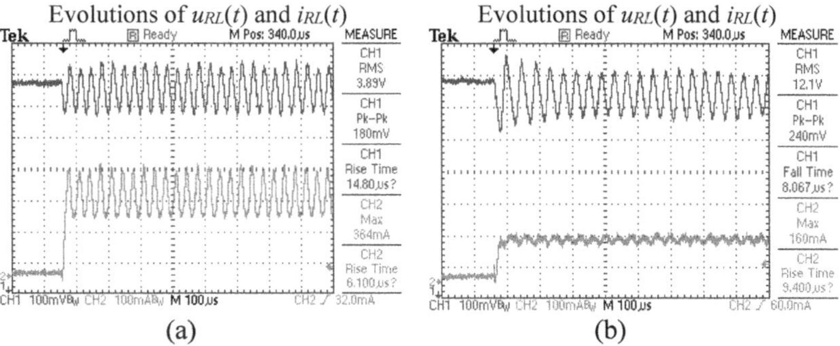

Figure 6.30. Closed-loop dynamics of voltage $u_{RL}(t)$ and current $i_{RL}(t)$ (top and bottom lines) in dc-dc regulator with $u=\tilde{\mathcal{H}}_{[0,\infty)}\left(4.47\int e dt+1.53e\right)$:

(a) For $u_{RL\text{ output}}$=3.96 V, output voltage u_{RL} and current i_{RL} oscillate at any load. There are limit cycles, voltage chattering and current ripple. For R_L=14 ohm, $\Delta u_{RL}=\pm0.09$ V, $\Delta i_{RL}=\pm0.129$ A;
(b) Specifying $u_{RL\text{ output}}$ =12.1 V, output voltage u_{RL} and current i_{RL} oscillate. For R_L=77.6 ohm, there are the voltage and current chattering $u_{RL}=12.1\pm0.11$ V and $i_{RL}=0.156\pm0.039$ A.

Minimal Complexity Continuous Bounded Control – Apply a Lyapunov stability and dynamic optimality paradigm. Consider a Lyapunov pair $\left\{V_{\mathbf{m}}(\mathbf{x})>0, \frac{dV_{\mathbf{m}}(\mathbf{x})}{dt}<0\right\}$, choosing $V_{\mathbf{m}}(\mathbf{x})$ such as control is a function of the tracking error $e(t)$.

Consider a continuous scalar function of all states $V_{\mathbf{m}}(\mathbf{x})$ as

$$V_{\mathbf{m}}(\mathbf{x})=k_{22}c(x_2)+k_{24}x_2x_e+k_{25}x_2e+\tfrac{1}{2}(k_{11}x_1^2+k_{33}x_3^2+k_{44}x_e^2+k_{55}e^2).$$

A physical limit on the duty ratio $0\le d_D\le 1$ is described by a continuous function $\Phi(\cdot)$, $\Phi\subseteq\mathcal{U}_\varepsilon$,

$$\Phi(x)=\frac{1}{1+e^{-a(x-b)}},\ x\equiv\frac{\partial V}{\partial\mathbf{x}},\ 0<\Phi<1,\ a\gg1,\ b\ll1,\ \Phi^{-1}(u)=\frac{1}{a}\left[-\ln\left(\frac{1-u}{u}\right)+ab\right],\ 0<u<1.$$

Note that the device-specific (a,b) are found to be a=100 and b=0.05.

Minimize the specified total derivative of the Lyapunov function

$$\frac{dV_{\mathbf{m}}}{dt}=-\tfrac{1}{2}\left(x_1^2+x_2^2+x_3^2+10x_e^2+10e^2+G\int\Phi^{-1}(u)du\right),\ \frac{dV_{\mathbf{m}}}{dt}<0.$$

A scalar Hamiltonian function is

$$H(x,u,\tfrac{\partial V_{\mathbf{m}}}{\partial \mathbf{x}})=\tfrac{1}{2}\left(x_1^2+x_2^2+x_3^2+10x_e^2+10e^2+G\int\Phi^{-1}(u)du\right)+\left(\frac{\partial V_{\mathbf{m}}}{\partial \mathbf{x}}\right)^T(\mathbf{Ax}+\mathbf{B}u)\cdot$$

Using $\frac{\partial}{\partial u}H(x,u,\frac{\partial V_{\mathbf{m}}}{\partial \mathbf{x}})=0$, we obtain a control law

$$u=-\Phi\left(G^{-1}\mathbf{B}^T\tfrac{\partial V_{\mathbf{m}}}{\partial \mathbf{x}}\right)=-\Phi\left(G^{-1}\tfrac{V_d}{L}\tfrac{\partial V_{\mathbf{m}}}{\partial x_2}\right)=\Phi\left(G^{-1}\tfrac{V_d}{L}\left(k_{24}\int edt+k_{25}e\right)\right),\ 0<u<1.$$

An asymmetric limit on the duty cycle $0\leq\mathcal{U}\leq1$ is described by a continuous function $\Phi\subseteq\mathcal{U}_\varepsilon$, $0<\Phi<1$ with device-specific (a,b), a=100, b=0.05. Used $\Phi(x)$ and $\Phi^{-1}(u)$ are consistent with MOSFET and comparator device physics, continuum voltage switching schemes, regulator transients and latencies in comparator, MOSFETs and ICs. Asymmetric limit and $\Phi(x)$ are illustrated in Figure 6.31.a.

Having found k_{ij}, we have $V_{\mathbf{m}}(\mathbf{x})>0$.

A bounded proportional-integral control law is

$$u=\Phi\left(3.16\int edt+0.307e\right).$$

Figures 6.31.b and c report experimental results. Fast dynamics, minimal overshoot, robustness, adequate current and voltage evolutions, minimal losses and high efficiency are guaranteed in a full operating envelope. The output voltage $u_{RL}(t)$ is stabilized for different specified voltages and loads. The settling time varies from 6×10^{-4} to 7.5×10^{-4} sec. Disturbance attenuation, voltage stabilization and robustness are ensured.

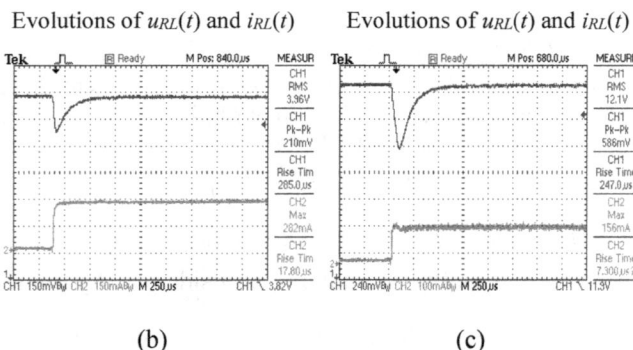

Asymmetric Bounds: Assumed Discontinuous Limit $0\leq\mathcal{U}\leq1$, and, Continuous $\Phi(x)=\dfrac{1}{1+e^{-a(x-b)}}$, $0<\Phi<1$

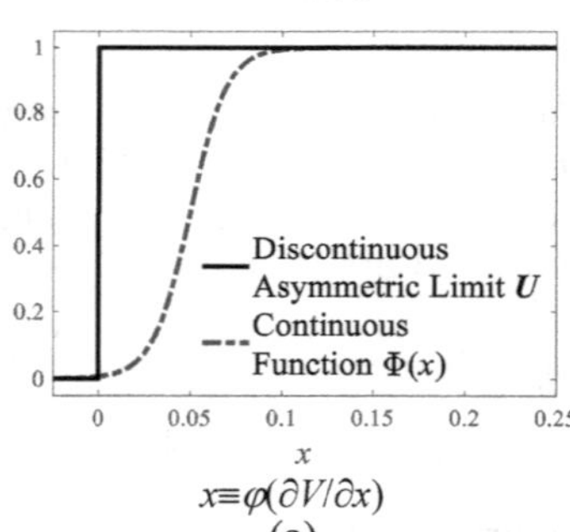

$x\equiv\varphi(\partial V/\partial x)$

Evolutions of $u_{RL}(t)$ and $i_{RL}(t)$ Evolutions of $u_{RL}(t)$ and $i_{RL}(t)$

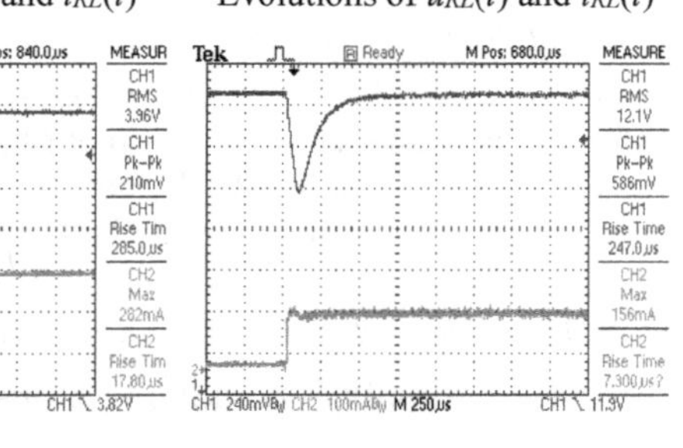

(a) (b) (c)

Figure 6.31. (a) Asymmetric limit $0\leq\mathcal{U}\leq1$ may assume a description by the right-continuous Heaviside unit step function $\mathcal{H}(x)=\begin{cases}1, & x>0\\0, & x\leq0\end{cases}$. Due to continuity of electrostatic device physics, continuous transients of comparators and MOSFETs, continuous switching in ICs, a limit $0\leq\mathcal{U}\leq1$ is described by $\Phi(x)=\dfrac{1}{1+e^{-a(x-b)}}$, $0<\Phi<1$, $\Phi\subseteq\mathcal{U}_\varepsilon$, $\Phi^{-1}(u)=\frac{1}{a}\left[-\ln\left(\frac{1-u}{u}\right)+ab\right]$, $0<u<1$, a=100, b=0.05;
(b) Closed-loop dc-dc regulator dynamics with $u=\Phi\left(3.16\int edt+0.307e\right)$: Evolutions of voltage $u_{RL}(t)$ and current $i_{RL}(t)$ (top and bottom lines) as the load R_L is applied. Output voltage is $u_{RL\ \text{output}}$ =3.96 V and i_{RL}=0.282 A for R_L=14 ohm;
(c) Closed-loop system transients: $u_{RL\ \text{output}}$ =12.1 V and i_{RL}=0.156 A, R_L=77.6 ohm.

6.11. Control of Discrete-Time Systems: Lyapunov Theory, Stability and Dynamic Optimization

For linear systems, dynamic optimization and solution of linear quadratic regulator problem were covered in section 5.10. System transients and control laws were found for (5.124) $x_{n+1} = A_n x_n + B_n u_n$. Nonlinear systems and control limits $u_{\min} \le u \le u_{\max}$ are investigated in this section. Control laws are synthesized, and, Lyapunov paradigm is applied to analyze, control and optimize discrete-time systems.

6.11.1. Lyapunov Stability Analysis of Discrete-Time Systems

Study discrete-time systems, described by difference equations.

Stability of a linear system

$$x_{n+1} = A_n x_n,\ n \ge 0, \tag{6.122}$$

is guaranteed if all eigenvalues of A_n have the magnitudes less than 1, $\forall|\lambda_i|<1$. For linear and nonlinear systems, stability and optimality are examined using a Lyapunov concept.

Theorem 6.2. A nonlinear discrete-time system

$$x_{n+1}=F(x_n),\ n\ge 0, \tag{6.123}$$

is asymptotically stable if there exists a positive definite scalar Lyapunov function of all states $V(x_n)$, and, the first difference $\Delta V(x_n)=V(x_{n+1})-V(x_n)$ is negative definite. If $V(x_n)>0$ and $\Delta V(x_n)<0$, states x_n evolve to a stable equilibrium $\bar{x}_n$, $x_n|_{n\to\infty} \to \bar{x}_n$, $\lim_{n\to\infty}\|x_n - \bar{x}_n\| = 0$. The sufficient conditions for asymptotic stability are given by a Lyapunov pair $\{V(x_n), \Delta V(x_n)\}$

$$\{V(x_n)>0,\ \Delta V(x_n)=[V(x_{n+1})-V(x_n)]<0\},\ \forall x_n \in X,\ \forall n. \tag{6.124}$$

■

Stability of Linear Systems – Theorem 6.2 is valid for linear and nonlinear systems. Study stability of a linear system (6.122) $x_{n+1} = A_n x_n$. Apply a quadratic function

$$V(x_n) = x_n^T K_n x_n,\ K_n>0. \tag{6.125}$$

The first difference is

$$\Delta V(x_n) = V(x_{n+1}) - V(x_n) = x_{n+1}^T K_{n+1} x_{n+1} - x_n^T K_n x_n = x_n^T A_n^T K_{n+1} A_n x_n - x_n^T K_n x_n. \tag{6.126}$$

Negative definiteness of $\Delta V(x_n)$ can be specified by a measure $\omega_x(x_n)$.

Define negative definiteness of $\Delta V(x_n)$ by applying a quadratic measure $\omega_x(x_n) = x_n^T Q_n x_n$ with a positive definite matrix $Q_n>0$. In particular,

$$\Delta V(x_n) \le -\omega_x(x_n),\ \Delta V(x_n) = -x_n^T Q_n x_n,\ \omega_x(x_n)>0,\ Q_n>0. \tag{6.127}$$

For $\Delta V(x_n)$, given by (6.126) and (6.127), a Lyapunov equation is

$$A_n^T K_n A_n - K_n = -Q_n. \tag{6.128}$$

An analytic closed-form solution for stability analysis is found. For a system, characterized by positive energy $\mathbb{E}$, matrix K_n is parameter-dependent. Using $V(x_n)\equiv\mathbb{E}$, one concludes that $K_n>0$. That is, for known (A_n,K_n), one finds Q_n and assesses $\Delta V(x_n)$.

Solution of Discrete Lyapunov Equation – An illustrative example in analysis of stability, solution of discrete Lyapunov equation and computing K_n are documented in Example 5.6 and Practice problem 5.4. Custom MATLAB codes are developed to solve equation (6.128), see Example 5.6.

Conjecture 6.6. A linear discrete-time system (6.122) $x_{n+1}=A_n x_n$ is stable, and, stability conditions given by a Lyapunov pair (6.124) are guaranteed if for a positive definite matrix Q_n, which implies negative definiteness of (6.127) $\Delta V(x_n)=-x_n^T Q_n x_n<0$, there exists a positive definite symmetric matrix K_n, which is a solution of the Lyapunov equation (6.128). With $K_n>0$, a quadratic function (6.125) is positive definite, $V(x_n)=x_n^T K_n x_n>0$. ■

Conjecture 6.7. Consider a linear discrete-time system (6.122) $x_{n+1}=A_n x_n$. For a given positive definite function (6.125) $V(x_n)=x_n^T K_n x_n$ with $K_n>0$, the sufficient conditions for stability (6.124) are guaranteed if the Lyapunov equation (6.128) yields a positive definite matrix Q_n, which imply negative definiteness of $\Delta V(x_n)=-x_n^T Q_n x_n$. ■

One quantitatively assesses system and Lyapunov pair (6.124) $\{V(x_n), \Delta V(x_n)\}$ evolutions, system energy $V(x_n)$, *admissible* energy transductions $\Delta V(x_n)$, stability margins, sensitivity to parameter variations, system robustness and other quantities. For linear systems, using the state transition matrix $\Phi[n]$, additional studies are performed. The relevant numeric studies are reported in Example 5.6.

Illustrative Studies 6.3. *Lyapunov Function Evolution*
For a linear system (6.122) $x_{n+1}=A_n x_n$, $x_n[0]=x_0$ with constant-coefficient parameters matrix A_n, the state transition matrix is $\Phi[n]=A_n^n$. Discrete system dynamics and state transition matrix were covered in details with examples in section 5.2.2.

Applying notations of this chapter, one has

$$x_n=\Phi[n]x_0=A_n^n x_0.$$

For an asymptotically stable system

$$\lim_{n\to\infty} x_n\to 0,\ \lim_{n\to\infty}\Phi[n]\to 0.$$

A system is stable if evolutions of $(x_n,\Phi[n])$ are bounded,

$$\|x_n\|\le\varepsilon,\ \|\Phi[n]\|\le M, \forall x_0,\ \forall n\ge 0.$$

For bounded $\Phi[n]$,

$$\|x_n\|=\|\Phi[n]x_0\|\le\|\Phi[n]\|\|x_0\|\le M\|x_0\|\le\varepsilon,\ \forall x_0,\ \forall n\ge 0.$$

For a positive definite Lyapunov function (6.125) $V(x_n)=x_n^T K_n x_n$ and negative definite first difference (6.127) $\Delta V(x_n)=-x_n^T Q_n x_n$, we have

$$\lim_{n\to\infty} V(x_n)\to 0,\ V(x_n)=x_n^T K_n x_n=x_0^T\Phi^T[n]K_n\Phi[n]x_0,$$

and, $$\lim_{n\to\infty}\Delta V(x_n)\to 0,\ \Delta V(x_n)=-x_n^T Q_n x_n=x_0^T\Phi^T[n]Q_n\Phi[n]x_0.$$ ■

Illustrative Studies 6.4. *Stability, Eigenvalues and Lyapunov Equation*
If $|\lambda_{\max}(A_n)|<1$, system is stable. Eigenvalues define evolutions of x_n. For stable systems,

$$\sum_{n=0}^{\infty}\|x_n\|^2\le c\|x_0\|^2\sum_{n=0}^{\infty}|\lambda_{\max}(A_n)|^n=c\|x_0\|^2\frac{1}{1-|\lambda_{\max}(A_n)|^n}<\infty, \forall x_0.$$

Consider a Lyapunov equation (6.128) $A_n^T K_n A_n-K_n=-Q_n$.

Using a cyclic permutation, $\text{tr}(A_n^T K_n A_n) = \text{tr}(A_n A_n^T K_n)$.

Apply the Cauchy-Schwarz inequality $\text{tr}(A_n^T K_n A_n) \le \text{tr}(A_n A_n^T)\text{tr}(K_n), \text{tr}(A_n A_n^T) = \text{tr}(A_n^T A_n)$.

The trace of a matrix is the sum of its eigenvalues. An upper bound for $\text{tr}(A_n^T A_n)$ is n times of the maximum eigenvalue of $A_n \in \mathbb{R}^{l\times l}$, e.g., $\text{tr}(A_n A_n^T) \le n|\lambda_{\max}(A_n)|^2$.

Hence, $\text{tr}(K_n)\left(1 - n|\lambda_{\max}(A_n)|^2\right) \le \text{tr}(Q_n)$.

We analyze system and energy governance and dependencies, assessed and evaluated by $(A_n,\Phi[n],K_n,Q_n)$, procured by a Lyapunov pair $\{V(x_n), \Delta V(x_n)\}$. ■

Illustrative Studies 6.5. *Nonlinear Systems and Lyapunov Stability*

Nonlinear systems models are found applying the total energy $\mathbb{E}$. Considering a nonlinear system (6.123) $x_{n+1}=F(x_n)$ with a nonlinear map $F(x_n)$, one designs $V(x_n)$ and selects $\omega_x(x_n)$.

Scalar Lyapunov functions can be found as

$V(x_n) = (x_n^{\langle d\rangle})^T K_n x_n^{\langle d\rangle}$, $V(x_n) = (x_n^{\langle d_1\rangle})^T K_n x_n^{\langle d_2\rangle}$, $K_n>0$, $d\in\mathbb{N}$, $(d_1,d_2)\in\mathbb{N}$.

Negative definiteness of $\Delta V(x_n)$ defines stability margins and system governance.

Specify $\Delta V(x_n)$ applying a positive definite measure $\omega_x(x_n)$ as

$\Delta V(x_n) \le -\omega_x(x_n)$, $\omega_x(x_n) = (x_n^{\langle q\rangle})^T Q_n x_n^{\langle q\rangle}$, $\omega_x(x_n)>0$, $Q_n>0$, $q\in\mathbb{N}$.

A Lyapunov pair $\{V(x_n)>0, \Delta V(x_n)<0\}$, $\Delta V(x_n)=[V(x_{n+1})-V(x_n)]$ yields stability analysis. ■

6.11.2. Stability and Optimal Control of Linear Discrete-Time Systems

Consider a linear system (5.124)

$$x_{n+1} = A_n x_n + B_n u_n. \tag{6.129}$$

Linear Quadratic Regulator – Section 5.10 covers design of control laws by minimizing a quadratic performance index (5.125)

$$J = \min_{x_n,u_n} \sum_{n=0}^{N-1} \left[x_n^T Q_n x_n + u_n^T G_n u_n \right], (Q_n, G_n) > 0. \tag{6.130}$$

A scalar Hamiltonian (5.126) $H\left(x_n, u_n, V(x_{n+1})\right) = x_n^T Q_n x_n + u_n^T G_n u_n + V(x_{n+1})$ and quadratic return function (5.127) $V(x_n) = x_n^T K_n x_n$, $K_n>0$ yield a control law (5.130)

$$u_n = -\left(G_n + B_n^T K_{n+1} B_n\right)^{-1} B_n^T K_{n+1} A_n x_n, \tag{6.131}$$

where K_{n+1} is a symmetric matrix, computed by solving (5.132)

$$-K_n + Q_n + A_n^T K_{n+1} A_n - A_n^T K_{n+1} B_n \left(G_n + B_n^T K_{n+1} B_n\right)^{-1} B_n^T K_{n+1} A_n = 0.$$

Lyapunov Stability and Optimality – Apply a Lyapunov paradigm to design control laws. Specify the negative definiteness of $\Delta V(x_n)$ using positive definite quadratic measures $(\omega_x(x_n),\omega_u(u_n))>0$, which define stability margins and system governance. Let

$\Delta V(x_n) \le -[\omega_x(x_n) + \omega_u(u_n)]$, $\omega_x(x_n) = x_n^T Q_n x_n$, $\omega_u(u_n) = u_n^T G_n u_n$, $(Q_n,G_n)>0$,

$$\Delta V(x_n) = -\left[x_n^T Q_n x_n + u_n^T G_n u_n \right],\ \Delta V(x_n)<0. \tag{6.132}$$

To find a closed-form solution of an optimization problem, an inequality problem is solved applying affine equality constraints and convex objective function.

Minimize $\Delta V(x_n)$ subject to (6.129) $x_{n+1} = A_n x_n + B_n u_n$.

For a quadratic function (6.125) $V(x_n)=x_n^TK_nx_n$, an explicit equation for $\Delta V(x_n)$ is

$$\begin{aligned}\Delta V(x_n)&=V(x_{n+1})-V(x_n)=(A_nx_n+B_nu_n)^TK_{n+1}(A_nx_n+B_nu_n)-x_n^TK_nx_n\\&=x_n^TA_n^TK_{n+1}A_nx_n+x_n^TA_n^TK_{n+1}B_nu_n+u_n^TB_n^TK_{n+1}A_nx_n+u_n^TB_n^TK_{n+1}B_nu_n-x_n^TK_nx_n,\end{aligned}\quad(6.133)$$

while the first difference $\Delta V(x_n)$ is specified by (6.132).

From (6.132) and (6.133), a scalar Hamiltonian function is

$$\begin{aligned}H(x_n,u_n)&=x_n^TQ_nx_n+u_n^TG_nu_n+x_n^TA_n^TK_{n+1}A_nx_n+x_n^TA_n^TK_{n+1}B_nu_n\\&\quad+u_n^TB_n^TK_{n+1}A_nx_n+u_n^TB_n^TK_{n+1}B_nu_n-x_n^TK_nx_n.\end{aligned}\quad(6.134)$$

The first differences ΔH, which is the stationarity condition to minimize $H(x_n,u_n)$, yields a control function

$$u_n=-\left(G_n+B_n^TK_{n+1}B_n\right)^{-1}B_n^TK_{n+1}A_nx_n\,.\quad(6.135)$$

From (6.134) and (6.135)

$$x_n^TK_nx_n=x_n^TQ_nx_n+x_n^TA_n^TK_{n+1}A_nx_n-x_n^TA_n^TK_{n+1}B_n\left(G_n+B_n^TK_{n+1}B_n\right)^{-1}B_n^TK_{n+1}A_nx\,.\quad(6.136)$$

A symmetric matrix K_{n+1} is computed solving an equation

$$-K_n+Q_n+A_n^TK_{n+1}A_n-A_n^TK_{n+1}B_n\left(G_n+B_n^TK_{n+1}B_n\right)^{-1}B_n^TK_{n+1}A_n=0\,.\quad(6.137)$$

From (6.135), for infinite horizon, $K_{n+1}\equiv K_n$,

$$u_n=-\left(G_n+B_n^TK_nB_n\right)^{-1}B_n^TK_nA_nx_n,\quad(6.138)$$

where a symmetric matrix K_n is computed by solving equation (6.137) with $K_{n+1}\equiv K_n$

$$-K_n+Q_n+A_n^TK_nA_n-A_n^TK_nB_n\left(G_n+B_n^TK_nB_n\right)^{-1}B_n^TK_nA_n=0,\ K_n=K_n^T\,.\quad(6.139)$$

We applied stability and optimality criteria (6.124) and (6.132) $\{V(x_n)>0,\ \Delta V(x_n)<0\}$. The derived results are equivalent to findings covered in section 5.10 when the quadratic performance indexes were minimized.

Conjecture 6.8. For a linear discrete-time system (6.129) $x_{n+1}=A_nx_n+B_nu_n$, a control law (6.138) $u_n=-\left(G_n+B_n^TK_nB_n\right)^{-1}B_n^TK_nA_nx_n$ is designed by specifying negative definiteness of the first difference for $V(x_n)$ as (6.132) $\Delta V(x_n)=V(x_{n+1})-V(x_n)$, $\Delta V(x_n)=-\left[x_n^TQ_nx_n+u_n^TG_nu_n\right]$, $(Q_n,G_n)>0$.

A symmetric positive definite matrix K_n is found by solving an equation (6.139), guaranteeing positive definiteness of the Lyapunov function (6.125) $V(x_n)=x_n^TK_nx_n$. Control law (6.138) stabilizes a closed-loop system because the stability criteria (6.124) $\{V(x_n)>0,\Delta V(x_n)<0\}$ are guaranteed. ■

Solution of Riccati and Lyapunov Equations – In equation (6.139), matrix $A_n^TK_nB_n\left(G_n+B_n^TK_nB_n\right)^{-1}B_n^TK_nA_n$ is invertable. Equation (6.139) is solved by forming the extended symplectic matrix $(\mathbf{H}-\lambda\mathbf{J})$, given by a pair of block matrices $(\mathbf{H},\mathbf{J})$ and generalized eigenvalues λ. Orthogonal eigenvalue-eigenvector decomposition algorithms with balancing yields $K_n\in\mathbb{R}^{n\times n}$. Alternatively, recursive algorithms are reported and illustrated in Example 6.52. Solutions of a Lyapunov equation (6.128) $A_n^TK_nA_n-K_n=-Q_n$ are documented in Example 5.6 and Practice Problem 5.4.

6.11.3. Optimization of Discrete-Time Systems With Control Limits

Systems With Control Limits – Consider a system with control limits. For continuous-time systems, section 6.7 reports design of constrained control laws. Control limits $\mathcal{U}_{\min}\le\mathcal{U}\le\mathcal{U}_{\max}$ are described by a real-valued bounded function $\Phi\subseteq\mathcal{U}_{\varepsilon}$ with range $(u_{\min},u_{\max})$ and domain $(-\infty, \infty)$. For Φ, there exist inverse functions Φ^{-1} with range $(-\infty,\infty)$ and domain $(u_{\min}, u_{\max})$.

Antiderivative and Antidifference – The indefinite integral of a function $f(u)$, or the antiderivative of $f(u)$, is a function $F(u)$, such that $F'(u)=f(u)$.One has $\int f(u)du=F(u)+C$, where $F(u)$ is an antiderivative of $f(u)$. Using discrete calculus, for a real-valued function $f(u)$, define an indefinite sum or *antidifference* of $f(u)$ as a function $F(u)$, such that $\Delta F(u)=F(u+1)-F(u)=f(u)$. The *antidifference* operator is Σ_u or Δ^{-1}, while the forward difference operator is Δ. The operator Σ_u is not the sum symbol. Expression $F(u)=\Sigma_u f(u)$ implies that $F(u)$ is an indefinite sum of $f(u)$.

The *antidifference* of inverse trigonometric and hyperbolic functions are

$$\Sigma_u \tan^{-1} au=\tfrac{i}{2}\ln\frac{\Gamma(u+\frac{i}{a})}{\Gamma(u-\frac{i}{a})}+C,\ \Sigma_u \tanh^{-1} au=\tfrac{1}{2}\ln\frac{\Gamma(u+\frac{1}{a})}{\Gamma(u-\frac{1}{a})}+C,$$

where $\Gamma(\cdot)$ is the incomplete gamma function.

We have $\int u_{\max}\tan^{-1}(au)du=u_{\max}\left[u\tan^{-1}(au)-\frac{1}{2a}\ln\left(a^2u^2+1\right)\right]$.

For a real-valued function $\Phi^{-1}(u_n)$, $\Phi\subseteq\mathcal{U}_{\varepsilon}$, $\int\Phi^{-1}(u)du\equiv\Sigma_u\Phi^{-1}(u_n)$.

Minimize a nonquadratic performance index

$$J=\min_{x_n,u_n\in U}\sum_{n=0}^{N-1}\left[x_n^TQ_nx_n-u_n^TB_n^TK_{n+1}B_nu_n+2\Sigma_u\left(\Phi^{-1}(u_n)\right)^TG_n\right],\ \begin{cases}x_n^TQ_nx_n>0,\ (Q_n,G_n)>0\\ \Sigma_u\left(\Phi^{-1}(u_n)\right)^TG_n>0\end{cases}\quad(6.140)$$

subject to

$$x_{n+1}=A_nx_n+B_nu_n,\ x_{n0}\in X_0,\ u_n\in U,\ u_{\min}\le u\le u_{\max},\ n\ge 0.\quad(6.141)$$

Performance index (6.140) is positive definite because

$$2\Sigma_u\left(\Phi^{-1}(u_n)\right)^TG_n>u_n^TB_n^TK_{n+1}B_nu_n,\ \forall x_n\in X(X_0,U),\ \forall u_n\in U.\quad(6.142)$$

Solution of dynamic optimization problems can be found on a class of quadratic return functions (6.125) $V(x_n)=x_n^TK_nx_n$.

From (6.125), (6.140) and (6.141), a scalar Hamiltonian is

$$H(x_n,u_n)=x_n^TQ_nx_n-u_n^TB_n^TK_{n+1}B_nu_n+2\Sigma_u\left(\Phi^{-1}(u_n)\right)^TG_n+\left(A_nx_n+B_nu_n\right)^TK_{n+1}\left(A_nx_n+B_nu_n\right).\quad(6.143)$$

Using the difference operator Δ, apply the stationarity condition $\Delta H(x_n,u_n)=0$. A constrained control law is

$$u_n=-\Phi\left(G_n^{-1}B_n^TK_{n+1}A_nx_n\right),\ u_n\in U.\quad(6.144)$$

From (6.143) and (6.144),

$$x_n^TK_nx_n=x_n^TQ_nx_n+x_n^TA_n^TK_{n+1}A_nx_n-2\Sigma_x\Phi^T\left(G_n^{-1}B_n^TK_{n+1}A_nx_n\right).\quad(6.145)$$

Real-valued functions $\Phi(\cdot)$ and $\Phi^{-1}(\cdot)$ lie in the first and third quadrants, and, $\Sigma_x\Phi(x_n)$ and $\Sigma_u\Phi^{-1}(u_n)$ exist. Therefore, equation (6.145) yields an equation to compute K_{n+1}, and, derive the feedback gains in control function (6.144).

Linear Approximation – To solve (6.145), apply the first-order Taylor polynomial approximation $\Sigma_x \Phi^T\left(G_n^{-1}B_n^T K_{n+1}A_n x_n\right) = k x_n^T A_n^T K_{n+1} B_n G_n^{-1} B_n^T K_{n+1} A_n x_n$, $k>0$.

A symmetric matrix K_{n+1} is found by solving

$$-K_n + Q_n + A_n^T K_{n+1} A_n - 2k A_n^T K_{n+1} B_n G_n^{-1} B_n^T K_{n+1} A_n = 0. \quad (6.146)$$

The closed-loop system (6.141) with (6.144) evolves in $X \times U$, $x_n \in X(X_0,U)$ and $u_n \in U$. A Lyapunov pair (6.124) $\{V(x_n)>0, \Delta V(x_n)<0\}$ yields a stability domain $S \subset \mathbb{R}^n$

$$S = \Big\{ \underset{\substack{x_{n+1}=A_n x_n + B_n u_n \\ u_n = -\Phi(G_n^{-1}B_n^T K_{n+1}A_n x_n)}}{x_n \in X : x_{n0} \in X_0,\ u_n \in U} \Big|\ \underset{V(x_n)=x_n^T K_n x_n \quad \Delta V(x_n)=V(x_{n+1})-V(x_n)}{V(0)=0,\ V(x_n)>0,\ \Delta V(x_n)<0}, \forall x_n \in X(X_0,U) \Big\}. \quad (6.147)$$

Stability is guaranteed if $X(X_0,U) \subseteq S$, $\forall x_n$, $n \geq 0$.

Minimizing a performance index (6.140) for $N \to \infty$, the bounded control law is

$$u_n = -\Phi\left(G_n^{-1} B_n^T K_n A_n x_n\right),\ u_n \in U. \quad (6.148)$$

A symmetric matrix K_n is computed by solving

$$-K_n + Q_n + A_n^T K_n A_n - 2k A_n^T K_n B_n G_n^{-1} B_n^T K_n A_n = 0. \quad (6.149)$$

Symmetric Control Limits – Physical limits $-\mathcal{U}_{\max} \leq \mathcal{U} \leq \mathcal{U}_{\max}$ imply designing control laws such that $-u_{\max} \leq u \leq u_{\max}$. Consider an infinitely differentiable, symmetric function of class C^∞. In particular, use $\Phi(x_n) = \boldsymbol{u}_{\max}\tanh(\boldsymbol{a}x_n)$, $\tanh(\boldsymbol{a}x_n) \to \pm 1$.

This $\Phi(\cdot)$ describes relays and saturation limits, and, $\Phi \subseteq \mathcal{U}_\varepsilon$.

Minimize a performance index

$$J = \min_{x_n, u_n \in U} \sum_{n=0}^{\infty} \left[x_n^T Q_n x_n - u_n^T B_n^T K_{n+1} B_n u_n + 2\Sigma_u \left(\boldsymbol{a}^{-1} \tanh^{-1}(\boldsymbol{u}_{\max}{}^{-1} u_n) \right)^T G_n \right] \quad (6.150)$$

subject to (6.141) $x_{n+1} = A_n x_n + B_n u_n$, $u_n \in U$.

Recall that $\Phi^{-1}(\cdot)$ exists, and, $\tanh^{-1} cu = \frac{1}{2}\ln\left(\frac{1+cu}{1-cu}\right)$.

Apply the stationarity condition on a Hamiltonian

$$H(x_n,u_n) = x_n^T Q_n x_n - u_n^T B_n^T K_{n+1} B_n u_n + 2\Sigma_u \left(\boldsymbol{a}^{-1} \tanh^{-1}(\boldsymbol{u}_{\max}{}^{-1} u_n) \right)^T G_n + \left(A_n x_n + B_n u_n \right)^T K_{n+1} \left(A_n x_n + B_n u_n \right). \quad (6.151)$$

A control law is found to be

$$u_n = -\boldsymbol{u}_{\max} \tanh\left(\boldsymbol{a} G_n^{-1} B_n^T K_n A_n x_n \right). \quad (6.152)$$

For $\Phi(x_n) = \boldsymbol{u}_{\max}\tanh(\boldsymbol{a}x_n)$, the control law has an asymptotes $\pm \boldsymbol{u}_{\max}$.

Approximation $\tanh(x) \approx x$ yields an equation to compute a matrix K_n

$$-K_n + Q_n + A_n^T K_n A_n - 2k A_n^T K_n B_n \boldsymbol{u}_{\max} \boldsymbol{a} G_n^{-1} B_n^T K_n A_n = 0. \quad (6.153)$$

Nonlinear Systems – Consider a nonlinear discrete-time system

$$x_{n+1} = F(x_n) + B(x_n) u_n,\ x_{n0} \in X_0,\ u_n \in U,\ u_{\min} \leq u_n \leq u_{\max},\ n \geq 0. \quad (6.154)$$

With control limits $\mathcal{U}_{\min} \leq \mathcal{U} \leq \mathcal{U}_{\max}$, described by $\Phi \subseteq \mathcal{U}_\varepsilon$, design a control law $u_n \in U$ minimizing

$$J = \min_{x_n, u_n \in U} \sum_{n=0}^{N-1} \left[x_n^T Q_n x_n - u_n^T B^T(x_n) K_{n+1} B(x_n) u_n + 2\Sigma_u \left(\Phi^{-1}(u_n) \right)^T G_n \right],\ Q_n \geq 0,\ G_n > 0, \quad (6.155)$$

$2\Sigma_u \left(\Phi^{-1}(u_n) \right)^T G_n > u_n^T B^T(x_n) K_{n+1} B(x_n) u_n$, $\forall x_n \in X(X_0,U)$, $\forall u_n \in U$.

For a nonlinear system (6.154), using the quadratic $V(x_n)$, we have

$$V(x_n)=x_n^T K_n x_n,\ V(x_{n+1})=x_{n+1}^T K_{n+1} x_{n+1}=\left(F(x_n)+B(x_n)u_n\right)^T K_{n+1}\left(F(x_n)+B(x_n)u_n\right).$$

The Hamiltonian is

$$H(x_n,u_n)=x_n^T Q_n x_n-u_n^T B^T(x_n)K_{n+1}B(x_n)u_n+2\Sigma_u\left(\Phi^{-1}(u_n)\right)^T G_n+\left(F(x_n)+B(x_n)u_n\right)^T K_{n+1}\left(F(x_n)+B(x_n)u_n\right). \tag{6.156}$$

A bounded control law is

$$u_n=-\Phi\left[G_n^{-1}B^T(x_n)K_{n+1}F(x_n)\right],\ u_n\in U. \tag{6.157}$$

From (6.156) and (6.157), we have

$$x_n^T K_n x_n=x_n^T Q_n x_n+F^T(x_n)K_{n+1}F(x_n)-2\Sigma_x\Phi^T\left(G_n^{-1}B^T(x_n)K_{n+1}F(x_n)\right). \tag{6.158}$$

One obtains a recurrence equation for K_{n+1}.

Investigate stability of the closed-loop system (6.154) with (6.157)

$$x_{n+1}=F(x_n)-B(x_n)\Phi\left[G_n^{-1}B^T(x_n)K_{n+1}F(x_n)\right],\ x_{n0}\in X_0,\ u_n\in U,\ u_{\min}\le u_n\le u_{\max},\ n\ge 0. \tag{6.159}$$

A domain of stability $S\subset\mathbb{R}^n$ is found by applying stability conditions defined by a Lyapunov pair (6.124). We have

$$S=\Big\{\underset{\substack{x_{n+1}=F(x_n)+B_n(x_n)u_n\\ u_n=-\Phi[G_n^{-1}B_n^T(x_n)K_{n+1}F(x_n)]}}{x_n\in X:\ x_{n0}\in X_0,\ u_n\in U}\Big|\ V(0)=0,\ \underset{V(x_n)=x_n^TK_nx_n}{V(x_n)>0},\ \underset{\Delta V(x_n)=V(x_{n+1})-V(x_n)}{\Delta V(x_n)<0},\ \forall x_n\in X(X_0,U)\Big\}. \tag{6.160}$$

For a closed-loop system (6.159), stability is guaranteed if $X(X_0,U)\subseteq S$.

Constrained Design Using Lyapunov Paradigm – Control problems are solved for linear and nonlinear discrete-time systems. As covered in section 6.11.2, negative definiteness of the first difference for $V(x_n)$, $\Delta V(x_n)=V(x_{n+1})-V(x_n)$ is specified using nonquadratic measures. Letting

$$\Delta V(x_n)=-\left[x_n^T Q_n x_n-u_n^T B^T(x_n)K_n B(x_n)u_n+2\Sigma_u\left(\Phi^{-1}(u_n)\right)^T G_n\right],$$

one finds a bounded control law and results, formulated by a Conjecture.

Conjecture 6.9. For a nonlinear discrete-time system (6.154) $x_{n+1}=F(x_n)+B(x_n)u_n$, a bounded control law (6.157) $u_n=-\Phi\left[G_n^{-1}B^T(x_n)K_nF(x_n)\right]$, $u_n\in U$, $u_{\min}\le u_n\le u_{\max}$ is designed by specifying negative definiteness of the first difference for $V(x_n)$ as $\Delta V(x_n)=-\left[x_n^T Q_n x_n-u_n^T B^T(x_n)K_n B(x_n)u_n+2\Sigma_u\left(\Phi^{-1}(u_n)\right)^T G_n\right]$. A functional equation

$$x_n^T K_n x_n=x_n^T Q_n x_n+F^T(x_n)K_nF(x_n)-2\Sigma_x\Phi^T\left(G_n^{-1}B^T(x_n)K_nF(x_n)\right),$$

yields an equation to compute a positive definite K_n, ensuring positive definiteness of Lyapunov function (6.125) $V(x_n)=x_n^T K_n x_n$. Control law (6.157) stabilizes a closed-loop system (6.154) with (6.157) in (6.160) $S\subset\mathbb{R}^n$, for which criteria imposed on a Lyapunov pair criteria (6.124) $\{V(x_n)>0,\ \Delta V(x_n)<0\}$, $\Delta V(x_n)=V(x_{n+1})-V(x_n)$ are guaranteed, $x_n\in X(X_0,U)\subseteq S$. ■

A domain of stability $S\subset\mathbb{R}^n$ depends on control bounds, initial conditions, etc. One may apply nonquadratic Lyapunov functions and $\omega_x(x_n)$. Tracking control problems are formulated and solved using tracking error dynamics, yielding control laws with state and tracking error feedback. Case studies in design of tracking control laws are documented in sections 6.15.1 and 6.15.4.

Example 6.52. *Bounded Control of a Fixed-Wing Aerial Vehicle*

Consider longitudinal dynamics of a fixed-wing aerial vehicle studied in section 6.8.2. The state variables (x_1,x_2,x_3,x_4) are the forward velocity V, angle of attack α, pitch rate q and pitch angle θ. The outputs are V and θ. Thrust u_t and elevator displacement u_E are controls.

Hence, $y=\begin{bmatrix} x_1 \\ x_4 \end{bmatrix}=\begin{bmatrix} V \\ \theta \end{bmatrix}$ and $u=\begin{bmatrix} u_1 \\ u_2 \end{bmatrix}=\begin{bmatrix} u_t \\ u_E \end{bmatrix}$.

Discretize the governing equations

$$\dot{x}=Ax+Bu,\quad \begin{bmatrix} \dot{x}_1 \\ \dot{x}_2 \\ \dot{x}_3 \\ \dot{x}_4 \end{bmatrix}=\begin{bmatrix} -0.15 & -1.3 & 0 & -9.81 \\ 0 & -7.4 & 0.96 & 0 \\ 0.1 & -5.6 & -2.4 & 0 \\ 0 & 0 & 1 & 0 \end{bmatrix}\begin{bmatrix} x_1 \\ x_2 \\ x_3 \\ x_4 \end{bmatrix}+\begin{bmatrix} 5.3 & -0.29 \\ 0 & 0.074 \\ 0 & 1.6 \\ 0 & 0 \end{bmatrix}\begin{bmatrix} u_1 \\ u_2 \end{bmatrix},\ y=\begin{bmatrix} 1 & 0 & 0 & 0 \\ 0 & 0 & 0 & 1 \end{bmatrix}\begin{bmatrix} x_1 \\ x_2 \\ x_3 \\ x_4 \end{bmatrix},\ x=\begin{bmatrix} x_1 \\ x_2 \\ x_3 \\ x_4 \end{bmatrix}=\begin{bmatrix} V \\ \alpha \\ q \\ \theta \end{bmatrix}.$$

System Discretization – For given (A,B,H,D), T_s=0.1 sec, matrices (A_n,B_n,H_n,D_n) are obtained by applying equations (5.19)

$A_n=e^{AT_s}$, $B_n=A^{-1}(e^{AT_s}-I)B$, $H_n=H$ and $D_n=D$.

We have

$x_{n+1}=A_nx_n+B_nu_n$, $y_n=H_nx_n$,

$$A_n=\begin{bmatrix} 0.9849 & -0.0831 & -0.0495 & -0.9736 \\ 0.00034 & 0.4619 & 0.0589 & -0.00012 \\ 0.007 & -0.344 & 0.7685 & -0.0045 \\ 0.00046 & -0.0204 & 0.0882 & 0.9998 \end{bmatrix},\quad B_n=\begin{bmatrix} 0.526 & -0.0319 \\ 0.000065 & 0.0108 \\ 0.0024 & 0.1395 \\ 0.000082 & 0.0073 \end{bmatrix}.$$

Linear Control – Solve (6.139) $-K_n+Q_n+A_n^TK_nA_n-A_n^TK_nB_n\left(G_n+B_n^TK_nB_n\right)^{-1}B_n^TK_nA_n=0$.

Let $Q_n=I$, $Q_n\in\mathbb{R}^{4\times4}$, $G_n=I$, $G_n\in\mathbb{R}^{2\times2}$, $G_n=\begin{bmatrix} 1 & 0 \\ 0 & 1 \end{bmatrix}$. We have $K_n=\begin{bmatrix} 2.387 & -0.132 & -0.236 & -2.78 \\ -0.132 & 3.429 & -2.945 & -7.9 \\ -0.236 & -2.945 & 5.503 & 12.83 \\ -2.78 & -7.9 & 12.83 & 54.45 \end{bmatrix}$.

A control law is

$$u_n=-\left(G_n+B_n^TK_{n+1}B_n\right)^{-1}B_n^TK_nA_nx_n,\quad \begin{bmatrix} u_{1n} \\ u_{2n} \end{bmatrix}=-\begin{bmatrix} 0.7412 & -0.0625 & -0.1383 & -1.517 \\ -0.0629 & -0.4623 & 0.7145 & 1.955 \end{bmatrix}\begin{bmatrix} x_{1n} \\ x_{2n} \\ x_{3n} \\ x_{4n} \end{bmatrix}.$$

A closed-loop system is stable, and, the eigenvalues are (0.906, 0.671, 0.569±j0.024).

Constrained Control – A bounded control law design problem is solved considering the symmetric saturation limits. For illustrative purposes and to demonstrate design, solve a stabilization problem.

Consider $-1\le\mathcal{U}_1\le1$ and $-1\le\mathcal{U}_2\le1$.

Physical limits are described as $\Phi_1=\tanh(x)$ and $\Phi_2=\tanh(x)$.

Minimizing a performance index (6.140) for $N\to\infty$, the control law is (6.148) $u_n=-\Phi\left(G_n^{-1}B_n^TK_nA_nx_n\right)$.

Solve (6.149) $-K_n+Q_n+A_n^TK_nA_n-2kA_n^TK_nB_nG_n^{-1}B_n^TK_nA_n=0$.

Let k=1, $Q_n=I$, $Q_n\in\mathbb{R}^{4\times 4}$, $G_n=I$, $G_n\in\mathbb{R}^{2\times 2}$. We have $K_n=\begin{bmatrix} 1.335 & -0.0674 & 0.0279 & -0.322 \\ -0.0674 & 2.323 & -1.416 & -2.998 \\ 0.0279 & -1.416 & 3.218 & 4.775 \\ -0.322 & -2.998 & 4.775 & 22.37 \end{bmatrix}$.

The bounded control law, which stabilizes a system, is

$$u_n=-\tanh\left(G_n^{-1}B_n^TK_nA_nx_n\right),\ \begin{bmatrix} u_{1n} \\ u_{2n} \end{bmatrix}=-\tanh\left(\begin{bmatrix} 0.692 & -0.0811 & -0.0333 & -0.8402 \\ -0.0367 & -0.2627 & 0.4215 & 0.8463 \end{bmatrix}\begin{bmatrix} x_{1n} \\ x_{2n} \\ x_{3n} \\ x_{4n} \end{bmatrix}\right).$$

MATLAB Solvers and Recursive Algorithms – Custom MATLAB codes are developed to solve equation (6.139) $-K_n+Q_n+A_n^TK_nA_n-A_n^TK_nB_n\left(G_n+B_n^TK_nB_n\right)^{-1}B_n^TK_nA_n=0$. Alternatively, the **dlqr** and **idare** MATLAB commands are used. Equation (6.149)

$$-K_n+Q_n+A_n^TK_nA_n-2kA_n^TK_nB_nG_n^{-1}B_n^TK_nA_n=0$$

can be solved by applying the Schur decomposition, recursive, iterative, numeric optimization and search algorithms. The recursive algorithm solves equations (6.139) and (6.149). Solutions are verified, and, the specified accuracy is achieved by robust numeric algorithms reported below with changes as appropriate.

```
clc; clear; close all;
A=[-0.15 -1.3 0 -9.81; 0 -7.4 0.96 0; 0.1 -5.6 -2.4 0; 0 0 1 0]; B=[5.3 -0.29; 0 0.074; 0 1.6; 0 0];
H=[1 0 0 0;0 0 0 1]; D=[0 0;0 0]; % Outputs: Forward velocity and pitch angle
Ts=0.1; An=expm(A*Ts); Bn=inv(A)*(An-eye(size(A)))*B;    % Discretization
k=1; Qn=1*eye(length(An)); Gn=1*eye(length(Bn(1,:)));
% LQR using the dlqr and idare commands
[KFd1,Kd1,Eigd1]=dlqr(An,Bn,Qn,Gn);  % dlqr command, solve equation (6.139)
[Kd2,KFd2,Eigd2]=idare(An,Bn,Qn,Gn); % idare command, solve equation (6.139)
   % Solve derived equation
X=Qn; small=false; acc=10^(-12); % Accuracy and solution error
while small==false
% Recursive solution
     X=Qn+An'*X*An-2*k*An'*X*Bn*inv(Gn)*Bn'*X*An;          % Bounded design, equation (6.149)
  %  X=Qn+An'*X*An-An'*X*Bn*inv(Gn+Bn'*X*Bn)*Bn'*X*An; % LQR design, Riccati equation (6.139)
% Accuracy and error analysis
     check=-X+Qn+An'*X*An-2*k*An'*X*Bn*inv(Gn)*Bn'*X*An;   % Bounded design, equation (6.149)
  % check=-X+Qn+An'*X*An-An'*X*Bn*inv(Gn+Bn'*X*Bn)*Bn'*X*An; % LQR design, equation (6.139)
if check(1,1)<acc && check(1,2)<acc && check(2,1)<acc && check(2,2)<acc
   small=true;
     end
        end
  %  Kn=X,  KFn=inv(Gn+Bn'*Kn*Bn)*Bn'*Kn*An % Kn and feedback gain matrix: LQR control (6.138)
     Kn=X,  KFn=inv(Gn)*Bn'*Kn*An % Kn and feedback gain matrix: Bounded control (6.148)
     Eign=eig(An-Bn*KFn); % Eigenvalues
% Verifying solvers and solutions
checkKd1=-Kd1+Qn+An'*Kd1*An-An'*Kd1*Bn*inv(Gn+Bn'*Kd1*Bn)*Bn'*Kd1*An; % dlqr, eq. (6.139)
checkKd2=-Kd2+Qn+An'*Kd2*An-An'*Kd2*Bn*inv(Gn+Bn'*Kd2*Bn)*Bn'*Kd2*An; % idare, eq. (6.139)
checkKn=-Kn+Qn+An'*Kn*An-2*k*An'*Kn*Bn*inv(Gn)*Bn'*Kn*An;          % Equation (6.149)
```

Transcendental, recurrence and difference equations may be solved using Newton's methods. To solve a system of p equations, compute roots of p functions $f:\mathbb{R}\to\mathbb{R}$ and a vector-valued function $F(K_n):\mathbb{R}^p\to\mathbb{R}^p$. Finding the Jacobian $J_F(K_n)$, one has $K_{n+1}=K_n-J_F^{-1}(K_n)F(K_n)$, and, $J_F(K_n)(K_{n+1}-K_n)=-F(K_n)$. The Newton, Newton-Fourier and quasi-Newton optimization methods, as well as nonlinear programming algorithms, are considered to compute $K_n\in\mathbb{R}^{n\times n}$. To solve (6.149), define a function $F=-K_n+Q_n+A_n^TK_nA_n-2kA_n^TK_nB_nG_n^{-1}B_n^TK_nA_n$ and Jacobian matrix.

The MATLAB code yields K_n by solving (6.139) and (6.149).

```
N_iteration=10000; n=length(An);
syms(sym('Kn',[n^2 1])); Kn=(sym('Kn',[n^2 1])); KN=reshape(Kn,[n n]);
% F(Kn)=-KN+Qn+An'*KN*An-An'*KN*Bn*inv(Gn+Bn'*KN*Bn)*Bn'*KN'*An; % Riccati equation (6.139)
  F(Kn)=-KN+Qn+An'*KN*An-2*k*An'*KN*Bn*inv(Gn)*Bn'*KN*An;      % Equation (6.149)
  F=reshape(F,[1 n^2]); J=jacobian(F,Kn);                        % Jacobian matrix
K0=reshape(Qn,[n^2 1]);                     % Initial values Kn=Qn
error=1e-5*ones([n^2,1]); errorK=error;     % Admissible error
for i=0:1:N_iteration
 while abs(errorK)>= error
   Fk0=double(subs(F,Kn,K0));
   JK0=double(subs(J,Kn,K0));  % Jacobian matrix
   KnF=K0-JK0\Fk0';            % Solution by Newton iteration
   errorK=K0-KnF;              % Computing error
   K0=KnF;
 end
end
E=reshape(subs(F,Kn,K0),[n n]);
% Kn=reshape(K0,[n n]), KFn=inv(Gn+Bn'*Kn*Bn)*Bn'*Kn*An % LQR: Kn and feedback gain matrix
Kn=reshape(K0,[n n]), KFn=inv(Gn)*Bn'*Kn*An % Bounded control (6.148): Kn and feedback matrix
```

■

6.12. Hierarchical Dynamic Networks in Analysis, Control and Coordination

Device and System Level Organization – System organization, as well as topics in *structural*, *physical* and *optimal* designs, were discussed and covered in previous chapters. Figures 1.1, 1.2 and 1.3, as well as 5.1, 5.2 and 5.3 report functional diagrams, system organization, and, system and hardware solutions. Analysis, design and optimization were performed for sensors, power electronics, actuators and servos, pertained to a system level hardware solutions. One conducts a modular design focusing on decoupling, correlation, correspondence, aggregation, coordination, interoperability and integration. Systems are decoupled into consistent, well-defined and interrelated interacting devices, components, modules and subsystems. One selects high-performance commercialized components. For selected hardware, analyses and aggregation are considered. Integration are achieved by means of *optimal modular design* and consistent modules selection. For example, in adequately designed systems, at system level, control problems may be approached by considering dominant dynamics, rigid-body mechanics, etc. Robots, manipulators, aerial vehicles and other systems were controlled assuming the *force* and *torque* control paradigms.

Practitioners acquire knowledge, skills, expertise, techniques and tools by means of enduring learning through focused activities. Practical knowledge and existing solutions lead to a deep understanding through experience, competence, learning, findings and best practices. One balances practical foundations and fundamentals, proven hardware schemes and practice with theoretical basics and interweaving designs. An integrated design is aimed to ensure functionality, operationability, complementarity, etc. Practical considerations are of a significant importance, and, multiple-level hardware-networked solutions exist, implemented and deployed.

Attempts to perform modular optimal and physical designs applying theoretical studies can be approached using hierarchical dynamic networks. Within well-defined organization, low-, medium- and high level features, decompositions and topological abstractions could be considered. Components, modules, systems, services, algorithms and protocols (estimation, interfacing, security, etc.) exhibit heterogeneities. Aggregation, information sharing, coordination and interoperability of components and systems are characterized by control principles, compliance and connectivity. Data rate, latencies, uncertainties and other descriptive features are considered within multi-physics domains and cyberspace.

Devising and examining system organization, the designer considers:

1. Components: Kinematics, electronics, actuators, sensors, MPUs, FPGAs, etc.;
2. Controlled and networked modules, data management system, etc.;
3. Algorithms and protocols, management system, data fusion, etc.;
4. Aggregated systems.

As an example, supervisory control and data acquisition (SCADA) systems are comprised of actuators with PWM amplifiers, distributed sensors, data communications links, peripheral modules, MPUs and other subsystems. The application-specific software and hardware for various multi-level SCADA systems with human-machine interface and *cloud* services have being commercialized in aerial, energy, manufacturing, robotic and other systems. One strengthens security, data quality and other features. Authentication, encryption-decryption and *cloud* services algorithms and protocols imply delays and latencies in data sharing, data fusion and spatiotemporal data management. Dynamic data driven applications, services and knowledge sharing tasks exhibit transients, and, dynamically evolve. While seamless integration and aggregation can be accomplished, governing and constitutive equations may be applied in spatiotemporal analyses.

Dynamic analyses, control schemes and optimization should be investigated examining integrated modules and services. For example, there are absolute limits on data rates, bandwidth, dynamic accuracy and other characteristics of sensors, transceivers and MPUs which cannot be exceeded. Despite heterogeneities, as illustrated in Figure 6.32.a, components are seamlessly integrated applying multi-physics domains and cyberspace considerations. One designs and implements control, networking and data management algorithms and protocols. Adaptive signal processing by MPUs and FPGAs supports asynchronous data fusion from heterogeneous MDOF sensors. Robust data sharing is ensured by means of high-throughput secure communication schemes. MPUs and FPGAs perform reconfigurable signal processing, pertained to control, data management, communication, security and other tasks. High assurance measurements, communication, processing, data fusion and control in networked learning-enabled systems support situational awareness, perception as well as descriptive, predictive and prescriptive data analytics. A cyclic design taxonomy implies a synergy of integrated hardware supporting learning and adaptation as reported in Figure 6.32.a. Advanced-technology components enable stability, dynamic range, precision, accuracy and robustness.

Mathematically, augmentation, analysis and optimization of interactive dynamic components may be approached using a hierarchical dynamic network paradigm. One investigates scalable organizations, and, studies application-compliant cyber-physical frameworks. Physical systems are controlled by MPUs and FPGAs using heterogeneous sensor measurements.

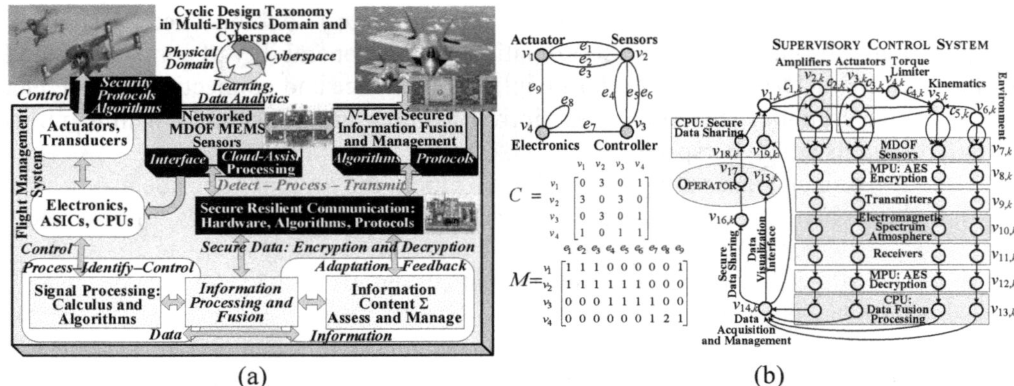

(a) (b)

Figure 6.32. (a) Seamless hardware and software integration within multi-physics domains and cyberspace: Adaptive learning-enable processing, secure communication and information sharing support *Detect* ↔ *Share* ↔ *Connect* ↔ *Assess* ↔ *Control* tasks in multi-level systems; (b) Conjectural graphs to analyze and visualize interacted system components.

For distributed sensor networks $\mathbf{S}=\bigcup_{k\in K} S_k$ with heterogeneous MDOF sensors $S_k \in K$, one evaluates and uses measurements by each MDOF sensor $S_i=\bigcup_{l\in L} s_l$, $s_i \in S_i$, $s_l \in L$, $s_{l,i} \cap s_{l,j}=\emptyset$, $i \neq j$. Descriptive deterministic and probabilistic models are derived to evaluate and empower data fusion and management. Adaptive learning-enable processing by MPUs should meet requirements on functionality, defined by the industry standards on compliant and complimentary hardware. Filtering, estimations and processing by MPUs yield latencies. In chapters 3 and 5, fixed-structure and reconfigurable adaptive filters $\mathcal{F}_{SC}(\boldsymbol{S},\boldsymbol{C}) \in \mathbf{C}_{\mathcal{F}}$ were designed to ensure accurate estimates, maximize data homogeneity, as well as to minimize data heterogeneity and information losses. Processing algorithms, implemented by MPUs or FPGAs P_i, $\mathbf{P}=\bigcup_{n\in N} P_n$, assess data from MDOF sensors S_l and controlled devices and modules D_h within an networks-of-network hardware organization. By means of communication and interfacing, resilient and secure high-throughput information sharing is performed by data acquisition and management systems. Data analytics and data mining are performed.

Graph Formulation – A time-invariant graph $\mathcal{G}$ is an ordered tuple (V,E,ϕ) with a nonempty set of vertices (nodes) V, a set of edges (links) E which is disjoint from V, and, an incidence function ϕ which associates with edges and vertices. For $V=\{v_1,\ldots,v_v\}$ and $E \subseteq V \times V$, the incidence and adjacency matrices are $M=[m_{ij}] \in \mathbb{R}^{v\times e}$ and $C=[c_{ij}] \in \mathbb{R}^{v\times v}$, where m_{ij} is the number of v_i and e_j occurrences; c_{ij} is the number of edges joining v_i and v_j; v and e are the numbers of vertices and edges. Any two vertices are connected by a unique path in a tree. Figure 6.32.b illustrates a representative directed a representative graph for a simplified system, such as robot, manipulator, aerial vehicle, etc. The correspondence of vertices and arcs to hardware and algorithmic solutions can be ensured, and, labeling may be performed in links. Applying graphs and sets formalisms, connectivity and aggregation may be studied in information-theoretic sense.

Example 6.53. Figure 6.32.b reports a graph with four nodes $V=\{v_1,v_2,v_3,v_4\}$ and nine links $E=\{e_1,e_2,e_3,e_4,e_5,e_6,e_7,e_8,e_9\}$. The incidence and adjacency matrices M and C are

$$M=\begin{matrix} & e_1 & e_2 & e_3 & e_4 & e_5 & e_6 & e_7 & e_8 & e_9 \\ v_1 & 1 & 1 & 1 & 0 & 0 & 0 & 0 & 0 & 1 \\ v_2 & 1 & 1 & 1 & 1 & 1 & 1 & 0 & 0 & 0 \\ v_3 & 0 & 0 & 0 & 1 & 1 & 1 & 1 & 0 & 0 \\ v_4 & 0 & 0 & 0 & 0 & 0 & 0 & 1 & 2 & 1 \end{matrix},\quad C=\begin{matrix} & v_1 & v_2 & v_3 & v_4 \\ v_1 & 0 & 3 & 0 & 1 \\ v_2 & 3 & 0 & 3 & 0 \\ v_3 & 0 & 3 & 0 & 1 \\ v_4 & 1 & 0 & 1 & 1 \end{matrix}.$$

Typifying an organization for a closed-loop transducer, consider a graph with six nodes, such as MPU (node 1), self-diagnostic MDOF sensors with self-monitoring and self-calibration (nodes 2, 3 and 4), PWM amplifier (node 5), transducer (node 6), and, corresponding edges. Using description of graph $\mathcal{G}(V,E,\phi)$ one may find the shortest path between nodes, distance, perform ranking and partitioning. These analyses have limited practicality, incompleteness and may be system-incompatible. For example, the shortest path distance between MPU and transducer is 2. However, the PWM amplifier is required. Using the MATLAB statement, the graph $\mathcal{G}(V,E,\phi)$ is visualized as reported in Figure 6.33.a.

```
V = [1 1 1 1 1 1 1 1 1 1 2 2 2 2 3 3 3 3 4 4 4 5 5 6]; E = [2 2 2 3 3 3 4 4 4 5 2 6 6 6 3 6 6 6 4 5 5 4 5 5];
F = [1 2 1 2 1 2 1 2 1 2 1 2 1 2 1 2 1 2 1 2 1 2 1 2];
G = graph(V,E,F); g = plot(G,'EdgeLabel',G.Edges.Weight,'linewidth',2);
```

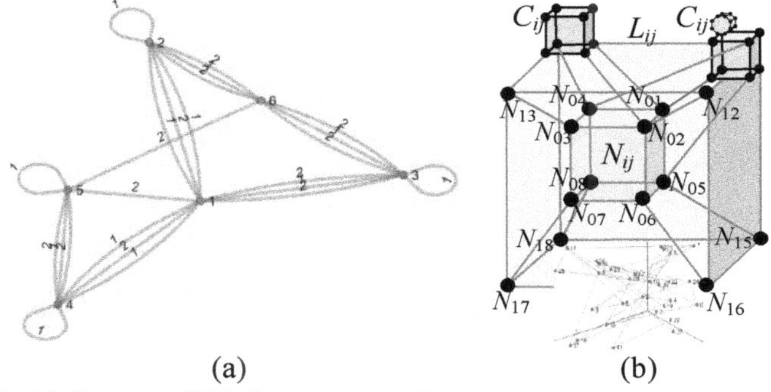

(a) (b)

Figure 6.33. (a) Graph $\mathcal{G}(V,E,\phi)$ with six nodes;
(b) Tesseract with nodes N_{ij}, cells C_{ij} and links L_{ij}. ■

Graph Models and Hierarchical Networks – In theoretical studies, graphs and networks are analyzed to model the data and information sharing, vulnerabilities, etc. The pertained analysis is covered in chapter 7. Within various services in adversarial environments, one may define the variables and interacting nodes. Using linear and metric temporal logic, the multi-objective graph models can be derived ensuring adequate representation and visualization. Using a finite set Q of states σ with initial and final states $\sigma_0\in Q$, $\sigma_f\in Q$, finite alphabet $a_{\mathcal{B}}$, transition function $\tau_{\mathcal{B}}$, $\tau_{\mathcal{B}}:Q\times a_{\mathcal{B}}\to 2^X$, and, acceptance condition $c_{\mathcal{B}}$, a generalized Büchi automata quintuple $\mathcal{B}=(Q,\sigma(\sigma_0,\sigma_f),a_{\mathcal{B}},\tau_{\mathcal{B}},c_{\mathcal{B}})$ yields directed graph structures within formal models. A propositional sequence is a sequence of states $(\sigma_0,\sigma_1,\ldots,\sigma_i,\sigma_{i+1},\ldots)\in Q$, $(\sigma,\sigma')\equiv(\sigma_i,\sigma_{i+1})\in\tau_{\mathcal{B}}$, $i\geq 0$. The Büchi automata is executed if $\mathcal{B}(\cdot)$ accepts a propositional sequences. The graph is synthesized using the model given by $\mathcal{B}(\cdot)$. For the Büchi model $\mathcal{M}(\mathcal{B})$, the correctness property is introduced to evaluate evolutions. The Markov statistical model and stochastic processes are considered as reported in chapter 7.

The *physical layer* nodes are the central processor (user's control station, flight management system, autonomous robot computer, etc.), MPUs, sensors, electronics, transducers, transceivers, etc. The *cyber layer* nodes are the servers, terminals, assets, query, host, external and internal firewalls, front- and back-end databases, antivirus, logins, service connections, event management, intrusion detection systems, shared resources, visualization and other entities. The *information layer* nodes are the data analytics, information management and fusion, information evaluation, and, other entities. Analysis, interactive visualization, control and other attributes can be examined. Dynamically evolving graph models can support analysis of processes and transitions, information processing, attack reachability, mitigation pattern, security schemes and system vulnerability, ensuring security by means of detection, risk mitigation and control as documented in chapter 7. Dynamic evolutions, adaptive reconfiguration and probabilistic analysis imply investigation of hierarchical topologies and dynamic networks $\mathcal{N}_t$.

Physical system organization and cyberspace layers are mapped by an interconnected *n*-dimensional-mesh network of connected modules. Each component interacts, and, nodes are connected in interconnected network of regular, semiregular and irregular *n*-polytopes. Depending on system organization and connectivity, one uses octahedron (polyhedron with 8 faces, 12 edges and 6 vertices), truncated tetrahedron (4 regular and 4 triangle faces, 18 edges and 12 vertices), hypercube and other polytopes. The tesseract with 16 nodes N_{ij} and 32 links L_{ij} is documented in Figure 6.33.b. Regular polygons $_p\{q\}_r$ are used in the graph theory, for which applying the Schläfli numbers $\{p,q,r\}$, $\begin{bmatrix} \mathrm{N}_0 & \mathrm{N}_{01} \\ \mathrm{N}_{10} & \mathrm{N}_1 \end{bmatrix} = \begin{bmatrix} \frac{g}{r} & r \\ p & \frac{g}{r} \end{bmatrix} = \begin{bmatrix} q & 2 \\ 2 & q \end{bmatrix}$, where g is the group order. Hierarchical networks are mapped by *n*-polytope, $n \geq 3$. A 4-polytope is a four-dimensional structure with vertices, edges, faces and polyhedron cells.

For example, a 3-polytype is a three-dimensional polytope with flat polygonal faces, straight edges and vertices $\begin{bmatrix} \mathrm{N}_0 & \mathrm{N}_{01} & \mathrm{N}_{02} \\ \mathrm{N}_{10} & \mathrm{N}_1 & \mathrm{N}_{12} \\ \mathrm{N}_{20} & \mathrm{N}_{21} & \mathrm{N}_2 \end{bmatrix} = \begin{bmatrix} \frac{g}{2q} & q & q \\ 2 & \frac{g}{4} & 2 \\ p & p & \frac{g}{2p} \end{bmatrix}$, $g = \frac{4}{4-(p-2)(q-2)} pq$.

Mapping System Organization and Topologies – The system organization, as a hierarchical network, is mapped by an *n*-dimensional regular polytope tabulated by the nondiagonal matrix $\begin{bmatrix} \mathrm{N}_0 & \mathrm{N}_{01} & \cdots & \mathrm{N}_{0n-1} \\ \mathrm{N}_{10} & \mathrm{N}_1 & \cdots & \mathrm{N}_{1n-1} \\ \vdots & \vdots & \ddots & \vdots \\ \mathrm{N}_{n-10} & \mathrm{N}_{n-11} & \cdots & \mathrm{N}_{n-1} \end{bmatrix}$ with the *i*- and *j*-faces, such as $\mathrm{N}_i\mathrm{N}_{ij}=\mathrm{N}_j\mathrm{N}_{ji}$, $i \neq j$, $0 \leq j < n$. Regular, semiregular and irregular *n*-polytopes describe complex organizations. For example, the tesseract with two 3-polytope is illustrated in Figures 6.33.b with nodes and links (N,L). Labeling of networks $\mathcal{N}_t$ with multiple components and modules are accomplished considering hierarchy, parallelism, mesh interconnect, redundancy, etc. Applying topological combinatorics, one defines the combinatorial structure of CPS as a partially ordered sets with specified dimensions. As illustrated in Figures 6.33.b, the pattern, structural permutations, boundary maps and topologically stratified spaces allow one to determine topological information such as homology and homeomorphism.

Dynamic Networks – Controlled networks $\mathcal{N}_t(t,N_t,C_t,L_t,\phi_t,\boldsymbol{x},\boldsymbol{y},\boldsymbol{u})$ are characterized by a time-varying topology $\mathcal{T}_t(t,N_t,C_t,L_t,\phi_t)$ and variables $(\boldsymbol{x},\boldsymbol{y},\boldsymbol{u})\in X\times Y\times U$. One considers physical and cyber states, pertained to physical processes, dynamic transitions, services, connections, event management, intrusions, encryption-decryption, etc. Networks $\mathcal{N}_t$ yield physics-consistent spatiotemporal dimensionality, supporting structural and physical designs, adaptive reconfiguration, reconfigurable connectivity assessment, etc. Aggregation, decoupling, high-fidelity description, dimensionality reduction, parameterization and node activities are investigated. Descriptive, predictive and prescriptive analyses are addressed. Spatiotemporal dynamics and topological evolution of high-dimensional $\mathcal{N}_t$ differ compared with time-invariant considerations. One considers controlled nodes N_t, dynamic activities in cells C_t, connectivity, dynamic interactions, data fusion, etc.

The aggregated topology $\mathcal{T}_t(t,N_t,L_t,\phi_t)$ may be reconfigured by solving the minimax optimization and adaptive reconfiguration problems. Dynamic evolutions are due to time-varying $\mathcal{T}_t$, as well as transitions and information sharing between nodes, cells and links $(N_{t_j},C_{t_k},L_{t_l})$. Dynamics of $(N_{t_j},C_{t_k},L_{t_l})$ are defined by components, interface, interconnect and communication evolutions for normal operation, as well as under uncertainties, electromagnetic spectrum and adversarial activities (jamming, spoofing, etc.), propagation delays, latencies, etc. Network synthesis and topology optimization problems are solved by analyzing $(\boldsymbol{x},\boldsymbol{y},\boldsymbol{u})$ in continuous and discrete time domains. For continuous-time nodes $N_{t_{j=1,\ldots,z}}$,

$$N_t=\begin{bmatrix}N_{t_1}\\ \vdots\\ N_{t_z}\end{bmatrix},\ \begin{bmatrix}\dot{x}_1=\sum_i \dot{x}_{1_i}\\ \vdots\\ \dot{x}_z=\sum_i \dot{x}_{z_i}\end{bmatrix}=\begin{bmatrix}A_1(\boldsymbol{x}_1)\\ \vdots\\ A_z(\boldsymbol{x}_z)\end{bmatrix}+\begin{bmatrix}B_1(\boldsymbol{x}_1,\boldsymbol{u}_1)\\ \vdots\\ B_z(\boldsymbol{x}_z,\boldsymbol{u}_z)\end{bmatrix}+\begin{bmatrix}\sum_{j,k,l}F_{1_{jkl}}(\boldsymbol{x}_{1_{jkl}},\boldsymbol{u}_{1_{jkl}})\\ \vdots\\ \sum_{j,k,l}F_{z_{jkl}}(\boldsymbol{x}_{z_{jkl}},\boldsymbol{u}_{z_{jkl}})\end{bmatrix}+\begin{bmatrix}\sum_{j,k,l}D_{1_{jkl}}(\boldsymbol{d}_{1_{jkl}})\\ \vdots\\ \sum_{j,k,l}D_{z_{jkl}}(\boldsymbol{d}_{z_{jkl}})\end{bmatrix},\ \begin{bmatrix}\boldsymbol{y}_1\\ \vdots\\ \boldsymbol{y}_w\end{bmatrix}=\begin{bmatrix}\sum_{j,k,l}H_{1_{jkl}}(\boldsymbol{x}_1,\boldsymbol{x}_{1_{jkl}})\\ \vdots\\ \sum_{j,k,l}H_{w_{jkl}}(\boldsymbol{x}_w,\boldsymbol{x}_{w_{jkl}})\end{bmatrix},\tag{6.161}$$

where $A_i(\cdot)$, $B_i(\cdot)$, $F_i(\cdot)$, $D_i(\cdot)$ and $H_i(\cdot)$ are the state-space model and node aggregation mappings pertained to states $\boldsymbol{x}_j=\Sigma_i x_{j_i}$, controls $\boldsymbol{u}_j=\Sigma_i u_{j_i}$, disturbances $\boldsymbol{d}_j=\Sigma_i d_{j_i}$ and outputs $\boldsymbol{y}_j=\Sigma_i y_{j_i}$.

For cells $C_{t_{k=1,\ldots,h}}$, illustrated in Figure 6.33.b, one has

$$C_t=\begin{bmatrix}C_{t_1}\\ \vdots\\ C_{t_h}\end{bmatrix},\ \begin{bmatrix}\dot{x}_1=\sum_i \dot{x}_{1_i}\\ \vdots\\ \dot{x}_h=\sum_i \dot{x}_{h_i}\end{bmatrix}=\begin{bmatrix}a_1(\boldsymbol{x}_1)\\ \vdots\\ a_h(\boldsymbol{x}_h)\end{bmatrix}+\begin{bmatrix}b_1(\boldsymbol{x}_1,\boldsymbol{u}_1)\\ \vdots\\ b_h(\boldsymbol{x}_h,\boldsymbol{u}_h)\end{bmatrix}+\begin{bmatrix}\sum_{j,k,l}f_{1_{jkl}}(\boldsymbol{x}_{1_{jkl}},\boldsymbol{u}_{1_{jkl}})\\ \vdots\\ \sum_{j,k,l}f_{h_{jkl}}(\boldsymbol{x}_{h_{jkl}},\boldsymbol{u}_{h_{jkl}})\end{bmatrix}+\begin{bmatrix}\sum_{j,k,l}d_{1_{jkl}}(\boldsymbol{d}_{1_{jkl}})\\ \vdots\\ \sum_{j,k,l}d_{h_{jkl}}(\boldsymbol{d}_{h_{jkl}})\end{bmatrix},\ \begin{bmatrix}\boldsymbol{y}_1\\ \vdots\\ \boldsymbol{y}_v\end{bmatrix}=\begin{bmatrix}\sum_{j,k,l}h_{1_{jkl}}(\boldsymbol{x}_1,\boldsymbol{x}_{1_{jkl}})\\ \vdots\\ \sum_{j,k,l}h_{v_{jkl}}(\boldsymbol{x}_v,\boldsymbol{x}_{v_{jkl}})\end{bmatrix}.\tag{6.162}$$

Sets of quantities and variable pairs are used to perform decentralized, aggregated and hierarchical controls and optimization by applying a system model

$$\mathcal{M}_t\equiv\mathcal{N}_t(t,N_t,C_t,L_t,\phi_t,\boldsymbol{x},\boldsymbol{y},\boldsymbol{u},A,B,F,D,H),\tag{6.163}$$

$$\dot{\boldsymbol{x}}(t)=\sum_{i,j,k,l}\underbrace{\phi_{i,j,k,l}^{(N_j,C_k,L_l)}(t)}_{\text{Aggregation and Topology Optimization}}\Big[\underbrace{\sum_{i,j,k,l}A_{j,k,l}^{(N_j,C_k,L_l)}(\boldsymbol{x}_{i_{j,k,l}})}_{\text{Structural Design }A_{j,k,l}^{(C_j,N_k,L_l)}}+\underbrace{\sum_{i,j,k,l}B_{j,k,l}^{(N_j,C_k,L_l)}(\boldsymbol{x}_{i_{j,k,l}},\boldsymbol{u}_{i_{j,k,l}})}_{\substack{\text{Structural Design }B_{k,l,m}^{(C_j,N_k,L_l)}\\ \text{Dyanamic Optimization and Design }\boldsymbol{u}_{i_{j,k,l}}=\Phi(\varphi(\boldsymbol{x}_{i_{j,k,l}}))}}\Big].$$

To find optimal topology and organization $\left(\phi_{i,j,k,l}^{(N_j,C_k,L_l)}, A_{j,k,l}^{(N_j,C_k,L_l)}, B_{j,k,l}^{(N_j,C_k,L_l)}\right)$, analyze data fusion and information management, design control laws $\boldsymbol{u}_{i_{j,k,l}} = \Phi\left(\varphi(\boldsymbol{x}_{i_{j,k,l}})\right)$, evaluate and optimize performance, etc.

One finds

$$\mathcal{N}_t^{\text{Optimal}}(t,N_t,C_t,L_t,\phi_t,\boldsymbol{x},\boldsymbol{y},\boldsymbol{u},A,B,F,D,H)\subset\mathcal{N}_t^{\text{Admissible}}. \quad (6.164)$$

While theoretical frameworks can be investigated, topological stability, connectivity, seamless data sharing, dynamic optimization and other problems are solved by applying a concurrent physical design, and, ensuring physical stability, optimality and control for components.

Optimal topologies are engineered as high-performance of the shelf hardware and software. In previous chapters, we analyze steady-state and dynamic governance of decoupled and aggregated nodes $N_{t_{j=1,\ldots,z}}$, cells $C_{t_{k=1,\ldots,h}}$ and links $L_{t_{l=1,\ldots,m}}$.

Example 6.54. Physical systems are designed such that physical controllability is guaranteed. Considering a simplified mathematical formulation, controllability and observability of dynamic networks $\mathcal{N}_t$ is usually simplified to conjectural descriptions. For example, as stated in Definitions 5.2 and 5.3, system $\dot{x}=Ax+Bu$ is controllable and observable if the controllability and observability matrices $C=[B\ AB\ \ldots\ A^{n-1}B]$, $C\in\mathbb{R}^{n\times n}$ and $O=[H\ HA\ \ldots\ HA^{n-1}]^T$, $O\in\mathbb{R}^{n\times n}$ have rank n. A linear system (6.161) is controllable and observable if the pairs (A,B) and (H,A) are controllable and observable. The controllability of nonlinear $\mathcal{N}_t$, described by (6.161), is studied using the Lie bracket operator $\left[\text{ad}_{\mathbf{F}}^k\mathbf{B}\right]=\left[\mathbf{F}\ldots[\mathbf{F},\mathbf{B}]\right]$. One applies $C=\left[\mathbf{B}\ldots[\text{ad}_{\mathbf{F}}^k\mathbf{B}]\right]$. The controllability is ensured if C spans q space with the rank q. ■

Multiple-Level Hierarchical Systems – Physical systems can be partitioned into subsystems, and, aggregated using hierarchical decoupling and aggregation. High fidelity is ensured by performing functional organization and connectivity assessment, multivariate decoupling, dimensionality and truncated dynamic mode analyses, etc. Structural decoupling and aggregation are critical. Design and implementation of centralized and decentralized control schemes may ensure optimality. Due to fast transients of microelectronic components and sensors, one considers pertained nonlinearities and dominant dynamics of decoupled and aggregated subsystems. Control laws, algorithms and protocols must be consistent with device physics and compliant with sensing and microelectronics hardware.

As discussed in chapter 1, performance and capabilities variables $(\boldsymbol{p},\boldsymbol{c})\in\mathcal{P}\times C$ are measurable and physically assessed, while the minimax optimization problem

$$J=\min_{\boldsymbol{p}\in\mathcal{P},\boldsymbol{c}\in C}\max W(\boldsymbol{p},\boldsymbol{c}),\ J:\mathcal{P}\times C\to\mathbb{R},\ \boldsymbol{p}=f_p(\boldsymbol{q},\boldsymbol{y},\boldsymbol{x},\boldsymbol{u}),\ \boldsymbol{c}=f_c(\boldsymbol{q},\boldsymbol{y},\boldsymbol{x},\boldsymbol{u}), \quad (6.165)$$

$$(\boldsymbol{q},\boldsymbol{y},\boldsymbol{x},\boldsymbol{u})\in Q\times Y\times X\times U,\ L_{\min}\le L(\boldsymbol{p},\boldsymbol{c})\le L_{\max},(l_p,l_c)\ge 0,$$

may not admit a solution due to incompleteness, uncertainties and complexity.

6.13. Physics and Essence of Control

To approach design and optimization of physical systems, we investigated mechanical and electromechanical systems, motion devices, sensors, power electronics and other components. The essence of optimization and control from design and hardware standpoints are studied. Control laws are designed and implemented by analog operational amplifiers, analog electronics, as well as using digital ASICs and MPUs. Optimal control laws are derived for physical systems. Distinct mathematical models with different level of hierarchy, consistency, details and fidelity are used.

Control variables u are physical quantities. From classical mechanics, u may imply the applied force F or torque T. Using the Newtonian and Lagrangian mechanics, we considered the *force* and *torque control* problems. To develop F and T, actuators are used. For electromagnetic, electrostatic, piezoelectric and other actuators, the applied voltage V and fed current i are considered as controls u. Actuators are controlled by using power amplifiers, PWM *controllers-drivers*, etc. The power transistors are switched (controlled) by the signal-level voltage u_c, which is the output of MOSFET *drivers*, comparator or ICs. One may continue this control path further. We focused on adequate control fidelity for devices and systems, while using consistent control hardware to implement control algorithms.

To control systems, analog and digital control laws are applied. Digital-to-analog and analog-to-digital converters support data processing, sensing and control schemes. The MPUs and FPGAs output analog, digital and PWM signals. Control laws are reported, such as PID, linear and nonlinear, *minimal complexity* and others implemented by regulators. Discontinuous control laws and relay switching have limited practicality due to abstract problem formulation under hypothetical assumptions. For example, it is impossible to develop applied force F and torque T with high frequency switching activity. Furthermore, the force and torque *limiters*, *absorbers* and *couplers* are used to reduce the abrupt force, torque, load and disturbances.

The *LC* and *LCL* filters are used to filter the PWM voltage applied to electromagnetic and electromechanical actuator windings. The switching frequency of transistors is limited, and, there is a continuous transient dynamics of switching in MOSFETs, comparators, etc. Due to device physics, the output PWM voltage waveforms are continuous, and, not trains of discontinuous pulses. Even if intended, one cannot generate high-frequency discontinuous voltage and current switching. The settling time of low-power and power transistors is from nanoseconds to microseconds. The designer always intents to ensure continuous *soft* switching of power transistors to minimize losses, ensure efficiency, minimize loads, etc. The *LC* output filters supply the adequate voltage to actuators, inverters and other loads, thereby minimizing current ripple.

Practical stabilizing and tracking control laws were designed which can be implemented by analog and digital controllers. Design consistency and hardware compliance are addressed, and, must be guaranteed.

6.14. Practice Problems

Practice Problem 6.1.
For a nonlinear system

$$\dot{x}_1 = -x_1 - x_1|x_1| - x_1^5 - x_2 - x_2^3 + x_1^2 u,$$
$$\dot{x}_2 = x_1,$$

study open-loop system stability, and, design a minimum-time control law by minimizing a functional

$$J = \min_{x,u} \int_{t_0}^{t_f} 1dt$$

subject to a system with limits $\mathcal{U}$=±1.

Stability Analysis – For a positive definite Lyapunov function

$$V_L = \tfrac{1}{2}x_1^2 + x_1x_2 + x_2^2,$$

the total derivative is

$$\frac{dV_L}{dt} = \frac{\partial V_L}{\partial x_1}\left(-x_1 - x_1|x_1| - x_1^5 - x_2 - x_2^3\right) + \frac{\partial V_L}{\partial x_2}x_1$$
$$= -x_1^2|x_1| - x_1^6 - x_1x_2^3 - x_1x_2|x_1| - x_1^5x_2 - x_2^2 - x_2^4.$$

The open-loop system is asymptotically stable because $V_L(x_1,x_2)>0$ and $dV_L/dt<0$, $\forall(x_1,x_2)\neq 0$.

Control Laws Design – Using calculus of variations, the switching curve and relay control can be derived if differential equations are separable and admit explicit solutions. For u=±1, assuming u=1 and u=–1, one cannot solve equations by separating variables. Apply the Hamilton optimization.

Minimizing a functional $J = \min_{x,u} \int_{t_0}^{t_f} 1dt$, subject to a systems, a scalar Hamiltonian function

$$H(x,u,\tfrac{\partial V}{\partial x}) = 1 + \frac{dV}{dx_1}\left(-x_1 - x_1|x_1| - x_1^5 - x_2 - x_2^3 + x_1^2 u\right) + \frac{dV}{dx_2}x_1 \cdot$$

Recalling that $u^2\equiv 1$. The stationarity condition yields a discontinuous control law

$$u = -\operatorname{sgn}\left(x_1^2 \frac{\partial V}{\partial x_1}\right).$$

To solve the Hamilton-Jacobi equation, approximate the sigmoid function by a continuous hyperbolic function $\Phi(x)=\tanh(ax)$, $-1<\Phi<1$, $a\gg 1$. That is, discontinuous limits $\mathcal{U}$=±1 are mapped by continuously differentiable function $\Phi\in C^k$, $k\geq 1$ with range (–1, 1), . Recall that physical systems exhibit continuous switching $\mathcal{U}\cong\pm 1$, and, using $\Phi(x)=\tanh(ax)$, $-1<\Phi<1$, $a\gg 1$, $\Phi\subseteq\mathcal{U}_\varepsilon$. Solution of an equation

$$\tanh^2\left(ax_1^2\frac{dV}{dx_1}\right) + \frac{dV}{dx_1}\left(-x_1 - x_1|x_1| - x_1^5 - x_2 - x_2^3 - x_1^2\tanh\left(ax_1^2\frac{dV}{dx_1}\right)\right) + \frac{dV}{dx_2}x_1 = 0, a\gg 1, \tanh(ax)\approx ax,$$

is approximated by a quadratic return function $V = \tfrac{1}{2}k_{11}x_1^2 + k_{12}x_1x_2 + \tfrac{1}{2}k_{22}x_2^2$.

Grouping the terms for x_1^2, x_1x_2 and x_1^6, we have three equations with three unknowns

$$-k_{11}+k_{12}=0,\ -k_{11}+k_{22}-k_{12}=0,\ a^2k_{11}^2 - ak_{11}^2 - k_{11} = 0.$$

Hence, $k_{11}=\dfrac{1}{a(a-1)}$, $k_{12}=k_{11}$ and $k_{22}=2k_{11}$, $(k_{11},k_{12},k_{22})>0$.

Selecting $a\gg 1$, and computes $(k_{11}, k_{12}, k_{22})>0$.

The continuous and discontinuous control laws are

$u=-\tanh\left[ax_1^2(k_{11}x_1+k_{12}x_2)\right]$, $u=-\mathrm{sgn}\left[x_1^2(k_{11}x_1+k_{12}x_2)\right]$.

The closed-loop systems are stable with designed control laws. The approximation error is solution of algebraic equation can be evaluated for a given X. One can be analyzed dynamics of the closed-loop systems, control switching activities, control energies, etc. ■

Practice Problem 6.2.

Consider a nonlinear electric drive with a dc servomotor, assuming that all parameters are ones. A nonlinear model is

$$\frac{di}{dt}=-i-i^3-\omega+u,\ \frac{d\omega}{dt}=i-\omega-\omega^3.$$

Use a positive definite Lyapunov function $V=\frac{1}{2}\left(i^2+\omega^2\right)>0$, consistent with the kinetic energy. The total derivative of $V(i,\omega)$ is

$$\frac{dV}{dt}=\frac{dV}{di}\frac{di}{dt}+\frac{dV}{d\omega}\frac{d\omega}{dt}=i\left(-i-i^3-\omega\right)+\omega\left(i-\omega-\omega^3\right)=-i^2-i^4-\omega^2-\omega^4<0.$$

The system is asymptotically stable because for $V(i,\omega)>0$, $dV(i,\omega)/dt<0$.

Surfaces for $V(i,\omega)$ and dV/dt are computed and documented in Figure 6.34.a.

```
x=linspace(-1,1,50); y=x; [X,Y]=meshgrid(x,y); V=(X.^2+Y.^2)./2 ;
figure(1); surf(x,y,V); xlabel('{\iti}','FontSize',20); ylabel('{\it\omega}','FontSize',20);
dV=-X.^2-X.^4-Y.^2-Y.^4;
figure(2); surf(x,y,dV); xlabel('{\iti}','FontSize',20); ylabel('{\it\omega}','FontSize',20);
```

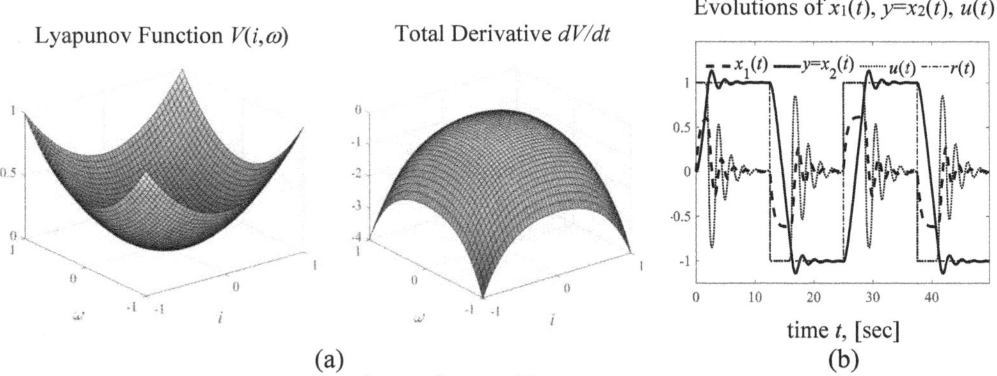

Figure 6.34. (a) Surfaces for $V=\frac{1}{2}\left(i^2+\omega^2\right)$ and $\frac{dV}{dt}=-i^2-i^4-\omega^2-\omega^4$;

(b) Closed-loop system dynamics with a constrained *minimal complexity* control law, $r(t)=\pm1$, Practice Problem 6.14.5. ■

Practice Problem 6.3.

Consider an axial topology hard disk drive actuator, described by

$$\frac{di}{dt}=-i-i^3-\omega+u,\ \frac{d\omega}{dt}=i-\omega-\omega^3-\theta-\theta^3,\ \frac{d\theta}{dt}=\omega.$$

Study the open-loop system stability using a Lyapunov function $V=\frac{1}{2}\left(i^2+\omega^2+2\omega\theta+\theta^2\right)>0$.

The total derivative of $V(i,\omega,\theta)$ is

$$\frac{dV}{dt}=\frac{dV}{di}\frac{di}{dt}+\frac{dV}{d\omega}\frac{d\omega}{dt}+\frac{dV}{d\theta}\frac{d\theta}{dt}=i\left(-i-i^3-\omega\right)+\left(\omega+\theta\right)\left(i-\omega-\omega^3-\theta-\theta^3\right)+\left(\omega+\theta\right)\omega$$

$$=-i^2-i^4-\omega^4-\omega\theta^3+i\theta-\omega\theta-\theta\omega^3-\theta^2-\theta^4<0.$$

The open-loop system is asymptotically stable because for $V(i,\omega,\theta)>0$, $dV(i,\omega,\theta)/dt<0$.■

Practice Problem 6.4.

Study stability of a system

$$\dot{x}_1(t)=-x_1+10x_2,$$

$$\dot{x}_2(t)=-10x_1-x_2^7.$$

Consider a scalar *Lyapunov* function $V(x_1,x_2)=\frac{1}{2}\left(x_1^2+x_2^2\right)>0$.

The total derivative is

$$\frac{dV(x_1,x_2)}{dt}=x_1\dot{x}_1+x_2\dot{x}_2=x_1(-x_1+10x_2)+x_2(-10x_1-x_2^7)=-x_1^2-x_2^8,\ \frac{dV(x_1,x_2)}{dt}<0.$$

The system is asymptotically stable. The quadratic positive definite function $V(x_1,x_2)=\frac{1}{2}\left(x_1^2+x_2^2\right)$ is the Lyapunov function. ■

Practice Problem 6.5.

Solve a tracking control problem for rotational motion of an aerial vehicle with a saturation limit on control $-1\le u\le 1$. Depart from a solution found Example 6.47. Our objective is to design a *minimal complexity* tracking control law using the tracking error $e(t)$, which is directly measured. The governing equations are

$$\dot{x}_1=-x_1-x_1\left|x_1\right|+u,$$

$$\dot{x}_2=x_1,\ y=x_2,$$

$$\dot{x}_e=r-y=r-x_2,$$

$$\dot{e}=-e+\dot{r}-\dot{x}_2=-e+\dot{r}-x_1.$$

To design a control law for the extended model $\dot{\mathbf{x}}(t)=\begin{bmatrix}\dot{x}(t)\\ \dot{x}_e(t)\\ \dot{e}(t)\end{bmatrix}=\mathbf{A}\mathbf{x}+A_r r+\mathbf{B}u$, a function is

$$V_{\mathbf{m}}(\mathbf{x})=k_{11}c(x_1)+k_{13}x_1x_e+k_{14}x_1e+\frac{1}{2}(k_{22}x_2^2+k_{33}x_e^2+k_{44}e^2),\ V_{\mathbf{m}}(\mathbf{x})>0.$$

With limits $-1\le u\le 1$, design a bounded control. A Lyapunov pair is $\{V_{\mathbf{m}}(\mathbf{x})>0,\frac{dV_{\mathbf{m}}}{dt}<0\}$.

Let the specified rate of change for $V_{\mathbf{m}}(\mathbf{x})$, which assesses optimality and stability, is

$$\frac{dV_{\mathbf{m}}(x,e)}{dt}\le-\frac{1}{2}\left(0.0001x_1^2+0.0001x_2^2+0.0001x_e^2+100e^2\right)-\int\tanh^{-1}u\,du.$$

That is, strengthen specifications are imposed on the tracking error $e(t)$. Apply a quadratic approximation $\int\tanh\left(\mathbf{B}^T\frac{\partial V_{\mathbf{m}}}{\partial\mathbf{x}}\right)d\left(\mathbf{B}^T\frac{\partial V_{\mathbf{m}}}{\partial\mathbf{x}}\right)=\ln\cosh\left(\mathbf{B}^T\frac{\partial V_{\mathbf{m}}}{\partial\mathbf{x}}\right)\approx\frac{1}{3}\left(\frac{\partial V_{\mathbf{m}}}{\partial x_1}\right)^2$, and, solve the resulting quadratic equations.

Grouping the terms one finds algebraic equations. For example, for x_e^2 and e^2, we have

$$-\tfrac{1}{3}k_{13}^2+\tfrac{1}{2}0.0001=0,\ -\tfrac{1}{3}k_{14}^2+k_{44}+\tfrac{1}{2}100=0.$$

We have, $k_{13}=-0.0122$ and $k_{14}=-7.07$.

Approximation $\int \tanh\left(\mathbf{B}^T\frac{\partial V_{\mathbf{m}}}{\partial \mathbf{x}}\right)d\left(\mathbf{B}^T\frac{\partial V_{\mathbf{m}}}{\partial \mathbf{x}}\right)=\ln\cosh\left(\mathbf{B}^T\frac{\partial V_{\mathbf{m}}}{\partial \mathbf{x}}\right)\approx\frac{1}{3}\left(\frac{\partial V_{\mathbf{m}}}{\partial x_1}\right)^2$ is satisfied with the state transitions $\mathbf{x}\in X$, chosen $V_{\mathrm{m}}(\mathbf{x})$ and computed (k_{ii},k_{ij}).

The *minimal complexity* bounded control law (6.114) is

$$u=\Phi\left(\varphi\left(\int edt,e\right)\right)=-\Phi\left(\mathbf{B}^T\frac{\partial V_{\mathbf{m}}}{\partial x_1}\right)=-\Phi\left(\frac{\partial V_{\mathbf{m}}}{\partial x_1}\right)=\tanh\left(0.0122\int edt+7.07e\right),\ -1<u<1.$$

The closed-loop system is simulated for reference $r(t)=\pm1$. Dynamics of (x_1,x_2) and $y=x_2$ are illustrated in Figure 6.34.b. ■

Practice Problem 6.6. *Tracking Control of a Fighter Aircraft With Control Limits*

In Section 5.11.2, tracking control laws were designed for a multirole fighter aircraft. Displacements of aerodynamic control surfaces are limited. Consider symmetric saturation limits $-u_{i\,\max}\le u_i\le u_{i\,\max}$ for the horizontal stabilizer δ_s, flaperon δ_f, canard δ_c and rudder δ_r.

We have $|\delta_s|\le0.5$ rad, $|\delta_f|\le0.5$ rad, $|\delta_c|\le0.6$ rad and $|\delta_r|\le0.6$ rad.

A linearized model $\dot{x}(t)=Ax+Bu$, see Section 5.11.2, with limits $-u_{i\,\max}\le u_i\le u_{i\,\max}$, is

$$\dot{x}=Ax+Bu,\ x=\begin{bmatrix}x_1\\x_2\\x_3\\x_4\\x_5\\x_6\\x_7\\x_8\\x_9\end{bmatrix}=\begin{bmatrix}V\\\alpha\\q\\\theta\\\beta\\p\\r\\\phi\\\psi\end{bmatrix},\ u=\begin{bmatrix}u_1\\u_2\\u_3\\u_4\end{bmatrix}=\begin{bmatrix}\delta_s\\\delta_f\\\delta_c\\\delta_r\end{bmatrix},\ |u_1|\le0.5\text{ rad},\ |u_2|\le0.5\text{ rad},\ |u_3|\le0.6\text{ rad},\ |u_4|\le0.6\text{ rad},$$

$$A=\begin{bmatrix}-0.0099&-9.81&0&9.81&0&0&0&0&0\\-0.00013&0.00054&1&-0.00054&-0.0252&0&0&0&0\\0&0&0&0&0&0.0212&0.0207&0&0\\0&0&1&0&0&0&0&-0.025&0\\0&0.0252&0&0&0.00054&0.01&-1&0&0\\0&0&0.0117&0&0&-0.00007&0.0118&0&0\\0&0&0.0234&0&0&0.0235&-0.00007&0&0\\0&0&0&0.025&0&1&0.025&0.00063&0\\0&0&0&0.00063&0&0&1&0.025&0\end{bmatrix},\ B=\begin{bmatrix}-0.0008&-0.0005&-0.0001&-0.0007\\0.093&0&-0.024&0\\5.3&0&1.9&0\\0&0&0&0\\0&0&0&0\\0.053&6.8&0&0.07\\0.081&0.07&0&8.2\\0&0&0&0\\0&0&0&0\end{bmatrix}.$$

The outputs are Euler angles (θ,ϕ,ψ). Hence, $y=[\theta,\ \phi,\ \psi]$.

The tracking errors are $e(t)=\begin{bmatrix}e_1\\e_2\\e_3\end{bmatrix}=\begin{bmatrix}r_1-y_1\\r_2-y_2\\r_3-y_3\end{bmatrix}=\begin{bmatrix}r_1-x_4\\r_2-x_8\\r_3-x_9\end{bmatrix}=\begin{bmatrix}r_1-\theta\\r_2-\phi\\r_3-\psi\end{bmatrix}$.

The governing equations for *error states* $x_e(t)$ and tracking error vector $e(t)$ are

$\dot{x}_e(t)=I_X(r-y)=I_X(r-Hx)$, $e(t)\equiv\dot{x}_e(t)$, $x_e\in\mathbb{R}^3$, $I_X\in\mathbb{R}^{3\times3}$, $I_X=I$,

$\dot{e}(t)=-I_Ee+\dot{r}-\dot{y}=-I_Ee+\dot{r}-HAx-HBu$, $e\in\mathbb{R}^3$, $I_E\in\mathbb{R}^{3\times3}$, $I_E=I$.

The extended model is

$$\dot{\mathbf{x}}(t)=\begin{bmatrix}\dot{x}(t)\\ \dot{x}_e(t)\\ \dot{e}(t)\end{bmatrix}=\mathbf{A}\mathbf{x}+A_r r+\mathbf{B}u=\begin{bmatrix}A & 0 & 0\\ -I_X H & 0 & 0\\ -HA & 0 & -I_E\end{bmatrix}\mathbf{x}+\begin{bmatrix}0 & 0\\ I_X & 0\\ 0 & I\end{bmatrix}\begin{bmatrix}r\\ \dot{r}\end{bmatrix}+\begin{bmatrix}B\\ 0\\ -HB\end{bmatrix}u,$$

$\mathbf{A}\in\mathbb{R}^{15\times 15}$, $A_r\in\mathbb{R}^{15\times 6}$, $\mathbf{B}\in\mathbb{R}^{15\times 4}$.

Describe saturation limits $-u_{i\max}\le u_i\le u_{i\max}$ as $\Phi_i=u_{i\max}\tanh(a_i x)$, $\forall a_i=1$.

Minimize a nonquadratic performance functional (6.67)

$$J=\min_{\mathbf{x},\,|u|\le \boldsymbol{u}_{\max}}\int_0^{\infty}\left[\tfrac{1}{2}\mathbf{x}^T Q\mathbf{x}+\int\left(\boldsymbol{a}^{-1}\tanh^{-1}(\boldsymbol{u}_{\max}{}^{-1}u)\right)^T G\,du\right]dt,\ \boldsymbol{u}_{\max}\in\mathbb{R}^{4\times 4},\ \boldsymbol{a}=I,\ \boldsymbol{a}\in\mathbb{R}^{4\times 4}$$

subject to a MIMO system.

Using the stationarity condition for Hamiltonian, the constrained control law (6.69) is

$$u=-\boldsymbol{u}_{\max}\tanh\left(\boldsymbol{a}G^{-1}\mathbf{B}^T\frac{\partial V}{\partial \mathbf{x}}\right),\ \boldsymbol{u}_{\max}=\begin{bmatrix}u_{1\max} & 0 & 0 & 0\\ 0 & u_{2\max} & 0 & 0\\ 0 & 0 & u_{3\max} & 0\\ 0 & 0 & 0 & u_{4\max}\end{bmatrix}.$$

Solution of (6.72)

$$\tfrac{1}{2}\mathbf{x}^T Q\mathbf{x}+\left(\frac{\partial V}{\partial \mathbf{x}}\right)^T\mathbf{A}\mathbf{x}-\int\left[\boldsymbol{u}_{\max}\tanh\left(\boldsymbol{a}G^{-1}\mathbf{B}^T\frac{\partial V}{\partial \mathbf{x}}\right)\right]^T d\left(\mathbf{B}^T\frac{\partial V}{\partial \mathbf{x}}\right)=0,$$

$$\int\left[\boldsymbol{u}_{\max}\tanh\left(\boldsymbol{a}G^{-1}\mathbf{B}^T\frac{\partial V}{d\mathbf{x}}\right)\right]^T d\left(\mathbf{B}^T\frac{\partial V}{d\mathbf{x}}\right)\approx\frac{9}{20}\frac{\partial V}{d\mathbf{x}}^T\mathbf{B}\boldsymbol{u}_{\max}\boldsymbol{a}G^{-1}\mathbf{B}^T\frac{\partial V}{d\mathbf{x}}$$

is found using the quadratic return function (6.73) $V(\mathbf{x})=\frac{1}{2}\mathbf{x}^T K\mathbf{x}$.

Let $q_{1,1}=q_{2,2}=q_{3,3}=q_{4,4}=q_{5,5}=q_{6,6}=q_{7,7}=q_{8,8}=q_{9,9}=0.01$ and $q_{10,10}=q_{11,11}=10$, $q_{12,12}=50$, $q_{13,13}=q_{14,14}=1$, $q_{15,15}=10$. For $Q\in\mathbb{R}^{15\times 15}$, $G=I$, $G\in\mathbb{R}^{4\times 4}$ and $\boldsymbol{a}=I$, $\boldsymbol{a}\in\mathbb{R}^{4\times 4}$.

The tracking control law is

$$u=-\boldsymbol{u}_{\max}\tanh(K_F\mathbf{x})=-\begin{bmatrix}u_{1\max} & 0 & 0 & 0\\ 0 & u_{2\max} & 0 & 0\\ 0 & 0 & u_{3\max} & 0\\ 0 & 0 & 0 & u_{4\max}\end{bmatrix}\tanh\left(K_F\begin{bmatrix}x\\ \int e_1 dt\\ \int e_2 dt\\ \int e_3 dt\\ e_1\\ e_2\\ e_3\end{bmatrix}\right),\ x=\begin{bmatrix}x_1\\ x_2\\ x_3\\ x_4\\ x_5\\ x_6\\ x_7\\ x_8\\ x_9\end{bmatrix},\ u=\begin{bmatrix}u_1\\ u_2\\ u_3\\ u_4\end{bmatrix},$$

and, the feedback gain matrix K_F is

$$K_F=\begin{bmatrix}-0.0464 & 3.728 & 1.68 & 0.267 & 1.997 & 0.0091 & 0.0136 & -0.025 & 2.037 & -4.335 & 0.0907 & 0.0194 & -0.0915 & -0.00075 & -0.0031\\ -0.005 & 0.202 & -0.0083 & -0.202 & 0.362 & 1.66 & 0.0033 & 4.076 & 0.3 & 0.0499 & -4.7 & 0.0336 & 0.00073 & -0.103 & -0.0002\\ 0.11 & -8.787 & 0.74 & 10.24 & -4.436 & -0.003 & -0.0021 & 0.0178 & -4.455 & -1.689 & -0.31 & -0.188 & -0.0327 & -0.00058 & -0.0011\\ -0.0023 & 0.0861 & -0.008 & -0.11 & 0.111 & 0.0025 & 1.803 & 0.0378 & 6.77 & 0.0258 & -0.0075 & -9.621 & 0.00041 & -0.00065 & -0.499\end{bmatrix}.$$

Figure 6.35 depicts evolutions of outputs (θ,ϕ,ψ) for references $r(t)=\begin{bmatrix}r_1(t)\\ r_2(t)\\ r_3(t)\end{bmatrix}=\begin{bmatrix}\theta_{\text{ref}}(t)\\ \phi_{\text{ref}}(t)\\ \psi_{\text{ref}}(t)\end{bmatrix}=\begin{bmatrix}\pm 0.4\text{ rad}\\ \mp 0.75\text{ rad}\\ \mp 1\text{ rad}\end{bmatrix}$. Coordinated maneuvers, flight path, adequate dynamics and stability are achieved despite limits on displacement of aerodynamic control surfaces $(\delta_s,\delta_f,\delta_c,\delta_r)$. Lateral and longitudinal decoupling, coordinated attitude control and other problems may be approached and solved.

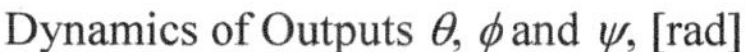

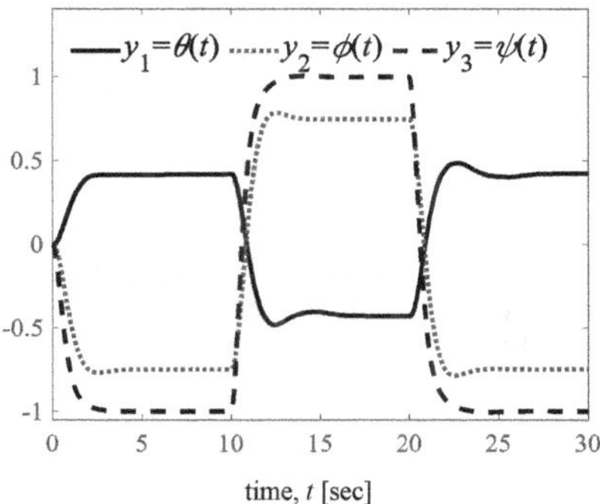

Figure 6.35. Aircraft outputs $y_1=x_4=\theta$, $y_2=x_8=\phi$, $y_3=x_9=\psi$ if $r(t)=\begin{bmatrix} r_1(t) \\ r_2(t) \\ r_3(t) \end{bmatrix}=\begin{bmatrix} \pm 0.4 \text{ rad} \\ \mp 0.75 \text{ rad} \\ \mp 1 \text{ rad} \end{bmatrix}$. ■

Practice Problem 6.7. *Nonlinear Analysis and Control of a Fighter Aircraft*

In Practice Problem 6.14.6, for a linear model, a bounded tracking control law was designed. Study a nonlinear coupled aerodynamics. Aerodynamic thrust vectoring is accomplished by steering the jet nozzle, airfoils which are mounted behind the nozzle, as well as exhaust vanes, thereby regulating $(F_X,F_Y,F_Z)_{\text{thrust}}$. Thrust vector control empowers aircraft controllability, agility and maneuverability by regulating the direction of thrust.

Simplify a problem by investigating multi-axis pitch-roll-yaw attitude control by means of steering aerodynamic control surfaces, which are stabilizers, flaperons and rudder. For a supersonic multirole fighter airplane, flying at 400 m/s, elevation 5000 m, near-zero Euler angles, consider a nonlinear model covered in section 2.4 with states, controls and outputs as

$$x=[\alpha, q, \theta, \beta, p, r, \phi, \psi],\ u=[\delta_s, \delta_f, \delta_r],\ y=[\theta, \phi, \psi].$$

Using equations of motion (2.52), at constant forward velocity V=400 m/s, a nonlinear model is

$$\dot{x}(t)=F(x)+Bu,\quad \begin{bmatrix} u_1 \\ u_2 \\ u_3 \end{bmatrix}=\begin{bmatrix} \delta_s \\ \delta_f \\ \delta_r \end{bmatrix},$$

$$\begin{bmatrix} \dot{\alpha}(t) \\ \dot{q}(t) \\ \dot{\theta}(t) \\ \dot{\beta}(t) \\ \dot{p}(t) \\ \dot{r}(t) \\ \dot{\phi}(t) \\ \dot{\psi}(t) \end{bmatrix}=\begin{bmatrix} q-(p\cos\alpha+r\sin\alpha)\tan\beta+\frac{g}{V\cos\beta}(\cos\alpha\cos\theta\cos\phi+\sin\alpha\sin\theta) \\ \frac{1}{I_{yy}}\left[(-I_{xx}+I_{zz})pr-I_{xz}(p^2-r^2)\right] \\ q\cos\phi-r\sin\phi \\ p\sin\alpha-r\cos\alpha+\frac{g}{V}(\cos\theta\cos\beta\sin\phi+\cos\alpha\sin\theta\sin\beta-\sin\alpha\cos\theta\sin\beta\cos\phi) \\ \frac{1}{I_{xx}I_{zz}-I_{xz}^2}\left[I_{xz}(I_{xx}-I_{yy}+I_{zz})qp-(I_{xz}^2-I_{yy}I_{zz}+I_{zz}^2)qr\right] \\ \frac{1}{I_{xx}I_{zz}-I_{xz}^2}\left[-(I_{xx}^2-I_{xx}I_{yy}+I_{xz}^2)qp+I_{xz}(I_{xx}-I_{yy}+I_{zz})qr\right] \\ p+q\tan\theta\sin\phi+r\tan\theta\cos\phi \\ q\sec\theta\sin\phi+r\sec\theta\cos\phi \end{bmatrix}+\begin{bmatrix} -0.025 & 0 & 0 \\ 0.57 & -0.09 & 0 \\ 0 & 0 & 0 \\ 0.0039 & -0.0062 & 0.01 \\ 0 & 2.4 & 0.06 \\ 0 & -0.007 & 5.8 \\ 0 & 0 & 0 \\ 0 & 0 & 0 \end{bmatrix}\begin{bmatrix} \delta_s \\ \delta_f \\ \delta_r \end{bmatrix},$$

where I_{xx}=2.7×10^4 kg-m^2, I_{xy}=0, I_{xz}= –3500 kg-m^2, I_{yy}=2.1×10^5 kg-m^2, I_{yz}=0 and I_{zz}=1.8×10^5 kg-m^2.

Linearize a nonlinear unstable model for a given equilibrium

$\bar{x}$ =[0.015 rad, 0.1 rad/s, 0.02 rad, 0, –0.1 rad/s, 0.1 rad/s, 0 rad, 0 rad].

For $\dot{x}=Ax+Bu$, find $A\in\mathbb{R}^{8\times 8}$. Linearization is performed using MATLAB as

```
syms V g Ixx Ixy Ixz Iyy Iyx Izz;syms xd1 xd2 xd3 xd4 xd5 xd6 xd7 xd8;syms x1 x2 x3 x4 x5 x6 x7 x8;
xd1=x2-x5*cos(x1)*tan(x4)-x6*sin(x1)*tan(x4)+(g/(V*cos(x4)))*(cos(x1)*cos(x3)*cos(x7)+sin(x1)*sin(x3));
xd2=(1/Iyy)*((-Ixx+Izz)*x5*x6-Ixz*(x5^2-x6^2));
xd3=x2*cos(x7)-x6*sin(x7);
xd4=x5*sin(x1)-x6*cos(x1)+(g/V)*(cos(x3)*sin(x4)*sin(x7)+cos(x1)*sin(x3)*sin(x4)-sin(x1)*cos(x3)*sin(x4)*cos(x7));
xd5=(1/(Ixx*Izz-Ixz^2))*(Ixz*(Ixx-Iyy+Izz)*x2*x5-(Ixz^2-Iyy*Izz+Izz^2)*x2*x6);
xd6=(1/(Ixx*Izz-Ixz^2))*(-(Ixx^2-Ixx*Iyy+Ixz^2)*x2*x5+Ixz*(Ixx-Iyy+Izz)*x2*x6);
xd7=x5+x2*tan(x3)*sin(x7)+x6*tan(x3)*cos(x7);
xd8=x2*sec(x3)*sin(x7)+x6*sec(x3)*cos(x7);
equations=[xd1 xd2 xd3 xd4 xd5 xd6 xd7 xd8];
x=[x1 x2 x3 x4 x5 x6 x7 x8]; % States
V=400; g=9.81; Ixx=2.7e4; Ixy=0; Ixz=-3500; Iyy=2.1e5; Iyx=0; Izz =1.8e5; % Constants
A_jacobian=jacobian(equations,x);
x1=0.015; x2=0.1; x3=0.02; x4=0; x5=-0.1; x6=0.1; x7=0; x8=0; % Equilibria
A=double(subs(A_jacobian)) % Matrix A
```

This yields
$$A=\begin{bmatrix} 0.00012 & 1 & -0.00012 & 0.0985 & 0 & 0 & 0 & 0 \\ 0 & 0 & 0 & 0 & 0.0695 & -0.076 & 0 & 0 \\ 0 & 1 & 0 & 0 & 0 & 0 & -0.1 & 0 \\ -0.0985 & 0 & 0 & 0.00012 & 0.015 & -1 & 0 & 0 \\ 0 & 0.111 & 0 & 0 & -0.00022 & 0.111 & 0 & 0 \\ 0 & -0.102 & 0 & 0 & 0.102 & 0.00022 & 0 & 0 \\ 0 & 0 & 0.1 & 0 & 1 & 0.02 & 0.002 & 0 \\ 0 & 0 & 0.002 & 0 & 0 & 1 & 0.1 & 0 \end{bmatrix}.$$

Linear Quadratic Regulator Problem – Design a tracking control law for linearized system by minimizing a quadratic performance functional

$$J=\min_{\mathbf{x},u}\int_0^{\infty}\tfrac{1}{2}\left(\mathbf{x}^T Q\mathbf{x}+u^T Gu\right)dt .$$

The system and tracking error dynamics with states (x,x_e,e) is described by $\dot{\mathbf{x}}(t)=\mathbf{A}\mathbf{x}+A_r r+\mathbf{B}u$, $y=Hx$, $\mathbf{x}\in\mathbb{R}^{14}$, $u\in\mathbb{R}^3$, $y\in\mathbb{R}^3$, $\mathbf{A}\in\mathbb{R}^{14\times 14}$, $A_r\in\mathbb{R}^{14\times 6}$, $\mathbf{B}\in\mathbb{R}^{14\times 3}$, $H\in\mathbb{R}^{3\times 8}$.

The weighting matrices (Q,G) affect closed-loop system dynamics. Let in $Q\in\mathbb{R}^{14\times 14}$ the weighting coefficients for the states $x(t)$ are $q_{1,1}=q_{2,2}=q_{3,3}=q_{4,4}=q_{5,5}=q_{6,6}=q_{7,7}=q_{8,8}=0.01$. For the tracking error states $(x_e(t),e(t))$, specify $q_{9,9}=q_{10,10}=q_{11,11}=q_{12,12}=q_{13,13}=10$ and $q_{14,14}=1$. Let $G=I$, $G\in\mathbb{R}^{3\times 3}$.

A tracking control law is $u=-K_F\mathbf{x}=-K_F\begin{bmatrix} x \\ \int e_1 dt \\ \int e_2 dt \\ \int e_3 dt \\ e_1 \\ e_2 \\ e_3 \end{bmatrix}$.

The feedback gain matrix K_F is computed to be

$$K_F=\begin{bmatrix} -0.0846 & 4.541 & 5.625 & -0.015 & 0.172 & -0.0173 & -0.148 & 0.221 & -3.151 & -0.178 & -0.193 & -0.328 & -0.0166 & -0.003 \\ 0.1097 & 0.0124 & -0.113 & -0.0865 & 1.923 & 0.014 & 3.926 & -0.246 & 0.17 & -3.154 & 0.148 & 0.0248 & -0.514 & 0.0029 \\ 0.0029 & -0.159 & -0.283 & 0.0885 & 0.0787 & 0.96 & 0.291 & 2.64 & 0.2 & -0.137 & -3.153 & 0.0275 & -0.0289 & -0.0723 \end{bmatrix}.$$

To assess aircraft agility, maneuverability and stability, nonlinear model is used in simulations with the designed control law. Figure 6.36.a depicts the outputs (θ,ϕ,ψ) for $r(t)=\begin{bmatrix} r_1(t)\\ r_2(t)\\ r_3(t)\end{bmatrix}=\begin{bmatrix}\theta_{\text{ref}}(t)\\ \phi_{\text{ref}}(t)\\ \psi_{\text{ref}}(t)\end{bmatrix}=\begin{bmatrix}\pm 0.5 \text{ rad}\\ \mp 0.5 \text{ rad}\\ \mp 0.25 \text{ rad}\end{bmatrix}$.

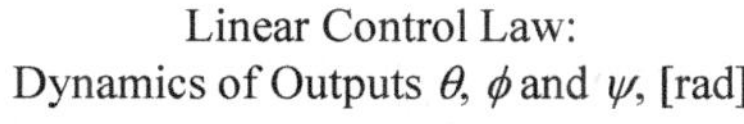

Linear Control Law:
Dynamics of Outputs θ, ϕ and ψ, [rad]

Bounded Control Law:
Dynamics of Outputs θ, ϕ and ψ, [rad]

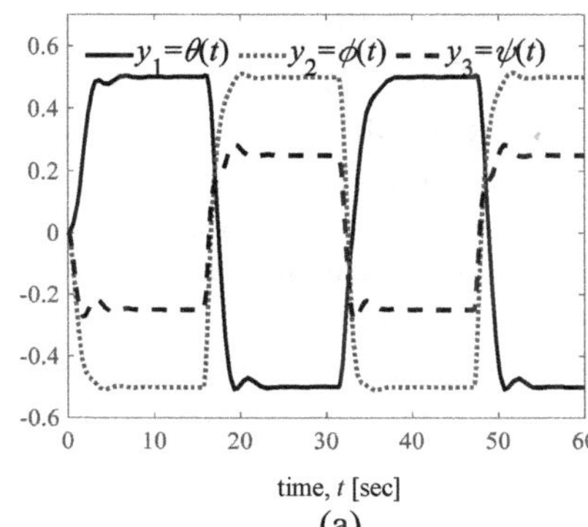

(a)

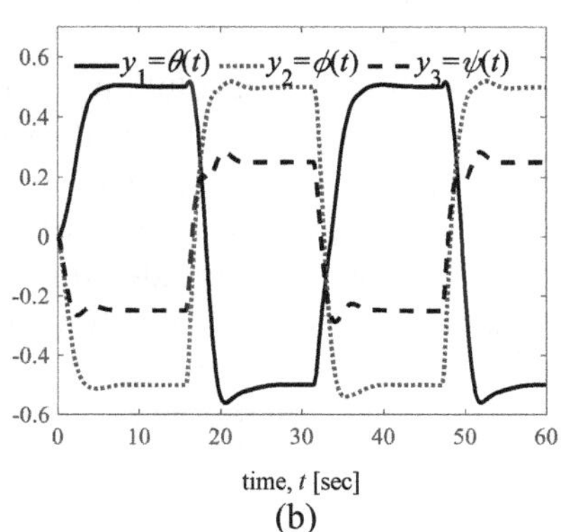

(b)

Figure 6.36. Simulation of a nonlinear system $\dot{x}(t)=F(x)+Bu$:
Output dynamics $y(t)=\begin{bmatrix} y_1(t)\\ y_2(t)\\ y_3(t)\end{bmatrix}=\begin{bmatrix}\theta(t)\\ \phi(t)\\ \psi(t)\end{bmatrix}$ for references $r(t)=\begin{bmatrix} r_1(t)\\ r_2(t)\\ r_3(t)\end{bmatrix}=\begin{bmatrix}\theta_{\text{ref}}(t)\\ \phi_{\text{ref}}(t)\\ \psi_{\text{ref}}(t)\end{bmatrix}=\begin{bmatrix}\pm 0.5 \text{ rad}\\ \mp 0.5 \text{ rad}\\ \mp 0.25 \text{ rad}\end{bmatrix}$:
(a) Nonlinear closed-loop system with a linear tracking control law $u=-G^{-1}\mathbf{B}^T K\mathbf{x}$;
(b) Nonlinear system with a constrained tracking control law $u=-\boldsymbol{u}_{\max}\tanh(\boldsymbol{a}G^{-1}\mathbf{B}^T K\mathbf{x})$.

Constrained Control Law Design – Displacements of aerodynamic control surfaces are limited. For the stabilizers, flaperons and rudder the mechanical limits $-\mathcal{U}_{i\max}\le\mathcal{U}_i\le\mathcal{U}_{i\max}$ are known. While $|\mathcal{U}_{i\min}|\neq|\mathcal{U}_{i\max}|$. assume a symmetry with $|\mathcal{U}_{i\min}|=|\mathcal{U}_{i\max}|$, and,

$|\delta_s|\le 0.4$ rad, $|\delta_f|\le 0.4$ rad, $|\delta_r|\le 0.5$ rad.

The symmetric saturation limits are described by $\Phi_i=u_{i\max}\tanh(a_i x)$, $\forall a_i=1$.

Minimize a nonquadratic performance functional (6.67)

$$J=\min_{\mathbf{x},\,|u|\le\boldsymbol{u}_{\max}}\int_0^\infty\left[\tfrac{1}{2}\mathbf{x}^T Q\mathbf{x}+\int\left(\boldsymbol{a}^{-1}\tanh^{-1}(\boldsymbol{u}_{\max}{}^{-1}u)\right)^T G\,du\right]dt,\ \boldsymbol{u}_{\max}\in\mathbb{R}^{3\times3},\ \boldsymbol{a}=I,\ \boldsymbol{a}\in\mathbb{R}^{3\times3}.$$

As reported in chapter 5 and section 6.8.1, for linear systems, positive definite weighting matrices (Q,G) with $(q_{ii},g_{jj})>0$ are usually chosen considering the expected or admissible variations of states and controls $(\tilde{x}_{i\max},\tilde{u}_{j\max})$, $q_{ii}=\dfrac{1}{\tilde{x}^2_{i\max}}$ and $g_{jj}=\dfrac{1}{\tilde{u}^2_{j\max}}$. In constrained control designs, weighing coefficients (q_{ii},g_{jj}) should be selected assessing $(\tilde{x}_{i\max},\tilde{u}^2_{j\max},\mathcal{U}_{\min}\le\mathcal{U}\le\mathcal{U}_{\max},\boldsymbol{u}_{\min}<\Phi<\boldsymbol{u}_{\max},\boldsymbol{a})$. The weighting coefficients, used solving the LQR problems, may be refined in constrained control law design. However, use the same (Q, G).

With $|u_1|\le 0.4$ rad, $|u_2|\le 0.4$ rad, $|u_3|\le 0.5$ rad, solving the constrained dynamic optimization problem, matrix $K\in\mathbb{R}^{14\times14}$ is computed. A constrained control law is

$$u = -\boldsymbol{u}_{\max} \tanh(\boldsymbol{a}G^{-1}\mathbf{B}^T K\mathbf{x}) = -\begin{bmatrix} u_{1\max} & 0 & 0 \\ 0 & u_{2\max} & 0 \\ 0 & 0 & u_{3\max} \end{bmatrix} \tanh\left(K_F \begin{bmatrix} x \\ \int e_1 dt \\ \int e_2 dt \\ \int e_3 dt \\ e_1 \\ e_2 \\ e_3 \end{bmatrix} \right).$$

Solution of (6.72) with (Q,G) reported above, yields

$$K_F = \begin{bmatrix} -0.161 & 13.3 & 13.1 & -0.0327 & 0.516 & -0.0543 & -0.513 & 0.698 & -6.092 & -0.389 & -0.485 & -0.476 & -0.0268 & -0.0058 \\ 0.217 & 0.0833 & -0.165 & -0.165 & 5.476 & 0.0447 & 8.78 & -0.613 & 0.362 & -6.103 & 0.349 & 0.0429 & -0.834 & 0.0054 \\ 0.0075 & -0.499 & -0.733 & 0.152 & 0.2327 & 2.36 & 0.623 & 5.332 & 0.452 & -0.285 & -5.451 & 0.049 & -0.052 & -0.11 \end{bmatrix}.$$

Figure 6.36.b depicts evolutions of Euler angles (θ,ϕ,ψ) for $r(t)$. A nonlinear aircraft model with a longitudinal-lateral mapping $F(x)$ and limits $-u_{i\max} \le u_i \le u_{i\max}$ is used to perform simulations. Coordinated maneuvers and flight paths are ensured in the considered flight envelope. While control limits affect aircraft capabilities, bounded control law ensures stability in considered flight envelope. System dynamics and stability margins are examined in a full operating flight envelope. Control law structure and feedback gains depend on optimization and design methods, decoupling schemes, linearization at specified equilibria $\bar{x}$, etc. Adaptive reconfigurable control and gain scheduling enhance flight envelopes.

Identification, Parameter Estimation and Reconfigurable Control – For a given model structure $(F(x),B(x))$, parameters can be estimated in near real time. One yields $\dot{x}(t)=F(x)+B(x)u$ or $\dot{x}(t)=Ax+Bu$. The proposed algorithms, covered in chapters 4 to 6, allow one to redesign and reconfigure control laws depending on flight path, conditions, missions, etc. There is a focus on dynamic characteristics and flying qualities of damaged aircraft to synthesize control laws to enable survivability, stability and controllability. If the airframe, wings and control surfaces are damaged, malfunction or faulty, the structure and parameters of governing equations are altered. The standard equations of motion with defined, fixed or variable structure can be applied. Identification and parameter estimation problems are solved in chapter 4. Adaptive flight control schemes should be developed for aircraft which may sustain mid-air collisions, aging and fatigue damages, defects, bird strike, turbulence, anti-aircraft warfare (gunfire, missiles damages and electromagnetic spectrum), etc. To redesign and reconfigure controllers, near real time identification and optimization may be performed by flight management systems.

Discretization and Digital Control Design – Discretization and design of digital control laws are performed. For given $(\mathbf{A},\mathbf{B},A_r,H,D)$ and sampling time T_s=0.01 sec, find (5.138)

$$\mathbf{x}_{n+1} = \mathbf{A}_n\mathbf{x}_n + A_{rn}r_n + \mathbf{B}_n u_n,\ \mathbf{x}_n = \begin{bmatrix} x_n \\ x_{en} \\ e_n \end{bmatrix},\ y_n = H_n x_n,\ n \ge 0.$$

Using $(\mathbf{A},\mathbf{B})$, matrices $(\mathbf{A}_n,\mathbf{B}_n)$ are found applying equations (5.19)

$$\mathbf{A}_n = e^{\mathbf{A}T_s},\ \mathbf{B}_n = \mathbf{A}^{-1}(e^{\mathbf{A}T_s} - I)\mathbf{B},\ H_n=H,\ D_n=D.$$

As reported in chapter 5, a discrete LQR is designed by minimizing a quadratic performance index

$$J = \min_{\mathbf{x}_n, u_n} \sum_{n=0}^{\infty} \left[\mathbf{x}_n^T Q_n \mathbf{x}_n + u_n^T G_n u_n \right].$$

An equation (5.137)

$-K_n + Q_n + \mathbf{A}_n^T K_n \mathbf{A}_n - \mathbf{A}_n^T K_n \mathbf{B}_n \left(G_n + \mathbf{B}_n^T K_n \mathbf{B}_n \right)^{-1} \mathbf{B}_n^T K_n \mathbf{A}_n = 0$, $Q_n=I$, $Q_n \in \mathbb{R}^{14\times 14}$, $G_n=I$, $G_n \in \mathbb{R}^{3\times 3}$

is solved, yielding K_n and feedback gain matrix in a control law

$$u_n = -\left(G_n + \mathbf{B}_n^T K_{n+1} \mathbf{B}_n \right)^{-1} \mathbf{B}_n^T K_n \mathbf{A}_n \mathbf{x}_n.$$

One finds an unconstrained control law for $Q_n=I$ and $G_n=I$ as

$$u_n = \begin{bmatrix} u_{1n} \\ u_{2n} \\ u_{3n} \end{bmatrix} = -\begin{bmatrix} -0.482 & 3.397 & 3.47 & -0.386 & 0.13 & -0.0157 & -0.12 & -0.261 & -0.975 & -0.17 & 0.03 & -0.053 & -0.0042 & -0.0037 \\ 0.935 & 0.058 & -0.961 & -0.427 & 1.675 & 0.01 & 2.135 & -0.49 & 0.165 & -0.963 & -0.0756 & 0.0053 & -0.09 & 0.00001 \\ -0.314 & -0.166 & 0.151 & 0.64 & 0.0626 & 1.304 & 0.13 & 2.786 & -0.0426 & 0.0695 & -0.958 & 0.0028 & -0.0028 & -0.0989 \end{bmatrix} x_n.$$

A constrained control problem is solved by minimizing a nonquadratic performance index

$$J = \min_{\mathbf{x}_n, u_n \in U} \sum_{n=0}^{\infty} \left[\mathbf{x}_n^T Q_n \mathbf{x}_n - u_n^T \mathbf{B}_n^T K_{n+1} \mathbf{B}_n u_n + 2\Sigma_u \left(\boldsymbol{a}^{-1} \tanh^{-1}(\boldsymbol{u}_{\max}{}^{-1} u_n) \right)^T G_n \right]$$

subject to $\mathbf{x}_{n+1} = \mathbf{A}_n \mathbf{x}_n + A_{rn} r_n + \mathbf{B}_n u_n$, $-u_{\max} \le u_n \le u_{\max}$, $u_n \in U$.

The symmetric saturation limits $-\mathcal{U}_{\max} \le \mathcal{U} \le \mathcal{U}_{\max}$ are described by

$\Phi(\mathbf{x}_n) = \boldsymbol{u}_{\max} \tanh(\boldsymbol{a}\mathbf{x}_n)$, $\Phi \subseteq \mathcal{U}_\varepsilon$.

The bounded control law (6.152) is

$$u_n = -\boldsymbol{u}_{\max} \tanh\left(\boldsymbol{a} G_n^{-1} \mathbf{B}_n^T K_n \mathbf{A}_n \mathbf{x}_n \right), \; \boldsymbol{u}_{\max} = \begin{bmatrix} u_{1\max} & 0 & 0 \\ 0 & u_{2\max} & 0 \\ 0 & 0 & u_{3\max} \end{bmatrix}, \; \boldsymbol{a} = \begin{bmatrix} 1 & 0 & 0 \\ 0 & 1 & 0 \\ 0 & 0 & 1 \end{bmatrix}.$$

Solving an equation (6.153)

$$-K_n + Q_n + \mathbf{A}_n^T K_n \mathbf{A}_n - 2k \mathbf{A}_n^T K_n \mathbf{B}_n \boldsymbol{u}_{\max} \boldsymbol{a} G_n^{-1} \mathbf{B}_n^T K_n \mathbf{A}_n = 0,$$

$k=1$, $\boldsymbol{a}=I$, $Q_n=I$, $Q_n \in \mathbb{R}^{14\times 14}$, $G_n=I$, $G_n \in \mathbb{R}^{3\times 3}$,

we find K_n and feedback gains. One has

$$u_n = \begin{bmatrix} u_{1n} \\ u_{2n} \\ u_{3n} \end{bmatrix} = -\begin{bmatrix} u_{1\max} & 0 & 0 \\ 0 & u_{2\max} & 0 \\ 0 & 0 & u_{3\max} \end{bmatrix} \tanh \begin{bmatrix} -0.347 & 2 & 2.339 & -0.275 & 0.075 & -0.0081 & -0.06 & -0.21 & -0.696 & -0.121 & 0.032 & -0.044 & -0.0034 & -0.0022 \\ 0.67 & 0.034 & -0.694 & -0.312 & 1.068 & 0.0039 & 1.475 & -0.345 & 0.118 & -0.695 & -0.059 & 0.0041 & -0.071 & -0.0004 \\ -0.237 & -0.09 & 0.146 & 0.469 & 0.0346 & 0.883 & 0.0811 & 2.014 & -0.042 & 0.053 & -0.704 & 0.0017 & -0.0019 & -0.076 \end{bmatrix} \mathbf{x}_n.$$

Algorithms and MATLAB codes to solve (6.53) are reported in Example 6.52. A closed-loop system is stable in the considered flight envelope. ■

Practice Problem 6.8. *Stability Analysis and Control: Van der Pol Equations and Nonlinear Systems*

Lyapunov Stability – Unsteady aerodynamics, elasticity, electromagnetics, electrostatics, heat transfer, hydrodynamics and other processes are described by nonlinear differential equations which admit instabilities in small. Solutions of van der Pol, Lotka-Volterra, Lorenz and other equations exhibit oscillatory and periodic dynamics. The aforementioned systems are unstable in small and stable in large. Assuming physical controllability, control schemes ensure asymptotic stability. While nonlinear differential equations may admit formal linearization, an overall consistency cannot be guaranteed.

In Example 2.22, the van der Pol oscillator was studied.
The differential equations are

$$\dot{x}_1 = x_2,$$

$$\dot{x}_2 = -x_1 + kx_2(1 - x_1^2),\ k>0.$$

Consider a positive definite function $V(x_1, x_2) = \frac{1}{2}\left(x_1^2 + x_2^2\right)$.

The total derivative is $\dfrac{dV(x_1, x_2)}{dt} = \dfrac{dV}{dx_1}\dfrac{dx_1}{dt} + \dfrac{dV}{dx_2}\dfrac{dx_2}{dt} = kx_2^2(1 - x_1^2)$.

For $V(x_1,x_2)>0$, $\dfrac{dV}{dt} \ge 0$ if $x_1^2 \le 1$, and, $\dfrac{dV}{dt} < 0$ if $x_1^2 > 1$. As stated in Conjecture 6.2, the system is unstable in small for $x_1^2 \le 1$, and, stable in large if $x_1^2 > 1$. States $x_1(t)$ and $x_2(t)$ exhibit periodic orbits.

For different classes of governing equations, which exhibit limit cycles, the Lyapunov stability analysis is applied. Consider the second-order equations

$$\dot{x}_1 = x_2,\ \dot{x}_2 = -x_1 + kx_2\left(1 - |x_1| - x_1^2\right),$$

and $$\dot{x}_1 = x_2,\ \dot{x}_2 = -x_1 + kx_2\left(1 - |x_1| - x_1^2 - |x_1^3| - x_1^4\right).$$

Using $V(x_1, x_2) = \frac{1}{2}x_1^2 + \frac{1}{2}x_2^2$, one finds X_δ where $\left.\dfrac{dV}{dt}\right|_{x\in X_\delta} > 0$, and, X_Δ where $\left.\dfrac{dV}{dt}\right|_{x\in X_\Delta} < 0$.

With $X_\delta \subset X_\Delta \in X$, systems are stable in large, and, exhibit limit cycles.

Solutions of (x_1,x_2) are bounded, such that $\|x(t) - x_e\| \le \varepsilon$, $\forall (t,x)\in T\times X$.

Control Problems – Consider a systems, described by differential equations

$$\dot{x}_1 = x_2,$$

$$\dot{x}_2 = -x_1 + kx_2(1 - x_1^2) + u,\ k>0.$$

The Lyapunov stability analysis established that the term kx_2 is destabilizing, while $-kx_2x_1^2$ guarantees stability in large. One should not perform feedback linearization canceling stabilizing term by means of $u_{\text{linearization}} = kx_2x_1^2$. Open-loop stability and stabilizing nonlinear map $-kx_2x_1^2$ imply that the design may be performed using $\dot{x}=Ax+Bu$, e.g., considering

$$\dot{x}_1 = x_2,$$

$$\dot{x}_2 = -x_1 + kx_2 + u,\ k>0.$$

Control laws can be designed by using a nonlinear model $\dot{x}=F(x)+Bu$, minimizing a functional, and, solving the Hamilton-Jacobi equation. For a quadratic return function

$$V(x_1, x_2) = \tfrac{1}{2}k_{11}x_1^2 + k_{12}x_1x_2 + \tfrac{1}{2}k_{22}x_2^2,\ (k_{11},k_{12},k_{22})>0,$$

one obtains a control law

$$u = -G^{-1}\frac{\partial V(x_1,x_2)}{\partial x_2} = -G^{-1}(k_{12}x_1 + k_{22}x_2),\ G>0,\ G^{-1}k_{22}>k.$$

One finds a feedback with a gain $G^{-1}k_{22}$ to compensate a destabilizing term kx_2, guaranteeing asymptotic stability of the closed-loop system. A range of nonlinear design and optimization problems are examined, including the Lyapunov concept to find a control law.

Third-Order Governing Equations – Investigate stability of systems

$$\dot{x}_1 = x_2,\ \dot{x}_2 = -x_1 + kx_2\left(1-|x_1|-x_1^2\right)+x_3,\ \dot{x}_3 = -x_1x_2 - x_3,$$

and

$$\dot{x}_1 = x_2,\ \dot{x}_2 = -x_1 + kx_2\left(1-|x_1|-x_1^2-|x_1^3|-x_1^4\right)+x_3,\ \dot{x}_3 = -x_1x_2 - x_3.$$

Apply $V(x_1,x_2,x_3)=\frac{1}{2}x_1^2+\frac{1}{2}x_2^2+\frac{1}{2}x_3^2$. We have $\left.\frac{dV}{dt}\right|_{x\in X_\delta}>0$ in X_δ, and, $\left.\frac{dV}{dt}\right|_{x\in X_\Delta}<0$ in X_Δ. Furthermore, $X_\delta\subset X_\Delta\in X$. Systems exhibits limit cycles, solutions (x_1,x_2,x_3) are bounded, $\|x(t)-x_e\|\le\varepsilon$, $\forall(t,x)\in T\times X$, and, systems are stable in large. ■

6.15. Case Studies in Analysis and Control

6.15.1. Inverted Arm on a Moving Platform

6.15.1.1. Mathematical Model

There are various pointing and steering systems used in various applications, such as high-precision servos, gimbals, robots, manipulators, etc. As educational studies and proof-of-concept platforms, investigate an inverted platforms within specific organization and hardware to:

1. Stabilize the inverted stand upward with angular displacement $\theta=\pi$;
2. Ensure platform repositioning in $x(t)$ direction and arm steering $\theta(t)$, or, upward arm pointing with $\theta=\pi$.

Applied and experimental studies are performed for a two-wheel platform illustrated in Figure 6.37.a. Stabilizing and tracking control laws are designed. Different controllers are implemented using the measured and observed states (x,v,θ,ω). The upward stabilization, arm pointing with $\theta(t)$ within allowed angular displacement $|\Delta\theta(t)|\le\varepsilon_\theta$, as well as platform repositioning in $x(t)$, are ensured for admissible initial conditions θ_0 and disturbances $d(t)$. The challenges are due to nonlinear unstable physical system, actuator limits on maximum force and torque (F,T), specified sensor accuracy and dynamic precision, sensing and processing latencies and delays, as well as other impediments. The considered system is open-loop unstable, and, system can be stabilized in an operating envelope within the maximum stand's angular displacement $|\theta(t)|\le\theta_{\max}$. An open-loop system is stable if the stand or arm points downward.

Figure 6.37.b documents that the implemented controller guarantees that the stand is stabilized and maintains upward pointing with a dynamic precision $|\Delta\theta(t)|\le\varepsilon_\theta$, which depends on the sensors accuracy, precision, linearity, sensors and ASICs errors and other factors. With PID controllers, the stand is unstable and falls to a horizontal position at $\pm\frac{1}{2}\pi$. Stable and unstable closed-loop system dynamics $\theta(t)$ are documented in Figures 6.37.b. With control laws, designed in section 6.15.1.3, the upward pointing of the stand is maintained. System is stable as illustrated by the central trace for the experimentally measured $\theta(t)$.

A four-wheel moving platform is illustrated in Figure 6.37.c. To avoid faulty functionality, in many applications, there are mechanical limits on linear and angular displacements $x_{\min}\le x\le x_{\max}$, $\theta_{\min}\le\theta\le\theta_{\max}$. The arm is at a stable equilibrium at the downward pointing with $\theta=0$ rad, and, $(F,T)=0$. An arm with length L rotates around a fixed pivot O at the cart, and, the center of mass m is at distance l. For an uniform rod of mass m and length L, with axis at the end, the moment of inertia is $J=\frac{1}{3}mL^2$. For a mass m at distance l, $J=\frac{1}{12}mL^2+ml^2$. For the uniform thin rod, $l=\frac{1}{2}L$ and $J=\frac{1}{3}mL^2$, a model with translational-rotational coupling is developed using the Lagrange equations.

Angular Displacement $\theta(t)$ [°]: Stand Maintains Upward Pointing With Designed Controller, and, Fall to Horizontal Position With a PID Controller

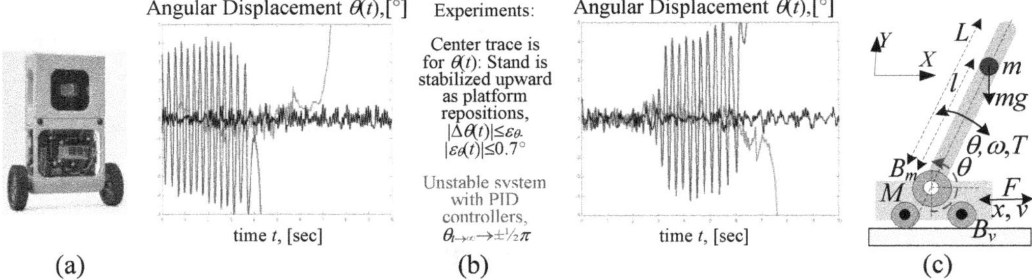

(a) (b) (c)

Figure 6.37. (a) Inverted stand: Two-wheel moving platform. The goal is to stabilize and maintain the upward pointing of the stand such as $|\Delta\theta(t)|\le\varepsilon_\theta$ during the platform translational repositioning on the surface in the x direction;
(b) Experimental studies: Nonlinear controller guarantees stability and upward pointing. Controllers, designed not considering physical limits yield unstable dynamics, and, stands falls to the horizontal position at $\pm\frac{1}{2}\pi$. With a nonlinear controller, the stand is stabilized upward $\theta=\pi$ with the dynamic error $|\Delta\theta(t)|\le\varepsilon_\theta$, $|\varepsilon_\theta(t)|\le 0.7°$, $\forall t$;
(c) Inverted arm on a four-wheel moving platform. A force F, developed by direct-drive servomotors in wheels is applied to move and reposition platform M in the $\pm x$ direction. To ensure arm pointing, steering and tracking, the arm is actuated by a direct-drive servomotor which develops torque T. The arm can be stabilized at $\theta=\pi$ and steered by regulating (F,T). The parameters are M, m, l, B_v and B_m. Physical limits on the force and toque (F,T) are $\mathcal{U}_{1\min}\le\mathcal{U}_1\le\mathcal{U}_{1\max}$ and $\mathcal{U}_{2\min}\le\mathcal{U}_2\le\mathcal{U}_{2\max}$. There are limits on an admissible angular displacement.

Inverted Arm on a Four-Wheel Moving Platform – An operating envelope $\{\theta_{\min}\le\theta\le\theta_{\max}, x_{\min}\le x\le x_{\max}\}$ depends on kinematics, control limits, mechanical limits, etc. Tracking of angular displacement $\theta(t)$ can be guaranteed if an adequate bidirectional torque $\pm T$ is developed by a servomotor. A direct drive wheel servo develops a torque, and, traction torque yields a bidirectional force $\pm F$. The outputs are (x,θ).

Within operating envelopes, the objectives are to accomplish:

1. Angular displacement steering of $\theta(t)$ by regulating (F,T) with pointing at $(\pi\pm\theta_{\text{steering}})$ rad;
2. Arm upward stabilization at $\theta=\pi$ within admissible displacement $(\pi\pm\varepsilon_\theta)$ rad by controlling the translational bidirectional motions $\pm x$ regulating F;
3. Platform repositioning $x(t)$ by regulating F, and, specified arm pointing.

A mathematical model is found by using the Lagrange equations of motion as documented in section 2.2.2. The kinetic, potential and dissipating energies for a cart M and arm m are

$$\Gamma=\Gamma_M+\Gamma_m=\tfrac{1}{2}M\dot{x}^2+\tfrac{1}{2}m\left(\dot{x}_m^2+\dot{y}_m^2\right)$$

$$=\tfrac{1}{2}M\dot{x}^2+\tfrac{1}{2}m\left[\dot{x}^2+2l\dot{x}\dot{\theta}\cos\theta+l^2\dot{\theta}^2\left(\cos^2\theta+\sin^2\theta\right)\right]=\tfrac{1}{2}(M+m)\dot{x}^2+\tfrac{1}{2}ml^2\dot{\theta}^2+ml\dot{x}\dot{\theta}\cos\theta,$$

$$\Pi=\Pi_M+\Pi_m=0+mgl(1-\cos\theta),\; D=D_M+D_m=\tfrac{1}{2}B_v\dot{x}^2+\tfrac{1}{2}B_m\dot{\theta}^2. \quad (6.163)$$

Using the *generalized coordinates* ($q_1=x$, $q_2=\theta$) and *generalized forces* ($Q_1=F$, $Q_2=T$), two second-order nonlinear differential equations are found by applying the Lagrange equations (2.16) $\frac{d}{dt}\left(\frac{\partial\Gamma}{\partial\dot{q}_i}\right)-\frac{\partial\Gamma}{\partial q_i}+\frac{\partial D}{\partial\dot{q}_i}+\frac{\partial\Pi}{\partial q_i}=Q_i$. Using (6.163), we have

$$(M+m)\ddot{x}+ml\ddot{\theta}\cos\theta-ml\dot{\theta}^2\sin\theta+B_v\dot{x}=F, \tag{6.164}$$
$$ml^2\ddot{\theta}+ml\ddot{x}\cos\theta+mgl\sin\theta+B_m\dot{\theta}=T.$$

One may find the resulting nonlinear differential equations in the Cauchy's form. Then, perform linearization using the Jacobian matrices (J,J_u) specifying the equilibria. Applying (6.8), find the governing equation $\dot{x}$=Ax+Bu for the state variations $\delta_x(t)$ in $X_\delta\subset X$ with inputs $\delta_u(t)$ in $U_\delta\subset U$,

$$\dot{\delta}_x(t)=\left.\frac{\partial F}{\partial x}\right|_{\substack{x=\bar{x}\\u=\bar{u}}}\delta_x+\left.\frac{\partial F}{\partial u}\right|_{\substack{x=\bar{x}\\u=\bar{u}}}\delta_u=J\big|_{\substack{x=\bar{x}\\u=\bar{u}}}\delta_x+J_u\big|_{\substack{x=\bar{x}\\u=\bar{u}}}\delta_u=\left.\begin{bmatrix}\frac{\partial f_1}{\partial x_1}&\cdots&\frac{\partial f_1}{\partial x_n}\\\vdots&\ddots&\vdots\\\frac{\partial f_n}{\partial x_1}&\cdots&\frac{\partial f_n}{\partial x_n}\end{bmatrix}\right|_{\substack{x=\bar{x}\\u=\bar{u}}}\delta_x+\left.\begin{bmatrix}\frac{\partial f_1}{\partial u_1}&\cdots&\frac{\partial f_1}{\partial u_m}\\\vdots&\ddots&\vdots\\\frac{\partial f_n}{\partial u_1}&\cdots&\frac{\partial f_n}{\partial u_m}\end{bmatrix}\right|_{\substack{x=\bar{x}\\u=\bar{u}}}\delta_u.$$
$$\dot{x}(t)=Ax+Bu$$

6.15.1.2. Inverted Pendulum on a Moving Cart: Stabilization and Control

Our goal is to stabilize arm upward at $\theta=\pi$ by controlling the cart motions $\pm x$ changing the force F, while T=0. The mathematical model is given by (6.164). For $Q_1=F\neq 0$ and $Q_2=T$=0, the Cauchy's form of differential equations are found using the MATLAB code

```
clear all; syms F T M m l g Bv Bm a v x ap om th % Symbols
% Lagrange equations
eqn1=(M+m)*a+m*l*ap*cos(th)-m*l*om*om*sin(th)+Bv*v==F;
eqn2=(m*l^2)*ap+m*l*a*cos(th)+m*g*l*sin(th)+Bm*om==0;
% Cauchy's equations for linear acceleration and angular acceleration
sol=solve([eqn1 eqn2], [a ap]); acceleration=sol.a; alpha=sol.ap;
pretty(acceleration),pretty(alpha)
```

The resulting equations are found as

$$\frac{dx}{dt}=v,$$
$$\frac{dv}{dt}=\frac{-lB_vv+B_m\omega\cos\theta+l^2m\omega^2\sin\theta+lmg\sin\theta\cos\theta+lF}{l\left(M+m-m\cos^2\theta\right)}, \tag{6.165}$$
$$\frac{d\theta}{dt}=\omega,$$
$$\frac{d\omega}{dt}=\frac{-B_m(M+m)\omega-(M+m)lmg\sin\theta-l^2m^2\omega^2\sin\theta\cos\theta+lB_vmv\cos\theta-Flm\cos\theta}{l^2m\left(M+m-m\cos^2\theta\right)}.$$

Continuous-Time System – Nonlinear model (6.165) is linearized at equilibrium

$$(\bar{x}_1,\bar{x}_2,\bar{x}_3,\bar{x}_4)=(\bar{x},\bar{v},\bar{\theta},\bar{\omega})=[0, 0, \pi, 0].$$

The parameters are M=0.5 kg, m=0.05 kg, l=0.2 m, $B_v=4\times10^{-6}$ N-sec/m and $B_m=4\times10^{-6}$ N-m-sec/rad. The linearized model is found using

```
M=0.5; m=0.05; l=0.2; g=9.81; Bv=4e-6; Bm=4e-6; % Parameters
x1=x; x2=v; x3=th; x4=om; % States
% Derivatives of states
dx1=x2; dx2=subs(acceleration);
dx3=x4; dx4=subs(alpha);
% Jacobian matrices
```

```
J= [diff(dx1,x1) diff(dx1,x2) diff(dx1,x3) diff(dx1,x4);
   diff(dx2,x1) diff(dx2,x2) diff(dx2,x3) diff(dx2,x4);
   diff(dx3,x1) diff(dx3,x2) diff(dx3,x3) diff(dx3,x4);
   diff(dx4,x1) diff(dx4,x2) diff(dx4,x3) diff(dx4,x4)];
Ju=[diff(dx1,F); diff(dx2,F); diff(dx3,F); diff(dx4,F)];
x=0; v=0; th=pi; om=0;  % Equilibria
% Linearized system dx/dt=Ax+Bu
A=double(subs(J)), B=double(subs(Ju)), H=[1 0 0 0]; D=0;
```

A linear model is $\dot{x}=\begin{bmatrix}0 & 1 & 0 & 0\\ 0 & -8\times10^{-6} & 0.981 & -4\times10^{-5}\\ 0 & 0 & 0 & 1\\ 0 & -4\times10^{-5} & 53.96 & -2.2\times10^{-3}\end{bmatrix}x+\begin{bmatrix}0\\2\\0\\10\end{bmatrix}u$, $y=Hx=[1\ \ 0\ \ 0\ \ 0]x$.

A constrained stabilization problem is solved considering a symmetric saturation limit on the actuator force F, $-0.5\le \mathcal{U}\le 0.5$ N. The limit is described by an infinitely-differentiable hyperbolic function

$u=\Phi(x)=u_{\max}\tanh(ax)$, $-0.5<u<0.5$ N, $u_{\max}=0.5$, $a=1$, $\Phi\subseteq\mathcal{U}_\varepsilon$.

Minimize a positive definite functional

$$J=\min_{x,u\in U}\int_0^\infty\left(\tfrac{1}{2}q_{11}x_1^2+\tfrac{1}{2}q_{22}x_2^2+\tfrac{1}{2}q_{33}x_3^2+\tfrac{1}{2}q_{44}x_4^2+G\int\tfrac{1}{a}\tanh^{-1}\left(\tfrac{1}{u_{\max}}u\right)du\right)dt$$

subject to linearized system with bounds on control.

The constrained control law is

$$u=-u_{\max}\tanh(aG^{-1}B^TKx)=-u_{\max}\tanh\left(K_F\begin{bmatrix}x_1\\x_2\\x_3\\x_4\end{bmatrix}\right),\ |u|\le 0.5\text{ N}.$$

An algebraic equation $\tfrac{1}{2}x^TQx+\dfrac{\partial V}{\partial x}^T Ax-\int u_{\max}\tanh\left(aG^{-1}B^T\dfrac{\partial V}{\partial x}\right)d\left(B^T\dfrac{\partial V}{\partial x}\right)=0$

is solved using the quadratic return function (6.66) $V(x)=\tfrac{1}{2}x^TKx$.

For the weighting coefficients $q_{11}=1000$, $q_{22}=1000$, $q_{33}=1000$, $q_{44}=1000$ and $G=1$,

$u=-u_{\max}\tanh\left(-54.77x_1-82.43x_2+356x_3+74.17x_4\right)$, $u_{\max}=0.5$, $-0.5<u<0.5$ N.

The closed-loop linearized system is simulated. The transient dynamics for (x_1, x_2, x_3, x_4) and bounded control function evolution are illustrated in Figure 6.38.a for initial conditions $(x,v,\theta,\omega)_0=[0\ 0\ 0.0872\ 0]$. System is stable, and, states $x(t)\equiv\delta_x(t)$ evolve to 0. Recall that the system was linearized at $(\bar{x}_1,\bar{x}_2,\bar{x}_3,\bar{x}_4)=(\bar{x},\bar{v},\bar{\theta},\bar{\omega})=[0, 0, \pi, 0]$.

That is, an arm points upward, pointing at $\theta=\pi$ rad.

We have $\lim_{t\to\infty}x_3(t)=0$, $x_3(t)\equiv\delta_\theta(t)$, and, $\lim_{t\to\infty}\theta(t)=\pi$ rad.

Consider a nonlinear system (6.165).

For upward pointing $\theta=\pi$, the constrained control law is

$u=-u_{\max}\tanh\left(-54.77x_1-82.43x_2+356(x_3-\pi)+74.17x_4\right)$, $u_{\max}=0.5$, $-0.5<u<0.5$ N.

Transient dynamics for the closed-loop nonlinear system are plotted in Figure 6.38.b. For $\theta_0=\pi+0.0872$ rad, arm steers upward at $\theta=\pi$, and, maintains upward pointing. A nonlinear system is stable. Due to control limits, stability depends on initial displacement, loads, disturbances, etc.

For $|\Delta\theta|\ge 0.0875$ rad, due to control bounds $|u|\le 0.5$ N, system becomes unstable. This implies that the arm descends to a mechanical limit $\theta_{\min}\le\theta\le\theta_{\max}$, or, points downward $\theta=0$ rad. Actuators should be chosen to ensure adequate torque or force.

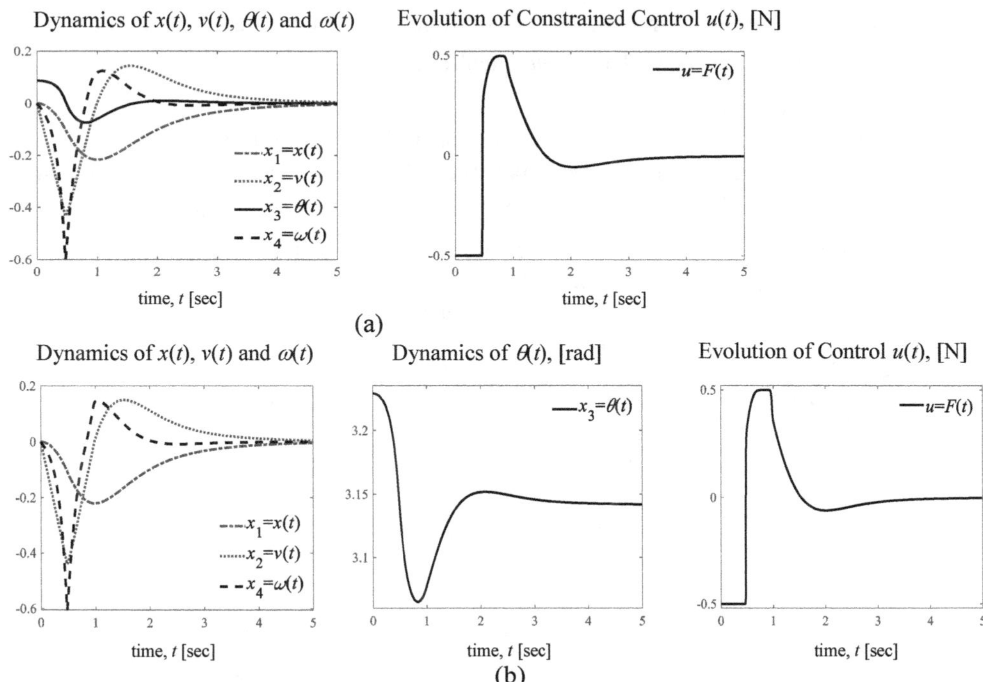

Figure 6.38. (a) Dynamics of the linearized system with initial displacement θ_0=0.0872 rad (θ_0=5°), when the equilibrium is $(\bar{x},\bar{v},\bar{\theta},\bar{\omega})$=[0 0 π 0]. Evolution of the constrained control $u(t)$=$F(t)$ is illustrated;

(b) Dynamics of a nonlinear closed-loop system for θ_0=π+0.0872 rad. Stability is guaranteed, and, arm is stabilized upward, θ=π rad. Evolutions of states $x(t)$ and control $u(t)$ are documented.

Discrete-Time System – A linearized continuous-time model

$$\dot{x}=\begin{bmatrix}0 & 1 & 0 & 0\\ 0 & -8\times10^{-6} & 0.981 & -4\times10^{-5}\\ 0 & 0 & 0 & 1\\ 0 & -4\times10^{-5} & 53.96 & -2.2\times10^{-3}\end{bmatrix}x+\begin{bmatrix}0\\2\\0\\10\end{bmatrix}u$$

is discretized with the sampling period T_s=0.001 sec.

Using the matrix exponential, A_n and B_n are found using (5.19). Hence $A_n=e^{AT_s}$, $B_n=A^{-1}(e^{AT_s}-I)B$, H_n=H and D_n=D.

One obtains

$$x_{n+1}=\begin{bmatrix}x_{1\ n+1}\\ x_{2\ n+1}\\ x_{3\ n+1}\\ x_{4\ n+1}\end{bmatrix}=A_nx_n+B_nu_n=\begin{bmatrix}1 & 1\times10^{-3} & 4.91\times10^{-7} & 1.43\times10^{-10}\\ 0 & 1 & 9.81\times10^{-4} & 4.51\times10^{-7}\\ 0 & -2\times10^{-11} & 1 & 1\times10^{-3}\\ 0 & -4\times10^{-8} & 5.4\times10^{-2} & 1\end{bmatrix}\begin{bmatrix}x_{1\ n}\\ x_{2\ n}\\ x_{3\ n}\\ x_{4\ n}\end{bmatrix}+\begin{bmatrix}0\\ 2\times10^{-3}\\ 5\times10^{-6}\\ 1\times10^{-2}\end{bmatrix}u_n,\ n\geq 0.$$

Minimize a nonquadratic performance index (6.150)

$$J=\min_{x_n,u_n\in U}\sum_{n=0}^{\infty}\left[x_n^TQ_nx_n-u_n^TB_n^TK_{n+1}B_nu_n+2G_n\textstyle\sum_u\left(\frac{1}{a}\tanh^{-1}(\frac{1}{u_{\max}}u_n)\right)\right],u_{\max}=0.5,\ a=1.$$

We find a control law (6.152) $u_n=-u_{\max}\tanh\left(aG_n^{-1}B_n^TK_nA_nx_n\right)$.

Solving (6.153) with the weighting coefficients q_{11n}=1, q_{22n}=1, q_{33n}=1, q_{44n}=1 and G_n=0.33, the positive definite matrix K_n and feedback gain matrix are found. We have

$$u_n = -u_{\max}\tanh\left(aG_n^{-1}B_n^T K_n A_n x_n\right) = -u_{\max}\tanh\left(-6.95x_{1n} - 11.74x_{2n} + 79.77x_{3n} + 11.06x_{4n}\right),$$

$u_{\max}$=0.5. $-0.5<u_n<0.5$ N.

Positive definiteness of the performance index is guaranteed, and, Lyapunov stability criteria are satisfied. Simulate a linearized closed-loop system with control law

$$x_{n+1}=A_n x_n+B_n u_n,\ u_n = -u_{\max}\tanh\left(-6.95x_{1n} - 11.74x_{2n} + 79.77x_{3n} + 11.06x_{4n}\right),\ |u_n|\le 0.5.$$

Transient dynamics for (x_{1n},x_{2n},x_{3n},x_{4n}) and evolution of u_n are illustrated in Figures 6.39.a. The initial conditions are x_{n0}=[0,0,0.0872,0]. System is stable, and, states x_n converge to 0.

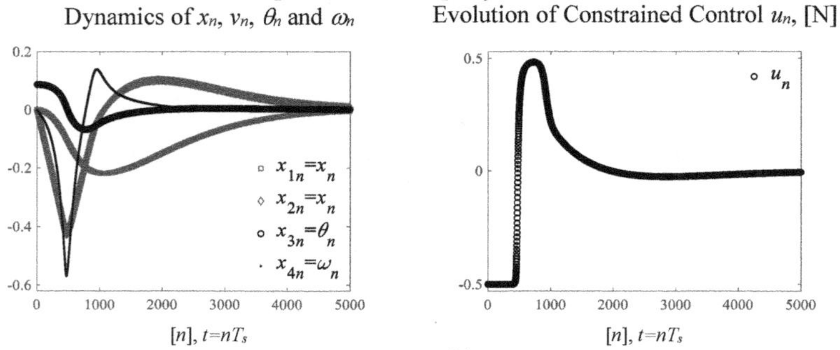

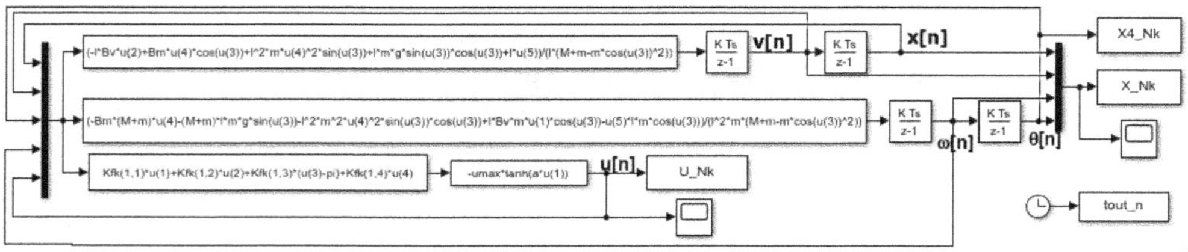

Figure 6.39. (a) Dynamics of the linearized discrete-time system with initial displacement θ_{n0}=0.0872 rad (θ_{n0}=5°). The evolution of the constrained force is reported, x_{n0}=[0, 0, 0.0872, 0]; (b) SIMULINK diagram to simulate a nonlinear discrete-time system with a constrained control u_n. Integrators are replaced by the Discrete-Time Integrator blocks which implement z-domain transfer function $G(z) = kT_s\frac{z^{-1}}{1-z^{-1}} = kT_s\frac{1}{z-1}$, corresponding to the forward Euler approximation. Numeric solutions are found at the fixed-step T_s;
(c) Dynamics of a nonlinear closed-loop system for $\theta_{n0}=\pi$+0.0872 rad. The arm points upward, $\theta_n=\pi$ rad, and, evolutions of states x_n and control u_n are documented.

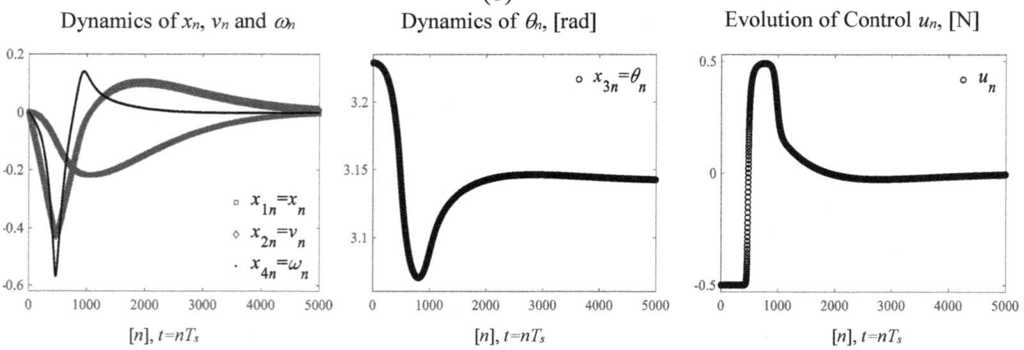

Nonlinear system (6.165) can be solved as a nonlinear discrete system using SIMULINK. The angular displacement $\theta_n=\pi$ corresponds to the upward arm pointing, which should be maintained despite disturbances.

The SIMULINK diagram is reported in Figure 6.39.b. Integration is approximated by the forward Euler equation, trapezoidal relationship, etc. For the forward Euler approximation $x_{n+1}=x_n+kT_s u_n$, $y_n=x_n$ with the integrator state, output and input tuple (x,y,u), the transfer function is $G(z)=kT_s\frac{z^{-1}}{1-z^{-1}}=kT_s\frac{1}{z-1}$. For the trapezoidal approximation, $G(z)=\frac{1}{2}T_s\frac{1+z^{-1}}{1-z^{-1}}=\frac{1}{2}T_s\frac{z+1}{z-1}$. The fixed-step numeric solution and solver are applied, and, T_s=0.001 sec. One may use the Discrete-Time Integrator block.

Transient dynamics for a nonlinear closed-loop system with a constrained control law $u_n=-u_{\max}\tanh\left(G_n^{-1}B_n^TK_nA_nx_n\right)=-u_{\max}\tanh\left(-6.95x_{1n}-11.74x_{2n}+79.77(x_{3n}-\pi)+11.06x_{4n}\right)$ are documented in Figures 6.39.c. For the displaced arm with $\theta_{n0}=\pi+0.0872$ rad, the arm points upward with $\theta_n=\pi$ rad.

A nonlinear closed-loop system is stable for $(\pi-0.0875)\le\theta_n\le(\pi+0.0875)$ rad.

6.15.1.3. Tracking Control: Continuous and Discrete Systems

To ensure arm pointing and cart repositioning, tracking control laws are designed. For a MIMO system, the outputs are $y_1=x$, $y_2=\theta$, while controls are $u_1=F$ and $u_2=T$.

The Cauchy form of nonlinear differential equations are found using (6.164). Linearization is performed using the Jacobian matrices (J,J_u). We have

$$\begin{aligned}
\frac{dx}{dt}&=v,\\
\frac{dv}{dt}&=\frac{-lB_vv+B_m\omega\cos\theta+l^2m\omega^2\sin\theta+lmg\sin\theta\cos\theta+lF-T\cos\theta}{l\left(M+m-m\cos^2\theta\right)},\\
\frac{d\theta}{dt}&=\omega,\\
\frac{d\omega}{dt}&=\frac{-B_m(M+m)\omega-(M+m)lmg\sin\theta-l^2m^2\omega^2\sin\theta\cos\theta+lB_vmv\cos\theta-Flm\cos\theta+(M+m)T}{l^2m\left(M+m-m\cos^2\theta\right)}.
\end{aligned}\tag{6.166}$$

The Cauchy's form of differential equations and linearized model at $(\bar{x}_1,\bar{x}_2,\bar{x}_3,\bar{x}_4)=(\bar{x},\bar{v},\bar{\theta},\bar{\omega})$=[0, 0, π, 0] with $(\bar{u}_1,\bar{u}_2)=(\bar{F},\bar{T})$ are found using a MATLAB code

```
clear all; syms F T M m l g Bv Bm a v x ap om th % Symbols
% Lagrange equations
eqn1=(M+m)*a+m*l*ap*cos(th)-m*l*om*om*sin(th)+Bv*v==F;
eqn2=(m*l^2)*ap+m*l*a*cos(th)+m*g*l*sin(th)+Bm*om==T;
% Cauchy's equations for linear acceleration and angular acceleration
sol=solve([eqn1 eqn2], [a ap]); acceleration=sol.a; alpha=sol.ap;
pretty(acceleration),pretty(alpha)
M=0.5; m=0.05; l=0.2; g=9.81; Bv=4e-6; Bm=4e-6; % Parameters
x1=x; x2=v; x3=th; x4=om; % States
% Derivative of each state
dx1=x2; dx2=subs(acceleration);
dx3=x4; dx4=subs(alpha);
% Jacobian matrices
```

```
J=[diff(dx1,x1) diff(dx1,x2) diff(dx1,x3) diff(dx1,x4);
   diff(dx2,x1) diff(dx2,x2) diff(dx2,x3) diff(dx2,x4);
   diff(dx3,x1) diff(dx3,x2) diff(dx3,x3) diff(dx3,x4);
   diff(dx4,x1) diff(dx4,x2) diff(dx4,x3) diff(dx4,x4)];
Ju=[diff(dx1,F) diff(dx1,T); diff(dx2,F) diff(dx2,T);...
    diff(dx3,F) diff(dx3,T); diff(dx4,F) diff(dx4,T)];
x=0; v=0; th=pi; om=0;  % Equilibrium
% Linearized system dx/dt=Ax+Bu
A=double(subs(J)), B=double(subs(Ju)), H=[1 0 0 0; 0 0 1 0]; D=[0 0; 0 0];
```

The linearized model is

$$\dot{x}=Ax+Bu=\begin{bmatrix}0 & 1 & 0 & 0\\ 0 & -8\times10^{-6} & 0.981 & -4\times10^{-5}\\ 0 & 0 & 0 & 1\\ 0 & -4\times10^{-5} & 53.96 & -2.2\times10^{-3}\end{bmatrix}x+\begin{bmatrix}0 & 0\\ 2 & 10\\ 0 & 0\\ 10 & 550\end{bmatrix}u,\ y=Hx=\begin{bmatrix}1 & 0 & 0 & 0\\ 0 & 0 & 1 & 0\end{bmatrix}\begin{bmatrix}x_1\\ x_2\\ x_3\\ x_4\end{bmatrix}.$$

The symmetric saturation limits on actuators force and torque are $-0.5\le\mathcal{U}_1\le0.5$ N, $-0.1\le\mathcal{U}_2\le0.1$ N-m.

The physical limits of servos are described by $\Phi_i\subseteq\mathcal{U}_\varepsilon$ as $\Phi_1=u_{1\max}\tanh(a_1x)$, $\Phi_2=u_{2\max}\tanh(a_2x)$, $u_{1\max}=0.5$, $u_{2\max}=0.1$, $a_1=10$, $a_2=1$.

Analytically design analog and discrete tracking control laws, such that $-0.5\le u_1\le0.5$ N, $-0.1\le u_2\le0.1$ N-m.

The tracking errors $e_1(t)=(r_1-y_1)$ and $e_2(t)=(r_2-y_2)$ are used to define the *error states* $x_e(t)$ and $e(t)$. The dynamics of $x_e(t)$ and $e(t)$ are governed as

$$\dot{x}_e(t)=I_X(r-y)=I_X(r-Hx),\ e(t)\equiv\dot{x}_e(t),\ x_e\in\mathbb{R}^2,$$

$$\dot{e}(t)=-I_Ee+\dot{r}-\dot{y}=-I_Ee+\dot{r}-HAx-HBu,\ e\in\mathbb{R}^2.$$

The extended model dynamics (6.66) with the extended vector $\mathbf{x}$ is

$$\dot{\mathbf{x}}(t)=\begin{bmatrix}\dot{x}(t)\\ \dot{x}_e(t)\\ \dot{e}(t)\end{bmatrix}=\mathbf{A}\mathbf{x}+A_rr+\mathbf{B}u=\begin{bmatrix}A & 0 & 0\\ -I_XH & 0 & 0\\ -HA & 0 & -I_E\end{bmatrix}\mathbf{x}+\begin{bmatrix}0 & 0\\ I_X & 0\\ 0 & I\end{bmatrix}\begin{bmatrix}r\\ \dot{r}\end{bmatrix}+\begin{bmatrix}B\\ 0\\ -HB\end{bmatrix}u\cdot$$

Hence,
$$\dot{\mathbf{x}}=\begin{bmatrix}0 & 1 & 0 & 0 & 0 & 0 & 0 & 0\\ 0 & -8\times10^{-6} & 0.981 & -4\times10^{-5} & 0 & 0 & 0 & 0\\ 0 & 0 & 0 & 1 & 0 & 0 & 0 & 0\\ 0 & -4\times10^{-5} & 53.96 & -2.2\times10^{-3} & 0 & 0 & 0 & 0\\ -1 & 0 & 0 & 0 & 0 & 0 & 0 & 0\\ 0 & 0 & -1 & 0 & 0 & 0 & 0 & 0\\ 0 & -1 & 0 & 0 & 0 & 0 & -1 & 0\\ 0 & 0 & 0 & -1 & 0 & 0 & 0 & -1\end{bmatrix}\mathbf{x}+\begin{bmatrix}0 & 0 & 0 & 0\\ 0 & 0 & 0 & 0\\ 0 & 0 & 0 & 0\\ 0 & 0 & 0 & 0\\ 1 & 0 & 0 & 0\\ 0 & 1 & 0 & 0\\ 0 & 0 & 1 & 0\\ 0 & 0 & 0 & 1\end{bmatrix}\begin{bmatrix}r_1\\ r_2\\ \dot{r}_1\\ \dot{r}_2\end{bmatrix}+\begin{bmatrix}0 & 0\\ 2 & 10\\ 0 & 0\\ 10 & 550\\ 0 & 0\\ 0 & 0\\ 0 & 0\\ 0 & 0\end{bmatrix}\begin{bmatrix}u_1\\ u_2\end{bmatrix}.$$

The physical limits on force and torque $-\mathcal{U}_{i\max}\le\mathcal{U}_i\le\mathcal{U}_{i\max}$ are described by $\Phi_i=u_{i\max}\tanh(a_ix)$.

Minimize a nonquadratic performance functional

$$J=\min_{\substack{\mathbf{x}\\ |u_1|\le u_{1\max},|u_2|\le u_{2\max}}}\int_0^\infty\left[\tfrac{1}{2}\mathbf{x}^TQ\mathbf{x}+\int\left(\boldsymbol{a}^{-1}\tanh^{-1}(\boldsymbol{u}_{\max}^{-1}u)\right)^TGdu\right]dt,\ \boldsymbol{u}_{\max}=\begin{bmatrix}u_{1\max} & 0\\ 0 & u_{2\max}\end{bmatrix},\ \boldsymbol{a}=\begin{bmatrix}a_1 & 0\\ 0 & a_2\end{bmatrix},$$

$u_{1\max}=0.5$, $u_{2\max}=0.1$, $a_1=1$, $a_2=1$.

The constrained tracking control law is

$$u=-\boldsymbol{u}_{\max}\tanh(\boldsymbol{a}G^{-1}\mathbf{B}^TK\mathbf{x})=-\begin{bmatrix}u_{1\max} & 0\\ 0 & u_{2\max}\end{bmatrix}\tanh\left(K_F\begin{bmatrix}x\\ \int e_1dt\\ \int e_2dt\\ e_1\\ e_2\end{bmatrix}\right),\ e(t)=\begin{bmatrix}e_1(t)\\ e_2(t)\end{bmatrix}=\begin{bmatrix}r_1-x\\ r_2-\theta\end{bmatrix}.$$

An equation $\tfrac{1}{2}\mathbf{x}^TQ\mathbf{x}+\left(\frac{\partial V}{d\mathbf{x}}\right)^T\mathbf{A}\mathbf{x}-\int\left[\boldsymbol{u}_{\max}\tanh\left(\boldsymbol{a}G^{-1}\mathbf{B}^T\frac{\partial V}{d\mathbf{x}}\right)\right]^Td\left(\mathbf{B}^T\frac{\partial V}{d\mathbf{x}}\right)=0$

is solved using $\int \left[\boldsymbol{u}_{\max} \tanh(\boldsymbol{a}G^{-1}\mathbf{B}^T \frac{\partial V}{d\mathbf{x}}) \right]^T d(\mathbf{B}^T \frac{\partial V}{d\mathbf{x}}) \approx \frac{1}{3} \frac{\partial V}{d\mathbf{x}}^T \mathbf{B}\boldsymbol{u}_{\max}\boldsymbol{a}G^{-1}\mathbf{B}^T \frac{\partial V}{d\mathbf{x}}$, $V(\mathbf{x}) = \frac{1}{2}\mathbf{x}^T K\mathbf{x}$. Let $q_{11}=1$, $q_{22}=1$, $q_{33}=1$, $q_{44}=1$, $q_{55}=10000$, $q_{66}=1\times10^6$, $q_{77}=1000$, $q_{88}=1000$, $G = \begin{bmatrix} 10 & 0 \\ 0 & 100 \end{bmatrix}$.

A tracking control law is

$$u = -\begin{bmatrix} u_{1\max} & 0 \\ 0 & u_{2\max} \end{bmatrix} \tanh\left(\begin{bmatrix} 37.14 & 11.29 & 5.87 & 0.027 & -54.33 & -69.75 & -1.4 & -0.032 \\ -3.41 & -1.04 & 37.03 & 1.48 & 4.93 & -384.15 & 0.12 & -0.18 \end{bmatrix} \begin{bmatrix} x \\ \int e_1 dt \\ \int e_2 dt \\ e_1 \\ e_2 \end{bmatrix} \right), \quad x = \begin{bmatrix} x_1 \\ x_2 \\ x_3 \\ x_4 \end{bmatrix} = \begin{bmatrix} x \\ v \\ \theta \\ \omega \end{bmatrix}.$$

Dynamics of the closed-loop system are reported in Figures 6.40 for references $r(t) = \begin{bmatrix} r_1(t) \\ r_2(t) \end{bmatrix} = \begin{bmatrix} x_{\text{ref}}(t) \\ \theta_{\text{ref}}(t) \end{bmatrix} = \begin{bmatrix} \pm 0.25 \text{ m} \\ \pm 0.2 \text{ rad} \end{bmatrix}$ and $r(t) = \begin{bmatrix} r_1(t) \\ r_2(t) \end{bmatrix} = \begin{bmatrix} x_{\text{ref}}(t) \\ \theta_{\text{ref}}(t) \end{bmatrix} = \begin{bmatrix} \mp 0.25 \text{ m} \\ \pm 0.2 \text{ rad} \end{bmatrix}$.
The outputs (x,θ) track references $r(t)$. An adequate performance is ensured despite control limits. Evolutions of $u_1(t)$ and $u_2(t)$ are depicted on Figures 6.40. Translational-rotational coupling and specifications affect definiteness of feedback gains. The closed-loop system is stable, and, adequate transients and tracking are achieved. Designer studies consistency of state and tracking error feedback, state dependencies, coupling, reference scheduling, measurements and observed states, hardware and algorithm complementarity, etc. *Minimal complexity* control laws are designed applying the Lyapunov theory, examining stability, studying feedback definiteness, etc.

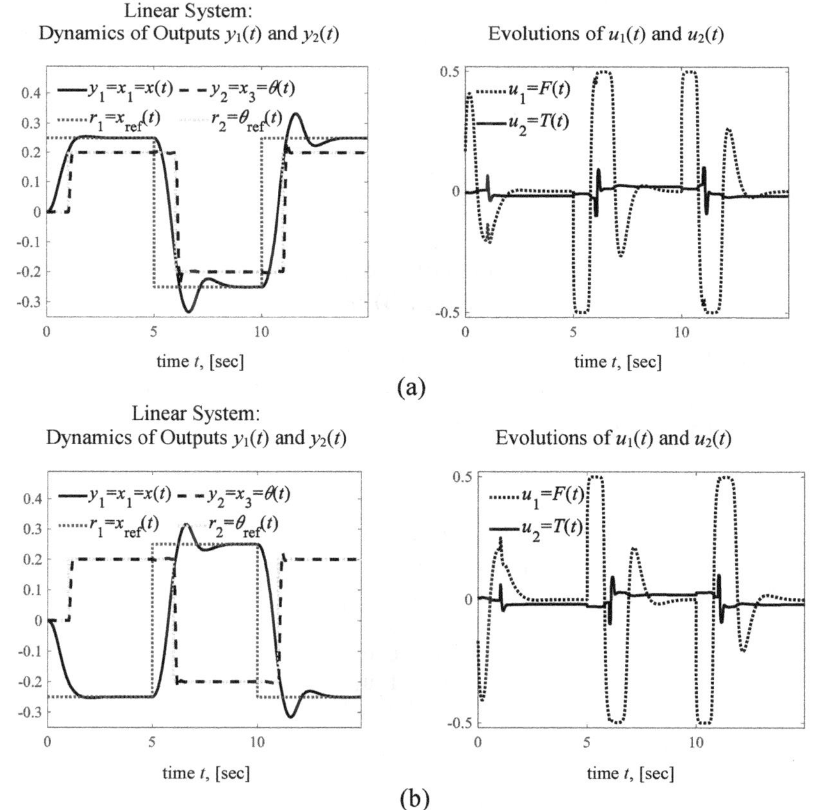

Figure 6.40. Linear system: Dynamics of outputs (y_1,y_2) and evolutions of controls (u_1,u_2) for references: (a) $r_1(t)=\pm0.25$ m, $r_2(t)=\pm0.2$ rad; (b) $r_1(t)=\mp 0.25$ m, $r_2(t)=\pm0.2$ rad.

For a nonlinear mathematical model (6.166), the constrained tracking control law $u=-\begin{bmatrix} u_{1\max} & 0 \\ 0 & u_{2\max} \end{bmatrix}\tanh(K_F\mathbf{x})$ is evaluated. Let the arm points upward, $\theta_0=\pi$. Outputs $y_1=x$ and $y_2=\theta$ track references $r(t)$ as documented in Figures 6.41. Controls (u_1,u_2) reach limits. Nonlinear system with limits on control is stable, and, tracking is achieved.

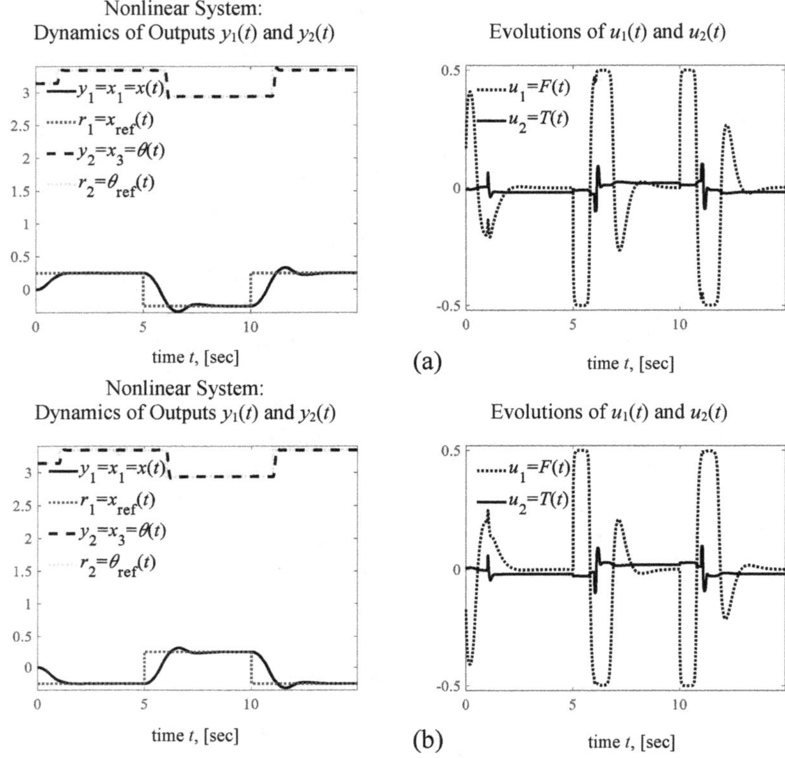

Figure 6.41. Nonlinear system: Dynamics of outputs (y_1,y_2) and evolutions of controls (u_1,u_2) for references: (a) $r_1(t)=\pm 0.25$ m, $r_2(t)=\pm 0.2$ rad; (b) $r_1(t)=\mp 0.25$ m, $r_2(t)=\pm 0.2$ rad.

Tracking control laws are designed for a discrete-time system. A continuous-time system $\dot{\mathbf{x}}(t)=\mathbf{A}\mathbf{x}+A_r r+\mathbf{B}u$ is discretized with $T_s=0.001$ sec.

Discretization yields

$$\mathbf{x}_{n+1}=\mathbf{A}_n\mathbf{x}_n+A_{rn}r_n+\mathbf{B}_n u_n,\ n\ge 0,$$

$$\mathbf{x}_{n+1}=\begin{bmatrix} 1 & 1\times10^{-3} & 4.9\times10^{-7} & 0 & 0 & 0 & 0 & 0 \\ 0 & 1 & 9.81\times10^{-4} & 4.51\times10^{-7} & 0 & 0 & 0 & 0 \\ 0 & 0 & 1 & 1\times10^{-3} & 0 & 0 & 0 & 0 \\ 0 & 0 & 5.4\times10^{-2} & 1 & 0 & 0 & 0 & 0 \\ -1\times10^{-3} & -5\times10^{-7} & 0 & 0 & 1 & 0 & 0 & 0 \\ 0 & 0 & -1\times10^{-3} & -5\times10^{-7} & 0 & 1 & 0 & 0 \\ 0 & -1\times10^{-3} & -4.9\times10^{-7} & 0 & 0 & 0 & 1 & 0 \\ 0 & 0 & -2.7\times10^{-5} & -1\times10^{-3} & 0 & 0 & 0 & 1 \end{bmatrix}\mathbf{x}+\begin{bmatrix} 0 & 0 \\ 0 & 0 \\ 0 & 0 \\ 0 & 0 \\ 1\times10^{-3} & 0 \\ 0 & 1\times10^{-3} \\ 0 & 0 \\ 0 & 0 \end{bmatrix}\begin{bmatrix} r_{1n} \\ r_{2n} \end{bmatrix}+\begin{bmatrix} 1\times10^{-6} & 5\times10^{-6} \\ 2\times10^{-3} & 1\times10^{-2} \\ 5\times10^{-6} & 2.75\times10^{-4} \\ 1\times10^{-2} & 0.55 \\ 0 & 0 \\ 0 & 0 \\ -1\times10^{-6} & -5\times10^{-6} \\ -5\times10^{-6} & -2.75\times10^{-4} \end{bmatrix}u_n.$$

Minimize the performance index (6.150)

$$J=\min_{\mathbf{x}_n,u_n\in U}\sum_{n=0}^{\infty}\left[\mathbf{x}_n^T Q_n\mathbf{x}_n-u_n^T\mathbf{B}_n^T K_{n+1}\mathbf{B}_n u_n+2\Sigma_u\left(\boldsymbol{a}^{-1}\tanh^{-1}(\boldsymbol{u}_{\max}^{-1}u_n)\right)^T G_n\right],$$

$u_{1\max}=0.5$, $u_{2\max}=0.1$, $a_1=1$, $a_2=1$.

We find a control law (6.152). Equation (6.153) is solved using the weighting coefficients q_{11n}=1, q_{22n}=1, q_{33n}=1, q_{44n}=1, q_{55n}=100, q_{66n}=100, q_{77n}=500, q_{88n}=10, G_{11n}=0.01 and G_{22n}=1.

A positive definite matrix K_n and feedback gain matrix K_{Fn} are

$$u_n = -\begin{bmatrix} u_{1\max} & 0 \\ 0 & u_{2\max} \end{bmatrix} \tanh\left(aG_n^{-1}B_n^T K_n A_n \mathbf{x}_n\right) = -\begin{bmatrix} u_{1\max} & 0 \\ 0 & u_{2\max} \end{bmatrix} \tanh\left(K_{Fn}\mathbf{x}_n\right),$$

$$K_{Fn} = \begin{bmatrix} 241.8 & 39.24 & 30.78 & 5.73 & -167 & -58.74 & -134.9 & -1.8 \\ -17.38 & -3.2 & 22.83 & 4.18 & 11.88 & -39.81 & 9.45 & -1.21 \end{bmatrix}.$$

Positive definiteness of the performance index is guaranteed. The closed-loop system is simulated. The transient dynamics for (y_{1n},y_{2n}) are illustrated in Figures 6.42 for $r_{1n}(t)$=±0.25 m and $r_{2n}(t)$=±0.2 rad. Tracking is achieved, and, platform repositions and arm steers as specified. In a linear model, there is overshoot for linear displacement $y_{1n}=x_{1n}=x_n$. A nonlinear model should be examined to design a controller with the objective to implement a controlled by a MPU.

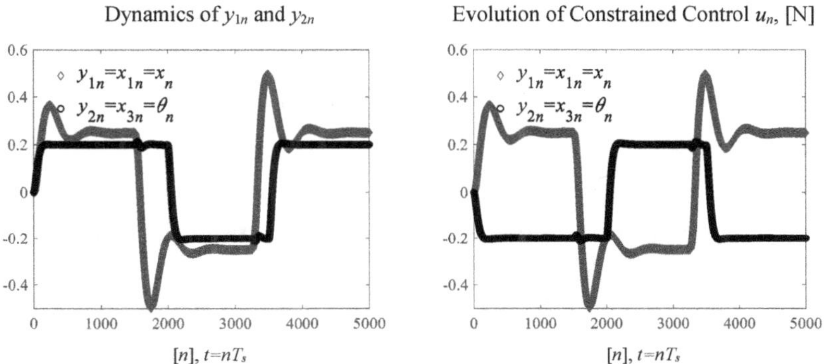

Figure 6.42. Linear discrete-time system: Dynamics of (y_{1n}, y_{2n}), r_{1n}=±0.25 m, r_{2n}=±0.2 rad.

For a nonlinear system with control limits, redesign a control law using the weighting coefficients q_{11n}=1, q_{22n}=1, q_{33n}=1, q_{44n}=1, q_{55n}=1000, q_{66n}=1000, q_{77n}=100, q_{88n}=1, G_{11n}=1 and G_{22n}=1. The control law with computed (K_n, K_{Fn}) is

$$u_n = -\begin{bmatrix} u_{1\max} & 0 \\ 0 & u_{2\max} \end{bmatrix} \tanh\left(aG_n^{-1}B_n^T K_n A_n \mathbf{x}_n\right) = -\begin{bmatrix} u_{1\max} & 0 \\ 0 & u_{2\max} \end{bmatrix} \tanh\left(K_{Fn}\mathbf{x}_n\right),$$

$$K_{Fn} = \begin{bmatrix} 37.77 & 11.54 & -0.072 & -0.132 & -54.96 & -1.24 & -1.42 & -0.0006 \\ -2.48 & -0.994 & 36.47 & 4.46 & 2.21 & -132 & 0.028 & -0.051 \end{bmatrix}.$$

To simulate a nonlinear system, a SIMULINK diagram is developed as illustrated in Figure 6.43.a. The arm points upward, $\theta_{n0}=\pi$. The outputs $y_{1n}=x_n$ and $y_{2n}=\theta_n$ track references $r_n = \begin{bmatrix} r_{1n} \\ r_{2n} \end{bmatrix} = \begin{bmatrix} \pm 0.25 \text{ m} \\ \pi \pm 0.2 \text{ rad} \end{bmatrix}$ and $r_n = \begin{bmatrix} r_{1n} \\ r_{2n} \end{bmatrix} = \begin{bmatrix} \mp 0.25 \text{ m} \\ \pi \pm 0.2 \text{ rad} \end{bmatrix}$ as depicted in Figures 6.43.b. The system is stable despite control limits, and, (u_{1n},u_{2n}) are reaching limits. To expand an operating envelope and ensure tracking in $\theta \in [0 \;\; 2\pi]$ rad, one should use an actuator which develops sufficient torque. The experimental studies are performed. The implemented tracking controller guarantees stability and disturbance rejection. The inverted stand is stabilized with dynamic error $|\theta(t)| \le \varepsilon_\theta$, $|\varepsilon_\theta(t)| \le 0.75°$ which is due to sensors accuracy, precision, linearity, dynamic range, noise, latencies, etc. The closed-loop system ensures repositioning, steering as well as angular pointing. Experimental results are reported in Figure 6.37.b.

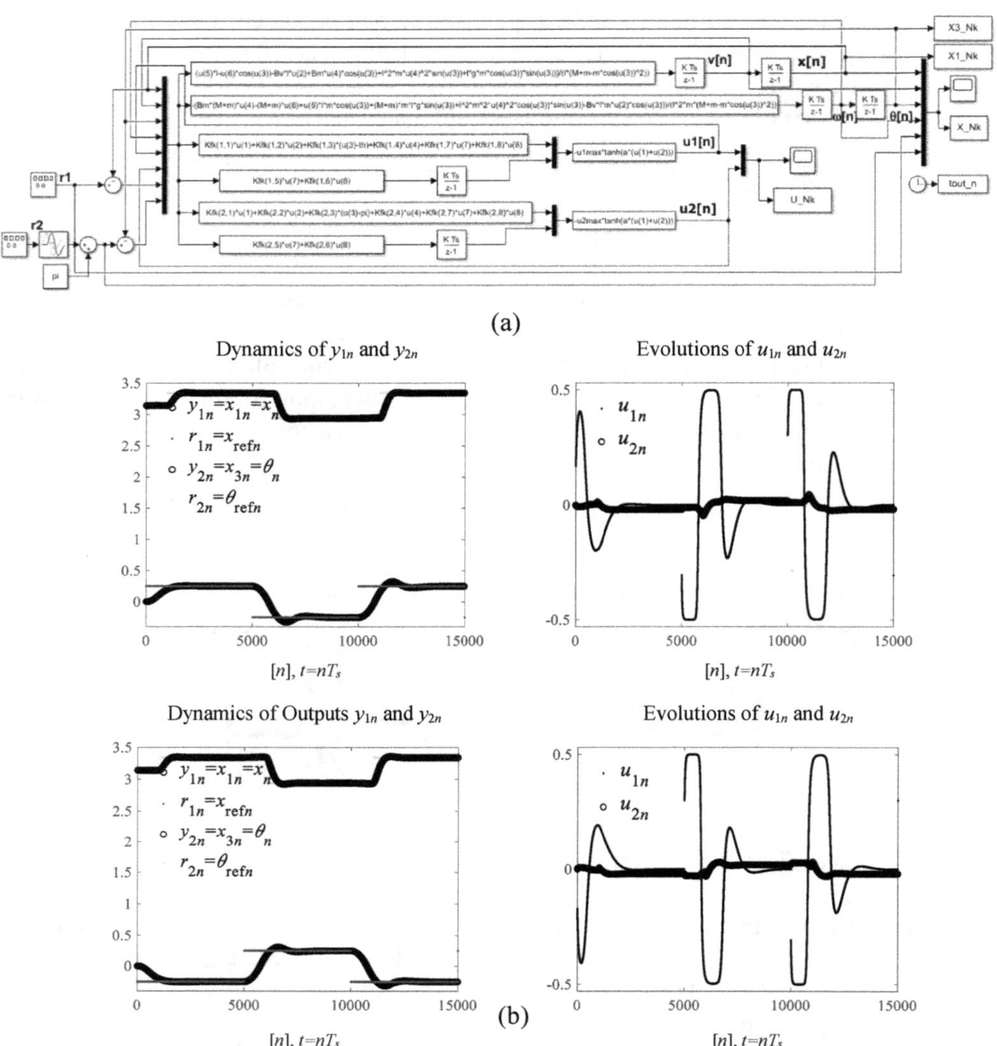

Figure 6.43. (a) SIMULINK diagram with the Discrete-Time Integrator block to simulate a nonlinear system with a tracking control law;

(b) Dynamics of outputs (y_{1n},y_{2n}) and evolutions of controls (u_{1n},u_{2n}), r_{1n}=±0.25 m,r_{2n}=π±0.2 rad, and, r_{1n}= ∓ 0.25 m, r_{2n}=π±0.2 rad.

6.15.2. Translational-Rotational Manipulator

6.15.2.1. Mathematical Models: Governing Equations and Linearization

Consider a robotic manipulator with two actuators to ensure the 360-degree angular pointing $-\pi \le \theta \le \pi$, and, translational repositioning $r_{\min} \le r \le r_{\max}$. A manipulator is illustrated in Figure 6.44. Simplified considerations are covered in [17]. Two servos develop the torque T and force F to regulate the angular and linear displacements (θ, r). The mechanical limits on angular displacement and linear arm displacement are $\theta_{\min} \le \theta \le \theta_{\max}$ and $r_{\min} \le r \le r_{\max}$. The shock absorber spring exerts a bidirectional force. The spring, characterized by constant k_s and *free length* r_0, absorb shock, prevents mechanical failures and enables stability. Tracking of angular and linear displacements must be accomplished. As illustrated in Figure 6.44, m_1 and m_2 are the masses, r and r_1 are the relative displacement of the end effector and base length, $r(t)=r_1+l(t)$, r_1=const. In the XY plane, the end effector position (x, y) depends on (θ, r).

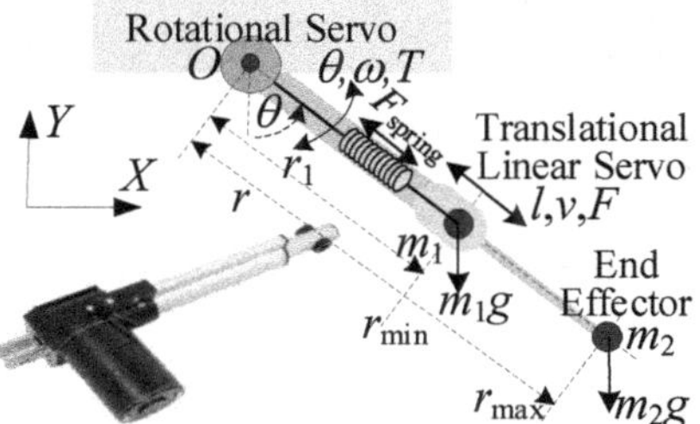

Figure 6.44. Two-degree-of-freedom manipulator with two servos, and, lamped-parameter schematics

The Lagrange equations of motion (2.16)

$$\frac{d}{dt}\left(\frac{\partial\Gamma}{\partial\dot{q}_i}\right)-\frac{\partial\Gamma}{\partial q_i}+\frac{\partial D}{\partial\dot{q}_i}+\frac{\partial\Pi}{\partial q_i}=Q_i,\quad \begin{bmatrix}q_1\\q_2\end{bmatrix}=\begin{bmatrix}\theta\\r\end{bmatrix},\quad \begin{bmatrix}Q_1\\Q_2\end{bmatrix}=\begin{bmatrix}T\\F\end{bmatrix}$$

yields nonlinear differential equations.

The *generalized coordinates* and *generalized forces* pairs (q_1,q_2) and (Q_1,Q_2) yields two second-order nonlinear differential equations. Apply the variables and parameters, shown in Figure 6.44. In the XY plane, we have

$$x_1=r_1\cos\theta,\ y_1=r_1\sin\theta,\ \dot{x}_1=-r_1\dot{\theta}\sin\theta,\ \dot{y}_1=r_1\dot{\theta}\cos\theta,$$

$$x_2=r\cos\theta,\ y_2=r\sin\theta,\ \dot{x}_2=-r\dot{\theta}\sin\theta+\dot{r}\cos\theta,\ \dot{y}_2=r\dot{\theta}\cos\theta+\dot{r}\sin\theta.$$

The kinetic energy is

$$\Gamma=\Gamma_1+\Gamma_2,\Gamma_1=\tfrac{1}{2}m_1v_1^2=\tfrac{1}{2}m_1\left(\dot{x}_1^2+\dot{y}_1^2\right)=\tfrac{1}{2}m_1r_1^2\dot{\theta}^2,\Gamma_2=\tfrac{1}{2}m_2v_2^2=\tfrac{1}{2}m_2\left(\dot{x}_2^2+\dot{y}_2^2\right)=\tfrac{1}{2}m_2\left(r^2\dot{\theta}^2+\dot{r}^2\right),$$

$$\Gamma=\tfrac{1}{2}m_1r_1^2\dot{\theta}^2+\tfrac{1}{2}m_2\left(r^2\dot{\theta}^2+\dot{r}^2\right)=\tfrac{1}{2}m_1r_1^2\dot{q}_1^2+\tfrac{1}{2}m_2\left(q_2^2\dot{q}_1^2+\dot{q}_2^2\right).$$

The total potential energy is a sum of potential energies of the first and second masses, as well as the spring potential energy

$$\Pi=\Pi_1+\Pi_2+\Pi_{\text{spring}},\ \Pi_1=m_1gh_1=m_1gr_1(1-\cos\theta),\ \Pi_2=m_2gh_2=m_2gr(1-\cos\theta),$$

$$\Pi_{\text{spring}}=\tfrac{1}{2}k_s(r-r_0)^2,\ \Pi=g(m_1r_1+m_2q_2)(1-\cos q_1)+\tfrac{1}{2}k_s(q_2-r_0)^2.$$

The total dissipating energy is $D=D_1+D_2=\tfrac{1}{2}B_m\dot{\theta}^2+\tfrac{1}{2}B_v\dot{r}^2=\tfrac{1}{2}B_m\dot{q}_1^2+\tfrac{1}{2}B_v\dot{q}_2^2$.

The Lagrange equations yield

$$\underbrace{m_1r_1^2\ddot{q}_1+m_2\left(q_2^2\ddot{q}_1+2q_2\dot{q}_1\dot{q}_2\right)}_{\frac{d}{dt}\frac{\partial\Gamma}{\partial\dot{q}_1}}-\underbrace{0}_{\frac{\partial\Gamma}{\partial q_1}}+\underbrace{B_m\dot{q}_1}_{\frac{\partial D}{\partial\dot{q}_1}}+\underbrace{g\left(m_1r_1+m_2q_2\right)\sin q_1}_{\frac{\partial\Pi}{\partial q_1}}=Q_1,$$

$$\underbrace{m_2\ddot{q}_2}_{\frac{d}{dt}\frac{\partial\Gamma}{\partial\dot{q}_2}}-\underbrace{m_2q_2\dot{q}_1^2}_{\frac{\partial\Gamma}{\partial q_2}}+\underbrace{B_v\dot{q}_2}_{\frac{\partial D}{\partial\dot{q}_2}}+\underbrace{m_2g(1-\cos q_1)+k_s(q_2-r_0)}_{\frac{\partial\Pi}{\partial q_2}}=Q_2.$$

Hence, $$\ddot{q}_1=\frac{-g(m_1r_1+m_2q_2)\sin q_1-2m_2q_2\dot{q}_1\dot{q}_2-B_m\dot{q}_1+Q_1}{m_1r_1^2+m_2q_2^2},$$

$$\ddot{q}_2=\frac{m_2q_2\dot{q}_1^2-m_2g(1-\cos q_1)-k_s(q_2-r_0)-B_v\dot{q}_2+Q_2}{m_2}.$$

A nonlinear equation in the state-space form is

$$\begin{bmatrix}\dot{x}_1\\ \dot{x}_2\\ \dot{x}_3\\ \dot{x}_4\end{bmatrix}=\begin{bmatrix}\dot{\theta}\\ \dot{r}\\ \dot{\omega}\\ \dot{v}\end{bmatrix}=\begin{bmatrix}x_3\\ x_4\\ \dfrac{-g(m_1r_1+m_2x_2)\sin x_1-2m_2x_2x_3x_4-B_mx_3+T}{m_1r_1^2+m_2x_2^2}\\ \dfrac{m_2x_2x_3^2-m_2g(1-\cos x_1)-k_s(x_2-r_0)-B_vx_4+F}{m_2}\end{bmatrix},\quad \begin{bmatrix}x_1\\ x_2\\ x_3\\ x_4\end{bmatrix}=\begin{bmatrix}\theta\\ r\\ \omega\\ v\end{bmatrix}. \tag{6.167}$$

The parameters are m_1=5 kg, m_2=2 kg, r_1=0.5 m, r_0=0.75 m, B_m=0.1 N-m-sec/rad, B_v=0.1 N-sec/m and k_s=25 N/m.

To perform linearization, specify the equilibrium at $\bar{x}=\begin{bmatrix}\bar{\theta}\\ \bar{r}\\ \bar{\omega}\\ \bar{v}\end{bmatrix}=\begin{bmatrix}\frac{\pi}{8}\\ 0.75\\ 0\\ 0\end{bmatrix}=\begin{bmatrix}0.3927\text{ rad}\\ 0.75\text{ m}\\ 0\\ 0\end{bmatrix}$.

For states equilibria $\bar{x}$, compute the corresponding $(\bar{T},\bar{F})$.

From $$\frac{\bar{T}-g(m_1r_1+m_2\bar{x}_2)\sin\bar{x}_1-B_m\bar{x}_3-2m_2\bar{x}_2\bar{x}_3\bar{x}_4}{m_1r_1^2+m_2\bar{x}_2^2}=0,$$

$$\frac{\bar{F}+m_2\bar{x}_2\bar{x}_3^2-m_2g(1-\cos\bar{x}_1)-k_s(\bar{x}_2-r_0)-B_v\bar{x}_4}{m_2}=0,$$

we find $\bar{u}=\begin{bmatrix}\bar{T}\\ \bar{F}\end{bmatrix}=\begin{bmatrix}15\text{ N-m}\\ 7.7435\text{ N}\end{bmatrix}$.

The Jacobian matrices are

$$J=\begin{bmatrix}\frac{\partial f_1}{\partial x_1}&\frac{\partial f_1}{\partial x_2}&\frac{\partial f_1}{\partial x_3}&\frac{\partial f_1}{\partial x_4}\\ \frac{\partial f_2}{\partial x_1}&\frac{\partial f_2}{\partial x_2}&\frac{\partial f_2}{\partial x_3}&\frac{\partial f_2}{\partial x_4}\\ \frac{\partial f_3}{\partial x_1}&\frac{\partial f_3}{\partial x_2}&\frac{\partial f_3}{\partial x_3}&\frac{\partial f_3}{\partial x_4}\\ \frac{\partial f_4}{\partial x_1}&\frac{\partial f_4}{\partial x_2}&\frac{\partial f_4}{\partial x_3}&\frac{\partial f_4}{\partial x_4}\end{bmatrix},\quad J_u=\begin{bmatrix}\frac{\partial f_1}{\partial u_1}&\frac{\partial f_1}{\partial u_2}\\ \frac{\partial f_2}{\partial u_1}&\frac{\partial f_2}{\partial u_2}\\ \frac{\partial f_3}{\partial u_1}&\frac{\partial f_3}{\partial u_2}\\ \frac{\partial f_4}{\partial u_1}&\frac{\partial f_4}{\partial u_2}\end{bmatrix}=\begin{bmatrix}0&0\\ 0&0\\ \frac{1}{m_1r_1^2+m_2x_2^2}&0\\ 0&\frac{1}{m_2}\end{bmatrix},$$

$$J=\begin{bmatrix}0&0&1&0\\ 0&0&0&1\\ -\frac{g(m_1r_1+m_2x_2)\cos x_1}{m_1r_1^2+m_2x_2^2}&\frac{-(m_1r_1^2+m_2x_2^2)m_2(g\sin x_1+2x_3x_4)+2m_2x_2\left(g(m_1r_1+m_2x_2)\sin x_1+B_m+2m_2x_2x_3x_4-T\right)}{\left(m_1r_1^2+m_2x_2^2\right)^2}&-\frac{2m_2x_2x_4+B_m}{m_1r_1^2+m_2x_2^2}&-\frac{2m_2x_2x_3}{m_1r_1^2+m_2x_2^2}\\ -g\sin x_1&x_3^2-\frac{k_s}{m_2}&2x_2x_3&-\frac{B_v}{m_2}\end{bmatrix}.$$

For given $(\bar{x},\bar{u})$ and derived $(\bar{T},\bar{F})$, $J=\begin{bmatrix}0&0&1&0\\ 0&0&0&1\\ -15.26&-3.16&-0.042&0\\ -3.75&-12.5&0&-0.05\end{bmatrix}$ and $J_u=\begin{bmatrix}0&0\\ 0&0\\ 0.42&0\\ 0&0.5\end{bmatrix}$.

The linearized model is $\dot{\delta}_x=A\delta_x+B\delta_u=\begin{bmatrix}0&0&1&0\\ 0&0&0&1\\ -15.26&-3.16&-0.042&0\\ -3.75&-12.5&0&-0.05\end{bmatrix}\delta_x+\begin{bmatrix}0&0\\ 0&0\\ 0.42&0\\ 0&0.5\end{bmatrix}\delta_u$.

Analytic solutions are verified using a MATLAB code

```
m1=5; m2=2; r1=0.5; r0=0.75; Bm=0.1; Bv=0.1; ks=25; g=9.81;  % Parameters
syms q1 q2 q3 q4 q1d q2d q3d q4d q1e q2e q3e q4e u1 u2 y1 y2 T F;
q1e=pi/8; q2e=0.75; q3e=0; q4e=0; qe=[q1e;q2e;q3e;q4e]; % Equilibrium states
q1s=q3; q2s=q4;  % States
q3s=(-g*(m1*r1+m2*q2)*sin(q1)-Bm*q3-2*m2*q2*q3*q4+T)/(m1*r1^2+m2*q2^2);
q4s=(m2*q2*q3^2-m2*g*(1-cos(q1))-ks*(q2-r0)-Bv*q4+F)/m2;
qd=[q1s; q2s; q3s; q4s];
u1=T; u2=F;  % Controls are T and F
q=[q1; q2; q3; q4]; u=[u1; u2];  % States and control vectors
% Solve for control inputs to obtain the equilibrium
feqe=subs(qd,q,qe); ue=solve(feqe(3)==0,feqe(4)==0);
ue=[double(ue.T);double(ue.F)];
% Jacobian matrices
J=jacobian(qd,q); Ju=jacobian(qd,u);
% State-space matrices for the linearized model
A=subs(J,q,qe); A=subs(A,u,ue); B=subs(Ju,q,qe);
A=double(A), B=double(B); H=[1 0 0 0; 0 1 0 0]; D=[0 0; 0 0];
 % Display the A and B matrices
fprintf([repmat('%6.2f',1,size(A,2)) '\n'],A')
fprintf([repmat('%6.2f',1,size(B,2)) '\n'],B')
```

6.15.2.2. Tracking Control Problem

Consider tracking of the angular and linear displacements (θ, r), which are the outputs. That is, $y_1=\theta$ and $y_2=r$. Controls are $u_1=T$ and $u_2=F$. There are symmetric saturation limits on the actuators' force and torque, $-50\le\mathcal{U}_1\le 50$ N-m and $-40\le\mathcal{U}_2\le 40$ N.

The saturation limits are described as

$\Phi_i=u_{i\,\max}\tanh(a_i x)$, $a_i=1$, $\Phi_i\subseteq\mathcal{U}_\varepsilon$.

Analytically design a tracking control law for a MIMO system

$\dot{x}=Ax+Bu$, $-u_{1\max}\le u_1\le u_{1\max}$, $-u_{2\max}\le u_2\le u_{2\max}$, $y_1=x_1=\theta$, $y_2=x_2=r$.

The tracking errors are

$e_1=r_1-y_1$, $e_2=r_2-y_2$.

The governing equations for the *error states* $x_e(t)$ and tracking error vector $e(t)$ are

$$\dot{x}_e(t)=I_X(r-y)=I_X(r-Hx),\ e(t)\equiv\dot{x}_e(t),\ \begin{bmatrix}\dot{x}_{e1}\\ \dot{x}_{e2}\end{bmatrix}=\begin{bmatrix}1&0\\0&1\end{bmatrix}\begin{bmatrix}e_1\\ e_2\end{bmatrix}=\begin{bmatrix}1&0\\0&1\end{bmatrix}\left(\begin{bmatrix}r_1\\ r_2\end{bmatrix}-Hx\right),\ \begin{bmatrix}e_1\\ e_2\end{bmatrix}\equiv\begin{bmatrix}\dot{x}_{e1}\\ \dot{x}_{e2}\end{bmatrix},$$

$$\dot{e}(t)=-I_E e+\dot{r}-\dot{y}=-I_E e+\dot{r}-HAx-HBu,\ e\in\mathbb{R}^2.$$

One finds (6.66) with the state vector $\mathbf{x}\in\mathbb{R}^8$ as

$$\dot{\mathbf{x}}=\mathbf{A}\mathbf{x}+A_r r+\mathbf{B}u$$

$$=\begin{bmatrix}0&0&1&0&0&0&0&0\\0&0&0&1&0&0&0&0\\-15.26&-3.16&-0.042&0&0&0&0&0\\-3.75&-12.5&0&-0.05&0&0&0&0\\-1&0&0&0&0&0&0&0\\0&-1&0&0&0&0&0&0\\0&0&-1&0&0&0&-1&0\\0&0&0&-1&0&0&0&-1\end{bmatrix}\mathbf{x}+\begin{bmatrix}0&0&0&0\\0&0&0&0\\0&0&0&0\\0&0&0&0\\1&0&0&0\\0&1&0&0\\0&0&1&0\\0&0&0&1\end{bmatrix}\begin{bmatrix}r_1\\ r_2\\ \dot{r}_1\\ \dot{r}_2\end{bmatrix}+\begin{bmatrix}0&0\\0&0\\0.42&0\\0&0.5\\0&0\\0&0\\0&0\\0&0\end{bmatrix}u,\ \mathbf{x}=\begin{bmatrix}\theta\\ r\\ \omega\\ v\\ \int e_1dt\\ \int e_2dt\\ e_1\\ e_2\end{bmatrix}.$$

Minimize a nonquadratic functional

$$J=\min_{\substack{\mathbf{x}\\ |u_1|\le u_{1\max}\\ |u_2|\le u_{2\max}}}\int_0^\infty\left[\tfrac{1}{2}\mathbf{x}^T Q\mathbf{x}+\int\left(\boldsymbol{a}^{-1}\tanh^{-1}(\boldsymbol{u}_{\max}^{-1}u)\right)^T G\,du\right]dt,\ a_1=1,\ a_2=1,\ u_{1\max}=50,\ u_{2\max}=40,$$

subject to the system dynamics.

The tracking control law is

$$u=-\boldsymbol{u}_{\max}\tanh(\boldsymbol{a}G^{-1}\mathbf{B}^{T}K\mathbf{x})=-\begin{bmatrix}u_{1\max} & 0\\ 0 & u_{2\max}\end{bmatrix}\tanh\left(K_F\begin{bmatrix}x\\ \int e_1dt\\ \int e_2dt\\ e_1\\ e_2\end{bmatrix}\right).$$

Solution of $\quad \frac{1}{2}\mathbf{x}^{T}Q\mathbf{x}+\left(\frac{\partial V}{d\mathbf{x}}\right)^{T}\mathbf{A}\mathbf{x}-\int\left[\boldsymbol{u}_{\max}\tanh\left(\boldsymbol{a}G^{-1}\mathbf{B}^{T}\frac{\partial V}{d\mathbf{x}}\right)\right]^{T}d\left(\mathbf{B}^{T}\frac{\partial V}{d\mathbf{x}}\right)=0,$

$$\int\left[\boldsymbol{u}_{\max}\tanh(\boldsymbol{a}G^{-1}\mathbf{B}^{T}\tfrac{\partial V}{d\mathbf{x}})\right]^{T}d(\mathbf{B}^{T}\tfrac{\partial V}{d\mathbf{x}})\approx\tfrac{1}{3}\tfrac{\partial V}{d\mathbf{x}}^{T}\mathbf{B}\boldsymbol{u}_{\max}\boldsymbol{a}G^{-1}\mathbf{B}^{T}\tfrac{\partial V}{d\mathbf{x}}$$

is approximated by a quadratic return function $V(\mathbf{x})=\frac{1}{2}\mathbf{x}^{T}K\mathbf{x}$.

Let q_{11}=1, q_{22}=1, q_{33}=1, q_{44}=1, q_{55}=100, q_{66}=1000, q_{77}=10, q_{88}=100, g_{11}=0.01, g_{22}=0.01.

A tracking control law is

$$u=-\begin{bmatrix}u_{1\max} & 0\\ 0 & u_{2\max}\end{bmatrix}\tanh\left(K_F\begin{bmatrix}x\\ \int e_1dt\\ \int e_2dt\\ e_1\\ e_2\end{bmatrix}\right),\ x=\begin{bmatrix}x_1\\ x_2\\ x_3\\ x_4\end{bmatrix}=\begin{bmatrix}x\\ v\\ \theta\\ \omega\end{bmatrix},\ e=\begin{bmatrix}e_1\\ e_2\end{bmatrix}=\begin{bmatrix}r_1-\theta\\ r_2-r\end{bmatrix},\qquad(6.168)$$

$$K_F=\begin{bmatrix}8.59 & -0.19 & 2.07 & -0.0067 & -17.32 & -0.075 & -0.51 & -0.0028\\ -0.3 & 23.53 & -0.008 & 2.75 & 0.027 & -61.24 & 0.0007 & -2.09\end{bmatrix}.$$

SIMULINK diagram is reported in Figure 6.45.a. Simulations are performed for random references r_1 and r_2. For linearized model, transients and evolutions are depicted in Figures 6.45.b.

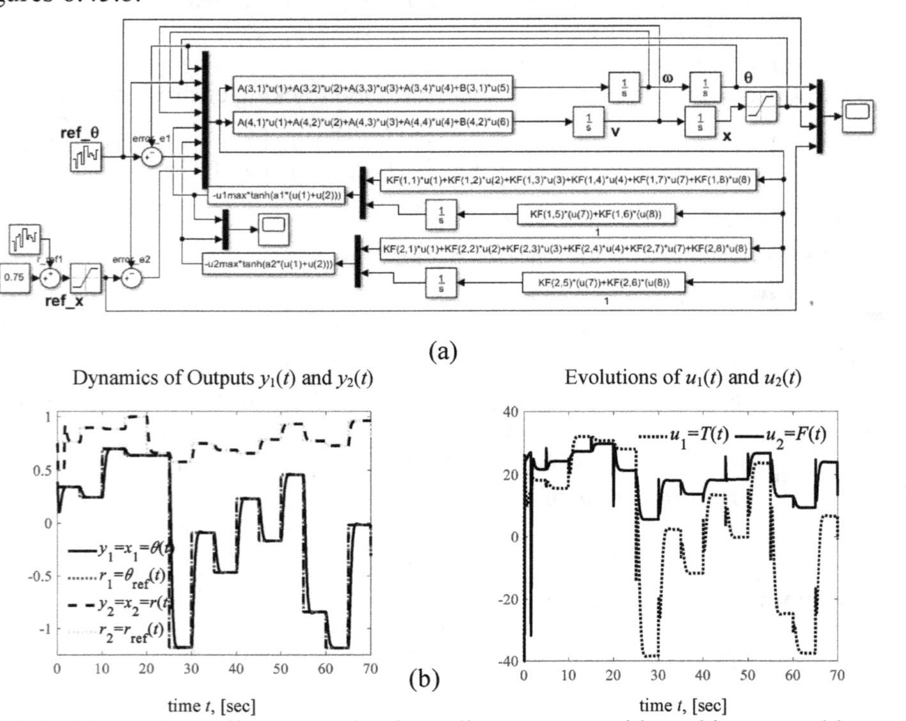

Figure 6.45. (a) SIMULINK diagram to simulate a linear system with tracking control law; (b) Dynamics of outputs (y_1,y_2) and control evolutions (u_1,u_2), $-\frac{1}{2}\pi<r_1<\frac{1}{2}\pi$ rad and $0.5\le r_2\le 1$ m.

For a nonlinear closed-loop MIMO system (6.167) with (6.168), the SIMULINK diagram is illustrated in Figure 6.46.a. The tracking is accomplished. Evolutions of outputs (θ, r) and controls (u_1, u_2) in an operating envelope with references $-\pi \le r_1 \le \pi$ rad, $0.5 \le r_2 \le 1$ m are investigated. Simulation results are documented in Figures 6.45.b. The closed-loop system is stable, and, tracking is achieved.

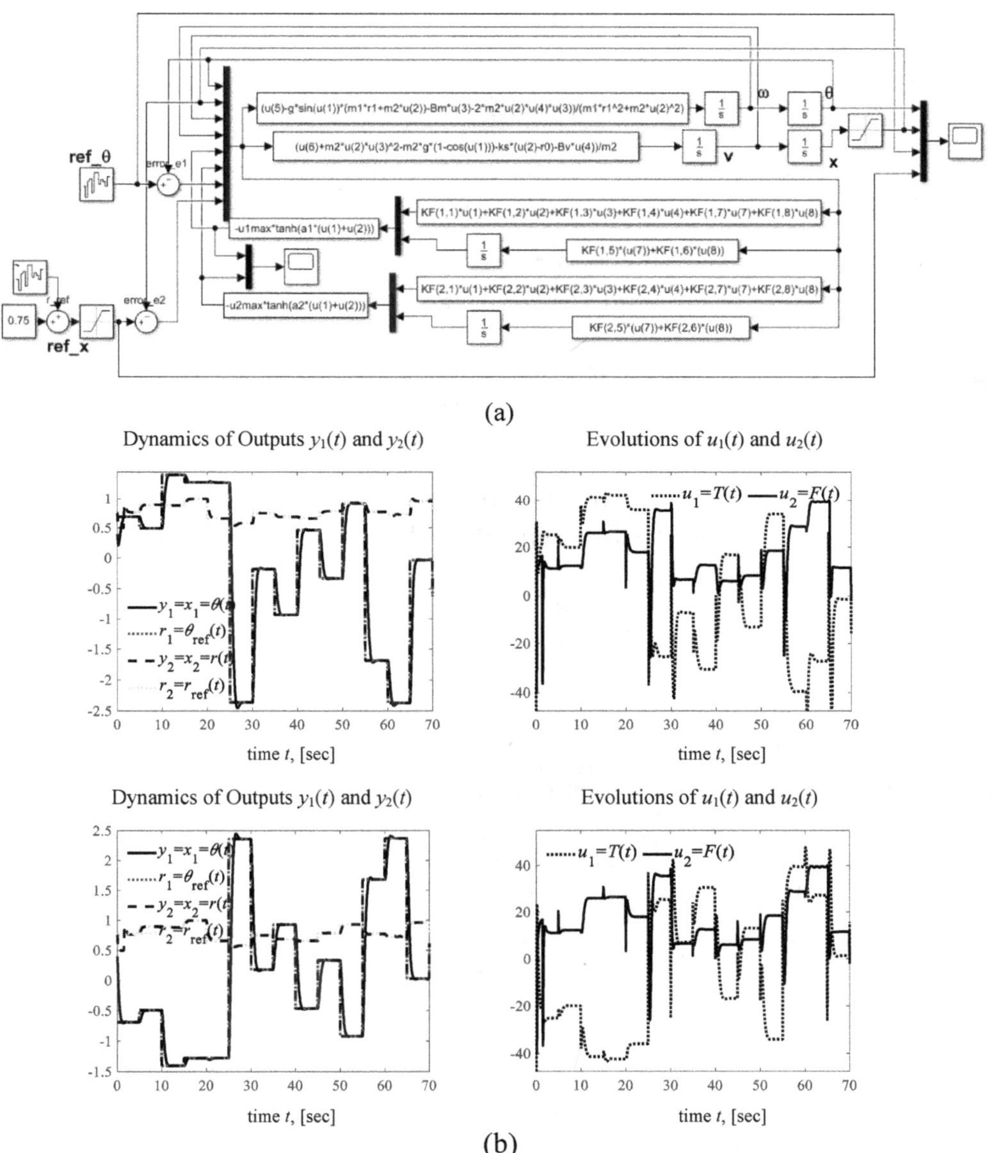

Figure 6.46. (a) SIMULINK diagram to simulate a nonlinear system with tracking control law; (b) Dynamics of outputs (y_1,y_2) and control evolutions (u_1,u_2), $-\pi \le r_1 \le \pi$ rad and $0.5 \le r_2 \le 1$ m.

6.15.3. Two-Link Robot Arm

Consider a two-degree-of-freedom kinematics with two joints, shown in Figure 6.47.a. The system is stable if pointing downward with equilibria $(\omega_{1e},\theta_{1e},\omega_{2e},\theta_{2e})=0$ for zero inputs. System is controlled by two actuators, which develop torques (T_1,T_2), applied to the first and second joints. These torques are controls (u_1,u_2), $u_1=T_1$ and $u_2=T_2$. The load torques (T_{1L},T_{2L}) are applied to the first and second links.

As documented in Example 2.8, using the Lagrange equations (2.16) $\frac{d}{dt}\left(\frac{\partial\Gamma}{\partial\dot{q}_i}\right)-\frac{\partial\Gamma}{\partial q_i}+\frac{\partial D}{\partial\dot{q}_i}+\frac{\partial\Pi}{\partial q_i}=Q_i$, $q_i=(q_1,q_2)$, $q_1=\theta_1$, $q_2=\theta_2$, the second-order nonlinear ordinary differential equations in non Cauchy's form are derived as

$$l_1\left[(m_1+m_2)l_1\ddot{\theta}_1+m_2l_2\cos(\theta_2-\theta_1)\ddot{\theta}_2-m_2l_2\sin(\theta_2-\theta_1)\dot{\theta}_2^2+(m_1+m_2)g\sin\theta_1\right]+B_{m1}\dot{\theta}_1=T_1+T_{L1}(t),$$

$$m_2l_2\left[l_2\ddot{\theta}_2+l_1\cos(\theta_2-\theta_1)\ddot{\theta}_1+l_1\sin(\theta_2-\theta_1)\dot{\theta}_1^2+g\sin\theta_2\right]+B_{m2}\dot{\theta}_2=T_2+T_{L2}(t).$$

The parameters are m_1=1 kg, m_2=0.5 kg, l_1=0.5 m, l_2=0.25 m, B_{m1}=0.1 N-m-sec/rad and B_{m2}=0.1 N-m-sec/rad.

Using non Cauchy's differential equations, simulate the open-loop stable system with initial conditions θ_{10}=1 rad, θ_{20}= –1 rad, ω_{10}=0 rad/sec and ω_{20}=0 rad/sec. System dynamics for $\omega_1(t)$, $\theta_1(t)$, $\omega_2(t)$ and $\theta_2(t)$, as well as SIMULINK diagram, are reported in Figures 6.47.b and c. The open-loop nonlinear system is stable with two masses pointing downward. The equilibrium is θ_{1e}=0 and θ_{2e}=0 for T_1=0 and T_2=0.

The Symbolic Math Toolbox supports finding a model in the Cauchy's form

$$\dot{x}=f(x,u),\quad x=\begin{bmatrix}x_1\\x_2\\x_3\\x_4\end{bmatrix}=\begin{bmatrix}\omega_1\\\theta_1\\\omega_2\\\theta_2\end{bmatrix},\quad u=\begin{bmatrix}u_1\\u_2\end{bmatrix}.$$

To derive Cauchy's form differential equations, the MATLAB code is

```
clear all; syms m1 m2 l1 l2 g Bm1 Bm2 a1 a2 o1 o2 t1 t2 u1 u2      % Lagrange equations
eqn1=l1*((m1+m2)*l1*a1+m2*l2*cos(t2-t1)*a2-m2*l2*sin(t2-t1)*o2^2 +(m1+m2)*g*sin(t1))+Bm1*o1==u1;
eqn2=m2*l2*(l2*a2+l1*cos(t2-t1)*a1+l1*sin(t2-t1)*o1^2+g*sin(t2))+Bm2*o2==u2;
% Cauchy's equations for alpha1 and alpha2
sol=solve([eqn1 eqn2], [a1 a2]); alpha1=sol.a1; alpha2 = sol.a2;
pretty(alpha1); pretty(alpha2); simplify(alpha1); simplify(alpha2);
```

The resulting equations of motion are

$$\dot{x}_1=\frac{1}{l_1^2l_2\left[m_1+m_2\left(1-\cos^2(x_2-x_4)\right)\right]}\Big[-l_1l_2^2m_2(x_1+x_3)x_3\sin(x_2-x_4)-l_1^2l_2m_2(x_1+x_3)x_1\sin(x_2-x_4)\cos(x_2-x_4)$$

$$-gl_1l_2\left[(m_1+m_2)\sin x_2-m_2\cos(x_2-x_4)\sin x_4\right]-B_{m1}l_2x_1+B_{m2}l_1x_3\cos(x_2-x_4)+l_2u_1-l_1\cos(x_2-x_4)u_2\Big],$$

$$\dot{x}_2=x_1,$$

$$\dot{x}_3=\frac{1}{l_1l_2^2m_2\left[m_1+m_2\left(1-\cos^2(x_2-x_4)\right)\right]}\Big[l_1^2l_2m_2x_1\left((m_1+m_2)x_1+m_1x_3\right)\sin(x_2-x_4)$$

$$+l_1l_2^2m_2^2(x_1+x_3)x_3\sin(x_2-x_4)\cos(x_2-x_4)$$

$$-gl_1l_2m_2\left[(m_1+m_2(1-\cos(x_2-x_4)))\sin x_4-m_1\cos(x_2-x_4)\sin x_2\right]$$

$$+B_{m1}l_2m_2x_1\cos(x_2-x_4)-B_{m2}l_1(m_1+m_2)x_3-l_2m_2\cos(x_2-x_4)u_1+l_1(m_1+m_2)u_2\Big],$$

$$\dot{x}_4=x_3.$$

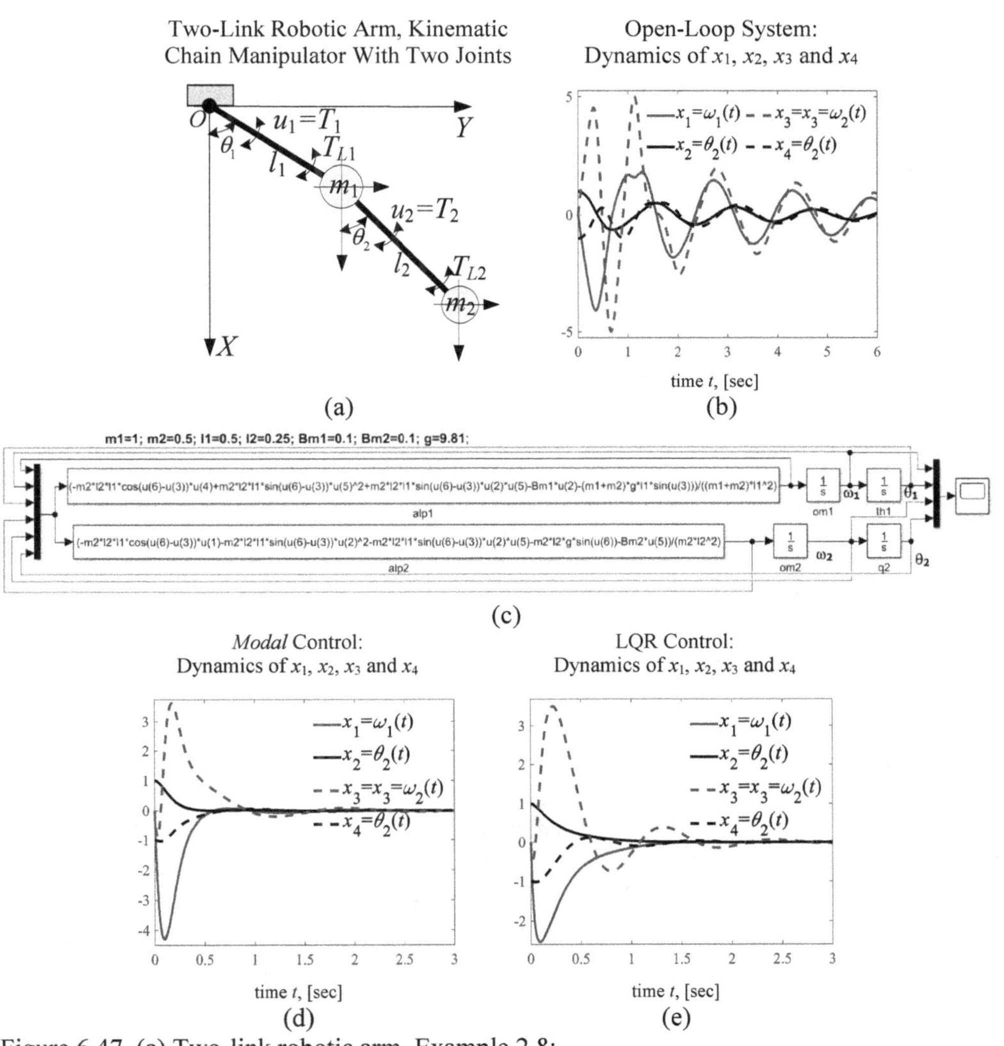

Figure 6.47. (a) Two-link robotic arm, Example 2.8;
(b) Open-loop nonlinear system dynamics for $x_1(t)=\omega_1(t)$, $x_2(t)=\theta_1(t)$, $x_3(t)=\omega_2(t)$ and $x_4(t)=\theta_2(t)$ if the initial conditions are $\omega_{10}=0$ rad/sec, $\theta_{10}=1$ rad, $\omega_{20}=0$ rad/sec, $\theta_{20}=-1$ rad. The torques are $T_1=0$ and $T_2=0$ N-m;
(c) SIMULINK diagram to simulate open-loop nonlinear system in non Cauchy's form;
(d) *Modal* control: Dynamics for $x_1(t)=\omega_1(t)$, $x_2(t)=\theta_1(t)$, $x_3(t)=\omega_2(t)$ and $x_4(t)=\theta_2(t)$;
(e) Closed-loop system dynamics with the stabilizing LQR control law: Dynamics for $x_1(t)$, $x_2(t)$, $x_3(t)$ and $x_4(t)$ if the initial conditions are $\omega_{10}=0$ rad/sec, $\theta_{10}=1$ rad, $\omega_{20}=0$ rad/sec, $\theta_{20}=-1$ rad.

Equations of Motion in Cauchy's Form and Linearization – To linearize a system, chose the state equilibria $\bar{x}=\begin{bmatrix}\bar{x}_1\\ \bar{x}_2\\ \bar{x}_3\\ \bar{x}_4\end{bmatrix}=\begin{bmatrix}0\\ -1\\ 0\\ 1\end{bmatrix}$. One has $\bar{u}=\begin{bmatrix}\bar{u}_1\\ \bar{u}_2\end{bmatrix}\neq 0$, and, $\bar{u}=\begin{bmatrix}\bar{u}_1\\ \bar{u}_2\end{bmatrix}=\begin{bmatrix}6.2\\ -1\end{bmatrix}$.

```
eqn3=alpha1==0; eqn4=alpha2==0;
m1=1; m2=0.5; l1=0.5; l2=0.25; g=9.81; Bm1=0.1; Bm2=0.1; % Parameters
t1=1; o1=0; t2=-1; o2=0; % States equilibrium
sol2=solve([eqn3 eqn4],[u1 u2]);
T1=double(subs(sol2.u1)), T2=double(subs(sol2.u2)) % Computing inputs at equilibria
```

The linearized state-space model

$\dot{x}=Ax+Bu$

is found using the MATLAB code

```
syms m1 m2 l1 l2 g Bm1 Bm2 a1 a2 o1 o2 t1 t2 u1 u2
m1=1; m2=0.5; l1=0.5; l2=0.25; g=9.81; Bm1=0.1; Bm2=0.1;
x1=o1; x2=t1; x3=o2; x4=t2; % States
% Derivative of states
dx1=subs(alpha1); dx2=x1; dx3=subs(alpha2); dx4=x3;
% Jacobian matrices
J=[diff(dx1, x1) diff(dx1, x2) diff(dx1, x3) diff(dx1, x4); diff(dx2, x1) diff(dx2, x2) diff(dx2, x3) diff(dx2, x4);
   diff(dx3, x1) diff(dx3, x2) diff(dx3, x3) diff(dx3, x4); diff(dx4, x1) diff(dx4, x2) diff(dx4, x3) diff(dx4, x4)];
Ju=[diff(dx1, u1) diff(dx1,u2); diff(dx2, u1) diff(dx2,u2);
    diff(dx3, u1) diff(dx3, u2); diff(dx4, u1) diff(dx4, u2)];
t1=1; o1=0; t2=-1; o2=0; u1=T1; u2=T2; % States and input equilibria, u1=6.1911; u2=-1.0319;
A=double(subs(J)), B=double(subs(Ju)), H=[0 1 0 1], D=[0] % Linearized system dx/dt=Ax+Bu
```

One has

$$\dot{x} = Ax + Bu, \quad \begin{bmatrix} \dot{x}_1 \\ \dot{x}_2 \\ \dot{x}_3 \\ \dot{x}_4 \end{bmatrix} = \begin{bmatrix} -0.283 & -11.25 & -0.2355 & -1.56 \\ 1 & 0 & 0 & 0 \\ -0.2355 & -9.364 & -3.396 & -22.5 \\ 0 & 0 & 1 & 0 \end{bmatrix} \begin{bmatrix} x_1 \\ x_2 \\ x_3 \\ x_4 \end{bmatrix} + \begin{bmatrix} 2.83 & 2.355 \\ 0 & 0 \\ 2.355 & 33.96 \\ 0 & 0 \end{bmatrix} \begin{bmatrix} u_1 \\ u_2 \end{bmatrix}, \quad \begin{bmatrix} x_1 \\ x_2 \\ x_3 \\ x_4 \end{bmatrix} = \begin{bmatrix} \delta_{\omega_1} \\ \delta_{\theta_1} \\ \delta_{\omega_2} \\ \delta_{\theta_2} \end{bmatrix}, \quad \begin{bmatrix} u_1 \\ u_2 \end{bmatrix} = \begin{bmatrix} \delta_{T_1} \\ \delta_{T_2} \end{bmatrix}.$$

The linearized system is stable. The eigenvalues of matrix A are

$\lambda_{1,2}=-0.1665\pm3.1946i$, $\lambda_{3,4}=-1.6731\pm4.5287i$.

Stabilizing Control: Modal and LQR Methods – Solve the stabilization problem using the *modal* and LQR control concepts.

The *modal* control law $u=-K_F x$

is found assigning eigenvalues to be $\lambda_1=-5$, $\lambda_2=-7.5$, $\lambda_3=-10$, $\lambda_4=-12.5$.

These λ_i define system transient, stability margins, settling time, etc.

A MATLAB code to find the feedback gain matrix K_F is

```
p=[-5 -7.5 -10 -12.5]; KF=place(A,B,p); % Desired eigenvalues
KF=place(A,B,p); KF  % Computing feedback matrix KF
Ac=A-B*KF; eig(Ac)  % Computing closed-loop system matrix and eigenvalues
```

A *modal* control law is $u=-K_F x,\ K_F = \begin{bmatrix} 6.685 & 24.95 & -1.357 & -9.74 \\ -0.544 & -2.65 & 0.497 & 2.02 \end{bmatrix}$.

The LQR problem is solved by minimizing a quadratic performance functional

$J = \min_{x,u} \int_0^\infty \frac{1}{2}\left(x^T Q x + u^T G u\right) dt$, $q_{11}=1$, $q_{22}=10$, $q_{33}=1$, $q_{44}=10$, $G = \begin{bmatrix} 0.01 & 0 \\ 0 & 0.01 \end{bmatrix}$.

Solve the algebraic equation $Q + A^T K + KA - KBG^{-1}B^T K = 0$, $K>0$.

Find the feedback gain matrix K_F using the **lqr** solver.

```
Q=1*eye(size(A)); Q(2,2)=10; Q(4,4)=10; G=[0.01 0; 0 0.01]; G=G; [KF,K,Eigenvalues]=lqr(A,B,Q,G)
Ac=A-B*KF; eig(Ac) % Computing closed-loop system matrix and eigenvalues
```

We obtain $u=-K_F x,\ K_F = \begin{bmatrix} 10.9 & 27.9 & -0.069 & 0.0105 \\ -0.076 & -0.0106 & 10 & 30.97 \end{bmatrix}$.

The closed-loop system eigenvalues are $\lambda_1=-3.16$, $\lambda_2=-3.26$, $\lambda_3=-25.93$, $\lambda_4=-341.3$.

The LQR control law ensures near-decoupling of control functions (u_1,u_2) with state feedback on the first and second joint states (x_1,x_2) and (x_3,x_4). The *modal* control couples the first and second joints with positive terms. It is difficult to derive adequate eigenvalues for MIMO systems. The LQR design yields consistent results.

Figures 6.47.d and e illustrate the closed-system nonlinear system dynamics for $x_1(t)$, $x_2(t)$, $x_3(t)$ and $x_4(t)$ if $x_0=[0, 1, 0, -1]$. The angular displacements converge to $(\theta_1,\theta_2)=0$ rad, which correspond to downward pointing.

The linearized system $\begin{bmatrix} x_1 \\ x_2 \\ x_3 \\ x_4 \end{bmatrix} = \begin{bmatrix} \delta_{\omega_1} \\ \delta_{\theta_1} \\ \delta_{\omega_2} \\ \delta_{\theta_2} \end{bmatrix}, \begin{bmatrix} u_1 \\ u_2 \end{bmatrix} = \begin{bmatrix} \delta_{T_1} \\ \delta_{T_2} \end{bmatrix}$ evolves within $\bar{x} = \begin{bmatrix} \bar{x}_1 \\ \bar{x}_2 \\ \bar{x}_3 \\ \bar{x}_4 \end{bmatrix} = \begin{bmatrix} 0 \\ -1 \\ 0 \\ 1 \end{bmatrix}, \bar{u} = \begin{bmatrix} \bar{u}_1 \\ \bar{u}_2 \end{bmatrix} = \begin{bmatrix} 6.2 \\ -1 \end{bmatrix}$.

A stabilizing *modal* control law and state *observer* for a linearized model are designed and investigated in Practice Problem 5.6. One may consider stabilizing and tracking controls assuming all states are measured or observed, as well as solve a *minimal complexity* control problem by applying the Lyapunov concept.

Nonlinear System and Tracking Control – Study a nonlinear system using physical system states $x(t)$ and controls $u(t)$. Investigate tracking using the PID control law with physical limits $-25 \le u_1 \le 25$ N-m and $-5 \le u_2 \le 5$ N-m.

Figures 6.48.a documents a SIMULINK diagram with a control law

$$u = \begin{bmatrix} u_1 \\ u_2 \end{bmatrix}, \quad \begin{matrix} u_1 = \underset{-25 \le u_1 \le 25}{\text{saturation}}\left(k_{p1}e_1 + \int k_{i1}e_1 dt + k_{d1}\dfrac{de_1}{dt}\right) \\ u_2 = \underset{-5 \le u_2 \le 5}{\text{saturation}}\left(k_{p2}e_2 + \int k_{i2}e_2 dt + k_{d2}\dfrac{de_2}{dt}\right) \end{matrix}, \quad \begin{bmatrix} e_1(t) \\ e_2(t) \end{bmatrix} = \begin{bmatrix} r_1(t) - \theta_1(t) \\ r_2(t) - \theta_2(t) \end{bmatrix}.$$

The feedback gains are $k_{p1}=250$, $k_{i1}=100$, $k_{d1}=10$ and $k_{p2}=125$, $k_{i2}=5$, $k_{d2}=2.5$. The output dynamics (y_1,y_2) for references $\begin{bmatrix} r_1(t) \\ r_2(t) \end{bmatrix} = \begin{bmatrix} \pm 0.25 \text{ rad} \\ \mp 0.5 \text{ rad} \end{bmatrix}$ and $\begin{bmatrix} r_1(t) \\ r_2(t) \end{bmatrix} = \begin{bmatrix} \pm 0.5 \text{ rad} \\ \mp 1 \text{ rad} \end{bmatrix}$ are reported in Figures 6.48.b. Tracking of $r_1(t)$ and $r_2(t)$ is ensured. The MIMO system is nonlinear with link coupling and saturation on controls (u_1,u_2), which affect dynamics, operating envelope, admissible references (r_1,r_2), etc.

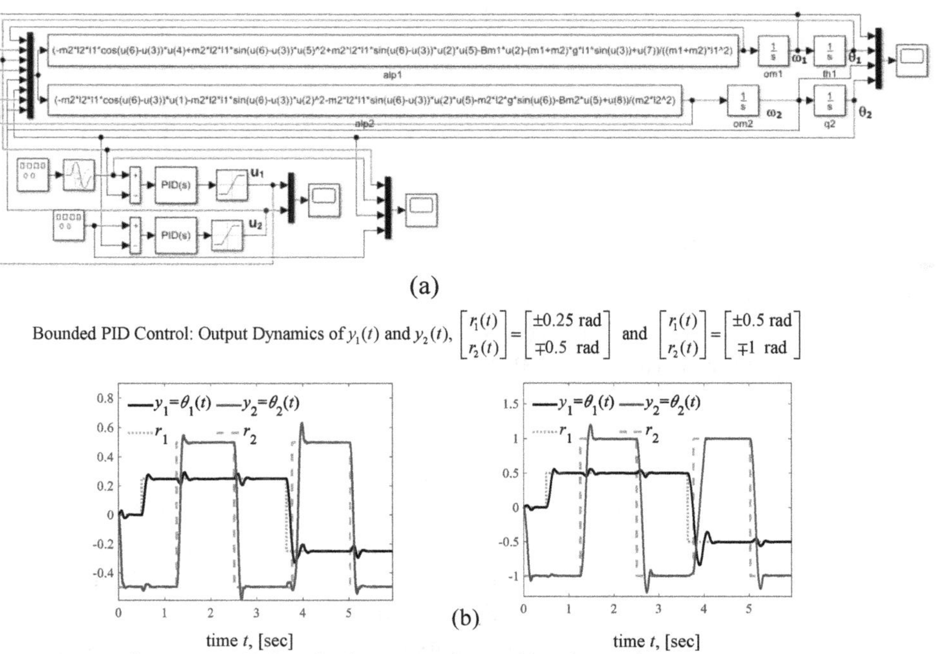

Figure 6.48. Nonlinear two-link robotic arm and transient dynamics:
(a) SIMULINK diagram of a system with bounded PID control, $k_{p1}=250$, $k_{i1}=100$, $k_{d1}=10$ and $k_{p2}=125$, $k_{i2}=5$, $k_{d2}=2.5$;
(b) Output dynamics (y_1,y_2) for references $r_1(t)$ and $r_2(t)$.

6.15.4. Three-Link Manipulator

Study a three-degree-of-freedom kinematics with three joints, shown in Figure 6.49.a. A model development problem was assigned in Homework Problem 2.1. The parameters are m_1=1 kg, m_2=0.75 kg, m_3=0.5 kg, l_1=0.25 m, l_2=0.25 m, l_3=0.25 m, B_{m1}=0.1 N-m-sec/rad, B_{m2}=0.1 N-m-sec/rad and B_{m3}=0.1 N-m-sec/rad. The system is stable if points downward. For example, $(\theta_{10},\theta_{20},\theta_{30})=0$ rad is a stable equilibrium for $\forall T_i=0$. In general, the admissible angular displacements are $-\pi \le (\theta_1,\theta_2,\theta_3) \le \pi$.

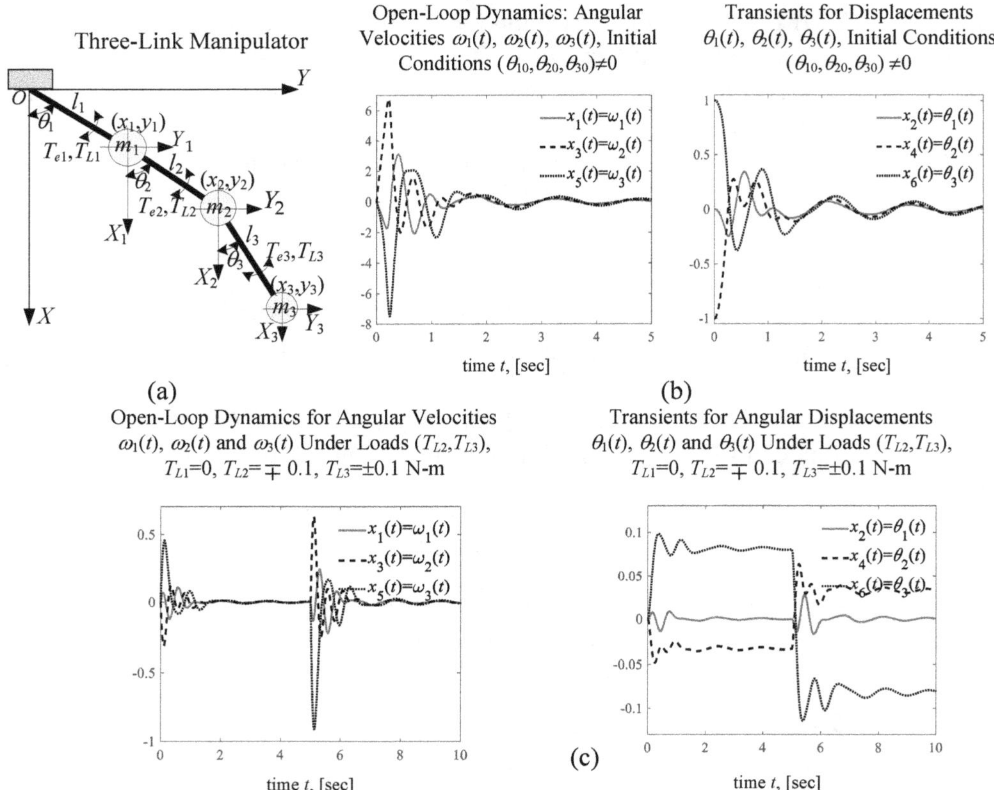

Figure 6.49. (a) Three-link robotic arm;
(b) Open-loop dynamics for $x_1(t)=\omega_1(t)$, $x_2(t)=\theta_1(t)$, $x_3(t)=\omega_2(t)$, $x_4(t)=\theta_2(t)$, $x_5(t)=\omega_3(t)$, $x_6(t)=\theta_3(t)$. Initial conditions are $(\omega_1,\omega_2,\omega_3)_0=0$ rad/s, $(\theta_1,\theta_2,\theta_3)_0=[0,-1, 1]$ rad. The loads are $\forall T_{Li}=0$ N-m;
(c) Open-loop dynamics if $(\omega_1,\omega_2,\omega_3)_0=0$ rad/s and $(\theta_1,\theta_2,\theta_3)_0=0$ rad. The load torques $T_{L1}=0$, $T_{L2}=\mp 0.1$ and $T_{L3}=\pm 0.1$ N-m are applied at f=0.1 Hz.

Mathematical Model – In the Lagrange equations (2.16) $\frac{d}{dt}\left(\frac{\partial \Gamma}{\partial \dot{q}_i}\right)-\frac{\partial \Gamma}{\partial q_i}+\frac{\partial D}{\partial \dot{q}_i}+\frac{\partial \Pi}{\partial q_i}=Q_i$, the angular displacements $(\theta_1,\theta_2,\theta_3)$ are the independent *generalized coordinates* (q_1,q_2,q_3). The *generalized forces* are the *net* torques, which are the actuators' torques (T_1,T_2,T_3), load torques (T_{L1},T_{L2},T_{L3}), and, the disturbance torques $(T_{\xi 1},T_{\xi 2},T_{\xi 3})$.

In the *xy* plane, the coordinates for mass points (m_1,m_2,m_3) are (x_1,y_1), (x_2,y_2) and (x_3,y_3),

$$x_1=l_1\cos\theta_1,\ y_1=l_1\sin\theta_1,\qquad x_2=l_1\cos\theta_1+l_2\cos\theta_2,\ y_2=l_1\sin\theta_1+l_2\sin\theta_2,$$

$$x_3=l_1\cos\theta_1+l_2\cos\theta_2+l_3\cos\theta_3,\ y_3=l_1\sin\theta_1+l_2\sin\theta_2+l_3\sin\theta_3.$$

The total kinetic, potential and dissipating energies (Γ,Π,D) are nonlinear functions of angular displacements ($\theta_1,\theta_2,\theta_3$) and accelerations $(\dot\theta_1,\dot\theta_2,\dot\theta_3)$. One has

$$\Gamma=\tfrac{1}{2}m_1v_1^2+\tfrac{1}{2}m_2v_2^2+\tfrac{1}{2}m_3v_3^2=\tfrac{1}{2}m_1\left(\dot x_1^2+\dot y_1^2\right)+\tfrac{1}{2}m_2\left(\dot x_2^2+\dot y_2^2\right)+\tfrac{1}{2}m_3\left(\dot x_3^2+\dot y_3^2\right)$$
$$=\tfrac{1}{2}(m_1+m_2+m_3)l_1^2\dot\theta_1^2+\tfrac{1}{2}(m_2+m_3)l_2^2\dot\theta_2^2+\tfrac{1}{2}m_3l_3^2\dot\theta_3^2+(m_2+m_3)l_1l_2\dot\theta_1\dot\theta_2\cos(\theta_2-\theta_1)$$
$$+m_3l_1l_3\dot\theta_1\dot\theta_3\cos(\theta_3-\theta_1)+m_3l_2l_3\dot\theta_2\dot\theta_3\cos(\theta_3-\theta_2),$$
$$\Pi=m_1gh_1+m_2gh_2+m_3gh_3=(m_1+m_2+m_3)gl_1(1-\cos\theta_1)+(m_2+m_3)gl_2(1-\cos\theta_2)+m_3gl_3(1-\cos\theta_3),$$
$$D=\tfrac{1}{2}B_{m1}\dot\theta_1^2+\tfrac{1}{2}B_{m2}\dot\theta_2^2+\tfrac{1}{2}B_{m3}\dot\theta_3^2.$$

For constant (m_1,m_2,m_3) and (l_1,l_2,l_3), one finds explicit expressions for $\dfrac{d}{dt}\left(\dfrac{\partial\Gamma}{\partial\dot q_i}\right)$, $\dfrac{\partial\Gamma}{\partial q_i}$, $\dfrac{\partial D}{\partial\dot q_i}$ and $\dfrac{\partial\Pi}{\partial q_i}$. The equations of motion, if $(T_{\xi1},T_{\xi2},T_{\xi3})=0$, are

$$l_1\Big[(m_1+m_2+m_3)l_1\ddot\theta_1+(m_2+m_3)l_2\cos(\theta_2-\theta_1)\ddot\theta_2+m_3l_3\cos(\theta_3-\theta_1)\ddot\theta_3-(m_2+m_3)l_2\sin(\theta_2-\theta_1)\dot\theta_2^2$$
$$-m_3l_3\sin(\theta_3-\theta_1)\dot\theta_3^2+(m_1+m_2+m_3)g\sin\theta_1\Big]+B_{m1}\dot\theta_1=T_1+T_{L1},$$
$$l_2\Big[(m_2+m_3)l_1\cos(\theta_2-\theta_1)\ddot\theta_1+(m_2+m_3)l_2\ddot\theta_2+m_3l_3\cos(\theta_3-\theta_2)\ddot\theta_3+(m_2+m_3)l_1\sin(\theta_2-\theta_1)\dot\theta_1^2$$
$$-m_3l_3\sin(\theta_3-\theta_2)\dot\theta_3^2+(m_2+m_3)g\sin\theta_2\Big]+B_{m2}\dot\theta_2=T_2+T_{L2},$$
$$l_3\Big[m_3l_1\cos(\theta_3-\theta_1)\ddot\theta_1+m_3l_2\cos(\theta_3-\theta_2)\ddot\theta_2+m_3l_3\ddot\theta_3+m_3l_1\sin(\theta_3-\theta_1)\dot\theta_1^2$$
$$+m_3l_2\sin(\theta_3-\theta_2)\dot\theta_2^2+m_3g\sin\theta_3\Big]+B_{m3}\dot\theta_3=T_3+T_{L3}.$$

Open-Loop Nonlinear System – Simulations are performed using the derived equations. For considered kinematics, stable in downward pointing, the transient dynamics are illustrated in Figures 6.49.b for initial conditions $\begin{bmatrix}\omega_{10}\\ \omega_{20}\\ \omega_{30}\end{bmatrix}=\begin{bmatrix}0\\0\\0\end{bmatrix}$ rad/s, $\begin{bmatrix}\theta_{10}\\ \theta_{20}\\ \theta_{30}\end{bmatrix}=\begin{bmatrix}0\\-1\\1\end{bmatrix}$ rad, and, for no-loads, $\forall T_{Li}=0$ N-m. For initial conditions $\forall\omega_{i0}=0$ rad/s, $\forall\theta_{i0}=0$ rad, and, applied loads $\begin{bmatrix}T_{L1}\\ T_{L2}\\ T_{L3}\end{bmatrix}=\begin{bmatrix}0\\ \mp0.1\\ \pm0.1\end{bmatrix}$ N-m at f=0.1 Hz, system dynamics are reported in Figures 6.49.c.

Torque Control – To ensure steering, actuators' torques (T_1,T_2,T_3) are applied to the first, second and third joints. For the *torque control* problem, these (T_1,T_2,T_3) are controls (u_1,u_2,u_3), and, $u_1=T_1$, $u_2=T_2$, $u_3=T_3$. To design control laws, find Cauchy's form of governing equations and linearize a nonlinear system, which is open-loop stable.

Cauchy's Form of Governing Equations and Linearization – Find Cauchy's form of state-space equations

$$\dot x=f(x,u,d),\quad x=\begin{bmatrix}x_1\\x_2\\x_3\\x_4\\x_5\\x_6\end{bmatrix}=\begin{bmatrix}\omega_1\\ \theta_1\\ \omega_2\\ \theta_2\\ \omega_3\\ \theta_3\end{bmatrix},\quad u=\begin{bmatrix}u_1\\u_2\\u_3\end{bmatrix},\quad d=\begin{bmatrix}T_{L1}\\T_{L2}\\T_{L3}\end{bmatrix}.$$

The results are found by using a MATLAB code

```
clear all; syms L1 m1 th1 w1 a1 B1 Ta1 TL1 L2 m2 th2 w2 a2 B2 Ta2 TL2 L3 m3 th3 w3 a3 B3 Ta3 TL3 g
eqn1 = (m1+m2+m3)*L1^2*a1 + (m2+m3)*L1*L2*cos(th2-th1)*a2 ...
  + m3*L1*L3*cos(th3-th1)*a3 - (m2+m3)*L1*L2*sin(th2-th1)*w2^2 ...
  - m3*L1*L3*sin(th3-th1)*w3^2 + B1*w1 + (m1+m2+m3)*g*L1*sin(th1) == Ta1 + TL1;
eqn2 = (m2+m3)*L1*L2*cos(th2-th1)*a1 + (m2+m3)*L2^2*a2 ...
  + m3*L2*L3*cos(th3-th2)*a3 + (m2+m3)*L1*L2*sin(th2-th1)*w1^2 ...
  - m3*L2*L3*sin(th3-th2)*w3^2 + B2*w2 + (m2+m3)*g*L2*sin(th2) == Ta2 + TL2;
eqn3 = m3*L1*L3*cos(th3-th1)*a1 + m3*L2*L3*cos(th3-th2)*a2 ...
  + m3*L3^2*a3 + m3*L1*L3*sin(th3-th1)*w1^2 ...
  + m3*L2*L3*sin(th3-th2)*w2^2 + B3*w3 + m3*g*L3*sin(th3) == Ta3 + TL3;
% Cauchy's equations for alpha1, alpha2 and alpha3
sol=solve([eqn1 eqn2 eqn3], [a1 a2 a3]); alpha1=sol.a1; alpha2 = sol.a2; alpha3 = sol.a3;
pretty(alpha1); pretty(alpha2); pretty(alpha3); simplify(alpha1); simplify(alpha2); simplify(alpha3);
```

The linearization is performed using a nonlinear system $\dot{x}=f(x,u,d)$. Sections 6.15.1, 6.15.2 and 6.15.3 illustrate the use of the MATLAB Symbolic Math Toolbox to find the Jacobian matrices. The model is linearized for chosen equilibrium $(\bar{\omega},\bar{\theta})$.

For $\begin{bmatrix}\bar{\omega}_1\\ \bar{\omega}_2\\ \bar{\omega}_3\end{bmatrix}=\begin{bmatrix}0\\0\\0\end{bmatrix}$ rad/s and $\begin{bmatrix}\bar{\theta}_1\\ \bar{\theta}_2\\ \bar{\theta}_3\end{bmatrix}=\begin{bmatrix}0\\-1\\1\end{bmatrix}$ rad, one has $\begin{bmatrix}\bar{u}_1\\ \bar{u}_2\\ \bar{u}_3\end{bmatrix}=\begin{bmatrix}\bar{T}_1\\ \bar{T}_2\\ \bar{T}_3\end{bmatrix}=\begin{bmatrix}0\\-2.58\\1\end{bmatrix}$ N-m.

The Jacobian mappings are found, and, the linearized equations of motion are

$$\dot{\delta}_x = A\delta_x + B\delta_u,\ A=\begin{bmatrix}-1.019 & -56.215 & 0.69 & 11.426 & 0.837 & 5.549\\ 1 & 0 & 0 & 0 & 0 & 0\\ 0.69 & 38.066 & -1.842 & -30.517 & -1.139 & -7.549\\ 0 & 0 & 1 & 0 & 0 & 0\\ 0.837 & 46.214 & -1.139 & -18.873 & -4.127 & -27.341\\ 0 & 0 & 0 & 0 & 1 & 0\end{bmatrix},\ B=\begin{bmatrix}10.187 & -6.898 & -8.375\\ 0 & 0 & 0\\ -6.898 & 18.424 & 11.394\\ 0 & 0 & 0\\ -8.375 & 11.394 & 41.267\\ 0 & 0 & 0\end{bmatrix}.$$

Despite inconsistencies, use notations $x\equiv\delta_x$ and $u\equiv\delta_u$.

Dynamics of the linearized system $\dot{x}=Ax+Bu$ is consistent with the nonlinear transitions for $-\tfrac{1}{2}\pi<(\theta_1,\theta_2,\theta_3)<\tfrac{1}{2}\pi$. The eigenvalues of matrix A are

$\lambda_{1,2}=-1.6933\pm j8.2566$, $\lambda_{3,4}=-0.6115\pm j4.28$, $\lambda_{5,6}=-1.1891\pm j4.362$.

Modal Control and Search Algorithm – Design a stabilizing *modal* control law

$u=-K_F x$

specifying eigenvalues for the closed-loop linear system

$\dot{x}=Ax+Bu,\ u=-K_F x$.

As discussed in section 5.4, the closed-loop system eigenvalues $\boldsymbol{\lambda}_F=(\lambda_{F1},\lambda_{F2},\lambda_{F3},\lambda_{F4},\lambda_{F5},\lambda_{F6})$ are chosen considering dynamic governance, settling time, stability margins, feedback gain definiteness, robustness under parameter variations, etc. Feedback gains $\forall k_{Fij}>0$ in K_F may be ensured by different eigenvalues. For specified range of eigenvalues $\Lambda_F^{\Lambda}\in[-10\quad -100]$, $\forall\boldsymbol{\lambda}_F\in\mathbb{R}$, we find

$\boldsymbol{\lambda}_F=[-10,-10,-15,-50,-70\ -70]$, $\boldsymbol{\lambda}_F=[-10,-10,-15,-55,-80,-80]$,
$\boldsymbol{\lambda}_F=[-10,-10,-15,-60,-60,-100]$,

as well as others $\boldsymbol{\lambda}_F$, yield $\forall k_{Fij}>0$. The search algorithm is used to find $\boldsymbol{\lambda}_F\in\mathbb{R}$ and feedback gains in a control law considering:

1. Specified $\boldsymbol{\lambda}_F\in\Lambda_F^{\Lambda}$, $\boldsymbol{\lambda}_F\in\mathbb{R}\times\mathbb{C}$ or $\forall\boldsymbol{\lambda}_F\in\mathbb{R}$;
2. *Admissible* gains $\forall k_{Fij}>0$ to ensure robustness and instabilities;
3. Matrix norms, $\min_{\boldsymbol{\lambda}_F\in\Lambda_F^{\Lambda}}\left\{\|K_F\|_1 : \forall k_{Fij}>0\right\}$, $\min_{\boldsymbol{\lambda}_F\in\Lambda_F^{\Lambda}}\left\{\|K_F\|_\infty : \forall k_{Fij}>0\right\}$.

Specify $\Lambda_F^A \in[-10 \quad -100]$, $\forall \boldsymbol{\lambda}_F \in \mathbb{R}$, $\forall k_{Fij}>0$, $\|K_F\|_1 \le 100$, $\|K_F\|_\infty \le 130$, $k_{F\max} \le 70$.

We find $\boldsymbol{\lambda}_F=[-10, -10, -15, -50, -15, -50]$.

The feedback gain matrix is $K_F=\begin{bmatrix} 8.375 & 66.33 & 2.372 & 27.57 & 1.581 & 21.91 \\ 2.596 & 22.78 & 3.237 & 31.24 & 0.98 & 18.18 \\ 1.001 & 8.615 & 0.044 & 3.744 & 1.025 & 9.399 \end{bmatrix}$

with $\forall k_{Fij}>0$, $\|K_F\|_1=97.749$, $\|K_F\|_\infty=128.14$ and $k_{F\max}=66.334$.

The stabilizing control $u=-K_F x$ is designed for the linearized system $\dot{\delta}_x = A\delta_x + B\delta_u$, $\delta_u=-K_F\delta_x$.

Nonlinear System With a Modal Control – Investigate a closed-loop nonlinear system. Simulations for a nonlinear system

$\dot{x}=f(x,u,d)$, $u=-K_F x$

are reported in Figures 6.50. The initial conditions are $\begin{bmatrix}\omega_{10}\\ \omega_{20}\\ \omega_{30}\end{bmatrix}=\begin{bmatrix}0\\0\\0\end{bmatrix}$ rad/s, $\begin{bmatrix}\theta_{10}\\ \theta_{20}\\ \theta_{30}\end{bmatrix}=\begin{bmatrix}0\\-1\\1\end{bmatrix}$ rad.

The loads torques $\begin{bmatrix}T_{L1}\\ T_{L2}\\ T_{L3}\end{bmatrix}=\begin{bmatrix}0\\ \mp 0.1\\ \pm 0.1\end{bmatrix}$ N-m are applied as steps at frequency f=0.5 Hz.

A closed-loop nonlinear system is stable, exhibits adequate dynamics, and, ensures disturbance rejection.

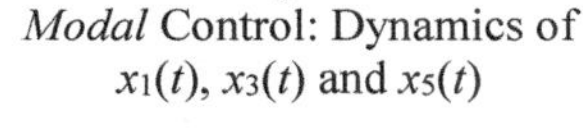
Modal Control: Dynamics of $x_1(t)$, $x_3(t)$ and $x_5(t)$

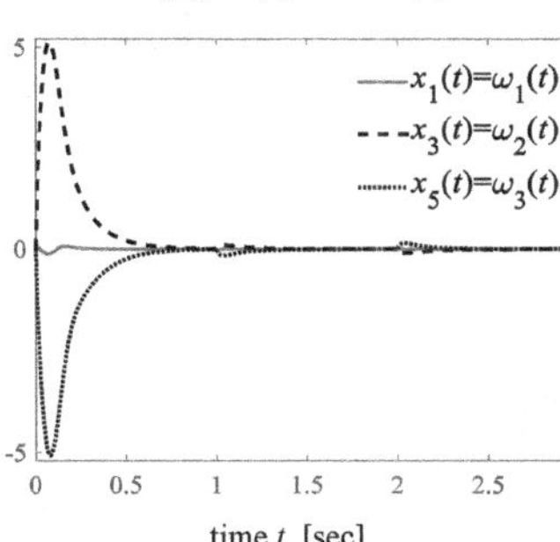

Modal Control: Dynamics of $x_2(t)$, $x_4(t)$ and $x_6(t)$

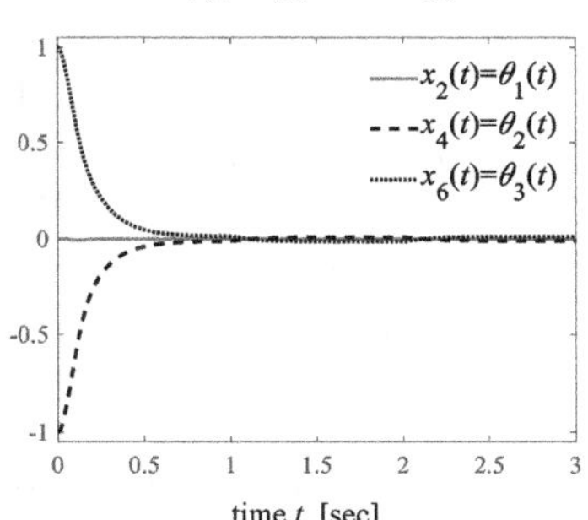

Figure 6.50. Nonlinear closed-loop system $\dot{x}=f(x,u,d)$, $u=-K_F x$: Dynamics if initial conditions are $(\omega_1,\omega_2,\omega_3)_0=0$ rad/s, $(\theta_1,\theta_2,\theta_3)_0=[0,-1,1]$ rad, and, the load torques $T_{L1}=0$, $T_{L2}=\mp 0.1$ and $T_{L3}=\pm 0.1$ N-m are applied, f=0.5 Hz.

Tracking Control Law – Applying the LQR concept, design the tracking control law.

Minimize a quadratic functional

$$J=\min_{\mathbf{x},u}\int_0^\infty \tfrac{1}{2}\left(\mathbf{x}^T Q\mathbf{x}+u^T G u\right)dt,\ \mathbf{x}\in\mathbb{R}^{12},\ u\in\mathbb{R}^3,$$

subject to the extended system dynamics

$\dot{x}=Ax+Bu$, $y_1=x_2=\theta_1$, $y_2=x_4=\theta_2$, $y_3=x_6=\theta_3$, $x\in\mathbb{R}^6$, $u\in\mathbb{R}^3$, $y\in\mathbb{R}^3$,

$$\dot{x}_e(t)=I_X(r-y)=I_X(r-Hx),\ e(t)\equiv\dot{x}_e(t),\ \begin{bmatrix}\dot{x}_{e1}\\ \dot{x}_{e2}\\ \dot{x}_{e3}\end{bmatrix}=\begin{bmatrix}1&0&0\\0&1&0\\0&0&1\end{bmatrix}\begin{bmatrix}e_1\\e_2\\e_3\end{bmatrix}=\begin{bmatrix}1&0&0\\0&1&0\\0&0&1\end{bmatrix}\left(\begin{bmatrix}r_1\\r_2\\r_3\end{bmatrix}-Hx\right),$$

$$\dot{e}(t)=-I_E e+\dot{r}-\dot{y}=-I_E e+\dot{r}-HAx-HBu,\ x_e\in\mathbb{R}^3,\ I_X=I,\ I_X\in\mathbb{R}^{3\times 3},\ e\in\mathbb{R}^3,\ I_E=I,\ I_E\in\mathbb{R}^{3\times 3}.$$

Having explicitly defined the governing equations for the MIMO system states, *error states* $x_e(t)$ and tracking error vector $e(t)$, we have

$$\dot{\mathbf{x}} = \mathbf{A}\mathbf{x} + A_r r + \mathbf{B}u,\ \mathbf{x} = \begin{bmatrix} x \\ \int e_1 dt \\ \int e_2 dt \\ \int e_3 dt \\ e_1 \\ e_2 \\ e_3 \end{bmatrix},\ x = \begin{bmatrix} x_1 \\ x_2 \\ x_3 \\ x_4 \\ x_5 \\ x_6 \end{bmatrix} = \begin{bmatrix} \omega_1 \\ \theta_1 \\ \omega_2 \\ \theta_2 \\ \omega_3 \\ \theta_3 \end{bmatrix},$$

$$\mathbf{A} = \begin{bmatrix} -1.019 & -56.215 & 0.69 & 11.426 & 0.837 & 5.549 & 0 & 0 & 0 & 0 & 0 & 0 \\ 1 & 0 & 0 & 0 & 0 & 0 & 0 & 0 & 0 & 0 & 0 & 0 \\ 0.69 & 38.066 & -1.842 & -30.517 & -1.139 & -7.549 & 0 & 0 & 0 & 0 & 0 & 0 \\ 0 & 0 & 1 & 0 & 0 & 0 & 0 & 0 & 0 & 0 & 0 & 0 \\ 0.837 & 46.214 & -1.139 & -18.873 & -4.127 & -27.341 & 0 & 0 & 0 & 0 & 0 & 0 \\ 0 & 0 & 0 & 0 & 1 & 0 & 0 & 0 & 0 & 0 & 0 & 0 \\ 0 & -1 & 0 & 0 & 0 & 0 & 0 & 0 & 0 & 0 & 0 & 0 \\ 0 & 0 & 0 & -1 & 0 & 0 & 0 & 0 & 0 & 0 & 0 & 0 \\ 0 & 0 & 0 & 0 & 0 & -1 & 0 & 0 & 0 & 0 & 0 & 0 \\ -1 & 0 & 0 & 0 & 0 & 0 & 0 & 0 & 0 & -1 & 0 & 0 \\ 0 & 0 & -1 & 0 & 0 & 0 & 0 & 0 & 0 & 0 & -1 & 0 \\ 0 & 0 & 0 & 0 & -1 & 0 & 0 & 0 & 0 & 0 & 0 & -1 \end{bmatrix},\ \mathbf{B} = \begin{bmatrix} 10.187 & -6.898 & -8.375 \\ 0 & 0 & 0 \\ -6.898 & 18.424 & 11.394 \\ 0 & 0 & 0 \\ -8.375 & 11.394 & 41.267 \\ 0 & 0 & 0 \\ 0 & 0 & 0 \\ 0 & 0 & 0 \\ 0 & 0 & 0 \\ 0 & 0 & 0 \\ 0 & 0 & 0 \\ 0 & 0 & 0 \end{bmatrix}.$$

Solution of the functional equation is satisfied by a continuous quadratic return function $V(\mathbf{x}) = \frac{1}{2}\mathbf{x}^T K\mathbf{x}$, $K=K^T$, $K>0$.

Solve an algebraic matrix equation $Q + \mathbf{A}^T K + K\mathbf{A} - K\mathbf{B}G^{-1}\mathbf{B}^T K = 0$.

For $q_{11}=0.01$, $q_{22}=1$, $q_{33}=0.01$, $q_{44}=1$, $q_{55}=0.01$, $q_{66}=1$, $q_{77}=100$, $q_{88}=100$, $q_{99}=100$, $q_{10,10}=1$, $q_{11,11}=1$, $q_{12,12}=1$, $G = \begin{bmatrix} 0.0001 & 0 & 0 \\ 0 & 0.0001 & 0 \\ 0 & 0 & 0.0001 \end{bmatrix}$, compute $K \in \mathbb{R}^{12\times 12}$ and feedback gain matrix $K_F \in \mathbb{R}^{3\times 12}$. A tracking control law is

$$u = -G^{-1}\mathbf{B}^T K\mathbf{x} = -K_F\mathbf{x},\ \mathbf{x}(t) = \begin{bmatrix} x(t) \\ x_e(t) \\ e(t) \end{bmatrix},\ \mathbf{x} = \begin{bmatrix} x \\ \int e_1 dt \\ \int e_2 dt \\ \int e_3 dt \\ e_1 \\ e_2 \\ e_3 \end{bmatrix},\ x = \begin{bmatrix} x_1 \\ x_2 \\ x_3 \\ x_4 \\ x_5 \\ x_6 \end{bmatrix} = \begin{bmatrix} \omega_1 \\ \theta_1 \\ \omega_2 \\ \theta_2 \\ \omega_3 \\ \theta_3 \end{bmatrix},\ e = \begin{bmatrix} e_1 \\ e_2 \\ e_3 \end{bmatrix} = \begin{bmatrix} r_1 - \theta_1 \\ r_2 - \theta_2 \\ r_3 - \theta_3 \end{bmatrix}.$$

The feedback gain matrix is

$$K_F = \begin{bmatrix} 12.46 & 202.61 & 0.751 & 3.497 & 0.309 & 1.452 & -1000 & 0.599 & 0.331 & -4.089 & 0.0173 & 0.0076 \\ 0.7648 & 3.743 & 11.38 & 201.24 & -0.25 & -1.227 & -0.599 & -1000 & -0.059 & 0.0124 & -4.11 & -0.0054 \\ 0.3167 & 1.586 & -0.252 & -1.25 & 10.51 & 198.01 & -0.331 & 0.059 & -1000 & 0.0049 & -0.0049 & -4.128 \end{bmatrix}.$$

The overall consistency is ensured. Simulate a closed-loop nonlinear system with the tracking control law

$$\dot{x} = f(x,u,d,r),\ u = -K_F\mathbf{x},\ \begin{bmatrix} T_{L1} \\ T_{L2} \\ T_{L3} \end{bmatrix} = \begin{bmatrix} 0 \\ \mp 0.1 \\ \pm 0.1 \end{bmatrix} \text{ N-m},\ f=1 \text{ Hz}.$$

The references are $\begin{bmatrix} r_1(t) \\ r_2(t) \\ r_3(t) \end{bmatrix} = \begin{bmatrix} \pm 0.25 \\ \mp 0.5 \\ \pm 0.75 \end{bmatrix}$ rad, f_1=0.15 Hz, f_2=0.175 Hz and f_3=0.2 Hz.

Results are documented in Figures 6.51. Tracking and disturbance rejection are ensured.

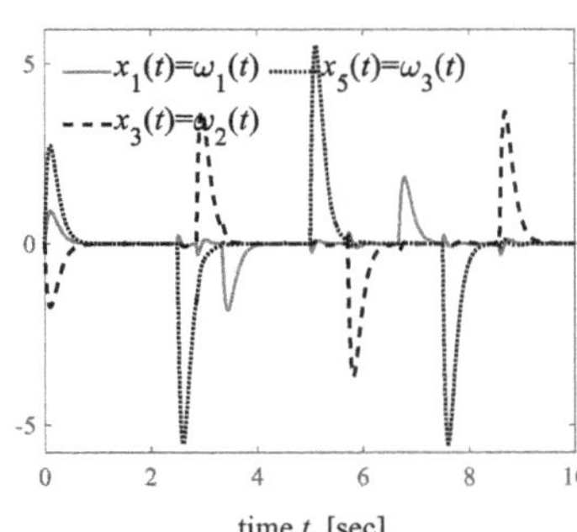

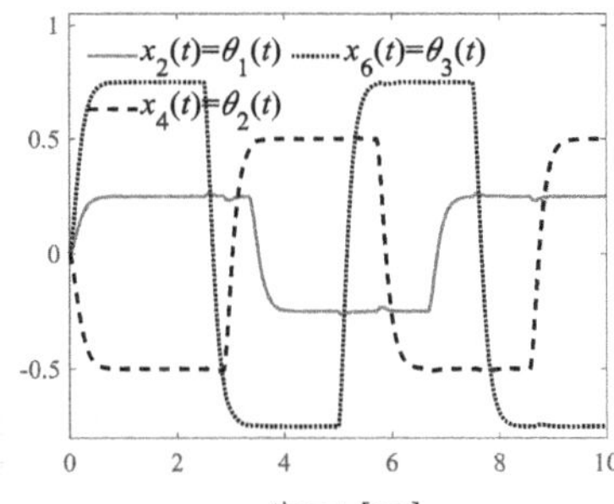

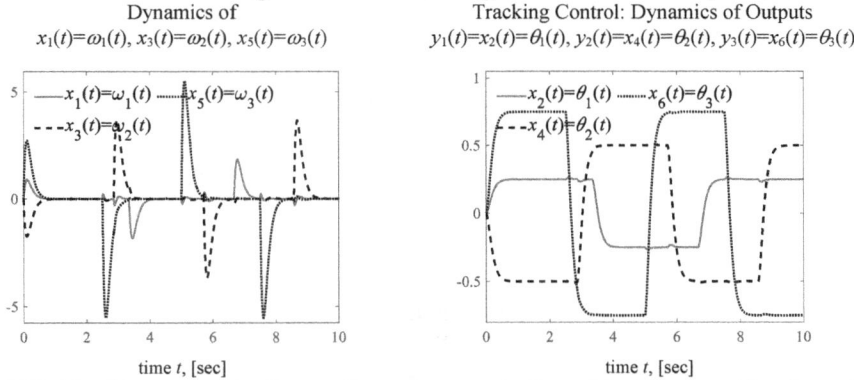

Figure 6.51. Tracking control of a nonlinear system $\dot{x}=f(x,u,d,r)$ with $u=-K_F\mathbf{x}$ for references $\begin{bmatrix} r_1(t) \\ r_2(t) \\ r_3(t) \end{bmatrix} = \begin{bmatrix} \pm 0.25 \\ \mp 0.5 \\ \pm 0.75 \end{bmatrix}$ rad, f_1=0.15 Hz, f_2=0.175 Hz, f_3=0.2 Hz. Load torques $\begin{bmatrix} T_{L1} \\ T_{L2} \\ T_{L3} \end{bmatrix} = \begin{bmatrix} 0 \\ \mp 0.1 \\ \pm 0.1 \end{bmatrix}$ N-m, f=1 Hz.

State Observer Design – Using results, reported in section 5.4, design the state *observer* to estimate not measured angular velocities $(\omega_1(t),\omega_2(t),\omega_3(t))$ using the directly measured angular displacements $(\theta_1(t),\theta_2(t),\theta_3(t))$ and inputs.

An equation for the state *observer* (5.50) is $\frac{d\hat{x}}{dt} = (A - K_E H)\hat{x} + Bu + K_E y$.

To observe $\begin{bmatrix} \hat{x}_1(t) \\ \hat{x}_3(t) \\ \hat{x}_5(t) \end{bmatrix} = \begin{bmatrix} \hat{\omega}_1(t) \\ \hat{\omega}_2(t) \\ \hat{\omega}_3(t) \end{bmatrix}$, specify the state *observer* eigenvalues $\boldsymbol{\lambda}_E \in \mathbb{R}$ within $\Lambda_E^{\Lambda} \in [-50\ \ -225]$. The objective is to ensure specified *observer* dynamics, positive definite gains $\forall k_{Eij}>0$, as well as avoid high gains by letting $\|K_E\|_1 \le 25000$, $\|K_E\|_\infty \le 25000$ and $k_{E\max} \le 17500$.

Using the optimization scheme, reported in the *State Observer and Modal Control Algorithm* 5.1, we find $\boldsymbol{\lambda}_E$=[–60, –75, –100, –150, –160, –210].

The *observer* gain matrix is $K_E = \begin{bmatrix} 16305.7 & 6017.9 & 1318.9 \\ 272.09 & 46.76 & 10.678 \\ 5026 & 13274.4 & 606.94 \\ 40.447 & 253.17 & 6.191 \\ 1237.6 & 778.69 & 10524.5 \\ 10.12 & 7.328 & 222.75 \end{bmatrix}$, $\forall k_{Eij}>0$,

and, $\|K_E\|_1$=22892, $\|K_E\|_\infty$=23642, $k_{E\max}$=16305.7.

The specifications are met. We solved a problem, postulated in section 5.4.

State Observer and Modal Control Algorithm 5.1 – For specified $(\boldsymbol{\lambda}_F \in \Lambda_F^{\Lambda}, \boldsymbol{\lambda}_E \in \Lambda_E^{\Lambda})$, $\forall(\boldsymbol{\lambda}_F,\boldsymbol{\lambda}_E)\in\mathbb{R}$, minimize $\min\limits_{\boldsymbol{\lambda}_F \in \Lambda_F^{\Lambda},\ \boldsymbol{\lambda}_E \in \Lambda_E^{\Lambda}} \left\{\|K_F\|_p, \|K_E\|_p\right\}$, $k_{Fij\min} \le k_{Fij} \le k_{Fij\max}$, $k_{Eij\min} \le k_{Eij} \le k_{Eij\max}$.

We guarantee asymptotic convergence of $(x(t), \hat{x}(t), \bar{e}(t))$ as $\begin{cases} \lim_{t\to\infty} x(t) \to x_e \\ \lim_{t\to\infty} \hat{x}(t) \to x, \lim_{t\to\infty} \bar{e}(t) \to 0 \end{cases}$, ensuring criteria imposed on the controller and *observer* gains (K_F,K_E).

Figures 6.52 illustrate dynamics of a nonlinear system $\dot{x}=f(x,u,d,r)$ with a tracking control law with the observed states

$$u = -G^{-1}\mathbf{B}^T K\mathbf{x} = -K_F\mathbf{x},\ \mathbf{x}(t) = \begin{bmatrix} \hat{x}(t) \\ x_e(t) \\ e(t) \end{bmatrix},\ \mathbf{x} = \begin{bmatrix} \hat{x} \\ \int e_1 dt \\ \int e_2 dt \\ \int e_3 dt \\ e_1 \\ e_2 \\ e_3 \end{bmatrix},\ \hat{x} = \begin{bmatrix} \hat{x}_1 \\ x_2 \\ \hat{x}_3 \\ x_4 \\ \hat{x}_5 \\ x_6 \end{bmatrix} = \begin{bmatrix} \hat{\omega}_1 \\ \theta_1 \\ \hat{\omega}_2 \\ \theta_2 \\ \hat{\omega}_3 \\ \theta_3 \end{bmatrix},\ e = \begin{bmatrix} e_1 \\ e_2 \\ e_3 \end{bmatrix} = \begin{bmatrix} r_1 - \theta_1 \\ r_2 - \theta_2 \\ r_3 - \theta_3 \end{bmatrix}.$$

The references $r(t)$ and disturbances $d(t)$ are as were used earlier. Results, reported in Figures 6.51 and 6.52, substantiate our findings and justify chosen eigenvalues and computed K_F and K_E. The *observer* simplifies hardware solutions, however, implies real-time *observer* implementation on MPU. Control laws and an *observer*, designed for a linear system, are adequate for a nonlinear system due to open-loop stability. If a system is under loads, disturbances and perturbations, these $d(t)\equiv(T_{Li}(t),T_{\xi i}(t))$ should be considered. Measurements of $(T_{Li}(t),d(t))$ is a challenging task. Consistency is ensured for *admissible* unmeasured $(T_{Li}(t),T_{\xi i}(t))$, for which one may assume d_i=0. For small $(T_{Li}(t),T_{\xi i}(t))$, the convergence is guaranteed, the observed states error is zero for d=0, and, the error is acceptable because $u(t)>d(t)$.

Closed-Loop System With an *Observer*: Dynamics of $x_1(t)=\omega_1(t)$, $x_3(t)=\omega_2(t)$, $x_5(t)=\omega_3(t)$

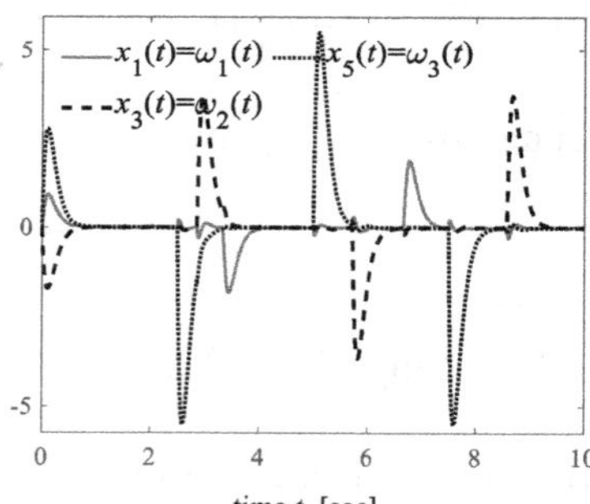

Tracking Control With an *Observer*: Output Dynamics of $y_1(t)=x_2(t)=\theta_1(t)$, $y_2(t)=x_4(t)=\theta_2(t)$, $y_3(t)=x_6(t)=\theta_3(t)$

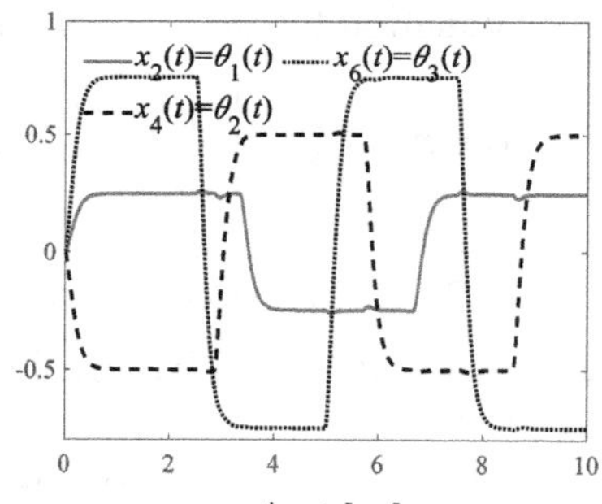

Figure 6.52. Control of a nonlinear closed-loop system $\dot{x}=f(x,u,d,r)$, $u=-K_F\mathbf{x}$ with an *observer* $\frac{d\hat{x}}{dt} = (A - K_E H)\hat{x} + Bu + K_E y$. Dynamics for references $\begin{bmatrix} r_1(t) \\ r_2(t) \\ r_3(t) \end{bmatrix} = \begin{bmatrix} \pm 0.25 \\ \mp 0.5 \\ \pm 0.75 \end{bmatrix}$ rad and loads $\begin{bmatrix} T_{L1} \\ T_{L2} \\ T_{L3} \end{bmatrix} = \begin{bmatrix} 0 \\ \mp 0.1 \\ \pm 0.1 \end{bmatrix}$ N-m.

Digital Control – To design a digital tracking control law, discretize a linearized system. For a continuous-time system $\dot{\mathbf{x}} = \mathbf{A}\mathbf{x} + A_r r + \mathbf{B}u,\ \mathbf{x} = \begin{bmatrix} x \\ \int e dt \\ e \end{bmatrix}$, find a discrete-time state-space model. Apply (5.19) and compute $\mathbf{A}_n = e^{\mathbf{A}T_s}$, $\mathbf{B}_n = \mathbf{A}^{-1}(e^{\mathbf{A}T_s} - I)\mathbf{B}$ for T_s=0.025 sec. We find

$$\mathbf{x}_{n+1} = \mathbf{A}_n\mathbf{x}_n + A_{rn}r_n + \mathbf{B}_n u_n,\ \mathbf{x}_n = \begin{bmatrix} x_n \\ x_{en} \\ e_n \end{bmatrix},\ H_n=H,\ D_n=D,\ n\geq 0,$$

$$\mathbf{A}_n=\begin{bmatrix} 0.9582 & -1.3587 & 0.0196 & 0.2683 & 0.0208 & 0.1273 & 0 & 0 & 0 & 0 & 0 & 0 \\ 0.0245 & 0.9828 & 0.0002 & 0.0034 & 0.0003 & 0.0016 & 0 & 0 & 0 & 0 & 0 & 0 \\ 0.0276 & 0.8939 & 0.9464 & -0.7332 & -0.0283 & -0.1725 & 0 & 0 & 0 & 0 & 0 & 0 \\ 0.0003 & 0.0114 & 0.02437 & 0.9907 & -0.0004 & -0.0022 & 0 & 0 & 0 & 0 & 0 & 0 \\ 0.0329 & 1.06 & -0.0316 & -0.4313 & 0.8948 & -0.6429 & 0 & 0 & 0 & 0 & 0 & 0 \\ 0.0004 & 0.0137 & -0.0004 & -0.0056 & 0.02369 & 0.9918 & 0 & 0 & 0 & 0 & 0 & 0 \\ -0.0003 & -0.0249 & 0 & 0 & 0 & 0 & 1 & 0 & 0 & 0 & 0 & 0 \\ 0 & 0 & -0.0003 & -0.0249 & 0 & 0 & 0 & 1 & 0 & 0 & 0 & 0 \\ 0 & -0.0001 & 0 & 0 & -0.0003 & -0.0249 & 0 & 0 & 1 & 0 & 0 & 0 \\ -0.0242 & 0.0171 & -0.0002 & -0.0034 & -0.0003 & -0.0016 & 0 & 0 & 0 & 0.975 & 0 & 0 \\ -0.0003 & -0.0113 & -0.0241 & 0.0092 & 0.0004 & 0.0022 & 0 & 0 & 0 & 0 & 0.975 & 0 \\ -0.0004 & -0.0135 & 0.0004 & 0.0055 & -0.0234 & 0.0081 & 0 & 0 & 0 & 0 & 0 & 0.975 \end{bmatrix},\ \mathbf{B}_n=\begin{bmatrix} 0.2462 & -0.162 & -0.1921 \\ 0.0031 & -0.0021 & -0.0025 \\ -0.162 & 0.4427 & 0.2604 \\ -0.0021 & 0.0056 & 0.0034 \\ -0.1921 & 0.2604 & 0.97029 \\ -0.0025 & 0.0034 & 0.0124 \\ 0 & 0 & 0 \\ 0 & 0 & 0 \\ 0 & 0 & 0 \\ -0.0031 & 0.0021 & 0.0025 \\ 0.0021 & -0.0056 & -0.0033 \\ 0.0025 & -0.0033 & -0.0123 \end{bmatrix}.$$

Minimize the performance index

$$J=\min_{\mathbf{x}_n,u_n}\sum_{n=0}^{\infty}\left[\mathbf{x}_n^TQ_n\mathbf{x}_n+u_n^TG_nu_n\right],\ Q_n\ge 0,\ G_n>0,$$

subject to a discrete-time system.

Using the quadratic return function $V(\mathbf{x}_n)=\mathbf{x}_n^TK_n\mathbf{x}_n$, $K_n>0$, a tracking control law is

$$u_n=-\left(G_n+\mathbf{B}_n^TK_n\mathbf{B}_n\right)^{-1}\mathbf{B}_n^TK_n\mathbf{A}_n\mathbf{x}_n,$$

where a symmetric matrix K_n is computed by solving

$$-K_n+Q_n+\mathbf{A}_n^TK_n\mathbf{A}_n-\mathbf{A}_n^TK_n\mathbf{B}_n\left(G_n+\mathbf{B}_n^TK_n\mathbf{B}_n\right)^{-1}\mathbf{B}_n^TK_n\mathbf{A}_n=0,\ K_n=K_n^T,\ K_n>0.$$

For $q_{11n}=0.01$, $q_{22n}=1$, $q_{33n}=0.01$, $q_{44n}=1$, $q_{55n}=0.01$, $q_{66n}=1$, $q_{77n}=100$, $q_{88n}=100$, $q_{99n}=100$, $q_{10,10\,n}=1$, $q_{11,11\,n}=1$, $q_{12,12\,n}=1$ and $G_n=\begin{bmatrix} 0.0001 & 0 & 0 \\ 0 & 0.0001 & 0 \\ 0 & 0 & 0.0001 \end{bmatrix}$, we have

$$u_n=-\left(G_n+\mathbf{B}_n^TK_n\mathbf{B}_n\right)^{-1}\mathbf{B}_n^TK_n\mathbf{A}_n\mathbf{x}_n=-K_{Fn}\mathbf{x}_n,$$

$$K_{Fn}=\begin{bmatrix} 5.703 & 73.88 & 1.584 & 20.621 & 0.717 & 9.788 & -350.06 & -89.32 & -43.62 & -1.411 & -0.353 & -0.17 \\ 1.586 & 20.69 & 3.419 & 47.49 & -0.626 & -8.899 & -89.28 & -219.21 & 40.06 & -0.358 & -0.883 & 0.162 \\ 0.717 & 9.814 & -0.626 & -8.904 & 1.44 & 21.36 & -43.32 & 40.04 & -99.18 & -0.175 & 0.162 & -0.4 \end{bmatrix}.$$

System dynamics are illustrated in Figures 6.53. Comparison of analog and digital control laws demonstrate similarity of transient dynamics and system evolutions.

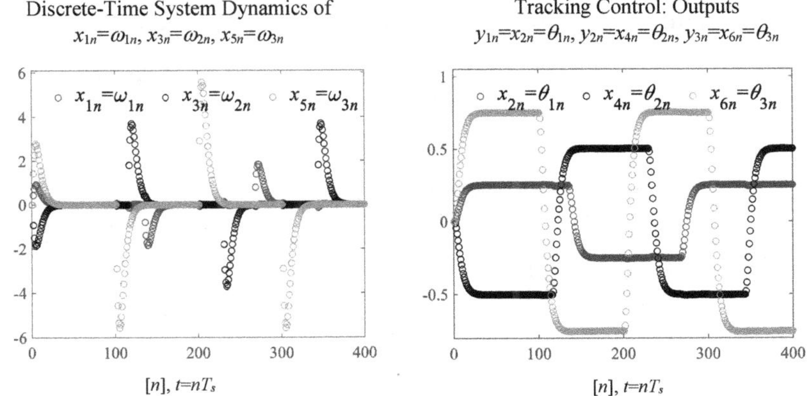

Figure 6.53. Dynamics of a discrete-time system with a control law $u_n=-K_F\mathbf{x}_n$, $\begin{bmatrix} r_{1n} \\ r_{2n} \\ r_{3n} \end{bmatrix}=\begin{bmatrix} \pm 0.25 \\ \mp 0.5 \\ \pm 0.75 \end{bmatrix}$ rad.

Constrained Tracking Control – Design a control law considering physical limits on torques (T_1,T_2,T_3), developed by actuators. Physical saturation limits $-\mathcal{U}_{i\,\max}\le\mathcal{U}_i\le\mathcal{U}_{i\,\max}$, $|\mathcal{U}_1|\le 4$ N-m, $|\mathcal{U}_2|\le 3$ N-m and $|\mathcal{U}_3|\le 2$ N-m are described using $u=\Phi(x)=\boldsymbol{u}_{\max}\tanh(\boldsymbol{a}x)$.

Minimize a nonquadratic functional

$$J=\min_{\substack{\mathbf{x}\\ |u_1|\le u_{1\max},|u_2|\le u_{2\max},|u_3|\le u_{3\max}}}\int_0^{\infty}\left[\tfrac{1}{2}\mathbf{x}^TQ\mathbf{x}+\int\left(\boldsymbol{a}^{-1}\tanh^{-1}(\boldsymbol{u}_{\max}^{-1}u)\right)^T G\,du\right]dt,\ u_{1\max}=4,\ u_{2\max}=3,\ u_{3\max}=2,$$

subject to the system dynamics. Here, $\boldsymbol{a}=I$, $\forall a_i=1$.

Design yields $u=-\boldsymbol{u}_{\max}\tanh(\boldsymbol{a}G^{-1}\mathbf{B}^TK\mathbf{x})=-\begin{bmatrix}u_{1\max}&0&0\\0&u_{2\max}&0\\0&0&u_{3\max}\end{bmatrix}\tanh\left(K_F\begin{bmatrix}x\\ \int e dt\\ e\end{bmatrix}\right)$.

Solution of $\tfrac{1}{2}\mathbf{x}^TQ\mathbf{x}+\left(\frac{\partial V}{\partial\mathbf{x}}\right)^T\mathbf{A}\mathbf{x}-\int\left[\boldsymbol{u}_{\max}\tanh\left(\boldsymbol{a}G^{-1}\mathbf{B}^T\frac{\partial V}{\partial\mathbf{x}}\right)\right]^T d\left(\mathbf{B}^T\frac{\partial V}{\partial\mathbf{x}}\right)=0$,

$$\int\left[\boldsymbol{u}_{\max}\tanh(\boldsymbol{a}G^{-1}\mathbf{B}^T\tfrac{\partial V}{d\mathbf{x}})\right]^T d(\mathbf{B}^T\tfrac{\partial V}{d\mathbf{x}})\approx\tfrac{1}{3}\frac{\partial V}{d\mathbf{x}}^T\mathbf{B}\boldsymbol{u}_{\max}\boldsymbol{a}G^{-1}\mathbf{B}^T\frac{\partial V}{d\mathbf{x}}$$

is approximated by a quadratic return function $V(\mathbf{x})=\tfrac{1}{2}\mathbf{x}^TK\mathbf{x}$.

Using the weighting matrix Q with $q_{11}=0.01$, $q_{22}=1$, $q_{33}=0.01$, $q_{44}=1$, $q_{55}=0.01$, $q_{66}=1$, $q_{77}=100$, $q_{88}=100$, $q_{99}=100$, $q_{10,10}=1$, $q_{11,11}=1$, $q_{12,12}=1$, and, $G=\begin{bmatrix}0.0001&0&0\\0&0.0001&0\\0&0&0.0001\end{bmatrix}$, compute $K\in\mathbb{R}^{12\times 12}$. The feedback gain matrix $K_F\in\mathbb{R}^{3\times 12}$ is computed.

A closed-loop nonlinear system is

$$\dot{x}=f(x,u,d,r),\ -u_{i\,\max}\le u_i\le u_{i\,\max},\ u=-\boldsymbol{u}_{\max}\tanh(K_F\mathbf{x})=-\begin{bmatrix}u_{1\max}&0&0\\0&u_{2\max}&0\\0&0&u_{3\max}\end{bmatrix}\tanh\left(K_F\begin{bmatrix}x\\ \int e dt\\ e\end{bmatrix}\right),$$

$$K_F=\begin{bmatrix}18.851&326.48&0.334&-4.99&-0.488&-14.25&-1630.37&44.91&80.87&-6.7&0.202&0.341\\1.2&11.92&15.53&282.14&0.26&8.78&-41.45&-1412.7&-51.26&-0.156&-5.82&-0.217\\0.88&12.52&-0.688&-9.76&12.04&227.87&-55.97&43.47&-1152.5&-0.225&0.175&-4.76\end{bmatrix}.$$

System transients and control evolutions are documented in Figures 6.54. The references are $\begin{bmatrix}r_1(t)\\r_2(t)\\r_3(t)\end{bmatrix}=\begin{bmatrix}\pm 0.25\\\mp 0.5\\\pm 0.75\end{bmatrix}$ rad, $f_1=0.15$ Hz, $f_2=0.175$ Hz, $f_3=0.2$ Hz. Loads $\begin{bmatrix}T_{L1}\\T_{L2}\\T_{L3}\end{bmatrix}=\begin{bmatrix}0\\\mp 0.1\\\pm 0.1\end{bmatrix}$ N-m are applied with $f=1$ Hz. Tracking and disturbance rejection are ensured.

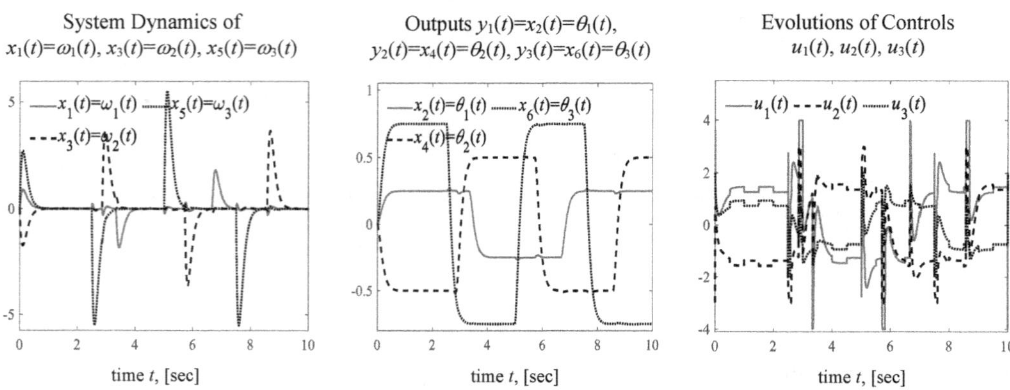

Figure 6.54. Tracking control of a nonlinear closed-loop system $\dot{x}=f(x,u,d,r)$, $-u_{i\,\max}\le u_i\le u_{i\,\max}$ under loads T_{Li}: Dynamics and tracking for references $r_i(t)$.

6.16. Homework Problems

Homework Problem 6.1.
Report specifications imposed on closed-loop automotive, aerial, electronic and/or robotic systems and their modules. Discuss physical controllability, control limits, hardware and software solutions. Report how to solve control problems for linear and nonlinear permanent magnet servos, studied in chapter 5 and 6. Consider different methods, such as transfer functions, Laplace transforms, *modal* control, LQR design, minimum-time control, calculus of variations, Hamiltonian optimization, Lyapunov paradigm, etc. How control limits affect system performance and stability? Report challenges in design and implementation of control laws.

Homework Problem 6.2.
Consider the following performance functionals

$$J=\min_{t,x,e,u}\int_0^\infty\left(t|e|+x^2+e^2+u^2\right)dt,\ J=\min_{t,e,u}\int_0^\infty\left(t|e|+1\right)dt,\ J=\min_{x,e,|u|\le 1}\int_0^\infty\left(x^2+e^4+\int\tanh u\,du\right)dt.$$

Explain how the specifications on settling time, tracking error, states and control are imposed and described by performance integrands. Use alternative integrands in the performance functionals. Note that the relay switching $u=\pm1$ implies $u^2\equiv1$, which yields

$$J=\min_{t,e,u}\int_0^\infty\left(t|e|+\underset{u=\pm1}{u^2}\right)dt\equiv\min_{t,e,u\in U}\int_0^\infty\left(t|e|+1\right)dt.$$

Explain the role of performance integrands in analytic design of control laws.

Homework Problem 6.3.
Consider an open-loop stable system

$$\dot x_1=-x_1-x_1^3+x_2,\ x_{10}=10,$$
$$\dot x_2=x_1-x_2^3+u,\ x_{20}=-10.$$

Problems to solve:

1. Minimizing a functional $J=\min_{x,u}\int_0^\infty\frac{1}{2}\left(x_1^2+2x_2^2+3x_2^4+u^2\right)dt$, derive a stabilizing control law. Apply a return function $V=\frac{1}{2}k_{11}x_1^2+k_{12}x_1x_2+\frac{1}{2}k_{22}x_2^2+\frac{1}{4}k_{44}x_2^4$ which approximates the solution of the algebraic Hamilton-Jacobi equation. Design and apply alternative nonquadratic return functions. Analyze the error in solution of the Hamilton-Jacobi functional equation;
2. Study the closed-loop system stability, and, design control laws using the Lyapunov theory. Design a *minimal complexity* control law using measured x_2;
3. Perform simulations of closed-loop systems for different control laws designed.

Homework Problem 6.4.
For a system with relay control function switching $u=\pm1$, consider the governing equation

$$\dot x_1(t)=x_2,$$
$$\dot x_2(t)=-x_1-x_1^3-x_1^5+u,\ -1\le u\le 1,\ u=\pm1.$$

Problems to solve:

1. Design an optimal control law using the calculus of variations, see Examples 6.16 and 6.18. For $u=\pm1$, study solutions for

$\frac{dx_2}{dx_1}=\frac{-x_1-x_1^3-x_1^5+1}{x_2}$ and $\frac{dx_2}{dx_1}=\frac{-x_1-x_1^3-x_1^5-1}{x_2}$;

2. Find the switching curve, and, derive a discontinuous switching control law;
3. Perform simulations for a closed-loop system;
4. Approximate discontinuous control on a class of continuous functions. Applying the Hamilton-Jacobi concept, design a control law by approximating a solution of the Hamilton-Jacobi equation by a positive definite return function;
5. With the derived controls, simulate the closed-loop systems in SIMULINK.

Homework Problem 6.5.

Let the system is modeled as

$$\dot{x}_1=-x_1-x_2-x_1^3-x_2^3+u,$$

$$\dot{x}_2=x_1,\ y=x_2.$$

Problems to solve:

1. Study an open-loop system stability;
2. Design a control law using the Lyapunov theory using the quadratic estimates on states and control. Simulate a closed-loop system in SIMULINK;
3. Let the reference is $r(t)=\pm 1$. By trial and error, obtain an adequate PID control law by varying the feedback gains. Report simulation results for different references $r(t)=\pm A$, A=const;
4. Design the stabilizing control laws minimizing performance functionals $J=\min_{x,u}\int_0^\infty\left(\omega_x(x)+\frac{1}{2}u^TGu\right)dt$ and $J=\min_{x,u}\int_0^\infty\left(\omega_x(x,\frac{\partial V}{\partial x})+\frac{1}{2}u^TGu\right)dt$, covered in section 6.3.2. With a choice of $\omega_x(x,\frac{\partial V}{\partial x})$, find an explicit solution of the Hamilton-Jacobi equation, satisfied by the quadratic return function. Perform simulations and analyze results;
5. Design stabilizing control laws by minimizing a performance functional $J=\min_{x,u}\int_0^\infty\frac{1}{2}\left[\omega^T(x)Q\omega(x)+\dot{\omega}^T(x)P\dot{\omega}(x)\right]dt$. As covered in section 6.9, consider and solve design problems if: (1) $\omega(x)=x$; (2) $\omega(x)=\tanh x$. Perform simulations and assess findings;
6. Design tracking control laws by minimizing performance a functional $J=\min_{\mathbf{x},u}\int_0^\infty\left(\omega_x(\mathbf{x})+\frac{1}{2}u^TGu\right)dt$. Perform simulations and analyze findings;
7. For a systems $\dot{x}_1=-x_1-x_1^3-x_2+u$, $\dot{x}_2=x_1$, consider a relay control $|u|\le 1$. Solve the constrained optimization problem by applying the calculus of variations (equations are not separable) as well by minimizing $J=\min_{t,u\in U}\int_{t_0}^{t_f}1dt$ using the Lyapunov paradigm. Approximate the control limit by $\Phi=\tanh(ax)$;
8. Evaluate and discuss different control laws designed. Make conclusions.

Homework Problem 6.6.

Consider a servo with a saturation limit on control

$$\dot{x}_1=-x_1-x_2-x_1^3-x_2^3+u,\ |u|\le 1,$$

$$\dot{x}_2=x_1,\ y=x_2.$$

Problems to solve:

1. Analytically solve the constrained optimization problem applying the Lyapunov stability theory if there is a saturation control limit. Design a consistent $V(x)$ and specify $dV(x)/dt$. Study dynamics of the closed-loop system due to initial conditions;
2. Approximate a saturation control limit by a hyperbolic tangent function $u=\Phi(x)$, $\Phi=\tanh(x)$, $|u|<1$. Minimize
$$J=\min_{x,u\in U}\int_0^\infty(\omega_x(x)+\underbrace{\int\tanh^{-1}u\,du}_{\omega_u(u)})dt \text{ and } J=\min_{x,u\in U}\int_0^\infty(\omega_x(x,\tfrac{\partial V}{\partial x})+\underbrace{\int\tanh^{-1}u\,du}_{\omega_u(u)})dt\,.$$
Design stabilizing control laws. Derive consistent integrands $\omega_x(x)$ and $\omega_x(x,\frac{\partial V}{\partial x})$ to approximate solutions of functional equations. Simulate closed-loop systems;
3. Solve the tracking control problem by minimize a functional $J=\min_{\mathbf{x},u\in U}\int_0^\infty(\omega_x(\mathbf{x})+\int\tanh^{-1}u\,du)dt$. The extended system is
$$\dot{\mathbf{x}}(t)=\begin{bmatrix}\dot x_1(t)\\ \dot x_2(t)\\ \dot x_e(t)\\ \dot e(t)\end{bmatrix}=\mathbf{F}(\mathbf{x})+\mathbf{A}\mathbf{x}+A_r r+\mathbf{B}u=\begin{bmatrix}-x_1^3-x_2^3\\0\\0\\0\end{bmatrix}+\begin{bmatrix}-1&-1&0&0\\1&0&0&0\\0&-1&0&0\\-1&0&0&-1\end{bmatrix}\begin{bmatrix}x_1\\x_2\\x_e\\e\end{bmatrix}+\begin{bmatrix}0\\0\\r\\\dot r\end{bmatrix}+\begin{bmatrix}1\\0\\0\\0\end{bmatrix}u,\ y=x_2.$$
Analyze closed-loop system dynamics for $r(t)=\pm1$;
4. Applying the Lyapunov stability theory, design a *minimal complexity* control for the tracking control problem if only the tracking error $e(t)$ is measured. Analytically solve unconstrained and constrained optimization problems for $\dot{\mathbf{x}}(t)=\mathbf{F}(\mathbf{x})+\mathbf{A}\mathbf{x}+A_r r+\mathbf{B}u$. For not bounded and bounded controls, consider
$$\frac{dV_{\mathbf{m}}}{dt}\le -x_1^2-x_2^2-10x_e^2-10e^2-u^2,\ \frac{dV_{\mathbf{m}}(\mathbf{x})}{dt}\le -x_1^2-x_2^2-10x_e^2-10e^2-\int\tanh^{-1}u\,du,$$
$V_{\mathbf{m}}(\mathbf{x})=k_{11}c(x_1)+k_{13}x_1x_e+k_{14}x_1e+\frac{1}{2}(k_{22}x_2^2+k_{33}x_e^2+k_{44}e^2)$;
5. Study dynamics of the closed-loop systems if $r(t)=\pm1$.

Homework Problem 6.7.

Using the Lyapunov theory, study stability of a system

$$\dot x_1(t)=-x_1+10x_2,$$
$$\dot x_2(t)=-10x_1+x_2+x_2^3+x_2^5-x_2^7.$$

Problems to solve:

1. Propose a consistent $V(x_1,x_2)$, and, find $dV(x_1,x_2)/dt$. Make a conclusion on the system stability;
2. Compute the surfaces for $V(x_1,x_2)$ and $dV(x_1,x_2)/dt$.

Homework Problem 6.8.

Let the governing equations for a system with saturation on control are

$$\dot x_1=-x_1-x_1^3+x_2,$$
$$\dot x_2=x_1-x_2^3+u,\ |u|\le1,\ y=x_1.$$

Problems to solve:

1. Examine open-loop system stability applying the Lyapunov concept. For a chosen $V(x_1,x_2)$, plot $V(x_1,x_2)$ and $dV(x_1,x_2)/dt$;
2. Specify $V(x_1,x_2)$ and $dV(x_1,x_2)/dt$, and, find a stabilizing control law. Simulate the closed-loop system using SIMULINK. Analyze transient dynamics for different initial conditions;
3. Design a tracking control law using a Lyapunov pair. Simulate and study the closed-loop system if the reference is $r(t)=\pm 1$;
4. Let the states (x_1,x_2) are not measured, or, cannot be used to implement the controller. Using the Lyapunov concept, solve the *minimal complexity* tracking control design problem finding the proportional-integral control law. Simulate the closed-loop system.

Homework Problem 6.9.

System is described by differential equations

$$\dot{x}_1(t) = -x_1 + 10x_2,$$

$$\dot{x}_2(t) = x_1 - x_2 + u.$$

Problems to solve:

1. Discretize system if the sampling time is T_s=0.01 sec. Use different discretization methods, and, verify the discretization results;
2. Analyze the open-loop system stability by solving the discrete Lyapunov equation. Examine the definiteness of a matrix K_n for a given $Q_n=I$. Examine the definiteness of a matrix Q_n for a given $K_n=I$;
3. Design a control law by minimizing a performance index $J = \min_{x_n, u_n} \sum_{n=0}^{\infty} \left(x_n^T \begin{bmatrix} 1 & 0 \\ 0 & 1 \end{bmatrix} x_n + u_n^2 \right)$. Simulate a closed-loop system;
4. Using the Lyapunov stability, prove that the closed-loop system is stabile;
5. Design a digital control law if $|u_n| \le 1$ by minimizing a nonquadratic performance index. Simulate a closed-loop system.

Homework Problem 6.10.

Study a two-link robot arm covered in section 6.15.3.

Problems to solve:

1. Linearize a model specifying the equilibrium as $\bar{x} = \begin{bmatrix} \bar{x}_1 \\ \bar{x}_2 \\ \bar{x}_3 \\ \bar{x}_4 \end{bmatrix} = \begin{bmatrix} 1 \\ 1 \\ -1 \\ -1 \end{bmatrix}$ with $\bar{u} = \begin{bmatrix} \bar{u}_1 \\ \bar{u}_2 \end{bmatrix} \neq 0$;
2. For the measured outputs (θ_1,θ_2) and controls (u_1,u_2), design the *modal* controller and state *observer* for $(\hat{\omega}_1, \hat{\omega}_2)$ as covered in sections 5.4, 6.15.3 and 6.15.4. Specify consistent eigenvalues, solve the search problem to design controller and *observer*, and, simulate a closed-loop system;
3. Design the stabilizing LQR control law. Simulate closed-loop linear and nonlinear system for initial conditions [–1, –1, 1, 1]. Study the LQR control law with the state *observer*, which yields $(\hat{\omega}_1, \hat{\omega}_2)$;

4. Solve the tracking control problem by minimizing $J = \min_{\substack{\mathbf{x} \\ |u_1| \le u_{1\max}, |u_2| \le u_{2\max}}} \int_0^\infty \left[\frac{1}{2}\mathbf{x}^T Q \mathbf{x} + \int \left(\tanh^{-1}(\boldsymbol{u}_{\max}{}^{-1} u)\right)^T G du \right] dt$, $u_{1\max}=25$, $u_{2\max}=5$, subject to the governing equation $\dot{\mathbf{x}} = \mathbf{A}\mathbf{x} + A_r r + \mathbf{B}u$;
5. Analyze the closed-loop system dynamics. Simulate and analyze closed-loop linear and nonlinear systems in SIMULINK for different references $r_1(t)=\pm A_1$ and $r_2(t)=\pm A_2$, $-\frac{1}{2}\pi < r_1 < \frac{1}{2}\pi$, $-\frac{1}{2}\pi < r_2 < \frac{1}{2}\pi$;
6. Using the linearized model, design the *minimal complexity* tracking control law applying the Lyapunov theory. The measured states which can be used in a control law are (θ_1, θ_2) and tracking error (e_1, e_2). Investigate a nonlinear system by performing simulations.

Homework Problem 6.11.

Study an advanced airframe multirole fighter airplane, covered in section 2.4, and, Practice Problems 6.14.6 and 6.14.7. Multi-axis pitch-roll-yaw attitude control is accomplished by deflecting aerodynamic control surfaces, e.g., stabilizers, flaperons, canard and rudder. For a fighter airplane flying at 270 m/s, elevation 5000 m, near-zero Euler angles, consider the state-space nonlinear model with variables $x(t)=[V,\alpha,q,\theta,\beta,p,r,\phi,\psi]$, outputs $y(t)=[\theta,\phi,\psi]$ and controls $u=[\delta_s,\delta_f,\delta_c,\delta_r]$. Use a nonlinear model, reported in section 2.4 and 5.11.2.

Problems to solve:

1. Consider an asymmetric aircraft with the products of inertia $I_{xy}\neq 0$, $I_{yx}\neq 0$, $I_{yz}\neq 0$ and $I_{zy}\neq 0$. Applying (2.46), and, find the Cauchy's form of differential equations for the angular rates (q,p,r). Note that matrix $B\in\mathbb{R}^{9\times 4}$ is reported in section 5.11.2.

$$\begin{bmatrix} L \\ M \\ N \end{bmatrix} = \begin{bmatrix} I_{xx} & -I_{xy} & -I_{xz} \\ -I_{yx} & I_{yy} & -I_{yz} \\ -I_{zx} & -I_{zy} & I_{zz} \end{bmatrix} \begin{bmatrix} \dot{p} \\ \dot{q} \\ \dot{r} \end{bmatrix} + \begin{bmatrix} p \\ q \\ r \end{bmatrix} \times \begin{bmatrix} I_{xx} & -I_{xy} & -I_{xz} \\ -I_{yx} & I_{yy} & -I_{yz} \\ -I_{zx} & -I_{zy} & I_{zz} \end{bmatrix} \begin{bmatrix} p \\ q \\ r \end{bmatrix}$$
$$= \begin{bmatrix} I_{xx}\dot{p} - I_{xy}\dot{q} - I_{xz}\dot{r} + (-I_{xz}p - I_{yz}q + I_{zz}r)q + (I_{xy}p - I_{yy}q + I_{yz}r)r \\ -I_{xy}\dot{p} + I_{yy}\dot{q} - I_{yz}\dot{r} + (I_{xz}p + I_{yz}q - I_{zz}r)p + (I_{xx}p - I_{xy}q - I_{xz}r)r \\ -I_{xz}\dot{p} - I_{yz}\dot{q} + I_{zz}\dot{r} + (-I_{xy}p + I_{yy}q - I_{yz}r)p + (-I_{xx}p + I_{xy}q + I_{xz}r)q \end{bmatrix},$$

$$B = \begin{bmatrix} -0.0008 & -0.0005 & -0.0001 & -0.0007 \\ 0.093 & 0 & -0.024 & 0 \\ 5.3 & 0 & 1.9 & 0 \\ 0 & 0 & 0 & 0 \\ 0 & 0 & 0 & 0 \\ 0.053 & 6.8 & 0 & 0.07 \\ 0.081 & 0.07 & 0 & 8.2 \\ 0 & 0 & 0 & 0 \\ 0 & 0 & 0 & 0 \end{bmatrix};$$

2. Using Cauchy's form of nine-order differential equations, linearize model at equilibrium
$\bar{x}$=[270 m/s, 0.01 rad, 0.025 rad/s, 0.025 rad, 0 rad, 0.025 rad/s, 0.025 rad/s, 0 rad, 0 rad].
Find $A\in\mathbb{R}^{9\times 9}$ in $\dot{x}=Ax+Bu$. Let m=28000 kg, $I_{xx}=2.75\times 10^4$ kg-m^2, I_{xy}=2500 kg-m^2, I_{xz}= −1280 kg-m^2, I_{yx}=3000 kg-m^2, $I_{yy}=2.51\times 10^5$ kg-m^2, I_{yz}=10000 kg-m^2, I_{zy}=5000 kg-m^2 and $I_{zz}=2.38\times 10^5$ kg-m^2, ρ=0.7364 kg/m^3, c_d=0.285 (at V=270 m/s) and A=4.9 m^2;
3. Design a tracking control law for a linear system minimizing a quadratic performance functional $J = \min_{\mathbf{x},u} \int_0^\infty \frac{1}{2}\left(\mathbf{x}^T Q \mathbf{x} + u^T G u\right) dt$ subject to the system and tracking error dynamics $\dot{\mathbf{x}}=\mathbf{A}\mathbf{x}+A_r r+\mathbf{B}u$, $\mathbf{x}\in\mathbb{R}^{15}$, $u\in\mathbb{R}^4$. The outputs are Euler angles $y=[\theta, \phi, \psi]$, $y\in\mathbb{R}^3$. Investigate how weighting matrices (Q, G) affect

eigenvalues, feedback gains and dynamics. Perform simulations of a closed-loop system for different $r(t)=\begin{bmatrix} r_1(t) \\ r_2(t) \\ r_3(t) \end{bmatrix}=\begin{bmatrix} \theta_{ref}(t) \\ \phi_{ref}(t) \\ \psi_{ref}(t) \end{bmatrix}$. Assess aircraft maneuverability;

4. Design a constrained tracking control law for linearized system minimizing $J=\min_{\mathbf{x},u\in U}\int_0^\infty\left[\frac{1}{2}\mathbf{x}^TQ\mathbf{x}+\int\left(\tanh^{-1}(\boldsymbol{u}_{\max}{}^{-1}u)\right)^T G du\right]dt$. Displacements of aerodynamic control surfaces are limited as $-u_{i\,\max}\le u_i\le u_{i\,\max}$, $|\delta_s|\le 0.5$ rad, $|\delta_f|\le 0.5$ rad, $|\delta_c|\le 0.6$ rad, and $|\delta_r|\le 0.6$ rad. Perform simulations of a closed-loop system for different $r(t)$. Study deflections of control surfaces. Evaluate the aircraft maneuvering capabilities assessed using the simulation results;
5. Consider a nonlinear model $\dot{x}=F(x)+Bu$, $-u_{i\,\max}\le u_i\le u_{i\,\max}$ with designed control laws. Perform simulations, evaluate transients, and, assess dynamics of a closed-loop system for different $r(t)$. Make conclusions on flight qualities;
6. Discretize a linear system finding $\mathbf{x}_{n+1}=\mathbf{A}_n\mathbf{x}_n+A_{rn}r_n+\mathbf{B}_nu_n$ if T_s=0.01 sec. Minimize a quadratic performance index $J=\min_{\mathbf{x}_n,u_n}\sum_{n=0}^\infty\left[\mathbf{x}_n^TQ_n\mathbf{x}_n+u_n^TG_nu_n\right]$. Design a linear tracking control laws if controls are not constrained. Perform simulations for linearized and nonlinear systems. Evaluate system dynamics for different r_n. For a nonlinear system, use the Discrete-Time Integrator block as reported in section 6.15.1. See SIMULINK diagrams in Figures 6.39.b and 6.43.a;
7. Design a control law if there are limits on deflection of aerodynamic control surfaces. Minimize a nonquadratic performance index $J=\min_{\mathbf{x}_n,u_n\in U}\sum_{n=0}^\infty\left[\mathbf{x}_n^TQ_n\mathbf{x}_n-u_n^T\mathbf{B}_n^TK_{n+1}\mathbf{B}_nu_n+2\sum_u\left(a^{-1}\tanh^{-1}(\boldsymbol{u}_{\max}{}^{-1}u_n)\right)^T G_n\right]$. Find a bounded control law and feedback gains for different (Q_n,G_n). Simulate linearized and nonlinear discrete systems for different r_n.

Homework Problem 6.12.

Study a path tracking control of a differential drive in a mobile robot. Many mobile robots and robotic platforms have two main wheels, each of which is rotated by a direct drive or geared electric servomotor. Passive wheels are placed in the front or rear ensuring directional and roll stability. Developing descriptive and constitutive equations, denote L is the distance between two wheels, and, R is the radius, see Figure 6.55. The linear velocities (v_R,v_L) depend on the angular wheel velocities (ω_R,ω_L). Assuming ideal traction, $v_R=R\omega_R$ and $v_L=R\omega_L$. The robot moves in the xy plane in the direction defined by the angular velocity of wheels, rotated clockwise and counterclockwise. Robot turns clockwise or counterclockwise if the wheels rotate in opposite directions. Define the body frame origin at the center of the axle between the wheels. A two-wheel robot schematics is depicted in Figure 6.55. No translation occurs if wheels rotate in opposite directions as $|\omega_R|=-|\omega_L|$ or $-|\omega_R|=|\omega_L|$.

For the considered kinematics, velocity V depends on the angular wheel velocities,

$$\dot{x}=\tfrac{1}{2}R(\omega_R+\omega_L)\cos\theta,\ \dot{y}=\tfrac{1}{2}R(\omega_R+\omega_L)\sin\theta,$$

$$\dot{\theta}=\frac{R}{L}(\omega_R-\omega_L).$$

If one wheel is fixed and other rotates, the robot exhibits circular motion depending on the kinematics, friction forces, traction, etc. In the *xy* plane, using the robot velocity V and angular velocity ω, path and trajectory are given as

$$\dot{x} = V\cos\theta,$$
$$\dot{y} = V\sin\theta,$$
$$\dot{\theta} = \omega.$$

One regulates the wheels angular velocities (ω_R, ω_L), which define (V, ω, θ) as

$$V = \sqrt{\dot{x}^2 + \dot{y}^2} = \tfrac{1}{2}R(\omega_R + \omega_L),\ \omega = \tfrac{R}{L}(\omega_R - \omega_L).$$

For specified (V,ω), $\omega_R = \frac{2V+L\omega}{2R}$ and $\omega_L = \frac{2V-L\omega}{2R}$.

In traction dc motors, the angular velocity (ω_R, ω_L) is controlled by changing the applied voltages (u_{a1}, u_{a2}), and, the PWM controllers-drivers are used as shown in Figure 6.55. The mobile robot output y vector could be $y=[x,y]$, $y=[V,\omega]$, $y=[V,\theta]$, etc. Problems to solve:

1. Consider the toque control, which implies $u = \begin{bmatrix} u_1 \\ u_2 \end{bmatrix} = \begin{bmatrix} T_1 \\ T_2 \end{bmatrix}$.

   The output vector is $y = \begin{bmatrix} V \\ \theta \end{bmatrix} = \begin{bmatrix} \frac{1}{2}R(\omega_R + \omega_L) \\ \theta \end{bmatrix}$.

   The governing equations are

   $$\frac{d\omega_R}{dt} = \tfrac{1}{J}(-B_m\omega_R - Rcmg + T_1),\ -5\le(T_1,T_2)\le 5 \text{ N-m},$$
   $$\frac{d\omega_L}{dt} = \tfrac{1}{J}(-B_m\omega_L - Rcmg + T_2),$$
   $$\frac{d\theta}{dt} = \tfrac{R}{L}(\omega_R - \omega_L).$$

   The parameters are R=0.1 m, L=0.2 m, B_m=0.001 N-m-s/rad, c=0.01, g=9.81 m/s^2, m=1 kg and J=0.01 kg-m^2. The rolling resistance force is $F=cmg$, where c is the rolling coefficient. Solve a tracking control problem by designing the bounded control law using the Hamilton-Jacobi optimization minimizing the performance functional $J = \min_{\mathbf{x},u\in U} \int_0^\infty \left[\tfrac{1}{2}\mathbf{x}^T Q\mathbf{x} + \int \left(\tanh^{-1}(\boldsymbol{u}_{\max}{}^{-1}u)\right)^T G du \right] dt$. Simulate a system if references are $r(t) = \begin{bmatrix} r_V(t) \\ r_\theta(t) \end{bmatrix} = \begin{bmatrix} e^{-t}\sin(5t) \\ \frac{1}{2}\pi\,\mathrm{sgn}[\cos(2.5t)] \end{bmatrix}$, $t\in[0\quad 5]$ sec;

2. Consider path tracking if the output vector is $y=[x, y]$. Using the governing and constitutive relationships, by trials and errors find a tracking PID control law $u=[T_1,T_2]$. Perform simulations and analyze robot dynamics. Let $r(t) = \begin{bmatrix} r_x(t) \\ r_y(t) \end{bmatrix} = \begin{bmatrix} e^{-t}\sin(2.5t) \\ e^{-0.5t}\cos(2.5t) \end{bmatrix}$, $t\in[0\quad 10]$ sec;

3. Using a *minimal complexity* system design, derive a bounded tracking control law considering $y = \begin{bmatrix} V \\ \theta \end{bmatrix} = \begin{bmatrix} \frac{1}{2}R(\omega_R + \omega_L) \\ \theta \end{bmatrix}$. Analyze closed-loop system dynamics if

$r(t)=\begin{bmatrix} r_V(t) \\ r_\theta(t) \end{bmatrix}=\begin{bmatrix} e^{-t}\sin(5t) \\ \pi\,\text{sgn}[\cos(2.5t)] \end{bmatrix}$. To regulate angular velocity of permanent magnet traction motors (ω_R,ω_L), one changes (u_{a1},u_{a2}) as reported in chapter 3. The control limits are $-50\le(u_{a1},u_{a2})\le 50$ V. The motor model is given by (3.19). Let r_a=3.15 ohm, L_a=0.0066 H and k_a=0.156 V-sec/rad. The armature currents (i_{a1}, i_{a2}) are not measured.

4\.

Governing Equations of Motion:

$$\frac{di_{a1}}{dt}=\frac{1}{L_a}\left(-r_a i_{a1}-k_a\omega_R+u_{a1}\right),$$

$$\frac{di_{a2}}{dt}=\frac{1}{L_a}\left(-r_a i_{a2}-k_a\omega_L+u_{a2}\right),$$

$$\frac{d\omega_R}{dt}=\frac{1}{J}(k_a i_{a1}-Rcmg-B_m\omega_R),$$

$$\frac{d\omega_L}{dt}=\frac{1}{J}(k_a i_{a2}-Rcmg-B_m\omega_L),$$

$$\frac{d\theta}{dt}=\frac{R}{L}(\omega_R-\omega_L).$$

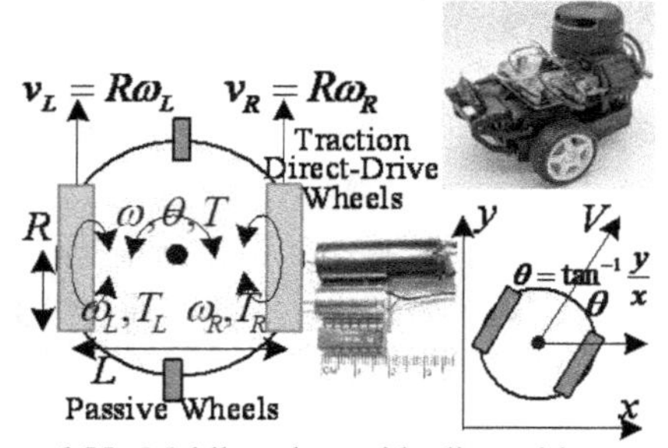

Figure 6.55. Mobile robot with direct-drive traction motors to rotate right and left wheels.

References

1. P. Hippe, *Windup in Control: Its Effects and Their Prevention*, Springer-Verlag, NY, 2006.
2. F. L. Lewis, D. Vrabie and V. L. Syrmos, *Optimal Control*, Wiley, NY, 2012.
3. S. E. Lyshevski, *Control Systems Theory With Engineering Applications*, Birkhauser, Boston, MA, 2001.
4. S. E. Lyshevski, *Mechatronics and Control of Electromechanical Systems*, CRC Press, Boca Raton, FL, 2017.
5. K. Ogata, *Discrete-Time Control Systems*, Prentice-Hall, Upper Saddle River, NJ, 1995.
6. A. Poznyak, *Advanced Mathematical Tools for Automatic Control Engineers*, Elsevier, Amsterdam-Boston, 2008.
7. R. A. DeCarlo, S. H. Zak and G. P. Matthews, "Variable structure control of nonlinear multivariable systems: A tutorial," *Proceedings of the IEEE,* vol. 76, no. 3, pp. 212-232, 1988.
8. S. V. Drakunov and V. I. Utkin, "Sliding mode control in dynamic systems," *Int. J. Control,* vol. 55, no. 4, pp. 1029-1037, 1992.
9. C. Edwards and S. Spurgeon, *Sliding Mode Control*, Taylor and Francis, Bristol, PA, 1998.
10. S. E. Lyshevski, "Sliding modes and soft switching control in dynamic systems," *Proc. American Control Conf.*, pp. 646-650, 2000.
11. S. E. Lyshevski, "Optimal control of nonlinear continuous-time systems: Design of bounded controllers via generalized nonquadratic functionals," *Proc. American Control Conf.,* vol. 1, pp. 205-209, 1998.
12. S. E. Lyshevski, "Role of performance functionals in control laws design," *Proc. American Control Conf.*, pp. 2400-2405, 2001.
13. S. E. Lyashevskiy, "Robust nonlinear control of uncertain systems with state and control constraints," *Proc. Conf. Decision and Control,* vol. 2, pp. 1670-1675, 1995.
14. S. E. Lyshevski, "Nonlinear discrete-time systems: Constrained optimization and application of nonquadratic costs," *Proc. American Control Conf.,* vol. 6, pp. 3699-3703, 1998.
15. A. M. Lyapunov, "Stability of motions," *Trans. Mathematic Society*, vol. 17, pp. 253-333, 1893.
16. R. E. Kalman and J. E. Bertram, "Control system analysis and design via the "second method" of Lyapunov. I. Continuous-time systems," *Trans. ASME J. Basic Eng.*, vol. 82, no. 2, pp. 371-393, 1960.
17. S. H. Zak, *Systems and Control*, Oxford University Press, NY, 2002.

7 Information Analysis and Analytics

7.1. Introduction

Controlled systems operate in physical domain and cyberspace. Cyberspace is defined as "a global domain within the information environment consisting of the interdependent network of information technology, infrastructures, and resident data, including the Internet, telecommunications networks, computer systems, and embedded processors and controllers" [1]. There are multiple data and information sources, assets and services within distributed networks in CPS. Physical and cyber vulnerabilities are investigated to guarantee resiliency, situational awareness, security and functionality. To enable resilience and attacks mitigation, controlled physical and information defensive measures to attacks, malicious actions, intrusions and disruptions are implemented prioritizing security, response strategies, etc.

Artificial Intelligence – Cognition and artificial intelligence assume representation of descriptive features by acquiring knowledge and exhibiting *learning* by means of perceiving, recognizing, conceiving and reasoning. There are debates on foundations, interpretation and implementation of explicit and implicit *knowledge generation* and *learning* schemes. Implicit *learning* assumes acquiring abstract knowledge on a system and environment without ascertain awareness on regularities by means of conjectural reasoning, fuzzy algorithms and artificial *intellect*. Explicit *learning* focuses on defined and consistent instructions, features, epistemic awareness and reasoning. Epistemic information applies considering situational awareness on propositional variables using definable explicit assessment and calculus which permit logics, reasoning and constructs.

Data and Information – Data and information are frequently used interchangeably and reciprocally. Measurements yield quantitative data on physical quantities. Spatiotemporal sensing of physical quantities from different sources, data fusion, data aggregation, data management, data governance and other tasks pertain to data analytics and information control. Different types of data is fused from sensors, streaming media, log files, applications, etc. Near real time analytics on diverse datasets facilitates investigation on fidelity, dimensionality reduction, decomposition and approximation using data from different sources. Incomplete and partially interpreted data and datasets lack informative features to acquire significance and attributes, and, insufficient to accomplish perceptive contextual and systematic content analyses. Data is aimed to ensure functionality, situational awareness, etc. One implements the following scheme

Measurements ⇔ Input Data ⇔ Data Processing ⇔ Output Information.

Define an information as an aggregated data and procured estimates, which yield perceptive features to accomplish descriptive, predictive and prescriptive analyses. Information measures, such as entropies, correlation, divergence and others may quantify dependencies, complexity, uncertainties, etc. Mutual information, divergences, perplexity and other information-theoretic measures may not always ensure uniqueness and practicality. Focusing on engineering science and design, consider pertained problems minimizing axiomatization. We focus on established analysis paradigms, optimization schemes, control solutions, algorithms and tools.

Data Quality, Data Integrity, Data Assurance and Information Security – In 2013, the International Organization for Standardization and the International Electrotechnical Commission released the ISO/IEC 27001 international standard on information security *Information Technology - Security Techniques - Information Security Management Systems - Requirements*. There are multiple control categories, such as assets management, access control, cryptography, physical security, communication security, etc. Problems on information security control, data security and risk analysis are addressed to enhance CPS capabilities. The *Security Event Manager*, *Security Information Management*, *Security Information and Event Management*, *Managed Security Service*, *Security as a Service System* and other platforms have being developed and implemented using developed algorithmic and schemes. The National Institute of Standards and Technology (NIST) reported the relevant solutions in the 2013 NIST Special Publication, revision 4, *Security and Privacy Controls for Federal Information Systems and Organizations*.

Current technologies ensure superior networking capabilities and unprecedented connectivity of mobile devices. There is a significant security risk of cellular networks. Information assurance and information sharing should guarantee secure multi-domain operation in various applications, services and systems, such as industrial control systems (ICS), supervisory control and data acquisition (SCADA) systems, surveillance and reconnaissance platforms, critical infrastructure systems, etc. To empower cybersecurity within information environments, effective methods and algorithms are under developments to minimize security risks and maximize resilience by reducing vulnerabilities, identifying threats, detecting attacks, etc. Focused efforts and systematic studies are aimed to advance information quality, empower security capabilities, enable resiliency, mitigate anomalous behavior, etc.

Critical Infrastructure Cybersecurity – Cybersecurity is a collection of preventive and proactive solutions, algorithms, tools and practices to protect servers, devices, communication and processing modules, networks and data from malicious attacks and unauthorized access. Numerous learning resources are available. Excellent books, guides and courses provide comprehensive coverage on best practices by reporting various solutions, studying practical problems and investigating case studies. These resources enable knowledge and skillset of cybersecurity practitioners. There are differences in design of industrial IoT platforms and CPS. One focuses on process- and application-centric cybersecurity, and, investigates information management schemes. Transformative findings are emerging in: (1) Applications security; (2) Distributed denial of service defense; (3) Mobile applications security; etc.

New and legacy systems depend on data quality, information security, as well as on resiliency to cyberattack and intrusions. The objective is to ensure overall functionality, guarantee safety, execute tasks and missions, etc. There are known and unknown threats which pose significant risks to critical infrastructure and information systems. Industrial control systems depend on information collection and sharing.

Cyber-physical systems are vulnerable to exploitation and attacks. There are different phases of intrusion, such as target analysis, persistence, exploits, etc. The NIST developed a *Framework for Improving Critical Infrastructure Cybersecurity* [2], and, new solutions and algorithms are studied. A range of solutions are stipulated and investigated using five core categories reported in see Figure 7.1.a, e.g., identify, protect, detect, respond and recover. The ad hoc manual expert's techniques, user-defined relationships, selected identifiers and security control schemes are implemented by means of attack prognostics with resiliency evaluation, mitigation and impact assessment. High-dimensional data-driven analyses, threat prioritization and security metrics support the NIST framework.

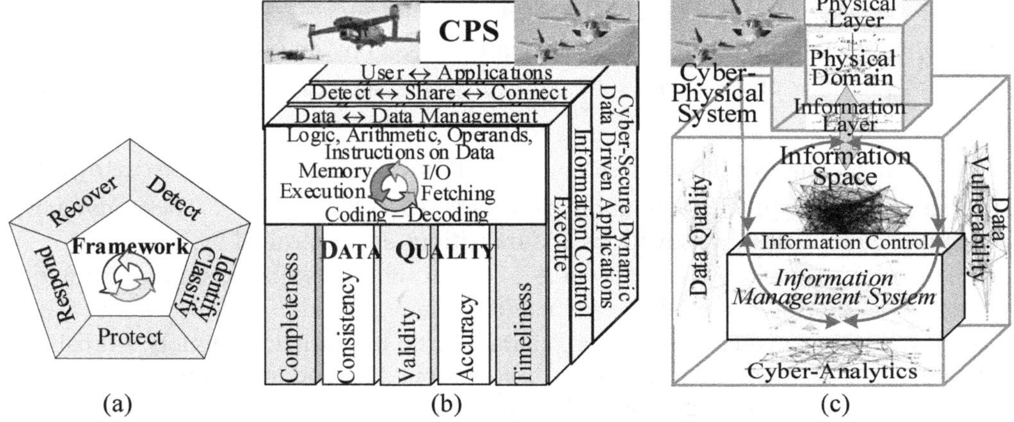

Figure 7.1. (a) Cybersecurity framework: Detect → Identify/Classify → Protect → Respond → Recover;

(b) Information sharing, information control and data quality within an information flow scheme: *Sensing → Sharing → Data Acquisition → Processing → Execution → Control → Adaptation and Reconfiguration*;

(c) System with an *Information Management System* to improve quality and utility of information: Interacting physical domain and information space.

Cybersecurity-Centric Data Quality and Physical Data Quality – Data quality, pertained to information security, is quantitatively characterized by the accuracy, completeness, consistency, conformity, integrity, relevance, validity, timeless and uniqueness, which may be considered as the outputs y^{cyber}. Data quality and information control taxonomy and schemes are illustrated in Figures 7.1.b and c. Using real-valued dynamic-probabilistic estimates on data quality homogeneity $\mathcal{M}_H(\cdot)$ and heterogeneity $\mathcal{M}_L(\cdot)$, one evaluates states and outputs $(\boldsymbol{x}^{\text{cyber}},\boldsymbol{y}^{\text{cyber}})$, finds control schemes $\boldsymbol{u}^{\text{cyber}}$, and, assesses measures $\mathcal{M}(\cdot)$. Outputs $\boldsymbol{y}^{\text{cyber}}$ and states $\boldsymbol{x}^{\text{cyber}}$ are frequently qualitative, and, not explicitly standardized. Challenges in defining and assessing a tuple $(\boldsymbol{y}^{\text{cyber}},\boldsymbol{x}^{\text{cyber}},\boldsymbol{u}^{\text{cyber}})$ and measures $\mathcal{M}(\cdot)$ are due to uncertainties in detectability, observability, controllability, etc. Mathematical formulation in attempts to solve axiomatized cyber-theoretical designs applying cybersecurity attributes are conjectural, while practical solutions exist.

Physical data quality, data measurements and processing were studied in previous chapters by covering matrix calculus, probability theory, stochastic processes, designing filters and state observers, etc. The designer aims to improve physical data quality. The quantitative and measurable outputs $\boldsymbol{y}^{\text{data}}$ are the accuracy, data rate, delays, latency, linearity, precision, range, resolution, sensitivity, etc. Analog and digital fixed-structure and adaptive filters were designed to maximize data fidelity. Considering the data fidelity as an output $\boldsymbol{y}^{\text{data}}$, the steady-state error (attenuation and frequency-dependent ripple), dynamic response (phase shift, latency, etc.) and other quantities can be characterized as the controlled states $\boldsymbol{x}^{\text{data}}$. In general, data states and controls $(\boldsymbol{x}^{\text{data}},\boldsymbol{u}^{\text{data}})$ describe data evolution, processing, data sharing, controlled data-driven applications, etc. The real-valued $\boldsymbol{x}^{\text{data}}$ are the channel capacity, data rate, bandwidth, communication losses, data losses, etc. In data transductions and sharing, multiphysics conversions, sensing, processing, peripheral interface, transceivers communication, logic operations and processing are controlled and reconfigured by $\boldsymbol{u}^{\text{data}}$.

Information Measures – For attributes $\boldsymbol{\mathcal{A}}$ and indicators $\boldsymbol{I}$, the $(\boldsymbol{y},\boldsymbol{x},\boldsymbol{u})$-dependent homogeneity $\mathcal{M}_{\mathcal{H}}(\cdot)$ and heterogeneity $\mathcal{M}_{\mathcal{L}}(\cdot)$ are assessed by $\mathcal{M}=(\mathcal{M}_{\mathcal{H}}(\cdot),\mathcal{M}_{\mathcal{L}}(\cdot))$.

A minimax problem $\boldsymbol{q}(\cdot)=\min\limits_{\substack{\mathcal{M}_{\mathcal{L}}\\ q\in Q\subseteq\mathcal{M}\times\mathbf{X}\times\mathbf{U}}}\max\limits_{\mathcal{M}_{\mathcal{H}}}\ \kappa(t,\mathcal{M},\boldsymbol{x},\boldsymbol{u})$ is described in section 7.3.

One evaluates system governance $\mathbb{M}(\cdot)$ by assessing $(\boldsymbol{y}(t),\boldsymbol{y}[n],\boldsymbol{x}(t),\boldsymbol{x}[n])$, data governance $(\boldsymbol{\mathcal{D}}(t),\boldsymbol{\mathcal{D}}[n])$, information processes $(\mathcal{J}(t),\mathcal{J}\ [n])$, etc. These dynamic-probabilistic processes, transductions and transitions are controlled by $(\boldsymbol{u}(t),\boldsymbol{u}[n])$.

Finite sequences S and strings s of codes, data, instructions and information processes have finite descriptive, computational and algorithmic complexities. These (S,s) may admit statistical description, characterized by statistical models. Probability distributions and other characteristics are computable and may found. Dynamic-probabilistic analysis of (S,s) may yield real-valued attributes, indicators and measures on information states $(\boldsymbol{x}^{\text{cyber}},\boldsymbol{x}^{\text{data}})$, controlled by $(\boldsymbol{u}^{\text{cyber}},\boldsymbol{u}^{\text{data}})$. These tasks are pertained to information analytics and control.

Measurable sequences of random X is defined on a probability space

$(\Omega,\mathscr{F},\mathbb{P})$,

where Ω is the sample space of outcomes, $\Omega\in\mathscr{F}$; $\mathscr{F}$ is the event space described by the σ-algebra (σ-filed) of measurable Ω, $\mathscr{F}\subseteq 2^{\Omega}$, $\varnothing\in\mathscr{F}$ with a countable evens $\cup_{i=1}^{N}A_i\in\mathscr{F}$; $\mathbb{P}$ is the probability on $\mathscr{F}$.

The probability space $([0,1],\mathscr{F}([0,1]),\mathbb{P})$ implies that the sample space Ω is the real interval $[0,1]$, $\mathscr{F}([0,1])$ is the Borel σ-algebra on $[0,1]$ with events $\{[0,a),0\leq a\leq 1\}$, $\mathscr{F}([0,1])=\sigma$(open sets on $[0,1]$), and, $\mathbb{P}(\cdot)$ is Lebesgue measure for all events $A\in\mathscr{F}$. For strings and sequences, apply a notation $X\equiv\mathsf{S}$ For an information process or state sequence, the statistical model $\mathscr{R}_X=(S,\mathcal{P})$ is a pair $(S,\mathcal{D})$. One finds $\mathscr{R}_X=(S,\mathcal{D})$. The probability of the event $A\in\mathscr{F}$ is $\mathbb{P}(A)\geq 0$. A real-valued $\mathbb{P}(\cdot)$ is defined on a set of events in a probability space, $\mathbb{P}:\mathscr{F}\to[0,\ 1]$, and, $\mathbb{P}(\Omega)=1$.

String – For a given a set X, a string over X is a finite ordered list of elements of X. If X is a set $\underline{X}=\{a,b,c\}$, then the strings over X are *aa*, *aaa*, *aaaa*, *aba*, *abba*, *abcc*, etc. Using a superscripts, $ab^3abc^2=abbbabcc$. The length of a string is its number of elements, e.g., 8 for ab^3abc^2.

Sequence of Random Variables – A sequence is an ordered list of elements. A sequence of random variables is a sequence of $X_j:\Omega\to\mathbb{R}$. Within a sample space, which consists of a finite number of elements, a random variable X is a mapping which specifies a real number to possible outcomes Ω_i, i=1,2,…,k. Hence, $X(\Omega_i)=x_i$. For a sequence of random $X_1,X_2,\ldots$, each X_j is a function from Ω to real numbers, $X_j(\Omega_i)=x_{ji}$, i=1,2,…,k.

Descriptive Statistics – For a random X, a statistical description $\mathcal{R}_X=(\mathcal{S},\mathcal{P})$ is a pair $(\mathcal{S},\mathcal{P})$, where $\mathcal{S}$ is the set of observations, and, $\mathcal{P}$ is a set of finite-dimensional probability distributions on $\mathcal{S}$. One finds a distribution $\mathcal{D}(\cdot)$, parameterized as $\mathcal{D}=\{p_X(x:\phi), \phi\in\Phi\}$, $\mathcal{D}=\{F_X(x:\phi), \phi\in\Phi\}$ or $\mathcal{D}=\{f_X(x:\phi), \phi\in\Phi\}$, where Φ is the set of parameters, $\Phi\in\mathbb{R}$. In general, processes are not stationary. For example, a stochastic process X is a Poisson process if it is characterized by a finite dimensional Poisson distribution for any finite partition. For a stationary process X, one finds a probability mass function $p_X(x)\geq 0$, and, $p_{X_i}(x)=p_{X_j}(x)$. For a Poisson process,

$$X\sim\text{Pois}(\lambda),\ p_X(k;\lambda)=\mathbf{P}(X=k)=\frac{\lambda^k e^{-\lambda}}{k!},\ F_X(k;\lambda)=e^{-\lambda}\sum_{i=0}^{\lfloor k\rfloor}\frac{\lambda^i}{i!},\ k=0,1,\ldots,\ \lambda=\mathbf{E}(X)=\text{var}(X),$$

where k is the number of occurrences; $\lfloor k \rfloor$ is the floor function which uses a real number k, and, outputs the greatest integer less than or equal to k.

Deterministic and Probabilistic Analyses – Strings pattern, signature and structure models are heterogeneous. One evaluates stationary and non-stationary random stochastic sequences, as well as investigates heterogeneous determinism and probabilistic descriptions of periodic, mixed, quasirandom, pseudorandom and random signals and sequences. Statistical distributions, descriptive characteristics and statistical parameters of $X(t)$ and $X[n]$ vary. The deterministic sequences, and, their models can be found applying pertained algorithms. For mixed processes, one decomposes stochastic and deterministic processes. Considering linear and nonlinear regressions $Y=X\beta+\varepsilon$, $Y=f(X,\varepsilon,\beta)$ for dependent and independent observed variables (Y,X,ε), (Y_i,X_i,ε_i), i=1,2,…,n, determine deterministic processes X and random processes $\varepsilon\sim\mathcal{D}(\cdot)$. For normal or Poisson distribution, one finds $\varepsilon\sim\mathcal{N}(\mu,\sigma)$ or $\varepsilon\sim\text{Pois}(\lambda)$. With identified deterministic processes, the centered stochastic process is analyzed. Using correlation, decompositions and other estimates, one concludes ether or not process is deterministic, mixed or completely nondeterministic. Due to heterogeneity of processes, hypotheses and assumed models, high fidelity analysis may not always be guaranteed.

Statistical Randomness, Determinism and Dependencies – Statistically random sequences assume disorder and unpredictability without regularities and patterns. Random and pseudorandom sequences admit probabilistic characterization. The Pearson chi-squared statistical independence test can be conducted, and, short length sequences may not meet the randomness assessment criteria. Sequences which exhibit patterns and dependences are not statistically random. These dependences and correlations may lead to feature detection, categorization, identification, classification and attributes evaluation. Spectral density estimation, nonparametric statistics, as well as other probabilistic and correlation analyses are applied to investigate statistical and deterministic dependences. Partitioned and shifted nonstationary superimposed sequences are studied.

7.2. Malware and Threats

Malware and Viruses – Cyber-physical systems are vulnerable to intelligence exploitation and cyberattacks. Sophistication of attacks have being increased, targeting systems by exploiting vulnerabilities and deficiencies s. A malicious virus code supports macros to execute it. Viruses may be embed and hidden in files, downloads, applications, images, audio and video files, etc. Antivirus programs are not effective against new viruses and embedded malware. It is difficult to detected polymorphic, oligomorphic, obfuscated and metamorphic viruses because codes morph, change signature and patterns, use random encryption schemes, etc. Viruses target Android, Linux, Windows and other fully-open source, partially-open source, as well as closed source operating systems. Malware is a malicious software. All viruses are malware, however, not all malware are viruses. The backdoors, bots, computer worms, Trojans, rootkits, scripts, spyware and adware are malware. Virus is a class of malware which morphs and propagates on its own without instructions. Malware uses a variety of mechanisms to infect hosts and processing devices by means of stealth schemes evading detection by antivirus software. Malware may infect and modify operating systems, modify or delete data, compromise or disable systems and networks, cause outages, etc. Heuristic signature-based malware detection and removal tools have being developed and commercialized, such as by AVG, Avast, Avira, Bitdefender, BullGuard, Kaspersky Lab, Malwarebytes, McAfee, Norton, ScanGuard, TotalAV and others. Antivirus and antimalware software are aimed to detect, classify and remove viruses, ransomware, spyware, rootkits, computer worms, spams, etc.

Antivirus software protects against threats and assesses vulnerabilities by:

1. Scanning files (background and full system scans) and comparing codes against malware database using matching signatures, patterns and modalities by means of data mining techniques;
2. Analyzing static and behavioral data and actions to detect and mitigate intrusions;
3. Identifying attacks, and, analyzing domains and services.

A *direct action virus* acts if a file is executed. A *resident virus* propagates into a device memory, and, executes when an operating system loads. The *memory resident viruses* allocate memory for themselves, and, may infect programs. A *non-resident virus* infects executable files even if they are not running. Metamorphic viruses are capable to refine their codes into temporary codes, edit the temporary representation, and then, translate changes back to original codes. Within streams, batch modes of scanning sequences, and, inspecting codes and searching strings to detect malware, one strives to ensure:

1. High detection probability and classification consistency;
2. Near real time capability in mitigation, removal and resilience.

Viruses can infect binary executables, such as executable and linkable file format, portable executable, DOS executable, etc. The virus is self-contained, and, operates independently affecting other files, codes, libraries and interpreters. The infected host continues files execution, and, spreads the self-propagating virus. There are attack strategies against host, memory, registers, buffers, data and files, while the host codes keep the same virtual addresses. Heuristic virus detection, search for virus signatures, patterns, modalities and other features may not always lead to desirable outcomes. Consider a set of q substrings $\mathbf{S}=\{\mathbf{s}_1,\mathbf{s}_2,\ldots,\mathbf{s}_{q-1},\mathbf{s}_q\}$. The complexity of search algorithms is excessive. With the maximum length of substring L, for a l-length $\mathbf{s}_i$, the Knuth-Morris-Pratt search algorithm complexity is $O(q(L+l))$. While small code viruses hide, ~1000 bytes stealth virus exhibit polymorphism, near-unbreakable encryption, packing and code compression, as well as self-replication, propagation and removal.

Antivirus Software and Tools – Different antivirus programs are installed to minimize probability to be affected by viruses and malicious software, called malware. Computer support specialists are highly knowledgeable on all aspects of applications, diagnostic software, operating systems, antivirus software, tools, etc. These activities are aimed to detect and remove viruses and malware ensuring seamless hardware and software operation, networking, routing, etc. There exist a broad range of malware detection, classification and virus removal software. Solutions are centered on malware analysis and detection, network security, web security, authentication, identity and access management, security as a service, security assessment, data loss prevention, cryptography, etc. While malware detection and removal may be often ensured, studies are required studying open information analysis problems, investigating alternative inroads, and, researching novel methodologies. Despite numerous resources and various solutions, exploratory findings and complimentary algorithms should be substantiated.

Hardware Security Features – Physical security of general-purpose and application-specific MPUs in networks, data management systems, sensors and controllers are enabled by cryptographic accelerator co-processors which perform computationally intensive cryptographic operations. Cryptographic accelerators implement symmetric and asymmetric encryption and decryption algorithms to deny unauthorized access and ensure secure hashing, supported by random number generation. Diagnostics, tamper detection and other mechanisms ensure physical security. Another hardware security feature is a *Trusted Execution Environment* which denies a memory assess or restricts runtime access using security control algorithms. Security breaches occur because security credentials and data (passwords, codes, cryptography keys, etc.) may reside in memory, from where an attacker could acquire them. The remote attacks could be through connectivity and *cloud* services. The *secure boot* combines the hardware security with firmware. A malware can be downloaded in a device, the device is rebooted, and, then, boots will run malware, which executes an attack, codes or exfiltrated data. The *secure boot* assesses and signs software images which run on embedded MPU. The aim is to stop untrustworthy codes execution. Once the trusted image is signed with a verified key, it is downloaded along with a signature.

Cyberattacks – There were known cyberattacks on ICS and SCADA systems, such as:

1. Duqu 2.0 malware which exploits zero-day vulnerabilities in host memory;
2. Stuxnet malware which caused inadequate operation of permanent magnet synchronous motors, regulated by controllers-drivers using Hall-effect sensors measurements;
3. Crash override malware which prompted inadequate operation of circuit breakers in electrical grids;

4. Jeep Cherokee remote hacking which wirelessly disabled control;
5. Heartbleed security bug in the OpenSSL cryptography library for the Transport Layer Security and Secure Sockets Layer protocols;
6. Stegomalware masks a malicious content and obscures signature detection.

Information security schemes are aimed to minimize security breaches, data alterations, unauthorized disclosure, disruptions, etc. Information security analysis implies assessment of data quality. The core services are authentication, domain name system, network time protocol, installation service, firewall, etc. Malware may assume and maintain control without the user knowledge. The basic input/output system (BIOS), which is a set of low-level routines, functions as a middleware which allows an operating system to interact with motherboard. The malware may erase, corrupt and replace the BIOS.

Cryptography and Information Security – Information assurance implies managing risks in information processing, storage and sharing with a focus on integrity and confidentiality. Information security is aimed to secure and protect data from unauthorized access, use, disclosure, disruption or modification. Cryptographic algorithms and encrypted data sharing support:

1. Data integrity, which implies information authenticity against improper information modification;
2. Data confidentiality, that is, preserve authorized restrictions on data access, sharing, use and disclosure;
3. Data availability, which implies ensuring timely and reliable access.

Information security is enabled by data encryption and decryption using the Advanced Encryption Standard (AES) [3-5]. We study secure data sharing, research communication under electromagnetic spectrum, as well as perform low-fidelity experimental studies within a *detect* ↔ *share* ↔ *connect* ↔ *assess* ↔ *evaluate* ↔ *manage* scheme in cyber-secure dynamic data-driven applications [6, 7]. Information security implies support, security management, control and reconfiguration of malware detection algorithms and encryption–decryption schemes. Robust algorithms should be developed to conduct assessment of information sharing, vulnerability and resiliency.

Arrays and Codes – There are different data types. Sequences and patterns of bits, crumb (two bits), nibble (four bits), bytes (eight bits), characters (multi-character symbols and words), integers, printable and non-printable ASCII characters (for example, 00, 01 and 02 denote null character, start of header, and, start of text) or text strings are used to characterize code signatures. Antivirus programs should detect hidden malware by searching and analyzing arrays and codes. Pattern recognition and similarity matching are applied to detect a signature with an objective to classify malware, including encrypted viruses. One researches intrusion detection schemes, investigates vulnerabilities, and, studies mitigation solutions. Heterogeneous topological chains and Markov models may ensure an adequate level of completeness.

Information assurance, resilience and adaptation to stealth malware are open problems due to:

1. Multiplicity of cyber threats, and, different types of coordinated distributed malicious attacks;
2. Distributed devices, network and connectivity, as well as *cloud* services vulnerability;
3. System complexity and multi-layer hierarchical organization with vulnerable MPUs;
4. Perplexing strings which are difficult to analyze.

Malware Detection and Cyber-Threat Analysis – Using identifiers, the signature-based detection may detect patterns of known malware, such as hash signatures, binary patterns, etc. There are the malware obfuscation techniques, such as:

1. Software packer executable compression;
2. Encryption by crypter to change a signature;
3. Morphic packing and crypting with advanced encrypting;
4. Dynamic-link library injection and anti-disassembly which prevent detection by signature-based techniques.

A crypter encrypts, obfuscates and refines malware to make it undetectable. A bitwise exclusive OR operation (XOR) additive cipher with a unique key yields immense number of encrypted sequences. Malware may be encrypted using the AES, RSA and other symmetric and asymmetric ciphers. The encryption key is embedded in the malware. Attempts to decrypt a file, which may be double-encrypted, without first removing the malware, may lead to subsequent encryption. Anti-malware and security tools may not restore encrypted files. The packers and crypters may not be detected by means of static analysis which uses not executed codes. Analysis of executed codes in memory may yield detection if memory can be assessed and codes are decrypted.

Semantic analysis, pattern recognition, data mining, code-structure analysis and signature-based classification can be performed on string of symbols, cryptic hash, binary patterns, etc. These schemes are aimed to assess and match patterns, as well as analyze signatures to detect malware. Classification of unspecified malware depends on feature extraction and learning on trained signatures. Heuristic analysis, vector models, support vector machines, Bayesian classifier, logistic regression, grid search, decision tree, clustering and machine learning algorithms are developed aiming to support malware detection using signatures, cross-validation and other schemes. For compromised devices, the attacker may alter or disable host-based intrusion detection capabilities.

Intrusion detection tools are not always ensure consistent analysis of malicious activities. Excessive data and feature dimensions, dataset incompleteness, features ranking ambiguity, scaling inconsistency, morphic packing, crypting obscurity, algorithms sensitivity, computational complexity, latency and other factors limit intrusion detection effectiveness. Advanced malware alters its signature to avoid detection. Complex patterns of unidentified and unknown zero-day malware may not be detected. Malware and intrusion detection systems are supported by:

1. Static- and behavior-based malware detection for defined signatures, patterns, modalities and types of known attacks and threats;

2. Anomaly detection using traffic, system calls, processing and functionality patterns;
3. Dynamic-probabilistic analysis, enabled by statistical models, heterogeneous Markov models and high-fidelity governing analysis.

Steganography – Steganography is concealing the embedded malicious codes. For example, image files contain data in a fixed size encrypted codes with the superimposed vectors which allow low-level decryption. It is challenging to detect malicious codes, macros, unsecure hyperlinks and other malicious content concealed in encrypted codes within partitioned and randomized blocks. If the content of binary strings or text are assessable or quantifiable, one analyzes finite-length sequence of strings.

Static- and behavior-based signature detection schemes, as well as conventional probabilistic concepts, are difficult to apply because classifiers and statistical models may not exist, vary, or, ambiguous due to a wide spectrum of distributed attacks, threats, malicious and adversarial activities, intrusion scenarios, etc. There have being endeavors to assess security metrics using postulated probabilities and selected scores in high-dimensional spaces for assumed or expected threats, attack patterns and vulnerability assessment schemes [8-16]. Machine learning, *knowledge* graphs and other concepts have being applied with moderate scalability and effectiveness [8-17]. There are challenges with simultaneous software-, hardware-, process- and system-aware attacks. Due to dependencies and variations, researchers depart from axiomatized probabilities, postulated statistical models and simplistic studies. Dynamic-probabilistic analysis empowers information security across multiple attack activities.

Electromagnetic Spectrum, Attacks and Detection – Different intrusions and detection schemes have being investigated, such as acoustic cryptoanalysis, side-channel attacks, side-band emission, power monitoring, etc.

The theoretical Shannon-Neumann-Landauer limit suggests the energy estimate required to write data in one binary memory bit as

$$E_{\min}=k_B T\ln 2,$$

where k_B is the Boltzmann constant, $k_B=1.380649\times10^{-23}$ J-K^{-1}; T is the absolute temperature.

At room temperature T=293K, a theoretical minimum energy per switching is $E_{\min}=2.804\times10^{-21}$ J. Study the theoretical switching limit. The Heisenberg uncertainty principle on the canonical measurable (t,E) is applied. An inequality on the canonical measurable energy and time is

$$\sigma_E\sigma_t\leq\tfrac{1}{2}\hbar,$$

where σ_E and σ_t are the standard-deviations; $\hbar$ is the reduced Plank constant, $\hbar=1.054577\times10^{-34}$ J-sec.

Consider a physical hardware. Assess energy estimates and absolute limits on measurements. There are billion transistors in 1 mm^2, and, the 4 nm technology node is expected by 2023. For FETs, fabricated using the 7 nm semiconductor pattern processing technology node, the switching energy is $\sim1\times10^{-16}$ J. The CPU clock frequency $f_{\text{clock}}=1/T_{\text{clock}}$, supported by ASICs, reaches 10 GHz.

The Heisenberg uncertainty principle $\sigma_E\sigma_t\leq\frac{1}{2}\hbar$ yields an absolute limit on the standard deviation of time σ_t for a given σ_E. For CMOS ASICs, one evaluates (σ_t,σ_E), finding that $\sigma_E\sigma_t \geq 1\times10^{-26}$ J-sec. Hence, the Heisenberg principle is met in testing, evaluation and characterization of circuits switching and functionality.

Measurements of radiated field emission and compatibility heterogeneity impose significant challenges. Ultra-low voltage, current and FETs capacitance yield ultra-low emitted field intensity, superimposed with multi-source noise. The thermal noise power is evaluated as $P=k_BTB$, where B is the bandwidth.

Investigating acoustic cryptoanalysis intrusion and detection premises, side-channel attacks, side-band emission and power monitoring, one assesses many factors and uncertainties, such as:

1. Technology benchmarks and fundamental limits of semiconductor technology, fabrication processes, metrology, device physics and digital system logic;
2. Ultra-low device switching energy, voltage and current;
3. Ultra-weak high-frequency field emission with a field spectrum at wide range of field frequencies;
4. Significant challenges in ultra-weak field measurements under interferences, distortions and multi-source noise;
5. Unidentifiable logic operations and instructions;
6. Unidentifiable transductions under assumed detection, and, unknown processing arithmetic;
7. Indistinguishable and undetectable switching, bit strings and character sequences.

Codes and Instructions – There are source code, object code, bytecode, machine code and microcode. Instructions and logic operations are defined by source codes, written in assembly or high-level languages. The source code is assembled, or, for higher-level languages, is compiled to object code. Assembly language has a direct correspondence between assembly code statements and machine instructions. The general-purpose programming languages do not have a direct mapping. The object code, generated by an assembler, compiler or translator, yields an executable bytecode which defines machine-level instructions. An interpreter analyzes the source code, and, converts commands to a machine code for 8-, 16-, 32- and 64-bit processors. In MPUs, operations and instructions are specified by a machine code, which is a sequence of 0s and 1s. The binary machine code is a sequence of machine instructions, and, code is executable by ASICs. Each instruction is a string of bits. The binary alphabet is $\mathcal{a}=\{0,1\}$. The bitwise operators are the logical AND, OR, exclusive OR, complement, right shift and left shift, denoted as &, |, ^, ~, << and >>. A binary operation on a non-empty set S is a mapping f, which assigns a uniquely determined element of S to each ordered pair of elements of S. The mapping is f:S×S→S with an image $a\circ b$, $(a,b)\in$S. The set S is said to be closed under the operation f, and, a binary operation is a tuple (S×S,S,f). The image of a function is the set of outputs, and, F is a function from domain X to codomain Y, $F{:}X\to Y$. The binary operations are denoted by special symbols, such as addition +, subtraction –, multiplication ·, division ÷, difference /, logic AND $\wedge$, logic OR $\vee$, union $\cup$, intersection $\cap$, oplus, ominus, otimes, oslash, min, max, defect and others.

Data driven applications imply various arithmetic and logic bitwise operations on integer operands, as well as instructions on input-output data during executions, coding-decoding, memory writing, addressing, memory access and other tasks. Bitwise operations on binary integers, control instructions, as well as register–register, register–memory and memory–memory ALU operations, support services, serial and parallel input-output (I/O) interface, etc. Operations are performed as instructions are fetch within a fetch–decode–execute cycle. Execution and data access are performed in multiple stages. Some instructions, data transfer, I/O data and data packets can be directly assessed from data registers, I/O ports, etc. Bytes, integers, strings, American Standard Code for Information Interchange (ASCII) codes, as well as the Universal Coded Character Set Transformation Format (UTF-8) variable-width character string sequences, may contain malicious codes and instructions.

Finite Alphabet and Strings – A string S is a finite and ordered sequence of characters such as letters, integers and spaces. Consider a finite alphabet $\mathcal{A}=\{a_1,a_2,\ldots,a_{l-1},a_l\}$. For a finite set of symbols (characters) in $\mathcal{A}$, a string S over $\mathcal{A}$ is a finite sequence of symbols from $\mathcal{A}$. The length of a string is the number of symbols.

For a binary alphabet $\mathcal{A}=\{0,1\}$

$\mathcal{A}^0=\{\varepsilon\}$, $\mathcal{A}^1=\{0,1\}$, $\mathcal{A}^2=\{00,01,10,11\}$, $\mathcal{A}^3=\{000,001,010,011,100,101,110,111\}$, etc.

Strings can be decomposed as crumb (two bits), nibble (four bits), bytes (eight bits), etc. A set of bit strings $\mathcal{A}^*$ is formed over $\mathcal{A}$, and, includes the empty strings. In particular, $\{0,1\}^*=\{\varepsilon,0,1,00,01,10,11,000,001,..\}$. A set of all strings $\mathcal{A}^*$ over an alphabet $\mathcal{A}$ is $\mathcal{A}^*=\mathcal{A}^0\cup\mathcal{A}^1\cup\mathcal{A}^2\cup\mathcal{A}^3\cup..$.

Alphabet and Language – Combinatorics and square-free morphism are used to investigate and characterize strings content, patterns, modalities and regularities. A set of natural numbers is $\mathbb{N}_0=\{0,1,2,...\}$. An alphabet $\mathcal{A}$ is a finite nonempty set, and, the elements are called *letters*. A *word w* over $\mathcal{A}$ is a finite sequence of *letters* from $\mathcal{A}$. The empty *word*, which is a sequence of length 0, is denoted as ε. For the empty string, $|\varepsilon|=0$, $\varepsilon\cdot\mathsf{S}=\mathsf{S}\varepsilon=\mathsf{S}$ and $\varepsilon^R=\varepsilon$, e.g., the reversal of the empty string yields an empty string with zero length.

A formal language $\mathcal{L}$ over an alphabet $\mathcal{A}$ is a subset of all strings $\mathcal{A}^*$, e.g., a set of *words* over $\mathcal{A}$, and, $\mathcal{A}^+=\mathcal{A}^*\backslash\{\varepsilon\}$. The propositional calculus establishes closure properties and operations on language families. A two-dimensional *word w*, with entries $w_{n,m}$, is a map from $\mathbb{N}^2$ to $\mathcal{A}$. A morphism between $\mathcal{A}^*$ and $\mathcal{B}^*$ is a map f:$\mathcal{A}^*\to\mathcal{B}^*$, such that $f(\varepsilon)=\varepsilon$ and $f(wu)=f(w)f(u)$ for *words* $(w,v)\in\mathcal{A}^*$. One may construct an infinite square-free *word* over $\mathcal{A}$ with three *letters*, and, there exists an infinite number of finite square-free *words*. No square-free word length $l\geq3$ occurs over two-letter alphabet $\mathcal{A}=\{a,b\}$. The ternary alphabet is $\mathcal{A}=\{0,1,2\}$. If $\mathcal{A}$ is an alphabet, and, $\mathcal{L}\subseteq\mathcal{A}^*$, then $\mathcal{L}$ is a formal language over $\mathcal{A}$. A language is a set of strings formed from some alphabet. For example, $\mathcal{L}(\emptyset)=_{\text{def}}\emptyset$ or $\mathcal{L}=_{\text{def}}\{000,001,010,011,100,101,110,111\}$. Here, ∅ denotes the empty language, e.g., a language over any alphabet.

Codes, Sequences and Strings – The hardware read-only microcode, which resides in high-speed memory, translates machine instructions, and, controls logic and data into sequences of circuit-level operations. MPUs operate on a predefined set of instructions by interpreting and executing instructions to perform logic and arithmetic operations. General-purpose MPUs consist of a central processing unit (CPU), coprocessors, arithmetic logic unit (ALU), control unit (CU), memory management unit (MMU), interrupts, input/output (I/O) interface, cache, etc. Specialized MPUs are used as graphical processing units (GPUs) and digital signal processing (DSP) processors, made by Advanced Micro Devices, Apple, Intel, Nvidia, Qualcomm, Texas Instruments, etc. These MPUs may be attacked by maliciously refining addresses and registers, as well as by modifying instructions, data and control paths, logic and memories (shared, flash, static random-access, external, etc.), interface, shared peripherals, etc. Different MPUs represent same operations and commands differently. Binary codes represent text as a sequence of binary 0s and 1s. Fixed-width binary codes use a set number of bits to represent each character in the text, while for variable-width binary codes, the number of bits may vary from character to character. There are 5-, 6-, 7-, 8-, 10-, 16- and 32-bit fixed-width binary codes. For example, the six-bit codes for logical AND, Add with Carry, SUB and Subtract with Borrow instructions may be 000000, 000100, 001010 and 000110. The common variable-width binary codes are UTF-8, UTF-16, etc.

A string of a byte (8 bits) is called a half word, two consecutive bytes (16 bits) are called a word, and, four bytes (32 bits) are called a double or long word. A quadword is 8 bytes, e.g., 64 bits. A binary string of eight bits can represent 256 possible values. One may analyze bit and byte sequences and strings. The user can assess the shared data and outputs. Considering combinatorial logic, arithmetic and memory architecture with micro-operations, micro-instructions and microcodes, one may not retrieve the CPU-specific built-in proprietary microcodes, such as hardware-level micro-instructions, microcode patches, etc. Integrated microelectronic devices and passive circuits implement logic gates, registers, multiplexers, converters and other ASICs.

The MPU performs arithmetic and bitwise logic operations on data by fetching instructtions from memory, executing them, and, fetching next instructions. Source codes and machine codes define operations, control, interfacing, memory, registers, addresses, etc. One may use disassemblers to disassemble machine codes to assembly language, and, decompilers for compiled codes for high-level languages. Machine codes may be compiled back to a specified programming language, which may not be an original language. One may not obtain the original source codes from machine codes.

Example 7.1. For the ASCII text **register**, the binary, hexadecimal and Base64 strings are '01110010 01100101 01100111 01101001 01110011 01110100 01100101 01110010', '72 65 67 69 73 74 65 72', and, 'cmVnaXN0ZXI='. ■

Code Disassembly – Disassembly uses a compiled executable, and, converts strings back to assembly language. While malicious executable may be disassembled, this does not yield the original source code, code structure, comments, etc. Security tools are the MPU specific. For example, the 32-bit Intel processors have specific machine instructions and eight general purpose registers, e.g. EAX, EBX, ECX, EDX, ESI, EDI, EBP and ESP. The 16-bit CPUs have AX, BX, CX, DX, SI, DI, BP and SP registers, while, for 8-bit CPUs, the registers are AH, AL, BH, BL, CH, CL, DH and DL. The *stack* is a temporary memory location for storing data, and, program instructions. Machine instructions are constantly *popping* and *pushing* data from and to the *stack*, and, to a register. The buffer overflows may overwrite the legitimate *stack* pointer to malicious one.

7.3. Networked Cyber-Physical Systems

Describe system governance using outputs $\boldsymbol{y}$, states $\boldsymbol{x}$, controls $\boldsymbol{u}$, disturbances $\boldsymbol{d}$ and perturbations $\boldsymbol{w}$ in cyber and physical domains. Physical and information quantities are:

1. Physical system variables $(\boldsymbol{y},\boldsymbol{x},\boldsymbol{d},\boldsymbol{w},\boldsymbol{u})^{\text{physical}}$;
2. Information and data variables $(\boldsymbol{y},\boldsymbol{x},\boldsymbol{d},\boldsymbol{w},\boldsymbol{u})^{\text{cyber}}$ and $(\boldsymbol{y},\boldsymbol{x},\boldsymbol{d},\boldsymbol{w},\boldsymbol{u})^{\text{data}}$.

While cybersecurity focuses on protecting of electronic and mobile devices against attacks, information security deals with protecting data availability, confidentiality, integrity, etc. Secure information sharing, information control and information management are considered. Physical, cyber and data domains in a system $\mathbb{M}(\mathbf{P}^{\text{physical}}\times\mathbf{C}^{\text{cyber}}\times\mathbf{D}^{\text{data}})$ are described and characterized by real-valued variables. Electromagnetic spectrum, interference, spoofing, electromagnetic spectrum and other malicious activities are considered as deterministic and stochastic disturbances and perturbations $(\boldsymbol{d}_a,\boldsymbol{w}_a)$. A union of measured, detected, observed and estimated $(\boldsymbol{y},\boldsymbol{x},\boldsymbol{d},\boldsymbol{w},\boldsymbol{u})$ describe system dynamics, information governance and data evolutions

$$\mathbb{M}(\mathbf{P}^{\text{physical}}\times\mathbf{C}^{\text{cyber}}\times\mathbf{D}^{\text{data}}):=\left\{\underset{\substack{\text{Outputs}\\ y\in\mathbf{Y}}}{\boldsymbol{y}=\begin{bmatrix}\boldsymbol{y}(t)\\ \boldsymbol{y}[n]\end{bmatrix}=\begin{bmatrix}\boldsymbol{y}^{\text{physical}}\\ \boldsymbol{y}^{\text{cyber}}\\ \boldsymbol{y}^{\text{data}}\end{bmatrix}},\ \underset{\substack{\text{States}\\ x\in\mathbf{X}}}{\boldsymbol{x}=\begin{bmatrix}\boldsymbol{x}(t)\\ \boldsymbol{x}[n]\end{bmatrix}=\begin{bmatrix}\boldsymbol{x}^{\text{physical}}\\ \boldsymbol{x}^{\text{cyber}}\\ \boldsymbol{x}^{\text{data}}\end{bmatrix}},\ \underset{\substack{\text{Controls}\\ u\in\mathbf{U}}}{\boldsymbol{u}=\begin{bmatrix}\boldsymbol{u}(t)\\ \boldsymbol{u}[n]\end{bmatrix}=\begin{bmatrix}\boldsymbol{u}^{\text{physical}}\\ \boldsymbol{u}^{\text{cyber}}\\ \boldsymbol{u}^{\text{data}}\end{bmatrix}},\ \underset{\substack{\text{Disturbances}\\ d\in\mathbf{D}}}{\boldsymbol{d}=\begin{bmatrix}\boldsymbol{d}(t)\\ \boldsymbol{d}[n]\end{bmatrix}},\ \underset{\substack{\text{Perturbations}\\ w\in\mathbf{W}}}{\boldsymbol{w}=\begin{bmatrix}\boldsymbol{w}(t)\\ \boldsymbol{w}[n]\end{bmatrix}}\right\},$$

$$\boldsymbol{y}=\begin{bmatrix}\boldsymbol{y}^{\text{physical}}\\ \boldsymbol{y}^{\text{cyber}}\\ \boldsymbol{y}^{\text{data}}\end{bmatrix}=\begin{bmatrix}\begin{bmatrix}\boldsymbol{y}(t)^{\text{physical}}\\ \boldsymbol{y}[n]^{\text{physical}}\end{bmatrix}\\ \begin{bmatrix}\boldsymbol{y}(t)^{\text{cyber}}\\ \boldsymbol{y}[n]^{\text{cyber}}\end{bmatrix}\\ \begin{bmatrix}\boldsymbol{y}(t)^{\text{data}}\\ \boldsymbol{y}[n]^{\text{data}}\end{bmatrix}\end{bmatrix},\ \boldsymbol{x}=\begin{bmatrix}\boldsymbol{x}^{\text{physical}}\\ \boldsymbol{x}^{\text{cyber}}\\ \boldsymbol{x}^{\text{data}}\end{bmatrix}=\begin{bmatrix}\begin{bmatrix}\boldsymbol{x}(t)^{\text{physical}}\\ \boldsymbol{x}[n]^{\text{physical}}\end{bmatrix}\\ \begin{bmatrix}\boldsymbol{x}(t)^{\text{cyber}}\\ \boldsymbol{x}[n]^{\text{cyber}}\end{bmatrix}\\ \begin{bmatrix}\boldsymbol{x}(t)^{\text{data}}\\ \boldsymbol{x}[n]^{\text{data}}\end{bmatrix}\end{bmatrix},\ \boldsymbol{u}=\begin{bmatrix}\boldsymbol{u}^{\text{physical}}\\ \boldsymbol{u}^{\text{cyber}}\\ \boldsymbol{u}^{\text{data}}\end{bmatrix}=\begin{bmatrix}\begin{bmatrix}\boldsymbol{u}(t)^{\text{physical}}\\ \boldsymbol{u}[n]^{\text{physical}}\end{bmatrix}\\ \begin{bmatrix}\boldsymbol{u}(t)^{\text{cyber}}\\ \boldsymbol{u}[n]^{\text{cyber}}\end{bmatrix}\\ \begin{bmatrix}\boldsymbol{u}(t)^{\text{data}}\\ \boldsymbol{u}[n]^{\text{data}}\end{bmatrix}\end{bmatrix},\tag{7.1}$$

$$\boldsymbol{d}=\begin{bmatrix}\boldsymbol{d}^{\text{physical}}\\ \boldsymbol{d}^{\text{cyber}}\\ \boldsymbol{d}^{\text{data}}\end{bmatrix}=\begin{bmatrix}\begin{bmatrix}\boldsymbol{d}(t)^{\text{physical}}\\ \boldsymbol{d}[n]^{\text{physical}}\end{bmatrix}\\ \begin{bmatrix}\boldsymbol{d}(t)^{\text{cyber}}\\ \boldsymbol{d}[n]^{\text{cyber}}\end{bmatrix}\\ \begin{bmatrix}\boldsymbol{d}(t)^{\text{data}}\\ \boldsymbol{d}[n]^{\text{data}}\end{bmatrix}\end{bmatrix},\ \boldsymbol{w}=\begin{bmatrix}\boldsymbol{w}^{\text{physical}}\\ \boldsymbol{w}^{\text{cyber}}\\ \boldsymbol{w}^{\text{data}}\end{bmatrix}=\begin{bmatrix}\begin{bmatrix}\boldsymbol{w}(t)^{\text{physical}}\\ \boldsymbol{w}[n]^{\text{physical}}\end{bmatrix}\\ \begin{bmatrix}\boldsymbol{w}(t)^{\text{cyber}}\\ \boldsymbol{w}[n]^{\text{cyber}}\end{bmatrix}\\ \begin{bmatrix}\boldsymbol{w}(t)^{\text{data}}\\ \boldsymbol{w}[n]^{\text{data}}\end{bmatrix}\end{bmatrix}.$$

Control of Cyber-Physical Systems – Control of physical systems have being explicitly studied. Using $(\boldsymbol{y},\boldsymbol{x},\boldsymbol{d},\boldsymbol{w},\boldsymbol{u})^{\text{physical}}$, designs have being performed, and, advanced-technology systems are deployed and broadly commercialized. Cybersecurity and data-driven information control are new paradigms, and, open-ended problems must be solved. To derive secure information management actions and controls $(\boldsymbol{u}^{\text{cyber}},\boldsymbol{u}^{\text{data}})$, one should measure or observe states $(\boldsymbol{x}^{\text{cyber}},\boldsymbol{x}^{\text{data}})$ and outputs $(\boldsymbol{y}^{\text{cyber}},\boldsymbol{y}^{\text{data}})$. Some states cannot be measured, and, there are uncertainties on estimations of $(\boldsymbol{x}^{\text{cyber}},\boldsymbol{x}^{\text{data}})$ and $(\boldsymbol{y}^{\text{cyber}},\boldsymbol{y}^{\text{data}})$. Evolutions and control $(\boldsymbol{y},\boldsymbol{x},\boldsymbol{u})^{\text{physical}}$ depend on information control, data management, etc. Cyber states $\boldsymbol{x}^{\text{cyber}}$ characterize software, middleware and application security and management, while $\boldsymbol{x}^{\text{data}}$ describe data governance. A vector $\boldsymbol{u}$ consists of physical, cyber, information and data controls, actions and commands. Information control is a very challenging problems. Section 7.5 reports a dynamic-probabilistic Turing machine (7.3) where the control vector $\boldsymbol{u}_{\text{T}}$ implies control of information, event and security, instructions and logic operations, etc.

Cyberattacks and malware are detected by examining codes and sequences, which affect logic and arithmetic operations, executed instructions, data access, data acquisition, data sharing, data retrieval, etc. One defines $(\boldsymbol{y},\boldsymbol{x},\boldsymbol{u})^{\text{cyber}}$ and examines states detectability, estimability and identifiability. Intrusion detection schemes are aimed to identify threats and malicious codes by searching binary arrays (bit and byte strings) which represent binary data, hexadecimal and character strings, text strings, encoded and decoded strings, printable and non-printable ASCII characters, character sequences in codes, data packets, etc. Information assurance, information security, data quality, resiliency and other features are described by $(\boldsymbol{y},\boldsymbol{x})^{\text{cyber}}$, and, controlled by $\boldsymbol{u}^{\text{cyber}}$. Data states and controls $(\boldsymbol{x},\boldsymbol{u})^{\text{data}}$ describe data evolution, logic, arithmetic and pertained calculus to execute processing, data sharing, controlled data-driven applications, etc. The real-valued $\boldsymbol{x}^{\text{data}}$ are the channel capacity, data rate, bandwidth, communication losses, data losses, etc. Physical data quality, controlled by $\boldsymbol{u}^{\text{data}}$, depends on physical data transductions, multiphysics conversion, sensing, processing, peripheral interface, transceivers communication, logic operations and processing. Data quality homogeneity and heterogeneity $(\mathcal{M}_{\mathcal{H}}(\cdot),\mathcal{M}_{\mathcal{L}}(\cdot))$ are found and characterized by finite dynamic-probabilistic measures, such as probability distributions, entropies, mutual information, information losses, etc. Some measures are used in communication [18].

Attributes, Indicators and Measures – To assess system functionality and security, apply ordered sets of explicitly defined multiple attributes $\boldsymbol{\mathcal{A}}$, composite indicators $\boldsymbol{I}$ and decision variables. Attributes and indicators $(\mathcal{A}_i,I_i)$ may be qualitative and quantitative. Functionality, normalcy and anomalies are assessed by descriptive states. Multi-objective analysis imply evaluation of high-dimensional multiple criteria on real-valued detectable and measurable $\boldsymbol{\mathcal{M}}=\sum_i \mathcal{M}_i$. Studies dynamic-probabilistic physical and information processes, transitions and transductions. For example, malware detection can be investigated by examining: (1) Semantics, signature and syntactic characteristics; (2) Occurrence and frequency of symbols and character sequences; (3) Combination of consecutive obfuscated strings and patterns; (4) Bit- and byte levels n-gram, and, subsets of n-grams; etc.

Using the real-valued $\mathcal{M}\equiv\varphi(\boldsymbol{\mathcal{A}},\boldsymbol{I},\boldsymbol{x},\boldsymbol{u})$, a minimax problem is

$$\boldsymbol{q}(\cdot)=\min_{\substack{\mathcal{M}_{\mathcal{L}}\\ q\in Q\subseteq\boldsymbol{\mathcal{M}}\times\mathbf{X}\times\mathbf{U}}}\max_{\mathcal{M}_{\mathcal{H}}}\ \kappa(t,\mathcal{M},\boldsymbol{x},\boldsymbol{u}),\ \boldsymbol{\mathcal{A}}=\begin{cases}\boldsymbol{\mathcal{A}}^{\text{physical}}\\ \boldsymbol{\mathcal{A}}^{\text{cyber}}\\ \boldsymbol{\mathcal{A}}^{\text{data}}\end{cases}=\begin{cases}\sum_i \mathcal{A}_i^{\text{physical}}\\ \sum_j \mathcal{A}_j^{\text{cyber}}\\ \sum_k \mathcal{A}_k^{\text{data}}\end{cases},\ \boldsymbol{I}=\begin{cases}\boldsymbol{I}^{\text{physical}}\\ \boldsymbol{I}^{\text{cyber}}\\ \boldsymbol{I}^{\text{data}}\end{cases}=\begin{cases}\sum_i I_i^{\text{physical}}\\ \sum_j I_j^{\text{cyber}}\\ \sum_k I_k^{\text{data}}\end{cases},\ \boldsymbol{\mathcal{M}}=\begin{cases}\boldsymbol{\mathcal{M}}^{\text{physical}}\\ \boldsymbol{\mathcal{M}}^{\text{cyber}}\\ \boldsymbol{\mathcal{M}}^{\text{data}}\end{cases}=\begin{cases}\sum_i \mathcal{M}_i^{\text{physical}}\\ \sum_j \mathcal{M}_j^{\text{cyber}}\\ \sum_k \mathcal{M}_k^{\text{data}}\end{cases},\tag{7.2}$$

where $\boldsymbol{q}(\cdot)$ is the vector of p-criterion functions, $\boldsymbol{q}\subseteq\mathbb{R}^{p}$; $\kappa(\cdot)$ is the nonlinear mapping.

The objective is to control and optimize dynamic-probabilistic processes, transductions and transitions in CPS $\mathbb{M}(\cdot)$ by finding $(\boldsymbol{u}(t),\boldsymbol{u}[n])$ to regulate system behavior $(\boldsymbol{y}(t),\boldsymbol{y}[n],\boldsymbol{x}(t),\boldsymbol{x}[n])$, data governance $(\boldsymbol{\mathcal{D}}(t),\boldsymbol{\mathcal{D}}[n])$, information processes $(\mathfrak{I}(t),\mathfrak{I}\,[n])$, etc.

Example 7.2. Analysis and Decision Spaces

Conducting high-dimensional analysis in decision spaces, define a p-criterion function $\boldsymbol{q}(\cdot)$ as well as nonlinear objective function $\kappa(t,\boldsymbol{\mathcal{M}},\boldsymbol{x},\boldsymbol{u})$ in a convex n-polytope $T\times\boldsymbol{\mathcal{M}}\times\mathbf{X}\times\mathbf{U}$. Examples of $\boldsymbol{q}^{\text{cyber}}(\cdot)$ are detection probability, false positive probability, prevention probability, cyberattack versus fault detection, etc. Cyberattack analysis is supported and visualized by descriptive polytopes with vertices, edges and facets, characterized by equalities and inequalities. For example, a 6-dimensional 2_{21} uniform 6-polytope with 27 vertices and 99 facets is constructed within the symmetry of the E_6-group, and, contains 27 5-orthoplexes and 72 5-simplexes. A 7-dimensional 3_{21} uniform 7-polytope with 56 vertices and 702 facets contains 126 3_{11} 6-orthoplexes and 576 6-simplexes. ■

Information security, data quality, normalcy and anomalies are assessed and asserted by $\mathcal{M}(\cdot)$. One may partition, factorize and correlate high-dimensional problems by considering a hierarchy of prioritized non-equally-weighted subproblems to calculate scores and estimates using nondominant and dominant events, governance, etc. Dimensionality reduction, low-rank factorization and correspondence analysis are investigated. If not all $(\boldsymbol{y},\boldsymbol{x},\boldsymbol{d},\boldsymbol{w})$ are detectable and measurable, estimates on $(\hat{\boldsymbol{y}},\hat{\boldsymbol{x}},\hat{\boldsymbol{d}},\hat{\boldsymbol{w}})$ may be found. For example, $\hat{\mathcal{M}}^{\text{cyber}}$ may be implicitly assessed by $\begin{bmatrix} \mathcal{M}^{\text{physical}} \\ \mathcal{M}^{\text{data}} \end{bmatrix}$.

Example 7.3. *Analysis of a Cyber-Physical Module in SCADA System*

Consider a module with a permanent magnet synchronous motor, PWM controller-driver with six power MOSFETs and control interface, Hall-effect sensors to measure rotor angular displacement, and, MPU. The vulnerability of a considered module and other electromechanical systems is well-known. The Win32/Stuxnet virus, revealed in the 2010, exploits deficiencies in the Microsoft Windows and Siemens SIMATIC WinCC interface. As documented in Practice Problem 7.11, this virus, which has three components (worm, link and rootkit), exploits and targets particular programmable logic controllers (PLCs) for electric drives. Stuxnet leads to inadequate data acquisition and inconsistent control in SCADA systems. Computationally-intensive algorithms for feature extraction and searches to detect and classify viruses may not yield desirable outcomes. Stealth viruses may not be detected. With obstacles and challenges in intrusion detection and data quality evaluation, one may assess an overall system normalcy using $\mathcal{M}(\boldsymbol{x},\boldsymbol{u})^{\text{physical}}$ and $\mathcal{M}(\boldsymbol{x},\boldsymbol{u})^{\text{data}}$.

Operation and control of synchronous motors and PWM controllers-drivers were covered in previous chapters. High frequency switching of six MOSFETs in a half-H bridge PWM driver are controlled by the MPU's PWM outputs. The balanced voltage set is implemented using the Hall-effect rotor displacement sensors measurements. With the MOSFETs switching frequency 100 kHz, the sensor measurements, data acquisition, processing and control are accomplished within microseconds. The MPU logic, instructions, binary logic gate level and arithmetic operations cannot be directly observed. The streaming data is stored in the RAM of the 32-bit MPUs. The data log is accomplished by using three pointers which define the start and end addresses of the memory and data indexes. The serial peripheral interface supports a serial 16-bit strings with high-speed bit-transfer rate.

The measured states $\boldsymbol{x}$ are the actuator variables, amplifier voltages and currents, MPU's PWM outputs, and, bit strings of the measurable peripherals data. Normalcy and key performance capabilities are explicitly defined by the p-criterion function $\boldsymbol{q}(\cdot)$.

Consider $\boldsymbol{\mathcal{A}} = \begin{cases} \sum_i \mathcal{A}_i^{\text{physical}} \\ \sum_k \mathcal{A}_k^{\text{data}} \end{cases}, \boldsymbol{I} = \begin{cases} \sum_i I_i^{\text{physical}} \\ \sum_k I_k^{\text{data}} \end{cases}$.

Define $\boldsymbol{\mathcal{M}} = \begin{cases} \sum_i \mathcal{M}_i^{\text{physical}} \\ \sum_k \mathcal{M}_k^{\text{data}} \end{cases}, \mathcal{M}_i^{\text{physical}} = \begin{cases} \mathcal{M}_{i\ \text{Actuator}}^{\text{physical}} \\ \mathcal{M}_{i\ \text{Amplifier}}^{\text{physical}} \\ \mathcal{M}_{i\ \text{Sensors}}^{\text{physical}} \\ \mathcal{M}_{i\ \text{MPU}}^{\text{physical}} \end{cases}, \mathcal{M}_k^{\text{data}} = \begin{cases} \mathcal{M}_{k\ \text{MPU data quality}}^{\text{data}} \\ \mathcal{M}_{k\ \text{Sensors data quality}}^{\text{data}} \end{cases}$.

The attributes $\mathcal{A}_i^{\text{physical}}$ are adaptiveness, complexity, connectivity, networking, modularity, cybersecurity compliance, etc. The indicators I_i^{physical} are controllability, detectability, observability, measurement fidelity, reconfigurability, robustness, etc. Algorithmic and physical controllability and observability are used. The measured and estimated $(\mathcal{M}_i^{\text{physical}}, \hat{\mathcal{M}}_i^{\text{physical}})$ are: (1) Physical normalcy defined by the motor and controller-driver (PWM amplifier) measured and estimated states within operating envelopes, such as (i_{as}, i_{bs}, i_{cs}), (u_{as}, u_{bs}, u_{cs}), (ω_r, θ_r) and others; (2) Normalcy defined by power, torque, torque ripple, efficiency η, losses, etc.; (3) Dynamic output error; (4) Tracking error; (5) Dynamic range, settling time and bandwidth; (6) Disturbance rejection; (7) Sensitivity to parameter variations; (8) Amplifier PWM voltage waveforms, harmonics and ripple.

Detected and estimated $(\mathcal{M}_k^{\text{data}}, \hat{\mathcal{M}}_k^{\text{data}})$, consistent with the physical and information-centric data quality, are: (1) Measurement fidelity and repeatability affected by denial of data, disruption or modifying data; (2) Algorithmic controller and filters data fidelity; (3) Measurement precision and information sensitivity, affected by electromagnetic interference, jamming, quantization (digitization) error, etc.; (4) Data delays and latency; (5) Data normalcy within an operating envelope.

Different $(\mathcal{M}_k^{\text{data}}, \hat{\mathcal{M}}_k^{\text{data}})$ can be used. For example, the information losses, bit error rate, bit error probability, environmental sensitivity, noise floor, sensor linearity and drift, etc. This yields sensing and data acquisition complexities, dimensionality increase, additional states to measure or estimate, etc. The measured and estimated variables $\boldsymbol{x} \equiv (x, \hat{x}, u)$, $\boldsymbol{x} \in \mathbb{R}^{15}$ are the Hall-effect sensor pulses, power source current, amplifier PWM voltages, phase voltages (u_{as}, u_{bs}, u_{cs}) with u_M defined by the controlled duty cycle u, phase currents (i_{as}, i_{bs}, i_{cs}), electromagnetic torque T_e, torque ripple ΔT_e, angular velocity ω_r, displacement θ_r, and, tracking error e_ω.

We assess $\boldsymbol{\mathcal{M}} = \begin{cases} \sum_i \mathcal{M}_i^{\text{physical}} \\ \sum_k \mathcal{M}_k^{\text{data}} \end{cases}, \boldsymbol{\mathcal{M}}^{\text{physical}} = \sum_i (\mathcal{M}_i^{\text{physical}}, \hat{\mathcal{M}}_i^{\text{physical}}), \boldsymbol{\mathcal{M}}^{\text{data}} = \sum_k (\mathcal{M}_k^{\text{data}}, \hat{\mathcal{M}}_k^{\text{data}})$.

Evaluate $\kappa(\boldsymbol{\mathcal{M}}, \boldsymbol{x}, \boldsymbol{u})$ by defining attributes $\mathcal{A}_j^{\text{cyber}}$ and indicators I_j^{cyber}. For example, the information assurance, information security, etc. The measures $\mathcal{M}_j^{\text{cyber}}$, applied in high-fidelity studies, could be: (1) Data breach probability; (2) Malware detection probability; (3) Resiliency to cyberattacks; (4) Information sharing security; (5) Cryptographic algorithm strength; etc. There are dependencies between $(\boldsymbol{\mathcal{A}}, \boldsymbol{I}, \boldsymbol{\mathcal{M}})$. While $\mathcal{M}_j^{\text{cyber}}$ are not evaluated, key $(\boldsymbol{\mathcal{A}}_j^{\text{cyber}}, I_j^{\text{cyber}}, \mathcal{M}_j^{\text{cyber}})$ may be implicitly asserted by

$$\boldsymbol{\mathcal{M}} = \begin{cases} \sum_i \mathcal{M}_i^{\text{physical}} \\ \sum_k \mathcal{M}_k^{\text{data}} \end{cases}, \boldsymbol{\mathcal{M}} \in \mathbb{R}^{13}.$$

The multidimensional mapping $\kappa(\mathcal{M},x,u)$ can be depicted as a convex polytope. A polyhedron P is characterized by vertices, edges and faces $(\mathcal{V},\mathcal{E},\mathcal{F})$. The Euler formula $(\mathcal{V}-\mathcal{E}+\mathcal{F})=2$ yields a description of graphs and skeletons for three-dimensional polytopes. For a regular polyhedron, use the Schläfli symbol $\{p,q\}$ with the number of sides of each face p, and, the number of faces meeting at each vertex q. Convexity and topological similarity are ensured for quasi-regular and nonuniform irregular polyhedrons, defined by dimensionality, degree of vertex configuration, skeleton, sequence of faces at vertexes, connections and dependencies.

In the closed-loop system, the proportional-integral control law regulates the amplitude u_M of phase voltages. The balanced phase voltages (u_{as},u_{bs},u_{cs}) are applied by measuring the angular displacement θ_r. That is, $u\equiv u_M$. A *minimal complexity* control law, designed using a Lyapunov pair $\left\{V(x)>0, \frac{dV}{dt}<0\right\}$, guarantees functionality and optimizes system dynamics. The fourth-order notch filters are designed and implemented. Having found physics-consistent control, filtering and data fusion, an information-centric control implies solution of a minimax problem (7.2). The $(\mathcal{M}_i^{\text{physical}},\hat{\mathcal{M}}_i^{\text{physical}})\in\mathbb{R}^8$ and $(\mathcal{M}_k^{\text{data}},\hat{\mathcal{M}}_k^{\text{data}})\in\mathbb{R}^5$ are normalized to be equally scaled.

Low-fidelity studies are conducted for a module with an optimal choice of devices and components, consistent data fusion, adequate processing algorithms, etc.

Simplify a minimax problem (7.2) $\bar{\boldsymbol{q}}(\cdot)=\min\limits_{\substack{\mathcal{M}_{\mathcal{L}} \\ q\in Q\subseteq \mathcal{M}\times\mathrm{X}\times\mathrm{U}}}\max\limits_{\mathcal{M}_{\mathcal{H}}}\ \kappa\left[t,\bar{\mathcal{M}}(\boldsymbol{x}),\bar{\boldsymbol{x}}\right],\ \boldsymbol{x}\equiv(x,\hat{x},u)\cdot$

Evaluate an optimized-by-design p-criterion function $\bar{\boldsymbol{q}}^*(\cdot)$ using normilized $\bar{\boldsymbol{q}}(\cdot)$ as

$$\bar{\boldsymbol{q}}\left[\bar{\mathcal{M}}(\boldsymbol{x}),\bar{\boldsymbol{x}}\right]=\frac{1}{N}\sum_{i,k}\left(\left|\bar{\mathcal{M}}_i^{\text{physical}}(\boldsymbol{x})\right|+\left|\bar{\mathcal{M}}_k^{\text{data}}(\boldsymbol{x})\right|\right),\ \boldsymbol{x}\equiv(x,\hat{x},u),\ \bar{\mathcal{M}}_i^{\text{physical}}=\frac{1}{\mathcal{M}_{i\max}^{\text{physical}}},\ \bar{\mathcal{M}}_k^{\text{data}}=\frac{1}{\mathcal{M}_{k\max}^{\text{data}}}.$$

For cross-tabulated measures $\bar{\mathcal{M}}=\begin{bmatrix}\bar{\mathcal{M}}_i^{\text{physical}}\in[\bar{\mathcal{M}}_{i\min}^{\text{physical}}\ \ \bar{\mathcal{M}}_{i\max}^{\text{physical}}],\ 0\le\bar{\mathcal{M}}_i^{\text{physical}}\le 1\\ \bar{\mathcal{M}}_k^{\text{data}}\in[\bar{\mathcal{M}}_{k\min}^{\text{data}}\ \ \bar{\mathcal{M}}_{k\max}^{\text{data}}],\ \ 0\le\bar{\mathcal{M}}_k^{\text{data}}\le 1\end{bmatrix}$ and normalized $\boldsymbol{x}\equiv(x,\hat{x},u)$, experimental results yield $\bar{\boldsymbol{q}}(\cdot)\in[0\ \ 1]$. The lower and upper bounds on time-varying $\mathcal{M}^{\text{physical}}(t)$ and $\mathcal{M}^{\text{data}}(t)$ exist.

Examples of $\bar{\mathcal{M}}^{\text{physical}}$ are:

1. $0\le\bar{\mathcal{M}}_{\text{normalcy}}^{\text{physical}}\le 1$ with $\sup\{\bar{\mathcal{M}}_{\text{normalcy}}^{\text{physical}}\}\approx 1\wedge\bar{\mathcal{M}}_{\text{normalcy}}^{\text{physical}}<1$;
2. $0\le\bar{\mathcal{M}}_{\text{sensitivity}}^{\text{physical}}\le 1$. If system is invariant to parameter variations, $\sup\{\bar{\mathcal{M}}_{\text{sensitivity}}^{\text{physical}}\}\approx 1\wedge\bar{\mathcal{M}}_{\text{sensitivity}}^{\text{physical}}<1$;
3. $0\le\bar{\mathcal{M}}_{\text{error}}^{\text{physical}}\le 1$ with $\sup\{\bar{\mathcal{M}}_{\text{error}}^{\text{physical}}\}\approx 1\wedge\bar{\mathcal{M}}_{\text{error}}^{\text{physical}}<1$ for zero tracking error.

For physical and multi-source information-centric data quality $\bar{\mathcal{M}}^{\text{data}}$, $0\le\bar{\mathcal{M}}_{\text{normalcy}}^{\text{data}}\le 1$, $0\le\bar{\mathcal{M}}_{\text{fidelity}}^{\text{data}}\le 1$, etc.

Under normal behavior, $\forall\bar{\mathcal{M}}_k^{\text{data}}\gg 0$, and, $\forall\sup\{\bar{\mathcal{M}}_k^{\text{data}}\}\approx 1\wedge\forall\bar{\mathcal{M}}_k^{\text{data}}<1$.

Experiments are conducted for step references $r(t)$=300±200 rad/sec, as well as for the sinusoidal references $r(t)$=300±200sin(ωt) rad/sec, $\omega\ge 10$ 1/sec, $t\in[0\ \ 25]$ sec.

The resulting convex $\bar{q}(\cdot)$ are shown in Figures 7.2.a using $\mathcal{M}=\begin{cases}\sum_i \mathcal{M}_i^{\text{physical}} \\ \sum_k \mathcal{M}_k^{\text{data}}\end{cases}$, $\mathcal{M}\in\mathbb{R}^{13}$ and $x\equiv(\bar{x},\bar{\tilde{x}},\bar{u})$, $x\in\mathbb{R}^{15}$. A multidimensional analysis of system performance and data quality yields a consistent assessment. The overall performance is better for step references $r(t)$ due to lower error, higher efficiency, etc. For time-varying $(\bar{\mathcal{M}}(t),\bar{x}(t))$ on $t\in[0\ \ 25]$ sec, the arithmetic means of $(\bar{\mathcal{M}},\bar{x})$ are depicted.

For anomalous operation, under faults, or, system affected by malware, key $\mathcal{M}$ differ. For example near 0, while for normal operation, these $\mathcal{M}$ are expected to be as $\begin{bmatrix}\bar{\mathcal{M}}_i^{\text{physical}}\in[\bar{\mathcal{M}}_{i\ \min}^{\text{physical}}\ \ \bar{\mathcal{M}}_{i\ \max}^{\text{physical}}] \\ \bar{\mathcal{M}}_k^{\text{data}}\in[\bar{\mathcal{M}}_{k\ \min}^{\text{data}}\ \ \bar{\mathcal{M}}_{k\ \max}^{\text{data}}]\end{bmatrix}$, $(\sup\{\forall\bar{\mathcal{M}}_i^{\text{physical}}\},\sup\{\forall\bar{\mathcal{M}}_k^{\text{data}}\})\approx 1\wedge(\forall\bar{\mathcal{M}}_i^{\text{physical}},\forall\bar{\mathcal{M}}_k^{\text{data}})<1$.

Abnormal conditions were tested, and, detected. Our analysis yields consistent assessment of an overall system normalcy not evaluating $\mathcal{M}^{\text{cyber}}$. Anomalies of time-varying $(\mathcal{M}^{\text{physical}}(t),\mathcal{M}^{\text{data}}(t))$ may indicate and reveal intrusion evidence or cyberattacks. Assessing $(\mathcal{M}^{\text{physical}}(t),\mathcal{M}^{\text{data}}(t))$ and $q(t)$ by a stand-alone MPU, control actions can be executed reconfiguration to ensure topology adaptation, resiliency and recovery.

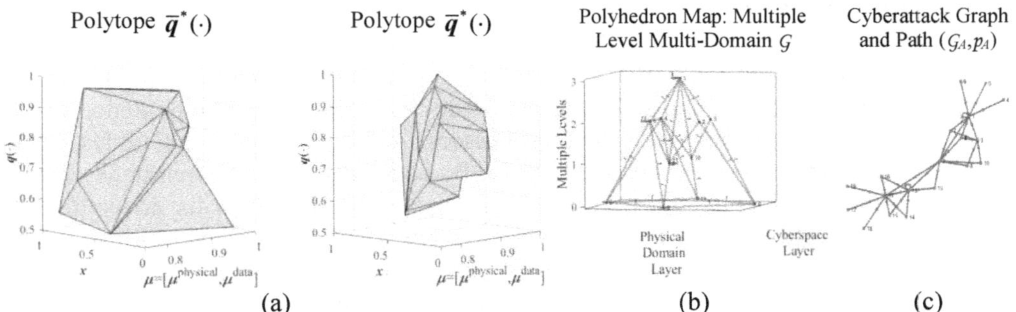

Figure 7.2. (a) Experimental findings for a closed-loop system, $r(t)$=300±200 rad/sec and $r(t)$=300±200sin(10t) rad/sec, $t\in[0\ 25]$ sec. Polytopes for multidimensional $\bar{q}^*[\bar{\mathcal{M}}(x),\bar{x}]$, $\bar{\mathcal{M}}\in\mathbb{R}^{13}$, $x\in\mathbb{R}^{15}$; (b) Multiple level graph $\mathcal{G}$ in physical and cyber domains, mapped by a polyhedron P, Example 7.4; (c) Cyberattack graph $\mathcal{G}_A$ with paths p_A, Example 7.4. ■

7.4. Graphs and Dynamic Networks

Advances in microelectronics, networking, computing, sensing and control empower CPS with additional capabilities and functionalities. However, CPS exhibit security and information vulnerabilities. It is critical to develop scalable cyber-analytic tools and methodologies to support design, characterization and implementation of secure information management and cybersecurity technologies.

Graphs and Networks – A graph $\mathcal{G}=(\mathcal{V},\mathcal{E})$ consists of a non-empty set of vertices (nodes) $\mathcal{V}=\{v_1,\ldots,v_n\}$ and a set of edges $\mathcal{E}=\{e_1,\ldots,e_m\}$. The structural topology is mapped by an adjacency matrix with elements which indicate whether pairs of vertices are adjacent or not. For an undirected graph with bidirectional edges, the descriptive incidence (0,1)-matrix is $A=[a_{ij}]$, $a_{ij}=\begin{cases}1 & \text{if } v_i\in e_j \\ 0 & \text{otherwise}\end{cases}$. Matrix A has a row for each vertex, and, a column for each edge, and, (v,e)=1 if a vertex v is incident upon edge e.

A subgraph $\mathcal{G}_0=(\mathcal{V}_0,\mathcal{E}_0)$ of $\mathcal{G}=(\mathcal{V},\mathcal{E})$ is a graph, such that $\mathcal{V}_0\subseteq\mathcal{V}$ and $\mathcal{E}_0\subseteq\mathcal{E}$. If $\mathcal{V}_0=\mathcal{V}$, then, $\mathcal{G}_0$ is a spanning subgraph of $\mathcal{G}$. Graphs admit l-dimensional topological manifolds. For example, an n-cube has 2^n vertices. In a complete directed graph $\mathcal{G}=(\mathcal{V},\mathcal{E})$, every pair of distinct vertices are connected by one edge. For a bipartite graph $\mathcal{G}=(\mathcal{V},\mathcal{E})$, vertices $\mathcal{V}$ are partitioned into two subsets $\mathcal{V}_1$ and $\mathcal{V}_2$, and, each edge in $\mathcal{G}$ connects vertexes $\mathcal{V}_1$ to vertexes $\mathcal{V}_2$. A supergraph is defined as $\mathcal{G}=(\mathcal{V},\mathcal{E})$, $\mathcal{V}=\{v_i,i\in I_v\}$, $\mathcal{E}=\{e_i,i\in I_e\wedge e_i\subseteq\mathcal{V}\wedge e_i\neq\emptyset\}$, where I_v and I_e are the index sets of the vertices and edges.

A network consists of a set of nodes (vertices) $\mathcal{V}$ and a set of links (connections) $\mathcal{L}$, which exhibit spatiotemporal dynamics and assume probabilistic characterization. In the n-node networks $\mathcal{N}=(t,\mathcal{V},\mathcal{L})$, the possible number of links is $m=n(n-1)/2$. There are similarities between $\mathcal{G}=(\mathcal{V},\mathcal{E})$ and $\mathcal{N}=(t,\mathcal{V},\mathcal{L})$ from combinatorial, spectral and topological analyses. Spatiotemporal dynamic-probabilistic consideration of $\mathcal{N}=(t,\mathcal{V},\mathcal{L})$ supports data-driven applications. Application of hierarchical networks $\mathcal{N}=(t,\mathcal{V},\mathcal{L})$ supports visualization of integrated physical components and hardware in *smart* electrical grids, autonomous swarms, etc. and software vulnerabilities, resilience, attacks, as well as data governance should be described. Multiple-level domains are mapped by networks $\mathcal{N}=\mathcal{N}^{\text{physical}}\cup\mathcal{N}^{\text{cyber}}\cup\mathcal{N}^{\text{data}}$.

Attack graphs $\mathcal{G}_A\subset\mathcal{G}$ and evidence graphs are considered in topological analyses of networked CPS. In $\mathcal{G}_A$, nodes represent states and events. The security predicates are the threats and vulnerability of devices and hosts. The evidence graphs procure intrusion, events, etc. Host processors and devices are the nodes, while dependencies between evidences are mapped by edges and links. Cyberattacks exploit vulnerabilities, analyzing preconditions and postconditions. Exploits are applied in sequences or parallel against a target module, aiming to explore vulnerabilities and alter states. As vulnerabilities are exploited, the postconditions are initiated. Each postcondition can be a precondition for exploits. Normalcy, anomalies and irregularities can be analyzed using attributes. Analytic and algorithmic solutions are developed to support security tasks. The attack path p_A and graph $\mathcal{G}_A$ may describe sequences of exploits, attack event sequences, detected and predicted anomalies, etc. A finite acyclic path of a sequence of states is $p_A=(s_0,s_1,\ldots,s_{f-1},s_f)$, where s_0 is an initial state. Any two adjacent states in p_A belong to a transition set τ_A, such that the execution of p_A leads to changes of a security property s_A. The typifying graph $\mathcal{G}$ is illustrated in Figure 7.2.b.

Finite-Dimensional Mapping – To visualize topologies and connected nodes, physical domain and cyberspace are mapped by a finite polyhedra. Multi-dimensional structures support interactive mathematical analysis of dependences, connectivity, information sharing, etc. For regular and irregular octahedron, dodecahedron, icosahedron or higher-dimensional polyhedral, one defines the symmetry group. Graphs with vertices and edges are applied to describe the polyhedra's skeleton. The skeleton of a convex polyhedron is a 3-connected planar graph. A regular high-dimensional polyhedron P is defined by low-dimensional polyhedra $P_i(\mathcal{V},\mathcal{E},\mathcal{F})$. A realization of P is a collection of vertices $\mathcal{V}$ in a finite Euclidean space $\mathbb{R}^n$ with a map $\beta:P_0\to\mathcal{V}$, and, action on each element of a group $\mathcal{A}(P)$ which induces an isometry of a finite $\mathcal{V}$. Aggregation is performed using unordered pair $\{i,j\}$ of $\mathcal{V}$ with edges $\mathcal{E}$, yielding a finite P [19-21].

Example 7.4.
A polyhedron P is characterized by vertices, edges and faces $(\mathcal{V},\mathcal{E},\mathcal{F})$. The irreducible polyhedron $\{3,6\}_{(b,0)}$ has $2b^2$ faces, $3b^2$ edges, and, b^2 vertices. Node interactions and processes are depicted by $\mathcal{G}$. Representative $\mathcal{G}$, as well as attack graph and path $(\mathcal{G}_A,p_A)$, are illustrated in Figures 7.2.b and c. ■

Quantitative statistical description and analysis may be conducted using high-dimensional MIMO dynamic, probabilistic and Markov models. A dynamic state-transition network $\mathcal{N}$ is comprised from nodes and links, characterized by outputs $\boldsymbol{y}$, states $\boldsymbol{x}$, controls $\boldsymbol{u}$, disturbances $\boldsymbol{d}$ and perturbations $\boldsymbol{w}$. Inputs $(\boldsymbol{u},\boldsymbol{d},\boldsymbol{w})$ govern dynamic transitions $\boldsymbol{\tau}_c$ for $(\boldsymbol{y}, \boldsymbol{x})$ and define security status $\boldsymbol{s}_S$. Dynamic transitions depend on attacks evolutions, changes in security status, etc. These $\boldsymbol{\tau}_c$ depend on preconditions of states and executions with post-conditions. Statistical models of processes and transitions for $(\boldsymbol{y},\boldsymbol{x},\boldsymbol{d},\boldsymbol{w})$ are described by probability distributions, and, $\mathcal{R}_X=(\mathcal{S},\mathcal{D})$. For nodes, one considers vulnerabilities, preconditions and postconditions. With assigned probabilities at each node, the likelihood may be estimated for the assumed hardware and software vulnerabilities. Probabilities of successful attacks and vulnerabilities are difficult to estimate for multi-vector attacks, different paths, risks, mean time to compromise, etc. Evidence subgraphs describe intrusion evidence and dependencies. Model checking algorithms depend on hypothesis and solutions of classification problems by estimating varying model coefficients, analyzing temporal logic, finding distributions, etc.

Descriptive, Predictive and Prescriptive Analytics – Systems are comprised from many components. There are multiple hierarchy, decentralization and control level, such as: (1) Component control, such as, interface and wireless modules, power management, PWM controllers-drivers with MPUs, programmable logic controllers, sensors with ASICs and MPUs, etc.; (2) Supervisory control; (3) Management system. The vulnerable components are MPUs, system-in-package processors, input-output devices, etc. Events, outputs and states are measured and estimated. Systems may be affected by attack instances, attack preconditions, attack postconditions, etc. One investigates anomalies, conducts threat analysis, examines system vulnerabilities, etc. Descriptive, predictive and prescriptive data analytics enable information analysis, data sharing and information control. The goal is to:

1. Investigate multi-domain designs;
2. Assess physical organization and vulnerabilities;
3. Analyze multi-level statistically-characterized descriptive models and dynamic;
4. Research and apply dynamic-probabilistic and statistical measures to detect abnormalities.

High-performance computing is critical to analyze interfacing, connectivity and information sharing. Resiliency and recovery should be ensured by hardware topological adaptation as well as by information-centric data-driven software- and hardware-defined reconfigurable schemes and algorithms.

7.5. Data Analytics, Factorization and Singular Value Decomposition

To perform data analytics, factorization schemes are applied. The singular value decomposition (SVD) is an eigendecomposition factorization applied in principal component analysis (PCA), low rank approximation, multi-dimensional scaling, factor analysis, classification, clustering, feature detection and extractions, regression analysis, etc. Analysis of high-dimensional spaces, complexity reduction, mapping reducibility and data visualization are supported by SVD. The SVD yields feature analysis of the Application Programming Interface (API) and Program Structure Interface (PSI), as well as characterizes data sets, string literals, etc. Study data, variable-length dynamic arrays and sequences.

Strings, Alphabet and Characters – Data structure pertains to data organization, storage and retrieval schemes to ensure access, control, processing and management. An array data structure, or array, is a composite data structure which consists of elements, e.g., characters, values, variables, etc. A string is a sequence of characters, as well as function, integer, hexadecimal and other literals. Strings $\mathsf{S}=\{\mathsf{s}_1,\mathsf{s}_2,\ldots,\mathsf{s}_{q-1},\mathsf{s}_q\}_i$ can be partitioned and analyzed using complimentary substrings $\mathsf{s}_{j,i}$. These $(\mathsf{S},\mathsf{s}_{j,i})$ of length $(l_\mathsf{S},l_\mathsf{s})$ are examined within the pertained alphabet $\mathcal{A}$. Text strings are sequences of characters, e.g., *words* or multi-character symbols. Each character is given by eight bits. A character, recognized and interpreted by the computer hardware and software, is a symbol, letter or number. Bit strings and byte strings represent non-textual binary data, and, the machine language instructions control CPU. Characters are converted into the bit strings. The finite sequences are called strings.

Semantic Indexing and Pattern Recognition – Malware polymorphism and metamorphism are of particular challenges. A virus can morph itself by encrypting its code, and, decrypts during execution. A polymorphic virus obfuscates decryption instructions using no operation k-byte instructions, instructions reordering, register reassignments, etc. Metamorphic viruses obfuscate codes by transposition, rearrangements, instruction substitutions, conditional jumps, etc. Signature and pattern matching can be studied using string segmentation, sequences partitioning and semantic instruction analysis. Hidden and fragmented sequences may be detected and characterized by finding probabilistic signatures and patterns with verification and validation. Perceiving probabilistic semantic indexing and statistical patterns may support malware detection capabilities. Data analytics is performed on $(\mathcal{A},X_a,\mathsf{S},\mathsf{S}=[\mathsf{s}_1,\ldots,\mathsf{s}_k]_i)$ assessing strings $X_a\in\mathbb{R}^{n\times m}$.

Probabilistic Turing Machine – – There are multiple controls, such as event control, assets management, access control, cryptography, physical security, communication security, etc. A non-deterministic Turing machine, pertained to information security, is an undecuple

$$\mathrm{T}=(X_\mathrm{T},x_\mathrm{T},\mathcal{A},X_a,\mathcal{O},\boldsymbol{\tau}_a,\mathbf{P}_\tau,c_a,p_a,\mathcal{A}_\tau,u_\mathrm{T}), \tag{7.3}$$

where X_T is the finite set of states x_T with initial and final states (x_{T0},x_{Tf}), $\forall x_T\in X_T$; X_a is a set of strings within an alphabet $\mathcal{A}$, $\mathcal{L}\subseteq\mathcal{A}^*$; $\mathcal{O}$ is the algebra, logic and calculus; $\boldsymbol{\tau}_a$ is the transition function; $\mathbf{P}_\tau$ is the transition probability of string-oriented computation,

$\mathbf{P}_{\tau}: X_T \times \mathcal{a}_{\tau} \times X_T^{'} \to [0, 1]$; c_a is a finite set of the state-alphabet combination; p_a is the distribution for X_a; $\mathcal{a}_{\tau}$ is the transition alphabet; $\boldsymbol{u}_{\mathrm{T}}$ is the control vector, such as event and security controls, instructions and logic operations, file transfer protocols, software uploads, security management, system management, service management and information control, $\boldsymbol{u}_{\mathrm{T}} \in \mathcal{O}$.

Example 7.5. *Strings and Characters*
One partitions and investigates the n strings and blocks as groups. Consider a string as an n-input sequence $\{X_j\}_{h\in[1,\ldots,n]}$. A function f describes the n-fold Cartesian product of $\{X_j\}$ to X_p

$$\prod_{j\in J} X_j = \left\{ f : J \to \bigcup_{j\in J} X_j \,\middle|\, f(j) \in X_j, \forall j \right\},\quad f : \times_{j=1}^{n} X_j \to X_p,$$

where J is the index set.

The Cartesian product of A and B is a set $A\times B=\{(a,b)|\ a\in A \wedge b\in B\}$. Hence, $A\times B$ contains all ordered pairs in which the first elements are chosen from A, and, the second elements are chosen from B.

Consider $A=\{c,d\}$ and $B=\{1,2,4,5\}$.

Hence, $A\times B=\{(c,1),(c,2),(c,4),(c,5),(d,1),(d,2),(d,4),(d,5)\}$,

while $B\times A=\{(1,c),(2,c),(4,c),(5,c),(1,d),(2,d),(4,d),(5,d)\}$.

The Cartesian product is not commutative, and, $A\times B\neq B\times A$. Note that, $A\times\emptyset=\emptyset$.

For m characters, one represents each string vector, and, obtains the number of occurrences of a character. Similarity, frequency, patterns and modalities of $(\mathbf{s}_i,\mathbf{s}_j)$ may be analyzed using the inner product $\mathbf{s}_i^T \mathbf{s}_j$. If the product is 0, the strings have no character in common, and, positive otherwise. ■

Sequences, Sets and Numbers – Deterministic, random, pseudorandom, quasirandom, stationary, non-stationary, periodic, aperiodic, partitioned and other data sequences $\{X_i(n), n\in\mathbb{R}\}$ are studied. For sets of natural numbers $\mathbf{N}$, integers $\mathbb{Z}$, rational numbers $\mathbb{Q}$, real and complex numbers ($\mathbb{R}$, $\mathbf{C}$), $\mathbf{N}\subset\mathbb{Z}\subset\mathbb{Q}\subset\mathbb{R}\subset\mathbf{C}$.

An integer $n\in\mathbb{Z}$ is any number in the infinite set $\mathbb{Z}=\{\ldots,-3,-2,-1,0,1,2,3,\ldots\}$. Positive integers $(1,2,3,\ldots)$ are within a finite countable set $\mathbb{Z}^+$, also denoted as $\mathbb{Z}_+$ and $\mathbb{Z}^>$. Negative integers $(\ldots,-3,-2,-1)$ are within $\mathbb{Z}^-$. A set $\mathbb{Z}^{0+}$ ($\mathbb{Z}^{\geq}$) maps positive integers including 0, and, a set $\mathbb{Z}^{0-}$ gives negative integers including 0. The set of natural numbers is $\mathbf{N}$, $\mathbf{N}=\mathbb{Z}^+$, and, $\mathbf{N}=\{1,2,3,\ldots\}$. For example, $\mathbf{N}=\{1,2,\ldots,m\}$ is a finite countable set of natural numbers from 1 to m. For $\mathbf{N}_0$, $\mathbf{N}_0=\{0,1,2,\ldots,m\}$.

For rational, irrational and complex numbers, $n\in\mathbb{Q}$, $n\in\mathbb{P}$ and $n\in\mathbf{C}$. For integers (p,q), consider the quotient p/q, $p\in\mathbb{Z}$, $q\in\mathbf{N}$. One may yield rational numbers $(p/q)\in\mathbb{Q}$, for example, the fractions are $1/3, -2/7$, etc. The example of irrational numbers are $\sqrt{2}$, $\sqrt[3]{5}$, π, Plank and reduced Plank constants h and $\hbar$, etc.

Singular Value Decomposition – Matrix operations, factorization, truncation, decomposition and other operations on real-valued $X\in\mathbb{R}^{n\times m}$ and complex $X\in\mathbb{C}^{n\times m}$ are performed by applying a unique decomposition.

For $X\in\mathbb{R}^{n\times m}$, the singular value decomposition and pertained relationships are

$$X=U\Sigma V^T,\ XX^T=U\Sigma\Sigma^T U^T,\ X^TX=V\Sigma^T\Sigma V^T, \tag{7.4}$$

where $U\in\mathbb{R}^{n\times n}$ and $V\in\mathbb{R}^{m\times m}$ are the real orthogonal matrices formed by left and right orthonormal column vectors, $U=[\boldsymbol{u}_1,\ldots,\boldsymbol{u}_n]$ and $V=[\boldsymbol{v}_1,\ldots,\boldsymbol{v}_m]$, $UU^T=U^TU=I\in\mathbb{R}^{n\times n}$, $VV^T=V^TV=I\in\mathbb{R}^{m\times m}$; $\Sigma\in\mathbb{R}^{n\times m}$ is the matrix with nonnegative singular values $\sigma_{i,i}$ on the diagonal, and, zeros off the diagonal, $\forall\sigma_{i,i}\geq 0$, $\forall\sigma_{i,j}=0$, and, complimentary null matrix.

The singular values $\sigma_{i,i}$ are computed using the eigenvalues λ_i of XX^T or X^TX, such that $\sigma_{11}\geq\sigma_{22}\geq\ldots\geq\sigma_{n-1,n-1}\geq\sigma_{n,n}$, $\sigma_{i,i}=\sqrt{\lambda_i}$.

The SVD (7.4) exists for any X. For strings and arrays, $X\equiv X_a$.

If $X\in\mathbb{C}^{n\times m}$, $X=U\Sigma V^*$, $U\in\mathbb{C}^{n\times n}$, $V\in\mathbb{C}^{m\times m}$, $\Sigma\in\mathbb{R}^{n\times m}$.

Here, $*$ denotes the complex conjugate transpose.

Singular Value Decomposition Tuple (U,Σ,V) – Find the eigenvalues and orthonormal eigenvectors for XX^T and X^TX. Compute the eigenvalues of XX^T or X^TX, and, arrange λ_i such that $\lambda_1\geq\lambda_2\geq\ldots\geq\lambda_{n-1}\geq\lambda_n$. The orthonormal linearly independent eigenvectors for XX^T are found, and, the columns $[\boldsymbol{u}_1,\ldots,\boldsymbol{u}_n]$ yield $U=[\boldsymbol{u}_1,\ldots,\boldsymbol{u}_n]$. The orthonormal linearly independent eigenvectors for X^TX result in the columns $[\boldsymbol{v}_1,\ldots,\boldsymbol{v}_m]$, and, $V=[\boldsymbol{v}_1,\ldots,\boldsymbol{v}_m]$. Columns of U are the left singular vectors of X, and, columns of V are the right singular vectors. The singular values $\sigma_{i,i}$ in Σ are the square roots of eigenvalues of XX^T or X^TX. The real-valued singular values $\sigma_{i,i}$ are the diagonal entries of the matrix Σ. Positive-definite $\sigma_{i,i}\geq 0$ are arranged in descending order, from largest to smallest. That is, the singular values $\sigma_{i,i}=\sqrt{\lambda_i}$ are ordered as $\sigma_{1,1}\geq\sigma_{2,2}\geq\ldots\geq\sigma_{n-1,n-1}\geq\sigma_{n,n}>0$.

Singular Value Decomposition Complexity – For $n\times m$ matrix, the SVD complexity is $O(n^3+n^2m+m^2n)$, $n>m$. Note that the complexity of two-dimensional fast Fourier transform for a square window of size $n\times n$ is $O(n\times n\times\log_2(n))$.

Principal Components – The eigenvectors of X^TX are the principal component directions of X. The first principal component direction has the following properties: (1) Eigenvector $\boldsymbol{v}_1$ associates with the largest eigenvalue λ_1 of X^TX; (2) Largest variance among all normalized linear combinations of the columns in X is $z_1=X\boldsymbol{v}_1$, and, z_1 is the first principal component of X. The second principal component direction, with the direction orthogonal to the first component which has the largest variance, is the eigenvector $\boldsymbol{v}_2$ which corresponds to the second largest eigenvalue λ_2 of X^TX. Factor analysis and PCA are applied to large data sets. The components are expected to be distinct from each other to be distinguished.

Strings, Sequences and Arrays: Randomness and Determinism – Dependencies, detectable correlations and distinguishable dependencies between variables are of interest. Data, strings and sequences X may not be random or pseudorandom. Discrepancies and uncertainties are minimized investigating a confidence interval, conducting null hypothesis significance testing, evaluating p-value probability, performing randomized tests, etc.

Eigenvalue-Eigenvector Problem – For square matrices $XX^T\in\mathbb{R}^{n\times n}$ and $X^TX\in\mathbb{R}^{m\times m}$, one finds (n,m) eigenvalues $(\boldsymbol{\lambda}_U,\boldsymbol{\lambda}_V)$ and eigenvectors $(\boldsymbol{u}_i,\boldsymbol{v}_i)$.

Eigenvectors $(\boldsymbol{u}_i,\boldsymbol{v}_i)$ correspond to λ_i, such that

$$(XX^T)\boldsymbol{u}_i=\lambda_i\boldsymbol{u}_i,\ (\lambda_iI_n-XX^T)\boldsymbol{u}_i=0,\ I_n\in\mathbb{R}^{n\times n},$$
$$(X^TX)\boldsymbol{v}_i=\lambda_i\boldsymbol{v}_i,\ (\lambda_iI_m-X^TX)\boldsymbol{v}_i=0,\ I_m\in\mathbb{R}^{m\times m}. \quad (7.5)$$

Eigenvalues – For a square matrix A, the eigenvalues λ_i are found by using the characteristic polynomial. Solution of $\det(A-\boldsymbol{\lambda}I)=0$ or $|A-\boldsymbol{\lambda}I|=0$ yields $\boldsymbol{\lambda}$. The spectral radius is $\rho(A)=\max\{|\boldsymbol{\lambda}|: \boldsymbol{\lambda}\in\boldsymbol{\lambda}(A)\}$.

For $X\in\mathbb{R}^{n\times m}$, matrix Σ has at most n or m non-zero $\sigma_{i,i}$ on the diagonal, and

$$X = U\Sigma V^T = \begin{bmatrix} \hat{U} & U_0 \end{bmatrix} \begin{bmatrix} \hat{\Sigma} \\ 0 \end{bmatrix} V^T = \hat{U}\hat{\Sigma}V^T,\ \hat{\Sigma} \in \mathbb{R}^{m\times m}, \tag{7.6}$$

where U_0 is the complementary matrix to $\hat{U}$.

Reduced and Truncated SVDs – Consider an $X\in\mathbb{R}^{n\times m}$ with rank r. For the reduced SVD, from (7.4), one has $X = U_r\Sigma_r V_r^T$, $U_r\in\mathbb{R}^{n\times r}$, $\Sigma_r\in\mathbb{R}^{r\times r}$ and $V_r\in\mathbb{R}^{r\times m}$.

The truncated SVD, solving (7.4) and (7.5), we have

$$X = U\Sigma V^T = \begin{bmatrix} U_t & U_\varepsilon & U_0 \end{bmatrix} \begin{bmatrix} \Sigma_t & 0 & 0 \\ 0 & \Sigma_\varepsilon & 0 \\ 0 & 0 & \Sigma_0 \end{bmatrix} \begin{bmatrix} V_t^T \\ V_\varepsilon^T \\ V_0^T \end{bmatrix},\ \Sigma_r = \begin{bmatrix} \Sigma_t & 0 \\ 0 & \Sigma_\varepsilon \end{bmatrix},\ \tilde{X} = U_t\Sigma_t V_t^T, \tag{7.7}$$

where (U_t,Σ_t,V_t), $(U_\varepsilon,\Sigma_\varepsilon,V_\varepsilon)$ and (U_0,Σ_0,V_0) are the matrices tuples; $(\Sigma_t,\Sigma_\varepsilon)$ are the square diagonal matrices with $\forall\sigma_{t\,i,i}>0$, $\forall\sigma_{\varepsilon\,i,i}>0$, $\sigma_{t\,i,i}>\sigma_{\varepsilon\,i,i}$ and $\forall\sigma_{i,j}=0$; Σ_0 is the null matrix.

The k largest singular values is used, and, $\Sigma_t\in\mathbb{R}^{k\times k}$ is truncated, with $0<k<r$. The corresponding left and right singular vectors are $U_t\in\mathbb{R}^{n\times k}$ and $V_r\in\mathbb{R}^{k\times m}$.

The Eckart-Young theorem states that the rank-k approximation of X, in a norm-p sense, is the rank-k SVD truncation as

$$\arg\min_{\tilde{X}\in M(k)} \left\|X-\tilde{X}\right\|_p,\ \tilde{X} = U_t\Sigma_t V_t^T, \tag{7.8}$$

where $M(k)$ is a set of rank-k matrices; Σ_t is the diagonal matrix which contains the leading $k\times k$ sub-block of Σ.

The first k components of the SVD for X (7.7) yield the low-rank matrix approximation of X as $\tilde{X} \equiv \tilde{U}\tilde{\Sigma}\tilde{V}^T$, evaluated using the p-norm (7.8). One applies the p-norm (p=1,2,…,∞), spectral norms and Frobenius norm

$$\|X\|_1 = \max_j \sum_{i=1}^{m} |X_{ij}|,\ j=1,\dots,n,\ \|X\|_2 = \sqrt{\max\lambda(X^TX)},\ \|X\|_F = \sqrt{\sum_{i=1}^{m}\sum_{j=1}^{n}|X_{ij}|^2}, \tag{7.9}$$

$$\|X\|_\infty = \max\left(\sum_{j=1}^{n}|X_{1j}|, \sum_{j=1}^{n}|X_{2j}|, \dots, \sum_{j=1}^{n}|X_{mj}|\right),\ \|X(r)\|_\Sigma = \frac{1}{\sum_i\sum_j |X_{ij}|}\sum_i\sum_j\left(|X_{ij}|-|X_{r_{ij}}|\right).$$

Pseudoinverse – The diagonal elements of $\hat{\Sigma} \in\mathbb{R}^{m\times m}$ are the singular values $\sigma_{i,i}\geq 0$. Matrix $\hat{\Sigma}$ is invertible, while Σ is not invertible.

From $X = U\Sigma V^T = \begin{bmatrix} \hat{U} & U_0 \end{bmatrix} \begin{bmatrix} \hat{\Sigma} & 0 \\ 0 & 0 \end{bmatrix} \begin{bmatrix} \hat{V}^T \\ V_0^T \end{bmatrix}$, the pseudoinverse X^{-1} is $X^{-1} = \hat{V}\hat{\Sigma}^{-1}\hat{U}^T$.

Covariance and Correlation – High-dimensional correlated data is analyzed using covariance and correlation. Arranging data as row vectors X_i, the resulting $X\in\mathbb{R}^{n\times m}$ yields features on observables.

Subtract the mean $\bar{x}$ with $\bar{x}_j = \frac{1}{n}\sum_{i=1}^{n} X_{ij}$ from $X\in\mathbb{R}^{n\times m}$. One finds the mean-zero

$$\hat{X} = (X - \bar{X}),\ \bar{X} = \begin{bmatrix} 1 \\ \vdots \\ 1 \end{bmatrix} \bar{x} = \mathbf{1}_n \bar{x}$$

with n samples in rows, and, m variables in columns. For the column-centered $X\in\mathbb{R}^{n\times m}$ with the number of samples n and the number of variables m, mean-centered $\hat{X}$ with column mean is subtracted and equal to zero, the symmetric square covariance matrix is

$$C = \frac{1}{n-1}\hat{X}^T\hat{X},\ C = \frac{1}{n-1}\sum_{i=1}^{n}(X_i - \bar{X})^T(X_i - \bar{X}),\ \hat{X} = (X - \bar{X}),\ \bar{X} = \begin{bmatrix} 1 \\ \vdots \\ 1 \end{bmatrix}\bar{x} = \mathbf{1}_n\bar{x},\ C\in\mathbb{R}^{m\times m}. \tag{7.10}$$

The diagonal entries of the covariance matrix $C\in\mathbb{R}^{m\times m}$ are the variances, and, off-diagonal entries are the covariances. That is, $c_{i,i}$ is the variance for x_i, while the off-diagonal $c_{i,j}$ is the covariance between x_i and x_j.

The covariance matrix $C\in\mathbb{R}^{m\times m}$ can be diagonalized as

$$C = V\Lambda V^T, \tag{7.11}$$

where V is the matrix of eigenvectors; Λ is the diagonal matrix of eigenvalues $\lambda_{i,i}$ in the decreasing order.

Principal Component Analysis – The data in an m-dimensional space $Y=XV$ is projected by m eigenvectors which correspond to the largest m eigenvalues as a new base. The principal components are found as the eigendecomposition of C.

The eigenvectors are called principal axes, or, principal directions. Projections of data on the principal axes are called principal components, or *scores*. The ith principal component is the ith column of XV. The principal components are uncorrelated to each other, but correlated with the data. The coordinates of the ith data point in a principal component space are the ith row of XV. The jth principal component is the jth column of XV.

The eigenvalues $\lambda_{i,i}$ depict the variances of the principal components.

In (7.4) $X=U\Sigma V^T$, U is the unitary matrix, columns of $U\Sigma$ are the principal components (*scores*), and, columns of V are the principal direction eigenvectors. We have

$$C = \frac{1}{n-1}V\Sigma U^T U\Sigma V^T = \frac{1}{n-1}V\Sigma^2 V^T,\ \Lambda = \frac{1}{n-1}\Sigma^2. \tag{7.12}$$

The right singular vectors V are principal directions, and, the singular values $\sigma_{i,i}$ are related to the eigenvalues of covariance matrix $\lambda_{i,i}$ as $\lambda_{i,i} = \frac{1}{n-1}\sigma_{i,i}^2$.

The principal components are

$$XV = U\Sigma V^T V = U\Sigma. \tag{7.13}$$

Multi-dimensional scaling yields representative estimates on similarity. The Euclidean, Chebychev, Hamming and other distances, which characterize X, are explicitly defined and can be computed.

Example 7.6. *Singular Value Decomposition Applied to Sequences*

Consider $X\in\mathbb{R}^{6\times 800}$ with 100 eight-bit in rows. With six columns, $n\leq m$. The pseudorandom uniformly distributed strings $\{0, 1\}^*$ are generated as

```
N=6; M=800; X=round(rand(N,M))
```

or

```
X=randi([0,1],[N,M])
```

Compute (U,Σ,V) and $(U,\hat{\Sigma},\hat{V})$, as well as the eigenvalues λ_U of the square matrix $U\in\mathbb{R}^{6\times 6}$. The columns of $U=[\boldsymbol{u}_1,\ldots,\boldsymbol{u}_6]$ and $V=[\boldsymbol{v}_1,\ldots,\boldsymbol{v}_{800}]$ are the orthonormal eigenvectors of XX^T and X^TX.

The *full* and reduced SVDs are computed using the statements **[U,S,V]=svd(X)** and **[Uhat,Shat,V]=svd(X,'econ')**.

For two pseudorandom uniformly distributed arrays $X=[X_1\ X_2\ X_3\ X_4\ X_5\ X_6]^T$, the singular value matrices are computed as

1. $\hat{\Sigma}=\begin{bmatrix}37.57&0&0&0&0&0\\0&14.82&0&0&0&0\\0&0&14.44&0&0&0\\0&0&0&14.14&0&0\\0&0&0&0&13.92&0\\0&0&0&0&0&13.53\end{bmatrix}$, $\lambda_U=(-0.958\pm0.287i,-0.217\pm0.976i,0.984\pm0.18i)$;

2. $\hat{\Sigma}=\begin{bmatrix}37.11&0&0&0&0&0\\0&14.72&0&0&0&0\\0&0&14.48&0&0&0\\0&0&0&14.14&0&0\\0&0&0&0&13.86&0\\0&0&0&0&0&13.42\end{bmatrix}$, $\lambda_U=(1,-1,-0.604\pm0.797i,\ 0.692\pm0.722i)$.

The computed eigenvalues are illustrated in Figures 7.3.a.

For different realization X, the computed (U,Σ,V) and $(U,\hat{\Sigma},\hat{V})$ are verified using (7.4) and (7.6), $X=U\Sigma V^T=U\begin{bmatrix}\hat{\Sigma}&0\end{bmatrix}\begin{bmatrix}\hat{V}^T\\0\end{bmatrix}=U\hat{\Sigma}\hat{V}^T$. ■

Example 7.7. Perturbed Pseudorandom Sequences: Feature Estimates

Statistical randomness, determinism and dependencies of arrays, sequences and strings are studied. Statistically random sequences assume disorder and unpredictability without regularities and patterns.

Consider a sequence $X=[X_1\ X_2]^T$ with normally distributed pseudorandom (X_1,X_2), perturbed by constants $(a_1,a_2)\in\mathbb{R}$.

Examine $X\in\mathbb{R}^{2\times 800}$ with two rows and 800 columns, $(a_1+X_1)\in\mathbb{R}^{1\times 800}$, $(a_2+X_2)\in\mathbb{R}^{1\times 800}$.

The mean values are found yielding $\bar{X}=\begin{bmatrix}\bar{X}_1\\\bar{X}_2\end{bmatrix}$. The covariance matrix $C\in\mathbb{R}^{2\times 2}$ is computed using (7.10) and (7.12). For two different realizations X, the covariance matrices are $C=\begin{bmatrix}0.478&0\\0&0.522\end{bmatrix}$ and $C=\begin{bmatrix}0.489&0\\0&0.511\end{bmatrix}$.

Finding the eigenvalues and eigenvectors (7.5), one computes variances. For two different sequences X, the principal components are computed and depicted to substantiate randomness of X, see Figures 7.3.b.

Sequences X may be random, not statistically random, deterministic, etc. Sequences X exhibit pattern, modalities and features due to dependences to detect, categorize, identify and classify. Consider perturbed sequences

$$X=\begin{bmatrix}\sin(n)+X_1\\2\cos(n)+X_2\end{bmatrix},\ (\sin(n)+X_1)\in\mathbb{R}^{1\times 800}\text{ and }(2\cos(n)+X_2)\in\mathbb{R}^{1\times 800}.$$

Figures 7.3.c depict the principal components for two different processes X. Randomness and periodicity are detected.

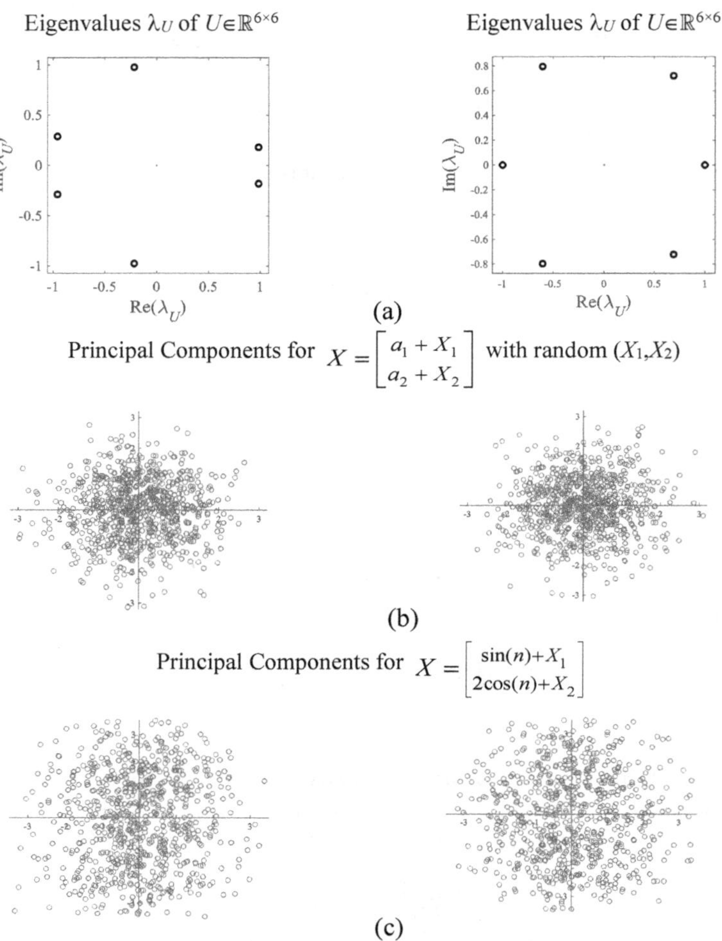

Figure 7.3. (a) Eigenvalues λ_U of the square matrix $U\in\mathbb{R}^{6\times6}$ for two arrays $X\in\mathbb{R}^{6\times800}$. The columns of U, $U=[\boldsymbol{u}_1,\ldots,\boldsymbol{u}_6]$ are the orthonormal eigenvectors for XX^T;

(b) Principal components for $X=\begin{bmatrix} a_1+X_1 \\ a_2+X_2 \end{bmatrix}$ with two pseudorandom (X_1,X_2), $X_i\in\mathbb{R}^{1\times800}$, $a_1=1$, $a_2=2$;

(c) Principal components for sequences $X=\begin{bmatrix} \sin(n)+X_1 \\ 2\cos(n)+X_2 \end{bmatrix}$ with periodic perturbations, $X_i\in\mathbb{R}^{1\times800}$. ■

Example 7.8. *Quantitative Features for Sequences*

Bit strings and sequences represent characters and integers. Study a sequence of 300 nibbles with four rows. Four bits in nibbles '0000' and '1111' yield 0 and 15.

For $X=[X_1\ \ X_2\ \ X_3\ \ X_4]^T$, $X\in\mathbb{R}^{4\times300}$, compute the SVD (7.4) $X=U\Sigma V^T$, and, perform PCA which implies finding eigenfeatures and multivariate statistics.

The random uniformly distributed binary numbers {0,1} and {0,1,2,…} are generated to construct rows X_i in X. The periodic X_j are also considered.

For example,

```
A=15; B=0.5; N=300; X1=round(A*rand(N,1)); X2=round(A*(B+B*sin(1:N).^3'));
```

Investigate three cases, computing the eigenfeatures for X:

1. Random bits in 300 nibbles '0000' to '1111' in four X_j, $X\in\mathbb{R}^{4\times300}$;

2. Bits in two rows X_i are random, while bits in other rows X_j are periodically varying integers;
3. Four rows X_j are periodically varying integers $X_1=15\sin(n)$, $X_2=15\sin^2(n)$, $X_3=15\cos(n)$ and $X_4=15\cos^2(n)$.

Figures 7.4 which depict (X_1,X_2,X_3,X_4). The component analysis is performed. The first and second principal components are reported in Figures 7.4 as bottom left plots. The first and second components describe ≥80% of quantitative features.

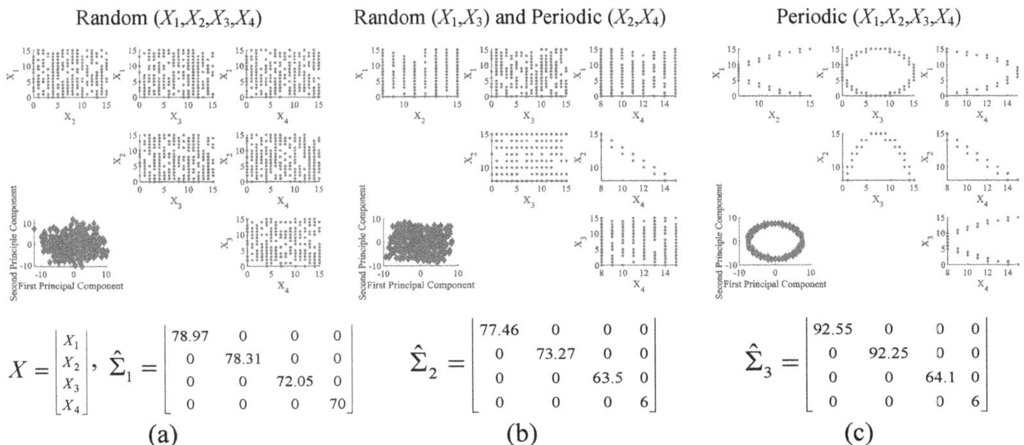

Figure 7.4. Nibbles '0000' to '1111' in (X_1,X_2,X_3,X_4), $X\in\mathbb{R}^{4\times300}$, and, component analysis:
(a) Random bits in X, $X\in\mathbb{R}^{4\times300}$, a singular value matrix $\hat{\Sigma}_1$ is reported;
(b) Bits in two rows X_i are random, while bits in other rows X_j are periodic, with $\hat{\Sigma}_2$;
(c) Four rows in $X=(X_1,X_2,X_3,X_4)$ are periodically varying integers, with $\hat{\Sigma}_3$. ■

Example 7.9. For $X_i\sim\mathcal{N}(\mu_i,\sigma_i^2)$, a sum of N statistically independent random variables $Z=\sum_{i=1}^{N}X_i$ are examined in Example 5.37 and Practice Problem 7.4. The uniformly distributed pseudorandom X are studied in Example 7.13 and Practice Problem 7.6.

Consider an array $X\in\mathbb{R}^{3200\times400}$, $\mathcal{A}=\{0,1\}$ with k independent and identically distributed uniform random variables on the interval [0,1], $X_i\sim\mathcal{U}_{[0,1]}$. Uniformly distributed pseudorandom binary numbers 0s and 1s for X_i are generated using the **rand** and **round** commands. Investigate X and regularized X_M. Using the diagonal-constant Toeplitz matrix $M\in\mathbb{R}^{400\times400}$, define X_M as with $\forall x_M=[0, 1]$ as

$$X_M=X\times M,\ X_M\in\mathbb{R}^{3200\times400},\ M\in\mathbb{R}^{400\times400},$$

$$M=\begin{vmatrix}1&1&0&0&0&\cdots&0&0&0&0&0\\1&1&1&0&0&\cdots&0&0&0&0&0\\0&1&1&1&0&\cdots&0&0&0&0&0\\0&0&1&1&1&\cdots&0&0&0&0&0\\\vdots&\vdots&\vdots&\vdots&\vdots&\ddots&\vdots&\vdots&\vdots&\vdots&\vdots\\0&0&0&0&0&\cdots&1&1&1&0&0\\0&0&0&0&0&\cdots&0&1&1&1&0\\0&0&0&0&0&\cdots&0&0&1&1&1\\0&0&0&0&0&\cdots&0&0&0&1&1\end{vmatrix},\ M_{2\times2}=\begin{bmatrix}1&1\\1&1\end{bmatrix},\ \forall(M_{i,i},M_{i,i\pm1})=1,\ \forall x_M=[0,1].$$

To ensure consistency, in the resulting X_M, $x_{Mi,j}>1$ are replaced by $x_{Mi,j}=1$. The eigenvalues and eigenvectors are found to examine the covariance and correlation in X

and X_M. The eigenvalues λ_i and corresponding eigenvectors $\boldsymbol{v}_i$ of the symmetric covariance matrices (7.10) $C\in\mathbb{R}^{400\times400}$ are calculated. The covariance matrix C yields the direction of variances. Symmetric non-diagonal square covariance matrix $C=[c_{ij}]$ represents correlations between variables. One may diagonalize C by changing basis using the eigenvalue decomposition. This may yield obscurities and interpretation inadequacies. The eigenvector with the largest eigenvalue yields the direction along with the sequence has maximum variance. The largest eigenvector corresponds to the principal component. In non-diagonal covariance matrix, the eigenvalues represent the variance along the principal components.

The Toeplitz matrix is generated using the **toeplitz** command. For two different X and corresponding $X_M=X\times M$, $\forall x_M=[0, 1]$, four largest principal eigenvectors mode evolutions of the covariance matrix are depicted in Figures 7.5.

The orthogonal eigenvectors of the distinct eigenvalues depict correlations and features observed for (X,X_M), and, $\boldsymbol{v}_1>\boldsymbol{v}_2>\boldsymbol{v}_3>\boldsymbol{v}_4$, $\boldsymbol{v}_1=\boldsymbol{v}_{\max}$. Non-stationary dependences and differences are detected and characterized. The eigenvalues yield variability of X and X_M in an orthogonal basis. The largest eigenvector of the covariance matrix points into the direction of the largest variance of X and X_M, while the magnitude of this eigenvector equals to the corresponding eigenvalue. The computing and assessment take ~0.06 sec.

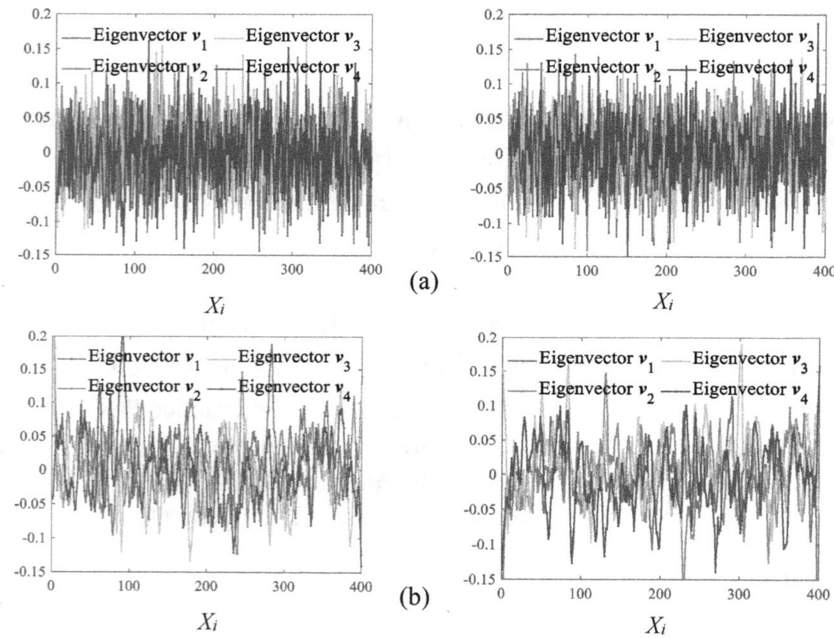

Figure 7.5. (a) Eigenvector modes for $X\in\mathbb{R}^{3200\times400}$, $X_i\sim\mathcal{U}_{[0,1]}$;

(b) Eigenvector modes for $X_M=X\times M$, $X_M\in\mathbb{R}^{3200\times400}$, $M\in\mathbb{R}^{400\times400}$, $M_{2\times2}=\begin{bmatrix}1 & 1\\ 1 & 1\end{bmatrix}$, $\forall m_{i,i}=1, \forall x_{M\,i,j}=[0, 1]$.■

Example 7.10. *Image Analysis*

Consider a 600×396×3 RGB image of the Mona Lisa oil painting, see Figure 7.6.a. In 8-bit-per-pixel images, each pixel has 8-bit strings for red, green and blue (RGB) colors. In three color channels, the pixel yields $2^8=256$ possible values. The pixel intensity is defined by an 8-bit integer giving a range of possible values from 0 to 255, with 0 is black, and, 255 is white.

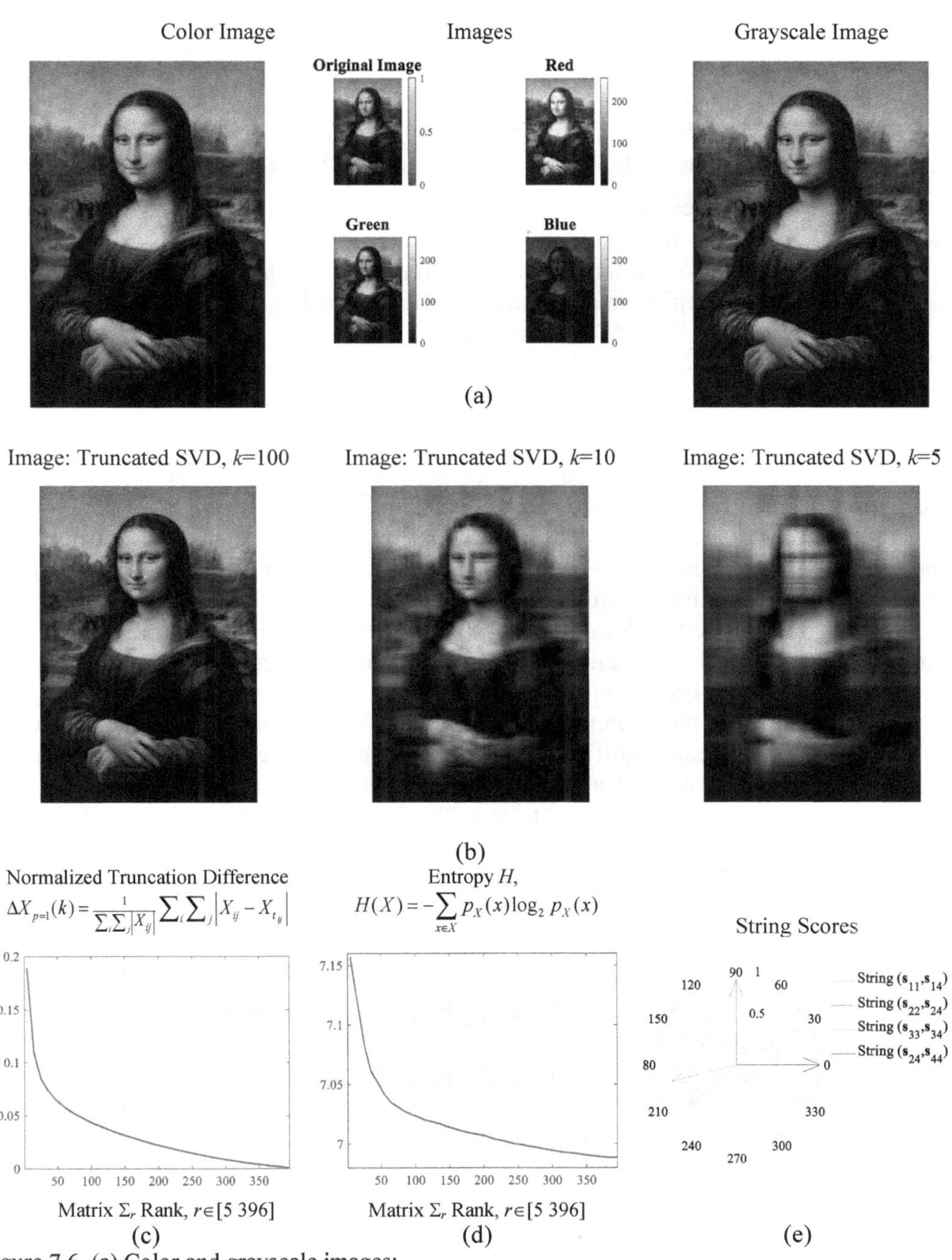

Figure 7.6. (a) Color and grayscale images;

(b) Compressed images using truncated SVDs (U_t,Σ_t,V_t) for k=100, k=10 and k=5. For k=10 and k=5, Σ_t=diag[3.606×10^4 1.09×10^4 6.949×10^3 5.726×10^3 4.022×10^3 3.425×10^3 2.747×10^3 2.205×10^3 2.089×10^3 1.871×10^3], and, Σ_t=diag[3.606×10^4 1.09×10^4 6.949×10^3 5.726×10^3 4.022×10^3];

(c) Normalized truncation difference of a full SVD and truncated rank-k SVD (U_t,Σ_t,V_t), $\Delta X_{p=1}(k) = \frac{1}{\sum_i \sum_j |X_{ij}|} \sum_i \sum_j \left| X_{ij} - \tilde{X}_{ij} \right|$, $k \in [5\ 396]$;

(d) Entropy $H(X)$ for image mapped by full $X=U\Sigma V^T$ and rank-k truncated SVD $\tilde{X} = U_t \Sigma_t V_t^T$, $k \in [5\ 396]$;

(e) Bit strings sematic indexing.

Figures 7.6.a illustrate the color image, red-, green- and blue-color images, as well as a grayscale image. For a RGB image, converted into the 600×396 grayscale image, compute a SVD (7.4)

$$X=U\Sigma V^T,\ U\in\mathbb{R}^{600\times 600},\ \Sigma\in\mathbb{R}^{600\times 396},\ V\in\mathbb{R}^{396\times 396}.$$

Study the dimensionality reduction using the truncated SVD (7.7) $\tilde{X}=U_t\Sigma_t V_t^T$

Finding a truncated SVD with a tuple (U_t,Σ_t,V_t), for k=100, k=10 and k=5, the compressed images are illustrated in Figures 7.6.b.

For (X,X_t), the quantitative estimates on the mismatch of X and X_t are evaluated. Figure 7.6.c depicts the normalized difference using (7.9)

$$\Delta X_{p=1}(k)=\frac{1}{\sum_i\sum_j\left|X_{ij}\right|}\sum_i\sum_j\left|X_{ij}-\tilde{X}_{ij}\right|$$

of the full SVD (7.4) and truncated SVD (7.7).

That is, we computed $X=U\Sigma V^T$ and rank-k truncated SVD $\tilde{X}=U_t\Sigma_t V_t^T$ for $k\in$[5 396].

Changes in the least significant bits may not be revealed, unless X is known or there are pertained features which can be detected.

Entropy measures the data complexity and uncertainties. The entropy $H(X)$ increases for the truncated SVDs pertained to the compressed images. The computed entropy $H(X)$ is reported in Figure 7.6.d.

As documented in Example 7.9, the eigenvector with the largest eigenvalue corresponds to the principal components, and, the direction has the maximum variance, defined by the largest eigenvector.

The malware may be embedded in image files. For example, malicious codes can be hidden in the least significant bits of bytes which represent the color pixels. The original bit-size is kept. Let a pixel is encoded as '11111111'. There are the least significant bits, reading left to right, and, '0011111' make no significant difference. Perform semantic indexing on the RGB image strings S with complimentary $\mathsf{s}_{i,j}$, $X_i \equiv X_i$ and $X_{i,j}\equiv \mathsf{s}_{i,j}$. The semantic indexing $\{\mathsf{s}_1,\mathsf{s}_2,\mathsf{s}_3\}_i$ to $\{\mathsf{s}_4\}_j$ are visualized as depicted in Figure 7.6.e. The score vectors with k elements in $(\mathsf{s}_{ij},\mathsf{s}_{nm})$ are depicted. ■

7.6. Statistical Models and Probabilistic Characteristics

The statistical models and descriptive probabilistic characteristics have being used in dynamic-probabilistic analysis and design of systems, such as transducers, sensors, transceivers, etc. Apply this analysis in data-driven information-centric studies. Source codes, instructions, text files, images, software and data may conceal malicious codes, malicious macros and other malicious content. It is challenging to detect malware in partitioned, randomized and encrypted blocks by decomposing strings and performing searches. New approaches and algorithms are needed to detect, identify and classify malware in obfuscated carrier strings. Embedded strings could be indistinguishable from data or pseudorandom codes.

Sequences and Strings Description – Finite sequences have finite descriptive, computational and algorithmic complexities. Finite strings and substrings (S,s) admit computable statistical descriptions and may yield descriptive attributes. The computed probabilistic characteristics of characterizable (S,s) are correlation, covariance, distributions, probabilities, entropies, entropy rates, entropy density, information losses, information rate, etc. The issues, pertained to descriptive and computational undecidability and complexity are partial observability, reducibility, incompleteness, etc. The finite statistical complexity implies the existence of computable information measures and statistical dependences.

Statistical Models, Probability Distributions and Descriptive Characteristics – A statistical model is a pair containing a sample space of observations $\mathcal{S}$ and a set of distributions $\mathcal{D}$ on $\mathcal{S}$. For a random X, consider a probability space $(\Omega,\mathcal{F},\mathbb{P})$ with the sample space Ω, event space $\mathcal{F}$ and probability space $\mathbb{P}$. Random $X:(\Omega,\mathcal{F})\rightarrow(\mathbb{R},\mathcal{S})$ in the measurable space $(\Omega,\mathcal{F})$ admits the cumulative distribution function (cdf) $F_X(x)$, $F_X(x):\mathbb{R}\rightarrow\mathbb{R}$, such as $X(F_X(x):\mathbb{R}\rightarrow[0,1])$, and, $F_X(x)=\mathbb{P}(X\leq x)$ determines the probabilities $\mathbb{P}(X\in[a,b])$ for $(a,b)\in\mathbb{R}$.

A probability distribution $\mathcal{D}$ is characterized by descriptive functions and parameters ϕ as $\mathcal{D}=\{F_X(x;\phi):\phi\in\Phi\}$, $\mathcal{D}=\{f_X(x;\phi):\phi\in\Phi\}$ or $\mathcal{D}=\{p_X(x;\phi):\phi\in\Phi\}$. For a random or pseudorandom X, there exists a probability distribution $\mathcal{D}(\cdot)$, characterized by a cdf $F_X(x)$, probability density function (pdf) $f_X(x)$, and, probability mass function (pmf) $p_X(x)$. The cdf uniquely defines the distribution. One finds a statistical model for X as (5.145) $\mathcal{R}_X=(\mathcal{S},\mathcal{P})$, $X\sim\mathcal{D}(\cdot)$. Continuous and discrete X are characterized by pdf $f_X(x)\geq0$ or pmf $p_X(x)\geq0$, respectfully. There exist the mixed continuous and discrete distributions. The singular probability distributions, such as the Cantor distribution, cannot be described as a combination of pdf and pmf. Such distribution has a cdf, described by a Hölder continuous Cantor function.

Processes and sequences $X\sim\mathcal{D}(\cdot)$ are characterized by cdfs, pdfs and pmfs. The real-valued $(F_X(x),f_X(x))$ or $(F_X(x),p_X(x))$ describe probabilistic signatures of strings, sequences, signals and processes. Statistical significance and statistical hypothesis testing are investigated. One finds and parameterizes $X\sim\mathcal{D}(\cdot)$ as $\mathcal{D}=\{F_X(x;\phi):\phi\in\Phi\}$, $\mathcal{D}=\{f_X(x;\phi):\phi\in\Phi\}$ or $\mathcal{D}=\{p_X(x;\phi):\phi\in\Phi\}$ on admissible probabilities $\mathbb{D}$, $\mathcal{D}\in\mathbb{D}$, where Φ is the set of model parameters.

Probability Distribution – For a random variable X_i with n outcomes $\{x_i: i=1,2,\ldots,n-1,n\}$, $X_i\equiv\mathbb{S}$, there exists a statistical model $\mathcal{R}_X=(\mathcal{S},\mathcal{P})$ with a probability distribution $\mathcal{D}\in\mathbb{D}$ under the hypotheses $h\in\mathcal{H}$. We find $(F_X(x),f_X(x))$, $f_X(x)=\dfrac{dF_X(x)}{dx}$, and, $(F_X(x),p_X(x))$.

Discretization or finding the first difference $\Delta F_X(x_k)$ yields finite $(p_X(x_k),F_X(x_k))$, $p_X(x_k)=\dfrac{f_X(x_k)}{\sum_{i=1}^{k}f_X(x_k)}$.

For continuous or discrete X, the problems are:

1. Find a histogram $h_X(x)$, normalize histogram $\bar{h}_X(x)$, $\sum_{x\in X}\bar{h}_X(x)=1$, and, define a probability distribution $\mathcal{D}\in\mathbb{D}$, characterized by a pdf $f_X(x)\geq0$ or pmf $p_X(x)\geq0$, as well as cdf $F_X(x)\geq0$, such as $\lim_{x\to-\infty}F_X(x)=0$, $\lim_{x\to\infty}F_X(x)=1$, $f_X(x)=\frac{d}{dx}F_X(x)$, $\int_{-\infty}^{\infty}f_X(x)dx=1$, $\sum_{x\in X}p_X(x)=1$;
2. Analyze covariance, variance, entropies, mutual information, Kullback-Leibler divergence, etc.

Any $X \equiv \mathsf{S}$ can be quantified by a distribution $X \sim \mathcal{D}(\cdot)$ under the hypothesis $\hbar \in \mathcal{H}$, and, characterized by pertained $(F_X(x), f_X(x))$, $(F_X(x), p_X(x))$, $\mathbb{E}[X]$, $H(X)$,etc. For example,

$$F_X(x) = \mathbb{P}(X \le x) = \sum_{x_i \le x} \mathbb{P}(X = x_i) = \sum_{x_i \le x} p_X(x_i),\ \mathbb{P}(X \in A) = \sum_{x \in A} p_X(x),$$

$$p_X(x) = \mathbb{P}(X = x) \text{ or } p_X(x_k) = \mathbb{P}(X = x_k),\ k = 1, 2, ..,$$

while the expected value $\mathbb{E}[X]$, which is called the expectation, statistical expectation or mean, is

$$\mathbb{E}[X] = \int_{-\infty}^{\infty} x f_X(x) dx \text{ or } \mathbb{E}[X] = \sum_i x_i p_X(x_i).$$

A probabilistic signature analysis may be used in descriptive analysis of data, information, malware detection and characterization, etc. The $F_X(x)$, $f_X(x)$ and $p_X(x)$ are parameterized by applying axioms (5.147)

$$\lim_{x \to -\infty} F_X(x) = 0,\ \lim_{x \to \infty} F_X(x) = 1,\ \int_{-\infty}^{\infty} f_X(x) dx = 1,\ \sum_{x \in X} p(x) = 1,$$

and, solving the minimization problem (5.149) using the l-norm

$$\min_{\substack{F_X \in (S, \mathcal{D}) \\ \mathcal{D} = \{F_X(x;\phi): \phi \in \Phi\}}} \int_{-\infty}^{\infty} \left\| F_{X_{F_X \in (S,\mathcal{D})}}(x) - F_{X_{\int_{-\infty}^{x} \tilde{\bar{h}}_X(y) dy}}(x) \right\|_l dx,\ \left\| F_{X_{F_X \in (S,\mathcal{D})}}(x) - F_{X_{\int_{-\infty}^{x} \tilde{\bar{h}}_X(y) dy}}(x) \right\|_l \le \varepsilon_F,\ 0 < \varepsilon_F \ll 1,$$

$$\min_{\substack{f_X \in (S, \mathcal{D}) \\ \mathcal{D} = \{f_X(x;\phi): \phi \in \Phi\}}} \int_{-\infty}^{\infty} \left\| f_{X_{f_X \in (S,\mathcal{D})}}(x) - f_{X_{\bar{h}_X(x)}}(x) \right\|_l dx,\ \left\| f_{X_{f_X \in (S,\mathcal{D})}}(x) - f_{X_{\bar{h}_X(x)}}(x) \right\|_l \le \varepsilon_f,\ 0 < \varepsilon_f \ll 1, \quad (7.14)$$

$$\min_{\substack{p_X \in (S, \mathcal{D}) \\ \mathcal{D} = \{p_X(x;\phi): \phi \in \Phi\}}} \int_{-\infty}^{\infty} \left\| p_{X_{p_X \in (S,\mathcal{D})}}(x) - p_{X_{\bar{h}_X(x)}}(x) \right\|_l dx,\ \left\| p_{X_{p_X(x) \in (S,\mathcal{P})}}(x) - p_{X_{\bar{h}_X(x)}}(x) \right\|_l \le \varepsilon_p,\ 0 < \varepsilon_p \ll 1.$$

Histogram – A histogram $h_X(x)$ is a plot, surface or manifold of values of a variable or variables versus the count of numbers observed. A histogram yields an observed distribution. Using a normalized histogram $\bar{h}_X(x)$, such as $\sum_{x \in X} \bar{h}_X(x) = 1$, one computes the pdf $f_X(x)$, $\int_{-\infty}^{\infty} f_X(x) dx = 1$, or pmf $p_X(x)$, $\sum_{x \in X} p(x) = 1$.

Cantor Distribution – For cdfs, $\inf_{x \in X} F_X(x) = 0$ and $\sup_{x \in X} F_X(x) = 1$. For a continuous probability distribution of a continuous random variable X, a cdf $F_X(x)$ is absolutely continuous. The continuous probability distributions are Cauchy, uniform, etc. For the Cantor distribution, the cdf is a Cantor function. This distribution does not admit the pdf and pmf, while the cdf is a Hölder continuous function. The Cantor singular distribution is not absolutely Lebesgue continuous, and, cannot be classified as discrete or absolutely continuous.

Example 7.11. *Density Functions: Probabilistic Characteristics For Continuous* (X, Y)

For continuous random (X, Y) with a joint pdf $f_{X,Y}(x, y)$, the conditional probability density function for X given Y is

$$f_{X|Y}(x|y) = \frac{f_{X,Y}(x,y)}{f_Y(y)},\ f_{X,Y}(x,y) > 0,\ f_Y(y) > 0,$$

and $$\mathbb{P}(a \le X \le b | Y = y) = \int_a^b f_{X|Y}(x|y) dx.$$

The conditional expectation of X given $Y = y$ is $\mathbb{E}(X | Y = y) = \int_{-\infty}^{\infty} x f_{X|Y}(x|y) dx$.

For an event Y=y, $\mathbf{E}(X|Y$=$y)$ is computed as $\mathbf{E}(X|Y=y)=\frac{1}{\mathbf{P}(Y=y)}\int_{Y=y}Xd\mathbf{P}$, $\mathbf{P}(Y$=$y)$>0.

If X and Y are independent

$$f_{X|Y}(x|y)=\frac{f_{X,Y}(x,y)}{f_Y(y)}=\frac{f_X(x)f_Y(y)}{f_Y(y)}=f_X(x),\mathbf{E}(X|Y=y)=\int_{-\infty}^{\infty}xf_{X|Y}(x|y)dx=\int_{-\infty}^{\infty}xf_X(x)dx=\mathbf{E}(X).$$

Consider the joint and marginal pdfs as

$$f_{X,Y}(x,y)=x^2y^4e^{-x^2-y^2},\ f_X(x)=ax^2e^{-x^2},\ f_Y(y)=by^4e^{-y^2},\ (a,b)>0.$$

Hence, X and Y are independent random variables.

For a pair (X,Y), having found the joint pdf $f_{X,Y}(x,y)$, depicted as a surface, the probability that (X,Y) are in a region $A\in\mathbb{R}^2$ is a volume under the surface $f_{X,Y}(x,y)$ within A.

That is, $\mathbf{P}((X,Y)\in A)=\iint_A f_{X,Y}(x,y)dxdy$.

The probability is analytically found or computed as

$$\mathbf{P}(a\le X\le b,c\le Y\le d)=\int_a^b\int_c^d f_{X,Y}(x,y)dxdy,\ \int_{-\infty}^{\infty}\int_{-\infty}^{\infty}f_{X,Y}(x,y)dxdy=1, f_{X,Y}(x,y)\ge 0,\forall(x,y).\ \blacksquare$$

Example 7.12. *Probability Density Function, Statistical Dependence and Independence*
For (X,Y), statistical models are found using observations which yield histograms. The normalized histograms $\bar{h}_X(x)$, $\bar{h}_Y(y)$, $\bar{h}_{X,Y}(x,y)$ are parameterized.

Let $\bar{h}_{X,Y}(x,y)$ implies a joint pdf as

$$f_{X,Y}(x,y)=\begin{cases}ax^2ye^{-(bx^2+cy^2)}, & 0\le x\le\infty,\ 0\le y\le\infty\\ 0, & \text{otherwise}\end{cases},\ (a,b,c)\in\mathbb{R},$$

which is parameterized using $\int_{-\infty}^{\infty}\int_{-\infty}^{\infty}f_{X,Y}(x,y)dxdy=1$, finding a=4.55, b=1 and c=1.

One has $f_{X,Y}(x,y)=\begin{cases}4.55x^2ye^{-(x^2+y^2)}, & 0\le x\le\infty,\ 0\le y\le\infty\\ 0, & \text{otherwise}\end{cases}.$

The marginal pdfs for independent (X,Y) are

$$f_X(x)=\int_{-\infty}^{\infty}f_{X,Y}(x,y)dy=4.55x^2e^{-x^2}\int_0^{\infty}ye^{-y^2}dy=2.275x^2e^{-x^2},$$

$$f_Y(y)=\int_{-\infty}^{\infty}f_{X,Y}(x,y)dx=4.55ye^{-y^2}\int_0^{\infty}x^2e^{-x^2}dx=\tfrac{4.55\sqrt{\pi}}{4}ye^{-y^2}=1.1375\sqrt{\pi}ye^{-y^2}.$$

Consider statistically dependent (X,Y). One finding $\bar{h}_{X,Y}(x,y)$, implying a joint pdf is

$$f_{X,Y}(x,y)=\begin{cases}ax^2ye^{-(bx^2+cxy+dy^2)}, & 0\le x\le\infty,\ 0\le y\le\infty\\ 0, & \text{otherwise}\end{cases},\ (a,b,c,d)\in\mathbb{R}.$$

The parametrization with $\int_{-\infty}^{\infty}\int_{-\infty}^{\infty}f_{X,Y}(x,y)dxdy=1$ yields

$$f_{X,Y}(x,y)=\begin{cases}\frac{32}{\sqrt{\pi}}x^2ye^{-(x^2+2xy+y^2)}, & 0\le x\le\infty,\ 0\le y\le\infty\\ 0, & \text{otherwise}\end{cases}.$$

The surfaces for $f_{X,Y}(x,y)$ are plotted in Figure 7.7.a.

Consider statistically dependent (X,Y) with $\bar{h}_{X,Y}(x,y)$ and multimodal pdf

$$f_{X,Y}(x,y)=0.5255\cos^2(2x^2+xy+2y^2)e^{-(x^2+xy+y^2)},\ \int_{-\infty}^{\infty}\int_{-\infty}^{\infty}f_{X,Y}(x,y)dxdy=1.$$

Figures 7.7.b depict surfaces for $f_{X,Y}(x,y)$.

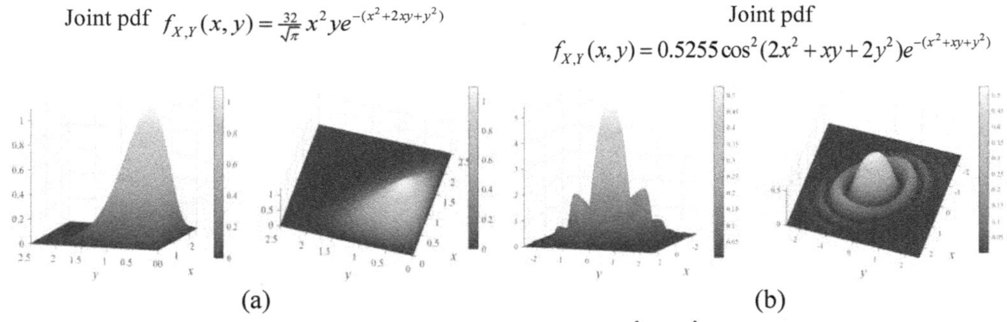

Figure 7.7. (a) Surfaces for a joint pdf $f_{X,Y}(x,y)=\frac{32}{\sqrt{\pi}}x^2ye^{-(x^2+2xy+y^2)}$, $0\le x\le\infty$, $0\le y\le\infty$;
(b) Surfaces for a multimodal pdf $f_{X,Y}(x,y)=0.5255\cos^2(2x^2+xy+2y^2)e^{-(x^2+xy+y^2)}$. ■

Probabilities and Expectations – For random (X,Y), probabilities are derived and numerically computed for specified $A\in\mathbb{R}^2$ as

$$\mathbf{P}((X,Y)\in A)=\iint_A f_{X,Y}(x,y)dxdy.$$

For a two-dimensional rectangle $A=\{(x,y)\colon a\le x\le b, c\le y\le d\}$, one has

$$\mathbf{P}((X,Y)\in A)=\int_a^b\int_c^d f_{X,Y}(x,y)dxdy.$$

For expectations, $\mathbf{E}[X]=\int_{-\infty}^{\infty}\int_{\infty}^{\infty} xf_{X,Y}(x,y)dxdy$, $\mathbf{E}[Y]=\int_{-\infty}^{\infty}\int_{\infty}^{\infty} yf_{X,Y}(x,y)dxdy$.

For a given $g(X,Y)$, the expectation is $\mathbf{E}[g(X,Y)]=\int_{-\infty}^{\infty}\int_{\infty}^{\infty} g(x,y)f_{X,Y}(x,y)dxdy$.

Conditional expectations are found using $f_{X|Y}(x|y)$ as

$$\mathbf{E}[Y|X=x]=\int_{-\infty}^{\infty} yf_{X|Y}(x|y)dy,\quad \mathbf{E}[X|Y=y]=\int_{-\infty}^{\infty} xf_{X|Y}(x|y)dx.$$

For random (X_1,X_2), jointly normal, a bivariate normal pdf is

$$f_{X_1,X_2}(x_1,x_2)=\frac{1}{2\pi\sigma_{11}\sigma_{22}\sqrt{1-\rho^2}}e^{-\frac{1}{2(1-\rho^2)}\left(\frac{(x_1-\mu_1)^2}{\sigma_{11}^2}-2\rho\frac{(x_1-\mu_1)(x_2-\mu_2)}{\sigma_{11}\sigma_{22}}+\frac{(x_2-\mu_2)^2}{\sigma_{22}^2}\right)},\quad \int_{x_1}\int_{x_2} f_{X_1,X_2}(x_1,x_2)dx_2dx_1=1$$

with variances $(\sigma_{11}^2,\sigma_{22}^2)$, mean $\boldsymbol{\mu}=\begin{bmatrix}\mu_1\\ \mu_2\end{bmatrix}$, correlation $\rho=\frac{\operatorname{cov}(X_1,X_2)}{\sigma_{11}\sigma_{22}}=\frac{\mathbf{E}[(X_1-\mu_1)(X_2-\mu_2)]}{\sigma_{11}\sigma_{22}}$,

and, a covariance matrix $\Sigma=\begin{bmatrix}\sigma_{11}^2 & \rho\sigma_{11}\sigma_{22}\\ \rho\sigma_{11}\sigma_{22} & \sigma_{22}^2\end{bmatrix}$.

Hence, $\mu_{X_1|X_2=x_2}=\mathbf{E}[X_1|X_2=x_2]=\mu_1+\rho\frac{\sigma_{11}}{\sigma_{22}}(x_2-\mu_2)$,

$$\mu_{X_2|X_1=x_1}=\mathbf{E}[X_2|X_1=x_1]=\int_{-\infty}^{\infty} x_2f_{X_1|X_2}(x_1|x_2)dx_2=\mu_2+\rho\frac{\sigma_{22}}{\sigma_{11}}(x_1-\mu_1).$$

Additional coverage is reported in Practice Problems 7.3 to 7.7.

Central Limit Theorem – Considers the independent and identically distributed random variables $\{X_1,X_2,\ldots,X_n\}$, $X_i\sim\mathcal{N}(\mu,\sigma^2)$ with expected value $\mathbf{E}[X_i]=\mu$, $\mu<\infty$ and finite variance $\operatorname{var}(X_i)=\sigma^2$, $0<\sigma^2<\infty$. Per the *central limit theorem*, the random variable

$$Z=\frac{X_1+X_2+\ldots+X_n-n\mu}{\sigma\sqrt{n}}=\frac{\bar{X}-n\mu}{\sigma/\sqrt{n}},\ \bar{X}=\frac{X_1+X_2+\ldots+X_n}{n},\ Z=\lim_{n\to\infty}\sqrt{n}\left(\tfrac{1}{\sigma}(\bar{X}-\mu)\right)$$

converges in distribution to the normal distribution as $n\to\infty$. Furthermore, the random $\sqrt{n}(\bar{X}-\mu)$, $n\to\infty$, yields a normal distribution $\sqrt{n}(\bar{X}-\mu)\xrightarrow{\mathcal{D}}\mathcal{N}(0,\sigma^2)$. For any real number z, $\lim_{n\to\infty}\mathbf{P}\left[\sqrt{n}(\bar{X}-\mu)\le z\right]=F_Z(\tfrac{1}{\sigma}z)$, where $F_Z(z)$ is the normal cdf evaluated at z.

Spectral Modalities and Fourier Transforms – Parametric and nonparametric estimates on signatures, pattern and modalities are applied to evaluate distinct features for arrays, sequences and bitstings in the frequency domain. Features of random and deterministic sequences are evaluated, and, illustrative results are documented in Practice Problems 7.9 and 7.10. Nonparametric statistical distributions can be investigated.

The autocorrelation of a real-valued $x(t)$ defines the self-similarity of sequence, and, $R_x(t_1,t_2)=\mathbf{E}[x(t_1)x(t_2)]$. For a stationary $x(t)$, autocorrelation depends only on time difference, $R_x(\tau)=\mathbf{E}[x(t)x(t+\tau)]$. The spectral density $S_x(f)$ of $x(t)$ is the Fourier transform of autocorrelation $R_x(\tau)$.

The Fourier transform exists, and, $x(t)$ is characterized in the frequency domain as

$$S_x(f)=\int_{-\infty}^{\infty}R_x(\tau)e^{-2j\pi f\tau}d\tau,\ S_x(f)\ge 0, \tag{7.15}$$

$$R_x(\tau)=\int_{-\infty}^{\infty}S_x(f)e^{2j\pi f\tau}df,\ R_x(\tau)=\mathbf{E}[x(t)x(t+\tau)],\ \mathbf{E}[X]=\int_{-\infty}^{\infty}xf_X(x)dx.$$

For two jointly stationary $x(t)$ and $y(t)$, the cross-power spectral density $S_{xy}(f)$ is found using the Fourier transform of the cross-correlation function $R_{xy}(\tau)$,

$$S_{xy}(f)=\int_{-\infty}^{\infty}R_{xy}(\tau)e^{-2j\pi f\tau}d\tau. \tag{7.16}$$

Discrete Sequences – Consider discrete bit-, byte- or character string sequences $x[n]$, which are absolutely summable $\sum_n|x[n]|<\infty$, $\sum_n|x[n]|^2<\infty$. Denote $\omega=2\pi fT$ and $X_{2\pi}(\omega)=TX(e^{j\omega})$. One finds the 2π-periodic discrete Fourier transform of aperiodic discrete sequences $x[n]$, defined over all integers $n\in\mathbb{Z}$

$$X(\omega)=X(e^{j\omega})=\mathcal{F}[x[n]]=\sum_{n=-\infty}^{\infty}x[n]e^{-j\omega n}, \tag{7.17}$$

$$X(\omega)=\mathrm{Re}\{X(\omega)\}+j\mathrm{Im}\{X(\omega)\}=|X(\omega)|e^{j\measuredangle X(\omega)},$$

where ω is the normalized frequency in $[\frac{\text{rad}}{\text{sample}}]$.

For $x[n]\overset{\mathcal{F}}{\leftrightarrow}X(\omega)$, the inverse Fourier transform is

$$x[n]=\mathcal{F}^{-1}[X(\omega)]=\frac{1}{2\pi}\int_{-\pi}^{\pi}X(e^{j\omega})e^{j\omega n}d\omega. \tag{7.18}$$

The linearity and Parseval relations are

$$a_1x_1[n]+a_2x_2[n]\overset{\mathcal{F}}{\leftrightarrow}a_1X_1(\omega)+a_2X_2(\omega),\ \sum_{n=-\infty}^{\infty}|x[n]|^2=\frac{1}{2\pi}\int_{-\pi}^{\pi}|X(\omega)|^2d\omega. \tag{7.19}$$

The Fourier transform frequencies form a *continuum*, which yields the frequency domain analysis of $x[n]$, $n\in\mathbb{Z}$.

For example, if $x[n]=a^nu[n]$, $|a|<1$, we have $X(e^{j\omega})=\sum_{n=0}^{\infty}a^ne^{-i\omega n}=\frac{1}{1-ae^{-j\omega}}$.

For a finite-length N-point discrete sequences $x[n]$, from (7.17) and (7.18)

$$X[k]=X(\omega)\big|_{\omega=\frac{2\pi k}{N}}=\sum_{n=0}^{N-1}x[n]e^{-j\frac{2\pi kn}{N}},\ k=0,1,2,\ldots,N-1, \tag{7.20}$$

$$x[n]=\frac{1}{N}\sum_{k=0}^{N-1}X[k]e^{j\frac{2\pi kn}{N}},\ n=0,1,2,\ldots,N-1. \tag{7.21}$$

Frequency spectrum and spectral modalities are described by a discrete Fourier transform. The periodogram is obtained by using the Fourier transform (7.20), and, finding the spectral density of sequences $x[n]$ as the squared magnitude

$$S_{xx}[k]=S_{xx}(\omega)\big|_{\omega=\frac{2\pi k}{N}}=\frac{1}{N}\left|X(e^{j\omega})\right|^2\Big|_{\omega=\frac{2\pi k}{N}}=\frac{1}{N}\left|X[k]\right|^2=\frac{1}{N}\left|\sum_{n=0}^{N-1}x[n]e^{-j\frac{2\pi kn}{N}}\right|^2,k=0,1,..,N-1. \tag{7.22}$$

The window $w[k]$ is applied by splitting $x[k]$ into overlapping segments.

Binary Sequences and Discrete Bernoulli Distribution – A binary sequence is a finite-length string within an alphabet $\mathcal{a}\in\{0,1\}$. A Bernoulli process is a sequence of independent, identically-distributed random $(X_1,X_2,\ldots,X_{k-1},X_k)$.

For the Bernoulli distribution $X\sim\mathcal{B}(p)$, let a discrete random variable X assumes the value 1 with probability p, and, 0 with probability $q=(1-p)$.

The probability that the random X is equal to a particular value x is denoted as $\mathbf{P}(X=x)$. We have, $\mathbf{P}(X=1)=p$, $\mathbf{P}(X=0)=(1-p)$, $0\leq p\leq 1$.

With the binary outcomes $x=1$ and $x=0$, the variance, cdf and pmf are

$$\sigma_X^2=\mathrm{var}(X)=\mathbf{E}\left[(X-\mathbf{E}[X])\cdot(X-\mathbf{E}[X])\right]=\mathbf{E}\left[(X-\mathbf{E}[X])^2\right]=pq=p(1-p), \tag{7.23}$$

$$F_X(x)=\begin{cases}0 & \text{for } x<0\\ (1-p) & \text{for } 0\leq x<1\\ 1 & \text{for } x\geq 1\end{cases},\quad p_X(x;p)=\mathbf{P}(X=x)=\begin{cases}(1-p) & \text{for } x=0\\ p & \text{for } x=1\\ 0 & \text{otherwise}\end{cases},$$

$$p_X(x;p)=p^x(1-p)^{1-x},\ x\in\{0,1\}.$$

For $X\sim\mathcal{B}(p)$, the entropy of mutually independent $(X_1,X_2,\ldots,X_{k-1},X_k)$ is

$$H_{\mathcal{B}}(X)=-p\log_2 p-(1-p)\log_2(1-p). \tag{7.24}$$

Plot for $H_{\mathcal{B}}(X)$ is illustrated in Figure 7.8.a.

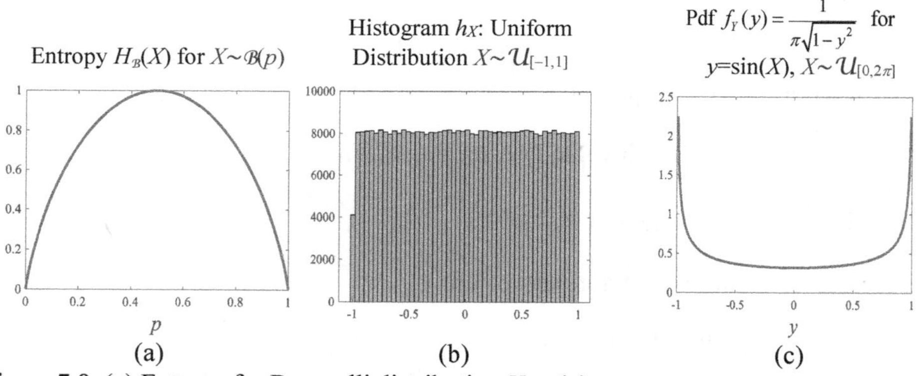

Figure 7.8. (a) Entropy for Bernoulli distribution $X\sim\mathcal{B}(p)$, $H_{\mathcal{B}}(X)=-p\log_2 p-(1-p)\log_2(1-p)$;
(b) Histograms: Uniform distribution $X\sim\mathcal{U}_{[-1,1]}$, $X\in\mathbb{R}^{1\times 400000}$, Example 7.13;
(c) Probability density function for $y=\sin(X)$, $X\sim\mathcal{U}_{[0,2\pi]}$, $f_Y(y)=\frac{1}{\pi\sqrt{1-y^2}}$, $y\in[-1,1]$, Example 7.13.

Binominal Distribution – For the binominal-distributed random variable $X \sim \mathcal{B}(n,p)$, X is the sum of independent Bernoulli processes $X \equiv X_{1\mathcal{B}} + \ldots + X_{n\mathcal{B}}$. It is assumed that there are two mutually two mutually exclusive outcomes, and, the probability of observing successes on a single trial is p. The cdf and pmf are

$$F_X(x;n,p) = \mathbf{P}(X \le x) = \sum_{i=0}^{\lfloor x \rfloor} \binom{n}{i} p^i (1-p)^{n-i},$$

$$p_X(x;n,p) = \mathbf{P}(X = x) = \binom{n}{x} p^x (1-p)^{n-x}, \quad \binom{n}{x} = \frac{n!}{x!(n-x)!}, \quad 0 \le p \le 1,$$

where $\lfloor x \rfloor$ is the greatest integer, $\lfloor x \rfloor \le x$, $x \in \{0,1,2,\ldots\}$.

Geometric Distribution – The geometric distribution describes the first event in an infinite sequence of independent Bernoulli processes $X_{j\mathcal{B}}$ with the success probability $p \in [0,1]$. The cdf and pmf are

$$F_X(x;p) = \mathbf{P}(X \le x) = 1 - (1-p)^x, \; p_X(x;p) = \mathbf{P}(X = x) = (1-p)^{x-1} p, \; x \in \{1,2,3,\ldots\}, 0 \le p \le 1.$$

Poison Distribution – The Poison discrete distribution yields the probability of a given number of events occurring in a fixed space if these events occur with a known constant mean rate and independently. Using the *rate* parameter $\lambda > 0$, the pmf is

$$p_X(x;\lambda) = \mathbf{P}(X = x) = \frac{\lambda^x}{x!} e^{-\lambda}, \; x \in \mathbb{N}^0, \; x \in \{0,1,2,\ldots\}, \; \lambda \in (0,\infty].$$

Discrete Uniform Distribution – The symmetric uniform probability distribution $X \sim \mathcal{U}_{[a,b]}$ is characterized by the maximum-entropy. In the interval $(a,b) \in \mathbb{Z}$, $b \ge a$,

$$p_X(x;a,b) = \frac{1}{b-a+1}, \; x \in \{a,\ldots,b].$$

Codes, data packets, images and text files are sequences of multiple-character characterizable strings. String sequences may exhibit unique and distinctive probabilistic characteristics. The strings $\mathsf{S} = \{\mathsf{s}_1, \mathsf{s}_2, \ldots, \mathsf{s}_{q-1}, \mathsf{s}_q\}$ and substrings s may be partitioned, grouped and analyzed. For bitstrings, $\mathcal{a} = \{0,1\}$. A string S of length l over $\mathcal{a}_{\mathsf{S}}$ may be partitioned to complimentary substrings $\mathsf{s}_{,j}$ of length g over $\mathcal{a}_{\mathsf{s}}$. Detectable and parameterizable distributions, computable probabilities and other quantities metrics can be asserted and evaluated to characterize $(\mathsf{S}, \mathsf{S}, \mathsf{s}, \mathsf{s}_{,j})$. For continuous processes and discrete sequences, different statistical models and statistical characteristics exist and applied, such as the Bayesian, Markov and Shannon analyses. One analyzes dynamic and statistically dependent signatures, patterns, modalities, similarity, homogeneity, etc. Malware sequences may be detected by applying indexing, analyzing cross-covariance and cross-correlation, evaluating probabilistic measures, etc.

Example 7.13. *Random and Deterministic Sequences*

Consider a random sequence $X \in \mathbb{R}^{1 \times 400000}$ with n outcomes. The histogram is shown in Figure 7.8.b, which implies a uniform distribution, $X \sim \mathcal{U}_{[-1,1]}$. The uniformly distributed pseudorandom X is generated by the MATLAB **rand** command which yields $X_{\mathcal{U}}$.

For the maximum-entropy uniform distribution $X\sim\mathcal{U}_{[a,b]}$ in the interval $[a,b]$ with $-\infty<a<b<\infty$, we have the cdf, pdf and entropy

$$F_X(x)=\begin{cases}0,\ x<a\\ \dfrac{x-a}{b-a},\ a\le x\le b\\ 1,\ x>b\end{cases},\quad f_X(x)=\begin{cases}\dfrac{1}{b-a},\ a\le x\le b\\ 0\quad\text{otherwise}\end{cases},\quad H(X)=\ln(b-a),\ x\in[a,b].$$

A random variable may be superimposed with a deterministic process. For example, $X=\frac{1}{2}[X_{\mathcal{U}}+\sin(n)]$, $n\in\mathbb{N}_0$. To ensure a descriptive statistics, deconstruct X, identify a periodic sequence, and, center the random process.

Consider $y=\sin(X)$, where a random variable X is uniformly distributed in $(0,2\pi)$, $X\in[-1,1]$.

We have $\mathbf{P}(Y\le y)=\mathbf{P}(\sin(X)\le y)=\dfrac{2\pi-2\cos^{-1}y}{2\pi}$, $f_Y(y)=\dfrac{d}{dy}\mathbf{P}(Y\le y)=\dfrac{1}{\pi\sqrt{1-y^2}}$.

The cdf and pdf are $F_Y(y)=\frac{1}{2}+\frac{1}{\pi}\sin^{-1}(y)$, $f_Y(y)=\frac{1}{\pi}\frac{d}{dy}\sin^{-1}(y)=\dfrac{1}{\pi\sqrt{1-y^2}}$, $-1<y<1$.

Thus, $f_Y(y)=\begin{cases}\dfrac{1}{\pi\sqrt{1-y^2}},\ -1<y<1\\ 0\quad\text{otherwise}\end{cases}$. The pdf $f_Y(y)$ is illustrated in Figure 7.8.c. ■

Covariance and Correlation – For a $k\times1$ vector $X=[X_1,X_2,\ldots,X_{k-1},X_k]^T$ with random X_i, the finite variance and expected value $\mathbf{E}[X_i]$ exist and computed. One has $\mu_X=\mathbf{E}[X]$.

The covariance matrices are

$$C_{X_iX_j}=\text{cov}[X_i,X_j]=\mathbf{E}\left[(X_i-\mathbf{E}[X_i])(X_j-\mathbf{E}[X_j])^T\right],\tag{7.25}$$

$$C_{XY}=\text{cov}[X,Y]=\mathbf{E}\left[(X-\mathbf{E}[X])(Y-\mathbf{E}[Y])^T\right]=\mathbf{E}[XY^T]-\mathbf{E}[X]\mathbf{E}[Y]^T,\ \mathbf{E}[X]=\tfrac{1}{k}\sum_{i=1}^{k}X_i,$$

$$C_X=\text{cov}[X,X]=\mathbf{E}\left[(X-\mathbf{E}[X])(X-\mathbf{E}[X])^T\right]=\mathbf{E}\left[(X-\mu_X)(X-\mu_X)^T\right]=\mathbf{E}\left[XX^T\right]-\mu_X\mu_X^T,$$

$$C_X=\begin{bmatrix}\mathbf{E}\left[(X_1-\mathbf{E}[X_1])(X_1-\mathbf{E}[X_1])\right] & \cdots & \mathbf{E}\left[(X_1-\mathbf{E}[X_1])(X_k-\mathbf{E}[X_k])\right]\\ \vdots & \ddots & \vdots\\ \mathbf{E}\left[(X_k-\mathbf{E}[X_k])(X_1-\mathbf{E}[X_1])\right] & \cdots & \mathbf{E}\left[(X_k-\mathbf{E}[X_k])(X_k-\mathbf{E}[X_k])\right]\end{bmatrix},\ C_X=C_X^T,\ C_X>0,$$

where $\mathbf{E}$ is the expected value, called the expectation, statistical expectation or mean of the argument.

The variance is

$$\sigma_X^2=\text{var}(X)=\mathbf{E}\left[(X-\mathbf{E}[X])\cdot(X-\mathbf{E}[X])\right]=\mathbf{E}\left[(X-\mathbf{E}[X])^2\right].\tag{7.26}$$

Consider partitioned data blocks (X,Y), where $X_i(n_i)$ are the rows of k samples $X=\begin{bmatrix}X_1(n_1) & \cdots & X_k(n_1)\\ \vdots & \ddots & \vdots\\ X_1(n_m) & \cdots & X_k(n_m)\end{bmatrix}$. The Pearson product-moment correlation matrix K_X is computed using normalized variables $\frac{1}{\sigma_{X_i}}X_i$. The element on the principal diagonal is the correlation of a random variable with itself, which is 1, while the off-diagonal elements are finite. One has

$$K_X=\operatorname{corr}[X]=\left(\operatorname{diag}(C_X)\right)^{-\frac{1}{2}}C_X\left(\operatorname{diag}(C_X)\right)^{-\frac{1}{2}}=\begin{bmatrix}1 & \frac{\mathbb{E}[(X_1-\mu_1)(X_2-\mu_2)]}{\sigma_{X_1}\sigma_{X_2}} & \cdots & \frac{\mathbb{E}[(X_1-\mu_1)(X_k-\mu_k)]}{\sigma_{X_1}\sigma_{X_k}}\\ \frac{\mathbb{E}[(X_2-\mu_2)(X_1-\mu_1)]}{\sigma_{X_2}\sigma_{X_1}} & 1 & \cdots & \frac{\mathbb{E}[(X_2-\mu_2)(X_k-\mu_k)]}{\sigma_{X_2}\sigma_{X_k}}\\ \vdots & \vdots & \ddots & \vdots\\ \frac{\mathbb{E}[(X_k-\mu_k)(X_1-\mu_1)]}{\sigma_{X_k}\sigma_{X_1}} & \frac{\mathbb{E}[(X_k-\mu_k)(X_2-\mu_2)]}{\sigma_{X_k}\sigma_{X_2}} & \cdots & 1\end{bmatrix}. \tag{7.27}$$

Two Random Variables – For (X_1,X_2), the covariance matrix and correlation coefficient are

$$C_{X_1X_2}=\operatorname{cov}(X_1,X_2)=\mathbb{E}\left[(X_1-\mu_{X_1})(X_2-\mu_{X_2})\right]=\mathbb{E}\left[X_1X_2\right]-\mu_{X_1}\mu_{X_2}, \tag{7.28}$$

$$\rho_{X_1X_2}=\frac{\operatorname{cov}(X_1,X_2)}{\sigma_{X_1}\sigma_{X_2}}=\frac{\operatorname{cov}(X_1,X_2)}{\sqrt{\operatorname{var}(X_1)}\sqrt{\operatorname{var}(X_2)}}.$$

Considering $(a_1+b_1X_1,a_2+b_2X_2)$, $C_{X_1X_2}=\operatorname{cov}(a_1+b_1X_1,a_2+b_2X_2)=b_1b_2\operatorname{cov}(X_1,X_2)$.

If (X_1,X_2) are independent, $C_{X_1X_2}=\operatorname{cov}(X_1,X_2)=0$.

The variances are $\operatorname{var}(X_1)=\operatorname{cov}(X_1,X_1)$ and $\operatorname{var}(X_2)=\operatorname{cov}(X_2,X_2)$.
For (aX_1+bX_2), one has $\operatorname{var}(aX_1+bX_2)=a^2\operatorname{var}(X_1)+b^2\operatorname{var}(X_2)+2ab\operatorname{cov}(X_1,X_2)$.
For independent (X_1,X_2), $\operatorname{var}(aX_1+bX_2)=a^2\operatorname{var}(X_1)+b^2\operatorname{var}(X_2)$.

Example 7.14. *Low-Fidelity Studies: Probabilistic Signature of Sequences*
Consider a Mona Lisa grayscale image used in Example 7.10. Study statistical characteristics and investigate probabilistic signatures of sequences X_i. Find histograms and probability functions for pixel intensity byte arrays X_i. For the 8-bit integers, there are 256 possible RGB and grayscale intensity values.

For a grayscale image, consider two rows X_1 and X_{250} in $X\in\mathbb{R}^{600\times396}$, and, find probability distributions $X_i\sim\mathcal{D}_i(\cdot)$, $\mathcal{D}_i\in\mathbf{D}$. The histograms are documented in Figures 7.9. The pmfs $p_X(x)$, which measure the [0,1]-valued probability, are used to calculate entropies, mutual information, etc.

The normalized histograms can be mapped by continuous pdfs $f_X(x)$, $x\in\mathbb{R}$. One can find $(p_X(x_k),F_X(x_k))$ using $(f_X(x),F_X(x))$, $f_X(x)=\frac{dF_X(x)}{dx}$, $x_k=0,1,2,\ldots$ or $x_k=0,\pm1,\pm2,\ldots$.

Discretization, or finding the first difference $\Delta F_X(x_k)$, yields finite $(p_X(x_k),F_X(x_k))$. For example, $p_X(x_k)=\frac{f_X(x_k)}{\sum_{i=1}^{k}f_X(x_k)}$.

For X_1, the normal and extreme value distributions are studied. From normalized $\bar{h}_{X_1}$, for $X\sim\mathcal{N}(\mu,\sigma^2)$ with

$$f_X(x;\mu,\sigma^2)=\frac{1}{\sqrt{2\pi}\sigma}e^{-\frac{(x-\mu)^2}{2\sigma^2}},\ F_X(x;\mu,\sigma^2))=\tfrac{1}{2}\left[1+\operatorname{erf}\left(\tfrac{1}{\sqrt{2}\sigma}(x-\mu)\right)\right],\ \mu\in\mathbb{R},\ \sigma>0,$$

using least-squares algorithm we find μ=103.06 and σ=8.579. The MATLAB command **fitdist** yields μ=102.66, σ=5.731.

For the extreme value distribution $X\sim\mathcal{EV}(\mu,\sigma)$,

$f_X(x;\mu,\sigma)=\frac{1}{\sigma}e^{\frac{x-\mu}{\sigma}}e^{-e^{\frac{x-\mu}{\sigma}}}$, μ=105, σ=7.656.

For the generalized extreme value distribution $X\sim\mathcal{GEV}(\mu,\sigma,k)$,

$f_X(x;\mu,\sigma,k)=\frac{1}{\sigma}e^{-(1+k\frac{x-\mu}{\sigma})^{-\frac{1}{k}}}\left(1+k\frac{1}{\sigma}(x-\mu)\right)^{-1-\frac{1}{k}}$, μ=105.269, σ=7.777, k=0.00105.

The histogram and pdfs are reported in Figures 7.9.a.

For X_i one may find multimodal histograms. Section 5.13 and Table 5.2 report multimodal distributions and parametric pdfs. Investigating $X_i\sim\mathcal{D}_i(\cdot)$, $\mathcal{D}_i\in\mathbb{D}$, parametric and nonparametric cdfs and pmfs can be found and computed. For a histogram pertained to X_{250}, shown in Figure 7.9.b, a nonparametric pmf is found applying a near-optimal, nonparametric, kernel-smoothed probability function. A plot for computed multimodal $p_X(x)$ is documented in Figure 7.9.b. We conducted low-fidelity studies by finding probabilistic signatures of sequences X_i, obtaining $X_i\sim\mathcal{D}_i(\cdot)$, $\mathcal{D}_i\in\mathbb{D}$.

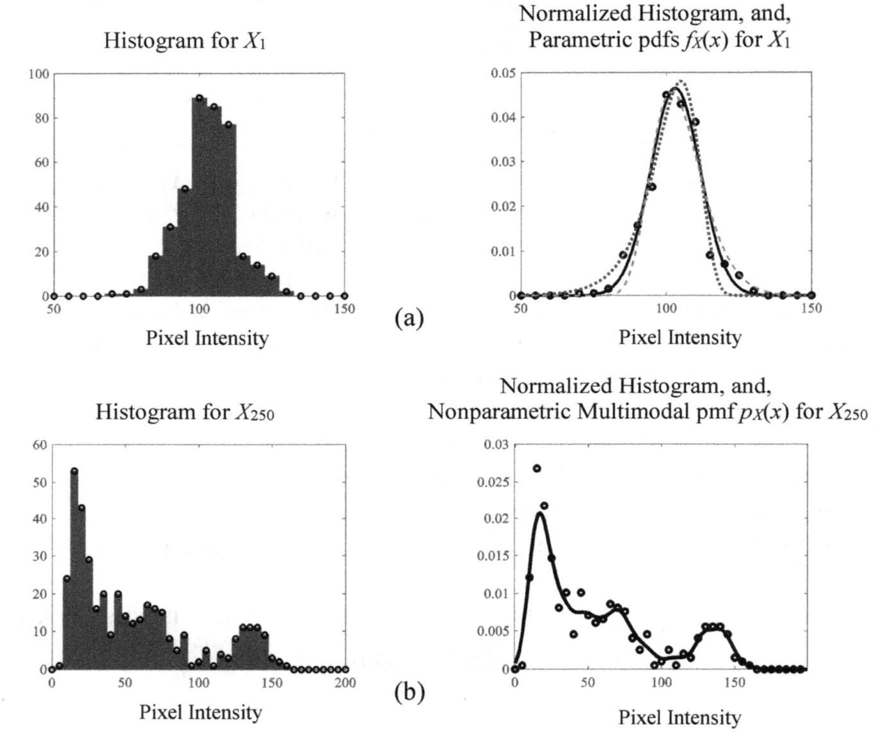

Figure 7.9. (a) Histogram for X_1 and parametric unimodal pdfs, $f_X(x)=\frac{1}{\sqrt{2\pi}\sigma}e^{-\frac{(x-\mu)^2}{2\sigma^2}}$, μ=103.06, σ=8.5794 (solid line), $f_X(x)=\frac{1}{\sigma}e^{\frac{x-\mu}{\sigma}}e^{-e^{\frac{x-\mu}{\sigma}}}$, μ=104.996, σ=7.6564 (dotted line), and, $f_X(x)=\frac{1}{\sigma}e^{-(1+k\frac{x-\mu}{\sigma})^{-\frac{1}{k}}}\left(1+k\frac{1}{\sigma}(x-\mu)\right)^{-1-\frac{1}{k}}$, μ=105.269, σ=7.777, k=0.00105 (dashed line);
(b) Histogram for X_{250}, and, nonparametric multimodal pmf $p_X(x)$.

Compute entropies for the red, green and blue colors, as well as for a grayscale image. Using (7.29) $H(X)=-\sum_{x\in X}p_X(x)\log_2 p_X(x)=\sum_{x\in X}p_X(x)\log_2\frac{1}{p_X(x)}=\mathbb{E}[\log_2\frac{1}{p_X(x)}]$, $H(X)\geq 0$, we have $H(X_{\text{red}})$=7.3481, $H(X_{\text{green}})$=6.9391, $H(X_{\text{blue}})$=5.851, $H(X_{\text{gray}})$=6.9887.

One may evaluate entropies for sequences X_i in bits and bytes, which can be found using decimal intensity values which vary from 0 to 255. Command **dec2bin** yields $X_{i\text{binary}}$=['00000001' '00000011' '00000111' ...]. Inconsistencies should be avoided as the entropies of bytes '00000001' and '100000000' are same.

Covariance and correlation of Xi are studied. The covariance matrix (7.25) is calculated as

X=double(Ig); C=cov(X)

Cross-covariance $C_{X_iX_j}$ (7.25) measures the similarity between X_i and shifted copies of X_i as a function of the samples lag, or, between distinct (X_i,X_j). Cross-covariance of sequences is evaluated for a grayscale image $X\in\mathbb{R}^{600\times 396}$. For X_3 and X_{103} consider (X_1,X_2,X_3,X_4,X_5) and $(X_{101},X_{102},X_{103},X_{104},X_{105})$. Plots for $C_{X_iX_j}$ are reported in Figures 7.10.a. For $(X_i\in\mathbb{R}^{1\times 396},X_j\in\mathbb{R}^{1\times 396})$, computed cross-correlations (7.27) $K_{X_iX_j}$ are illustrated in Figures 7.10.b.

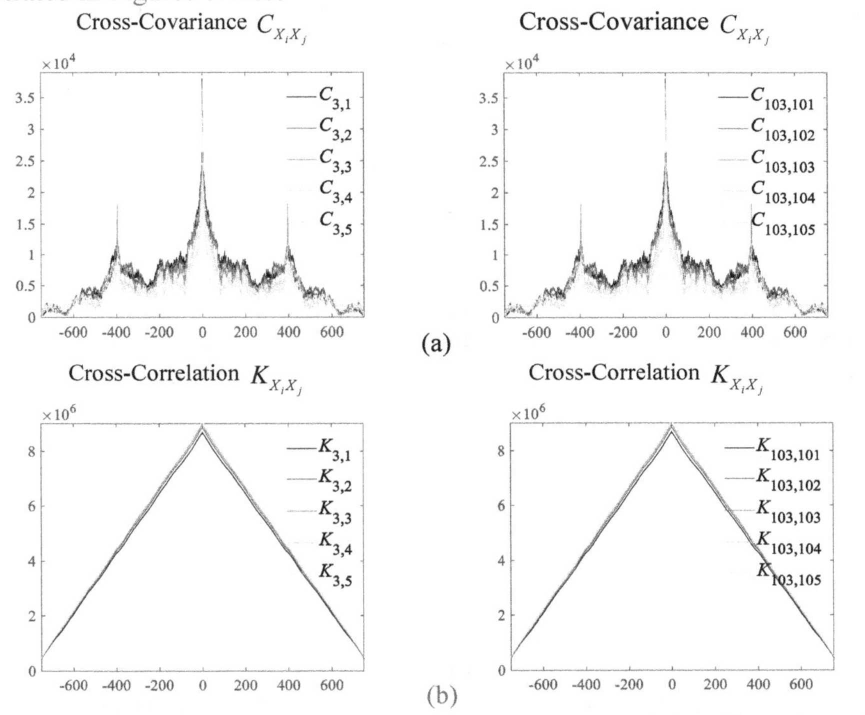

Figure 7.10. (a) Cross-covariance $C_{X_iX_j}$ of (X_i,X_j); (b) Cross-correlation $K_{X_iX_j}$ of (X_i,X_j). ■

Example 7.15. *Analysis of Sequences: Statistical Dependences*

Let $X=[X_1,X_2,X_3,X_4,X_5]^T\in\mathbb{R}^{5\times 800}$. The cross-covariance $C_{X_iX_j}$ and cross-correlation $K_{X_iX_j}$ of discrete stationary random sequences with $\mathcal{a}=\{0,1\}$ are computed and evaluated.

Consider $X=[X_1,X_2,X_3,X_4,X_5]\in\mathbb{R}^{5\times 800}$, $\forall x_i\in[0\ \ 1]$:

1. Stationary pseudorandom sequence
$X=[X_{1_{\mathcal{U},\,a=\{0,1\}}},X_{2_{\mathcal{U},\,a=\{0,1\}}},X_{3_{\mathcal{U},\,a=\{0,1\}}},X_{4_{\mathcal{U},\,a=\{0,1\}}},X_{5_{\mathcal{U},\,a=\{0,1\}}}]$, $X\in\mathbb{R}^{5\times 800}$;
2. Periodic sequences $X=[X_1,X_2,X_3,X_4,X_5]$, $X\in\mathbb{R}^{5\times 800}$, $\forall x_i\in[0\ \ 1]$,
$X_1=|\sin(n)|$, $X_2=|\cos(n)|$, $X_3=|\sin(2n)|$, $X_4=|\sin(n)\cos(n)|$, $X_5=|\sin(2n)\cos(2n)|$;

3. Sequence $X=[X_1,X_2,X_3,X_4,X_5]$, $X\in\mathbb{R}^{5\times800}$, where $X_i\in\mathbb{R}^{1\times800}$ is a sum of stationary pseudorandom $X_{i_{\mathcal{U},a=\{0,1\}}}$ and real-valued periodic $X_{i\text{periodic}}$ sequences, $X_i=(X_{i_{\mathcal{U},a=\{0,1\}}}+X_{i_{\text{periodic}}})_{i=1,2,3,4,5}$, $\forall x_i\in[0\ \ 1]$, $\mathbb{E}[x]=\mathbb{E}[x_{\mathcal{U}}(n)]+\mathbb{E}[x_{\text{periodic}}(n)]$.

   For uniformly distributed pseudorandom $X_{i_{\mathcal{U},a=\{0,1\}}}$, rounded periodic and

mixed sequences X_i with $X_{i_{\text{periodic}}}$, cross-covariance $C_{X_iX_j}$ and cross-correlation $K_{X_iX_j}$ are examined. For $X\in\mathbb{R}^{5\times800}$, $\forall x_i\in[0\ \ 1]$, the quantitative features on periodicity, patterns and modalities are revealed as illustrated in Figures 7.11. For stationary and non-stationary random sequences, deterministic processes and mixed sequences, statistical characteristics and dependencies can be analyzed.

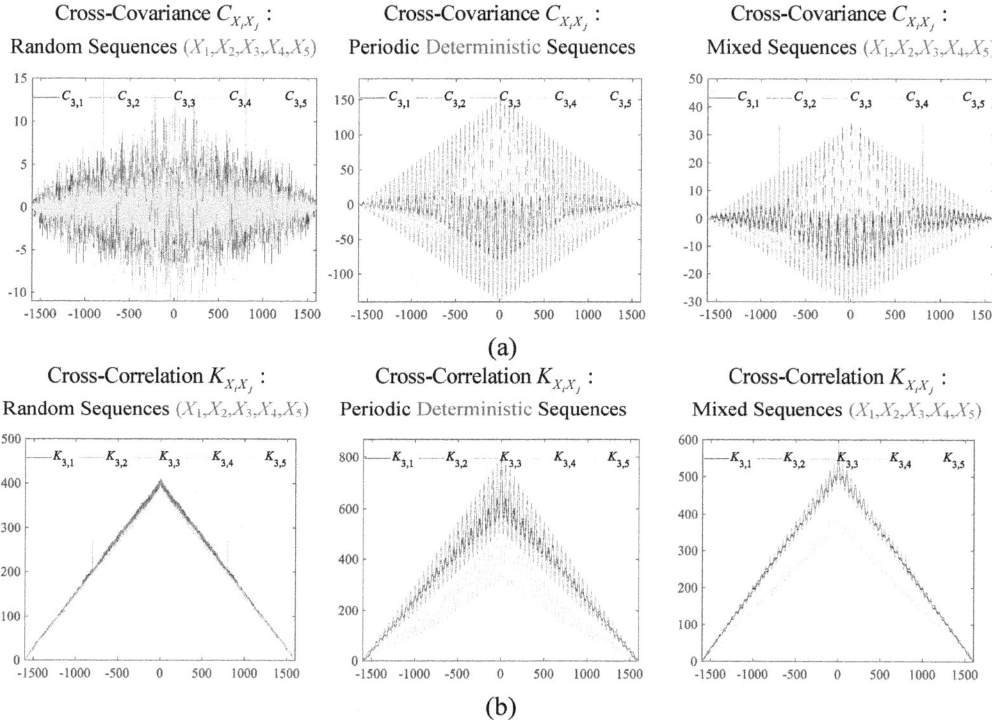

Figure 7.11. (a) Cross-covariance $C_{X_iX_j}$ for: Random $X\in\mathbb{R}^{5\times800}$, $a=\{0,1\}$; Periodic $X\in\mathbb{R}^{5\times800}$; Mixed $X\in\mathbb{R}^{5\times800}$ with $X_i=(X_{i_{\mathcal{U},a=\{0,1\}}}+X_{i_{\text{periodic}}})_{i=1,2,3,4,5}$, $\forall x_i\in[0\ \ 1]$;

(b) Cross-correlation $K_{X_iX_j}$ for random, periodic and mixed $X\in\mathbb{R}^{5\times800}$. ■

7.7. Sequences, Entropies and Information Measures

To approach problems in information control and information management, investigate statistical models, assess probabilistic measures, conduct heterogeneous Markov analysis, as well as adaptively optimize data-driven applications. Analysis and optimization of quantitative dynamic-probabilistic measures empower information sharing, evaluation of strings and sequences, as well as support analyses of data quality and information security. As documented in Figure 7.12.a, using arrays, data packets and sequences, an information management system may able to detect intrusions, classify malware, evaluate assets, as well as analyze cyber threats and attacks. Exploratory solutions are under developments to enable information control, assurance and security. Due to heterogeneity, multi-vector attacks, asynchronous processes, uncertainties and other factors, for the postulated attack and threats, the estimated probabilities are difficult to acquire. Dynamic-probabilistic studies, statistical hypothesis testing and verification are applied.

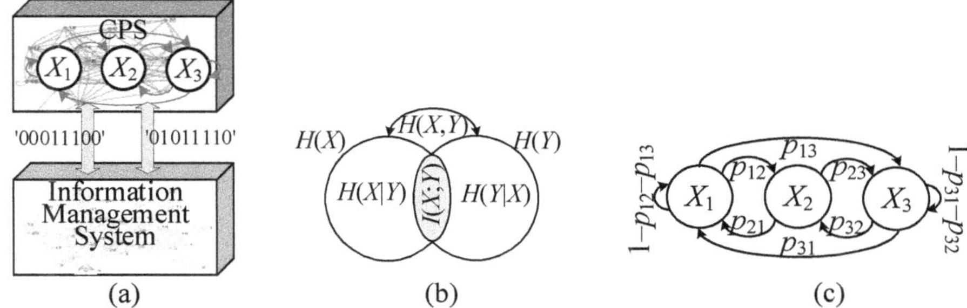

Figure 7.12. (a) Cyber-physical system with an *Information Management System*; (b) Venn diagram depicts information measures, such as entropies and mutual information $I(X;Y)$; (c) Descriptive three-state Markov chain.A transition probability matrix is $P=\begin{bmatrix} 1-p_{12}-p_{13} & p_{12} & p_{13} \\ p_{21} & 0 & p_{23} \\ p_{31} & p_{32} & 1-p_{31}-p_{32} \end{bmatrix}$.

Entropies – For a random X_i with n outcomes $\{x_i: i=1,2,\ldots,n-1,n\}$, $X_i\equiv\mathsf{S}$, let the probability distribution $\mathcal{D}(\cdot)$ and pmf $p_X(x_i)$ are found. The entropy $H(X)$ assesses complexity, randomness and uncertainties of discrete X. For (X,Y), entropy $H(X)$, joint entropy $H(X,Y)$ and conditional entropy $H(Y|X)$ are [18]

$$H(X)=-\sum_{x\in X}p_X(x)\log_2 p_X(x)=\sum_{x\in X}p_X(x)\log_2\frac{1}{p_X(x)}=\mathbb{E}[\log_2\frac{1}{p_X(x)}], H(X)\geq 0, p_X(x)\geq 0, \sum_{x\in X}p_X(x)=1,$$

$$H(X,Y)=H(X)+H(Y|X)=H(Y)+H(X|Y),\ H(X,Y)=\sum_{x\in X,y\in Y}p_{X,Y}(x,y)\log_2\frac{1}{p_{X,Y}(x,y)},\quad (7.29)$$

$$H(Y|X)=-\sum_{x\in X,y\in Y}p_{X,Y}(x,y)\log_2 p_{Y|X}(y|x)=\sum_{x\in X,y\in Y}p_{X,Y}(x,y)\log_2\frac{p_X(x)}{p_{X,Y}(x,y)},$$

where $p_X(x)$ and $p_Y(y)$ are the marginal pmfs of X and Y, $p_X(x)\geq 0$, $p_Y(y)\geq 0$; $p_{X,Y}(x,y)$ and $p_{X|Y}(x|y)$ are the joint and conditional pmfs, $p_{X,Y}(x,y)=\mathbb{P}(X=x \text{ and } Y=y)$, $p_{X,Y}(x,y)\geq 0$, $p_{X|Y}(x|y)=\mathbb{P}(X=x|Y=y)=\frac{p_{X,Y}(x,y)}{p_Y(y)}$, $p_{X|Y}(x|y)\geq 0$.

The conditional entropy $H(Y|X)$ is a measure which quantifies and describes the outcome of a random variable Y given known random X. For a random Y, conditioned on the event $\{X=x\}$, which is a function of a conditional pmf $p_{Y|X}(y|x)$

$$H(Y|X=x)=\sum_{x\in X,y\in Y}p_{Y|X}(y|x)\log_2\frac{1}{p_{Y|X}(y|x)}=\sum_{x\in X,y\in Y}p_{X,Y}(x,y)\log_2\frac{1}{p_{Y|X}(y|x)}.\quad (7.30)$$

The chain rules for probabilities and entropy are

$$p_{XYZ}(x,y,z)=p_{Z|XY}(z|x,y)p_{Y|X}(y|x)p_X(x),\quad (7.31)$$

$$H(X,Y,Z)=H(Z|X,Y)+H(Y|X)+H(X),\ H(X_{1:n})=\sum_{i=1}^{n}H(X_i|X_{1:i-1}).$$

Mutual Information – The mutual information expresses the average amount of information that system obtains on x by observing y. The conditional entropy of X, given Y, is $H(X|Y)\equiv H(X,Y)-H(Y)$, $H(X|Y)=-\sum_{x\in X,y\in Y}\mathbf{P}(X=a,Y=b)\log_2\mathbf{P}(X=a|Y=b)$.

One has

$$H(X|Y)=\sum_{x\in X,y\in Y}p_Y(y)H(X|Y=y),\ 0\leq H(X|Y)\leq H(X). \tag{7.32}$$

A mutual information $I(X;Y)$ defines mutual statistical dependences between X and Y. Applying the input-output entropies

$$I(X;Y)=I(Y;X)=H(X)-H(X|Y)=H(Y)-H(Y|X)=H(X)+H(Y)-H(X,Y), \tag{7.33}$$

$$I(X;Y)\geq 0,\ 0\leq I(X;Y)\leq \min[H(X),H(Y)],$$

$$I(X;Y)=\sum_{x\in X,y\in Y}p_{X,Y}(x,y)\log_2\frac{p_{X,Y}(x,y)}{p_X(x)p_Y(y)}=\sum_{x\in X,y\in Y}p_{X,Y}(x,y)\log_2\frac{p_{X|Y}(x|y)}{p_X(x)}$$

$$=\sum_{x\in X,y\in Y}p_{X,Y}(x,y)\log_2\frac{p_{Y|X}(y|x)}{p_Y(y)}=\mathbf{E}_{p_{X,Y}(x,y)}\left[\log_2\frac{p_{Y|X}(y|x)}{p_Y(y)}\right].$$

Randomness yields high entropy. The uniform probability distribution yields maximum entropy. The conditional entropy $H(Y|X)$ measures the uncertainty of Y given known X. The mutual information $I(X;Y)=H(X)-H(X|Y)=H(Y)-H(Y|X)$ yields the reduction of uncertainty of X or Y due to known Y or X.

The conditional mutual information, between X and Y, given Z, is

$$I(X;Y|Z)=H(X|Z)-H(X|Y,Z)=H(X|Z)+H(Y|Z)-H(X,Y|Z),\ H(X|Y)\geq H(X|Y,Z), \tag{7.34}$$

$$I(X;Y|Z)=\sum_{x,y,z}p_{XYZ}(x,y,z)\log_2\frac{p_{XYZ}(x,y,z)}{p_{X|Z}(x|z)p_{Y|Z}(y|z)p_Z(z)}=\sum_{x,y,z}p_{XYZ}(x,y,z)\log_2\frac{p_{XY|Z}(x,y|z)}{p_{X|Z}(x|z)p_{Y|Z}(y|z)}$$

$$=\mathbf{E}_{p_{XYZ}(x,y,z)}\left[\log_2\frac{p_{XY|Z}(x,y|z)}{p_{X|Z}(x|z)p_{Y|Z}(y|z)}\right],$$

$I(X;Y|Z)+I(Y;Z)=I(Y;(X,Z))$, $I(X;Y|Z)\geq 0$.

The chain rule for mutual information is

$$I(X_1,X_2,X_3;Y)=I(X_1;Y)+I(X_2;Y|X_1)+I(X_3;Y|X_1,X_2),\ I(X_{1:n};Y)=\sum_{i=1}^{n}I\left(X_i;Y|X_{1:i-1}\right). \tag{7.35}$$

The relationships between individual entropies ($H(X)$,$H(Y)$), joint entropy $H(X,Y)$, conditional entropies ($H(X|Y)$,$H(Y|X)$) and mutual information $I(X;Y)$ are illustrated by a Venn diagram, depicted in Figure 7.12.b. Figure 7.12.c documents a time invariant Markov chain for the set of states $X=\{X_1,X_2,X_3\}$ with a transition probability matrix $P=\begin{bmatrix}1-p_{12}-p_{13} & p_{12} & p_{13}\\ p_{21} & 0 & p_{23}\\ p_{31} & p_{32} & 1-p_{31}-p_{32}\end{bmatrix}$. Probabilities, entropies and mutual information can be investigated in information management systems.

String Entropy – Consider binary strings $\{0,1\}^*=\{\varepsilon,0,1,00,01,10,11,000,001,\ldots\}^*$ of an arbitrary finite length over an alphabet $\mathcal{A}=\{0,1\}$. For the Bernoulli distribution $X\sim\mathcal{B}(p)$, for countable bitstrings $\{0,1\}^*$, the binary alphabet-specific entropy is

$$H_{\{0,1\}^*}(X)=\sum_{n=0}^{N}\left[-p_n\log_2 p_n-(1-p_n)\log_2(1-p_n)\right]. \tag{7.36}$$

Consistent measures should be used. The Shannon entropy (7.29) $H(X)=-\sum_{x\in X}p_X(x)\log_2 p_X(x)$ for the bitstrings '11000000', '00011000' and '00000011' yields the same $H(X)$, $H(X)=0.8113$. Probabilistic measures should be consistent.

Example 7.16. *Mutual Information and Statistical Dependences*
Let (X,Y) is a bivariate random vector with a joint pmf $p_{X,Y}(x,y)$ and marginal pmfs $(p_X(x),p_Y(y))$. For any x such that $\mathbf{P}(X=x)=p_X(x)>0$, the conditional pmf of Y, given $X=x$, is a function of y, denoted as $p_{Y|X}(y|x)$, is

$$p_{Y|X}(y|x)=\mathbf{P}(Y=y|X=x)=\frac{p_{X,Y}(x,y)}{p_X(x)}.$$

For a bivariate random (X,Y) with a joint pmf $p_{X,Y}(x,y)$, the X and Y are independent random variables if and only if for every $x\in\mathbb{R}$ and $y\in\mathbb{R}$, $p_{X,Y}(x,y)=p_X(x)p_Y(y)$.

If X and Y are independent, $H(X|Y)=H(X)$.

From (7.33), $I(X;Y)=I(Y;X)=H(X)-H(X|Y)=H(Y)-H(Y|X)=0$.

Finding $\log_2\frac{p_{X,Y}(x,y)}{p_X(x)p_Y(y)}=\log_2 1=0$, one concludes that $I(X;Y)=0$. ■

Example 7.17. *Continuous and Discrete Sequences, Entropies and Information Rate*
For a continuous random variable X, denote the differential entropy as $\mathcal{H}(X)$. For a discrete random variable, the Shannon entropy is $H(X)$. Consider $Y=X$. A continuous random variable, which exists over an interval of infinite positive length, admits an infinite number of possible outcomes x. This implies an infinite information, and,

$\mathcal{H}(X)\to\infty$, $\mathcal{H}(X|X)=-\infty$, $\mathcal{H}(Y|X)=-\infty$ and $I_{\mathcal{H}}(X;Y)=\infty$.

For $X=(X_1,X_2)$ in $\Omega\in\mathbb{R}^2$, let $X\sim\mathcal{N}(\boldsymbol{\mu},\Sigma)$. The pdfs for $X_1\sim\mathcal{N}(\mu_1,\sigma_{11}^2)$ and $X_2\sim\mathcal{N}(\mu_2,\sigma_{22}^2)$ are

$$f_{X_1}(x_1)=\frac{1}{\sqrt{2\pi}\sigma_{11}}e^{-\frac{(x_1-\mu_1)^2}{2\sigma_{11}^2}},\ f_{X_2}(x_2)=\frac{1}{\sqrt{2\pi}\sigma_{22}}e^{-\frac{(x_2-\mu_2)^2}{2\sigma_{22}^2}}.$$

The entropies are $\mathcal{H}(X_1)=\frac{1}{2}\log_2\left(2\pi e\sigma_{11}^2\right)$ and $\mathcal{H}(X_2)=\frac{1}{2}\log_2\left(2\pi e\sigma_{22}^2\right)$.

These $\mathcal{H}(X_i)$ may admit negative definiteness, depending on σ_{ii}^2.

For a bivariate normal distribution with correlation ρ,

$$I_{\mathcal{H}}(X_1;X_2)=\mathcal{H}(X_1)+\mathcal{H}(X_2)-\mathcal{H}(X_1,X_2)=-\tfrac{1}{2}\log_2\left(1-\rho^2\right).$$

Discrepancies arises for the continuous variables and sequences. For example, $\mathcal{H}(X)<0$ if $2\pi e\sigma^2<1$.

Using quantitative probabilistic measures for discrete sequences, from (7.29) and (7.33), $H(X)\ge 0$, $I(X;Y)\ge 0$.

For two discrete random sequences $\{X_n,Y_n\}$, with the direct information flow $X_n\to Y_n$, from (7.33),

$I(X_n;Y_n)=H(X_n)+H(Y_n)-H(X_n,Y_n)$, $0\le I(X_n;Y_n)\le\min[H(X_n),H(Y_n)]$.

The mutual information rate for (X_n,Y_n) is

$$\hat{I}_{X_n\to Y_n}(X_n;Y_n)=\limsup_{n\to\infty}\tfrac{1}{n}I(X_n;Y_n).$$

For a finite alphabet, one estimates the information capacity (information rate) $\hat{I}_{X_n\to Y_n}(X_n;Y_n)$ to evaluate data sharing between two sources of information defined by data sequences in dynamic networks. ■

Relative Entropy – The relative entropy, which is the Kullback-Leibler divergence, measures a difference between two probability distributions over the same discrete random variable, characterized by pmfs $p(x)>0$ and $q(x)>0$ for any x in X. The divergence $D(P\|Q)$ from two distributions $P(x)$ to $Q(x)$, derived from observations, is

$$\begin{aligned} D(P\|Q) &= \sum_{x\in X} p(x)\log_2 \tfrac{p(x)}{q(x)} = -\sum_{x\in X} p(x)\log_2 \tfrac{q(x)}{p(x)} \\ &= -\sum_{x\in X} p(x)\log_2 q(x) + \sum_{x\in X} p(x)\log_2 p(x) = H(P,Q) - H(P), \end{aligned} \quad (7.37)$$

where $H(P,Q)$ is the cross-entropy loss function, $H(P,Q) = -\sum_{x\in X} p(x)\log_2 q(x)$.

The Kullback–Leibler divergence measures the information loss. A probability distribution $Q(x)$ is used to verify a probability distribution $P(x)$, verifying hypotheses $\hbar_0$ against $\hbar_1$, etc. A statistical expectation of the logarithm likelihood ratio in (7.37) is

$$D(P\|Q) = \mathbf{E}\left[\log_2 \tfrac{p(x)}{q(x)}\right].$$

The relative entropy is nonnegative, $D(P\|Q)\geq 0$.

Furthermore, $D(P\|Q)=0$ if and only if distributions are same, $P(x)=Q(x)$.

The studied measures provide systematic metrics to examine networks and Markov chains, as well as statistically characterize descriptive models.

Example 7.18. *Kullback-Leibler Divergence, Entropies and Relative Entropy*
Study the Kullback-Leibler divergence $D(P\|Q) = \int_X p(x)\log_2 \frac{p(x)}{q(x)}dx$.

Let the pdfs are $p(x) = \frac{1}{\sqrt{2\pi}\sigma_1} e^{-\frac{(x-\mu_1)^2}{2\sigma_1^2}}$ and $q(x) = \frac{1}{\sqrt{2\pi}\sigma_2} e^{-\frac{(x-\mu_2)^2}{2\sigma_2^2}}$.

From $\mathrm{var}(x) = \sigma^2 = \langle x^2\rangle - \langle x\rangle^2$ and $\langle x^2\rangle = \sigma^2 + \mu^2$,

$$D(P\|Q) = -\int p(x)\log_2 q(x)dx + \int p(x)\log_2 p(x)dx = \tfrac{1}{2}\left[\tfrac{(\mu_2-\mu_1)^2}{\sigma_2^2} + \tfrac{\sigma_1^2}{\sigma_2^2} - \log_2 \tfrac{\sigma_1^2}{\sigma_2^2} - 1\right].$$

If $\mu_1=\mu_2$ and $\sigma_1=\sigma_2$, we have $D(P\|Q)=0$.

For X=[1 2 3 4 5 6 7 8 9 10 11 12 13 14 15], let

$$P = \left[\tfrac{1}{20}\ \tfrac{1}{20}\ \tfrac{1}{10}\ \tfrac{1}{10}\ \tfrac{1}{10}\ \tfrac{1}{10}\ \tfrac{1}{10}\ \tfrac{1}{20}\ \tfrac{1}{20}\ \tfrac{1}{20}\ \tfrac{1}{20}\ \tfrac{1}{20}\ \tfrac{1}{15}\ \tfrac{1}{30}\ \tfrac{1}{20}\right],$$

$$Q = \left[\tfrac{1}{10}\ \tfrac{1}{10}\ \tfrac{1}{10}\ \tfrac{1}{20}\ \tfrac{1}{15}\ \tfrac{1}{30}\ \tfrac{1}{30}\ \tfrac{1}{15}\ \tfrac{1}{10}\ \tfrac{1}{10}\ \tfrac{1}{20}\ \tfrac{1}{30}\ \tfrac{1}{15}\ \tfrac{1}{15}\ \tfrac{1}{30}\right].$$

The entropies are $H(P)=3.8138$ and $H(Q)=3.7892$.

The divergences from P to Q, and, from Q to P, are different, $D(P\|Q)\neq D(Q\|P)$.

We have $D(P\|Q)=0.2799$ and $D(Q\|P)=0.2607$.

The cross-entropy loss function is $H(P,Q)=4.0937$. ■

Conditional Entropy and Information Losses – The conditional entropy $H(X|Y=y)$ of a random X, conditional on a particular realization y of Y, defines the expected conditional information content with respect to both X and Y.

One has $H(X)\geq H(X|Y)$.

We obtain $H(X|Y)=H(X)$ if X and Y are independent, yielding $I(X;Y)=0$.

As illustrated in Figure 7.12.b, the joint entropy (7.29) $H(X,Y)=H(X)+H(Y|X)=H(Y)+H(X|Y)$, defines the uncertainty associated with a set of variables. The conditional entropy $H(X|Y)$ characterizes the average loss of information

$$\mathcal{L}_I = H(X|Y),\ H(X|Y) = -\sum_{x\in X, y\in Y} p_{X,Y}(x,y)\log_2 p_{X|Y}(x|y),\ H(X)\geq H(X|Y). \quad (7.38)$$

This quantitative estimate may be used to examine information processing.

For a stationary time-invariant Markov process, the conditional entropy information rate $\hat{H}(X)$ is

$$\hat{H}(X) = \lim_{n\to\infty} H\left(X_n \middle| X_{n-1},\ldots,X_1\right). \tag{7.39}$$

If $(X_1,X_2,\ldots,X_n)$ are independent and identically distributed random variables, the entropy rate per symbol is

$$H(X) = \lim_{n\to\infty} \tfrac{1}{n} H\left(X_1,X_2,\ldots,X_n\right) = \lim \tfrac{1}{n} nH\left(X_1\right) = H\left(X_1\right). \tag{7.40}$$

For independent $(X_1,X_2,\ldots,X_n)$, not identically distributed, $H(X) = \lim_{n\to\infty} \tfrac{1}{n} \sum_{i=1}^{n} H\left(X_i\right)$. The entropy rates characterize sequences and estimate complexity.

Markov Chain and Entropies – Consider a Markov chain with a finite number of states $X=\{x_1,x_2,\ldots,x_n\}$ with a transition probability matrix $P\in\mathbb{R}^{n\times n}$ with $P_{i,j}$ for $(i,j)\in\{1,2,\ldots,n\}$, and, a stationary distribution π.

If the chain starts with its stationary distribution, the marginal distribution of all states at any time are the stationary distribution $\pi=\pi_i(x_i)$. Assuming irreducibility, the stationary distribution is unique if exists, and, its existence is due to recurrence of all states. If the system is in state x_i, then transitions to state x_k is characterized by a transition probability $p_{i,k}$. The transition from x_i to x_k, $x_i\to x_k$ is random y_i which satisfy $\{p_{i,1},p_{i,2},\ldots,p_{i,n}\}$, and, the entropy is $H\left(Y_i\right) = -\sum_{k=1}^{n} p_{i,k} \log p_{i,k}$.

The entropies of the Markov chain within a single step, and, z steps, starting at x_i, are the a single- and z-step entropies

$$H^{(1)} = \sum_{i=1}^{n} \pi_i H_i^{(1)} = -\sum_{i=1}^{n}\sum_{k=1}^{n} \pi_i p_{i,k} \log p_{i,k}, \; H^{(z)} = \sum_{i=1}^{n} \pi_i H_i^{(z)}, \; H^{(z)} = zH^{(1)}. \tag{7.40}$$

7.8. Dynamic-Probabilistic Analysis and Information Security Control

Fixed and variable length byte sequences, sequences of characters, strings, executables, memory maps, system calls traces, normalcy and anomalies have being investigated to manage information and data controls, find signatures, detect malware, acquire attack events, analyze security attributes, etc. Information management and malware detection schemes analyze strings, sequences, arrays and codes using static, dynamic and probabilistic estimates and measures. Information security controls are open problems.

For a known malware, the exiting algorithms yield different detection and classification capabilities depending on many factors, while unknown viruses may not be detected [8-17]. The security control is aimed to detect attacks and minimize security risks. It is important to characterize, describe, extract and systematically evaluate data behavior and governance.

The feature vectors may be assessed using classifiers based on identifiers, attributes $\mathcal{A}$, composite indicators $\boldsymbol{I}$ and measures $\mathcal{M}$. System functionality is assessed by estimating quantitative measures on dynamic evolutions, data management, information sharing, etc. Information losses, data heterogeneity, information homogeneity and other metrics are searched, extracted, assessed and evaluated. Applied spatiotemporal models, systematic probability analysis and classification algorithms should ensure near real time evaluation. Consider a tuple $(\mathcal{A},\boldsymbol{I},\mathcal{M})$. Evaluation of vulnerabilityand risk assessment should yield prioritization, reconfigurable defense strategies, adaptive security actions, etc.

System and Information Governance, and, Control – Governance of a system $\mathbf{M}(\mathbf{P}^{\text{physical}}\times\mathbf{C}^{\text{cyber}}\times\mathbf{D}^{\text{data}})$ is studied, and, descriptive organization-consistent hierarchical hybrid models are described as a network $\mathcal{N}\equiv\mathcal{N}^{\text{physical}}\cup\mathcal{N}^{\text{cyber}}\cup\mathcal{N}^{\text{data}}$ with statistically-characterized dynamic nodes (modules, components and devices), which are affected by actions, events and activities. Analog and digital outputs $y=\begin{bmatrix} y(t) \\ y[n] \end{bmatrix}$, states $x=\begin{bmatrix} x(t) \\ x[n] \end{bmatrix}$ and controls $u=\begin{bmatrix} u(t) \\ u[n] \end{bmatrix}$ are unions of real-valued physical, cyber and data variables. Distortions and disturbances $d=\begin{bmatrix} d(t) \\ d[n] \end{bmatrix}$, as well as perturbations $w=\begin{bmatrix} w(t) \\ w[n] \end{bmatrix}$, are considered.

The $(\boldsymbol{y},\boldsymbol{x},\boldsymbol{u})$ are affected by the electromagnetic spectrum, malicious tampering, data modification, as well as malicious control actions and activities $(\boldsymbol{d_a},\boldsymbol{w_a})$.

One has $\boldsymbol{d}=\begin{bmatrix} d(t) \\ d_a(t) \\ d[n] \\ d_a[n] \end{bmatrix}$ and $\boldsymbol{w}=\begin{bmatrix} w(t) \\ w_a(t) \\ w[n] \\ w_a[n] \end{bmatrix}$.

Dynamic-Probabilistic Models – Physical states $\boldsymbol{x}^{\text{physical}}$ map system dynamics, $\boldsymbol{x}^{\text{cyber}}$ quantitatively characterize information governance under cyberattacks, and, $\boldsymbol{x}^{\text{data}}$ describe data governance under uncertainties, heterogeneities, electromagnetic spectrum, etc. These $(\boldsymbol{x}^{\text{cyber}},\boldsymbol{u}^{\text{cyber}})$ and $(\boldsymbol{x}^{\text{data}},\boldsymbol{u}^{\text{data}})$ describe governance, transitions and control of dynamic information and data evolutions, sharing, processing (coding, encoding, encrypting, decrypting, fetching, retreating, etc.), logic and arithmetic operations, etc. System hierarchy and organization affect vulnerabilities and threats. High-fidelity and dimensionality analyses, parameterization, topological evolution and dynamics are examined by using descriptive states $\boldsymbol{x}$, controls $\boldsymbol{u}$, disturbances $\boldsymbol{d}$ and perturbations $\boldsymbol{w}$.

Physical systems are affected by information management and malicious activities. One has

$$\begin{bmatrix} \dot{x}(t) \\ \boldsymbol{x}_{n+1} \end{bmatrix}=\begin{bmatrix} F_t(t,\ \underbrace{\boldsymbol{x},\boldsymbol{d},\boldsymbol{w},}_{\substack{\text{States, Distortions} \\ \text{and Perturbations}}}\ \underbrace{\boldsymbol{d}_a,\ \boldsymbol{w}_a}_{\substack{\text{Cyber Attacks} \\ \text{Malware Actions} \\ \text{Electromagnetic Spectrum}}},\ \underbrace{\boldsymbol{u}}_{\text{Control}}) \\ F_n(n,\ \underbrace{\boldsymbol{x}_n,\boldsymbol{d}_n,\boldsymbol{w}_n,}_{\substack{\text{States, Distortions} \\ \text{and Perturbations}}}\ \underbrace{\boldsymbol{d}_{a_n},\ \boldsymbol{w}_{a_n}}_{\substack{\text{Cyber Attacks} \\ \text{Malware Actions} \\ \text{Electromagnetic Spectrum}}},\ \underbrace{\boldsymbol{u}_n}_{\text{Control}}) \end{bmatrix},\ y=H(\boldsymbol{x}).$$

High-fidelity analysis implies consideration of nonlinear dynamic-probabilistic models, dynamic networks $\mathcal{N}$, heterogeneous Markov chains, etc. Stochastically-perturbed dynamic models support and enable analysis of probabilistic finite-state machines, finite-state automata, Markov models, etc. Analysis in physical domain and information space are addressed by using high-dimensional topologically-evolving graphs $\mathcal{G}$ and dynamic networks $\mathcal{N}$. Descriptive probabilistic models are $\mathcal{R}_X=(\mathcal{S},\mathcal{D})$.

Considering physical processes, services, data sharing, asynchronous information processing, electromagnetic spectrum and potential cyberattacks, statistically-characterized model is

$$\mathcal{G}\equiv\boldsymbol{\phi}_{\mathcal{G}}(\mathcal{V},\mathcal{E},A^{(M,C)}),\ \mathcal{N}\equiv\boldsymbol{\phi}_{\mathcal{M}}(\mathcal{G},\boldsymbol{x},\boldsymbol{u},\boldsymbol{d},\boldsymbol{w},\boldsymbol{\mu},F_t(\cdot),F_n(\cdot),(\mathcal{S},\mathcal{P})),$$

$$\mathcal{N}\equiv\mathcal{N}^{\text{physical}}\cup\mathcal{N}^{\text{cyber}}\cup\mathcal{N}^{\text{data}}, \tag{7.42}$$

Statistical Models $\mathcal{R}_X=(\mathcal{S},\mathcal{D}),\ \mathcal{R}_{a^*}=(\mathcal{S}_{a^*},\mathcal{D}_{a^*})$,

$$\mathbf{M}_{\mathcal{N}} := \begin{bmatrix}\dot{\boldsymbol{x}}(t)\\ \boldsymbol{x}_{n+1}\\ \boldsymbol{x}_{\mathrm{T}_{n+1}}\\ \boldsymbol{x}_{\mathcal{M}_{n+1}}\end{bmatrix} = \begin{bmatrix} \overbrace{\underbrace{\sum_{i,j}\mathrm{P}_{ij}^{\mathcal{N}}F_{ij}\left(\begin{bmatrix}x(t)\\ x_n\\ x_{\mathrm{T}_n}\\ x_{\mathcal{M}_n}\end{bmatrix},\begin{bmatrix}u(t)\\ u_n\\ u_{\mathrm{T}_n}\end{bmatrix},\begin{bmatrix}d(t)\\ d_a(t)\\ d_n\\ d_{a_n}\end{bmatrix},\begin{bmatrix}w(t)\\ w_a(t)\\ w_n\\ w_{a_n}\end{bmatrix}\right)}_{\substack{\text{Fixed System Structure:}\\ \text{System Control and Information Transitions}}}+\underbrace{\sum_{k,l}\mathrm{P}_{kl}^{\mathcal{N}}F_{kl}\left(\begin{bmatrix}x(t)\\ x_n\\ x_{\mathrm{T}_n}\\ x_{\mathcal{M}_n}\end{bmatrix},\begin{bmatrix}u(t)\\ u_n\\ u_{\mathrm{T}_n}\end{bmatrix},\begin{bmatrix}d(t)\\ d_a(t)\\ d_n\\ d_{a_n}\end{bmatrix},\begin{bmatrix}w(t)\\ w_a(t)\\ w_n\\ w_{a_n}\end{bmatrix}\right)}_{\substack{\text{Reconfigurable Adaptation:}\\ \text{Optimization, Resiliency and Security}}}}^{\text{System and Information Governance}} \\ F_{\mathrm{T}}\left(\begin{bmatrix}x_n\\ x_{\mathrm{T}_n}\end{bmatrix},\begin{bmatrix}u_n\\ u_{\mathrm{T}_n}\end{bmatrix},\begin{bmatrix}d_n\\ d_{a_n}\end{bmatrix},\begin{bmatrix}w_n\\ w_{a_n}\end{bmatrix},\boldsymbol{a},X_{\boldsymbol{a}},\mathcal{O},\boldsymbol{\tau}_{\boldsymbol{a}},\mathbf{P}_{\boldsymbol{\tau}},\boldsymbol{c}_{\boldsymbol{a}},\boldsymbol{a}_{\boldsymbol{\tau}}\right) \\ F_{\mathcal{M}}\left(\begin{bmatrix}x(t)\\ x_n\\ x_{\mathrm{T}_n}\\ x_{\mathcal{M}_n}\end{bmatrix},\begin{bmatrix}u(t)\\ u_n\\ u_{\mathrm{T}_n}\end{bmatrix},\boldsymbol{a}_T\right)\end{bmatrix}.$$

$$y=H(\boldsymbol{x}),\ y=\begin{bmatrix}y(t)\\ y_n\\ y_{\mathrm{T}_n}\end{bmatrix}$$

Reconfigurable adaptive control and optimization schemes are sought on a class of admissible solutions considering organization, reconfigurability, intrusion detection, etc. Functionality, compliance, topological stability, connectivity and data sharing should be ensured by means of adequate component selection, secure control and information processing schemes, reconfigurable data management, state measurements, events detection, etc.

Example 7.19. *Bayesian Statistics*

The chain rule for random events A_i is

$$\mathbf{P}(A_n\cap\ldots\cap A_1)=\mathbf{P}(A_n\,|\,A_{n-1}\cap\ldots\cap A_1)\cdot\mathbf{P}(A_{n-1}\cap\ldots\cap A_1)=\prod_{i=1}^{n}\mathbf{P}(A_i|\bigcap_{j=1}^{i-1}A_j)\cdot$$

For random variables $(X_1,\ldots,X_n)$, to derive the joint distribution, the conditional probability is

$$\mathbf{P}(X_n\cap\ldots\cap X_1)=\mathbf{P}(X_n\,|\,X_{n-1},\ldots,X_1)\cdot\mathbf{P}(X_{n-1},\ldots,A_1),\ \mathbf{P}(\bigcap_{i=1}^{n}X_i)=\prod_{i=1}^{n}\mathbf{P}(X_i|\bigcap_{j=1}^{i-1}X_j)\cdot$$

The conditional probability of A_1, given A_2 occurs, is $\mathbf{P}(A_1\,|\,A_2)=\dfrac{\mathbf{P}(A_1\cap A_2)}{\mathbf{P}(A_2)}$, $\mathbf{P}(A_2)>0$.

For the probability that both A_1 and A_2 occur $\mathbf{P}(A_1\cap A_2)$, the multiplication rule is $\mathbf{P}(A_1\cap A_2)=\mathbf{P}(A_1\,|\,A_2)\mathbf{P}(A_2)$. If (A_1,A_2) are independent, $\mathbf{P}(A_1\cap A_2)=\mathbf{P}(A_1)\mathbf{P}(A_2)$ and $\mathbf{P}(A_1|A_2)=\mathbf{P}(A_1)$. For mutually exclusive (A_1,A_2), $\mathbf{P}(A_1\cap A_2)=0$. For disjoint (A_1,A_2), $A_1\cap A_2=\varnothing$, and, $\mathbf{P}(A_1|A_2)=0$.

It is difficult to calculate $\mathbf{P}(A_2)$. To estimate $\mathbf{P}(A_1|A_2)\propto\mathbf{P}(A_2|A_1)\mathbf{P}(A_1)$, a product of the prior probability and likelihood can be applied. This hypothesis may not ensure adequacy, and, likelihood may not yield consistent evaluation. If $\mathbf{P}(A_1)=0$, $\mathbf{P}(A_2)=0$,

$\mathbf{P}(A_2|A_1)$ and $\mathbf{P}(A_1|A_2)$ are not defined. It is difficult to estimate $\mathbf{P}(A_1 \cap A_2)$. For random variables (X_1, X_2) which are nondegenerate with the pdf $f_{X_1,X_2}(x_1,x_2)$

$$\mathbf{P}(X_1 \in A_1 \mid X_2 \in A_2) = \frac{\int_{x_2 \in A_2}\int_{x_1 \in A_1} f_{X_1,X_2}(x_1,x_2)dx_1dx_2}{\int_{x_2 \in \mathbb{R}}\int_{x_1 \in \mathbb{R}} f_{X_1,X_2}(x_1,x_2)dx_1dx_2}, \mathbf{P}(X_1 \in A_1 \mid X_2 = x_{2_0}) = \frac{\int_{x_1 \in A_1} f_{X_1,X_2}(x_1,x_{2_0})dx_1}{\int_{x_1 \in \mathbb{R}} f_{X_1,X_2}(x_1,x_{2_0})dx_1},$$

where $X_2 = x_{2_0}$ is the single-point.

Consider an event $\mathbf{A}$ and a set of pairwise disjoint measurable events A_i whose union is an entire sample space Ω. One has $\mathbf{P}(\mathbf{A}) = \sum_{i=1}^{n} \mathbf{P}(\mathbf{A} \cap A_i) = \sum_{i=1}^{n} \mathbf{P}(\mathbf{A}|A_i)\mathbf{P}(A_i)$.

The conditional probability of an event $\mathbf{A}$, given events A_i occurred with degrees a_i, is

$$\mathbf{P}(\mathbf{A} \cap A_1 \equiv a_1, \ldots, A_n \equiv a_n) = \sum_{i=1}^{n} a_i \mathbf{P}(\mathbf{A} \cap A_i).$$

Consider a finite probability space Ω with elements $(\omega_1, \ldots, \omega_n)$. For a subset A of the sample space Ω with probability $\mathbf{P}(A)=p$, the probability that A occurs in k of N trials is

$$\mathbf{P}(A)\big|_{0 \le k \le N} = \binom{N}{k} \cdot p^k (1-p)^{N-k},$$

where k is the integer, $0 \le k \le N$, $N>0$.

Figures 7.13.a reports $(\mathbf{A}, A_1, A_2, A_3, A_4, A_5)$ in a sample space Ω, as well as a graph $\mathcal{G}(\mathbf{A} \cap A_1)$. The conditional probabilities are

$\mathbf{P}(\mathbf{A}|A_1) = \frac{0.04}{0.04+0.06} = 0.4$, $\mathbf{P}(\mathbf{A}|A_2)=1$, $\mathbf{P}(\mathbf{A}|A_3)=0$, $\mathbf{P}(\mathbf{A}|A_4) = \frac{0.06}{0.06+0.04} = 0.6$, $\mathbf{P}(\mathbf{A}|A_5)=0$.

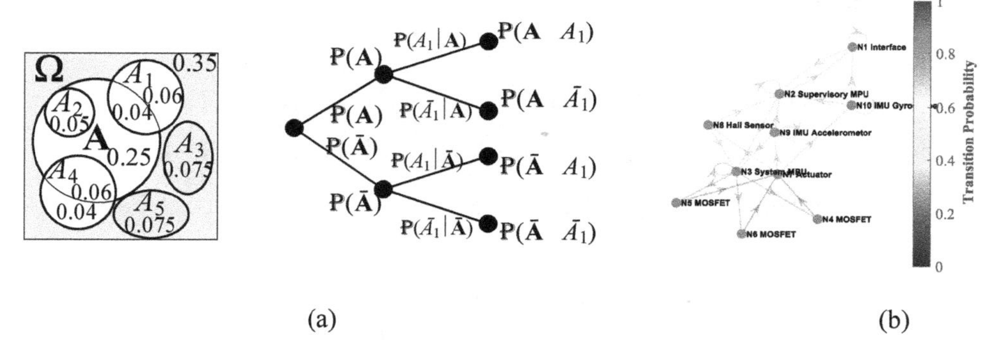

Figure 7.13. Sample space Ω and $\mathcal{G}(\mathbf{A} \cap A_1)$;
(b) Transition probabilities in a physical system $p_{ij} \ge 0$, Example 7.21. ■

Büchi Automata – Evidence subgraphs, graphs and networks describe intrusion evidence, electromagnetic spectrum dependencies, threats and anomalies within a pertained analysis. The *model checking* algorithms apply hypothesis, analyze temporal logic, and, solve estimation problems with varying coefficients and reconfigurable model structures. The Büchi automata quintuple is

$$\mathcal{B}=(Q, \sigma(\sigma_0, \sigma_f), a_{\mathcal{B}}, \tau_{\mathcal{B}}, c_{\mathcal{B}}), \tag{7.43}$$

where Q is a finite set of states σ with initial and final states (σ_0, σ_f); $a_{\mathcal{B}}$ is a fine alphabet set; $\tau_{\mathcal{B}}$ is the transition function; $c_{\mathcal{B}}$ is a finite set of the admissible acceptance conditions.

Transition Matrix and Markov Chain – Cyber-physical systems should ensure measure–manage–and–control capabilities in dynamic data-driven applications by means of information collection, sharing, processing and control. Within information paths in systems (7.42) $\mathbf{M}_{\mathcal{N}}$, consider input-output transductions $(R,D,X,U)\rightarrow Y$. Discrete X_n, $n\geq 0$ on $\mathbf{X}$ is a collection of the real-valued random $X_n\in\mathbf{X}$ defined on a probability space $(\Omega,\mathcal{F},\mathbf{P})$. Evaluate the transition matrix $P=[p_{ij}]$, $p_{ij}\geq 0$ on a family of events $\mathcal{F}$ in Ω. Finite-dimensional distributions of processes are analyzed, and, probabilities of events are $\mathbf{P}\{X_0=i_0,\ldots i_k,X_n=i_n\}$, $(i_0,\ldots,i_k)\in\mathbf{X}$, $n\geq 0$.
Two X_i and X_j, defined on same or different probability spaces, are characterized. Stochastic X on $\mathbf{X}$ is a homogeneous Markov chain if for all states $(i,j)\in\mathbf{X}$ at $n\geq 0$

$$\mathbf{P}\{X_{n+1}=j|\ X_n=i,\ X_{n-1}=i_{n-1},\ \ldots,\ X_0=i_0\}\equiv\mathbf{P}\{X_{n+1}=j|\ X_n=i\}=p_{ij},\ \forall(n,i,j),\qquad(7.44)$$

where p_{ij} is the Markov chain transitions probability from state i to state j.

For a one-step transition probability, $p_{ij}(n)=\mathbf{P}\{X_{n+1}=j|\ X_n=i\}$, $p_{ij}\in[0,\ 1]$, $\forall(i,j)$.
Define the *m-step* transition probabilities from n to $(n+m)$ as

$$p_{ij}(n,n+m)=\mathbf{P}\{X_{n+m}=j|\ X_n=i\}.\qquad(7.45)$$

The Markov transition probabilities satisfy

$$\sum\nolimits_j p_{ij}(n)=1,\ p_{ij}(n)\geq 0,\ (i,j)\in T\times\mathbf{X}.\qquad(7.46)$$

A one-step transition probability matrix of the chain is $P=[p_{ij}]$, $(i,j)\in\mathbf{X}$.

At any n, the next state X_{n+1} is conditionally independent of the past $(X_0,\ldots,X_{n-1})$, given the present state X_n. Hence, next state is dependent on the past and present only through the present state. An unconditional distribution at n is

$$\pi_i(n)=\mathbf{P}\{X_n=i\},\ i\in\mathbf{X}.\qquad(7.47)$$

The row vector of P is $\pi_i(n)=[\pi_0(n)\ \pi_i(n),\pi_1(n),\ldots,\pi_i(n),\ldots]$.

A finite-state discrete Markov chains is a quintuple

$$\mathcal{M}=(X_{\mathcal{M}},x_{\mathcal{M}}(x_{\mathcal{M}0},x_{\mathcal{M}f}),\mathbf{P},p_{\mathcal{M}},\mathfrak{a}_T),\qquad(7.48)$$

where $X_{\mathcal{M}}$ is the finite set of states $x_{\mathcal{M}}$ with initial and final states $(x_{\mathcal{M}0},x_{\mathcal{M}f})$, and, transitions between states are defined by probabilities and admissible alphabet; $\mathbf{P}$ is the transition probability, $\mathbf{P}: X_{\mathcal{M}}\times\mathfrak{a}_T\times X'_{\mathcal{M}}\rightarrow[0,1]$, $\sum_{(x_{\mathcal{M}},a)\in X_{\mathcal{M}}\times\mathfrak{a}_T} p_{ij}(x_{\mathcal{M}},\boldsymbol{a},x'_{\mathcal{M}})=1$, $\forall x_{\mathcal{M}}\in X_{\mathcal{M}}$, $\forall\boldsymbol{a}\in\mathfrak{a}_T$; $p_{\mathcal{M}}$ is the distribution on $X_{\mathcal{M}}$; $\mathfrak{a}_T$ is the transition alphabet.

Optimization, Control and Information Governance – A system $\mathbf{M}(\mathbf{P}^{\text{physical}}\times\mathbf{C}^{\text{cyber}}\times\mathbf{D}^{\text{data}})$ with descriptive Turing machine $\mathbf{T}(\cdot)$ (7.3), system governance (7.42), statistical model $\mathcal{R}_X=(\mathcal{S},\mathcal{D})$, Markov model $\mathcal{M}$ (7.48) with a finite path and $\mathbf{P}^{\mathcal{M}(\mathfrak{a}_T)}$, admits a descriptive analyses, dynamic-probabilistic descriptions and *model checking*.

For $\mathbf{M}_{\mathcal{N}\cup\mathrm{T}\cup\mathcal{M}}$, investigating a minimax (7.2) problem, consider

$$\boldsymbol{q}(\cdot)=\min_{\substack{\mathcal{M}_{\mathcal{L}}\\ q\in Q\subseteq\mathcal{M}\times\mathbf{X}\times\mathbf{U}}}\max_{\mathcal{M}_{\mathcal{H}}}\ \kappa(t,\mathcal{M},\boldsymbol{x},\boldsymbol{u}),\ \boldsymbol{q}\subseteq\mathbb{R}^p,\qquad(7.49)$$

Dynamic Network $\mathcal{N} \equiv \mathcal{N}^{\text{physical}} \cup \mathcal{N}^{\text{cyber}} \cup \mathcal{N}^{\text{data}}$ with $[\boldsymbol{y}, \boldsymbol{x}, \boldsymbol{u}, \boldsymbol{d}, \boldsymbol{w}]$,

Turing Machine $\mathrm{T} = (X_\mathrm{T}, x_\mathrm{T}, \mathcal{A}, X_a, \mathcal{O}, \tau_a, \mathbf{P}_\tau, c_a, p_a, \mathcal{A}_\tau, u_\mathrm{T})$, $\mathbf{P}_\tau : X_T \times \mathcal{A}_\tau \times X_T' \to [0, 1]$,

Markov Model $\mathcal{M} = (X_\mathcal{M}, x_\mathcal{M}, \mathbf{P}, p_\mathcal{M}, \mathcal{A}_T)$, $\mathbf{P} : X_\mathcal{M} \times \mathcal{A}_T \times X_\mathcal{M}' \to [0, 1]$,

Dynamic-Probabilistic Model $\mathcal{R}_X = (\mathcal{S}, \mathcal{D})$, $\mathcal{R}_{a^*} = (\mathcal{S}_{a^*}, \mathcal{D}_{a^*})$,

$$\mathbf{M}_{\mathcal{N}\cup\mathrm{T}\cup\mathcal{M}} := \begin{bmatrix} \dot{x}(t) \\ x_{n+1} \\ x_{\mathrm{T}_{n+1}} \\ x_{\mathcal{M}_{n+1}} \end{bmatrix} = \left[\begin{array}{c} \overbrace{\underbrace{\sum_{i,j} \mathrm{P}_{ij}^{\mathcal{N}} F_{ij} \left(\begin{bmatrix} x(t) \\ x_n \\ x_{\mathrm{T}_n} \\ x_{\mathcal{M}_n} \end{bmatrix}, \begin{bmatrix} u(t) \\ u_n \\ u_{\mathrm{T}_n} \end{bmatrix}, \begin{bmatrix} d(t) \\ d_a(t) \\ d_n \\ d_{a_n} \end{bmatrix}, \begin{bmatrix} w(t) \\ w_a(t) \\ w_n \\ w_{a_n} \end{bmatrix} \right)}_{\substack{\text{Fixed System Structure:} \\ \text{System Control and Information Transitions}}} + \underbrace{\sum_{k,l} \mathrm{P}_{kl}^{\mathcal{N}} F_{kl} \left(\begin{bmatrix} x(t) \\ x_n \\ x_{\mathrm{T}_n} \\ x_{\mathcal{M}_n} \end{bmatrix}, \begin{bmatrix} u(t) \\ u_n \\ u_{\mathrm{T}_n} \end{bmatrix}, \begin{bmatrix} d(t) \\ d_a(t) \\ d_n \\ d_{a_n} \end{bmatrix}, \begin{bmatrix} w(t) \\ w_a(t) \\ w_n \\ w_{a_n} \end{bmatrix} \right)}_{\substack{\text{Reconfigurable Adaptation:} \\ \text{Optimization, Resiliency and Security}}}}^{\text{System and Information Governance}} \\ F_\mathrm{T}\left(\begin{bmatrix} x_n \\ x_{\mathrm{T}_n} \end{bmatrix}, \begin{bmatrix} u_n \\ u_{\mathrm{T}_n} \end{bmatrix}, \begin{bmatrix} d_n \\ d_{a_n} \end{bmatrix}, \begin{bmatrix} w_n \\ w_{a_n} \end{bmatrix}, \mathcal{A}, X_a, \mathcal{O}, \tau_a, \mathbf{P}_\tau, c_a, \mathcal{A}_\tau \right) \\ F_\mathcal{M}\left(\begin{bmatrix} x(t) \\ x_n \\ x_{\mathrm{T}_n} \\ x_{\mathcal{M}_n} \end{bmatrix}, \begin{bmatrix} u(t) \\ u_n \\ u_{\mathrm{T}_n} \end{bmatrix}, \mathcal{A}_T \right) \end{array} \right].$$

$$y = H(x),\ y = \begin{bmatrix} y(t) \\ y_n \\ y_{\mathrm{T}_n} \end{bmatrix}$$

Example 7.20. *Quantified Prognostics*

Quantitative estimates, systematic metrics, attack probabilities, risk assessment and vulnerabilities were studied in [8-16]. The probability of successful attacks and vulnerabilities are difficult to estimate considering multi-vector attacks, different paths, mean time to compromise, etc. There are subjectivity on attack likelihood estimates, assumptions to characterize risks, deficiencies of estimations, intrusion detection inadequacy, etc. For systems and modules, one examines attack types ($\mathcal{T}_1, \ldots, \mathcal{T}_n$), probabilities $\mathbf{P}_i(\mathcal{T}_i)$, vulnerability $\mathbf{V}_i(\mathcal{T}_i)$, etc. Probabilistic risk and vulnerability assessments may be investigated using quantified prognostics on attack probabilities $\mathbf{P}_i \times \mathbf{V}_i$. This approach can be tested and evaluated considering the exposure to attacks of sensor networks, distributed computing, etc. ■

Example 7.21. *Markov Model and Analysis For a Cyber-Physical Module*

Consider a permanent-magnet synchronous motor, PWM controller-driver, Hall-effect sensors, MPU, transceiver, as well as inertial MDOF sensors to evaluate vertical and lateral motions, vibrations, asymmetry, etc. These modules are vulnerable to cyberattacks and electromagnetic spectrum. These servos have being used in aerial, automotive, robotic and other systems. Within a formal description (7.49), information processing and sharing are performed within the following scheme

Servo Control ↔ Sensing ↔ Data Control ↔ Information Management ↔ Information Sharing ↔ Evaluation ↔ Execution.

Conducting low-fidelity studies for $\mathcal{N} \equiv \mathcal{N}^{\text{physical}} \cup \mathcal{N}^{\text{data}}$.

Consider a system with the following nodes:

Node 1 – Interface and communication;

Node 2 – Supervisory management system;

Node 3 – C2000 MPU with 32-bit core C28x CPU which controls a synchronous motor by regulating the MPU's PWM outputs which drive the power MOSFETs of an amplifier;

Nodes 4, 5 and 6 – Power MOSFET stages;
Node 7 – Three-phase permanent magnet synchronous motor;
Node 8 – Hall-effect sensors;
Nodes 9 and 10 – System-in-package accelerometer and gyroscope sensors.

Experimental studies are performed. Physical and data states are measured. Transition probabilities are evaluated to compute a matrix $P=[p_{ij}]$, $p_{ij}\geq 0$. For finite-state state transductions, the transition probabilities between nodes $(i,j)\in\mathbf{X}$ represent the Markov chain with states $x_{\mathcal{M}}$. The probabilities p_{ij} are estimated, and, depicted in Figure 7.13.b. Connectivity, reachability and accessibility of $(i,j)\in\mathbf{X}$ are considered studying services, interface, information sharing and control. An adequate organization ensures measurability and accessibility of states, such as $i\rightarrow j$, $i\leftarrow j$ and $i\leftrightarrow j$. For the asynchronous sensors data rate from 2000 to 8000 Hz, probabilistic measures and distributions are found.

Finding the mutual information (7.35) $I(X_{1:n};Y)=\sum_{i=1}^{n} I\left(X_i;Y|X_{1:i-1}\right)$, estimations of data homogeneity $\mathcal{M}_{\mathcal{H}}(\cdot)$ and data heterogeneity $\mathcal{M}_{\mathcal{L}}(\cdot)$ may take seconds. To evaluate data integrity and data uniformity in near real time, one may reduce complexity, as well as evaluate the lower and upper bounds on $I(X_{1:n};Y)$ for critical components. ■

Example 7.22. *Axiomatizable Logic and Propositional Calculus*
For $(\mathcal{A},\boldsymbol{I},\mathcal{M})$, using the computable probabilistic measures $\mathcal{M}(\cdot)$, consider a sets $A_{\mathcal{A}}\times A_I\times A_{\mathcal{M}}$, $A=\bigcup_i A_i$. For $U\subseteq V\subseteq W$, one may define the V which stipulates true if W_i is in U, while false if W_j is not in U, $\phi_V^U \equiv \bigwedge_{W_i\in U} W_k \wedge \bigwedge_{W_i\in V\setminus U} \neg W_i$.

Apply logical operations, such as conjunction $\wedge$, disjunction $\vee$, exclusive disjunction $\oplus$, negation $\neg$, contradiction $\perp$, etc. Probabilistic reasoning is accomplished using the probability calculus considering dependencies, exclusiveness, relevance, sufficiency of measures, estimates, conditional logic, etc. For example, $A\Rightarrow A_1\vee A_2\wedge(\neg A_3)$ and $A\Rightarrow A_1\vee(\neg A_2)\wedge A_3$ yield truth tables. Consider a complete set of countable propositions $\Sigma\subseteq\text{Prop}(A)$ with a descriptive mappings $\mathbf{\Sigma}$ which describes system status, activities, etc.

Applying the ordered $\bigcup_i A_i$, perform computably axiomatizable operations on $A\cup\mathbf{\Sigma}$ by using a logic algorithm $\mathcal{A}$ yielding a decidable D

$$\{\mathcal{N},\mathcal{M},A\cup\mathbf{M}_{\mathcal{N}\cup\mathcal{M}},\mathcal{L}\}\rightarrow\text{D}$$

where $\mathcal{L}$ are the logic operations, such as $\mathcal{L}\equiv[\wedge,\vee,\oplus,\neg,\perp,=,\exists,\forall]$.

Finite- and infinite-valued logic can be applied. Finite-valued Łukasiewicz logic or Gödel logic can be simplified to three-, four- and multiple-valued logic. For the Łukasiewicz logic, strong and weak conjunctions $(\&,\wedge)$ are defined by the truth degree functions $(u\&v):=\max\{0,(u+v-1)\}$, $(u\wedge v):=\min\{u,v\}$, negation connective $\neg u:=(1-u)$, as well as an implication connective $(u\rightarrow v):=\min\{1,(1-u+v)\}$.

The logical matrix in the interval [0, 1] is $L_m=\left\{\frac{k}{m-1}\,\middle|\, 0\leq k\leq(m-1)\right\}$. ■

7.9. Data Encryption and Decryption Algorithms

Basic and applied research in data encryption and decryption is critical for multi-domain IoT and CPS. Wireless communication is enabled by dynamic reconfiguration with software-defined capabilities to ensure resiliency and secure data sharing. There are different cryptographic algorithms, such as:

1. AES symmetric key block cipher, 128-, 192- and 256-bit key length, 128-bit block size [3-5];
2. Blowfish symmetric key block cipher, 32- to 448-bit key length, 64-bit block size;
3. Data Encryption Standard;
4. Rivest-Shamir-Adleman asymmetric cryptographic algorithm; etc.

Information security and robust information sharing are critical for ICS, SCADA and other platforms which should resiliently operate in adverse environments and electromagnetic spectrum. Strength, encryption and decryption time, memory used, algorithmic operations, complexity, and other factors are considered. There are advantages of the AES algorithm. Examine encryption and decryption algorithms, as well as investigate a secure wireless communication to advance CPS capabilities [22-28]. The AES algorithms are implemented using MPUs and FPGAs [25, 26, 29]. The Galois field calculus is applied to analyze, optimize and design reconfigurable secure information sharing schemes [24, 25, 30-33]. These results are applicable in multi-domain wireless, free-space optical, radio-frequency and hybrid communication. Computationally-efficient algorithms are empowered by data-intensive processing and high-performance computing.

High data rate secure information sharing depends on progress in cross-cutting engineering science. Fundamental, applied and experimental results advance synergetic system-level integration of cryptography and algorithms. Algorithms, processor architecture and processor datapath are analyzed and optimized by applying Galois field operations. Cryptographic encryption and decryption schemes and computationally-effective algorithms are designed using the Galois field calculus. Front-end microelectronics supports communication schemes, processing and algorithms, yielding a high-throughput and resilient *connect – secure sharing – data management – applications*. Affordability, enhanced functionality, compliance, integration, modularity and multiple modalities are empowered by advanced-technology processors and transceivers in multi-level hierarchical distributed secure information sharing architectures.

Joan Daemen and Vincent Rijmen developed a symmetric-key block cipher AES algorithm, used by industry and government for classified data sharing [3, 4]. To ensure authentication, microcontrollers, FPGAs, desktops, tablets and phones have a unique identification number (UID), which is usually a 256-bit AES key, integrated into a secure coprocessor as device manufactured. This AES key is protected against physical attacks, and, cannot be read by software or firmware. Encryption is performed by using a randomly generated key, itself encrypted with a key derived from the UID.

The AES uses a fixed 128-bit block ciphers with cryptographic 128-, 192- or 256-bit key. The algorithm has 10 rounds for 128-bit keys, 12 rounds for 192-bit keys, and, 14 rounds for 256-bit keys. The substitution-permutation operations are expressed using a finite Galois field, describing encryption and decryption schemes to optimize pipelined parallel processing architectures for compact implementation on MPUs and

FPGAs. Applying the *encryption operation* $\mathcal{E}$ with an encryption key $\mathcal{K}$ on a *plaintext* $\mathcal{T}$, the encryption yields a *ciphertext* $C=\mathcal{E}_{\mathcal{K}}(\mathcal{T})$. The decryption yields $\mathcal{T}=\mathcal{D}_{\mathcal{K}}(\mathcal{E}_{\mathcal{K}}(\mathcal{T}))=C$, where $\mathcal{D}$ is the *decryption operation*. These operations

$$C=\mathcal{E}_{\mathcal{K}}(\mathcal{T}),\ \mathcal{T}=\mathcal{D}_{\mathcal{K}}(\mathcal{E}_{\mathcal{K}}(\mathcal{T}))=C, \qquad (7.50)$$

are described by the Galois field $GF(p^m)$, $p\geq 2$, and, implemented using MPUs or FPGAs. The ASICs can be designed and verified applying $GF(p^m)$. The Galois field $GF(p^m)$ is described by polynomials of degree $(m-1)$ over the field $\mathbf{Z}_p$. In polynomial

$$p(x)=a_{m-1}x^{m-1}+\ldots+a_1x_1+a_0x_0, \qquad (7.51)$$

the coefficients a_i take values in the set $\{0,1,\ldots,p-1\}$.

The polynomial addition, multiplication, inverse and other operations are described in $GF(p^m)$. For Boolean algebra and binary circuits [34], $p=2$. For $GF(2^m)$, the degree of polynomials d_p is $d_p\leq(m-1)$, and, the elements are the m-bit strings. Each bit corresponds to the polynomial coefficient.

Example 7.23. The $GF(2^3)$ has eight elements $\{0, 1, x, x+1, x^2, x^2+1, x^2+x, x^2+x+1\}$. Here, $(x+1)$ is $0x^2+1x+1$, or a bit string '011', while $(x^2+x)=1x^2+1x+0$, or '110'.

Hence, a field $GF(2^3)$ has eight elements, yielding 3-bit numbers.

The $g(x)=x^3+x+1$ as an irreducible polynomial, and,

$g(x)=x^3+x+1\equiv[1\ 0\ 1\ 1]=11_{\text{decimal}}$.

Numbers 3 ('11') and 5 ('101') correspond to the polynomials $(x+1)$ and (x^2+1). Sum of 3 and 5 is (x^2+x) because $1+1=0$ mod 2. Hence, the sum is 6. To compute the product of 5 and 6, multiply $(x^2+1)(x^2+x)$. Dividing by (x^3+x+1) mod 2, we have a remainder $(x+1)$. Hence, the product is 3. ■

Example 7.24. Applying modulo 2 arithmetic, performed by digital operations on binary numbers, addition and subtraction are performed using an exclusive OR (XOR) operation on the corresponding binary digits of each operand

$-1\equiv 1$ mod 2, $0\pm 0\equiv 0$ mod 2, $1\pm 0\equiv 1$ mod 2, and, $1\pm 1\equiv 0$ mod 2.

Using the XOR operations, $0\oplus 0=0$, $1\oplus 0=1$ and $1\oplus 1=0$.

For binary polynomials, addition is a bit-by-bit XOR operation.

For example, $(x^2+x+1)+(x+1)=x^2+2x+2$, where $2\equiv 0$ mod 2. Hence, $(x^2+x+1)+(x+1)=x^2$ mod 2.

One has $111\oplus 011=100$, where 100 is the bit string representation of x^2. ■

Multiplication of binary polynomials is performed using the bit-shift and XOR operations. The product of polynomials $A(x)=\sum_{i=0}^{n}a_ix^i$ and $B(x)=\sum_{j=0}^{m}b_jx^j$ is

$$A(x)\cdot B(x)=\sum_{k=0}^{n+m}c_kx^k,\ c_k=\sum_{\substack{k=i+j\\0\leq i\leq n\\0\leq j\leq m}}a_ib_j,\ (a_i,b_j,c_k)\in GF(p^m). \qquad (7.52)$$

Example 7.25. Consider multiplication of binary polynomials using the bit-shift and XOR operations. For $(x^2+x+1)\cdot(x^2+1)=x^4+x^3+2x^2+x+1$, e.g., x^4+x^3+x+1 mod 2.

One has $111\times 101=11100\oplus 111=11011$, which is a bit string representation of (x^4+x^3+x+1).

Modulo 2 division is performed by subtracting the denominator from the leading enumerator, and, computing the modulo 2 subtraction until the end. ■

Example 7.26. One may reduce a modulo of irreducible polynomials. Consider an irreducible polynomial (x^3+x+1). We have $(x^4+x^3+x+1)\equiv(x^2+x)$ mod (x^3+x+1).

The bit-string representation of (x^4+x^3+x+1) and (x^3+x+1) are '11011' and '1011', where the degrees of the irreducible polynomials are 4 and 3. The reduction starts by shifting the irreducible polynomial 1011 one bit left, yielding 10110. We obtain $11011\oplus10110=1101$ with a degree 3.

Applying the XOR operation, $1101\oplus1011=0110$, one finds (x^2+x). ■

Example 7.27. The $GF(2^m)$ operations and calculations are algorithmically supported and implemented by MPUs.

In MATLAB, the **gf(X,m)** commands creates a Galois field array from X in $GF(2^m)$, $1\leq m\leq16$. The elements of X are integers.

The command **gf(X,m,P)** generates a Galois field array from X using a primitive polynomial P to define the field. Primitive polynomials are the irreducible polynomials.

A polynomial $x^8+x^4+x^3+x^2+1$, or ('x^8+x^4+x^3+x^2+1'), is [1 0 0 0 1 1 1 0 1]. The polynomial division is performed by using the **deconv** command. For example,

```
p1=gf( [1 1 1 0 1 1 1] ); p2=gf( [1 0 1 1] );
[q,r]=deconv(p1,p2)  % Deconvolution is the polynomial division
```

yields

```
q = GF(2) array.
Array elements = 1  1  0  0
r = GF(2) array.
Array elements = 0  0  0  0  0  1  1
```

■

Data structures, arithmetic circuit graphs and physical circuit design are described by the Galois field. The logic operations on $x=(GF(2^m), [0,1])$ are applied. In AES, the key-alternating block cipher operations consist of bytes substitutions, rows shifting, columns mixing and round key adding. These operations use the blocks of data as arrays, as well as matrices and mappings of specific sizes. The physical implementation on MPUs and FPGAs are performed. The AES operations are described by the finite Galois field $GF(2^8)$ with 256 distinct polynomials over the Galois field. All operations are described in $GF(2^8)$ using the minimal irreducible binary polynomial of degree eight $p(x)=x^8+x^4+x^3+x+1$, or, x^8 mod (x^4+x^3+x+1). The function, which represents the 8-bit pattern $b_7b_6b_5b_4b_3b_2b_1b_0$ is

$$f(x)=b_7x^7+b_6x^6+b_5x^5+b_4x^4+b_3x^3+b_2x^2+b_1x+b_0. \tag{7.53}$$

Example 7.28. For '10000011', one has $\sum_i b_ix^i = x^8 + x + 1$. ■

AES Algorithm – The byte, given as bits '$b_7b_6b_5b_4b_3b_2b_1b_0$', is expressed as a polynomial with coefficients in {0,1}. Polynomial addition, subtraction and multiplication are realized as bitwise XOR operation ⊕. Multiplication of polynomials in mix columns operation involves modulo operations with an irreducible polynomial of degree 8 in the hexadecimal system. In an AES algorithm, for an N-bit key, a 4×4 matrix $A(a_{i,j})$, $0\leq(i,j)\leq3$ with $a_{i,j}$ is formed. Each entry in $A(a_{i,j})$ is a byte '$b_7b_6b_5b_4b_3b_2b_1b_0$'. The byte substitution operation requires finding the multiplicative inverse in $GF(2^8)$, and, carrying out the affine transformations in $GF(2^8)$ on each byte.

For a vector $x=[x_0x_1 \ldots x_7x_8]$, whose elements are the binary representations, the operation is $A_{8\times8}\cdot x+b$, where b is a vector of coefficients.

A 128-bit state space is a vector in $GF(2^8)^{16}$. The arithmetic addition by XOR, performs multiplication and multiplicative inverse of each field element in $GF(2^8)$. The shift row operations is performed finding A', and, columns mixing. Using the key, states are computed using the pertained operations [3, 4]. Byte substitution, rows shifting, columns mixing, round key adding and round operations are [24, 30-33]

$$s_{i,j} = S[a_{i,j}] = \sum_{k=0}^{7} c_k \cdot a_{i,j}^{-2^k} + c_8, \tag{7.54}$$

$$r_{i,j} = s_{j,i+j \bmod 4},\ m_{i,j} = \sum_{k=0}^{3} v_{i+k \bmod 4} \cdot r_{k,j},$$

$$d_{i,j} = k_{i,j} + m_{i,j},\ d_{i,j} = k_{i,j} + \sum_{k=0}^{3} v_{i+k \bmod 4} \cdot \left[\sum_{l=0}^{7} c_l \cdot a_{i,i+j \bmod 4}^{-2^l} + c_8 \right],$$

where $s_{i,j}$, $r_{i,j}$, $m_{i,j}$ and $d_{i,j}$ are the (i,j) output state of the *SubBytes*, *ShiftRows*, *MixColumns*, and, *AddRoundKey* and *Round* operations; c_8 is the vector of constants.

Cryptographic Processors – Decryption is performed using *InvShiftRows*, *InvSubBytes*, *AddRoundKey* and *InvMixColumns* operations [3, 4]. The instruction sets for the AES algorithms are developed for desktops, laptops, mobile and wearable devices, IoT platforms, etc. The Galois field calculus and encryption/decryption operations (7.54) are used to develop, verify and implement algorithms and protocols on a physical hardware, as well as to design and optimize cryptographic processors. Formal graph-based designs of arithmetic circuits can be accomplished by using polynomial. The circuit description can be verified and translated into the hardware description language codes ensuring design flow. Encryption schemes and algorithms are implemented by low-power cryptographic MPUs and FPGAs [25, 26, 35]. The elliptic-curve cryptography yield symmetric and asymmetric encryption scheme thereby enhancing ciphertext security and empowering information security.

Authenticated Encryption and Confidentiality – In wireless communication, the counter mode encryption (counter with cipher block chaining message authentication code CBC-MAC) ensures authentication and confidentiality. A block cipher Galois counter mode (GCM) supports authenticated encryption which uses universal hashing over a Galois field. High speed, low latency and hundreds of Gbps throughput is achieved due to algorithmic efficiency, high-performance computing and parallelism. In contrast, many authentication and confidentiality algorithms (counter with CBC-MAC, encrypt-then-authenticate-then-translate, one-key message authentication, etc.) cannot be pipelined and difficult to parallelized.

Apply the minimal irreducible binary polynomial of degree 128 in $GF(2^{128})$, $p(x)=x^{128}+x^7+2x^2+x+1$. The authentication tag is constructed by feeding blocks of data into the function

$$\text{GHASH}(H,A,C)=X_{m+n+1}, \tag{7.55}$$

and, encrypting the result. In (7.55), H is the 128-bit string hash key encrypted using the block cipher; A is the authenticated and unencrypted data; C is the ciphertext; m and n are the indexes for 128-bit blocks in A and C.

The variable X_i for $i=0, ..., m+n+1$, $X \in GF(2^{128})$ is

$$X_i=\begin{cases}0, & i=0\\ (X_{i-1}\oplus A_i)\cdot H, & i=1,\dots,m-1\\ (X_{m-1}\oplus(A_m^*\|0_{128-l_A}))\cdot H, & i=m\\ (X_{i-1}\oplus C_{i-m})\cdot H, & i=m+1,\dots,m+n-1\\ (X_{m+n-1}\oplus(C_n^*\|0_{128-l_C}))\cdot H, & i=m+n\\ (X_{m+n}\oplus(A_{(64)}\|C_{(64)}))\cdot H, & i=m+n+1\end{cases}, \quad (7.56)$$

where l_A and l_C are the bit length of the final blocks of A and C; $\|$ denotes concatenation of bit strings; A_{64} and C_{64} are the 64-bit representations of the bit lengths of A and C.

The authenticated decryption is similar to encryption. Encryption and decryption are implemented by data-intensive-computing algorithms. The XOR and multiplication over $GF(2^{128})$ support computing of the 128-bit string hash key $H\in GF(2^{128})$.

AES Implementation and Cryptographic Digital ASIC Design – Processing-intensive AES algorithms are implemented using the ARM processors. Optimized datapath and computationally-efficient processing are implemented. A compact implementation of (7.54)-(7.56) operations, expressed in $GF(2^8)$, $GF(2^{128})$ and $GF(2^m)$, yield optimal AES-coprocessor architecture and datapath. The ASICs are designed using $GF(2^2)$ and $GF(2)$ primitives. The circuit design and verifications are reported in [30, 33]. The Galois field calculus and encryption/decryption operations support the hardware architecture design, implementation of algorithms, physical design with vector parallelism (64-bit registers and a vector pipelined processor which performs 32-bit XOR instructions), and, physical verification. Arithmetic and processing operations are pre-computed and stored as tables. For example, the Intel® Core™ processors support the Intel® AES New Instructions AES-NI, and, the Intel® Core™ multi-core processors are optimized for 32-bit applications within 32-bit operating systems. The FPGA implementation enables floating-point operation, multi-processing, parallelism, pipelining, high-performance computing, etc.

7.10. Secure Data Sharing: Experimental Studies

The AES algorithms are implemented and tested for the inverted platform, examined in section 6.15.1. The MPU-6500 with Digital Motion Processor outputs 16-bit data sequences for the linear accelerations and angular rates using the I^{2}C at 400 kHz and serial peripheral interface (SPI) at 1 MHz. Using the angular rates ($\omega_\phi[n],\omega_\theta[n],\omega_\psi[n]$), the Euler angles ($\hat{\phi}[n],\hat{\theta}[n],\hat{\psi}[n]$) are estimated by a MPU.

The data, including $\begin{bmatrix}\omega_\phi[n],\omega_\theta[n],\omega_\psi[n]\\ \hat{\phi}[n],\hat{\theta}[n],\hat{\psi}[n]\end{bmatrix}$, is encrypted, transmitted, decrypted, fused and monitored ensuring human-robot interaction and end-to-end functionality. Figures 7.14 report encryption, communication and decryption. The AES-128, 192 and 256 algorithms are implemented and tested. The key exchange, encryption and decryption take 0.25, 0.37 and 0.49 msec. High throughput parallel processing and optimized algorithms, implemented by FPGAs, maximize data rates.

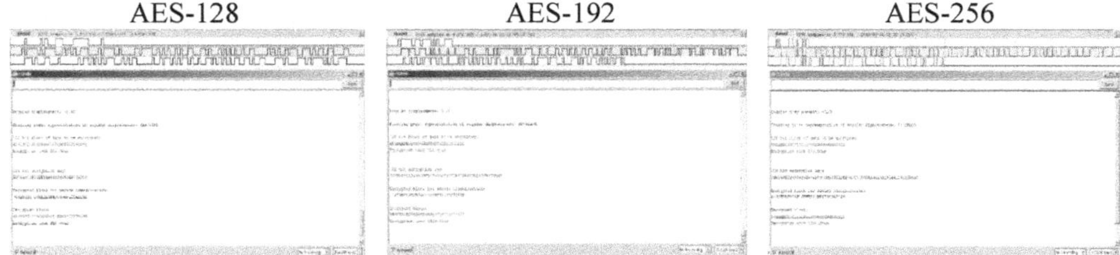

Figure 7.14. Secure communication and data sharing in dynamic data-driven applications: AES-128, 192 and 256 algorithms with randomly generated key. Data is encrypted, transmitted and decrypted.

The electromagnetic spectrum affects measurements, connectivity and data sharing. Figures 7.15 document secure information sharing and fusion for the (x,y,z)-orientation with no interference and under adverse electromagnetic spectrum. The TDK InvenSense MPU-9250 and Bosch Sensortec BNO055 system-in-package modules integrate triaxial accelerometer, gyroscope and magnetometer. Direct sensor measurements are discretized and processed by a 32-bit ARM Cortex MPU. The BNO055 outputs 14-bit resolution accelerations, 16-bit resolution angular rates, magnetic field density, as well as inertial and orientation estimates.

The MPU-9250 and BNO055, installed on the quadrotor helicopter, are used to conduct experimental studies. For example, the estimated UAV orientation $(\hat{x}_n, \hat{y}_n, \hat{z}_n)$ is encrypted using the AES-128 algorithm, transmitted and decrypted.

Study a long-range, low-power communication capabilities at 12 km for an UAV with a 5.8 GHz quad multiband microstrip patch bidirectional antenna (31 dBm transmission power, 16 and 12 dB transmitter and receiver gains). Data is transmitted to a ground station. For long range communication, transmission power, antenna gain and sensitivity should be optimized, adjusted and reconfigured. In critical applications, validity, redundancy, accuracy, error detection and error correction are addressed.

Due to low signal strength, low signal-to-noise ratio, high bandwidth, high gain, –96 dBm receiver sensitivity, there are interference, distortions and bit synchronization error. The error rate and bit error probability may be studied and characterized. Redundant information sharing ensures acceptable bit error rate. Under adverse electromagnetic spectrum, there are data gaps, distortions and errors. The estimated and fused $(\hat{x}_n, \hat{y}_n, \hat{z}_n)$ are reported in Figures 7.15.

Figures 7.16 illustrate data measurements by the solid-state sensors which measure temperature, humidity and heat index. To ensure the desired accuracy and resolution, the direct analog measurements by sensors are digitized by 16-bit ADCs. The MPU processes measurements. Sensors are installed on the UAV. Encryption by AES-128, wireless communication by a transceiver with 5.8 GHz quad multiband microstrip patch antenna, and, decryption are performed. Data is measured at 100 meter altitude flight, processed by MPUs, transmitted, and, fused into an information management system at 10 km. Figures 7.16 document data streams, which illustrate data packets at the transmitter and receiver. High-power electromagnetic interference causes heterogeneity and information losses.

Descriptive univariate, bivariate and multivariate analyses can be accomplished. For example, the data incompleteness analysis can be accomplished by using the signal strength, radiated power, receiver sensitivity, delays (latencies) and other variables. Statistical model $\mathscr{R}_X=(\mathcal{S},\mathcal{D})$ may be derived, supporting probabilistic assessment of data incompleteness and undecidability supporting dynamic data driven control, optimization and applications.

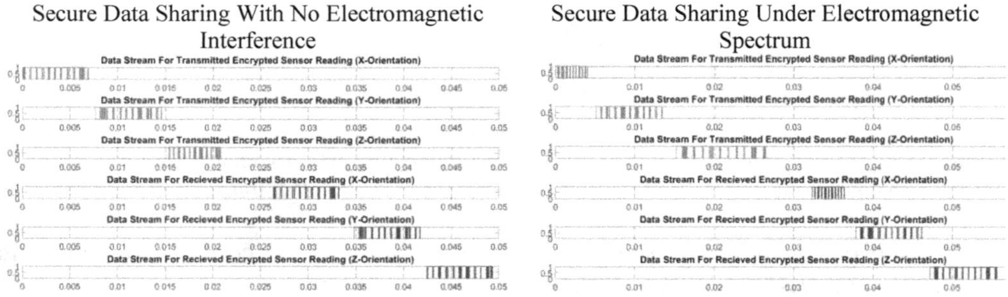

Secure Data Sharing With No Electromagnetic Interference

Secure Data Sharing Under Electromagnetic Spectrum

Figure 7.15. Estimated and fused UAV orientation $(\hat{x}_n, \hat{y}_n, \hat{z}_n)$. Controlled information sharing, processing protocols, applications, wireless communication, synchronization and resiliency are assessed by evaluating multiple binary data packets. Physical quantities are measured, discretized by ADCs, filtered by digital filters, processed, transmitted and evaluated as *Data Measured* → *Data Processed and Encrypted* → *Data Shared* (wireless communication) → *Data Decrypted.* Data streams are variable-size binary data packets of *N* bits, divided as sensor ID, location ID, timestamp and numeric value (16 and 32 bit). Each cipher encrypts and decrypts data in blocks of 128 bits using cryptographic keys of 128 bits.

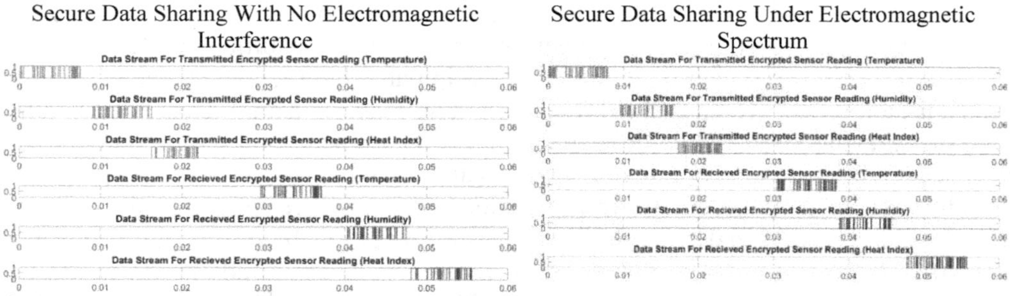

Secure Data Sharing With No Electromagnetic Interference

Secure Data Sharing Under Electromagnetic Spectrum

Figure 7.16. Encrypted and decrypted variable-size binary data packets for measured temperature, humidity and heat index. Timing and synchronization of transmitted data packets are established by timestamps.

Studies were conducted using computationally-efficient algorithms. Resiliency, data security and information control are tested and substantiated for distinct platforms over different environments. The reported findings empower system-consistent, application-compliant and data-centric solutions.

The key advantages are:

1. Modularity and scalability;
2. Resiliency and dynamic adaptiveness;
3. Information conformity, completeness and validity;
4. Overall functionality.

For diverse CPS, proven tools and algorithmic solutions enable acceptance and transfer of findings and technologies to high-fidelity operational environment in critical applications.

7.11. Practice Problems

Practice Problem* 7.1. *Singular Value Decomposition and Pseudoinverse

Find the SVD (7.4) $A=U\Sigma V^T$ for a matrix

$$A=\begin{bmatrix}3 & 1 & 1\\ -1 & 3 & 1\end{bmatrix},\ A\in\mathbb{R}^{2\times 3}.$$

The eigenvalue-eigenvector problem is solved for $AA^T\in\mathbb{R}^{2\times 2}$ and $A^TA\in\mathbb{R}^{3\times 3}$ to find a tuple (U,Σ,V).

For $AA^T=\begin{bmatrix}3 & 1 & 1\\ -1 & 3 & 1\end{bmatrix}\begin{bmatrix}3 & -1\\ 1 & 3\\ 1 & 1\end{bmatrix}=\begin{bmatrix}11 & 1\\ 1 & 11\end{bmatrix}$,

the characteristic polynomial is

$$p(\lambda)=\det(\lambda I-AA^T)=\det\begin{bmatrix}\lambda-11 & -1\\ -1 & \lambda-11\end{bmatrix}=\lambda^2-22\lambda+120\cdot$$

The characteristic polynomial yields the characteristic equation $\lambda^2-22\lambda+120=0$.

The eigenvalues are $\lambda_1=12$ and $\lambda_2=10$.

Using the trace and determinant of AA^T, $\mathrm{tr}(AA^T)=\sum_i\lambda_i$, $\det(AA^T)=\prod_i\lambda_i$.

We have $\mathrm{tr}(AA^T)=\lambda_1+\lambda_2=22$ and $\det(AA^T)=\lambda_1\lambda_2=120$.

Using (7.5),

$(AA^T)\boldsymbol{u}_i=\lambda_i\boldsymbol{u}_i$, $(\lambda_iI-AA^T)\boldsymbol{u}_i=0$,

obtain the linearly independent eigenvectors.

Solve $(\lambda_iI-AA^T)\boldsymbol{u}_i=0$, $\begin{bmatrix}\lambda_i-11 & -1\\ -1 & \lambda_i-11\end{bmatrix}\begin{bmatrix}u_{1_i}\\ u_{2_i}\end{bmatrix}=\begin{bmatrix}0\\ 0\end{bmatrix}$, $i=1,2$, $\lambda_1=12$, $\lambda_2=10$.

The linearly independent eigenvectors are $\boldsymbol{u}_{1_{\lambda=12}}=\begin{bmatrix}1\\ 1\end{bmatrix}$, $\boldsymbol{u}_{2_{\lambda=10}}=\begin{bmatrix}1\\ -1\end{bmatrix}$.

From $[\boldsymbol{u}_1,\boldsymbol{u}_2]$, find orthonormal eigenvectors, and, $U=\frac{1}{\sqrt{2}}\begin{bmatrix}1 & 1\\ 1 & -1\end{bmatrix}$.

For $A^TA=\begin{bmatrix}3 & -1\\ 1 & 3\\ 1 & 1\end{bmatrix}\begin{bmatrix}3 & 1 & 1\\ -1 & 3 & 1\end{bmatrix}=\begin{bmatrix}10 & 0 & 2\\ 0 & 10 & 4\\ 2 & 4 & 2\end{bmatrix}$,

the characteristic polynomial is

$$p(\lambda)=\det(\lambda I-A^TA)=\det\begin{bmatrix}\lambda-10 & 0 & -2\\ 0 & \lambda-10 & -4\\ -2 & -4 & \lambda-2\end{bmatrix}=\lambda(\lambda^2-22\lambda+120)\cdot$$

One finds $\lambda_1=12$, $\lambda_2=10$ and $\lambda_3=0$.

From (7.5) $(A^TA)\boldsymbol{v}_i=\lambda_i\boldsymbol{v}_i$, solve $(\lambda_iI-A^TA)\boldsymbol{v}_i=0$ to find the eigenvectors.

From $\begin{bmatrix}\lambda_i-10 & 0 & -2\\ 0 & \lambda_i-10 & -4\\ -2 & -4 & \lambda_i-2\end{bmatrix}\begin{bmatrix}v_{1_i}\\ v_{2_i}\\ v_{3_i}\end{bmatrix}=\begin{bmatrix}0\\ 0\\ 0\end{bmatrix}$, $\lambda_1=12$, $\lambda_2=10$, $\lambda_3=0$,

we find the linearly independent eigenvectors $\boldsymbol{v}_{1_{\lambda=12}}=\begin{bmatrix}1\\ 2\\ 1\end{bmatrix}$, $\boldsymbol{v}_{2_{\lambda=10}}=\begin{bmatrix}2\\ -1\\ 0\end{bmatrix}$, $\boldsymbol{v}_{3_{\lambda=0}}=\begin{bmatrix}1\\ 2\\ -5\end{bmatrix}$.

Computed $[\boldsymbol{v}_1,\boldsymbol{v}_2,\boldsymbol{v}_3]$ yield the orthonormal linearly independent eigenvectors, and

$$V=\begin{bmatrix}\frac{1}{\sqrt{6}} & \frac{2}{\sqrt{5}} & \frac{1}{\sqrt{30}}\\ \frac{2}{\sqrt{6}} & -\frac{1}{\sqrt{5}} & \frac{2}{\sqrt{30}}\\ \frac{1}{\sqrt{6}} & 0 & -\frac{5}{\sqrt{30}}\end{bmatrix}.$$

One has $UU^T=U^TU=I\in\mathbb{R}^{2\times 2}$, $VV^T=V^TV=I\in\mathbb{R}^{3\times 3}$.
The eigenvalues for AA^T are $\lambda_1=12$ and $\lambda_2=10$, $\lambda_1>\lambda_2$.
The singular values are $\sigma_{i,i}=\sqrt{\lambda_i}$, $\sigma_{1,1}>\sigma_{2,2}>0$.
Hence, $\sigma_{1,1}=\sqrt{12}$, $\sigma_{2,2}=\sqrt{10}$.

The singular matrix $\Sigma\in\mathbb{R}^{2\times 3}$ is $\Sigma=\begin{bmatrix}\sqrt{12} & 0 & 0\\ 0 & \sqrt{10} & 0\end{bmatrix}$.

Results are verified using computed (U,Σ,V)

$$A=U\Sigma V^T=\begin{bmatrix}\frac{1}{\sqrt{2}} & \frac{1}{\sqrt{2}}\\ \frac{1}{\sqrt{2}} & -\frac{1}{\sqrt{2}}\end{bmatrix}\begin{bmatrix}\sqrt{12} & 0 & 0\\ 0 & \sqrt{10} & 0\end{bmatrix}\begin{bmatrix}\frac{1}{\sqrt{6}} & \frac{2}{\sqrt{6}} & \frac{1}{\sqrt{6}}\\ \frac{2}{\sqrt{5}} & -\frac{1}{\sqrt{5}} & 0\\ \frac{1}{\sqrt{30}} & \frac{2}{\sqrt{30}} & -\frac{5}{\sqrt{30}}\end{bmatrix}=\begin{bmatrix}3 & 1 & 1\\ -1 & 3 & 1\end{bmatrix}.$$

The inverse of A is $A^{-1}=V\Sigma^{-1}U^T$, where $\Sigma^{-1}=\begin{bmatrix}\frac{1}{\sqrt{12}} & 0\\ 0 & \frac{1}{\sqrt{10}}\\ 0 & 0\end{bmatrix}$.

We compute $A^{-1}=\begin{bmatrix}\frac{17}{60} & -\frac{7}{60}\\ \frac{1}{15} & \frac{4}{15}\\ \frac{1}{12} & \frac{1}{12}\end{bmatrix}$. One has $AA^{-1}=\begin{bmatrix}3 & 1 & 1\\ -1 & 3 & 1\end{bmatrix}\begin{bmatrix}\frac{17}{60} & -\frac{7}{60}\\ \frac{1}{15} & \frac{4}{15}\\ \frac{1}{12} & \frac{1}{12}\end{bmatrix}=\begin{bmatrix}1 & 0\\ 0 & 1\end{bmatrix}$.

Hence, $AA^{-1}=I\in\mathbb{R}^{2\times 2}$.
However, $A^{-1}A\neq I$. ■

Practice Problem 7.2. SVD, Correlation, and, Unitary Transformations

Consider a sequence $X\in\mathbb{R}^{n\times m}$. Using the centered mean zero $\hat{X}=(X-\bar{X})$, $\bar{X}=\mathbf{1}_n\bar{x}$, the covariance matrix $C\in\mathbb{R}^{m\times m}$ is given as (7.10), $C=\frac{1}{n-1}\hat{X}^T\hat{X}$.

The explicit expressions for correlations are XX^T and X^TX. Using (7.4), $X^TX=V\Sigma^T\Sigma V^T$, $XX^T=U\Sigma\Sigma^TU^T$.

Columns of the square matrices $U\in\mathbb{R}^{n\times n}$ and $V\in\mathbb{R}^{m\times m}$ are the ordered orthonormal linearly independent eigenvectors of XX^T and X^TX.

The largest eigenvectors in $U=[\boldsymbol{u}_1,\ldots,\boldsymbol{u}_n]$ yield dominant correlations among columns of X. The columns of $V=[\boldsymbol{v}_1,\ldots,\boldsymbol{v}_m]$ represent correlation among rows of X.

For a left unitary transformation $X{:}Y=PX$, the correlation is $Y^TY=X^TP^TPX=X^TX$.
The projected data has the same eigendecomposition, resulting in V_X and Σ_X.
From $U_Y=PU_X$, $\Sigma_X=\Sigma_Y$ and $V_X=V_Y$, one finds $Y=PX=PU_X\Sigma_XV_X^T$.
For a right unitary transformation $X{:}Y=XP^T$, one has

$Y^T Y = PX^T XP^T = PV_X \Sigma_X^T V_X^T P^T$.

This yields the eigendecomposition $Y^T YPV_X = PV_X \Sigma_X^2$.

From $U_Y = YPV_X \Sigma_X^{-1} = XV_X \Sigma_X^{-1} = U_X$, $\Sigma_Y = \Sigma_X$, $V_Y = PV_X$,

$Y = XP^T = U_X \Sigma_X V_X^T P^T$. ■

Practice Problem 7.3. Sum of Statistically Independent Random Variables

Consider a sum of N statistically independent random variables X_i, $Z = \sum_{i=1}^{N} X_i$.

If Z is a sum of X_i with continuous pdfs, the pdf $f_Z(z)$ is a convolution of pdfs f_{X_i}

$$f_Z(z) = \left(f_{X_1}(x_1) * ... * f_{X_N}(x_N)\right)(z),$$

where $*$ is the convolution symbol.

For example, $\left(f_{X_1}(x_1) * f_{X_2}(x_2)\right)(z) = \int_{-\infty}^{+\infty} f_{X_2}(z - x_1) f_{X_1}(x_1) dx_1 = \int_{-\infty}^{+\infty} f_{X_1}(z - x_2) f_{X_2}(x_2) dx_2$.

For a Gaussian distribution, $f_{X_i}(x_i; \mu_i, \sigma_i^2) = \frac{1}{\sqrt{2\pi}\sigma_i} e^{-\frac{(x_i - \mu_i)^2}{2\sigma_i^2}}$, $\mu_i \in \mathbb{R}$, $\sigma_i > 0$.

Consider statistically independent random X_1 and X_2.

For $Z = X_1 + X_2$, we have $Z \sim \mathcal{N}(\mu, \sigma^2)$, $f_Z(z; \mu_z, \sigma_z^2) = \frac{1}{\sqrt{2\pi}\sqrt{\sigma_{11}^2 + \sigma_{22}^2}} e^{-\frac{[z - (\mu_1 + \mu_2)]^2}{2(\sigma_{11}^2 + \sigma_{22}^2)}}$.

For $Z = X_1 + X_2$, $Z \sim \mathcal{N}(\mu_1 + \mu_2, \sigma_{11}^2 + \sigma_{22}^2, \rho)$.

For $Z = X_1 - X_2$, $Z \sim \mathcal{N}(\mu_1 - \mu_2, \sigma_{11}^2 + \sigma_{22}^2, \rho)$.

Consider $Z = \sum_{i=1}^{N} X_i$.

The mean value is the sum of the individual means, $\mu_z = \sum_{i=1}^{N} \mu_i$, $\mu_i \in \mathbb{R}$.

The variance of Z is the sum of the individual variances, $\sigma_z^2 = \sum_{i=1}^{N} \sigma_i^2$, $\forall \sigma_i > 0$. ■

Practice Problem 7.4. Uniform Distribution

Consider $X = [X_1, X_2, \ldots, X_{k-1}, X_k]$ with k independent and identically distributed uniform random variables on the interval [0,1], $X_i \sim \mathcal{U}_{[0,1]}$.

For $Z = X_1 \cdot X_2 \cdot \ldots \cdot X_{k-1} \cdot X_k$, $f_Z(z) = \begin{cases} \frac{(-1)^{k-1}}{(k-1)!} (\ln z)^{k-1}, 0 \le z \le 1 \\ 0 \quad \text{otherwise} \end{cases}$.

For $Z = \prod_{i=1}^{k} X_i$, $X_i \sim \mathcal{U}_{[0,1]}$, plots for $f_Z(z)$ are documented in Figure 7.17.b.

One has $f_{X_1,\ldots,X_k}(x_1, \ldots, x_k) = \frac{\partial^k F_{X_1,\ldots,X_k}(x_1, \ldots, x_k)}{\partial x_1 \ldots \partial x_k}$, $\int_{x_k} \ldots \int_{x_1} f_{X_1,\ldots,X_k}(x_1, \ldots, x_k) dx_k \ldots dx_1 = 1$.

For independent uniform random variables $X = [X_1, X_2, \ldots, X_{k-1}, X_k]$, study $Z = \prod_{i=1}^{k} X_i$, $X_i \sim \mathcal{U}_{[a_i, b_i]}$, $-\infty < a_i < b_i < \infty$, $x_i \in [a_i, b_i]$.

The pdf and cdf are $f_{X_i}(x_i) = \begin{cases} \frac{1}{b_i - a_i} & \text{for } x_i \in [a_i, b_i] \\ 0 & \text{otherwise} \end{cases}$, $F_{X_i}(x_i) = \begin{cases} 0 & \text{for } x_i < a_i \\ \frac{x_i - a_i}{b_i - a_i} & \text{for } x_i \in [a_i, b_i] \\ 1 & \text{for } x_i > b_i \end{cases}$.

One finds $f_Z(z)$ using the limits. ■

7.12. Exploration Problems

Exploration Problem* 7.1. *Bivariate Probability Distribution

For an m-tuple of real $X=[X_1,\ldots,X_m]$ in $\Omega\in\mathbb{R}^m$, consider $X\sim\mathcal{N}(\boldsymbol{\mu},\Sigma)$.

The m-variate pdf is $f_X(\mathbf{x})=\left(\frac{1}{2\pi}\right)^{m/2}|\Sigma|^{-1/2}e^{-\frac{1}{2}[\mathbf{x}-\boldsymbol{\mu}]^T\Sigma^{-1}[\mathbf{x}-\boldsymbol{\mu}]}$.

For $X_1\sim\mathcal{N}(\mu_1,\sigma_{11}^2)$ and $X_2\sim\mathcal{N}(\mu_2,\sigma_{22}^2)$,

$$f_{X_1}(x_1)=\frac{1}{\sigma_{11}\sqrt{2\pi}}e^{-\frac{(x_1-\mu_1)^2}{2\sigma_{11}^2}},\ \mathbf{E}(X_1)\equiv\mu_1,\ \mathrm{var}(X_1)\equiv\sigma_{11}^2,$$

$$f_{X_2}(x_2)=\frac{1}{\sigma_{22}\sqrt{2\pi}}e^{-\frac{(x_2-\mu_2)^2}{2\sigma_{22}^2}},\ \mathbf{E}(X_2)\equiv\mu_2,\ \mathrm{var}(X_2)\equiv\sigma_{22}^2.$$

Marginal, joint and conditional pdfs can be experimentally obtained from histograms. Random (X_1,X_2) may be statistically dependent or independent. Function $f_{X_1,X_2}(x_1,x_2):\mathbb{R}^2\to\mathbb{R}$ is a joint pdf of random jointly continuous X_1 and X_2,

$$f_{X_1,X_2}(x_1,x_2)=\frac{\partial^2F_{X_1,X_2}(x_1,x_2)}{\partial x_1\partial x_2},\ \int_{-\infty}^{\infty}\int_{-\infty}^{\infty}f_{X_1,X_2}(x_1,x_2)dx_1dx_2=1.$$

For $A\in\mathbb{R}^2$, $\mathbf{P}\left((X_1,X_2)\in A\right)=\iint_A f_{X_1,X_2}(x_1,x_2)dx_1dx_2$.

The marginal pdfs are

$$f_{X_1}(x_1)=\int_{-\infty}^{\infty}f_{X_1,X_2}(x_1,x_2)dx_2,\ f_{X_2}(x_2)=\int_{-\infty}^{\infty}f_{X_1,X_2}(x_1,x_2)dx_1.$$

For a conditional distribution, the conditional density X_2, given $X_1=x_1$, is

$$f_{X_2|X_1}(x_2|x_1)=\frac{f_{X_1,X_2}(x_1,x_2)}{f_{X_1}(x_1)}.$$

A conditional distribution of X_1, given $X_2=x_2$, is denoted as $f_{X_1|X_2}(x_1|x_2)$.

One has $f_{X_1,X_2}(x_1,x_2)=f_{X_2|X_1}(x_2|x_1)f_{X_1}(x_1)=f_{X_1|X_2}(x_1|x_2)f_{X_2}(x_2)$.

Hence, $f_{X_1|X_2}(x_1|x_2)=\dfrac{f_{X_1,X_2}(x_1,x_2)}{f_{X_2}(x_2)}$, $f_{X_2|X_1}(x_2|x_1)=\dfrac{f_{X_1,X_2}(x_1,x_2)}{f_{X_1}(x_1)}$.

For statistically independent random variable,

$f_{X_1,X_2}(x_1,x_2)=f_{X_1}(x_1)f_{X_2}(x_2)$.

Study $X_1\sim\mathcal{N}(\mu_1,\sigma_{11}^2)$ and $X_2\sim\mathcal{N}(\mu_2,\sigma_{22}^2)$.

A covariance is

$\mathrm{cov}[X_1,X_2]=\mathbf{E}[(X_1-\mathbf{E}[X_1])(X_2-\mathbf{E}[X_2])]=\mathbf{E}[X_1X_2]-\mathbf{E}[X_1]\mathbf{E}[X_2]$.

For (X_1,X_2), one obtains variances $(\sigma_{11}^2,\sigma_{22}^2)$, mean $\boldsymbol{\mu}=\begin{bmatrix}\mu_1\\ \mu_2\end{bmatrix}$ and correlation

$$\rho=\frac{\mathrm{cov}(X_1,X_2)}{\sigma_{11}\sigma_{22}}=\frac{\mathbf{E}[(X_1-\mu_1)(X_2-\mu_2)]}{\sigma_{11}\sigma_{22}}.$$

Consider a bivariate normal distribution for (X_1,X_2).

Let a covariance matrix is $\Sigma = \begin{bmatrix} \sigma_{11}^2 & \rho\sigma_{11}\sigma_{22} \\ \rho\sigma_{11}\sigma_{22} & \sigma_{22}^2 \end{bmatrix}$.

One has

$$f_{X_1,X_2}(x_1,x_2) = \frac{1}{2\pi\sigma_{11}\sigma_{22}\sqrt{1-\rho^2}} e^{-\frac{1}{2(1-\rho^2)}\left(\frac{(x_1-\mu_1)^2}{\sigma_{11}^2} - 2\rho\frac{(x_1-\mu_1)(x_2-\mu_2)}{\sigma_{11}\sigma_{22}} + \frac{(x_2-\mu_2)^2}{\sigma_{22}^2}\right)}, \int_{x_1}\int_{x_2} f_{X_1,X_2}(x_1,x_2)dx_2dx_1 = 1.$$

The conditional pdf of X_2, given $X_1=x_1$, is

$$f_{X_2|X_1}(x_2|x_1) = \frac{f_{X_1,X_2}(x_1,x_2)}{f_{X_1}(x_1)} = \frac{1}{\sqrt{2\pi}\sigma_{22}\sqrt{1-\rho^2}} e^{-\frac{1}{2\sigma_{22}^2(1-\rho^2)}\left[x_2-\left(\mu_2+\rho\frac{\sigma_{22}}{\sigma_{11}}(x_1-\mu_1)\right)\right]^2},$$

and $\mu_{X_2|X_1=x_1} = \mathbf{E}[X_2|X_1=x_1] = \int_{-\infty}^{\infty} x_2 f_{X_1|X_2}(x_1|x_2)dx_2 = \mu_2 + \rho\frac{\sigma_{22}}{\sigma_{11}}(x_1-\mu_1)$,

$\mathrm{var}[X_2|X_1=x_1] = \sigma_{X_2|X_1=x_1}^2 = \sigma_{22}^2(1-\rho^2)$.

The conditional probability density function of X_1, given $X_2=x_2$, is

$$f_{X_1|X_2}(x_1|x_2) = \frac{f_{X_1,X_2}(x_1,x_2)}{f_{X_2}(x_2)} = \frac{1}{\sqrt{2\pi}\sigma_{11}\sqrt{1-\rho^2}} e^{-\frac{1}{2\sigma_{11}^2(1-\rho^2)}\left[x_1-\left(\mu_1+\rho\frac{\sigma_{11}}{\sigma_{22}}(x_2-\mu_2)\right)\right]^2},$$

and $\mu_{X_1|X_2=x_2} = \mathbf{E}[X_1|X_2=x_2] = \mu_1 + \rho\frac{\sigma_{11}}{\sigma_{22}}(x_2-\mu_2)$,

$\mathrm{var}[X_1|X_2=x_2] = \sigma_{X_1|X_2=x_2}^2 = \sigma_{11}^2(1-\rho^2)$.

For $\mu_1=0$, $\mu_2=0$, $\sigma_{11}=0.25$, $\sigma_{22}=0.375$ and $\rho=0.5$, pdfs $f_{X_1,X_2}(x_1,x_2)$ and $f_{X_2|X_1}(x_2|x_1)$ are depicted in Figure 7.17.a.

For statistically independent X_1 and X_2,

$f_{X_1,X_2}(x_1,x_2) = f_{X_1}(x_1)f_{X_2}(x_2)$.

Consider a joint pdf $f_{X_1,X_2}(x_1,x_2) = \frac{1}{2\pi}e^{-\frac{1}{2}(x_1^2+x_2^2)}$.

Marginal distributions for statistically independent (X_1,X_2) are

$$f_{X_1}(x_1) = \int_{-\infty}^{\infty} f_{X_1,X_2}(x_1,x_2)dx_2 = \int_{-\infty}^{\infty}\frac{1}{2\pi}e^{-\frac{1}{2}(x_1^2+x_2^2)}dx_2 = \frac{1}{\sqrt{2\pi}}e^{-\frac{1}{2}x_1^2}\int_{-\infty}^{\infty}\frac{1}{\sqrt{2\pi}}e^{-\frac{1}{2}x_2^2}dx_2 = \frac{1}{\sqrt{2\pi}}e^{-\frac{1}{2}x_1^2},$$

$$f_{X_2}(x_2) = \int_{-\infty}^{\infty} f_{X_1,X_2}(x_1,x_2)dx_1 = \int_{-\infty}^{\infty}\frac{1}{2\pi}e^{-\frac{1}{2}(x_1^2+x_2^2)}dx_1 = \frac{1}{\sqrt{2\pi}}e^{-\frac{1}{2}x_2^2}\int_{-\infty}^{\infty}\frac{1}{\sqrt{2\pi}}e^{-\frac{1}{2}x_1^2}dx_1 = \frac{1}{\sqrt{2\pi}}e^{-\frac{1}{2}x_2^2}.$$

Recall that $\int_{-\infty}^{\infty}\frac{1}{\sqrt{2\pi}}e^{-\frac{1}{2}x_1^2}dx_1 = 1$ and $\int_{-\infty}^{\infty}\frac{1}{\sqrt{2\pi}}e^{-\frac{1}{2}x_2^2}dx_2 = 1$.

For $X\sim\mathcal{N}(\mathbf{\mu},\Sigma)$, $X\sim\mathcal{N}(\mu_1,\mu_2,\sigma_{11}^2,\sigma_{22}^2,\rho)$, $\rho=0$, the marginal distributions are

$$f_{X_1}(x_1) = \int_{-\infty}^{\infty} f_{X_1,X_2}(x_1,x_2)dx_2 = \frac{1}{\sqrt{2\pi}\sigma_{11}}e^{-\frac{(x_1-\mu_1)^2}{2\sigma_{11}^2}},$$

$$f_{X_2}(x_2) = \int_{-\infty}^{\infty} f_{X_1,X_2}(x_1,x_2)dx_1 = \frac{1}{\sqrt{2\pi}\sigma_{22}}e^{-\frac{(x_2-\mu_2)^2}{2\sigma_{22}^2}}.$$

Marginal pdfs can be used to find joint pdfs when the random variables are statistically independent.

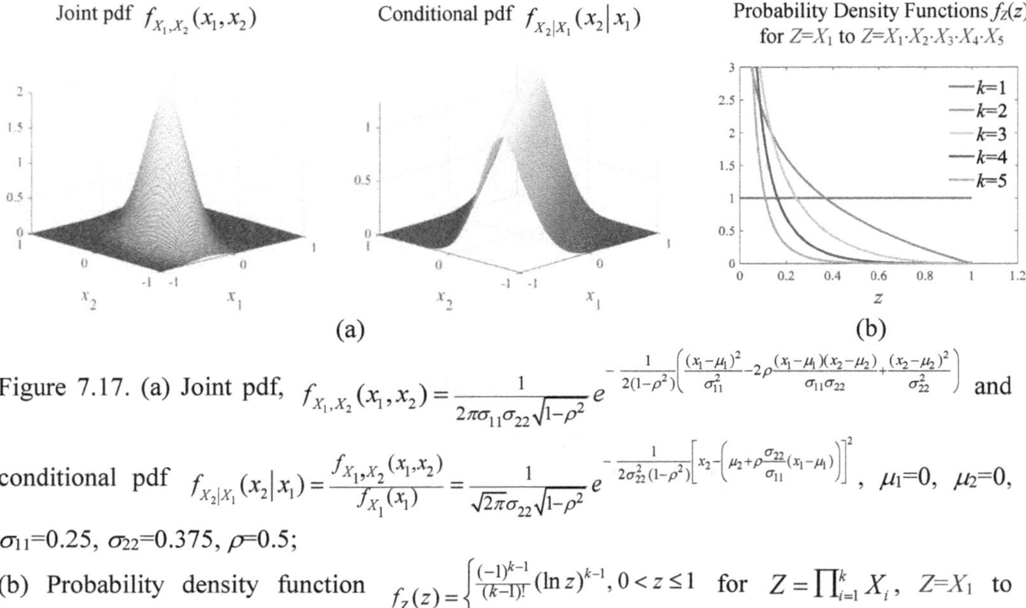

Figure 7.17. (a) Joint pdf, $f_{X_1,X_2}(x_1,x_2)=\frac{1}{2\pi\sigma_{11}\sigma_{22}\sqrt{1-\rho^2}}e^{-\frac{1}{2(1-\rho^2)}\left(\frac{(x_1-\mu_1)^2}{\sigma_{11}^2}-2\rho\frac{(x_1-\mu_1)(x_2-\mu_2)}{\sigma_{11}\sigma_{22}}+\frac{(x_2-\mu_2)^2}{\sigma_{22}^2}\right)}$ and conditional pdf $f_{X_2|X_1}(x_2|x_1)=\frac{f_{X_1,X_2}(x_1,x_2)}{f_{X_1}(x_1)}=\frac{1}{\sqrt{2\pi}\sigma_{22}\sqrt{1-\rho^2}}e^{-\frac{1}{2\sigma_{22}^2(1-\rho^2)}\left[x_2-\left(\mu_2+\rho\frac{\sigma_{22}}{\sigma_{11}}(x_1-\mu_1)\right)\right]^2}$, $\mu_1=0$, $\mu_2=0$, $\sigma_{11}=0.25$, $\sigma_{22}=0.375$, $\rho=0.5$;
(b) Probability density function $f_Z(z)=\begin{cases}\frac{(-1)^{k-1}}{(k-1)!}(\ln z)^{k-1}, 0<z\le 1\\ 0 \quad \text{otherwise}\end{cases}$ for $Z=\prod_{i=1}^{k}X_i$, $Z=X_1$ to $Z=X_1{\cdot}X_2{\cdot}X_3{\cdot}X_4{\cdot}X_5$, $X_i\sim\mathcal{U}_{[0,1]}$, Practice Problem 7.6. ■

Exploration Problem 7.2. Product of Correlated Random Variables
Consider jointly continuous independent (X_1,X_2).

For $X_1\sim\mathcal{N}(\mu_1,\sigma_{11}^2)$ and $X_2\sim\mathcal{N}(\mu_2,\sigma_{22}^2)$, $f_{X_1}(x_1)=\frac{1}{\sqrt{2\pi}\sigma_{11}}e^{-\frac{(x_1-\mu_1)^2}{2\sigma_{11}^2}}$, $f_{X_2}(x_2)=\frac{1}{\sqrt{2\pi}\sigma_{22}}e^{-\frac{(x_2-\mu_2)^2}{2\sigma_{22}^2}}$.

Find the pdf $f_X(x)$ for $X=X_1X_2$.

Using $e^x=\sum_{k=0}^{\infty}\frac{1}{k!}x^k$ and binominal theorem $e^x=\lim_{n\to\infty}\left(1+\frac{1}{n}x\right)^n$, one finds [36, 37]

$$f_X(x)=\frac{1}{\pi}e^{-\frac{1}{2(1-\rho^2)}\left(\frac{\mu_1^2}{\sigma_{11}^2}-\frac{2\rho(x+\mu_1\mu_2)}{\sigma_{11}\sigma_{22}}+\frac{\mu_2^2}{\sigma_{22}^2}\right)}$$

$$\times\sum_{n=0}^{\infty}\sum_{m=0}^{2n}\frac{\sigma_{11}^{m-n-1}}{(2n)!(1-\rho^2)^{2n+\frac{1}{2}}\sigma_{22}^{m-n+1}}x^{2n-m}|x|^{m-n}\left[{2n \atop m}\right]\left(\frac{\mu_1}{\sigma_{11}^2}-\frac{\rho\mu_2}{\sigma_{11}\sigma_{22}}\right)^m\left(\frac{\mu_2}{\sigma_{22}^2}-\frac{\rho\mu_1}{\sigma_{11}\sigma_{22}}\right)^{2n-m}\mathcal{K}_{m-n}\left(\frac{1}{(1-\rho^2)\sigma_{11}\sigma_{22}}|x|\right),$$

where $\mathcal{K}_h(\cdot)$ is the modified Bessel function of the second kind and order h.

If $\mu_1=\mu_2=0$ and $\rho\neq 0$, $f_X(x)=\frac{1}{\pi\sqrt{1-\rho^2}\sigma_{11}\sigma_{22}}e^{-\frac{\rho}{(1-\rho^2)\sigma_{11}\sigma_{22}}x}\mathcal{K}_0\left(\frac{1}{(1-\rho^2)\sigma_{11}\sigma_{22}}|x|\right)$.

For statistically independent uncorrelated X_1 and X_2, $\rho=0$. Hence,

$$f_X(x)=\frac{1}{\pi}e^{-\frac{\mu_1^2}{2\sigma_{11}^2}-\frac{\mu_2^2}{2\sigma_{22}^2}}\sum_{n=0}^{\infty}\sum_{m=0}^{2n}\frac{\mu_1^m\mu_2^{2n-m}}{(2n)!\sigma_{11}^{n+m+1}\sigma_{22}^{3n-m+1}}x^{2n-m}|x|^{m-n}\left[{2n \atop m}\right]\mathcal{K}_{m-n}\left(\frac{1}{\sigma_{11}\sigma_{22}}|x|\right).$$

If $\mu_1=\mu_2=0$, $f_X(x)=\frac{1}{\pi\sigma_{11}\sigma_{22}}\mathcal{K}_0\left(\frac{1}{\sigma_{11}\sigma_{22}}|x|\right)$.

The Bessel function of the second kind $\mathcal{Y}_\alpha(x)$ is

$$\mathcal{Y}_h(x)=\frac{\mathcal{J}_{-h}(x)\cos(h\pi)-\mathcal{J}_h(x)}{\sin(h\pi)},\ \mathcal{J}_h(x)=\sum_{m=0}^{\infty}\frac{1}{m!\Gamma(m+h+1)}\left(\tfrac{1}{2}x\right)^{2m+h},$$

where $\mathcal{J}_h(x)$ is the Bessel function of the first kind; Γ is the gamma function; h is the integer or multivalued function with singularity at zero.

Various approximations of the Bessel functions for $h\geq 0$ are applied.

For h=0, $\mathcal{J}_h(x)\approx\sqrt{\frac{2}{\pi x}}\cos\left(x-(\frac{1}{2}h+\frac{1}{4})\pi\right)$, $\mathcal{Y}_h(x)\approx\sqrt{\frac{2}{\pi x}}\sin\left(x-(\frac{1}{2}h+\frac{1}{4})\pi\right)$.

The modified Bessel function of the second kind $\mathcal{K}_h(x)$ is

$$\mathcal{K}_h(x)=\frac{\pi}{2\sin(h\pi)}\left[\mathcal{J}_{-h}(x)-\mathcal{J}_h(x)\right],\ \mathcal{J}_h(x)=\sum_{m=0}^{\infty}\frac{1}{m!\Gamma(m+h+1)}\left(\tfrac{1}{2}x\right)^{2m+h}.$$

The approximation is $\mathcal{K}_h(x)\approx\sqrt{\frac{\pi}{2x}}e^{-x}\left(1+\frac{4h^2-1}{8x}+\frac{(4h^2-1)(4h^2-9)}{2!(8x)^2}+...\right)$. ■

Exploration Problem 7.3. Nonparametric and Parametric Bivariate Distributions

Bivariate and high-dimensional unimodal, bimodal and multimodal distributions are studied. Examine a joint pdf $f_{X_1,X_2}(x_1,x_2):\mathbb{R}^2\to\mathbb{R}$ of random jointly continuous (X_1,X_2). Probability distributions are found, and, cdfs and pdfs are parameterized using histograms $h_{X_1,X_2}(x_1,x_2)$. Histograms are normalized, and, $\int_{-\infty}^{\infty}\int_{-\infty}^{\infty}f_{X_1,X_2}(x_1,x_2)dx_2dx_1=1$.

Consider experimental data. In particular, study noise in the MDOF acceleration and angular velocity (rate) channels. We found the corresponding histograms $h_{X_1,X_2}(x_1,x_2)$ for $(X_1,X_2)\in\mathbb{R}^{2\times 1000}$, documented in Figures 7.18 for (X_1,X_2). Investigating parametric distributions $(X_1,X_2)\sim\mathcal{D}(\cdot)$ on admissible distribution classes $\mathcal{D}\in\mathbf{D}$. Assume near-uniform distribution $(X_1,X_2)\sim\mathcal{D}_{\mathcal{U}_{[0,1]}}$ and near-normal normal distribution $(X_1,X_2)\sim\mathcal{D}_{\mathcal{N}}(\boldsymbol{\mu},\boldsymbol{\Sigma})$. The cdfs, pdfs, moment-generating functions and characteristic functions can be found.

Histograms $h_{X_1,X_2}(x_1,x_2)$ are interpolated as $\tilde{h}^{\text{spline}}_{X_1,X_2}(x_1,x_2)$. Figures 7.18.a illustrate $h_{X_1,X_2}(x_1,x_2)$, $(x_1,x_2)\in[0,1]$ and $\bar{h}_{X_1,X_2}(x_1,x_2)$ for the first data set $(X_1,X_2)\in\mathbb{R}^{2\times 1000}$.

A uniform distribution $(X_1,X_2)\sim\mathcal{D}_{\mathcal{U}_{[0,1]}}$ implies $(X_1,X_2)\sim\mathcal{U}_{[0,1]}$,

$$f_{X_1,X_2}(x_1,x_2)=\begin{cases}1, & 0\leq(x_1,x_2)\leq 1\\ 0, & \text{otherwise}\end{cases},\ \int_{-\infty}^{\infty}\int_{-\infty}^{\infty}f_{X_1,X_2}(x_1,x_2)dx_2dx_1=1.$$

For a given pseudorandom (X_1,X_2), distribution $(X_1,X_2)\sim\mathcal{U}_{[0,1]}$ are found. Parametric joint pdfs $f_{X_1,X_2}(x_1,x_2):\mathbb{R}^2\to\mathbb{R}$ are computed for $h_{X_1,X_2}(x_1,x_2)$ and $\bar{h}_{X_1,X_2}(x_1,x_2)$. Apply

$$f_{X_1,X_2}(x_1,x_2)=\begin{cases}p_0+p_1e^{-p_2(x_1-p_3)^2+p_4(x_1-p_3)(x_2-p_5)-p_6(x_2-p_5)^2}+\sum\limits_{n,m,l,q}p_{n,m}\sin(np_lx_1)\sin(mp_qx_2), & 0\leq(x_1,x_2)\leq 1\\ 0, \text{ otherwise} & \forall p_i\in\mathbb{R}.\end{cases}$$

Truncate a pdf to $f_{X_1,X_2}(x_1,x_2)=\begin{cases}p_0+p_{1,1}\sin(p_1x_1)\sin(p_2x_2), & 0\leq(x_1,x_2)\leq 1\\ 0, \text{ otherwise}\end{cases}$, $\forall p_i\in\mathbb{R}$.

Computed p_i are reported in annotations in Figures 7.18.a.

For the second data set (X_1,X_2), a histogram $h_{X_1,X_2}(x_1,x_2)$ is illustrated in Figure 7.18.b. Consider a bivariate normal distribution $(X_1,X_2)\sim\mathcal{D}_{\mathcal{N}}(\boldsymbol{\mu},\boldsymbol{\Sigma})$ with

$$f_{X_1,X_2}(x_1,x_2)=\frac{1}{2\pi\sigma_{11}\sigma_{22}\sqrt{1-\rho^2}}e^{-\frac{1}{2(1-\rho^2)}\left(\frac{(x_1-\mu_1)^2}{\sigma_{11}^2}-2\rho\frac{(x_1-\mu_1)(x_2-\mu_2)}{\sigma_{11}\sigma_{22}}+\frac{(x_2-\mu_2)^2}{\sigma_{22}^2}\right)}.$$

Apply $f_{X_1,X_2}(x_1,x_2)=\frac{1}{2\pi p_3p_4\sqrt{1-p_5^2}}e^{-\frac{1}{2(1-p_5^2)}\left(\frac{(x_1-p_1)^2}{p_3^2}-2p_5\frac{(x_1-p_1)(x_2-p_2)}{p_3p_4}+\frac{(x_2-p_2)^2}{p_4^2}\right)}$, $\forall p_i\in\mathbb{R}$.

For unknown p_i, derive $\begin{bmatrix}p_1\\p_2\end{bmatrix}=\begin{bmatrix}\hat{\mu}_1\\\hat{\mu}_2\end{bmatrix}$, $\begin{bmatrix}p_3\\p_4\end{bmatrix}=\begin{bmatrix}\hat{\sigma}_{11}\\\hat{\sigma}_{22}\end{bmatrix}$ and $p_5=\hat{\rho}$, thereby parameterizing the pdf. Computed p_i are reported in annotations in Figures 7.18.b.

Finally, for random (X_1,X_2) with $h_{X_1,X_2}(x_1,x_2)$, illustrated in Figure 7.18.b, study $(X_1,X_2)\sim\mathcal{D}(\cdot)$ with

$$f_{X_1,X_2}(x_1,x_2)=p_1e^{-p_2(x_1-p_3)^2+p_4(x_1-p_3)(x_2-p_5)-p_6(x_2-p_5)^2},\ \forall p_i\in\mathbb{R}.$$

The unknown p_i are computed. Surfaces for $f_{X_1,X_2}(x_1,x_2)$ are identical for $(X_1,X_2)\sim\mathcal{D}_\mathcal{N}(\boldsymbol{\mu},\boldsymbol{\Sigma})$ and $(X_1,X_2)\sim\mathcal{D}(p)$. Convergence of an algorithm and overall consistency are ensured.

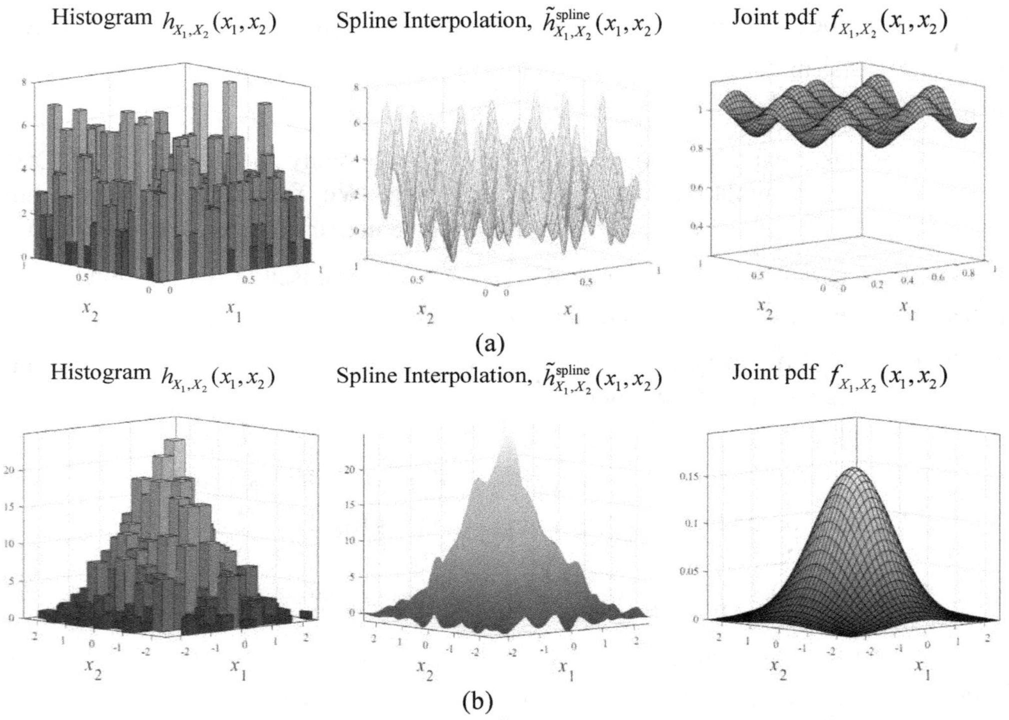

Figure 7.18. (a) Histograms $h_{X_1,X_2}(x_1,x_2)$ and $\tilde{h}^{\text{spline}}_{X_1,X_2}(x_1,x_2)$ for $(X_1,X_2)\in\mathbb{R}^{2\times1000}$. The surface for a parameterized joint pdf $f_{X_1,X_2}(x_1,x_2)=\begin{cases}p_0+p_{1,1}\sin(p_1x_1)\sin(p_2x_2), & 0\le(x_1,x_2)\le1\\ 0, & \text{otherwise}\end{cases}$ with derived p_0=0.9999, $p_{1,1}$=0.12, p_1=11.82, p_2=13.99;
(b) For $(X_1,X_2)\in\mathbb{R}^{2\times1000}$, histogram $h_{X_1,X_2}(x_1,x_2)$ and $\tilde{h}^{\text{spline}}_{X_1,X_2}(x_1,x_2)$ imply a bivariate near-normal distribution $(X_1,X_2)\sim\mathcal{D}_\mathcal{N}(\boldsymbol{\mu},\boldsymbol{\Sigma})$. For $f_{X_1,X_2}(x_1,x_2)=\frac{1}{2\pi p_3p_4\sqrt{1-p_5^2}}e^{-\frac{1}{2(1-p_5^2)}\left(\frac{(x_1-p_1)^2}{p_3^2}-2p_5\frac{(x_1-p_1)(x_2-p_2)}{p_3p_4}+\frac{(x_2-p_2)^2}{p_4^2}\right)}$, $p_1\equiv\mu_1$= –0.0673, $p_2\equiv\mu_2$=0.011, $p_3\equiv\sigma_{11}$=0.9174, $p_4\equiv\sigma_{22}$=1.082, $p_5\equiv\rho$=0.0291. For parameterized $f_{X_1,X_2}(x_1,x_2)=p_1e^{-p_2(x_1-p_3)^2+p_4(x_1-p_3)(x_2-p_5)-p_6(x_2-p_5)^2}$, p_1=0.161, p_2=0.551, p_3= –0.0674, p_4=0.0599, p_5=0.0112, p_6=0.466. ■

Exploration Problem 7.4. Pearson Correlation and Statistical Significance

The Pearson product-moment correlation matrix K_X is given by (7.27). The correlation coefficient ρ measures linear dependences between variables, $-1\le\rho\le1$.

The Pearson correlation coefficient for a pair of random (X_i,X_j) is

$$\rho_{X_iX_j}=\frac{\text{cov}(X_i,X_j)}{\sigma_{X_i}\sigma_{X_j}}=\frac{\text{cov}(X_i,X_j)}{\sqrt{\text{var}(X_i)}\sqrt{\text{var}(X_j)}}=\frac{\mathbb{E}\left[(X_i-\mathbb{E}[X_i])(X_j-\mathbb{E}[X_j])^T\right]}{\sigma_{X_i}\sigma_{X_j}}.$$

For a discrete sequence (X_1,X_2), the Pearson correlation coefficient is

$$r_{x_1x_2}=\frac{\sum_{i=1}^{n}\left(x_{1_i}-\bar{x}_1\right)\left(x_{2_i}-\bar{x}_2\right)}{\sqrt{\sum_{i=1}^{n}(x_{1_i}-\bar{x}_1)^2}\sqrt{\sum_{i=1}^{n}(x_{2_i}-\bar{x}_2)^2}}.$$

For an observed test-statistic value t from a distribution $\mathcal{D}$ for a random X, the p-value $0\le p\le1$ yields a probability estimate of observing t if the null hypothesis $\mathcal{H}_0$ is true.

For two-sided and symmetric case, $p=2\min\{\mathbb{P}(\mathcal{D}\ge t|\ \mathcal{H}_0),\ \mathbb{P}(\mathcal{D}\le t|\ \mathcal{H}_0)\}$, $p=\mathbb{P}(|\mathcal{D}|\ge t|\ \mathcal{H}_0)$.

Small p-value $p\ll1$ implies statistical significance. The p-value is conditional on validity of the null hypothesis. Other hypothesis and estimates are applied depending on the exhibited distribution.

Analyze grayscale and RGB Mona Lisa images, considered in Examples 7.10 and 7.14. For $X_g\in\mathbb{R}^{600\times396}$ and $X_{RGB}\in\mathbb{R}^{600\times396\times3}$, study datasets using double precision for (X_g,X_{RGB}). For $m=1,\ldots,396$, consider $X=[X_1,X_2,X_3,X_4]_m$, $X\in\mathbb{R}^{396\times4}$. Assume linearity and normal distribution.

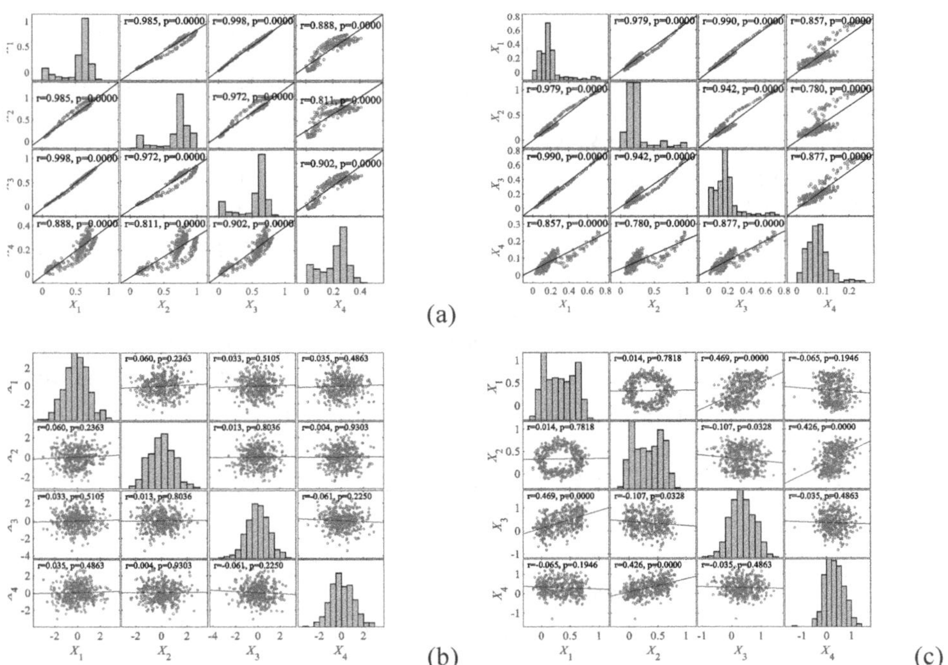

Figure 7.19. Histograms and Pearson correlations for $X=[X_1,X_2,X_3,X_4]$, $X\in\mathbb{R}^{396\times4}$;
(a) Mona Lisa image $X=[X_1,X_2,X_3,X_4]_m$, $m=100$ and $m=200$;
(b) Dataset $X=[X_1,X_2,X_3,X_4]$, $X\in\mathbb{R}^{396\times4}$ with pseudorandom normally distributed $X_i\sim\mathcal{N}(0,1)$;
(c) Dataset $X=[X_1,X_2,X_3,X_4]$, $X\in\mathbb{R}^{396\times4}$, $X_1=⅓[1(n)+\sin(n)+X_{1\mathcal{N}}(n)]$, $X_2=⅓[1(n)+\cos(n)+X_{2\mathcal{N}}(n)]$, $X_3=⅓[1(n)+\sin^3(n)+X_{3\mathcal{N}}(n)]$, $X_4=⅓[1(n)+\cos^3(n)+X_{4\mathcal{N}}(n)]$, $X_{i\mathcal{N}}\sim\mathcal{N}(0,1)$.

The histograms for X_i are at the chart diagonal, while the correlation coefficients $-1 \le r_{ij} \le 1$ for (X_i, X_j) are off diagonal. The least-squares regressions are found. The null hypothesis is tested. The null hypothesis is met if the p-value is small, which implies a statistical significance between (X_i, X_j). Histograms, Pearson correlation coefficients r_{ij} and p-values are reported in Figures 7.19.a.

Investigate pseudorandom $X=[X_1, X_2, X_3, X_4]$, $X \in \mathbb{R}^{396\times 4}$ with independent and identically distributed pseudorandom normally distributed $X_i \sim \mathcal{N}(0,1)$. Figure 7.19.b documents histograms, Pearson correlation coefficients r_{ij}, as well as statistical significance, evaluated by p_{ij}. Study and report results for periodic sequences X_i

$X=[X_1, X_2, X_3, X_4]$, $X \in \mathbb{R}^{396\times 4}$,

$X_1 = \frac{1}{3}[1(n)+\sin(n)+X_{1\mathcal{N}}(n)]$, $X_2 = \frac{1}{3}[1(n)+\cos(n)+X_{2\mathcal{N}}(n)]$,

$X_3 = \frac{1}{3}[1(n)+\sin^3(n)+X_{3\mathcal{N}}(n)]$, $X_4 = \frac{1}{3}[1(n)+\cos^3(n)+X_{4\mathcal{N}}(n)]$, $X_{i\mathcal{N}} \sim \mathcal{N}(0,1)$.

Histograms, computed (r_{ij}, p_{ij}) and periodicity are reported in Figure 7.19.c. ■

***Exploration Problem* 7.5. *Frequency Analysis of Strings: Binary Alphabet* $\mathcal{A}$=\{0,1\}**

Let a binary array X consists of four sequences,

$$X_{\mathcal{U},\mathcal{A}=\{0,1\}} = [X_1, X_2, X_3, X_4]_{\mathcal{U},\mathcal{A}=\{0,1\}},\ \mathcal{A}=\{0,1\}.$$

Study the rounded uniformly distributed pseudorandom $X_{i_{\mathcal{U},\mathcal{A}=\{0,1\}}} \sim \mathcal{U}$, $X_i \in \mathbb{R}^{1\times 800}$, which is weakly perturbed by deterministic X_{id}. Examine

$X=[X_1, X_2, X_3, X_4]$, $X_i = X_{i_{\mathcal{U},\mathcal{A}=\{0,1\}}} + X_{i_d} = X_{i_{\mathcal{U},\mathcal{A}=\{0,1\}}} + 0.025\sin(\omega_i n)$, $x_i \in [0\ \ 1]$, ω_i=1,2,3,4 1/s.

Nonparametric estimates of the spectral modality are computed for two datasets, illustrated in Figures 7.20,

$X_{(1)}=[X_1, X_2, X_3, X_4]_{(1)}$, $X_{(2)}=[X_1, X_2, X_3, X_4]_{(2)}$.

The descriptive features for X are observed as depicted in Figures 7.20 using the spectral density. As depicted in Figure 7.20, one detects distorted sequences and classifies perturbed bitstrings using peaks of $S_X(\omega)$, occurred at frequencies ω_i. ■

Exploration e Problem* 7.6. *Spectral Analysis for a Color Image

Estimates and descriptive features of sequences in the frequency domain are analyzed applying spectral analysis. Investigate red, green and blue channels of the RGB image $X \in \mathbb{R}^{600\times 396\times 3}$, studied in Examples 7.10 and 7.14. For an array X, the spectral density $S_X(\omega)$ is computed for columns j=100 and j=300. Figures 7.21 report $S_X(\omega)$ for:

1. The RGB image sequences $X_j = X_{i,j}|_{i=R,G,B} = [X_R, X_G, X_B]_j$, j=100 and j=300;
2. Weakly perturb $X_{i,j}$ by adding distortions
   $$X_{i_d} = \begin{cases} 5\,\mathrm{sgn}(\sin(\omega_i n)) & \text{if } \sin(\omega_i n) > 0 \\ 0 & \text{if } \sin(\omega_i n) \le 0 \end{cases},\ \omega_i\text{=1,2,3 1/sec},\ x_{i,j} \in [0\ \ 256];$$
3. Weakly perturb $X_{i,j}$ by adding rounded uniformly distributed pseudorandom $X_{i,d_{\mathcal{U},\mathcal{A}=\{0,1\}}}$, $\forall x_{i,j} \in [0\ \ 256]$.

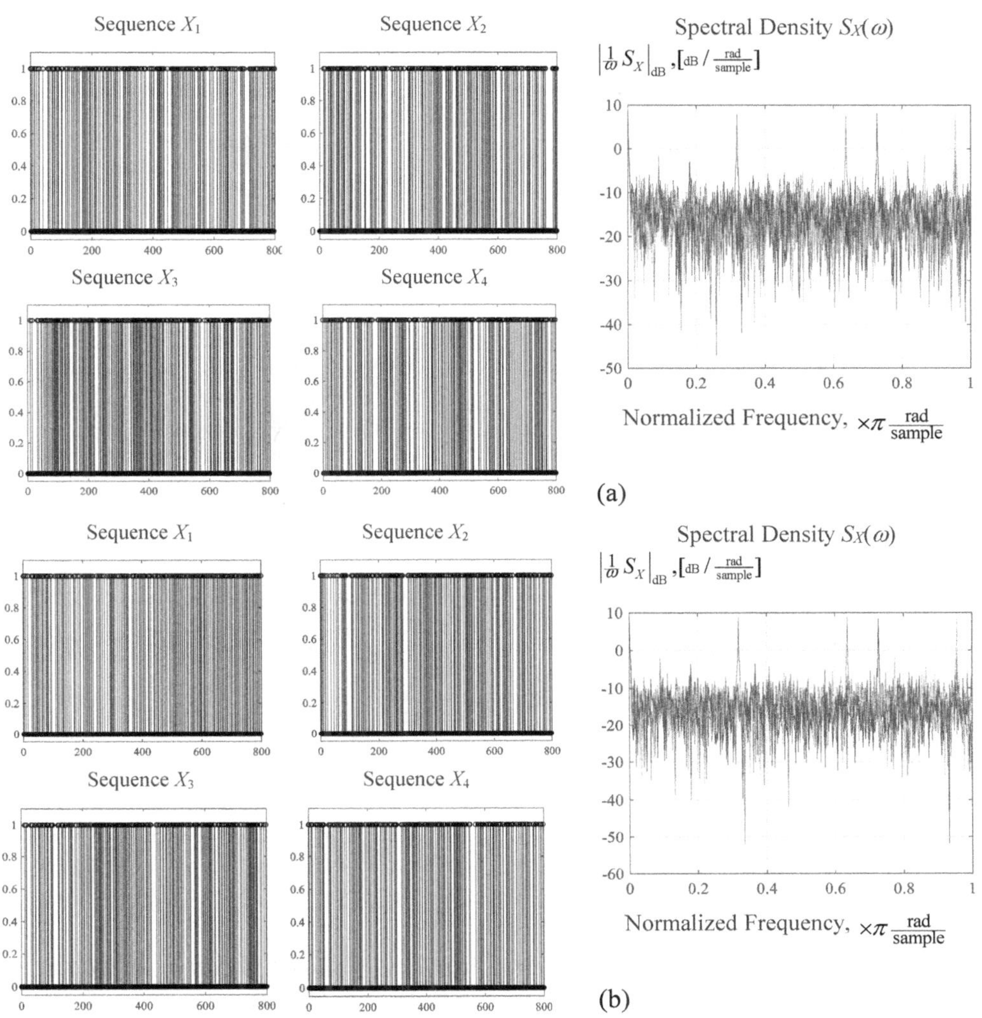

Figure 7.20. Spectral modality analysis evaluating spectral density $S_X(\omega)$:
(a) Array $X_{(1)}=[X_1,X_2,X_3,X_4]_{(1)}$; (b) Array $X_{(2)}=[X_1,X_2,X_3,X_4]_{(2)}$.

Figures 7.21 depict $S_X(\omega)$ for columns j=100 and j=300 of $[X_R,X_G,X_B]_j$.

Three peaks at ω_i are observed in $S_X(\omega)$ which imply weak periodic distortions. The random distortions change covariance and correlation matrices, variance, mutual information and other quantitative statistical characteristics which can be applied to evaluate and detect embedded sequences and codes.

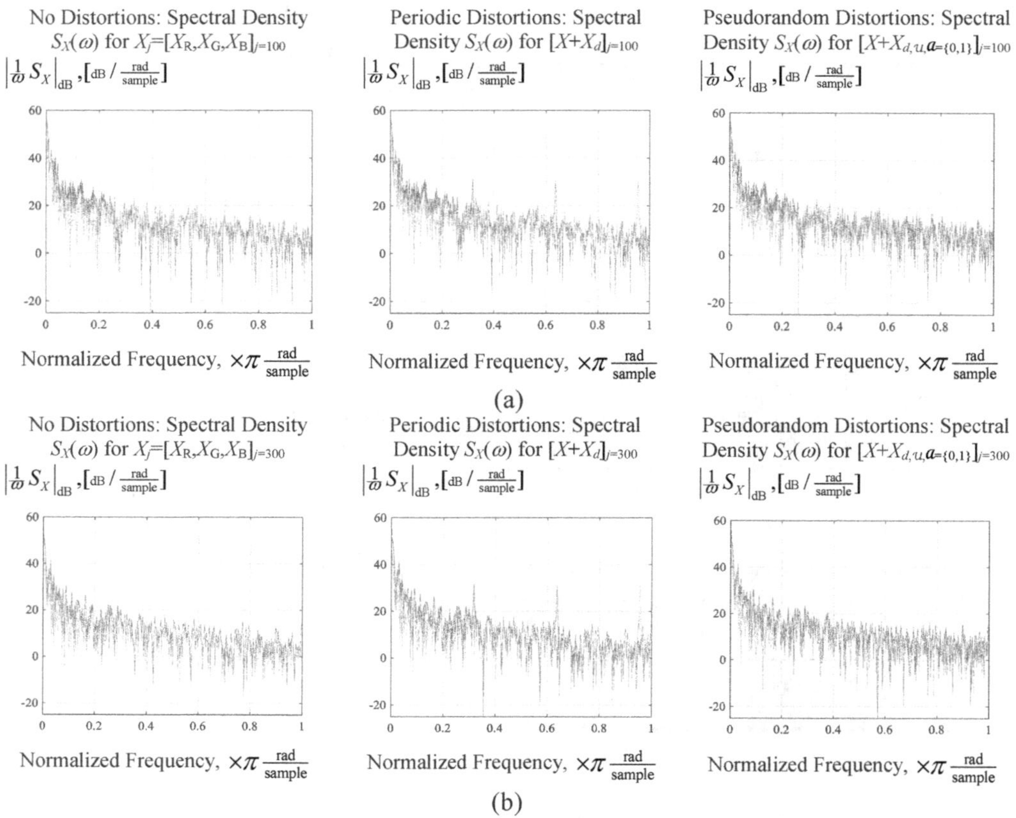

Figure 7.21. (a) Spectral density $S_X(\omega)$ for $X_j=[X_R,X_G,X_B]_{j=100}$, and, perturbed X_j by weak periodic distortions $[X+X_d]_{j=100}$ and $[X+X_{d,\mathcal{U},\boldsymbol{a}=\{0,1\}}]_{j=100}$;
(b) Spectral density $S_X(\omega)$ for $X_j=[X_R,X_G,X_B]_{j=300}$, $[X+X_d]_{j=300}$ and $[X+X_{d,\mathcal{U},\boldsymbol{a}=\{0,1\}}]_{j=300}$. ■

Exploration Problem 7.7. Computer Vision, Machine Vision and Pattern Recognition
Foundations of computer vision, machine vision and pattern recognition have being rapidly developed. Machine learning and convolutional neural networks (CNNs) support the aforementioned research areas. Innovative technologies were developed. Computer vision tasks imply assessing an information on objects. In contrast, machine vision should identify specific distinct or critical features of an object, and, process pertained data from electromagnetic spectrum detection, to information fusion and management, and, processing by ASICs to ensure feature and image recognition.

Two- and three-dimensional spaces and Cartesian coordinate geometry mappings are used. In $\mathbf{r}\in\mathbb{R}^3$, consider coordinates relative to a fixed coordinate frame $\mathbf{X}(r_X,r_Y,r_Z)$. The prospective image of an object by a camera is expressed using transformations between homogeneous coordinates, $\mathbf{I}(x,y)=DCP\mathcal{R}\mathbf{X}(r_X,r_Y,r_Z)$, where D and C are the depth and calibration mappings, which depend on the rotational symmetry; P is the projection matrix; $\mathcal{R}$ is the coordinate transformation mapping.

The CMOS image active-pixel sensors array, comprised of CMOS photodiodes, measures the multispectral radiation. An electronic shutter with ASICs post-processing yield a temporal image $I(x,y)$ as a two-dimensional array, defined on a compact region Ω of a two-dimensional surface, $I:\Omega\subset\mathbb{R}^2\to\mathbb{R}_+$, $(x,y)\mapsto I(x,y)$. The domain

Ω and range $\mathbb{R}_+$ are discrete, and, $\mathbb{R}_+$ is within the interval $\mathbb{N}_0=\{0,1,2,...,m\}$. The 8-,12-, 14- and 16-bit modes are used. For example, $\Omega=[1,640]\times[1,640]\subset\mathbb{N}^2$, and, for a 8-bit-mode, $[0, 255]\subset\mathbb{N}_0$.

The measured radiation is processed, and, a microprocessor implements a two-dimensional discrete Fourier transform (DFT). Frequency domain analysis supports filters design, image compression (JPEG, PNG and others formats) and other tasks. For audio and videos compressions, the fast Fourier transform (FFT) are used. Note that DFT is a mathematical operation, while FFT is an implementation algorithm or a family of algorithms.

In CNN models, two-dimensional arrays $(x,y)\mapsto I(x,y)$ are used. To extract features, eigenvector operators and multidimensional factorization spaces can be applied to ensure information-dense data analytics, dimensionality reduction, image reconstruction, etc. One strives to ensure high fidelity and accuracy (probability of recognition and localization), high mean average precision (mAP), robustness to nefarious data, minimal latency, etc. Calculus on eigenvector operators empowers descriptive analytics.

The *full*, rank-*k truncated*, and, *conformable partitioned* SVDs exist. For an image, we have $I(x,y)=U\Sigma V^T\approx U_t\Sigma_t V_t^T$.

For dataset mages $D_j(x,y)$, pertained to classes and subclasses (C_n,c_m), $\forall C_n\in\mathbf{C}$, the descriptive features in the eigenvector domain can be found, yielding

$$D_j(x,y)=U_j^{C_n}\Sigma_j^{C_n}(V_j^{C_n})^T\to\mathcal{F}_j^{C_n}.$$

For $(I_i(x,y),D_j(x,y))$, one may perform rotation, scaling and other transformations, yielding $\rho(I_i(x,y))$ and $\varphi(D_j(x,y))$. Without loss of generality, use

$$I_i(x,y)\equiv\rho(I_i(x,y)),\ D_j(x,y)\equiv\varphi(D_j(x,y)).$$

In general, nonlinear topological mapping between geometric primitives $\rho(\cdot)$ and $\varphi(\cdot)$ should be considered. Hence, $(I_i(x,y),\ D_j(x,y))$ are explicitly characterized by the SVD factorizations and pertained features

$$I_i(x,y)=U_i\Sigma_iV_i^T\to\mathcal{F}_i,\ D_j(x,y)=U_j^{C_n}\Sigma_j^{C_n}(V_j^{C_n})^T\to\mathcal{F}_j^{C_n},$$

yielding $(\mathcal{F}_i,\ \mathcal{F}_j^{C_n(c_m)})$. These $(\mathcal{F}_i,\ \mathcal{F}_j^{C_n(c_m)})$ support feature extraction, matching and pattern recognition. A pattern recognition algorithm is

$$\begin{cases}\text{Dataset}, D_j(x,y)=U_j^{C_n}\Sigma_j^{C_n}(V_j^{C_n})^T\to\mathcal{F}_j^{C_n}\\ \text{Object},\ I_i(x,y)=U_i\Sigma_iV_i^T\to\mathcal{F}_i\end{cases},\ \begin{cases}\text{Object}\in C_n\subseteq\mathbf{C},\mathcal{F}_i^{I_i(x,y)}\triangleq\mathcal{F}_j^{C_n}\ \text{if}\,(U_i,\Sigma_i,V_i)\cong(U_j^{C_n},\Sigma_j^{C_n},V_j^{C_n})\\ \text{Object}\notin C_n\subseteq\mathbf{C},\mathcal{F}_i^{I_i(x,y)}\neq\mathcal{F}_j^{C_n}\ \text{if}\,(U_i,\Sigma_i,V_i)\neq(U_j^{C_n},\Sigma_j^{C_n},V_j^{C_n})\end{cases}.$$

Images are described and uniquely characterized by the *full*, rank-*k truncated*, and, *conformable partitioned* SVDs with corresponding (U,Σ,V), computed for any image quality. The information-dense eigenfactorization is topologically simple, admits compact metric space and invariant measures, allows dimensionality (rank) reconfigurability and reduction, and, yields explicit descriptive features. SVDs can be computed for:

1. Near-singular, ill-conditioned and badly scaled grayscale, RGB, multispectral and hyperspectral images $I(x,y)$ with any number of spectral bands and channels;
2. Sparse and ragged matrices with zero-valued, distorted blocks and uncertainties.

Consider a scaling. Multiplying a spatial localized map X, $X=U\Sigma V^T$ with (U,Σ,V) by a scalar $a>0$, aX, yields the same SVD eigenvectors (U,V) for aX, while all singular values

σ_{ii} are multiplied by a, and, one finds $a\Sigma$, $aX=U(a\Sigma)V^T$.

By using computed tuples (U,Σ,V) and $(U^{C_n},\Sigma^{C_n},V^{C_n})$, one able to:

1. Find principal components (*scores*) and principal directions, e.g., accomplish principal component analysis and independent component analysis;
2. Analyze non-stationary dependences and differences;
3. Compute correlation K_X, covariance C and cross-covariance, which yield localization and dependences;
4. Perform spectral analytics and image segmentation;
5. Accomplish sequence mining and distortions correction;
6. Evaluate geometric contour, surface topology, directional derivatives, gradients $(\frac{\partial}{\partial x}I(x,y),\frac{\partial}{\partial y}I(x,y))$, etc.

Singular Value Decomposition Complexity – For $n\times m$ matrix, the SVD complexity is $O(n^3+n^2m+nm^2)$. The complexity of two-dimensional FFT for a square window of size $n\times n$ is $O(n\times n\times\log_2(n))$.

Pattern Recognition and Cross-Correlation – Matching image blocks is a fundamental problem. Cross correlation can be performed with block rotation, translation and scaling. The SVD factorization supports invariant cross-correlation, large angle rotation, robust scaling, and, similarity analysis. The mismatch of the *full* SVD $X_{I(x,y)}$, rank-k truncated SVD $X_t=U_t\Sigma_tV_t^T$, as well as $D(x,y)=U^{C_n}\Sigma^{C_n}(V^{C_n})^T$ are computed using the normalized measures. For the p-norm,

$$\Delta X_p(C_n)=\frac{1}{\sum_i\sum_j\|X_{ij}\|_p}\sum_i\sum_j\left\|X_{ij}-\mathcal{F}_{ij}^{C_n}\right\|_p,\ \Delta X_{p=1}(C_n)=\frac{1}{\sum_i\sum_j|X_{ij}|}\sum_i\sum_j\left|X_{ij}-\mathcal{F}_{ij}^{C_n}\right|,\ p=1.$$

Detection Transformer – The *detection transformer* (DETR) is developed by the Facebook AI Research group. In computer vision applications, this deep learning model is aimed to ensure end-to-end object detection. The neural network *transformer* architecture enables detection of dependencies between elements in data sets. Departing from region proposal networks in object recognition, a direct prediction is accomplished by using bounding boxes and classes performing *learning* on spatial features and fixed object queries for $I(x,y)$. The objective is to extract high-level features using object queries on spatial arrays. The *self-attention* allows parallel predictions, not implemented by traditional CNNs. A *bipartite matching* is implemented by assessing similarity predicted bounding boxes with ground truth objects.

Exploration Problem 7.8. Low-Fidelity Studies: Stuxnet Malware Sequences

Malware Detection – Probabilistic characteristics of binary sequences, codes, portable executable header, byte strings and n-gram may be extracted to assess quantitative features on known malware, semantic meaning, concealed malicious content, as well as hidden, encrypted and polymorphic codes. Variable length vectors may represent semantic meaning of codes and executable, yielding possibly interpretable detection. Lateral semantic indexing supports analysis. Antivirus programs are not effective against new viruses. A worm executes routines, a link file executes propagated copies of the worm, and, a rootkit hides all malicious files and processes to prevent detection. Polymorphic, oligomorphic, obfuscated and metamorphic codes are difficult to detect.

Signature-based detection cannot identify all possible attacks and malware, while the anomaly-based detection may exhibit false positives. Analysis of data, system governance and anomalies supports evaluation of security, resiliency, information sharing, traffic, etc. Dynamic-probabilistic statistical measures support analysis of codes and normalcy by analyzing data and system governance in $\mathbb{M}(\mathbf{P}^{\text{physical}}\times\mathbb{C}^{\text{cyber}}\times\mathbb{D}^{\text{data}})$. Analysis of anomalies in physical processes, caused by malware, may yield indirect indication of intrusion. An anomaly-based approach centers on behavior patterns and functionality to analyze security, resiliency and potentially detect intrusions.

Information Security Control – Reconfigurable control of information security and security management are aimed to derive feedback and adaptive schemes to deter and counteract malware actions, minimize security risks, as well as to protect information and physical components. Data quality, intrusions risk and other security attributes are of interest. Information security control schemes should minimize consequence of attacks and intrusions, counteract security incidents, enable security awareness, etc. Detection algorithms are intended to identify and characterize cyberattacks and incidents in progress, while control actions should prevent unauthorized attempts, reconfigure system, as well as empower recovery and normalcy.

Stuxnet Virus – The Win32/Stuxnet virus exploits deficiencies in the Microsoft Windows and Siemens SIMATIC WinCC interface in electric drives [38-41]. This processes- and functionality-centric virus has three components, e.g., worm, link and rootkit. Stuxnet leads to inadequate data acquisitionand control by affecting data acquisition and management. The MPU controls a *brushless* DC motor. The worm executes routines of the attack. It uses vulnerabilities to propagate and execute routines in networked modules. It implements a Microsoft Remote Procedure Call, which is distributed client-server program, to execute communication between affected systems and Siemens SIMATIC WinCC SCADA systems. The link file automatically executes the propagated copies of the worm for automatic execution. The rootkit hides malicious files and processes. There are different Stuxnet variants with precursors. The encrypted source codes, written in C/C++ and Assembly, are not available.

Typified characterizable binary sequences are devised by reverse engineering [38-41]. For illustrative purposes, convert fragments of a malware binary codes into bytes, and, yield sequences X. Histograms are found for two codes (X_1,X_2) which define:

1. Addresses, stacks and registers declarations;
2. Malware rootkit.

All affected systems use the Universally Unique Identifier UUID 000204e1-0000-0000-c000-000000000046 which is targeted by Stuxnet to identify, communicate, connect with remote servers and non-malicious and malicious URLs, exchange IP address and domain, update and execute. The worm is a self-replicating malware which duplicates itself to spread to uninfected devices once it compromised a target system, The worm searches for the legitimate link file X.DLL. For the Siemens WinCC, and, as the X.DLL is found, it is modifies adding malicious functions.

Consider bit strings of codes. One byte represents an ASCII character, while two bytes correspond to the UTF-16 character. Strings and encoded sequences are mapped by a finite countable set of natural numbers $\mathbb{N}_0=\{0,1,2,...,m\}$. For two codes with 481 and 18936 ASCII characters, two sequences to characterize are

$$(X_1,X_2),\ X_1\in\mathbb{R}^{1\times 481},\ X_2\in\mathbb{R}^{1\times 18936}.$$

Find nonparametric statistical distributions, histograms and cdfs. Encode sequences (X_1,X_2) by the XOR cipher, yielding $X_{1\text{XOR}}\in\mathbb{R}^{1\times 1545}$ and $X_{2\text{XOR}}\in\mathbb{R}^{1\times 56804}$. For sequences $(X_{1\text{XOR}},X_{2\text{XOR}})$, histograms and spline interpolations are depicted in Figures

7.22. As described in section 7.6, the Cantor singular distribution is not absolutely Lebesgue continuous, and, cannot be classified as discrete, continuous or mixed. It admits a cdf, which is a Hölder continuous Cantor function. In low-fidelity demonstrative studies, probabilistic signatures and descriptive statistics are found using the normalized histograms $\bar{h}_X^{\text{normalized}}(x)$, $\sum_{x\in X}\bar{h}_X^{\text{normalized}}(x)=1$.

The histograms $h_X(x)$ may be interpolated to be $\tilde{h}_X^{\text{interpolated}}(x)$, yielding cdfs $F_X(x)$ and pdfs. To analyze sequence and code signatures, and, detect embedded in codes characters, find and analyze the distributions, histograms $h_X(x)$, cross-covariance, correlation matrices, and, spectral characteristics. For X_2 and $X_{2\text{XOR}}$, the spectral density $S_X(\omega)$ are illustrated in Figures 7.23.

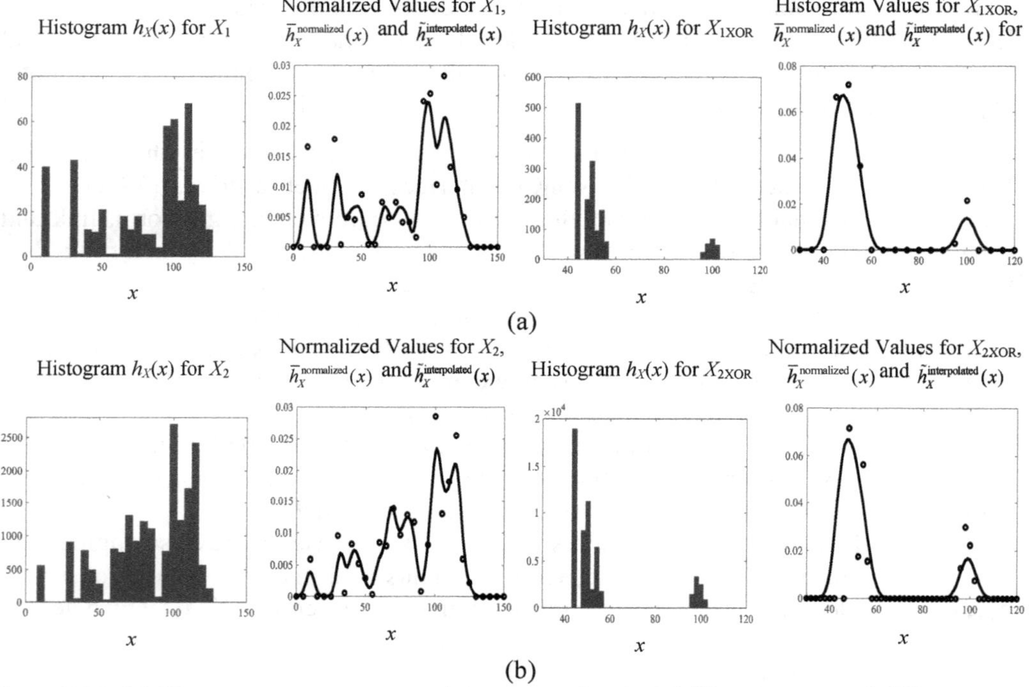

Figure 7.22. (a) Histograms $\bar{h}_X^{\text{normalized}}(x)$ and $\tilde{h}_X^{\text{interpolated}}(x)$ for $X_1\in\mathbb{R}^{1\times 481}$ and $X_{1\text{XOR}}\in\mathbb{R}^{1\times 1545}$; (b) Histograms $\bar{h}_X^{\text{normalized}}(x)$ and $\tilde{h}_X^{\text{interpolated}}(x)$ for $X_2\in\mathbb{R}^{1\times 18936}$ and $X_{2\text{XOR}}\in\mathbb{R}^{1\times 56804}$.

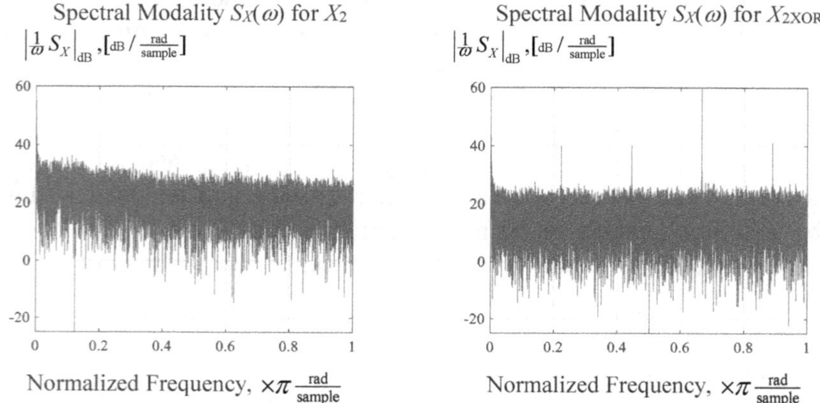

Figure 7.23. Spectral modalities $S_X(\omega)$ for $X_2\in\mathbb{R}^{1\times 18936}$ and $X_{2\text{XOR}}\in\mathbb{R}^{1\times 56804}$. ∎

References

1. *Strategy for Operations in the Information Environment*, US Department of Defense, Washington, D.C., 2016.
2. *Framework for Improving Critical Infrastructure Cybersecurity*, National Institute of Standards and Technology, U.S. Department of Commerce, 2018.
3. *Advanced Encryption Standard (AES)*, Federal Information Processing Standard Publication 197, 2001.
4. M. Dworkin, *Recommendation for Block Cipher Modes of Operation*, *NIST Special Publication* 800-38G, 2016.
5. S. Gueron, "Intel® advanced encryption standard (AES) new instructions set," Intel White Paper 323641-001, Rev. 3.0, pp. 1-81, 2010.
6. E. P. Blasch, R. Breton, P. Valin and E. Bosse, "User information fusion decision making analysis with the C-OODA model," *Proc. Conf. Information Fusion*, pp. 1-8, 2011.
7. S. E. Lyshevski, A. Aved and P. Morrone, "Information-centric cyberattack analysis and spatiotemporal networks applied to cyber-physical systems," *Proc. IEEE Microwave Theory and Techniques in Wireless Communication Workshop*, vol. 1, pp. 172-177, 2020.
8. J.-H. Cho, S. Xu, P. M. Hurley, M. Mackay, T. Benjamin and M. Beaumont, "STRAM: Measuring the trustworthiness of computer-based systems," *J. ACM Computing Surveys*, vol. 51, no. 6, Article 128, pp. 1-47, 2019.
9. M. Christodorescu, S. Jha, S. A. Seshia, D. Song and R. E. Bryant, "Semantic-aware malware detection," *Proc. Symp. Security and Privacy*, 2005.
10. J. Clark and S. Banik, "Building contemporary and efficient static models for malware detection," *Proc. SouthestCon Conf*, 2020.
11. M. Husák, J. Komárková, E. Bou-Harb and P. Celeda, "Survey on attack projection, prediction, and forecasting in cyber security," *IEEE Communication Surveys and Tutorials*, vol. 21, no. 1, pp. 640-660, 2019.
12. H. S. Lallie, K. Debattista and J. Bal, "A review of attack graph and attack tree visual syntax in cyber security," *Computer Science Review*, vol. 35, pp. 1-41, 2020.
13. A. Moser, C. Kruegel and E. Kirda, "Limits of static analysis for malware detection," *Proc. Computer Security Applications Conf.*, pp. 421-430, 2007.
14. W. Ren, T. Yu, T. Yardley and K. Nahrstedt, "CAPTAR: Causual-polytree-based anomaly reasoning for SCADA networks," *Proc. Conf. Communications, Control, and Computing Technologies for Smart Grids*, 2019.
15. C.-W. Ten, G. Manimaran and C.-C. Liu, "Cybersecurity for critical infrastructure: Attack and defense modeling," *IEEE Trans. Systems, Man, and Cybernetics*, vol. 40, no. 4, pp. 853-865, 2010.
16. C.-W. Ten, C.-C. Liu and G. Manimaran, "Vulnerability assessment of cybersecurity for SCADA systems," *IEEE Trans. Power Systems*, vol. 23, no. 4, pp. 1836-1846, 2008.
17. I. Khokhlov, M. Perez and L. Reznik, "Machin learning in anomaly detection: Example of colluded applications attack in Android devices", *Proc. Conf. Machine Learning and Applications*, pp. 1328-1333, 2019.
18. C. E. Shannon, "A mathematical theory of communications", *Bell Systems Technical Journal*, vol. 27, no. 3, pp. 379-423, no. 4, 623-656, 1948.
19. H. Burgiel and D. Stanton, "Realizations of regular abstract polyhedra of types {3, 6} and {6, 3}," *Discrete and Computational Geometry*, vol. 24, pp. 241-255, 2000.
20. P. McMullen, "Realizations of regular polytopes," *Aequationes Mathematicae*, vol. 37, pp. 38-56, 1989.
21. P. McMullen and E. Schulte, "The mix of a regular polytype with a face," *Annals of Combinatorics*, vol. 6, pp. 77-86, 2002.

22. D. Canright, "A very compact S-box for AES", *Proc. Workshop on Cryptographic Hardware and Embedded Systems*, pp. 441-455, 2005.
23. L. Dong, N. Wu and X. Zhang, "Low-power state machine design for AES encryption coprocessor", *Proc. Conf. Engineers and Computer Scientists*, pp. 1-4, 2015.
24. L. Herlihy, E. Golen, L. Reznik and S. E. Lyshevski, "Secure communication and signal processing in inertial navigation systems," *Proc. IEEE Conf. Electronics and Nanotechnologies*, pp. 414-419, 2017.
25. M. Kalimoldaev, S. Tynymbayev, M. Magzom, D. Tananova and S. E. Lyshevski, "FPGA implementation of encryption algorithms based on residual polynomials," *Proc. IEEE Conf. Electronics and Nanotechnology*, pp. 631-637, 2020.
26. S. Lemsitzer, J. Wolkerstorfer, N. Felber and M. Braendli, Multi-gigabit GCM-AES architecture optimized for FPGAs, Ed. P. Paillier and I. Verbauwhede, *Cryptographic Hardware and Embedded Systems, Lecture Notes in Computer Science*, 4727, pp. 227-238, 2007.
27. J. Ma, X. Chen, R. Xu and J. Shi, "Implementation and evaluation of different parallel designs of AES Using CUDA," Proc. *IEEE Int. Conf. Data Science in Cyberspace*, pp. 606-614, 2017.
28. I. Puchades, P. P. Markopoulos, B. J. Landi and S. E. Lyshevski, "Planar carbon nanotube antennas and data-centric secure communication," *Proc. IEEE Conf. Electronics and Nanotechnologies*, pp. 643-647, 2018.
29. D.-H. Bui, D. Puschini, S. Bacles-Min, E. Beigné and X.-T. Tran, "Ultra low-power and low-energy 32-bit datapath AES architecture for IoT applications", *Proc. Conf. IC Design and Technology*, pp. 1-4, 2016.
30. N. Homma, K. Saito and T. Aoki, "A formal approach to designing cryptographic processors based on GF(2^m) arithmetic circuit," *IEEE Trans. Information Forensics and Security*, vol. 7, no. 1, pp. 3-13, 2012.
31. D. McGrew, "Efficient authentication of large, dynamic data sets using Galois/counter mode (GCM)," *Proc. IEEE Int. Security in Storage Workshop*, pp. 1-6, 2005.
32. R. Ueno, S. Morioka, N. Miura, K. Matsuda, M. Nagata, S. Bhasin, Y. Mathieu, T. Graba, J.-L. Danger and N. Homma, "High throughput/gate AES hardware architecture based on datapath compression," *IEEE Trans. Computers*, vol. 69, no. 4, pp. 534-544, 2020.
33. R. Ueno, N. Homma, Y. Sugawara and T. Aoki, "Formal approach for verifying Galois field arithmetic circuits of higher degrees," *IEEE Trans. Computers*, vol. 66, no. 3, pp. 431-442, 2017.
34. S. Yanushkevich, V. Shmerko and S. E. Lyshevski, *Computer Arithmentics for Nanoelectronics*, CRC Press, Boca Raton, FL, 2009.
35. K. Wilkinson, *Application Note: Using Encryption and Authentication to Secure an UltraScale/UltraScale + FPGA Bitstream*, XILINX, 2018.
36. G. Cui, X. Yu, S. Iommelli and L. Kong, "Exact distribution for the product of two correlated Gaussian random variables," *IEEE Trans. Processing Letters*, vol. 23, no. 11, pp. 1662-1666, 2016.
37. M. K. Simon, *Probability Distributions Involving Gaussian Random Variables: A Handbook for Engineers and Scientists*, Springer, NY, 2006.
38. D. Kushner,"The real story of Stuxnet," *IEEE Spectrum*, vol.50, issue 3, pp.49-53,2013.
39. S. Karnouskos, "Stuxnet worm impact on industrial cyber-physical system security," *Proc. Industrial Electronics Conf.*, pp. 4490-4494, 2011.
40. T. Miyachi, H. Narita, H. Yamada and H. Furuta, "Myth and reality on control system security revealed by Stuxnet," *Proc. SICE Conf.*, pp. 1537-1540, 2011.
41. B. Bakić, M. Milić, I. Antović, D. Savić and T. Stojanović, "10 years since Stuxnet: What have we learned from this mysterious computer software worm," *Proc. Conf. Information Technology*, pp. 1537-1540, 2021.

Index

E

F

G

H

I

K

L

M

N

O

P

Q

R

For Product Safety Concerns and Information please contact our EU representative GPSR@taylorandfrancis.com
Taylor & Francis Verlag GmbH, Kaufingerstraße 24, 80331 München, Germany

www.ingramcontent.com/pod-product-compliance
Lightning Source LLC
Chambersburg PA
CBHW080347220326
41598CB00030B/4632

* 9 7 8 1 0 3 2 5 2 6 7 8 2 *